APA ENGINEERED WOOD HANDBOOK

Thomas G. Williamson, PE, Editor

McGRAW-HILL

New York Chicago San Francisco Lisbon London
Madrid Mexico City Milan New Delhi San Juan Seoul
Singapore Sydney Toronto

Cataloging-in-Publication Data is on file with the Library of Congress.

McGraw-Hill 

A Division of The ***McGraw-Hill*** *Companies*

1 2 3 4 5 6 7 8 9 0 DOC/DOC 0 7 6 5 4 3 2 1

ISBN 0-07-136029-8

The sponsoring editor for this book was Larry Hager and the production supervisor was Sherri Souffrance. It was set in Times Roman by Pro-Image Corporation.

Printed and bound by R. R. Donnelley & Sons Company.

McGraw-Hill books are available at special quantity discounts to use as premiums and sales promotions, or for use in corporate training programs. For more information, please write to the Director of Special Sales, Professional Publishing, McGraw-Hill, Two Penn Plaza, New York, NY 10121-2298. Or contact your local bookstore.

This book is printed on recycled, acid-free paper containing a minimum of 50% recycled de-inked fiber.

CONTENTS

CONTRIBUTORS

William A. Baker, P.E. *Manager of Codes and Engineering, TSD*
Zhaozhen Bao, PhD *Associate Scientist, TSD*
Richard D. Carlson *Senior Scientist, TSD*
H. Fulton Desler *Senior Engineer, TSD*
Edward L. Keith, P.E. *Senior Engineer, TSD*
Zeno A. Martin, P.E. *Associate Engineer, TSD*
John Rose *Retired Senior Engineer, TSD*
Sheldon Q. Shi, PhD *Associate Scientist, TSD*
Thomas D. Skaggs, PhD, P.E. *Senior Engineer, TSD*
Borjen Yeh, PhD, P.E. *Manager, Research and Development, TSD*
Steven C. Zylkowksi *Director, Engineered Wood Systems*

PREFACE

Over the past six decades, *APA-The Engineered Wood Association,* has developed a large volume of technical information associated with the design and use of glued engineered wood composite products in building construction applications. These products include plywood, oriented strand board (OSB), glued laminated timber (glulam), laminated veneer lumber (LVL) and pre-fabricated wood I-joists, plus other specialty products. Until now, this technical information has only been available to design professionals and others through a wide array of APA and other industry publications. The purpose of this Handbook is to bring this material all together in one reference document.

Glued engineered wood composites have taken on added importance over the last decade as the wood products industry has moved toward the manufacture of more environmentally friendly products. All of the products discussed in this handbook can be produced from second- and third-growth managed forestlands. Also, they permit the use of a wide variety of species and species mixes, which further expands the available wood fiber resource.

Chapter 1 provides a basic introduction to the structure and properties of wood as an engineering material. Subsequent chapters discuss the unique design and construction characteristics of each glued engineered wood composite product.

Chapter 3 explains how wood structural panel products can be combined to create box beams, stressed skin panels and structural insulated panels. Chapter 7 describes how wood framed shear walls and diaphragms can be used to resist the lateral forces induced by wind and seismic events—technology which is essential to the design of any wood framed building. And since any building design is only as good as the connection detailing, Chapter 8 provides a comprehensive review of fasteners and connections typically used with glued composite products.

Some of the questions most commonly posed to the APA technical staff relate to product protection—how to provide maximum durability with minimum maintenance. Chapter 9 discusses treatments and finishes, Chapter 10 describes fire- and noise-related systems, and Chapter 12 focuses on designing and detailing for permanence. These chapters provide the most up to date industry recommendations on these topics.

The next technology breakthrough on the horizon will combine glued wood composite products with other technologies, such as fiber-reinforced polymers. These products of the future, commonly called advanced engineered wood composites, are discussed in Chapter 11.

Each chapter has been written by one of the scientists or engineers in the Technical Services Division (TSD) of APA. Collectively these professionals represent an unequalled wealth of knowledge related to glued engineered wood composites. A sincere thanks is extended to each of these dedicated APA staff members, who did most of their writing in off-work hours. In addition, a special thanks is extended to Marilyn LeMoine, Director of the Market Communications Division of APA, and her dedicated staff, especially Mike Martin and Mary Trodden, who helped generate all of the tables and artwork used in this handbook.

Thomas G. Williamson, Editor

APA ENGINEERED WOOD HANDBOOK

CHAPTER ONE
INTRODUCTION TO WOOD AS AN ENGINEERING MATERIAL

Steven Zylkowski
Director, Engineered Wood Systems

1.1 BASIC STRUCTURE OF WOOD

The unique characteristics and abundant supply of wood have made it the most desirable building material throughout history. Manufacturing technologies based on the modern understanding of wood have led to a family of engineered wood products that optimize properties to meet the specific needs of design professionals. While there are many types of technically advanced wood products such as machine stress-rated (MSR) lumber and metal-plated connected wood trusses that are considered to be part of the broad definition of engineered wood products, they have not been included in this Handbook. Information on these and other non-glued wood products may be found in other manuals and literature sources. For the purposes of this Handbook, engineered wood products are defined as products manufactured from various forms of wood fiber bonded together with water-resistant adhesives. Engineered wood products are intended for structural applications and include such products as structural plywood, oriented strand board (OSB), glued-laminated timber, laminated veneer lumber (LVL), and wood I-joists. These products are also known as wood composites.

As an organic material derived from trees, wood has a cellular structure composed of longitudinally arranged fibers. This directional aspect of wood imparts directional properties that are well recognized by wood engineers and manufacturers of wood products. While the properties of engineered wood products are largely determined by manufacturing processes that change the configuration of the wood fibers, basic wood characteristics are primary to the end product. Wood characteristics are determined by many factors, such as species, growing conditions, and wood quality.

1.1.1 Softwoods and Hardwoods

One fundamental characterization of wood is based upon whether the wood is from a hardwood or softwood tree. Hardwood, or deciduous, trees are those that lose their leaves during the fall. Softwoods, or coniferous, trees are those that have needles that typically remain green throughout the year. Classification of hardwood and softwood is not based upon the hardness or density of the wood; there are many examples of low-density hardwoods such as basswood or aspen and dense softwoods such as southern pines. Rather, the classification is based upon the taxonomy of the tree. Table 1.1 lists some of the prevalent commercial hardwood and softwood wood species in North America.

1.1.2 Earlywood and Latewood Formation

Trees grow at different rates throughout the year, resulting in growth rings in the wood. During favorable growing conditions, such as during the spring in temperate climates, trees grow at a faster rate, resulting in lower density fibers. This wood fiber appears as lighter areas in the wood. This portion of the wood is known as earlywood or springwood. As growth rates slow, the wood fibers develop thicker cell walls, resulting in denser fibers that appear as the darker portion of the growth ring. This portion of a growth ring is known as latewood or summerwood. Figure 1.1 depicts the earlywood and latewood portions of a growth ring.

TABLE 1.1 Prevalent Wood Species of North America

Hardwoods	Softwoods
Alder, red	Cedar, eastern red
Ash, white	Cedar, western red
Aspen	Cedar, yellow
Basswood, American	Douglas fir, coast type
Beech, American	Fir, balsam
Birch, paper	Fir, grand
Cherry, black	Fir, noble
Cottonwood	Fir, Pacific silver
Elm, American	Fir, white
Hackberry	Hemlock, eastern
Maple, silver	Hemlock, western
Maple, sugar	Larch, western
Oak, northern red	Pine, loblolly
Oak, southern red	Pine, lodgepole
Oak, white	Pine, longleaf
Sweetgum	Pine, ponderosa
Sycamore, American	Pine, red
Tupelo, black	Pine, shortleaf
Tupelo, swamp	Pine, sugar
Walnut, black	Pine, western white
Yellow-poplar	Redwood
	Spruce, black
	Spruce, Engelmann
	Spruce, Sitka

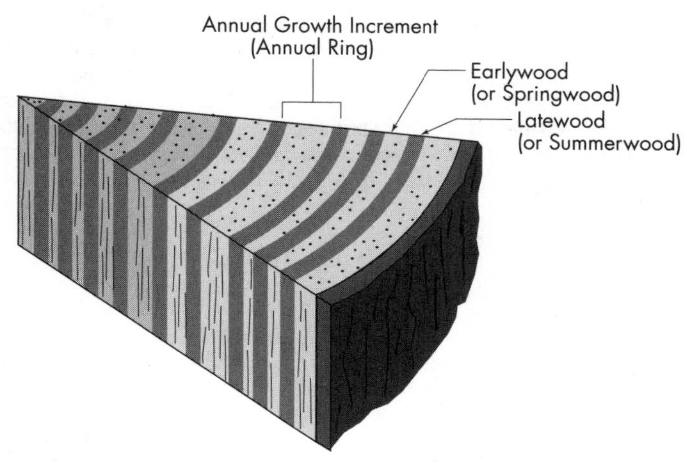

FIGURE 1.1 Earlywood and latewood.

1.1.3 Sapwood and Heartwood

The inner portion of a cross section of a tree trunk often displays a variation in color compared to the outer portion of the trunk. The lighter colored outer portion of the cross section is the sapwood. As shown in Fig. 1.2, the darker inner portion is the heartwood. The width of the sapwood and the color of the heartwood various considerably between different tree species. The sapwood portion of a stem is the newer portion and is used by the tree for conduction of water and nutrients for

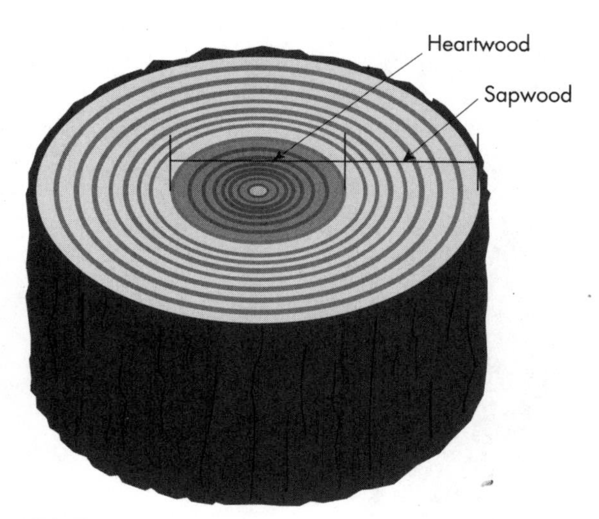

FIGURE 1.2 Sapwood and heartwood.

growth. As trees age, the center portion of the stem can collect excess nutrients that metabolize into various extractives that discolor the wood. These extractives can include waxes, oils, resins, fats, and tannins, along with aromatic and coloring materials. The color and characteristics of the heartwood is critical to woods used for decorative uses such as furniture, but less critical for engineered wood uses. Most properties of sapwood and heartwood are identical, except that the heartwood of some species is resistant to decay fungi as discussed further in Section 1.5 and Chapter 9.

1.1.4 Anisotrophy

The appearance of wood, as well as its properties, are significantly influenced by the surface orientation relative to its location in the tree stem. As shown in Fig. 1.3, wood has different orientations relative to the growth rings and longitudinal fiber arrangement. The cross section is perpendicular to the longitudinal direction of the fibers. The surface from the center of the stem outward is the radial surface. The outer surface, parallel to the growth rings, is called the tangential surface. Wood properties are significantly influenced by the direction relative to the fiber and growth ring orientation.

1.1.5 Chemical Makeup of Wood

From the standpoint of basic chemical elements, wood is primarily composed of carbon, hydrogen, and oxygen, as shown in Table 1.2. The basic chemical elements

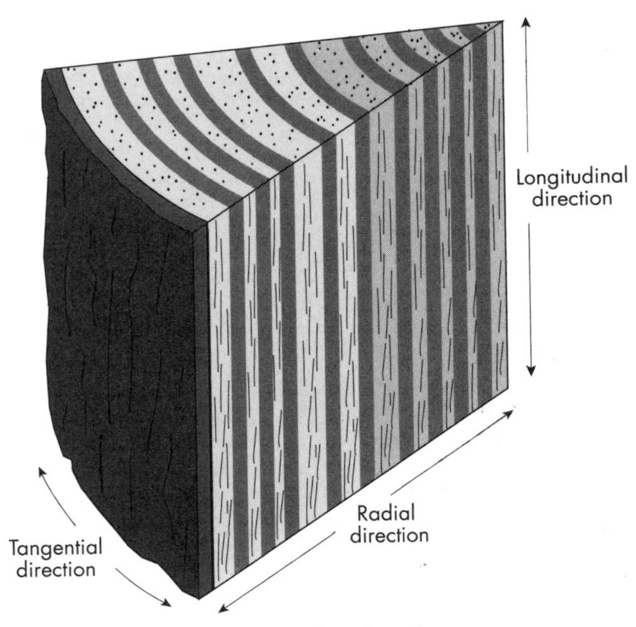

FIGURE 1.3 Directional orientation of wood.

TABLE 1.2 Basic Chemical Composition of Wood

Element	% of dry weight
Carbon	49
Hydrogen	6
Oxygen	44
Nitrogen	slight amount
Ash[a]	0.2–1.0

[a] What remains of wood after complete combustion in the presence of abundant oxygen.

of wood are incorporated into a number of organic compounds. The primary organic compounds are cellulose, hemicellulose, and lignin. These three compounds account for almost all of the extractive-free dry weight of wood. On average, proportions of cellulose, hemicellulose, and lignin differ slightly between hardwood and softwood species as shown in Table 1.3.

1.2 RESOURCE AND THE ENVIRONMENT

1.2.1 Distribution of Forests Throughout the World

The northern hemisphere contains mostly softwood timberlands and the southern hemisphere mostly hardwoods. As shown in Table 1.4, excerpted from Ref. 1, North America contains a large source of the world's softwood forests.

North America is nearly 21% forestland. As shown in Fig. 1.4, this percentage has been fairly constant since the 1920s. Each year the amount of timber cut is less than the growth of standing timber, as shown in Fig. 1.5 (from Ref. 1).

1.2.2 Volume of Engineered Wood Products from North America

The United States and Canada are major manufacturers and exporters of engineered wood products. The producers of these products in North America are among the most advanced in the world. Engineered wood products have been readily adopted by the construction industry in North America due to an overall familiarity and preference for wood. Table 1.5 (from Ref. 2) shows the volume of plywood, OSB, I-joists, LVL, and glulam produced by North American producers as well as the percentage of worldwide production.

TABLE 1.3 Organic Compounds in Wood (% of Oven-Dry Weight)

	Cellulose	Hemicellulose	Lignin
Hardwoods	40–44	15–35	18–25
Softwoods	40–44	20–32	25–35

TABLE 1.4 Distribution of Forests Throughout the World

Region	Softwood or coniferous forests		Hardwood or deciduous forests		Combined softwood and hardwood forests	
	Land area	%	Land area	%	Land area	%
North America	400	30.5	230	13.4	630	20.8
Central America	20	1.5	10	2.3	60	2.0
South America	10	0.8	550	32.0	560	18.5
Africa	2	0.2	188	10.9	190	6.3
Europe	107	8.2	74	4.3	181	6.0
CIS[a]	697	53.0	233	13.6	930	30.6
Asia	65	5.0	335	19.5	400	13.2
Oceania	11	0.8	69	4.0	80	2.6
Total World	1,312	100.0	1,719	100.0	3,031	100.0

[a] CIS—Confederation of Independent States of the former Soviet Union.

Note: The totals for combined softwood and hardwood forests do not always add up because no breakdowns have been given for areas in Europe, and the Confederation of Independent States is excluded by law from exploitation.

Source: R. Sedjo and K. Lyon, *The Long Term Adequacy of World Timber Supply,* Resources for the Future, Washington, DC, 1990. Excerpted from ref. 1.

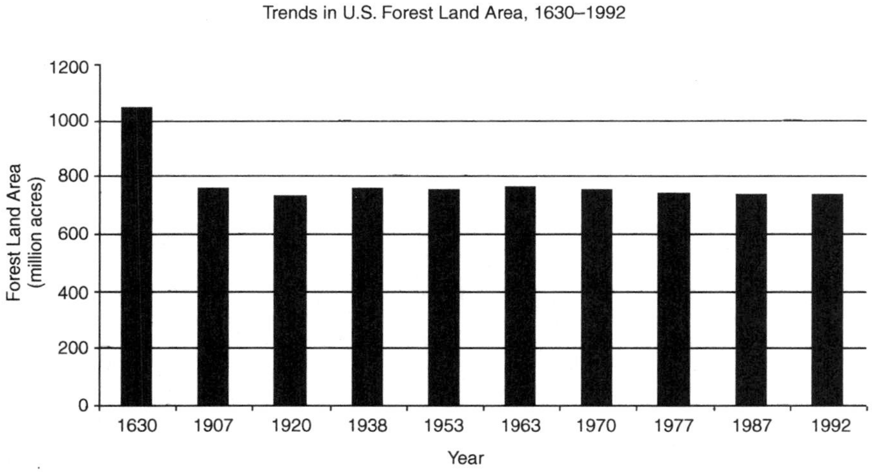

Trends in U.S. Forest Land Area, 1630–1992

FIGURE 1.4 Percent forestland in the U.S. from ref. 1.

1.2.3 Environmental Advantages of Wood Construction

As construction professionals have become increasingly interested in the environmental impacts of construction materials, the preference for wood products has increased since they are an excellent environmental choice. An environmental study was conducted by the ATHENA Sustainable Materials Institute for the Canadian Wood Council[3] to compare the environmental impact of constructing a house using

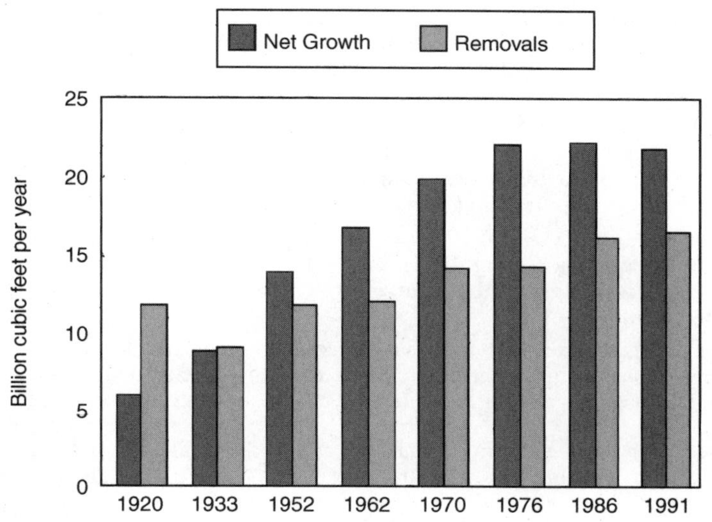

FIGURE 1.5 Harvest ratio of forest land in the U.S.

TABLE 1.5 2000 Volume of Engineered Wood Products from North America

Product	Units	U.S.	Canada
Structural plywood	10^6 ft^2-⅜ in. basis	17,475	2,200
OSB	10^6 ft^2-⅜ in. basis	11,910	8,740
LVL	10^6 ft^3	44.4	4.4
I-Joist	10^6 ft	693	173
Glulam	10^6 bd ft	356	21

wood framing, steel framing, and concrete. The study used life-cycle analysis to assess the environmental effects at all stages of the product's life, including resource procurement, manufacturing, on-site construction, building service life, and decommissioning at the end of the useful life of the building.

The study evaluated a typical 2400 ft^2 house designed for the Toronto, Ontario, Canada market. The wood-designed house was framed with lumber, wood I-joists for the floor, and wood roof trusses. The steel house used light-gage steel members for wall and floor framing. The concrete house used insulated concrete forms (ICF) for walls and a composite floor system with open web steel joists and concrete slab. The study considered the following environmental aspects.

- *Embodied energy* measures the total amount of direct and indirect energy used to extract, manufacture, transport, and install the construction materials. It includes potential energy contained in raw or feedstock materials, such as natural gas used in the production of resins.
- *Global warming potential* is a reference measurement using carbon dioxide as a common reference for global warming or greenhouse gas effects. All greenhouse

gases are referred to as having a "CO_2 equivalence effect." While greenhouse gas emissions are largely a function of energy combustion, some products also emit greenhouse gases during processing of raw materials, such as during the calcination of limestone during the production of cement.

- *Air and water toxicity* indices represent the human health effects of substances emitted during the various stages of the life cycle of the materials. The commonly accepted measure is the critical volume method, used to estimate the volume of ambient air or water required to dilute contaminants to acceptable levels.
- *Weighted resource use* is the sum of the weighted resource requirements for all products used in each design. This can be thought of as "ecologically weighted kilograms," which reflect the relative ecological carrying capacity effects of extracting resources.
- *Solid waste* is reported on a mass basis according to general life-cycle conventions that tend to favor lighter materials. Solid waste is more related to building practices than materials and careful planning can significantly reduce waste.

Table 1.6 shows the environmental measure of the design house built with each type of construction material. Construction with wood uses less energy, represents less global warming potential, has fewer impacts on air and water, and represents less weighted resource use. The results of this study clearly demonstrate that wood systems have fewer environmental impacts than the other construction systems used in the study.

1.3 PHYSICAL PROPERTIES OF WOOD

The wide spectrum of wood species provides a diverse range of properties. Wood properties are largely a result of several primary influences, such as wood species, grade, and moisture content. Other factors that influence wood's utility for certain applications are permeability, decay resistance, thermal properties, chemical resistance, and shrinkage and expansion characteristics. Engineered wood products rely on a combination of underlying wood properties and manufacturing techniques to optimize the desirable characteristics of wood and minimize undesirable characteristics. This section provides a review of the most important physical characteristic of wood as they affect engineered wood products.

TABLE 1.6 Environmental Impacts from Various Building Systems

Athena study results			
	Wood	Steel	Concrete
Embodied energy, GJ	255	389	562
Global warming potential, kg CO_2 equivalent	62,183	76,453	93,573
Air toxicity, critical volume measurement	3,236	5,628	6,971
Water toxicity, critical volume measurement	407,787	1,413,784	876,189
Weighted resource use, kg	121,804	138,501	234,996
Solid waste, kg	10,746	8,897	14,056

1.3.1 Density and Specific Gravity of Wood

The density, or weight per volume, of wood varies considerably between and within wood species. Density of wood is always reported in combination with its moisture content due to the significant influence moisture content has on overall density. To standardize comparisons between species or products, it is common to report the specific gravity, or density relative to that of water, on the basis of oven dry weight of wood and volume at a specified moisture content.

In general, the greater the density or specific gravity of wood, the greater its strength, expansion and shrinkage, and thermal conductivity. Table 1.7 reports the specific gravity of common commercial species in North America. The relationship between specific gravity and mechanical properties can be represented by the equation

$$P = kG^n \tag{1.1}$$

where P = a mechanical property
k and n = constants that depend upon the specific mechanical property and species
G = specific gravity

The constants that describe the relationship between properties and specific gravity are presented in Table 1.8 (from Ref. 4).

1.3.2 Moisture and Wood

Similar to other organic materials, wood is hygroscopic in that it absorbs or loses moisture to reach equilibrium with the surrounding environment. Wood can naturally hold large quantities of water. Understanding the effects of water in wood is important because it influences many properties of wood and engineered wood products.

The measure of water in wood is called the *moisture content* and is reported as the weight of water per weight of oven-dry, or moisture-free, wood. As can be seen in Table 1.9, the moisture content of freshly cut wood can exceed 100% because the weight of water in a given volume of wood can exceed the weight of the oven-dry wood.

With a cellular structure, wood can hold water both in the cell cavity and in the cell wall itself. Water is held in the cavity as liquid, or *free*, water. Water held in the cell walls is chemically bound water. Cell walls can chemically hold about 30% moisture. The term *fiber saturation point* describes the conceptual point where the cell walls are saturated while no water exists in the cavity. The fiber saturation point is important because moisture changes involving the chemically held water result in a different behavior of the wood product than with the loss of free water in the cell cavities. Section 1.3.3 describes how moisture changes in chemically held water lead to changes in strength properties and the dimensions of the wood, respectively. As discussed in Chapter 9, the threshold for wood decay is exceeded when the moisture content exceeds the fiber saturation point of the wood.

When wood is acclimated to be in equilibrium with ambient air conditions, yet protected from direct wetting, the moisture content of the wood will be below the fiber saturation point. Under this condition, the moisture content of wood is a

TABLE 1.7 Specific Gravity of Common Commercial Wood Species

Species	Specific gravity
Softwoods	
Douglas fir, coast type	0.45
Fir, grand	0.35
Fir, noble	0.37
Fir, Pacific silver	0.39
Fir, white	0.37
Hemlock, western	0.42
Larch, western	0.48
Pine, loblolly	0.47
Pine, longleaf	0.54
Pine, shortleaf	0.47
Cedar, eastern red	0.46
Cedar, western red	0.31
Fir, balsam	0.32
Hemlock, eastern	0.39
Pine, lodgepole	0.39
Pine, ponderosa	0.39
Pine, red	0.42
Pine, sugar	0.34
Pine, western white	0.35
Redwood	0.39
Spruce, black	0.38
Spruce, Engelmann	0.33
Spruce, Sitka	0.38
Hardwoods	
Alder, red	0.38
Ash, white	0.54
Basswood, American	0.32
Beech, American	0.57
Birch, paper	0.48
Cottonwood, eastern	0.37
Elm, American	0.46
Hackberry	0.49
Maple, sugar	0.57
Maple, silver	0.44
Oak, northern red	0.56
Oak, southern red	0.53
Oak, white	0.60
Sycamore, American	0.46
Sweetgum	0.46
Tupelo, black	0.47
Yellow-poplar	0.40

TABLE 1.8 Mechanical Properties as a Function of Specific Gravity for Wood at 12% Moisture Content

Property[a]	Specific gravity–strength relationship	
	Softwoods	Hardwoods
Static bending		
MOE (lb/in.^{2})	24,760 $G^{1.01}$	24,850 $G^{0.13}$
MOE ($\times 10^6$ lb/in.^{2})	2.97 $G^{0.84}$	2.39 $G^{0.7}$
WML (in.-lbf/in.^{3})	25.9 $G^{1.34}$	31.8 $G^{1.54}$
Impact bending (lbf)	77.7 $G^{1.39}$	95.1 $G^{1.65}$
Compression parallel (lb/in.^{2})	13,590 $G^{0.97}$	11,030 $G^{0.89}$
Compression perpendicular (lb/in.^{2})	2,390 $G^{1.57}$	3,130 $G^{2.09}$
Shear parallel (lb/in.^{2})	2,410 $G^{0.85}$	3,170 $G^{1.13}$
Tension perpendicular (lb/in.^{2})	870 $G^{1.11}$	1,460 $G^{1.3}$
Side hardness (lbf)	1,930 $G^{1.5}$	3,440 $G^{2.09}$

[a] Compression parallel to grain is maximum crushing strength; compression perpendicular to grain is fiber stress at proportional limit. MOR is modulus of rupture; MOE, modulus of elasticity; and WML, work to maximum load. For green wood, use specific gravity based on oven-dry weight and green volume; for dry wood, use specific gravity based on oven-dry weight and volume at 12% moisture.

TABLE 1.9 Average Moisture Content of Freshly Cut Wood

Hardwoods	Moisture content (%)		Softwoods	Moisture content (%)	
	Heartwood	Sapwood		Heartwood	Sapwood
Alder, red	—	97	Cedar, eastern red	33	—
Ash, white	46	44	Cedar, western red	58	249
Aspen	95	113	Cedar, yellow	32	166
Basswood, American	81	133	Douglas fir, coast type	37	115
Beech, American	55	72	Fir, balsam	88	173
Birch, paper	89	72	Fir, grand	91	136
Cherry, black	58	—	Fir, noble	34	115
Cottonwood	162	146	Fir, Pacific silver	55	164
Elm, American	95	92	Fir, white	98	160
Hackberry	61	65	Hemlock, eastern	97	119
Maple, silver	58	97	Hemlock, western	85	170
Maple, sugar	65	72	Larch, western	54	119
Oak, northern red	80	69	Pine, loblolly	33	110
Oak, southern red	83	75	Pine, lodgepole	41	120
Oak, white	64	78	Pine, longleaf	31	106
Sweetgum	79	137	Pine, ponderosa	40	148
Sycamore, American	114	130	Pine, red	32	134
Tupelo, black	87	115	Pine, shortleaf	32	122
Tupelo, swamp	101	108	Pine, sugar	98	219
Walnut, black	90	73	Pine, western white	62	148
Yellow-poplar	83	106	Redwood	86	210
			Spruce, black	52	113
			Spruce, Engelmann	51	173
			Spruce, Sitka	41	142

function of relative humidity and to a much smaller degree the temperature. Table 1.10 provides the equilibrium moisture content of wood at given relative humidity and temperatures.

Engineered wood products and most lumber products are dried to a relatively low moisture content as part of the manufacturing process. One objective of drying wood during manufacturing is to lower the moisture to conditions near that of in-service conditions. Drying of wood used for engineered wood products is also necessary for effective bonding using most wood adhesives.

Wood products in service are always undergoing changes in moisture content, as products respond to exposure to seasonal and short-term weather changes. To a lesser degree, the equilibrium moisture content of engineered wood products is also somewhat dependent upon the manufacturing process. At a specific relative humidity, the moisture content of products such as glued laminated timber is similar to the moisture content of the lumber constituents. Products that involve elevated temperatures during manufacturing, such as plywood and OSB, have reduced equilibrium moisture contents compared to wood at the same relative humidity, as shown in Table 1.11.

1.3.3 Movement of Wood and Engineered Wood Products

Wood shrinks and expands when the moisture content changes while below the fiber saturation point. As chemically held moisture in the cell wall is lost, the wood shrinks. As moisture is added to the cell wall, the wood expands. The amount of wood shrinkage or expansion is highly dependent upon the direction relative to the wood grain. As previously described in Section 1.1.4, the three principal grain orientations are a function of the growth of fibers around the perimeter of the tree stem. The expansion/shrinkage rate longitudinal to the grain is between 0.10% and 0.20% from green to oven dry and is therefore considered negligible in most cases. The expansion/shrinkage across the grain, that is, in the radial and tangential directions, is a function of wood species and other variables. Table 1.12 presents the shrinkage rate of common commercial species as the moisture changes from the green (freshly cut) condition to oven dry. The rate of shrinkage or expansion is

TABLE 1.10 Equilibrium Moisture Content of Wood (%)

°F	10%	20%	30%	40%	50%	60%	70%	80%	90%
40	2.6	4.6	6.3	7.9	9.5	11.3	13.5	16.5	21.0
50	2.6	4.6	6.3	7.9	9.5	11.2	13.4	16.4	20.9
60	2.5	4.6	6.2	7.8	9.4	11.1	13.3	16.2	20.7
70	2.5	4.5	6.2	7.7	9.2	11.0	13.1	16.0	20.5
80	2.4	4.4	6.1	7.6	9.1	10.8	12.9	15.7	20.2
90	2.3	4.3	5.9	7.4	8.9	10.5	12.6	15.4	19.8
100	2.3	4.2	5.8	7.2	8.7	10.3	12.3	15.1	19.5
110	2.2	4.0	5.6	7.0	8.4	10.0	12.0	14.7	19.1
120	2.1	3.9	5.4	6.8	8.2	9.7	11.7	14.4	18.6
130	2.0	3.7	5.2	6.6	7.9	9.4	11.3	14.0	18.2
140	1.9	3.6	5.0	6.3	7.7	9.1	11.0	13.6	17.7
150	1.8	3.4	4.8	6.1	7.4	8.8	10.6	13.1	17.2

TABLE 1.11 Equilibrium Moisture Content of Engineered Wood Products

	Moisture content (%)		
Relative humidity	Solid wood[a]	Plywood	OSB
10	2.5	1.2	0.8
20	4.5	2.8	1.0
30	6.2	4.6	2.0
40	7.7	5.8	3.6
50	9.2	7.0	5.2
60	11.0	8.4	6.3
70	13.1	11.1	8.9
80	16.0	15.3	13.1
90	20.5	19.4	17.2

[a] From *Agriculture Handbook* No. 72 by U.S. Forest Products Laboratory.

TABLE 1.12 Shrinkage Rate of Common Commercial Wood Species, %, from Green to Oven-Dry Moisture Content

Hardwoods species	Radial	Tangential	Volumetric
Alder, red	4.4	7.3	12.6
Ash, white	4.9	7.8	13.3
Aspen, quaking	3.5	6.7	11.5
Basswood, American	6.6	9.3	15.8
Beech, American	5.5	11.9	17.2
Birch, paper	6.3	8.6	16.2
Cherry, black	3.7	7.1	11.5
Cottonwood, eastern	3.9	9.2	13.9
Elm, American	4.2	9.5	14.6
Hackberry	4.8	8.9	13.8
Maple, silver	3.0	7.2	12.0
Maple, sugar	4.8	9.9	14.7
Oak, northern red	4.0	8.6	13.7
Oak, southern red	4.7	11.3	16.1
Oak, white	4.4	8.8	12.7
Sweetgum	5.3	10.2	15.8
Sycamore, American	5.0	8.4	14.1
Tupelo, black	5.1	8.7	14.4
Walnut, black	5.5	7.8	12.8
Yellow-poplar	4.6	8.2	12.7

From ref. 4.

affected by other variables. Generally there is greater shrinkage with wood with higher density.

Engineered wood products that cross-laminate wood elements, such as plywood and OSB, have expansion/shrinkage properties that are much lower than the cross-grain expansion/shrinkage rates of wood. This is due to the restraining affect that the longitudinal direction imposes upon the cross-grain movement.

The in-service moisture content of wood products used in construction depends upon the ambient relative humidity of the locale. The resultant moisture content of wood products exposed to ambient outdoor conditions, but protected from direct wetting, is presented in a U.S. Forest Service publication.[5]

1.3.4 Thermal Properties

In some applications the thermal properties of wood become important. Thermal properties of wood include thermal conduction, expansion, combustion temperature, and thermal influence of structural properties.

Conduction. Thermal conductivity is a measure of the rate of heat flow through a material subjected to a temperature change. Wood has low thermal conductivity compared to most other structural building materials. The thermal conductivity of wood products is affected by a number of variables such as density, moisture content, and grain direction. On average, thermal conductivity along the grain is about 1.8 times as high as across the grain. The thermal conductivity of some commercial wood species is presented in Table 1.13.

The thermal conductivity of plywood and OSB is slightly higher than solid wood of similar density. Researchers theorize that interstitial voids in plywood and OSB provide increased thermal resistance.[6] Table 1.14 presents thermal conductivity values for some plywood and OSB panels.

Expansion. Like most solid materials, dry wood expands with increase in temperature. The coefficient of thermal expansion is approximately 1.7×10^{-6} to 2.5×10^{-6} in./in. per °F along the grain. The thermal expansion in the radial and tangential directions is approximated by the following equations:

$$\alpha_r = (18G + 5.5) \times 10^{-6} \text{ per °F} \tag{1.2}$$

$$\alpha_t = (18G + 10.2) \times 10^{-6} \text{ per °F} \tag{1.3}$$

where α_r and α_t are the radial and tangential coefficient of thermal expansion, respectively, and G is the oven-dry specific gravity of the wood.

Using the above equations, OSB or plywood with 60% or less of the cross section of the panel with grain in the along direction would have a coefficient of thermal expansion of approximately 3.4×10^{-6} in./in./°F.

Under normal circumstances, wood expansion from thermal affects is offset by shrinkage as the wood product loses moisture. So for wood at normal moisture levels, net dimensional changes will generally be negative after prolonged heating because the shrinkage from moisture reduction more than offsets the thermal expansion.

TABLE 1.13 Thermal Conductivity of Selected Hardwoods and Softwoods

Species	Specific gravity	Conductivity (W/m · K (Btu · in./hr · ft² · °F)) Oven-dry	12% MC
Hardwoods			
Ash, white	0.63	0.41 (0.98)	0.17 (1.2)
Aspen, quaking	0.40	0.10 (0.67)	0.12 (0.80)
Basswood, American	0.38	0.092 (0.64)	0.11 (0.77)
Beech, American	0.68	0.15 (1.0)	0.18 (1.3)
Cherry, black	0.53	0.12 (0.84)	0.15 (1.3)
Cottonwood, eastern	0.43	0.10 (0.71)	0.12 (0.85)
Elm, American	0.54	0.12 (0.86)	0.15 (1.0)
Hackberry	0.57	0.13 (0.90)	0.16 (1.1)
Maple, silver	0.50	0.12 (0.80)	0.14 (0.97)
Maple, sugar	0.66	0.15 (1.0)	0.18 (1.2)
Oak, northern red	0.65	0.14 (1.0)	0.18 (1.2)
Oak, southern red	0.62	0.14 (0.96)	0.17 (1.2)
Oak, white	0.72	0.16 (1.1)	0.19 (1.3)
Sweetgum	0.55	0.13 (0.87)	0.15 (1.1)
Sycamore, American	0.54	0.12 (0.86)	0.15 (1.0)
Tupelo, black	0.54	0.12 (0.86)	0.15 (1.0)
Yellow-poplar	0.46	0.11 (0.75)	0.13 (0.90)
Softwoods			
Cedar, eastern red	0.48	0.11 (0.77)	0.14 (0.94)
Cedar, western red	0.33	0.083 (0.57)	0.10 (0.68)
Douglas fir, coast	0.51	0.12 (0.82)	0.14 (0.99)
Fir, balsam	0.37	0.090 (0.63)	0.11 (0.75)
Fir, white	0.41	0.10 (0.68)	0.12 (0.82)
Hemlock, eastern	0.42	0.10 (0.69)	0.12 (0.84)
Hemlock, western	0.48	0.11 (0.77)	0.14 (0.94)
Larch, western	0.56	0.13 (0.88)	0.15 (1.1)
Pine, loblolly	0.54	0.12 (0.86)	0.15 (1.0)
Pine, lodgepole	0.43	0.10 (0.71)	0.12 (0.85)
Pine, ponderosa	0.42	0.10 (0.69)	0.12 (0.84)
Pine, red	0.46	0.11 (0.75)	0.13 (0.90)
Pine, shortleaf	0.54	0.12 (0.82)	0.15 (1.0)
Pine, sugar	0.37	0.090 (0.63)	0.11 (0.75)
Pine, western white	0.40	0.10 (0.67)	0.12 (0.80)
Redwood	0.10	0.12 (0.82)	0.12 (0.82)
Spruce, black	0.43	0.10 (0.71)	0.12 (0.85)
Spruce, Engelmann	0.37	0.090 (0.63)	0.11 (0.75)
Spruce, Sitka	0.42	0.10 (0.69)	0.12 (0.84)

From ref. 4.

TABLE 1.14 Thermal Conductivity of Plywood and OSB Panels

Panel	Btu · in./hr · ft^2 · °F
7/16 in. aspen OSB	0.57
23/32 in. aspen OSB	0.60
3/8 in. Douglas fir plywood	0.43
3/4 in. Douglas fir plywood	0.66

Combustion. Wood will burn when exposed to extreme heat. The exact thermal degradation process depends on the rate of heating, temperature, presence of air, and presence of an ignition source. Wood decomposes into volatiles and char when heated beyond 212°F. The general sequence of thermal degradation is as follows:

- At temperatures of 230–302°F, wood will char over time, and if the heat and small amounts of volatiles are not dissipated, there is a possibility of combustion.
- At temperatures of 302–392°F charring takes place at a greater rate. Volatiles produced at this temperature are not readily ignited by a flame source, but chance of combustion is present if the heat is not dissipated.
- At temperatures of 392–536°F the formation of char is rapid and spontaneous combustion is probable.
- At temperatures of 536°F and greater spontaneous combustion will occur rapidly.

Chapter 10 provides an extensive discussion on the fire performance of engineered wood products.

1.4 MECHANICAL PROPERTIES

1.4.1 Species

The properties of engineered wood products are a function of the basic wood properties as well as the manufacturing techniques used to produce the products. Other chapters of this book present the structural properties and design values for engineered wood products. This section presents some basic wood properties and variables that influence the properties of wood. Although there is a wide variety of wood species, the number of commercially used wood species is limited due to commercial viability and availability (see Table 1.1). Many of the differences in wood properties are a function of the density or specific gravity of the wood. Table 1.15 presents the mechanical properties of clear wood at 12% moisture content for many commercially available species in North America.

Within a given species there is considerable variation in mechanical properties due to factors such as local growing conditions, soil type, elevation, and forestry practices. Furthermore, knots and grain angle influence the properties of the final wood products and are not considered in these clear wood properties. Design values for lumber products are presented in the *National Design Specification for Wood Construction*[8] and take these factors into consideration. Later chapters of this book present design values for engineered wood products.

TABLE 1.15 Average Mechanical Properties of Clear Wood

Species	Modulus of rupture (psi)	Modulus of elasticity (1000's psi)	Compression parallel to grain (psi)	Shear strength (psi)	Compression perpendicular to grain (psi)	Specific gravity
Softwoods						
Douglas fir, coast type	7665	1560	3784	904	700	0.45
Fir, grand	5839	1250	2939	739	475	0.35
Fir, noble	6169	1380	3013	802	478	0.37
Fir, Pacific silver	6410	1420	3142	746	414	0.39
Fir, white	5854	1161	2902	756	491	0.37
Hemlock, western	6637	1307	3364	864	457	0.42
Larch, western	7652	1458	3756	869	676	0.48
Pine, loblolly	7300	1402	3511	863	661	0.47
Pine, longleaf	8538	1586	4321	1041	804	0.54
Pine, shortleaf	7435	1388	3527	905	573	0.47
Cedar, eastern red	030	649	3570	1008	700	0.46
Cedar, western red	184	939	2774	771	244	0.31
Fir, balsam	517	1251	2631	662	187	0.32
Hemlock, eastern	420	1073	3080	848	359	0.39
Pine, lodgepole	490	1076	2610	685	252	0.39
Pine, ponderosa	130	997	2450	704	282	0.39
Pine, red	820	1281	2730	686	259	0.42
Pine, sugar	893	1032	2459	718	214	0.34
Pine, western white	688	1193	2434	677	192	0.35
Redwood	500	1177	4210	803	424	0.39
Spruce, black	118	1382	2836	739	242	0.38
Spruce, Engelmann	705	1029	2180	637	197	0.33
Spruce, Sitka	660	1230	2670	757	279	0.38
Hardwoods						
Alder, red	540	1167	2960	770	250	0.38
Ash, white	500	1436	3990	1354	667	0.54
Basswood, American	960	1038	2220	599	170	0.32
Beech, American	570	1381	3550	1288	544	0.57
Birch, paper	380	1170	2360	836	273	0.48
Cottonwood, eastern	260	1013	2280	682	196	0.37
Elm, American	190	1114	2910	1002	355	0.46
Hackberry	480	954	2650	1070	399	0.49
Maple, sugar	480	1546	4020	1465	645	0.57
Maple, silver	820	943	2490	1053	369	0.44
Oak, northern red	300	1353	3440	1214	614	0.56
Oak, southern red	920	1141	3030	934	547	0.53
Oak, white	300	1246	3560	1249	671	0.60
Sycamore, American	470	1065	2920	996	365	0.46
Sweetgum	110	1201	3040	991	367	0.46
Tupelo, black	040	1031	3040	1098	485	0.47
Yellow-poplar	950	1222	2660	792	269	0.40

1.4.2 Effects of Moisture on Mechanical Properties

As discussed in Section 1.3.3, wood is a hygroscopic material that absorbs or loses moisture in response to ambient moisture conditions. Changes in moisture content below the fiber saturation point results in changes to mechanical properties. A decrease in moisture content will generally increase mechanical properties; an increase in moisture content will lead to a decrease in mechanical properties. As covered in later chapters, design values for engineered wood products include moisture content adjustments because it is an important design parameter. An equation to estimate the effects of changes in moisture content on clear wood properties is available in the *Wood Handbook*.[4] Table 1.16 presents approximate changes in certain mechanical properties for a 1% change in moisture content. The relation that describes the effect of moisture on mechanical properties when below the fiber saturation point is:

$$P_m = P_{12} \times (P_{12}/P_{grn})^{(12-M)/(M_p-12)} \tag{1.4}$$

where P_m = the mechanical property at moisture content m.
P_{12} = the mechanical property at 12% moisture content (see Table 1.15).
P_{grn} = the mechanical property at the green condition (values are available in the *Wood Handbook*[4]).
M = the moisture content of interest.
M_p = the moisture content corresponding to the threshold at which lowing the moisture content effects the strength. This point is slightly below the fiber saturation point. Table 1.17 presents M_p values for some species. For other species, assume M_p = 25%.

1.4.3 Effect of Temperature on Mechanical Properties

Although the strength of wood products can vary with temperatures, the effects in most structural applications are small enough such that they are negligible. How-

TABLE 1.16 Approximate Percentage Increase (or Decrease) in Mechanical Properties for 1% Decrease (or Increase) in Moisture Content

Property	% change per 1% change in moisture content
Static bending	
Fiber stress at proportional limit	4
Modulus of elasticity	2
Compression parallel to grain	
Fiber stress at proportional limit	5
Maximum crushing strength	6
Compression perpendicular to grain	
Fiber stress at proportional limit	5.5
Shear parallel to grain	
Maximum shearing strength	3
Tension perpendicular to grain	
Maximum tensile strength	1.5

TABLE 1.17 Values of M_p for Calculating Moisture Effect on Strength

Species	M_p (%)
Ash, white	24
Birch, yellow	27
Douglas fir	24
Hemlock, western	28
Larch, western	28
Pine, loblolly	21
Pine, longleaf	21
Redwood	21
Spruce, Sitka	27

ever, exposure to extremely high temperature can weaken wood in two regards. First, wood has lower strength at elevated temperatures. Furthermore, elevated temperature can lead to irreversible strength degradation. The effects of elevated temperature are compounded when the wood is also at high moisture content. The *Wood Handbook*[4] provides a detailed discussion on the compounded relation between temperature and moisture effects on wood strength. Table 1.18 presents data on the reversible effect of temperature on wood at various moisture contents.

1.4.4 Time Effects on Wood Strength Properties

The load that a wood product will support continuously over a long period of time is less than the load it will carry over a short duration. For example, when continuously loaded for 10 years, a wood product will carry about 60% of the load that it will carry in the 5 to 10 minutes it typically takes to test the strength of the wood product in a laboratory. Considerable research has been conducted to evaluate the time–strength relationship for wood products. Early research on clear, straight-grain wood specimens led to development of widely recognized load duration factors that have been incorporated into building codes. Additional research has evaluated the suitability of this load duration relationship for other engineered wood products and found that it applies to these other wood products as well.

The allowable stress design method most commonly used for wood products is based on a 10-year duration of load. Published design stresses for wood products take this time effect into account and include a reduction of design stresses so that the value is applicable for loads of 10-year duration. Figure 1.6 presents the relationship between strength and load duration.

If a given design anticipates continuous or combined intermittent loading exceeding 10 years, the published design stress is reduced. Conversely, if the anticipated load duration is less than 10 years, the design stress can be increased. Table 1.19 presents the load duration adjustments applicable for allowable stress design. These factors are applicable for design with the engineered wood products discussed in this book. Note that load duration factors apply to strength properties only, but not to stiffness properties.

TABLE 1.18 Approximate Middle-Trend Effects of Temperature on Mechanical Properties of Clear Wood at Various Moisture Conditions

Property	Moisture condition (%)	Relative change in mechanical property from 20°C (68°F) at −50°C (−58°F) (%)	Relative change in mechanical property from 20°C (68°F) at +50°C (+122°F) (%)
MOE parallel to grain	0	+11	−6
	12	+17	−7
	>FSP	+50	—
MOE perpendicular to grain	6	—	−20
	12	—	−35
	≥20	—	−38
Shear modulus	>FSP	—	+25
Bending strength	≤4	+18	−10
	11–15	+35	−20
	18–20	+60	−25
	>FSP	+110	−25
Tensile strength parallel to grain	0–12	—	−4
Compressive strength parallel to grain	0	+20	−10
	12–45	+50	−25
Shear strength parallel to grain	>FSP	—	−25
Tensile strength perpendicular to grain	4–6	—	−10
	11–16	—	−20
	≥18	—	−30
Compressive strength perpendicular to grain at proportional limit	≥10	—	−35

1.4.5 Effect of Grain Angle on Strength Properties

As mentioned earlier, the anisotropic nature of wood is such that the mechanical properties of wood are much higher parallel to grain than at other angles. The grain or fiber direction of a wood product may not be parallel to the axis of the product for a number of reasons. First, the piece may not have been cut parallel, for a number of reasons, such as crook in the log. Second, the natural grain direction can be variable within a log. Many engineered wood products take advantage of cross lamination in manufacturing to impart desirable properties into the product. For example, the cross laminating of plies in plywood or strands in OSB provides for desirable stabilization of shrinkage and expansion in the two primary panel directions as well as desirable cross-panel strength properties.

The relationship between grain angle and properties has been empirically described by the Hankinson equation[4]:

$$N = \frac{PQ}{P \sin^n\theta + Q \cos^n\theta} \tag{1.5}$$

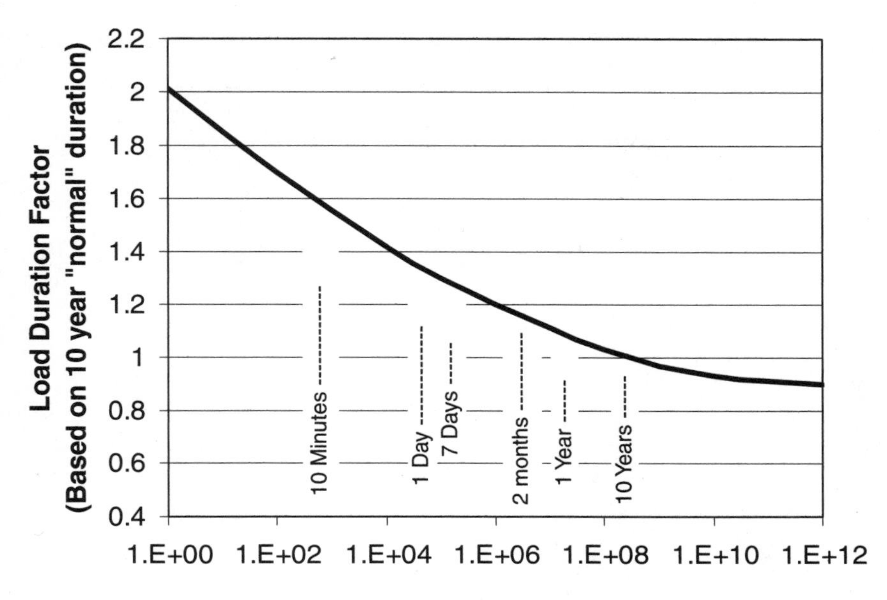

FIGURE 1.6 Relationship between load duration and wood strength.

TABLE 1.19 Load Duration Adjustments

Load duration	C_D	Typical design loads
Permanent	0.9	Dead load
Ten years	1.0	Occupancy live load
Two months	1.15	Snow load
Seven days	1.25	Construction load
Ten minutes	1.6	Wind/earthquake load
Impact	2.0	Impact load

where N = strength at angle θ from fiber direction
P = strength parallel to grain
Q = strength perpendicular to grain
n = empirically determined constant

Table 1.20 provides values for n and associated ratios of Q/P to be used in the Hankinson equation. Figure 1.7 presents the effect of grain angle using the Hankinson equation.

1.4.6 Effect of Chemicals on Wood Products

The behavior of wood in chemical environments depends upon factors such as the pH and the wood species. Chemical effects on the strength of wood occur by two

TABLE 1.20 Constants for Calculating Effect of Grain Angle on Strength of Wood

Property	n	Q/P
Tensile strength	1.5–2	0.04–0.07
Compression strength	2–2.5	0.03–0.40
Bending strength	1.5–2	0.04–0.10
Modulus of elasticity	2	0.04–0.12
Toughness	1.5–2	0.06–0.10

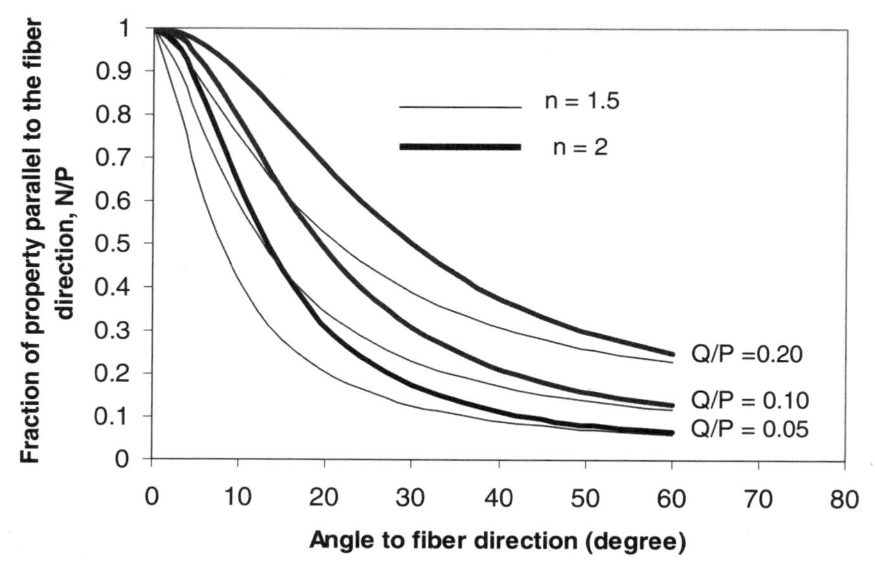

FIGURE 1.7 Effect of grain angle on wood strength.

general types of action. One involves swelling similar to moisture response and is almost completely reversible. The other type is nonreversible and involves the chemical interaction with the wood.

Chemicals such as alcohol and other polar solvents swell dry wood and cause a strength reduction proportional to the volumetric change due to swelling. Non-polar liquids such as petroleum hydrocarbons, often used as carriers for wood preservatives, have negligible swelling effect and therefore no effect on strength.

The second type of chemical interaction is what causes permanent changes and is caused by acids, alkalis, and salts. The chemical attack reduces the strength in proportion to the reduction in the cross-sectional area of the wood member. For purposes of this discussion, chemicals and their effect on wood can be grouped into classes as follows:

- *Inorganic acids:* The rate of strength loss due to attack on the hemicellulose and cellulose polymers depends upon the pH and temperature.
- *Organic acids:* These acids, such as formic, acetic, propionic, and lactic acid, do not hydrolyze wood rapidly because the pH of these acids in solution is about the same as wood (around 3 to 6).
- *Alkalis:* Alkaline chemicals such as sodium, calcium, and magnesium hydroxide will swell wood and react with hemicelluloses and cellulose and will dissolve the lignin. Wood is seldom used in contact with these solutions, and a pH greater than 11 at room temperature will lead to rapid strength loss. Alkalis at pH greater then 9.5 will lead to strength loss over time or at elevated temperatures.
- *Salts:* As with acids and alkalis, the effect of salts can usually be predicted based upon the pH and temperature. The acid salts can be considered as weak acids and will not have a rapid effect on wood unless the temperature is high. Most neutral salts will not react with wood. Alkaline salts are harmful to wood and behave similar to alkalis.

Because wood can withstand mild acid conditions, it has an important advantage in structures exposed to corrosive environments. Wood is much more resistant to degradation than mild steel and cast iron under these conditions.

1.5 DURABILITY OF ENGINEERED WOOD PRODUCTS

Wood construction has a long history of successful use. There are numerous examples of structures that are centuries old. However, like all other building materials, wood is subject to deterioration by natural elements. Wood is subject to decay from fungi and to attack by certain insects.

Decay from wood-destroying fungi is the most costly form of damage to wood construction. As with other forms of organic material, wood is subject to fungal attack when the wood moisture content reaches or exceeds the fiber saturation point described in Section 1.3.2. For practical purposes, wood is at risk of decay when the moisture content exceeds approximately 20% for a prolonged period of time. When properly used in construction, wood moisture content will be below this threshold for fungal growth. Wood decay may occur as a result of improper design, installation, or maintenance of the structure. Chapter 12 provides a detailed discussion of design and installation procedures to ensure that the construction will not be susceptible to high levels of moisture and associated fungal decay.

Of the insects that may attack wood, termites are the most destructive. Figure 1.8 presents a map showing the risk of termite attack. Chapter 12 provides construction techniques to assure longevity of wood structures in termite-prone areas.

Some buildings will naturally involve high moisture conditions that are favorable to decay fungi or locations prone to termite attack. In these cases, the use of preservative-treated wood may be necessary. Preservative treatments involve the use of certain chemicals that are impregnated into the wood product by a treating company as a postmanufacturing process in accordance with standard developed by the American Wood-Preserver's Association. Chapter 9 discusses the use of treated wood for construction. Although the use of treated wood is the most practical way to design for decay-prone applications, there are some woods that provide natural resistance to decay fungi and insects. Table 1.21 lists the natural decay resistance of the heartwood of various species.

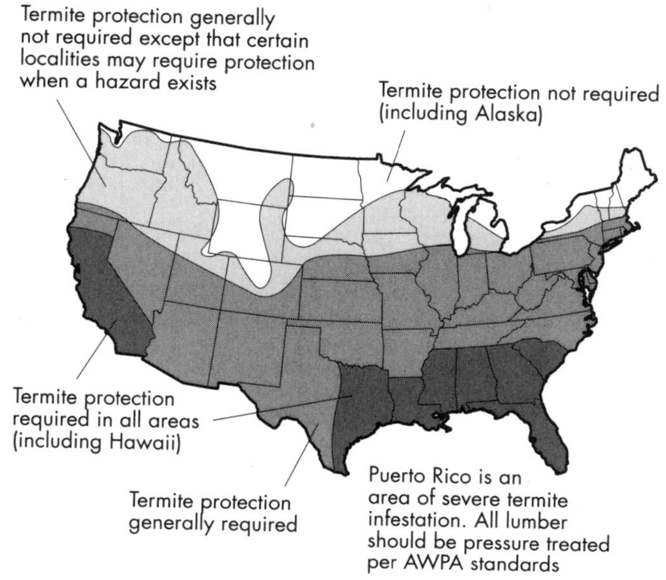

FIGURE 1.8 Termite risk.

TABLE 1.21 Grouping of Some Domestic Woods According to Average Heartwood Decay Resistance

Resistant or very resistant	Moderately resistant	Slightly or nonresistant
Cedar, eastern red	Douglas-fir	Alder, red
Cedar, northern white	Larch, western	Aspens
Cedar, yellow	Pine, eastern white	Basswood
Cherry, black	Pine, longleaf, old growth	Birches
Oak, white	Pine, slash, old growth	Cottonwood
Redwood	Redwood, young growth	Elms
Walnut, black		Firs, true
		Hemlocks
		Maples
		Pines (other than those listed elsewhere)
		Sweetgum
		Sycamore
		Willows
		Yellow poplar

1.6 SUMMARY

Engineered wood products combine the time-proven properties of wood with modern manufacturing techniques to optimize the performance for the building professional. The study of wood as an engineering material has led to a thorough understanding of the advantages and limitations of wood and engineered wood products. As a result, the North American construction industry is reliant on the vast volumes of engineered wood products produced annually.

1.7 REFERENCES

1. Bowyer, J. L., and R. L. Smith, 1998. *The Nature of Wood and Wood Products,* University of Minnesota, Forest Products Management Institute, Minneapolis, MN, 1998.
2. APA—The Engineered Wood Association and Southern Forest Products Association, *Market Outlook, Lumber, Structural Panels and Engineered Wood Products.* Tacoma, WA, 2000.
3. Canadian Wood Council, *Life Cycle Analysis for Residential Buildings—Case Study No. 5,* Ottawa, ON, 2000.
4. U.S. Forest Products Laboratory, *Wood Handbook—Wood as an Engineering Material,* Madison, WI, 1999.
5. *Equilibrium Moisture Content of Wood in Outdoor Locations in the United States and Worldwide,* USDA FPL RN-268, Forest Products Laboratory, Madison, WI.
6. Kamke, F., "Effects of wood-based panel characteristics on thermal conductivity," *Forest Products Journal,* Vol. 39, No. 5, 1987.
7. *Plywood in Hostile Environments,* APA Research Report 132, APA—The Engineered Wood Association, Tacoma, WA, 1975.
8. *National Design Specification for Wood Construction,* American Forest and Paper Association, Washington, DC, 1997.
9. Faherty, K. F., and T. G. Williamson, *Wood Engineering and Construction Handbook,* McGraw-Hill, New York, 1995.
10. American Society of Civil Engineers, *Evaluation, Maintenance and Upgrading of Wood Structures,* New York, 1982.
11. *APA Technical Note, Controlling Decay in Wood Construction,* Form No. R495, APA—The Engineered Wood Association, Tacoma, WA, 1998.
12. *APA Technical Note, Termite Protection for Wood Framed Construction,* Form No. K830, APA—The Engineered Wood Association, Tacoma, WA, 1987.

CHAPTER TWO
WOOD STRUCTURAL PANELS

William A. Baker, P.E.
Manager, Codes and Engineering, TSD

2.1 INTRODUCTION

The variety of wood structural panels available today was born out of necessity—a response to changes in wood resources, manufacturing, and construction trends. A wood structural panel, also referred to as a structural-use panel, is a panel product composed primarily of wood, which, in its commodity end use, is essentially dependent upon certain mechanical and/or physical properties for successful performance in service. Such a product is identified in a manner clearly conveying its intended end use. Today, wood structural panels include all-veneer plywood, composite panels containing a combination of veneer and wood-based material, and mat-formed panels such as oriented strand board.

In the early days of plywood manufacture, every mill worked with the same species and technology. Manufacturing techniques didn't vary much from mill to mill. To produce panels under prescriptive standards, a mill used wood of a certain species, peeled it to veneer of a prescribed thickness, then glued the veneers together in a prescribed manner using approved adhesives.

As technology changed, mills started using a broader range of species and different manufacturing techniques. With the development of U.S. Product Standard PS 1-66 for Softwood Plywood—Construction and Industrial,[1] three existing plywood standards were combined into one. And, for the first time, span ratings were incorporated into the standard. The span rating concept would later be used as a basis for the development of performance standards.

At the same time, there was growing concern over efficient use of forest resources. Working in cooperation with the U.S. Forest Service, the American Plywood Association (APA) (now *APA—The Engineered Wood Association*) tested panels manufactured with a core of compressed wood strands and traditional wood veneer on the face and back for use in structural applications. By using cores of wood strands, manufacturers were able to make more efficient use of the wood resource and use a broader range of species. Today, these panels are called composite panels or COM-PLY.®

In the course of the research on composite panels, performance standards were developed that led to a system of performance rated panels. Soon, manufacturers were making wood structural panels composed entirely of wood strands. Most current production of these panels, intended for use in structural applications, is referred to as oriented strand board (OSB).

2.1.1 Plywood

Plywood is the original wood structural panel. It is composed of thin sheets of veneer, or plies, arranged in layers to form a panel. Plywood always has an odd number of layers, each one consisting of one or more plies, or veneers.

In plywood manufacture, a log is turned on a lathe and a long knife blade peels the veneer. The veneers are clipped to a suitable width, dried, graded, and repaired if necessary. Next the veneers are laid up in cross-laminated layers. Sometimes a layer will consist of two or more plies with the grain running in the same direction, but there will always be an odd number of layers, with the face layers typically having the grain oriented parallel to the long dimension of the panel.

Adhesive is applied to the veneers that are to be laid up. Laid-up veneers are then put in a hot press, where they are bonded to form panels.

Wood is strongest along its grain, and shrinks and swells most across the grain. By alternation of grain direction between adjacent layers, strength and stiffness in both directions are maximized, and shrinking and swelling are minimized in each direction.

2.1.2 Oriented Strand Board

Panels manufactured of compressed wood wafers or strands have been marketed with such names as waferboard and oriented strand board. Today, virtually all mat-formed wood structural panels are manufactured with oriented strands or oriented wafers, and are commonly called oriented strand board (OSB).

OSB is composed of compressed strands arranged in layers (usually three to five) oriented at right angles to one another and bonded under heat and pressure with a waterproof and boil-proof adhesive. The orientation of layers achieves the same advantages of cross-laminated veneers in plywood. Since wood is stronger along the grain, the cross-lamination distributes wood's natural strength in both directions of the panel. Whether a panel is composed of strands or wafers, most manufacturers orient the material to achieve maximum performance.

Most OSB sheathing panels have a non-skid surface on one side for safety on the construction site, particularly when used as sheathing on pitched roofs.

2.1.3 Composite Panels

COM-PLY is an APA product name for composite panels that are manufactured by bonding layers of wood fibers between wood veneer. By combining reconstituted wood fibers with conventional veneer, COM-PLY panels allow for more efficient resource use while retaining the wood grain appearance on the panel face and back.

COM-PLY panels are manufactured in a three- or five-layer arrangement. A three-layer panel has a wood fiber core and veneer for face and back. The five-layer panel has a wood veneer crossband in the center and veneer on the face and back. When manufactured in a one-step pressing operation, voids in the veneers are filled automatically by the reconstituted wood particles or strands as the panel is pressed in the bonding process.

2.2 GROWTH OF THE INDUSTRY

The North American structural panel industry began in Portland, Oregon, when Portland Manufacturing Company, a small wooden box factory, experimented with laminated veneers for an exhibit at the 1905 World's Fair. Door manufacturers placed orders for the new product to make door panels, and others used it to make trunks and drawer bottoms. The laminated product became known as plywood, and by 1933 softwood plywood production had grown to 390 million ft^2 (all panel production is reported on an equivalent ⅜-in. thickness basis) (345,000 m^3) when the Douglas Fir Plywood Association was chartered. Shortly thereafter, the association began to establish uniform grading rules and helped manufacturers improve product quality. Through the World War II period, the uses for plywood were still mostly for industrial or manufactured products and military uses, including landing craft, ammunition boxes, and field tables. Plywood was promoted for residential construction in the 1940s and 1950s to meet the growing demand for housing. North American softwood plywood production reached 9.4 billion ft^2 (8,000,000 m^3) in 1960.

In 1964, plywood production expanded to the U.S. South and large plants were built. By the 1960s, sheathing for residential construction was clearly the largest plywood use, consuming just under 50% of production. The repair and remodeling and the nonresidential building markets were also growing, and plywood was much less dependent on industrial markets. By 1980, North American plywood production reached 18.5 billion ft^2 (16,000,000 m^3).

By 1980, waferboard and OSB were being manufactured according to a structural panel standard promulgated by the American Plywood Association. Because it could be made thinner and lighter than waferboard, OSB became the product of choice for construction sheathing. Both OSB and softwood plywood grew in what became known as the structural panel industry. By 1990, North American structural panel production totaled 30.9 billion ft^2 (27,000,000 m^3)—23.2 billion (20,000,000 m^3) of plywood and 7.7 billion (7,000,000 m^3) of OSB.

Throughout the 1990s, environmental pressures locked up millions of acres of productive forestland that plywood manufacturers had relied upon. In addition, the cost of producing OSB was less than that for plywood and the structural panel industry quickly shifted to building OSB mills to meet growing demand. By 1999, total structural panel industry production reached 40.2 billion ft^2 (35,000,000 m^3)—20.0 billion (17,000,000 m^3) of plywood and 20.2 billion (18,000,000 m^3) of OSB. Continued structural panel growth is expected in building construction markets as well as for industrial uses. The outlook is for about 42 billion ft^2 (37,000,000 m^3) of industry production by 2005, as shown in Fig. 2.1, which also shows the historic growth of the wood structural panel industry.

2.3 SELECTING PANELS

Wood structural panels are selected according to a number of key attributes. These attributes are identified in the qualified inspection and testing agency trademarks found on the panels. Examples of APA trademarks are shown in Fig. 2.2 and further explained in the paragraphs that follow.

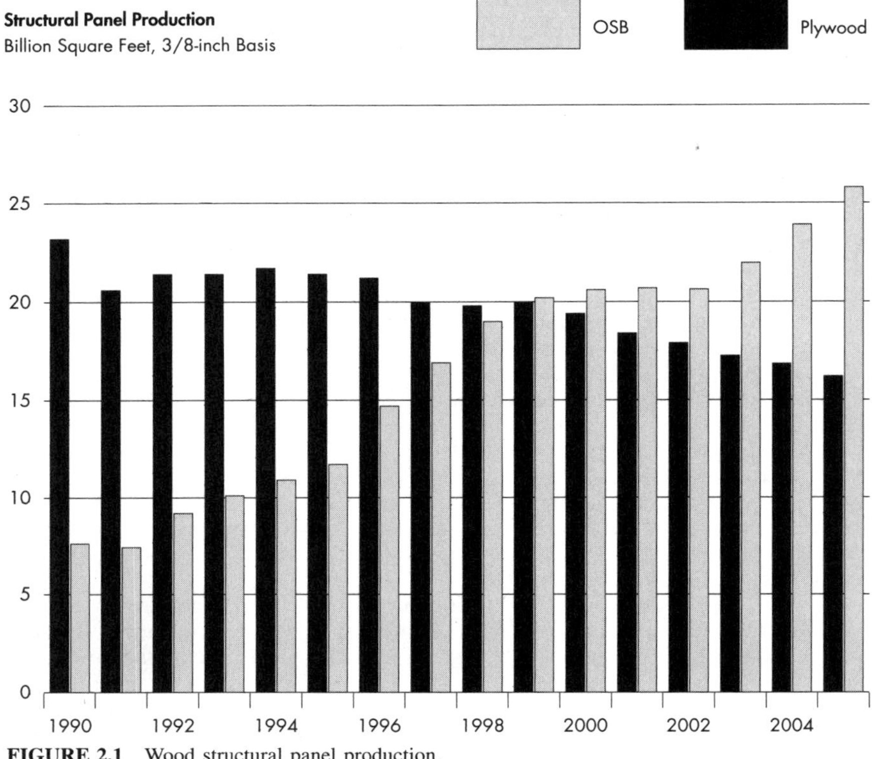

FIGURE 2.1 Wood structural panel production.

2.3.1 Standards

Manufacturing standards for wood structural panels are primarily of two types: prescriptive and performance based. Traditionally, plywood standards have been of the prescriptive type. The standard provides a recipe for panel layup, specifying the species of veneer and the number, thickness, and orientation of plies that are required to achieve panels of the desired nominal thickness and strength. A more recent development for wood structural panels is that of performance-based standards. Such standards are blind to actual panel construction, but do specify performance levels required for common end uses. Performance standards permitted the introduction of OSB into the construction market, since mat-formed panels (panels laid up in a mat rather than by stacking veneers) don't lend themselves to prescriptive layups.

Another distinction between standards is whether they are consensus-based or proprietary. Consensus-based standards are developed following a prescribed set of rules that provide for input and/or review by people of varying interests following one of several recognized procedures. Other standards are of a proprietary nature and may be developed by a single company or industry on a less formal basis. Sometimes proprietary standards become the forerunners of consensus standards.

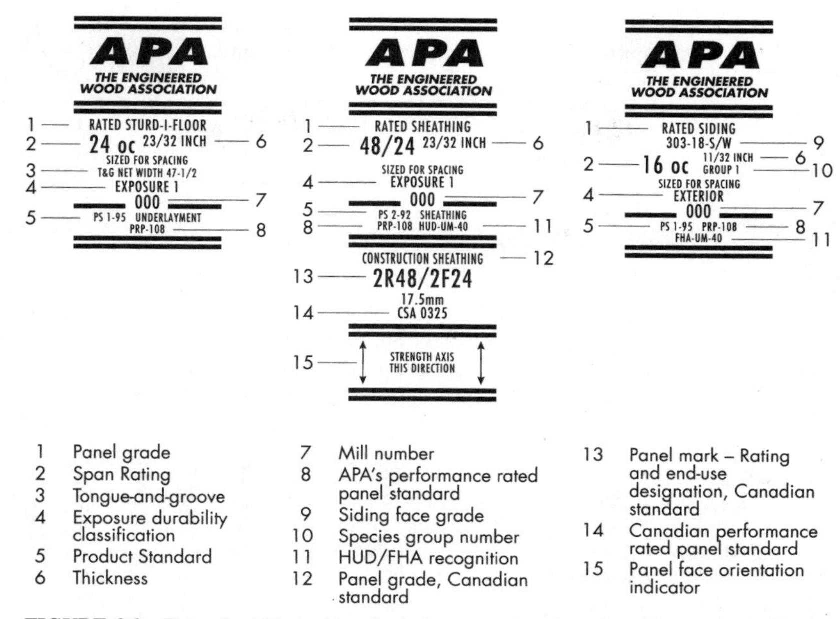

FIGURE 2.2 Example APA trademarks (other agency trademarks will contain similar information).

This was the case with APA's proprietary standard PRP-108, *Performance Standards and Policies for Structural-Use Panels,*[3] which became the foundation for the consensus-based Voluntary Product Standard PS 2, which was developed to achieve broader recognition of performance standards for wood structural panels.

Voluntary Product Standard PS 1. Voluntary Product Standard PS 1, *Construction and Industrial Plywood,*[1] is a consensus standard that originated in 1966 when it combined several preceding Commercial Standards, each covering a different species of plywood. It is often referred to as a prescriptive standard, although in the 1983 version performance-based provisions were added as an alternative method of qualifying sheathing and single-floor grades of plywood for span ratings. PS 1 continues to offer only prescriptive provisions for other panel grades, such as a variety of sanded plywood grades.

Voluntary Product Standard PS 2. Voluntary Product Standard PS 2,[2] *Performance Standard for Wood-Based Structural-Use Panels,* was promulgated in 1992 as the first consensus-based performance standard for wood structural panels. PS 2 is not limited to plywood, but is used extensively for all wood-based structural panel types. It covers sheathing and single-floor grades only, and includes performance criteria, a qualification policy, and test methods. As mentioned earlier, PS 2 is modeled after APA's performance standard, PRP-108, and most panels qualified under one also meet the other. Wood structural panels manufactured in conformance with PS 1 and PS 2 are recognized in all model building codes and most local codes in the United States.

Proprietary Standards. Two or three proprietary performance standards for wood structural panels are currently being used. The prototype standard, however, is APA PRP-108, *Performance Standards and Policies for Structural-Use Panels*. The APA standard includes performance provisions for sheathing and single-floor grades, but also includes provisions for siding. Although PRP-108, promulgated in 1980, is quite mature, it remains in effect to take advantage of technical developments more expeditiously than would be possible with the rather time-consuming consensus process required by PS 2.

2.3.2 Veneer

Wood veneer is at the heart of a plywood panel, but veneer is also an important component of a COM-PLY panel. Whether the product is plywood or COM-PLY, the veneer used is classified according to species group and grade requirements of PS 1.

Species Groups. Plywood can be manufactured from over 70 species of wood (see Table 2.1). These species are divided on the basis of strength and stiffness into five Groups under PS 1. Strongest species are in Group 1; the next strongest in Group 2, and so on. The Group number that appears in the trademark on panels—primarily sanded grades—is based on the species used for face and back veneers. Where face and back veneers are not from the same species Group, the higher Group number (the lower strength species) is used, except for sanded panels ⅜ in. (9.5 mm) thick or less and Decorative panels of any thickness. These are identified by face species because they are chosen primarily for appearance and used in applications where structural integrity is not critical. Sanded panels greater than ⅜ in. (9.5 mm) are identified by face species if C or D grade backs are at least ⅛ in. (3 mm) and are no more than one species Group number higher. Some species are used widely in plywood manufacture; others rarely. The specifier should check local availability if a particular species is desired.

Grades. Veneer grades define veneer appearance in terms of natural unrepaired growth characteristics and allowable number and size of repairs that may be made during manufacture (see Table 2.2). The highest quality commonly available veneer grade is A. The minimum grade of veneer permitted in Exterior plywood is C-grade. D-grade veneer is used in panels intended for interior use or applications protected from long-term exposure to weather.

2.3.3 Panel Grades

Wood structural panel grades are generally identified in terms of the veneer grade used on the face and back of the panel (e.g., A-B, B-C), or by a name suggesting the panel's intended end use (e.g., APA Rated Sheathing, APA Rated Sturd-I-Floor). See Table 2.3. Unsanded and touch-sanded panels, and panels with B-grade or better veneer on one side only, usually carry the trademark of a qualified inspection and testing agency (such as APA) on the panel back. Panels with both sides of B-grade or better veneer, or with special overlaid surfaces (such as High Density Overlay), usually carry the trademark on the panel edge.

TABLE 2.1 Classification of Species

Group 1	Group 2		Group 3	Group 4	Group 5
Apitong[a,b]	Cedar, Port Orford	Maple, black	Alder, red	Aspen	Basswood
Beech, American	Cypress	Mengkulang[a]	Birch, paper	bigtooth	Poplar, balsam
Birch	Douglas fir 2[c]	Meranti, red[a,d]	Cedar, Alaska	quaking	
sweet	Fir	Mersawa[a]	Fir, subalpine	Cativo	
yellow	balsam	Pine	Hemlock, eastern	Cedar	
Douglas fir 1[c]	California red	pond	Maple, bigleaf	incense	
Kapur[a]	grand	red	Pine	western red	
Keruing[a,b]	noble	Virginia	jack	Cottonwood	
Larch, western	Pacific silver	western white	lodgepole	eastern	
Maple, sugar	white	Spruce	ponderosa	black (western poplar)	
Pine	Hemlock, western	black	spruce	Pine	
Caribbean	Lauan	red	Redwood	Eastern white	
Ocote	Almon	Sitka	Spruce	Sugar	
Pine, southern	Bagtikan	Sweetgum	Englemann		
loblolly	Mayapis	Tamarack	white		
longleaf	Red Lauan	Yellow-poplar			
shortleaf	Tangile				
slash	White Lauan				
Tanoak					

[a] Each of these names represents a trade group of woods consisting of a number of closely related species.

[b] Species from the genus Dipterocarpus marketed collectively: apitong if originating in the Philippines, keruing if originating in Malaysia or Indonesia.

[c] Douglas fir from trees grown in the states of Washington, Oregon, California, Idaho, Montana, and Wyoming and the Canadian provinces of Alberta and British Columbia shall be classed as Douglas fir No. 1. Douglas fir from trees grown in the states of Nevada, Utah, Colorado, Arizona, and New Mexico shall be classed as Douglas fir No. 2.

[d] Red meranti shall be limited to species having a specific gravity of 0.41 or more based on green volume and oven-dry weight.

TABLE 2.2 Veneer Grades

Grade	Description
A	Smooth, paintable. Not more than 18 neatly made repairs, boat, sled, or router type, and parallel to grain, permitted. Wood or synthetic repairs permitted. May be used for natural finish in less demanding applications.
B	Solid surface. Shims, sled or router repairs, and tight knots to 1 in. across grain permitted. Wood or synthetic repairs permitted. Some minor splits permitted.
C Plugged	Improved C veneer with splits limited to ⅛ in. width and knotholes or other open defects limited to ¼ × ½ in. Wood or synthetic repairs permitted. Admits some broken grain.
C	Tight knots to 1½ in. Knotholes to 1 in. across grain and some to 1½ in. if total width of knots and knotholes is within specified limits. Synthetic or wood repairs. Discoloration and sanding defects that do not impair strength permitted. Limited splits allowed. Stitching permitted.
D	Knots and knotholes to 2½ in. width across grain and ½ in. larger within specified limits. Limited splits are permitted. Stitching permitted. Limited to exposure 1 or interior panels.

Note: 1 in. = 25.4 mm.

Unsanded. Sheathing panels are unsanded since a smooth surface is not a requirement of their intended end use for subfloor, roof, and wall applications. Sheathing panels are classified by span ratings, which identify the maximum recommended support spacings for specific end uses. Design capacities provided in Section 2.6.4 are on the basis of span ratings.

Structural I sheathing panels meet the requirements of sheathing grades as well as enhanced requirements associated with use in panelized roof systems, diaphragms, and shear walls (e.g., increased cross-panel strength and stiffness and racking shear resistance).

Touch Sanded. Underlayment, Single Floor, C-D Plugged, and C-C Plugged grades require only touch sanding for sizing to make the panel thickness more uniform. Panels rated for single-floor (combination subfloor-underlayment) applications are usually manufactured with tongue-and-groove (T&G) edge profiles and are classified by span ratings. Panel span ratings identify the maximum recommended support spacings for floors. Design capacities provided in Section 2.6.4 are on the basis of span ratings. Other thinner panels intended for separate underlayment applications (Underlayment or C-C Plugged) are identified with a species Group number but no span rating.

Sanded. Plywood panels with B-grade or better veneer faces are always sanded smooth in manufacture to fulfill the requirements of their intended end use—applications such as cabinets, shelving, furniture, and built-ins. Sanded grades are classed according to nominal thickness and the species group of the faces, and design capacities provided in Section 2.6.4 are on that basis.

Overlaid. High Density Overlay (HDO) and Medium Density Overlay (MDO) plywood may or may not have sanded faces, depending on whether the overlay is applied at the same time the panel is pressed (one-step) or after the panel is pressed (two-step). For purposes of assigning design capacities provided in Section 2.6.4, HDO and MDO panels are assumed to be sanded (two-step).

TABLE 2.3 Guide to Panel Use

Panel grade	Description and use	Common nominal thickness	Panel construction		
			OSB	COM-PLY	Plywood
APA Rated Sheathing EXP 1	Unsanded sheathing grade for wall, roof, subflooring, and industrial applications such as pallets and for engineering design with proper capacities.	5/16, 3/8 15/32, 1/2 19/32, 5/8 23/32, 3/4	Yes	Yes	Yes
APA Structural I Rated Sheathing EXP 1	Panel grades to use where shear and cross-panel strength properties are of maximum importance.	19/32, 5/8 23/32, 3/4	Yes	Yes	Yes
APA Rated Sturd-I-Floor EXP 1	Combination subfloor-underlayment. Provides smooth surface for application of carpet and pad. Possesses high concentrated and impact load resistance during construction and occupancy. Touch-sanded. Available with tongue-and-groove edges.	19/32, 5/8 23/32, 3/4 7/8, 1 1 3/32, 1 1/8	Yes	Yes	Yes
APA Underlayment EXP 1	For underlayment under carpet and pad. Touch-sanded. Available with tongue-and-groove edges.	1/4 11/32, 3/8 15/32, 1/2 19/32, 5/8 23/32, 3/4	No	No	Yes

TABLE 2.3 Guide to Panel Use (*Continued*)

Panel grade	Description and use	Common nominal thickness	Panel construction		
			OSB	COM-PLY	Plywood
APA C-C Plugged EXT	For underlayment, refrigerated or controlled atmosphere storage rooms, open soffits, and other similar applications where continuous or severe moisture may be present. Touch-sanded. Available with tongue-and-groove edges.	½ 19/32, 5/8 23/32, ¾	No	No	Yes
APA sanded grades EXP 1 or EXT	Generally applied where a high quality surface is required. Includes APA A-A, A-B, A-C, A-D, B-B, B-C, and B-D grades	¼, 11/32, 3/8 15/32, ½ 19/32, 5/8 23/32, ¾	No	No	Yes
APA Marine EXT	Superior Exterior plywood made only with Douglas fir or western larch. Special solid-core construction. Available with MDO or HDO face. Ideal for boat hull construction.	¼, 11/32, 3/8 15/32, ½ 19/32, 5/8 23/32, ¾	No	No	Yes

Note: 1 in. = 25.4 mm.

Rough Sawn. Plywood panels with rough-sawn faces are a special case applicable mostly to siding grades. Panel grades with rough-sawn faces are decorative and are usually not associated with engineered design, although racking shear resistance values are usually provided in the building codes.

2.3.4 Bond Classifications

Wood structural panels may be produced in four bond classifications: Exterior, Exposure 1, Exposure 2, and Interior. The bond classification relates to adhesive bond and thus to structural integrity of the panel. By far the predominant bond classifications are Exposure 1 and Exterior. Therefore, design capacities provided herein are on that basis.

Bond classification relates to moisture resistance of the glue bond and *does not* relate to fungal decay resistance of the panel. Fungal decay of wood products may occur when the moisture content exceeds approximately 20% for an extended period. Prevention of fungal decay is a function of proper design to prevent prolonged exposure to moisture, of material specification, of construction, and of maintenance of the structure.

Aesthetic (nonstructural) attributes of panels may be compromised to some degree by exposure to weather. Panel surfaces may become uneven and irregular under prolonged moisture exposure. Panels should be allowed to dry, and panel joints and surfaces may need to be sanded before some finish materials are applied.

Exterior. Exterior panels have a fully waterproof and boil-proof bond and are designed for applications subject to long-term exposure to the weather or moisture.

Exposure 1. Exposure 1 panels have a fully waterproof and boil-proof bond and are designed for applications where temporary exposure to weather due to construction delays or high humidity or other conditions of similar severity may be expected. Exposure 1 panels are made with the same exterior adhesives used in exterior panels. However, because other compositional factors may affect bond performance, only Exterior panels should be used for long-term exposure to the weather. Exposure 1 panels may, however, be used where exposure to the outdoors is on the underside only, such as at roof overhangs. Appearance characteristics of the panel grade should be considered.

C-D Exposure 1 plywood, sometimes called CDX in the trade, is occasionally mistaken for an Exterior panel and erroneously used in applications for which it does not possess the required resistance to weather. CDX should only be used for applications as outlined above.

Other Classifications. Although seldom produced, panels identified as Exposure 2 are intended for protected construction applications where potential for conditions of high humidity exists.

Panels identified as Interior and that lack further glueline information in their trademarks are manufactured with interior glue and are intended for interior applications only. Panels classed Interior were commonplace prior to the 1970s, but are not commonly produced today.

2.3.5 Span Ratings

Sheathing, Single Floor, and Siding grades carry numbers in their trademarks called span ratings. These denote the maximum recommended center-to-center spacing of supports, in inches, over which the panels should be placed in construction applications. Except for Siding panels, the span rating applies when the long panel dimension or strength axis is across supports, unless the strength axis is otherwise identified. The span rating of Siding panels applies when installed vertically.

Sheathing. The span rating on Sheathing grade panels appears as two numbers separated by a slash, such as 32/16 or 48/24. The left-hand number denotes the maximum recommended spacing of supports when the panel is used for roof sheathing with the long dimension or strength axis of the panel across three or more supports (two spans). The right-hand number indicates the maximum recommended spacing of supports when the panel is used for subflooring with the long dimension or strength axis of the panel across three or more supports. A panel marked 32/16, for example, may be used for roof sheathing over supports up to 32 in. (800 mm) on center (o.c.) or for subflooring over supports up to 16 in. (400 mm) on center.

Certain of the roof sheathing maximum spans are dependent upon panel edge support. See Section 2.4.3.

Sheathing panels rated for use only as wall sheathing are usually identified as either Wall-24 or Wall-16. The numerical index (24 or 16) corresponds to the maximum wall stud spacing. Wall sheathing panels are performance tested with the secondary axis (usually the short dimension of panel) spanning across supports, or studs. For this reason, wall sheathing panels may be applied with either the strength axis or secondary axis across supports.

Single Floor. The span rating on Single Floor grade panels appears as a single number. Single Floor panels are designed specifically for Single Floor (combined subfloor-underlayment) applications under carpet and pad and are manufactured with span ratings of 16, 20, 24, 32, and 48 oc. The span ratings for Single Floor panels, like those Sheathing grade, are based on application of the panel with the long dimension or strength axis across three or more supports.

Siding. Siding is available with span ratings of 16 oc and 24 oc. Span-rated panels and lap siding may be used direct to studs or over nonstructural wall sheathing (single-wall construction) or over nailable panel or lumber sheathing (double-wall construction). Panels and lap siding with a span rating of 16 oc may be applied direct to studs spaced 16 in. (400 mm) on center. Panels and lap siding bearing a span rating of 24 oc may be used direct to studs 24 in. (600 mm) on center. All Siding panels may be applied horizontally direct to studs 16 or 24 in. (400 or 600 mm) on center, provided horizontal joints are blocked. When used over nailable structural sheathing, the span rating of Siding panels refers to the maximum recommended spacing of vertical rows of nails rather than to stud spacing.

2.4 CONSTRUCTION APPLICATIONS

Building code provisions and APA installation recommendations for wood structural panels in construction applications are prescriptive in nature. The basis for many of the recommendations is the long history of satisfactory performance, but application testing has had the effect of periodically confirming many of these long-standing recommendations. The load-span tables in the codes and in this section are based directly on minimum test criteria of the performance standards for structural-use panels. Since in order to qualify for trademarks, wood structural panels must meet or exceed the specified criteria, these load-span tables tend to be conservative. For those occasional cases where sheathing-type applications must be engineered, or for other engineered panel applications, design capacities are given in Section 2.6.4.

2.4.1 Floors

Wood structural panel floors are typically installed with panel strength axis (usually the long panel dimension) across supports. Although this is not necessarily true for floors of manufactured homes, there are generally no provisions in building codes for installation of floors with panel strength axis along supports.

Single Floor. Single Floor grade is a span-rated product designed specifically for use in single-layer floor construction beneath carpet and pad. It is manufactured in conformance with PS 2[2] or PS 1[1] (where it is called Underlayment grade or C-C Plugged Exterior grade) or proprietary performance standards such as APA's PRP-108[3] (where it is called Rated Sturd-I-Floor). Panels are manufactured with span ratings of 16, 20, 24, 32, and 48 oc. These span ratings assume use of the panel continuous over two or more spans with the long dimension or strength axis across supports. The span rating applies when the long panel dimension is across supports unless the strength axis is otherwise identified.

Many builders glue-nail Single Floor panels, though panels may also be nailed only. Application recommendations for both methods are given in Table 2.4. (See below, APA Glued Floor System, for more detailed gluing recommendations.) Smooth panel faces and tongue-and-groove (T&G) edges should be protected from damage prior to and during application. Recommended live loads are given in Table 2.5.

Although Single Floor grade is suitable for direct application of carpet and pad, an additional thin layer of underlayment is recommended under tile, sheet flooring, or fully adhered carpet. This added layer restores a smooth surface over panels that may have been scuffed or roughened during construction, or over panels that may not have received a sufficiently sanded surface. Glued T&G edges are recommended under thin floor coverings to assure snug joints.

If the floor has become wet during construction, it should be allowed to dry before application of finish floor, including carpet, underlayment, hardwood flooring, and ceramic tile. After it is dry, the floor should be checked for flatness, especially at joints.

When floor members are dry, fasteners should be flush with or below the surface of the Single Floor panels just prior to installation of thin floor coverings. Fasteners should be set if green framing will present nail-popping problems upon drying. Do not fill nail holes. Fill and thoroughly sand edge joints (this step may not be necessary under some carpet and structural flooring products—check recommendations of flooring manufacturer). Fill any other damaged or open areas, such as splits, and sand all surface roughness.

APA Glued Floor System. The APA Glued Floor System is based on thoroughly tested gluing techniques and field-applied construction adhesives that firmly and permanently secure a layer of wood structural panels to wood joists. The glue bond

TABLE 2.4 Single-Floor Grade[a]

Span rating (maximum joist spacing) (in.)	Panel thickness[b] (in.)	Fastening: glue-nailed[c]			Fastening: nailed only		
			Maximum spacing (in.)			Maximum spacing (in.)	
		Nail size and type	Supported panel edges[g]	Intermediate supports	Nail size and type	Supported panel edges[g]	Intermediate supports
16	19/32, 5/8	6d ring- or screw-shank[d]	12	12	6d ring- or screw-shank	6	12
20	19/32, 5/8	6d ring- or screw-shank[d]	12	12	6d ring- or screw-shank	6	12
24	23/32, 3/4	6d ring- or screw-shank[d]	12	12	6d ring- or screw-shank	6	12
	7/8	8d ring- or screw-shank[d]	6	12	8d ring- or screw-shank	6	12
32	7/8	8d ring- or screw-shank[d]	6	12	8d ring- or screw-shank	6	12
48	1 3/32, 1 1/8	8d ring- or screw-shank[e]	6	[f]	8d ring- or screw-shank[e]	6	[f]

Note: 1 in. = 25.4 mm.

[a] Special conditions may impose heavy traffic and concentrated loads that require construction in excess of the minimums shown.

[b] Panels in a given thickness may be manufactured in more than one span rating. Panels with a span rating greater than the actual joist spacing may be substituted for panels of the same thickness with a span rating matching the actual joist spacing. For example, 19/32 in. thick Single Floor 20 oc may be substituted for 19/32 in. thick Single Floor 16 oc over joists 16 in. on center.

[c] Use only adhesives conforming to APA Specification AFG-01 or ASTM D3498, applied in accordance with the manufacturer's recommendations. If OSB panels with sealed surfaces and edges are to be used, use only solvent-based glues; check with panel manufacturer.

[d] 8d common nails may be substituted if ring- or screw-shank nails are not available.

[e] 10d common nails may be substituted with 1 1/8 in. panels if supports are well seasoned.

[f] Space nails maximum 6 in. for 48 in. spans and 12 in. for 32 in. spans.

[g] Fasten panels 3/8 in. from panel edges.

TABLE 2.5 Recommended Uniform Floor Live Loads for Single Floor and Sheathing with Panel Strength Axis Perpendicular to Supports

Single Floor span rating	Sheathing span rating	Minimum panel thickness (in.)	Maximum span (in.)	Allowable live loads (psf)[a] Joist spacing (in.)						
				12	16	20	24	32	40	48
16 oc	24/16, 32/16	7/16[c]	16	185	100					
20 oc	40/20	19/32, 5/8	20	270	150	100				
24 oc	48/24	23/32, 3/4	24	430	240	160	100			
32 oc	60/32[b]	7/8	32		430	295	185	100		
48 oc		1 3/32, 1 1/8	48			460	290	160	100	55

Note: 1 in. = 25.4 mm; 1 psf = 47.88 N/m^2.
[a] 10 psf dead load assumed. Live load deflection limit is $\ell/360$.
[b] Check with supplier for availability.
[c] 7/16 in. is not a permitted thickness of Single Floor.

is so strong that floor and joists behave like integral T-beam units. Floor stiffness is increased appreciably over conventional construction, particularly when tongue-and-groove joints are glued. Gluing also helps eliminate squeaks, floor vibration, bounce, and nail-popping.

The system is normally built with span-rated Single Floor panels (Fig. 2.3), although double-layer floors are also applicable. In both cases, single-floor and subflooring panels should be installed continuous over two or more spans with the long dimension or strength axis across supports.

Tongue-and-groove panels are highly recommended for single-floor construction. Before each panel is placed, a line of glue is applied to the joists with a caulking gun. The panel T&G joint should also be glued, although less heavily to avoid squeeze-out. If square-edge panels are used, edges must be supported between joists with 2 × 4 in. (38 × 89 mm) blocking. Glue panels to blocking to minimize squeaks. The blocking is not required when structural finish flooring, such as wood strip flooring, is to be applied, or if a separate underlayment layer is installed.

Only adhesives conforming with Performance Specification AFG-01,[4] developed by APA, or with ASTM D3498,[5] are recommended for use with the glued floor system. A number of brands meeting this specification are available from building supply dealers. If OSB panels with sealed surfaces and edges are to be used, use only solvent-based glues; check with panel manufacturer. The specific application recommendations of the glue manufacturer should be followed.

Subfloor. In addition to single-floor (combination subfloor-underlayment) applications, there is, of course, the traditional double-layer system consisting of a separate layer each of subfloor and underlayment or other structural flooring. The subfloor component is shown in Fig. 2.4.

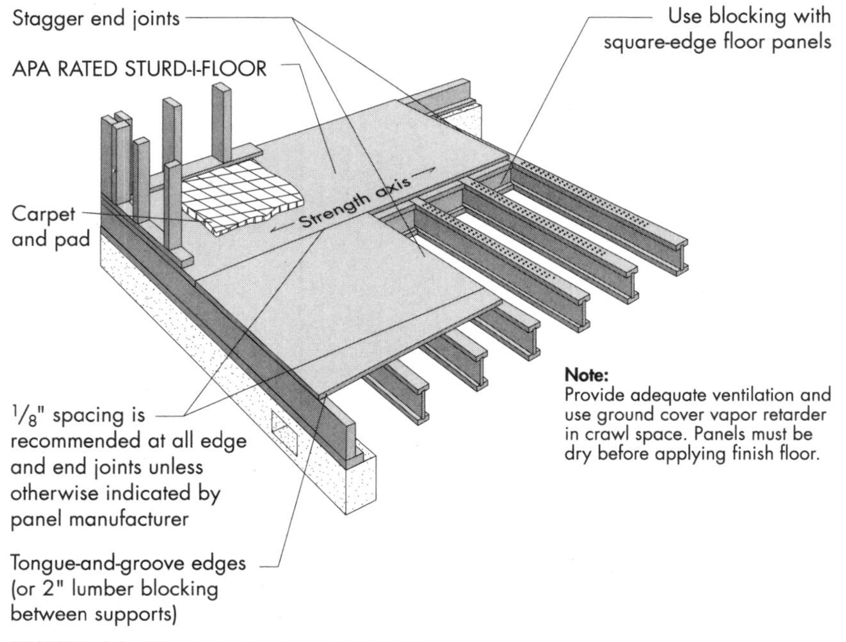

FIGURE 2.3 Wood structural panel single floor.

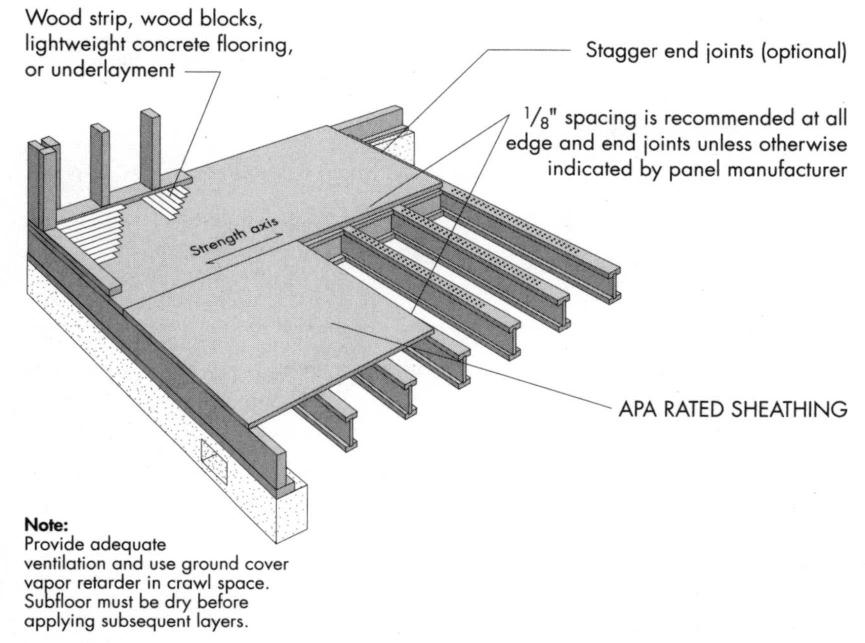

FIGURE 2.4 Wood structural panel subfloor.

The limiting factor in the design of floors is deflection under concentrated loads at panel edges. The span ratings in Table 2.6 apply to Sheathing grades only, and are the minimum recommended for the spans indicated. The spans assume panels continuous over two or more spans with the long dimension or strength axis across supports. Sheathing grade is manufactured in conformance with PS 2[2] or with PS 1[1] (where it is called C-D Exposure 1 or C-C Exterior grade) or proprietary performance standards such as APA's PRP-108[3] (where it is called Rated Sheathing). The span rating in the trademark applies when the long panel dimension is across supports, unless the strength axis is otherwise identified.

Recommended live loads are given in Table 2.5. Spans are limited to the values shown because of the possible effect of concentrated loads.

Nailing recommendations for subfloor panels are given in Table 2.6. Panel subflooring may also be glued for added stiffness and to reduce squeaks using nailing recommendations in Table 2.4.

Long edges should be tongue and groove or supported with blocking unless:

1. A separate underlayment layer is installed with its joints offset from those in the subfloor. The minimum thickness of underlayment should be ¼ in. (6.5 mm) for subfloors on spans up to 24 in. (600 mm), and 11⁄32 in. (8.5 mm) or thicker panels on spans longer than 24 in. (600 mm).
2. A minimum of 1½ in. (38 mm) of lightweight concrete is applied over the panels.
3. ¾ in. (19 mm) wood strip flooring is installed over the subfloor perpendicular to the unsupported edge.

TABLE 2.6 Panel Subflooring[a,b]

Panel span rating	Panel thickness (in.)	Maximum span (in.)	Nail size and type[e]	Maximum nail spacing (in.)	
				Supported panel edges[g]	Intermediate supports
24/16	7/16	16	6d common	6	12
32/16	15/32, 1/2	16	8d common[c]	6	12
40/20	19/32, 5/8	20[d]	8d common	6	12
48/24	23/32, 3/4	24	8d common	6	12
60/32[f]	7/8	32	8d common	6	12

Note: 1 in. = 25.4 mm.
[a] For subfloor recommendations under ceramic tile or under gypsum concrete, contact manufacturer of tile or floor topping.
[b] Single Floor grade may be substituted when the span rating is equal to or greater than tabulated maximum span.
[c] 6d common nail permitted if panel is ½ in. or thinner.
[d] Span may be 24 in. if a minimum 1½ in. of lightweight concrete is applied over panels.
[e] Other code-approved fasteners may be used.
[f] Check with supplier for availability.
[g] Fasteners shall be located ⅜ in. from panel edges.

If the floor has become wet during construction, it should be allowed to dry before application of finish floor, including underlayment, hardwood flooring, and ceramic tile. After it is dry, the floor should be checked for flatness, especially at joints.

In some nonresidential buildings, or in high-traffic common areas of multifamily buildings, the greater traffic and/or heavier concentrated loads may require construction in excess of the minimums given. Where joists are 16 in. (400 mm) on center, for example, panels with a span rating of 40/20 or 48/24 will give additional stiffness. For beams or joists 24 or 32 in. (600 or 800 mm) on center, 1⅛ in. (28.5 mm) panels provide additional stiffness.

Underlayment. In double-layer floor systems, smooth underlayment panels are applied as a second layer over the structural subfloor (see Fig. 2.5). Underlayment grades of plywood have a solid, touch-sanded surface for direct application of carpet and pad. Underlayment grade plywood panels are manufactured in conformance with PS 1.[1] For areas to be covered with resilient floor covering, underlayment panels with sanded face should be specified, or certain other grades as noted in Table 2.7. Special inner-ply construction of Underlayment resists dents and punctures from concentrated loads. Applied as recommended, plywood underlayment is also dimensionally stable and eliminates excessive swelling and subsequent buckling or humps around nails.

Always protect plywood underlayment against physical damage or water prior to application. However, allow panels to equalize to atmospheric conditions by standing individual panels on edge for several days before installation.

Install plywood underlayment immediately before laying the finish floor. For maximum stiffness, place face grain across supports. End and edge joints of underlayment panels should be offset by at least 2 in. from joints of subfloor panels, and should not coincide with framing below.

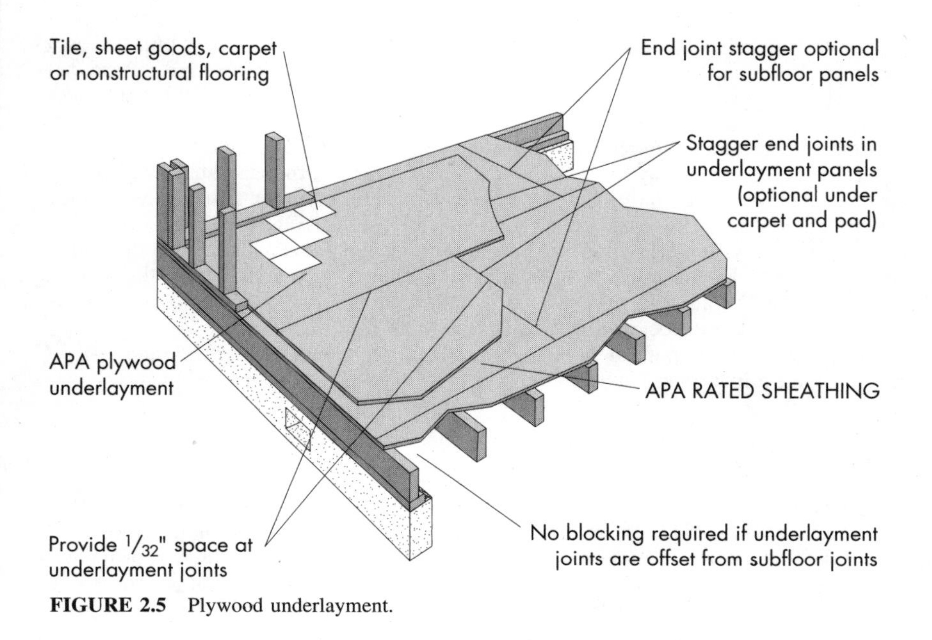

FIGURE 2.5 Plywood underlayment.

TABLE 2.7 Plywood Underlayment[c]

Plywood grades[a]	Application	Minimum plywood thickness (in.)	Fastener size and type	Maximum fastener spacing (in.)[a] Panel edges[d]	Intermediate
Underlayment, C-C Plugged EXT, or Single Floor (19⁄32 in. or thicker)	Over smooth subfloor	1⁄4	3d × 1¼-in. ring-shank nails[b] min. 12½ gage (0.099 in.) shank dia.	3	6 each way
	Over lumber subfloor or uneven surfaces	11⁄32		6	8 each way

Note: 1 in. = 25.4 mm.

[a] In areas to be finished with resilient floor coverings such as tile or sheet vinyl, or with fully adhered carpet, specify Underlayment, C-C Plugged or veneer-faced Single Floor with sanded face. Underlayment A-C, Underlayment B-C, Marine EXT or sanded plywood grades marked "plugged crossbands under face," "plugged crossbands (or core)," "plugged inner plies," or "meets underlayment requirements" may also be used under resilient floor coverings.

[b] Use 4d × 1½ in. ring-shank nails, minimum 12½ gage (0.099 in.) shank diameter, for underlayment panels 19⁄32 in. to ¾ in. thick.

[c] For underlayment recommendations under ceramic tile, contact manufacturer of tile.

[d] Fasten panels ⅜ in. from panel edges.

[e] Fasteners for five-ply plywood underlayment panels and for panels greater than ½ in. thick may be spaced 6 in. on center at edges and 12 in. each way intermediate.

Begin fastening at one edge next to a preceding panel. Ensuring that the panel is uniformly flat, continue by fully fastening toward opposite edge. Make sure fasteners are flush with or just slightly below surface of Underlayment just prior to installation of resilient floor coverings such as tile, or sheet vinyl (see Table 2.7 for underlayment recommendations for thin flooring products). Fill and thoroughly sand edge joints (this step may not be necessary under some carpet and structural flooring products—check recommendations of flooring manufacturer). Fill any other damaged or open areas, such as splits, and sand all surface roughness.

The plywood underlayment needed to bridge an uneven floor will depend on roughness and loads applied. Although a minimum 11⁄32 in. (8.5 mm) thickness is recommended, 1⁄4 in. (6.5 mm) plywood underlayment may also be acceptable over smooth subfloors, especially in remodeling work (see Table 2.7).

C-D Plugged is not an adequate substitute for underlayment grade or C-C Plugged Exterior grade since it does not ensure equivalent dent resistance.

2.4.2 Exterior Walls

Engineered wood walls consist of wood structural panel siding and/or sheathing over wall studs. Such wall systems provide shear and racking strength (more fully discussed in Chapter 8). In addition, wood structural panel sheathing provides a solid base for virtually every kind of exterior finish material, such as wood or vinyl siding, stucco, or brick veneer.

Siding. Neither PS 1[1] nor PS 2[2] includes a grade specifically for siding applications; however, such a grade is provided in APA's proprietary performance standard PRP-108.[3] Proprietary OSB and COM-PLY siding products, as well as plywood, have been qualified under this standard. Further, APA developed a specification in the 1960s for plywood, termed *APA 303 Siding Manufacturing Specification,*[6] which has become very well known over the years. The primary features of wood structural panel siding include an Exterior bond classification and a textured or otherwise treated surface for appearance. Wood structural panel siding products may be applied direct to studs (single-wall construction) or over nailable sheathing (double-wall construction).

Single Wall. Single-wall construction (called by APA the APA Sturd-I-Wall system) consists of Siding (panel or lap) applied direct to studs or over nonstructural sheathing such as fiberboard, gypsum, or rigid foam insulation. Nonstructural sheathing is defined as sheathing not recognized by building codes as meeting both bending and racking strength requirements.

In single-wall construction, a single layer of panel siding, since it is strong and resistant to racking, eliminates the cost of installing separate structural sheathing or diagonal wall bracing. Panel sidings are normally installed vertically, but may also be placed horizontally (long dimension across supports) if horizontal joints are blocked. Maximum stud spacings for both applications are given in Table 2.8. Gluing of siding to framing is not recommended.

See Fig. 2.6 for panel siding installation recommendations for the single-wall system.

All panel siding edges in single-wall construction should be backed with framing or blocking. Use nonstaining, noncorrosive nails as described in Table 2.8 to prevent staining the siding.

Where siding is to be applied at an angle, it should only be installed over nailable sheathing.

TABLE 2.8 Single-Wall Construction (Recommendations apply to siding direct to studs and over nonstructural sheathing)

	Siding description[a]	Nominal thickness (in.) or span rating	Max. stud. spacing (in.)		Nail size (use nonstaining box, siding or casing nails)[b,c]	Max. nail spacing[c] (in.)	
			Long dimension vertical	Long dimension horizontal		Panel edges[h]	Intermediate supports
Panel siding	MDO	11/32 and 3/8	16	24	6d for siding 1/2 in. thick or less; 8d for thicker siding	6[d]	12[f]
		15/32 and thicker	24	24			
	Siding	16 oc (including T1-11)	16	16[g]			
		24 oc	24	24			
Lap siding	Siding	16 oc	—	16	6d for siding 1/2 in. thick or less; 8d for thicker siding	16 along bottom edge	—
		24 oc	—	24		24 along bottom edge	—

Note: 1 in. = 25.4 mm. Galvanized fasteners may react under wet conditions with the natural extractives of some wood species and may cause staining if left unfinished. Such staining can be minimized if the siding is finished in accordance with APA recommendations, or if the roof overhang protects the siding from direct exposure to moisture and weathering.

[a] For veneered siding, such as APA 303 Siding, recommendations apply to all species groups.

[b] If panel siding is applied over foam insulation sheathing, use next regular nail size. If lap siding is installed over rigid foam insulation sheathing up to 1 in. thick, use 10d (3 in.) nails for 3/8 in. or 7/16 in. siding, 12d (3 1/4 in.) nails for 15/32 in. or 1/2 in. siding, and 16d (3 1/2 in.) nails for 19/32 in. or thicker siding. Use nonstaining box nails for siding installed over foam insulation sheathing.

[c] Hot-dipped or hot-tumbled galvanized steel nails are recommended for most siding applications. For best performance, stainless steel nails or aluminum nails should be considered. APA tests also show that electrically or mechanically galvanized steel nails appear satisfactory when plating meets or exceeds thickness requirements of ASTM A641 Class 2 coatings and is further protected by yellow chromate coating.

[d] For braced wall section with 11/32 in. or 3/8 in. panel siding applied horizontally over studs 24 in. o.c. space nails 3 in. o.c. along panel edges.

[e] Recommendations of siding manufacturer may vary.

[f] Where basic wind speed exceeds 80 mph, nails attaching siding to intermediate studs within 10% of the width of the narrow side from wall corners shall be spaced 6 in. o.c.

[g] Stud spacing may be 24 in. o.c. for veneer-faced siding panels.

[h] Fasteners shall be located 3/8 in. from panel edges.

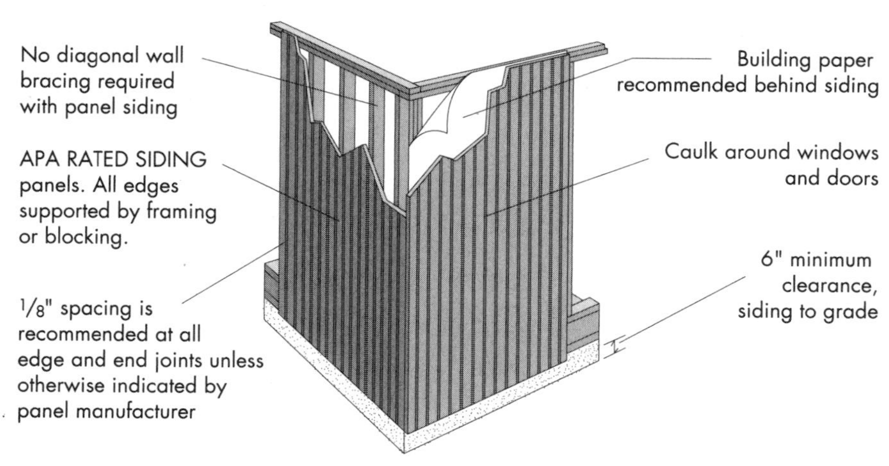

FIGURE 2.6 Wood structural panel siding.

Double Wall. Wood structural panel sheathing meets building code wall sheathing requirements for bending and racking strength without let-in corner bracing. Even when fiberboard or other nonstructural sheathing is used, Sheathing grade corner panels of the same thickness can eliminate let-in bracing. Installation recommendations are given in Fig. 2.7. Gluing of wall sheathing to framing is not recom-

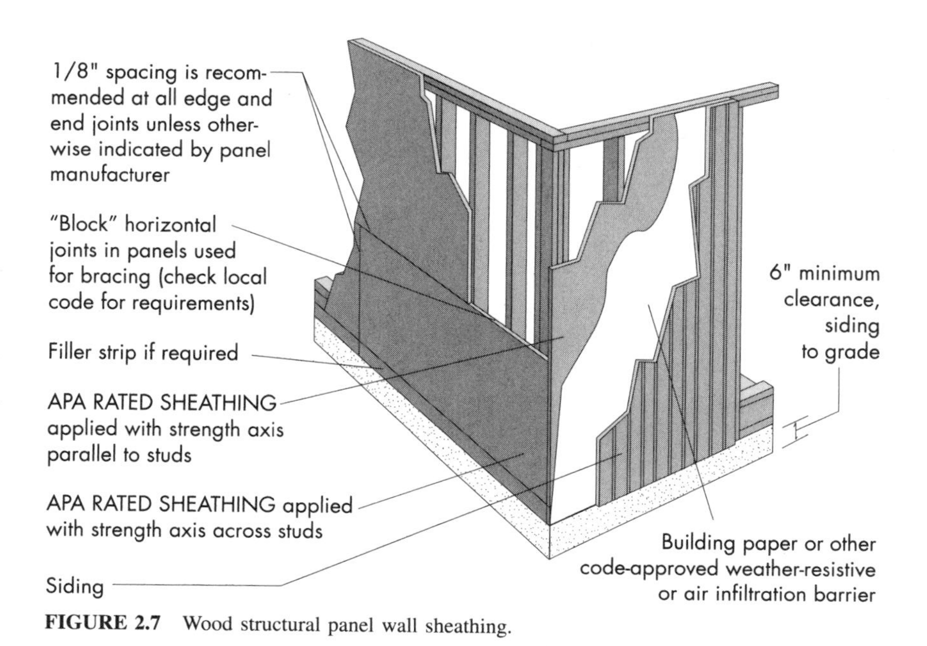

FIGURE 2.7 Wood structural panel wall sheathing.

mended, except when recommended by the adhesive manufacturer for wall sheathing that already has been permanently protected by siding.

Span and fastening recommendations for structural panel wall sheathing are given in Table 2.9. Increased nail schedules may be required for engineered shear walls (see Chapter 8). Recommended wall sheathing spans with brick veneer or masonry are the same as those for panel sheathing.

The recommendations in Table 2.10 for panel and lap siding apply to siding installed over nailable sheathing. Unless otherwise indicated in the local building code, nailable sheathing includes:

1. Nominal 1 in. (19 mm) boards with studs 16 or 24 in. (400 or 600 mm) o.c.
2. Wood structural panel sheathing with roof span rating of 24 in. (600 mm) or greater installed with strength axis either parallel or perpendicular to studs 16 or 24 in. (400 or 600 mm) o.c. (except three-ply plywood panels must be applied with strength axis across studs when studs are spaced 24 in. [600 mm] o.c). Check local building codes for blocking requirements between studs for braced or engineered shear wall segments, when wall sheathing is installed horizontally across studs.
3. Wood structural panel sheathing with roof span rating less than 24 in. (600 mm) installed with strength axis either parallel or perpendicular to studs 16 in. (400 mm) o.c. (except plywood panels ⅜ in. [9.5 mm] thick or less must be applied with strength axis across studs). Check local building codes for blocking requirements between studs for braced or engineered shear wall segments, when wall sheathing is installed horizontally across studs.

Lap siding joints, if staggered, and panel siding joints may occur away from studs when applied over nailable sheathing.

Greater stiffness is recommended for wall sheathing when stucco is to be applied. To increase stiffness, apply the long panel dimension or strength axis across studs. Blocking or a plywood cleat is recommended at horizontal joints. Blocking is required for braced wall sections or shear wall applications. For panel recommendations applied horizontally or vertically, see Table 2.11 and Fig. 2.8.

TABLE 2.9 Wall Sheathing[a] (Panels continuous over two or more spans)

			Maximum nail spacing (in.)	
Panel span rating	Maximum stud spacing (in.)	Nail size[b]	Supported panel edges[c]	Intermediate supports
12/0, 16/0, 20/0 or Wall-16 oc	16	6d for panels ½ in. thick or less; 8d for thicker panels	6	12
24/0, 24/16, 32/16 or Wall-24 oc	24			

Note: 1 in. = 25.4 mm.
[a] See requirements for nailable panel sheathing when exterior covering is to be nailed to sheathing.
[b] Common, smooth, annular, spiral-thread, or galvanized box.
[c] Fasteners shall be located ⅜ in. from panel edges.

TABLE 2.10 Siding over Nailable Sheathing **(For siding over types of nonstructural sheathing, see single-wall recommendations)**

	Siding description[a]	Nominal thickness (in.) or span rating	Max. spacing of vertical rows of nails (in.)		Nail size (use nonstaining box, siding or casing nails)[b,c]	Max. nail spacing[c] (in.)	
			Long dimension vertical	Long dimension horizontal		Panel edges[d]	Intermediate supports
Panel siding	MDO	11/32 & 3/8	16	24	6d for siding 1/2 in. thick or less; 8d for thicker siding	6	12
		15/32 and thicker	24	24			
	Siding	16 oc (including APA T1-11)	16	24			
		24 oc	24	24			
Lap siding	MDO	11/32 and thicker	—	—	6d for siding 1/2 in. thick or less; 8d for thicker siding	8 along bottom edge	—
	Siding	11/32 and thicker, or 16 oc or 24 oc	—	—			

Note: 1 in. = 25.4 mm. Galvanized fasteners may react under wet conditions with the natural extractives of some wood species and may cause staining if left unfinished. Such staining can be minimized if the siding is finished in accordance with APA recommendations, or if the roof overhang protects the siding from direct exposure to moisture and weathering.

[a] For veneered siding, such as APA 303 Siding, recommendations apply to all species groups.

[b] Hot-dipped or hot-tumbled galvanized steel nails are recommended for most siding applications. For best performance, stainless steel nails or aluminum nails should be considered. APA tests also show that electrically or mechanically galvanized steel nails appear satisfactory when plating meets or exceeds thickness requirements of ASTM A641 Class 2 coatings and is further protected by yellow chromate coating.

[c] Recommendations of siding manufacturer may vary.

[d] Fasten panels 3/8 in. from panel edges..

TABLE 2.11 Recommended Thickness and Span Rating for Panel Wall Sheathing for Stucco Exterior Finish

Stud spacing (in.)	Panel orientation[a]	Sheating[c] Minimum thickness (in.)	Minimum span rating
16	Horizontal[b]	3/8	24/0
	Vertical	15/32,[d] 1/2[d]	32/16
24	Horizontal[b]	7/16	24/16
	Vertical	19/32,[d] 5/8[d]	40/20

Note: 1 in. = 25.4 mm.

[a] Strength axis (long panel dimension) perpendicular to studs for horizontal application; or parallel to studs for vertical application.

[b] Blocking recommended between studs along horizontal panel joints.

[c] Recommendations apply to all-veneer plywood, oriented strand board (OSB), or composite (APA COM-PLY) panels except as noted.

[d] OSB or five-ply/five-layer plywood.

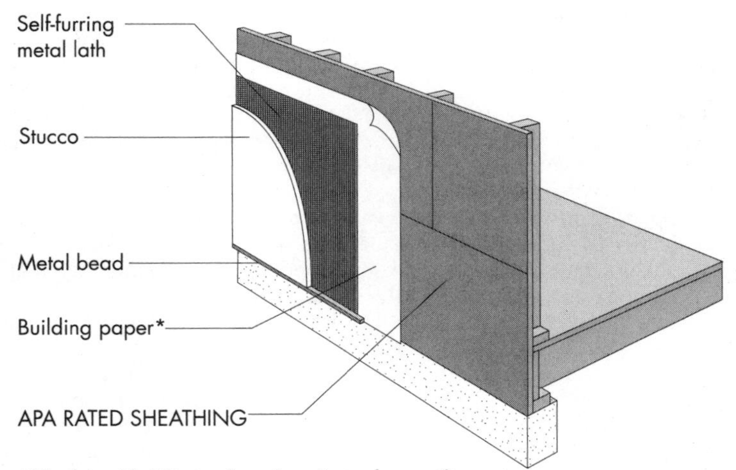

*Check local building code and applicator for specific requirements.
Note:
Uniform Building Code requires two layers of grade D paper for stucco over wood-based sheathing.

FIGURE 2.8 Stucco over wood structural panel wall sheathing.

2.4.3 Roofs

Wood structural panel roof sheathing is typically installed with panel strength axis (usually the long panel dimension) across supports. Relative to the sheathing, this is termed a conventional roof application. Some roof systems, however, are designed to apply the panel strength axis parallel to supports. Typical among these is the panelized roof system.

Gluing of roof sheathing to framing is not recommended, except when recommended by the adhesive manufacturer for roof sheathing that already has been permanently protected by roofing.

Conventional Framing. Recommendations for roof sheathing assume that panels are continuous over two or more spans with the long dimension or strength axis oriented across supports. Requirements for uniform load deflection limits are typically 1/180 of span under live load plus dead load and 1/240 under live load only. Special conditions, such as heavy concentrated loads, may require constructions in excess of these minimums, or allowable live loads may have to be decreased for dead loads greater than 10 psf (480 N/m^2), such as for some tile roofs. The span rating in the trademark applies when the long panel dimension is across supports unless the strength axis is otherwise identified. Allowable loads are given in Table 2.12.

Good performance of built-up, single-ply, or modified bitumen roofing applied on low-slope roofs requires a stiffer deck than does prepared roofing applied on pitched roofs. Although span-rated panels used as roof sheathing at maximum span are adequate structurally, an upgraded system is recommended for low slope roofs. Table 2.13 provides recommended maximum spans for low slope roof decks. Recommended live loads can be determined from Table 2.12, and minimum fastener requirements are given in Table 2.14. See Fig. 2.9 for installation recommendations. Increased nail schedules may be required in high-wind zones (see Table 2.17).

When support spacing exceeds the maximum length of an unsupported edge (see Table 2.12), provide adequate blocking, tongue-and-groove edges, or other edge support such as panel clips. Some types of panel clips, in addition to edge support, automatically ensure recommended panel spacing. When required, use one panel clip per span, except use two clips for 48 in. (1200 mm) or longer spans.

Panelized Framing. In panelized, or preframed, wood roof construction used in commercial applications, wood structural panels are attached to 2 × 4 or 2 × 6 (38 × 89 or 38 × 140 mm) dimension lumber subpurlins, or stiffeners, typically spaced 24 in. (600 mm) on center. The stiffeners are attached to secondary wood framing members, referred to as purlins, which are usually at 8 ft (2400 mm) on center to match the typical panel length. This work is done on the ground level and only one or two workers are needed to complete the assembly sequence on the roof. This minimizes the potential for falls and increases safety on the jobsite.

The entire preframed panelized unit is then lifted into position at the roof level using high-lift capacity forklifts. The purlins are attached to the primary glulam beams using preengineered metal hangers. The free edge of the wood decking for each panelized unit is nailed to the framing edge of the previously placed unit. Preframed panel ends attached to the main glulam beams complete the assembly. These preframed roof sections speed the erection process and add strength, dimensional stability, and high diaphragm capacity to the roof.

For the sheathing, unsanded 4 × 8 ft (1200 × 2400 mm) APA panels with stiffeners preframed at 16 or 24 in. (400 or 600 mm) on center (Fig. 2.10) are common. The strength axis of the panel typically runs parallel to supports. Stiffeners and roof purlins provide support for all panel edges. Minimum nailing requirements for preframed panels are the same as for conventionally applied roof sheathing.

In preframed panels 8 × 8 ft (2400 × 2400 mm) or larger, the long panel dimension may run either parallel or perpendicular to stiffeners spaced 16 or 24 in. (400 or 600 mm) on center. Placing the long dimension across supports may require edge support such as panel clips or cleats between stiffeners at midspan in accord-

TABLE 2.12 Recommended Roof Uniform Live Loads for Sheathing and Single Floor Panels with Strength Axis Perpendicular to Supports[d]

Panel span rating	Minimum panel thickness (in.)	Maximum span (in.)		Allowable live loads (psf)[c] — Spacing of supports center-to-center (in.)							
		With edge support[a]	Without edge support	12	16	20	24	32	40	48	60
Sheathing panels[c]											
12/0	5/16	12	12	30							
16/0	5/16	16	16	70	30						
20/0	5/16	20	20	120	50	30					
24/0	3/8	24	20[b]	190	100	60	30				
24/16	7/16	24	24	190	100	65	40				
32/16	15/32, 1/2	32	28	325	180	120	70	30			
40/20	19/32, 5/8	40	32	—	305	205	130	60	30		
48/24	23/32, 3/4	48	36	—	—	280	175	95	45	35	
60/32[f]	7/8	60	48	—	—	—	305	165	100	70	35
Single Floor panels[e]											
16 oc	19/32, 5/8	24	24	185	100	65	40				
20 oc	19/32, 5/8	32	32	270	150	100	60	30			
24 oc	23/32, 3/4	48	36	—	240	160	100	50	30	25	
32 oc	7/8	48	40	—	—	295	185	100	60	40	
48 oc	1 3/32, 1 1/8	60	48	—	—	—	290	160	100	65	40

Note: 1 in. = 25.4 mm; 1 psf = 47.88 N/m^2.

[a] Tongue-and-groove edges, panel edge clips (one midway between each support, except two equally spaced between supports 48 in. on center or greater), lumber blocking, or other.

[b] 24 in. for 15/32 in. and 1/2 in. panels.

[c] 10 psf dead load assumed.

[d] Applies to panels 24 in. or wider applied over two or more spans.

[e] Also applies to C-C Plugged grade plywood.

[f] Check with supplier for availability.

TABLE 2.13 Recommended Maximum Spans for APA Panel Roof Decks for Low-Slope Roofs[a] (Panel strength axis perpendicular to supports and continuous over two or more spans)

Grade	Minimum nominal panel thickness (in.)	Minimum span rating	Maximum span (in.)	Panel clips per span[b] (number)
Sheathing	15/32	32/16	24	1
	19/32	40/20	32	1
	23/32	48/24	48	2
	7/8	60/32[c]	60	2
Single Floor	19/32	20 oc	24	1
	23/32	24 oc	32	1
	7/8	32 oc	48	2

Note: 1 in. = 25.4 mm.
[a] Low-slope roofs are applicable to built-up, single-ply, and modified bitumen roofing systems. *For guaranteed or warranted roofs contact membrane manufacturer for acceptable deck.*
[b] Edge support may also be provided by tongue-and-groove edges or solid blocking.
[c] Check with supplier for availability.

TABLE 2.14 Recommended Minimum Fastening Schedule for Panel Roof Sheathing (Increased nail schedules may be required in high-wind zones)

Panel thickness[b] (in).	Nailing[c,d]		
		Maximum spacing (in.)	
	Size	Supported panel edges[e]	Intermediate
5/16–1	8d	6	12[a]
1 1/8	8d or 10d	6	12[a]

Note: 1 in. = 25.4 mm.
[a] For spans 48 in. or greater, space nails 6 in. at all supports.
[b] For stapling asphalt shingles to 5/16 in. and thicker panels, use staples with a 15/16 in. minimum crown width and a 1 in. leg length. Space according to shingle manufacturer's recommendations.
[c] Use common smooth or deformed shank nails with panels to 1 in. thick. For 1 1/8 in. panels, use 8d ring- or screw-shank or 10d common smooth-shank nails.
[d] Other code-approved fasteners may be used.
[e] Fasteners shall be located 3/8 in. from panel edges.

ance with Table 2.12. Recommendations in Table 2.15 are based on strength axis of the panel parallel to supports. Deflection limits are 1/180 of the span for total load and 1/240 of the span for live load only. See Table 2.16 for design information on stiffeners for preframed panels. Nailing requirements for preframed panels are the same as for roof sheathing.

Wind Uplift Considerations

Fastening. This section provides recommended nailing schedules for wood structural panel roof sheathing—plywood, COM-PLY, and OSB. These schedules were established to provide resistance to wind uplift pressure, with particular emphasis on high-wind exposures.

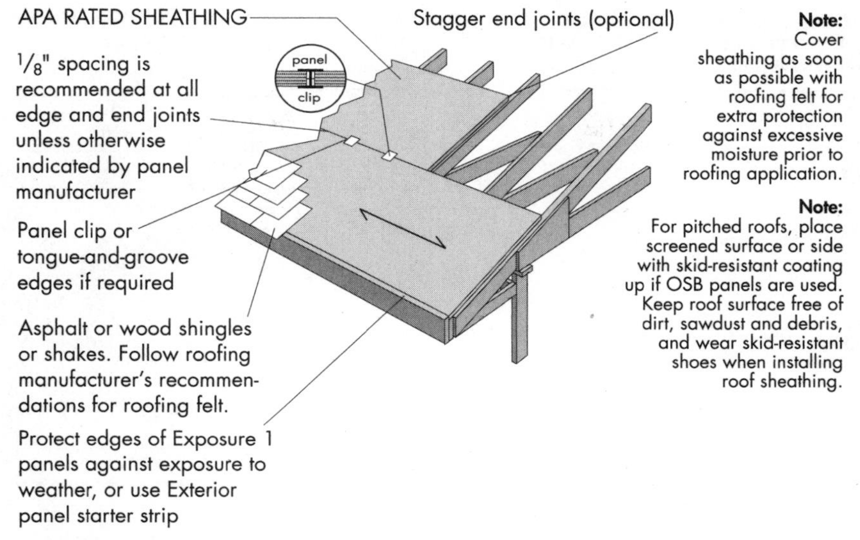

FIGURE 2.9 Wood structural panel roof sheathing.

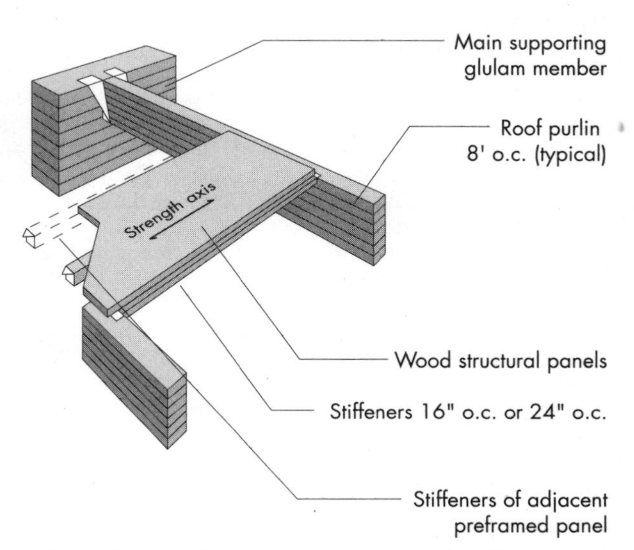

FIGURE 2.10 Preframed wood structural panel roof sheathing (4 × 8 ft panels with strength axis parallel to supports).

TABLE 2.15 Recommended Roof Loads (psf) for Sheathing Panels with Strength Axis Parallel to Supports[e,f] (OSB, composite and five-ply/five-layer plywood panels unless otherwise noted)

Panel grade	Thickness (in.)	Span rating	Max. span (in.)	Load at maximum span	
				Live	Total
Structural I Sheathing	7⁄16	24/0, 24/16	24[d]	20	30
	15⁄32	32/16	24	35[a]	45[a]
	½	32/16	24	40[a]	50[a]
	19⁄32, 5⁄8	40/20	24	70	80
	23⁄32, ¾	48/24	24	90	100
Sheathing	7⁄16[b]	24/0, 24/16	16	40	50
	15⁄32[b]	32/16	24[d]	20	25
	½[b]	24/0, 32/16	24[d]	25	30
	19⁄32	40/20	24	40[c]	50[c]
	5⁄8	32/16, 40/20	24	45[c]	55[c]
	23⁄32, ¾	40/20, 48/24	24	60[c]	65[c]

Note: 1 in. = 25.4 mm; 1 psf = 47.88 N/m^2.
[a] For four-ply plywood marked PS 1, reduce load by 15 psf.
[b] Composite panels must be 19⁄32 in. or thicker.
[c] For composite and four-ply plywood panels, reduce load by 15 psf.
[d] Solid blocking recommended at panel ends for 24 in. span.
[e] *For guaranteed or warranted roofs, contact membrane manufacturer for acceptable deck.*
[f] Provide edge support.

Recommendations were developed through computer analysis and verified by full-scale laboratory testing. Wet as well as dry specimens of plywood and OSB panels were tested for full-panel withdrawal under uniform pressure. The results of this testing were compared with wind loads calculated in accordance with the design provisions of ASCE 7-88, *Minimum Design Loads for Buildings and Other Structures*.[7]

The fastening schedules presented in Table 2.17 reflect the differences in wind uplift pressures that may be anticipated over various portions of roof systems. Higher pressures at eaves, corners, ridges, and gable ends require more restrictive schedules than at interior portions of the roof system. For this reason, fastening schedules may have different requirements for each of the three roof fastening zones illustrated in Fig. 2.11.

Three fastening schedules are provided in Table 2.17 for roof applications with framing spaced at 24 in. (600 mm) on center or less. These schedules assume the use of wood structural panels 5⁄8 in. (16 mm) thick or less and are appropriate for buildings with a mean roof height of up to 35 ft (10.5 m). All fasteners listed in the tables are minimum 8d common nails with smooth or ring shanks, depending on the basic wind speed and fastener location. All recommendations are based on the use of full-length nails meeting the requirements of Federal Specification FF-N-105B[8] or ASTM F1667.[9]

The three schedules provided give nailing recommendations for basic uplift, intermediate uplift, and high-wind uplift conditions as follows:

Basic uplift: The basic uplift fastening schedule is appropriate for buildings located in areas where the basic wind speed, as determined by your local build-

TABLE 2.16 Stiffener Load-Span Tables for Preframed Panel Roof Decks

Douglas fir-larch		Allowable roof live load (psf)[a]											
		Select Structural			No. 1 and Btr			No. 1			No. 2		
Center-to center purlin spacing[b] (ft)	Stiffener size and spacing (in.)	Defl.[c]	Strength[d] 1.15	Strength[d] 1.25	Defl.[c]	Strength[d] 1.15	Strength[d] 1.25	Defl.[c]	Strength[d] 1.15	Strength[d] 1.25	Defl.[c]	Strength[d] 1.15	Strength[d] 1.25
	2×4@16	37	67	73	35	51	57	33	41	46	31	36	40
	2×4@24	23	41	46	21	31	34	19	24	27	18	21	23
8	2×6@16	144	154	168	136	121	133	129	99	109	121	88	97
	2×6@24	96	99	109	91	78	85	86	63	69	81	56	61
	2×6@32	72	61	68	68	47	52	64	38	42	61	33	37

Southern pine		Allowable roof life load (psf)[a]											
		Select Structural			No. 1 Dense			No. 1			No. 2		
Center-to center purlin spacing[b] (ft)	Stiffener size and spacing (in.)	Defl.[c]	Strength[d] 1.15	Strength[d] 1.25	Defl.[c]	Strength[d] 1.15	Strength[d] 1.25	Defl.[c]	Strength[d] 1.15	Strength[d] 1.25	Defl.[c]	Strength[d] 1.15	Strength[d] 1.25
	2×4@16	35	87	96	35	58	64	33	53	59	31	41	46
	2×4@24	21	55	60	21	35	39	19	32	36	18	24	27
8	2×6@16	136	205	223	136	137	150	129	129	141	121	95	104
	2×6@24	91	133	146	91	88	97	86	83	91	81	60	66
	2×6@32	68	83	91	68	54	60	64	50	56	61	36	40

Note: 1 in. = 25.4 mm; 1 psf = 47.88 N/m^2.

[a] Final allowable load is the lesser of the loads as determined by deflection and stress.

[b] Actual span of stiffeners taken as 3½ in. less than center-to-center spacing of purlins.

[c] Deflection limitations: Span/240 under live load only; Span/180 under total load, assuming a dead load of 10 psf.

[d] Loads limited by stress are based on two conditions of duration of load: 2 months, such as for snow (1.15); and 7 days (1.25); includes effects of 10 psf dead load.

TABLE 2.17 Roof Sheathing Fastening Schedule

Region	Nails	Panel location	Roof fastening zone 1	2	3
			Fastening schedule (in. on center)		
High-wind uplift	8d common	Panel edges[a]	6	6	4[b]
		Panel field	6	6	6[b]
Intermediate uplift	8d common	Panel edges[a]	6	6	4
		Panel field	12	6	6
Basic Uplift	8d common	Panel edges[a]	6	6	6
		Panel field	12	12	12

Note: 1 in. = 25.4 mm.
[a] Edge spacing also applies over roof framing at gable-end walls.
[b] Use 8d ring-shank nails in this zone if mean roof height is greater than 25 ft.

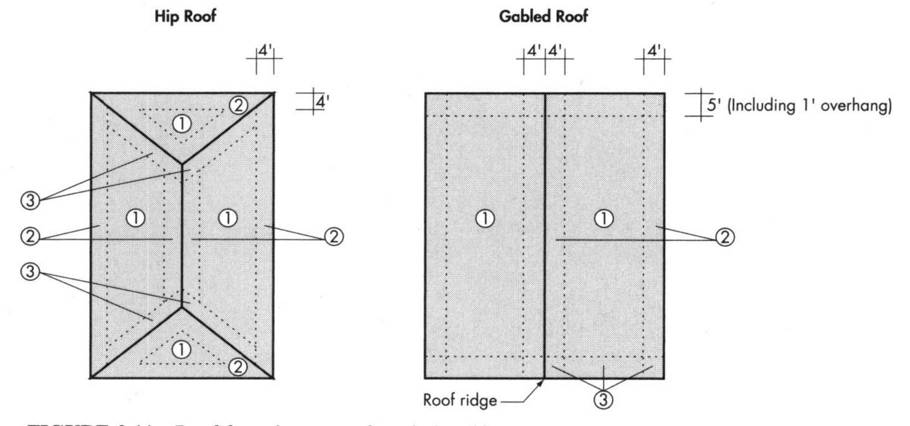

FIGURE 2.11 Roof fastening zones for wind uplift (zones shown indicate areas of the roof with different fastening requirements and should not be confused with ASCE 7 pressure coefficient zones).

ing department, is 80 mi/h (128 km/h) (fastest mile wind speed) or less. These areas are normally included under the prescriptive sections of building codes. As such, the nailing schedule is the familiar 6 in. (150 mm) on center at supported panel edges, including at gable-end walls, and 12 in. (300 mm) on center over intermediate panel supports. Note, however, that minimum 8d nails are recommended for all panels ⅝ in. (16 mm) thick or less. Former APA minimum fastening recommendations included the use of 6d nails for panels ½ in. (12.5 mm) thick or less.

Intermediate uplift: The intermediate uplift fastening schedule is appropriate in inland areas with a basic wind speed above 80 mph (128 km/h) (fastest mile wind speed) and below the basic wind speeds for which the high-wind uplift schedule is recommended.

High-wind uplift: The schedule for high-wind uplift is appropriate for all hurricane oceanline regions (Atlantic and Gulf of Mexico coastal areas). In addition, this schedule should be considered for the transition zone between hurricane oceanline and inland regions. The paragraphs below provide assistance in determining at which basic wind speed (for inland regions) the high-wind uplift schedule is recommended. Contact your local building department for basic wind speed used for design in your area.

For conditions that are not addressed by these general guidelines, such as the "special wind regions" identified in ASCE 7,[7] engineered design is recommended. To determine the basic wind speed at which the high-wind uplift fastening schedule is recommended for a specific structure in an inland region, consider the following:

1. The ability of a roof sheathing panel to resist high winds is directly related to how well it is secured to the roof framing. The type and number of fasteners required for a specific application is obviously an important consideration. Another important consideration is the wood species of the roof framing members into which the sheathing fasteners are driven. Wood of more dense species, such as Douglas fir and southern pine, provides greater nail withdrawal resistance and significantly improves the performance of the sheathing nailing. As shown in Table 2.18, if less dense species, such as spruce-pine-fir or hem-fir are specified and used, the high-wind uplift schedule is recommended at lower basic wind speeds than if the denser species are used.
2. Another consideration relates to the condition of the building envelope during the high wind event. If the building envelope remains intact during the storm, the destructive forces of the wind are considerably less than experienced if a large window, sliding glass door, or garage door is breached or if there are permanent openings. Breaching of the building envelope can be prevented by the use of impact-resistant glazing or shutters.

Generally speaking, well-designed and installed shutter systems are intended to keep the building envelope intact during high wind conditions. In addition to main-

TABLE 2.18 Basic Wind speeds[a,b] for Which the High-Wind Uplift Schedule Is Recommended for Inland Regions

Wood species of roof framing	Building envelope intact (shutters or impact-resistant glazing)	Building envelope breached
Hemlock, eastern spruce, hem-fir, white pine, northern pine or spruce-pine-fir (specific gravity between 0.42 and 0.49)	100 mph or greater	90 mph or greater
Douglas fir or southern pine, (specific gravity between 0.50 and 0.55)	110 mph or greater	100 mph or greater

Note: 1 mph = 1.6 km/h.
[a] Fastest mile wind speed.
[b] Contact your local building department for basic wind speed used for design in your area.

taining the building envelope intact and lowering the wind forces on the structure, shutters also serve to protect the interior of the building from water damage caused by failed doors and glass. Simple, do-it-yourself shutter designs[10] are available from APA—The Engineered Wood Association. Also, many more sophisticated shutter products have become available since Hurricane Andrew hit South Florida in 1992. As can be seen from Table 2.18, the high-wind uplift schedule is recommended at lower basic wind speeds when there is a possibility that the envelope may be breached by breakage or by permanent openings.

Rated Assemblies. Wind resistance of a structure largely determines extended coverage endorsement (ECE) and is an important factor in determining total insurance costs. Underwriters Laboratories (U.L.) and Factory Mutual Research Corporation (FMRC) rate roof systems for wind resistance, based on their performance in a wind uplift test. Many fire-rated wood roof assemblies can also qualify for wind uplift ratings. Systems meeting U.L. requirements are assigned a semi-wind-resistive classification (Class 30 or 60) or fully wind-resistive classification (Class 90).

Two plywood roof systems with hot-mopped built-up roofing over a mechanically fastened roofing base sheet are qualified for fully wind-resistive ratings (Class 90). One of these systems, U.L. Construction No. NM519,[11] is illustrated in Fig. 2.12. It uses 15⁄32 in. (12 mm) APA Rated Sheathing Exposure 1 marked PS 1[1] (untreated CDX plywood), installed across nominal 2 in. (38 mm) wood joists spaced 24 in. (600 mm) oc. For a fully wind-resistive rating (Class 90), the three-

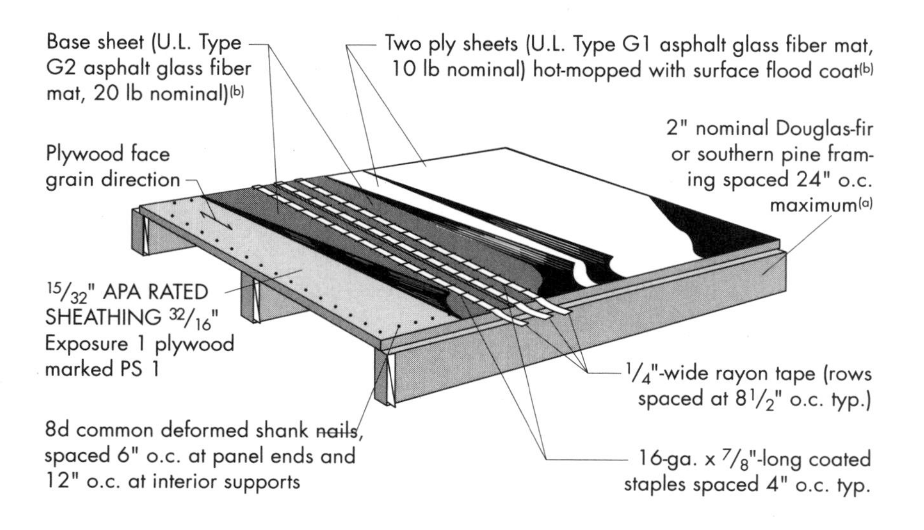

FIGURE 2.12 Fully wind-resistive roof assembly—U.L. Class 90 (NM519).

ply built-up roofing consists of a fiberglass mat base sheet (U.L. Type G2) that is mechanically fastened to the plywood roof deck at lapped edges and along three intermediate rows with a staple/tape system, and two plies of fiberglass mat ply sheets (U.L. Type G1) that are hot-mopped to the base sheet.

The second system is U.L. Construction No. NM520,[11] a panelized roof deck of $^{15}/_{32}$ in. (12 mm) APA Rated Sheathing Exposure 1 marked PS 1[1] (untreated CDX plywood). The panels are installed parallel to 2 × 4 (38 × 89 mm) joists spaced 24 in. (600 mm) oc, which span 8 ft (2400 mm) between purlins framed into glulam beams (Fig. 2.13). For a fully wind-resistive (Class 90), the three-ply built-up roofing is installed as described above for NM519[11] construction. If the roofing base sheet is fastened to the plywood roof deck at lapped edges and along two intermediate rows with a staple/tape system, the roofing system qualifies for a semi-wind-resistive rating (Class 60).

2.4.4 Code Provisions

Recommendations given in the preceding sections for construction applications are consistent with provisions given in the model building codes in the United States,

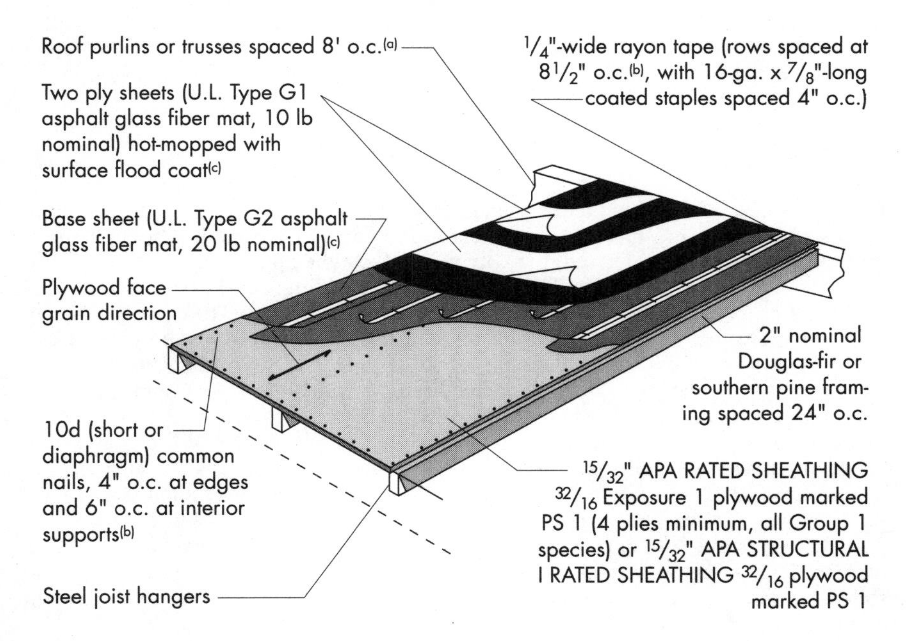

(a) Trusses or I-joists used for purlins must have chords or flanges of 1 3/4" minimum depth for plywood deck nailing.

(b) For semi-wind-resistive assemblies (Class 60), plywood deck nailing spaced 6" o.c. at all supports and roofing base sheet attached with rayon tape rows spaced 11 1/3" o.c.

(c) Install roofing base and ply sheets with roll direction parallel to plywood face grain direction.

FIGURE 2.13 Fully wind-resistive roof assembly—U.L. Class 90 (NM520).

with the exception of stucco wall sheathing recommendations, which are not specifically addressed in the codes. However, most of the preceding information has been expanded compared to the code provisions, to be more useful to designers.

The general recommendations apply primarily to conventional or nonengineered construction, but can also be considered conservative for engineered construction. On the other hand, for engineered construction, codes contain provisions for acceptance of engineering calculations, and mechanical properties given later in this chapter may be used. In many cases, calculations using mechanical properties in this chapter will lead to higher design capacities for sheathing. This is because the general recommendations are based on minimum structural requirements or criteria of the performance standards, while the mechanical properties are based on actual characteristics of panels qualified under the performance standards. Since it would be difficult to manufacture a truly "minimum" panel with regard to all properties, most panel characteristics actually exceed requirements of the standards.

2.5 SPECIAL CONSIDERATIONS

Nearly every special consideration for wood construction relates to either moisture or fire. Fire considerations are discussed in Chapter 10. Basic moisture properties of wood are discussed in Chapter 1, and moisture control is addressed in Chapter 12. This section will emphasize design for panel movement due to moisture.

2.5.1 Panel Buckling

Buckling of wood structural panel sheathing such as plywood and OSB occasionally results when high-moisture conditions cause the panels to expand. Although structural properties are not affected, the waviness affects the appearance and may cause concerns about serviceability. The potential for buckling can be significantly reduced by understanding the factors that contribute to buckling risk and providing for the natural increase in panel dimensions that results from moisture exposure.

The tendency of expansion to cause buckling is related to mechanical and physical properties of the panel, natural variability of wood, and installation techniques. Mechanical properties such as panel stiffness are important for resisting the stresses that develop as the panel tries to expand. The physical properties of the panel, such as the orientation of veneers or strands, will influence the panel's dimensional response to moisture conditions. Installation practices such as panel edge spacing are important to minimize the build-up of stresses that can cause buckling.

Laboratory and field experience indicate that certain types of installation involve increased buckling risks that merit special attention. When one or more of the following factors are present, additional techniques should be considered to help assure best performance:

- Shear wall or diaphragm applications with panels applied with strength axis parallel to supports and edge nail spacing 4 in. (100 mm) oc or closer
- Use of three-ply plywood panels with the face grain parallel to supports (i.e., walls)
- Use of oversized panels which are larger than 4 × 8 ft (1200 × 2400 mm)

These applications can be high-risk because the tight nailing schedule reduces the effectiveness of the panel edge gap in absorbing the panel expansion; the low panel stiffness direction spans between the supports; and/or the oversize panel dimension allows panel expansion to build up over a longer length.

For these applications, the following techniques help offset the increased buckling risk.

Panel Edge Spacing. Additional attention to edge spacing is required due to the increased buckling risk. Normal edge spacing recommendations for sheathing, ⅛ in. (3 mm) gap at edges and ends, may be insufficient. For example, for oversized panels, consider increasing the panel gaps at edges (length parallel to strength axis marked on the panel) to ¼ in. (6.5 mm). This can be accomplished either by increasing the framing module or by specifying a special size cut from the panel manufacturer. Such special cut panels are denoted with edge gapping recommendations on the panels. In applications where high-density nailing schedules are followed, such as diaphragms, edge gapping will not be very effective.

APA's recommendations for spacing are designed to mitigate panel buckling. After panel installation, the panel gap will naturally close as a result of panel expansion due to moisture absorption. The absence of a gap during later inspection may be indicative of gap closure, rather than an absence of a gap during installation. Whether or not a gap is present immediately prior to roofing, if the deck flatness is acceptable, roofing may generally proceed.

Panel Nailing. To allow for expansion of panels if subjected to job-site wetting, the following nailing sequence should be considered where nail spacing 4 in. (100 mm) o.c. or closer is specified:

- Temporarily nail panels with a nail spacing of 12 in. (300 mm) o.c. at ends, edges, and intermediate supports (rather than at the specified shear wall or diaphragm schedule) during the framing phase of construction. For temporary nailing, use nail size specified. With this lighter nailing schedule, resultant panel expansion is more readily absorbed by the panel edge gaps.
- Complete final nailing immediately prior to covering with siding or roofing or after panels have been acclimated to job-site moisture conditions.

2.5.2 Temporary Expansion Joints

If wood structural panels are exposed to moisture or humidity during construction of buildings with large, continuous floor or roof decks, panel expansion may accumulate through the framing.

All wood products absorb moisture from or give up moisture to the environment until they reach a moisture content in equilibrium with their surroundings. Wood structural panels have good dimensional stability because the tendency of individual veneers or strands to swell or shrink is greatly restricted by the adjacent veneers or strands in the panel.

In typical sheathing applications, relative humidity might vary between 40% and 80%, with corresponding equilibrium moisture content of wood structural panels ranging between 6% and 14%. Total dimensional change of an unrestrained 48 × 96 in. (1200 × 2400 mm) panel exposed to this range of conditions typically averages ⅛ in. (3 mm) in length and width. If the panel gets wet during construction, dimensional change could be slightly greater. Recommended spacing of ⅛ in.

(3 mm) at ends and edges of floor and roof deck panels will absorb some or most of this expansion.

However, such dimensional change in installed panels typically is reduced due to partial restraint by fasteners and framing. Field experience indicates that there can be net overall expansion of floor or roof decks that reflects the combined effects of panel expansion as absorbed by the spacing at panel edges and ends, and restraint afforded by panel fasteners and framing.

Floor panels are interconnected by bottom plates of exterior and interior walls that typically are nailed to the floor, or through the floor to the floor framing. Also, floor panels are often nail-glued to floor framing for added floor stiffness and to minimize or eliminate floor squeaks. Either or both of these situations may partially offset the effectiveness of the recommended spacing at panel edges and ends, resulting in accumulation of panel expansion along the length or width of the building.

For example, in an 80 ft (24 m) long building, if net overall expansion of 0.05% occurs in the floor deck during construction, an increase in building length of ½ in. (12.5 mm) or ¼ in. (6.5 mm) at each end may result. If this expansion occurs on the first floor with a concrete or masonry foundation below, the rim or band joists might be displaced out-of-plumb by ¼ in. (6.5 mm), which typically could be accommodated without problem. If this expansion occurs on the second floor of a multistory building (assuming an on-grade concrete slab for the first floor), the top end of the first story walls theoretically might be displaced out-of-plumb by ¼ in. (6.5 mm), which typically would not be noticeable. However, if the building is 160 ft (49 m) or 240 ft (73 m) long, the overall expansion could be two or three times as much, and out-of-plumb rim joists or end (and interior) walls would be noticeable. In multistory buildings, walls would be plumb at the building's midlength or midwidth, but wall displacement (out-of-plumb) would gradually increase to a maximum at the exterior walls. The squareness of door or window openings also might be affected, both in interior and exterior walls.

Designers and contractors can minimize displacement by incorporating temporary expansion joints in floors of buildings with wood- or steel-framed walls when the building plan dimension (length or width) exceeds 80 ft (24 m). Such joints for floors might consist of an extra-wide spacing gap, such as ¾ in. (19 mm), between panel ends at the desired expansion joint intervals. Panel ends can be supported on adjacent doubled floor joists and not nailed to them until later, to allow for floor expansion. Also, it is important to ensure that wall bottom plates do not extend across the expansion joint. After the building is closed in, fastening of the floor panels can be completed, and a filler piece or nonshrink grout can be installed to fill the gap between panels, where necessary. For shear walls or braced wall panels, a short lumber bottom plate filler block and doubler could be added later between studs, to splice the bottom plate of walls over the expansion joint. See Fig. 2.14 for a possible construction detail for incorporating an expansion joint in floors; other effective expansion joint details also may be used.

Minimizing exposure to moisture during construction can reduce expansion of floor panels. If rain (or snow) occurs during construction and there are areas of the floor that are subject to water ponding, such as when water is trapped on the floor by bottom plates of walls, drill drainage holes through the floor to allow the water to escape. These holes can be patched later with glued wood dowels or grout, and backer plates cut from wood structural panels that are screw-glued to the underside of the floor panels, or with sheet metal patches on top of the floor.

In the construction of large roof decks with wood structural panels fastened to trusses or rafters, sheath 80 ft (24 m) sections, omitting a roof sheathing panel (in

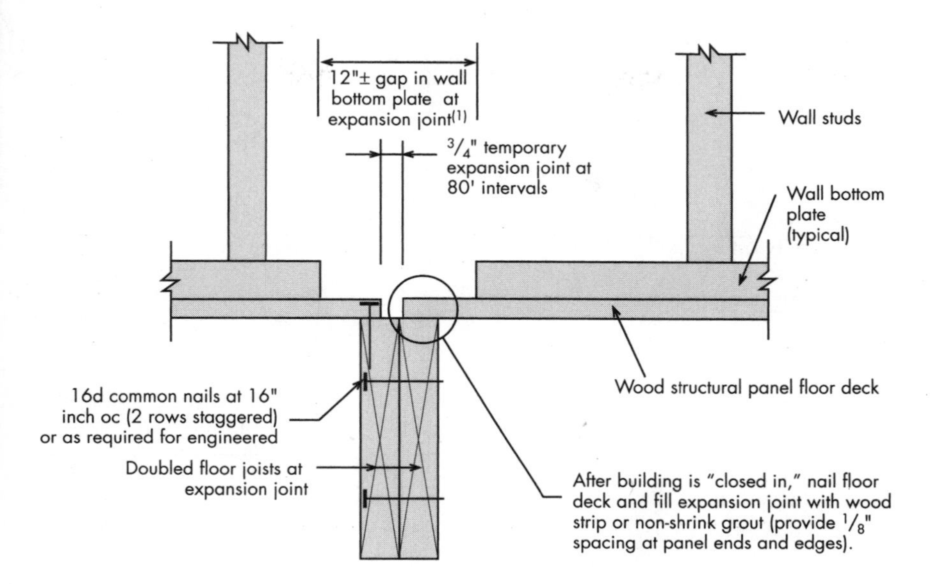

FIGURE 2.14 Temporary expansion joint detail for floors.

each course of sheathing) between sections. This provides effective temporary expansion joints (see Fig. 2.15). Then the installation can be completed with fill-in panels, cut to size as necessary. The roof deck should be covered with roofing underlayment as soon as possible for protection against excessive moisture prior to roofing application. On large roof decks, installation of roofing underlayment and roofing can be scheduled in sections to avoid exposing the entire expanse of roof deck to weathering during construction.

Designers or contractors may choose to omit temporary expansion joints in large buildings, based on their individual experience as affected by materials they choose and the environment and techniques of construction. For example, when large buildings are constructed in warm, dry regions or in summer months where moisture is not likely to occur during construction, the need for expansion joints is less. However, if expansion joints are not incorporated in the design or construction of large buildings, it is done with the understanding that the designer or contractor may face the potential risk of structural modifications or repairs if problems occur later. Although problems are relatively few, incorporating temporary expansion joints in such large buildings is recommended as good construction practice.

2.6 MECHANICAL PROPERTIES

Wood structural panels can typically be incorporated into construction projects without the need for engineering design of the panels themselves. They lend themselves to tabular and descriptive presentation of design recommendations and pro-

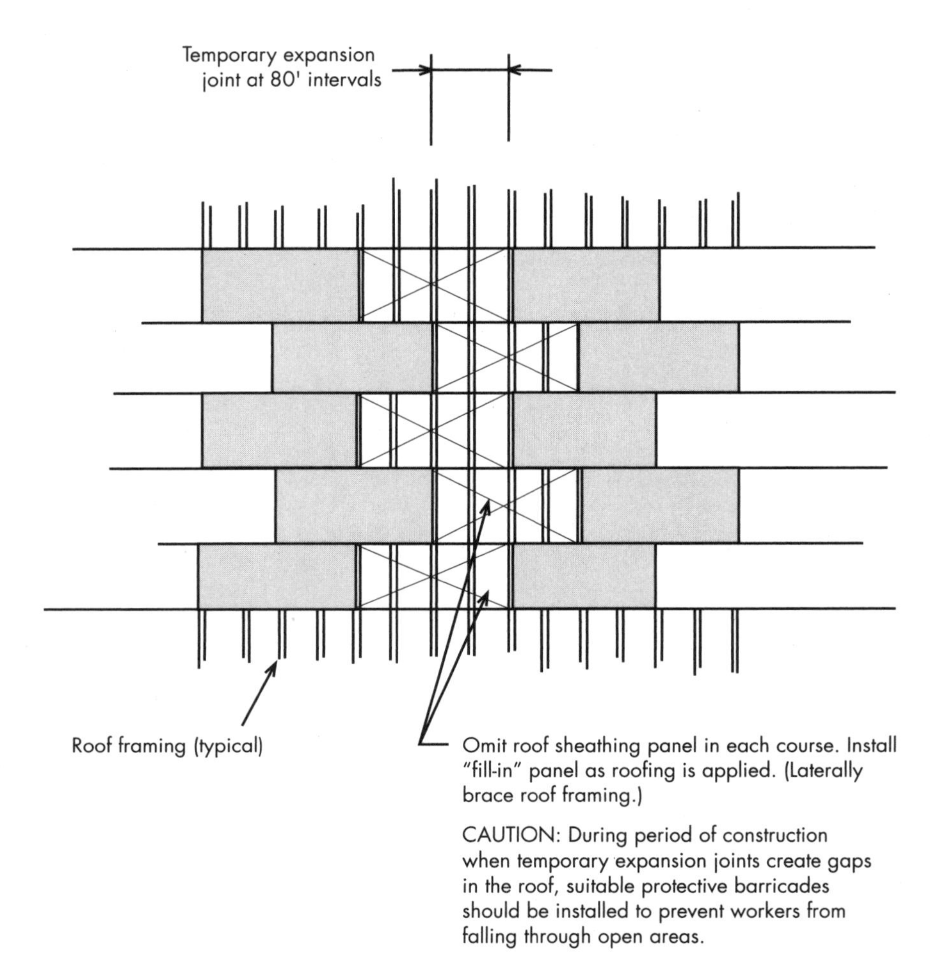

FIGURE 2.15 Temporary expansion joint detail for roofs.

visions. Occasionally, however, there is a need to engineer panel applications that call for panel properties or capacities; or it may be necessary to evaluate specific panel constructions that yield superior mechanical properties compared to those that are the basis for general use recommendations.

2.6.1 Strength Axis

A feature of most wood structural panel types, primarily plywood and OSB, is that there is a strength axis associated with their manufacture. The layered construction of both products, in which layers are oriented 90° from one another, creates dissimilar properties in the two principal directions. This is illustrated in Fig. 2.16. The orientation of the face and back layer determines the direction of the strength axis.

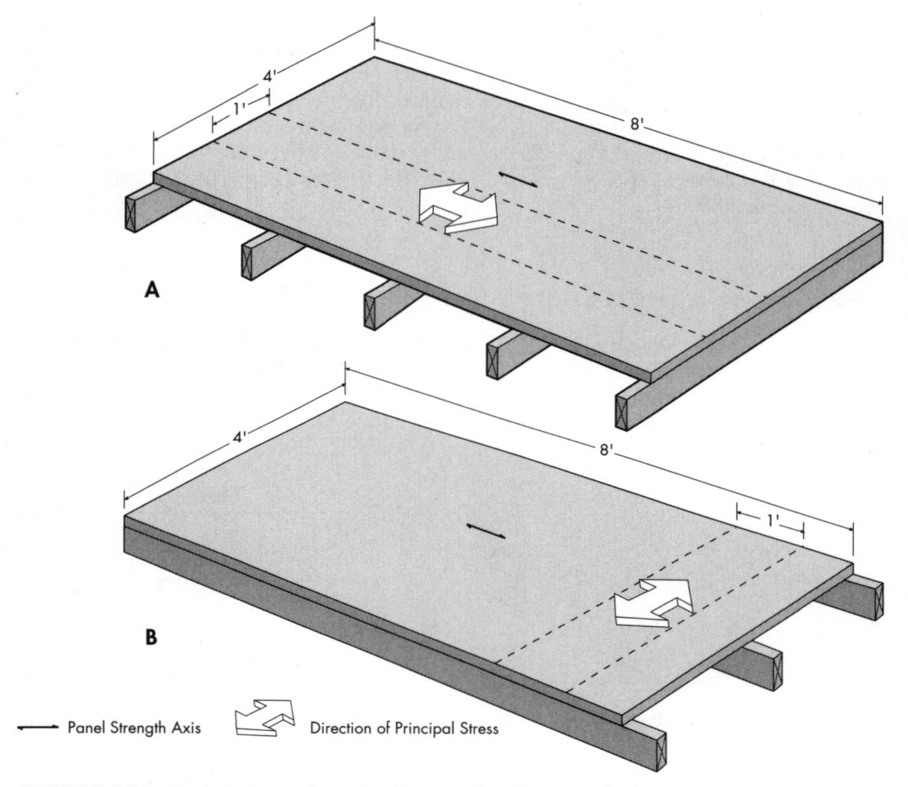

FIGURE 2.16 Typical plywood panel with strength axis perpendicular to or across supports (*A*) and parallel to supports (*B*) (note representative portion of panel used in calculation of section capacities for stress parallel (*A*) or perpendicular (*B*) to the strength axis).

The panel strength axis is typically in the long panel direction; that is, the panel is typically stronger and stiffer along the panel length than across the panel width. Specification of panel orientation, then, can be stated as "strength axis is perpendicular (or parallel) to supports" or sometimes, "stress is parallel (or perpendicular) to strength axis." In the case of plywood or composite panels, the strength axis is sometimes referred to as the face grain direction.

2.6.2 Panel Construction

Plywood mills may use different layups for the same panel thickness and span rating to make optimum use of raw materials. Design calculations must take into account the direction in which the stresses will be imposed in the panel. If stresses can be expected in both directions, then both the parallel and perpendicular directions should be checked. For this reason, tabulated capacities are given for both directions.

Capacities parallel to the face grain of plywood are based on a panel construction that gives minimum values in that direction (see Fig. 2.17). Capacities perpendicular to the face grain are usually based on a different panel construction that gives minimum values in that direction. Both values, therefore, are conservative. Capacities given for the two directions are not necessarily for the same panel construction.

Similar layers occur also in OSB manufacture. However, the layers are not defined and therefore cannot be specified. For this reason, ply-layer options are not tabulated for OSB.

2.6.3 Properties and Stresses

Plywood properties have traditionally been separately tabulated as section properties and design stresses. These are, of course, multiplied together to obtain a capacity. In many cases the resulting capacity will be quite conservative. Design stresses are conservatively developed, taking into account grade factors and manufacturing fac-

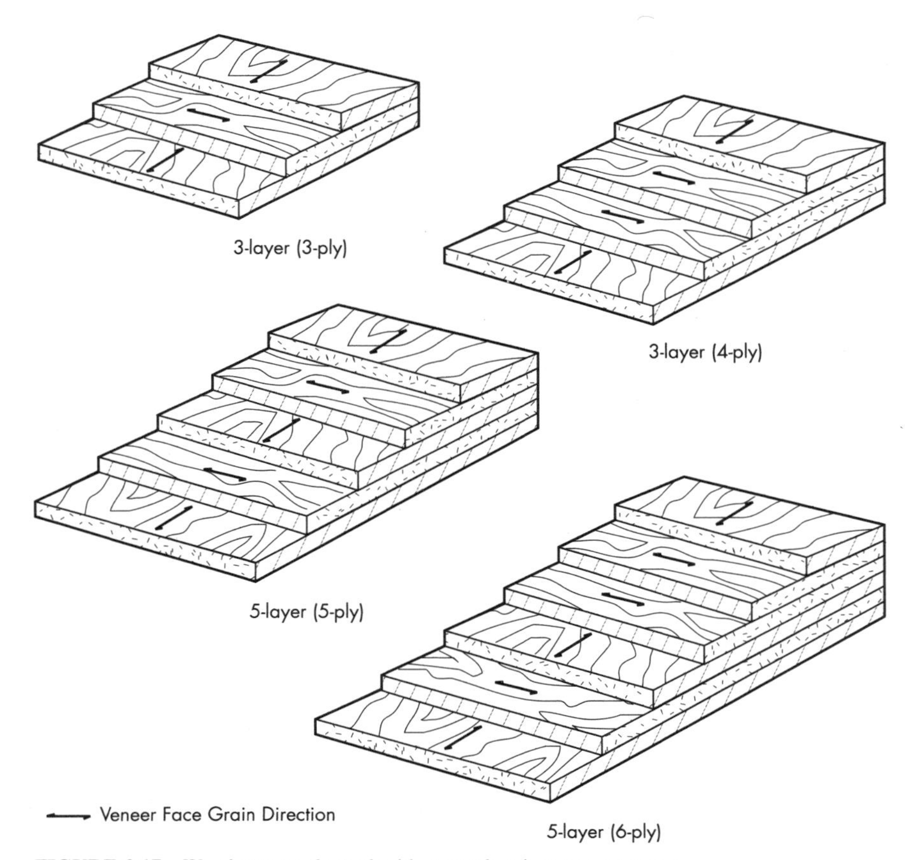

FIGURE 2.17 Wood structural panel with strength axis across supports.

tors, and then the data are statistically manipulated such that they represent the low end of possible values. The stress is then further adjusted by a load factor or, as some call it, a factor of safety.

At the same time, section properties are developed for virtually all possible layup combinations of veneer thickness and species. The lowest property value for a given panel thickness or span rating is then chosen for tabulation. The resulting capacity combines two already conservative values. In the 1990s this procedure was largely replaced by direct publication of panel capacities.

2.6.4 Capacities

Panel design capacities listed in Tables 2.19 through 2.22 are furnished courtesy of APA—The Engineered Wood Association and are minimum for grade and span

TABLE 2.19 Panel Dry Design Bending Stiffness and Strength Capacities

	Bending stiffness, EI (lbf-in²/ft width)		Bending strength, F_bS (lbf-in/ft width)	
	Along	Across	Along	Across
Span rating		Rated Sheathing		
24/0	60,000	3,600	250	54
24/16	78,000	5,200	320	64
32/16	115,000	8,100	370	92
40/20	225,000	18,000	625	150
48/24	400,000	29,500	845	225
Span rating		Rated Sturd-I-Floor		
16oc	150,000	11,000	415	100
20oc	210,000	13,000	480	140
24oc	300,000	26,000	640	215
32oc	650,000	75,000	870	380
48oc	1,150,000	160,000	1,600	680
Nominal thickness		Sanded plywood		
1/4	15,000	700	97	14
11/32	34,000	1,750	155	26
3/8	49,000	2,750	205	37
15/32	120,000	11,000	355	110
1/2	140,000	15,500	390	145
19/32	205,000	37,500	520	225
5/8	230,000	48,500	560	270
23/32	320,000	90,500	645	380
3/4	355,000	115,000	680	470
7/8	500,000	185,000	850	650
1	760,000	330,000	1,100	975
1 1/8	985,000	490,000	1,350	1,250

Note: 1 in. = 25.4 mm; 1 lbf-in²/ft width = 9.415×10^{-3} N · m²/m width; 1 lbf-in/ft width = 0.371 N · m/m width.

TABLE 2.19A Adjustments to Flexural Design Capacities Based on Panel Grade and Constructions, C_g

	Strength axis[a]			
	Perpendicular to supports		Parallel to supports	
	Other	Structural I	Other	Structural I
Panel bending stiffness, *EI*				
Rated panels				
3-ply plywood	1.1	1.1	1.0	1.5
4-ply plywood, COM-PLY	1.1	1.1	2.2	3.3
5-ply plywood[b]	1.1	1.1	3.1	5.2
OSB	1.0	1.0	3.1	5.2
Sanded plywood				
A-A, A-C	1.0	1.0	1.0	1.4
Marine	1.0	1.0	1.4	1.4
Other	1.0	1.0	1.0	1.4
Panel bending strength, $F_b S$				
Rated panels				
3-ply plywood	1.0	1.0	1.0	1.3
4-ply plywood	1.1	1.1	1.2	1.7
COM-PLY	1.2	1.2	1.2	1.7
5-ply plywood[b] OSB	1.2	1.2	1.8	2.8
Sanded plywood				
A-A, A-C	1.2	1.3	1.2	1.7
Marine	1.1	1.1	1.4	1.4
Other	1.0	1.1	1.0	1.4

[a] The strength axis is the long panel dimension unless otherwise identified.
[b] Adjustments apply to plywood with five or more layers; for five-ply/three-layer plywood, use adjustments for four-ply.

TABLE 2.20 Panel Dry Design Axial Stiffness, Tension, and Compression Capacities

	Axial stiffness, EA (lbf/ft width)		Tension, F_tA (lbf/ft width)		Compression, F_cA (lbf/ft width)	
	Along	Across	Along	Across	Along	Across
Span rating			Rated Sheathing			
24/0	3,350,000	2,900,000	2,300	600	2,850	2,500
24/16	3,800,000	2,900,000	2,600	990	3,250	2,500
32/16	4,150,000	3,600,000	2,800	1,250	3,550	3,100
40/20	5,000,000	4,600,000	2,900	1,600	4,200	4,000
48/24	5,850,000	5,000,000	4,000	1,950	5,000	4,800
Span rating			Rated Sturd-I-Floor			
16oc	4,500,000	4,200,000	2,600	1,450	4,000	3,600
20oc	5,000,000	4,600,000	2,900	1,600	4,200	4,000
24oc	5,850,000	5,000,000	3,350	1,950	5,000	4,800
32oc	7,500,000	7,300,000	4,000	2,500	6,300	6,200
48oc	15,000,000	14,600,000	5,600	5,000	13,000	12,500
Nominal thickness			Sanded plywood			
1/4	1,800,000	625,000	1,650	550	1,550	550
11/32	1,800,000	750,000	1,650	700	1,550	650
3/8	2,350,000	1,150,000	2,150	1,050	2,000	950
15/32	3,500,000	2,150,000	3,200	2,000	3,000	1,850
1/2	3,500,000	2,250,000	3,200	2,050	3,000	1,900
19/32	4,350,000	2,500,000	4,000	2,300	3,750	2,150
5/8	4,450,000	2,750,000	4,100	2,500	3,800	2,350
23/32	5,100,000	3,150,000	4,650	2,850	4,350	2,650
3/4	5,200,000	3,750,000	4,750	3,450	4,450	3,200
7/8	5,300,000	4,750,000	4,850	4,350	4,550	4,100
1	6,700,000	5,700,000	6,150	5,200	5,750	4,850
1 1/8	6,950,000	5,700,000	6,350	5,250	5,950	4,900

Note: 1 in. = 25.4 mm; 1 lbf/ft width = 14.59 N/m width.

TABLE 2.20A Adjustments to Axial Design Capacities Based on Panel Grade and Constructions, C_g

	Strength axis[a]			
	Along the loading direction		Across the loading direction	
	Other	Structural I	Other	Structural I
Panel axial stiffness, EA				
Rated panels				
3-ply plywood, COM-PLY	1.0	1.0	1.0	1.0
4-ply plywood	1.0	1.0	1.0	1.0
5-ply plywood[b], OSB	1.0	1.0	1.0	1.0
Sanded plywood				
A-A, A-C	1.0	1.0	1.0	1.8
Marine	1.0	1.0	1.8	1.8
Other	1.0	1.0	1.0	1.8
Panel tension, F_tA				
Rated panels				
3-ply plywood, COM-PLY	1.0	1.0	1.0	1.0
4-ply plywood	1.0	1.0	1.0	1.0
5-ply plywood[b]	1.3	1.3	1.3	1.3
OSB	1.0	1.0	1.3	1.3
Sanded plywood				
A-A, A-C	1.2	1.2	1.2	2.1
Marine	1.0	1.0	1.8	1.8
Other	1.0	1.0	1.0	1.8
Panel compression, F_cA				
Rated panels				
3-ply plywood, COM-PLY	1.0	1.0	1.0	1.0
4-ply plywood	1.5	1.5	1.5	1.5
5-ply plywood[b]	1.5	1.5	1.5	1.5
OSB	1.0	1.0	1.0	1.0
Sanded plywood				
A-A, A-C	1.1	1.1	1.1	2.0
Marine	1.0	1.0	1.8	1.8
Other	1.0	1.0	1.0	1.8

[a] The strength axis is the long panel dimension unless otherwise identified.
[b] Adjustments apply to plywood with five or more layers; for five-ply/three-layer plywood, use adjustments for four-ply.

TABLE 2.21 Panel Dry Shear Capacities in the Plane

	Shear in the plane, $Fs(Ib/Q)$ (lbf/ft width)	
	Along	Across
Span rating	Rated Sheathing	
24/0	165	105
24/16	190	105
32/16	210	130
40/20	265	165
48/24	340	190
Span rating	Rated Sturd-I-Floor	
16oc	225	145
20oc	265	170
24oc	340	195
32oc	400	280
48oc	600	450
Nominal thickness	Sanded plywood	
1/4	97	97
11/32	135	125
3/8	150	170
15/32	195	115
1/2	215	130
19/32	265	135
5/8	280	150
23/32	315	185
3/4	325	195
7/8	385	245
1	425	340
1 1/8	475	405

Note: 1 in. = 25.4 mm; 1 lbf/ft width = 14.59 N/m width.

TABLE 2.21A Adjustments to Shear Capacities in the Plane Based on Panel Grade and Constructions, C_g

	Strength axis[a]			
	Perpendicular to supports		Parallel to supports	
	Other	Structural I	Other	Structural I
Panel shear in the plane, $Fs(Ib/Q)$				
Rated panels				
3-ply plywood	1.0	1.4	2.8	5.2
4-ply plywood	1.0	1.4	3.9	7.9
5-ply plywood[b]	1.1	1.6	1.0	1.4
OSB, COM-PLY	1.0	1.0	1.0	1.0
Sanded plywood				
A-A, A-C	1.0	1.3	1.0	1.5
Marine	1.3	1.3	1.5	1.5
Other	1.0	1.3	1.0	1.5

[a] The strength axis is the long panel dimension unless otherwise identified.

[b] Adjustments apply to plywood with five or more layers; for five-ply/three-layer plywood, use adjustments for four-ply.

TABLE 2.22 Panel Dry Rigidity and Shear Capacities through the Thickness

	Rigidity through the thickness, $G_v t_v$ (lbf/in. of panel depth)		Shear through the thickness, $F_v t_v$ (lbf/in. of shear-resisting panel length)	
	Along	Across	Along	Across
Span rating		Rated Sheathing		
24/0	25,000	25,000	53	53
24/16	27,000	27,000	57	57
32/16	27,000	27,000	62	62
40/20	28,500	28,500	68	68
48/24	31,000	31,000	75	75
Span rating		Rated Sturd-I-Floor		
16oc	27,000	27,000	58	58
20oc	28,000	28,000	67	67
24oc	30,000	30,000	74	74
32oc	36,000	36,000	80	80
48oc	50,500	50,500	105	105
Nominal thickness		Sanded plywood		
1/4	22,000	22,000	43	43
11/32	23,500	23,500	45	45
3/8	23,500	23,500	46	46
15/32	34,500	34,500	67	67
1/2	35,000	35,000	68	68
19/32	44,500	44,500	87	87
5/8	45,000	45,000	88	88
23/32	46,000	46,000	90	90
3/4	46,500	46,500	91	91
7/8	48,000	48,000	94	94
1	67,000	67,000	130	130
1 1/8	68,500	68,500	135	135

Note: 1 in. = 25.4 mm; 1 lbf/in. = 0.175 N/mm.

TABLE 2.22A Adjustments to Design Capacities Based on Panel Grade and Constructions, C_g

	Strength axis[a]			
	Perpendicular to supports		Parallel to supports	
	Other	Structural I	Other	Structural I
Panel rigidity through the thickness, $G_v t_v$				
Rated panels				
3-ply plywood	1.0	1.3	1.0	1.3
4-ply plywood, COM-PLY	1.3	1.7	1.3	1.7
5-ply plywood[b]	1.5	1.7	1.5	1.7
OSB	3.1	3.1	3.1	3.1
Sanded plywood				
A-A, A-C	1.0	1.3	1.0	1.3
Marine	1.3	1.3	1.3	1.3
Other	1.0	1.3	1.0	1.3
Panel shear through the thickness, $F_v t_v$				
Rated panels				
3-ply plywood	1.0	1.3	1.0	1.3
4-ply plywood, COM-PLY	1.3	1.7	1.3	1.7
5-ply plywood[b]	1.5	2.0	1.5	2.0
OSB	2.9	2.9	2.9	2.9
Sanded plywood				
A-A, A-C	1.0	1.3	1.0	1.3
Marine	1.3	1.3	1.3	1.3
Other	1.0	1.3	1.0	1.3

[a] The strength axis is the long panel dimension unless otherwise identified.
[b] Adjustments apply to plywood with five or more layers; for five-ply/three-layer plywood, use adjustments for four-ply.

rating. Tables 2.19A through 2.22A list allowable increases in capacity for specific panel constructions and grades. The tabulated capacities and adjustments are based on data from tests of panels bearing the APA trademark. To take advantage of these capacities and adjustments, the specifier must ensure that the correct panel span rating is used in the final construction.

Panel Flexure (Flat Panel Bending). Panel design capacities reported in Table 2.19 are based on flat panel bending as measured by testing according to the principles of ASTM D3043[13] Method C (large panel testing) (see Fig. 2.18).

Stiffness (EI). Panel bending stiffness is the capacity to resist deflection and is represented in bending equations as EI. The E is the modulus of elasticity of the material and the I is the moment of inertia of the cross section. Units of EI are lb-in.^{2}/ft of panel width.

Strength (F_bS). Allowable bending strength capacity is the design maximum moment, represented in bending equations as F_bS. Terms are the allowable extreme fiber stress of the material (F_b) and the section modulus (S). Units of F_bS are lb-in./ft of panel width.

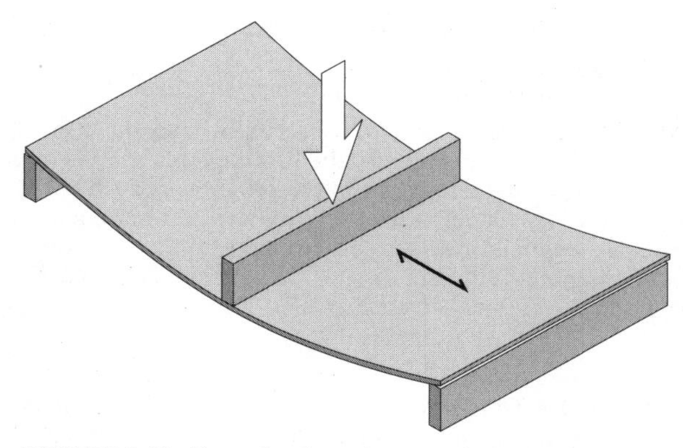

FIGURE 2.18 Example of wood structural panel in bending.

Panel Axial Strength

Tension (F_tA). Allowable tension capacities are reported in Table 2.20 based on testing according to the principles of ASTM D3500[14] Method B. Tension capacity is given as F_tA, where F_t is the allowable tensile stress of the material and A is the area of the cross section. Units of F_tA are lb/ft of panel width.

Compression (F_cA). Allowable compression capacities are reported in Table 2.20 based on testing according to the principles of ASTM D3501[15] Method B. Compressive properties are generally influenced by buckling; however, this effect was eliminated by restraining edges of specimens during testing. Compression capacity is given as F_cA, where F_c is the allowable compression stress of the material and A is the area of the cross section. Units of F_cA are lb/ft of panel width. Axial compression strength is illustrated in Fig. 2.19.

Panel Axial Stiffness (EA). Panel axial stiffness is reported in Table 2.21 based on testing according to the principles of ASTM D3501[15] Method B. Axial stiffness is the capacity to resist axial strain and is represented by EA. The E is the axial

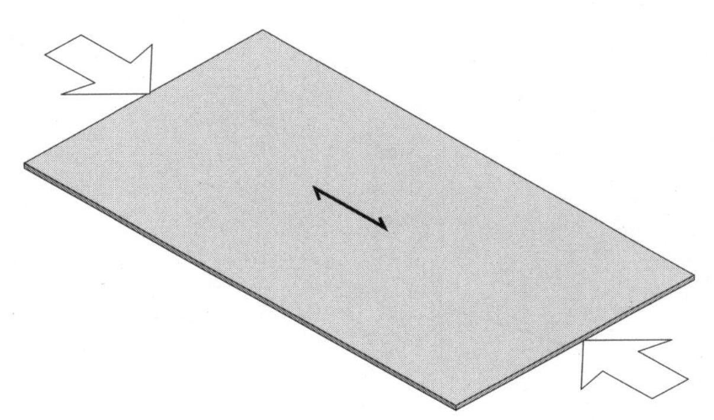

FIGURE 2.19 Wood structural panel with axial compression load in the plane of the panel.

modulus of elasticity of the material and A is the area of the cross section. Units of EA are lb/ft of panel width.

Shear in the Plane of the Panel (Fs[Ib/Q]). Allowable shear in the plane of the panel (or interlaminar shear, sometimes called rolling shear in plywood) is reported in Table 2.21 based on testing according to the principles of ASTM D2718.[16] Shear strength in the plane of the panel is the capacity to resist horizontal shear breaking loads when loads are applied or developed on opposite faces of the panel, as in flat panel bending (see Figure 2.20). The term F_s is the allowable material stress, while lb/Q is the panel cross-sectional shear constant. The units of $F_s(Ib/Q)$ are lb/ft of panel width.

Panel Shear through the Thickness. Panel shear through the thickness capacities are reported based on testing according to the principles of ASTM D2719[17] (see Figure 2.20).

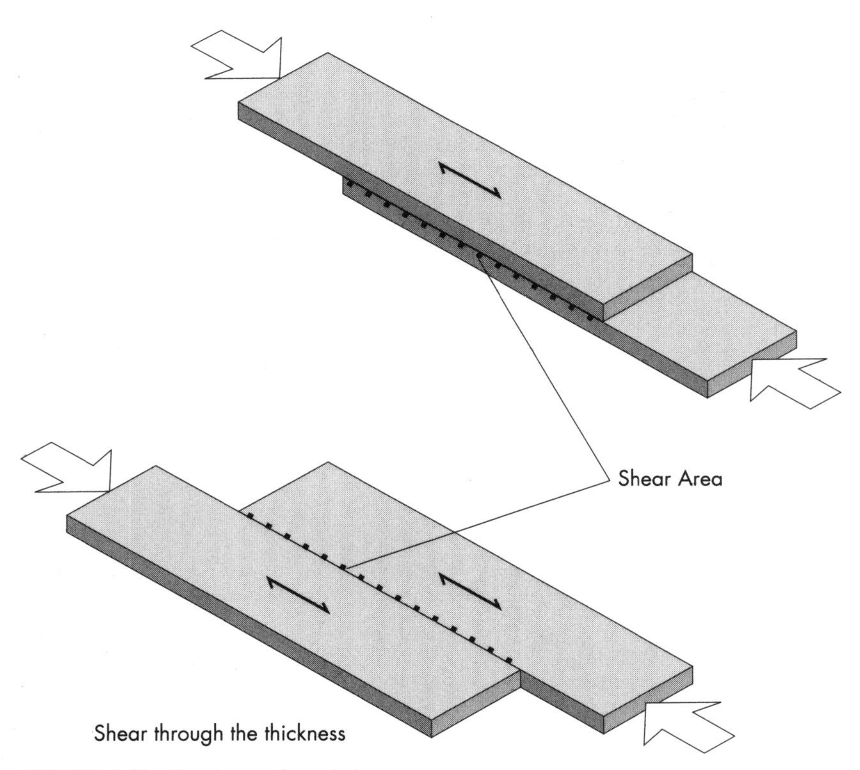

FIGURE 2.20 Two types of panel shear.

Panel Shear through the Thickness ($F_v t_v$). Allowable shear through the thickness is the capacity to resist horizontal shear breaking loads when loads are applied or developed on opposite edges of the panel, such as in an I-beam, and is reported in Table 2.22 (see Fig. 2.20). Where additional support is not provided to prevent buckling, design capacities in Table 2.22 are limited to sections 2 ft or less in depth. Deeper sections may require additional reductions. The term F_v is the allowable stress of the material, while t_v is the effective panel thickness for shear. The units of $F_v t_v$ are lb/in. of shear-resisting panel length.

Panel Rigidity through the Thickness ($G_v t_v$). Panel rigidity is reported in Table 2.22 and is the capacity to resist deformation when under shear-through-the-thickness stress. Rigidity is represented by $G_v t_v$, where G_v is the modulus of rigidity and t_v is the effective panel thickness for shear. The units of $G_v t_v$ are lb/in. of panel depth (for vertical applications). Multiplication of $G_v t_v$ by panel depth gives GA, used by designers for some applications.

2.6.5 Adjustments

Panel design capacities may be adjusted as required under the following provisions.

Panel Grade and Construction. Panel design capacities listed in Tables 2.19 through 2.22 are minimum for grade and span rating. Tables 2.19A, 2.20A, 2.21A, and 2.22A list allowable increases in capacity (C_G) for specific panel constructions and grades. To take advantage of these capacities and adjustments, the specifier must ensure that the correct panel is used in the final construction.

Duration of Load (DOL). Design capacities listed are based on normal duration of load as traditionally used for solid wood in accordance with U.S. Forest Products Laboratory Report R-1916[18] and successfully used for plywood for approximately 40 years. Adjustment factors for strength capacities are:

Time under load	DOL adjustment factor (C_D)
Permanent	0.90
Normal	1.00
Two months	1.15
Seven days	1.25
Wind or Earthquake	1.60[a]

[a] Check local building code.

Creep. Wood-based panels under constant load will creep (deflection will increase) over time. For typical construction applications, panels are not normally under constant load and, accordingly, creep need not be considered in design. When panels will sustain permanent loads that will stress the product to one-half or more of its design strength capacity, allowance should be made for creep. Limited data indicates that under such conditions, creep may be taken into account in deflection calculations by applying the applicable following adjustment factor (C_C) to panel stiffness, *EI*:

Moisture condition	Creep adjustment factor (C_c) for permanent loads: Plywood	OSB
Dry	½	½
16% m.c. or greater	½	⅙

See below, Service Moisture Conditions, for additional adjustments related to service moisture conditions, which for *EI* are cumulative with the adjustment for creep.

Service Moisture Conditions. Design capacities apply to panels under moisture conditions that are continuously dry in service; that is, where equilibrium moisture content is less than 16%. Adjustment factors for conditions where the panel moisture content in service is expected to be 16% or greater are as follows:

	Moisture content adjustment factor (C_m)
Strength (F_bS, F_tA, F_cA, $F_s(Ib/Q)$, F_vt_v)	0.75
Stiffness (EI, EA, G_vt_v)	0.85

Pressure Treatment

Preservative Treatment. Capacities given in this document apply, without adjustment, to plywood pressure-impregnated with preservative chemicals and redried in accordance with American Wood Preservers Association (AWPA) Standard C-9.[19] Due to the absence of applicable treating industry standards, OSB and COM-PLY panels are not currently recommended for applications requiring pressure-preservative treating.

Fire-Retardant Treatment. Discussion in this book does not apply to fire-retardant-treated structural panels. However, some general information on fire-retardant treated plywood roof sheathing is available in a report from APA—The Engineered Wood Association.[20] For fire-retardant-treated panels, all capacities and end-use conditions shall be in accordance with the recommendations of the company providing the treating and redrying service.

Panel Size. Strength capacity in bending and tension are appropriate for panels 24 in. (600 mm) or greater in width. For panels less than 24 in. (600 mm) in width used in applications where failure could endanger human life, the following adjustment shall be made (x is the width, or dimension perpendicular to the applied stress):

- When x is 24 in. (600 mm) or greater, then

$$C_s = 1.00$$

- When x is a minimum of 8 in. (200 mm) to a maximum of 24 in. (600 mm), then

$$C_s = 0.25 + 0.0313x$$

- When x is less than or equal to 8 in. (200 mm), then

$$C_s = 0.50$$

Single strips less than 8 in. (200 mm) wide used in stressed applications shall be chosen such that they are relatively free of surface defects.

2.6.6 Section Properties

Where required, geometric cross-sectional properties may be calculated by assuming a uniform rectangular cross section in conjunction with nominal panel thickness given in Table 2.23. Computed rectangular (geometric) properties on a per-foot-of-panel-width basis are provided in Table 2.24.

Similarly, where design stress is required, design capacity may be divided by the applicable rectangular section property in Table 2.24.

2.6.7 Uniform Load Computations

Computation of uniform load capacity of wood structural panels shall be as outlined in this section for such applications as roofs, floors, and walls. The design capacities are subject to adjustment as specified earlier in this chapter.

Three basic span conditions are presented for computing uniform load capacities of wood structural panels. For normal framing practice and a standard panel size (4 × 8 ft [1200 × 2400 mm]), the APA has used the following assumptions in computing recommendations for load-span tables. When the panel strength axis is across (perpendicular to) the supports, the three-span condition is assumed for support spacing up to and including 32 in. (800 mm). The two-span condition is assumed for support spacing greater than 32 in. (800 mm).

When the panel strength axis is placed parallel to the supports, the three-span condition is assumed for support spacing up to and including 16 in. (400 mm), the two-span condition is assumed when the support spacing is greater than 16 in. (400 mm) up to 24 in. (600 mm), and a single span is assumed for spans greater than 24 in. (600 mm).

Two in. nominal (38 mm) lumber framing is assumed for support spacings less than 48 in. (1200 mm). Four in. nominal (89 mm) lumber framing is assumed for support spacing of 48 in. (1200 mm) or greater.

TABLE 2.23 Nominal Thickness by Span Rating (The nominal thickness is given. The predominant thickness for each span rating is highlighted in **bold** type)

	Nominal thickness (in.)										
Span rating	3/8	7/16	15/32	1/2	19/32	5/8	23/32	3/4	7/8	1	1-1/8
APA Rated Sheathing											
24/0	**0.375**	0.437	0.469	0.500							
24/16		**0.437**	0.469	0.500							
32/16			**0.469**	0.500	0.594	0.625					
40/20					**0.594**	0.625	0.719	0.750			
48/24							**0.719**	0.750	0.875		
APA Rated Sturd-I-Floor											
16 oc					**0.594**	0.625					
20 oc					**0.594**	0.625					
24 oc							**0.719**	0.750			
32 oc									**0.875**	1.000	
48 oc											**1.125**

Note: 1 in. = 25.4 mm.

TABLE 2.24 Panel Section Properties[a]

Nominal panel thickness	Approximate weight[b] (psf)	Nominal thickness t (in.)	Area A (in.^{2}/ft)	Moment of Inertia I (in./ft)	Section modulus S (in.^{3}/ft)	Statical moment Q (in.^{3}/ft)	Shear constant lb/Q (in.^{2}/ft)
3/8 in.	1.1	0.375	4.500	0.053	0.281	0.211	3.000
7/16 in.	1.3	0.437	5.250	0.084	0.383	0.287	3.500
15/32 in.	1.4	0.469	5.625	0.103	0.440	0.330	3.750
1/2 in.	1.5	0.500	6.000	0.125	0.500	0.375	4.000
19/32 in.	1.8	0.594	7.125	0.209	0.705	0.529	4.750
5/8 in.	1.9	0.625	7.500	0.244	0.781	0.586	5.000
23/32 in.	2.2	0.719	8.625	0.371	1.033	0.775	5.750
3/4 in.	2.3	0.750	9.000	0.422	1.125	0.844	6.000
7/8 in.	2.6	0.875	10.500	0.670	1.531	1.148	7.000
1 in.	3.0	1.000	12.000	1.000	2.000	1.500	8.000
1-1/8 in.	3.3	1.125	13.500	1.424	2.531	1.898	9.000

Note: 1 in. = 25.4 mm; 1 psf = 4.88 kg/m^2; 1 in.^{2}/ft width = 2116.67 mm^2/m width; 1 in.^{3}/ft width = 53763 mm^3/m width; 1 in.^{4}/ft width = 1.3656×106 mm^4/m width.

[a] Properties based on rectangular cross section of 1 ft width.

[b] Approximate plywood weight for calculating actual dead loads. For OSB and COM-PLY panels, increase tabulated weights by 10%.

The equations presented in this section are standard beam formulas altered to accept the mixed units noted. These formulas are provided for computing uniform loads on wood structural panels over conventional lumber framing. Because it is assumed that no blocking is used, the formulas are for one-way beam action, rather than two-way plate action. The resulting loads are assumed to be applied to full-sized panels in standard sheathing-type applications. Loads are for the panels only, and in no way account for the design of the framing supports. Further consideration should be given to concentrated loads, in compliance with local building codes and with maximum span recommendations of APA—The Engineered Wood Association.

Uniform Loads Based on Bending Strength. The following formulas shall be used for computing loads based on design bending strength capacity (F_bS):

- For a single span:

$$w_b = \frac{95\ F_bS}{\ell_1^2}$$

- For a two-span condition:

$$w_b = \frac{96\ F_bS}{\ell_1^2}$$

- For a three-span condition:

$$w_b = \frac{120\ F_bS}{\ell_1^2}$$

where w_b = uniform load based on bending strength (psf)
F_bS = design bending strength capacity (lb-in./ft)
ℓ_1 = span (in., center-to-center of supports)

Uniform Loads Based on Shear Strength. The following formulas shall be used for computing loads based on design shear strength capacity ($F_s[Ib/Q]$):

- For a single span:

$$w_s = \frac{24\ F_s(Ib/Q)}{\ell_2}$$

- For a two-span condition:

$$w_s = \frac{19.2\ F_s(Ib/Q)}{\ell_2}$$

- For a three-span condition:

$$w_s = \frac{20\ F_s(Ib/Q)}{\ell_2}$$

where w_s = uniform load based on shear strength (psf)
$F_s(Ib/Q)$ = design shear strength capacity (lb/ft)
ℓ_2 = clear span (in., center-to-center of supports minus support width)

Uniform Loads Based on Deflection Requirements. The following formulas shall be used for computing deflection under uniform load, or allowable loads based on deflection requirements.

- For a single span:

$$\Delta = \frac{w\ell_3^4}{921.6\ EI}$$

- For a two-span condition:

$$\Delta = \frac{w\ell_3^4}{2220\ EI}$$

- For a three-span condition:

$$\Delta = \frac{w\ell_3^4}{1743\ EI}$$

where Δ = deflection (in.)
w = uniform load (psf)
EI = design bending stiffness capacity (lb-in.^{2}/ft)
ℓ_3 = clear span + SW (in.)
SW = support-width factor, equal to 0.25 in. (6.5 mm) for 2 in. nominal (38 mm) lumber framing and 0.625 in. (16 mm) for 4 in. nominal (89 mm) lumber framing (for additional information on this factor see *APA Research Report 120*).

Uniform Load. For uniform load based on a deflection requirement, compute bending deflection with a uniform load (w) equal to one psf. The allowable uniform load based on the allowable deflection is then computed as:

$$w_d = \frac{\Delta_{\text{all.}}}{\Delta}$$

where w_d = uniform load based on deflection (psf)
$\Delta_{\text{all.}}$ = allowable deflection (in.)

2.6.8 Design Examples Showing Use of Capacity Tables

Note: In these examples, panel type is selected for illustrative purposes. Normally specification is by grade and span rating without regard to panel type, and calculations should assume the lowest adjustments (C_G) applicable to available constructions as given in Table 2.30 for the specified span rating.

Example 1—Conventional Roof. A four-ply plywood panel trademarked APA Rated Sturd-I-Floor 24 oc with tongue-and-groove edges was inadvertently installed over 4 in. nominal (89 mm) roof supports 48 in. (1200 mm) on center. The long dimension (strength axis) of the panel was placed perpendicular to supports. The local building code requires that the panel support a 25 psf (1200 N/m^2) snow load.

Bending Strength. From Table 2.19, a Rated Sturd-I-Floor 24 oc panel with stress applied parallel to the strength axis (long panel dimension perpendicular to supports) has a bending strength capacity (F_bS) of 640 lb-in./ft (237 N · m/m). This capacity is adjusted by a C_G factor of 1.1 as shown in Table 2.19A for four-ply plywood, and by a duration-of-load factor (C_D) of 1.15 (see above, Duration of Load). From Section 2.6.7, a two-span condition is assumed.

$$W_b = \frac{96\ F_bS}{\ell_1^2}$$

$$= \frac{96 \times (640 \times 1.1 \times 1.15)}{48^2}$$

$$= 34 \text{ psf } [1628 \text{ N/m}^2]$$

Shear Strength in the Plane. From Table 2.21, a Rated Sturd-I-Floor 24 oc panel with stress applied parallel to the strength axis has shear strength in the plane ($F_s[Ib/Q]$) of 340 lb/ft (4964 N/m). This capacity is adjusted by a C_G factor of 1.0 for four-ply plywood, and by a duration-of-load factor (C_D) of 1.15 (see above),

$$W_s = \frac{19.2\ F_s(Ib/Q)}{\ell_2}$$

$$= \frac{192\ (340 \times 1.0 \times 1.15)}{(48 - 3.5)}$$

$$= 169 \text{ psf } [8092 \text{ N/m}^2]$$

Bending Stiffness. From Table 2.19, a Rated Sturd-I-Floor 24 oc panel with stress applied parallel to the strength axis has a dry stiffness capacity (EI) of 300,000 lb-in.^{2}/ft (2825 N · m^2/m). This capacity is adjusted by a C_G factor of 1.1 for four-ply plywood as shown in Table 2.19A. The deflection limit for live load is $\ell/240$.

$$\Delta = \frac{w\ell_3^4}{2{,}220\ EI}$$

$$= \frac{1.0\ (48 - 3.5 + 0.625)^{-4}}{2{,}220 \times (300{,}000 \times 1.1)}$$

$$= 5.66 \times 10^{-3} \text{ in.}$$

$$W_d = \frac{\Delta_{\text{all.}}}{\Delta} = \frac{48/240}{5.66 \times 10^{-3}} = 35 \text{ psf}(1676 \text{ N/m}^2)$$

Bending strength controls (provides the lowest capacity) for this application. The bending strength capacity of 34 psf (1628 N/m^2) represents total load, from which dead load is subtracted to arrive at live load capacity. The bending stiffness capacity of 35 psf (1676 N/m^2) represents live load only. Here, if dead load (panel weight plus roofing) is no more than 9 psf (431 N/m^2), the 25 psf (1200 N/m^2) snow load capacity is achieved. The tongue-and-groove edges provide required edge supports.

Example 2—Panelized Roof. An oriented strand board (OSB) panel, APA Structural I Rated Sheathing 32/16 is to be used in a panelized roof system over 2 in. nominal (38 mm) framing members 24 in. (600 mm) on center. The long panel dimension (strength axis) of the panel will be placed parallel to supports.

Bending Strength. From Table 2.19, a Rated Sheathing 32/16 panel with stress applied perpendicular to strength axis (long panel dimension parallel to supports) has a bending strength capacity (F_bS) equal to 92 lb-in./ft (34.1 N · m/m). This capacity is adjusted by a C_G factor of 2.8 for OSB structural I (Table 2.19A), and by a duration-of-load factor (C_D) of 1.15 (see above, Duration of Load, under Section 2.6.5). This duration-of-load factor is normally associated with snow loads for roof structures. From Section 2.6.7, a two-span condition is assumed.

$$W_b = \frac{96\ F_bS}{\ell_1^2}$$

$$= \frac{96\ (92 \times 2.8 \times 1.15)}{24^2} = 49 \text{ psf } (2346 \text{ N/m}^2)$$

Shear Strength in the Plane. From Table 2.21, a Rated Sheathing 32/16 panel with stress applied perpendicular to strength axis has shear strength in the plane ($F_s[Ib/Q]$) of 130 lb/ft (1898 N/m). This capacity is adjusted by a C_G factor of 1.0 for OSB structural I (Table 2.21A) and by a duration-of-load factor (C_D) of 1.15 (see above, Duration of Load, under Section 2.6.5).

$$W_s = \frac{19.2\ F_s(Ib/Q)}{\ell_2}$$

$$= \frac{19.2\ (130 \times 1.0 \times 1.15)}{24 - 1.5)}$$

$$= 128 \text{ psf } [6129 \text{ N/m}^2]$$

Bending Stiffness. From Table 2.19, a Rated Sheathing 32/16 panel with stress applied perpendicular to strength axis has a dry stiffness capacity (EI) of 8100 lb-in.^{2}/ft (76.3 N · m^2/m). This capacity is adjusted by a C_G factor of 5.2 for OSB Structural I as shown in Table 2.19A. The deflection limit for live load is l/240.

$$\Delta = \frac{w\ell_3^4}{2220\ EI} = \frac{1.0\ (24 - 1.5 + 0.25)^4}{2220 \times (8100 \times 5.2)} = 2.865 \times 10^{-3} \text{ in.}$$

$$W_d = \frac{\Delta_{\text{all.}}}{\Delta} = \frac{24/240}{2.865 \times 10^{-3}} = 35 \text{ psf } (1676 \text{ N/m}^2)$$

Example 3—Floor. A COM-PLY panel marked APA Rated Sturd-I-Floor 24 oc is to be used in a floor system over supports 24 in. (600 mm) on center. The panels will be placed with the long panel dimension (strength axis) perpendicular to supports. Supports are 2 in. nominal (38 mm) framing members. The capacity of the panel will be computed based on bending strength, shear strength in the plane, and bending stiffness.

Bending Strength. From Table 2.19, a Rated Sturd-I-Floor 24 oc panel with stress applied parallel to the strength axis (long panel dimension perpendicular to supports) has a bending strength capacity (F_bS) of 640 lb-in./ft (237 N · m/m). This capacity is adjusted by a C_G factor of 1.2 as shown in Table 2.19A for COM-PLY. From Section 2.6.7, a three-span condition is assumed.

$$W_b = \frac{120\ F_bS}{\ell_1^2} = \frac{120 \times (640 \times 1.2)}{24^2} = 160 \text{ psf } (7661 \text{ N/m}^2)$$

Shear Strength in the Plane. From Table 2.21, a rated Sturd-I-Floor 24 oc panel with stress applied parallel to the strength axis has shear strength in the plane ($F_s[Ib/Q]$) equal to 340 lb/ft (4964 N/m). This capacity is adjusted by a C_G factor of 1.0 as shown in Table 2.21A for COM-PLY.

$$W_s = \frac{20\ F_s(Ib/Q)}{\ell^2}$$

$$= \frac{20\ (340 \times 1.0)}{24 - 1.5)}$$

$$= 302 \text{ psf } (14{,}460 \text{ N/m}^2)$$

Bending Stiffness. From Table 2.19, a Rated Sturd-I-Floor 24 oc panel with stress applied parallel to the strength axis has a dry stiffness capacity (*EI*) of 300,000 lb-in.^{2}/ft (2825 N · m^2/m). This capacity is adjusted by a C_G factor of 1.1 as shown in Table 2.19A for COM-PLY. The deflection limit for live load is $\ell/360$.

$$\Delta = \frac{w\ell_3^4}{1743\ EI} = \frac{1.0\ (24 - 1.5 + 0.25)^4}{1743 \times (300{,}000 \times 1.1)} = 4.657 \times 10^{-4} \text{ in.}$$

$$W_d = \frac{\Delta_{\text{all.}}}{\Delta} = \frac{24/360}{4.657 \times 10^{-4}} = 143 \text{ psf } (6847 \text{ N/m}^2)$$

While the above calculations would indicate that this Sturd-I-Floor construction has a live load capacity of 143 psf (6847 N/m^2) (limited by bending stiffness), it is important to note that some structural panel applications are not controlled by uniform load. Residential floors, commonly designed for 40 psf (1900 N/m^2) live load, are a good example. The calculated allowable load is greatly in excess of the typical design load. This excess does not mean that floor spans for Sturd-I-Floor can be increased, but only that there is considerable reserve strength and stiffness for uniform loads. Recommended maximum spans for wood structural panel floors are based on deflection under concentrated loads, how the floor feels to passing foot traffic, and other subjective factors which relate to user acceptance. Always check the maximum floor and roof spans for wood structural panels before making a final selection for these applications.

2.6.9 Load-Span Tables

The following load-span tables are furnished courtesy of APA—The Engineered Wood Association and apply to wood structural panels qualified and manufactured in accordance with APA PRP-108, PS 1 and PS 2. Loads are tabulated based on capacities given in Section 2.6.4 and are provided for applications where the panel strength axis is applied perpendicular to supports and parallel to supports. For each combination of span and span rating or thickness, loads are given for deflections of L/360, L/240, L/180, and maximum loads controlled by bending and shear capacity. The load-span tables were generated using the methodology given in Section 2.6.7.

The values given in Tables 2.25 and 2.26 represent the maximum allowable loads without regard to panel type. Table 2.27 is based on sanded plywood conforming to PS 1. These values may be adjusted for panel type or for sanded plywood grade using Table 2.28. Once the allowable loads have been adjusted for panel type, they should be further adjusted for application conditions using Table 2.29.

TABLE 2.25 Uniform Loads (psf) on APA Rated Sheathing (Multispan, normal duration of load, dry conditions)

		Span, center-to-center supports (in.)												
		Strength axis perpendicular to supports										Strength axis parallel to supports		
Span rating	Load governed by	12	16	19.2	24	30	32	36	40	48	60	12	16	24
24/0	$L/360$	261	98	54	26	13	10	9				16	6	
	$L/240$	392	147	81	39	19	16	14				23	9	
	$L/180$	522	196	107	52	26	21	18				31	12	
	Bending	208	117	81	52	33	29	19				45	25	
	Shear	314	228	186	147	116	108	92				200	125	
24/16	$L/360$	339	128	70	34	17	14	12	9			23	9	
	$L/240$	509	191	105	51	25	20	18	13			34	13	
	$L/180$	679	255	140	68	33	27	24	17			45	17	
	Bending	267	150	104	67	43	38	24	19			53	30	
	Shear	362	262	215	169	133	125	106	95			200	145	
32/16	$L/360$	500	188	103	50	24	20	18	13			35	13	4
	$L/240$	750	282	154	75	37	30	26	19			53	20	7
	$L/180$	1001	376	206	100	49	40	35	25			70	27	9
	Bending	308	173	120	77	49	43	27	22			77	43	15
	Shear	400	290	237	187	147	138	117	105			248	179	111
40/20	$L/360$	979	368	201	98	48	39	34	25	16		78	29	10
	$L/240$	1468	552	302	146	72	58	51	37	24		117	44	15
	$L/180$	1958	736	403	195	96	78	69	49	32		157	59	20
	Bending	521	293	203	130	83	73	46	38	26		125	70	25
	Shear	505	366	299	236	186	174	147	132	114		314	228	141
48/24	$L/360$	1740	655	358	174	85	69	61	44	29	14	128	48	16
	$L/240$	2610	982	537	260	128	104	91	66	43	21	193	72	24
	$L/180$	3480	1309	716	347	170	139	122	88	57	28	257	97	33
	Bending	704	396	275	176	113	99	63	51	35	23	188	105	38
	Shear	648	469	384	302	239	223	189	170	147	116	362	262	162

Note: 1 in. = 25.4 mm; 1 psf = 47.88 N/m^2.

TABLE 2.26 Uniform Loads (psf) on APA Rated Sturd-I-Floor (Multispan, normal duration of load, dry conditions)

Span rating	Load governed by	Span, center-to-center supports (in.)												
		Strength axis perpendicular to supports										Strength axis parallel to supports		
		12	16	19.2	24	30	32	36	40	48	60	12	16	24
16 oc	$L/360$	653	245	134	65	32	26	23	16	11		48	18	6
	$L/240$	979	368	201	98	48	39	34	25	16		72	27	9
	$L/180$	1305	491	269	130	64	52	46	33	21		96	36	12
	Bending	346	195	135	86	55	49	31	25	17		83	47	17
	Shear	429	310	254	200	158	148	125	112	97		276	200	124
20 oc	$L/360$	914	344	188	91	45	36	32	23	15		57	21	7
	$L/240$	1370	516	282	137	67	55	48	34	22		85	32	11
	$L/180$	1827	687	376	182	89	73	64	46	30		113	43	14
	Bending	400	225	156	100	64	56	36	29	20		117	66	23
	Shear	505	366	299	236	186	174	147	132	114		324	234	145
24 oc	$L/360$	1305	491	269	130	64	52	46	33	21	10	113	43	14
	$L/240$	1958	736	403	195	96	78	69	49	32	16	170	64	22
	$L/180$	2610	982	537	260	128	104	91	66	43	21	226	85	29
	Bending	533	300	208	133	85	75	47	38	27	17	179	101	36
	Shear	648	469	384	302	239	223	189	170	147	116	371	269	166
32 oc	$L/360$	2828	1064	582	282	138	113	99	71	46	23	326	123	41
	$L/240$	4242	1596	873	473	207	169	148	107	70	34	489	184	62
	$L/180$	5656	2128	1164	564	276	225	198	142	93	45	653	245	83
	Bending	725	408	283	181	116	102	64	52	36	23	317	178	63
	Shear	762	552	452	356	281	262	223	199	173	136	533	386	239
48 oc	$L/360$	5003	1882	1030	499	244	199	175	126	82	40	696	262	88
	$L/240$	7505	2823	1545	748	367	299	263	189	123	60	1044	393	133
	$L/180$	10006	3764	2060	998	489	399	350	252	164	80	1392	524	177
	Bending	1333	750	521	333	213	188	119	96	67	43	567	319	113
	Shear	1143	828	678	533	421	393	334	299	259	204	857	621	384

Note: 1 in. = 25.4 mm; 1 psf = 47.88 N/m^2.

TABLE 2.27 Uniform Loads (psf) on APA Group 1 Sanded Panels (Multispan, normal duration of load, dry conditions)

Nom thick (in.)	Load governed by	Span, center-to-center supports (in.)												
		Strength axis perpendicular to supports										Strength axis parallel to supports		
		12	16	19.2	24	30	32	36	40	48	60	12	16	24
11/32	$L/360$	148	56	30	15	7	6	5				8	3	1
	$L/240$	222	83	46	22	11	9	8				11	4	1
	$L/180$	296	111	61	29	14	12	10				15	6	2
	Bending	129	73	50	32	21	18	11				22	12	4
	Shear	276	200	164	129	102	95	81				257	186	115
3/8	$L/360$	213	80	44	21	10	8	7				12	5	2
	$L/240$	320	120	66	32	16	13	11				18	7	2
	$L/180$	426	160	88	43	21	17	15				24	9	3
	Bending	171	96	67	43	27	24	15				31	17	6
	Shear	314	228	186	147	116	108	92				352	255	158
15/32	$L/360$	522	196	107	52	26	21	18	13	9		48	18	6
	$L/240$	783	295	161	78	38	31	27	20	13		72	27	9
	$L/180$	1,044	393	215	104	51	42	37	26	17		96	36	12
	Bending	296	166	116	74	47	42	26	21	15		92	52	18
	Shear	419	303	249	196	154	144	122	110	95		248	179	111
1/2	$L/360$	609	229	125	61	30	24	21	15	10		67	25	9
	$L/240$	914	344	188	91	45	36	32	23	15		101	38	13
	$L/180$	1,218	458	251	121	60	49	43	31	20		135	51	17
	Bending	325	183	127	81	52	46	29	23	16		121	68	24
	Shear	448	324	266	209	165	154	131	117	101		276	200	124
19/32	$L/360$	892	336	184	89	44	36	31	22	15	7	163	61	21
	$L/240$	1,338	503	275	133	65	53	47	34	22	11	245	92	31
	$L/180$	1,784	671	367	178	87	71	62	45	29	14	326	123	41
	Bending	433	244	169	108	69	61	39	31	22	14	188	105	38
	Shear	552	400	328	258	204	190	161	145	125	99	286	207	128
5/8	$L/360$	1,001	376	206	100	49	40	35	25	16	8	211	79	27
	$L/240$	1,501	565	309	150	73	60	53	38	25	12	317	119	40
	$L/180$	2,001	753	412	200	98	80	70	50	33	16	422	159	54
	Bending	467	263	182	117	75	66	41	34	23	15	225	127	45
	Shear	590	428	350	276	218	203	173	155	134	105	314	228	141

TABLE 2.27 Uniform Loads (psf) on APA Group 1 Sanded Panels (Multispan, normal duration of load, dry conditions) (*Continued*)

Nom thick (in.)	Load governed by	Span, center-to-center supports (in.)												
		Strength axis perpendicular to supports										Strength axis parallel to supports		
		12	16	19.2	24	30	32	36	40	48	60	12	16	24
23/32	$L/360$	1,392	524	287	139	68	55	49	35	23	11	394	148	50
	$L/240$	2,088	786	430	208	102	83	73	53	34	17	591	222	75
	$L/180$	2,784	1,047	573	278	136	111	97	70	46	22	787	296	100
	Bending	538	302	210	134	86	76	48	39	27	17	317	178	63
	Shear	667	483	395	311	246	230	195	175	151	119	381	276	171
3/4	$L/360$	1,544	581	318	154	75	62	54	39	25	12	500	188	64
	$L/240$	2,317	871	477	231	113	92	81	58	38	19	750	282	95
	$L/180$	3,089	1,162	636	308	151	123	108	78	51	25	1,001	376	127
	Bending	567	319	221	142	91	80	50	41	28	18	392	220	78
	Shear	686	497	407	320	253	236	200	180	155	122	410	297	183
7/8	$L/360$	2,175	818	448	217	106	87	76	55	36	17	805	303	102
	$L/240$	3,263	1,227	672	325	159	130	114	82	54	26	1,207	454	153
	$L/180$	4,351	1,637	895	434	213	173	152	109	71	35	1,610	606	204
	Bending	708	398	277	177	113	100	63	51	35	23	542	305	108
	Shear	810	586	480	378	298	279	237	212	183	144	514	372	230
1	$L/360$	3,306	1,244	681	330	162	132	116	83	54	26	1,436	540	182
	$L/240$	4,960	1,866	1,021	495	242	198	174	125	81	40	2,154	810	273
	$L/180$	6,613	2,488	1,361	659	323	263	231	166	109	53	2,871	1,080	365
	Bending	917	516	358	229	147	129	81	66	46	29	813	457	163
	Shear	895	648	531	418	330	308	262	234	203	160	714	517	320
1⅛	$L/360$	4,285	1,612	882	427	209	171	150	108	70	34	2,132	802	271
	$L/240$	6,428	2,418	1,323	641	314	256	225	162	105	51	3,198	1,203	406
	$L/180$	8,571	3,224	1,764	855	419	341	300	216	141	68	4,264	1,604	541
	Bending	1,125	633	439	281	180	158	100	81	56	36	1,042	586	208
	Shear	1,000	724	593	467	368	344	292	262	227	178	848	614	380

Note: 1 in. = 25.4 mm; 1 psf = 47.88 N/m^2.

TABLE 2.28 Adjustments to Allowable Load Capacities Based on Panel Grade and Constructions, C_g

	Strength axis[a]			
	Perpendicular to supports		Parallel to supports	
	Other	Structural I	Other	Structural I
	Stiffness (L/360, L/240, L/180)			
Rated panels				
3-ply plywood	1.1	1.1	1.0	1.5
4-ply plywood, COM-PLY	1.1	1.1	2.2	3.3
5-ply plywood[b]	1.1	1.1	3.1	5.2
OSB	1.0	1.0	3.1	5.2
Sanded plywood				
A-A, A-C	1.0	1.0	1.0	1.4
Marine	1.0	1.0	1.4	1.4
Other	1.0	1.0	1.0	1.4
	Bending			
Rated panels				
3-ply plywood	1.0	1.0	1.0	1.3
4-ply plywood	1.1	1.1	1.2	1.7
COM-PLY	1.2	1.2	1.2	1.7
5-ply plywood,[b] OSB	1.2	1.2	1.8	2.8
Sanded plywood				
A-A, A-C	1.2	1.3	1.2	1.7
Marine	1.1	1.1	1.4	1.4
Other	1.0	1.1	1.0	1.4
	Shear			
Rated panels				
3-ply plywood	1.0	1.4	2.8	5.2
4-ply plywood	1.0	1.4	3.9	7.9
5-ply plywood[b]	1.1	1.6	1.0	1.4
OSB, COM-PLY	1.0	1.0	1.0	1.0
Sanded plywood				
A-A, A-C	1.0	1.3	1.0	1.4
Marine	1.3	1.3	1.4	1.4
Other	1.0	1.3	1.0	1.4

[a] The strength axis is the long panel dimension unless otherwise identified.

[b] Adjustments apply to plywood with five or more layers; for five-ply/three-layer plywood, use adjustments for four-ply.

TABLE 2.29 Application Adjustment Factors

Duration of load, C_D (applies to bending and shear only)	
Permanent load (over 10 years)	0.90
Two months, as for snow	1.15
Seven days	1.25
Wind or earthquake	1.60[a]
Impact	2.00
Span adjustment	
Two-span to 1-span	
Deflection	0.42
Bending	1.00
Shear	1.25
Three-span to 1-span	
Deflection	0.53
Bending	0.80
Shear	1.20
Wet or damp locations, C_M (moisture content 16% or more)	
Deflection	0.85
Bending	0.75
Shear	0.75

[a] Check local building code.

TABLE 2.30 Typical APA Panel Constructions[a]

	Plywood				
Span rating	3-ply	4-ply	5-ply[b]	COM-PLY	OSB
APA Rated Sheathing					
24/0	X				X
24/16					X
32/16	X	X	X		X
40/20	X	X	X		X
48/24		X	X		X
APA Rated Sturd-I-Floor					
16 oc					
20 oc		X	X	X	X
24 oc		X	X	X	X
32 oc			X	X	X
48 oc			X	X	X

[a] Constructions listed may not be available in every area. Check with suppliers concerning availability.
[b] Applies to plywood with five or more layers.

2.6.10 Design Examples Showing Use of Load-Span Tables

To assist in ascertaining the availability of a specific panel type, Table 2.30 has been developed by APA.

Example 1—Floor. Find the allowable uniform floor load for APA Rated Sheathing 32/16, plywood, when applied at its rated span. From Table 2.30 it can be seen that 32/16 plywood sheathing is available in three-, four- or five-ply. Since actual construction may not be known during design, assume the most conservative values. Table 2.28 indicates the most conservative values for plywood are those for three-ply. Assume 10 psf (479 N/m^2) dead load, and panel strength axis across supports 16 in. (400 mm) oc. Unless stated otherwise, assume floor deflection criteria to be $L/360$ under live load and $L/240$ under total load.

Note: In these examples, panel type is selected for illustrative purposes. Often specification is by grade and span rating without regard to panel type, and calculations should assume the lowest adjustments (Table 2.28) applicable to typical constructions (Table 2.30) for the specified span rating.

From Tables 2.25 and 2.28:

For APA Rated Sheathing Panels with Strength Axis Across Supports

Load governed by	Load (psf)		Adjustment for panel grade and construction, C_G		Adjusted load (psf)
$L/360$	188	×	1.1	=	207
$L/240$	282	×	1.1	=	310
$L/180$	376	×	1.1	=	414
Bending	173	×	1.0	=	173
Shear	290	×	1.0	=	290

Note: 1 psf = 47.88 N/m^2.

Allowable total load for floors is the least of loads for $L/240$, bending and shear.

Allowable total load is 173 psf (8283 N/m^2).

Live load is the lesser of the load for $L/360$, and total load, as determined above, minus dead load.

$$L/360 = 207 \text{ psf } (9911 \text{ N/m}^2)$$

Total load − dead load = 173 − 10 = 163 psf (7804 N/m^2)

Allowable live load = 163 psf (7804 N/m^2), or 165 psf (7900 N/m^2) (rounded to nearest 5 psf (239 N/m^2)).

Note: Do not increase span beyond the floor span rating even though the allowable uniform live load greatly exceeds the 40 psf (1900 N/m^2) design live load normally used for floors. Recommended maximum span reflects performance under concentrated and impact loads in addition to uniform load.

Example 2—Conventional Roof. Find the allowable snow load for APA Rated Sturd-I-Floor 24 oc, OSB, when the panel is used as roof sheathing with the strength axis across supports spaced 32 in. (800 mm) o.c. In question are several panels in the one-span condition. Deflection criteria are $L/240$ under live load only and $L/180$ under total load. Assuming a two-month duration of load for snow, allowable loads for bending and shear may be increased 15%. Assume that 10 psf (479 N/m²) dead load is supported by the Sturd-I-Floor.

From Tables 2.26, 2.28, and 2.29:

For APA Rated Sturd-I-Floor Panels with Strength Axis Across Supports

Load governed by	Load (psf)		Adjustment for panel grade and construction, C_G		Adjustment for duration of load, C_D		Adjustment for span		Adjusted load (psf)
$L/360$	52	×	1.0			×	0.53	=	28
$L/240$	78	×	1.0			×	0.53	=	41
$L/180$	104	×	1.0			×	0.53	=	55
Bending	75	×	1.2	×	1.15	×	0.80	=	83
Shear	223	×	1.0	×	1.15	×	1.20	=	308

Note: 1 psf = 47.88 N/m².

Allowable total load is the lesser of the load for $L/180$, bending and shear. Allowable total load is 55 psf (2633 N/m²).

Live load is the lesser of the load for $L/240$, and total load as determined above minus dead load.

$$L/240 = 41 \text{ psf } (1963 \text{ N/m}^2)$$

$$\text{Total load} - \text{dead load} = 55 - 10 = 45 \text{ psf } (2155 \text{ N/m}^2)$$

In this case, live load is governed by deflection of $L/240$: allowable live load = 41 psf (1963 N/m²), or 40 psf (1900 N/m²) (rounded to nearest 5 psf [239 N/m²]).

Example 3—Soil Pressure. Find allowable soil pressure on 23/32 in. APA B-C Group 1 EXT if supports are 16 in. (400 mm) o.c. Face grain is across supports. Deflection need not be considered. Assume soil pressure is permanent load.

From Tables 2.27 and 2.28:

For APA Group 1 Sanded Panels with Strength Axis across Supports

Load governed by	Load (psf)		Adjustment for panel grade and construction, C_G		Adjustment for duration of load, C_D		Adjustment for moisture		Adjusted load (psf)
$L/360$	524	×	1.0			×	0.85	=	445
$L/240$	786	×	1.0			×	0.85	=	668
$L/180$	1047	×	1.0			×	0.85	=	890
Bending	302	×	1.0	×	0.90	×	0.75	=	204
Shear	483	×	1.0	×	0.90	×	0.75	=	326

Note: 1 psf = 47.88 N/m².

Allowable load = 204 psf (9768 N/m^2), or 205 psf (9815 N/m^2) (rounded to nearest 5 psf (239 N/m^2)).

2.7 PHYSICAL PROPERTIES

The physical properties of wood as a material are discussed in Chapter 1. The purpose of this section is to discuss the effect of basic physical properties on as-built construction using wood structural panels.

2.7.1 Thermal

Heat has a number of important effects on wood structural panels. Temperature affects the equilibrium moisture content and the rate of adsorption and desorption. At temperatures above 100°F (38°C) the equilibrium moisture content will decrease. As temperatures are lowered, the strength and stiffness of wood increase.

Elevated Temperature. Capacities in Tables 2.19 through 2.22 apply at temperatures of 70°F (21°C) and lower. Wood structural panel parts of buildings should not be exposed to temperatures above 200°F (93°C) for more than very brief periods. However, between 70°F (21°C) and 200°F (93°C) adjustments to capacity generally do not need to be made, because the need for adjustment of dry capacities depends upon whether moisture content will remain in the 12–15% range or whether the panel will dry to lower moisture contents as a result of the increase in temperature. If drying occurs, as is usually the case, the increase in strength due to drying can offset the loss in strength due to elevated temperature. For instance, temperatures of up to 150°F (66°C) or higher do occur under roof coverings of buildings on hot days, but they are accompanied by moisture content reductions that offset the strength loss, so that high temperatures are not considered in the design of roof structures. To maintain a moisture content of 12% at 150°F [66°C], sustained relative humidity of around 80% would be required. The designer needs to exercise judgment in determining whether high temperature and moisture content occur simultaneously and the corresponding need for temperature adjustment of capacities.

Cryogenic Temperatures. Investigations of wood in low temperatures, down to −300°F (−184°C), have shown mechanical strength to increase. The increase is up to three times the property measured at room temperature, depending on the strength property and moisture content. This increase is consistent with other materials that exhibit increased resistance to changes in form as the temperature drops.

Cycling of freezing and thawing does not seem to affect the properties of the wood itself but may reduce the strength of some fastenings by as much as 10%. In practical applications of wood products the increase in strength due to exposure to subnormal temperature will tend to offset strength losses caused by other factors.

On the basis of available test information, published capacities for wood structural panels are considered applicable at temperatures down to −300°F (−184°C).

Coefficient of Linear Thermal Expansion. The thermal expansion of wood is much smaller than swelling due to absorption of water. Because of this, thermal expansion can be neglected in cases where wood is subject to considerable swelling and shrinking. It may be of importance only where the wood is used in assemblies with other materials.

The thermal expansion of wood is quite small and requires exacting techniques for its measurement. The effect of temperature on plywood dimensions is related to the percentage of panel thickness in plies having grain perpendicular to the direction of expansion or contraction. The average coefficient of linear thermal expansion is about 3.4×10^{-6} in./in. per degree F (6.1×10^{-6} mm/mm per degree C) for a plywood panel with 60% of the plies or less running perpendicular to the face. The coefficient of thermal expansion for panel thickness is approximately 16×10^{-6} in./in. per degree F (29×10^{-6} mm/mm per degree C).

Thermal Resistance. For most wood structural panel applications, the important thermal quality is resistance, or insulating effectiveness. Although thermal conductivity or resistance varies somewhat depending on wood species, for most practical purposes it is neither necessary nor feasible to determine the actual species makeup of the panel. For determining the overall coefficient of heat transmission (U-value) of a construction assembly, APA publications use $k = 0.80$ Btu · in./(h · ft^2 · °F) (0.115 W/(m · °C)] for softwood, as listed by the American Society of Heating, Refrigerating and Air Conditioning Engineers (ASHRAE).[12] Use of this single value simplifies computations and produces only insignificant differences in resulting design heat losses.

Table 2.31 shows thermal resistance, R, for several panel thicknesses, based on $k = 0.80$ Btu · in./(h · ft^2 · °F) [0.115 W/(m · °C)]. Thermal resistance represents the ability of the material to retard heat flow and is the reciprocal of k, adjusted for actual material thickness.

Thermal Degradation and Ignition Point. There are a few applications where extremes in temperature may be encountered, such as solar collectors, heating ducts, and other industrial applications. From an appearance standpoint, unprotected plywood should not be used when temperatures exceed 200°F (93°C). Between 70°F (21°C) and 200°F (93°C) strength loss is recovered when temperature is reduced. At temperatures above 200°F (93°C) plywood undergoes slow thermal decompo-

TABLE 2.31 Thermal Resistance

Panel thickness, in.	Thermal resistance, R
1/4	0.31
5/16	0.39
3/8	0.47
7/16	0.55
15/32	0.59
1/2	0.62
19/32	0.74
5/8	0.78
23/32	0.90
3/4	0.94
7/8	1.09
1	1.25
1 1/8	1.41

Note: 1 in. = 25.4 mm; R (ft^2 · h · °F/Btu) = 0.176 m^2 · °K/W.

sition, which permanently reduces its strength. Exposure to sustained temperatures higher than 200°F (93°C) will result in charring and weight loss. Using plywood in applications involving periodic exposure to temperatures from 200°F (93°C) to 302°F (150°C) should be based on the amount of exposure and the amount of decomposition that can be tolerated without impairing the serviceability of the panel.

A number of attempts have been made to measure a definite ignition temperature of wood, with little success. A specific temperature is hard to define, due to the fact that there are so many contributing factors, such as size and shape of the material, air circulation, rate of heating, moisture content of the wood, and so on. Estimates range from 510°F (265°C) to 932°F (500°C), but no value should be accepted as an absolute.

2.7.2 Water Vapor Permeability

Water vapor permeance is the rate of water vapor transmission through a material at a given vapor pressure gradient between the two surfaces. The accepted unit of permeance is the perm, given as one grain/ft^2/hr/in. of mercury vapor pressure. A grain is a unit of weight (1 lb (453.6 g) = 7000 grains).

The permeability of plywood is different from solid wood in several ways. The veneers from which plywood is made generally contain lathe checks from the manufacturing process. These small cracks provide pathways for materials to pass when entering through the panel edge.

When permeability is measured through the panel thickness, a number of variables will affect the actual flow rate. The anatomy of the species, consistency of the glueline, numbers of void spaces, and growth characteristics all combine to determine permeability. Due to natural variability, tables of water vapor permeance should be used only as best estimates.

Research at the National Institute of Science and Technology has shown that the water vapor permeance is very sensitive to the relative humidity gradients. For example, at 50% humidity the water vapor permeance of plywood is approximately 1 perm (57.45 ng/(s · m^2 · Pa)), but the water vapor permeance may be increased by a factor of 10 when the humidity is increased to 90%. Similar results are reported for an OSB siding product that had been coated with a latex paint.

Values for water vapor permeance are given in Table 2.32 for typical grades of plywood and OSB with and without various finishes and overlays. The data for this table were collected from testing in accordance with the dry-cup method given in ASTM Standard C355[21] or E96[22] (supersedes C355). All materials tested were selected to be representative of the plywood, OSB, and overlay as produced at the time of testing.

2.7.3 Dimensional Stability

When wood structural panels are exposed to a constant relative humidity they will eventually reach an equilibrium moisture content, the moisture content when it has reached equilibrium with the surrounding atmosphere. The equilibrium moisture content (EMC) of panels is highly dependent on relative humidity, but essentially independent of temperature between 32°F (0°C) and 100°F (38°C). The typical relationship between relative humidity and EMC of wood structural panels is given in Table 2.33.

TABLE 2.32 Water Vapor Permeance

Product	Water vapor permeance	
	Perms[a]	g/hr · m² · mmHg
Exterior plywood ⅜ in.[b]	0.8	0.024
Exterior plywood ⅜ in.[c]	0.2	0.006
Exterior MDO plywood ⅜ in. (overlaid 1 side)	0.3	0.008
Exterior MDO plywood ½ in. (overlaid 2 sides)	0.2	0.006
Exterior HDO plywood ½ in. & ⅝ in. (overlaid 2 sides)	0.1	0.003
OSB 7⁄16 in.	0.9	0.025
OSB 15⁄32 in., ½ in.	0.7	0.019
OSB 19⁄32 in., ⅝ in.	0.7	0.020
OSB 23⁄32 in., ¾ in.	0.5	0.013

Note: 1 in. = 25.4 mm; 1 perm = 57.45 ng/(s · m² · Pa).
[a] A "perm" is 1 grain/hr/ft²/in. of mercury vapor pressure.
[b] Range of seven species: 0.5–1.4.
[c] Surface finished with one coat exterior oilprimer plus two coats exterior finish paint.

TABLE 2.33 Relative Humidity (RH) Related to Equilibrium Moisture Content of Panels (EMC) at 75°F

RH (%)	EMC (%)	
	Plywood	OSB
10	1	1
20	3	1
30	5	2
40	6	4
50	7	5
60	8	6
70	11	9
80	15	13
90	19	17

Wood structural panels generally exhibit greater dimensional stability than other wood-based building products. It is a well-known fact that when wood loses or gains water it shrinks and swells. The shrinking of solid wood along the grain with changes in moisture content is about 1⁄20 to 1⁄40 of that across the grain. The tendency of individual plywood veneers to swell or shrink crosswise, therefore, is greatly restricted by the relative longitudinal stability of the adjacent plies.

The average coefficient of hygroscopic expansion or contraction in length and width for plywood panels is about 0.0002 in./in. (0.0002 mm/mm) for each 10% change in equilibrium relative humidity. The total change from oven dry to fiber saturation averages about 0.2%.

Expansion of a plywood panel that is free to move consists of not only uniform swelling across the width and length, but also additional swelling at the panel edges. Edge swelling is independent of panel size, and varies with the thickness of veneers having grain perpendicular to the direction of expansion. For identical veneer thicknesses, this expansion is about twice as great for the face ply as for the inner plies. For panels of balanced construction, edge swelling could reach a maximum of about 0.002 in. (0.05 mm) at each edge for each 10% increase in equilibrium relative humidity above 40%.

The average coefficient of hygroscopic expansion in thickness is about 0.01 in./in. (0.01 mm/mm) of original thickness for each 10% change in equilibrium relative humidity.

In normal conditions of dry use, relative humidity may vary between 40 and 80%, with corresponding equilibrium moisture contents ranging from 6–14%. Total dimensional changes of a 48 × 96 in. (1200 × 2400 mm) panel exposed to this change in conditions may be expected to average about 0.05 in. (1.25 mm) across the width and 0.09 in. (2.3 mm) along the length. Certain plywood constructions (combinations of veneer thickness) exhibit greater resistance to warping than others. The ideal construction has about 50% of the veneers running in each direction and is assembled in a balanced manner about the central plane. As a general rule, the greater the number of plies, the better the stability. For most plywood applications a minimum space of ⅛ in. (3 mm) should be provided at panel ends and edges to allow for expansion due to pickup in moisture content. When spacing recommendations are not followed, there is increased possibility of unsightly appearance due to panel buckling. See Section 2.5.2.

2.8 STORAGE AND HANDLING

Like all building materials, wood structural panels should be properly stored, handled, and installed to ensure superior in-service performance. Protect the edges and ends of panels, especially tongue-and-groove and shiplap-edged panels. Place panels to be moved by forklift on pallets or bunks when received to avoid damage by fork tines.

Panels to be transported on open truck beds should be covered with standard tarpaulins. For open railcar shipment, use lumber wrap to avoid extended weather exposure.

Store panels whenever possible under a roof, especially if they won't be used soon after they are received. Keep sanded and other appearance grades away from open doorways, and weight down the top panel in a stack to help avoid any possible warpage from humidity. If moisture absorption is expected, cut steel banding on panel bundles to prevent edge damage.

Panels to be stored outside should be stacked on a level platform supported by 4 × 4 (89 × 89 mm) stringers or other blocking, as illustrated in Fig. 2.21. Never leave panels or the platform in direct contact with the ground. Use at least three full-width supports along the 8 ft length of the panel—one centered and the others 12–16 in. (300–400 mm) from each end.

Cover the stack loosely with plastic sheets or tarps. Anchor the covering at the top of the stack, but keep it open and away from the sides and bottom to assure good ventilation. Tight coverings prevent air circulation and, when exposed to sunlight, create a greenhouse effect, which may encourage mold formation.

Build platform of cull panel and scrap lumber 4x4s for stacking panels.

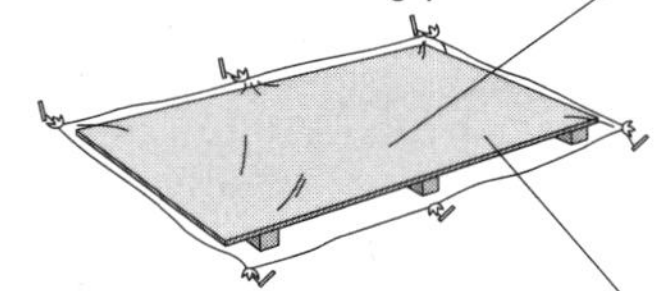

Stretch plastic film over platform to block passage of ground moisture.

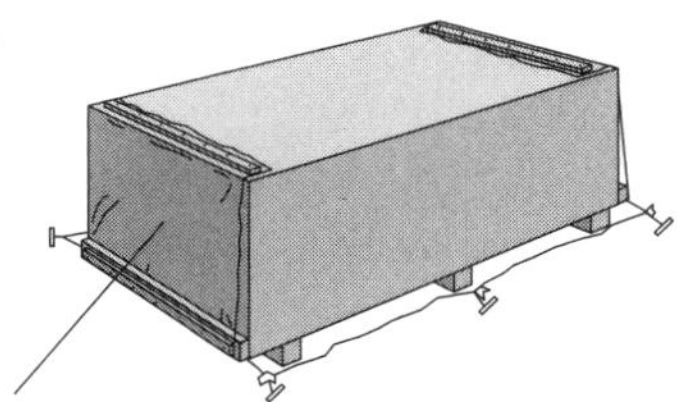

Nail film to top panel and drape over ends for protection against driving rain. Weight lower end with 2x4.

Lay two 2x4s on top of stack.
Pad corners with rags.

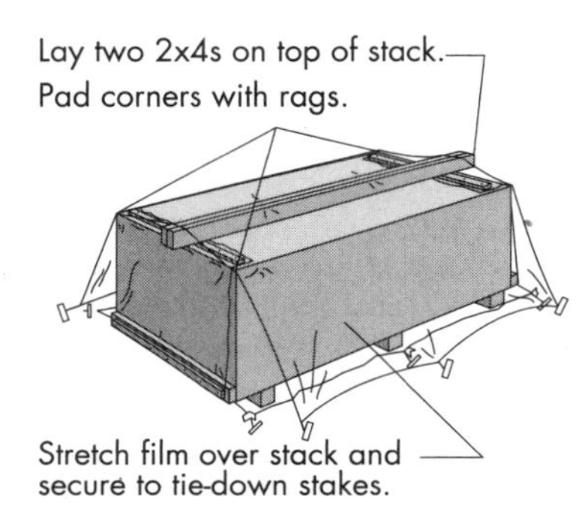

Stretch film over stack and secure to tie-down stakes.

FIGURE 2.21 Proper storage of wood structural panels.

2.9 REFERENCES

1. *Construction and Industrial Plywood,* Voluntary Product Standard PS 1, U.S. Department of Commerce, Washington, DC.
2. *Performance Standard for Wood-Based Structural-Use Panels,* Voluntary Product Standard PS 2, U.S. Department of Commerce, Washington, DC.
3. *Performance Standards and Policies for Structural-Use Panels,* PRP-108, APA—The Engineered Wood Association.
4. *Adhesives for Field-Gluing Plywood to Wood Framing,* Specification AFG-01, APA—The Engineered Wood Association.
5. *Specification for Adhesives for Field-Gluing Plywood to Lumber Framing for Floor Systems,* ASTM D 3498, American Society for Testing and Materials, Philadelphia.
6. *303 Siding Manufacturing Specifications,* Form B840, APA—The Engineered Wood Association, Tacoma, WA.

7. *Minimum Design Loads for Buildings and Other Structures,* ASCE 7, American Society of Civil Engineers, Reston, VA.
8. *Wire, Cut and Wrought Nails, Staples and Spikes,* Federal Specification FF-N-105B, General Services Administration, Washington, DC.
9. *Specification for Driven Fasteners: Nails, Spikes and Staples,* ASTM F 1667, American Society for Testing and Materials, Philadelphia.
10. *Hurricane Shutter Designs,* Form T450, APA—The Engineered Wood Association, Tacoma, WA.
11. *Roofing Materials and Systems,* Underwriters Laboratories, Inc., Northbrook, IL.
12. *Handbook—Fundamentals,* American Society of Heating, Refrigerating and Air-Conditioning Engineers, Atlanta, GA.
13. *Methods of Testing Structural Panels in Flexure,* ASTM D 3043, American Society for Testing and Materials, Philadelphia.
14. *Test Method for Structural Panels in Tension,* ASTM D 3500, American Society for Testing and Materials, Philadelphia.
15. *Test Method for Testing Structural Panels in Compression,* ASTM D 3501, American Society for Testing and Materials, Philadelphia.
16. *Test Method for Structural Panels in Planar Shear (Rolling Shear),* ASTM D 2718, American Society for Testing and Materials, Philadelphia.
17. *Test Method for Structural Panels in Shear Through-the-Thickness,* ASTM D 2719, American Society for Testing and Materials, Philadelphia.
18. *Relation of Strength of Wood to Duration of Load,* Report FPL R-1916, USDA Forest Products Laboratory, Madison, WI.
19. *Plywood—Preservative Treatment by Pressure Process,* AWPA Standard C9, American Wood-Preservers' Association, Granbury, TX.
20. *Fire-Retardant-Treated Plywood Roof Sheathing—General Information,* Form SPE-1007, APA—The Engineered Wood Association, Tacoma, WA.
21. *Test Methods for Water Vapor Transmission of Thick Materials,* ASTM C 355 (replaced by ASTM E 96), American Society for Testing and Materials, Philadelphia.
22. *Test Methods for Water Vapor Transmission of Materials,* ASTM E 96, American Society for Testing and Materials, Philadelphia.

CHAPTER THREE
WOOD STRUCTURAL PANELS IN STRUCTURAL COMPONENTS

H. Fulton Desler
Senior Engineer, TSD

*3.1 INTRODUCTION**

The structural components included in this chapter are box beams, stressed-skin panels, and structural insulated panels (SIPs). These are composed of a combination of wood structural panels and lumber, or structural composite lumber, although flanges, stiffeners, and splices may also be composed of wood structural panels with little or no sawn lumber products required.

Components made of wood structural panels and lumber are often major structural members, which depend on the glued or mechanically fastened joints to combine the separate pieces into an efficient structural unit capable of carrying the design loads. Materials in these components may be stressed to an appreciably higher level than in nonengineered construction.

Since improperly designed or fabricated components could constitute a hazard to life safety and property damage, it is strongly recommended that they be designed by qualified design professionals, using recognized design and fabrication methods, and that adequate quality control be maintained during manufacture.

To ensure that such quality control has been carefully maintained, it is recommended that the services of an independent third-party testing agency be employed. A requirement that each unit bear the trademark of an approved agency will ensure adequate independent inspection.

Working design capacities for wood structural panels are given in Chapter 2. References are also made to the American Forest and Paper Association publication *National Design Specification for Wood Construction* (NDS)[1] for other wood products.

Presentation of these specific design methods is not intended to preclude further innovation. Therefore, where adequate test data are available, the design provisions

**Caution on the use of equations:* Metric equivalents are frequently given in this chapter. Many of the equations contain constants or variables that are intended to permit the use of mixed units and may make these equations incapable of being used by directly substituting equivalent metric units.

may be appropriately modified. If they are modified, any such change should be noted when cross-referencing the design procedure to those presented in this Handbook.

Quality of workmanship and the conditions under which wood structural panels are used vary widely. Because the authors and publisher of this Handbook have no control over those elements, they cannot accept responsibility for wood structural panel or lumber performance or designs as actually constructed.

3.1.1 Growth of Industry and History

There is little if any actual box beam industry in the United States. These components are often built on-site, for residential construction, using nails. Nailed box beams are most-often used by do-it-yourselfers since the time spent in fabrication has less value to the builder than the cost of buying a ready-made alternative, such as a glulam or LVL beam.

Factory-built beams may be fastened together mechanically with nails or staples or with a structural adhesive such as resorcinol. Best estimates put the size of the industry at about 2,000,000 ft^2 (3⁄8 in. [9.5 mm] basis). Because of the availability of alternative engineered wood products, the use of wood structural panels in box beams is not expected to grow.

Of the different fabricated-component industries utilizing wood structural panels, the structural insulated panel (SIP) industry is the major user. Here the usage is in the neighborhood of 100,000,000 ft^2 (3⁄8 in. [9.5 mm] basis) and is growing steadily as designers and users recognize the benefits of using SIPs. SIPs are used in both residential and nonresidential construction. They are built to specification in a factory for rapid installation at the job site.

Stressed-skin panels are assemblies that have wood structural panel faces and backs with framing lumber or ribs in between. No statistics on utilization of wood structural panels for use in stressed-skin panels are available. Their use has declined in recent years, but they are still occasionally used in floors and roofs in the manufactured housing industry.

3.2 DESIGN OF GLUED PANEL LUMBER BOX BEAMS

3.2.1 General

This design method applies only to box beams with joints glued with structural adhesive. Design of mechanically fastened box beams is covered in Section 3.3. The primary difference in analysis between the two methods of fastening box beam components together is in the analysis of rolling shear stresses. With glued beam components, rolling shear must be considered in the design. With mechanically fastened box-beam components, planar (rolling) shear is seldom a consideration.

Beam Behavior. In wood structural panel box beams, the lumber flanges carry most of the bending, and one or more panel webs carry the shear. Joints between them transfer stresses between components.

Vertical stiffeners set between flanges distribute concentrated loads and resist web buckling. Deflection resulting from shear is usually significant and must be added to the bending deflection. Lateral restraint is often required to maintain stability. End joints in flange laminations and webs may require splicing.

Shape. Loads, spans, and allowable stresses, as well as desired appearance, determine the beam proportions. The depth and cross section may be varied along the length of the beam to fit design requirements, provided the resisting moments and shears at all sections are adequate. Typical cross sections are shown in Fig. 3.1.

3.2.2 Design Considerations

Design Loads. Live loads are typically those that are caused by objects moved into the structure after it is completed, including the occupants and their equipment and possessions. Snow, wind, and earthquake are special cases of live loads. The design live loads should not be less than required by the governing building code. Dead loads are those that will remain in place for 10 years or more. Dead load is the actual weight of the members and the permanent elements it supports. Allowance should be made for any temporary erection (construction) loads, or moving concentrated loads, such as cranes.

Allowable Working Capacities. Working capacities are determined as described in Chapter 2, with due regard for duration of loading. For symmetrical sections, the design should be based on the allowable stress in axial tension or compression, whichever is less. When butt joints occur in the tension flange, the design should be based on 0.8 of the allowable tensile stress.

Values for compression and tension parallel with lumber grain depend on species, grade, number of laminations, slope of scarf joints, and moisture condition. Values are applied as outlined in Chapter 2.

Allowable stresses for stress-grade lumber flanges shall not exceed those given in the latest edition of the NDS.[1] Allowable stress level at any point in the flanges must be determined based on the number of laminations continuous at that point. Any lamination with a butt joint within 10 times the lamination thickness of the point under investigation is considered discontinuous.

Allowable Deflection. Deflection should not exceed that allowed by the applicable building code. Maximum deflections recommended, shown in Table 3.1, are the proportions of the span, L, in inches.

More severe limitations may be required for special conditions, such as for supporting vibrating machinery, long spans, or beams over large glass windows or sliding doors.

Camber. Camber may be provided opposite to the direction of anticipated deflection for purposes of appearance or utility. It will have no effect on strength or actual stiffness.

Where roof and floor beams are cambered, a recommended amount is 1.5 times the deflection due to dead load only. This will provide a nearly level beam under conditions of long-term dead load application.

Additional camber may be introduced as desired to provide for drainage or appearance. Members used in low-slope roof applications must be designed to prevent ponding of water. This may be done either by cambering or by providing slope or stiffness such that ponding will not occur.

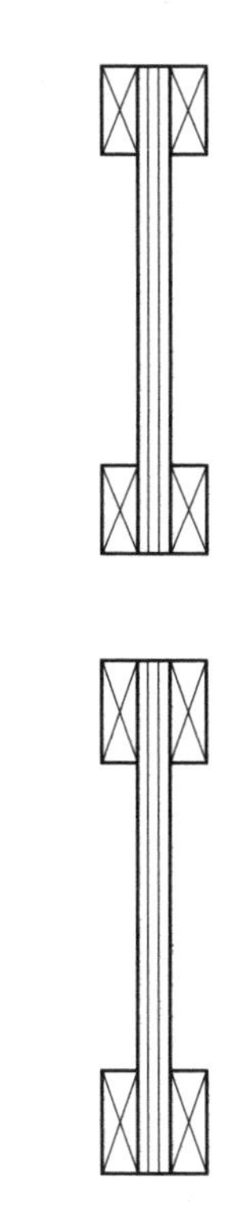

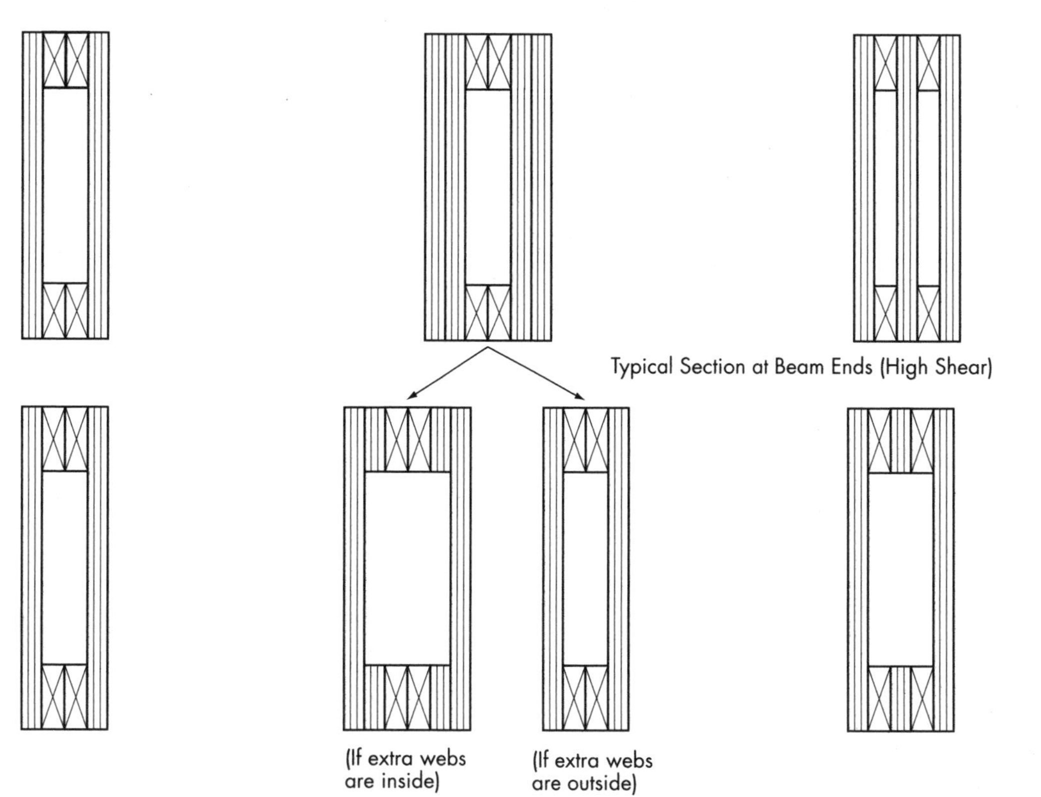

FIGURE 3.1 Typical beam sections.

TABLE 3.1 Standard Allowable Deflections for Beams

Floor beams	
Live load only	$L/360$
Dead plus live load	$L/240$
Roof beams	
Live load only	$L/240$
Dead plus live load	$L/180$

3.2.3 Trial Section

The first step in the actual design of a wood-structural-panel box beam is the selection of a trial section. Suitable beam depths vary somewhat, ranging generally from ⅛ to 1/12 of the span (although ratios up to 1/22 have been successfully used). The depth should ordinarily be equal to an available width of wood structural panel, such that waste is minimized. As a rule, the flange depth should be equal to at least four times the adjoining wood-structural-panel web thickness in order to have sufficient contact area between the flange and web for gluing.

Selection from Table. Table 3.2 lists preliminary bending and shear capacities for typical glued box beams with two webs.

First, determine the design requirements in terms of maximum moment and shear. A cross section that meets the design requirements may then be selected directly from the table. Tabular maximums, however, may also be subject to a number of adjustments based on duration of load, allowable flange tension stress (grade of lumber), and web thickness and grade. Note that further adjustment will be necessary when butt joints are allowed in lumber flanges. For example, it may be necessary to add a lamination to those shown in the table. The final design must then take into account provisions of Section 3.6.6.

Lumber Flanges

Symmetrical Sections. Symmetrical cross sections are generally used in wood-structural-panel beams for several practical reasons. These practical considerations usually outweigh the savings in material that theoretically can be achieved with unsymmetrical sections.

The design stresses for flanges are those for allowable stress in axial tension and axial compression. With symmetrical sections, the lower of these allowable stresses will limit the flange design. The equations in this section assume a symmetrical section.

Allowance for Surfacing. To allow for resurfacing of flange laminations for gluing, each lamination should be considered ⅛ in. (3 mm) smaller in dimension perpendicular to gluing surfaces (1/16 in. per surface) (2 mm) than its standard, net lumber size.

Beams should be designed for an actual depth, h, slightly less than their nominal depth, to allow for resurfacing, which may occur for the sake of appearance or uniformity of depth.

Actual depth of beams under 24 in. (610 mm) deep should be considered ⅜ in. (10 mm) less than nominal; for beams 24 in. (610 mm) and deeper, actual depth

TABLE 3.2 Preliminary Maximum Moments and Shears for Glued Box Beams

Preliminary selection estimates[a]

Table basis

DOL = 1.00

Panel webs

Thickness = 15⁄32 in. Span rating = 32/16

Grade = Rated Sheathing—Structural I

EA = 4,150,000 lb/ft of width, each web

$F_v t_v$ = 62 lb/in., each web

$F_s(\text{Ib}/Q)$ = 210 lb/in., each web

Lumber flanges

Species and Grade: Douglas fir-larch Select Structural

F_t = 1,000 psi, unadjusted for size factor

E = 1,900,000 psi

Beam depth, in.	Flange		Max. moment, ft-lb			Max. shear, lb		Stiffness, lb-in.²
	Lams	Size	M_f	M_w	M_t	V_h	V_s	EI
12	1	2 × 4	3,563	512	4,076	1,055	2,091	405,359,633
12	2	2 × 4	7,126	512	7,639	1,068	1,857	720,168,009
12	3	2 × 4	10,690	512	11,202	1,073	1,779	1,034,976,385
16	1	2 × 4	5,657	926	6,583	1,486	3,105	891,656,921
16	2	2 × 4	11,314	926	12,240	1,519	2,722	1,563,438,930
16	3	2 × 4	16,971	926	17,897	1,532	2,594	2,235,220,938
16	1	2 × 6	5,863	802	6,665	1,369	4,253	1,023,152,754
16	2	2 × 6	11,725	802	12,527	1,379	3,796	1,826,430,596
16	3	2 × 6	17,588	802	18,390	1,383	3,644	2,629,708,438
20	1	2 × 4	7,826	1,460	9,286	1,913	4,213	1,602,875,041
20	2	2 × 4	15,652	1,460	17,112	1,969	3,640	2,770,093,183
20	3	2 × 4	23,477	1,460	24,938	1,992	3,449	3,937,311,325
20	1	2 × 6	8,639	1,266	9,905	1,798	5,758	1,922,470,875
20	2	2 × 6	17,279	1,266	18,545	1,823	5,106	3,409,284,850
20	3	2 × 6	25,918	1,266	27,184	1,833	4,888	4,896,098,825
20	1	2 × 8	8,632	1,168	9,800	1,700	6,979	2,044,959,026
20	2	2 × 8	17,264	1,168	18,432	1,709	6,235	3,654,261,152
20	3	2 × 8	25,896	1,168	27,064	1,713	5,987	5,263,563,278
24	1	2 × 4	9,831	2,094	11,925	2,322	5,301	2,503,820,790
24	2	2 × 4	19,662	2,094	21,756	2,404	4,509	4,259,611,285
24	3	2 × 4	29,493	2,094	31,587	2,440	4,245	6,015,401,780
24	1	2 × 6	11,389	1,815	13,204	2,220	7,271	3,095,116,623
24	2	2 × 6	22,779	1,815	24,594	2,264	6,392	5,442,202,951
24	3	2 × 6	34,168	1,815	35,983	2,282	6,099	7,789,289,280
24	1	2 × 8	11,820	1,675	13,495	2,114	8,800	3,386,764,149
24	2	2 × 8	23,639	1,675	25,315	2,138	7,828	6,025,498,003
24	3	2 × 8	35,459	1,675	37,134	2,147	7,504	8,664,231,858
24	1	2 × 10	11,452	1,536	12,988	2,010	10,458	3,537,200,608
24	2	2 × 10	22,905	1,536	24,440	2,018	9,352	6,326,370,920
24	3	2 × 10	34,357	1,536	35,893	2,022	8,984	9,115,541,233
30	2	2 × 4	26,229	3,300	29,529	3,061	6,010	7,360,192,535
30	3	2 × 4	39,343	3,300	42,643	3,119	5,607	10,300,425,217
30	4	2 × 4	52,457	3,300	55,757	3,151	5,406	13,240,657,899
30	2	2 × 6	31,666	2,860	34,526	2,940	8,549	9,671,384,201
30	3	2 × 6	47,498	2,860	50,358	2,973	8,113	13,767,212,717

TABLE 3.2 Preliminary Maximum Moments and Shears for Glued Box Beams (*Continued*)

Beam depth, in.	Flange		Max. moment, ft-lb			Max. shear, lb		Stiffness, lb-in.^{2}
	Lams	Size	M_f	M_w	M_t	V_h	V_s	EI
30	4	2 × 6	63,331	2,860	66,191	2,991	7,895	17,863,041,233
30	2	2 × 8	34,101	2,640	36,741	2,812	10,514	11,036,469,878
30	3	2 × 8	51,151	2,640	53,791	2,832	10,044	15,814,841,233
30	4	2 × 8	68,202	2,640	70,842	2,843	9,809	20,593,212,587
30	2	2 × 10	34,397	2,420	36,817	2,669	12,543	11,995,692,795
30	3	2 × 10	51,595	2,420	54,015	2,680	12,027	17,253,675,608
30	4	2 × 10	68,793	2,420	71,213	2,686	11,769	22,511,658,420
30	2	2 × 12	32,693	2,200	34,893	2,549	14,454	12,474,215,712
30	3	2 × 12	49,039	2,200	51,239	2,554	13,882	17,971,459,983
30	4	2 × 12	65,385	2,200	67,585	2,556	13,596	23,468,704,253
36	2	2 × 4	32,842	4,779	37,621	3,706	7,603	11,439,373,785
36	3	2 × 4	49,262	4,779	54,041	3,787	7,032	15,869,711,155
36	4	2 × 4	65,683	4,779	70,462	3,834	6,746	20,300,048,524
36	2	2 × 6	40,721	4,142	44,862	3,611	10,810	15,255,365,451
36	3	2 × 6	61,081	4,142	65,223	3,661	10,201	21,593,698,655
36	4	2 × 6	81,441	4,142	85,583	3,690	9,896	27,932,031,858
36	2	2 × 8	44,930	3,823	48,753	3,489	13,336	17,731,416,753
36	3	2 × 8	67,395	3,823	71,218	3,523	12,690	25,307,775,608
36	4	2 × 8	89,860	3,823	93,684	3,542	12,366	32,884,134,462
36	2	2 × 10	46,605	3,505	50,110	3,341	15,955	19,725,189,670
36	3	2 × 10	69,908	3,505	73,412	3,362	15,260	28,298,434,983
36	4	2 × 10	93,210	3,505	96,715	3,374	14,912	36,871,680,295
36	2	2 × 12	45,487	3,186	48,673	3,201	18,362	20,987,462,587
36	3	2 × 12	68,231	3,186	71,417	3,213	17,610	30,191,844,358
36	4	2 × 12	90,975	3,186	94,161	3,220	17,234	39,396,226,128
42	2	2 × 6	49,871	5,660	55,531	4,274	13,168	22,268,846,701
42	3	2 × 6	74,806	5,660	80,467	4,344	12,356	31,343,447,092
42	4	2 × 6	99,742	5,660	105,402	4,384	11,950	40,418,047,483
42	2	2 × 8	55,968	5,225	61,193	4,164	16,269	26,185,038,628
42	3	2 × 8	83,952	5,225	89,177	4,214	15,416	37,217,734,983
42	4	2 × 8	111,936	5,225	117,161	4,241	14,989	48,250,431,337
42	2	2 × 10	59,220	4,789	64,009	4,018	19,516	29,589,561,545
42	3	2 × 10	88,830	4,789	93,619	4,052	18,610	42,324,519,358
42	4	2 × 10	118,440	4,789	123,229	4,071	18,157	55,059,477,170
42	2	2 × 12	58,956	4,354	63,310	3,871	22,498	32,011,784,462
42	3	2 × 12	88,434	4,354	92,788	3,894	21,533	45,957,853,733
42	4	2 × 12	117,912	4,354	122,266	3,906	21,050	59,903,923,003
48	2	2 × 6	59,080	7,415	66,495	4,929	15,619	30,786,527,951
48	3	2 × 6	88,621	7,415	96,036	5,020	14,575	43,091,158,030
48	4	2 × 6	118,161	7,415	125,576	5,073	14,052	55,395,788,108
48	2	2 × 8	67,135	6,845	73,980	4,833	19,304	36,472,035,503
48	3	2 × 8	100,703	6,845	107,547	4,901	18,214	51,619,419,358
48	4	2 × 8	134,270	6,845	141,115	4,939	17,669	66,766,803,212
48	2	2 × 10	72,087	6,274	78,361	4,695	23,197	41,663,508,420
48	3	2 × 10	108,130	6,274	114,404	4,744	22,050	59,406,628,733
48	4	2 × 10	144,173	6,274	150,447	4,770	21,477	77,149,749,045
48	2	2 × 12	72,843	5,704	78,547	4,548	26,792	45,621,881,337
48	3	2 × 12	109,265	5,704	114,969	4,582	25,583	65,344,188,108
48	4	2 × 12	145,686	5,704	151,390	4,600	24,979	85,066,494,878

TABLE 3.2 Preliminary Maximum Moments and Shears for Glued Box Beams (*Continued*)

[a] Bases and adjustments:

1. Basis: normal duration of load (C_D): 1.00
   Adjustments:
   0.90 for permanent load (over 50 years)
   1.15 for 2 months, as for snow
   1.25 for 7 days
   1.6 for 10 minutes, as for wind or earthquake
   2.00 for impact
2. Basis: F_t of flange = 1000 psi, (6,895 kPa) corrected by C_F (Douglas fir-larch select structural, 1997 NDS)[1]
   $2 \times 4 = 1{,}000 \times 1.5 = 1{,}500$ psi
   $2 \times 6 = 1{,}000 \times 1.3 = 1{,}300$ psi
   $2 \times 8 = 1{,}000 \times 1.2 = 1{,}200$ psi
   $2 \times 10 = 1{,}000 \times 1.1 = 1{,}100$ psi
   $2 \times 12 = 1{,}000 \times 1.0 = 1{,}000$ psi
   Adjustment: $\frac{F_t}{1000}$ for other tabulated tension stresses
   (C_F in numerator and denominator cancel when flanges are same width.)
3. Basis: one web effective in bending because web joints are assumed to be unspliced.
   Adjustment: 2.0 for web splices

should be considered ½ in. (13 mm) less than nominal. This resurfacing also results in reduced flange dimensions.

Bending Moment—Symmetrical Sections. In a symmetrical section allowable bending moment may be calculated by the formula

$$M = F'_t S_T \tag{3.1}$$

where M = allowable bending moment (lb-in. or lb-ft)
F'_t = allowable controlling working tensile stress of the flange lumber after all permissible/required adjustments (psi)

$$S_T = \frac{I_T}{c} \tag{3.2}$$

where S = section modulus of beam cross section
I_T = total moment of inertia of beam cross section
c = distance from beam neutral axis to outermost fiber

Unsymmetrical Sections. When the cross section is not symmetrical about its center, the resisting moment may be calculated as above, except that the distance from the neutral axis to the extreme fiber of each flange is used in place of the value $0.5h$ and the moment of inertia is calculated with due regard for the location of the neutral axis. The location of the neutral axis is computed based on the total cross section, without reduction for butt joints.

Net Moment of Inertia (I_n). The net moment of inertia, I_n, is the sum of I of the flanges plus the sum of all effective web material. I_n is used in determining the beam's allowable bending moment and deflections.

Wood Structural Panel Webs. When calculating moment of inertia of the wood structural panel webs, consider only effective material parallel to the span. The effective thickness, t', is 1⁄12 of the appropriate area, A. The effective area, A, is derived from the axial stiffness, AEC_G, of the panel. AEC_G is obtained from Chapter 2, where it is in units of lb/ft of panel width. Dividing by E gives the designer the panel area in in.^{2}/ft of width. Dividing by E_{lumber} gives the effective area of the panel transformed by the E of the lumber. (For a more detailed discussion of transformed sections see below, Transformed Section, under Section 3.4.1.) Dividing this by 12 in./ft of width gives the effective thickness, t', of panel transformed by E_{lumber}. The equation is:

$$t' = \frac{AEC_G}{12 \times E_{\text{lumber}}} \tag{3.3}$$

for tension and compression

where AE = panel stiffness capacity, from Chapter 2
C_G = adjustment to stiffness capacity, from Chapter 2
E_{lumber} = modulus of elasticity for lumber flanges

Butt joints in wood structural panel webs are usually spliced to transmit shear only, with a splice plate only as deep as the clear distance between flanges. If such butt joints in webs are staggered 24 in. (610 mm) or more, only one web need be disregarded in computing moment of inertia for bending stress. When unequal web thicknesses are used, use the most critical condition for computing I_n, unless the location of butt joints is specified in the design. For joints closer than 24 in. (610 mm), the contribution of the webs should be neglected in computing I_n.

When webs are spliced full-depth to carry direct flange stresses, they may all be included in computing the moment of inertia from allowable section capacities as given in Chapter 2.

Flange Lumber. Butt joints in the lumber flanges are required by the Fabrication Specification (Section 3.2.4) to be spaced at least 30 times the lamination thickness in adjoining laminations. Adjoining laminations refers to multiple-ply lumber flanges that are in direct face-to-face contact such as shown in Fig. 3.1. Butt joints in the lumber flanges are required to be spaced at least 10 times the lamination thickness in nonadjoining laminations (where web material separates multiple flanges), if not otherwise stipulated in the design. Ignore any panel material between laminations.

If butt-joint location is not otherwise stipulated by the designer, the net moment of inertia of flanges in which butt joints occur may be calculated by ignoring one lamination and 10% of the two adjoining laminations. The effective area of such adjoining laminations shall be computed by multiplying their gross area by the percentages in Table 3.3. Butt joints spaced closer than $10t$ (t = lamination thickness) shall be considered as occurring in the same section.

Wood Structural Panel Webs. Webs are primarily stressed in shear through their thickness, although they may also carry bending moment, if individual panels are properly spliced to transmit both types of stresses. In addition, sufficient contact

TABLE 3.3 Effective Area of Flange Laminations

Butt joint Spacing (t = lamination thickness)	Effective factor
$30t$	90%
$20t$	80%
$10t$	60%

area with the flanges must be provided to transmit the stresses between web and flange.

The number and thickness of the webs may be varied along the beam length in proportion to the shear requirements considering both shear through the panel thickness at the neutral axis and planar/rolling shear between flange and web. Where web joints result in a change of beam thickness, wood structural panel or lumber shims may be glued to the flanges to maintain beam width as required for appearance or for gluing pressure.

When the depth of a beam is tapered, the net vertical component of the direct forces in the flanges should be considered in determining the net shear to be resisted by the webs and the flange-web joints. This vertical component may add to or subtract from the external shear. It is equal to M/L_1, where M is the bending moment acting on the section and L_1 is the horizontal distance from the section to the intersection of the flange centerlines.

Horizontal Shear. The allowable horizontal shear on a section can be calculated by the following formula:

$$V_h = \frac{F_v t_v' C_G I_T N}{Q_T} \tag{3.4}$$

where V_h = allowable total horizontal shear (through the panel thickness) on section (lb).

$F_v t_v'$ = allowable shear capacity through the panel thickness (lb/in.), as given in Chapter 2, with adjustments such as duration of load, if applicable. Note that per Plywood Design Specification, Section 3.9.1,[3] F_v, and by extension $F_v t_v$, can commonly be increased by 19% for plywood and 33% for marine-grade plywood.

C_G = adjustment factor, depending on panel type, as given in Chapter 2

I_T = total moment of inertia of all flanges and webs about the neutral axis regardless of any butt joints (in.^{4}).

N = number of webs effective in shear (typically the same as the number of webs).

Q_T = statical moment about the neutral axis of all flanges and webs, regardless of any butt joints, lying above (or below) the neutral axis (in.^{3}).

End Joints for Tension or Bending. End joints across the face grain shall be considered capable of transmitting the following stresses parallel with the face plies (normal duration of load).

Scarf Joints and Finger Joints. Scarf joints 1 in 8 or flatter shall be considered as transmitting full allowable stress in tension or flexure. Scarf joints 1 in 5 shall be considered as transmitting 75% of the allowable stress. Scarf joints steeper than 1 in 5 shall not be used. Finger joints are acceptable, at design levels supported by adequate test data.

Butt Joints. When backed with a glued wood-structural-panel splice plate on one side having its strength axis perpendicular to the joint, the same width as the panels spliced, of a grade and span rating the same as the panel itself, joints may be considered capable of transmitting tensile or flexural stresses as in Table 3.4 (normal duration of loading). Splices are to be at least 14 in. long on each side of the joint. With adequately supported test data, it may be possible to make splices shorter. Mated faces of glued joints must be clean and free of oils and waxes prior to application of the adhesive.

TABLE 3.4 Panel Butt Joints—Tension or Flexure (Minimum splice length = 16 in. on each side of joint)

		Maximum stress $\frac{F_t A C_G}{A}$ psi							
		Regular grades				Structural I grades			
Span rating	Thickness (in.)	3 ply	4 ply	5 ply	OSB	3 ply	4 ply	5 ply	OSB
24/0	3/8	510	—	—	510	510	—	—	665
24/16	7/16	—	—	—	495	—	—	—	645
32/16	15/32	500	500	645	500	500	500	645	645
40/20	19/32	405	405	530	405	—	—	530	530
48/24	23/32	—	465	605	465	—	—	605	605
16 oc	19/32	365	365	475	365	—	—	475	475
20 oc	19/32	405	405	530	405	—	—	530	530
24 oc	23/32	—	390	505	390	—	—	505	505

End Joints for Compression. End joints across the face grain may be considered as transmitting 100% of the compressive strength of the panels joined when conforming to the requirements of this section (normal duration of load).

End Joints for Shear

Scarf Joints and Finger Joints. Scarf joints along or across the face grain, with slope of 1 in 8 or flatter, may be designed for 100% of the shear strength of the panels joined. Finger joints are acceptable, at design levels supported by adequate test data.

Butt Joints. Butt joints, along or across the face grain, may be designed for 100% of the strength of the panels joined when backed with a glued plywood splice plate on one side, no thinner than the panel itself, of a grade and species group equal to the plywood spliced, and of a length equal to at least 12 times the panel thickness.

Shear strength may be taken proportionately for shorter splice-plate lengths.

Combination of Stresses. Joints subject to more than one type of stress (for example, tension and shear), or to a stress reversal (for example, tension and compression), shall be designed for the most severe case.

Permissible Alternative Joints. Other types of glued joints, such as tongue-and-groove joints, or those backed with lumber framing, may be used at stress levels demonstrated by acceptable tests.

A 2 in. wide (50 mm) nominal stiffener alone may be used as a shear-splice plate when the web is 24 in. (610 mm) deep or less and is no thicker than 3/8 in. (10 mm) or carries no more shear than would be allowed on a 3/8 in. (10 mm) panel.

Holes in Webs. Holes in webs should be avoided if possible. If they are required, they should be located in areas of low shear, with proper consideration for the shear capacity of the remaining section. It is good practice to avoid sharp corners, and to use a wood-structural-panel "doubler" in the area of the hole (i.e. laminate another layer of wood structural panel around the hole).

Flange-Web Joints. Joints between flanges and webs at any section must be designed to transfer the shear acting along that section. Stresses are transferred wholly by glue, not by a combination of glue with mechanical fasteners.

Beams with One or Two Webs. The allowable flange-web shear on a glued, symmetrical two-web section in which only one face of each web contacts the flange, or on an I section, may be calculated by Eq. (3.5).

$$V_s = \frac{2F'_s dI_T}{Q_{\text{flange}}} \tag{3.5}$$

where V_s = allowable total shear on the section (lb) (N)

$F'_s = \dfrac{F_s \dfrac{\text{Ib}}{Q_{\text{web}}} C_G C_D,\ \text{etc.}}{\dfrac{\text{Ib}}{Q_{\text{web}}}}$ = Planar/rolling shear stress (psi). F'_s is adjusted per Chapter 2, with the 50% reduction for shear concentration where appropriate.

d = flange depth (in.)

I_T = total moment of inertia about the neutral axis of all effective material, regardless of any butt joints (in.^{4}).

$F_s \dfrac{\text{Ib}}{Q_{\text{web}}} C_g$ = Panel shear in the plane capacity (lb/ft of width) (N/m), with

$\dfrac{\text{Ib}}{Q_{\text{web}}}$ = shear constant from Chapter 2 (in.^{2}/ft).

Q_{flange} = statical moment about the neutral axis of all effective material in the upper (or lower) flanges, regardless of any butt joints (in.^{3}).

Beams with More than Two Webs. For purposes of designing the flange-to-web glued joint, maximum flange-web shear on beams with more than two webs may be computed using the assumption that the horizontal shear stress is equal in all webs. For calculations, flanges are then broken down into areas tributary to each web, and flange-web shear figured separately for each contact surface. Tributary areas are generally assigned such that the first moment (Q) of the area tributary to each web is proportional to the thickness of the webs.

For a beam in which the center web is less than twice the thickness of an outer web, the maximum stress occurs on the outside web, and allowable shear is given by the following formula.

$$V_s = \frac{F'_s dI_T}{Q_{\text{flange}}} \times \frac{\sum F_s \dfrac{\text{Ib}}{Q_{\text{Total}}}}{F_s \dfrac{\text{Ib}}{Q_{\text{Outer}}}} \tag{3.6}$$

Other notations are as previously described in this section.

Deflection. The deflection of wood-structural-panel beams may be taken as the sum of the calculated deflections due to bending and to shear. It should generally not exceed the values given in Table 3.1.

The bending deflection (Δ_b) may be calculated by conventional engineering formulas, with due regard to loading conditions and fixity of supports. Deflection due to several simultaneously applied loads may be calculated separately and added (superposition of deflection due to individual load cases).

Total deflection may then be obtained by one of the following methods. If the approximate method indicates that total deflection governs the design, a check may be made using the refined method.

- *Approximate method:* The approximate deflection (Δ_A) of simply supported, uniformly loaded wood-structural-panel beams may be found by multiplying the bending deflection (Δ_b) by a shear deflection factor, S_F, depending on the span-depth ratio, to account for shear deflection. The bending deflection is found by conventional formulas, using the elastic modulus of the flange lumber tabulated in the NDS and the moment of inertia of all effective material in the section, regardless of any butt joints. The shear-deflection factors (S_F) given in Table 3.5 may then be applied to the bending deflection, with interpolation permitted.
- *Refined method:* The total deflection (Δ_R) may be calculated by separately computing the bending deflection (Δ_b) and shear deflection (Δ_s) and adding the two.

Bending Deflection. In calculating the bending deflection, the tabulated elastic modulus (E) of the flange lumber may be increased by 3% over the values tabulated in the NDS[1] ($E' = 1.03E$) to obtain a "true" modulus of elasticity. The moment of inertia used for computing the bending deflection is I_T, the moment of inertia of the effective material in the section, regardless of butt joints.

Shear Deflection. The shear deflection for simple beams is shown in Fig. 3.2 and may be calculated using the formula.

$$\Delta_s = \frac{KC}{AG} \tag{3.7}$$

where Δ_s = shear deflection (in.)
K = a constant determined by the beam cross section, shown in Fig. 3.2
C = a coefficient depending on the manner of loading, shown in Fig. 3.2
$A = A_{\text{flange}} + A_{\text{web}}$ = cross-sectional area of the beam (in.^{2})

When calculating area of wood structural panel webs for shear, use

$$(A_{\text{web}}) = \left[\frac{(N_{\text{webs}})(G_v t_v C_G)}{G_{\text{lumber}}}\right] h \tag{3.8}$$

where G_{lumber} = shear modulus of the webs (psi).

If deflection is critical for loading conditions other than those shown in Fig. 3.2, refer to Ref. 2.

TABLE 3.5 Shear Factor S_F

Span/depth ratio	Factor S_F
10	1.5
15	1.2
20	1.0

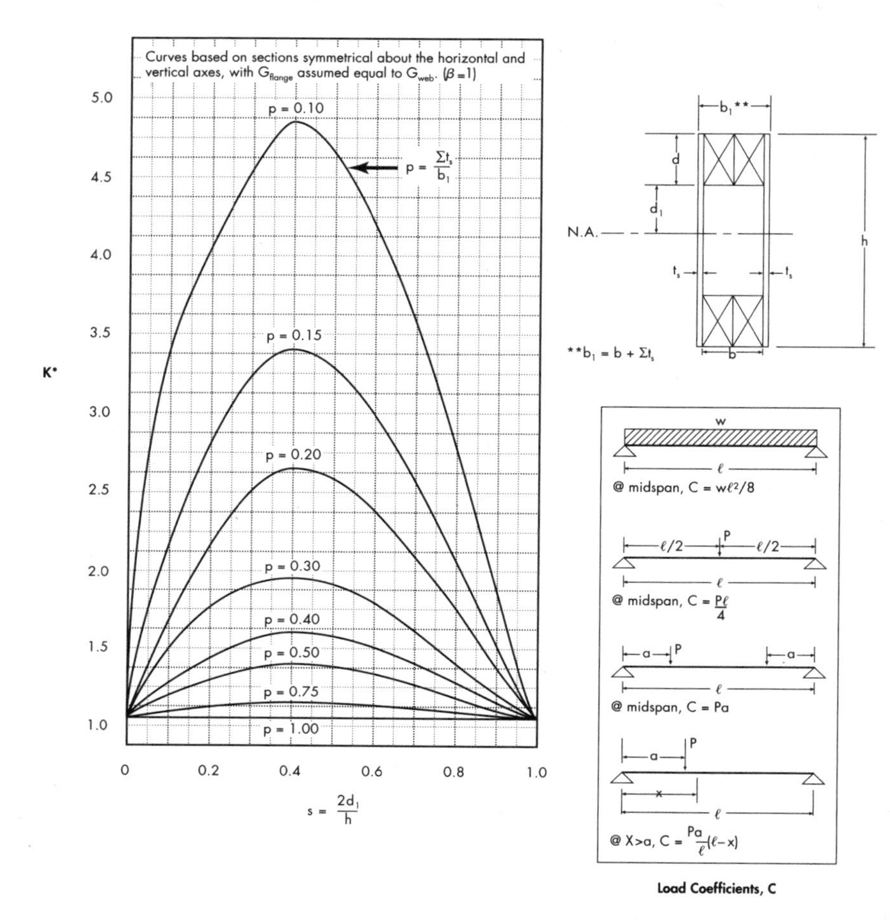

$$*K = \frac{\frac{9}{2}\left[\frac{1}{P}(1-s)+s\right]\left\{\frac{1}{P^2}\left[\frac{s^5}{2}-s^3+\frac{s}{2}\right]+\frac{1}{P}\left[-s^5\left(\frac{3}{30\beta}+\frac{2}{3}\right)+s^3\left(\frac{1}{3\beta}+\frac{2}{3}\right)-\frac{s}{2\beta}+\frac{8}{30\beta}\right]+\frac{8s^5}{30}\right\}}{\left[\frac{1}{P}\left(1-s^3\right)+s^3\right]^2}$$

where $\beta = \frac{G_{flange}}{G_{web}}$

Section Constant, K

FIGURE 3.2 Section constant and load coefficients for shear deflection equation.

Stiffeners

Bearing Stiffeners. Lumber bearing stiffeners are required over reactions and where other heavy concentrated loads occur, to distribute such loads into the beam. They should fit accurately against the flanges, and the webs should be securely attached to them.

Bearing Stiffeners at Ends of Beams. Bearing stiffeners at ends of beams should be the same width as the lumber flange at that section. Their dimension parallel to the beam span should not be less than that given by the following two considerations.

- *Compressive strength:* The thickness of stiffener must be at least equal to x in the following equation.

$$x = \frac{P}{F_{c\perp} b} \tag{3.9}$$

where x = thickness (in.) of stiffener (parallel to beam span)
P = concentrated load or reaction (lb)
$F_{c\perp}$ = allowable stress in compression perpendicular to grain for the flange lumber (psi)
b = flange width (in.)

- *Rolling, or planar shear:* For beams with one or two webs, the thickness of stiffeners must be at least equal to x in the following equation. For beams with more than two webs, the rolling (also called planar) shear stress will be less likely to govern.

$$x = \frac{P}{2hF_s} \tag{3.10}$$

where P = concentrated Load (lb)
h = depth of beam (in.)
F_s = allowable wood structural panel planar/rolling shear stress (psi)

Bearing Stiffeners Not at Ends of Beam. For bearing stiffeners not at ends of beam, factors given in 1997 NDS[1] Section 2.3.10 may be applied to $F_{c\perp}$.

Intermediate Stiffeners. Intermediate stiffeners are required to stabilize the flanges, space them accurately during fabrication, reinforce the webs in shear and prevent their buckling, and serve as backing for gluing of web splice plates where prespliced or scarfed webs are not used. Such stiffeners are usually of 2 in. (50 mm) dimension lumber and are equal in width to the lumber flange between webs, allowing for splice plates, if any.

Intermediate stiffeners spaced 48 in. o.c. (1220 mm) or less on centers will develop all, or nearly all, the shear strength of a beam of normal proportions.

Lateral Stability. Deep, narrow beams, particularly those used on long spans, may require definite lateral restraint to prevent buckling. The ratio of the total moment of inertia of all effective material about the horizontal neutral axis to that about the vertical axis will determine the minimum lateral support required, as given in Table 3.6.

3.2.4 Fabrication of Glued Wood Structural Panel Lumber Beams

General. This specification covers the fabrication of glued wood structural panel lumber beams, in which flanges are stress-graded lumber or glulam, and webs are trademarked wood structural panels.

TABLE 3.6 Provisions for Lateral Bracing

$\frac{\Sigma I_x}{\Sigma I_y}$	Provision for lateral bracing
Up to 5	None required
5 to 10	Ends held in position at bottom flanges at supports
10 to 20	Beams held in line at ends (both top and bottom flanges restrained from horizontal movement in planes perpendicular to beam axis)
20 to 30	One edge (either top or bottom) held in line
30 to 40	Beam restrained by bridging or other bracing at intervals of not more than 8 ft (2440 mm)
More than 40	Compression flanges fully restrained and forced to deflect in a vertical plane, as with a well-fastened joist and sheathing, or stressed-skin panel system

Wood structural panel/lumber beams should be designed by a qualified design professional in accordance with the latest edition of this Handbook, using the method set forth in Sections 3.2 and 3.3 of this chapter. Other design methods may be employed, provided they are supported by adequate test data.

Wood structural panel/lumber beams shall be fabricated and assembled in accordance with engineering drawings and specifications, except that minimum requirements herein shall be observed.

Materials

Wood Structural Panels. Wood structural panels shall conform to the latest edition of Voluntary Product Standard PS 1, *Construction and Industrial Plywood,*[4] (PS 1) or Voluntary Product Standard PS 2, *Performance Standard for Wood-Based Structural-Use Panels*[5] (PS 2). Each original panel shall bear the trademark of an approved quality assurance agency. Any precut wood structural panels shall be accompanied by an affidavit from the precutter certifying that each original panel was of the specified type and grade and carried the trademark of an approved quality assurance agency.

At the time of gluing, the wood structural panel shall be conditioned to a moisture content between 7% and 16%. Pieces to be assembled into a single beam shall be selected for moisture content to conform to Assembly, below.

Surfaces of wood structural panels to be glued shall be clean and free from oil, dust, wax, paper tape, and other material that would be detrimental to satisfactory gluing. Medium-density overlaid surfaces shall not be relied on for a structural glue bond.

Lumber. Grades shall be in accordance with current lumber grading rules. Knotholes up to the same size as the sound and tight knots specified for the grade by the grading rules may be permitted. When lumber is resawn, it shall be regraded based on the new size. Lumber for stiffeners shall be 2 in. (50 mm) minimum nominal thickness and of a grade equal to that of the flanges, except for extra stiffeners used only to supply pressure behind splice plates.

At the time of gluing, the lumber shall be conditioned to a moisture content between 7% and 16%. Pieces to be assembled into a single beam shall be selected for moisture content to conform to Assembly, below.

Surfaces of lumber to be glued shall be clean and free from oil, dust, and other foreign matter that would be detrimental to satisfactory gluing. Each piece of lumber shall be machine finished, but not sanded, to a smooth surface with a maximum allowable variation of 1⁄32 in. (1 mm) in the surface to be glued. Warp, twist, cup, or other characteristics that would prevent intimate contact of mating glued surfaces shall not be permitted.

Glue. Glue shall be of the type specified by the designer for anticipated exposure conditions.

Interior-type glue shall conform to ASTM Specification D3024 or D4689. Exterior-type glue shall conform to ASTM Specification D2559.

Mixing; spreading; storage, pot, and working life; and assembly time, temperature, and pressure shall be in accordance with the manufacturers' recommendations.

Fabrication

Webs. Only plywood wood structural panels shall be scarf or finger jointed. Scarf and finger joints shall be glued under pressure and over their full contact area and shall meet the requirements of Section 5.9 of PS 1.[4] In addition, no core gap shall intersect the sloped surface of the joint.

Unless otherwise noted in the design, butt joints in wood-structural-panel webs shall be backed with wood-structural-panel shear-splice plates centered over the joint and glued over their full contact area. The plate shall extend to within 1⁄4 in. (6 mm) of each flange on the inside of the beam and shall be at least equal in thickness to the web being spliced. Strength axis of the splice plate shall be parallel to that of the web. Length of the plate shall be at least 12 times the web thickness.

Surfaces of high-density overlaid wood structural panel to be glued shall be roughened, as by a light sanding, before gluing.

Framing. Scarf and finger joints may be used in flange lumber, provided the joints are as required for the grade and stress used in the design. Knots or knotholes in the end joints shall be limited to those permitted by the lumber grade, but in any case shall not exceed 1⁄4 the nominal width of the piece. Scarf slopes shall not be steeper than 1 in 8 in the tension flange, or 1 in 5 in the compression flange.

The edges of the framing members to which the wood structural panel webs are to be glued shall be surfaced prior to assembly to provide a maximum variation in depth of 1⁄16 in. (2 mm) for all members in a beam. (Allow for actual thickness of any splice plates superimposed on stiffeners.)

Assembly. The range of moisture content of the various pieces assembled into a single beam shall not exceed 5%.

All side-grain wood joints at flanges and stiffeners shall be glued over their full contact area.

Scarf and finger joints in stress-grade lumber flanges shall be well scattered throughout. Unless otherwise specified, they shall not be spaced closer than 16 times the lamination thickness in adjoining laminations, measured from center to center. (Ignore wood structural panels between laminations.) In flanges of three or fewer laminations, only one joint shall be allowed at any one cross section; in flanges of four or more laminations, two joints may be allowed at the same cross section.

Unless otherwise specified, butt joints in lumber flanges shall be spaced at least 30 times the lamination thickness in adjoining laminations and at least 10 times the lamination thickness in nonadjoining laminations. (Ignore wood structural panels between laminations.) No butt joints shall be allowed in portions of beams intended for mechanical splices or other stressed connections, unless specifically covered in the design.

Stiffeners shall be placed as shown in the design, but in any case, they shall be spaced not to exceed 4 ft (1220 mm) on center, and at reactions and other concentrated load points. Stiffeners shall be held in tight contact with the flanges by positive lateral pressure during assembly.

Unless otherwise specified by the designer, web butt joints shall be staggered at least 24 in. (610 mm). When glued during assembly, web splice plates shall be backed with one or more lumber stiffeners accurately machined in width to obtain adequate pressure. Where the design calls for the stiffeners to act as the web splices, web butt joints shall be located over the center of the stiffener, within 1⁄16 in. (2 mm), and webs shall be glued to the stiffener.

Where two adjacent webs are used, their contacting surfaces shall be glued together over the full flange and stiffener area. Wood structural panel webs shall be glued to framing members over their full contact area, using means that will provide close contact and substantially uniform pressure. Where clamping or other positive mechanical means are used, as required where webs are enclosed both sides with lumber laminations or where flanges are being glued simultaneously with the beam assembly, the pressure on the net framing area shall be sufficient to provide adequate contact and ensure good glue bond with 100–150 psi (690–1,035 kPa) on the net glued area recommended and shall be uniformly distributed by caul plates, beams, or other effective means. Where webs enclose a lumber flange or another web, nail gluing may be used in place of mechanical pressure methods. Nail sizes and spacings shown in the following schedule are suggested as a guide.

Nail sizes shall be:

- At least 4d (4-penny common) for wood structural panels up to 3⁄8 in. (10 mm) thick
- At least 6d (6-penny common) for 1⁄2–7⁄8 in. (12.7–22 mm) wood structural panels
- At least 8d (8-penny common) for 1 inch to 11⁄8 in. (25–30 mm) wood structural panels

Nail spacing shall not exceed:

- 3 in. (75 mm) along the flanges for wood structural panels through 3⁄8 in. (10 mm)
- 4 in. (100 mm) for wood structural panels 1⁄2 in. (10 mm) and thicker

Lines of nails shall be set in 3⁄4 in. (19 mm) from the lumber edge

- Two lines shall be used for 4 in. (100 mm) nominal flange lumber
- Three lines for lumber 6, 8, and 10 in. (150, 200, and 255 mm) nominal
- Four lines for 12 in. (305 mm) nominal

Application of pressure or nailing may start at any point but shall progress to an end or ends. In any case, it shall be the responsibility of the fabricator to produce a continuous glue bond that meets or exceeds applicable specifications.

Unless otherwise specified, width of beams shall equal, within ± 1⁄16 in. (2 mm), the sum of the lumber and wood structural panel dimensions, allowing for resurfacing. The net flange dimension in the plane of the laminations shall be no more than 1⁄4 in. (6 mm) less than the standard surfaced lumber width. To allow for resurfacing for finish appearance and uniformity, actual beam depth may be up to ⅜ in. (10 mm) less than nominal for beams up to 24 in. (635 mm) deep, ½ in. (12.7 mm) less for beams 24 in. (635 mm) and deeper, with tolerances of −⅛ in. (3 mm) and +¼ in. (6 mm). Length of beam shall be as specified ±¼ in. (6 mm).

Identification. Each member shall be identified by the appropriate trademark of an independent inspection and testing agency, legibly applied to be clearly visible. Locate trademark approximately 2 ft (610 mm) from either end, except appearance of installed beam shall be considered. If the strength of one flange is different from that of the other, the top flange shall be clearly marked on the outside surface of the finished beam.

3.2.5 Glued Box Beam Design Example

This example is intended for use as a general guide. Review of those sections pertinent to your specific design is recommended before proceeding.

Preliminary considerations as to the grade of wood structural panel and lumber to be used for a given design should include a check on availability. Where full exterior durability is not required for the wood structural panel, the wood structural panel may be specified exposure 1 (interior panel with exterior glue), generally permitting the use of higher allowable wood structural panel shear stresses.

Problem

Design a 28 ft (8.5 m) simple-span roof beam to support a total uniform load of 290 plf (265 N/m). Maximum design depth for the beam is 24 in. (610 mm). Allowable deflection under total load = $L/240$. Panel strength axis parallel to beam length. Duration of load = 1.15 as for snow loads.

Trial Section

$$\text{Acting moment, } M = \frac{wL^2}{8} \tag{3.11}$$

$$M = \frac{290 \times 28^2}{8} = 28{,}420 \text{ lb-ft} = 341{,}040 \text{ lb-in. (38,500 N-m)}$$

Try flanges consisting of two Douglas fir-larch No. 1 and Better 2 × 4s with two unspliced webs of 23⁄32-in. (18.3 mm) 48/24 OSB rated sheathing (Fig. 3.3).

Trial 1

Check Bending Strength

$$M_{\text{available}} = M_{\text{flange}} + M_{\text{webs}} = F_t S_{\text{flanges}} + F_t S_{\text{webs}} \tag{3.12}$$

$F_t = F_t C_F$ = allowable tensile stress in flanges (NDS[1] tabulated value)

$= 800 \times 1.5 = 1200$ psi (8.27 MPa)

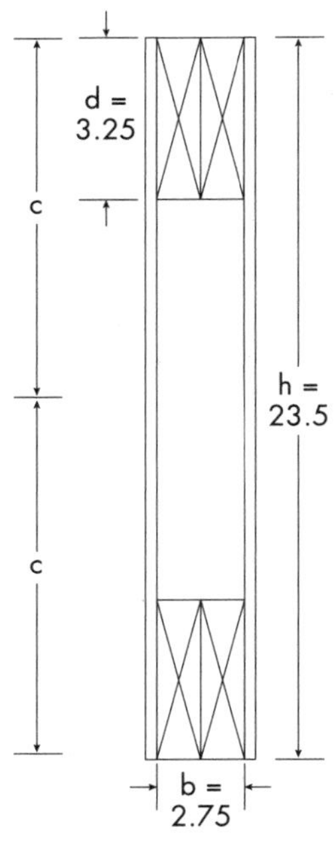

FIGURE 3.3 Assumed Section: Trial 1.

$$S_{\text{flanges}} + S_{\text{webs}} = \frac{I_{\text{flange}}}{c} + \frac{I_{\text{web}}}{c} \tag{3.13}$$

$$I_{\text{flange}} = \frac{bh^3}{12} = \frac{b}{12}(h^3 - d_1^3) \tag{3.14}$$

h = net depth of beam = 24 in. − 0.5 in. = 23.5 in. (595 mm)

d = net depth of flanges = 3.5 in. − $\left(\frac{0.5}{2}\right)$ = 3.25 in. (85 mm)

d_1 = net beam depth − 2(d) = 23.5 − 2(3.25) = 17 in. (430 mm)

b = net width of flanges = 2 × (1.50 in. − 0.125 in.) = 2.75 in. (70 mm)

$$I_{\text{flange}} = \frac{2.75}{12}[(23.5)^3 - (17)^3] = 1848.2 \text{ in.}^4 \ (769.28 \times 10^6 \text{ mm}^4)$$

$$I_{\text{web}} = \frac{bh^3}{12} = \frac{t'_{\text{web}}h^3}{12} \tag{3.15}$$

$$b_{\text{web}} = t'_{\text{web}} = \frac{EA_{\text{web}}C_G}{12E_{\text{flange}}} \tag{3.16}$$

where t'_{web} = effective thickness of a web
$EA_{web}C_G$ = tabulated axial stiffness for 23⁄32 in. 48/24 OSB rated sheathing = 5,850,000 × 1.0 = 5,850,000 lb/ft of width (85.40 kN/m)
E_{flange} = 1,800,000 psi (12.4 GPa)

Calculating the effective thickness of a web by transforming the web into lumber, EA_{web}/E_{flange}, and dividing by 12 in./ft of width,

$$t_{web} = \frac{5{,}850{,}000(1.0)}{12(1{,}800{,}000)} = 0.271 \text{ in. (7 mm)}$$

$$I_{web} = \left(\frac{(0.271)(23.5)^3}{12}\right) = 293.1 \text{ in.}^4 \ (122 \times 10^6 \text{ mm}^4)$$

Note that only a single web is used in calculating bending strength because the webs are not spliced for bending in this example.

Find

$$S = \frac{I}{c}$$

where c = distance from neutral axis to outermost fiber

$$= \frac{23.5}{2} = 11.75 \text{ in. (298 mm)}$$

$$S_{flange} = \frac{1848.2}{11.75} = 157.3 \text{ in.}^3 \ (2578 \times 10^3 \text{ mm}^3)$$

$$S_{web} = \frac{293.1}{11.75} = 24.9 \text{ in.}^3 \ (408 \times 10^3 \text{ mm}^3)$$

Section moment capacity:

$$(F'_tS)_{total} = F_t(C_D)S_{Total} = 1200 \times 1.15 \times (157.3 + 24.9) = 251{,}436 \text{ lb-in.}$$

Converting to lb-ft:

$$F_t(C_D)S_{Total} = \frac{251{,}436}{12}$$

$$= 20{,}953 \text{ lb-ft (28,408 N-m)} < 28{,}420 \text{ lb-ft (38,532 N-m)} \quad \text{OK}$$

Trial 1 Summary

Section Properties (transformed)

	I_{net} (in.^{4})	I_{gross} (in.^{4})	S_{net} (in.^{3})	Moment capacity (lb-ft)
Flanges	1848.2	1848.2	157.3	—
Webs	293.1	586.2	24.9	—
Total	2141.3	2434.4	182.2	20,953

Because this is considerably less moment capacity than required, the beam strength must be increased. Because bending controls the design and the beam depth is limited, full-depth web splices may be considered. It is often more convenient, however, to increase the number of flange laminations and/or specify a higher-grade flange lumber.

For convenience, refer to Table 3.2. This table gives the moment, shear and stiffness capacities of the most commonly used glued box beams. The table C_D is 1.0, so the moment and shear capacities may be multiplied by 1.15 for this example problem. The sheathing capacities are based on 15⁄32 in. (12 mm) 32/16 rated sheathing, as opposed to 23⁄32 in. (18 mm) 48/24 rated sheathing in the previous example. The capacities will be altered when different thicknesses and types of panels and flanges are used.

Trial 2

Select a 24-in. deep (610 mm) member with two 2 × 6 Douglas fir select structural flanges. This time let the webs be 15⁄32 in. (12 mm) five-ply 32/16 Structural I plywood with face grain parallel to the flanges. The span is 28 ft. The M_{total} capacity (tabulated value × C_D) is 24,594 × 1.15 = 28,283, (38,347 N-m), which is within 1% of the desired moment capacity for the 28 ft span without considering any contribution by the webs. The method of hand calculation of the moment capacity is the same as already shown.

Assume trial section as shown in Fig. 3.4.

Before calculating the moment of inertia (I) and the statical moment (Q) for a given trial section, the probable location of butt joints (if any) in both the web and flange members must be determined and adjustments applied. For this example, consider scarf joints of 1:12 slope for both the tension and compression flanges and butt joints in the wood-structural-panel webs staggered 24 in. (635 mm).

Where the beam design is controlled by horizontal shear, possible revisions include a specification of thicker wood structural panels, use of Structural I sheathing (such as in this example), or the addition of web member(s) to the end quarter sections of the beam.

Where flange-web (planar/rolling) shear controls the design, Structural I webs should be considered. In addition, greater flange-web area may be required.

Check Bending Strength

$$M_{\text{available}} = M_{\text{flange}} + M_{\text{webs}} = F_t S_{\text{flanges}} + F_t S_{\text{webs}}$$

Tabulated E for lumber flanges = 1,900,000 psi (13.1 GPa)

$F_t C_F$ = allowable tensile stress in flanges = 1000 × 1.3 = 1300 psi (8.96 MPa)

where F_t = tabulated tensile stress (psi)
C_F = size factor (NDS[1])

Moment of Inertia (I) of Beam Components

$$I_T = \left(\frac{bh^3}{12}\right)_{\text{flanges}} + \left(\frac{bh^3}{12}\right)_{\text{webs}} \tag{3.17}$$

$$I = \frac{b}{12}(h^3 - d_1^3)$$

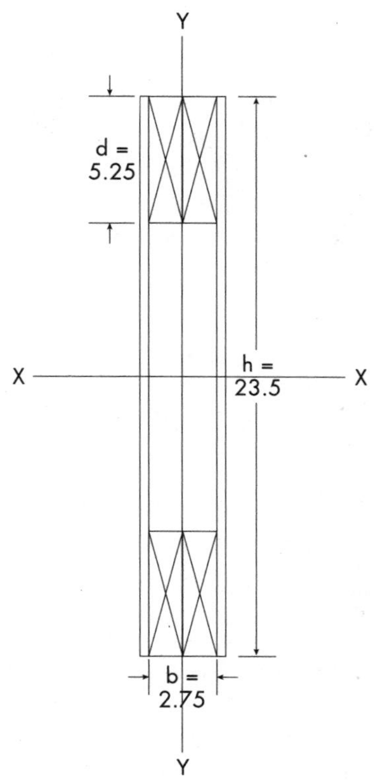

FIGURE 3.4 Assumed Section: Trial 2.

where b = net width of flanges = 2 × (1.50 in. − 0.125 in.) = 2.75 in. (70 mm)
h = Net depth of beam = 24 in. − 0.5 in. = 23.5 in. (597 mm)
d = Net depth of flanges = 5.5 in. − $\left(\frac{0.5}{2}\right)$ = 5.25 in. (133 mm)

$$d_1 = 23.5 - 2(5.25) = 13 \text{ in. } (330 \text{ mm})$$

$$I_{\text{flange}} = \frac{2.75}{12}\left[(23.5)^3 - (13)^3\right] = 2470.6 \text{ in.}^4 \ (1028 \times 10^6 \text{ mm}^4)$$

$$t_{\text{web}} = \frac{EA_{\text{web}}C_G}{12E_{\text{lumber}}}$$

$$t_{\text{web}} = \frac{4{,}150{,}000(1.0)}{12(1{,}900{,}000)} = 0.182 \text{ in. } (5 \text{ mm})$$

$$I_{\text{web}} = \left(\frac{(0.182)(23.5)^2}{12}\right) = 196.8 \text{ in.}^4 \ (81.91 \times 10^6 \text{ mm}^4)$$

Section Modulus of Beam Components

$$S = \frac{I}{c}$$

$$c = \text{distance from neutral axis to outermost fiber}$$

$$= \frac{D}{2} = \frac{23.5}{2} = 11.75 \text{ in. (298 mm)}$$

$$S_{\text{flange}} = \frac{2470.6}{11.75} = 210.3 \text{ in.}^3 \ (3446 \times 10^3 \text{ mm}^3)$$

$$S_{\text{web}} = \frac{196.8}{11.75} = 16.8 \text{ in.}^3 \ (275 \times 10^3 \text{ mm}^3)$$

Section Moment Capacity

$$M_T = F_t'S = (F_t'S)_{\text{total}}$$

$$F_t'(C_D)S = 1300 \times 1.15 \times (210.3 + 16.8)$$

$$= 339{,}514 \text{ lb-in./ft (125.85 kN-m/ft)}$$

$$F_t(C_D)S = \frac{339{,}514}{12}$$

$$= 28{,}292 \text{ lb-ft (38,359 N-m)} \cong 28{,}420 \text{ lb-ft (38,532 N-m)} \qquad \text{OK}$$

Note that the use of five-ply, structural I did not improve the bending capacity. The primary benefit from using structural I will be when the beam capacity is controlled by planar/rolling shear, F_s, or shear through the thickness, F_v.

Check Horizontal or Shear-Through-the-Thickness and Planar Shear Capacity

$$V_{h \text{ req.}} = \frac{wL}{2} \tag{3.18}$$

$$V_{h \text{ req.}} = \frac{290(28)}{2} = 4060 \text{ lb (18.06 kN)}$$

$$V_h = \frac{F_v t_v' C_G I_T N}{Q_T} \tag{3.19}$$

where $F_v t_v C_G$ = panel shear-through-the-thickness capacity = 62(2.0) = 124 lb/in. (21.7 N/mm)

$F_v t_v' C_G$ = $F_v t_v C_G$ adjusted for C_D and adjustment for adhesive on only one side of wood structural panel web and glued area ≥30% of beam depth

N = number of webs

$$F_v t_v' C_G = F_v t_v C_G (C_D)(F_A) = 124(1.15)(1.19)$$

$$= 169.7 \text{ lb-in. (29.72 N-mm)}$$

where F_A = allowable increase for continuous glued edge framing parallel to face grain per PDS 3.8.1.[3]

$$I_T = I_{\text{flange}} + 2I_{\text{web}} = 2470.6 \text{ in.}^4 + 2(196.8)$$

$$= 2864.2 \text{ in.}^4 \ (1192 \times 10^6 \text{ mm}^4)$$

First moment of beam (Q):

$$Q_{\text{flange}} = (A_{\text{flange}}) \left(\frac{h}{2} - \frac{d}{2}\right) \tag{3.20}$$

d = depth of flange (in.)

$$Q_{\text{flange}} = 5.25(2.75) \left(\frac{23.5}{2} - \frac{5.25}{2}\right) = 131.7 \text{ in.}^3 \ (2158 \times 10^3 \text{ mm}^3)$$

$$Q_{\text{web}} = (A_{\text{web}}) \left(\frac{h}{2} - \frac{c}{2}\right) \tag{3.21}$$

c = distance from neutral axis to outermost fiber

$$Q_{\text{web}} = 23.5(0.182) = \left(\frac{23.5}{2} - \frac{11.75}{2}\right) = 25.1 \text{ in.}^3 \ (411 \times 10^3 \text{ mm}^3)$$

$$Q_T = 131.7 + 25.1 = 156.9 \text{ in.}^3 \ (2571 \times 10^3 \text{ mm}^3)$$

$$V_h = \frac{169.7(2864.2)^2}{156.9} = 6196 \text{ lb} > 4060 \text{ lb} \qquad \text{OK}$$

Check Panel Shear in the Plane (Rolling Shear)

$$V_{s,\ \text{req}} = 4060 \text{ lb (18.06 kN)}$$

$$V_s = \frac{2F_s d I_T}{Q_{\text{flange}}} C_D(0.5)$$

Note: the basic planar/rolling-shear stress Eq. (3.5), is reduced 0.50 = 50% for wood structural panel beam flange-web shear design (Net Moment of Inertia, above, under Section 3.2.3).

$$F_s = \frac{F_s(\text{Ib}/Q)C_G}{\text{Ib}/Q}$$

$$F_s = \frac{210(1.6)}{3.750} = 89.6 \text{ psi (618 kPa)}$$

$$d = 5.25 \text{ in. (133 mm)}$$

$$V_s = \frac{(2)(89.6)(1.15)(0.5)(5.25)(2864.2)}{131.7}$$

$$= 11{,}764.8 \text{ lb (52.33 kN)} > 4{,}060 \text{ lb (18.06 kN)} \qquad \text{OK}$$

Check Beam Deflection

Both the approximate and the Refined methods for determining deflection are illustrated below. As a rule, if the deflection calculates to near the allowable limit using the approximate method, recalculate by the refined method before altering the trial conditions.

Check Deflection—Δ_A—Approximate Method

$$\Delta_A = \text{total deflection allowed} = \frac{L}{240} = \frac{28(12)}{240} = 1.4 \text{ in. (36 mm)}$$

$$S_F = \text{shear factor (Table 3.5)}$$

$$\text{Span-to-depth ratio} = \frac{\text{span}}{\text{depth}} = \frac{28(12)}{24} = 14$$

Interpolating from Table 3.5, $S_F = 1.26$

$$\Delta_T = \frac{5wL^4}{384EI_T} S_F \tag{3.23}$$

$$\Delta_T = \frac{5(290)(28)^4(12)^3}{384(1{,}900{,}000)(2864.2)} (1.26)$$

$$= 0.929 \text{ in. (24 mm)} < 1.4 \text{ in. (36 mm)} \qquad \text{OK}$$

Check Deflection—Refined Method

Even though the approximate deflection is only ⅔ of the allowable deflection, the following calculation is provided as an example of calculating a more precise deflection:

$$\Delta_R = \frac{5wL^4}{384E'_{\text{lumber}}\, I_T} + \frac{KC}{A_T G_{\text{lumber}}} \tag{3.24}$$

$$I_T = I_{\text{flange}} + I_{\text{webs}} = 2864.2 \text{ in.}^4 \text{ } (1192 \times 10^6 \text{ mm}^4)$$

$$E' = 1{,}900{,}000(1.03) = 1{,}957{,}000 \text{ psi (13,493 GPa)}$$

$$K = 2.21$$

$$C = M_T = 339{,}514 \text{ lb-in. (38.36 kN-m)}$$

$$A_T = A_{\text{flanges}} + A_{\text{webs}}$$

$$t_{\text{transformed webs}} = 0.182 \text{ in. (5 mm)}$$

$$A_T = N_{\text{flanges}}\,(bd) + \left[\frac{(N_{\text{webs}})(G_v t_v C_G)}{G_{\text{flanges}}}\right] h \tag{3.25}$$

$$C_v t_v C_G = (27{,}000)(1.7)$$

$$= 45{,}900 \text{ lb/in. (8,038 N/mm) for five-ply Structural I plywood}$$

$$G_{\text{lumber}} = E_{\text{flanges}}(0.06) = 1{,}900{,}000(0.06) = 114{,}00 \text{ psi (786 MPa)}$$

$$A_T = 2(2.75)(5.25) + \left[\frac{2(27{,}000)(1.7)}{114{,}000}\right](23.5) = 47.80 \text{ in.}^2 \text{ } (30.84 \times 10^3 \text{ mm}^2)$$

$$\Delta_R = \frac{5(290)(28)^4(12)^3}{384(1{,}957{,}000)(2864.2)} + \frac{2.21(341{,}040)}{47.80(114{,}000)}$$

$$= 0.8538 \text{ in. (22 mm)} < 1.4 \text{ in. (36 mm)} \qquad \text{OK}$$

If deflection controls the design, beam stiffness may be increased by increasing the beam depth and/or, with less pronounced results, the width. Note that structural I provides the maximum effective area for a given panel thickness, making it the stiffest grade for the webs. Also, consider using lumber with a higher E.

Trial 2 Summary

Section Properties (transformed)

	I_{net} (in.^{4})	I_{gross} (in.^{4})	S_{net} (in.^{3})	Q_{gross} (in.^{3})	Moment capacity (lb-ft)
Flanges	2470.6	2470.6	210.3	131.7	—
Webs	196.8	393.6	16.8	25.1	—
Total	2667.4	2864.2	227.1	156.8	28,292

Check for Bearing Stiffeners

$$P = \text{end reaction} = \frac{wL}{2} = 4{,}060 \text{ lb (18,060 N)}$$

$$F_{c\perp} = 625 \text{ psi (4,300 kPa)}$$

Stiffener thickness required for compression at bearing ends:

$$x = \frac{P}{F_{c\perp}\, b} \tag{3.26}$$

$$x = \frac{4{,}060}{(625)(2.75)} = 2.36 \text{ in. (60 mm)}$$

Use double 2 × 4s at ends.

Stiffener thickness required for planar/rolling shear at bearing ends:

$$x = \frac{P}{2hF'_s} \tag{3.27}$$

$$F'_s = 89.6C_D = 89.6(1.15) = 103 \text{ psi (710 kPa)}$$

$$x = \frac{4{,}060}{2 \times 23.5 \times 103} = 0.84 \text{ in. (44 mm)}$$

double 2 × 4s are OK

Check Lateral Stability

$$\text{Stability index} = S_i = \frac{\sum I_x}{\sum I_y} \tag{3.28}$$

$$I_x = 2667.4 \text{ in.}^4 \ (943.8 \times 10^6 \text{ mm}^4)$$

$$\sum I_y = I_y \text{ (flanges)} + I_y \text{ (webs)}$$

$$I_y \text{ (flanges)} = 2 \times \frac{db^3}{12} = \frac{2 \times 5.25 \times 2.75^3}{12} = 18.2 \text{ in.}^4 \ (7.58 \times 10^6 \text{ mm}^4)$$

$$\text{Parallel axis theorem:} \qquad I = \bar{I} + AD^2 \tag{3.29}$$

$$I_y \text{ (webs)} = 2\ [\bar{I}_{\text{web}} + A_{\text{web}}D^2]$$

$$= 2\left[\left(\frac{EIC_G}{E_{\text{lumber}}} \times \frac{h}{12}\right) + \left(\frac{EAG_G}{E_{\text{lumber}}} \times \frac{h}{12}\right)\left(\frac{b+t}{2}\right)^2\right] I_{y \text{ webs}} \tag{3.30}$$

I_y (webs)

$$= 2\left\{\left[\left(\frac{115{,}000(1.1)}{1{,}900{,}000}\right)\left(\frac{23.5}{12}\right)\right] + \left[\left(\frac{4{,}150{,}000(1.0)}{1{,}900{,}000}\right)\left(\frac{23.5}{12}\right)\left(\frac{2.75 + 0.4688}{2}\right)^2\right]\right\}$$

$$= 22.4 \text{ in.}^4 \ (9.32 \times 10^6 \text{ mm}^4)$$

$$I_{y \text{ total}} = I_{\text{flanges}} + I_{\text{webs}} = 18.2 + 22.4 = 40.6 \text{ in.}^4 \ (16.9 \times 10^6 \text{ mm}^4)$$

$$S_i = \frac{2864.2}{40.6} = 70.5 \text{ in.}^3 \ (1155 \times 10^3 \text{ mm}^3)$$

For this design example, the compression flange should be fully restrained as indicated in Table 3.6 since the $\Sigma I_x/\Sigma I_y$ ratio exceeds 40. This can usually be achieved by sheathing and/or by ceiling material.

As final steps in the overall design, review the structural adequacy of beam supports and develop beam connection details.

3.3 NAILED BOX BEAM DESIGN

There are several differences between glued beam design methods and nailed (stapled) box beam design. One difference is that the nailed connections between the flanges and the webs are not rigid, as with the glued connections. This permits an indeterminate amount of slip between the flanges and the webs. To maintain a conservative design, therefore, the webs are not typically considered in calculating maximum moment capacity. In nailed design, the flanges are assumed to carry all tensile and compressive stresses and the webs carry only shear.

Another difference is that planar/rolling shear does not enter into the calculations. Consequently, glued area need not be maximized and the flanges may be oriented either flatwise or edgewise in the beam. In addition, the flanges need not be planed down to smaller dimensions to accommodate optimal glue thickness and curing pressure. This saves some labor and gives a larger flange cross-sectional area to work with. Some of this labor savings is lost, however, because nailed beams require a considerable amount of nailing.

A more efficient beam can be designed with flatwise orientation of the flange lumber because this orientation maximizes the distance of the flange lumber from the neutral axis of the beam (Fig. 3.5). In addition, a beam with a single flatwise-oriented flange on the top and bottom gives more depth of penetration to accommodate maximum design capacity from each fastener. Oriented vertically, a single flange may not be thick enough to develop full fastener lateral (shear) capacity.

Nails may be spaced along the flange according to the amount of shear that must be transferred at each point. In a uniformly loaded simple span beam, nailing will be heaviest at the ends and least through the middle. The design capacities of the nails may be adjusted for C_D. Care must be taken to avoid splitting the lumber. Staples have less tendency to split lumber and may be spaced much more closely, without splitting, than nails. To simplify construction, consider spacing nails to develop the required maximum shear transfer in the outer ¼ of beam length and at twice that spacing through the center ½ of beam length. Maximum fastener spacing shall be 6 in. o.c. (150 mm).

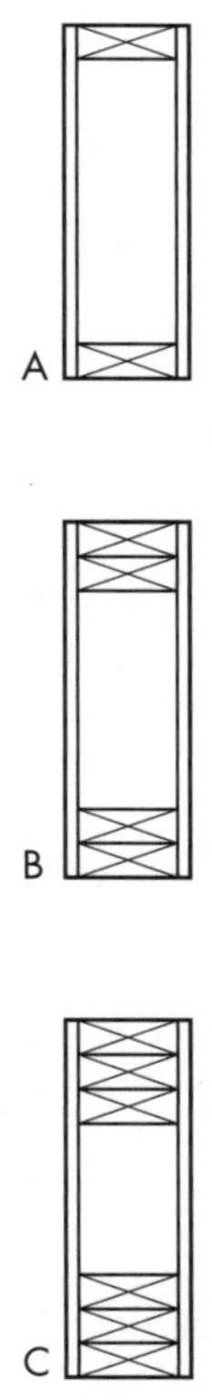

FIGURE 3.5 Cross sections for nailed box beam design example.

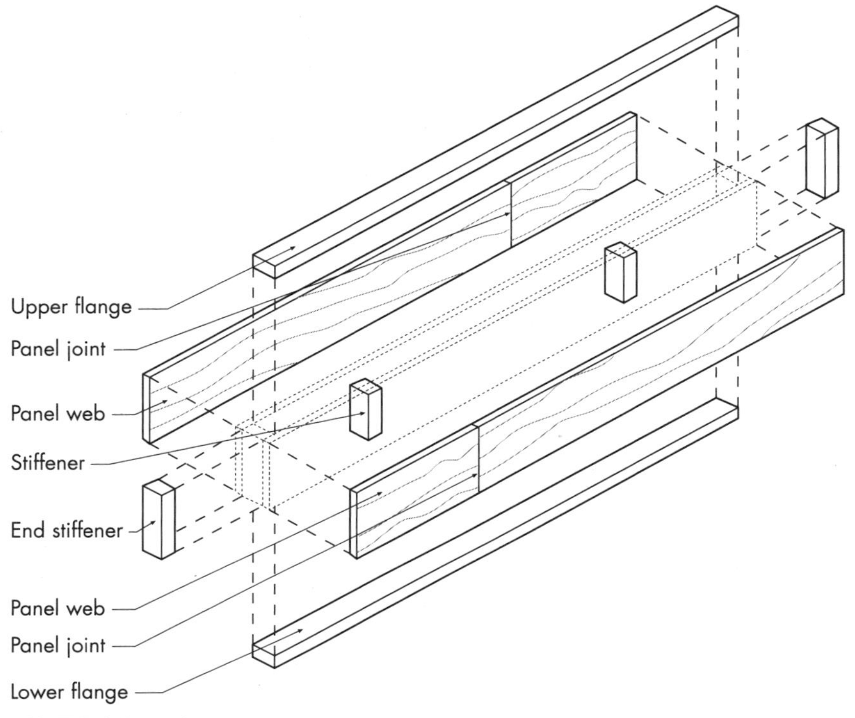

FIGURE 3.6 Basic components.

Glue-bond quality, so easily compromised under field-assembly conditions, is not an issue with nailed box beam designs. Nailed box beams may, therefore, be field assembled and consistently achieve full design capacity. They should be designed by a design professional, but third-party inspection is not generally required, provided that the beam is assembled according to engineered drawings (Fig. 3.6).

3.3.1 Nailed Box Beam Design Example

Using the design example from Trial 2 of the glued wood structural panel lumber beam above, design a 28 ft long (8.53 m) nailed box beam with 32/16 rated, $^{15}/_{32}$ in. (12 mm) five-ply Structural I plywood webs. The strength axis of the wood structural panel (plywood in this case) is oriented parallel to the beam span and the continuous double 2 × 6 Douglas fir select structural flanges are oriented flatwise for maximum efficiency (Fig. 3.7). Determine the maximum moment capacity, F_tS, of the beam, the maximum shear capacity, and the nailing requirements. The lateral design capacity for 8d common nails is 74 lb/nail (329 N) (unadjusted for C_D). Assume one row of nails in each flange lamination.

Check Bending Strength

$$M_{\text{available}} = M_{\text{flanges}} = F_t'S_{\text{flanges}}$$

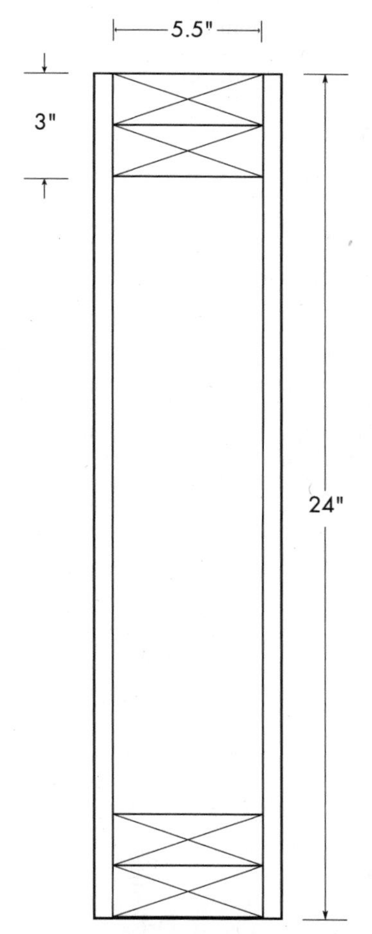

FIGURE 3.7 Nailed box beam design example.

where $F_{y\ \text{flanges}} = F_t(C_F) = 1000(1.3)\ 1300$ psi (8.96 MPa)

$F'_{t\ \text{flanges}} = F_t C_f C_D = 1300(1.15) = 1495$ psi (10.31 MPa)

$S_{\text{flanges}} = \dfrac{I_{\text{flanges}}}{c}$ (in.^{3})

$$I_{\text{flanges}} = \frac{b[h^3 - (h - 2d)^3)]}{12} \text{ (in.}^4\text{)} \tag{3.31}$$

c = distance from neutral axis to outermost fiber $= \dfrac{h}{2} = \dfrac{24}{2}$

= 12 in. (305 mm)

$$d = \text{depth of flange (in.)}$$

$$I_{\text{flanges}} = 5.5 \left\{ \frac{24^3 - [24 - (2)(3)]^3}{12} \right\} = 3663 \text{ in.}^4 \ (1524 \times 10^6 \text{ mm}^4)$$

$$S_{\text{flanges}} = \frac{I}{c} = \frac{3663}{12} = 305.25 \text{ in.}^3 \ (5002 \times 10^3 \text{ mm}^3)$$

$$M_{\text{flanges}} = F_t' S_{\text{flanges}} \ 1495(305.25) = 456{,}349 \text{ lb-in.} \ (51.56 \text{ kN-m})$$

$$M_{\text{flanges}} = \frac{456{,}349}{12} = 38{,}029 \text{ lb-ft} \ (51.560 \text{ kN-m})$$

This moment is nearly 99% of the 38,500 lb-ft (52.199 kN-m) calculated as required in the glued beam example above. As a practical matter, this is acceptable since the webs are not counted as contributing to bending strength, while, in fact, they do contribute something toward bending strength. The degree of web contribution is simply difficult to quantify due to nail slip/deformation.

Check Web Shear

$$V_{h \text{ req.}} = \frac{wL}{2} = \frac{290(28)}{2} = 4060 \text{ lb} \ (18{,}060 \text{ N})$$

$$V_h = \frac{F_v t_v C_G' I_T N}{Q_T} \tag{3.32}$$

$$F_v t_v C_G = 62(2) = 124 \text{ lb/in.} \ (21.7 \text{ N/mm})$$

$$F_v t_v C_G' = 124(C_D) = 124(1.15) = 142.6 \text{ lb/in.} \ (24.97 \text{ N/mm})$$

$$I_T = I_{\text{flange}} + I_{\text{web}}$$

$$I_{\text{webs}} = \frac{Nbh^3}{12} \tag{3.33}$$

$$N = \text{no. of webs} = 2$$

$$b = t_{\text{webs}} = \frac{EAC_G}{12E_{\text{lumber}}}. \tag{3.34}$$

$$\text{Tabulated } E_{\text{flange}} = 1{,}900{,}000 \text{ psi} \ (13.1 \text{ GPa})$$

$$b = \frac{(4{,}150{,}000)(1.0)}{12(1{,}900{,}000)} = 0.182 \text{ in.} \ (4.6 \text{ mm})$$

$$I_{\text{webs}} = \frac{Nbh^3}{12} = \frac{2(0.182)(24)^3}{12} = 419.3 \text{ in.}^4 \ (174.5 \times 10^6 \text{ mm}^4)$$

$$h = \text{flange height (beam depth)} = 24 \text{ in.} \ (610 \text{ mm})$$

$$I_T = 3663 + 419.3 = 4082.3 \text{ in.}^4 \ (6.802 \times 10^6 \text{ mm}^4)$$

$$Q_T = Q_{\text{flange}} + Q_{\text{webs}} \quad (3.35)$$

$$Q_{\text{flange}} = A\left(\frac{h}{2} - \frac{d}{2}\right) \quad (3.36)$$

$$Q_{\text{flange}} = 3(5.5)\left(\frac{24}{2} - \frac{3}{2}\right) = 173.25 \text{ in.}^3 \ (2839 \times 10^3 \text{ mm}^3)$$

$$Q_{\text{web}} = \frac{A}{2}\left(\frac{h}{2} - \frac{h}{4}\right) \quad (3.37)$$

$$Q_{\text{web}} = \frac{(0.182)(24)}{2}\left(\frac{24}{2} - \frac{24}{4}\right) = 13.1 \text{ in.}^3 \ (214 \times 10^3 \text{ mm}^3)$$

$$Q_{\text{webs}} = NQ_{\text{web}} \quad (3.38)$$

$$Q_{\text{webs}} = 2(13.1) = 26.2 \text{ in.}^3 \ (429 \times 10^3 \text{ mm}^3)$$

$$Q_T = 173.25 + 26.2 = 199.4 \text{ in.}^3 \ (3268 \times 10^3 \text{ mm}^3)$$

$$V_h = \frac{F_v t_v C'_G I_T N}{Q_T} = \frac{(142.6)(4082.3)(2)}{199.4}$$

$$= 5839 \text{ lb } (25.97 \text{ kN}) > 4060 \text{ lb } (18.06 \text{ kN}) \qquad \text{OK}$$

Determine Nail Spacing—Assume Two Rows (R) of Nails

$$V_n = \text{nail shear} = \frac{F_n C_D I_T N R}{s Q_{\text{flanges}}} \quad (3.39)$$

$$s = \text{nail spacing (in.)}$$

$$s = \frac{F_n C_D I_T N R}{V_n Q_{\text{flanges}}} \quad (3.40)$$

$$F_n = 74 \text{ lb/nail } (323 \text{ N/nail})$$

$$R = \text{no. of rows of nails} = 2$$

$$V_{n,\text{ req.}} = V_{h \text{ req.}} = 4060 \text{ lb/ft } (59.25 \text{ N/mm})$$

$$s = \frac{74(1.15)(4082.3)(2)(2)}{(4060)(173.3)} = 1.98 \text{ in. } (50 \text{ mm})$$

Space two rows of 8d common nails 2 in. o.c. (50 mm) in the outer ¼ (7 ft) (2.1 m) to develop the required maximum shear capacity. Space nails at 4 in. o.c. (102 mm) along the inner ½ (14 ft) (4.3 m) of beam. Stagger nails when used at 2 in. o.c. (51 mm) to minimize splitting.

Bearing Capacity and Lateral Stability. Check for bearing stiffeners and lateral stability according to Wood Structural Panel Webs, above, under Section 3.2.3, and Table 3.6.

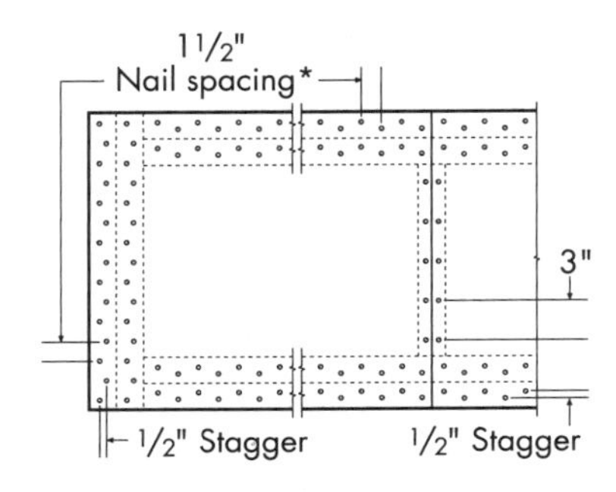

*When end stiffeners extend through the beam, nail spacing is the same as for flanges, except space nails 1 in. on center when double end stiffeners are used in beams with three members per flange (cross-section C of Figure 3.5). When end stiffeners are inserted between flanges, nails may be spaced 3 inches on center.

FIGURE 3.8 Side view of nailed box beam nailing pattern.

3.4 DESIGN AND FABRICATION OF WOOD STRUCTURAL PANEL/STRESSED-SKIN PANELS

Flat panels with stressed wood structural panel skins and spaced lumber stringers act like a series of built-up I-beams, with the wood structural panel skins taking most of the bending stresses as well as performing a sheathing function and the lumber stringers taking shear stresses.

Since stressed-skin panels are usually relatively shallow, any shear deformation between skins and webs will contribute materially to deflection. For maximum stiffness, therefore, a rigid connection is required between the wood structural panel and the lumber. Thus, all panels considered in this design method are assumed to be assembled with glue.

Although it is possible to use laminated or scarf-jointed members for the stringers of stressed-skin panels, such panels are usually restricted to single-lamination stringers. The maximum length of lumber available, therefore, generally determines panel maximum length.

Headers (at the ends of the panel) and blocking (within the panel) serve to align the stringers, back up splice plates, support skin edges, and help to distribute concentrated loads. They may be omitted in some cases, but should always be used when stressed-skin panels are applied with their stringers horizontal on a sloping roof. Without headers and blocking, panels so applied may tend to assume a parallelogram cross section.

Panels with top and bottom skins, as shown in Fig. 3.9, are most common. One-sided panels, as shown in Fig. 3.10, are also used, especially when special ceiling treatment is desired, or when no ceiling is required. A variation of the single-skin panel, with lumber strips on the bottom of the stringers, as shown in Fig. 3.11, is called a T-flange panel.

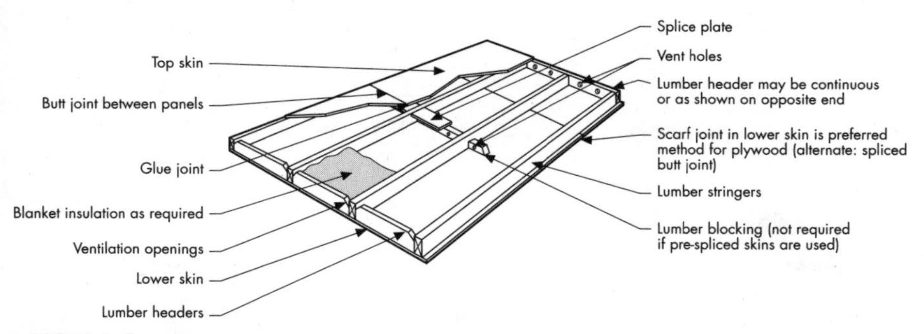

FIGURE 3.9 Typical two-sided stressed-skin panel.

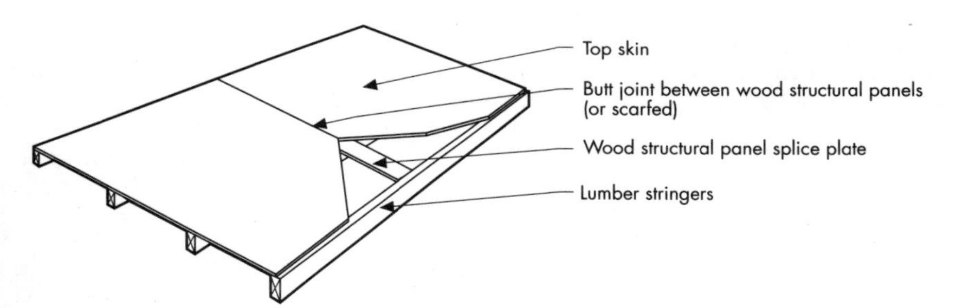

FIGURE 3.10 Typical one-sided stressed-skin panel.

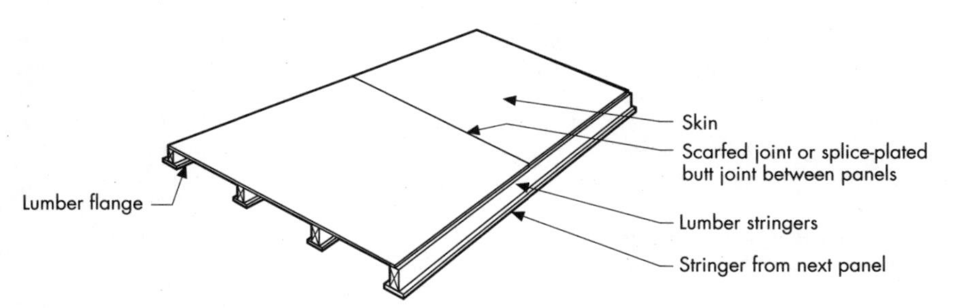

FIGURE 3.11 Typical T-flange stressed-skin panel.

Two-sided panels do not require bridging. Stringers in one-sided panels may be bridged, as are joists of the same depth in conventional construction.

Owing to the high strength of wood structural panels, calculations will often indicate that a thin bottom skin is structurally sufficient. There is some possibility, however, of a slight bow when ¼ in. (6 mm) bottom skins are used with strength axis parallel to stringers on 16 in. centers (405 mm). Such a bow, although of no importance structurally, may be undesirable from an appearance standpoint. For stringers so spaced, therefore, ⅜ in. (8 mm) wood structural panels are the minimum recommended.

A number of decorative plywood surfaces are adaptable for panels where the bottom skin will serve as a ceiling for a habitable room. The design of panels using such decorative plywood must, of course, allow for the special properties of the product. See Materials, above, under Section 3.2.4, for a general discussion of the wood components used in the fabrication of SIPs.

3.4.1 Design Considerations

Transformed Section. In calculating section properties for stressed-skin panels, the designer must take into account the composite nature of the unit. Unless all materials in the panel have similar moduli of elasticity, some method must be employed to make allowance for the differences in MOE.

Different moduli of elasticity may be reconciled by the use of a transformed, or effective, section. The transformed section approach is common to structural design of composite sections. It consists of transforming the actual section into one of equivalent strength and stiffness, as if composed of a single material. Where a portion of a section is to be transformed to another material, its actual area must be multiplied by the ratio of its MOE to that of the other material.

$$\left(\frac{E_1}{E_2}\right) = n \tag{3.42}$$

thus arriving at an effective area of the new material.

For instance, assume a stressed-skin panel with the modulus of elasticity of the stringers half that of the skins. Properties of the section could be stated in terms of those of a transformed section having the modulus of elasticity of the skins and calculated as if the stringers were only half as wide as they actually are.

Design Loads. The design live loads must not be less than required by the governing building regulations. Allowance must be made for any temporary erection loads or moving concentrated loads. Roof panels must be designed to resist uplift due to wind load, combined with internal pressure developed by wind through openings in the sidewalls, minus the dead load of the panels and roofing. Lateral loads, which develop diaphragm action, may require special consideration, particularly to fastenings between panels and to framing.

Allowable Working Stresses and Capacities. Wood structural panel working capacities are determined as described in Chapter 2.

Wood Structural Panels in Stressed-Skin Panels

Effective Sections. For all panels, whether containing butt joints or not, deflection and shear stresses are based on the gross section of all material having its strength axis parallel with the direction of principal stress.

All wood structural panels and all lumber having their strength axis parallel to the direction of stress may be considered effective in resisting bending stress, except when butt-jointed and except as reduced for **b** distance below, under Section 3.4.2.

The best method for splicing skins, from both the structural and appearance standpoint, is to scarf joint plywood to the desired length.

End Joints for Tension or Bending. End joints across the face grain shall be considered capable of transmitting the following stresses parallel with the face plies.

Scarf Joints and Finger Joints. Scarf joints 1 in 8 or flatter shall be considered as transmitting full allowable stress in tension or flexure. Scarf joints 1 in 5 shall be considered as transmitting 75% of the allowable stress. Scarf joints steeper than 1 in 5 shall not be used. Finger joints are acceptable, at design levels supported by adequate test data. OSB panels shall not be scarf or finger jointed.

Butt Joints. When backed with a glued wood structural panel splice plate on one side, centered over butt joint, and having its strength axis perpendicular to the joint, and of a grade equal to the wood structural panel spliced, and being no thinner than the panel itself, joints may be considered capable of transmitting tensile or flexural stresses as in Table 3.4 (normal duration of loading). Strength may be taken proportionately for shorter splice-plate lengths.

End Joints for Compression. End joints across the face grain may be considered as transmitting 100% of the compressive strength of the panels joined when conforming to the requirements of this section (normal duration of load).

Scarf Joints and Finger Joints. Slope no steeper than 1 in 5.

Butt Joints. Spliced as in Table 3.4 and with the splice lengths tabulated therein. Strength maybe taken proportionately for shorter splice-plate lengths.

End Joints for Shear

Scarf Joints and Finger Joints. OSB panels shall not be scarf or finger jointed. Scarf joints in plywood panels, along or across the face grain, with slope of 1 in 8 or flatter, may be designed for 100% of the shear strength of the panels joined. Finger joints in plywood panels are acceptable, at design levels supported by adequate test data.

Butt Joints. Butt joints, along or across the face grain, may be designed for 100% of the strength of the panels joined when backed with a glued wood structural panel splice plate on one side, no thinner than the panel itself, of a grade and species group equal to the wood structural panel spliced, and of a length equal to at least 12 times the panel thickness.

Strength may be taken proportionately for shorter splice-plate lengths.

Combination of Stresses. Joints subject to more than one type of stress (for example, tension and shear), or to a stress reversal (for example, tension and compression), shall be designed for the most severe case.

Permissible Alternative Joints. Other types of glued joints, such as tongue-and-groove joints or those backed with lumber framing, may be used at stress levels demonstrated by acceptable tests.

Splice plates are used only between the stringers, and consequently there is a certain percentage of the panel, which is not spliced. When calculating the strength of the panel at the splices, only the portion of the skin, which is actually spliced, should be considered effective. Width and thickness of plates should not be reduced.

Allowable Deflection. Deflection must not exceed that allowed by the applicable building code. Maximum recommended deflections are given in Table 3.1. More severe limitations may be required by special conditions such as the installation of marble or ceramic tile.

Camber. Camber may be provided opposite to the direction of anticipated deflection for purposes of appearance or utility. It will have no effect on strength or actual stiffness.

Where panels are cambered, typically in floors, a recommended minimum amount is 1.5 times the deflection due to dead load only. This will provide a nearly level panel under conditions of long-term application of dead load.

Additional camber may be introduced as desired to provide for drainage or appearance. Roof panels must be designed to prevent ponding of water. Ponding may be prevented either by cambering or by providing slope or stiffness such that it will not occur.

Continuous and Cantilevered Spans. Stressed-skin panels may be cantilevered and/or continuous over interior supports. Stresses can be calculated by normal engineering formulas for multiple-span application or for cantilever action. Negative moments, of course, require that the top skin be adequately spliced for tension and the bottom skin be thick enough to resist applied compressive stresses.

Connections. Connections of panels to supporting members must resist uplift as well as downward loads. Connections between stressed-skin panels must transfer concentrated loads between sections without excessive differential deflection. Connections should also be detailed to restrain slight bowing, which sometimes results from moisture changes.

To avoid bowing, which can be caused by slight expansion of panels, it is advisable not to drive them tightly together. A suitable allowance is about 1/16 in. (1.5 mm) at the side of 4 ft wide (1.22 m) panels, and perhaps 1/4 in. (6 mm) at the ends of 20 ft long (6.1 m) panels.

Diaphragms. A roof, wall, or floor designed as a diaphragm may require more fastening than would ordinarily be needed simply to attach the panels for resisting gravity loads.

The glue bond between the stringers and wood structural panel skins must transfer shear due to loads in the plane of the panel, as well as shear due to loads perpendicular to the stressed-skin panels. These shear stresses will be additive and the resultant stress should be checked against the allowable planar/rolling shear stress of the wood structural panels, appropriately adjusted for duration of load. When analyzing the shear stresses, appropriate load combinations should be selected since the diaphragm load will generally not occur at the same time as full, normal-to-the-surface, vertical dead load and live load.

The allowable diaphragm load may be limited by the glueline stresses, particularly if one-sided or T-flange stressed-skin panels are used.

3.4.2 Flexural Panel Design

Due to the structural efficiency possible with stressed-skin panels, whereby relatively shallow panels prove adequate for strength, the design is likely to be controlled by the allowable deflection. The first aspect of the assumed section to be checked, therefore, will be deflection. Moment will be checked next, and shear last, since it is least likely to control. Shear, however, will sometimes govern when one or both skins are thick and the span is short.

Trial Section. Stressed-skin panels are designed by a trial-and-error method. A trial section must first be assumed and then checked for its ability to do the job intended. The entire 4 ft wide (1220 mm) panel is usually designed as a unit, in order to allow for edge conditions. The equations in the following sections are based on 4 ft wide (1220 mm) panels. They will require adjustment for any other panel width.

b ***Distance.*** In some cases the whole 4 ft (1.22 m) width of the skins cannot be considered effective, since there is a tendency for thin wood structural panels to dish toward the neutral axis of the panel when stringers are widely spaced. Basic spacing, usually referred to as the **b** distance, is used to consider this tendency. The **b** distance represents the maximum amount of the skin that may be considered to act in conjunction with the stringer for bending stress calculations. Table 3.7 provides **b** distances. Panels in which the clear distance between longitudinal members exceeds 2**b** for both covers should not be considered as having stressed skins.

One-Sided Panels

Panels over 3 in. (75 mm) Deep. One-sided stressed-skin panels, except the shallow ones covered in this section, are designed just as are two-sided panels. There is, of course, no bottom skin to be taken into account in calculating moment of inertia, and the resisting moment based on the bottom of the panel uses the allowable bending stress value for the stringers.

A variation on the one-sided panel sometimes called a T-flange panel, is illustrated in Fig. 3.11. The lumber web and lumber bottom flange may be considered integral for design purposes when they are glued together. When determining load governed by a 1 × 4 or 2 × 4 T-flange, use the allowable bending stress of the flange lumber. The allowable tensile stress of the flange lumber should be used for 1 × 6 and 2 × 6 T-flanges. The designer should use engineering judgment when selecting the allowable stress for other size T-flanges.

Panels Less Than 3 in. (75 mm) Deep. In one-sided stressed-skin panels less than 3 in. (75 mm) deep, the tendency for the skins to dish is more serious than in other panels. This factor should be taken into account in their design (Fig. 3.14). When designing such shallow sections, a value of 0.5**b** should be used in place of the full **b** value given in Table 3.7.

TABLE 3.7 Basic Spacing, **b**, for Various Wood Structural Panel Thicknesses

	Basic spacing, **b** (in.)			
	Strength axis parallel to stringers		Strength axis perpendicular to stringers	
Wood structural panel thickness (in.)	3, 4, 5-ply 3-layer plywood	5 ply 5-layer plywood and OSB	3, 4, 5 ply 3-layer plywood and OSB	5 ply 5-layer plywood
Unsanded panels				
3/8	14	—	17	—
15/32, 1/2	18	22	21	27
19/32, 5/8	23	28	22	31
23/32, 3/4	31	32	29	31
Touch-sanded panels				
1/2	19	24	21	27
19/32, 5/8	26	28	24	29
23/32, 3/4	32	34	28	36
1, 1 1/8	—	55	—	55

Neutral Axis and Moment of Inertia

Allowance for Surfacing. To allow for resurfacing of lumber members for gluing, they should be considered 1⁄16, in. (1.5 mm) smaller in dimension perpendicular to each gluing surface than their standard lumber size. Stringers for two-sided panels, which have two gluing surfaces, should then be considered 1⁄8 in. (3 mm) smaller than their standard lumber size.

Gross and Net Sections. Allowable bending stresses are determined by applying suitable reduction factors to the stresses obtained at ultimate. Allowable deflections are arbitrarily set, and applied to the behavior of the panel in its working range. Dishing of the skins at high load is discussed in Section 3.3.2. Because of it, a smaller net section may be effective at ultimate loads than in the working range. For thin skins and wide stringer spacings, therefore, this net moment of inertia for bending, I_n, may be less than the gross moment of inertia for deflection and shear, I_g (Fig. 3.13).

Calculation Method. The usual system for calculating the neutral axis and moment of inertia involves taking moments about the plane of the bottom of the panel. Use the actual resurfaced cross-sectional area of the lumber members, modified as appropriate by the transformed section ratio, n, Eq. (3.41), also considering applicable **b** distances.

Deflection—Other Than Shallow One-Sided Sections. The transformed width of wood structural panel is used for neutral-axis and moment-of-inertia calculations for deflection of the panels in this section. It is, of course, entirely possible to convert the lumber stringers to an MOE that matches that of the panels, but it is probably easier to work with the lumber as a basis unless both the top and bottom skins are the same. This includes panels like the one shown in Fig. 3.12, which does not have a stringer at each edge. The resultant moment of inertia, calculated from the transformed section, is called the gross moment of inertia, or I_g.

Shallow One-Sided Sections. As mentioned above, under One-Sided Panels, for one-sided sections less than 3 in. (75 mm) deep, the gross width may be used as the basis from which to compute the transformed section's moment of inertia only when the clear distance between stringers is less than **b** (Table 3.7). When the clear distance is greater than **b**, the flange width for stiffness calculations must be taken as the sum of the stringer width plus a distance equal to 0.25**b** on each side of each stringer (except, of course, for exterior stringers where only part of that distance is available). This jump, from physical width, where spacing between stringers is **b** to slightly over half the gross width where spacing is greater than **b**, is in accordance with test results (Fig. 3.13).

Deflection

Stiffness. The *EI* (stiffness capacity) of the stressed-skin panel (skins plus stringers) will be the stiffness capacity for bending deflection alone. It will not include shear deflection.

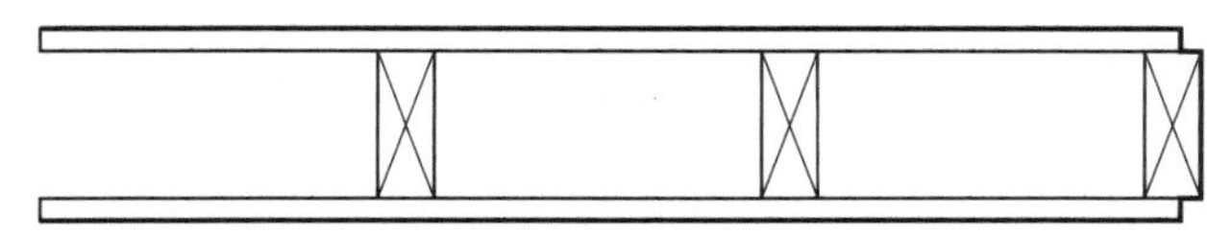

FIGURE 3.12 Total panel width effective (without reference to **b** distance).

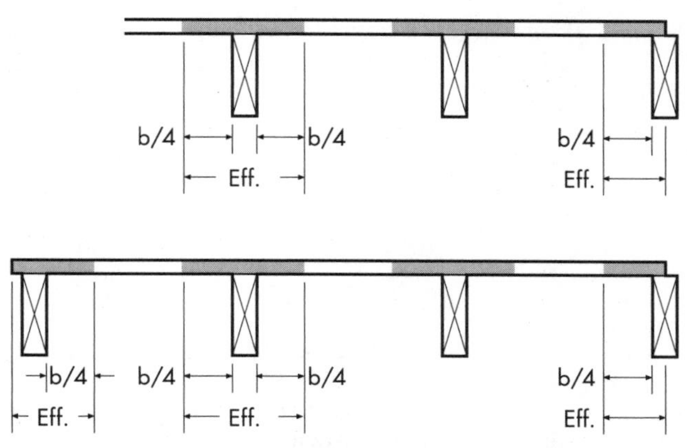

FIGURE 3.13 Effective width is sum of widths shown shaded.

In dealing with the stiffness of the panel, *EI*, begin with the skin stiffness capacity from Chapter 2, EAC_G. It reflects the composite nature of the panel and the fact that the skins of stressed-skin panels are typically in combined axial tension and/or compression about the stressed-skin-panel axis, rather than about the axis of the wood structural panel. The *E* can be obtained by dividing the skin's axial stiffness, *EA*, by the appropriate area, *A*, for the skin being considered.

Shear deflection should be considered separately for stressed-skin panels, whereas for most timber design it is taken into account by adjusting the modulus of elasticity. The MOE values listed for timber materials are not actually MOE values for pure bending but have been adjusted to account for shear deflection. Listed values for wood structural panels include a 10% allowance, and lumber values include a 3% allowance for shear deflection. These factors are close approximations for most applications, but are excessive for most stressed-skin panels. These MOE values are often referred to as the apparent MOE.

Bending deflection must be calculated using a true value for bending modulus of elasticity and not the published apparent MOE. A value for shear deflection, computed separately, is then added to the first figure representing the moment deflection. The simple-span shear deflection for uniform (also quarter-point) loading is given by Eq. (3.42).

$$\Delta_s = \frac{1.8PL}{A_{T,\text{stringers}} G_{\text{stringers}}} \tag{3.42}$$

where Δ_s = shear deflection (in.)
P = total load on panel (lb)
L = span length (ft)
$A_{T,\text{ stringers}}$ = actual total cross-sectional area of all stringers (in.^{2})
$G_{\text{stringers}}$ = modulus of rigidity of stringers (psi)

G may be taken as 0.06 of the true bending modulus of elasticity of stringers. This equation gives the shear deflection for uniform loading or quarter-point loading. The shear deflection for a single concentrated load at the center is double this amount; the shear deflection of a cantilever with a uniformly distributed load is four times this amount.

Allowable Load Combining equations for bending and shear deflections for a uniformly loaded simple-span panel, the allowable load based on deflection is then given by Eq. (3.43).

$$W_{\Delta} = \frac{1}{CL[(7.5L^2/EI_g) + (0.6/A_T G)]} \tag{3.43}$$

where w_{Δ} = allowable load based on deflection (psf).
C = factor for allowable deflection, usually 360 for floors, 240 for roofs.
L = span length (ft).
EI_g = stiffness of the stressed-skin panel assembly using the "true" E (lb-in.^{2}).
A_T = actual total cross-sectional area of all stringers, and T-flanges if applicable (in.^{2}).
G = modulus of rigidity of stringers (psi). G may be taken as 0.06 of the true bending modulus of elasticity of stringers.

Note that if the allowable-deflection factor, C, is based on live load only, this equation will yield allowable live load, to which dead load may be added. The constants shown in the above equation result from collecting constants for the simple-span beam equations, appropriately adjusted for panel width and conversion of units. For instance,

$$7.5 = \frac{5}{384} \times 4 \times \frac{1}{12} \times 1728$$

Top Skin Deflection. In addition to computing the deflection of the whole entire stressed-skin panel acting as a unit, the designer must also check the deflection of the top skin between stringers. Sections are usually selected such that deflection only must be checked for this top skin, but for many applications, moment and shear should also be investigated. For two-sided panels, this skin will be a fixed-end beam (because the panel is rigidly glued to the outside stringers) for which the equation is:

$$\Delta = \frac{4wL_1^4}{4608\ EI_{top}C_G} \tag{3.44}$$

where Δ = deflection (in.)
w = load (psf)
L_1 = clear span between stringers (in.)
$EI_{top}\ C_G$ = modulus of elasticity for top skin (psi)
I = moment of inertia (in direction perpendicular to stringers) of 1 ft (305 mm) width of top skin (in.^{4})

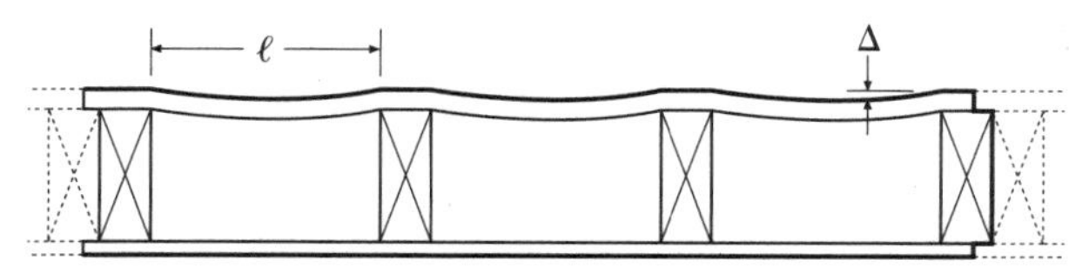

FIGURE 3.14 Top-skin deflection.

For one-sided panels with four stringers (Fig. 3.13), the skin will act like a three-span beam, with the following equation:

$$\Delta = \frac{wL_1^4}{1743\ EI_{\text{top}}C_G} \tag{3.45}$$

where all symbols are as above for the refined method that includes shear deflection of the wood structural panel.

In most stressed-skin panels the strength axis will be parallel to supports and it will therefore be necessary to use in these equations the correct stiffness capacity value, EIC_G, for the perpendicular-to-strength-axis direction. This value for $EI_\perp$ can be obtained in Chapter 2. Tabulated values of EIC_G are typically given per foot of panel width.

Bending Moment

Other Than Shallow One-Sided Sections. As mentioned above, under **b** Distance, for bending considerations only, it is sometimes necessary to calculate beginning with the reduced physical section provided by using the **b** distance to arrive at effective skin widths. If the clear distance between stringers is less than **b**, the effective width of skin is equal to the full panel width. Neutral axis and moment of inertia are then as calculated above for deflection. If the clear distance is greater than **b**, the effective width of skins must be reduced, as if the unshaded areas shown in Fig. 3.13 were not part of the skin. The effective panel width (Fig. 3.15) equals the sum of the widths of the stringers plus a portion of the skin extending a distance equal to 0.5**b** each side of each stringer (except, of course, for outside stringers where only part of that distance is available). The resultant moment of inertia is called the net moment of inertia, or I_n.

Shallow One-Sided Sections. The neutral axis and moment of inertia used for shallow, one-sided panels are the same as those used above for deflection; that is, using 0.25**b** instead of 0.5**b** when stringer spacing is greater than **b** (see Fig. 3.13).

Allowable Stresses. As the spacing between framing members is increased, the allowable load is reduced in two ways. The first has been accounted for in considering the **b** distance. The second involves reductions in allowable stresses. The allowable stresses, in both tension and compression, for the grade of wood structural panel used, are reduced in accordance with Fig. 3.16.

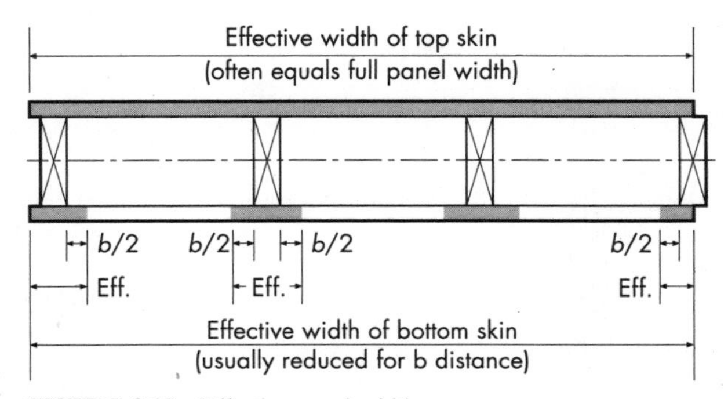

FIGURE 3.15 Effective panel widths.

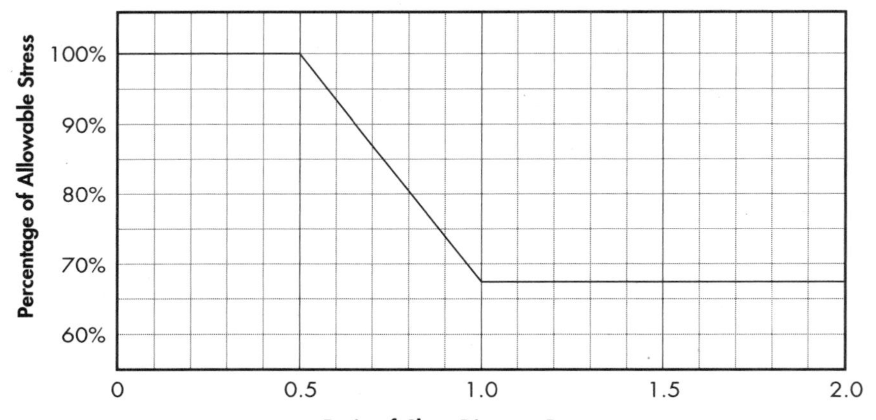

FIGURE 3.16 Stress reduction factor for framing member spacing.

This reduction is to provide against buckling of the skins. It applies to working stresses in both tension and compression parallel to strength axis, but not to rolling-shear stresses.

Allowable Load. Most stressed-skin panels with spaced stringers are not symmetrical about the neutral axis. For such unbalanced sections, it will be necessary to calculate the allowable load in bending as determined both by the top skin and by the bottom. The lower of these values will then govern, unless there is a splice in an area of high moment. If there is, an additional check on the splice will be required.

For uniform loads, the general equation

$$M = \frac{FI}{c} \tag{3.46}$$

reduces to the following:

$$w_b = \frac{8\left(\dfrac{FAC_G}{A}\right)(EI_n)}{48cL^2E_{\text{lumber}}} \tag{3.47}$$

where w_b = allowable load based on bending (psf)
FAC_G = allowable stress (either tension or compression adjusted as appropriate for C_D, etc.)
EI_n = stiffness of the transformed section (lb-in.^{2})
c = distance from neutral axis to extreme tension or compression fiber (in.)
L = span (ft)
E_{lumber} = modulus of elasticity for the lumber stringers (psi)

Splice-Plate Design. Resisting tension (or compression) in the skins resulting from the bending moment must be resisted by the splice plate. For splice plates, the allowable stress is applied over only the width of skin actually covered by splice plates. The gross I is used for calculating this splice-plate resisting moment, since full width of skins is effective at this point, due to the stiffening effect of splice plate and blocking. The following equation is valid for a splice plate at the point of maximum moment.

$$w_p = \frac{8\left(\frac{FAC_G}{A}\right)(EI_g)}{48cL^2E_{\text{lumber}}} \tag{3.48}$$

where w_p = allowable load based on splice-place stress (psf)
FAC_G = allowable splice-plate stress (F_t or F_c) multiplied by the proportion of the width actually spliced (psi)
EI_g = stiffness factor, discussed above under Deflection, for deflection (lb-in.^{2}) and other symbols are as above.

If the splice is in an area of high moment and is found to control the design, often the best thing to do is to change the location of the splice to an area of lower stress.

Planar/Rolling Shear

Location of Critical Stress. When the wood structural panel skin has its strength axis parallel to the longitudinal framing members, as is usually the case, the critical shear plane lies within the wood structural panel. With plywood, it is at the glueline, between the inner face ply and the adjacent perpendicular ply at A or B of Fig. 3.17.

OSB, however, has no clearly defined shear plane. As a practical matter, the actual ply thicknesses of plywood are seldom available to the designer and are subject to change by the manufacturers without notification. As a simplification, therefore, the critical interior shear plane is assumed to be 0.10 in. (2.5 mm) from the glued face of the panel, regardless of panel composition. This will typically give conservative results and if additional planar/rolling shear capacity must be found, other wood structural panel combinations must be evaluated.

When the strength axis of the skin is perpendicular to the framing members, the critical planar/rolling-shear plane lies between the skin's inner face and the framing member (at A' or B' for plywood). The standard equation for shear is

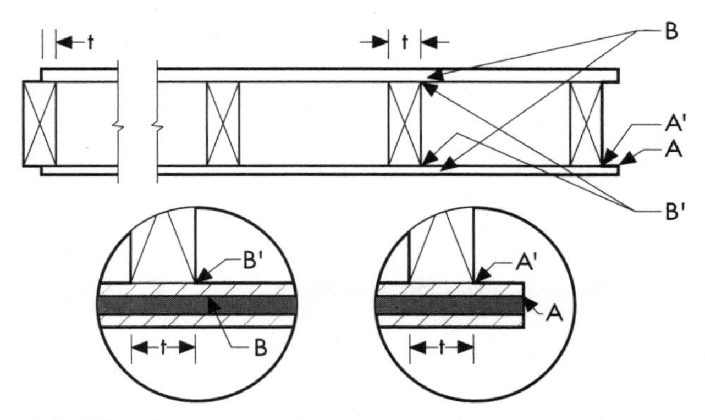

FIGURE 3.17 Location of critical planar/rolling shear stress.

$$f_v = \frac{QV}{\text{Ib}} \tag{3.49}$$

where F_s = allowable rolling/planar shear stress (lb-in.^{2})
Q = statical moment of area of skin (in.^{3}/ft of width)
V = shear (lb)
I = moment of inertia of 1 ft width of skin or stringer (in.^{4})
b = effective width of skin (in.)

An unbalanced (different thickness or type of panel on each side) stressed-skin panel has two values for Q—one for the thicker skin and one for the thinner. Since the Q of the thicker skin is usually larger than that of the thinner skin, it is generally sufficient to compute planar/rolling shear only for the thicker skin.

Transformed Section. In calculating Q, use of the transformed section is required. The approach used here, however, is typically different. The Q calculated is for that area outside the shear-critical plane, not the Q for the entire skin. Since skins are typically applied with strength axis parallel to the stringers, in plywood the skin's actual shear-critical plane is at the first glueline away from the stringer. In a five-ply plywood skin, for instance, the actual Q is calculated for the outer four plies. As a practical matter, however, for plywood and OSB, the internal shear-critical plane may be taken as 0.1 in. from the glued face of the panel.

If the value for Q is for the transformed section, the designer does not convert the allowable stresses back to those of an untransformed section. If the designer chooses to work with the full physical dimensions of the section, the calculated allowable stresses for planar/rolling shear must be converted back to those of the full-width panel by the use of n. This may seem backward, but the reason is that in the equation for calculating w_s, the ratio between I_g and Q_s is based on I of the transformed section. If both Q and I are based on the transformed section, the ratio is one-to-one. If Q is based on the full physical width of the panels, the ratio $n = E_1/E_2$ must be used to correct for the use of the transformed moment of inertia, I, with the untransformed Q.

The distance, d_s, by which this area must be multiplied is not presented directly, but it is easy to obtain. It is the distance from the centroid of the area outside the assumed shear plane to the neutral axis of the stressed-skin panel. Note that the neutral axis is as determined above under Neutral Axis and Moment of Inertia. To obtain d_s, subtract from the total distance between N.A. and outside of panel, c, the distance between outside of panel and centroid of transformed area outside the critical plane for rolling shear, $\bar{Y}$. In equation form,

$$\bar{Y} = c - \left(\frac{t - 0.1 \text{ in.}}{2}\right) \tag{3.50}$$

where t = panel thickness (in.)

Statical Moment. The Q_s, or statical moment for rolling shear, is then given by the following equation:

$$Q_s = Ad_s \tag{3.51}$$

Allowable Capacity. It is convenient to compute a value for the sum of the allowable shear stress times the applicable shear width, $\Sigma F_s t$, over all joints.

$$F_s = \frac{F_s\,(\text{Ib}/Q)C_G}{\text{Ib}/Q} \tag{3.52}$$

Note that due to stress concentrations, the allowable stress in rolling/planar shear for exterior stringers is only half the stress for interior stringers. This reduction applies to exterior stringers whose clear distance to the panel edge is less than half the clear distance between stringers.

Allowable Load. The allowable uniform load on a simple-span stressed-skin panel as determined by rolling shear can be expressed by the following equation:

$$w_s = \frac{2(EI_g) \sum F_s t}{4Q_s L(E_{\text{stringer}})} \tag{3.53}$$

where w_s = allowable load based on rolling shear stress (psf)

EI_g = stiffness factor of transformed section (lb-in.^{2})

$\Sigma F_s t$ = sum of the glueline widths over each stringer, each multiplied by its applicable allowable rolling-shear stress (lb-in.)

Q_s = the first moment (about the neutral axis) of the area outside of the critical rolling-shear plane, full 4 ft (1220 mm) panel width (in.^{3})

L = length of the stressed-skin panel (ft)

E_{stringer} = modulus of elasticity of the lumber (psi)

Horizontal Shear

Statical Moment. The Q to be used in the equation for horizontal shear includes the statical moment of all material above (or below) the neutral axis. Thus, the value of Q_s obtained for rolling shear for the top skin cannot be used since it will not include the stringers nor will it include the bottom 0.10 in. (2.5 mm) of the top skin.

Again, the neutral axis of the transformed section is used, to allow for differences in modulus of elasticity, in calculating effective I. The value of Q may be of either the transformed panel widths or the full physical panel widths. Since the whole depth of the wood structural panel is involved, the area, per foot of width, is that listed in the Chapter 2. The d distance is from the N.A. to the middepth of the skin. If the transformed width is used to calculate area, n need not be applied in finding Q_{skin}, but the effectiveness of the full-width skin must be transformed to be compatible with the stringer. Thus, the Q_v of the panel is given in the following equation:

$$Q_v = Q_{\text{stringers}} + \frac{E_{\text{skins}}}{E_{\text{stringers}}} \times Q_{\text{skin}} \tag{3.54}$$

where

$$\frac{E_{\text{skin}}}{E_{\text{stringers}}} = n$$

Allowable Shear Stress. Allowable horizontal shear stresses for stress-grade lumber stringers are given in the NDS.[1]

Allowable Load. The equation for allowable uniform load on a stressed-skin panel is as follows:

$$w_v = \frac{2(EI_g) \sum F_v t}{4Q_v LE_{\text{stringers}}} \tag{3.55}$$

where w_v = allowable load based on horizontal shear stress (psf)
EI_g = stiffness factor, discussed above under Deflection, Eq. (3.43) (lb-in.^{2})
F_v = allowable horizontal shear stress in the lumber stringers (psi)
t = stringer width (in.)
Q_v = first moment about the neutral axis of all material above (or below) the neutral axis of the 4 ft wide panel (in.^{3})
L = length of the panel (ft)
$E_{\text{stringers}}$ = modulus of elasticity of stringers (psi)

Final Allowable Load. The final allowable load on the panel is the lowest of the figures obtained above under Deflection, Bending Moment, and Planar/Rolling Shear.

3.4.3 Wall Panel Design

Stressed-skin panels used for walls, or other applications where they are loaded in axial compression, can be designed in accordance with standard procedures. In practice, when these panels are used for walls, their thickness is usually determined by appearance, acoustics, or insulation requirements, and sometimes by the necessity for wind resistance, but seldom by their actual load-bearing capacity as columns.

Vertical-Load Formula. Few practical end joints for stressed-skin panels will provide any appreciable degree of fixity. Under vertical load, therefore, panels will behave as pin-ended columns. The pin-ended-column equations reduce to:

$$P_a = \frac{3.619EI_g}{144L^2} P_a \quad \text{or} \quad F_c A \text{ (whichever is less)} \tag{3.56}$$

where P_a = allowable axial load on the panel (lb)
E = appropriate modulus of elasticity (see below) (psi)
I_g = transformed moment of inertia of panel about neutral axis (in.^{4})
L = unsupported vertical height of panel (ft)
F_c = allowable compressive stress (axial) for wood-structural-panel skins (psi)
A = total transformed vertical-grain material of stringers and skins (in.^{2})

Combined Bending and Axial Load. When designing wall panels subject to wind loads, or any other panels where bending and axial stresses are both present, the usual combined-load equation,

$$\left(\frac{P/A}{F_a}\right) + \left(\frac{M/S}{F_b}\right) \leq 1.0 \qquad (3.57)$$

which reduces to the following:

$$\left(\frac{P}{P_a}\right) + \left(\frac{M/S}{F_c}\right) \leq 1.0$$

where P = total allowable axial load on the panel with combined loading (lb)
P_a = allowable axial load on the panel (lb)
M = total allowable bending moment on the stressed-skin panel with combined loading (lb-in.)
S = section modulus of stressed-skin panel (compression side) = I_n/c (in.^{3})
F_c = allowable stress in compression parallel to strength axis, from Chapter 2.

Values for P_a and F_c can be adjusted for duration of load. See Chapter 1.

3.4.4 Fabrication of Wood Structural Panel Stressed-Skin Panels

The provision of Section 3.2.4 shall be followed except as modified by the provisions of this section.

Fabrication

Skins. Slope of scarf and finger joints in plywood skins shall not be steeper than 1 in 8 in the tension skin and 1 in 5 in the compression skin. Scarf and finger joints shall be glued under pressure over their full contact area, and shall meet the requirements of PS 1[4] or PS 2.[5] In addition, the aggregate width of all knots and knotholes falling wholly within the critical section of plywood skins shall be not more than 10 in. (254 mm) on each face of the jointed panel for a 4 ft wide (1220 mm) panel, and proportionately for other widths. The critical section for a plywood scarf joint shall be defined as a 12 in. wide (305 mm) strip, 6 in. (152 mm) on each side of the joint in the panel face, extending across the width of the panel. OSB shall not be scarf jointed.

Butt joints in wood structural panel skins shall be backed with wood structural panel splice plates centered over the joint and glued over their full contact area. Splice plates shall be at least equal in thickness to the skin, except that minimum thickness shall be 15⁄32 in. (11.9 mm) if nail glued. Minimum splice plate lengths, strength axis parallel with that of the skin, shall be as follows (unless otherwise called for in the design). See Table 3.8.

Surfaces of high-density overlaid plywood to be glued shall be roughened, as by a light sanding, before gluing.

Assembly. Wood structural panel skins shall be glued to framing members over their full contact area, using means that will provide close contact and substantially uniform pressure. Where clamping or other positive mechanical means are used, the pressure on the net framing area shall be sufficient to provide adequate contact

TABLE 3.8 Minimum Splice Plate Dimensions

Skin thickness	Splice plate length
¼ in. (6.4 mm)	6 in. (150 mm)
5/16 in. (7.9 mm)	8 in. (205 mm)
⅜ in. (9.5 mm) sanded	10 in. (255 mm)
⅜ in. (9.5 mm) unsanded	12 in. (305 mm)
15/32 in. and ½ in. (11.9–12.7 mm)	14 in. (355 mm)
19/32 in.–¾ in. (15–19 mm)	16 in. (405 mm)

and ensure good glue bond (100–150 psi [690–1035 kPa]) on the net glued area is recommended), and shall be uniformly distributed by caul plates, beams, or other effective means. In place of mechanical pressure methods, nail gluing may be used.

Where a tongue-and-groove-type stressed-skin panel edge joint is specified (and not otherwise detailed), the longitudinal framing member forming the tongue shall be of at least 2 in. (50 mm) nominal width, set out ¾ in. (19 mm) ± 1/16 in. (1.5 mm) from the wood structural panel edge. Edges of the tongue shall be eased to provide a flat area at least ⅜ in. (9.5 mm) wide. Any corresponding framing member forming the base of the groove shall be set back ¼ in. (6.4 mm) to 1 inch (25 mm) more than the amount by which the tongue protrudes. One skin may be cut back slightly to provide a tight fit for the opposite skin.

Unless otherwise specified, panel length and width shall be accurate within ± ⅛ in. (3 mm). Panel edges shall be straight within 1/16 in. (1.5 mm) for an 8 ft (2440 mm) length and proportionately for other lengths. Panels in the same group shall not vary in thickness by more than 1/16 in. (1.5 mm), nor differ from design thickness by more than ⅛ in. (3 mm). Panels shall be square, as measured on the diagonals, within ⅛ in. (3 mm) for a 4 ft (1220 mm) wide panel and proportionately for other widths. Panels shall lie flat at all points within ¼ in. (6.4 mm) for 4 ft (1220 mm) × 8 ft (2440 mm) panels, and proportionately for other sizes. Panel edge cross sections shall be square within 1/16 inch (1.5 mm) for constructions with lumber stringers 4 in. (100 mm) deep, and proportionately for other sizes.

Insulation, vapor-barrier materials, and ventilation shall be provided as specified in the design. Such materials shall be securely fastened in the assembly in such a way that they cannot interfere in the process of gluing the wood structural panel skins, or with the ventilation pattern. When ventilation is specified, panels shall be vented through blocking and headers on the cool side (cooler climates) of the insulation. Provision shall be made to line up the vent holes within and between panels. Stringers shall not be notched for ventilation, unless so specified.

3.4.5 Identification

Each member shall be identified by the appropriate trademark of an independent inspection and testing agency, legibly applied to be clearly visible. Locate trademark approximately 2 ft from either end, except appearance of the installed panel shall be considered.

3.4.6 Stressed-Skin Panel Design Example

Since this example is intended for use as a general guide to illustrate the principles provided in this handbook, a review of those sections pertinent to your specific design is recommended before proceeding.

Preliminary considerations as to the grade of wood structural panel and lumber to be used for a given design should include a check on availability. Where full exterior durability is not required for the wood structural panel, Exposure 1 wood structural panels may be specified.

Problem

Design a stressed skin floor panel for a 14 ft (4.3 m) span.

Live load: 40 psf (1.92 kPa)

Dead load: 10 psf (0.48 kPa)

Deflection limitation: $L/360$ under live load

2×6 Stringers

$E_{\text{stringer}} = 1{,}800{,}000$ psi (12.4 GPa)

Trial Section

Try the section shown in Figs. 3.18 and 3.19.

Top skin: $^{19}\!/_{32}$ in. (15.1 mm) combination subfloor/underlayment 20 o.c. EXP 1 marked PS 1 or PS 2 five-ply plywood panel.

$$A_{\text{top}} = 7.125 \text{ in.}^2/\text{ft } (15.1 \text{ mm}^2/\text{mm})$$

$$I = 0.209 \text{ in.}^4/\text{ft } (0.442 \text{ mm}^2/\text{mm}) \text{ (from Chapter 2)}$$

Bottom skin: $^3\!/_8$ in. (9.5 mm) three-ply three-layer panel of plywood rated sheathing 24/0 exposure 1 marked PS 1[4] or PS 2.[5]

$$A_{\text{bottom}} = 4.50 \text{ in.}^2/\text{ft } (9.53 \text{ mm}^2/\text{mm})$$

$$I = 0.053 \text{ in.}^4/\text{ft } (72.38 \text{ mm}^4/\text{mm})$$

$$\text{Clear distance between stringers} = \frac{48 - (3 \times 1.5) - 1 - 0.75}{3}$$

$$= 13.9 \text{ in. } (345 \text{ mm})$$

$$\text{Total splice plate width (three plates)} = 3(13.9 - 2(0.25)) = 40.2 \text{ in. } (1020 \text{ mm})$$

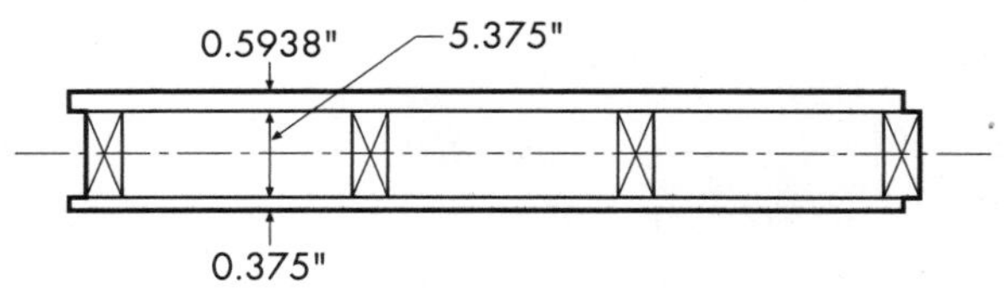

FIGURE 3.18 Cross section for deflection determination.

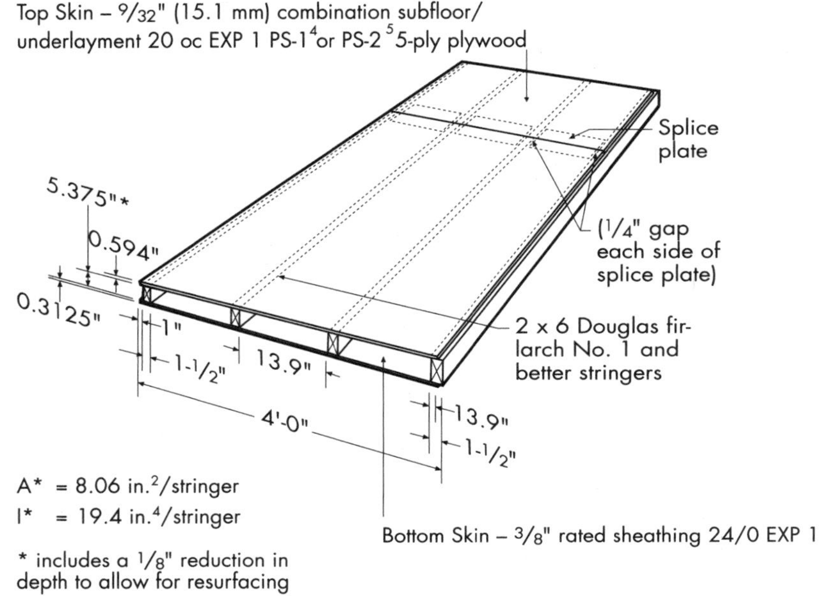

FIGURE 3.19 Trial section—panel layout.

Basic Spacing

From Table 3.7, for 19⁄32 in. (15.1 mm) five-ply five-layer touch-sanded wood structural panels, strength axis parallel to stringers, **b** = 28 in. (710 mm). For 3⁄8 in. (9.5 mm) three-ply unsanded wood structural panels, **b** = 14 in. (356 mm)

Note that for calculations involving section properties and allowable capacities, the final answer seldom has more than two significant figures. More significant figures are often used in the design example to facilitate continuity of calculations.

Calculation of Neutral Axis for Deflection

The neutral axis is determined by summing moments about any given point, using the transformed areas.

$$\Sigma M = 0 = M_{\text{top}} + M_{\text{stringer}} + M_{\text{bottom}}$$

$$E_{\text{top}} = \frac{EA_{\text{top}}C_G}{A_{\text{top}}} = \frac{5{,}000{,}000(1.0)}{7.125} = 701{,}754 \text{ psi } (4.838 \text{ GPa})$$

$$E'_{\text{top}} = E(1.10) = 701{,}754(1.10) = 771{,}930 \text{ psi } (5.32 \text{ GPa})$$

$$E_{\text{bottom}} = \frac{EA_{\text{bottom}}C_G}{A_{\text{bottom}}} = \frac{3{,}350{,}000(1.0)}{4.50} = 744{,}444 \text{ psi } (5.133 \text{ GPa})$$

$$E'_{\text{bottom}} = E(1.10) = 744{,}444(1.10) = 818{,}889 \text{ psi } (5.346 \text{ GPa})$$

$$E_{\text{stringers}} = 1{,}8000{,}000(1.03) = 1{,}854{,}000 \text{ psi } (12.78 \text{ GPa})$$

$$n_{\text{top}} = \frac{771{,}930}{1{,}854{,}000} = 0.4164$$

$$n_{\text{bottom}} = \frac{818{,}889}{1{,}854{,}000} = 0.4417$$

Transformed width = panel width × n

$$w_{t,\ \text{top}} = 48(0.4164) = 19.99 \text{ in. (508 mm)}$$

$$w_{e,\ \text{bottom}} = 48(0.4417) = 21.20 \text{ in. (538 mm)}$$

$$A_{\text{top}} = tb_e = 0.5938(19.99) = 11.8691 \text{ in.}^2 \ (7.657 \times 10^3 \text{ mm}^2)$$

$$A_{\text{stringer}} = tb_e = 5.375(1.5) = 32.25 \text{ in.}^2 \ (20.81 \times 10^3 \text{ mm}^2)$$

$$A_{\text{bottom}} = tb_e = 0.375(21.20) = 7.95 \text{ in.}^2 \ (5.13 \times 10^3 \text{ mm}^2)$$

Summary: Neutral-Axis Calculation for Deflection:
$\Sigma M = 0 = M_{\text{top}} + M_{\text{stringer}} + M_{\text{bottom}}$

Item	$A = tw_e$ (in.^{2})	d (in.)	Ad (in.^{3})
Top skin	11.8754	6.0469	71.7773
Stringer	32.25	3.0625	98.7656
Bottom skin	7.95	0.1875	1.406
Total	52.075	—	172.0335

$$\bar{y} = \frac{\Sigma Ad}{\Sigma A} = \frac{172.0335}{52.075} = 3.304 \text{ in. (84 mm)}$$

Determine Stiffness

Parallel axis theorem

$$I = \bar{I} + AD^2$$

$$\bar{I} = \frac{bh^2}{12}$$

$$\bar{I}_{\text{top}} = \frac{19.99(0.5938)^3}{12} = 0.3488 \text{ in.}^4 \ (0.1452 \times 10^6 \text{ mm}^4)$$

$$\bar{I}_{\text{stringers}} = 4\left[\frac{1.5(5.375)^3}{12}\right] = 77.6436 \text{ in.}^4 \ (32.318 \times 10^6 \text{ mm}^4)$$

$$\bar{I}_{\text{bottom}} = \frac{21.20(0.375)^3}{12} = 0.0932 \text{ in.}^4 \ (0.0388 \times 10^6 \text{ mm}^4)$$

$$D_{\text{top}} = 3.304 - \left(\frac{0.5938}{2}\right) = 2.7371 \text{ in. (70 mm)}$$

$$D_{\text{stringers}} = 3.304 - 0.375 - \left(\frac{5.375}{2}\right) = 0.2415 \text{ in. (6 mm)}$$

$$D_{\text{bottom}} = 3.304 - \left(\frac{0.375}{2}\right) = 3.1165 \text{ in. (79 mm)}$$

Summary: Stiffness Calculations for Deflection

Item	$\bar{I}_{\text{item}}$ (in.^{4})	A_{item} (in.^{2})	D (in.)	D^2 (in.^{2})	$\bar{I} + AD^2$ (in.^{4})
Top Skin	0.3488	11.88	2.7371	7.4917	89.35
Lumber	77.6436	32.25	0.2415	0.0583	79.52
Bottom skin	0.0932	7.95	3.1165	9.7126	77.31
Total	—	52.07	—	—	246.18

(Fig. 3.20)

Stiffness = $EI_g = E_{\text{stringers}} I_g$ = 1,854,000(246.18) = 456,400,000 (lb-in.^{2}) (1,310 kN-m^2) per 4 ft wide (1220 mm) stressed-skin panel section.

Allowable Load Based on Deflection

Using Eq. 3.43,

$$w_\Delta = \frac{1}{CL\left[\dfrac{7.5L^2}{EI_g} + \dfrac{0.6}{A_T G}\right]} \quad \text{(includes shear deflection from Eq. [3.42])}$$

$$G = 0.06 \times 1{,}854{,}000 = 111{,}200 \text{ psi (765 mPa)}$$

$$= \frac{1}{360 \times 14\left[\dfrac{7.5 \times 14^2}{456{,}400{,}000} + \dfrac{0.6}{52.1 \times 111{,}200}\right]}$$

$$= \frac{1}{5040\,[0.0000032 + 0.0000001036]} = 59.68 \text{ psf (2.86 kPa)}$$

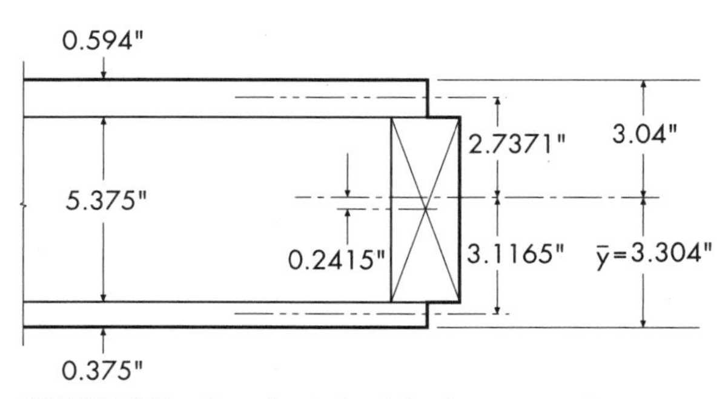

FIGURE 3.20 Neutral axis for deflection cross section.

$$w_{\Delta LL} = 59.7 \text{ psf } (2.86 \text{ kPa}) > \text{live load} = 40 \text{ psf } (1.92 \text{ kPa}) \qquad \text{OK}$$

For purposes of comparing with other allowable load figures, add 10 psf (0.48 kPa) for dead load and change C to 240. Then,

$$w_{\Delta TL} = 89.5 \text{ psf } (2.95 \text{ kPa}) > \text{total load} = 50 \text{ psf } (2.39 \text{ kPa}) \qquad \text{OK}$$

Top-Skin Deflection

$$\text{Allowable live load deflection} = \frac{L}{360} = \frac{13.9}{360} = 0.0386 \text{ in. } (1 \text{ mm})$$

From Eq. (3.44),

$$\Delta_{LL} = \frac{wL_1^4}{4608(EIC_{G,\ \text{top}})}$$

where EI = stiffness of panel in direction perpendicular to stringes (psi)
L_1 = clear span between stringers (in.)

$$= \frac{40(13.9)^4}{4608(13{,}000)(3.1)}$$

$$= 0.0080 \text{ in. } (0.2 \text{ mm}) < 0.0386 \text{ in. } (1 \text{ mm}) \qquad \text{OK}$$

Allowable Stresses

Top-Skin Allowable Compressive Stress

$$\text{Basic } F_cAC_G \text{ from Chapter 2} = 4200(1.5) = 6300 \text{ lb/ft } (91.94 \text{ N/mm})$$

From Fig. 3.16, ratio of clear distance between stringer clear distance and basic spacing, **b**:

$$\frac{\text{clear dist.}}{\mathbf{b}} = \frac{13.9}{28} = 0.4964 < 0.5$$

therefore 100% allowable stress may be used.

Bottom Skin Allowable Tensile Stress

From Chapter 2,

$$\text{Basic } F_tAC_G = 2300(1.0) = 2300 \text{ lb/ft } (33.57 \text{ N/mm})$$

$$\frac{\text{clear dist.}}{\mathbf{b}} = \frac{13.9}{14} = 0.9929 \text{ in. } (25 \text{ mm})$$

From Fig. 3.16,

$$\text{stress reduction factor} = 66.7\%$$

Therefore,

$$F_tAC_G = 2300(0.667) = 1534 \text{ lb/ft } (22.39 \text{ N/mm})$$

Because only 67% of the bottom skin can be considered effective for bending (above, **b** Distance), the neutral axis must be recalculated using the transformed effective width.

$$\text{Bottom skin effective width for bending} = 21.20 \text{ in. } (0.667)$$

$$= 14.1 \text{ in. } (358 \text{ mm}) \text{ (Fig. 3.21)}$$

Calculation of Neutral Axis for Bending

$$\text{Transformed width} = (\text{panel width}) \times (n)$$

$$w_{t,\,\text{top}} = 19.99 \text{ in. } (508 \text{ mm})$$

$$w_{e,\,\text{bottom}} = 14.1 \text{ in. } (358 \text{ mm})$$

$$A_{\text{top}} = tb_e = 11.8691 \; (7.657 \times 10^3 \text{ mm}^2)$$

$$A_{\text{stringer}} = tb_e = 32.25 \text{ in.}^2 \; (20.81 \times 10^3 \text{ mm}^2)$$

$$A_{\text{bottom}} = tb_e = 0.375(14.1) = 5.2875 \text{ in.}^2 \; (3.411 \times 10^3 \text{ mm}^2)$$

Summary: Neutral-Axis Calculation for Bending:
$\Sigma M = 0 = M_{\text{top}} + M_{\text{stringer}} + M_{\text{bottom}}$

Item	$A = tw_e$ (in.^{2})	d (in.)	Ad (in.^{3})
Top skin	11.8754	6.0469	71.7773
Stringer	32.25	3.0625	98.7656
Bottom skin	5.2875	0.1875	0.9914
Total	49.4129	—	171.5343

$$\bar{y} = \frac{\Sigma Ad}{\Sigma A} = \frac{171.5343}{49.4129} = 3.4714 \text{ in. } (88 \text{ mm})$$

Determine Stiffness

Parallel axis theorem

$$I = \bar{I} + AD^2$$

$$\bar{I} = \frac{bh^3}{12}$$

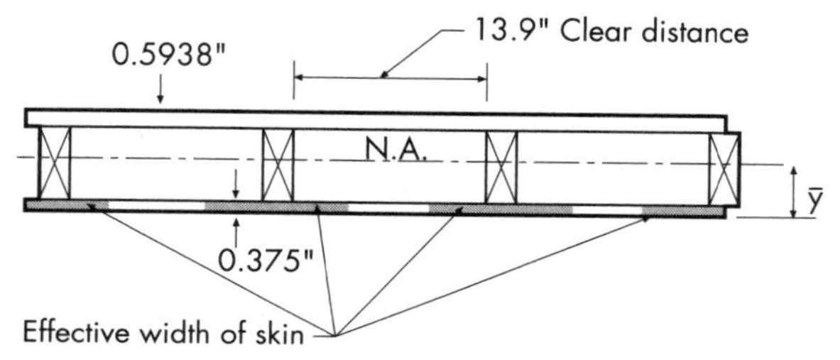

FIGURE 3.21 Cross section for neutral axis determination for bending calculations.

$$\bar{I}_{\text{top}} = 0.3488 \text{ in.}^4 \ (0.145 \times 10^6 \text{ mm}^4)$$

$$\bar{I}_{\text{stringers}} = 77.6436 \text{ in.}^4 \ (32.32 \times 10^6 \text{ mm}^4)$$

$$\bar{I}_{\text{bottom}} = \frac{14.1(0.375)^3}{12} = 0.0620 \text{ in.}^4 \ (25.8 \times 10^3 \text{ mm}^4)$$

$$D_{\text{top}} = 2.8724 - \left(\frac{0.5938}{2}\right) = 2.5755 \text{ in. (65 mm)}$$

$$D_{\text{stringers}} = 3.4714 - 0.375 - \left(\frac{5.375}{2}\right) = 0.4089 \text{ in. (10 mm)}$$

$$D_{\text{bottom}} = 3.4714 - \left(\frac{0.375}{2}\right) = 3.2839 \text{ in. (83 mm)}$$

Summary: Stiffness Calculations for Bending

Item	$\bar{I}_{\text{item}}$ (in.^{4})	A_{item} (in.^{2})	D (in.)	D^2 (in.^{2})	$\bar{I} + AD^2$ (in.^{4})
Top skin	0.3488	11.88	2.5755	6.6331	79.15
Lumber	77.6436	32.25	0.4089	0.1672	83.0358
Bottom skin	0.0620	5.2875	3.2839	10.7840	57.0824
Total	—	49.4175	—	—	219.27

(Fig. 3.22)

$$\text{Stiffness} = EI_g = E_{\text{stringers}} I_g$$
$$= 1{,}854{,}000(219.27)$$
$$= 406{,}500{,}000 \text{ (lb-in.}^2\text{) (1166 kN-m}^2\text{)}$$

per 4 ft wide (1220 mm) stressed-skin panel section.

Allowable Load Based on Bending Stress

Allowable load if top skin controls, from Eq. (3.47):

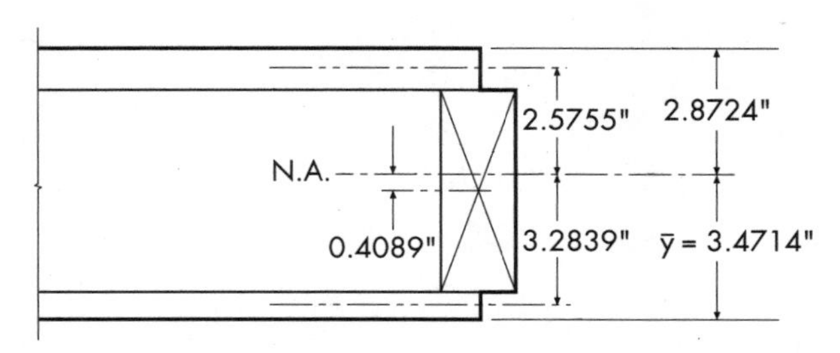

FIGURE 3.22 Neutral axis for bending.

$$w_{b,\ \text{top}} = \frac{8(F_c A C_G / A)(EI_n)}{48\ cL^2\ E_{\text{stringer}}}$$

$$w_b = \frac{8(6300/7.125)(406{,}500{,}000)}{48(2.8724)(14)^2(1{,}854{,}000)} = 57.3924 \text{ psf (2.75 kPa)} \quad \text{(transformed)}$$

$$n_{\text{top}} = 0.4164$$

$$w_{b,\ \text{top}} = \frac{57.3924}{0.4164} = 137.8 \text{ psf (6.598 kPa)}$$

Allowable load if bottom skin controls, from Eq. (3.46),

$$w_{b,\ \text{bottom}} = \frac{8(F_t A C_G / A)(EI_n)}{48\ cL^2\ E_{\text{stringer}}}$$

$$= \frac{8(2300/4.5)(406{,}500{,}00)}{(48)(3.4174)(14)^2(1{,}854{,}000)}$$

$$= 27.8845 \text{ psf (1.34 kPa)} \quad \text{(transformed)}$$

$$n_{\text{bottom}} = 0.4417$$

$$w_{b,\ \text{bottom}} = \frac{27.8845}{0.4417} = 63.1 \text{ psf (3.02 kPa)} > 50 \text{ psf (2.39 kPa)} \qquad \text{OK}$$

Skin Splices

Tension Splice
From Eq. (3.51),

$$w_p = \frac{8(F_t A C_G / A)(EI_n)}{48\ cL^2\ E_{\text{stringer}}}$$

$$\text{Allowable stress}_{\text{top}} = F_c = \frac{4200(1.5)}{7.125} = 884.21 \text{ psi (6.096 MPa)}$$

The splice plate is to be 13.4 in. (340 mm) wide.
The ratio of splice-plate width to total skin width is

$$\text{Ratio} = \frac{3 \times 13.4}{48} = 0.8375$$

$$\text{Allowable stress} = \frac{F_t A C_G}{A}(\text{ratio}) = \frac{2300(1.0)}{4.5}(0.8375) = 428.1 \text{ psi (2.952 MPa)}$$

$$w_{p,\ \text{bottom}} = \frac{8(428.1)(456{,}400{,}000)}{48\ (3.304)(14)^2(1{,}854{,}000)} = 27.1 \text{ psf (186.8 kPa)}$$

$$n_{\text{bottom}} = 0.4417$$

$$w_{p,\ \text{bottom}} = \frac{27.1}{0.4417} = 61.4 \text{ psf (2.94 kPa)} > 50 \text{ psf (2.39 kPa)} \qquad \text{OK}$$

Compression Splice

OK by inspection because it is thicker (19⁄32 inch vs. 3⁄8 in.) (15 mm vs. 9.5 mm). Calculation, however, by the same approach as above, will yield a capacity of 122.46 psf (5.86 kPa).

Statical Moment for Rolling Shear

$$Q = Ad$$

As discussed above under Planar/Rolling Shear, the location of the critical shear plane is assumed to be 0.10 in. (2.5 mm) from the glued connection. The distance from the neutral axis (*NA*) to the assumed shear-critical plane is called $\overline{Y}_{cr}$ in this example.

From Fig. 3.20, and detailed in Fig. 3.23

$$\overline{Y}_{cr\ \text{skin, top}} = 3.04\ \text{in.} - 0.5938\ \text{in.} + 0.1\ \text{in.} + \frac{0.5938 - 0.1}{2}$$

$$= 2.7931\ \text{in. (71 mm)}$$

$$A'_{\text{skin, top}} = (0.5938 - 0.1)\ (19.99)$$

$$= 9.8711\ \text{in.}^2\ (\text{mm}^2)\ \text{for a 4 ft wide (1220 mm) planel, and}$$

$$Q_{\text{skin, top}} = (9.8711)(2.7931) = 27.5710\ \text{in.}^3\ (16.387 \times 10^3\ \text{mm}^3)$$

$$\overline{Y}_{cr\ \text{skin, bottom}} = 3.304 - 0.375 + \left(\frac{0.375 - 0.1}{2}\right) = 3.1665\ \text{in. (80 mm)}$$

$$A'_{\text{skin, bottom}} = (0.375 - 0.1)(21.2) = 5.83\ \text{in.}^2\ (3.76 \times 10^3\ \text{mm}^2)$$

$$Q_{\text{skin, bottom}} = (5.83)(3.1665) = 18.4602\ \text{in.}^3\ (302.51 \times 10^3\ \text{mm}^3)$$

Allowable Planar/Rolling Shear Capacity, F_s

$$F_s = \frac{F_s\ (\text{Ib}/Q)C_G}{\text{Ib}/Q} \tag{3.58}$$

$$\text{Allowable } F_s = \frac{265(1.1)}{4.75} = 61.4\ \text{psi (423 kPa) at interior stringers}$$

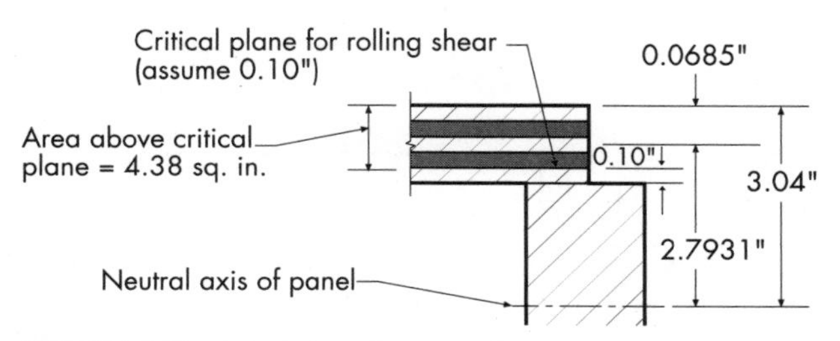

FIGURE 3.23 Calculation of Q, top skin.

$$\text{Allowable } F_s = \frac{61.4}{2} \text{ psi} = 30.7 \text{ psi (212 kPa) at exterior stringers}$$

As stated above under Allowable Capacity, the allowable stress at the end stringers is only half that at the interior stringers.

$$F_s t = F_s t_{\text{left}} + F_s t_{\text{middle}} + F_s t_{\text{right}}$$

$$\Sigma F_s t_{\text{top}} = \left[\left(\frac{61.4}{2}\right)(1.5) + [(2)(61.4)(1.5)] + 0.75\left(\frac{61.4}{2}\right)\right]$$

$$F_s t_{\text{top}} = 46.05 + 184.2 + 23$$

$$\Sigma F_s t_{\text{top}} = 253 \text{ lb/in. (N/mm)}$$

$$F_{s,\text{ bottom}} \frac{165(1.1)}{3} = 55 \text{ psi (379 kPa)}$$

$$\Sigma F_s t_{\text{bottom}} = \left[\left(\frac{55}{2}\right)(1.5) + [(2)(55)(1.5)] + 0.75\left(\frac{55}{2}\right)\right]$$

$$= 226.8 \text{ lb/in. (39.72 N/mm)}$$

Allowable Load Based on Rolling Shear (Fig. 3.24)
From Eq. (3.53),

$$w_{s,\text{ top}} = \frac{2(EI_g)\,\Sigma F_{s,\text{ top}} t}{4Q_{s,\text{ top}}\, LE_{\text{stringers}}}$$

$$w_{s,\text{ top}} = \frac{2(456{,}400{,}000)(253)}{4(27.5710)(14)(1{,}854{,}000)} = 80.6763 \text{ psf (3.86 kPa)}$$

$$n_{\text{top}} = 0.4164$$

$$w_{s,\text{ top}} = \frac{80.6763}{0.4164} = 193.7 \text{ psf (kPa)} > 50 \text{ psf (2.39 kPa)} \qquad \text{OK}$$

$$w_{s,\text{ bottom}} = \frac{2(456{,}400{,}000)(226.875)}{4(18.4602)(14)(1{,}854{,}000)} = 108.2 \text{ psf (5.18 kPa)}$$

$$n_{\text{bottom}} = 0.4417$$

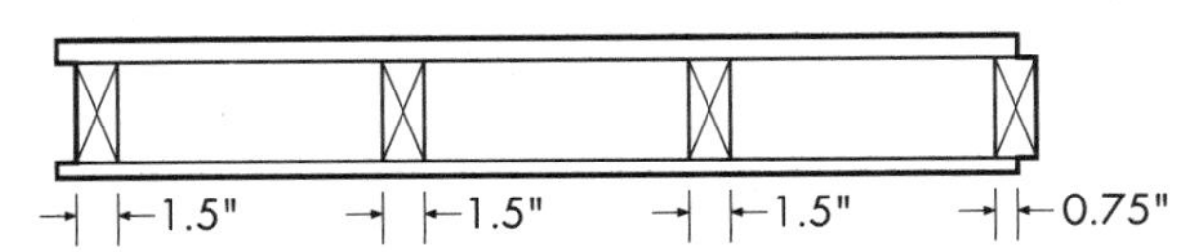

FIGURE 3.24 Effective width for shear.

$$\frac{108.2}{0.4417} = 245 \text{ psf (11.7 kPa)} > 50 \text{ psf (2.39 kPa)} \qquad \text{OK}$$

Top governs at 193.7 psf (9.27 kPa)

Statical Moment for Horizontal Shear (Fig. 3.25)
From Eq. (3.54),

$$Q_v = Q_{\text{stringers}} + \frac{E_{\text{skin}}}{E_{\text{stringers}}} \times Q_{\text{skin}} = Q_{\text{stringers}} + n_{\text{top}} (Q_{\text{skin}})$$

$$A_{\text{stringers, above } NA} = = 14.6760 \text{ in.}^2 \ (9.468 \times 10^3 \text{ mm}^2)$$

$$d_{\text{stringers}} = \frac{(5.375/2) - 0.2415}{2} = 1.223 \text{ in. (31 mm)}$$

$$Q_{\text{stringers}} = 14.6760(1.2230) = 17.9487 \text{ in.}^3 \ (294.13 \times 10^3 \text{ mm}^3)$$

$$A_{\text{skin}} = 0.5938(48) = 28.5024 \text{ in.}^2 \ (18.389 \times 10^3 \text{ mm}^2)$$

$$d_{\text{skin}} = 3.04 - \frac{0.5938}{2} = 2.7431 \text{ in. (70 mm)}$$

$$Q_{\text{skin}} = A_{\text{skin}} d_{\text{skin}} = 28.50241 \ (2.7431) = 78.1849 \text{ in.}^3 \ (1281.2 \times 10^3 \text{ mm}^3)$$

$$Q_v = 17.9487 + 0.4164(78.1849) = 50.5 \text{ in.}^3 \ (827 \times 10^3 \text{ mm}^3)$$

Allowable Load Based on Horizontal Shear
From NDS,[1]

$$F_v = 95 \text{ psi (655 kPa)} \quad \text{for Douglas fir-larch stringers}$$

From Eq. (3.58),

$$w_v = \frac{2(EI_g)F_v t}{4Q_v L E_{\text{stringers}}}$$

$$= \frac{2(456{,}400{,}000)(95)(4)(1.5)}{4(50.5)(14)(1{,}854{,}000)} = 99.22 \text{ psf (4.756 kPa)}$$

$$w_v = 99.2 \text{ psf (4.75 kPa)} > 50 \text{ psf (2.39 kPa)} \qquad \text{OK}$$

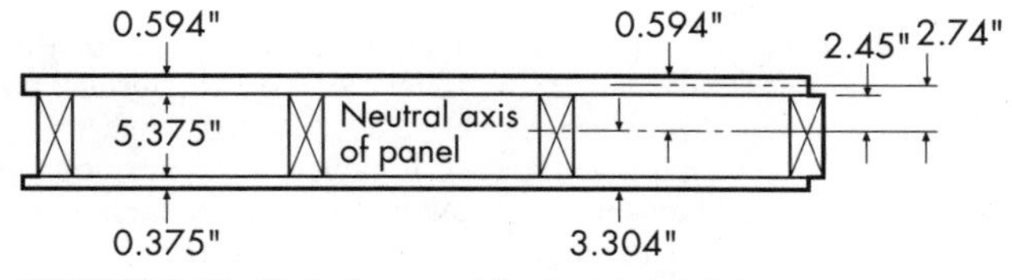

FIGURE 3.25 Statical moment for horizontal shear.

Final Allowable Loads Summary

$$w_{\Delta,\ 50\ \mathrm{psf}} = 89.5\ \mathrm{psf}$$

$$w_{b,\ \mathrm{bottom}} = 63.2\ \mathrm{psf}$$

$$w_{b,\ \mathrm{top}} = 138\ \mathrm{psf}$$

$$w_{p,\ \mathrm{top}} = 122.5\ \mathrm{psf}$$

$$w_{p,\ \mathrm{bottom}} = 61.4\ \mathrm{psf}$$

$$w_{s,\ \mathrm{top}} = 193.7\ \mathrm{psf}$$

$$w_{s,\ \mathrm{bottom}} = 245\ \mathrm{psf}$$

$$w_{v} = 99.2\ \mathrm{psf}$$

Allowable design load is 61.4 psf (2.84 kPa) and is controlled by the bottom-skin splice plate. Since this is greater than the 50 psf (2.39 kPa) required, the panel is OK as designed. Note that, in this case, the allowable load on the panel could be increased slightly by moving the splice plate away from the mid-span of the panel.

3.5 DESIGN AND FABRICATION OF WOOD STRUCTURAL INSULATED PANELS

3.5.1 General

A structural insulated panel (SIP) is an assembly consisting of a lightweight core securely laminated between two relatively thin, strong facings.

Load Direction. This section presents a method for design of structural insulated panels for horizontal, vertical, or combined loadings. The method may be used for panels with only one type of loading by eliminating the equations that do not apply.

Panel Behavior. Axial forces in a structural insulated panel are carried by compression in the facings, stabilized by the core material against buckling; bending moments are resisted by an internal couple composed of forces in the facings; shearing forces are resisted by the core.

Materials. Plywood and OSB serve as ideal materials for the facings of structural insulated panels. They are strong, light in weight, easily finished, dimensionally stable, and easily repaired if damaged. A variety of core materials may be used with wood structural panels to complete the SIP. Among these are polystyrene foams, polyurethane foams, and even paper honeycombs. Besides resistance to shearing forces, for some applications such as exterior wall panels and roof panels the core should possess high resistance to heat and vapor transmission. The designer should consider the suitability of the core material to his application. Factors to consider include resistance to degradation by heat, age, and moisture, and compatibility with glues, etc.

Bond between Faces and Core. The bond between the faces or skins and the core is extremely important since the structural performance of the assembly depends on its integrity. Several types of bond may be acceptable. For instance, glues may be used or, and in the case of some of the foam materials, direct adhesion of the foam to the faces during expansion may be used. In exterior wall panels, the bond should be waterproof. The combination of core material and bond should be such as not to creep excessively under the long-term loads and temperatures anticipated. Emulsion polymers and moisture-cured urethanes are most commonly used for this purpose. Isocyanate and reactive hot-melt adhesives are used occasionally.

3.5.2 Panel Design for Combined Loading

All equations that follow have been adjusted with constants so that substitution of values with dimensions given in the legends will produce answers with the desired dimension.

In general, the structural design of a SIP may be compared to that of an I-beam. Faces carry the compressive and tensile stresses, and the core resists shear. This core should be thick enough to space the faces so that they provide bending stiffness. The core also supports the faces against buckling. The composite structure must be checked for possible buckling as a column, and for wrinkling or buckling of the skins. Bending stiffness (EI) of the core (parallel to faces) is assumed to be insignificant. Tests have proven this assumption correct for panels with face thicknesses up to 15% of the total panel thickness (Fig. 3.26).

Note: This design method assumes a distribution of stresses based on principles of mechanics and has been verified by testing. Stress distributions around supports should be checked to confirm that stresses in components are within allowable limits. Verification by testing may be advisable in some cases.

Trial Section. Since in many cases the core thickness will be determined by requirements for insulation rather than for strength, the suggested design method is to select a given construction and to check it for all possible modes of failure under the design loads. Also see the note above.

Calculate Approximate Panel Area Required. Having the design loads w (uniform load in psf) and P (concentrated load in pounds), the required panel area is determined by

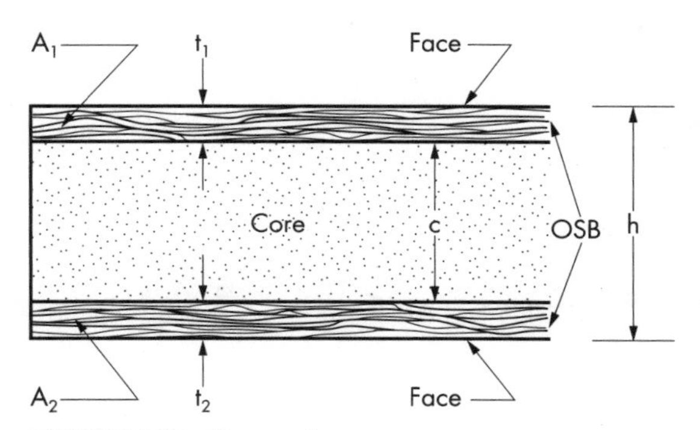

FIGURE 3.26 Core section.

$$A_1 + A_2 = \frac{P}{F_c} \tag{3.59}$$

where

$$F_c = \frac{F_c A}{A}$$

Determine Core Thickness. For simply supported end conditions, the maximum bending moment (lb-in. per ft panel width) (N-m per meter of panel width) is

$$M = \frac{12wL^2}{8} = 1.5wL^2 \tag{3.60}$$

To obtain the approximate core thickness required, assume that $A_1 = A_2 = A$. Then the section modulus

$$S = \frac{2I}{h} = \frac{A(h + c)^2}{4h} \tag{3.61}$$

This equation can be most easily solved by assuming values of c. The value of c thus obtained usually will be too low for insulation and deflection purposes. A minimum c of 3-½ in. (89 mm) is suggested for load-bearing wall panels subjected to wind loads.

With trial values of c, A_1, A_2, and h, the panel may be checked for all possible modes of failure (Fig. 3.27).

Find Neutral Axis

$$\bar{y} = \frac{A_1\left(h - \frac{t_1}{2}\right) + A_2\left(\frac{t_2}{2}\right)}{A_1 + A_2} \tag{3.62}$$

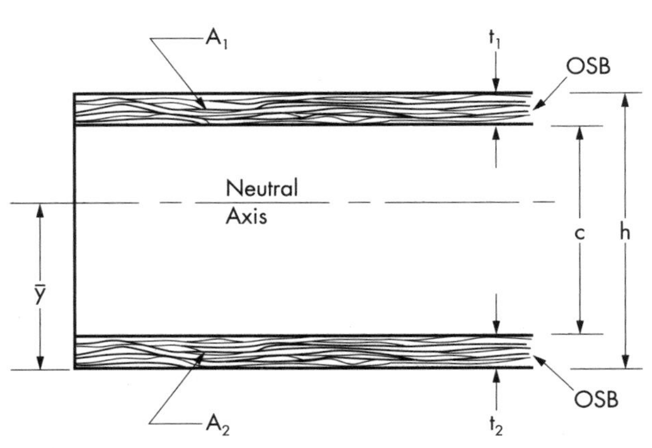

FIGURE 3.27 Neutral axis.

Find Moment of Inertia and Section Modulus

$$I = \frac{A_1A_2(h + c)^2}{4(A_1 + A_2)} \tag{3.63}$$

$$S_1 = \frac{I}{h - \bar{y}}, \; S_2 = \frac{I}{\bar{y}}$$

Determine Column Buckling Load

$$P_{cr} = \frac{\pi^2 EI}{(12L)^2 \left[1 + \dfrac{\pi^2 EI}{(12L)^2 \times 6(h + c)G_c}\right]} \tag{3.64}$$

Compute Approximate Skin Stress at Buckling. For foam or balsa core,

$$C_{cr} = 0.5 \sqrt[3]{EE_cG_c} \tag{3.65}$$

where C_{cr} = theoretical skin stress at buckling (wrinkling).
For honeycomb core,

$$C_{cr} = 0.82 \sqrt[3]{\frac{E_c t}{E_c}} \tag{3.66}$$

Note: Allowable should be approximately ⅓ C_{cr} for design. Skins are assumed to be flat.

Find Deflection. Deflection due to transverse loading only is equal to

$$\Delta = \Delta_b + \Delta_s = \frac{5wL^4 \times 1728}{384EI} + \frac{wL^2}{4(h + c)G_c} \tag{3.67}$$

where L = span (ft)
Δ_T = total deflection (in.)
w = uniform load (psf)
Δ_b = deflection due to bending (in.)
Δ_s = deflection due to shear (in.)

Total deflection including the effects of axial load is approximately equal to

$$\Delta_{max} = \frac{\Delta_T}{1 - (P/P_{cr})} \tag{3.68}$$

The deflection is usually limited to $L/240$, $L/360$, or some other predetermined allowable amount, such as shown in Table 3.1.

Calculate Maximum Bending Stress. This stress includes the bending due to the axial load acting through the initial transverse load deflection and is equal to

$$f_{b\,max} = \frac{1.5wL^2 + P\Delta_{max}}{S_1} \tag{3.69}$$

Determine Maximum Combined Stress. This stress will occur at midlength or midheight of the panel. It is the sum of the axial stress and the compressive bending stress in the concave side of the panel.

$$f_{c\ \max} = \frac{P}{A_1 + A_2} + f_{b\ \max} \tag{3.70}$$

$f_{c\ \max}$ must be less than F_c, and less than ⅓ C_{cr}.

Compute Shear Stress. The calculated shear stress f_v must be no more than the allowable shear stress, F_v.

$$f_v = \frac{wL}{(h + c)12} \leq F_v \tag{3.71}$$

In addition to the structural design, numerous architectural details must also be considered, such as connections, joint details, and finishes. They are not covered in this design method.

3.5.3 Fabrication of Wood Structural Insulated Panels

This specification covers the fabrication of glued SIP panels, with preformed cores of paper honeycomb, foamed plastic, or other material of demonstrated strength and durability for which accepted shear strength, shear modulus, compressive strength, and compressive modulus values are available. The core may be used with or without auxiliary wood members.

SIP panels should be designed by a qualified design professional using the method set forth in Section 3.5.2. Other design methods may be employed, provided they are supported by adequate test data.

SIPs shall be fabricated and assembled in accordance with engineering drawings and specifications, except that minimum requirements herein shall be observed.

There are wide variations in quality of workmanship and in the conditions under which wood structural panels are used. Because the authors and publisher have no control over those elements, they cannot accept responsibility for panel performance or designs as actually constructed.

Materials. Materials shall conform to the provisions of Sections 3.2.4 and 3.4.4, as applicable, and with modifications as follows.

Glue. Glue shall be of the type specified by the designer for anticipated exposure conditions and must be compatible with the core material being used.

Interior-type glue shall conform to ASTM Specification D3024. Exterior-type glue shall conform to ASTM Specification D2559.

Mixing; spreading; storage, pot, and working life; and assembly time and temperature shall be in accordance with the manufacturers' recommendations for the specific core and facing materials used.

Core Material. Properties of foamed plastic and other materials shall be in accordance with the design and compatible with the glue being used.

Paper honeycomb shall be preexpanded, resin-impregnated Kraft of specified paper weight, resin content, and cell size.

Fabrication

Framing. In order to provide gluing pressure for the core material and auxiliary framing members simultaneously, wood-framing members shall be surfaced so that they are shallower than the core. The difference in depth shall be as given by the following equation:

$$\Delta_{fc} = \frac{cp}{E_c} \tag{3.72}$$

where Δ_{fc} = amount by which framing lumber is shallower than core material (in.)
c = core depth (in.)
p = desired pressure on core (psi)
E_c = modulus of elasticity of core material perpendicular to skin

Assembly. The range of moisture content of the various pieces assembled into a single sandwich panel shall not exceed 5%.

Wood structural panel skins shall be glued to auxiliary framing members and core material over their full contact area, using mechanical pressure gluing in a press. The pressure on the net framing area shall be sufficient to provide adequate contact and ensure good glue bond, and shall be uniformly distributed by caul plates, beams, or other effective means. A pressure of 100–150 psi (689–1034 kPa) on the net glued area is recommended for wood-to-wood joints. A pressure equal to from 40–60% of the compressive yield strength of the core is suggested.

Application of pressure may start at any point, but shall progress to an end or ends. In any case, it shall be the responsibility of the fabricator to produce a continuous glue bond that meets or exceeds applicable specifications.

Where a tongue-and-groove-type panel edge joint is specified (and not otherwise detailed), the longitudinal framing member forming the tongue shall be of at least 2 in. (51 mm) nominal width, set out ¾ in. ± 1⁄16 in. (19 mm ± 1.6 mm) from the wood structural panel edge. Corners of the tongue shall be eased to facilitate ease of assembly, but tongue must provide a flat area at least ⅜ in. (9.5 mm) wide. Any corresponding framing member forming the base of the groove shall be set back ¼ in. 1 in. (6.4–25 mm) more than the amount by which the tongue protrudes. One skin may be cut back slightly to provide a tight fit for the opposite skin.

Unless otherwise specified, panels shall be trademarked by a recognized quality-assurance agency as complying with the provisions of PS 1[4] or PS 2.[5] In addition, panels shall lie flat at all points within ¼ in. (6.4 mm) for 4 × 8 ft (1220 × 2440 mm) panels, and proportionately for other sizes. Panel edge cross sections shall be square within 1⁄16 in. (1.6 mm) for constructions with lumber stringers 4 in. (102 mm) deep and proportionately for other sizes.

3.5.4 Identification

Each member shall be identified by the appropriate trademark of an independent inspection and testing agency, legibly applied to be clearly visible. Locate trademark approximately 2 ft from either end, except appearance of installed panel shall be considered.

3.5.5 Structural Insulated Panel Design Example

Since this example is intended for use as a general guide to illustrate the principles provided in the Handbook, a review of those sections pertinent to a specific design is recommended before proceeding.

Preliminary considerations as to the grade of wood structural panel to be used for a given design should include a check on availability.

Problem: Design of Exterior Bearing Wall

Application: Residential Exterior Bearing Wall

Height (L)	8 ft (2440 mm)
Axial load	600 lb/ft (8.76 kN/m) (due to snow load)
Normal load	20 psf (0.96 kPa) (due to wind load)
Deflection limitation	L/240 = 0.40 in. (10 mm)
Face materials	7⁄16 in. (11 mm) 24/16 OSB
Core material	Commercially available 1 lb/ft^3 expanded polystyrene

Material Properties

Wood Structural Panel Design Values

$E = \dfrac{EAC_G}{A} = \dfrac{3{,}800{,}000(1.0)(1.10)}{5.25} = 796{,}300$ psi (5.490 GPa) (increase of 10% over usually published value, since shear deflection will be computed separately)

$$F_c = \frac{F_c A C_G}{A} = \frac{3{,}250(1.0)}{5.25} = 619 \text{ psi (4.27 MPa) (normal duration of load)}$$

$$F_t = \frac{F_t A C_G}{A} = \frac{2{,}600(1.0)}{5.25} = 495.2 \text{ psi (3.41 MPa) (normal duration of load)}$$

$$F_b = \frac{F_b S C_G}{S} = \frac{370(1.2)}{0.383} = 1159 \text{ psi (7.99 MPa) (normal duration of load)}$$

For snow loads F_c and F_t stresses may be increased 15%. A 60% increase over normal-duration values may be taken for wind loadings.

Core Material Design Values

$$E_c = 200 \text{ psi (1.38 MPa)}$$

$$G_c = 300 \text{ psi (2.07 MPa)}$$

$$F_v = 6 \text{ psi (41.4 kPa) (from manufacturer's information)}$$

Note: Manufacturer's F_v is often an ultimate value and should be reduced by a factor of approximately three for design.

Trial Section

Find approximate wood structural panel area required from Eq. (3.59),

$$A_1 + A_2 = \frac{P}{F_c} = \frac{600}{619(1.15)} = 0.8429 \text{ in.}^2\text{/ft (1.78 mm}^2\text{/mm)}$$

Check 7⁄16 in. (11.1 mm) Exposure 1 OSB both sides.

$$\text{Area for tension or compression} = 0.4375(12)(2)$$

$$= 10.50 \text{ in.}^2\text{/ft. (22.2 mm}^2\text{/mm)}$$

$$> 0.8429 \text{ in.}^2\text{/ft (1.78 mm}^2\text{/mm)} \quad \text{OK}$$

Determine Core Thickness

Assume that, for insulation purposes, a minimum core thickness of 3.5 in. (89 mm) is required.

Then

$$c = 3.5 \text{ in. (89 mm)}$$

$$h = 3.5 + (2 \times 0.4375) = 4.375 \text{ in. (111 mm)}$$

Determine Neutral Axis

Because of the symmetrical construction, the neutral axis is at the center of the panel (Fig. 3.28).

Calculate I and S

In this, and nearly all cases, the MOE of the core material is very low relative to that of the skins. For that reason, the contribution to the SIP stiffness is negligible. Checking this assumption:

$$\frac{E_{\text{core}}}{E_{\text{panel}}} = \frac{200}{796{,}300} = 0.0003 = 0.03\%$$

Ignoring the contribution of the core is therefore justified.

Using parallel axis theorem, $I = \bar{I} + AD^2$, $\bar{I} = 0$, and simplifying, from Eq. (3.63),

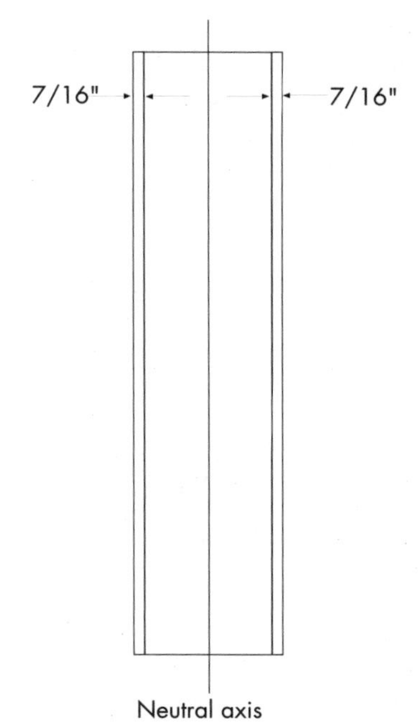

FIGURE 3.28 Example problem cross section.

$$I = \frac{A_1 A_2 (h + c)^2}{4(A_1 + A_2)}$$

$$I = \frac{[(0.4375)(12)]^2(4.375 + 3.5)^2}{4(5.25 + 5.25)} = 40.6978 \text{ in.}^4 \; (16.94 \times 10^6 \text{ mm}^4)$$

$$S = S_1 = S_2 = \frac{I}{y} = \frac{40.6978}{2.1875} = 18.6047 \text{ in.}^3/\text{ft} \; (1000 \text{ mm}^3/\text{mm})$$

Evaluate Column Buckling

From Eq. (3.64),

$$P_{cr} = \frac{\pi^2 EI}{(12L)^2 \left[1 + \dfrac{\pi^2 EI}{(12L)^2 \times 6(h + c)G_c}\right]}$$

$$= \frac{\pi^2(796{,}300)(40.6978)}{96^2 \left[1 + \dfrac{\pi^2(796{,}300)(40.6978)}{96^2(6)(4.375 + 3.5)(300)}\right]}$$

$$= 10{,}064 \text{ lb/ft } (146.9 \text{ N/mm}) > 600 \text{ lb/ft } (8.8 \text{ N/mm}) \qquad \text{OK}$$

Check for Skin Buckling

From Eq. (3.65),

$$C_{cr} = 0.5 \sqrt[3]{EE_c G_c}$$

$$C_{cr} = 0.5 \sqrt[3]{796{,}300(200)(300)} = 1{,}814.3 \text{ lb/ft } (26.48 \text{ N/mm})$$

Desired safety factor = 3

$$\frac{C_{cr}}{3} = \frac{1814}{3} = 605 \text{ lb/ft } (88.3 \text{ N/mm})$$

Deflection

From Eq. (3.67),

$$\Delta = \Delta_b + \Delta_s = \frac{5wL^4 \times 1728}{384EI} + \frac{sL^2}{4(h + c)G_c}$$

$$= \frac{5(20)(8)^4 \times 1728}{(384)(796{,}000)(40.6978)} + \frac{20(8)^2}{4(4.375 + 3.5)(300)}$$

$$\Delta = 0.0569 + 0.1354 = 0.1923 \text{ in. } (5 \text{ mm}) < 0.2667 \text{ in. } (7 \text{ mm}) \qquad \text{OK}$$

From Eq. (3.68),

$$\Delta_{max} = \frac{\Delta}{1 - (P/P_{cr})}$$

$$= \frac{0.1923}{1 - (600/10{,}064)} = 0.2045 \text{ in. } (5 \text{ mm}) < 0.40 \text{ in. } (10 \text{ mm}) \qquad \text{OK}$$

Check Bending Stress
From Eq. (3.69),

$$f_{b\ \max} = \frac{1.5wL^2 + P\Delta_{\max}}{S_1} C_D$$

$$S_1 = S_2 = 18.6047 \text{ in.}^3 \ (304.9 \times 10^3 \text{ mm}^3)$$

$$F_{b\ \max} = \frac{1.5(20)(8)^2 + (600)(0.2045)}{18.6047}(1.60) = 175.7 \text{ psi } (1.21 \text{ MPa})$$

Combined Stress
From Eq. (3.70),

$$f_{c\ \max} = \frac{P}{A_1 + A_2} + f_{b\ \max}$$

$$f_{c\ \max} = \frac{600}{2(5.25)} + 175.7 = 233 \text{ psi}(1.61 \text{ mPa}) < \frac{1}{3} C_{cr}$$

$$= 605 \text{ psi } (4.17 \text{ mPa}) \text{ and}$$

$$< F_c = 619 \times 1.15 = 712 \text{ psi } (4.19 \text{ mPa}) \qquad \text{OK}$$

Check Shear Stress
From Eq. (3.71),

$$f_v = \frac{wL}{(h + c)12} = \frac{20(8)}{(4.375 + 3.5)(12)}$$

$$= 1.69 \text{ psi } (11.7 \text{ Pa}) < \frac{F_y}{3} = \frac{6}{3} = 2 \text{ psi } (13.8 \text{ kPa}) \qquad \text{OK}$$

Conclusion

The proposed panel has met all structural requirements.

Note: Support conditions are not analyzed because there is no recognized method for analysis of SIPs under the various support conditions. If the designer thinks such evaluation may prove critical, testing is the standard means of determining acceptability.

3.6 LUMBER FOR PANEL/LUMBER COMPONENTS

3.6.1 Lumber: Allowable Working Stresses and Capacities

Lumber working stresses are as follows. Lumber for use in wood-structural-panel structural assemblies shall fall into one of the following two categories:

Stress-Grade Lumber. Stress-grade lumber is defined in this Specification as lumber conforming with standard stress-grading rules, and so identified by a qualified grading agency, but not subject to the additional restrictions imposed on glued-laminated lumber. Even if laminated, it is still defined as stress-grade lumber.

Allowable stresses and modifications thereof shall be as defined in the latest edition of *National Design Specification for Wood Construction*[1] (NDS), American Forest and Paper Association. Stress-grade lumber does not qualify for structural glued laminated timber stresses, regardless of the number of laminations.

Structural Glued Laminated Timber. Allowable stresses for structural glued laminated timber and modifications thereof shall be as defined in the latest edition of *Standard Specifications for Structural Glued Laminated Timber of Softwood Species*, AITC 117,[6] or with approved Engineered Wood Systems (EWS) glued laminated timber combinations.

3.6.2 Number of Laminations for Determining Allowable Stress Level

The number of laminations to be used in determining the allowable stress level of a laminated member shall include all lumber laminations of appropriate grade that are subjected to the principal stress, but shall not include wood structural panel webs or shims within the member. Lumber shims, if appropriately graded, may be grouped to equal or exceed the lamination thickness, and the group considered as a lamination. Similarly, in a member where laminations are ripped diagonally (as in some folded-plate chords), ripped portions of laminations may be paired to equal or exceed the full lamination width, and the pair considered as a lamination.

3.6.3 Number of Laminations for Resisting Stress

All laminations, including webs and shims, may be considered as resisting stress with due consideration for grade and end joints.

3.6.4 Adjustments for Service Moisture Conditions

Allowable stresses for stress-grade lumber shall be modified for in-use moisture content of the lumber as set forth in the latest edition of *National Design Specification for Wood Construction.*[1]

Allowable stresses for structural glued laminated timber shall be modified for in-use moisture content as set forth in the latest edition of *Standard Specifications for Structural Glued Laminated Timber of Softwood Species*, AITC 117,[6] or with approved engineered wood systems glued laminated timber combinations.

3.6.5 Allowance for Surfacing

In applying stresses, actual sizes of finished lumber shall be used, including any necessary allowance for resurfacing.

Shear Deflection. When the shear deflection of an assembly having lumber flanges (such as a plywood beam) is calculated separately and added to the bending deflection, the elastic modulus of the lumber flanges may be increased 3% in calculating bending deflection.

Values for modulus of elasticity have been derived from tests that involve both bending and shear deflection, while such built-up assemblies as stressed-skin panels and box beams have such low shear stresses in the flanges that shear deflection in the flanges may be ignored. For these assemblies, therefore, the usual modulus of elasticity of the flange material may be increased to restore the portion of the deflection that is ordinarily caused by shear.

3.6.6 Glued Lumber End Joints

End Joints in Structural Glued Laminated Timber. In structural glued laminated timber, end joints shall comply with the requirements of ANSI A190.1 for *Structural Glued Laminated Timber.*[7] Allowable stresses shall be those of the latest edition of *Standard Specifications for Structural Glued Laminated Timber of Softwood Species*, AITC 117,[6] or with approved Engineered Wood Systems glued laminated timber combinations.

Scarf and Finger Joints in Stress-Grade Lumber

Members Stressed Principally in Axial Tension. Slope of scarf joints shall not be steeper than 1 in 8 for dry conditions of use, nor 1 in 10 for wet conditions of use. They may then be considered to carry the full allowable tensile stress of the members glued. Finger joints are acceptable, at design levels supported by adequate test data.

Members Stressed Principally in Axial Compression. Slope of scarf joints shall not be steeper than 1 in 5 for dry conditions of use, nor 1 in 10 for wet conditions of use. They may then be considered to carry the full allowable compressive stress of the members glued. Finger joints are acceptable, at design levels supported by adequate test data.

Butt Joints. Butt joints may be used in stress-grade lumber tension and compression members, in which case the effective cross-sectional area shall be computed by subtracting from the cross-sectional area the area of all laminations containing butt joints at a single cross section. In addition, laminations adjoining (actually touching) those containing butt joints and themselves containing butt joints, shall be considered only partially effective if the spacing in adjoining laminations is less than 50 times the lamination thickness. The effective area of such adjoining laminations shall be computed by multiplying their gross area by the percentages given in Table 3.2.

Tension. For butt joints in tension members or tension portions of members, the appropriate allowable stress in tension shall be multiplied by 0.8 at sections containing the joints.

Compression Members—End Grain Bearing Requirements for Butt Joints. Members in compression may be butted and spliced, provided there is adequate lateral support and the end cuts are accurately squared and parallel and maintained in tight contact.

Allowable Stresses. Allowable stresses for bearing on end grain shall be as determined from the latest edition of the *National Design Specification for Wood Construction*, American Forest and Paper Association.[1]

3.7 REFERENCES

1. *National Design Specification for Wood Construction*, National Forest and Paper Association, Washington, DC, 1997.
2. Orosz, I., *Simplified Method for Calculation Shear Deflections of Beams*, U.S.D.A. Forest Service Research Note FPL-0210, U.S. Department of Agriculture, Forest Service Products Laboratory, Madison, WI, 1970.
3. *Plywood Design Specification*, Form No. Y510, APA—The Engineered Wood Association, Tacoma, WA, September 1998.
4. *Voluntary Product Standard PS 1-95,* Construction and Industrial Plywood, Office of Standards Services, National Institute of Standards and Technology, Gaithersburg, MD, 1995.
5. *Voluntary Product Standard PS 2-92,* Performance Standard for Wood-Based Structural-Use Panels, Office of Standards Services, National Institute of Standards and Technology, Gaithersburg, MD, 1992.
6. *Standard Specifications for Structural Glued Laminated Timber of Softwood Species,* AITC 117, American Institute of Timber Construction, Englewood, CO.
7. *Structural Glued Laminated Timber*, American National Standard ANSI A190.1, American National Standards Institute.

3.8 ADDITIONAL READING

Design Capacities of APA Performance Rated Structural-Use Panels, APA Technical Note N375B, APA—The Engineered Wood Association, Tacoma, WA, June 1995.

Plywood Design Specification, Supplement 2, Design and Fabrication of Glued Plywood-Lumber Beams, APA—The Engineered Wood Association, Tacoma, WA, November 1998.

Plywood Design Specification, Supplement 3, Design and Fabrication of Plywood Stressed-Skin Panels, APA—The Engineered Wood Association, Tacoma, WA, April 1996.

Plywood Design Specification, Supplement 4, Design and Fabrication of Plywood Sandwich Panels, APA—The Engineered Wood Association, Tacoma, WA, Septemberr 1993.

National Design Specification for Wood Construction, Supplement, National Forest and Paper Association, Washington, DC 1997.

CHAPTER FOUR
STRUCTURAL GLUED LAMINATED TIMBER (GLULAM)

Borjen Yeh, P.E., PhD
Manager, Research and Development, TSD

4.1 INTRODUCTION

Structural glued laminated timber (glulam) is a structural member glued up from suitably selected and prepared pieces of laminating lumber or "laminations" either in a straight or curved form with the grain of all pieces parallel to the longitudinal axis of the member. Glued laminated timber members are produced in laminating plants by gluing together dry lumber, normally of 2 in. or 1 in. nominal thickness, under controlled conditions of temperature and pressure. Members with a wide variety of sizes, profiles, and lengths can be produced for superior characteristics of strength, serviceability, and appearance. Glued laminated timber beams are manufactured with the strongest laminations on the bottom and top of the beam, where greatest tension and compression stresses occur in bending. This allows a more efficient use of the lumber resource by placing higher grade lumber in zones that have higher stresses and lumber with less structural quality in lower-stressed zones.

Glued laminated timber is manufactured from several softwood species, primarily Douglas fir-larch, southern pine, hem-fir, spruce-pine-fir, eastern spruce, western woods, Alaska cedar, Durango pine, and California redwood. Several hardwood species, including red oak, red maple, and yellow poplar, are also used. Standard glued laminated timber sizes are given in Section 4.1.2. Any length, subject to the maximum length permitted by transportation and handling restrictions, is available.

Glued laminated timber is typically manufactured with kiln-dry lumber having a maximum moisture content at the time of fabrication of 16%. As a result, the allowable design stresses for glued laminated timber are higher than dry (moisture content of 19% or less) or green lumber. The use of kiln-dry laminating lumber also means that the moisture content of glued laminated timber is relatively uniform throughout the member, unlike green sawn timbers, which may have widely varying moisture contents within a given member. This use of uniformly dry lumber gives glued laminated timber excellent dimensional stability. Thus, a glued laminated timber member will not undergo the dimensional changes normally associated with larger solid-sawn green timbers, and will remain straight and true in cross-section. A dry glued laminated timber is also less susceptible to the checking and splitting that is often associated with green timbers.

Glued laminated timber is one of the most versatile of the family of glued engineered wood products and is used in applications ranging from concealed beams and headers in residential construction to structures with large open spaces (see Figs. 4.1 and 4.2). Glued laminated timber has greater strength and stiffness than comparable dimensional lumber. Pound for pound, it is stronger than steel. Because of their composition, large glued laminated timber members can be manufactured from smaller trees harvested from second- and third-growth forests and plantations. With glued laminated timber, the designer and builder can continue to enjoy the strength and versatility of large wood members without relying on the old growth-dependent solid-sawn timbers.

FIGURE 4.1 Glulam beam supports second floor I-joist construction.

FIGURE 4.2 Disney ICE rink in Anaheim, California, features glulam arches curved to a 75-ft radius to form the ice center's roof systems.

4.1.1 100 Years of Glued Laminated Timber

In terms of current needs to optimize products from a carefully managed timber resource, glued laminated timber is one of the most resource-efficient approaches to wood building products. It is an engineered product manufactured to meet the most demanding structural requirements. But glued laminated timber is not a new product. The first patents for glued laminated timber were issued in Switzerland and Germany in the late 1890s. A 1906 German patent signaled the true beginning of glued laminated timber construction. One of the first glued laminated timber structures erected in the United States was a research laboratory at the USDA Forest Products Laboratory in Madison, Wisconsin. The structure was erected in 1934 and is still in service today.

A significant development in the glued laminated timber industry was the introduction of fully water-resistant phenol-resorcinol adhesives in 1942. This allowed glued laminated timber to be used in exposed exterior environments without concern of glueline degradation. The first U.S. manufacturing standard for glued laminated timber was Commercial Standard CS253-63, which was published by the Department of Commerce in 1963. The most recent standard is ANSI/AITC A190.1-92,[1] which took effect in 1993.

4.1.2 Typical Sizes

Individual pieces of laminating lumber used in glued laminated timber manufacturing are typically 1⅜ in. thick for southern pine and 1½ in. thick for Western species, although other thicknesses may also be used. Glued laminated timber products typically range in net widths from 2½ to 10¾ in., although virtually any member width can be custom produced.

Glued laminated timber is available in both custom and stock sizes. Stock beams are manufactured in commonly used dimensions and cut to length when the beam is ordered from a distributor or dealer. Typical stock beam widths include 3⅛, 3½, 5⅛, 5½, and 6¾ in., which meet the requirements for most residential construction applications. Where long spans, unusually heavy loads, or other circumstances control design, custom members are typically specified. Custom members are available in virtually any size and shape that may be required to meet the design conditions. Some of the common custom shapes that are available include curved beams, pitched and curved beams, radial arches and tudor arches (see Figs. 4.3 and 4.4).

4.1.3 Common Uses

Glued laminated timber has a reputation for being used in striking applications such as vaulted ceilings and other designs with soaring open spaces. In churches, schools, restaurants, and other commercial buildings, glued laminated timber is often specified for its beauty as well as its strength for good reason.

Glued laminated timber has the classic natural wood appearance that holds a timeless appeal. Aesthetics aside, there are many other applications where the strength and durability of glued laminated timber beams make them the ideal structural choice. Typical uses range from simple purlins, ridge beams, floor beams, and cantilevered beams to complete commercial roof systems. In some instances, warehouse and distribution centers with roof areas exceeding 1 million ft^2 have been constructed using glued laminated timber framing. In large open spaces, glued laminated timber beams can span more than 100 ft.

FIGURE 4.3 Two-lane highway bridge in Colorado using glulam radial arches.

One of the greatest advantages of glued laminated timber is that it can be manufactured in a wide range of shapes, sizes, and configurations. In addition to straight prismatic sections, beams can also be produced in a variety of tapered configurations, such as single-tapered, double-tapered, and off-centered ridges. Curved shapes range from a simple curved beam to a pitched and tapered curved beam to a complex arch configuration. Spans using glued laminated timber arches are virtually unlimited. For example, in reticulated glued laminated timber framed dome structures, arches can span more than 500 ft.

Glued laminated timber trusses also take many shapes including simple pitched trusses, complicated scissors configurations, and long-span bowstring trusses with curved upper chords. When designed as space frames, glued laminated timber truss systems can create great clear spans for auditoriums, gymnasiums, and other applications requiring large open floor areas. When manufactured with waterproof phenol resorcinol adhesives, glued laminated timber products can be fully exposed to the environment, provided they are properly pressure-preservative treated. Exposed applications include utility poles and cross-arms, marinas, docks, and other waterfront structures and bridges.

Bridges represent a growing market for glued laminated timber in pedestrian and light vehicular applications for stream and roadway crossings. Glued laminated timber is also used in secondary highway bridge designs ranging from straight girders to soaring arches. And the railroads are finding glued laminated timber to be a viable structural product for use in their heavily loaded bridge structures.

In all of these uses, the strength and stiffness of glued laminated timber give builders and designers more design versatility than they have with other structural products. And, these advantages come at a cost that is competitive with other structural systems. Table 4.1 lists economical spans for selected timber framing systems

FIGURE 4.4 This 236,000 ft^2 potash storage building in Portland, Oregon, features glulam arches.

using glued laminated timber members in buildings. This table may be used for preliminary design purposes to determine the economical span ranges for the selected framing systems. However, all systems require a more extensive analysis for final design.

4.1.4 Availability

Glued laminated timber members are available in both custom and stock sizes. Custom beams are manufactured to the specifications of a specific project, while stock beams are made in common dimensions, shipped to distribution yards, and cut to length when the beam is ordered. Stock beams are available in virtually every major metropolitan area. Although glued laminated timber members can be custom fabricated to provide a nearly infinite variety of forms and sizes, the best economy is generally realized by using standard-size members. When in doubt, the designer is advised to check with the glued laminated timber suppliers or manufacturers concerning the availability of a specific size glued laminated timber members prior to specification. The following trade associations are available for technical assistance:

TABLE 4.1 Economical Spans for Glulam Framing Systems

Type of framing system	Economical spans (ft)
Roof	
Simple-span beams	
Straight or slightly cambered	10–100
Tapered, double tapered-pitched, or curved	25–105
Cantilevered beams (main span)	up to 90
Continuous beams (interior spans)	10–50
Girders	40–100
Three-hinged arches	
Gothic	40–100
Tudor	40–140
A-frame	20–100
Three-centered, parabolic, or radial	40–250
Two-hinged arches	
Radial or parabolic	50–200
Trusses (four- or more ply chords)	
Flat or parallel chord	50–150
Triangular or pitched	50–150
Bowstring (continuous chord)	50–200
Trusses (two- or three-ply chords)	
Flat or parallel chord	20–75
Triangular or pitched	20–75
Tied arches	50–200
Dome structures	200–500+
Floor	
Simple span beams	10–40
Continuous beams (individual spans)	10–40
Headers	
Windows and doors	<10
Garage doors	9–18

APA—The Engineered Wood Association and
Engineered Wood Systems (EWS), a related corporation of APA
7011 South 19th Street
Tacoma, WA 98466
Phone: (253) 565-6600
Fax: (253) 565-7265

American Institute of Timber Construction
7012 South Revere Parkway, Suite 140
Englewood, CO 80112
Phone: (303) 792-9559
Fax: (303) 792-0669

4.2 GROWTH OF INDUSTRY

Table 4.2 shows the recent history of the glued laminated timber production in North America. Glued laminated timber production is expected to increase steadily in the years to come, as shown in Fig. 4.5. New-generation 30F beams with higher shear strengths are being introduced. A new family of I-joist compatible products is beginning to make market inroads. In addition, fiber-reinforced technology should be more widely available in the near future to help glued laminated timber, and wood construction in general, penetrate the commercial building market. All of these product innovations will be important to the future growth of this industry.

Approximately one-half of the glued laminated timber produced in the United States goes to new residential and remodeling uses. The next largest segment is the nonresidential market, as shown in Fig. 4.6.

4.3 STANDARDS

4.3.1 ANSI/AITC A190.1

ANSI/AITC A190.1, the *American National Standard for Structural Glued Laminated Timber,*[1] is a national consensus standard for glued laminated timber manufacturing. Detailed manufacturing requirements for glued laminated timber are documented in ANSI/AITC A190.1. This standard is recognized in the U.S. model building codes, and a construction specification for glued laminated timber should include reference to this standard.

4.3.2 ASTM D3737

ASTM D 3737, *Standard Practice for Establishing Stresses for Structural Glued Laminated Timber,*[2] provides a consensus approach in deriving allowable properties for glued laminated timber manufactured in accordance with ANSI A190.1. In the U.S. glued laminated timber industry, two computer programs developed by the major trade associations, APA—The Engineered Wood Association and American Institute for Timber Construction (AITC) are recognized by the model building codes as an alternative to the procedures given in ASTM D3737 for establishing

TABLE 4.2 U.S. and Canada Glued Laminated Timber Production

	1993	1994	1995	1996	1997	1998	1999
	(Million board ft)						
U.S. Production	241.0	266.1	282.0	309.0	300.0	287.2	315.8
Canada Production	NA[a]	NA[a]	13.0	13.0	15.0	13.0	15.2
North America Total	–	–	295.0	322.0	315.0	300.2	331.0

[a] NA, not available; data collection started in 1995.
Source: APA—The Engineered Wood Association (April 2000).

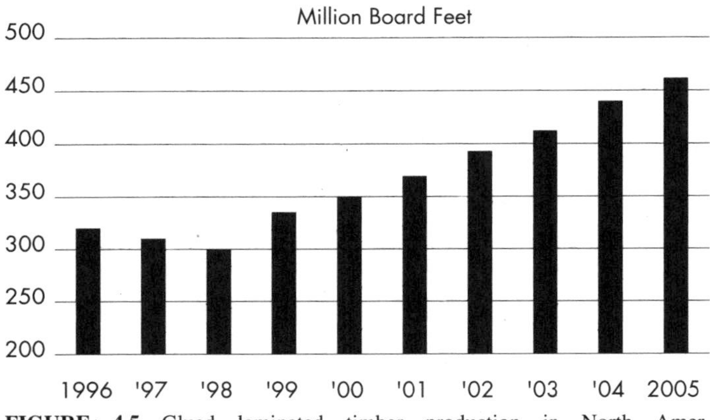

FIGURE 4.5 Glued laminated timber production in North America. *(Forecast for 2001 and beyond by APA.)*

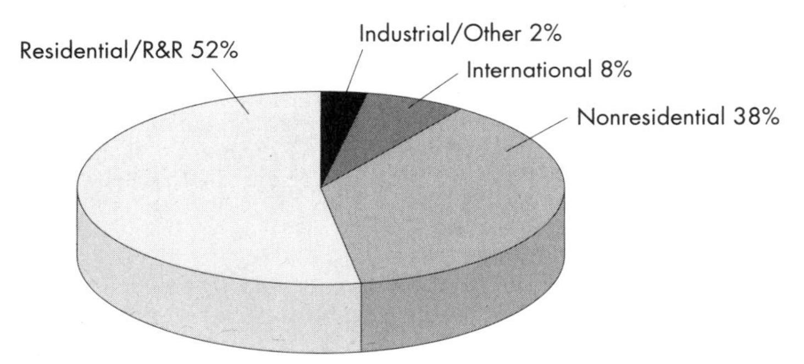

FIGURE 4.6 Glued laminated timber end use in the United States. *(2000 production volume per APA.)*

design properties for glued laminated timber. These associations share the database required for their computer programs on generic laminating lumber grades.

4.4 MECHANICAL PROPERTIES

4.4.1 Lay-up Principles

The laminating process used in glued laminated timber manufacturing results in a random dispersion of strength-reducing growth characteristics, such as knots and slope of grain, of lumber throughout the glued laminated timber member. Consequently, glued laminated timber has higher mechanical properties with a lower variability than sawn lumber products of comparable sizes. For example, the coefficient of variation for the modulus of elasticity (E) of glued laminated timber is published as 10%, which is equal to or lower than any other wood product.

Single- and Multiple-Grade Lay-ups. Glued laminated timber can be manufactured using a single grade or multiple grades of lumber, depending on the intended use. A mixed-species glued laminated timber member is also possible. When the member is intended to be primarily loaded either axially or in bending with the loads acting parallel to the wide faces of the laminations, a single-grade combination is recommended. On the other hand, a multiple-grade combination provides better cost-effectiveness when the member is primarily loaded in bending due to loads applied perpendicular to the wide faces of the laminations.

On a multiple-grade combination, a glued laminated timber member can be produced as either a balanced or unbalanced combination, depending on the geometrical arrangement of the laminations about the middepth of the member. This is further explained below.

Balanced and Unbalanced Lay-ups. Glued laminated timber may be manufactured as unbalanced or balanced members, as shown in Fig. 4.7. The most critical zone of a glued laminated timber bending member with respect to controlling strength is the outermost tension zone. In unbalanced beams, the quality of lumber used on the tension side of the beam is higher than the lumber used on the corresponding compression side, allowing a more efficient use of the timber resource. Therefore, unbalanced beams have different bending stresses assigned to the compression and tension zones and must be installed accordingly. To ensure proper installation of unbalanced beams, the top of the beam is clearly stamped with the word "TOP" (see Fig. 4.8).

While the unbalanced combination is primarily for use in simple-span applications, it could also be used for short-cantilever applications (cantilever less than approximately 20% of the back span) or for continuous-span applications when the design is controlled by shear or deflection. If members are inadvertently installed in an improper orientation, i.e., upside down, the allowable bending stress for the compression zone stressed in tension should be used. In this case, the controlling bending stress and the capacity of the beam in this orientation shall be checked to determine if they are still adequate to carry the design loads.

Balanced members are symmetrical in lumber quality about the midheight. Balanced beams are used in applications such as cantilevers or continuous spans, where

UNBALANCED	BALANCED
No. 2D	Tension Lam
No. 2	No. 1
No. 2	No. 2
No. 3	No. 3
No. 3	No. 3
No. 3	No. 3
No. 2	No. 2
No. 1	No. 1
Tension Lam	Tension Lam

FIGURE 4.7 Unbalanced and balanced lay-up combinations.

FIGURE 4.8 Glued laminated timber with a "TOP" stamp.

either the top or bottom of the member may be stressed in tension due to service loads. They can also be used in single-span applications, although an unbalanced beam is more efficient for this use.

Visually Graded and E-Rated Lay-ups. Allowable design properties are a key factor in designing glued laminated timber. Bending members are typically specified on the basis of the maximum allowable bending stress of the member. For example, a 24F designation indicates a member with an allowable bending stress of 2400 psi. Similarly, a 20F designation refers to a member with an allowable bending stress of 2000 psi. These different stress levels are achieved by varying the percentages and grade of higher-quality lumber in the beam lay-up. Use of different species may also result in different stress designations.

To identify whether the lumber used in the beam is visually or mechanically graded, the stress combination also includes a second set of designations. For example, for an unbalanced 24F lay-up using visually graded lumber, the lay-up designation may be identified as a 24F-V4. The "V" indicates that the lay-up uses visually graded lumber. ("E" is used for mechanically graded lumber, which is sorted by the modulus of elasticity (MOE) of the laminating lumber.) The number "4" further identifies a specific combination of lumber used to which a full set of design stresses such as horizontal shear, MOE, etc. are assigned. Figure 4.9 shows a typical trademark for a glued laminated timber beam.

Horizontally and Vertically Laminated Lay-ups. Glued laminated timber beams are typically installed with the wide face of the laminations perpendicular to the applied load, as shown in Fig. 4.10. These are commonly referred to as horizontally

APA EWS

1 — B IND — 7

EWS Y117 — 4

EWS 24F-1.8E — 6

WW — 5

2 — MILL 0000

ANSI A190.1-1992 — 3

(1) Indicates structural use: B-Simple span bending member. C-Compression member. T-Tension member. CB-Continuous or cantilevered span bending member.

(2) Mill number.

(3) Identification of ANSI Standard A190.1, Structural Glued Laminated Timber. ANSI A190.1 is the American National Standard for glulam beams.

(4) Applicable laminating specification.

(5) Western woods (see note 6).

(6) Structural grade designation. The APA EWS 24F-1.8E designation is a glulam grade commonly used in residential applications. Combining a group of six layup combinations made with Douglas-fir-Larch, Spruce-Pine-Fir, southern pine, and/or Hem fir, this grade provides strength (allowable bending stress of 2,400 psi and allowable shear stress of 195 psi) and stiffness (modulus of elasticity of 1.8 x 10^6 psi) needed for typical residential applications, while greatly simplifying the design specification.

(7) Designation of appearance grade. INDUSTRIAL, ARCHITECTURAL, PREMIUM, or FRAMING.

FIGURE 4.9 Typical trademark.

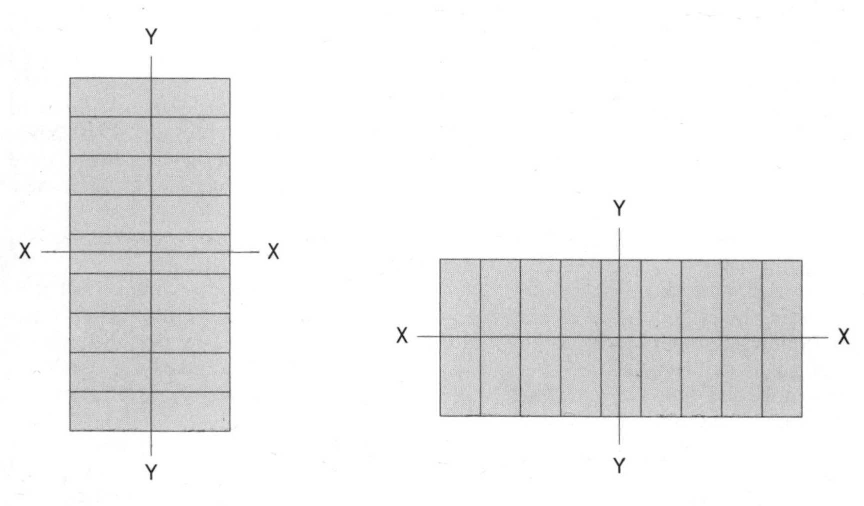

FIGURE 4.10 Horizontally and vertically laminated lay-ups with loading in the X–X and Y–Y axes.

laminated members. If this same member is rotated 90° such that the load is applied parallel to the wide face of the laminations, it is considered to be a vertically laminated member. Glued laminated timber members have different tabulated stress properties, depending on whether the member is used in a horizontal or vertical orientation.

4.4.2 Allowable Stresses and Lay-up Combinations

Many different species of lumber can be used to produce glued laminated timber. In addition, a wide range of grades of both visually graded and mechanically graded lumber can be used in the manufacture of glued laminated timber. This wide variety of available species and grades results in numerous options for the producers to combine species and grades to create a wide array of glued laminated timber lay-up combinations.

For some lay-up combinations, the use of different species within the same member is permitted. This is done when it is desirable to use a lower-strength species in the core of a glued laminated timber and a higher-strength species in the outer zones. However, it should be cautioned that when mixed species are used, they may result in an appearance that may not be suitable for an exposed application as the species will typically have different coloration and visual characteristics.

Glued laminated timber lay-up combinations can be developed based on the desired allowable properties and available lumber resources. Once the lumber resources are identified, the allowable stresses for glued laminated timber are determined in accordance with the principles of ASTM D3737 or using recognized computer software.

Bending or Axially Loaded Members. In addition to being able to produce virtually any size or shape of structural member, the laminating process used in glued laminated timber manufacturing also permits the manufacturer to optimize the use of the available wood fiber resource by selecting and positioning the lumber based on the stresses it will be subjected to in-service. For example, for members stressed primarily in bending, a graded lay-up of lumber is used throughout the depth of the beam with the highest-quality laminations used in the outer zones of the beam where the bending stresses are highest with lower-quality laminations being used in zones subjected to lower bending stresses. Lay-up combinations for members stressed primarily in bending about the X–X axis (see Fig. 4.10) are provided in Table 4.3. These members may range in cross-section from straight rectangular beams to pitched and tapered curved beams.

Although permitted, for members primarily stressed in axial loading or in bending about the Y–Y axis (also see Fig. 4.10), the use of glued laminated timber combinations given in Table 4.3 is not efficient. In such cases, the designer should select glued laminated timber combinations from Table 4.4. Similarly, glued laminated timber combinations in Table 4.3 are inefficiently utilized if the primary use is not bending about the X–X axis.

It should be noted that Tables 4.3 and 4.4 tabulate the lay-up combinations based on species, whether the combination is for a balanced or unbalanced layup and whether the lumber used is visually or mechanically graded as signified by a V (visual) or E (E-rated or mechanically graded). The allowable properties given in Tables 4.3 and 4.4 should be used in conjunction with the dimensions provided in Section 4.4.4.

The values for allowable properties given in Tables 4.3 and 4.4 are based on use under normal duration of load (10 years) and dry conditions (less than 16%

TABLE 4.3 Design Values for Structural Glued-Laminated Timber Intended Primarily for Members Stressed in Bending Due to Loads Applied Perpendicular to Wide Faces of the Laminations for Normal Duration of Load and Dry Conditions of Use[1,2,3]

		Bending about X-X Axis							Bending about Y-Y Axis							Axially Loaded		
		Loaded Perpendicular to Wide Faces of Laminations							Loaded Parallel to Wide Faces of Laminations									
		Extreme Fiber in Bending F_{bx}		Compression Perpendicular to Grain $F_{c\perp x}$[9]									Shear Parallel to Grain F_{vy}					
Combination Symbol[4]	Species-Outer Lams/Core Lams[5]	Tension Zone Stressed in Tension[6,7] psi	Compression Zone Stressed in Tension[8] psi	Tension Face psi	Compression Face psi	Shear Parallel to Grain F_{vx} psi		Modulus of Elasticity E_x 10^6 psi	Extreme Fiber in Bending[10] F_{by} psi	Compression Perpendicular to Grain $F_{c\perp y}$[9] psi	Shear Parallel to Grain F_{vy} psi		(For members with multiple piece lams which are not edge glued)[11] psi		Modulus of Elasticity E_y 10^6 psi	Tension Parallel to Grain F_t psi	Compression Parallel to Grain F_c psi	Modulus of Elasticity E 10^6 psi
1	2	3	4	5	6	7[19]	8[20]	9	10	11	12[19]	13[20]	14[19]	15[20]	16	17	18	19
VISUALLY GRADED WESTERN SPECIES																		
The following combinations are NOT BALANCED and are intended primarily for simple-span applications.																		
EWS 16F-V3	DF/DF	1,600	1,150	560[12,13]	560	240	190	1.5	1,450	560	210	165	105	85	1.5	950	1,550	1.5
EWS 20F-V4	DF/DF	2,000	1,450	590[12,13]	560[12]	240	190	1.6	1,450	560	210	165	105	85	1.6	1,000	1,550	1.6
EWS 20F-V12	AC/AC	2,000	1,400	560	560	240	190	1.5	1,200	470	210	165	105	85	1.4	900	1,500	1.4
EWS 24F-V4	DF/DF	2,400	1,850	650	650	240	190	1.8	1,500	560	210	165	105	85	1.6	1,100	1,650	1.6
EWS 24F-V5	DF/HF	2,400	1,600	650	650	195	155	1.7	1,350	375	180	140	90	70	1.5	1,100	1,450	1.5
EWS 24F-V5M1	DF/SPF	2,400	1,600	650	650	195	155	1.8	1,350	375	180	140	90	70	1.5	1,050	1,450	1.5
The following combination is NOT BALANCED and shall be applicable to members having a depth of 27 inches or less.																		
EWS 24F-V5M2	DF/HF	2,400	1,600	650	650	195	155	1.8	1,350	375	180	140	90	70	1.5	1,100	1,450	1.5
The following combination is NOT BALANCED and shall be applicable to members having a depth of 24 inches or less.																		
EWS 24F-V5M3	DF/HF	2,400	1,600	650	650	195	155	1.8	1,350	375	180	140	90	70	1.5	1,100	1,450	1.5
The following combinations are NOT BALANCED and are intended for straight or slightly cambered members for dry use and industrial appearance[14], and shall be applicable to members having a depth of 9 to 20 laminations.																		
EWS 24F-V4M1	DF/DF	2,400	1,850	650	650	240	190	1.8	1,500	560	210	165	105	85	1.6	1,100	1,650	1.6
EWS 24F-V4M2	DF/DF	2,400	1,850	650	650	200	160	1.8	1,500	560	210	165	105	85	1.6	1,100	1,650	1.6
The following combinations are BALANCED and are intended for members continuous or cantilevered over supports and provide equal capacity in both positive and negative bending.																		
EWS 20F-V8	DF/DF	2,000	2,000	590[12,13]	590[12,13]	240	190	1.7	1,450	560	210	165	105	85	1.6	1,000	1,600	1.6
EWS 20F-V13	AC/AC	2,000	2,000	560	560	240	190	1.5	1,250	470	210	165	105	85	1.5	925	1,550	1.5
EWS 24F-V8	DF/DF	2,400	2,400	650	650	240	190	1.8	1,450	560	210	165	105	85	1.6	1,100	1,650	1.6
EWS 24F-V10	DF/HF	2,400	2,400	650	650	195	155	1.8	1,400	375	180	140	90	70	1.6	1,150	1,600	1.6
The following combinations are BALANCED and are intended for straight or slightly cambered members for dry use and industrial appearance[14], and shall be applicable to members having a depth of 9 to 20 laminations.																		
EWS 24F-V8M1	DF/DF	2,400	2,400	650	650	240	190	1.8	1,450	560	210	165	105	85	1.6	1,100	1,650	1.6
EWS 24F-V8M2	DF/DF	2,400	2,400	650	650	200	160	1.8	1,450	560	210	165	105	85	1.6	1,100	1,650	1.6
E-RATED WESTERN SPECIES																		
The following combinations are NOT BALANCED and are intended primarily for simple-span applications.																		
EWS 20F-E8	ES/ES	2,000	1,300	450	450	180	145	1.5	1,400	300	160	125	80	65	1.4	800	1,000	1.4
EWS 24F-E15M1	HF/HF	2,400	1,600	500	500	195	155	1.8	1,300	375	170	135	85	70	1.5	950	1200	1.5
EWS 24F-E/CSP3	CSP[21]/CSP	2,400	1,550	560	650	195	160	1.6	1,200	470	175	140	90	70	1.5	900	1,750	1.5
Wet-use factors[2]		0.8		0.53		0.875		0.833	0.8	0.53	0.875				0.833	0.8	0.73	0.833

TABLE 4.3 Design Values for Structural Glued-Laminated Timber Intended Primarily for Members Stressed in Bending Due to Loads Applied Perpendicular to Wide Faces of the Laminations for Normal Duration of Load and Dry Conditions of Use[1,2,3] *(Continued)*

		Bending about X-X Axis							Bending about Y-Y Axis							Axially Loaded		
		Loaded Perpendicular to Wide Faces of Laminations							Loaded Parallel to Wide Faces of Laminations									
		Extreme Fiber in Bending F_{bx}		Compression Perpendicular to Grain $F_{c\perp x}$[9]									Shear Parallel to Grain F_{vy} (For members with multiple piece lams which are not edge glued)[11]					
Combination Symbol[4]	Species-Outer Lams/Core Lams[5]	Tension Zone Stressed in Tension[6,7] psi	Compression Zone Stressed in Tension[8] psi	Tension Face psi	Compression Face psi	Shear Parallel to Grain F_{vx} psi		Modulus of Elasticity E_x 10^6 psi	Extreme Fiber in Bending[10] F_{by} psi	Compression Perpendicular to Grain $F_{c\perp y}$[9] psi	Shear Parallel to Grain F_{vy} psi		psi		Modulus of Elasticity E_y 10^6 psi	Tension Parallel to Grain F_t psi	Compression Parallel to Grain F_c psi	Modulus of Elasticity E 10^6 psi
1	2	3	4	5	6	7[19]	8[20]	9	10	11	12[19]	13[20]	14[19]	15[20]	16	17	18	19
E-RATED WESTERN SPECIES																		
The following combinations are **NOT BALANCED** and are intended primarily for simple-span applications.																		
EWS 24F-E/CSP4	CSP[21]/CSP	2,400	1,700	560	650	195	160	1.8	1,400	470	180	145	90	75	1.6	1,150	1,900	1.6
EWS 24F-E/ES1	ES/ES	2,400	1,700	560	560	180	145	1.7	1,100	300	160	125	80	65	1.5	1,050	1,150	1.5
EWS 24F-E/SPF3	SPF[21]/SPF	2,400	1,550	560	650	195	160	1.6	1,200	470	175	140	90	70	1.5	900	1,750	1.5
EWS 24F-E/SPF4	SPF[21]/SPF	2,400	1,700	560	650	195	160	1.8	1,400	470	180	145	90	75	1.6	1,150	1,900	1.6
The following combination is **NOT BALANCED** and shall be applicable to members having a depth of 9-1/2, 11-7/8, 14, and 16 inches.																		
EWS 26F-E/DF1	DF/DF	2,600	1,950[16]	650	650	240	190	2.0	1,850	560	210	165	105	85	1.8	1,400	1,800	1.8
The following combinations are **BALANCED** and are intended for members continuous or cantilevered over supports and provide equal capacity in both positive and negative bending.																		
EWS 20F-E8M1	ES/ES	2,000	2,000	450	450	180	145	1.5	1,400	300	160	125	80	65	1.4	800	1,000	1.4
EWS 24F-E/CSP1	CSP/CSP	2,400	2,400	560	560	195	160	1.6	1,150	470	170	140	85	70	1.5	900	1,800	1.5
EWS 24F-E/CSP2	CSP/CSP	2,400	2,400	560	560	195	160	1.8	1,500	470	170	140	85	70	1.6	1,150	2,000	1.6
EWS 24F-E/SPF1	SPF/SPF	2,400	2,400	560	560	195	160	1.6	1,150	470	170	140	85	70	1.5	900	1,800	1.5
EWS 24F-E/SPF2	SPF/SPF	2,400	2,400	560	560	195	160	1.8	1,500	470	170	140	85	70	1.6	1,150	2,000	1.6
The following combination is **BALANCED** and shall be applicable to members having a depth of 7-1/2, 9, 9-1/2, 11-7/8, and 14 inches, and a width of 1-1/2 through 3-1/2 inches.																		
EWS 20F-E/SPF1	SPF/SPF	2,000	2,000	425	425	195	160	1.5	875	425	170	140	85	70	1.4	425	1,100	1.4
The following combination is **BALANCED** and shall be applicable to members having a depth of 9-1/2, 11-7/8, 14, and 16 inches.																		
EWS 26F-E /DF1M1	DF/DF	2,600	2,600	650	650	240	190	2.0	1,850	560	210	165	105	85	1.8	1,400	1,800	1.8
VISUALLY GRADED/E-RATED WESTERN SPECIES OR SOUTHERN PINE																		
The following combination is **NOT BALANCED** and is intended primarily for simple-span header applications.																		
EWS 24F-1.8E Glulam Header[13]	WS, SP /WS, SP	2,400	1,600	500	500	195	155	1.8	1,300	375	180	140	90	70	1.5	950	1,200	1.5
VISUALLY GRADED SOUTHERN PINE																		
The following combinations are **NOT BALANCED** and are intended primarily for simple-span applications.																		
EWS 24F-V1	SP/SP	2,400	1,750	740	650[12]	270	240	1.7	1,500	650	235	210	120	105	1.5	1,100	1,350	1.5
EWS 24F-V3	SP/SP	2,400	1,950	740	740	270	240	1.8	1,600	650	235	210	120	105	1.6	1,150	1,700	1.6
EWS 26F-V1	SP/SP	2,600	1,950	740	740	270	240	1.8	1,900	650	235	210	120	105	1.6	1,150	1,600	1.6
Wet-use factors[2]		0.8		0.53		0.875		0.833	0.8	0.53	0.875				0.833	0.8	0.73	0.833

Combination Symbol[4]	Species- Outer Lams/ Core Lams[5]	Bending about X-X Axis							Bending about Y-Y Axis							Axially Loaded		
		Loaded Perpendicular to Wide Faces of Laminations							Loaded Parallel to Wide Faces of Laminations									
		Extreme Fiber in Bending F_{bx}		Compression Perpendicular to Grain $F_{c\perp x}$[9]									Shear Parallel to Grain F_{vy} (For members with multiple piece lams which are not edge glued)[11]					
		Tension Zone Stressed in Tension[6,7] psi	Compression Zone Stressed in Tension[8] psi	Tension Face psi	Compression Face psi	Shear Parallel to Grain F_{vx} psi		Modulus of Elasticity E_x 10^6 psi	Extreme Fiber in Bending[10] F_{by} psi	Compression Perpendicular to Grain $F_{c\perp y}$[9] psi	Shear Parallel to Grain F_{vy} psi		psi		Modulus of Elasticity E_y 10^6 psi	Tension Parallel to Grain F_t psi	Compression Parallel to Grain F_c psi	Modulus of Elasticity E 10^6 psi
1	2	3	4	5	6	7[19]	8[20]	9	10	11	12[19]	13[20]	14[19]	15[20]	16	17	18	19
VISUALLY GRADED SOUTHERN PINE																		
The following combinations are **NOT BALANCED** and are intended primarily for simple-span applications.																		
EWS 26F-V2	SP/SP	2,600	2,100	740	740	270	240	1.9	2,200	740	235	210	120	105	1.8	1,200	1,650	1.8
EWS 26F-V3	SP/SP	2,600	2,100	740	740	270	240	1.9	2,100	650	235	210	120	105	1.8	1,150	1,600	1.8
The following combinations are **NOT BALANCED** and are intended for straight or slightly cambered members for dry use and industrial appearance[14], and shall be applicable to members having a depth of 9 to 20 laminations.																		
EWS 24F-V3M1	SP/SP	2,400	1,950	740	740	270	240	1.8	1,600	650	235	210	120	105	1.6	1,150	1,700	1.6
EWS 24F-V3M2	SP/SP	2,400	1,950	740	740	225	200	1.8	1,600	650	235	210	120	105	1.6	1,150	1,700	1.6
EWS 26F-V3M1	SP/SP	2,600	2,100	740	740	270	240	1.9	2,100	650	235	210	120	105	1.8	1,150	1,600	1.8
EWS 26F-V3M2	SP/SP	2,600	2,100	740	740	225	200	1.9	2,100	650	235	210	120	105	1.8	1,150	1,600	1.8
The following combinations are **BALANCED** and are intended for members continuous or cantilevered over supports and provide equal capacity in both positive and negative bending.																		
EWS 24F-V5	SP/SP	2,400	2,400	740	740	270	240	1.7	1,600	650	235	210	120	105	1.5	1,150	1,700	1.5
EWS 24F-V5M1	SP/SP	2,400	2,400	740	740	270	240	1.8	1,600	650	235	210	120	105	1.5	1,150	1,700	1.5
EWS 26F-V4	SP/SP	2,600	2,600	740	740	270	240	1.9	2,100	650	235	210	120	105	1.8	1,150	1,600	1.8
The following combinations are **BALANCED** and are intended for straight or slightly cambered members for dry use and industrial appearance[14], and shall be applicable to members having a depth of 9 to 20 laminations.																		
EWS 24F-V5M2	SP/SP	2,400	2,400	740	740	270	240	1.8	1,600	650	235	210	120	105	1.5	1,150	1,700	1.5
EWS 24F-V5M3	SP/SP	2,400	2,400	740	740	225	200	1.8	1,600	650	235	210	120	105	1.5	1,150	1,700	1.5
EWS 26F-V4M1	SP/SP	2,600	2,600	740	740	270	240	1.9	2,100	650	235	210	120	105	1.8	1,150	1,600	1.8
EWS 26F-V4M2	SP/SP	2,600	2,600	740	740	225	200	1.9	2,100	650	235	210	120	105	1.8	1,150	1,600	1.8
E-RATED SOUTHERN PINE																		
The following combinations are **NOT BALANCED** and are intended primarily for simple-span applications.																		
EWS 28F-E1	SP/SP	2,800	2,300	740	740	270	240	2.0	1,600	650	235	210	120	105	1.7	1,300	1,850	1.7
EWS 28F-E1M1	SP/SP	2,800	2,300	740	740	270	240	2.1	1,600	650	235	210	120	105	1.7	1,300	1,850	1.7
The following combinations are **BALANCED** and are intended for members continuous or cantilevered over supports and provides equal capacity in both positive and negative bending.																		
EWS 28F-E2	SP/SP	2,800	2,800	740	740	270	240	2.0	1,600	650	235	210	120	105	1.7	1,300	1,850	1.7
EWS 28F-E2M1	SP/SP	2,800	2,800	740	740	270	240	2.1	1,600	650	235	210	120	105	1.7	1,300	1,850	1.7
The following combination is **NOT BALANCED** and shall be applicable to members having a nominal width of 6 inches or less.																		
EWS 30F-E1	SP/SP	3,000	2,400	740	740	270	240	2.0	1,750	650	235	210	120	105	1.7	1,250	1,750	1.7
Wet-use factors[2]		0.8		0.53		0.875		0.833	0.8	0.53	0.875				0.833	0.8	0.73	0.833

TABLE 4.3 Design Values for Structural Glued-Laminated Timber Intended Primarily for Members Stressed in Bending Due to Loads Applied Perpendicular to Wide Faces of the Laminations for Normal Duration of Load and Dry Conditions of Use[1,2,3] (*Continued*)

Combination Symbol[4]	Species-Outer Lams/Core Lams[5]	Bending about X-X Axis							Bending about Y-Y Axis								Axially Loaded		
		Loaded Perpendicular to Wide Faces of Laminations							Loaded Parallel to Wide Faces of Laminations										
		Extreme Fiber in Bending F_{bx}		Compression Perpendicular to Grain $F_{c\perp x}$[9]									Shear Parallel to Grain F_{vy} (For members with multiple piece lams which are not edge glued)[11]						
		Tension Zone Stressed in Tension[6,7] psi	Compression Zone Stressed in Tension[8] psi	Tension Face psi	Compression Face psi	Shear Parallel to Grain F_{vx} psi		Modulus of Elasticity E_x 10^6 psi	Extreme Fiber in Bending[10] F_{by} psi	Compression Perpendicular to Grain $F_{c\perp y}$[9] psi	Shear Parallel to Grain F_{vy} psi		psi		Modulus of Elasticity E_y 10^6 psi	Tension Parallel to Grain F_t psi	Compression Parallel to Grain F_c psi	Modulus of Elasticity E 10^6 psi
1	2	3	4	5	6	7[19]	8[20]	9	10	11	12[19]	13[20]	14[19]	15[20]	16	17	18	19
E-RATED SOUTHERN PINE																		
The following combination is **NOT BALANCED** and shall be applicable to members having a nominal width of 6 inches or less, and a depth of 18 inches or less.																		
EWS 30F-E1M1	SP/SP	3,000	2,400	740	740	270	240	2.1	1,750	650	235	210	120	105	1.7	1,250	1,750	1.7
The following combination is **BALANCED** and shall be applicable to members having a nominal width of 6 inches or less.																		
EWS 30F-E2	SP/SP	3,000	3,000	740	740	270	240	2.0	1,750	650	235	210	120	105	1.7	1,250	1,750	1.7
The following combination is **BALANCED** and shall be applicable to members having a nominal width of 6 inches or less, and a depth of 18 inches or less.																		
EWS 30F-E2M1	SP/SP	3,000	3,000	740	740	270	240	2.1	1,750	650	235	210	120	105	1.7	1,250	1,750	1.7
The following combination is **BALANCED** and shall be applicable to members having a depth of 16 inches or less.																		
EWS 30F-E2M2	DF/SP	3,000	3,000	650[22]	650[22]	270	240	2.1	1,750	650	235	210	120	105	1.7	1,250	1,750	1.7
The following combination is **BALANCED** and shall be applicable to members having a depth of 18 inches or less.																		
EWS 30F-E2M3	DF/SP	3,000	3,000	510[22]	510[22]	270	240	2.1	1,750	650	235	210	120	105	1.7	1,250	1,750	1.7
E-RATED DURANGO PINE																		
The following combinations are **NOT BALANCED** and are intended primarily for simple-span applications.																		
EWS 16F-E1[17]	DP/DP	1,600[15]	1,300[15]	650[12]	650[12]	270	240	1.6	1,500	650	235	210	120	105	1.5	1,050	1,550	1.5
EWS 20F-E1[17]	DP/DP	2,000[15]	1,550[15]	650[12]	650[12]	270	240	1.7	1,600	650	235	210	120	105	1.5	1,050	1,550	1.5
EWS 24F-E1[17]	DP/DP	2,400[15]	1,750[15]	740	650[12]	270	240	1.8	1,600	650	235	210	120	105	1.6	1,100	1,550	1.6
E-RATED DURANGO PINE																		
The following combinations are **BALANCED** and are intended for members continuous or cantilevered over supports and provide equal capacity in both positive and negative bending.																		
EWS 16F-E3[17]	DP/DP	1,600[15]	1,600[15]	650[12]	650[12]	270	240	1.6	1,700	650	235	210	120	105	1.5	1,100	1,600	1.5
EWS 20F-E3[17]	DP/DP	2,000[15]	2,000[15]	650[12]	650[12]	270	240	1.7	1,800	650	235	210	120	105	1.5	1,150	1,650	1.5
EWS 24F-E4[17]	DP/DP	2,400[15]	2,400[15]	740	740	270	240	1.8	2,000	650	235	210	120	105	1.6	1,250	1,700	1.6
Wet-use factors[2]		0.8		0.53		0.875		0.833	0.8	0.53	0.875				0.833	0.8	0.73	0.833

[1]The combinations in this table are applicable to members consisting of 4 or more laminations and are intended primarily for members stressed in bending due to loads applied perpendicular to the wide faces of the laminations. Design values are tabulated, however, for loading both perpendicular and parallel to the wide faces of the laminations. For combinations and design values applicable to members loaded primarily axially or parallel to the wide faces of the laminations, see Table 4-4. For members of 2 or 3 laminations, see Table 4-4.

[2]The tabulated design values are for dry conditions of use. For wet conditions of use, multiply the tabulated values by the factors shown at the end of the table.

[3]The tabulated design values are for normal duration of loading. For other durations of loading, see applicable building code.

[4]The combination symbols relate to a specific combination of grades and species in Table 4-5 that will provide the design values shown for the combination. The first two numbers in the combination symbol correspond to the design value in bending shown in Column 3. The letter in the combination symbol (either a "V" or an "E") indicates whether the combination is made from visually graded (V) or E-rated (E) lumber in the outer zones.

[5]The symbols used for species are DF = Douglas fir-larch, HF = Hem-fir, AC = Alaska cedar, ES = Eastern spruce, SPF = Spruce-pine-fir, CSP = Canadian spruce-pine, SP = Southern pine, and DP = Durango pine.

[6]The tabulated design values in bending, F_{bx}, are based on members 5-1/8 inches (130 mm) in width by 12 inches (305 mm) in depth by 21 feet (6,400 mm) in length. For members with a larger volume, F_{bx} shall be multiplied by a volume factor, C_v, determined in accordance with applicable building code.

[7]The design values in bending about the X-X axis, F_{bx}, in this column for bending

lamination(s). See Column 3 in Table 4-5 for the special tension lamination(s) provisions.

[8]Design values in this column are for extreme fiber stress in bending when the member is loaded such that the compression zone laminations are subjected to tensile stresses.

[9]The compression perpendicular to grain design values in this table are not subject to the duration of load modifications.

[10]The values of F_{by} were calculated based on members 12 inches in depth (bending about Y-Y axis). When the depth is less than 12 inches, the values of F_{by} shall be permitted to be increased by multiplying by the following factors:

All Species	Beam Depth, inches				
	10-3/4 or 10-1/2	8-3/4 or 8-1/2	6-3/4	5-1/8 or 5	3-1/8 or 3
Factor	1.01	1.04	1.07	1.10	1.16

For members with depth greater than 12 inches, the value of F_{by} shall be reduced by applying the size factor, $(12/d)^{1/9}$, where d is the beam depth in inches.

[11]These values for shear parallel to grain, F_{vy}, apply to members manufactured using multiple piece laminations with unbonded edge joints. For members manufactured using single piece laminations or using multiple piece laminations with bonded edge joints, the shear parallel to grain values in Columns 12 and 13 apply. The values in this column do not apply to members with 5, 7 or 9 laminations when unbonded edge joints occur in alternate laminations at mid-depth of the member with no edge joints in adjacent laminations and the outside lamination contains unbonded edge joints. The value in this column shall be reduced by 16 percent for members containing 5 laminations when unbonded edge joints occur in each lamination forming a staggered pattern in the member with no edge joint closer than 1 inch to the mid-depth of the member.

[12]Where specified by the designer, this value shall be permitted to be increased to 650 psi for Douglas fir-Larch, or 740 psi for southern pine or Durango pine by providing in the bearing area at least one dense 2-inch nominal thickness lamination of Douglas fir-larch for western species combinations, southern pine for southern pine combinations, or Durango pine for Durango pine combinations. These dense laminations shall be backed by a medium grain lamination of the same species.

[13]For bending members greater than 15 inches in depth, the design value for compression stress perpendicular to grain is 650 psi on the tension face.

[14]When containing wane, these combinations shall be used in dry conditions only. In this case, wet-use factors shall not be applied. Because of the wane, these combinations are assigned for an industrial appearance grade. If wane is omitted, these restrictions shall not apply.

[15]This tabulated design value shall be multiplied by a volume factor, C_v, based on the same volume factor applicable to Southern pine structural glued laminated timber in accordance with Footnote No. 6 to this table.

[16]This tabulated value is permitted to be increased to 2,200 psi for beam depths less than 16 in.

[17]For connection design, use the specific gravity of 0.52 for Durango pine based on oven-dry weight and volume.

[18]This combination shall be manufactured from either EWS 24F-V4/WS, EWS 24F-V5M1/WS, EWS 24F-V5M2/WS, EWS 24F-V5M3/WS, EWS 24F-E15M1/WS, EWS 24F-E/SPF4, or EWS 24F-V3/SP, and is intended primarily for use in header applications. Design values given for this combination are based on the minimum values for all combinations mentioned above. For connection design, use the specific gravity of 0.42 for Spruce-pine-fir based on oven-dry weight and volume.

[19]These tabulated shear values are applicable to prismatic glulam members subjected to typical static and transient dead, live, snow, wind and earthquake loadings but excluding impact or cyclic loadings such as may occur in bridges or crane rail applications. These values allow for checking up to 10% of the glulam width in the shear critical zone.

[20]These tabulated shear values are applicable to non-prismatic glulam members subjected to typical static and transient loadings and all members subjected to impact or cyclic loads such as may occur in bridges or crane rail applications. These values allow for checking up to 10% of the glulam width in the shear critical zone.

[21]L2D/DF laminations are used in the outer compression zone. See Table 4-5 for detailed grade requirements.

[22]The allowable compressive stress perpendicular to grain of the beam shall be permitted to be increased to the published allowable compressive stress perpendicular to grain of the outermost laminated veneer lumber.

TABLE 4.4 Design Values for Structural Glued-Laminated Timber for Normal Duration of Load and Dry Conditions of Use[1,2,3]

					Axially Loaded			Bending about Y-Y Axis, Loaded Parallel to Wide Faces of Laminations											Bending about X-X Axis, Loaded Perpendicular to Wide Faces of Laminations			
					Tension Parallel to Grain F_t	Compression Parallel to Grain F_c		Extreme Fiber in Bending F_{by}[7]			Shear Parallel to Grain F_{vy}								Extreme Fiber in Bending F_{bx}[9]		Shear Parallel to Grain, F_{vx}[13]	
Comb. Symbol	Species[4]	Grade[5]	Modulus of Elasticity E 10^6 psi	Compression Perpendicular to Grain $F_{c\perp}$[6] psi	2 or More Lams psi	4 or More Lams psi	2 or 3 Lams psi	4 or More Lams psi	3 Lams psi	2 Lams psi	4 or More Lams (For members with multi-lams)[8] psi		4 or More Lams psi		3 Lams psi		2 Lams psi		2 Lams to 15 in. Deep psi	4 or More Lams psi	2 or More Lams psi	
1	2	3	4	5	6	7	8	9	10	11	12[19]	13[20]	14[19]	15[20]	16[19]	17[20]	18[19]	19[20]	20[10]	21[11,12]	22[19]	23[20]
VISUALLY GRADED WESTERN SPECIES																						
EWS 1	DF	L3	1.5	560[14]	900	1,550	1,200	1,450	1,250	1,000	105	85	210	165	200	160	180	145	1,250	1,500	240	190
EWS 2	DF	L2	1.7	560[14]	1,250	1,900	1,600	1,800	1,600	1,300	105	85	210	165	200	160	180	145	1,700	2,000	240	190
EWS 3	DF	L2D	1.8	650	1,450	2,300	1,850	2,100	1,850	1,550	105	85	210	165	200	160	180	145	2,000	2,300	240	190
EWS 5	DF	L1	2.0	650	1,600	2,400	2,100	2,400	2,100	1,800	105	85	210	165	200	160	180	145	2,200	2,400	240	190
EWS 22	WW[16]	L3	1.0[15]	255	525	850	675	800	700	550	80	60	155	120	145	115	130	105	725	850	175	140
EWS 70	AC	L2	1.4	470	1,000	1,450	1,550	1,250	1,100	925	105	85	210	165	200	160	180	145	1,350	1,550	240	190
VISUALLY GRADED SOUTHERN PINE																						
EWS 47	SP	N2M[17]	1.4	650[14]	1,200	1,900	1,150	1,750	1,550	1,300	120	105	235	210	225	200	205	180	1,400	1,600	270	240
EWS 48	SP	N2D[17]	1.7	740	1,400	2,200	1,350	2,000	1,800	1,500	120	105	235	210	225	200	205	180	1,600	1,900	270	240
EWS 49	SP	N1M[17]	1.7	650[14]	1,350	2,100	1,450	1,950	1,750	1,500	120	105	235	210	225	200	205	180	1,800	2,100	270	240
EWS 50	SP	N1D[17]	1.9	740	1,550	2,300	1,700	2,300	2,100	1,750	120	105	235	210	225	200	205	180	2,100	2,400	270	240
The following combination is intended for dry use and industrial appearance[18], and shall be applicable to members having a depth of 9 to 20 laminations.																						
EWS 49M1	SP	N1M[17]	1.7	650[14]	1,350	2,100	1,450	1,950	1,750	1,500	120	105	235	210	225	200	205	180	1,800	2,100	270	240
The following combination is intended for dry use and industrial appearance[18], and shall be applicable to members having a depth of 4 to 20 laminations.																						
EWS 49M2	SP	N1M[17]	1.7	650[14]	1,350	2,100	1,450	1,950	1,750	1,500	120	105	235	210	225	200	205	180	1,800	2,100	225	200
Wet-use factors[2]			0.833	0.53	0.8	0.73		0.8			0.875								0.8		0.875	

[1]The combinations in this table are intended primarily for members loaded either axially or in bending with the loads acting parallel to the wide faces of the laminations. Design values for bending due to loading applied perpendicular to the wide faces of the laminations are also included, however, the combinations in Table 4-3 are usually better suited for this condition of loading. The design values for bending about the X-X axis (F_{bx}) shown in Column 20 are for members from 2 laminations to 15 inches deep without tension laminations. Design values approximately 15 percent higher for members with 4 or more laminations are shown in Column 21. These higher design values, however, require special tension lamination(s). See Columns 4 through 6 in Table 4-6 for the tension lamination(s) provisions.

[2]The tabulated design values are for dry conditions of use. For wet conditions of use, multiply the tabulated values by the factors shown at the end of the table.

[3]The tabulated design values are for normal duration of loading. For other durations of loading, see applicable building code.

[4]The symbols used for species are DF = Douglas fir-larch, WW = softwood species, and SP = southern pine.

[5]Grade designations are given in Footnote No. 8 of Table 4-5.

[6]The compression perpendicular to grain design values in this table are not subject to the duration of load modifications.

[7]Footnote No. 10 of Table 4-3 also applies.

[8]These values for shear parallel to grain, F_{vy}, apply to members manufactured using multiple piece laminations with unbonded edge joints. For members using single piece laminations or using multiple piece laminations with bonded edge joints, the shear parallel to grain values tabulated in Columns 14 though 19 apply. The values in Columns 12 and 13 do not apply to members with 5, 7 or 9 laminations when unbonded edge joints occur in alternate laminations at mid-depth of the member with no edge joints in adjacent laminations and the outside laminations contains unbonded edge joints. The values in Columns 12 and 13 shall be reduced by 16 percent for members containing 5 laminations when unbonded edge joints occur in each lamination forming a staggered pattern in the member with no edge joint closer than 1 inch to the mid-depth of the member.

[9]Footnote No. 6 of Table 4-3 also applies.

[10]These design values are for members of from 2 laminations to 15 inches in depth without tension laminations.

[11]These design values are for members of 4 or more laminations in depth and require special tension laminations. See Columns 4 through 6 in Table 4-6 for the tension lamination(s) provisions. When these values are used for the design and the member is specified by combination symbol, the designer shall also specify the required design value in bending.

[12]When special tension laminations are not used, the design values in bending about the X-X axis (F_{bx}) shall be multiplied by 0.75 for bending members over 15 inches deep. For bending members 15 inches and less in depth, use the design values in Column 20.

[13]Laminations shall not contain wane.

[14]When tension laminations are used to obtain the design value for F_{bx} shown in Column 21 (see Footnote No. 11), the compression perpendicular to grain value

$F_{c\perp}$, for the tension face is permitted to be increased to 650 psi for Douglas fir-larch and 740 psi for southern pine and Durango pine because the tension laminations are required to be dense.

[15]The modulus of elasticity (E) shall be 900,000 psi when the following softwood species (WW) are used: western cedars, western cedars (North), white woods (western woods) and California redwood - open grain.

[16]When the following softwood species (WW) are used, the shear value parallel to grain (F_{vy}) specified in Columns 12 and 13 shall be reduced 5 psi, the shear value parallel to grain (F_{vy}) specified in Columns 14 through 19 shall be reduced 10 psi, and the shear value parallel to grain (F_{vx}) specified in Columns 22 and 23 shall be reduced 10 psi: coast Sitka spruce, coast species, western white pine and eastern white pine.

[17]Combinations EWS 47, EWS 48, EWS 49 and EWS 50 have more restrictive slope of grain requirements than the basic slope of grain of the grades of lumber used in order to obtain higher tension parallel to grain values and design values in bending when loaded perpendicular to the wide faces of the laminations. All laminations shall not contain a slope of grain less restrictive than 1:14 for EWS 47, EWS 48, and EWS 50, and 1:16 for EWS 49. Where these slope of grain requirements are not met, the design values specified in the following table for the applicable slope of grain apply. The design bending values, F_{bx}, specified in Column 5 of the table are applicable to members with 2 or more laminations up to 15 inches in depth without tension laminations, and the values specified in Column 6 of the table are for members with 4 or more laminations with tension laminations.

Slope of Grain	Comb. No.	Tension Parallel to Grain, F_t	Comp. Parallel to Grain, F_c		Bending about the X-X Axis, F_{bx}		Bending about the Y-Y Axis, F_{by}		
		2 or More Lams psi	2 or 3 Lams psi	4 or More Lams psi	2 Lams to 15 in. psi	4 or More Lams psi	2 Lams psi	3 Lams psi	4 or More Lams psi
1		2	3	4	5	6	7	8	9
1:12	EWS 47	1,200	1,150	1,900	1,400	1,600	1,300	1,550	1,750
	EWS 48	1,400	1,350	2,200	1,600	1,900	1,500	1,800	2,000
	EWS 49	1,300	1,450	1,900	1,750	2,100	1,500	1,750	1,950
	EWS 50	1,550	1,700	2,200	2,100	2,400	1,750	2,100	2,300
1:10	EWS 47	1,150	1,450	1,700	1,550	1,850	1,500	1,750	1,850
	EWS 48	1,150	1,450	1,700	1,550	1,850	1,500	1,750	1,850
	EWS 49	1,150	1,450	1,700	1,550	1,850	1,500	1,750	1,850
	EWS 50	1,350	1,700	2,000	1,800	2,100	1,750	2,100	2,100
1:8	EWS 47	1,000	1,150	1,500	1,350	1,600	1,300	1,550	1,600
	EWS 48	1,150	1,350	1,750	1,600	1,850	1,500	1,800	1,850
	EWS 49	---	---	---	---	---	---	---	---
	EWS 50	---	---	---	---	---	---	---	---

[18]When containing wane, these combinations shall be used in dry conditions only. In this case, wet-use factors shall not be applied. Because of the wane, these combinations are assigned for an industrial appearance grade. If wane is omitted, these restrictions shall not apply.

[19]These tabulated shear values are applicable to prismatic glulam members subjected to typical static and transient dead, live, snow, wind and earthquake loadings but excluding impact or cyclic loadings such as may occur in bridges or crane rail applications. These values allow for checking up to 10% of the glulam width in the shear critical zone.

[20]These tabulated shear values are applicable to non-prismatic glulam members subjected to typical static and transient loadings and all members subjected to impact or cyclic loads such as may occur in bridges or crane rail applications. These values allow for checking up to 10% of the glulam width in the shear critical zone.

moisture content). When used under other conditions, see Section 4.4.3 for adjustment factors. It is important to note that the allowable bending stresses given in Table 4.3 are based on members loaded as simple beams. When glued laminated timber is used in continuous or cantilevered beams, the allowable bending stresses given in column 4 of Table 4.3 should be used for the design of stress reversal (when compression zone is stressed in tension).

Tables 4.5 and 4.6 provide the grade requirements for the laminations used in manufacturing the glued laminated timber listed in Tables 4.3 and 4.4, respectively. In addition to the lay-up combinations tabulated in Tables 4.3–4.6, the glulam industry periodically evaluates the use of new layup combinations and stresses based on the use of a computer program such as the one identified as Glulam Allowable Properties (GAP). The GAP program is based on the provisions of ASTM D3737 and has been verified by extensive laboratory testing of full-sized glued laminated timber beams at the APA Research Center in Tacoma, Washington, and at other laboratories throughout North America. As these new special lay-ups are evaluated and approved, they are added to the code evaluation reports as part of the periodic reexamination process.

Combined Stress Members. If the member is going to be subjected to high bending stresses as well as axial stresses such as occur in arches or beam-columns, a bending member combination as tabulated in Table 4.3 is typically the most efficient. Tapered beams or pitched and tapered curved beams are special configurations that are also specified using Table 4.3 bending member combinations.

Straight-Tapered End Cuts on the Compression Face. Straight-tapered end cuts on the top of a beam are sometimes used to improve drainage, to provide extra head for downspouts and scuppers, to facilitate discharge of water, and to reduce the height of the wall. Table 4.7 provides allowable stresses and mean moduli of elasticity for glued laminated timber with straight-tapered end cuts on the compression face. The allowable stresses are provided for bending, F_b, and compression perpendicular to grain, $F_{c\perp}$, and replace the allowable values provided in Table 4.3 when tapered end cut members are used.

Radial Tension and Compression. When a curved member is loaded in bending, radial stresses are induced. When the bending moment is in the direction that tends to decrease the curvature or increase the radius, the radial stress is in radial tension, F_{rt}. On the other hand, when the bending moment is in the direction that tends to increase the curvature or decrease the radius, the radial stress is in radial compression, F_{rc}. Table 4.8 provides allowable radial tensile stresses for glued laminated timber. These values are subject to adjustments for duration of load and wet conditions of use (16% moisture content or higher). If the adjusted value is exceeded, appropriate mechanical reinforcements shall be used to resist all applied radial tensile stresses. The maximum moisture content of the laminations shall not exceed 12% at the time of the reinforcement manufacturing.

The allowable radial compressive stress has been traditionally limited to the design value in compression perpendicular to grain, $F_{c\perp}$, of the grade and species being used. Also given in Table 4.8 are allowable radial compressive stresses for glued laminated timber. These allowable radial compressive stresses are not subject to the adjustments for the duration of load, but shall be adjusted for wet conditions of use when appropriate.

TABLE 4.5 Grade Requirements for Members Stressed Principally in Bending and Loaded Perpendicular to the Wide Faces of Laminations[1,2]

Combination Symbol	Depth of Member	Tension Lam[3]	Minimum Grade of Laminations[4,5,6,7]									
			Percent/Grade/Species Each Zone[8]					Percent/Slope of Grain[9]				
			Outer Tension Zone	Inner Tension Zone	Core	Inner Comp. Zone	Outer Comp. Zone	Outer Tension Zone	Inner Tension Zone	Core	Inner Comp. Zone	Outer Comp. Zone
1	2	3	4	5	6	7	8	9	10	11	12	13
VISUALLY GRADED WESTERN SPECIES (WS)												
The following combinations are NOT BALANCED and are intended primarily for simple-span applications.												
EWS	< 12 in.	--	10%L2D/DF	--	L3/DF	--	L3/DF[10]	--	--	--	--	--
16F-V3/WS	12 to 15 in.	--	10%L2D/DF	--	L3/DF	--	L3/DF[10]	--	--	--	--	--
	< 12 in.	302-20	10%L2/DF[10,11]	--	L3/DF	--	L3/DF[10]	--	--	--	--	--
	12 to 15 in.	302-20	10%L2/DF[10,11]	--	L3/DF	--	L3/DF[10]	--	--	--	--	--
	> 15 in.	302-20	L3/DF[10,11]	--	L3/DF	--	L3/DF[10,11]	--	--	--	--	--
EWS	< 12 in.	--	15%L1CL/DF[10]	15%L2/DF	L3/DF	5%L2/DF	10%L2/DF[10]	10% 1:14	--	--	--	--
20F-V4/WS	12 to 15 in.	--	20%L1CL/DF[10]	25%L2/DF	L3/DF	10%L2/DF	10%L2D/DF	10% 1:14	--	--	--	--
	< 12 in.	302-20	10%L1CL/DF[10]	--	L3/DF	--	10%L2/DF[10]	--	--	--	--	--
	12 to 15 in.	302-20	10%L1CL/DF[10]	--	L3/DF	--	10%L2/DF[10]	--	--	--	--	--
	> 15 in.	302-22	5%L1CL/DF[10,11]	10%L2/DF	L3/DF	--	5%L2/DF[10]	5% 1:14	--	--	--	--
EWS	< 12 in.	302-20	15%L1D/AC	10%L2/AC	L3/AC	15%L2/AC	10%L1D/AC	--	--	--	--	--
20F-V12/WS	12 to 15 in.	302-22	15%L1D/AC	10%L2/AC	L3/AC	15%L2/AC	10%L1D/AC	--	--	--	--	--
	> 15 in.	302-24	10%L1S/AC	10%L1D/AC	L3/AC	10%L2/AC	10%L1D/AC	5% 1:16	--	--	--	--
EWS	< 12 in.	302-20	15%L1/DF	15%L2/DF	L3/DF	10%L2/DF	10%L2D/DF	--	--	--	--	--
24F-V4/WS	12 to 15 in.	302-22	15%L1/DF	15%L2/DF	L3/DF	10%L2/DF	10%L2D/DF	--	--	--	--	--
	> 15 in.	302-24	10%L1/DF	10%L2/DF	L3/DF	10%L2/DF	10%L2D/DF	5% 1:16	--	--	--	--
	< 12 in.	302-20	15%L1/DF	15%L2/DF	L3/DF	10%L2/DF	10%L1/DF	--	--	--	--	--
	12 to 15 in.	302-22	15%L1/DF	15%L2/DF	L3/DF	10%L2/DF	10%L1/DF	--	--	--	--	--
	> 15 in.	302-24	10%L1/DF	10%L2/DF	L3/DF	10%L2/DF	10%L1/DF	5% 1:16	--	--	--	--
EWS	< 12 in.	302-20	20%L1/DF	20%L1/HF	L3/HF	20%L2/HF	20%L2D/DF	--	--	--	--	--
24F-V5/WS	12 to 15 in.	302-22	20%L1/DF	20%L1/HF	L3/HF	20%L2/HF	20%L2D/DF	--	--	--	--	--
	> 15 in.	302-24	15%L1/DF	20%L1/HF	L3/HF	10%L2/HF	10%L2D/DF	5% 1:16	--	--	--	--
EWS	4 lams	302-20	30%L1/DF	--	1.4E/SPF	--	10%L2D/DF	5% 1:16	--	--	--	--
24F-V5M1	5 lams to < 12 in.	302-20	10%L1/DF	10%1.8E/SPF or 10%L2/DF	1.4E/SPF	10%1.8E/SPF or 10%L2/DF	10%L2D/DF	5% 1:16	--	--	--	--
1.8E 1050Ft /WS	12 to 15 in.	302-22	15%L1/DF	10%1.8E/SPF or 10%L2/DF	1.4E/SPF	10%1.8E/SPF or 10%L2/DF	10%L2D/DF	5% 1:16	--	--	--	--
	> 15 in.	302-24	15%L1/DF	10%1.8E/SPF or 10%L2/DF	1.4E/SPF	10%1.8E/SPF or 10%L2/DF	10%L2D/DF	5% 1:16	--	--	--	--
The following combination is NOT BALANCED and shall be applicable to members having a depth of 27 inches or less.												
EWS	< 12 in.	302-20	20%L1/DF	20%L1/HF	L3/HF	20%L2/HF	20%L2D/DF	5% 1:14	--	--	--	--
24F-V5M2	12 to 15 in.	302-22	20%L1/DF	20%L1/HF	L3/HF	20%L2/HF	20%L2D/DF	5% 1:16	--	--	--	--
1.8E/WS	> 15 to 27 in.	302-24	15%L1/DF	20%L1/HF	L3/HF	10%L2/HF	10%L2D/DF	5% 1:16	--	--	--	--
The following combination is NOT BALANCED and shall be applicable to members having a depth of 24 inches or less.												
EWS	< 12 in.	302-20	15%L1/DF	20%L2/HF	L3/HF	20%L2/HF	10%L2D/DF	5% 1:16	--	--	--	--
24F-V5M3	12 to 15 in.	302-22	15%L1/DF	20%L2/HF	L3/HF	20%L2/HF	15%L2D/DF	5% 1:16	--	--	--	--
1.8E/WS	15 to 24 in.	302-24	15%L1/DF	20%L2/HF	L3/HF	20%L2/HF	10%L2D/DF	5% 1:16	--	--	--	--

TABLE 4.5 Grade Requirements for Members Stressed Principally in Bending and Loaded Perpendicular to the Wide Faces of Laminations[1,2] *(Continued)*

Combination Symbol	Depth of Member	Tension Lam[3]	Minimum Grade of Laminations[4,5,6,7]									
			Percent/Grade/Species Each Zone[8]					Percent/Slope of Grain[9]				
			Outer Tension Zone	Inner Tension Zone	Core	Inner Comp. Zone	Outer Comp. Zone	Outer Tension Zone	Inner Tension Zone	Core	Inner Comp. Zone	Outer Comp. Zone
1	2	3	4	5	6	7	8	9	10	11	12	13
VISUALLY GRADED WESTERN SPECIES (WS)												
The following combinations are NOT BALANCED and are intended for straight or slightly cambered members for dry use and industrial appearance, and shall be applicable to members having a depth of 9 to 20 laminations.												
EWS 24F-V4M1[15] /WS	9 lams to ≤ 15 in.	302-22	15%L1/DF	15%L2/DF	L3/DF	10%L2/DF	10%L2D/DF or 10% L1/DF	--	--	--	--	--
	> 15 in. to 20 lams	302-24	10%L1/DF	10%L2/DF	L3/DF	10%L2/DF	10%L2D/DF or 10%L1/DF	5% 1:16	--	--	--	--
EWS 24F-V4M2[16] 200Fvx/WS	9 lams to ≤ 15 in.	302-22	15%L1/DF	15%L2/DF	L3/DF	10%L2/DF	10%L2D/DF or 10% L1/DF	--	--	--	--	--
	> 15 in. to 20 lams	302-24	10%L1/DF	10%L2/DF	L3/DF	10%L2/DF	10%L2D/DF or 10%L1/DF	5% 1:16	--	--	--	--
The following combinations are BALANCED and are intended for members continuous or cantilevered over supports and provide equal capacity in both positive and negative bending (Footnote No. 3 applies both top and bottom of bending members).												
EWS 20F-V8/WS	< 12 in.	--	15%L1CL/DF[10]	20%L2/DF	L3/DF	20%L2/DF	15%L1CL/DF[10]	10% 1:14	--	--	--	10% 1:14
	12 to 15 in.	--	15%L1CL/DF[10]	20%L2/DF	L3/DF	20%L2/DF	15%L1CL/DF[10]	10% 1:14	--	--	--	10% 1:14
	< 12 in.	302-20	10%L1CL/DF[10]	--	L3/DF	--	10%L1CL/DF[10]	--	--	--	--	--
	12 to 15 in.	302-20	10%L1CL/DF[10]	--	L3/DF	--	10%L1CL/DF[10]	--	--	--	--	--
	> 15 in.	302-22	5%L1CL/DF[10,11]	5%L2/DF	L3/DF	5%L2/DF	5%L1CL/DF[10,11]	--	--	--	--	--
EWS 20F-V13/WS	< 12 in.	302-20	15%L1D/AC	10%L2/AC	L3/AC	10%L2/AC	15%L1D/AC	--	--	--	--	--
	12 to 15 in.	302-22	15%L1D/AC	10%L2/AC	L3/AC	10%L2/AC	15%L1D/AC	--	--	--	--	--
	> 15 in.	302-24	10%L1S/AC	10%L1D/AC	L3/AC	10%L1D/AC	10%L1S/AC	5% 1:16	--	--	--	5% 1:16
EWS 24F-V8/WS	< 12 in.	302-20	10%L1/DF	10%L2/DF	L3/DF	10%L2/DF	10%L1/DF	--	--	--	--	--
	12 to 15 in.	302-22	10%L1/DF	10%L2D/DF	L3/DF	10%L2D/DF	10%L1/DF	--	--	--	--	--
	> 15 in.	302-24	10%L1/DF	5%L2/DF	L3/DF	5%L2/DF	10%L1/DF	5% 1:16	--	--	--	5% 1:16
EWS 24F-V10/WS	< 12 in.	302-20	20%L1/DF	10%L2/HF	L3/HF	10%L2/HF	20%L1/DF	--	--	--	--	--
	12 to 15 in.	302-22	20%L1/DF	10%L2/HF	L3/HF	10%L2/HF	20%L1/DF	--	--	--	--	--
	> 15 in.	302-24	15%L1/DF	15%L2/HF	L3/HF	15%L2/HF	15%L1/DF	5% 1:16	--	--	--	5% 1:16
The following combinations are BALANCED and are intended for straight or slightly cambered members for dry use and industrial appearance, and shall be applicable to members having a depth of 9 to 20 laminations.												
EWS 24F-V8M1[15]/WS	9 lams to ≤ 15 in.	302-22	10%L1/DF	10%L2D/DF	L3/DF	10%L2D/DF	10%L1/DF	--	--	--	--	--
	> 15 in. to 20 lams	302-24	10%L1/DF	5%L2/DF	L3/DF	5%L2/DF	10%L1/DF	5% 1:16	--	--	--	5% 1:16
EWS 24F-V8 M2[16] 200Fvx/WS	9 lams to ≤ 15 in.	302-22	10%L1/DF	10%L2D/DF	L3/DF	10%L2D/DF	10%L1/DF	--	--	--	--	--
	> 15 in. to 20 lams	302-24	10%L1/DF	5%L2/DF	L3/DF	5%L2/DF	10%L1/DF	5% 1:16	--	--	--	5% 1:16
E-RATED WESTERN SPECIES (WS)												
The following combinations are NOT BALANCED and are intended primarily for simple-span applications.												
EWS 20F-E8/WS	< 12 in.	302-20	10%B/ES	15%C4/ES	D/ES	15%D4/ES	10%C6/ES	--	--	--	--	--
	12 to 13-1/2 in.	302-22	10%B/ES	15%C4/ES	D/ES	15%D4/ES	10%C6/ES	--	--	--	--	--
	> 13-1/2 to 19-1/2 in.	302-22	20%B/ES	10%C4/ES	D/ES	10%D4/ES	15%C6/ES	--	--	--	--	--
	> 19-1/2 to 24 in.	302-24	15%B/ES	10%C4/ES	D/ES	15%D4/ES	10%C6/ES	--	--	--	--	--
	> 24 in.	302-24	15%B/ES	10%C4/ES	D/ES	10%D4/ES	15%C6/ES	--	--	--	--	--

Combination Symbol	Depth of Member	Tension Lam[3]	Minimum Grade of Laminations[4,5,6,7]									
			Percent/Grade/Species Each Zone[8]					Percent/Slope of Grain[9]				
			Outer Tension Zone	Inner Tension Zone	Core	Inner Comp. Zone	Outer Comp. Zone	Outer Tension Zone	Inner Tension Zone	Core	Inner Comp. Zone	Outer Comp. Zone
1	2	3	4	5	6	7	8	9	10	11	12	13
E-RATED WESTERN SPECIES (WS)												
The following combinations are **NOT BALANCED** and are intended primarily for simple-span applications.												
EWS	< 12 in.	302-20[17]	10%2.1E/HF	10%1.9E/HF	L3/HF	10%1.9E/HF	10%2.0E/HF	1/6	1/2	1/2	1/2	1/3
24F-E15M1	12 to 15 in.	302-22[17]	10%2.1E/HF	10%1.9E/HF	L3/HF	10%1.9E/HF	10%2.0E/HF	1/6	1/2	1/2	1/2	1/3
/WS	> 15 in.	302-24	10%2.1E/HF	10%1.9E/HF	L3/HF	10%1.9E/HF	10%2.0E/HF	1/6	1/2	1/2	1/2	1/3
EWS 24F-E	≤ 10-1/2 in.	Special	20%2.0E/CSP	5%1.8E/CSP	1.4E/CSP	5%1.8E/CSP	20%L2D/DF	1/6	1/3	1/2	1/3	--
/CSP3	> 10-1/2 in.	rules[14]	5%2.0E/CSP	10%1.8E/CSP	1.4E/CSP	10%1.8E/CSP	5%L2D/DF	1/6	1/3	1/2	1/3	--
EWS 24F-E	≤ 12 in.	Special	20%2.0E/CSP	10%1.8E/CSP	1.4E/CSP	10%1.8E/CSP	20%L2D/DF	1/6	1/3	1/2	1/3	--
/CSP4	> 12 in.	rules[14]	20%2.0E/CSP	10%1.8E/CSP	1.4E/CSP	10%1.8E/CSP	20%L2D/DF	1/6	1/3	1/2	1/3	--
EWS 24F-E	4 lams	302-20	5%1.9E/ES	--	B/ES	--	5%1.9E/ES	1/6	--	--	--	1/6
/ES1	5 lams to 10-1/2 in.	302-20	25%1.9E/ES	10%C4/ES	D/ES	10%C4/ES	15%B/ES	1/6	--	--	--	--
	12 to 15 in.	302-22	25%1.9E/ES	10%C4/ES	D/ES	10%C4/ES	20%B/ES	1/6	--	--	--	--
	> 15 in.	302-24	25%1.9E/ES	10%C4/ES	D/ES	10%C4/ES	20%B/ES	1/6	--	--	--	--
EWS 24F-E	≤ 10-1/2 in.	Special	20%2.0E/SPF	5%1.8E/SPF	1.4E/SPF	5%1.8E/SPF	20%L2D/DF	1/6	1/3	1/2	1/3	--
/SPF3	> 10-1/2 in.	rules[14]	5%2.0E/SPF	10%1.8E/SPF	1.4E/SPF	10%1.8E/SPF	5%L2D/DF	1/6	1/3	1/2	1/3	--
EWS 24F-E	≤ 12 in.	Special	20%2.0E/SPF	10%1.8E/SPF	1.4E/SPF	10%1.8E/SPF	20%L2D/DF	1/6	1/3	1/2	1/3	--
/SPF4	> 12 in.	rules[14]	20%2.0E/SPF	10%1.8E/SPF	1.4E/SPF	10%1.8E/SPF	20%L2D/DF	1/6	1/3	1/2	1/3	--
The following combination is **NOT BALANCED** and shall be applicable to members having a depth of 9-1/2, 11-7/8, 14, and 16 inches.												
EWS 26F-E /DF1	9-1/2 in.	302-22	1-2.3E/DF + 1-L1/DF	1-L2/DF	1-L3/DF	1-L2/DF	2-L1/DF	1/6	--	--	--	--
	11-7/8 in.	302-24	2-2.3E/DF + 1-L1/DF	1-L2/DF	1-L3/DF	1-L2/DF	2-L1/DF	1/6	--	--	--	--
	14 in.	302-24	2-2.3E/DF + 1- L1/DF	1-L2/DF	3-L3/DF	1-L2/DF	2-L1/DF	1/6	--	--	--	--
	16 in.	302-26	3-2.3E/DF + 1- L1/DF	1-L2/DF	3-L3/DF	1-L2/DF	2-L1/DF	1/6	--	--	--	--
The following combinations are **BALANCED** and are intended for members continuous or cantilevered over supports and provide equal capacity in both positive and negative bending (Footnote No. 3 applies both top and bottom of bending members).												
EWS	< 12 in.	302-20	10%B/ES	15%C4/ES	D/ES	15%C4/ES	10%B/ES	--	--	--	--	--
20F-E8M1	12 to 13-1/2 in.	302-22	10%B/ES	15%C4/ES	D/ES	15%C4/ES	10%B/ES	--	--	--	--	--
/WS	> 13-1/2 to 19-1/2 in.	302-22	20%B/ES	10%C4/ES	D/ES	10%C4/ES	20%B/ES	--	--	--	--	--
	> 19-1/2 to 24 in.	302-24	15%B/ES	10%C4/ES	D/ES	10%C4/ES	15%B/ES	--	--	--	--	--
	> 24 in.	302-24	15%B/ES	10%C4/ES	D/ES	10%C4/ES	15%B/ES	--	--	--	--	--
EWS 24F-E	≤ 10-1/2 in.	Special	20%2.0E/CSP	5%1.8E/CSP	1.4E/CSP	5%1.8E/CSP	20%2.0E/CSP	1/6	1/3	1/2	1/3	1/6
/CSP1	> 10-1/2 in.	rules[14]	5%2.0E/CSP	10%1.8E/CSP	1.4E/CSP	10%1.8E/CSP	5%2.0E/CSP	1/6	1/3	1/2	1/3	1/6
EWS 24F-E	≤ 12 in.	Special	20%2.0E/CSP	10%1.8E/CSP	1.4E/CSP	10%1.8E/CSP	20%2.0E/CSP	1/6	1/3	1/2	1/3	1/6
/CSP2	> 12 in.	rules[14]	20%2.0E/CSP	10%1.8E/CSP	1.4E/CSP	10%1.8E/CSP	20%2.0E/CSP	1/6	1/3	1/2	1/3	1/6
EWS 24F-E	≤ 10-1/2 in.	Special	20%2.0E/SPF	5%1.8E/SPF	1.4E/SPF	5%1.8E/SPF	20%2.0E/SPF	1/6	1/3	1/2	1/3	1/6
/SPF1	> 10-1/2 in.	rules[14]	5%2.0E/SPF	10%1.8E/SPF	1.4E/SPF	10%1.8E/SPF	5%2.0E/SPF	1/6	1/3	1/2	1/3	1/6

TABLE 4.5 Grade Requirements for Members Stressed Principally in Bending and Loaded Perpendicular to the Wide Faces of Laminations[1,2] (*Continued*)

Combination Symbol	Depth of Member	Tension Lam[3]	Minimum Grade of Laminations[4,5,6,7]									
			Percent/Grade/Species Each Zone[8]					Percent/Slope of Grain[9]				
			Outer Tension Zone	Inner Tension Zone	Core	Inner Comp. Zone	Outer Comp. Zone	Outer Tension Zone	Inner Tension Zone	Core	Inner Comp. Zone	Outer Comp. Zone
1	2	3	4	5	6	7	8	9	10	11	12	13
E-RATED WESTERN SPECIES (WS)												
The following combination is BALANCED and is intended for members continuous or cantilevered over supports and provide equal capacity in both positive and negative bending (Footnote No. 3 applies both top and bottom of bending members).												
EWS 24F-E /SPF2	≤ 12 in.	Special rules[14]	20%2.0E/SPF	10%1.8E/SPF	1.4E/SPF	10%1.8E/SPF	20%2.0E/SPF	1/6	1/3	1/2	1/3	1/6
	> 12 in.		20%2.0E/SPF	10%1.8E/SPF	1.4E/SPF	10%1.8E/SPF	20%2.0E/SPF	1/6	1/3	1/2	1/3	1/6
The following combination is BALANCED and shall be applicable to members having a depth of 7-1/2, 9, 9-1/2, 11-7/8, and 14 inches, and a width of 1-1/2 through 3-1/2 inches.												
EWS 20F-E[19] /SPF1	7-1/2 in.	--	1-B/SPF	1-C4/SPF	6-D/SPF	1-C4/SPF	1-B/SPF	--	--	--	--	--
	9 in.	--	1-B/SPF	1-C4/SPF	8-D/SPF	1-C4/SPF	1-B/SPF	--	--	--	--	--
	9-1/2 in.	--	2-B/SPF	1-C4/SPF	7-D/SPF	1-C4/SPF	2-B/SPF	--	--	--	--	--
	11-7/8 in.	--	2-B/SPF	2-C4/SPF	8-D/SPF	2-C4/SPF	2-B/SPF	--	--	--	--	--
	14 in.	--	3-B/SPF	2-C4/SPF	9-D/SPF	2-C4/SPF	3-B/SPF	--	--	--	--	--
The following combination is BALANCED and shall be applicable to members having a depth of 9-1/2, 11-7/8, 14, and 16 inches.												
EWS 26F-E /DF1M1	9-1/2 in.	302-22	1-2.3E/DF + 1-L1/DF	1-L2/DF	1-L3/DF	1-L2/DF	1-2.3E/DF + 1-L1/DF	1/6	--	--	--	1/6
	11-7/8 in.	302-24	2-2.3E/DF + 1-L1/DF	--	2-L2/DF	--	2-2.3E/DF + 1-L1/DF	1/6	--	--	--	1/6
	14 in.	302-24	2-2.3E/DF + 1-L1/DF	1-L2/DF	2-L3/DF	1-L2/DF	2-2.3E/DF + 1-L1/DF	1/6	--	--	--	1/6
	16 in.	302-26	3-2.3E/DF + 1-L1/DF	1-L2/DF	1-L3/DF	1-L2/DF	3-2.3E/DF + 1-L1/DF	1/6	--	--	--	1/6
VISUALLY GRADED SOUTHERN PINE (SP)												
The following combinations are NOT BALANCED and are intended primarily for simple-span applications.												
EWS 24F-V1/SP	< 12 in.	302-20[13]	10%N1D/SP	10%N2D/SP	N3M/SP	10%N2D/SP	10%N1D/SP	10% 1:14	10% 1:8	1:8	10% 1:8	10% 1:10
	12 to 15 in.	302-22	15%N1D/SP	15%N2M/SP	N3M/SP	15%N2M/SP	15%N1M/SP[10]	5% 1:14 + 10% 1:10	15% 1:8	1:8	15% 1:8	5% 1:12 + 10% 1:10
	> 15 in.	302-24	15%N1D/SP	15%N2M/SP	N3M/SP	15%N2M/SP	15%N1M/SP[10]	5% 1:16 + 5% 1:12 + 5% 1:10	15% 1:8	1:8	15% 1:8	5% 1:14 + 5% 1:12 + 5% 1:10
EWS 24F-V3/SP	< 12 in.	302-20[13]	10%N1D/SP	15%N2D/SP	N2M/SP	10%N2D/SP	10%N1D/SP	10% 1:14	15% 1:8	1:8	10% 1:8	10% 1:10
	12 to 15 in.	302-22	10%N1D/SP	15%N2D/SP	N2M/SP	10%N2D/SP	10%N1D/SP	10% 1:14	15% 1:8	1:8	10% 1:8	10% 1:12
	> 15 in.	302-24	10%N1D/SP	15%N2D/SP	N2M/SP	10%N2D/SP	10%N1D/SP	5% 1:14 + 5% 1:12	15% 1:8	1:8	10% 1:8	5% 1:12 + 5% 1:10
EWS 26F-V1/SP	7 lams to < 12 in.	302-22	10%N1D/SP	15%N1D/SP	N1M/SP	10%N1D/SP	10%N1D/SP	10% 1:14	15% 1:8	1:8	10% 1:8	10% 1:10
	12 to 15 in.	302-24	10%N1D/SP	15%N1D/SP	N2D/SP	10%N1D/SP	10%N1D/SP	10% 1:14	15% 1:8	1:8	10% 1:8	10% 1:10
	> 15 in.	302-26	10%N1D/SP	15%N2D/SP	N2M/SP	10%N2D/SP	10%N1D/SP	10% 1:14	15% 1:8	1:8	10% 1:8	10% 1:12
EWS 26F-V2/SP	7 lams to < 12 in.	302-22	10%N1D/SP	15%N1D/SP	N2D/SP	10%N1D/SP	10%N1D/SP	10% 1:14	15% 1:8	1:8	10% 1:8	10% 1:10
	12 to 15 in.	302-24	10%N1D/SP	15%N1D/SP	N2D/SP	10%N1D/SP	10%N1D/SP	10% 1:14	15% 1:8	1:8	10% 1:8	10% 1:10
	> 15 in.	302-26	10%N1D/SP	15%N1D/SP	N2D/SP	15%N1D/SP	10%N1D/SP	10% 1:14	15% 1:8	1:8	15% 1:8	10% 1:12

Combination Symbol	Depth of Member	Tension Lam[3]	Minimum Grade of Laminations[4,5,6,7]									
			Percent/Grade/Species Each Zone[8]					Percent/Slope of Grain[9]				
			Outer Tension Zone	Inner Tension Zone	Core	Inner Comp. Zone	Outer Comp. Zone	Outer Tension Zone	Inner Tension Zone	Core	Inner Comp. Zone	Outer Comp. Zone
1	2	3	4	5	6	7	8	9	10	11	12	13
VISUALLY GRADED SOUTHERN PINE (SP)												
The following combination is <u>NOT BALANCED</u> and is intended primarily for simple-span applications.												
EWS 26F-V3/SP	7 lams to < 12 in. 12 to 15 in. > 15 in.	302-22 302-24 302-26	10%N1D/SP 10%N1D/SP 10%N1D/SP	15%N1D/SP 15%N1D/SP 15%N1D/SP	N1M/SP N1M/SP N1M/SP	10%N1D/SP 10%N1D/SP 15%N1D/SP	10%N1D/SP 10%N1D/SP 10%N1D/SP	10% 1:14 10% 1:14 10% 1:14	15% 1:8 15% 1:8 15% 1:8	1:8 1:8 1:8	10% 1:8 10% 1:8 15% 1:8	10% 1:10 10% 1:10 10% 1:12
The following combinations are <u>NOT BALANCED</u> and are intended for straight or slightly cambered members for dry use and industrial appearance, and shall be applicable to members having a depth of 9 to 20 laminations.												
EWS 24F-V3M1[15] /SP	9 lams to ≤ 15 in. > 15 in. to 20 lams	302-22 302-24	10%N1D/SP 10%N1D/SP	15%N2D/SP 15%N2D/SP	N2M/SP N2M/SP	10%N2D/SP 10%N2D/SP	10%N1D/SP 10%N1D/SP	10% 1:14 5% 1:14 +5% 1:12	15% 1:8 15% 1:8	1:8 1:8	10% 1:8 10% 1:8	10% 1:12 5% 1:12 +5% 1:10
EWS 24F-V3M2[16] 225Fvx/SP	9 lams to ≤ 15 in. > 15 in. to 20 lams	302-20[13] 302-24	10%N1D/SP 10%N1D/SP	15%N2D/SP 15%N2D/SP	N2M/SP N2M/SP	10%N2D/SP 10%N2D/SP	10%N1D/SP 10%N1D/SP	10% 1:14 5% 1:14 +5% 1:12	15% 1:8 15% 1:8	1:8 1:8	10% 1:8 10% 1:8	10% 1:10 5% 1:12 +5% 1:10
EWS 26F-V3M1[15]/SP	9 lams to ≤ 15 in. > 15 in. to 20 lams	302-24 302-26	10%N1D/SP 10%N1D/SP	15%N1D/SP 15%N1D/SP	N1M/SP N1M/SP	10%N1D/SP 15%N1D/SP	10%N1D/SP 10%N1D/SP	10% 1:14 10% 1:14	15% 1:8 15% 1:8	1:8 1:8	10% 1:8 15% 1:8	10% 1:10 10% 1:12
EWS 26F-V3M2[16] 225Fvx/SP	9 lams to ≤ 15 in. > 15 in. to 20 lams	302-24 302-26	10%N1D/SP 10%N1D/SP	15%N1D/SP 15%N1D/SP	N1M/SP N1M/SP	10%N1D/SP 15%N1D/SP	10%N1D/SP 10%N1D/SP	10% 1:14 10% 1:14	15% 1:8 15% 1:8	1:8 1:8	10% 1:8 15% 1:8	10% 1:10 10% 1:12
The following combinations are <u>BALANCED</u> and are intended for members continuous or cantilevered over supports and provide equal capacity in both positive and negative bending (Footnote No. 3 applies both top and bottom of bending members).												
EWS 24F-V5/SP	< 12 in. 12 to 15 in. > 15 in.	302-20[13] 302-22 302-24	10%N1D/SP 10%N1D/SP 10%N1D/SP	5%N2D/SP 5%N2D/SP 5%N2D/SP	N2M/SP N2M/SP N2M/SP	5%N2D/SP 5%N2D/SP 5%N2D/SP	10%N1D/SP 10%N1D/SP 10%N1D/SP	10% 1:14 10% 1:15 5% 1:16 +5% 1:12	5% 1:8 5% 1:10 5% 1:10	1:8 1:8 1:8	5% 1:8 5% 1:10 5% 1:10	10% 1:14 10% 1:15 5% 1:16 +5% 1:12
EWS 24F-V5M1 1.8E/SP	< 12 in. 12 to 15 in. > 15 in.	302-20 302-22 302-24	10%N1D/SP 10%N1D/SP 10%N1D/SP	15%N2D/SP 15%N2D/SP 15%N2D/SP	N2M/SP N2M/SP N2M/SP	15%N2D/SP 15%N2D/SP 15%N2D/SP	10%N1D/SP 10%N1D/SP 10%N1D/SP	5% 1:14 +5% 1:10 5% 1:16 +5% 1:10 5% 1:16 +5% 1:12	15% 1:8 15% 1:8 15% 1:8	1:8 1:8 1:8	15% 1:8 15% 1:8 15% 1:8	5% 1:14 +5% 1:10 5% 1:16 +5% 1:10 5% 1:16 +5% 1:12
EWS 26F-V4/SP	7 lams to < 12 in. 12 to 15 in. > 15 in.	302-22 302-24 302-26	10%N1D/SP 10%N1D/SP 10%N1D/SP	15%N1D/SP 15%N1D/SP 15%N1D/SP	N1M or N2D/SP N1M or N2D/SP N1M or N2D/SP	15%N1D/SP 15%N1D/SP 15%N1D/SP	10%N1D/SP 10%N1D/SP 10%N1D/SP	10% 1:14 10% 1:14 10% 1:14	15% 1:8 15% 1:8 15% 1:8	1:8 1:8 1:8	15% 1:8 15% 1:8 15% 1:8	10% 1:14 10% 1:14 10% 1:14
The following combinations are <u>BALANCED</u> and are intended for straight or slightly cambered members for dry use and industrial appearance, and shall be applicable to members having a depth of 9 to 20 laminations.												
EWS 24F-V5M2[15] 1.8E/SP	9 lams to ≤ 15 in. > 15 in. to 20 lams	302-22 302-24	10%N1D/SP 10%N1D/SP	15%N2D/SP 15%N2D/SP	N2M/SP N2M/SP	15%N2D/SP 15%N2D/SP	10%N1D/SP 10%N1D/SP	5% 1:16 +5% 1:10 5% 1:16 +5% 1:12	15% 1:8 15% 1:8	1:8 1:8	15% 1:8 15% 1:8	5% 1:16 +5% 1:10 5% 1:16 +5% 1:12

TABLE 4.5 Grade Requirements for Members Stressed Principally in Bending and Loaded Perpendicular to the Wide Faces of Laminations[1,2] *(Continued)*

Combination Symbol	Depth of Member	Tension Lam[3]	Minimum Grade of Laminations[4,5,6,7]									
			Percent/Grade/Species Each Zone[8]					Percent/Slope of Grain[9]				
			Outer Tension Zone	Inner Tension Zone	Core	Inner Comp. Zone	Outer Comp. Zone	Outer Tension Zone	Inner Tension Zone	Core	Inner Comp. Zone	Outer Comp. Zone
1	2	3	4	5	6	7	8	9	10	11	12	13
VISUALLY GRADED SOUTHERN PINE (SP)												
The following combinations are BALANCED and are intended for straight or slightly cambered members for dry use and industrial appearance, and shall be applicable to members having a depth of 9 to 20 laminations.												
EWS 24F-V5M3[16] 1.8E 225Fvx/SP	9 lams to ≤ 15 in.	302-22	10%N1D/SP	15%N2D/SP	N2M/SP	15%N2D/SP	10%N1D/SP	5% 1:16 + 5% 1:10	15% 1:8	1:8	15% 1:8	5% 1:16 + 5% 1:10
	> 15 in. to 20 lams	302-24	10%N1D/SP	15%N2D/SP	N2M/SP	15%N2D/SP	10%N1D/SP	5% 1:16 + 5% 1:12	15% 1:8	1:8	15% 1:8	5% 1:16 + 5% 1:12
EWS 26F-V4M1[15] /SP	9 lams to ≤ 15 in.	302-24	10%N1D/SP	15%N1D/SP	N1M or N2D/SP	15%N1D/SP	10%N1D/SP	10% 1:14	15% 1:8	1:8	15% 1:8	10% 1:14
	> 15 in. to 20 lams	302-26	10%N1D/SP	15%N1D/SP	N1M or N2D/SP	15%N1D/SP	10%N1D/SP	10% 1:14	15% 1:8	1:8	15% 1:8	10% 1:14
EWS 26F-V4M2[16] 225Fvx/SP	9 lams to ≤ 15 in.	302-24	10%N1D/SP	15%N1D/SP	N1M or N2D/SP	15%N1D/SP	10%N1D/SP	10% 1:14	15% 1:8	1:8	15% 1:8	10% 1:14
	> 15 in. to 20 lams	302-26	10%N1D/SP	15%N1D/SP	N1M or N2D/SP	15%N1D/SP	10%N1D/SP	10% 1:14	15% 1:8	1:8	15% 1:8	10% 1:14
E-RATED SOUTHERN PINE (SP)												
The following combinations are NOT BALANCED and are intended primarily for simple-span applications.												
EWS 28F-E1[12]/SP	≤ 13-3/4 in.	302-24[12]	10%2.3E[12] + 10% N1D 2.3E[12]/SP	10%N1D[12] /SP	N2M/SP	10%N1D[12] /SP	10%N1D 2.3E[12]/SP	1/3 & 10% 1:12 + 10% 1:12	10% 1:12	1:8	10% 1:12	10% 1:12
	> 13-3/4 in.	302-24[12]	5%2.3E[12] + 5% N1D 2.3E[12]/SP	15%N1D[12] /SP	N2M/SP	15%N1D[12] /SP	10%N1D 2.3E[12]/SP	1/5 & 5% 1:16 + 5% 1:12	15% 1:12	1:8	15% 1:12	10% 1:12
EWS 28F-E1M1[12] /SP	≤ 13-3/4 in.	302-24[12]	10%2.3E[12] + 10% N1D 2.3E/SP	10%N1D[12] /SP	N2M/SP	15%N1D[12] /SP	10%N1D 2.3E[12]/SP	1/3 & 10% 1:12 + 10% 1:12	10% 1:12	1:8	15% 1:12	10% 1:12
	> 13-3/4 in.	302-24[12]	5%2.3E[12] + 5% N1D 2.3E[12]/SP	15%N1D[12] /SP	N2M/SP	15%N1D[12] /SP	15%N1D 2.3E[12]/SP	1/5 & 5% 1:16 + 5% 1:12	15% 1:12	1:8	15% 1:12	15% 1:12
The following combination is BALANCED and is intended for members continuous or cantilevered over supports and provides equal capacity in both positive and negative bending (Footnote No. 3 applies both top and bottom of bending members).												
EWS 28F-E2[12]/SP	≤ 13-3/4 in.	302-24[12]	10%2.3E[12] + 10% N1D 2.3E[12]/SP	10%N1D[12] /SP	N2M/SP	10%N1D[12] /SP	10%2.3E[12] + 10% N1D 2.3E[12]/SP	1/3 & 10% 1:12 + 10% 1:12	10% 1:12	1:8	10% 1:12	1/3 &10% 1:12 + 10% 1:12
	> 13-3/4 in.	302-24[12]	5%2.3E[12] + 5% N1D 2.3E[12]/SP	15%N1D[12] /SP	N2M/SP	15%N1D[12] /SP	5%2.3E[12] + 5% N1D 2.3E[12]/SP	1/5 & 5% 1:16 + 5% 1:12	15% 1:12	1:8	15% 1:12	1/5 &5% 1:16 + 5% 1:12

Combination Symbol	Depth of Member	Tension Lam[3]	Minimum Grade of Laminations[4,5,6,7]									
			Percent/Grade/Species Each Zone[8]					Percent/Slope of Grain[9]				
			Outer Tension Zone	Inner Tension Zone	Core	Inner Comp. Zone	Outer Comp. Zone	Outer Tension Zone	Inner Tension Zone	Core	Inner Comp. Zone	Outer Comp. Zone
1	2	3	4	5	6	7	8	9	10	11	12	13
E-RATED SOUTHERN PINE (SP)												
The following combination is **BALANCED** and is intended for members continuous or cantilevered over supports and provides equal capacity in both positive and negative bending (Footnote No. 3 applies both top and bottom of bending members).												
EWS 28F-E2M1[12] /SP	≤ 13-3/4 in.	302-24[12]	10%2.3E[12] + 10% N1D 2.3E/SP	10%N1D[12] /SP	N2M/SP	10%N1D[12] /SP	10%2.3E[12] + 10% N1D 2.3E/SP	1/3 & 10% 1:12 + 10% 1:12	10% 1:12	1:8	10% 1:12	1/3 &10% 1:12 + 10% 1:12
	> 13-3/4 in.	302-24[12]	5%2.3E[12] + 10% N1D 2.3E/SP	15%N1D[12] /SP	N2M/SP	15%N1D[12] /SP	5%2.3E[12] + 10% N1D 2.3E/SP	1/5 & 5% 1:16 + 10% 1:12	15% 1:12	1:8	15% 1:12	1/5 &5% 1:16 + 10% 1:12
The following combination is **NOT BALANCED** and shall be applicable to members having a nominal width of 6 inches or less.												
EWS 30F-E1[12]/SP	≤ 13-3/4 in.	302-24[12]	10%2.3E[12] + 10% N1D 2.3E[12]/SP	10%N1D[12] /SP	N2M/SP	20%N1D[12] /SP	10%N1D 2.3E[12]/SP	1/3 & 10% 1:12 + 10% 1:12	10% 1:12	1:8	20% 1:12	10% 1:12
	> 13-3/4 in.	302-24[12]	5%2.3E[12] + 5% N1D 2.3E[12]/SP	15%N1D[12] /SP	N2M/SP	15%N1D[12] /SP	15%N1D 2.3E[12]/SP	1/5 & 5% 1:16 + 5% 1:12	15% 1:12	1:8	15% 1:12	15% 1:12
The following combination is **NOT BALANCED** and shall be applicable to members having a nominal width of 6 inches or less, and a depth of 18 inches or less.												
EWS 30F-E1M1[12] /SP	≤ 13-3/4 in.	302-24[12]	10%2.3E[12] + 10% N1D 2.3E[12]/SP	10%N1D[12] /SP	N2M/SP	20%N1D[12] /SP	10%N1D 2.3E[12]/SP	1/3 & 10% 1:12 + 10% 1:12	10% 1:12	1:8	20% 1:12	10% 1:12
	> 13-3/4 in. to 18 in.	302-24[12]	5%2.3E[12] + 5% N1D 2.3E[12]/SP	15%N1D[12] /SP	N2M/SP	15%N1D[12] /SP	15%N1D 2.3E[12]/SP	1/5 & 5% 1:16 + 5% 1:12	15% 1:12	1:8	15% 1:12	15% 1:12
The following combination is **BALANCED** and shall be applicable to members having a nominal width of 6 inches or less.												
EWS 30F-E2[12]/SP	≤ 13-3/4 in.	302-24[12]	10%2.3E[12] + 10% N1D 2.3E[12]/SP	10%N1D[12] /SP	N2M/SP	10%N1D[12] /SP	10%2.3E[12] + 10% N1D 2.3E[12]/SP	1/3 & 10% 1:12 + 10% 1:12	10% 1:12	1:8	10% 1:12	1/3 &10% 1:12 + 10% 1:12
	> 13-3/4 in.	302-24[12]	5%2.3E[12] + 5%N1D 2.3E[12]/SP	15%N1D[12] /SP	N2M/SP	15%N1D[12] /SP	5%2.3E[12] + 5% N1D 2.3E[12]/SP	1/5 & 5% 1:16 + 5% 1:12	15% 1:12	1:8	15% 1:12	1/5 & 5% 1:16 + 5% 1:12
The following combination is **BALANCED** and shall be applicable to members having a nominal width of 6 inches or less, and a depth of 18 inches or less.												
EWS 30F-E2M1[12] /SP	≤ 13-3/4 in.	302-24[12]	10%2.3E[12] + 10% N1D 2.3E[12]/SP	10%N1D[12] /SP	N2M/SP	10%N1D[12] /SP	10%2.3E[12] + 10% N1D 2.3E[12]/SP	1/3 & 10% 1:12 + 10% 1:12	10% 1:12	1:8	10% 1:12	1/3 &10% 1:12 + 10% 1:12
	> 13-3/4 in. to 18 in.	302-24[12]	5%2.3E[12] + 10% N1D 2.3E[12]/SP	15%N1D[12] /SP	N2M/SP	15%N1D[12] /SP	5%2.3E[12] + 10% N1D 2.3E[12]/SP	1/5 & 5% 1:16 + 10% 1:12	15% 1:12	1:8	15% 1:12	1/5 & 5% 1:16 + 10% 1:12

TABLE 4.5 Grade Requirements for Members Stressed Principally in Bending and Loaded Perpendicular to the Wide Faces of Laminations[1,2] *(Continued)*

Combination Symbol	Depth of Member	Tension Lam[3]	Minimum Grade of Laminations[4,5,6,7] — Percent/Grade/Species Each Zone[8]: Outer Tension Zone	Inner Tension Zone	Core	Inner Comp. Zone	Outer Comp. Zone	Percent/Slope of Grain[9]: Outer Tension Zone	Inner Tension Zone	Core	Inner Comp. Zone	Outer Comp. Zone
1	2	3	4	5	6	7	8	9	10	11	12	13
E-RATED SOUTHERN PINE (SP)												
The following combinations is BALANCED and shall be applicable to members having a depth of 16 inches or less.												
EWS 30F-E2M2 2.1E 650F_{cp}/SP	9-1/2 in.	--	1-1.50 in. 2.4E LVL[20]	1-1.30 in. N1D2.3E/SP	3-1.30 in. N2M/SP	1-1.30 in. N1D2.3E/SP	1-1.50 in. 2.4E LVL[20]	--	1:12	1:8	1:12	--
	11-7/8 in.	--	1-1.75 in. 2.4E LVL[20]	1-1.40 in. N1D2.3E/SP	4-1.40 in. N2M/SP	1-1.40 in. N1D2.3E/SP	1-1.75 in. 2.4E LVL[20]	--	1:12	1:8	1:12	--
	14 in.	--	1-1.50 in. 2.4E LVL[20]	1-1.50 in. N1D2.3E/SP[21]	6-1.34 in. N2M/SP	1-1.50 in. N1D2.3E/SP[21]	1-1.50 in. 2.4E LVL[20]	--	1:12	1:8	1:12	--
	16 in.	--	1-1.75 in. 2.4E LVL[20]	1-1.50 in. N1D2.3E/SP[21]	7-1.36 in. N2M/SP	1-1.50 in. N1D2.3E/SP[21]	1-1.75 in. 2.4E LVL[20]	--	1:12	1:8	1:12	--
The following combinations is BALANCED and shall be applicable to members having a depth of 18 inches or less.												
EWS 30F-E2M3 2.1E 510F_{cp}/SP	9-1/2 in.	--	1-1.375 in. 2.4E LVL[22]	1-1.35 in. N1D2.3E/SP	3-1.35 in. N2M/SP	1-1.35 in. N1D2.3E/SP	1-1.375 in. 2.4E LVL[22]	--	1:12	1:8	1:12	--
	11-7/8 in.	--	1-1.75 in. 2.4E LVL[22]	1-1.40 in. N1D2.3E/SP	4-1.40 in. N2M/SP	1-1.40 in. N1D2.3E/SP	1-1.75 in. 2.4E LVL[22]	--	1:12	1:8	1:12	--
	14 in.	--	1-1.75 in. 2.4E LVL[22]	1-1.31 in. N1D2.3E/SP	6-1.31 in. N2M/SP	1-1.31 in. N1D2.3E/SP	1-1.75 in. 2.4E LVL[22]	--	1:12	1:8	1:12	--
	16 in.	--	1-1.75 in. 2.4E LVL[21]	1-1.50 in. N1D2.3E/SP	7-1.36 in. N2M/SP	1-1.50 in. N1D2.3E/SP	1-1.75 in. 2.4E LVL[22]	--	1:12	1:8	1:12	--
	18 in.	--	1-1.75 in. 2.4E LVL[22]	2-1.32 in. N1D2.3E/SP	7-1.32 in. N2M/SP	2-1.32 in. N1D2.3E/SP	1-1.75 in. 2.4E LVL[22]	--	1:12	1:8	1:12	--
E-RATED DURANGO PINE (DP)												
The following combinations are NOT BALANCED and are intended primarily for simple-span applications.												
EWS 16F-E1/DP	< 12 in.	--	10%2.1E/DP	--	N2M/DP	--	10%1.9E/DP[18]	1/6	--	--	--	1/2
	12 to 15 in.	--	10%2.1E/DP	--	N2M/DP	--	10%1.9E/DP[18]	1/6	--	--	--	1/2
	< 12 in.	302-20	10%1.9E/DP[18]	--	N2M/DP	--	10%1.9E/DP[18]	1/6	--	--	--	1/2
	12 to 15 in.	302-20	10%1.9E/DP[18]	--	N2M/DP	--	10%1.9E/DP[18]	1/6	--	--	--	1/2
	> 15 in.	302-20	10%1.9E/DP[18]	--	N2M/DP	--	5%1.9E/DP[18]	1/6	--	--	--	1/2
EWS 20F-E1/DP	< 12 in.	--	20%2.1E/DP	--	N2M/DP	--	10%2.1E/DP	1/6	--	--	--	1/2
	12 to 15 in.	--	15%2.1E/DP	15%1.9E/DP	N2M/DP	--	10%2.1E/DP	1/6	1/6	--	--	1/2
	< 12 in.	302-20	10%2.1E/DP	--	N2M/DP	--	10%1.9E/DP[18]	1/6	--	--	--	1/2
	12 to 15 in.	302-20	10%2.1E/DP	10%1.9E/DP	N2M/DP	--	10%1.9E/DP[18]	1/6	1/6	--	--	1/2
	> 15 in. to 19-1/2 in.	302-20	5%2.1E/DP	5%1.9E/DP	N2M/DP	--	5%1.9E/DP[18]	1/6	1/6	--	--	1/2
	> 19-1/2 in.	302-22	15%1.9E/DP[18]	--	N2M/DP	--	15%1.9E/DP[18]	1/6	--	--	--	1/2
EWS 24F-E1/DP	< 12 in.	302-20	10%2.1E/DP	20%1.9E/DP	N2M/DP	--	10%2.1E/DP	1/6	1/6	--	--	1/2
	12 to 15 in.	302-22	10%2.1E/DP	20%1.9E/DP	N2M/DP	--	10%2.1E/DP	1/6	1/6	--	--	1/2
	> 15 in.	302-24	10%2.1E/DP	10%1.9E/DP	N2M/DP	10%1.9E/DP	10%1.9E/DP[18]	1/6	1/6	--	1/2	1/2

Combination Symbol	Depth of Member	Tension Lam[3]	Minimum Grade of Laminations[4,5,6,7]									
			Percent/Grade/Species Each Zone[8]					Percent/Slope of Grain[9]				
			Outer Tension Zone	Inner Tension Zone	Core	Inner Comp. Zone	Outer Comp. Zone	Outer Tension Zone	Inner Tension Zone	Core	Inner Comp. Zone	Outer Comp. Zone
1	2	3	4	5	6	7	8	9	10	11	12	13
E-RATED DURANGO PINE (DP)												
The following combinations are BALANCED and are intended for members continuous or cantilevered over supports and provide equal capacity in both positive and negative bending (Footnote No. 3 applies both top and bottom of bending members).												
EWS 16F-E3/DP	< 12 in.	--	10%1.9E/DP[10]	--	N2M/DP	--	10%1.9E/DP[10]	1/6	--	--	--	1/6
	12 to 15 in.	--	10%1.9E/DP[10]	--	N2M/DP	--	10%1.9E/DP[10]	1/6	--	--	--	1/6
	> 15 in.	302-20	10%1.9E/DP[10]	--	N2M/DP	--	10%1.9E/DP[10]	1/6	--	--	--	1/6
EWS 20F-E3/DP	< 12 in.	--	10%2.1E/DP	10%1.9E/DP	N2M/DP	10%1.9E/DP	10%2.1E/DP	1/6	1/6	--	1/6	1/6
	12 to 15 in.	--	10%2.1E/DP	10%1.9E/DP	N2M/DP	10%1.9E/DP	10%2.1E/DP	1/6	1/6	--	1/6	1/6
	< 12 in.	302-20	10%1.9E/DP[10]	10%1.9E/DP	N2M/DP	10%1.9E/DP	10%1.9E/DP[10]	1/6	1/6	--	1/6	1/6
	12 to 15 in.	302-20	10%1.9E/DP[10]	10%1.9E/DP	N2M/DP	10%1.9E/DP	10%1.9E/DP[10]	1/6	1/6	--	1/6	1/6
	> 15 in. to 19-1/2 in.	302-20	5%2.1E/DP	5%1.9E/DP	N2M/DP	5%1.9E/DP	5%2.1E/DP	1/6	1/6	--	1/6	1/6
	> 19-1/2 in.	302-22	10%1.9E/DP[10]	5%1.9E/DP	N2M/DP	5%1.9E/DP	10%1.9E/DP[10]	1/6	1/6	--	1/6	1/6
EWS 24F-E4/DP	< 12 in.	302-20	10%2.1E/DP	20%1.9E/DP	N2M/DP	20%1.9E/DP	10%2.1E/DP	1/6	1/6	--	1/6	1/6
	12 to 15 in.	302-22	10%2.1E/DP	20%1.9E/DP	N2M/DP	20%1.9E/DP	10%2.1E/DP	1/6	1/6	--	1/6	1/6
	> 15 in..	302-24	10%2.1E/DP	10%1.9E/DP	N2M/DP	10%1.9E/DP	10%2.1E/DP	1/6	1/6	--	1/6	1/6

[1]The combinations in this table are primarily applicable to members stressed in bending due to a load applied perpendicular to the wide faces of the laminations.

[2]The combinations are applicable to arches, compression members, tension members and bending members. For bending members, Footnote No. 3 applies. All combinations are applicable to members with four or more laminations. The tension lamination requirements in Footnote No. 3 do not apply to arches, compression members or tension members.

[3]In addition to the grade requirements tabulated for the outer tension zone, the grading restrictions as contained in AITC 302-24, 302-22 and 302-20 tension lamination requirements are applicable to the outer 5 percent of the total depth of bending members. These tension lamination requirements are shown in Column 3. E-rated tension laminations conforming to C1.3, C2.3 or C3.3 of Annex C in AITC 117--Manufacturing are permitted to be used with visually graded combinations provided Douglas fir-larch and southern pine laminations have a 2.0E and hem-fir laminations have a 1.8E. For EWS 20F-V4 and EWS 20F-V8 WS from 4 lams to 15 inches in depth, the required 302-20 tension laminations used with L1/DF outer tension zone laminations need not be upgraded to 1:14 slope of grain. Where 302-20 tension laminations are used in conjunction with L2, L2D or L3 outer tension zone laminations, the required slope of grain is 1:12. The 302- tension lamination required for some bending members is permitted to be omitted provided Footnote No. 7 to Table 4-3 applies. This reduction does not apply to arches which do not require special tension laminations.

[4]Percent values are based on the total depth of the member. All fractional numbers of laminations shall be rounded upward to the next whole number. For the inner tension and compression zones, the resulting excess of percentage resulting from rounding upward of the outer zone is permitted to be subtracted from the inner zone requirements. The actual depth of the member shall be used to determine the tension lamination requirements from Column 3. In no case shall the tension lamination requirements in Footnote No. 3 be less than 5 percent of the total depth of the member in inches.

[5]The following substitutions of E-Rated Douglas fir-larch lumber are permitted for EWS 24F-V4 and EWS 24F-V8 WS with at least 7 laminations in depth:

Douglas Fir-Larch Visual Grade	Member Location	Douglas Fir-Larch E-Rated Substitution Grades
L1	Tension Side	2.2E-2 (2.2E-3 in outer tension zone)
	Comp. Side	2.2E-2, 2.0E-3
L2D	Tension Side	2.2E-2 (2.2E-3 in outer tension zone), 2.0E-3
	Comp. Side	1.9E-2*
L2	Tension Side	1.9E-6, 2.0E-3
	Comp. Side	1.9E-2

* 7- and 8-lam beams shall be 1.9E-6. The substitution of 1.9E-2 for an L2D does not apply to the outermost compression lam.

[6]The combinations in this table have been established based on procedures given in ASTM D 3737 as modified by subsequent research.

[7]Where specified to have an extreme fiber in bending stress on the compression side which results in tension on the compression (top) side greater than the value given in Column 4, Table 4-3 (except for balanced combinations but not exceeding 200 psi higher than the value in Column 4) tension zone end joint spacing restrictions shall be applied to both the tension and compression zones.

[8]Grade designations are as follows:

Visually Graded - Western Species
- L1 is L1 laminating grade (dense for Douglas fir-larch and Douglas fir south)
- L1D is L1 dense laminating grade for hem-fir and Alaska cedar
- L1S is a special grade of Alaska cedar
- L1CL is L1 close grain laminating grade
- L2D is L2 dense laminating grade (dense)
- L2 is L2 laminating grade (medium grade)
- L3 is L3 laminating grade (medium grade for Douglas fir-larch, Douglas fir south and hem-fir)

Visually Graded - Southern Pine or Durango Pine
- N1D is No. 1 dense structural joists and planks or structural light framing grade or No. 1 boards both graded as dense
- N2D is No. 2 dense structural joists and planks or structural light framing grade or No. 2 boards both graded as dense
- N1M is No. 1 structural joists and planks or structural light framing grade or No. 1 boards both with medium grain rate of growth
- N2M is No. 2 structural joists and planks or structural light framing grade or No. 2 boards with a medium grain rate of growth
- N3M is No. 3 structural joists and planks or structural light framing grade or No. 3 boards both with a medium grain rate of growth

E-Rated Grades - All Species (examples)
- 2.0E-6 has 2.0E with 1/6 edge characteristic
- 1.8E-3 has 1.8E with 1/3 edge characteristic
- 1.4E-2 has 1.4E with 1/2 edge characteristic

E-Rated Grades - Eastern Spruce
- B/ES has a minimum long-span E of 1.55×10^6 psi
- C6/ES has a minimum long-span E of 1.6×10^6 psi
- C4/ES has a minimum long-span E of 1.4×10^6 psi
- D4/ES has a minimum long-span E of 1.4×10^6 psi
- D/ES has no minimum long-span E requirement

These Eastern Spruce referenced herein shall apply to the following species grown in the United States or Canada: White spruce, Black spruce, and Red spruce. In addition to the minimum long-span E given above, these laminating lumber shall be graded in accordance with the requirements in CSA Standard O122.

[9]Where slope of grain is not tabulated, it shall be the slope of grain required for the grade. Slope of grain is not specified for E-rated lumber except for tension laminations, but slope near the ends of the piece shall not be steeper than slopes of grain in the remainder of the piece.

[10]When required to have 650 psi compression perpendicular to grain design value for Douglas fir-larch or 740 psi for southern pine, at least one 2-inch nominal thickness lamination of dense Douglas fir-larch for western species, or dense southern pine for southern pine species, shall be used in place of the tabulated lamination in the bearing area, provided the next inner lamination is medium grain Douglas fir-larch or southern pine.

[11]For Western species combinations EWS 16F-V3, EWS 20F-V4, and EWS 20F-V8, the next inner 5 percent of the outermost tension laminations are required to be L2D/DF for the same conditions indicated by Footnote No. 3. The excess of percentage of the outer 5 percent required by Footnote No. 3 resulting from rounding upward is permitted to be subtracted from the next inner 5 percent required by this footnote.

TABLE 4.5 Grade Requirements for Members Stressed Principally in Bending and Loaded Perpendicular to the Wide Faces of Laminations[1,2] (*Continued*)

[12]For the manufacture of 28F and 30F (limited to a nominal beam width of 6" or less) southern pine members, quality control procedures for daily QC monitoring of the average and minimum MOE of the E-rated grades shall be established. End joints for the tension laminations shall be qualified at 1.67 x 3,000 = 5,010 psi. Following initial qualification, daily QC shall be maintained through the use of a statistical process control methodology. The visually-graded and E-rated laminations shall meet the following requirements:

Grade	Grade Requirements
2.3E-5&16 Tension Lam	Must meet all requirements for 302-24 tension lam; Average MOE ≥ 2.3 x 10^6 psi with no piece less than 1.96 x 10^6 psi; Edge characteristics ≤ 20%; Centerline characteristics ≤ 25%; Slope of grain ≤ 1/16; Both ends shall be dense.
2.3E-3&12 Tension Lam	Must meet all requirements for 302-24 tension lam; Average MOE ≥ 2.3 x 10^6 psi with no piece less than 1.96 x 10^6 psi; Edge characteristics ≤ 33%; Centerline characteristics ≤ 33%; Slope of grain ≤ 1/12; Both ends shall be dense.
N1D 2.3E	Average MOE ≥ 2.3 x 10^6 psi with no piece less than 1.96 x 10^6 psi; Slope of grain ≤ 1/12.
N1D	Average MOE ≥ 2.0 x 10^6 psi with no piece less than 1.67 x 10^6 psi; Slope of grain ≤ 1/12.

[13]The 302 grade tension laminations are included in C1, C2, and C3 of AITC 117–Manufacturing. When used in the indicated depth range with this combination, the laminating lumber shall have a slope of grain not steeper than that shown in Column 9 for the outer tension zone. This footnote applies to the 302-20 requirement of EWS 24F-V1, EWS 24F-V3, and EWS 24F-V5 southern pine.

[14]In addition, the 2.0E lam material shall be visually graded in accordance with provisions of Section C14.2, AITC 117-Manufacturing (alternate provisions for 302 tension lam) with the exception that general slope of grain restrictions are not applicable.

[15]This combination contains wane. Therefore, it shall be used in dry conditions only and the wet-use factors shall not be applied. Wane lumber is allowed for use with the following restrictions: (a) Maximum wane is 1/6 of the finished member width; (b) No wane is allowed for the outer top and bottom lams; (c) No wane is allowed for the 302 tension lams; (d) No wane is allowed in the central 40% of the member depth; (e) Maximum wane is 1/2 the lam thickness for No. 1 or L1 and 2/3 the thickness for No. 2 or L2; (f) Wane is allowed only on one side of the finished member; and (g) The first interior lam from the top or bottom shall have the wane located away from the outside lam.

[16]This combination contains wane. Therefore, it shall be used in dry conditions only and the wet-use factors shall not be applied. Wane lumber is allowed for use with the following restrictions: (a) Maximum wane is 1/6 of the finished member width; (b) No wane is allowed for the outer top and bottom lams; (c) No wane is allowed for the 302 tension lams; (d) Maximum wane is 1/2 the lam thickness for No. 1 or L1 and 2/3 the thickness for No. 2 or L2; (e) Wane is allowed only on one side of the finished member; and (f) The first interior lam from the top or bottom shall have the wane located away from the outside lam.

[17]The outer 5% tension lamination(s) shall have a slope of grain not steeper than 1:16.

[18]Where specified to have 650 psi compression perpendicular to grain design value for Douglas fir-Larch, or 740 psi for southern pine or Durango pine, at least one 2-inch nominal thickness lamination of Douglas fir-larch for western species combinations, southern pine for southern pine combinations, and Durango pine for Durango pine combinations having a modulus of elasticity (*E*) value 200,000 psi higher than the specified E values shall be used in the bearing area provided the next inner lamination is at least 1.9E modulus of elasticity.

[19]The layup requires the use of 1x laminations with a maximum nominal width of 4 inches. Only one ripping is permitted to achieve the specific beam width without varying the basic grade requirements of the full-width laminating lumber.

[20]2.4E LVL used in this layup combination is laminated veneer lumber with a minimum average long-span E (flatwise) of 2.4 x 10^6 psi and a characteristic tensile strength (5th percentile with 75% confidence) of 5,400 psi. The allowable compressive stress perpendicular to grain of the LVL shall not be less than 650 psi.

[21]The N1D2.3E/SP is permitted to be replaced by L1D2.3E/DF in this layup combination, which shall meet all requirements for dense L1/DF. In addition, the L1D2.3E/DF shall have a minimum average long-span E of 2.3 × 10^6 psi with no piece less than 1.96 × 10^6 psi. The slope of grain shall be no steeper than 1:14.

[22]2.4E LVL used in this layup is laminated veneer lumber with a minimum average long-span E (flatwise) of 2.4 x 10^6 psi and a characteristic tensile strength (5th percentile with 75% confidence) of 6,400 psi. The allowable compressive stress perpendicular to grain of the LVL shall not be less than 510 psi.

TABLE 4.6 Grade Requirements for Members with Two or More Laminations Stressed Principally in Bending Parallel to the Wide Faces of the Laminations[1,2,3,4]

Combination Number	Minimum Grade of Laminations[5]	Species	Tension Laminations Required[6]			Slope of Grain
			4 lams to < 12 in. deep	12 in. to 15 in. deep	> 15 in. deep	
Visually Graded Western Species (WS)						
EWS 1	L3	DF	302-20	302-20	302-20	1:8
EWS 2	L2	DF	302-20	302-20	302-20	1:12
EWS 3	L2D	DF	302-20	302-22	302-24	1:12
EWS 5	L1	DF	302-20	302-22	302-24	1:14
EWS 22[7]	L3	WW	302-20	302-20	302-20	1:8
EWS 70	L2	AC	302-20	302-20	302-20	1:12
Visually Graded Southern Pine (SP)						
EWS 47	N2M	SP	302-20	302-20	302-20	1:14[8]
EWS 48	N2D	SP	302-20	302-20	302-20	1:14[8]
EWS 49	N1M	SP	302-20	302-20	302-22	1:16[8]
EWS 50	N1D	SP	302-20	302-22	302-24	1:14[8]
The following combination is intended for dry use and industrial appearance[9], and shall be applicable to members having a depth of 9 to 20 laminations.						
EWS 49M1[10]	N1M[10]	SP	Not permitted	302-20	302-22	1:16[8]
The following combination is intended for dry use and industrial appearance[9], and shall be applicable to members having a depth of 4 to 20 laminations.						
EWS 49M2[11] 225Fvx/SP	N1M[11]	SP	302-20	302-20	302-22	1:16[8]

[1]The tension laminations required by this table are to be used only when the design values for bending about the x-x axis (F_{bx}) exceed those listed in Column 16 of Table 4-4, but are not greater than those listed in Column 17. These tension laminations are permitted to be omitted for bending members over 15 inches deep provided the design value in Column 17 is multiplied by 0.75. This reduction does not apply to arches which do not require special tension laminations.

[2]The tabulated combinations in this table are primarily intended for members loaded axially or in bending with the loads acting parallel to the wide faces of the laminations. The combinations are also permitted to be used as bending members loaded perpendicular to the wide faces of the laminations; however, combinations in Table 4-5 for four or more laminations are usually better suited for this condition of loading.

[3]It is not intended that these combinations be used for deep bending members which are loaded perpendicular to the wide faces of the laminations. If higher design values in bending about the x-x axis are specified for these combinations that those contained in Table 4-4, Column 16, see Footnote No. 1.

[4]The allowable wane permitted in some grades in Table 4-5 is not allowed for combinations in this table.

[5]Grade designations are the same as Footnote No. 8 to Table 4-5.

[6]The outer 5 percent of laminations on the tension side of bending members shall be replaced with the tension lamination shown in this table. Percent values are based on the total depth of members. Laminations of different thicknesses shall be permitted to be used in the same member provided that the total thickness of tension lamination(s) equals or exceeds 5 percent of the depth.

[7]The following species or species groups shall be permitted to be used for softwood species (WW) provided the design values in modulus of elasticity (E) in Column 4 of Table 4-4 are reduced by 100,000 psi: Western cedars, Western cedars (North), White woods (Western woods) and California redwood - open grain.

The following species or species groups shall be permitted to be used for softwood species (WW) provided the design values in shear parallel to grain in Columns 13, 14, and 15 (F_{vy}), and Column 18 (F_{vx}) of Table 4-4 are reduced by 10 psi and the design values in shear parallel to grain in Column 12 (F_{vy}) of Table 4-4 are reduced by 5 psi: coast Sitka spruce, coast species, western white pine, and eastern white pine (north).

[8]When the designer uses the reduced stresses as indicated in Footnote No. 1 to Table 4-4 and so specifies or else specifies only a given stress and the manufacturer selects the combination to use, the slope of grain for the combination shall be permitted to be as follows, depending on the design value specified (see Table 4-4):

Combination	Slope of Grain
EWS 47 and 48	1:8, 1:10 or 1:12
EWS 49, 50, 50M1, and 50M2	1:10 or 1:12

[9]Footnote No. 14 to Table 4-3 applies.

[10]Footnote No. 15 to Table 4-5 applies.

[11]Footnote No. 16 to Table 4-5 applies.

TABLE 4.7 Allowable Properties and Moduli of Elasticity for Glued Laminated Timber with Tapered Cuts on Compression Face

Comb. symbol	F_{bx},[a] psi	$F_{c\perp x}$,[b] psi	E_x,[a] 10^6 psi	Comb. symbol	F_{bx},[a] psi	$F_{c\perp x}$,[b] psi	E_x,[a] 10^6 psi
Visually graded western species				Visually graded southern pine			
16F-V3	1,600	560	1.5	20F-V2	1,600	650	1.5
16F-V6	1,900	560	1.5	20F-V5	1,550	650	1.5
20F-V4	1,600	560	1.4	24F-V3	2,300	650	1.6
20F-V8	2,000	560	1.6	24F-V5	2,300	650	1.6
20F-V9	1,900	375	1.5	26F-V1	2,300	650	1.6
20F-V10	1,950	375	1.5	26F-V2	2,400	740	1.8
20F-V12	1,850	470	1.3	26F-V4	2,400	650	1.8
24F-V4	2,200	560	1.7				
24F-V5	2,100	375	1.6				
24F-V8	2,200	560	1.7				
24F-V10	2,000	375	1.6				
24F-1.8E	2,000	375	1.6				
E-rated western species				E-rated southern pine			
20F-E1	1,400	255	1.3	24F-E1	2,200	650	1.7
20F-E8	1,600	300	1.3	24F-E4	2,200	650	1.7
20F-E/SPF1	1,600	470	1.3	28F-E1	2,600	650	1.8
24F-E10	2,300	560	1.8	28F-E2	2,600	650	1.8
24F-E11	2,000	375	1.6	30F-E1	2,600	650	1.8
24F-E14	2,300	560	1.8	30F-E2	2,600	650	1.8
24F-E15	1,800	375	1.5				
24F-E18	2,100	560	1.7				
24F-E20	1,800	350	1.5				
24F-E/ES1	2,000	300	1.5				
26F-E/DF1M1	2,100	560	1.8				

[a] Value is applicable to members that have up to one-half the depth on the compression side removed by taper cutting. Value is for dry conditions of use and 12 in. or less in depth.
[b] Design value in compression perpendicular to grain for the core laminations of the combination.

TABLE 4.8 Allowable Radial Stresses[a]

Species	Radial tension (F_{rt}), psi: Wind and earthquake	Radial tension (F_{rt}), psi: Other Loading	Radial compression (F_{rc}), psi
Alaska cedar	63	15	470
California redwood	42	42	315
Canadian spruce pine	53	15	560
Douglas fir-larch	55	15	560
Douglas fir-south	55	15	560
Eastern spruce	48	15	300
Hem-fir	52	15	375
Softwood species (WW)	47	15	255
Southern pine	67	67	650

[a] The wet-use factor, C_M, is equal to 0.875 for radial tensile stresses and 0.53 for radial compressive stresses.

4.4.3 Adjustment Factors

The adjustment factors provided in this section are for nonreference end-use conditions and material modification effects. These factors shall be used to modify the allowable properties when one or more of the specific end-use or material modification conditions fall outside the limits of the reference conditions given in this section.

Load-Duration Factor, C_D. Allowable properties tabulated in Tables 4.3 and 4.4 apply to normal load duration, which means a structural member is subject to full design loads for a cumulative duration of approximately 10 years. For other cumulative duration of the full design loads, the allowable properties, except for modulus of elasticity and compression perpendicular to grain, shall be adjusted by the load duration factor given in Table 4.9.

Wet Service Factor, C_M. Design values provided in this section are applicable to dry use conditions of glued laminated timber (moisture content in service is less than 16%, as in most covered structures) and its connections. When glued laminated timber is exposed to wet service conditions, the adjustment factors given in Tables 4.3 and 4.4 apply.

Temperature Factor, C_t. The temperature factor, C_t, shall be applied when glued laminated timber is exposed to a sustained elevated temperature ranging from 100–150°F. When the equilibrium moisture content of a glued laminated timber member exceeds the reference condition limitation during sustained elevated temperature exposure, both the temperature and wet service (moisture) factors shall be applied. When the equilibrium moisture content of glued laminated timber falls within the limits of the reference conditions during sustained exposure to elevated temperatures, only the temperature factor shall be applied. The temperature factors are given in Table 4.10.

Preservative Treatment. Whenever practical, the moisture content of permanent structural wood members should be kept below 20%. If this is not feasible, then preservative treatment may be required unless the heartwood of a naturally decay-

TABLE 4.9 Load-Duration Factor for Glued Laminated Timber, C_D

Load duration	C_D	Typical design loads
Permanent	0.9	Dead Load
Ten years	1.0	Occupancy live load
Two months	1.15	Snow load
Seven days	1.25	Construction load
Ten minutes	1.6[b]	Wind/earthquake load
Impact[a]	2.0	Impact load

[a] The impact load duration factor shall not apply to glued laminated timber members treated with waterborne preservatives to a heavy retention required for marine exposure, nor to members pressured treated with fire retardant chemicals.

[b] Check applicable building code for the appropriate load duration factor subject to earthquake loading.

TABLE 4.10 Temperature Factor for Glued Laminated Timber Exposed to Sustained Elevated Temperature, C_t

Design values	In-service moisture conditions[a]	C_t		
		$T \leq 100°F$	$100 < T \leq 125$	$125 < T \leq 150$
F_t, E	Dry or wet	1.0	0.9	0.9
F_b, F_v, F_c and $F_{c\perp}$	Dry	1.0	0.8	0.7
	Wet	1.0	0.7	0.5

[a] In-service conditions are defined in Section 4.4.3.

resistant species such as redwood, Port Orford cedar, or Alaska yellow cedar is used.

Most preservative chemicals used today do not significantly alter the strength properties of structural wood products. However, the method of pre- and postconditioning, as well as the treatment method itself, may weaken the wood. For preservative treatment methods with AWPA accepted manufacturing control, as listed in Table 4.11, the effect on strength degradation of glued laminated timber is negligible. For more information concerning preservative treatment of glued laminated timber, refer to Chapter 9 of this handbook and to APA EWS Technical Note S580, *Preservative Treatment of Glulam Beams.*[3]

Fire-Retardant Treatment. Effects of fire-retardant treatments on allowable design stresses shall be considered for wood products treated with pressure-impregnated fire retardants. The glued laminated timber industry does not recommend the use of fire-retardant treatments with glued laminated timber. Specific adjustment factors for fire retardants used in conjunction with glued laminated timber shall be obtained from the company providing the treatment services, and the glued laminated timber manufacturer accepts no responsibility for any structural glued laminated timber that is fire retardant treated.

Beam Stability Factor, C_L. Allowable bending stresses of glued laminated timber shall be adjusted by the beam stability factor, C_L, whenever applicable Refer to the *National Design Specification for Wood Construction*[6] (NDS), published by American Forest and Paper Association (AF&PA), for the determination of an appropriate

TABLE 4.11 Preservative Treatment Effect on Glued Laminated Timber

No adjustment is required when glued laminated timber is preservative-treated using the following American Wood Preservers' Association Standards

Designation	Title
C1-88	All Timber Products—Preservative Treatment by Pressure Processes
C14-89	Wood for Highway Construction—Preservative Treatment by Pressure Processes
C15-88	Wood for Commercial—Residential Construction—Preservative Treatment by Pressure Processes
C28-93	Standard for Preservative Treatment of Structural Glued Laminated Members and Laminations before Gluing of Southern Pine, Pacific Coast Douglas-fir, Hem Fir, and Western Hemlock by Pressure Processes

beam stability factor. It is important to note that C_L is not accumulative with the volume factor, C_v, given below under Volume Factor, for the design of glued laminated timber.

Column Stability Factor, C_P. Allowable values for compression parallel to grain of glued laminated timber are affected by the dimensions and modulus of elasticity. Refer to the NDS or Section 4.7.2 for the determination of an appropriate column stability factor.

Volume Factor, C_v. Allowable bending stresses of glued laminated timber are affected by geometry and size. Generally, larger sizes have a correspondingly lower allowable bending stress than smaller members. To account for this behavior, a volume factor, C_v, which is the product of beam width, depth, and length shall be applied. C_v shall not exceed 1.0 and is computed as follows:

$$C_v = \left(\frac{5.125}{b}\right)^p \left(\frac{12}{d}\right)^p \left(\frac{21}{\ell}\right)^p \leq 1.0 \tag{4.1}$$

where b = width of bending member being checked (in.). For multiple-piece width lay-ups, b = width of widest piece in the lay-up. For practical purposes, b is assumed to be $\leq$10.75 in.

d = depth of bending member being checked (in.)

ℓ = length of bending member being checked between points of zero moment (ft)

p = as defined in Table 4.12

Table 4.12 provides the exponent values for use with the volume effect factor. Separate exponent values are given for western species, hardwoods, and southern pine. No volume adjustment is required for properties other than allowable bending stresses.

Curvature Factor, C_c. The curvature factor, C_c, is used to adjust the allowable bending stresses of curved glued laminated timber members only. It takes into account the difference in extreme outer fiber stress between a curved member and a straight prismatic member, as well as any residual stresses that may remain in a lamination that has been bent to the stated curvature. However, the curvature factor, C_c, shall not be applied to the allowable bending stress in the straight portion of a member, regardless of curvature in other portions. Also, this factor is not applicable to cambered glued laminated timber members or in the design of pitched and tapered curved glued laminated timber members. The curvature factor, C_c, shall be calculated in accordance with the following equation:

$$C_c = 1 - 2000\left(\frac{t}{R}\right)^2 \tag{4.2}$$

TABLE 4.12 Exponents for Volume Factor Equation

	Exponent		
Exponent symbol	Western species	Southern pine	Hardwoods
p	0.10	0.05	0.10

TABLE 4.13 Flat Use Factor,[a] C_{fu}

Member dimensions parallel to wide faces of laminations	C_{fu}
10¾ or 10½	1.01
8¾ or 8½	1.04
6¾	1.07
5⅛ or 5	1.10
3⅛ or 3	1.16
2½	1.19

[a] Values for C_{fu} are rounded values from the equation $(12/d)^{1/9}$ where d is the dimension of the wide faces of the laminations in inches.

where t = thickness of lamination (in.)
R = radius of curvature of inside face of lamination (in.)
$t/R \leq 1/100$ for hardwoods and southern pine
$t/R \leq 1/125$ for other species

Flat Use Factor, C_{fu}. Allowable bending stresses of glued laminated timber shall be adjusted by the flat use factor, C_{fu}, when loaded in bending parallel to wide faces of the laminations (the y–y axis). The allowable bending stresses for loads applied parallel to the wide faces of the laminations, F_{by}, as given in Tables 4.3 and 4.4 of this chapter, are based on members with laminations 12 in. wide. For members with laminations less than 12 in. wide, the tabulated F_{by} values shall be adjusted by a flat use factor, C_{fu}, as listed in Table 4.13. When the width of the laminations is greater than 12 in., as may occur in members with multiple-piece laminations, C_{fu} shall be obtained by use of the equation given in footnote (a) to Table 4.13.

4.4.4 Section Properties

Net section properties for both western species and southern pine glued laminated timber are given in Section 4.7.1. Note that the plane of the glueline is in the X–X direction. Further, the width of glued laminated timber is in the X–X direction and its depth is in the Y–Y direction. The thickness of each lamination for western species and southern pine glued laminated timber members is based on 1½ and 1⅜ in., respectively, which are typical for these species. However, other lamination thicknesses may be used in glued laminated timber manufacturing, and the availability should be verified prior to design.

4.5 PHYSICAL PROPERTIES

4.5.1 General

This section contains information concerning physical properties of glued laminated timber members.

4.5.2 Specific Gravity

Table 4.14 provides specific gravity values for some of the most common wood species used to manufacture glued laminated timber. These values are used in determining various physical and connection properties. Further, weight factors are provided at four moisture contents. When the net cross-sectional area (in.^{2}) is multiplied by the appropriate weight factor, it provides the weight of the glued laminated timber member per linear foot of length. For other moisture contents, the tabulated weight factors can be interpolated or extrapolated.

Glued laminated timber members may be manufactured using different species in different portions of the cross section. In this case the weight of the glued laminated timber may be computed by the sum of the products of the cross-sectional area and the weight factor for each species.

4.5.3 Dimensional Changes Due to Moisture

Due to the hygroscopic nature of wood, it changes dimensions as its moisture content is altered below the fiber saturation point. For most species the longitudinal shrinkage of normal wood drying from fiber saturation point to oven-dry condition is approximately 0.1–2%. However, certain atypical types of wood may exhibit excessive longitudinal shrinkage, and these types should be avoided in use where longitudinal stability is important.

The change in radial (R), tangential (T), and volumetric (V) dimensions is computed as:

$$X = X_0(\Delta\text{MC})e_{\text{ME}} \tag{4.3}$$

where X_0 = initial dimension or volume
X = new dimension or volume
e_{ME} = coefficient of moisture expansion (in./in./%MC for linear expansion, in.^{3}/in.^{3}/%MC for volumetric expansion), as given in Table 4.15, and

TABLE 4.14 Average Specific Gravity and Weight Factor

Species combination	Specific gravity[a]	Weight factor[b]			
		12%	15%	19%	25%
California redwood (close grain)	0.44	0.195	0.198	0.202	0.208
Douglas fir-larch	0.50	0.235	0.238	0.242	0.248
Douglas fir (south)	0.46	0.221	0.225	0.229	0.235
Eastern spruce	0.41	0.191	0.194	0.198	0.203
Hem-fir	0.43	0.195	0.198	0.202	0.208
Red maple	0.58	0.261	0.264	0.268	0.274
Red oak	0.67	0.307	0.310	0.314	0.319
Southern pine	0.55	0.252	0.255	0.259	0.265
Spruce-pine-fir (north)	0.42	0.195	0.198	0.202	0.208
Yellow poplar	0.43	0.213	0.216	0.220	0.226

[a] Specific gravity is based on weight and volume when oven-dry.
[b] Weight factor shall be multiplied by net cross-sectional area in in.^{2} to obtain weight in pounds per lineal foot.

TABLE 4.15 Coefficient of Moisture Expansion, e_{ME}, and Fiber Saturation Point (FSP) for Solid Woods

Species	e_{ME} Radial (in./in./%)	e_{ME} Tangential (in./in./%)	e_{ME} Volumetric ($in.^3/in.^3/\%$)	FSP (%)
Alaska cedar	0.0010	0.0021	0.0033	28
Douglas fir-larch	0.0018	0.0033	0.0050	28
Englemann spruce	0.0013	0.0024	0.0037	30
Redwood	0.0012	0.0022	0.0032	22
Red oak	0.0017	0.0038	0.0063	30
Southern pine	0.0020	0.0030	0.0047	26
Western hemlock	0.0015	0.0028	0.0044	28
Yellow poplar	0.0015	0.0026	0.0041	31

ΔMC = moisture content change (%), as defined as follows:

$$\Delta MC = M - M_0 \tag{4.4}$$

where M_0 = initial moisture content % ($M_0 \leq$ FSP)
M = new moisture content % ($M \leq$ FSP)
FSP = fiber saturation point, also given in Table 4.15 for selected species

For more information concerning the effects of moisture changes on glued laminated timber, refer to APA EWS Technical Note Y260, *Dimensional Changes in Structural Glued laminated Timber.*[4]

4.5.4 Dimensional Changes Due to Temperature

The thermal expansion of solid wood, including glued laminated timber, is computed by the relationship:

$$X = X_0(\Delta T)e_{TE} \tag{4.5}$$

where X_0 = reference dimension at T_0
X = computed dimension at T
e_{TE} = coefficient of thermal expansion (in./in./°F), see Table 4.16
ΔT = temperature change (°F), defined as follows:

$$\Delta T = T - T_0 \tag{4.6}$$

where T_0 = reference temperature (°F), $-60°F \leq T_0 \leq 130°F$
T = new temperature (°F), $-60°F \leq T_0 \leq 130°F$

The coefficient of thermal expansion of oven-dry wood parallel to grain appears to be independent of specific gravity and species. In tests of both hardwoods and softwoods, the parallel-to-grain values have ranged from about 1.7×10^{-6} to 2.5×10^{-6} per °F.

The linear expansion coefficients across the grain (radial and tangential) are proportional to the density of wood. These coefficients are about 5–10 times greater

TABLE 4.16 Coefficient of Thermal Expansion, e_{TE}, for Solid Woods

	e_{TE}	
Species	Radial (10^{-6} in./in./°F)	Tangential (10^{-6} in./in./°F)
California redwood	13	18
Douglas fir-larch[a]	15	19
Douglas fir (south)	14	19
Eastern spruce	13	18
Hem-fir[a]	13	18
Red oak	18	22
Southern pine	15	20
Spruce-pine-fir	13	18
Yellow poplar	14	18

[a] Also applies when species name includes the designation "North."

than the parallel-to-grain coefficients. The radial and tangential thermal expansion coefficients for oven-dry wood in the oven-dry specific gravity (SG) range of about 0.1–0.8 can be approximated by the following equations.

Radial:

$$e_{TE} = [18(SG) + 5.5](10^{-6}\text{in./in./°F}) \tag{4.7}$$

Tangential:

$$e_{TE} = [18(SG) + 10.2](10^{-6}\text{in./in./°F}) \tag{4.8}$$

where SG = specific gravity, provided in Table 4.14

Table 4.16 provides the numerical values for e_{TE} for the most commonly used commercial species or species groups.

Wood that contains moisture reacts to varying temperature differently than does dry wood. When moist wood is heated, it tends to expand because of normal thermal expansion and to shrink because of loss in moisture content. Unless the wood is very dry initially (perhaps 3 or 4% MC or less), the shrinkage due to moisture loss on heating will be greater than the thermal expansion. As a result, the net dimensional change on heating will be negative. Wood at intermediate moisture levels (about 8–20%) will expand when first heated, then gradually shrink to a volume smaller than the initial volume, as the wood gradually loses water while in the heated condition.

Even in the longitudinal (grain) direction, where dimensional change due to moisture change is very small, such changes will still predominate over corresponding dimensional changes due to thermal expansion unless the wood is very dry initially. For wood at usual moisture levels, net dimensional changes will generally be negative after prolonged heating.

Computation of actual changes in dimensions can be accomplished by determining the equilibrium moisture content of wood at the temperature value and relative humidity of interest. Then the relative dimensional changes due to temperature change alone and moisture content change alone are computed. By combining these two changes the final dimension of lumber and timber can be established.

4.6 SPECIAL CONSIDERATIONS

4.6.1 Appearance Classifications

Lumber grade characteristics, such as open knots or other voids, and the appearance considerations inherent in the glued laminated timber manufacturing process itself may be cosmetically altered based on the desired finished member appearance for any layup combination. Glued laminated timber members are typically produced in four appearance classifications, premium, architectural, industrial, and framing. Premium and architectural beams are higher in appearance qualities and are surfaced for a smooth finish ready for staining or painting. Industrial classification beams are normally used in concealed applications or in construction where appearance is not important. Framing classification beams are typically used for headers and other concealed applications in residential construction.

Industry recommendations for the finished appearance of glued laminated timber have traditionally identified three grades: premium, architectural, and industrial. All of these appearance grades permit the individual laminations used in the glued laminated timber layup to possess the natural growth characteristics of the laminating lumber grades specified. In order to account for the visual impact of these natural growth characteristics, cosmetic repairs may be made to the finished member as required to meet the desired finished appearance.

However, the use of the term appearance *grades* associated with these nonstructural cosmetic classifications often leads to confusion in the marketplace. Designers and users often think of appearance grades in terms of structural distinctions, when in fact structural properties for a given glued laminated timber layup are independent of appearance considerations. As an example, a premium appearance classification may be specified based on the presumption that the member will have better structural characteristics than an architectural or industrial classification when, in fact, all three classifications will have identical strength characteristics if manufactured to the same layup combination. To minimize this confusion, these different appearance characteristics are more appropriate to be identified as appearance classifications than grades. APA EWS Technical Note Y110, *Glued Laminated Timber Appearance Classifications for Construction Applications,*[5] provides detailed information on these appearance classifications.

4.6.2 Camber

Introduction. One of the most important design qualities of wood framing is its resilience under load. Wood's unique stiffness makes wood floors comfortable to walk, work, and stand on. These qualities also contribute greatly to wood's excellent damping characteristics, thus giving wood-framed structures the ability to absorb impact loadings.

The size and span capabilities of sawn lumber beams are generally limited by the physical characteristics of the timber supply. Consequently, stiffness, or deflection under load, at the relatively short spans commonly used with sawn lumber beams is seldom a governing factor in the design of these wood structural elements. With the availability of glued laminated timber (glued laminated timber), it is possible for architects and engineers to design wood members in large sizes and long lengths.

Design professionals recognize that for long spans, design is often controlled by deflection limits rather than by beam strength. One way to reduce the adverse aesthetic effect and structural significance of beam deflection is through the use of camber. Camber is an initial curvature built into a fabricated member such as a glued laminated timber which is opposite in direction to the calculated deflection which will occur under gravity loads. The use of camber in glued laminated timber beams also gives the designer the ability to negate the possible adverse effects of long-term deflection or creep that may occur with wood members.

While this section provides a discussion of camber, it is important to recognize that building codes do not require this technique for deflection control. It is only one option available to the designer. Other design techniques for providing deflection control include specifying beams of greater stiffness than otherwise would be required, increasing slope for roof drainage to minimize possible effects of ponding, or limiting spans such that deflection is not a controlling design consideration.

In some situations, especially in construction applications where the framing elements are rarely subjected to specified design live loads and the deflection due to dead loads is minimal, camber may not be necessary. In such applications, camber designed into a beam may never relax. In floor construction this can cause a permanent, unwanted crown and, in multistory applications it may increase the difficulty of framing successive floors during construction. As a general rule, if noncambered sawn lumber beams have traditionally performed well in a given application, then cambering is probably not required when a glued laminated timber having equivalent or greater stiffness is substituted. Residential garage door headers and floor beams are good examples of uses where introducing camber may not be necessary.

Controlling Deflection. Building code limits on beam deflection were first imposed in the early 1900s to prevent the cracking of brittle plaster ceilings in habitable units. The deflections determined for this serviceability requirement remain as the basis for the allowable deflections listed in building codes today.

Building code requirements aside, the appearance of a sagging floor or roof beam does not instill structural confidence in the eye of the beholder. While a sagging beam may have sufficient strength to support its design loads, it may be perceived as being underdesigned and may adversely affect the acceptance of the overall building design. Glued laminated timber is generally the only engineered wood beam that can be cambered.

In addition to the aesthetic reasons mentioned above, there is a structural reason for limiting deflection in large flat roofs. This is to prevent a phenomenon called ponding, which may occur when water collects in a depression in a flat roof. As water builds up in a depression, the load on the supporting members increases. This causes additional deflection that permits more water to accumulate. Once initiated, this cycle can continue until a roof failure occurs. Ponding is more prevalent in areas where roofs are designed for low live-to-dead-load ratios. Roofs designed for snow loads are typically stiffer and have a greater ability to resist ponding than similar roofs designed only for relatively light construction loads.

A general guideline for determining if provisions to prevent ponding are necessary is to compute the deflection of the member in question under a uniform load of 5 psf. If the resulting deflection is equal to or greater than ½ in., this is an indication that ponding is a potential problem and beam camber combined with other positive means to insure drainage should be used. This simplified procedure should not be substituted for a detailed engineering analysis to control ponding of low slope or flat roofs.

Camber Recommendations. Calculated beam deflection is used to assess compliance with specifications and the building code and to specify beam camber. This calculated deflection is measured from a plane connecting the beam supports. When camber is built into the beam, the built-in deflection is above this plane.

The code-specified deflection limit may be measured from the plane connecting the supports to a point below that plane. When camber is not required to prevent ponding, a cambered beam, just as one without camber, may be permitted to deflect below the plane connecting the supports by the amount permitted in the code. This is acceptable provided that the beam is not overstressed and adjacent construction (such as plaster walls or ceilings), machinery, and objects connected to or resting on the beam are not adversely affected by this deflection.

For most roof beam applications, the glued laminated timber industry recommends the use of 1½ times the calculated dead load deflection to arrive at camber specifications. This amount of camber is generally sufficient to allow the beam to deflect back to near level after many years under sustained dead load. For floor beams, the recommended camber is 1.0 times the calculated dead load deflection. However, for residential construction, floor beams are typically not cambered due to the relatively short spans and the need to minimize unevenness in the floor construction.

In addition to cambering for deflection induced by dead load, roof beams in simple-span applications may also be cambered to provide the industry recommended slope of ¼ in./ft to ensure proper drainage. When this is done, the required camber for slope is calculated and added to the dead-load camber described above. This becomes the total camber, which can be built into the beam during fabrication.

Camber Tolerances. Because of the variability of mechanical and physical properties in all wood products and manufacturing associated tolerances, a cambered glued laminated timber beam will not always be manufactured with exactly the camber specified. In addition, moisture changes can also affect the camber of any specific beam. For these reasons, a certain degree of variability associated with any target camber is allowed and should be expected. ANSI A190.1 gives the tolerances for camber or straightness of a glued laminated timber as: "Up to 20 feet, the tolerance is plus or minus ¼ inch. Over 20 feet, increase tolerance by ⅛ inch per additional 20 feet or fraction thereof, but not to exceed ¾ inch." This variability in camber is generally so small with respect to beam length that it is undetectable to the human eye and has little or no effect on beam performance.

The camber for beams is often specified as inches of camber. It may also be specified as a radius of curvature. Thus specified, the offsets at various points along a beam of any length can be represented by a single number. The circular arc formed by this method closely approximates the actual deflected curvature of the beam under load. The beam is thus fabricated with a built-in mirror image of the expected deflected curvature.

Table 4.17 is a precalculated deflection/radius-of-curvature table used by fabricators. This table can be used to specify a camber radius based on a given span and calculated deflection. See below, under Stock Beam Camber, for camber recommendations for residential or stock beam applications.

Calculating Beam Camber. As an alternative to using Table 4.17 to specify camber, the following formula may be used to calculate the approximate radius of curvature, given the beam span and camber desired.

TABLE 4.17 Camber (in.) Based on Radius and Beam Span (ft)[a]

	Radius, ft											
Span, ft	1000	1200	1400	1600	1800	2000	2500	3000	3500	4000	4500	5000
10	0.15	0.13	0.11	0.09	0.08	0.08	0.06	0.05	0.04	0.04	0.03	0.03
12	0.22	0.18	0.15	0.14	0.12	0.11	0.09	0.07	0.06	0.05	0.05	0.04
14	0.29	0.25	0.21	0.18	0.16	0.15	0.12	0.10	0.08	0.07	0.07	0.06
16	0.38	0.32	0.27	0.24	0.21	0.19	0.15	0.13	0.11	0.10	0.09	0.08
18	0.49	0.41	0.35	0.30	0.27	0.24	0.19	0.16	0.14	0.12	0.11	0.10
20	0.60	0.50	0.43	0.38	0.33	0.30	0.24	0.20	0.17	0.15	0.13	0.12
22	0.73	0.61	0.52	0.45	0.40	0.36	0.29	0.24	0.21	0.18	0.16	0.15
24	0.86	0.72	0.62	0.54	0.48	0.43	0.35	0.29	0.25	0.22	0.19	0.17
26	1.01	0.85	0.72	0.63	0.56	0.51	0.41	0.34	0.29	0.25	0.23	0.20
28	1.18	0.98	0.84	0.74	0.65	0.59	0.47	0.39	0.34	0.29	0.26	0.24
30	1.35	1.13	0.96	0.84	0.75	0.68	0.54	0.45	0.39	0.34	0.30	0.27
32	1.54	1.28	1.10	0.96	0.85	0.77	0.61	0.51	0.44	0.38	0.34	0.31
34	1.73	1.45	1.24	1.08	0.96	0.87	0.69	0.58	0.50	0.43	0.39	0.35
36	1.94	1.62	1.39	1.22	1.08	0.97	0.78	0.65	0.56	0.49	0.43	0.39
38	2.17	1.81	1.55	1.35	1.20	1.08	0.87	0.72	0.62	0.54	0.48	0.43
40	2.40	2.00	1.71	1.50	1.33	1.20	0.96	0.80	0.69	0.60	0.53	0.48
42	2.65	2.21	1.89	1.65	1.47	1.32	1.06	0.88	0.76	0.66	0.59	0.53
44	2.90	2.42	2.07	1.82	1.61	1.45	1.16	0.97	0.83	0.73	0.65	0.58
46	3.17	2.65	2.27	1.98	1.76	1.59	1.27	1.06	0.91	0.79	0.71	0.63
48	3.46	2.88	2.47	2.16	1.92	1.73	1.38	1.15	0.99	0.86	0.77	0.69
50	3.75	3.13	2.68	2.34	2.08	1.88	1.50	1.25	1.07	0.94	0.83	0.75
52	4.06	3.38	2.90	2.54	2.25	2.03	1.62	1.35	1.16	1.01	0.90	0.81
54	4.37	3.65	3.12	2.73	2.43	2.19	1.75	1.46	1.25	1.09	0.97	0.87
56	4.70	3.92	3.36	2.94	2.61	2.35	1.88	1.57	1.34	1.18	1.05	0.94
58	5.05	4.21	3.60	3.15	2.80	2.52	2.02	1.68	1.44	1.26	1.12	1.01
60	5.40	4.50	3.86	3.38	3.00	2.70	2.16	1.80	1.54	1.35	1.20	1.08
62	5.77	4.81	4.12	3.60	3.20	2.88	2.31	1.92	1.65	1.44	1.28	1.15
64	6.15	5.12	4.39	3.84	3.41	3.07	2.46	2.05	1.76	1.54	1.37	1.23
66	6.54	5.45	4.67	4.08	3.63	3.27	2.61	2.18	1.87	1.63	1.45	1.31
68	6.94	5.78	4.96	4.34	3.85	3.47	2.77	2.31	1.98	1.73	1.54	1.39
70	7.35	6.13	5.25	4.59	4.08	3.68	2.94	2.45	2.10	1.84	1.63	1.47

[a] Industry-accepted manufacturing tolerance on camber is ± ¼ in. for lengths up to 20 ft with an increase in tolerance of ⅛ in. per additional 20 ft of length or fraction thereof, but not to exceed ¾ in.

$$R = \frac{3L^2}{2\Delta} \tag{4.9}$$

where R = approximate radius of curvature (ft)
L = span (ft)
Δ = desired camber (in.)

See Fig. 4.11 for a graphic representation of beam camber parameters.

An example of calculating the radius of curvature to specify beam camber is given as follows:

Calculate the required camber for a glued laminated timber roof beam for the following design conditions:

Glued laminated timber beam size = 6¾ × 27 in.

$E = 1.6 \times 10^6$ psi

$I = 11{,}072$ in.^{4}

Span = 48 ft

Dead load = 120 lb/ft

Beam weight = 44 lb/ft

$$\text{Dead load deflection} = \frac{5wL^4 1728}{384EI}$$

w = dead load + beam weight (lb/ft)

L = beam span (ft)

E = modulus of elasticity (psi)

I = moment of inertia (in.^{4})

$$\text{Dead load deflection} = \frac{5(120 + 44)(48^4(1728)}{384(1.6 \times 10^6)(11{,}072)} = 1.11 \text{ in.}$$

Recommended camber = 1.5 × 1.11 = 1.66 in.

Radius of curvature (R) required: from Table 4.17 by interpolation R_{2100} = 1.66 in. for 48 ft span or by calculation:

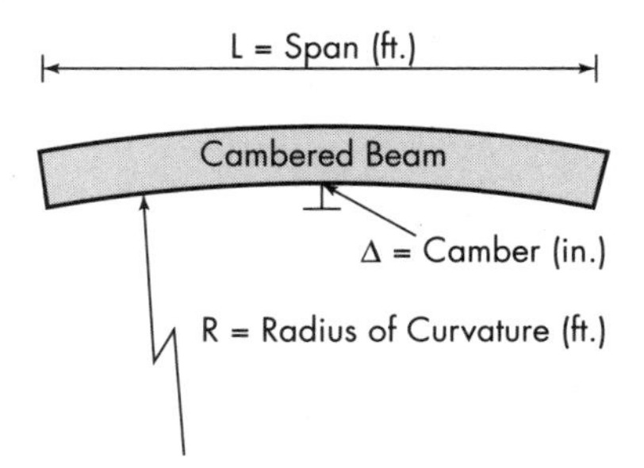

FIGURE 4.11 Beam camber parameters.

$$R = \frac{3(48)^2}{2(1.66)} = 2082 \text{ ft}$$

Note that a camber radius of 2000 ft would provide a camber of 1.73 in. and should be considered for this case.

Cantilevered Beams. The cambering of cantilevered beams is a subject that is often misunderstood. While it is theoretically possible to specify camber for each span section of single and double cantilevered beams, it is, in most cases, not practical. If the camber does not relax, a beam with an S-shaped camber for a cantilever (or W-shaped camber in the case of a double cantilever) can severely impact roof drainage and can even cause ponding over the outside supports. Ponding over the supports can lead to bearing or support failures and subsequent roof leaks. For most typical cantilever length to main span length conditions, the preferred method is to camber only the span length between the supports, leaving the cantilevered portion uncambered. This will generally ensure good drainage of the roof system. If unusually long cantilever lengths are used, a further analysis of member deflection and camber is warranted. Based on this analysis, the designer may choose to specify individual camber values for both the interior beam span and the cantilever span.

Stock Beam Camber. While the ability to precamber glued laminated timber beams is a distinct advantage for many longer-span applications, it can be a disadvantage in applications where too much camber can cause job-site framing problems. This can happen when a stock beam for residential construction is specified with too much camber. Stock beams are glued laminated timbers that are manufactured for inventory by a stocking distributor or dealer. Theses are typically supplied to the distributor in long lengths and cut to length as ordered by the contractor or end user.

Most stock beams are used in residential construction in combination with other wood products such as dimension lumber or I-joists, which are not cambered. Therefore, it is important that stock beams be supplied with the right amount of camber to be compatible with other noncambered framing members.

In order to eliminate the effect of having too much camber, glued laminated timber manufacturers typically supply stock beams with either zero camber or with a camber radius of 3500 ft or flatter. As shown in Table 4.17, a camber radius of 3500 ft provides an actual camber of ¼ in. for a span of 24 ft. Since most residential floor beam and header applications have spans of less than 24 ft, camber will typically be less than ¼ in. and this amount of camber should be virtually unnoticeable.

As previously noted, the manufacturing tolerance for camber is plus or minus ¼ in. for beams up to 20 ft in length and increases by ⅛ in. per additional 20 ft of length or fraction thereof, not to exceed ¾ in.

The amount of camber manufactured into stock beams is often dependent on the geographic region of the market and can vary from region to region. Since the degree of camber in stock beams may be almost imperceptible, it is very important that contractors be aware of the "TOP" stamp (see Fig. 4.8), which is placed on the compression side of the beam by the manufacturer to ensure proper installation and performance of these members.

4.6.3 Fire

Introduction. Glued laminated timber beams and columns provide architectural warmth and beauty along with structural strength and natural fire resistance. In the presence of fire, the outer portion of a glued laminated timber member becomes charred. This layer of charred wood then functions as an insulator, helping to protect the undamaged interior of the member from the heat. The rate of advancement of this insulating char layer into the remaining, undamaged portion of the member has been measured (approximately 0.025 in./minute) and forms the theoretical basis of the equations used to predict fire endurance. Tests on loaded beams and columns have confirmed the validity of the equations in predicting their load-carrying ability under fire conditions, and the method has been recognized by all U.S. model building codes.

Fires do not normally start in structural framing, but rather in the building's contents. These fires generally reach temperatures of between 1290°F and 1650°F. Glued laminated timber members perform very well under these conditions. Unprotected steel members typically suffer severe buckling and twisting during fires, often collapsing catastrophically.

Wood ignites at about 480°F, but charring may begin as low as 300°F. Wood typically chars at $\frac{1}{40}$ in./minute. Thus, after half an hour's fire exposure, only the outer $\frac{3}{4}$ in. of the glued laminated timber will be damaged. Char insulates a wood member and hence raises the temperature it can withstand. Most of the cross section will remain intact, and the member will continue supporting loads during a typical building fire.

It is important to note that neither building materials nor building features nor detection and fire extinguishing equipment alone can provide adequate safety from fire in buildings. To ensure a safe structure in the event of fire, authorities base fire and building code requirements on research and testing, as well as fire histories. Chapter 10 provides a more detailed discussion of designing for fire safety.

The model building codes and the International Building Code (IBC) classify heavy timber as a specific type of construction and give minimum sizes for roof and floor beams. The requirements set out for heavy timber construction in the codes do not constitute one-hour fire resistance. However, procedures are available to estimate the glued laminated timber size required for projects in which one-hour fire resistance is required. The minimum depths for selected glued laminated timber sizes that can be adopted for one-hour fire ratings are given in Tables 4.21 for glued laminated timber beams.

To achieve a one-hour fire rating for beams whose dimensions qualify them for this rating, the basic layup must be modified—one core lamination must be removed from the center and the tension face augmented with the addition of a tension lamination.

Design Methodology. Calculation of the ability of a glued laminated timber beam or column to resist fire for up to one hour is accepted in the 1997 Uniform Building Code (UBC), Volume 1, Section 703.3 and UBC Volume 3, Section 7.727, and in the 1997 Standard Building Code, Section 709.6.3. The method is also recognized in Section 6.1 of *Guidelines for Determining Fire Resistance Ratings of Building Elements,* published by Building Officials and Code Administrators International, Inc. (BOCA), and Section 720.6.3 of the 2000 International Building Code (IBC). The equations apply to members with fire on three or four sides.

Beams:

$$\text{Fire on 3 sides} \quad t = 2.54ZB\left[4 - \frac{B}{D}\right] \tag{4.10}$$

$$\text{Fire on 4 sides} \quad t = 2.54ZB\left[4 - \frac{2B}{D}\right] \tag{4.11}$$

Columns:

$$\text{Fire on 3 sides} \quad t = 2.54ZB\left[3 - \frac{B}{2D}\right] \tag{4.12}$$

$$\text{Fire on 4 sides} \quad t = 2.54ZB\left[3 - \frac{B}{D}\right] \tag{4.13}$$

where t = fire resistance (minutes)
Z = partial load compensation factor (see Fig. 4.12), which is a function of applied load to design capacity
B = the breadth or width of a beam or the smaller dimension of a column (in.) (see Fig. 4.13)
D = the depth of a beam or the larger dimension of a column (in.) (see Fig. 4.13)

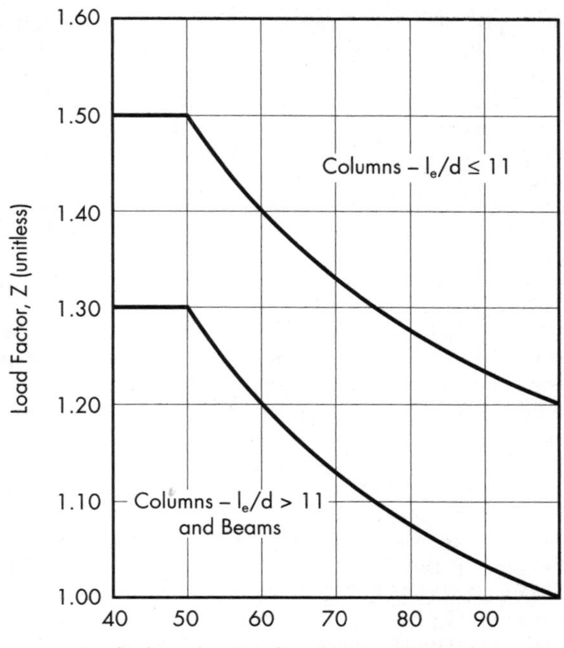

FIGURE 4.12 Factor Z as a percentage of design capacity.

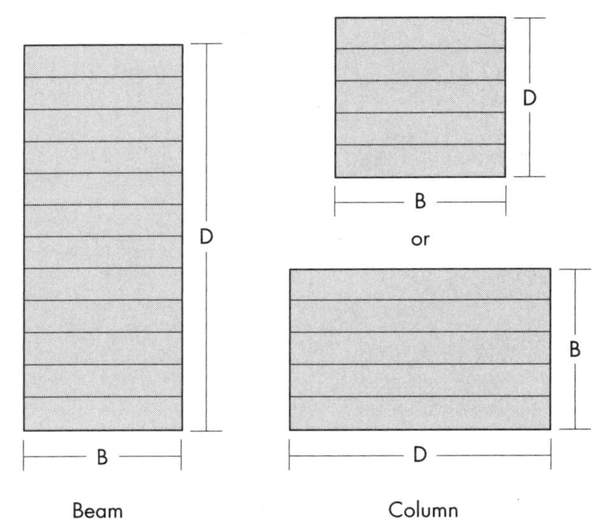

FIGURE 4.13 Dimensions for calculation of fire resistance, where B is always the lead dimension.

These equations apply to glued laminated timbers with a minimum nominal size of 6 × 6 in. or more before exposure to fire. Equations (4.10) and (4.12) are accurate only when the smallest dimension (B) is the side not exposed to the fire. When a beam or column is partially recessed into a wall, floor, or ceiling, the full dimension of the member, including the portion of the column recessed into the wall, floor or ceiling may be used in the calculations to obtain the maximum calculated fire resistance (the Standard Building Code requires the use of the full column dimension in calculating fire endurance).

Equations (4.12) and (4.13) are slightly altered from the way they appear in the U.S. building codes. The dimensions B and D are reversed to maintain consistency and clarity of notation on glued laminated timber beams and columns (see Fig. 4.13).

Tables 4.18, 4.19, and 4.20 show the minimum dimensions of a glued laminated timber member that will provide 100% design capacity and one-hour fire protection. These tables have been generated using Eqs. (4.10)–(4.13).

Beams and columns with dimensions less than those shown in Tables 4.18–4.20, but at least 6 × 6 in. nominal size, may meet the requirements for a one-hour fire resistance when the member is overdesigned for the applied load. This principle is demonstrated in the design examples that follow.

Specifying a One-Hour Fire-Resistant Glued Laminated Timber. Tension laminations of glued laminated timber beams are always positioned as the outermost laminations of the beam subjected to maximum tension stresses, and in a fire, the outermost fibers in a wood member are the first to be damaged. For this reason, when a one-hour rating is required for a glued laminated timber beam, the designer should specify one additional tension lamination in place of a core lamination (see Fig. 4.14) and the glued laminated timber should be marked "Fire-rated one-hour" by the manufacturer. For a balanced beam lay-up, an additional tension lamination

TABLE 4.18 Minimum Depths at Which Selected Beam Sizes Can Be Adopted for One-Hour Fire Ratings[a]

Member type	Beam									
Fire exposure	Fire three sides					Fire four sides				
Beam width (in.)[b]	6¾	8½	8¾	10½	10¾	6¾	8½	8¾	10½	10¾
Minimum depth (in.): 1½ in. thick laminations	13½	–	7½	–	6	27	–	13½	–	12
Minimum depth (in.): 1⅜ in. thick laminations	13⅜	6⅞	–	6⅞	–	27½	13¾	–	12⅜	–

[a] Assuming a load factor of 1.0 (design loads are equal to the capacity of the member). The minimum depths may be reduced when the design loads are less than the member capacity.

[b] Glued laminated timber members having a net width of 8½ in. or 10½ in. are typically manufactured using 1⅜ in. thick laminations. Glued laminated timber members having a net width of 8¾ in. or 10¾ in. are typically manufactured using 1½ in. thick laminations.

TABLE 4.19 Minimum Depths at Which Selected Column Sizes Can Be Adopted for One-Hour Fire Ratings with Fire Exposure from Three Sides[a]

Member type	Column							
Fire exposure	Fire four sides*							
$K_e l/d$ condition	≤11				>11			
Column width (in.)	8½	8¾	10½	10¾	8½	8¾	10½	10¾
Minimum depth (in.): 1½ in. thick laminations	–	9	–	7½	–	15	–	10½
Minimum depth (in.): 1⅜ in. thick laminations	8¼	–	8¼	–	19¼	–	9⅝	–

*Minimum dimensions are only valid when the unexposed side of the column is the smaller side.

[a] Assuming a load factor of 1.0 (design loads are equal to the capacity of the member). The minimum depths may be reduced when the design loads are less than the member capacity.

[b] Glued laminated timber members having a net width of 8½ in. or 10½ in. are typically manufactured using 1⅜ in. thick laminations. Glued laminated timber members having a net width of 8¾ in. or 10¾ in. are typically manufactured using 1½ in. thick laminations.

TABLE 4.20 Minimum Depths at Which Selected Column Sizes Can Be Adopted for One-Hour Fire Ratings with Fire Exposure from Four Sides[a]

Member type	Column							
Fire exposure	Fire four sides							
$K_e l/d$ condition	≤11				>11			
Column width (in.)	8½	8¾	10½	10¾	8½	8¾	10½	10¾
Minimum depth (in.): 1½ in. thick laminations	–	12	–	10½	–	30	–	13½
Minimum depth (in.): 1⅜ in. thick laminations	12⅜	–	9⅝	–	38½	–	13¾	–

[a] Assuming a load factor of 1.0 (design loads are equal to the capacity of the member). The minimum depths may be reduced when the design loads are less than the member capacity.

[b] Glued laminated timber members having a net width of 8½ or 10½ are typically manufactured using 1⅜ in. thick laminations. Glued laminated timber members having a net width of 8¾ in. or 10¾ in. are typically manufactured using 1½ in. thick laminations.

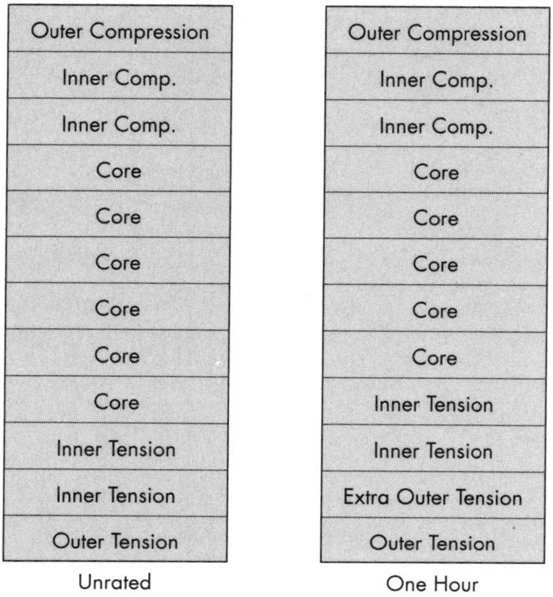

Unrated	One Hour
Outer Compression	Outer Compression
Inner Comp.	Inner Comp.
Inner Comp.	Inner Comp.
Core	Core
Core	Core
Core	Core
Core	Core
Core	Core
Core	Inner Tension
Inner Tension	Inner Tension
Inner Tension	Extra Outer Tension
Outer Tension	Outer Tension

FIGURE 4.14 Typical unbalanced glued laminated timber beam lay-ups for unrated and one-hour fire resistance (one extra tension lamination added for one-hour resistance).

should be added to both outer zones. An additional tension lamination is not required for columns and arches.

Fasteners. Because metal fasteners conduct heat directly into the member, exposed fasteners must be given rated protection from fire that is equivalent to that expected of the member. For a one-hour rating, sufficient wood, gypsum wallboard, or other material must be applied to protect the exposed portions of the fasteners for one hour. This may be 1½ in. (38 mm) of wood, ⅝ in. (16 mm) type X gypsum board or other approved material. Example details can be found in Figs. 4.15–4.20.

Design Examples for One-Hour Fire Rating

Beam Example Assume a simply supported roof beam is to span 30 ft, carry 240 lb/ft of total load (dead load plus snow load), and be used in a dry service condition. It is continuously supported along its compression side and will have three sides exposed to fire. A one-hour rating is required. The beam used will be a 24F-V4/DF (Douglas fir) with the following allowable design stresses:

$$F_b = 2400 \text{ psi}$$

$$E = 1.8 \times 10^6 \text{ psi}$$

$$F_v = 240 \text{ psi}$$

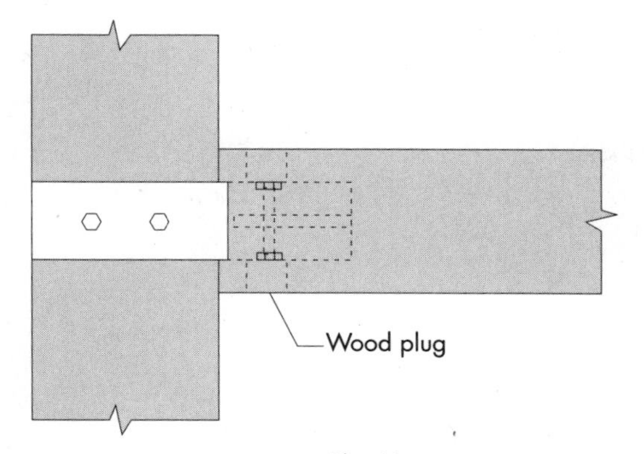

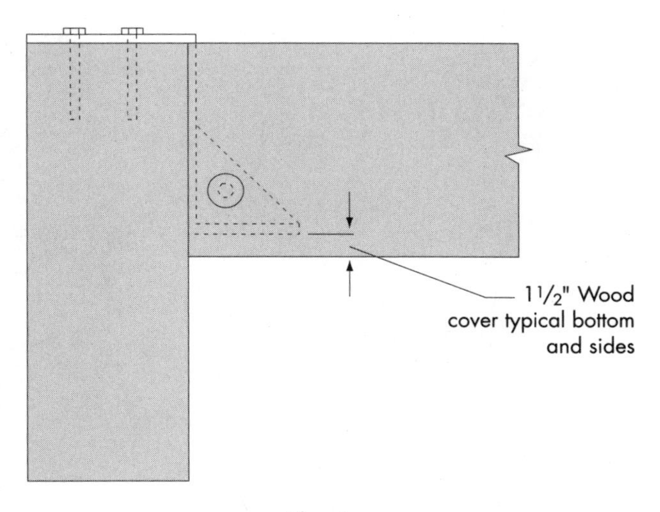

FIGURE 4.15 Beam-to-girder concealed connection.

What size glued laminated timber beam should be used?

Use an initial estimate of the dead weight of the beam as 25 lb/ft. Total design load = 240 + 25 = 265 lb/ft.

From Table 4.25, select a 5⅛ × 13½ beam with total capacity of 266 lb/ft, which is greater than 265 lb/ft.

Determining the actual beam depth that will continue to carry the design load for one hour is aided by the use of Fig. 4.21. From this graph, the range of depths that might be practical to use can by anywhere from approximately 12–32 in. Obviously, this beam must be deeper than 13½ in. as the beam in this example is stressed to approximately 100% of design capacity.

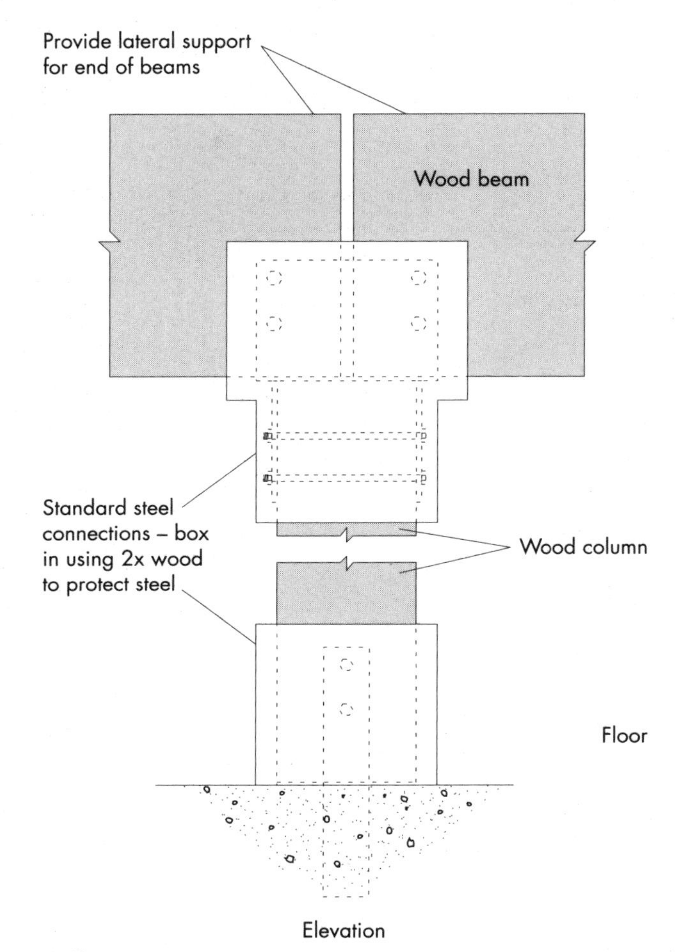

FIGURE 4.16 Covered column connections.

All of these 5⅛ in. wide beams will retain 50–60% of design after one hour. An initial estimate of a percentage that corresponds with a practical beam depth is 55%.

$$\text{Approximate } S_{\text{Required}} \frac{1.42}{0.55} = 258 \text{ in.}^3$$

$$\text{Approximate } d_{\text{Required}} = \sqrt{\frac{6S}{b}} = \sqrt{\frac{6(258)}{5.125}} = 17.4 \text{ in.}$$

Try an 18 in. deep beam with $S = 277$ in.^{3}

Determine if this beam will have sufficient strength left after one hour of fire exposure to continue to carry the design load by determining the ratio of applied moment to design flexural capacity. Beam size will also have to be checked for shear and deflection.

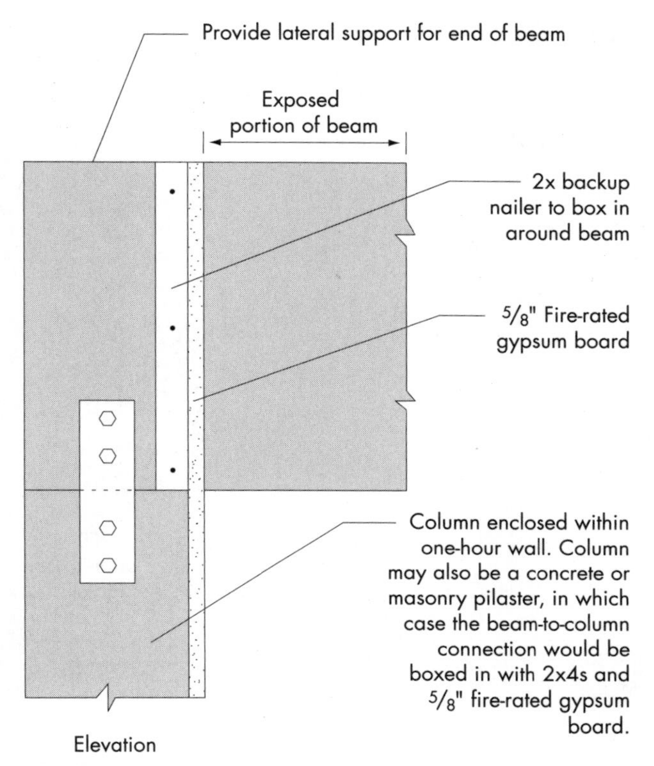

FIGURE 4.17 Beam-to-column connection when the connection is not exposed to fire.

Determine F'_b

$$\text{Volume factor} = C_v = \left(\frac{21}{L}\right)^{1/10}\left(\frac{12}{d}\right)^{1/10}\left(\frac{5.125}{b}\right)^{1/10}$$

$$= C_v = \left(\frac{21}{30}\right)^{1/10}\left(\frac{12}{18}\right)^{1/10}\left(\frac{5.125}{5.125}\right)^{1/10} = 0.9266$$

$$F'_b = F_b C_D C_D = (2400)(1.15)(0.9266) = 2557 \text{ lb/in.}^2$$

Determine f_b:

Calculate beam weight using 35 lb/ft³

$$\text{Beam} = \frac{(5.125)(18)(12)}{12^3}(35) = 22.4 \text{ lb/ft}$$

$$M_{\text{Applied}} = \frac{wL^2}{8} = \frac{(240 + 22.4)\ 30^2}{8} = 29{,}520 \text{ ft-lb} = 354{,}240 \text{ in.-lb}$$

$$f_b = \frac{M}{S} = \frac{354{,}240}{277} = 1279 \text{ psi}$$

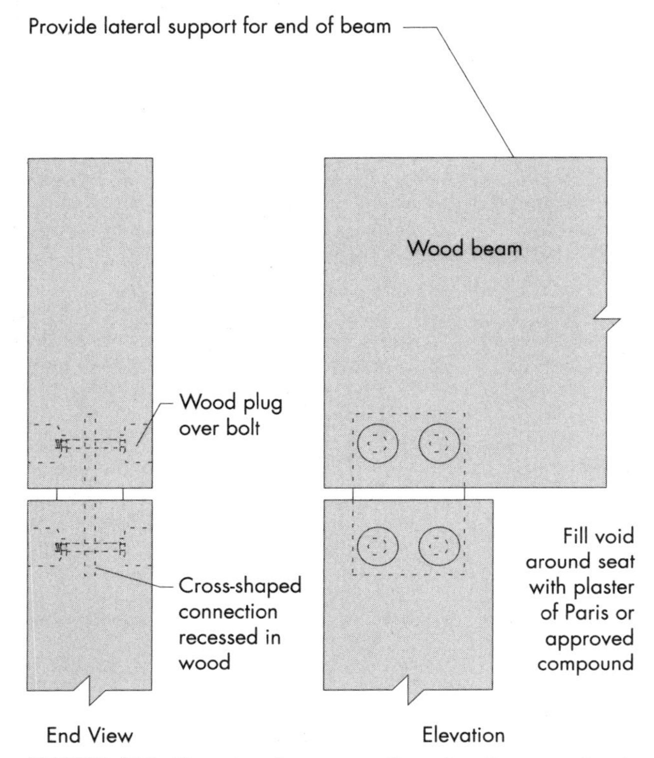

FIGURE 4.18 Beam-to-column connection when the connection is exposed to fire and appearance is a factor.

Check the ratio of applied moment to flexural capacity

$$\frac{f_b}{F'_b} = \frac{1279}{2557} = 0.500 < 55\% \rightarrow \text{OK}$$

Check fire endurance:

From Fig. 4.12, for a beam loaded to 50% of capacity, Z is approximately 1.3. Using Eq. (4.10):

$$t = 2.54(1.3)(5.125)\left(4 - \frac{5.125}{18.0}\right) = 62.9 \text{ minutes} > 60 \rightarrow \text{OK}$$

This beam has a moment capacity that is significantly greater than is needed if the one-hour fire resistance is not a requirement. In some cases, a wider beam may be required to satisfy a beam depth limitation while still meeting the one-hour fire resistance requirement. For instance, any 6¾ in. wide beam that is 13½ in. deep, and has the extra tension lamination, will carry 100% of its design load after one hour of fire exposure on three sides (Table 4.18 and Fig. 4.21).

The designer will also need to confirm that the design shear and deflection values for the trial beam size are less than 50% of these capacities. When specifying the beam,

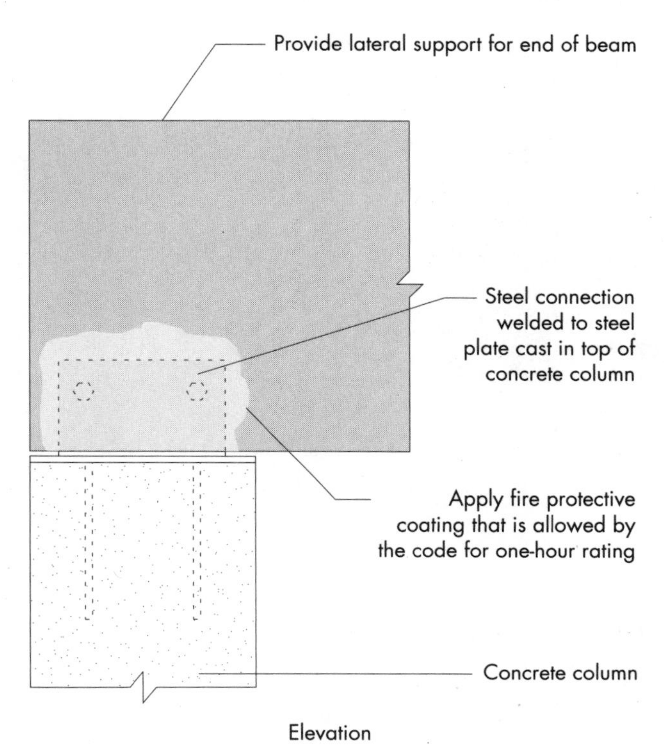

FIGURE 4-19 Beam-to-column connection when the connection is exposed to fire and appearance is not a factor.

advise the manufacturer to eliminate one core lamination and substitute one additional tension lamination (Fig. 4.14) and mark the beam "Fire-rated one-hour."

Column Example An existing building is to be remodeled with a change of occupancy requiring that the glued laminated timber columns meet a one-hour fire resistance. The existing glued laminated timber column is 22 ft high, measures 8¾ in. wide by 10½ in. deep, and will remain dry in service. It supports a concentrated total floor load ($DL + LL$) of 50,000 lb (C_D = 1.0) applied concentrically to the top of the column. The column is not subjected to any lateral loads. From the original specifications, the glued laminated timber is a Douglas fir combination 5. Determine if the column is adequate to carry the imposed axial load for one hour with fire on four sides and how long it can be expected to carry the applied load. If it is not adequate, determine what size column is required.

To answer these questions, the total load capacity of the column must be determined along with the percentage of the total load capacity used by the applied load and the partial-load compensation factor, Z.

Determine load capacity:

$$d = 8.75 \text{ in.}$$

$$A = 8.75(10.5) = 91.9 \text{ in.}^2$$

$$C_D = 1.0 \text{ for } DL \text{ plus floor } LL$$

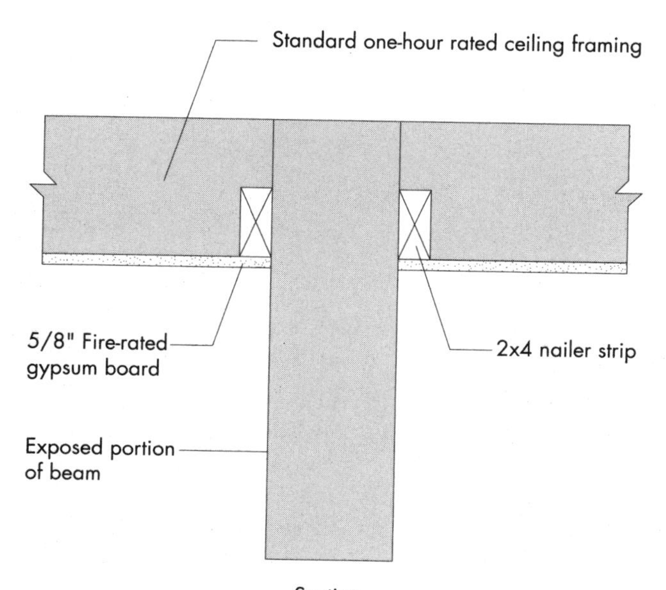

FIGURE 4.20 Connections for ceiling construction.

$$E = E' = 2{,}000{,}000 \text{ psi}$$

$$F_c = 2400 \text{ psi}$$

$$l = 22(12) = 264 \text{ in.}$$

$$K_e = 1.0 \text{ (column is assumed to be pinned at both ends)}$$

$$l_e = lK_e = 264(1.0) = 264 \text{ in.}$$

$$\frac{l_e}{d} = 264/8.75 = 30.17$$

$$K_{cE} = 0.418$$

$$c = 0.9$$

where E = tabulated modulus of elasticity (psi)
F_c = tabulated compression design value parallel to grain (psi)
A = area of cross section (in.^{2})
d = least dimension being evaluated for potential buckling (in.)
L = column length (ft)
l = length of column (in.)
l_e = lK_e = effective length of column (in.)
K_e = buckling length coefficient for compression members
c = coefficient that depends on member type (0.9 for glued laminated timber)
K_{cE} = coefficient that depends on the coefficient of variation of the member (0.418 for glued laminated timber)
F_c^* = tabulated compression design value multiplied by all applicable adjustment factors except C_p (psi)

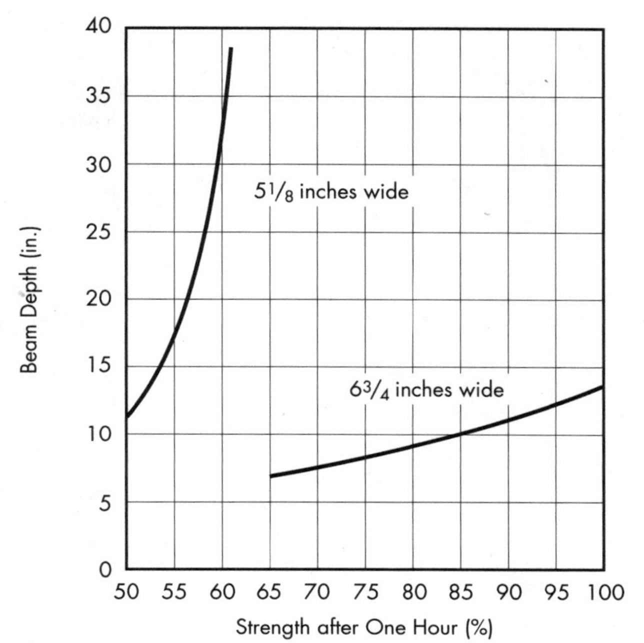

FIGURE 4.21 Beam—Fire three sides.

$$F_{cE} = \frac{K_{cE}E'}{\left(\frac{l_e}{d}\right)^2} = \frac{0.418(2{,}000{,}000)}{30.17^2} = 918 \text{ psi}$$

$$\frac{F_{cE}}{F_c^*} = \frac{918}{2400} = 0.382$$

$$C_p = \frac{1 + \frac{F_{cE}}{F_c^*}}{2c} - \sqrt{\left[\frac{1 + \frac{F_{cE}}{F_c^*}}{2c}\right] - \frac{\left(\frac{F_{cE}}{F_c^*}\right)}{c}}$$

$$= \frac{1 + 0.382}{2(0.9)} - \sqrt{\left[\frac{1 + 0.382}{2(0.9)}\right]^2 - \frac{0.382}{0.9}} = 0.362$$

$$F_c' = F_c^* C_p = \text{allowable compressive stress (psi)}$$

$$F_c' = 0.362(2400) = 869 \text{ psi}$$

$$\text{Axial load capacity} = AF_c' = 91.9(869)$$

$$= 79\ 861 \text{ lb} > 50\ 000 \rightarrow \text{OK for design axial load}$$

Check fire endurance based on ratio of applied load to design capacity:

$$\frac{\text{Applied load}}{\text{Design capacity}} = \frac{50{,}000}{79{,}861} = 0.626 = 62.6\%$$

From Fig. 4.24, for a column 8¾ in. wide and $l_e/d > 11$, 62.6% corresponds to about a 12 in. depth, which is greater than the existing column's depth of 10½ in. The existing column will therefore not carry the applied load for one hour. To check, calculate the fire endurance.

From Fig, 4.12, for a column with $l_e/d > 11$ and the load at 62.6% of capacity, Z is approximately 1.18.

Using Eq. (4.13):

$$B = 8.75 \text{ in.}$$

$$D = 10.5 \text{ in.}$$

$$t = 2.54ZB\left[3 - \frac{B}{D}\right] = 2.54(1.18)(8.75)\left[3 - \frac{8.75}{10.5}\right] = 57 \text{ minutes} < 60 \rightarrow \text{NG}$$

The existing column is inadequate to meet the one-hour fire-resistance requirement even though it is adequate to carry the applied load in occupancies not requiring a one-hour fire rating.

Determine column size necessary to carry the design load and meet the one-hour requirement.

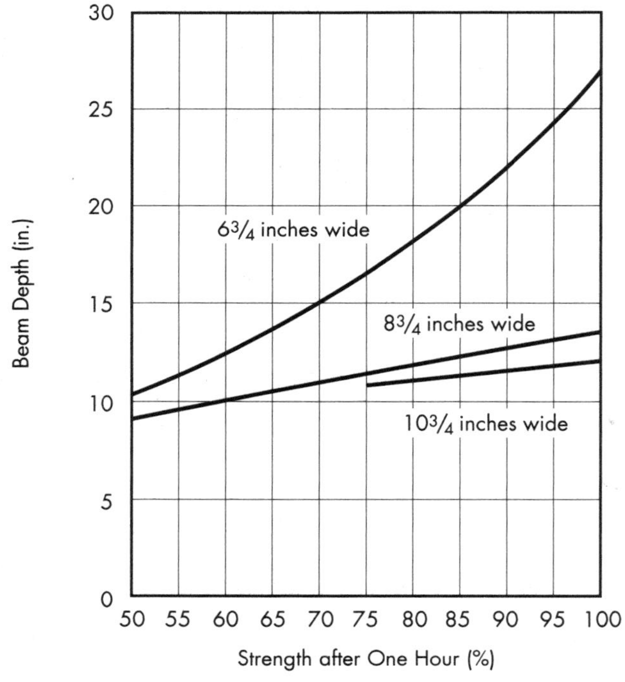

FIGURE 4.22 Beam—Fire four sides.

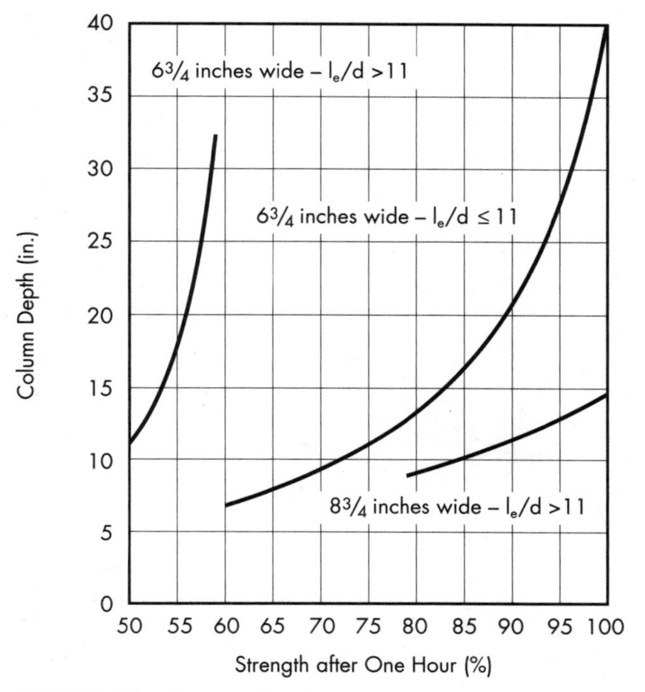

FIGURE 4.23 Column—Fire three sides.

Using Fig. 4.24 as a guide, try a 10¾ in. × 10½ in. glued laminated timber, Douglas fir combination 5. Another trial option would be 8¾ × 13½ in.

$$A = 10.75(10.5) = 112.875 \text{ in.}^2$$

$$l_e/d = 264/10.5 = 25.14$$

$$F_{cE} = \frac{K_{cE}E'}{\left(\frac{l_e}{d}\right)^2} = \frac{0.418(2{,}000{,}000)}{25.14^2} = 1323 \text{ psi}$$

$$F_c^* = 2400 \text{ psi}$$

$$\frac{F_{cE}}{F_c^*} = \frac{1323}{2400} = 0.5513$$

$$C_p = \frac{1 + \frac{F_{cE}}{F_c^*}}{2c} - \sqrt{\left[\frac{1 + \frac{F_{cE}}{F_c^*}}{2c}\right] - \left(\frac{\frac{F_{cE}}{F_c^*}}{c}\right)}$$

$$= \frac{1 + 0.553}{2(0.9)} - \sqrt{\left[\frac{1 + 0.553}{2(0.9)}\right]^2 - \frac{0.5513}{0.9}} = 0.5010$$

$$F_c' = 0.5010(2400) = 1202 \text{ psi}$$

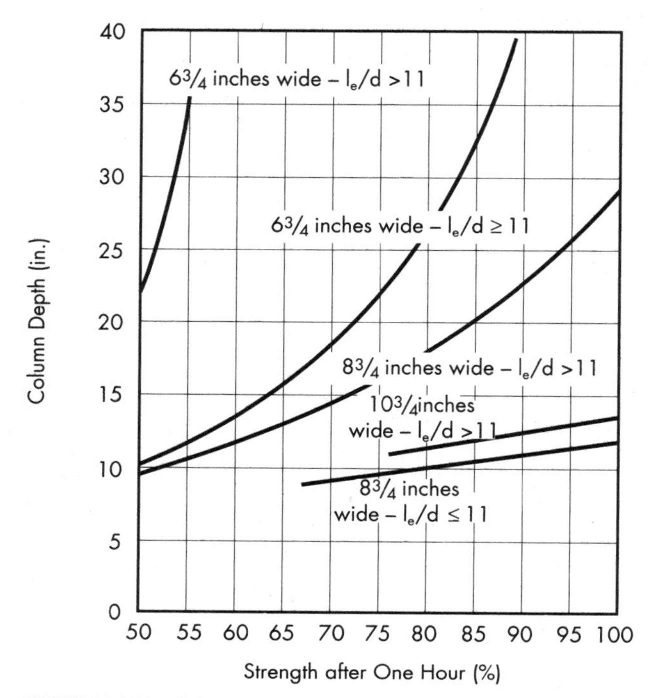

FIGURE 4.24 Column—Fire four sides.

$$\text{Axial load capacity} = AF'_c = 112.875(1202) = 135{,}676 \text{ lb} \rightarrow >50{,}000 \text{ lb}$$

Check the fire endurance:

$$\frac{\text{Applied load}}{\text{Maximum capacity}} = \frac{50{,}000}{135{,}676} = 0.368 = 36.8\%$$

Z, from Fig. 4.12, is 1.3.
Using Eq. (4.13):

$$B = 10.5 \text{ in.}$$

$$D = 10.75 \text{ in.}$$

$$t = 2.54ZB\left[3 - \frac{B}{D}\right] = 2.54(1.3)(10.5)\left[3 - \frac{10.5}{10.75}\right]$$

$$= 70 \text{ minutes} > 60 \rightarrow \text{OK}$$

Summary. As shown by the preceding examples, glued laminated timber members can be designed to provide a one-hour fire rating when required. Tables 4.18, 4.19, and 4.20 provide basic minimum dimensions for one-hour fire-rated glued laminated timber beams and columns when the applied load represents 100% of the member design capacity. Figures 4.21–4.24 provide estimated sizes of beams and columns that will satisfy a requirement for a one-hour fire rating when the member is loaded to less than 100% of capacity.

4.6.4 Moisture and Decay

As evidenced by buildings worldwide, wood construction can provide centuries of service life. However, as a natural, organic material, wood is susceptible to degradation by organisms under certain conditions. Further discussions of the cause and prevention of decay fungi are provided in Chapters 1, 9, and 12 of this Handbook.

4.6.5 Preservative Treatments of Glued Laminated Timber

Although glued laminated timbers do not require preservative treatment for most uses, certain applications may present environmental conditions conducive to decay or insect or marine borer attack. Conditions that favor such attack are the presence of oxygen and moisture (20% or greater moisture content of the wood) accompanied by temperatures ranging from 50 to 90°F. Decay progresses more slowly at temperatures outside this range and virtually ceases at temperatures below 35 or above 100°F. These hazards are typically controlled through recognized design principles and construction techniques such as use of overhangs, flashings, ventilation, and proper joint connection details. Elimination of potential decay hazards through effective design detailing is the preferred method of controlling decay. When hazards of decay or insect attack cannot be avoided, glued laminated timbers must be preservative treated. Examples of construction where such hazards may exist include direct exposure to weather, ground contact (including direct contact with concrete foundations and footings), contact with fresh water or seawater, and exposure to excessive condensation. Detailed descriptions of preservative treatments for glued laminated timber are covered in Chapter 9 of this Handbook and in APA EWS Technical Note S580, *Preservative Treatment of Glulam Beams.*[3]

4.6.6 Connection Details for Glued Laminated Timber

Introduction. Proper connection details are important to the structural performance and serviceability of any timber-framed structure. While this is true for solid sawn as well as glued laminated (glued laminated timber) timbers, the larger sizes and longer spans made possible with glued laminated timber components make the proper detailing of connections even more critical. Careful consideration of moisture-related expansion and contraction characteristics of wood is essential in detailing glued laminated timber connections to prevent inducing tension perpendicular-to-grain stresses. Connections must be designed to transfer design loads to and from the structural glued laminated timber member without causing localized stress concentrations that may initiate failure at the connection.

It is also important to design connections to isolate all wood members from potential sources of excessive moisture. In addition to accentuating any connection problems related to expansion or contraction of the wood due to moisture cycling, equilibrium moisture content in excess of approximately 20% may promote the growth of decay-causing organisms in untreated wood.

Structural Effects of Shrinkage and Improper Detailing. Wood expands and contracts as a result of changes in its internal moisture content. While expansion in the direction parallel to grain in a wood member is minimal, dimensional change in the direction perpendicular to grain can be significant and must be considered in connection design and detailing. A 24 in. deep beam can decrease in depth through shrinkage by approximately ⅛ in. as it changes from 12 to 8% in equilibrium moisture content. In designing connections for glued laminated timber members it is important to design and detail the connection such that the member's shrinkage is not restrained. If restrained, shrinkage of the beam can cause tension perpendicular-to-grain stresses to develop in the member at the connection. If these stresses exceed the capacity of the member, they may cause the glued laminated timber to split parallel to the grain. Once a tension-splitting failure has occurred in a member, its shear and bending capacity are greatly reduced.

In addition to the moisture-induced tension perpendicular-to-grain failures discussed above, similar failures can result from a number of different incorrect connection design details. Improper beam notching, eccentric (out-of-plane) loading of truss connections and loading beams from the tension side can induce internal moments and tension perpendicular-to-grain stresses.

Effects of Moisture Accumulation. Because most connections occur at the ends of beams where the wood end grain is exposed, it is critical that these connections be designed to prevent moisture accumulation. This can usually be accomplished by detailing drain holes or slots in box-type connectors and by maintaining a gap of at least ½ in. between the wood and concrete or masonry construction. Because most connections require the exposure of end grain due to fastener penetration, even those connections that occur away from beam ends must be considered potential decay locations. Field studies have shown that any metal connectors or parts of connectors that are placed in the cold zone of the building (that area outside of the building insulation envelope) can become condensation points for ambient moisture. This moisture has ready access to the inside of the beam through fasteners and exposed end grain. A few examples of these kinds of fasteners are saddle-type hangers, cantilever beam hinges, and beam-to-column connectors.

Connection Examples. Figure 4.25a–t illustrates various connection types common to the framing of glued laminated timber. These illustrations show correct connection details along with examples of common incorrect details and a discussion of the failures that may occur due to the incorrect detailing. While the figures are not all-inclusive, they are provided as a tool to illustrate the principles discussed in the preceding section. Reviewing the examples with these principles in mind will enable the designer to detail proper connections.

While the details in this section address serviceability concerns associated with glued laminated timber connection detailing, it is important to emphasize that all connection details must effectively transfer the design loads imposed on the structure and that all designs be in accordance with accepted engineering practice. There are a number of manufacturers of preengineered metal connectors which have been specifically designed for use in glued laminated timber framing, and it is recommended that these connectors be used whenever possible.

Summary. The details given in this section have been provided to illustrate both the correct and the incorrect manner of making a connection involving glued laminated timbers. These details emphasize seven basic principles, which if followed will lead to efficient, durable and structurally sound connections:

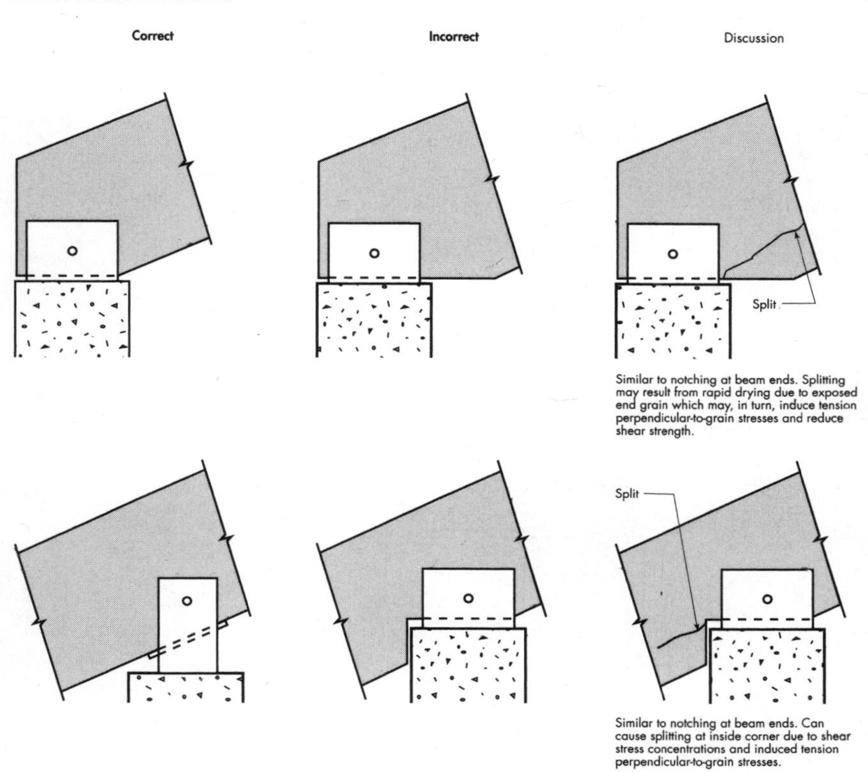

FIGURE 4.25 (*a*) Connection details.

1. Transfer loads in compression bearing whenever possible.
2. Allow for dimensional changes in the glued laminated timber due to potential in-service moisture cycling.
3. Avoid the use of details that induce tension perpendicular-to-grain stresses in the member.
4. Avoid moisture entrapment at connections.
5. Do not place the glued laminated timber in direct contact with masonry or concrete.
6. Avoid eccentricity in joint details.
7. Minimize exposure of end grain.

4.6.7 Allowable Depth-to-Width Ratios

The NDS[6] has, since the 1944 edition, provided prescriptive lateral bracing recommendations for wood bending members having varying depth-to-width (d/b)

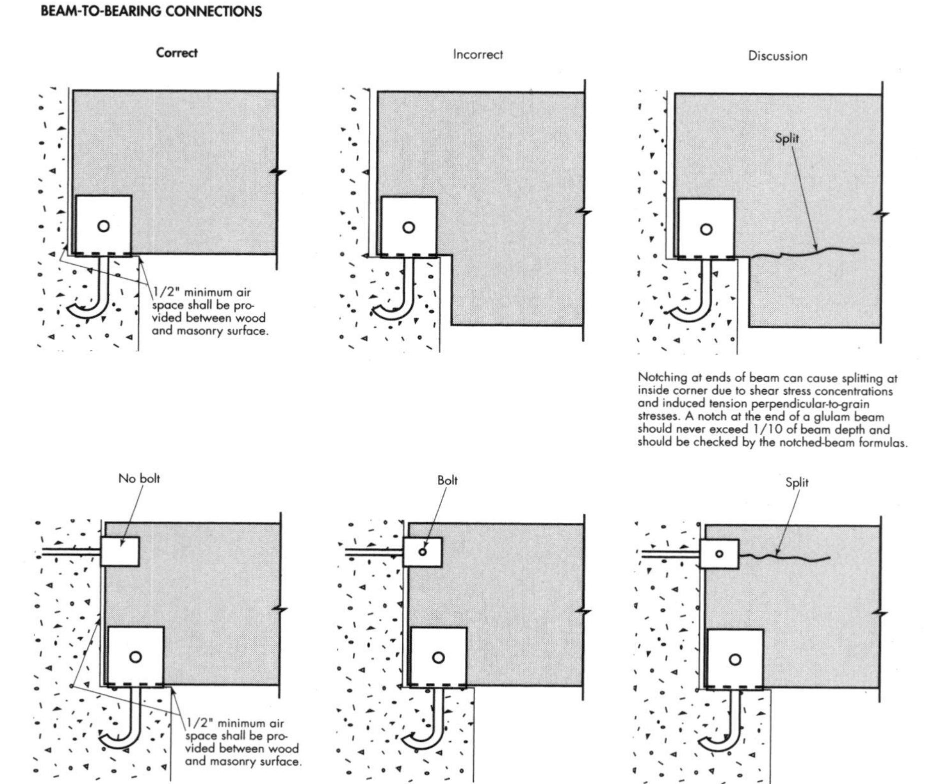

FIGURE 4.25 (*b*) Connection details. (*Continued*)

ratios. Current guidelines for the lower and upper limits of bracing requirements are shown below, with bracing requirements for intermediate d/b ratios falling somewhere between these.

$d/b \leq 2$	no lateral bracing required
$6 < d/b \leq 7$	lateral restraint required at the ends to prevent torsion and lateral displacement and both edges must be laterally restrained over their full lengths

While intended for sawn lumber members, these guidelines are often interpreted as being applicable to other rectangular-shaped engineered wood products such as glued laminated timber. Also, since no mention is made of d/b ratios greater than 7, it is often assumed that the maximum d/b ratio for a wood bending member is

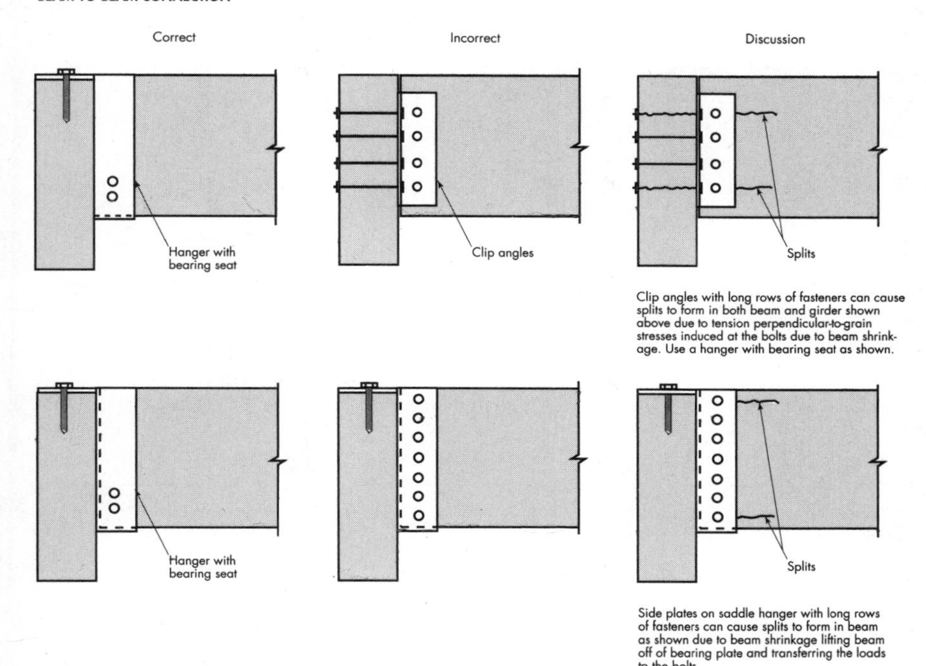

FIGURE 4.25 (*c*) Connection details. (*Continued*)

7 when, in fact, a wood bending member having any d/b ratio can be designed using the beam stability calculations provided in the NDS.

Equations, experimentally verified, for calculating critical lateral buckling loads in rectangular shaped wood beams have been available since 1931. Research conducted at the University of British Columbia in the 1960s[7,8] led to the adoption of provisions in the 1968 edition of the NDS for determining the adequacy of lateral support for glued laminated timber bending members. One of their test beams was a glued laminated timber beam with a cross section of $3\frac{1}{4} \times 66\frac{5}{8}$ in. (d/b ratio = 20.5). While this represented an extreme example of a high d/b ratio, no lateral buckling of this beam was observed at failure when the compression edge was fully restrained laterally.

One of the advantages of glued laminated timber is that it can be manufactured in virtually any size or shape and one of the most efficient configurations is a narrow deep bending member. Since the early 1980s, one of the most popular forms of glued laminated timber has been resawn beams used as purlins in panelized roof construction. These members are typically $2\frac{1}{2}$ in. in width with depths up to $28\frac{1}{2}$ in. with a d/b ratio of 11.4. An emerging technology uses high-strength fiber-reinforced plastic to reinforce the tension zone of glued laminated timber beams, with resulting d/b ratios of 10 or greater.

When the narrow, deep resawn glued laminated timber purlins were introduced, concern was expressed by structural engineers regarding their inherent lateral sta-

(d)

BEAM-TO-BEAM CONNECTION

Correct | **Incorrect** | **Discussion**

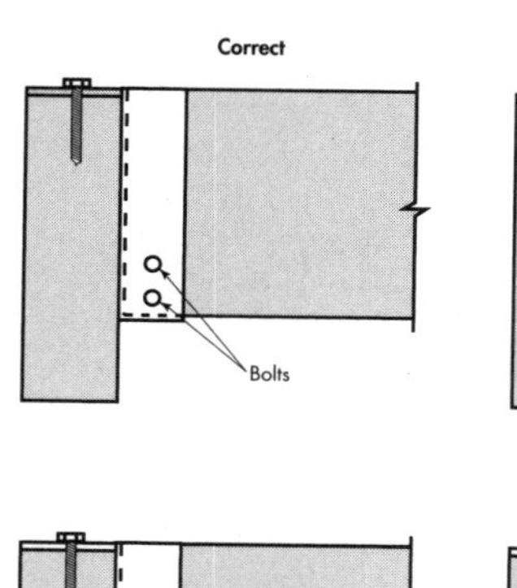

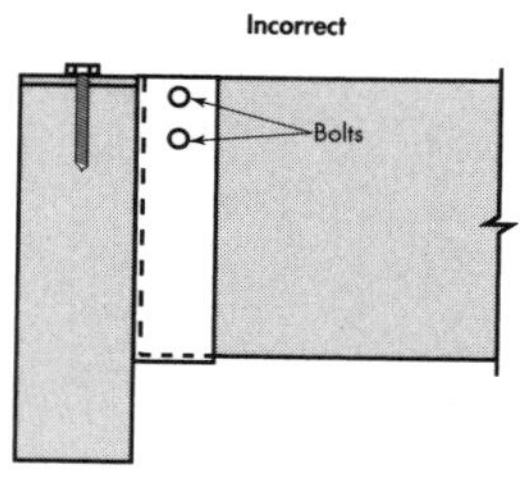

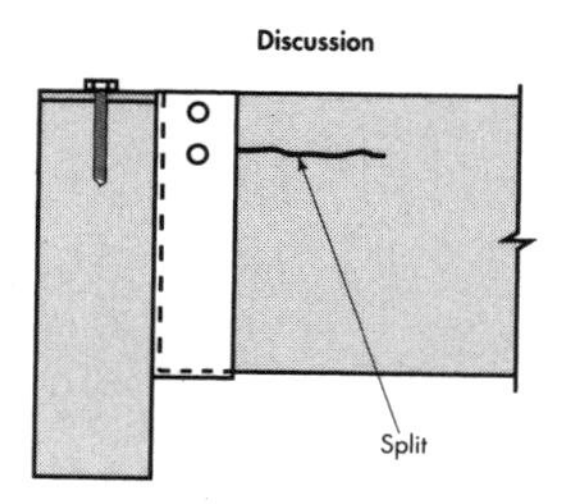

Shrinkage of supported beam causes bearing load to transfer from beam saddle to bolts. This can cause splitting of beam.

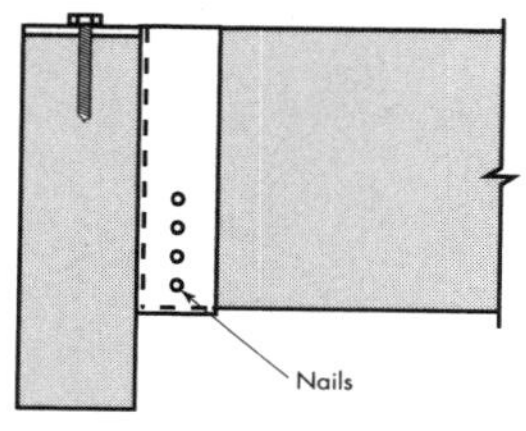

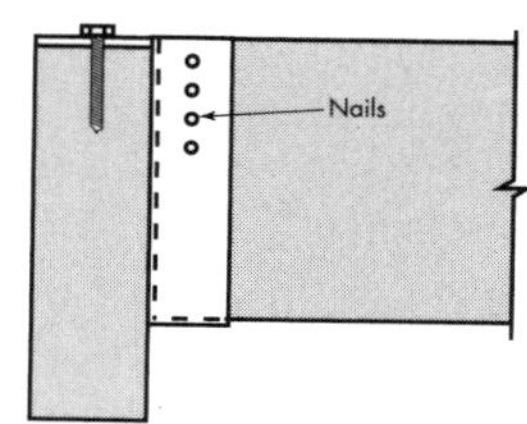

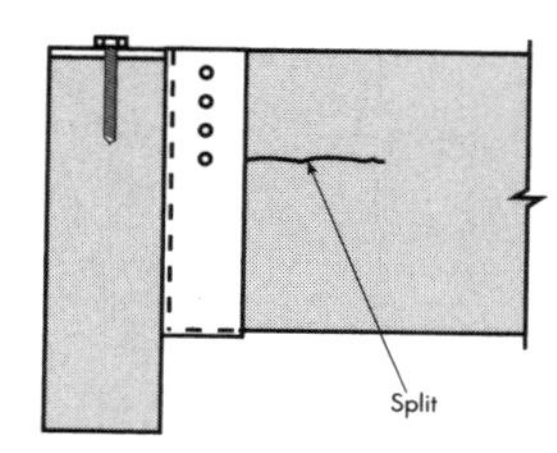

Shrinkage of supported beam causes bearing load to transfer from beam saddle to nail group. This can cause splitting of beam.

(e)

BEAM-TO-BEAM CONNECTION

Correct | **Incorrect** | **Discussion**

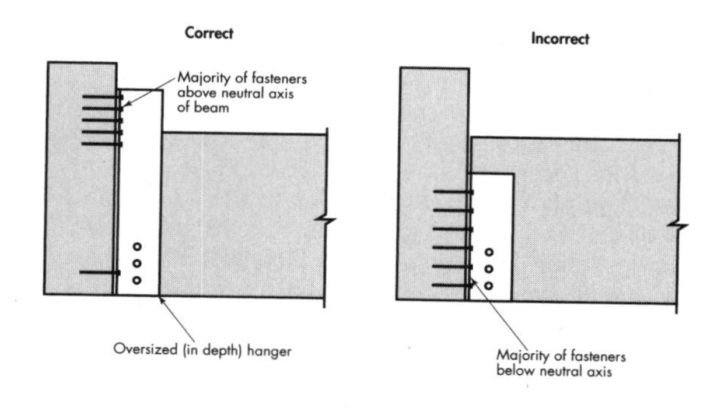

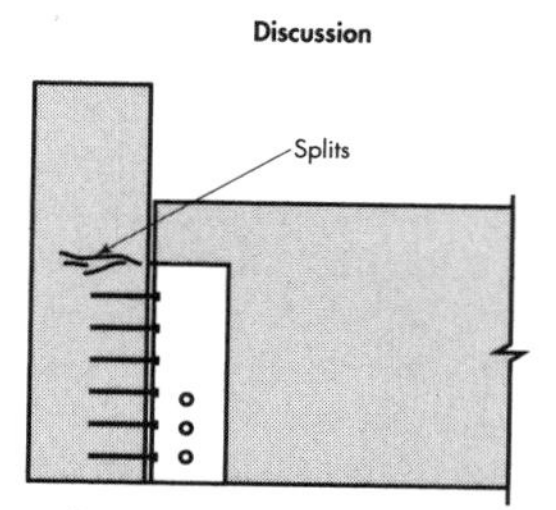

Application of load via fasteners below the neutral axis can cause a tension-perpendicular-to-grain failure in the beam. Location of majority of fasteners above neutral axis or use of top-mounted hanger will minimize the possibility of splitting of the beam. Note that when face-mounted hangers are used, oversized (in depth) hangers may be required to place majority of fasteners above neutral axis.

FIGURE 4.25 (*d*)(*e*) Connection details. (*Continued*)

(f)

BEAM-TO-BEARING CONNECTION – FIRE CUT

Correct

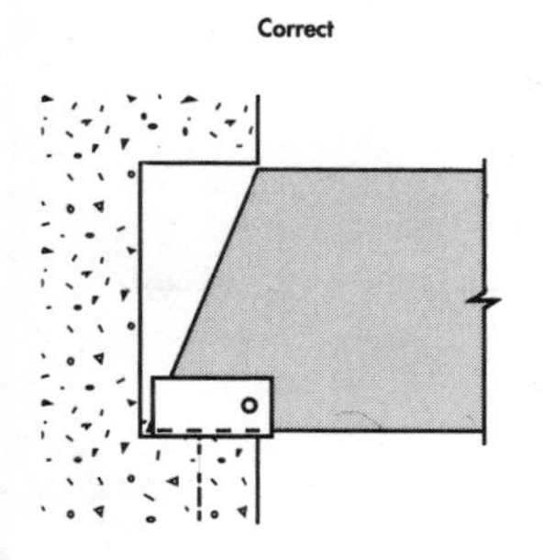

Incorrect

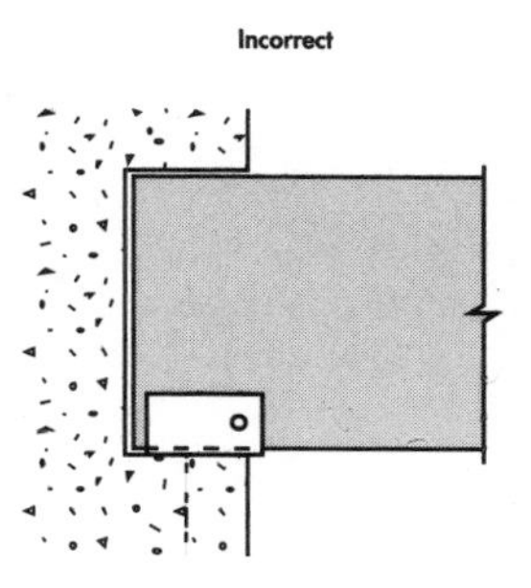

Discussion

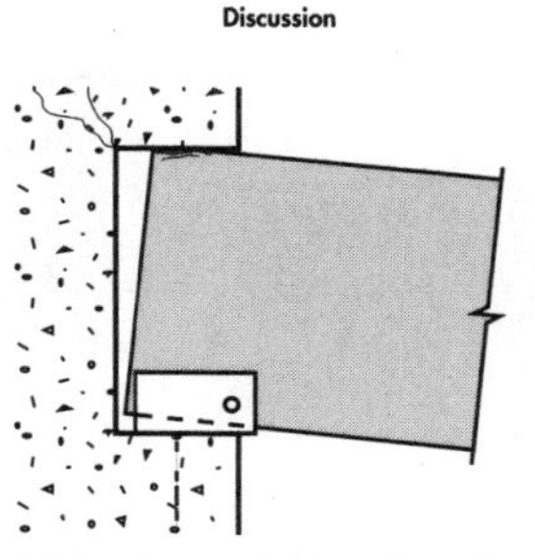

Deflection of square end-cut beams during a fire can cause structural damage to bearing wall. While such a failure is unlikely due to the excellent performance of heavy timber construction during fires, such detailing is prudent.

(g)

BEAM-TO-BEAM CONNECTIONS – SEMI-CONCEALED USING FISH PLATES

Correct

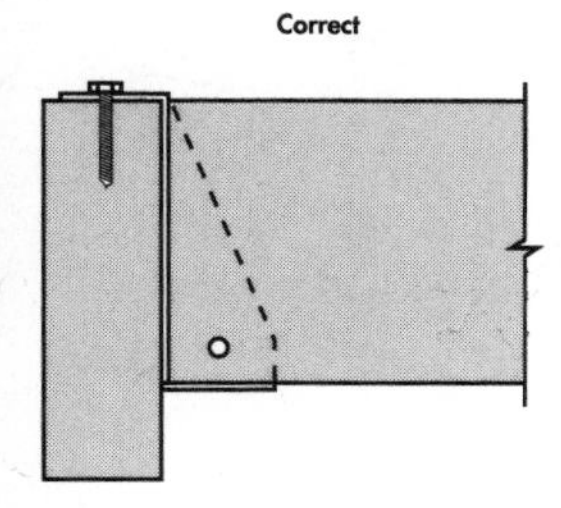

Incorrect

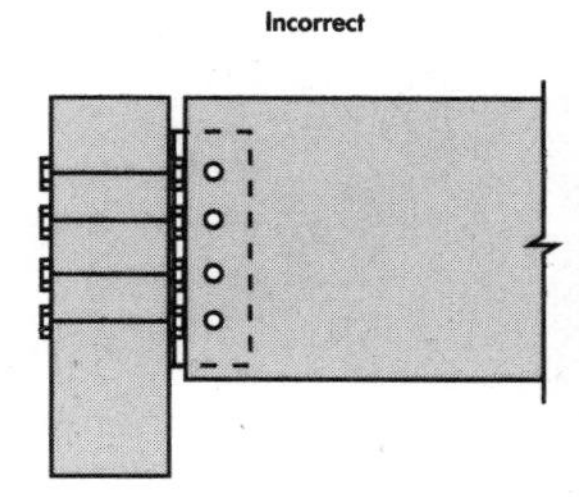

Discussion

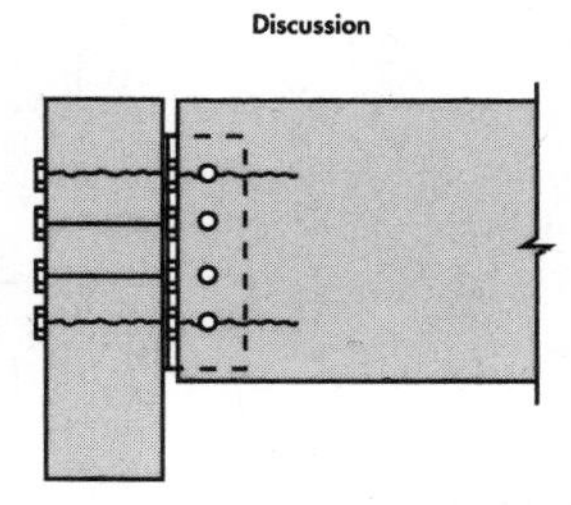

Concealed fish plate with long row of fasteners can cause splits to form in both beam and girder as shown above. Use a fish plate with bearing seat as shown to the left.

(h)

HEAVY CONCENTRATED LOADS SUSPENDED FROM BEAM

Correct

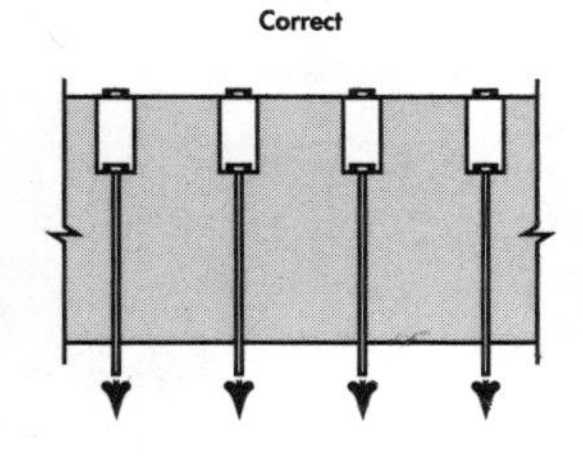

Incorrect

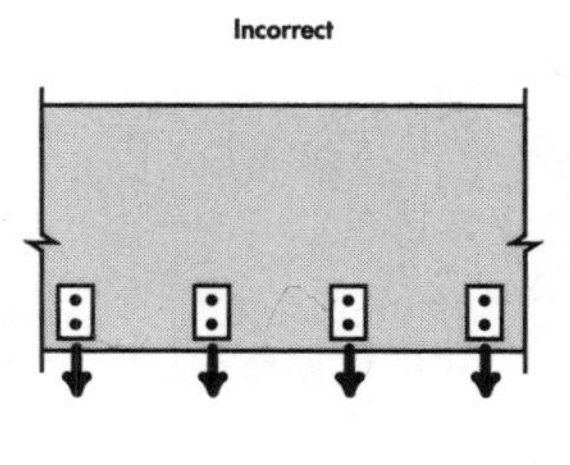

Discussion

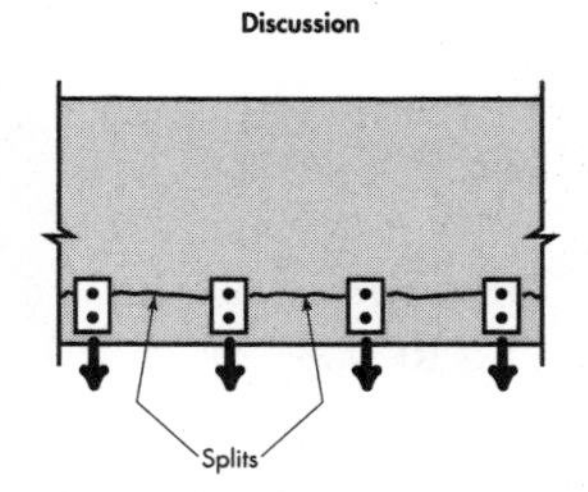

Heavy concentrated loads such as heating and air conditioning units, crane rails or main framing members suspended from the bottom of beams induce tension perpendicular-to-grain stresses and may cause splits as shown.

This is not intended to apply to light loads such as from 2x – joists attached to the main beam with light gauge nail-on metal hangers.

FIGURE 4.25 (*f–h*) Connection details. (*Continued*)

(i)

NOTCH IN BEAM OVER COLUMN

Correct

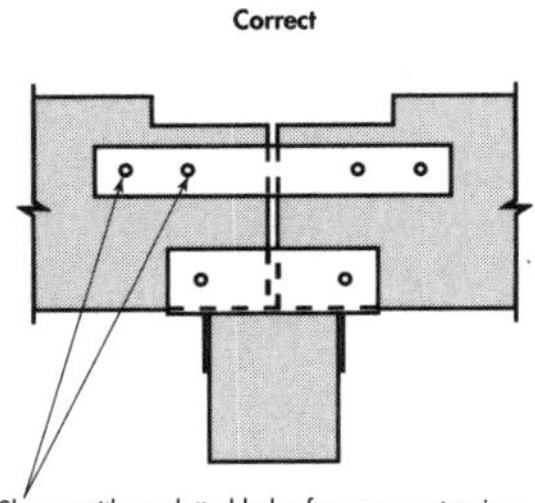

Shown with no slotted holes for use as a tension tie. Design must insure no excessive rotation of beams under load.

If used as a lateral support plate only, slotted holes may be used with no further restrictions on beam rotation required.

Incorrect

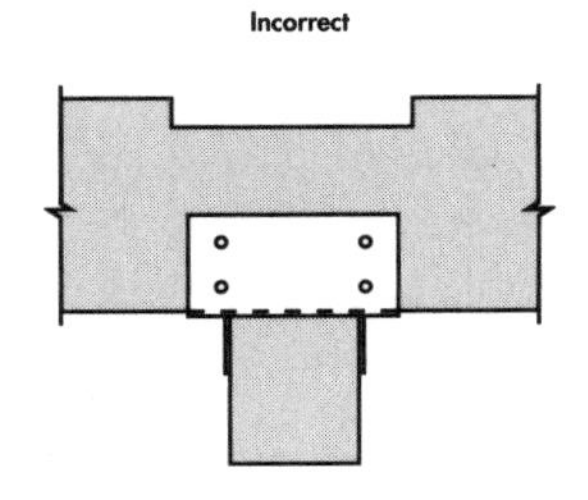

Discussion

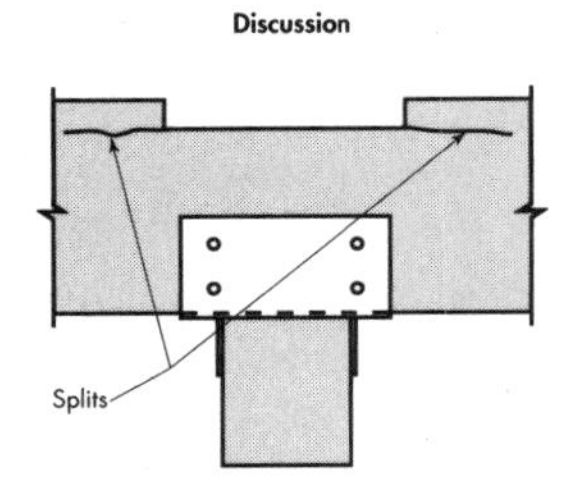

A notch in the top of a continuous beam over a center support occurs in the tension zone of the beam, greatly reducing its capacity. Design as two simply supported beams if top notch is required.

(j)

CANTILEVER BEAM CONNECTION – INDEPENDENT TENSION TIE

Correct

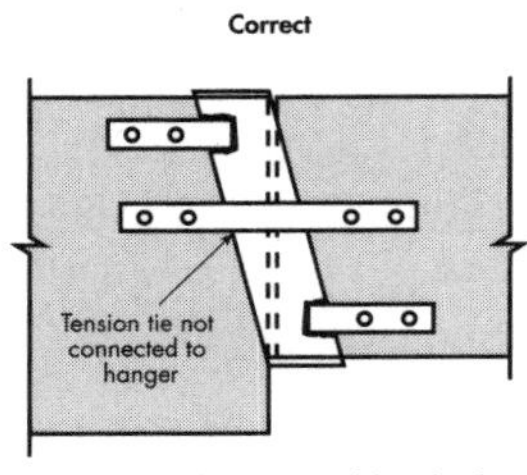

The relative vertical positioning of the side tabs shown in this detail is very important to minimize the possibility of splitting along the axis of these tabs due to beam shrinkage.

Incorrect

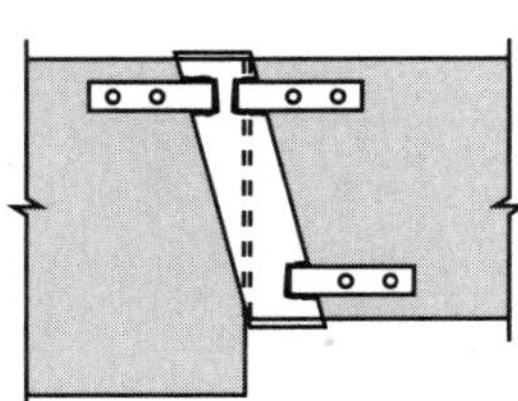

Discussion

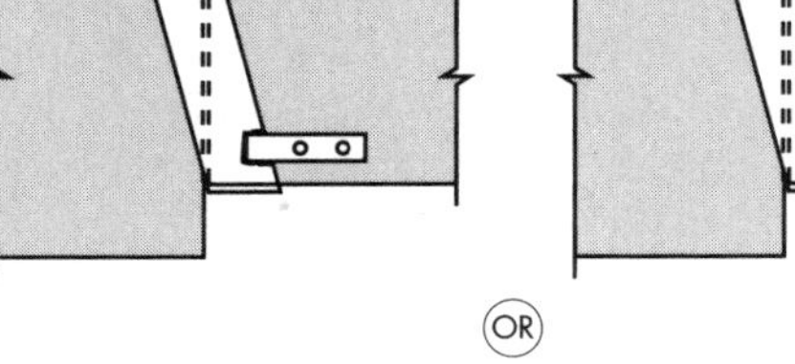

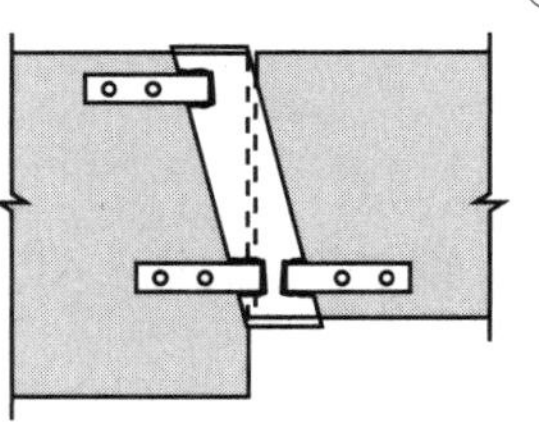

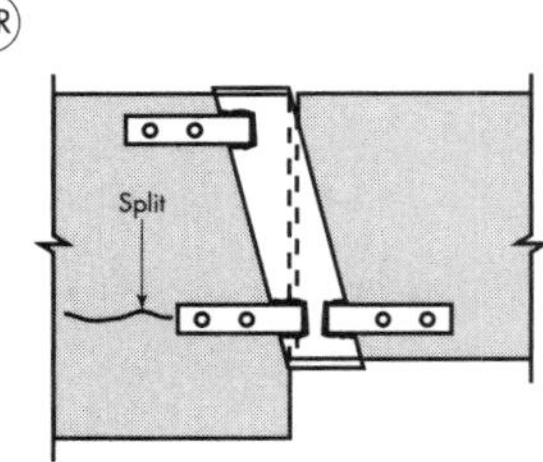

An integral tension-tie connection can cause tension perpendicular-to-grain stress to develop due to beam shrinkage. This can happen regard-less of the location of the integral tension tie connector. If a tension connection is required, a separate connector may be used as shown in the figure to the left. This tie is not welded to the beam hanger.

FIGURE 4.25 *(i)(j)* Connection details. *(Continued)*

(k)

CANTILEVER BEAM CONNECTION – WELDED TIE TENSION

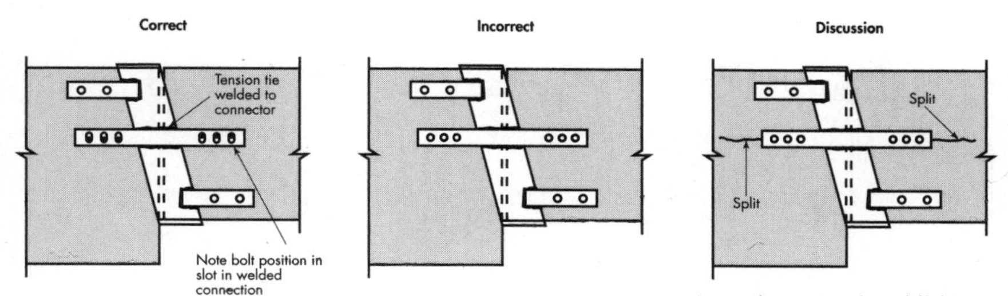

An integral tension tie can be used if holes in tie are vertically slotted and tie attachment bolts are placed as shown to allow motion of bolt in slot due to shrinkage of timber elements. If movement is not allowed at this location, tension perpendicular-to-grain stresses may develop in both members and cause splitting.

(l)

CANTILEVER BEAM CONNECTION – NO TENSION TIE

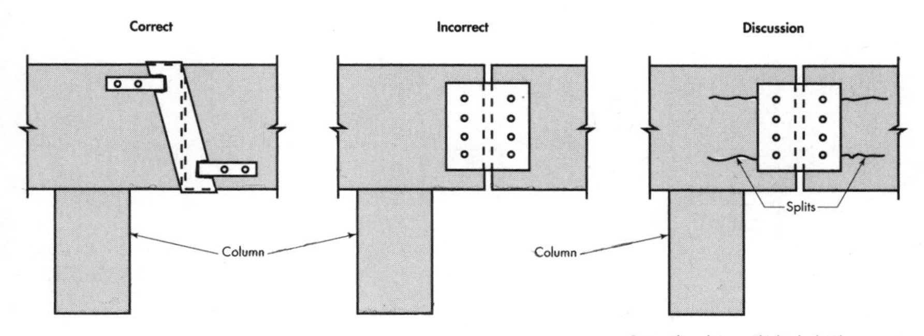

Deep splice plates applied to both sides can cause splitting of both members if members shrink. Sideplates resist this shrinkage and may induce tension perpendicular-to-grain stresses which may in turn cause splits.

FIGURE 4.25 (*k*)(*l*) Connection details. (*Continued*)

bility. To address these concerns, the glued laminated timber industry sponsored a series of full-scale bending tests of narrow, deep glued laminated timber purlins at Oregon State University in 1989. This involved testing a series of glued laminated timber beams having a span of 40 ft with d/b ratios of 9 and 11.4, which were representative of the resawn purlins being used in the field. These tests were conducted with varying degrees of lateral bracing of the top or compression edge of the beams. This bracing ranged from a fully braced compression edge to unbraced lengths of 15 ft, 0 in.

As anticipated, the purlins with minimal lateral bracing (unbraced length = 15 ft, 0 in.) did undergo varying degrees of lateral buckling, typically buckling in an S-shaped curve between points of lateral support. However, the failure load in all cases exceeded the values predicted by applying the NDS lateral stability analysis.

(m)

CANTILEVER BEAM CONNECTION – NO TENSION TIE

Correct | Incorrect | Discussion

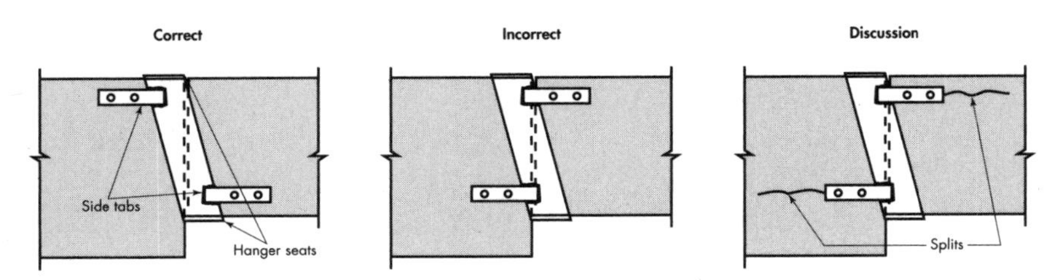

With side tabs inverted, glulam beam shrinkage shifts load from hanger seats to side tabs. This is likely to induce tension perpendicular-to-grain stresses which can lead to the development of splits and beam failure.

(n)

BEAM TO COLUMN – U-BRACKET – WOOD OR PIPE COLUMN

Correct | Incorrect | Discussion

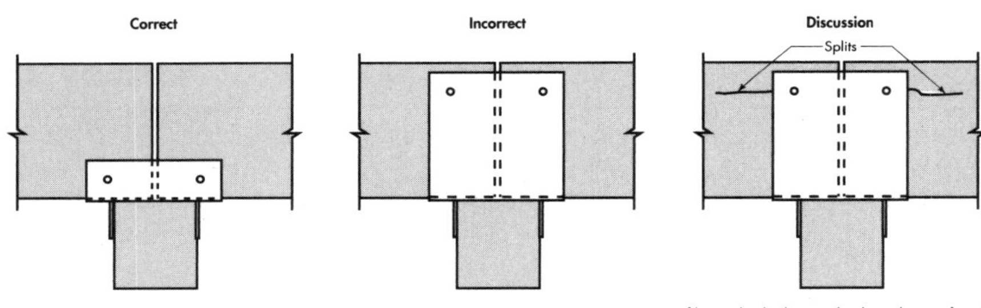

If beam shrinks, bearing load may be transferred to bolts. This can cause splitting of beam. This detail also restrains beam rotation due to deflection under loading which can also cause splitting.

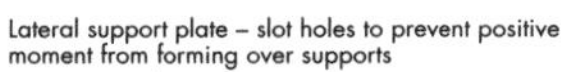

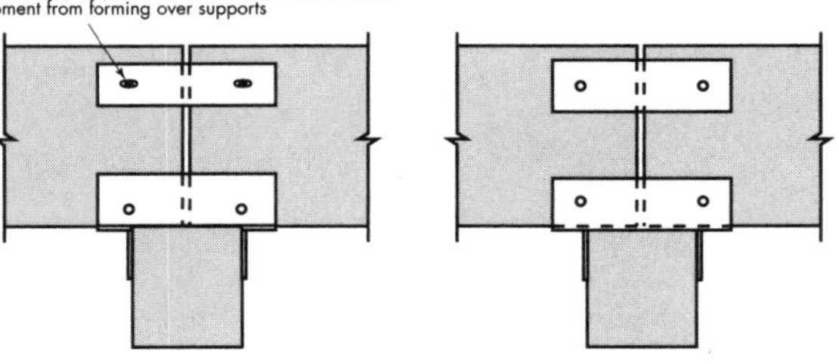

Splits

Rotation of the beams under loading can cause splitting at the tension tie plate unless slotted.

FIGURE 4.25 (*m*)(*n*) Connection details. (*Continued*)

(o)

BEAM TO COLUMN – T-BRACKET

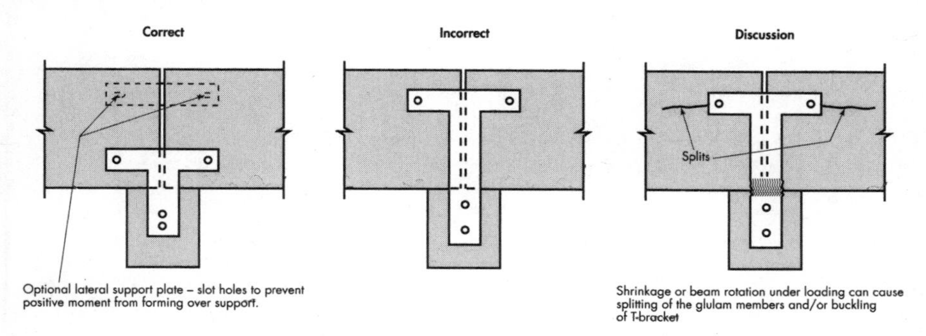

(p)

BEAM TO COLUMN – WITH TOP LATERAL SUPPORT PLATE

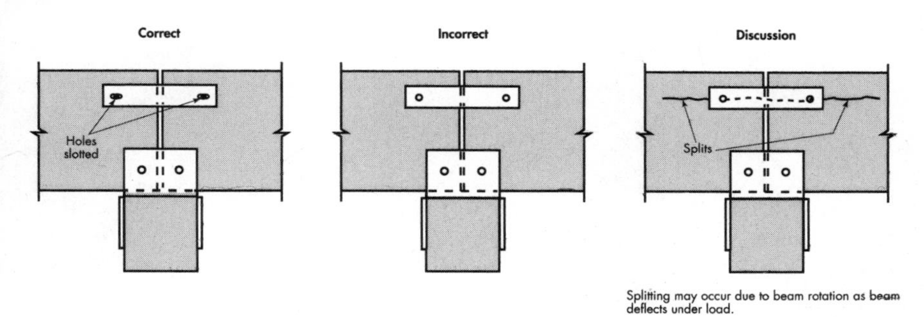

(q)

WOOD COLUMN TO CONCRETE BASE

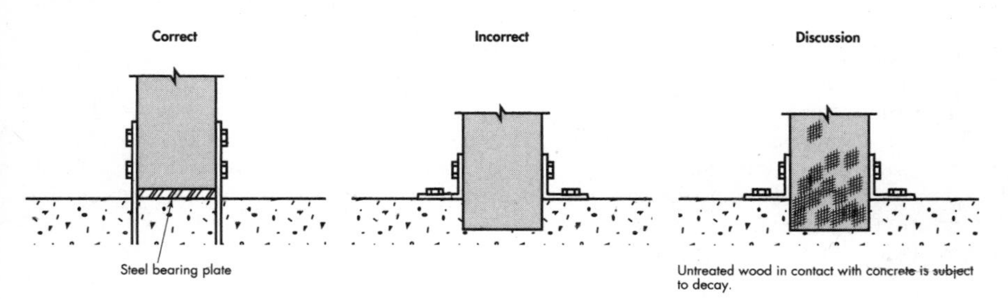

FIGURE 4.25 (*o–q*) Connection details. (*Continued*)

(r)

BEAM IN BENT HANGER

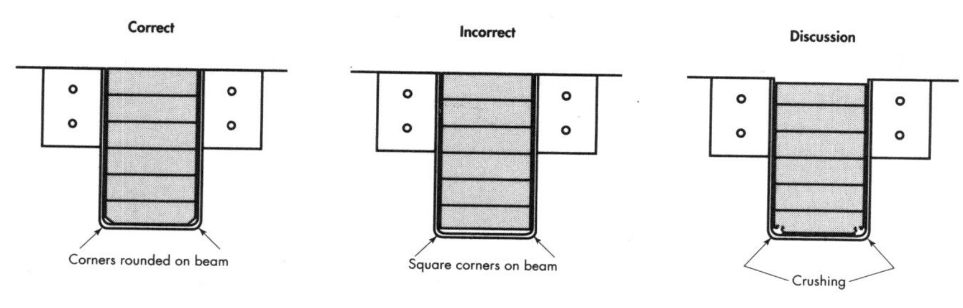

(s)

GLULAM ARCH TO FOUNDATION

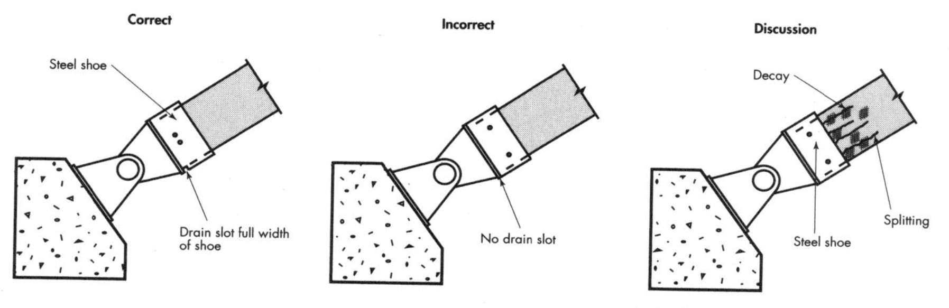

FIGURE 4.25 (*r*)(*s*) Connection details. (*Continued*)

In conclusion, if a glued laminated timber beam has the compression edge fully braced as would occur by the normal attachment of sheathing or decking with mechanical fasteners, there is no need to limit the d/b ratio of these members to 7. If the compression edge is not braced, the member must be checked in accordance with the lateral stability provisions of the NDS but this procedure does not require any specific limit on the beam d/b ratio. As a practical handling and transportation limitation, it is recommended that an upper limit d/b ratio of 12 be applied for glued laminated timber beams regardless of the bracing used.

4.6.8 Checking of Glued Laminated Timber

Introduction. Wood swells or shrinks as it gains or loses moisture in response to changing relative humidity and temperature in the surrounding environment. If the moisture differential between the core and surface of the wood becomes great

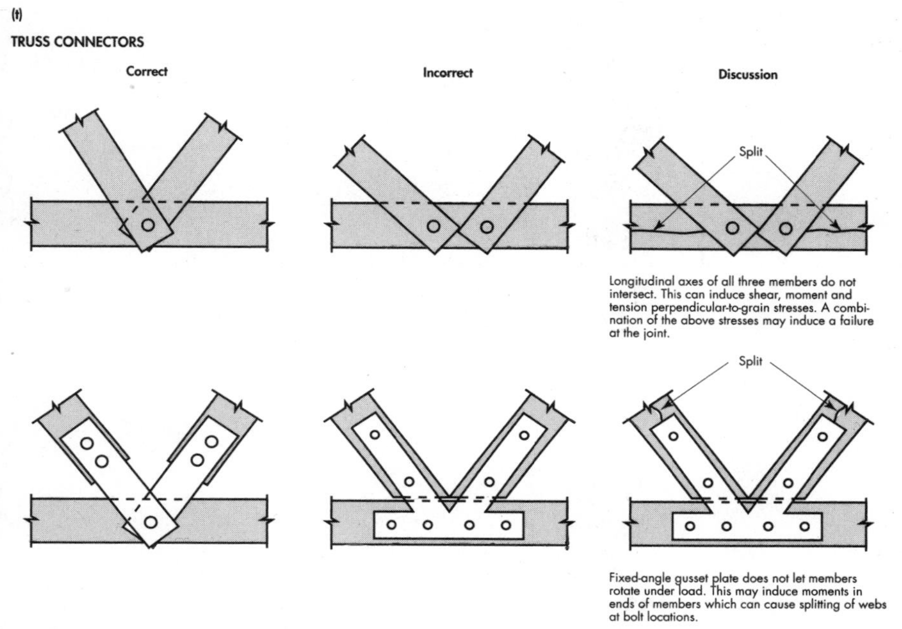

FIGURE 4.25 (*t*) Connection details. (*Continued*)

enough, the internal stresses associated with these dimensional changes can induce significant tensile stresses perpendicular to the grain. These stresses can result in seasoning checks developing across annual growth rings. This phenomenon is particularly pronounced in large sawn lumber beams.

In glued laminated timber (glued laminated timber), individual laminations are relatively thin (normally 1½ in.) and therefore dry more uniformly. Also, the maximum moisture content of each lamination is typically limited to 16% or less for gluing purposes. As a result, seasoning checks occur less frequently in glued laminated timber than in sawn lumber.

It is important to distinguish seasoning checks in glued laminated timbers from delamination, which results from inadequate glue bond between laminations. The key to identifying seasoning checks is the presence of torn wood fibers in the separation along the grain. Delamination is represented by a smooth surface without torn wood fibers. Delamination is seldom observed in glued laminated timber members produced in accordance with the ANSI A190.1 standard for glued laminated timber. While delamination could occur at any of the gluelines between laminations, seasoning checks are often concentrated along the first glueline adjacent to an outer lamination (see Fig. 4.26) but may occur anywhere on the surface of the member. Seasoning checks located at or near the first glueline are typically a result of environmental exposure to the large surface areas of the outer laminations.

Evaluating Check Size. For glued laminated timber bending members, the presence of seasoning checks generally affects only the horizontal shear capacity and usually is not significant outside the shear-critical zone. Bending and tensile strengths are virtually unaffected. The shear-critical zones, as shown in Fig. 4.27,

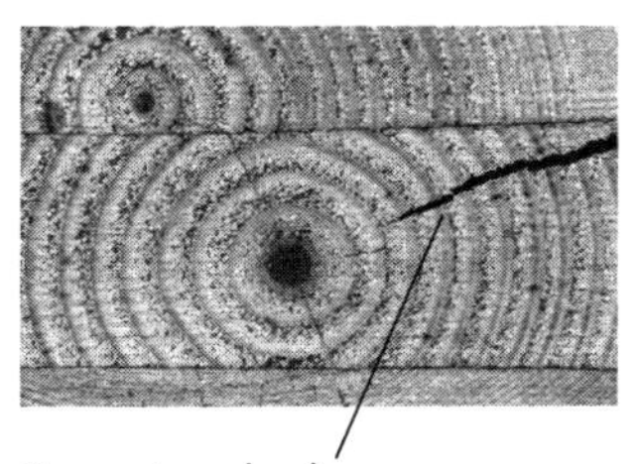

FIGURE 4.26 Glued laminated timber beam cross section illustrating checking.

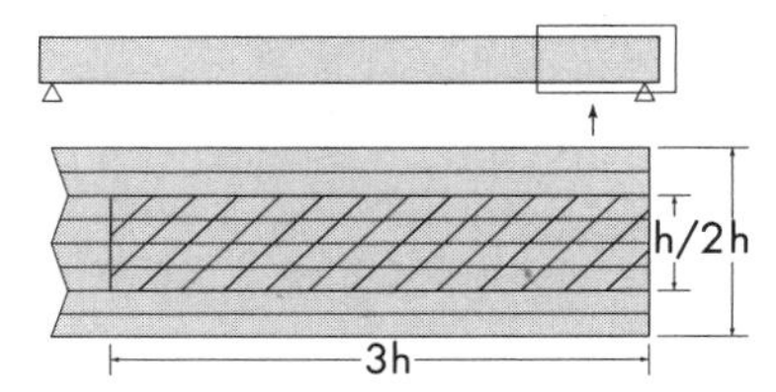

FIGURE 4.27 Location of the shear-critical zone at end of balanced glued laminated timber beam.

are typically defined as the areas at both ends of a simply supported timber beam within a distance from each end equal to 3 times the beam depth and within the middle ½ depth of the beam. For continuous beams, the shear-critical zone also includes a similar area over interior supports.

The published shear stresses are applicable to glued laminated timber members having a rectangular cross section and subjected to typical building design loads of dead, live, snow, wind, or earthquake. In establishing the published values, both APA and AITC Technical Advisory Committees applied a reduction of 10% to the actual values as determined by testing to account for some degree of checking of beams in service. This permits seasoning checks up to 10% of the beam width to occur without requiring an engineering analysis of the effects of the checks, even when they occur in the shear critical zone.

It is noted that if the designer wants to make the assumption that no seasoning checks will occur in the shear-critical zone of the beam, then the published horizontal shear stresses can be increased by 10%. Such an assumption would require a subsequent engineering analysis of any seasoning checks that occur in the shear-critical zone in accordance with the procedures that follow.

There are two types of checks: side checks and end checks (or splits) as shown in Fig. 4.28. Side checks are the separation of wood on either side of the beam and are measured as the average depth of separation perpendicular to the face on which it appears. End checks occur across the full width of the beam at the ends and are measured as the average length of penetration on each face from the end. The allowable check size when located in the shear-critical zone of a glued laminated timber beam may be calculated using Eqs. (4.14) and (4.15). Results of the

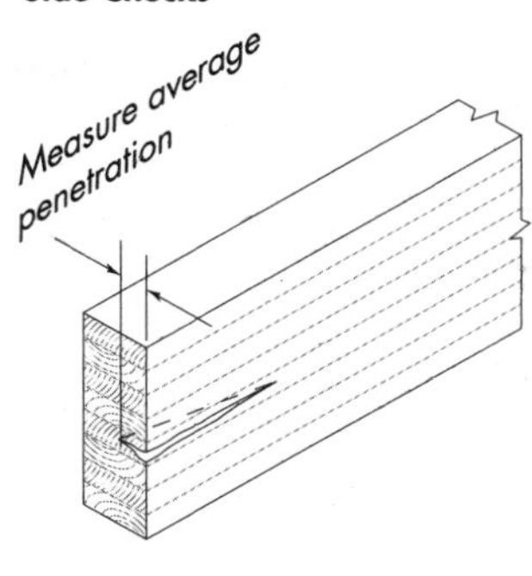

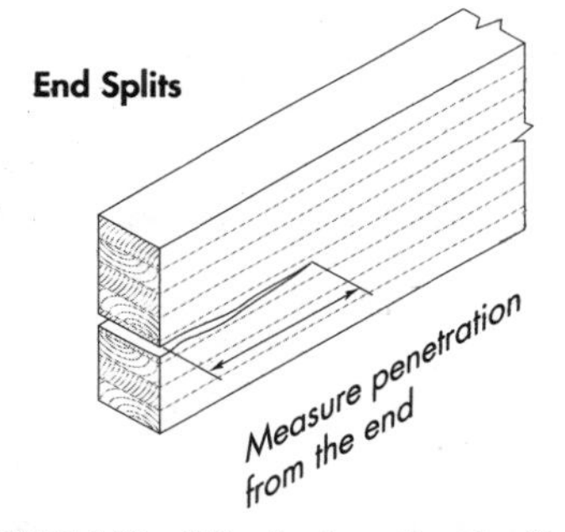

FIGURE 4.28 Side checks and end splits.

calculations for the allowable check size for various beam widths are given in Table 4.21.

$$\text{Allowable side check size (depth)} = \text{glulam beam width} \times 0.10 \qquad (4.14)$$

$$\text{Allowable end check size (length)} = \text{glulam beam width} \times 0.10 \times 3 \qquad (4.15)$$

When checks appear outside the shear-critical zone, the allowable check size may be determined in linear proportion to the shear stress distribution in the beam.

TABLE 4.21 Allowable Check Size Inside the Shear-Critical Zone

Beam width (in.)	Allowable side checks (depth) (in.)	Allowable end checks (length) (in.)
$2\frac{1}{2}$	$\frac{1}{4}$	$\frac{3}{4}$
$3\frac{1}{8}$	$\frac{3}{8}$	1
$3\frac{1}{2}$	$\frac{3}{8}$	1
$5\frac{1}{8}$	$\frac{1}{2}$	$1\frac{1}{2}$
$6\frac{3}{4}$	$\frac{5}{8}$	2
$8\frac{3}{4}$	$\frac{7}{8}$	$2\frac{5}{8}$
$10\frac{3}{4}$	$1\frac{1}{8}$	$3\frac{1}{4}$

In general, the allowable check size increases as its distance from middepth of the beam increases. For a typical glued laminated timber beam such as 24F-V4/DF or 24F-V3/SP, Eqs. (4.16) and (4.17) may be used to determine the allowable size of checks located outside the shear-critical zone. Table 4.22 shows results of these calculations for various beam widths. Refer to the report titled *Shear Stress in Two Wood Beams over Wood Block Supports,* published by U.S. Forest Products Laboratory (Report No. 2249),[9] for more detailed information.

$$\text{Allowable side check size (depth)} = C \times \text{glulam beam width} \tag{4.16}$$

$$\text{Allowable end check size (length)} = C \times \text{glulam beam width} \times 3 \tag{4.17}$$

where $C = 3.6\ (y/h) - 0.8 \leq 0.85$
h = glued laminated timber beam depth (height) (in.), and
y = distance away from middepth of the beam, $0.25h \leq y \leq 0.50h$ (in.)

The following examples illustrate the principles of determining the effect of checks on glued laminated timber performance.

Example 1 A side check with an average depth of 2.0 in. is located near the end and 1.5 in. up from the bottom face of a 5⅛ × 24 in. 24F-V4/DF glued laminated timber beam (between the outer tension and adjacent lamination). The beam had no other strength-reducing characteristics at the check location. Does this beam remain structurally sound?

solution

Step 1
Determine the location of the check as a percentage of the beam depth, y/h:

$$y/h = \frac{(24/2) - 1.5}{24} = 0.4375$$

Therefore, the check is located outside the shear-critical zone, $y/h > 0.25$.

Step 2
According to Eq. (4.16):

TABLE 4.22 Allowable Check Size Outside the Shear-Critical Zone

Beam width (in.)	Allowable side checks (depth) (in.)				Allowable end checks (length) (in.)			
	y/h							
	0.30	0.35	0.40	0.45	0.30	0.35	0.40	0.45
2½	¾	1⅛	1⅝	2	2⅛	3½	4¾	6⅛
3⅛	⅞	1½	2	2⅝	2⅝	4⅜	6	7¾
3½	1	1⅝	2¼	2⅞	3	4⅞	6¾	8⅝
5⅛	1⅜	2⅜	3¼	4¼	4¼	7⅛	9⅞	12⅝
6¾	1⅞	3⅛	4⅜	5½	5⅝	9⅜	13	16⅝
8¾	2½	4	5⅝	7⅛	7⅜	12⅛	16¾	21½
10¾	3	5	6⅞	8⅞	9	14⅞	20⅝	26½

$$C = 3.6 \times 0.4375 - 0.8 = 0.775 < 0.85$$

$$\text{Allowable check size} = 0.775 \times 5.125 = 3.97 \text{ in.} > 2.0 \text{ in.} \rightarrow \text{OK}$$

Or, interpolating from Table 4.22, the allowable side check $\cong$ 4.0 in.

Example 2 The same situation as Example 1 except the check is located at 7.5 in. up from bottom face of the beam.

solution

Step 1

Determine the location of the check as a percentage of the beam depth, y/h:

$$y/h = \frac{(24/2) - 7.5}{24} = 0.1875$$

Therefore, the check is located inside the shear-critical zone, $y/h \leq 0.25$

Step 2

According to Eq. (4.14):

$$\text{Allowable check size} = 5.125 \times 0.10 = 0.51 \text{ in.} < 2.0 \text{ in.} \rightarrow \text{NG}$$

Or, interpolating from Table 4.21, the allowable side check $\approx$ 0.5 in. Therefore, it is necessary to reexamine carefully the adequacy of shear capacity based on the actual loading conditions and the net dimensions of the beam after allowing for a check of 2.0 in. in depth.

Discussion. Allowable check sizes discussed above are applicable to glued laminated timber used as bending members (beams). Checks occurring on the soffit face of a bending member loaded in the direction perpendicular to the wide face of the laminations should be of no structural concern as long as they run parallel to the grain of the wood. For compression members (columns), checks are generally not structurally significant unless they develop into a split, thereby increasing the length-to-depth (l/d) ratio. In such a case, the load-carrying capacity of the column should be reduced based on the new l/d ratio. A detailed evaluation performed by a design professional knowledgeable in timber design may be needed to ensure the integrity of the structural members.

It is important to note that even though a check located outside the shear-critical zone can be relatively large without jeopardizing the structural safety of the glued laminated timber beam, such checking may be an indication of excessive humidity and temperature conditions in the surrounding environment. Therefore, the cause of the excessive checking should be identified and corrected to prevent further progression that might ultimately lead to a check that is structurally significant. See APA EWS Technical Note R465, *Checking in Glued Laminated Timber,*[10] for a general discussion of checking in glued laminated timber.

4.6.9 Filed Notching and Drilling of Glued Laminated Timber

Introduction. Glued laminated timber beams are highly engineered components manufactured from specially selected and positioned lumber laminations of varying strength and stiffness properties. These fabricated beams are often specified where the span, strength or stiffness requirements for a specific application call for a highly engineered wood product. As such, glued laminated timber beams are generally designed for and used in applications where they will be highly stressed under design loads. For this reason, field modifications such as notching, tapering, or drilling not shown on the design or shop drawings should be avoided, and never done without a thorough understanding of their effects on the structural integrity of the member.

As mentioned above, glued laminated timber beams are fabricated using selected lumber laminations of various strength, stiffness and species types. The beams are assembled with the lower-grade material in the middepth of the beam and higher-grade material positioned on the top and bottom. The highest grade of material is used as the outermost laminations on the tension side of the beam. Because of this, any drilling, dapping, or notching that takes place in these outermost tension laminations has a twofold effect on the strength and serviceability of the beam. First, such modifications reduce the section of the beam. Second, they remove wood fiber from the laminations having the highest strength. Given this, along with the fact that stress concentrations occur in such areas created by notching or drilling, it is easy to see why a cautious approach to field modifications is vital.

This section provides recommendations for field notching, tapering, and drilling glued laminated timber beams. Beams illustrated herein are assumed to be simple span and are shown with the compression side up. All equations and notching guidelines are presented using the same assumptions. For continuous or cantilevered beams, this information can be used, but only with extreme caution and only after careful analysis based on sound engineering judgment.

It is important to understand that improperly made field notches or holes may reduce the capacity of a properly designed member to the point where a structural failure may occur. The effects of any field notching or drilling should be checked by a designer competent in engineered timber design.

Notching. Notching of bending members should be avoided whenever possible, especially on the tension side of the member (see Fig. 4.29). Tension-side notching of glued laminated timber beams is not permitted except at member ends and then

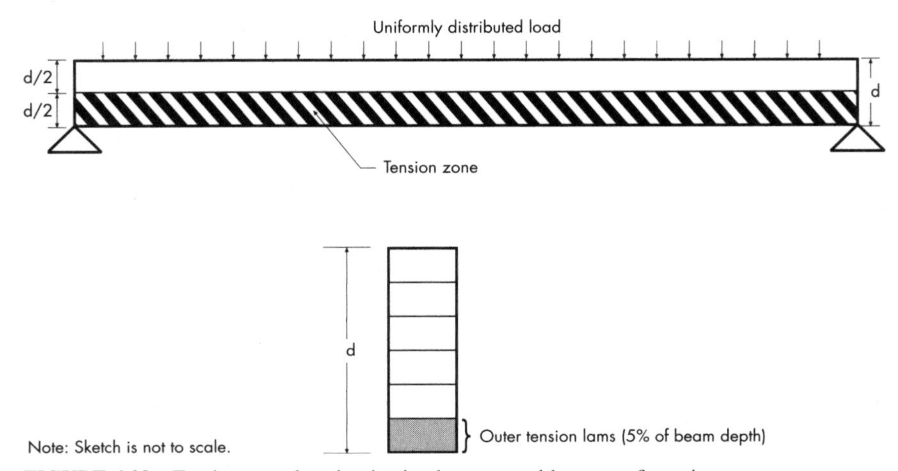

FIGURE 4.29 Tension zone location in simply supported beam configuration.

only under specific conditions. The notching of a bending member on the tension side results in a decrease in strength caused by stress concentrations that develop around the notch as well as a reduction of the area resisting the bending and shear forces. Such notches induce perpendicular-to-grain tension stresses, which, in conjunction with horizontal shear forces, can cause splitting along the grain, typically starting at the inside corner of the notch. Stress concentrations due to notches can be reduced by using a gradually tapered notch configuration in lieu of a square-cornered notch. Rounding the square corner of a notch with a radius of approximately ½ in. is also recommended to reduce stress concentrations in these areas.

For square-cornered notches occurring at the ends of beams on the tension side, the designer may consider the use of reinforcement such as full-threaded lag screws to resist the tendency to split at the notch (see Fig. 4.30). A number of design methodologies exist for sizing such screws. The design methodology selected and subsequent fabrication details are the responsibility of the designer/engineer of record. If lag screws are used, lead holes shall be predrilled in accordance with accepted practice.

Where glued laminated timber members are notched at the ends for bearing over a support, the notch depth shall not exceed ⅒ of the beam depth or 3 inches, whichever is less. Figure 4.31f is provided to assist in evaluating the associated reductions to beam strength due to notching on the tension side. For notches on the compression side, a less severe condition exists and equations for the analysis of the effects of these notches are also given in Fig. 4.31. The equations given are empirical in nature and were developed for the conditions shown.

Because this guideline is limited to single-span, simply supported beams, the notches shown in Fig. 4.31 occur in areas of high shear and effectively zero moment. For this reason, the design equations given are shear equations. In situations where compression-side notches extend into areas of significant moment, the bending capacity of the beam should also be checked using the remaining section of the beam and the appropriate allowable stresses for those laminations remaining at the notch location.

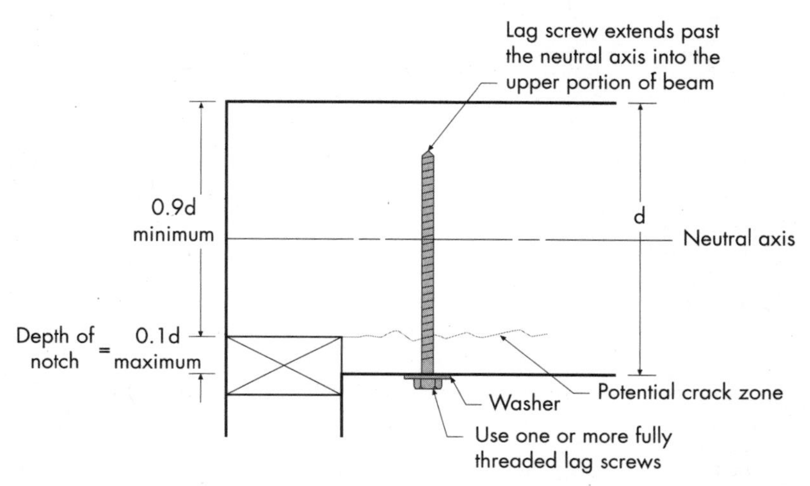

FIGURE 4.30 Recommended reinforcement on a tension-side notch.

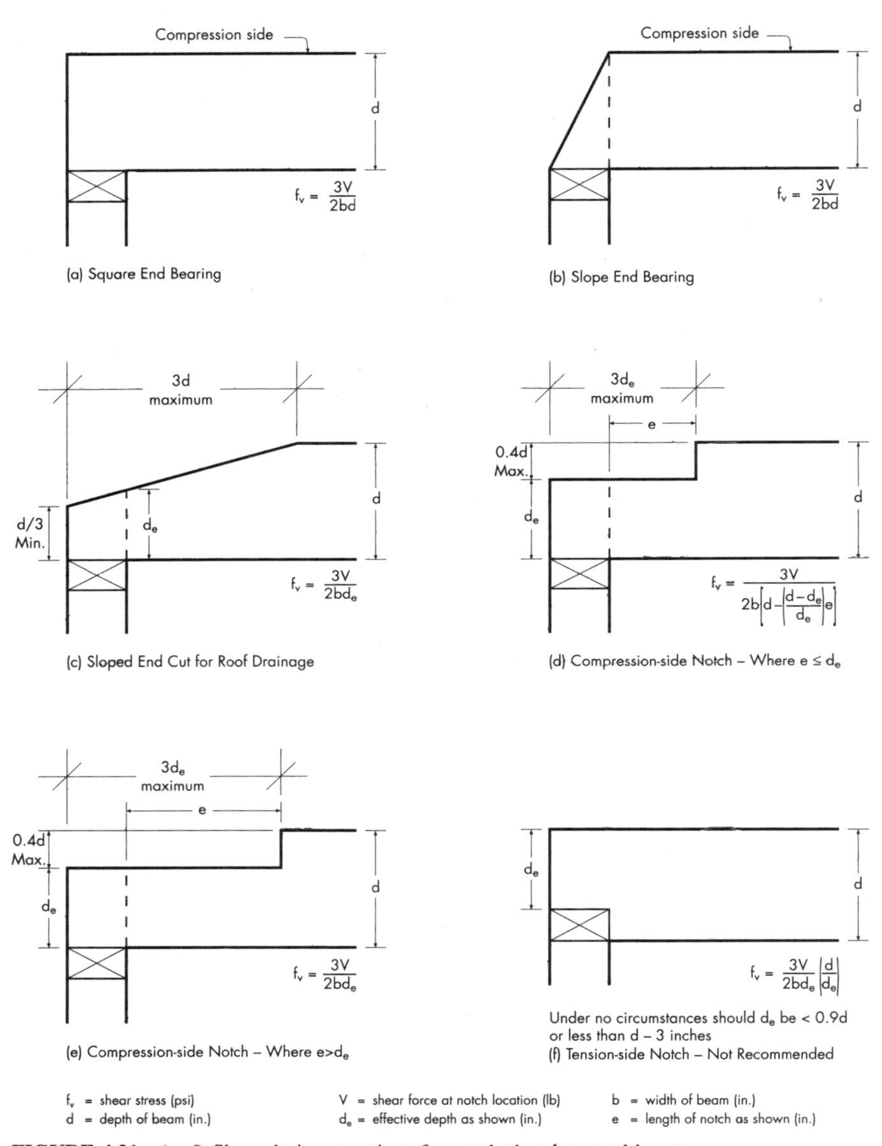

FIGURE 4.31 (*a–f*) Shear design equations for notched and tapered beams.

When it becomes necessary to cut a small notch in the top of a glued laminated timber (in the compression zone) to provide passage for small-diameter pipe or conduit, this cut should be made in areas of the beam stressed to less than 50% of the design bending stress. The net section in these areas should be checked for shear and bending stresses to ensure adequate performance.

All field notches should be accurately cut. Avoid overcutting at the root of the notch. Drilling a pilot hole in the member at the interior corner location of a notch as a stop point for the saw blade provides both a rounded corner and minimizes overcutting at the corner.

Holes

Horizontal Holes. Like notches, holes in a glued laminated timber beam remove wood fiber, thus reducing the net area of the beam at the hole location, and they introduce stress concentrations. These effects cause a reduction in the capacity of the beam in the area of the penetration. For this reason, horizontal holes in glued laminated timbers are limited in size and location to maintain the structural integrity of the beam. Figure 4.32 shows the zones of a uniformly loaded, simply supported beam where the field drilling of holes may be considered. These noncritical zones are located in portions of the beam stressed to less than 50% of design bending stress and less than 50% of design shear stress. For beams of more complex loading or other than simple spans, similar diagrams may be developed.

Field-drilled horizontal holes should be used for access only and should not be used as attachment points for brackets or other load bearing hardware unless specifically designed as such by the engineer or designer. Examples of access holes include those used for the passage of wires, electrical conduit, small diameter sprinkler pipes, fiberoptic cables, and other small, lightweight materials.

Field-drilled horizontal holes should be spaced at least 2 ft on center in the noncritical zones, as defined in Fig. 4.32. Under no circumstances should holes be drilled in the outer four laminations of a flexural member on the tension side in the zone or zones stressed to 50% or greater of the design stress. Holes with a diameter larger than the thickness of one lamination (typically 1½ in.) or ⅒ of the beam depth, whichever is smaller, are not recommended. However, for glued laminated timber members that have been oversized for architectural reasons, this limit may be increased based on an engineering analysis.

Regardless of the hole location, the net section of the beam remaining should be checked for flexure and horizontal shear. In addition, holes drilled horizontally through the member should be positioned and sized with the understanding that the beam will deflect over a period of time under in-service loading conditions. This deflection could cause distress to supported equipment or piping unless properly considered.

Vertical Holes. As a rule of thumb, vertical holes drilled through the depth of a glued laminated timber beam cause a reduction in the capacity at that location directly proportional to the ratio of 1½ times the diameter of the hole to the width

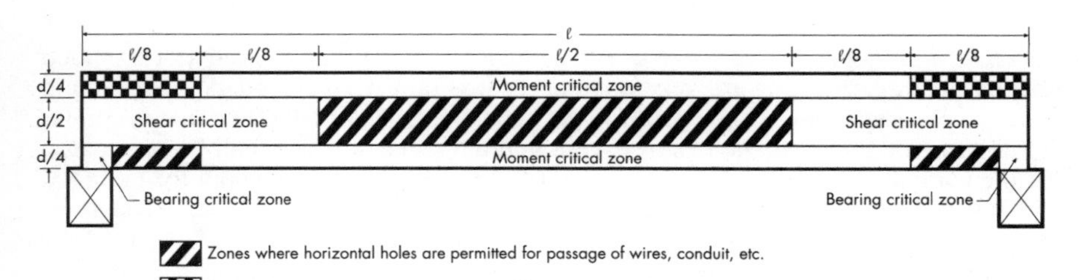

FIGURE 4.32 Small horizontal hole locations for uniformly loaded, simply supported beam.

of the beam. For example, a 1 in. hole drilled in a 6 in. wide beam would reduce the capacity of the beam at that section by approximately $(1 \times 1\frac{1}{2})/6 = \frac{1}{4}$. For this reason, when it is necessary to drill vertical holes through a glued laminated timber member, the holes should be positioned in areas of the member that are stressed to less than 50% of design in bending. In a simply supported, uniformly loaded beam, this area would be located from the end of the beam inward approximately ⅛ of the beam span. In all cases, use a drill guide to minimize wandering of the bit as it passes through knots or material of varying density and to ensure a true alignment of the hole through the depth of the beam.

Holes for Support of Suspended Equipment. Heavy equipment or piping suspended from glued laminated timber beams should be attached such that load is applied to the top of the members to avoid introducing tension perpendicular-to-grain stresses. Any horizontal holes required for support of significant weight, such as suspended heating and cooling units or main water lines, must be located above the neutral axis of the member and in a zone stressed to less than 50% of the design flexural stress (see Fig. 4.32). Fasteners supporting light loads such as light fixtures must be placed at least four laminations away from the tension face of the member. The design capacity of the beam should be checked for all such loads to ensure proper performance.

Protection of Field-Cut Notches and Holes. Frequently, glued laminated timber beams are provided from the manufacturer with the ends sealed by a protective coating. This sealer is applied to the end grain of the glued laminated timber beams to retard the migration of moisture in and out of the beam ends during transit and jobsite storage. Field-cutting a notch in the end of a beam can change the moisture-absorption characteristics of the beam at the notch location. This can result in seasoning checks or even localized splitting developing at the root of the notch. To minimize this possibility, all notches should be sealed immediately after cutting using a water-repellent sealer. Sealing other field-cut locations as well as field-drilled holes is also recommended. These sealers can be applied by brush, swab, or roller, or with a spray gun.

4.6.10 Design of Curved Bending Members

Curved Beams. In addition to straight or tapered straight members, glued laminated timber can be manufactured into a variety of shapes or forms. The most common simple-span curved beams include curved beams with constant cross section, curved beams with constant cross section and mechanically attached haunch, curved beams with constant curvature, and curved beams with constant cross section and framed haunch. Detailed description and design of these curved bending members can be found in Reference 11, chap. 4.

Pitched and Tapered Curved Beams. Pitched and tapered curved (PTC) glued laminated timber beams are very popular as structural roof members when a long interior clear span and sloping roof is desired. The design of PTC beams is similar to that of straight prismatic beams except that the radial stresses shall be considered in the curved portion and the distribution of the bending stresses about the assumed neutral axis is different from that in a prismatic member. Reference 11, chap. 4 provides detailed design procedures for the PTC beams.

Arches. Glued laminated timber can be designed as arches with a variety of shapes and structural characteristics. Three-hinged arches are generally more economical than two-hinged arches due to the elimination of the moment splice required for most two-hinged arches. Reference 11, chap. 9 gives detailed description and design procedures for glued laminated timber arches.

4.7 DESIGN TABLES

4.7.1 Beams

Introduction. Glued laminated timber beams are used in a wide range of applications in both commercial and residential construction. The tables in this section provide recommended preliminary design loads for two of the most common glued laminated timber beam applications: roofs and floors. The recommendations apply to glued laminated timber beams meeting the requirements of ANSI A190.1.

The tables included in this section include values for section properties and capacities and allowable loads for simple-span and cantilevered beams. The tables are based on an allowable bending stress of F_b = 2400 psi for both Douglas fir and southern pine.

Section Properties and Design Capacities. Tables 4.23 and 4.29 provide section properties and design capacities for two commonly used species of glued laminated timber beams under dry-use conditions. Bending moment and shear capacities are based on a normal (10-year) duration of load. Dimensions shown are net sizes, and capacities are based on loading perpendicular to the wide faces of the laminations; that is, bending about the *x–x* axis of the beam as shown in Fig. 4.10. Final design should include a complete analysis, including bearing stresses and lateral stability.

See Design Examples 1 and 4 in this section for examples of preliminary design using glued laminated timber beam section capacities from Tables 4.23 and 4.29.

Allowable Loads for Simple-Span Glued Laminated Timber Beams. Tables 4.24, 4.25, 4.30, and 4.31 provide allowable loads for glued laminated timber beams used as simple span roof members in snow load areas (DOL factor = 1.15) and for nonsnow loads (DOL factor = 1.25). Tables 4.26 and 4.32 provide similar information for floor members (DOL factor = 1.25). The tables can be used to size such members for preliminary design. Final design should include a complete analysis, including bearing stresses and lateral stability.

See Design Examples 2 and 3 in this section for examples of preliminary design using glued laminated timber beam load-span tables.

Allowable Loads for Cantilevered Glued Laminated Timber Roof Beams. Tables 4.27, 4.28, 4.33, and 4.34 are for preliminary design of cantilevered roof beams. The tables are based on balanced (fully loaded) as well as unbalanced loading. They do *not* include deflection criteria limitations. Final designs should include deflection requirements per the applicable building code, in addition to the bending and shear strength assessments incorporated in these tables. Final design should include a complete analysis, including bearing stresses and lateral stability.

A minimum roof slope of ¼ in. per foot in addition to specified camber is recommended to help avoid ponding of water on the roof.

The cantilever beam tables presented are applicable to balanced lay-ups, such as 24F-V8 for Douglas fir and 24F-V5 for southern pine, for three different systems. See Fig. 4.33 for details of the following cantilever systems:

SYSTEM 1

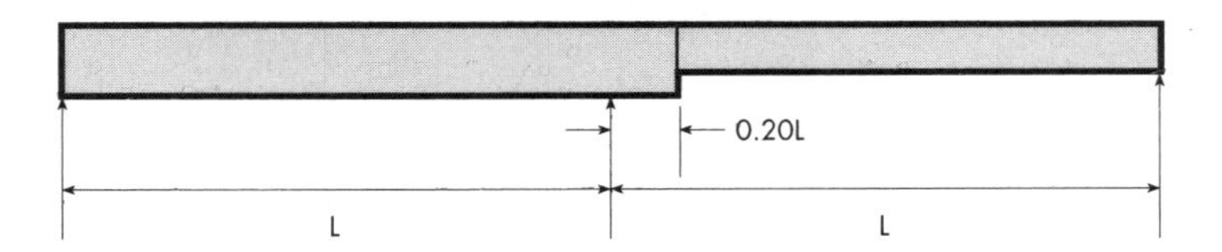

SYSTEM 2

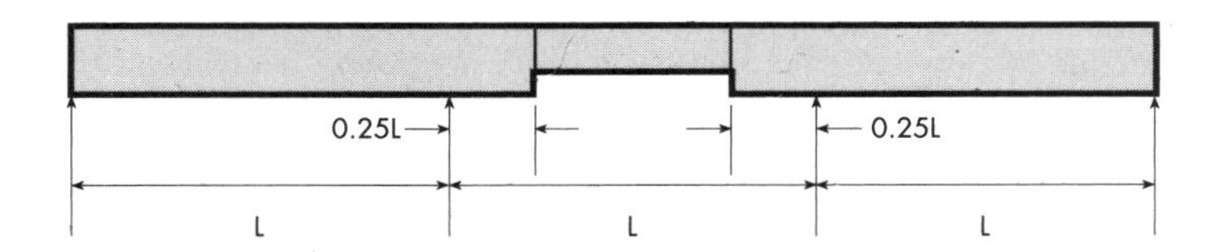

SYSTEM 3

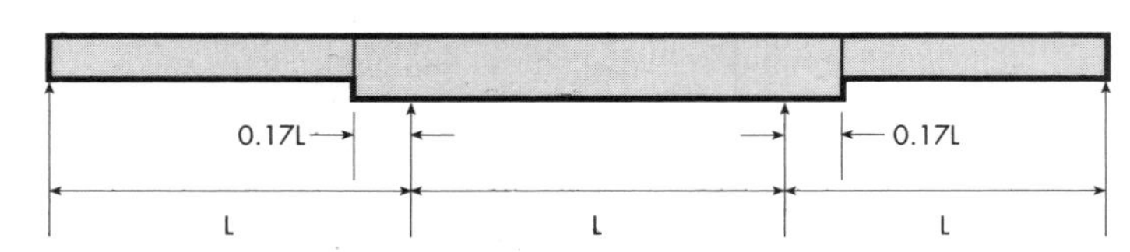

FIGURE 4.33 Typical cantilever beam systems.

System 1 is a two-equal-span cantilever system with the cantilevered beam extending past the center support by approximately 0.20 × the span, or 0.20*L*. Its overall length is therefore 1.2*L*, and the suspended beam's length is 0.8*L*.

System 2 is a three-equal-span cantilever system, with each of the two outer cantilevered beams extending past the center support into the middle span by 0.25*L*. Their length is therefore 1.25*L*, and the interior suspended beam's length is 0.5*L*.

System 3 is also a three-equal-span cantilever system, but the two outer span beams are suspended from the interior, double cantilevered beam, which extends past its two supports by approximately 0.17*L*. Its length is 1.34*L*, and the suspended beams are 0.83*L* each.

The following are additional notes that apply to Tables 4.27, 4.28, 4.33, and 4.34:

1. Span = spacing of column supports for cantilevered beams.
2. Load-duration factor = as noted.
3. Cantilevered beam lay-up = balanced.
4. Deflection has not been considered.
5. Service condition = dry.
6. Tabulated values represent total loads and have taken the dead weight of the beam into account (assumed 35 lb/ft^3 for Douglas fir and 36 lb/ft^3 for southern pine). Live load is assumed to be 0.6 × total load for purposes of checking strength under full unbalanced live load.

7. Volume factor is included.
8. Light (unshaded) areas limited by bending strength; dark-shaded areas limited by shear strength.

Design Examples

Design Example 1: Low-Slope Roof Design Using Section Capacities
Given:

24 ft span, 24 ft wide tributary area
Live load = 30 psf (snow); duration of load = 1.15
Dead load = 10 psf (actual)
Allowable total load deflection = $L/180$
Allowable live load deflection = $L/240$
Use 24F Douglas fir glued laminated timber

Then:

$$\text{Load, } w = (30 + 10)(24) = 960 \text{ lb/ft to glued laminated timber}$$

$$\text{Max. moment} = \frac{wL^2}{8} = \frac{960 \times 24^2}{8} = 69{,}120 \text{ lb/ft}$$

$$\text{Max. shear} = \frac{wL}{2} = \frac{960 \times 24}{2} = 11{,}520 \text{ lb}$$

Design:

From Table 4.23, try 5⅛ × 21 (weight = 26 lb/ft)
Total load = 960 + 26 = 986 lb/ft
From Eq. (4.1), volume factor = 0.9330
Design moment capacity = 75 338 × 0.9330 × 1.15 = 80,834 lb-ft

$$69\ 120 \times \frac{986}{960} = 70{,}992 \text{ lb-ft} < 80{,}834 \text{ lb-ft—OK}$$

Design shear capacity = 17,220 × 1.15 = 19 803 lb

(For shear design, neglect all loads within a distance from supports equal to the depth of the beam.)

$$\text{Deflection, } TL = \frac{5wL^4}{384EI} = \frac{5 \times 986 \times 24^4 \times 1728}{384 \times 7119 \times 10^6}$$
$$= 1.03 \text{ in.} = L/279 < L/180\text{—OK}$$

$$\text{Deflection, } LL = \frac{30 \times 24}{986} = 1.03 = 0.75 \text{ in.} = L/383 < L/240\text{—OK}$$

Design Example 2: Low-Slope Roof Design Using Load-Span Tables
Given:

24 ft span, 24 ft wide tributary area
Live load = 30 psf (snow); duration of load = 1.15
Dead load = 10 psf (actual)

TABLE 4.23 Section Properties and Design Capacities for Douglas Fir Glulam

3-1/8-INCH WIDTH

Depth (in.)	6	7-1/2	9	10-1/2	12	13-1/2	15	16-1/2	18	19-1/2	21	22-1/2	24	25-1/2	27
Beam Weight (lb/ft)	4.6	5.7	6.8	8.0	9.1	10.3	11.4	12.5	13.7	14.8	16.0	17.1	18.2	19.4	20.5
A (in.^{2})	18.8	23.4	28.1	32.8	37.5	42.2	46.9	51.6	56.3	60.9	65.6	70.3	75.0	79.7	84.4
S (in.^{3})	19	29	42	57	75	95	117	142	169	198	230	264	300	339	380
I (in.^{4})	56	110	190	301	450	641	879	1170	1519	1931	2412	2966	3600	4318	5126
EI (106 lb-in.^{2})	101	198	342	543	810	1153	1582	2106	2734	3476	4341	5339	6480	7773	9226
Moment Capacity (lb-ft)	3750	5859	8438	11484	15000	18984	23438	28359	33750	39609	45938	52734	60000	67734	75938
Shear Capacity (lb)	3000	3750	4500	5250	6000	6750	7500	8250	9000	9750	10500	11250	12000	12750	13500

3-1/2-INCH WIDTH

Depth (in.)	6	7-1/2	9	10-1/2	12	13-1/2	15	16-1/2	18	19-1/2	21	22-1/2	24	25-1/2	27
Beam Weight (lb/ft)	5.1	6.4	7.7	8.9	10.2	11.5	12.8	14.0	15.3	16.6	17.9	19.1	20.4	21.7	23.0
A (in.^{2})	21.0	26.3	31.5	36.8	42.0	47.3	52.5	57.8	63.0	68.3	73.5	78.8	84.0	89.3	94.5
S (in.^{3})	21	33	47	64	84	106	131	159	189	222	257	295	336	379	425
I (in.^{4})	63	123	213	338	504	718	984	1310	1701	2163	2701	3322	4032	4836	5741
EI (106 lb-in.^{2})	113	221	383	608	907	1292	1772	2358	3062	3893	4862	5980	7258	8705	10334
Moment Capacity (lb-ft)	4200	6563	9450	12863	16800	21263	26250	31763	37800	44363	51450	59063	67200	75863	85050
Shear Capacity (lb)	3360	4200	5040	5880	6720	7560	8400	9240	10080	10920	11760	12600	13440	14280	15120

5-1/8-INCH WIDTH

Depth (in.)	12	13-1/2	15	16-1/2	18	19-1/2	21	22-1/2	24	25-1/2	27	28-1/2	30	31-1/2	33
Beam Weight (lb/ft)	14.9	16.8	18.7	20.6	22.4	24.3	26.2	28.0	29.9	31.8	33.6	35.5	37.4	39.2	41.1
A (in.^{2})	61.5	69.2	76.9	84.6	92.3	99.9	107.6	115.3	123.0	130.7	138.4	146.1	153.8	161.4	169.1
S (in.^{3})	123	156	192	233	277	325	377	432	492	555	623	694	769	848	930
I (in.^{4})	738	1051	1441	1919	2491	3167	3955	4865	5904	7082	8406	9887	11531	13349	15348
EI (106 lb-in.^{2})	1328	1891	2595	3453	4483	5700	7119	8757	10627	12747	15131	17796	20756	24028	27627
Moment Capacity (lb-ft)	24600	31134	38438	46509	55350	64959	75338	86484	98400	111084	124538	138759	153750	169509	186038
Shear Capacity (lb)	9840	11070	12300	13530	14760	15990	17220	18450	19680	20910	22140	23370	24600	25830	27060

5-1/2-INCH WIDTH

Depth (in.)	12	13-1/2	15	16-1/2	18	19-1/2	21	22-1/2	24	25-1/2	27	28-1/2	30	31-1/2	33
Beam Weight (lb/ft)	16.0	18.0	20.1	22.1	24.1	26.1	28.1	30.1	32.1	34.1	36.1	38.1	40.1	42.1	44.1
A (in.^{2})	66.0	74.3	82.5	90.8	99.0	107.3	115.5	123.8	132.0	140.3	148.5	156.8	165.0	173.3	181.5
S (in.^{3})	132	167	206	250	297	349	404	464	528	596	668	745	825	910	998
I (in.^{4})	792	1128	1547	2059	2673	3398	4245	5221	6336	7600	9021	10610	12375	14326	16471
EI (106 lb-in.^{2})	1426	2030	2784	3706	4811	6117	7640	9397	11405	13680	16238	19098	22275	25786	29648
Moment Capacity (lb-ft)	26400	33413	41250	49913	59400	69713	80850	92813	105600	119213	133650	148913	165000	181913	199650
Shear Capacity (lb)	10560	11880	13200	14520	15840	17160	18480	19800	21120	22440	23760	25080	26400	27720	29040

6-3/4-INCH WIDTH

Depth (in.)	18	19-1/2	21	22-1/2	24	25-1/2	27	28-1/2	30	31-1/2	33	34-1/2	36	37-1/2	39
Beam Weight (lb/ft)	29.5	32.0	34.5	36.9	39.4	41.8	44.3	46.8	49.2	51.7	54.1	56.6	59.1	61.5	64.0
A (in.^{2})	121.5	131.6	141.8	151.9	162.0	172.1	182.3	192.4	202.5	212.6	222.8	232.9	243.0	253.1	263.3
S (in.^{3})	365	428	496	570	648	732	820	914	1013	1116	1225	1339	1458	1582	1711
I (in.^{4})	3281	4171	5209	6407	7776	9327	11072	13021	15188	17581	20215	23098	26244	29663	33367
EI (106 lb-in.^{2})	5905	7508	9377	11533	13997	16789	19929	23438	27338	31647	36386	41577	47239	53394	60060
Moment Capacity (lb-ft)	72900	85556	99225	113906	129600	146306	164025	182756	202500	223256	245025	267806	291600	316406	342225
Shear Capacity (lb)	19440	21060	22680	24300	25920	27540	29160	30780	32400	34020	35640	37260	38880	40500	42120

8-3/4-INCH WIDTH

Depth (in.)	24	25-1/2	27	28-1/2	30	31-1/2	33	34-1/2	36	37-1/2	39	40-1/2	42	43-1/2	45
Beam Weight (lb/ft)	51.0	54.2	57.4	60.6	63.8	67.0	70.2	73.4	76.6	79.8	82.9	86.1	89.3	92.5	95.7
A (in.^{2})	210.0	223.1	236.3	249.4	262.5	275.6	288.8	301.9	315.0	328.1	341.3	354.4	367.5	380.6	393.8
S (in.^{3})	840	948	1063	1185	1313	1447	1588	1736	1890	2051	2218	2392	2573	2760	2953
I (in.^{4})	10080	12091	14352	16880	19688	22791	26204	29942	34020	38452	43253	48439	54023	60020	66445
EI (106 lb-in.^{2})	18144	21763	25834	30383	35438	41023	47167	53896	61236	69214	77856	87190	97241	108036	119602
Moment Capacity (lb-ft)	168000	189656	212625	236906	262500	289406	317625	347156	378000	410156	443625	478406	514500	551906	590625
Shear Capacity (lb)	33600	35700	37800	39900	42000	44100	46200	48300	50400	52500	54600	56700	58800	60900	63000

Notes:

(1) Beam weight is based on density of 35 pcf.

(2) Moment capacity must be adjusted for volume effect. The volume factor is given in Section 4.4.3.

(3) Moment and shear capacities are based on a normal (10 years) duration of load and should be adjusted for the design duration of load per the applicable building code.

TABLE 4.24 Allowable Loads for Simple-Span Douglas Fir Glulam Roof Beams (lb/ft)—Non-snow Loads

3-1/8-INCH WIDTH

Depth (in.)	SPAN (ft) 8	10	12	14	16	18	20	22	24	26	28	30	32	34	36	38	40	42	44	46	48
6	581	295	169	105	69	—	—	—	—	—	—	—	—	—	—	—	—	—	—	—	—
7-1/2	910	580	333	208	137	95	68	—	—	—	—	—	—	—	—	—	—	—	—	—	—
9	1312	837	579	362	240	167	120	88	66	51	—	—	—	—	—	—	—	—	—	—	—
10-1/2	1786	1140	790	578	385	268	193	143	108	84	65	52	—	—	—	—	—	—	—	—	—
12	2335	1491	1033	756	577	402	291	216	164	127	100	80	64	52	—	—	—	—	—	—	—
13-1/2	2925	1888	1308	958	731	576	417	311	237	184	145	116	94	77	63	52	—	—	—	—	—
15	3398	2332	1616	1184	904	712	575	429	328	255	202	162	132	108	89	74	62	52	—	—	—
16-1/2	3916	2823	1957	1434	1095	863	696	573	439	342	272	219	178	146	121	101	85	72	61	52	—
18	4486	3201	2330	1708	1305	1028	830	684	570	447	355	286	234	192	160	134	113	96	81	70	60
19-1/2	5117	3596	2736	2006	1532	1208	975	801	664	559	454	367	299	247	206	173	146	124	106	91	78
21	5817	4023	3072	2328	1778	1402	1131	923	766	645	550	460	377	311	260	218	185	158	135	116	100
22-1/2	6601	4483	3392	2673	2043	1611	1290	1053	874	736	628	541	466	385	322	271	230	196	169	145	126
24	7482	4982	3732	2982	2326	1825	1459	1191	989	834	711	612	533	467	393	332	282	241	207	179	155
25-1/2	8481	5524	4094	3250	2627	2049	1639	1338	1111	936	799	688	599	525	463	400	340	291	251	217	189
27	9622	6116	4479	3532	2914	2286	1828	1493	1240	1045	891	768	668	586	517	460	407	348	300	260	227

3-1/2-INCH WIDTH

Depth (in.)	SPAN (ft) 8	10	12	14	16	18	20	22	24	26	28	30	32	34	36	38	40	42	44	46	48
6	651	331	189	117	77	53	—	—	—	—	—	—	—	—	—	—	—	—	—	—	—
7-1/2	1019	650	373	233	154	106	76	55	—	—	—	—	—	—	—	—	—	—	—	—	—
9	1469	937	649	406	269	187	134	99	74	57	—	—	—	—	—	—	—	—	—	—	—
10-1/2	2001	1277	884	647	431	300	216	160	121	94	73	58	—	—	—	—	—	—	—	—	—
12	2615	1670	1156	847	646	451	326	242	184	143	112	89	72	58	—	—	—	—	—	—	—
13-1/2	3275	2115	1465	1073	819	645	467	348	265	206	163	130	105	86	71	58	—	—	—	—	—
15	3805	2612	1810	1327	1013	797	643	480	367	286	226	182	147	121	100	83	69	58	—	—	—
16-1/2	4386	3162	2192	1606	1227	966	780	642	491	384	304	245	199	164	136	113	95	80	68	58	—
18	5025	3585	2610	1913	1461	1151	930	760	631	501	398	321	262	216	179	150	126	107	91	78	67
19-1/2	5731	4028	3064	2247	1716	1353	1086	886	736	619	509	411	335	277	231	194	164	139	119	102	88
21	6515	4505	3441	2607	1992	1566	1252	1022	848	714	609	516	422	349	291	245	207	177	151	130	112
22-1/2	7393	5021	3799	2994	2288	1787	1428	1166	968	815	695	599	520	432	361	304	258	220	189	163	141
24	8380	5580	4180	3340	2594	2021	1616	1319	1095	923	787	678	589	517	440	371	316	270	232	201	174
25-1/2	9498	6187	4585	3640	2912	2269	1814	1481	1230	1037	884	762	663	581	513	448	381	326	281	243	212
27	10777	6850	5017	3956	3247	2531	2024	1653	1373	1157	987	851	740	649	573	509	454	390	336	292	254

5-1/8-INCH WIDTH	SPAN (ft)																				
Depth (in.)	8	10	12	14	16	18	20	22	24	26	28	30	32	34	36	38	40	42	44	46	48
12	3829	2445	1693	1240	946	660	477	355	270	209	164	131	105	85	69	57	—	—	—	—	—
13-1/2	4796	3097	2145	1572	1199	944	684	509	389	302	238	191	154	126	103	85	71	59	—	—	—
15	5572	3825	2651	1942	1483	1160	926	703	537	419	332	266	216	177	146	121	101	85	72	60	51
16-1/2	6422	4630	3209	2352	1788	1392	1111	906	720	562	446	358	292	240	199	166	139	118	100	85	72
18	7358	5249	3821	2802	2111	1644	1313	1071	888	733	583	470	383	316	262	220	185	157	134	114	98
19-1/2	8391	5898	4487	3264	2460	1915	1530	1248	1036	872	743	601	491	405	338	284	240	204	174	149	128
21	9541	6597	5039	3759	2833	2207	1763	1439	1194	1005	857	738	618	511	426	358	303	259	221	191	165
22-1/2	10825	7352	5563	4287	3232	2518	2012	1642	1363	1148	979	843	732	632	528	445	377	322	277	239	207
24	12270	8170	6120	4848	3655	2848	2277	1858	1543	1300	1108	954	830	727	641	544	462	395	340	294	255
25-1/2	13908	9060	6713	5330	4103	3197	2556	2087	1733	1460	1245	1073	933	817	721	641	558	478	412	356	310
27	15781	10030	7346	5793	4576	3566	2851	2328	1934	1629	1390	1198	1042	913	806	716	639	572	493	427	372
28-1/2	17941	11093	8023	6281	5073	3953	3162	2582	2145	1807	1542	1329	1156	1014	895	795	710	638	575	506	441
30	20463	12263	8748	6796	5554	4360	3487	2848	2366	1994	1701	1467	1276	1119	988	878	785	705	636	576	519
31-1/2	23443	13555	9527	7341	5968	4785	3827	3126	2597	2189	1868	1611	1402	1230	1086	965	863	775	699	633	576
33	27019	14992	10367	7918	6402	5229	4183	3417	2839	2394	2043	1762	1533	1345	1188	1056	944	848	765	694	631

TABLE 4.24 Allowable Loads for Simple-Span Douglas Fir Glulam Roof Beams (lb/ft)—Non-snow Loads (*Continued*)

5-1/2-INCH WIDTH	SPAN (ft)																				
Depth (in.)	8	10	12	14	16	18	20	22	24	26	28	30	32	34	36	38	40	42	44	46	48
12	4109	2624	1817	1331	1015	708	512	381	290	224	176	140	113	91	74	61	—	—	—	—	—
13-1/2	5147	3323	2302	1687	1287	1010	734	547	417	324	256	205	165	135	111	92	76	63	53	—	—
15	5980	4105	2845	2085	1588	1235	986	755	577	449	356	286	232	190	157	130	109	91	77	65	55
16-1/2	6892	4969	3444	2524	1905	1483	1184	965	772	603	478	385	313	257	213	178	150	126	107	91	77
18	7896	5633	4101	2985	2249	1751	1399	1141	946	787	625	504	411	339	281	236	199	168	143	122	105
19-1/2	9006	6329	4815	3478	2621	2041	1631	1330	1104	929	791	645	527	435	362	304	257	219	187	160	138
21	10239	7080	5407	4005	3019	2351	1879	1533	1272	1071	913	786	663	548	457	384	326	277	238	205	177
22-1/2	11617	7890	5970	4568	3444	2683	2144	1750	1453	1223	1043	898	780	678	567	477	405	346	297	256	222
24	13168	8768	6568	5166	3895	3034	2426	1980	1644	1385	1180	1017	884	774	683	584	496	424	365	315	273
25-1/2	14926	9722	7205	5720	4372	3407	2724	2224	1847	1556	1326	1143	994	871	769	682	599	513	442	382	332
27	16935	10764	7884	6217	4876	3800	3038	2481	2060	1736	1481	1276	1110	973	859	763	681	611	529	458	399
28-1/2	19254	11905	8610	6740	5405	4212	3369	2751	2285	1926	1643	1416	1232	1080	953	847	757	679	612	543	474
30	21960	13160	9388	7293	5960	4646	3716	3034	2521	2125	1813	1563	1360	1192	1053	936	836	751	677	613	557
31-1/2	25158	14547	10225	7878	6404	5099	4078	3331	2768	2333	1991	1716	1494	1310	1157	1028	919	825	745	675	613
33	28996	16089	11125	8497	6870	5572	4457	3641	3025	2550	2176	1877	1633	1433	1266	1125	1006	904	815	739	672

6-3/4-INCH WIDTH	SPAN (ft)																				
Depth (in.)	10	12	14	16	18	20	22	24	26	28	30	32	34	36	38	40	42	44	46	48	50
18	6913	4972	3589	2704	2105	1681	1371	1137	957	767	618	504	416	345	289	244	207	176	150	129	110
19-1/2	7768	5791	4181	3151	2453	1960	1599	1326	1116	951	792	647	534	445	373	316	268	229	197	169	146
21	8689	6636	4815	3629	2826	2259	1843	1529	1287	1097	944	813	672	561	472	400	341	292	251	217	188
22-1/2	9683	7327	5492	4140	3225	2577	2103	1746	1470	1253	1079	937	821	696	586	497	424	364	314	272	236
24	10761	8061	6211	4682	3648	2916	2380	1976	1664	1419	1222	1062	930	821	716	609	520	447	387	336	292
25-1/2	11932	8842	6971	5256	4095	3274	2673	2219	1870	1594	1373	1194	1046	923	820	732	630	542	469	408	356
27	13210	9676	7629	5862	4568	3652	2982	2476	2086	1779	1533	1333	1169	1031	916	818	734	649	562	490	428
28-1/2	14610	10567	8272	6498	5064	4050	3307	2747	2314	1974	1702	1480	1297	1145	1017	909	816	735	666	581	509
30	16151	11522	8951	7166	5585	4467	3647	3030	2554	2178	1878	1634	1433	1265	1124	1004	902	813	736	669	599
31-1/2	17854	12548	9668	7860	6130	4903	4004	3327	2804	2392	2063	1795	1574	1390	1235	1104	991	894	810	736	672
33	19746	13654	10428	8432	6698	5358	4376	3636	3065	2616	2256	1963	1722	1521	1352	1208	1085	979	887	807	736
34-1/2	21861	14847	11234	9031	7291	5833	4764	3959	3338	2849	2457	2138	1876	1657	1473	1317	1183	1068	968	880	803
36	24241	16141	12091	9661	7907	6326	5168	4295	3621	3091	2666	2321	2036	1799	1600	1430	1285	1160	1052	957	873
37-1/2	26938	17547	13003	10323	8547	6839	5587	4644	3916	3342	2883	2510	2203	1947	1731	1548	1391	1256	1139	1036	946
39	30022	19081	13976	11020	9093	7370	6022	5005	4221	3603	3109	2707	2375	2099	1867	1670	1501	1356	1229	1118	1021

8-3/4-INCH WIDTH	SPAN (ft)																				
Depth (in.)	12	14	16	18	20	22	24	26	28	30	32	34	36	38	40	42	44	46	48	50	52
24	10449	7844	5913	4606	3682	3005	2494	2101	1790	1542	1340	1174	1035	919	789	675	580	501	435	379	331
25-1/2	11462	8804	6638	5172	4134	3375	2802	2360	2012	1733	1507	1320	1165	1034	923	816	703	608	529	462	404
27	12543	9817	7403	5768	4612	3765	3126	2634	2246	1935	1683	1475	1301	1156	1032	926	834	729	635	555	487
28-1/2	13698	10723	8206	6395	5114	4175	3468	2922	2492	2148	1868	1637	1445	1284	1146	1029	927	839	753	660	580
30	14936	11603	9049	7052	5640	4605	3826	3224	2750	2370	2062	1808	1596	1418	1267	1137	1025	928	843	769	683
31-1/2	16266	12533	9931	7741	6191	5056	4200	3540	3020	2604	2265	1986	1754	1559	1393	1251	1128	1021	928	847	774
33	17699	13518	10852	8459	6766	5526	4591	3870	3302	2847	2478	2173	1919	1706	1524	1369	1235	1119	1017	928	849
34-1/2	19247	14563	11707	9207	7365	6016	4999	4214	3596	3101	2699	2368	2091	1859	1662	1493	1347	1220	1110	1013	927
36	20923	15673	12523	9986	7988	6526	5423	4572	3902	3365	2929	2570	2271	2019	1805	1621	1463	1326	1206	1101	1008
37-1/2	22746	16856	13382	10794	8636	7055	5863	4944	4220	3640	3168	2780	2457	2184	1953	1755	1584	1436	1307	1193	1092
39	24735	18117	14285	11632	9307	7604	6320	5329	4549	3924	3417	2998	2650	2356	2107	1894	1710	1550	1411	1288	1179
40-1/2	26914	19466	15238	12500	10002	8172	6793	5728	4890	4219	3673	3224	2850	2535	2267	2038	1840	1668	1518	1387	1270
42	29311	20911	16244	13274	10720	8759	7282	6141	5243	4524	3939	3458	3056	2719	2432	2187	1975	1791	1630	1489	1364
43-1/2	31960	22463	17307	14070	11462	9366	7787	6568	5608	4839	4214	3699	3270	2909	2603	2340	2114	1917	1745	1594	1461
45	34904	24135	18434	14904	12228	9992	8308	7008	5984	5164	4497	3948	3491	3106	2779	2499	2257	2048	1864	1703	1561

Notes:

(1) Span = simply supported beam.

(2) Maximum deflection = L/180 under total load. Other deflection limits may apply.

(3) Service condition = dry.

(4) Tabulated values represent total loads and have taken the dead weight of the beam (assumed 35 pcf) into account.

(5) Sufficient bearing length shall be provided at supports.

(6) Maximum beam shear is located at a distance from the supports equal to the depth of the beam.

(7) Light (unshaded) areas limited by deflection; medium-shaded areas limited by bending strength; dark-shaded areas limited by shear strength.

TABLE 4.25 Allowable Loads for Simple-Span Douglas Fir Glulam Roof Beams (lb/ft)—Snow Loads

3-1/8-INCH WIDTH	SPAN (ft)																				
Depth (in.)	8	10	12	14	16	18	20	22	24	26	28	30	32	34	36	38	40	42	44	46	48
6	535	295	169	105	69	—	—	—	—	—	—	—	—	—	—	—	—	—	—	—	—
7-1/2	837	533	333	208	137	95	68	—	—	—	—	—	—	—	—	—	—	—	—	—	—
9	1206	769	532	362	240	167	120	88	66	51	—	—	—	—	—	—	—	—	—	—	—
10-1/2	1643	1049	726	531	385	268	193	143	108	84	65	52	—	—	—	—	—	—	—	—	—
12	2147	1371	949	695	530	402	291	216	164	127	100	80	64	52	—	—	—	—	—	—	—
13-1/2	2690	1736	1203	881	672	529	417	311	237	184	145	116	94	77	63	52	—	—	—	—	—
15	3125	2145	1486	1089	831	654	528	429	328	255	202	162	132	108	89	74	62	52	—	—	—
16-1/2	3602	2597	1799	1319	1007	793	640	527	439	342	272	219	178	146	121	101	85	72	61	52	—
18	4126	2943	2143	1571	1199	945	763	628	523	440	355	286	234	192	160	134	113	96	81	70	60
19-1/2	4706	3307	2516	1844	1409	1110	896	735	610	513	437	367	299	247	206	173	146	124	106	91	78
21	5351	3699	2825	2140	1635	1288	1039	848	703	592	504	434	377	311	260	218	185	158	135	116	100
22-1/2	6071	4123	3119	2458	1878	1480	1186	967	803	676	576	496	431	378	322	271	230	196	169	145	126
24	6882	4582	3432	2742	2138	1678	1341	1095	909	765	652	562	488	428	377	332	282	241	207	179	155
25-1/2	7801	5081	3765	2988	2415	1884	1506	1229	1021	860	733	632	549	481	425	377	336	291	251	217	189
27	8851	5625	4119	3248	2679	2101	1680	1372	1139	960	818	705	613	537	474	421	376	338	300	260	227

3-1/2-INCH WIDTH	SPAN (ft)																				
Depth (in.)	8	10	12	14	16	18	20	22	24	26	28	30	32	34	36	38	40	42	44	46	48
6	599	331	189	117	77	53	—	—	—	—	—	—	—	—	—	—	—	—	—	—	—
7-1/2	937	597	373	233	154	106	76	55	—	—	—	—	—	—	—	—	—	—	—	—	—
9	1351	862	596	406	269	187	134	99	74	57	—	—	—	—	—	—	—	—	—	—	—
10-1/2	1840	1174	813	595	431	300	216	160	121	94	73	58	—	—	—	—	—	—	—	—	—
12	2405	1535	1063	778	594	451	326	242	184	143	112	89	72	58	—	—	—	—	—	—	—
13-1/2	3013	1945	1347	987	753	592	467	348	265	206	163	130	105	86	71	58	—	—	—	—	—
15	3500	2402	1664	1219	931	733	591	480	367	286	226	182	147	121	100	83	69	58	—	—	—
16-1/2	4034	2908	2015	1477	1127	888	717	590	490	384	304	245	199	164	136	113	95	80	68	58	—
18	4621	3297	2400	1759	1343	1058	854	698	579	487	398	321	262	216	179	150	126	107	91	78	67
19-1/2	5271	3704	2818	2066	1578	1243	998	814	675	568	484	411	335	277	231	194	164	139	119	102	88
21	5993	4143	3164	2397	1831	1440	1150	938	779	655	558	481	417	349	291	245	207	177	151	130	112
22-1/2	6800	4618	3494	2753	2103	1642	1313	1071	889	748	638	549	477	418	361	304	258	220	189	163	141
24	7708	5132	3844	3071	2385	1858	1485	1212	1006	847	722	622	541	474	418	371	316	270	232	201	174
25-1/2	8737	5690	4216	3347	2677	2086	1668	1361	1130	952	812	699	608	533	470	417	372	326	281	243	212
27	9913	6300	4614	3638	2986	2326	1860	1519	1261	1062	906	781	679	595	525	466	416	373	336	292	254

5-1/8-INCH WIDTH	SPAN (ft)																				
Depth (in.)	8	10	12	14	16	18	20	22	24	26	28	30	32	34	36	38	40	42	44	46	48
12	3521	2248	1557	1140	869	660	477	355	270	209	164	131	105	85	69	57	—	—	—	—	—
13-1/2	4411	2848	1972	1445	1102	867	684	509	389	302	238	191	154	126	103	85	71	59	—	—	—
15	5125	3518	2437	1786	1363	1065	850	693	537	419	332	266	216	177	146	121	101	85	72	60	51
16-1/2	5907	4258	2951	2163	1643	1279	1021	832	689	562	446	358	292	240	199	166	139	118	100	85	72
18	6767	4827	3514	2576	1940	1510	1206	983	815	686	583	470	383	316	262	220	185	157	134	114	98
19-1/2	7718	5424	4126	3001	2261	1760	1406	1146	951	800	681	586	491	405	338	284	240	204	174	149	128
21	8775	6067	4633	3456	2605	2028	1620	1322	1097	923	786	677	587	511	426	358	303	259	221	191	165
22-1/2	9957	6762	5116	3942	2971	2314	1849	1509	1252	1054	898	773	672	588	518	445	377	322	277	239	207
24	11286	7514	5628	4458	3360	2618	2092	1707	1417	1193	1017	876	761	666	588	521	462	395	340	294	255
25-1/2	12793	8332	6174	4901	3773	2939	2349	1917	1592	1341	1143	984	856	750	661	587	524	470	412	356	310
27	14516	9225	6756	5327	4207	3278	2621	2139	1776	1496	1276	1099	956	837	739	656	585	525	473	427	372
28-1/2	16503	10203	7378	5775	4664	3634	2906	2372	1970	1660	1416	1220	1061	930	820	729	651	584	526	476	432
30	18823	11279	8045	6249	5106	4008	3205	2617	2174	1832	1562	1346	1171	1027	906	805	719	645	582	527	478
31-1/2	21564	12468	8762	6750	5487	4399	3518	2873	2387	2011	1716	1479	1287	1128	996	885	791	710	640	579	527
33	24854	13790	9534	7281	5886	4807	3845	3140	2609	2199	1876	1617	1407	1234	1090	968	865	777	701	635	577

TABLE 4.25 Allowable Loads for Simple-Span Douglas Fir Glulam Roof Beams (lb/ft)—Snow Loads (*Continued*)

5-1/2-INCH WIDTH	SPAN (ft)																				
Depth (in.)	8	10	12	14	16	18	20	22	24	26	28	30	32	34	36	38	40	42	44	46	48
12	3779	2413	1671	1223	933	708	512	381	290	224	176	140	113	91	74	61	—	—	—	—	—
13-1/2	4734	3056	2117	1550	1183	927	734	547	417	324	256	205	165	135	111	92	76	63	53	—	—
15	5500	3775	2615	1916	1459	1135	906	738	577	449	356	286	232	190	157	130	109	91	77	65	55
16-1/2	6339	4570	3167	2321	1751	1362	1088	886	735	603	478	385	313	257	213	178	150	126	107	91	77
18	7262	5181	3771	2744	2067	1609	1285	1048	869	730	622	504	411	339	281	236	199	168	143	122	105
19-1/2	8283	5821	4428	3197	2409	1875	1498	1222	1013	852	726	624	527	435	362	304	257	219	187	160	138
21	9417	6511	4972	3683	2775	2161	1726	1408	1168	983	837	721	626	548	457	384	326	277	238	205	177
22-1/2	10685	7256	5490	4200	3166	2466	1970	1607	1334	1123	957	824	715	626	552	477	405	346	297	256	222
24	12112	8064	6040	4750	3581	2789	2229	1819	1510	1271	1083	933	811	710	626	555	495	424	365	315	273
25-1/2	13729	8942	6626	5259	4020	3132	2503	2043	1696	1428	1218	1049	912	798	704	625	558	500	442	382	332
27	15578	9900	7250	5716	4483	3493	2792	2279	1893	1594	1359	1171	1018	892	787	699	624	559	504	456	399
28-1/2	17711	10949	7918	6198	4970	3872	3096	2528	2099	1769	1508	1300	1130	990	874	776	693	622	560	507	460
30	20200	12104	8634	6707	5480	4271	3415	2788	2316	1951	1664	1435	1248	1094	965	858	766	687	620	561	509
31-1/2	23142	13380	9403	7244	5889	4687	3749	3061	2543	2143	1828	1576	1371	1202	1061	943	842	756	682	617	561
33	26673	14799	10232	7814	6317	5123	4097	3346	2780	2343	1999	1723	1499	1315	1161	1032	922	828	747	676	615

6-3/4-INCH WIDTH	SPAN (ft)																				
Depth (in.)	10	12	14	16	18	20	22	24	26	28	30	32	34	36	38	40	42	44	46	48	50
18	6358	4572	3299	2485	1934	1544	1259	1044	878	747	618	504	416	345	289	244	207	176	150	129	110
19-1/2	7144	5325	3844	2896	2254	1801	1468	1218	1024	872	750	647	534	445	373	316	268	229	197	169	146
21	7991	6102	4427	3336	2598	2075	1692	1404	1182	1006	866	752	658	561	472	400	341	292	251	217	188
22-1/2	8905	6738	5050	3806	2964	2368	1932	1603	1349	1150	990	859	752	663	586	497	424	364	314	272	236
24	9897	7413	5711	4305	3353	2679	2186	1815	1528	1302	1121	974	853	752	667	595	520	447	387	336	292
25-1/2	10974	8131	6410	4833	3764	3009	2456	2038	1717	1463	1260	1095	959	846	751	670	600	541	469	408	356
27	12150	8898	7015	5389	4199	3356	2740	2275	1916	1633	1407	1223	1072	945	839	749	672	605	547	490	428
28-1/2	13438	9718	7607	5975	4655	3722	3038	2523	2125	1812	1562	1358	1190	1050	932	832	747	673	609	553	503
30	14855	10596	8231	6589	5134	4105	3352	2784	2345	2000	1724	1499	1314	1160	1030	920	825	744	673	612	557
31-1/2	16421	11540	8891	7227	5635	4506	3680	3056	2576	2197	1894	1647	1444	1275	1132	1012	908	819	741	673	614
33	18162	12557	9590	7753	6158	4925	4022	3341	2816	2402	2071	1802	1580	1395	1239	1107	994	897	812	738	673
34-1/2	20108	13655	10331	8304	6703	5361	4379	3638	3066	2616	2256	1963	1721	1520	1351	1207	1084	978	886	805	734
36	22297	14845	11119	8883	7270	5815	4750	3947	3327	2839	2448	2130	1869	1651	1467	1311	1178	1063	963	875	799
37-1/2	24778	16138	11958	9492	7859	6287	5135	4267	3597	3070	2648	2304	2022	1786	1588	1419	1275	1151	1043	948	865
39	27615	17550	12853	10133	8360	6775	5535	4600	3878	3310	2855	2485	2180	1926	1713	1531	1376	1242	1126	1024	934

8-3/4-INCH WIDTH	SPAN (ft)																				
Depth (in.)	12	14	16	18	20	22	24	26	28	30	32	34	36	38	40	42	44	46	48	50	52
24	9609	7212	5436	4234	3383	2760	2291	1928	1643	1415	1229	1076	948	841	750	672	580	501	435	379	331
25-1/2	10541	8095	6103	4753	3799	3100	2573	2167	1847	1590	1382	1210	1067	947	845	757	682	608	529	462	404
27	11535	9027	6806	5302	4238	3459	2872	2418	2062	1776	1543	1352	1193	1058	945	847	763	690	626	555	487
28-1/2	12597	9860	7545	5878	4700	3836	3185	2683	2288	1971	1713	1501	1325	1176	1050	942	848	767	697	634	579
30	13736	10670	8320	6483	5184	4232	3514	2961	2525	2176	1892	1658	1463	1299	1160	1041	938	849	771	702	642
31-1/2	14960	11525	9132	7116	5690	4646	3859	3251	2773	2390	2079	1822	1608	1429	1276	1145	1032	934	849	774	707
33	16278	12431	9978	7777	6219	5078	4218	3555	3032	2614	2274	1994	1760	1564	1397	1254	1131	1024	930	848	775
34-1/2	17701	13392	10765	8465	6770	5529	4593	3871	3303	2847	2477	2172	1918	1704	1523	1367	1233	1117	1015	926	847
36	19243	14413	11515	9181	7343	5997	4983	4200	3584	3090	2689	2358	2083	1851	1654	1486	1340	1214	1104	1007	921
37-1/2	20920	15501	12305	9924	7938	6484	5388	4542	3876	3342	2909	2551	2254	2003	1791	1608	1451	1315	1196	1091	998
39	22750	16661	13136	10695	8556	6989	5808	4896	4179	3604	3137	2752	2431	2161	1932	1736	1567	1420	1291	1178	1079
40-1/2	24754	17901	14012	11493	9195	7511	6242	5263	4492	3875	3373	2959	2615	2325	2079	1868	1686	1528	1390	1269	1162
42	26959	19231	14937	12205	9855	8052	6692	5643	4817	4155	3617	3174	2805	2494	2230	2004	1810	1640	1492	1363	1248
43-1/2	29396	20659	15915	12937	10538	8610	7156	6035	5152	4444	3869	3396	3001	2669	2387	2146	1937	1756	1598	1459	1337
45	32104	22197	16951	13704	11242	9185	7635	6439	5497	4743	4130	3625	3204	2850	2549	2291	2069	1876	1708	1559	1429

Notes:

(1) Span = simply supported beam.

(2) Maximum deflection = L/180 under total load. Other deflection limits may apply.

(3) Service condition = dry.

(4) Tabulated values represent total loads and have taken the dead weight of the beam (assumed 35 pcf) into account.

(5) Sufficient bearing length shall be provided at supports.

(6) Maximum beam shear is located at a distance from the supports equal to the depth of the beam.

(7) Light (unshaded) areas limited by deflection; medium-shaded areas limited by bending strength; dark-shaded areas limited by shear strength.

TABLE 4.26 Allowable Loads for Simple-Span Douglas Fir Glulam Floor Beams (lb/ft)

3-1/8-INCH WIDTH	SPAN (ft)																				
Depth (in.)	8	10	12	14	16	18	20	22	24	26	28	30	32	34	36	38	40	42	44	46	48
6	362	183	104	64	—	—	—	—	—	—	—	—	—	—	—	—	—	—	—	—	—
7-1/2	710	361	206	128	84	57	—	—	—	—	—	—	—	—	—	—	—	—	—	—	—
9	1048	626	359	224	148	102	72	53	—	—	—	—	—	—	—	—	—	—	—	—	—
10-1/2	1428	911	574	358	237	164	118	86	65	—	—	—	—	—	—	—	—	—	—	—	—
12	1866	1191	824	538	357	248	178	132	99	76	59	—	—	—	—	—	—	—	—	—	—
13-1/2	2338	1508	1044	765	511	356	257	190	144	111	87	69	55	—	—	—	—	—	—	—	—
15	2716	1864	1291	945	704	491	355	264	201	155	122	97	78	63	51	—	—	—	—	—	—
16-1/2	3130	2256	1563	1145	874	656	475	354	270	209	165	132	106	87	71	59	—	—	—	—	—
18	3586	2558	1861	1364	1041	820	619	462	353	274	217	174	141	115	95	79	65	55	—	—	—
19-1/2	4090	2874	2186	1602	1223	963	777	590	451	351	278	224	182	149	123	102	86	72	61	51	—
21	4651	3215	2455	1859	1420	1118	901	735	566	441	350	282	229	189	156	131	110	93	78	67	57
22-1/2	5277	3583	2710	2135	1631	1285	1029	839	696	545	433	349	285	234	195	163	137	116	99	84	72
24	5982	3982	2982	2382	1857	1457	1164	950	788	663	528	426	348	287	239	200	169	144	123	105	90
25-1/2	6781	4415	3271	2596	2097	1636	1307	1067	885	745	635	514	420	347	289	243	206	175	150	129	111
27	7694	4889	3579	2822	2327	1824	1458	1190	988	832	709	611	501	414	346	291	246	210	180	155	134

3-1/2-INCH WIDTH	SPAN (ft)																				
Depth (in.)	8	10	12	14	16	18	20	22	24	26	28	30	32	34	36	38	40	42	44	46	48
6	405	205	116	71	—	—	—	—	—	—	—	—	—	—	—	—	—	—	—	—	—
7-1/2	795	404	231	143	94	64	—	—	—	—	—	—	—	—	—	—	—	—	—	—	—
9	1174	701	403	251	165	114	81	59	—	—	—	—	—	—	—	—	—	—	—	—	—
10-1/2	1599	1020	642	401	266	184	132	97	72	55	—	—	—	—	—	—	—	—	—	—	—
12	2090	1334	923	602	400	278	200	148	111	85	66	52	—	—	—	—	—	—	—	—	—
13-1/2	2618	1690	1170	856	573	399	288	213	162	125	97	77	62	—	—	—	—	—	—	—	—
15	3042	2087	1446	1059	788	550	397	295	225	174	137	109	87	71	58	—	—	—	—	—	—
16-1/2	3506	2527	1751	1282	979	735	532	396	302	234	185	148	119	97	80	66	54	—	—	—	—
18	4017	2865	2085	1528	1166	918	693	517	395	307	243	195	158	129	106	88	73	61	51	—	—
19-1/2	4581	3219	2448	1794	1370	1079	866	660	505	394	312	250	203	167	138	115	96	81	68	57	—
21	5209	3601	2749	2082	1590	1249	998	814	633	494	392	316	257	211	175	146	123	104	88	75	64
22-1/2	5910	4013	3035	2392	1827	1426	1139	929	771	611	485	391	319	263	218	183	154	130	111	95	81
24	6700	4460	3340	2668	2071	1613	1289	1051	872	734	592	477	390	322	268	225	190	161	137	118	101
25-1/2	7594	4945	3663	2908	2325	1811	1447	1181	980	825	703	575	470	388	324	272	230	196	168	144	124
27	8617	5475	4009	3160	2593	2020	1614	1317	1094	921	785	676	561	464	387	326	276	235	202	174	150

5-1/8-INCH WIDTH									SPAN (ft)											
Depth (in.)	8	10	12	14	16	18	20	22	24	26	28	30	32	34	36	38	40	42	44	46	48
12	3060	1953	1352	882	586	407	293	216	163	125	97	76	60	—	—	—	—	—	—	—	—
13-1/2	3834	2474	1713	1254	838	584	421	312	237	182	143	113	90	72	58	—	—	—	—	—	—
15	4454	3056	2117	1550	1154	805	582	433	329	255	200	159	128	104	84	69	56	—	—	—	—
16-1/2	5134	3700	2563	1878	1426	1076	779	580	442	343	271	216	175	142	117	96	79	66	55	—	—
18	5882	4195	3053	2237	1684	1310	1015	757	578	450	356	285	231	189	156	129	107	90	75	63	53
19-1/2	6708	4713	3585	2606	1963	1527	1219	967	739	576	457	367	298	244	202	168	141	118	100	84	71
21	7627	5272	4026	3002	2261	1760	1406	1146	928	724	574	462	376	309	256	214	180	152	129	109	93
22-1/2	8654	5876	4445	3424	2580	2008	1604	1308	1085	895	711	573	467	385	320	267	225	191	162	139	119
24	9810	6530	4890	3873	2918	2272	1815	1481	1228	1034	867	699	571	471	392	329	278	236	201	172	148
25-1/2	11120	7241	5364	4257	3276	2551	2039	1663	1380	1162	990	843	689	569	474	398	337	287	245	211	182
27	12618	8017	5870	4627	3654	2846	2274	1856	1540	1297	1105	951	821	679	567	477	404	345	295	254	220
28-1/2	14346	8867	6411	5017	4051	3156	2522	2058	1709	1439	1226	1056	918	803	671	565	479	409	351	303	262
30	16363	9803	6991	5429	4435	3480	2782	2271	1885	1588	1354	1166	1013	888	783	663	563	481	414	358	310
31-1/2	18746	10837	7614	5865	4766	3820	3054	2493	2070	1744	1487	1281	1114	976	861	764	656	561	483	418	363
33	21607	11986	8285	6326	5113	4175	3338	2725	2263	1907	1626	1401	1218	1068	942	837	747	649	559	484	421

TABLE 4.26 Allowable Loads for Simple-Span Douglas Fir Glulam Floor Beams (lb/ft)

5-1/2-INCH WIDTH	SPAN (ft)																				
Depth (in.)	8	10	12	14	16	18	20	22	24	26	28	30	32	34	36	38	40	42	44	46	48
12	3284	2096	1451	946	628	437	314	232	175	134	104	82	65	51	—	—	—	—	—	—	—
13-1/2	4114	2655	1838	1346	900	626	452	335	254	196	153	121	97	78	63	50	—	—	—	—	—
15	4780	3280	2272	1664	1239	864	624	464	353	273	215	171	137	111	90	74	61	—	—	—	—
16-1/2	5509	3971	2751	2015	1520	1155	836	622	474	368	291	232	187	153	125	103	85	71	59	—	—
18	6312	4502	3276	2383	1795	1396	1090	813	620	483	382	306	248	203	167	138	115	96	81	67	57
19-1/2	7199	5058	3847	2777	2091	1627	1299	1038	793	618	490	393	320	262	217	180	151	127	107	90	76
21	8185	5658	4320	3199	2410	1875	1498	1221	995	777	616	496	404	332	275	230	193	163	138	117	100
22-1/2	9288	6306	4770	3649	2749	2140	1709	1394	1156	960	763	614	501	413	343	287	242	205	174	149	127
24	10528	7008	5248	4126	3110	2421	1934	1578	1309	1101	930	750	612	505	421	353	298	253	216	185	159
25-1/2	11934	7771	5757	4569	3491	2719	2172	1772	1470	1238	1054	904	739	610	509	428	362	308	263	226	195
27	13541	8604	6300	4966	3893	3032	2423	1977	1641	1381	1177	1014	881	729	608	512	434	370	317	273	236
28-1/2	15396	9516	6881	5385	4317	3362	2687	2193	1820	1533	1306	1125	978	856	720	606	515	439	377	325	282
30	17560	10520	7503	5827	4760	3708	2964	2419	2009	1692	1442	1242	1080	946	834	712	604	517	444	384	333
31-1/2	20118	11629	8171	6294	5115	4071	3254	2656	2206	1858	1584	1365	1186	1040	917	814	704	602	518	448	390
33	23188	12863	8891	6789	5487	4449	3557	2904	2411	2031	1732	1493	1298	1137	1004	891	796	697	600	520	452

6-3/4-INCH WIDTH	SPAN (ft)																				
Depth (in.)	10	12	14	16	18	20	22	24	26	28	30	32	34	36	38	40	42	44	46	48	50
18	5525	3972	2865	2157	1678	1337	997	761	593	469	375	304	249	205	170	141	118	99	83	69	58
19-1/2	6208	4626	3338	2514	1956	1562	1272	974	759	601	483	392	322	266	221	185	156	131	111	94	79
21	6944	5302	3845	2897	2254	1800	1467	1216	954	757	609	495	407	338	282	237	200	169	144	123	104
22-1/2	7739	5854	4386	3305	2572	2054	1675	1389	1169	936	754	615	506	421	352	297	251	214	183	156	134
24	8601	6441	4961	3738	2910	2325	1896	1573	1323	1127	921	752	620	516	433	366	310	265	227	195	168
25-1/2	9537	7065	5569	4197	3268	2611	2130	1767	1487	1267	1090	907	749	625	525	444	378	323	278	239	207
27	10559	7732	6095	4681	3645	2913	2376	1972	1660	1415	1218	1058	895	747	628	532	454	389	335	289	251
28-1/2	11679	8444	6608	5189	4042	3230	2636	2188	1842	1570	1352	1175	1029	884	744	631	539	463	399	346	300
30	12911	9208	7151	5723	4458	3563	2908	2414	2033	1733	1493	1297	1136	1002	873	742	634	545	471	409	356
31-1/2	14273	10028	7724	6278	4893	3912	3193	2651	2233	1904	1640	1425	1249	1102	978	864	739	636	550	478	417
33	15786	10912	8332	6734	5348	4276	3490	2898	2442	2082	1794	1560	1367	1206	1071	956	855	737	638	555	485
34-1/2	17478	11867	8976	7214	5821	4655	3800	3156	2659	2268	1954	1699	1489	1315	1167	1042	935	843	734	640	559
36	19381	12901	9661	7717	6314	5049	4123	3424	2885	2461	2121	1845	1617	1428	1268	1132	1016	916	829	732	641
37-1/2	21538	14025	10390	8246	6826	5459	4457	3703	3120	2662	2294	1996	1750	1545	1373	1226	1101	993	899	816	729
39	24005	15252	11168	8803	7261	5883	4804	3991	3364	2870	2474	2152	1888	1667	1481	1323	1188	1072	970	882	804

8-3/4-INCH WIDTH	SPAN (ft)																				
Depth (in.)	12	14	16	18	20	22	24	26	28	30	32	34	36	38	40	42	44	46	48	50	52
24	8349	6265	4720	3675	2935	2394	1985	1670	1422	1193	974	804	669	561	474	402	343	294	253	218	188
25-1/2	9159	7032	5300	4126	3297	2689	2231	1877	1599	1376	1176	971	810	680	575	490	419	360	310	268	232
27	10023	7842	5911	4603	3678	3000	2490	2096	1785	1537	1335	1160	968	814	690	588	504	434	375	325	283
28-1/2	10946	8566	6553	5104	4079	3328	2762	2325	1981	1706	1482	1298	1144	965	819	699	600	517	448	390	340
30	11936	9270	7227	5629	4499	3672	3048	2566	2187	1884	1637	1434	1264	1122	962	822	707	610	530	461	403
31-1/2	13000	10013	7932	6179	4939	4031	3347	2819	2403	2070	1799	1576	1390	1234	1101	958	825	713	620	541	473
33	14145	10800	8668	6753	5399	4407	3659	3082	2628	2264	1968	1724	1521	1351	1205	1081	955	827	720	629	551
34-1/2	15383	11636	9351	7351	5877	4798	3984	3357	2862	2466	2144	1879	1658	1473	1315	1180	1063	952	829	725	636
36	16723	12523	10003	7973	6375	5205	4323	3642	3106	2677	2328	2041	1801	1600	1428	1282	1155	1046	949	831	730
37-1/2	18181	13469	10689	8619	6893	5628	4675	3939	3360	2896	2519	2208	1949	1732	1547	1388	1252	1133	1029	938	832
39	19772	14477	11412	9289	7429	6066	5039	4247	3623	3123	2717	2382	2103	1869	1669	1499	1351	1224	1112	1014	927
40-1/2	21514	15555	12173	9983	7984	6520	5417	4566	3895	3358	2922	2562	2262	2010	1796	1613	1455	1318	1198	1092	991
42	23431	16711	12977	10602	8558	6990	5807	4895	4177	3601	3134	2748	2427	2157	1928	1731	1562	1415	1286	1173	1073
43-1/2	25550	17952	13827	11238	9151	7475	6211	5236	4468	3853	3353	2941	2598	2309	2064	1854	1673	1515	1378	1257	1150
45	27904	19289	14728	11904	9763	7975	6627	5587	4768	4112	3579	3139	2773	2465	2204	1980	1787	1619	1472	1344	1230

Notes:

(1) Span = simply supported beam.

(2) Maximum deflection = L/360 under live load, based on live/total load = 0.8. Where additional stiffness is desired or for other live/total load ratios, design for deflection must be modified per requirements.

(3) Service condition = dry.

(4) Tabulated values represent total loads based on live/total load = 0.8 and have taken the dead weight of the beam (assumed 35 pcf) into account.

(5) Sufficient bearing length shall be provided at supports.

(6) Maximum beam shear is located at a distance from the supports equal to the depth of the beam.

(7) Light (unshaded) areas limited by deflection; medium-shaded areas limited by bending strength; dark-shaded areas limited by shear strength.

TABLE 4.27 Allowable Loads for Cantilevered Douglas Fir Glulam Roof Beams (lb/ft)—Non-snow Loads

5-1/8-INCH WIDTH	SPAN (ft)																	
Depth (in.)	44			48			52			56			60			64		
	sys 1	sys 2	sys 3	sys 1	sys 2	sys 3	sys 1	sys 2	sys 3	sys 1	sys 2	sys 3	sys 1	sys 2	sys 3	sys 1	sys 2	sys 3
24	495	489	585	407	402	482	340	335	403	286	283	341	244	241	291	209	206	250
25-1/2	557	550	658	459	453	543	383	378	454	323	319	384	275	272	328	236	233	282
27	623	615	735	513	507	607	428	423	508	362	357	430	309	305	367	265	262	316
28-1/2	692	684	816	570	563	674	477	471	564	403	398	478	344	339	409	296	292	352
30	764	755	902	630	623	745	527	521	624	446	440	529	381	376	452	328	324	390
31-1/2	840	830	991	693	685	819	580	573	686	491	485	582	419	414	498	361	357	430
33	920	909	1085	759	750	897	635	628	751	538	531	637	460	454	546	396	391	471
34-1/2	1003	991	1182	828	818	977	693	685	819	587	580	695	502	496	596	433	428	515
36	1089	1076	1283	899	889	1061	753	744	890	638	631	755	546	540	648	471	465	560

6-3/4-INCH WIDTH	SPAN (ft)																	
Depth (in.)	44			48			52			56			60			64		
	sys 1	sys 2	sys 3	sys 1	sys 2	sys 3	sys 1	sys 2	sys 3	sys 1	sys 2	sys 3	sys 1	sys 2	sys 3	sys 1	sys 2	sys 3
24	633	625	748	521	514	617	434	429	515	366	361	435	311	307	371	267	263	319
25-1/2	713	704	842	587	579	694	489	483	580	413	408	491	351	347	419	302	298	360
27	797	787	941	656	648	776	548	541	649	462	457	549	394	389	469	339	334	404
28-1/2	885	875	1045	729	721	863	609	602	722	515	508	611	439	434	522	378	373	450
30	978	966	1154	806	797	953	674	666	798	570	563	676	486	480	578	418	413	499
31-1/2	1075	1063	1269	887	877	1048	742	733	878	628	620	744	536	529	637	461	456	549
33	1177	1163	1388	971	960	1147	813	803	961	688	679	815	588	580	698	506	500	602
34-1/2	1283	1268	1513	1059	1047	1251	887	876	1048	751	742	889	642	634	762	553	546	658
36	1393	1377	1642	1151	1137	1358	964	952	1139	816	806	966	698	690	828	602	595	716
37-1/2	1508	1490	1777	1246	1231	1470	1044	1031	1233	884	874	1047	757	748	897	653	645	776
39	1627	1608	1917	1345	1329	1586	1127	1113	1331	955	944	1130	818	808	969	706	697	838
40-1/2	1750	1730	2058	1447	1430	1706	1213	1198	1432	1028	1016	1216	881	870	1043	761	751	903
42	1878	1856	2134	1553	1534	1831	1302	1286	1537	1104	1091	1305	946	935	1120	817	807	969
43-1/2	2010	1986	2210	1662	1643	1959	1394	1377	1645	1183	1169	1398	1014	1001	1200	876	865	1039
45	2146	2121	2286	1775	1754	2090	1489	1471	1757	1264	1249	1493	1083	1070	1282	937	925	1110
46-1/2	2286	2259	2363	1891	1869	2159	1587	1568	1872	1347	1331	1591	1155	1141	1366	999	987	1184
48	2376	2402	2439	2011	1988	2229	1688	1668	1991	1433	1416	1693	1229	1215	1454	1064	1051	1259

8-3/4-INCH WIDTH	SPAN (ft)																	
Depth (in.)	44			48			52			56			60			64		
	sys 1	sys 2	sys 3	sys 1	sys 2	sys 3	sys 1	sys 2	sys 3	sys 1	sys 2	sys 3	sys 1	sys 2	sys 3	sys 1	sys 2	sys 3
36	1758	1737	2073	1452	1434	1714	1215	1201	1437	1029	1017	1219	880	869	1044	759	749	902
37-1/2	1903	1880	2243	1572	1553	1855	1316	1300	1556	1115	1102	1320	954	942	1131	823	813	978
39	2053	2029	2419	1696	1676	2001	1421	1404	1679	1204	1190	1425	1031	1018	1222	890	879	1056
40-1/2	2209	2183	2602	1825	1804	2153	1530	1511	1807	1297	1281	1534	1110	1097	1315	959	947	1138
42	2370	2342	2766	1959	1936	2310	1642	1623	1939	1393	1376	1647	1193	1178	1412	1030	1018	1222
43-1/2	2536	2506	2865	2097	2072	2472	1758	1737	2076	1492	1474	1763	1278	1262	1513	1104	1091	1309
45	2708	2676	2964	2240	2213	2640	1878	1856	2217	1594	1575	1883	1366	1349	1616	1181	1166	1399
46-1/2	2885	2851	3063	2386	2358	2799	2002	1978	2362	1699	1679	2007	1457	1439	1723	1259	1244	1492
48	3067	3031	3161	2538	2508	2889	2129	2104	2512	1808	1786	2135	1550	1531	1833	1341	1324	1588
49-1/2	3176	3211	3260	2694	2662	2980	2261	2234	2666	1920	1897	2267	1646	1627	1947	1424	1407	1687
51	3272	3308	3359	2854	2820	3070	2395	2367	2825	2035	2010	2402	1746	1725	2063	1511	1492	1788
52-1/2	3368	3405	3458	3018	2983	3160	2534	2504	2909	2153	2127	2541	1847	1825	2183	1599	1580	1892
54	3465	3502	3556	3166	3150	3251	2676	2645	2992	2274	2247	2684	1952	1928	2306	1690	1669	1999
55-1/2	3561	3600	3655	3254	3290	3341	2822	2789	3075	2399	2370	2830	2059	2034	2432	1783	1762	2109
57	3657	3697	3754	3342	3379	3431	2972	2937	3158	2526	2496	2924	2169	2143	2562	1879	1856	2222
58-1/2	3753	3794	3853	3430	3468	3521	3125	3088	3241	2657	2625	3001	2282	2254	2694	1977	1953	2337
60	3850	3892	3952	3518	3557	3612	3238	3244	3324	2791	2758	3078	2397	2369	2830	2077	2052	2455

See Fig. 4.33 for notes and description of cantilever systems.

TABLE 4.28 Allowable Loads for Cantilevered Douglas Fir Glulam Roof Beams (lb/ft)—Snow Loads

5-1/8-INCH WIDTH	SPAN (ft)																	
Depth (in.)	44			48			52			56			60			64		
	sys 1	sys 2	sys 3	sys 1	sys 2	sys 3	sys 1	sys 2	sys 3	sys 1	sys 2	sys 3	sys 1	sys 2	sys 3	sys 1	sys 2	sys 3
24	453	447	536	372	368	441	310	306	368	261	258	311	222	219	265	190	187	228
25-1/2	510	504	603	419	414	497	350	345	415	295	291	351	251	247	299	215	212	257
27	570	563	674	469	464	556	392	387	464	330	326	393	281	278	335	241	238	288
28-1/2	634	626	748	522	516	617	436	430	516	368	363	437	313	309	373	269	266	321
30	700	692	827	577	570	682	482	476	571	407	402	483	347	343	413	298	295	356
31-1/2	770	761	909	635	627	750	531	524	628	448	443	532	383	378	455	329	325	392
33	843	833	994	695	687	822	581	574	688	492	486	583	420	415	499	361	357	430
34-1/2	919	908	1084	758	749	896	634	627	750	537	530	636	459	453	545	395	390	470
36	998	986	1177	824	814	973	690	681	815	584	577	691	499	493	592	430	425	511

6-3/4-INCH WIDTH	SPAN (ft)																	
Depth (in.)	44			48			52			56			60			64		
	sys 1	sys 2	sys 3	sys 1	sys 2	sys 3	sys 1	sys 2	sys 3	sys 1	sys 2	sys 3	sys 1	sys 2	sys 3	sys 1	sys 2	sys 3
24	579	572	685	476	470	564	396	391	471	333	329	397	283	279	338	242	239	290
25-1/2	652	644	771	536	530	635	447	441	531	376	372	448	320	316	382	274	271	328
27	729	721	862	600	593	711	500	494	594	422	417	502	359	354	428	308	304	368
28-1/2	811	801	958	667	659	790	557	550	660	470	464	558	400	395	477	344	339	410
30	896	885	1058	738	729	873	616	609	730	520	514	618	444	438	528	381	376	455
31-1/2	985	973	1163	812	802	960	678	670	804	573	566	680	489	483	582	420	415	501
33	1079	1066	1273	889	879	1051	743	734	880	628	621	745	536	530	638	462	456	550
34-1/2	1176	1162	1387	970	958	1146	811	801	960	686	678	814	586	579	696	505	498	601
36	1277	1262	1506	1054	1042	1245	882	871	1043	746	737	884	638	630	757	549	542	654
37-1/2	1383	1366	1630	1141	1128	1348	955	944	1130	809	799	958	691	683	820	596	589	709
39	1492	1474	1759	1232	1217	1454	1032	1019	1219	874	863	1034	747	738	886	644	636	766
40-1/2	1605	1586	1888	1326	1310	1565	1110	1097	1312	941	929	1114	805	795	954	695	686	825
42	1722	1702	1958	1423	1406	1679	1192	1178	1408	1010	998	1196	865	854	1025	747	737	886
43-1/2	1843	1822	2028	1523	1505	1797	1277	1261	1508	1082	1069	1280	927	916	1098	800	790	950
45	1968	1945	2098	1627	1608	1917	1364	1348	1610	1157	1143	1368	991	979	1173	856	845	1015
46-1/2	2097	2072	2167	1734	1714	1980	1454	1437	1716	1233	1219	1458	1057	1044	1251	913	902	1083
48	2179	2203	2237	1844	1822	2044	1547	1528	1825	1312	1297	1551	1125	1111	1331	972	960	1152

8-3/4-INCH WIDTH	SPAN (ft)																	
Depth (in.)	44			48			52			56			60			64		
	sys 1	sys 2	sys 3	sys 1	sys 2	sys 3	sys 1	sys 2	sys 3	sys 1	sys 2	sys 3	sys 1	sys 2	sys 3	sys 1	sys 2	sys 3
36	1611	1592	1901	1329	1314	1571	1112	1098	1316	941	929	1115	803	793	954	692	683	824
37-1/2	1744	1724	2057	1440	1422	1700	1205	1190	1425	1019	1007	1208	871	860	1034	751	741	893
39	1882	1860	2219	1554	1536	1835	1301	1285	1538	1101	1088	1304	942	930	1117	812	802	965
40-1/2	2025	2001	2387	1673	1653	1974	1400	1384	1655	1186	1172	1404	1015	1002	1203	875	864	1040
42	2173	2147	2538	1795	1774	2118	1504	1486	1777	1274	1259	1508	1090	1077	1292	941	929	1117
43-1/2	2326	2298	2628	1922	1899	2267	1610	1591	1902	1365	1348	1615	1168	1154	1384	1008	996	1197
45	2483	2454	2719	2053	2028	2421	1720	1700	2032	1459	1441	1725	1249	1234	1479	1079	1065	1280
46-1/2	2646	2615	2810	2188	2162	2567	1834	1812	2165	1555	1537	1839	1332	1316	1578	1151	1137	1365
48	2814	2781	2900	2327	2299	2650	1951	1928	2303	1655	1635	1956	1418	1401	1679	1225	1210	1453
49-1/2	2913	2945	2991	2470	2441	2733	2071	2047	2444	1758	1737	2077	1506	1488	1783	1302	1286	1543
51	3002	3035	3082	2617	2586	2816	2195	2169	2590	1863	1841	2201	1597	1578	1890	1381	1364	1636
52-1/2	3090	3124	3172	2768	2735	2898	2322	2295	2667	1972	1948	2329	1691	1670	2000	1462	1444	1732
54	3178	3213	3263	2904	2889	2981	2453	2424	2743	2083	2058	2460	1786	1765	2112	1546	1527	1830
55-1/2	3267	3302	3353	2985	3017	3064	2587	2556	2819	2197	2171	2594	1885	1862	2228	1631	1611	1931
57	3355	3392	3444	3065	3099	3147	2724	2692	2896	2314	2287	2680	1986	1962	2347	1719	1698	2034
58-1/2	3443	3481	3535	3146	3180	3230	2865	2831	2972	2434	2405	2751	2089	2064	2469	1809	1787	2140
60	3531	3570	3625	3227	3262	3313	2969	2974	3048	2557	2527	2821	2195	2169	2593	1901	1878	2248

See Fig. 4.33 for notes and description of cantilever systems.

TABLE 4.29 Section Properties and Design Capacities for Southern Pine Glulam

3-INCH WIDTH															
Depth (in.)	**6-7/8**	**8-1/4**	**9-5/8**	**11**	**12-3/8**	**13-3/4**	**15-1/8**	**16-1/2**	**17-7/8**	**19-1/4**	**20-5/8**	**22**	**23-3/8**	**24-3/4**	**26-1/8**
Beam Weight (lb/ft)	5.2	6.2	7.2	8.3	9.3	10.3	11.3	12.4	13.4	14.4	15.5	16.5	17.5	18.6	19.6
A ($in.^2$)	20.6	24.8	28.9	33.0	37.1	41.3	45.4	49.5	53.6	57.8	61.9	66.0	70.1	74.3	78.4
S ($in.^3$)	24	34	46	61	77	95	114	136	160	185	213	242	273	306	341
I ($in.^4$)	81	140	223	333	474	650	865	1123	1428	1783	2193	2662	3193	3790	4458
EI (106 $lb\text{-}in.^2$)	146	253	401	599	853	1170	1557	2021	2570	3210	3948	4792	5747	6822	8024
Moment Capacity (lb-ft)	4727	6806	9264	12100	15314	18906	22877	27225	31952	37056	42539	48400	54639	61256	68252
Shear Capacity (lb)	3713	4455	5198	5940	6683	7425	8168	8910	9653	10395	11138	11880	12623	13365	14108

3-1/2-INCH WIDTH															
Depth (in.)	**6-7/8**	**8-1/4**	**9-5/8**	**11**	**12-3/8**	**13-3/4**	**15-1/8**	**16-1/2**	**17-7/8**	**19-1/4**	**20-5/8**	**22**	**23-3/8**	**24-3/4**	**26-1/8**
Beam Weight (lb/ft)	6.0	7.2	8.4	9.6	10.8	12.0	13.2	14.4	15.6	16.8	18.0	19.3	20.5	21.7	22.9
A ($in.^2$)	24.1	28.9	33.7	38.5	43.3	48.1	52.9	57.8	62.6	67.4	72.2	77.0	81.8	86.6	91.4
S ($in.^3$)	28	40	54	71	89	110	133	159	186	216	248	282	319	357	398
I ($in.^4$)	95	164	260	388	553	758	1009	1310	1666	2081	2559	3106	3725	4422	5201
EI (106 $lb\text{-}in.^2$)	171	295	468	699	995	1365	1817	2358	2998	3745	4606	5590	6705	7959	9361
Moment Capacity (lb-ft)	5514	7941	10808	14117	17866	22057	26689	31763	37277	43232	49629	56467	63746	71466	79627
Shear Capacity (lb)	4331	5198	6064	6930	7796	8663	9529	10395	11261	12128	12994	13860	14726	15593	16459

5-INCH WIDTH															
Depth (in.)	**12-3/8**	**13-3/4**	**15-1/8**	**16-1/2**	**17-7/8**	**19-1/4**	**20-5/8**	**22**	**23-3/8**	**24-3/4**	**26-1/8**	**27-1/2**	**28-7/8**	**30-1/4**	**31-5/8**
Beam Weight (lb/ft)	15.5	17.2	18.9	20.6	22.3	24.1	25.8	27.5	29.2	30.9	32.7	34.4	36.1	37.8	39.5
A ($in.^2$)	61.9	68.8	75.6	82.5	89.4	96.3	103.1	110.0	116.9	123.8	130.6	137.5	144.4	151.3	158.1
S ($in.^3$)	128	158	191	227	266	309	354	403	455	510	569	630	695	763	833
I ($in.^4$)	790	1083	1442	1872	2380	2972	3656	4437	5322	6317	7429	8665	10031	11534	13179
EI (106 $lb\text{-}in.^2$)	1421	1950	2595	3369	4284	5350	6580	7986	9579	11371	13373	15598	18056	20760	23722
Moment Capacity (lb-ft)	25523	31510	38128	45375	53253	61760	70898	80667	91065	102094	113753	126042	138961	152510	166690
Shear Capacity (lb)	11138	12375	13613	14850	16088	17325	18563	19800	21038	22275	23513	24750	25988	27225	28463

5-1/2-INCH WIDTH

Depth (in.)	12-3/8	13-3/4	15-1/8	16-1/2	17-7/8	19-1/4	20-5/8	22	23-3/8	24-3/4	26-1/8	27-1/2	28-7/8	30-1/4	31-5/8
Beam Weight (lb/ft)	17.0	18.9	20.8	22.7	24.6	26.5	28.4	30.3	32.1	34.0	35.9	37.8	39.7	41.6	43.5
A (in.^{2})	68.1	75.6	83.2	90.8	98.3	105.9	113.4	121.0	128.6	136.1	143.7	151.3	158.8	166.4	173.9
S (in.^{3})	140	173	210	250	293	340	390	444	501	562	626	693	764	839	917
I (in.^{4})	869	1191	1586	2059	2618	3269	4021	4880	5854	6949	8172	9532	11034	12687	14497
EI (106 lb-in.^{2})	1563	2145	2855	3706	4712	5885	7238	8785	10537	12508	14710	17157	19862	22837	26094
Moment Capacity (lb-ft)	28076	34661	41940	49913	58578	67936	77988	88733	100172	112303	125128	138646	152857	167761	183359
Shear Capacity (lb)	12251	13613	14974	16335	17696	19058	20419	21780	23141	24503	25864	27225	28586	29948	31309

6-3/4-INCH WIDTH

Depth (in.)	17-7/8	19-1/4	20-5/8	22	23-3/8	24-3/4	26-1/8	27-1/2	28-7/8	30-1/4	31-5/8	33	34-3/8	35-3/4	37-1/8
Beam Weight (lb/ft)	30.2	32.5	34.8	37.1	39.4	41.8	44.1	46.4	48.7	51.0	53.4	55.7	58.0	60.3	62.6
A (in.^{2})	120.7	129.9	139.2	148.5	157.8	167.1	176.3	185.6	194.9	204.2	213.5	222.8	232.0	241.3	250.6
S (in.^{3})	359	417	479	545	615	689	768	851	938	1029	1125	1225	1329	1438	1551
I (in.^{4})	3213	4012	4935	5990	7184	8528	10030	11698	13542	15570	17792	20215	22848	25701	28782
EI (106 lb-in.^{2})	5783	7222	8883	10781	12932	15350	18054	21057	24376	28027	32025	36386	41127	46262	51808
Moment Capacity (lb-ft)	71891	83377	95713	108900	122938	137827	153566	170156	187597	205889	225032	245025	265869	287564	310110
Shear Capacity (lb)	21718	23389	25059	26730	28401	30071	31742	33413	35083	36754	38424	40095	41766	43436	45107

8-1/2-INCH WIDTH

Depth (in.)	24-3/4	26-1/8	27-1/2	28-7/8	30-1/4	31-5/8	33	34-3/8	35-3/4	37-1/8	38-1/2	39-7/8	41-1/4	42-5/8	44
Beam Weight (lb/ft)	52.6	55.5	58.4	61.4	64.3	67.2	70.1	73.0	76.0	78.9	81.8	84.7	87.7	90.6	93.5
A (in.^{2})	210.4	222.1	233.8	245.4	257.1	268.8	280.5	292.2	303.9	315.6	327.3	338.9	350.6	362.3	374.0
S (in.^{3})	868	967	1071	1181	1296	1417	1543	1674	1811	1953	2100	2253	2411	2574	2743
I (in.^{4})	10739	12630	14731	17053	19607	22404	25455	28772	32364	36244	40422	44910	49718	54857	60339
EI (106 lb-in.^{2})	19330	22734	26516	30696	35293	40328	45820	51789	58256	65239	72760	80837	89492	98742	108610
Moment Capacity (lb-ft)	173559	193379	214271	236234	259268	283373	308550	334798	362118	390509	419971	450504	482109	514786	548533
Shear Capacity (lb)	37868	39971	42075	44179	46283	48386	50490	52594	54698	56801	58905	61009	63113	65216	67320

Notes:

(1) Beam weight is based on density of 36 pcf.

(2) Moment capacity must be adjusted for volume effect. The volume factor is given in Section 4.4.3.

(3) Moment and shear capacities are based on a normal (10 years) duration of load and should be adjusted for the design duration of load per the applicable building code.

TABLE 4.30 Allowable Loads for Simple-Span Southern Pine Glulam Roof Beams (lb/ft)—Non-snow Loads

3-INCH WIDTH	SPAN (ft)																				
Depth (in.)	8	10	12	14	16	18	20	22	24	26	28	30	32	34	36	38	40	42	44	46	48
6-7/8	733	428	246	153	101	69	—	—	—	—	—	—	—	—	—	—	—	—	—	—	—
8-1/4	1057	674	427	267	177	122	87	64	—	—	—	—	—	—	—	—	—	—	—	—	—
9-5/8	1440	919	636	426	283	197	141	104	79	60	—	—	—	—	—	—	—	—	—	—	—
11	1882	1202	832	609	425	296	214	158	120	93	73	57	—	—	—	—	—	—	—	—	—
12-3/8	2384	1522	1054	772	589	424	307	228	174	134	106	84	68	55	—	—	—	—	—	—	—
13-3/4	2944	1880	1303	954	728	573	423	315	240	187	148	118	95	78	64	53	—	—	—	—	—
15-1/8	3563	2276	1577	1156	882	695	561	422	322	251	199	160	129	106	88	73	61	51	—	—	—
16-1/2	4230	2710	1878	1377	1051	828	668	550	421	328	260	209	170	140	116	97	81	68	58	—	—
17-7/8	4793	3182	2205	1617	1235	973	785	647	537	420	333	269	219	180	150	125	106	89	76	65	55
19-1/4	5409	3691	2559	1876	1433	1129	912	751	627	527	419	338	276	228	189	159	134	114	97	83	72
20-5/8	6087	4227	2939	2155	1646	1297	1048	861	718	607	517	418	342	282	235	198	167	142	122	105	90
22	6837	4673	3345	2453	1874	1477	1192	978	815	689	590	509	417	345	288	242	205	175	150	129	112
23-3/8	7671	5152	3777	2770	2117	1669	1343	1101	919	777	665	575	502	416	347	293	249	212	182	157	136
24-3/4	8604	5669	4224	3107	2374	1869	1502	1232	1028	870	744	644	562	494	415	350	297	254	219	189	164
26-1/8	9654	6227	4593	3463	2646	2078	1670	1370	1143	967	828	716	625	550	487	414	352	301	259	225	195

3-1/2-INCH WIDTH	SPAN (ft)																				
Depth (in.)	8	10	12	14	16	18	20	22	24	26	28	30	32	34	36	38	40	42	44	46	48
6-7/8	856	499	287	178	117	81	57	—	—	—	—	—	—	—	—	—	—	—	—	—	—
8-1/4	1234	787	498	311	206	143	102	75	56	—	—	—	—	—	—	—	—	—	—	—	—
9-5/8	1680	1072	742	497	330	229	165	122	92	70	55	—	—	—	—	—	—	—	—	—	—
11	2196	1402	971	711	496	345	249	185	140	108	85	67	54	—	—	—	—	—	—	—	—
12-3/8	2781	1776	1230	901	687	495	358	266	202	157	123	98	79	64	52	—	—	—	—	—	—
13-3/4	3434	2194	1520	1113	850	669	493	368	280	218	172	138	111	91	75	62	51	—	—	—	—
15-1/8	4157	2656	1840	1348	1029	811	654	492	376	293	232	186	151	124	102	85	71	59	—	—	—
16-1/2	4936	3162	2191	1606	1226	966	780	642	491	383	304	244	199	163	135	113	95	80	68	57	—
17-7/8	5592	3712	2573	1886	1440	1135	916	752	627	490	389	313	255	210	175	146	123	104	89	76	65
19-1/4	6311	4306	2985	2189	1672	1317	1062	870	725	613	489	394	322	265	221	185	157	133	113	97	83
20-5/8	7102	4932	3428	2514	1921	1513	1216	997	831	703	601	487	398	329	274	231	195	166	142	122	105
22	7977	5452	3902	2862	2186	1717	1380	1132	944	798	683	590	486	402	336	283	240	204	175	151	131
23-3/8	8950	6011	4406	3232	2468	1934	1554	1275	1063	899	770	665	580	485	405	342	290	248	213	184	159
24-3/4	10038	6613	4928	3625	2760	2163	1739	1426	1190	1007	862	745	650	572	484	408	347	297	255	221	192
26-1/8	11263	7265	5359	4040	3068	2405	1933	1586	1323	1120	959	829	724	636	563	483	411	352	303	262	228

5-INCH WIDTH	SPAN (ft)																				
Depth (in.)	8	10	12	14	16	18	20	22	24	26	28	30	32	34	36	38	40	42	44	46	48
12-3/8	3973	2537	1757	1287	982	707	511	380	289	224	176	141	113	92	75	61	50	—	—	—	—
13-3/4	4906	3134	2171	1590	1214	955	705	525	401	311	246	197	159	130	107	88	73	61	51	—	—
15-1/8	5939	3794	2629	1926	1470	1155	927	703	537	419	331	266	216	177	146	121	101	85	71	60	51
16-1/2	7051	4517	3130	2294	1750	1370	1100	901	701	547	434	349	284	233	193	161	135	114	97	82	70
17-7/8	7988	5303	3676	2695	2047	1603	1288	1055	879	700	556	448	365	301	250	209	176	149	127	108	92
19-1/4	9015	6152	4265	3120	2367	1854	1489	1221	1017	860	698	563	460	379	316	265	224	190	162	139	119
20-5/8	10145	7046	4898	3571	2710	2123	1706	1398	1166	985	843	696	569	470	392	330	279	237	203	175	151
22	11396	7788	5568	4052	3075	2410	1936	1588	1324	1119	958	828	695	575	480	404	342	292	250	216	186
23-3/8	12786	8587	6269	4562	3463	2714	2181	1789	1492	1261	1080	933	814	693	579	488	414	354	304	262	227
24-3/4	14340	9448	7010	5102	3873	3035	2440	2001	1669	1412	1209	1045	912	802	691	583	495	424	365	315	274
26-1/8	16090	10379	7655	5671	4305	3375	2713	2225	1857	1571	1345	1163	1015	892	790	689	586	502	432	374	326
27-1/2	18075	11389	8308	6269	4760	3731	3000	2461	2054	1738	1488	1287	1123	988	875	780	688	589	508	440	384
28-7/8	20346	12488	9003	6897	5237	4105	3301	2709	2260	1913	1638	1417	1237	1088	964	859	770	686	592	514	448
30-1/4	22969	13689	9744	7553	5735	4497	3616	2968	2477	2096	1795	1554	1356	1193	1057	942	845	761	684	594	518
31-5/8	26006	15007	10535	8112	6257	4906	3945	3238	2703	2288	1960	1696	1481	1303	1155	1029	923	831	752	683	596

TABLE 4.30 Allowable Loads for Simple-Span Southern Pine Glulam Roof Beams (lb/ft)—Non-snow Loads (*Continued*)

5-1/2-INCH WIDTH	SPAN (ft)																				
Depth (in.)	8	10	12	14	16	18	20	22	24	26	28	30	32	34	36	38	40	42	44	46	48
12-3/8	4370	2791	1933	1415	1080	777	562	418	318	247	194	155	124	101	82	67	55	—	—	—	—
13-3/4	5397	3447	2388	1750	1335	1048	775	578	441	343	271	216	175	143	117	97	80	67	56	—	—
15-1/8	6532	4173	2892	2119	1615	1264	1015	774	591	460	364	292	237	194	160	133	111	93	78	66	56
16-1/2	7756	4969	3443	2524	1916	1500	1204	986	772	602	478	384	312	257	213	177	149	126	106	90	77
17-7/8	8787	5833	4043	2955	2241	1755	1409	1155	962	770	611	492	401	331	275	230	194	164	139	119	102
19-1/4	9917	6767	4691	3416	2592	2030	1630	1336	1114	941	768	619	506	417	347	291	246	209	178	153	131
20-5/8	11160	7750	5373	3910	2967	2324	1867	1531	1276	1079	923	766	626	517	431	362	307	261	223	192	166
22	12535	8567	6096	4436	3366	2638	2120	1738	1449	1225	1048	906	764	632	528	444	376	321	275	237	205
23-3/8	14064	9446	6863	4995	3791	2971	2387	1958	1633	1381	1182	1022	891	762	637	537	456	389	334	289	250
24-3/4	15774	10393	7674	5585	4240	3323	2671	2191	1827	1546	1323	1144	998	877	760	641	545	466	401	347	301
26-1/8	17699	11417	8421	6208	4713	3694	2970	2436	2032	1719	1472	1273	1111	977	865	758	645	552	476	412	358
27-1/2	19883	12528	9139	6863	5211	4085	3284	2694	2248	1902	1629	1409	1230	1081	958	853	757	648	559	484	422
28-7/8	22381	13737	9903	7550	5733	4494	3614	2965	2474	2094	1793	1551	1354	1191	1055	940	843	755	651	565	492
30-1/4	25266	15058	10718	8269	6279	4923	3959	3249	2711	2295	1965	1701	1485	1306	1157	1031	924	832	753	654	570
31-5/8	28500	16507	11588	8923	6849	5371	4319	3545	2958	2504	2145	1856	1621	1426	1264	1127	1010	910	823	748	656

6-3/4-INCH WIDTH	SPAN (ft)																				
Depth (in.)	10	12	14	16	18	20	22	24	26	28	30	32	34	36	38	40	42	44	46	48	50
17-7/8	7159	4934	3589	2722	2132	1712	1403	1169	945	750	604	493	406	337	282	238	201	171	146	125	107
19-1/4	8303	5703	4149	3148	2466	1980	1623	1353	1143	942	760	621	512	426	358	302	256	219	187	161	139
20-5/8	9501	6527	4749	3603	2823	2268	1859	1550	1310	1120	940	768	635	529	445	376	320	274	236	203	176
22	10514	7405	5388	4089	3204	2574	2111	1760	1488	1273	1100	938	776	648	545	462	394	338	291	252	218
23-3/8	11592	8336	6067	4605	3608	2900	2378	1983	1677	1435	1241	1082	935	782	659	559	478	410	354	307	267
24-3/4	12755	9321	6784	5150	4036	3244	2661	2219	1877	1607	1389	1212	1065	933	787	669	572	492	426	370	322
26-1/8	14011	10335	7541	5725	4487	3607	2959	2468	2088	1788	1546	1349	1186	1050	931	792	678	584	505	440	384
27-1/2	15375	11216	8337	6329	4962	3989	3273	2730	2310	1978	1711	1493	1313	1163	1036	928	796	686	595	518	453
28-7/8	16859	12154	9171	6963	5459	4389	3602	3005	2543	2178	1884	1645	1447	1281	1142	1023	921	799	693	604	529
30-1/4	18480	13154	10044	7627	5980	4808	3946	3293	2787	2387	2065	1803	1586	1405	1253	1122	1011	914	802	700	613
31-5/8	20259	14222	10951	8320	6524	5246	4305	3593	3041	2605	2255	1969	1732	1535	1368	1226	1105	999	908	805	706
33	22219	15365	11737	9042	7090	5702	4680	3907	3307	2833	2452	2141	1884	1670	1489	1335	1202	1088	988	901	807
34-3/8	24390	16593	12566	9793	7680	6177	5070	4233	3583	3070	2657	2321	2043	1811	1615	1448	1304	1180	1073	978	895
35-3/4	26807	17913	13443	10574	8293	6670	5476	4571	3870	3316	2871	2508	2208	1957	1745	1565	1410	1277	1160	1058	968
37-1/8	29516	19338	14372	11384	8928	7182	5896	4923	4168	3572	3093	2702	2378	2109	1881	1687	1520	1376	1251	1141	1045

8-1/2-INCH WIDTH	SPAN (ft)																				
Depth (in.)	12	14	16	18	20	22	24	26	28	30	32	34	36	38	40	42	44	46	48	50	52
24-3/4	11603	8445	6410	5024	4038	3312	2762	2336	1999	1729	1508	1326	1173	991	842	720	620	536	465	406	355
26-1/8	12896	9387	7126	5585	4489	3683	3072	2599	2225	1924	1679	1476	1307	1164	997	854	735	637	554	483	424
27-1/2	14124	10377	7878	6176	4965	4073	3398	2875	2462	2129	1858	1634	1447	1289	1155	1002	864	749	652	570	500
28-7/8	15305	11416	8667	6795	5463	4483	3740	3165	2710	2345	2047	1800	1594	1421	1273	1146	1006	873	761	666	585
30-1/4	16564	12503	9493	7443	5985	4911	4098	3468	2970	2570	2244	1974	1749	1558	1396	1257	1137	1010	881	772	679
31-5/8	17909	13638	10356	8120	6530	5359	4472	3785	3242	2806	2450	2156	1910	1702	1526	1374	1243	1129	1013	889	783
33	19349	14780	11255	8825	7097	5825	4862	4116	3526	3052	2665	2345	2078	1853	1661	1496	1354	1230	1121	1016	895
34-3/8	20895	15824	12190	9560	7688	6311	5268	4460	3821	3307	2888	2542	2253	2009	1801	1623	1469	1334	1217	1113	1018
35-3/4	22557	16928	13162	10322	8302	6815	5690	4817	4127	3573	3121	2747	2435	2172	1947	1755	1588	1443	1316	1205	1106
37-1/8	24352	18098	14170	11113	8939	7339	6127	5188	4445	3849	3362	2960	2624	2340	2099	1892	1712	1556	1420	1299	1193
38-1/2	26294	19337	15214	11933	9599	7881	6580	5572	4775	4134	3612	3180	2820	2515	2256	2033	1841	1674	1527	1398	1283
39-7/8	28402	20655	16220	12781	10282	8442	7049	5969	5116	4430	3871	3408	3022	2696	2419	2180	1974	1795	1638	1500	1377
41-1/4	30699	22057	17203	13657	10987	9022	7534	6380	5469	4736	4138	3644	3232	2883	2587	2332	2112	1921	1753	1605	1474
42-5/8	33211	23553	18237	14562	11716	9620	8034	6804	5833	5051	4414	3888	3448	3077	2760	2489	2254	2050	1871	1714	1574
44	35699	25152	19326	15495	12467	10238	8550	7242	6208	5377	4699	4139	3671	3276	2940	2651	2401	2184	1994	1826	1678

Notes:

(1) Span = simply supported beam.

(2) Maximum deflection = L/180 under total load. Other deflection limits may apply.

(3) Service condition = dry.

(4) Tabulated values represent total loads and have taken the dead weight of the beam (assumed 36 pcf) into account.

(5) Sufficient bearing length shall be provided at supports.

(6) Maximum beam shear is located at a distance from the supports equal to the depth of the beam.

(7) Light (unshaded) areas limited by deflection; medium-shaded areas limited by bending strength; dark-shaded areas limited by shear strength.

TABLE 4.31 Allowable Loads for Simple-Span Southern Pine Glulam Roof Beams (lb/ft)—Snow Loads

3-INCH WIDTH	SPAN (ft)																				
Depth (in.)	8	10	12	14	16	18	20	22	24	26	28	30	32	34	36	38	40	42	44	46	48
6-7/8	674	428	246	153	101	69	—	—	—	—	—	—	—	—	—	—	—	—	—	—	—
8-1/4	972	620	427	267	177	122	87	64	—	—	—	—	—	—	—	—	—	—	—	—	—
9-5/8	1324	845	585	426	283	197	141	104	79	60	—	—	—	—	—	—	—	—	—	—	—
11	1731	1105	765	560	425	296	214	158	120	93	73	57	—	—	—	—	—	—	—	—	—
12-3/8	2192	1400	969	710	541	424	307	228	174	134	106	84	68	55	—	—	—	—	—	—	—
13-3/4	2707	1729	1198	877	669	527	423	315	240	187	148	118	95	78	64	53	—	—	—	—	—
15-1/8	3277	2093	1450	1062	811	638	515	422	322	251	199	160	129	106	88	73	61	51	—	—	—
16-1/2	3891	2492	1727	1266	966	761	614	505	421	328	260	209	170	140	116	97	81	68	58	—	—
17-7/8	4408	2926	2028	1486	1135	894	721	594	497	420	333	269	219	180	150	125	106	89	76	65	55
19-1/4	4975	3395	2353	1725	1317	1038	838	690	575	486	416	338	276	228	189	159	134	114	97	83	72
20-5/8	5599	3888	2702	1981	1513	1192	963	791	659	557	476	412	342	282	235	198	167	142	122	105	90
22	6289	4298	3076	2255	1723	1358	1095	898	749	633	541	468	408	345	288	242	205	175	150	129	112
23-3/8	7056	4739	3473	2547	1946	1534	1234	1012	844	713	610	528	460	404	347	293	249	212	182	157	136
24-3/4	7914	5214	3885	2857	2183	1718	1381	1132	944	799	683	591	515	453	401	350	297	254	219	189	164
26-1/8	8880	5728	4224	3184	2433	1910	1535	1259	1050	888	760	657	574	504	446	397	352	301	259	225	195

3-1/2-INCH WIDTH	SPAN (ft)																				
Depth (in.)	8	10	12	14	16	18	20	22	24	26	28	30	32	34	36	38	40	42	44	46	48
6-7/8	787	499	287	178	117	81	57	—	—	—	—	—	—	—	—	—	—	—	—	—	—
8-1/4	1134	723	498	311	206	143	102	75	56	—	—	—	—	—	—	—	—	—	—	—	—
9-5/8	1545	986	682	497	330	229	165	122	92	70	55	—	—	—	—	—	—	—	—	—	—
11	2020	1289	892	653	496	345	249	185	140	108	85	67	54	—	—	—	—	—	—	—	—
12-3/8	2557	1633	1131	828	631	495	358	266	202	157	123	98	79	64	52	—	—	—	—	—	—
13-3/4	3159	2017	1397	1023	781	614	493	368	280	218	172	138	111	91	75	62	51	—	—	—	—
15-1/8	3823	2442	1692	1240	946	745	601	492	376	293	232	186	151	124	102	85	71	59	—	—	—
16-1/2	4540	2908	2015	1476	1127	887	716	589	491	383	304	244	199	163	135	113	95	80	68	57	—
17-7/8	5143	3414	2366	1734	1324	1043	842	691	575	486	389	313	255	210	175	146	123	104	89	76	65
19-1/4	5804	3961	2745	2012	1537	1211	975	799	666	563	481	394	322	265	221	185	157	133	113	97	83
20-5/8	6532	4536	3153	2311	1765	1391	1117	916	763	645	551	476	398	329	274	231	195	166	142	122	105
22	7337	5014	3588	2631	2010	1578	1268	1040	867	733	627	541	472	402	336	283	240	204	175	151	131
23-3/8	8232	5528	4052	2972	2269	1778	1428	1171	977	826	706	611	532	468	405	342	290	248	213	184	159
24-3/4	9233	6083	4532	3333	2538	1989	1598	1311	1093	924	791	684	596	524	464	408	347	297	255	221	192
26-1/8	10360	6682	4928	3715	2821	2211	1777	1458	1216	1028	880	761	664	584	517	460	411	352	303	262	228

5-INCH WIDTH	SPAN (ft)																				
Depth (in.)	8	10	12	14	16	18	20	22	24	26	28	30	32	34	36	38	40	42	44	46	48
12-3/8	3654	2333	1615	1183	902	707	511	380	289	224	176	141	113	92	75	61	50	—	—	—	—
13-3/4	4512	2882	1996	1462	1115	878	705	525	401	311	246	197	159	130	107	88	73	61	51	—	—
15-1/8	5462	3489	2417	1771	1351	1061	851	697	537	419	331	266	216	177	146	121	101	85	71	60	51
16-1/2	6485	4154	2878	2109	1608	1259	1010	827	689	547	434	349	284	233	193	161	135	114	97	82	70
17-7/8	7347	4877	3380	2477	1882	1473	1183	969	807	681	556	448	365	301	250	209	176	149	127	108	92
19-1/4	8292	5658	3922	2869	2176	1704	1368	1121	934	789	675	563	460	379	316	265	224	190	162	139	119
20-5/8	9332	6480	4504	3284	2491	1951	1567	1284	1070	905	773	668	569	470	392	330	279	237	203	175	151
22	10482	7163	5121	3726	2827	2215	1779	1458	1216	1028	879	759	662	575	480	404	342	292	250	216	186
23-3/8	11760	7898	5765	4195	3183	2494	2004	1643	1370	1158	991	856	747	656	579	488	414	354	304	262	227
24-3/4	13190	8689	6446	4691	3561	2790	2242	1839	1533	1297	1109	959	836	735	650	579	495	424	365	315	274
26-1/8	14800	9546	7040	5215	3958	3102	2493	2045	1705	1442	1234	1067	931	818	724	645	577	502	432	374	326
27-1/2	16627	10475	7641	5765	4376	3430	2757	2262	1887	1596	1366	1181	1031	906	802	714	640	576	508	440	384
28-7/8	18716	11486	8280	6342	4815	3774	3034	2489	2077	1757	1504	1301	1135	998	884	787	705	635	574	514	448
30-1/4	21129	12591	8961	6946	5274	4134	3324	2727	2275	1925	1649	1426	1245	1095	970	864	774	697	630	572	518
31-5/8	23922	13803	9689	7460	5753	4510	3626	2976	2483	2101	1800	1557	1359	1196	1059	944	846	761	689	625	570

TABLE 4.31 Allowable Loads for Simple-Span Southern Pine Glulam Roof Beams (lb/ft)—Snow Loads

5-1/2-INCH WIDTH	SPAN (ft)																				
Depth (in.)	8	10	12	14	16	18	20	22	24	26	28	30	32	34	36	38	40	42	44	46	48
12-3/8	4019	2566	1777	1301	992	777	562	418	318	247	194	155	124	101	82	67	55	—	—	—	—
13-3/4	4964	3170	2196	1608	1227	963	772	578	441	343	271	216	175	143	117	97	80	67	56	—	—
15-1/8	6008	3838	2659	1948	1484	1161	932	763	591	460	364	292	237	194	160	133	111	93	78	66	56
16-1/2	7134	4569	3166	2320	1761	1378	1106	906	754	602	478	384	312	257	213	177	149	126	106	90	77
17-7/8	8082	5365	3718	2716	2060	1613	1295	1061	883	746	611	492	401	331	275	230	194	164	139	119	102
19-1/4	9121	6224	4314	3140	2382	1865	1498	1227	1023	864	738	619	506	417	347	291	246	209	178	153	131
20-5/8	10265	7128	4941	3595	2727	2136	1716	1406	1172	990	847	731	626	517	431	362	307	261	223	192	166
22	11530	7879	5606	4079	3095	2424	1948	1597	1331	1125	962	831	724	632	528	444	376	321	275	237	205
23-3/8	12936	8687	6311	4592	3485	2730	2194	1799	1500	1268	1085	937	817	718	635	537	456	389	334	289	250
24-3/4	14509	9558	7057	5136	3898	3054	2454	2013	1678	1419	1214	1050	915	804	712	634	545	466	401	347	301
26-1/8	16280	10500	7744	5709	4333	3396	2729	2238	1867	1579	1351	1168	1019	896	793	706	632	552	476	412	358
27-1/2	18289	11522	8405	6311	4791	3755	3018	2476	2065	1747	1495	1293	1128	992	878	782	700	630	559	484	422
28-7/8	20587	12635	9108	6943	5271	4132	3321	2725	2273	1923	1646	1424	1243	1093	968	862	772	695	628	565	492
30-1/4	23242	13850	9857	7604	5773	4526	3639	2985	2491	2108	1805	1561	1363	1198	1061	946	847	763	689	626	570
31-5/8	26216	15183	10658	8206	6298	4938	3970	3258	2718	2300	1970	1704	1488	1309	1159	1033	926	833	754	684	623

6-3/4-INCH WIDTH	SPAN (ft)																				
Depth (in.)	10	12	14	16	18	20	22	24	26	28	30	32	34	36	38	40	42	44	46	48	50
17-7/8	6584	4537	3299	2502	1959	1572	1288	1073	906	750	604	493	406	337	282	238	201	171	146	125	107
19-1/4	7636	5244	3815	2893	2266	1819	1491	1242	1049	897	760	621	512	426	358	302	256	219	187	161	139
20-5/8	8738	6002	4366	3312	2594	2084	1708	1423	1202	1028	888	768	635	529	445	376	320	274	236	203	176
22	9670	6809	4954	3759	2945	2365	1939	1616	1366	1168	1009	880	772	648	545	462	394	338	291	252	218
23-3/8	10662	7666	5578	4233	3317	2665	2185	1821	1540	1317	1138	992	872	771	659	559	478	410	354	307	267
24-3/4	11731	8572	6238	4734	3710	2981	2445	2038	1724	1475	1275	1112	977	864	769	669	572	492	426	370	322
26-1/8	12887	9504	6934	5263	4125	3315	2719	2267	1918	1641	1419	1238	1088	963	857	767	678	584	505	440	384
27-1/2	14141	10315	7666	5819	4561	3666	3007	2508	2122	1816	1570	1370	1205	1066	949	850	765	686	595	518	453
28-7/8	15506	11178	8434	6402	5018	4034	3310	2761	2336	2000	1729	1509	1327	1175	1047	937	843	762	692	604	529
30-1/4	16998	12097	9237	7013	5497	4420	3626	3025	2560	2192	1896	1655	1455	1289	1148	1029	926	837	760	692	613
31-5/8	18634	13080	10071	7650	5997	4822	3957	3302	2794	2393	2070	1807	1589	1408	1254	1124	1012	915	831	757	692
33	20437	14132	10794	8314	6519	5242	4301	3590	3038	2602	2251	1965	1729	1532	1365	1224	1102	996	905	825	754
34-3/8	22434	15261	11556	9005	7061	5678	4660	3889	3292	2820	2440	2131	1875	1661	1481	1327	1195	1081	982	895	819
35-3/4	24658	16475	12363	9723	7625	6132	5033	4201	3556	3046	2636	2302	2026	1795	1601	1435	1293	1170	1062	969	886
37-1/8	27149	17786	13217	10468	8209	6602	5419	4524	3830	3281	2840	2480	2183	1935	1725	1547	1394	1261	1146	1045	956

8-1/2-INCH WIDTH	SPAN (ft)																				
Depth (in.)	12	14	16	18	20	22	24	26	28	30	32	34	36	38	40	42	44	46	48	50	52
24-3/4	10670	7765	5893	4618	3710	3043	2537	2145	1835	1586	1383	1215	1075	957	842	720	620	536	465	406	355
26-1/8	11860	8631	6551	5134	4126	3384	2822	2386	2042	1766	1540	1353	1198	1066	954	854	735	637	554	483	424
27-1/2	12990	9542	7243	5677	4563	3743	3122	2640	2260	1954	1705	1499	1327	1181	1058	951	859	749	652	570	500
28-7/8	14076	10498	7969	6246	5021	4119	3436	2907	2488	2152	1878	1651	1462	1302	1166	1049	948	860	761	666	585
30-1/4	15234	11497	8729	6842	5501	4513	3765	3186	2728	2359	2059	1811	1604	1429	1280	1152	1041	945	861	772	679
31-5/8	16471	12541	9522	7465	6002	4925	4109	3477	2978	2576	2248	1978	1752	1561	1398	1259	1138	1033	941	860	783
33	17796	13592	10349	8114	6524	5354	4468	3781	3238	2802	2446	2152	1906	1699	1522	1371	1240	1126	1026	938	860
34-3/8	19217	14553	11209	8789	7067	5800	4841	4097	3509	3037	2651	2333	2067	1842	1651	1487	1345	1222	1114	1018	934
35-3/4	20747	15568	12103	9490	7632	6264	5228	4426	3791	3281	2865	2521	2234	1992	1785	1608	1455	1322	1205	1102	1011
37-1/8	22397	16643	13030	10218	8218	6745	5630	4766	4084	3535	3087	2717	2408	2147	1925	1734	1569	1426	1300	1189	1091
38-1/2	24184	17784	13991	10972	8825	7244	6047	5120	4387	3797	3316	2919	2587	2307	2069	1864	1687	1533	1398	1279	1174
39-7/8	26123	18996	14916	11752	9453	7760	6478	5485	4700	4069	3554	3129	2774	2474	2218	1999	1810	1645	1500	1373	1260
41-1/4	28236	20286	15820	12558	10101	8293	6924	5863	5024	4350	3800	3346	2966	2646	2373	2139	1936	1760	1606	1470	1349
42-5/8	30547	21661	16771	13390	10771	8843	7384	6253	5359	4640	4054	3569	3165	2823	2532	2283	2067	1879	1714	1569	1441
44	32836	23132	17772	14248	11462	9411	7858	6655	5704	4939	4316	3800	3370	3006	2697	2431	2202	2002	1827	1673	1536

Notes:

(1) Span = simply supported beam.

(2) Maximum deflection = L/180 under total load. Other deflection limits may apply.

(3) Service condition = dry.

(4) Tabulated values represent total loads and have taken the dead weight of the beam (assumed 36 pcf) into account.

(5) Sufficient bearing length shall be provided at supports.

(6) Maximum beam shear is located at a distance from the supports equal to the depth of the beam.

(7) Light (unshaded) areas limited by deflection; medium-shaded areas limited by bending strength; dark-shaded areas limited by shear strength.

TABLE 4.32 Allowable Loads for Simple-Span Southern Pine Glulam Floor Beams (lb/ft)

3-INCH WIDTH	SPAN (ft)																				
Depth (in.)	8	10	12	14	16	18	20	22	24	26	28	30	32	34	36	38	40	42	44	46	48
6-7/8	524	266	152	94	61	—	—	—	—	—	—	—	—	—	—	—	—	—	—	—	—
8-1/4	845	462	265	164	108	74	52	—	—	—	—	—	—	—	—	—	—	—	—	—	—
9-5/8	1151	734	423	264	174	120	86	63	—	—	—	—	—	—	—	—	—	—	—	—	—
11	1504	960	634	396	263	182	130	96	72	55	—	—	—	—	—	—	—	—	—	—	—
12-3/8	1905	1216	842	566	376	262	188	139	105	81	63	—	—	—	—	—	—	—	—	—	—
13-3/4	2353	1502	1040	761	519	361	260	193	146	113	88	70	56	—	—	—	—	—	—	—	—
15-1/8	2848	1819	1260	922	693	483	349	259	197	153	120	95	77	62	50	—	—	—	—	—	—
16-1/2	3382	2166	1500	1099	838	630	456	339	258	201	158	126	102	83	68	56	—	—	—	—	—
17-7/8	3832	2543	1762	1291	985	776	582	434	331	257	203	163	132	108	89	73	61	51	—	—	—
19-1/4	4324	2950	2044	1498	1144	901	727	544	416	324	256	206	167	137	113	94	78	66	55	—	—
20-5/8	4867	3379	2348	1721	1314	1035	835	671	513	401	318	255	208	171	141	118	99	83	70	60	51
22	5467	3735	2672	1959	1496	1179	950	779	625	488	388	312	254	209	174	145	122	103	88	75	64
23-3/8	6134	4118	3018	2213	1690	1332	1071	878	731	588	467	377	307	253	211	176	149	126	107	92	79
24-3/4	6880	4531	3376	2482	1896	1491	1198	982	819	692	557	449	367	303	252	212	179	152	130	111	96
26-1/8	7719	4978	3671	2766	2113	1658	1332	1092	911	770	657	531	434	358	299	251	213	181	155	133	115

3-1/2-INCH WIDTH	SPAN (ft)																				
Depth (in.)	8	10	12	14	16	18	20	22	24	26	28	30	32	34	36	38	40	42	44	46	48
6-7/8	611	310	177	109	71	—	—	—	—	—	—	—	—	—	—	—	—	—	—	—	—
8-1/4	985	539	309	192	126	86	61	—	—	—	—	—	—	—	—	—	—	—	—	—	—
9-5/8	1343	856	493	308	203	140	100	73	54	—	—	—	—	—	—	—	—	—	—	—	—
11	1755	1120	739	462	306	212	152	112	84	64	—	—	—	—	—	—	—	—	—	—	—
12-3/8	2222	1418	982	661	439	305	219	162	122	94	73	57	—	—	—	—	—	—	—	—	—
13-3/4	2745	1753	1213	888	605	421	304	225	171	132	103	82	65	52	—	—	—	—	—	—	—
15-1/8	3323	2122	1470	1076	808	564	407	303	230	178	140	111	89	72	59	—	—	—	—	—	—
16-1/2	3946	2527	1750	1282	978	734	531	396	301	234	185	147	119	97	79	65	54	—	—	—	—
17-7/8	4470	2967	2055	1506	1149	905	678	506	386	300	237	190	154	126	103	86	71	59	—	—	—
19-1/4	5045	3442	2385	1748	1334	1051	846	634	485	378	299	240	195	160	132	110	92	77	65	54	—
20-5/8	5678	3942	2739	2008	1533	1207	969	783	599	467	371	298	242	199	165	137	115	97	82	70	59
22	6378	4358	3118	2286	1745	1370	1100	902	730	570	452	364	297	244	203	169	143	120	102	87	74
23-3/8	7156	4805	3521	2581	1970	1543	1239	1016	847	686	545	439	358	295	246	206	174	147	125	107	92
24-3/4	8026	5286	3938	2895	2204	1726	1387	1137	948	801	650	524	428	353	294	247	209	177	151	130	112
26-1/8	9006	5808	4282	3227	2450	1920	1542	1264	1054	891	762	619	506	418	349	293	248	211	181	155	134

5-INCH WIDTH	SPAN (ft)																				
Depth (in.)	8	10	12	14	16	18	20	22	24	26	28	30	32	34	36	38	40	42	44	46	48
12-3/8	3175	2026	1403	944	627	436	314	232	175	134	104	82	65	51	—	—	—	—	—	—	—
13-3/4	3922	2504	1733	1269	864	602	434	322	244	188	147	117	93	75	60	—	—	—	—	—	—
15-1/8	4747	3031	2099	1537	1154	805	582	432	329	255	200	159	128	103	84	69	56	—	—	—	—
16-1/2	5637	3609	2500	1831	1396	1049	759	565	431	334	264	210	170	138	113	93	77	64	53	—	—
17-7/8	6386	4238	2936	2151	1633	1278	969	723	551	429	339	271	220	179	148	122	102	85	71	59	—
19-1/4	7207	4917	3407	2491	1889	1479	1187	906	693	540	427	343	278	228	188	156	131	110	92	78	66
20-5/8	8111	5631	3913	2852	2163	1693	1359	1114	856	668	529	426	346	284	235	196	165	139	117	99	84
22	9111	6225	4449	3236	2455	1922	1543	1265	1042	814	646	520	424	349	289	242	204	172	146	124	106
23-3/8	10223	6864	5009	3644	2764	2165	1739	1425	1187	980	779	628	512	422	351	294	248	210	179	153	131
24-3/4	11466	7552	5601	4075	3092	2422	1946	1595	1329	1123	928	749	612	505	420	353	298	253	216	185	159
26-1/8	12866	8296	6118	4530	3438	2693	2164	1774	1479	1250	1069	885	723	597	498	419	354	302	258	222	191
27-1/2	14453	9104	6640	5009	3801	2978	2393	1962	1636	1383	1183	1023	847	701	585	492	417	355	305	262	227
28-7/8	16270	9983	7195	5510	4182	3277	2634	2160	1801	1523	1303	1127	983	815	681	573	486	415	356	307	266
30-1/4	18368	10944	7787	6035	4581	3590	2885	2367	1974	1669	1429	1235	1078	940	786	663	563	481	414	357	310
31-5/8	20797	11997	8420	6482	4997	3917	3148	2583	2154	1822	1560	1349	1177	1035	902	761	647	553	476	412	358

TABLE 4.32 Allowable Loads for Simple-Span Southern Pine Glulam Floor Beams (lb/ft)
(*Continued*)

5-1/2-INCH WIDTH	SPAN (ft)																				
Depth (in.)	8	10	12	14	16	18	20	22	24	26	28	30	32	34	36	38	40	42	44	46	48
12-3/8	3492	2229	1543	1038	690	479	345	255	192	148	115	90	71	57	—	—	—	—	—	—	—
13-3/4	4314	2754	1907	1396	951	662	478	354	268	207	162	128	102	82	66	53	—	—	—	—	—
15-1/8	5222	3334	2309	1691	1270	886	640	476	362	280	220	175	141	114	93	76	62	51	—	—	—
16-1/2	6200	3970	2750	2015	1528	1154	835	622	474	368	290	231	187	152	124	102	85	70	58	—	—
17-7/8	7025	4662	3230	2359	1788	1399	1066	795	607	472	373	299	242	197	162	134	112	93	78	65	54
19-1/4	7928	5408	3748	2727	2068	1619	1299	997	762	594	470	377	306	251	207	172	144	121	101	85	72
20-5/8	8922	6194	4293	3122	2368	1854	1488	1219	941	734	582	468	381	313	259	216	181	153	129	109	93
22	10022	6848	4871	3543	2687	2104	1690	1384	1147	895	711	572	466	384	318	266	224	189	161	137	117
23-3/8	11245	7550	5484	3989	3026	2370	1903	1560	1300	1078	857	691	563	464	386	323	273	231	197	168	144
24-3/4	12612	8307	6132	4461	3385	2652	2130	1746	1455	1230	1021	824	673	555	462	388	328	279	238	204	175
26-1/8	14152	9126	6730	4959	3763	2948	2369	1942	1619	1368	1170	973	795	657	548	461	390	332	284	244	210
27-1/2	15899	10014	7304	5483	4161	3260	2620	2148	1791	1514	1295	1120	932	771	643	541	459	391	335	289	249
28-7/8	17897	10982	7915	6032	4578	3588	2883	2364	1971	1667	1427	1233	1075	896	749	631	535	457	392	338	293
30-1/4	20205	12038	8566	6607	5015	3930	3159	2591	2161	1827	1564	1352	1179	1034	865	729	619	529	455	393	341
31-5/8	22791	13197	9262	7130	5471	4288	3446	2827	2358	1995	1707	1476	1288	1132	992	837	712	609	524	453	393

6-3/4-INCH WIDTH	SPAN (ft)																				
Depth (in.)	10	12	14	16	18	20	22	24	26	28	30	32	34	36	38	40	42	44	46	48	50
17-7/8	5721	3941	2865	2172	1699	1308	976	744	579	458	366	297	242	199	165	137	114	96	80	67	56
19-1/4	6636	4556	3313	2512	1966	1578	1224	935	728	577	463	376	308	254	211	176	148	125	105	88	75
20-5/8	7594	5215	3792	2876	2251	1807	1480	1155	901	715	574	467	384	318	265	222	187	158	134	114	97
22	8404	5916	4303	3264	2556	2052	1681	1401	1099	872	702	572	471	391	327	275	232	197	168	143	123
23-3/8	9266	6661	4846	3676	2879	2312	1895	1579	1323	1051	847	691	570	474	397	335	284	242	207	177	152
24-3/4	10195	7449	5419	4111	3221	2587	2120	1767	1493	1253	1011	826	681	568	476	402	342	292	250	215	186
26-1/8	11200	8259	6024	4571	3581	2877	2358	1966	1662	1421	1194	976	807	672	565	478	407	348	299	258	223
27-1/2	12291	8964	6660	5054	3960	3182	2609	2175	1839	1573	1360	1144	946	789	664	563	480	411	354	306	266
28-7/8	13477	9714	7327	5561	4358	3502	2871	2394	2025	1732	1498	1306	1100	919	774	657	561	481	415	359	312
30-1/4	14774	10513	8025	6091	4774	3836	3146	2624	2219	1899	1642	1432	1259	1061	895	760	649	558	482	418	364
31-5/8	16197	11367	8750	6645	5208	4186	3434	2864	2422	2074	1793	1564	1375	1217	1027	873	747	643	556	483	421
33	17764	12281	9378	7222	5661	4551	3733	3114	2634	2255	1950	1702	1496	1325	1172	997	854	735	637	554	483
34-3/8	19501	13263	10041	7823	6133	4930	4045	3374	2855	2444	2114	1845	1623	1437	1280	1132	970	836	724	631	551
35-3/4	21434	14319	10742	8447	6622	5324	4368	3645	3084	2641	2285	1994	1754	1553	1384	1240	1096	945	820	714	625
37-1/8	23600	15458	11485	9095	7130	5733	4704	3926	3322	2845	2461	2149	1890	1674	1492	1337	1204	1064	923	805	705

8-1/2-INCH WIDTH	SPAN (ft)																				
Depth (in.)	12	14	16	18	20	22	24	26	28	30	32	34	36	38	40	42	44	46	48	50	52
24-3/4	9272	6745	5117	4008	3220	2639	2199	1858	1578	1273	1040	858	715	600	507	431	368	315	271	234	202
26-1/8	10306	7498	5689	4457	3580	2935	2447	2068	1769	1504	1229	1016	847	712	602	513	439	377	325	281	244
27-1/2	11288	8290	6291	4929	3960	3247	2707	2288	1958	1692	1440	1191	994	836	709	604	518	446	386	334	291
28-7/8	12232	9120	6922	5424	4358	3574	2980	2520	2156	1863	1625	1385	1157	975	827	706	606	523	453	393	343
30-1/4	13238	9989	7582	5942	4775	3916	3266	2762	2364	2043	1782	1566	1337	1127	957	818	703	607	527	459	401
31-5/8	14314	10897	8271	6483	5210	4274	3564	3015	2580	2231	1946	1711	1514	1294	1100	941	809	700	608	530	464
33	15465	11810	8990	7046	5664	4646	3876	3279	2807	2427	2118	1862	1648	1468	1256	1075	926	802	697	609	533
35-3/4	16701	12645	9738	7633	6136	5034	4200	3553	3042	2631	2296	2019	1788	1593	1425	1221	1053	912	794	694	609
34-3/8	18031	13528	10514	8243	6627	5437	4536	3838	3287	2843	2481	2183	1933	1722	1543	1380	1190	1032	900	787	691
37-1/8	19466	14462	11320	8875	7136	5855	4886	4134	3541	3063	2674	2352	2083	1856	1663	1497	1339	1162	1014	888	780
38-1/2	21018	15454	12155	9530	7663	6288	5248	4441	3804	3291	2873	2528	2239	1996	1788	1610	1456	1302	1137	996	876
39-7/8	22705	16507	12959	10208	8209	6737	5622	4759	4076	3527	3080	2710	2401	2140	1918	1727	1563	1419	1269	1113	980
41-1/4	24542	17628	13745	10908	8772	7200	6009	5087	4357	3771	3293	2898	2568	2289	2052	1848	1672	1519	1385	1238	1091
42-5/8	26551	18824	14572	11632	9354	7678	6409	5425	4648	4023	3513	3092	2740	2443	2190	1973	1785	1622	1479	1353	1210
44	28541	20103	15442	12377	9955	8171	6821	5775	4948	4283	3740	3292	2918	2602	2333	2102	1902	1729	1576	1442	1324

Notes:

(1) Span = simply supported beam.

(2) Maximum deflection = L/360 under live load, based on live/total load = 0.8. Where additional stiffness is desired or for other live/total load ratios, design for deflection must be modified per requirements.

(3) Service condition = dry.

(4) Tabulated values represent total loads based on live/total load = 0.8 and have taken the dead weight of the beam (assumed 36 pcf) into account.

(5) Sufficient bearing length shall be provided at supports.

(6) Maximum beam shear is located at a distance from the supports equal to the depth of the beam.

(7) Light (unshaded) areas limited by deflection; medium-shaded areas limited by bending strength; dark-shaded areas limited by shear strength.

TABLE 4.33 Allowable Loads for Cantilevered Southern Pine Glulam Roof Beams (lb/ft)—Non-snow Loads

5-INCH WIDTH	SPAN (ft)																	
Depth (in.)	44			48			52			56			60			64		
	sys 1	sys 2	sys 3	sys 1	sys 2	sys 3	sys 1	sys 2	sys 3	sys 1	sys 2	sys 3	sys 1	sys 2	sys 3	sys 1	sys 2	sys 3
24-3/4	551	545	649	456	451	538	382	378	451	324	320	383	277	274	329	239	236	284
26-1/8	614	607	722	509	503	599	427	422	503	362	358	428	310	306	367	267	264	318
27-1/2	681	673	800	564	557	664	473	468	558	402	397	475	344	340	407	297	294	353
28-7/8	750	742	882	622	615	732	522	516	616	444	438	524	380	376	450	329	325	390
30-1/4	823	814	967	683	675	803	574	567	676	487	482	575	418	413	494	362	357	428
31-5/8	899	889	1056	746	737	877	627	620	739	533	527	629	458	452	541	396	391	469
33	979	968	1149	812	803	955	683	675	804	581	574	685	499	493	589	432	427	511
34-3/8	1062	1050	1246	881	871	1036	741	733	872	631	623	743	542	536	640	470	464	555
35-3/4	1148	1135	1347	953	942	1120	802	793	944	683	675	804	587	580	692	509	502	601

6-3/4-INCH WIDTH	SPAN (ft)																	
Depth (in.)	44			48			52			56			60			64		
	sys 1	sys 2	sys 3	sys 1	sys 2	sys 3	sys 1	sys 2	sys 3	sys 1	sys 2	sys 3	sys 1	sys 2	sys 3	sys 1	sys 2	sys 3
24-3/4	733	724	862	606	599	714	508	502	600	431	425	509	368	364	437	317	313	377
26-1/8	816	807	960	676	668	796	567	560	669	481	475	568	411	406	488	355	351	422
27-1/2	904	894	1063	749	740	882	629	621	742	534	527	630	457	452	541	395	390	468
28-7/8	997	986	1172	826	817	972	694	686	818	589	582	696	505	499	598	436	431	517
30-1/4	1094	1081	1285	907	896	1067	762	753	898	647	640	764	555	549	657	480	474	569
31-5/8	1195	1182	1404	991	980	1166	833	824	981	708	700	836	608	601	718	526	520	623
33	1301	1286	1528	1079	1067	1269	908	897	1069	772	763	910	663	655	783	574	567	679
34-3/8	1411	1395	1657	1171	1158	1377	985	974	1159	838	828	988	720	711	850	624	616	737
35-3/4	1526	1508	1791	1267	1252	1488	1066	1053	1254	907	896	1069	779	770	920	675	667	798
37-1/8	1645	1626	1930	1366	1350	1604	1149	1136	1352	979	967	1153	841	831	992	729	721	862
38-1/2	1768	1747	2074	1468	1451	1724	1236	1222	1454	1053	1040	1240	905	895	1067	785	776	927
39-7/8	1895	1873	2223	1575	1556	1849	1326	1311	1559	1130	1116	1330	972	960	1145	843	833	995
41-1/4	2027	2004	2364	1684	1665	1977	1419	1403	1668	1209	1195	1423	1041	1028	1226	903	892	1065
42-5/8	2163	2138	2443	1798	1777	2110	1515	1497	1780	1291	1276	1519	1112	1098	1309	965	953	1138
44	2303	2277	2522	1915	1893	2247	1614	1595	1896	1376	1360	1618	1185	1171	1395	1029	1017	1213
45-3/8	2448	2420	2601	2036	2012	2378	1716	1696	2015	1463	1446	1721	1260	1246	1484	1095	1082	1290
46-3/4	2597	2568	2680	2160	2135	2450	1821	1800	2139	1553	1535	1826	1338	1322	1575	1163	1149	1370

8-1/2-INCH WIDTH	SPAN (ft)																	
Depth (in.)	44			48			52			56			60			64		
	sys 1	sys 2	sys 3	sys 1	sys 2	sys 3	sys 1	sys 2	sys 3	sys 1	sys 2	sys 3	sys 1	sys 2	sys 3	sys 1	sys 2	sys 3
37-1/8	2046	2023	2401	1699	1679	1996	1430	1413	1682	1217	1203	1434	1046	1034	1234	907	896	1072
38-1/2	2200	2174	2581	1827	1806	2146	1538	1520	1809	1310	1294	1542	1126	1113	1328	976	965	1153
39-7/8	2358	2331	2766	1959	1936	2300	1650	1631	1939	1405	1389	1654	1209	1194	1425	1048	1036	1238
41-1/4	2522	2493	2958	2096	2072	2461	1765	1745	2075	1504	1487	1770	1294	1279	1525	1123	1110	1325
42-5/8	2692	2661	3076	2237	2211	2626	1885	1863	2215	1606	1588	1890	1383	1366	1629	1200	1186	1416
44	2866	2834	3176	2383	2355	2796	2008	1985	2359	1712	1692	2013	1474	1456	1736	1280	1264	1509
45-3/8	3046	3012	3275	2533	2504	2972	2135	2110	2508	1821	1799	2141	1568	1549	1846	1362	1345	1605
46-3/4	3232	3195	3374	2688	2657	3085	2266	2240	2661	1933	1910	2272	1665	1645	1959	1446	1429	1704
48-1/8	3384	3384	3473	2847	2814	3175	2401	2373	2819	2048	2024	2407	1764	1744	2076	1533	1515	1806
49-1/2	3481	3518	3573	3010	2976	3266	2539	2510	2981	2166	2141	2546	1867	1845	2196	1622	1603	1911
50-7/8	3577	3616	3672	3179	3142	3357	2681	2650	3090	2288	2262	2688	1972	1949	2319	1714	1694	2019
52-1/4	3674	3714	3771	3351	3313	3448	2827	2795	3174	2413	2385	2835	2080	2056	2446	1809	1787	2129
53-5/8	3771	3812	3870	3447	3485	3538	2977	2943	3257	2541	2512	2985	2191	2166	2576	1905	1883	2243
55	3868	3909	3970	3535	3574	3629	3130	3095	3341	2673	2642	3094	2305	2278	2709	2005	1981	2359
56-3/8	3964	4007	4069	3624	3663	3720	3288	3250	3424	2807	2775	3171	2421	2393	2846	2106	2082	2478
57-3/4	4061	4105	4168	3712	3753	3811	3417	3409	3508	2945	2911	3249	2541	2511	2986	2211	2185	2600
59-1/8	4158	4203	4267	3801	3842	3901	3499	3537	3592	3086	3051	3326	2663	2632	3096	2317	2290	2725

See Fig. 4.33 for notes and description of cantilever systems.

TABLE 4.34 Allowable Loads for Cantilevered Southern Pine Glulam Roof Beams (lb/ft)—Snow Loads

5-INCH WIDTH	SPAN (ft)																	
Depth (in.)	44			48			52			56			60			64		
	sys 1	sys 2	sys 3	sys 1	sys 2	sys 3	sys 1	sys 2	sys 3	sys 1	sys 2	sys 3	sys 1	sys 2	sys 3	sys 1	sys 2	sys 3
24-3/4	505	499	594	417	412	492	349	345	413	296	292	350	253	250	300	218	215	259
26-1/8	563	556	662	465	460	548	390	385	461	330	326	391	283	279	335	243	240	290
27-1/2	623	616	733	516	510	608	433	428	511	367	362	434	314	310	372	271	267	322
28-7/8	687	679	808	569	562	670	478	472	563	405	400	479	347	343	411	300	296	356
30-1/4	754	746	887	625	618	736	525	518	619	445	440	526	382	377	452	330	326	391
31-5/8	824	815	969	683	675	804	574	567	676	487	482	575	418	413	494	361	357	428
33	897	887	1054	744	735	875	625	618	737	531	525	627	456	450	539	394	389	467
34-3/8	973	962	1143	807	798	949	679	671	799	577	570	681	495	489	585	429	423	507
35-3/4	1053	1040	1236	873	863	1027	734	726	864	625	617	736	536	530	633	464	459	549

6-3/4-INCH WIDTH	SPAN (ft)																	
Depth (in.)	44			48			52			56			60			64		
	sys 1	sys 2	sys 3	sys 1	sys 2	sys 3	sys 1	sys 2	sys 3	sys 1	sys 2	sys 3	sys 1	sys 2	sys 3	sys 1	sys 2	sys 3
24-3/4	671	663	790	554	548	654	464	458	548	393	388	465	335	331	398	289	285	344
26-1/8	747	739	880	618	611	729	518	512	612	439	433	519	375	370	445	323	319	384
27-1/2	828	819	975	685	677	808	575	568	678	487	481	576	417	412	494	359	355	427
28-7/8	913	903	1074	756	747	891	634	627	749	538	532	636	461	455	546	398	393	472
30-1/4	1002	991	1178	830	821	978	697	689	822	592	584	699	507	501	600	438	432	519
31-5/8	1096	1083	1287	908	897	1068	762	753	899	647	640	764	555	548	657	480	474	569
33	1193	1179	1401	989	977	1163	831	821	979	706	697	833	605	598	716	523	517	620
34-3/8	1294	1279	1519	1073	1060	1262	902	891	1062	766	757	904	658	650	777	569	562	674
35-3/4	1399	1383	1643	1160	1147	1364	976	964	1149	830	820	978	712	704	841	617	609	730
37-1/8	1508	1491	1770	1251	1237	1471	1053	1040	1239	895	885	1055	769	760	908	666	658	788
38-1/2	1621	1602	1903	1346	1330	1581	1132	1119	1332	963	952	1135	828	818	977	717	709	848
39-7/8	1738	1718	2040	1443	1427	1695	1215	1200	1429	1034	1022	1218	889	878	1048	770	761	910
41-1/4	1859	1838	2170	1544	1526	1814	1300	1285	1529	1107	1094	1303	952	940	1122	825	815	975
42-5/8	1984	1962	2242	1648	1629	1936	1388	1372	1632	1182	1168	1392	1017	1005	1199	882	871	1041
44	2113	2089	2314	1756	1736	2062	1479	1462	1738	1260	1245	1483	1084	1071	1278	940	929	1110
45-3/8	2246	2221	2387	1867	1845	2181	1573	1554	1848	1340	1325	1577	1153	1140	1359	1001	989	1181
46-3/4	2383	2356	2459	1981	1958	2247	1669	1650	1961	1423	1406	1674	1225	1210	1442	1063	1051	1254

8-1/2-INCH WIDTH	SPAN (ft)																	
Depth (in.)	44			48			52			56			60			64		
	sys 1	sys 2	sys 3	sys 1	sys 2	sys 3	sys 1	sys 2	sys 3	sys 1	sys 2	sys 3	sys 1	sys 2	sys 3	sys 1	sys 2	sys 3
37-1/8	1876	1855	2203	1557	1539	1830	1309	1294	1541	1114	1100	1313	956	945	1129	828	818	980
38-1/2	2017	1994	2368	1674	1655	1967	1408	1392	1657	1198	1184	1412	1030	1017	1215	892	881	1054
39-7/8	2163	2138	2538	1796	1775	2110	1511	1493	1778	1286	1271	1515	1105	1092	1304	958	946	1132
41-1/4	2313	2287	2714	1921	1899	2257	1617	1598	1902	1377	1361	1621	1184	1170	1396	1026	1014	1212
42-5/8	2469	2441	2823	2051	2027	2409	1727	1707	2030	1471	1453	1731	1265	1250	1491	1097	1084	1295
44	2630	2599	2914	2185	2160	2565	1840	1819	2163	1567	1549	1845	1348	1332	1589	1170	1156	1381
45-3/8	2795	2763	3005	2323	2296	2727	1957	1934	2299	1667	1648	1962	1435	1418	1690	1245	1230	1469
46-3/4	2965	2932	3096	2465	2436	2830	2077	2053	2440	1770	1749	2082	1523	1506	1795	1322	1307	1560
48-1/8	3105	3105	3187	2611	2581	2913	2200	2175	2585	1876	1854	2206	1615	1596	1902	1402	1385	1653
49-1/2	3194	3229	3278	2761	2730	2996	2327	2301	2734	1985	1961	2334	1709	1689	2012	1484	1467	1750
50-7/8	3283	3318	3370	2916	2882	3080	2458	2430	2835	2096	2072	2465	1806	1784	2125	1568	1550	1848
52-1/4	3371	3408	3461	3074	3039	3163	2592	2562	2911	2211	2185	2599	1905	1883	2242	1655	1636	1950
53-5/8	3460	3498	3552	3162	3197	3246	2730	2698	2988	2329	2302	2737	2007	1983	2361	1744	1723	2054
55	3549	3587	3643	3243	3279	3329	2871	2838	3064	2450	2421	2837	2111	2086	2483	1835	1813	2161
56-3/8	3637	3677	3734	3324	3361	3413	3015	2980	3141	2573	2543	2908	2218	2192	2609	1928	1906	2270
57-3/4	3726	3767	3825	3405	3443	3496	3134	3127	3218	2700	2669	2979	2328	2301	2737	2024	2000	2382
59-1/8	3815	3856	3916	3487	3525	3579	3209	3244	3294	2829	2797	3050	2440	2411	2838	2122	2097	2497

See Fig. 4.33 for notes and description of cantilever systems.

Maximum deflection under total load = $L/180$

Use 24F southern pine glued laminated timber

Then:

Total applied load, $w = (30 + 10)(24) = 960$ lb/ft excluding beam weight

From Table 4.31 for 24 ft span, select

$3 \times 26\frac{1}{8}$ ($w = 1050$ lb/ft)

or $3\frac{1}{2} \times 23\frac{3}{8}$ ($w = 977$ lb/ft)

or $5 \times 20\frac{5}{8}$ ($w = 1070$ lb/ft)

or $5\frac{1}{2} \times 19\frac{1}{4}$ ($w = 1023$ lb/ft)

or $6\frac{3}{4} \times 17\frac{7}{8}$ ($w = 1073$ lb/ft)

Note that the beam weight has been included in the table.

Design Example 3 Panelized Roof Design Using Load-Span Tables

A warehouse/office building is to be 85×180 ft. It has a flat roof with a minimum slope of ¼:12. The design live load (non-snow load) is the minimum required by the Uniform Building Code, with a duration of load factor of 1.25. Assume design dead load = 8 psf. It is desired to minimize the number of interior columns.

Assume three 60 ft bays (equals 180 ft) and two 42.5 ft bays (equals 85 ft) requiring two interior columns.

Main Beam Design—Option 1

Try system 3 (double cantilever) with three 60 ft bays. The tributary area for each cantilever beam's main span is $60 \times 42.5 = 2550$ ft^2. The suspended beam's tributary area is $0.83 \times 60 \times 42.5 = 2117$ ft^2. Per Table 23-C (method 1) of the Uniform Building Code (UBC), the minimum design live load is 12 psf for tributary areas greater than 600 ft^2 per beam. Therefore, the design live load for these beams is $12 \times 42.5 =$ 510 lb/ft and the design total load, excluding beam weight, is $(12 + 8) \times 42.5 = 850$ lb/ft.

Assume 24F-V8 Douglas fir glued laminated timber with $F_b = 2400$ psi and $E =$ 1,800,000 psi for the main cantilever beam. From Table 4.27, a double-cantilever beam (system 3) with 60 ft span, 6¾ in. wide and 37½ in. deep, can carry 897 lb/ft Note that the beam weight has been included in the table—OK.

From Table 4.24, a simple-span 24F-V4 Douglas-fir glued laminated timber beam 50 ft (0.83×60) long, 6¾ in. wide, and 36 in. deep can carry 873 lb/ft. Note that the beam weight has been included in the table—OK.

Option 2

Try system 2 (single cantilever with suspended center beam) with three 60 ft bays. Loads are the same as for Option 1, since all members carry more than 600 ft^2 of tributary area.

From Table 4.27, a single cantilever beam (system 2) with a 60 ft main span, 6¾ in. wide and 40½ in. deep, can carry 870 lb/ft. Note that the beam weight has been included in the table—OK.

From Table 4.24, a simple-span beam 30 ft ($2 \times 0.25 \times 60$) long, 5⅛ in. wide and 24 in. deep can carry 954 lb/ft. Note that the beam weight has been included in the table—OK.

Note: A 6¾ × 21 in. beam can carry 944 lb/ft, and it is also OK, but its area of 142 in.^{2} is greater than the area of the 5⅛ × 24 beam (123 in.^{2}), suggesting it may be less economical.

The two options can then be compared by beam volume, which will typically indicate the most economical option.

Beam Volume for Option 1

$$\frac{6.75 \times 37.5}{144}(1 + 2 \times 0.17)\,60 + \frac{2(6.75 \times 36)}{144}(0.83 \times 60) = 309.4 \text{ ft}^3$$

Beam Volume for Option 2

$$\frac{2(6.75 \times 24)}{144}(1 + 0.25)\,60 + \frac{(5.125 \times 24)}{144}[(1 - 2 \times 0.25) \times 60] = 310.4 \text{ ft}^3$$

For this example, the beam volumes are approximately equal and the final selection is the designer's option.

Secondary Beam Design

Secondary beams, all perpendicular to the main beams and all simple span, are spaced at 8 ft on center, as is typical with a panelized system panel deck. For a non-panelized system, they could be at some greater spacing, such as 20 ft on center, with subpurlins between these members at a closer on center spacing.

The secondary beams have a simple span of approximately 42 ft.

Assume secondary beams 8 ft on center. The tributary area is 42 × 8 = 336 ft^2. Per method 1 of UBC. Table 23-C, the design live load is 16 psf. Total load, excluding beam weight, is 8(16 + 8) = 192 lb/ft. From Table 4.24, a simple-span beam 42 ft long, 3⅛ in. wide, and 22½ in. deep can carry 196 lb/ft—OK.

Other types of framing members, such as solid-sawn lumber, wood I-joists, and wood trusses, can also be used as secondary beams, depending on the span and loading conditions.

A comparison of material costs will provide guidance as to their relative economies. In addition, hardware (hanger) requirements, as well as any labor differences, need to be considered in order to obtain a complete economic comparison of the systems.

Design Example 4 Floor Design Using Section Capacities

Given:

Two span-continuous beam with spans of 23.25 ft and 19.25 ft. Beams spaced at 10 ft on center.

Floor Live load = 125 psf (light storage); duration of load = 1.0

Dead load = 10 psf (actual)

Allowable total load deflection = $L/240$

Allowable live load deflection = $L/360$

Beam depth limited to 24 in. or less, due to height restrictions

Use 24F-V5 southern pine glued laminated timber

Then:

Assume beam weight of 36 lb/ft

Live load, w_l = 125 × 10 = 1250 lb/ft

Dead load, w_d = (10 × 10) + 36 = 136 lb/ft

Total load, w_t = 1250 + 136 = 1386 lb/ft

Max. moment, fully loaded, M = 80 312 lb/ft, at interior reaction

Max. moment, unbalanced loading, M_u = 69,790 lb/ft at approximately 10 ft from the outer support of the 23.25 ft span

Max. shear, fully loaded, V = 16,795 lb at 24 in. away from the interior reaction, in the 23.25 ft span

Max. shear, unbalanced loading, V_u = 15,544 lb

Max. reaction, R = 37,079 lb at interior support

Design:

From Table 4.29, a 3½ in. wide beam would exceed the depth limitation, based on shear requirements.

Try a 5 in. wide × 23⅜ in. deep beam. (For purposes of the volume factor, the moment capacity span is the distance between points of zero moment and is approximately 20 ft.) From Tables 4.29 and Eq. (4.1), the allowable moment capacity = 91,065 × 0.9708 = 88,406 lb-ft > 80,312 lb-ft. The actual beam weight of 29.2 lb/ft is less than the assumed 36 lb/ft—OK.

The allowable compression perpendicular to grain, $F_{c\perp}$ = 740 psi. Minimum bearing length at interior support = 37,079/(740 × 5) = 10 in. Revised design shear, V = 16,867 lb at 23⅜ in. away from the face of the interior support <21,038 lb—OK.

Max. deflection: total load on longer span, dead load only on shorter span = 0.66 in. = $L/425 < L/240$ → OK.

Max. deflection: live load on longer span = 0.62 in. = $L/454 < L/360$—OK.

4.7.2 Columns

Introduction. While glued laminated timbers are typically used as some type of bending member, they are also ideally suited for use as columns. Because glued laminated timber is manufactured with dry lumber having a maximum moisture content at the time of fabrication of 16%, it has excellent dimensional stability. Thus, a glued laminated timber column will not undergo the dimensional changes normally associated with larger solid-sawn sections, which are typically supplied as green timbers. Glued laminated timber will remain straight and true in cross-section. Since glued laminated timber is manufactured with dry lumber, it is also less susceptible to checking and splitting, which often occur with green timbers, and it has better fastener holding capacities.

Member Sizes. Like other glued laminated timber shapes, columns can be manufactured in virtually any cross-sectional size and length required. However, since they are manufactured using dimension lumber, specifying glued laminated timber column in the typical widths as shown by Table 4.35 will ensure maximum efficiency of the resource and product availability.

The depths of glued laminated timber columns are normally specified in multiples of 1½ in. for western species and 1⅜ in. for southern pine. Examples of column sizes are given in Table 4.36 to show the use of typical glued laminated timber width and depth size multiples.

Another advantage of glued laminated timber is that any length can be supplied, thereby eliminating the need for costly splices to create long-length columns for multistory applications or high open areas. Availability of specific cross-section dimensions and lengths should be verified with the supplier or manufacturer.

TABLE 4.35 Typical Glulam Column Widths

Nominal Width	4*	6*	8	10	12
Western species	3-1/8	5-1/8	6-3/4	8-3/4	10-3/4
Southern pine	3	5	6-3/4	8-1/2	10-1/2

*For the 4-inch and 6-inch nominal widths, glulam may also be available in 3-1/2" and 5-1/2" widths respectively. These "full-width" members correspond to the dimensions of 2x4 and 2x6 framing lumber and are supplied with "hit or miss" surfacing which is only acceptable for concealed applications. For additional information on the appearance characteristics of glulam, see EWS Technical Note Y110,[5] Appearance Classifications for Glued Laminated Timber.

TABLE 4.36 Typical Glulam Column Sizes

Nominal Size	6x6	8x8	8x10	10x10	10x12
Western species	5-1/8x6	6-3/4x7-1/2	6-3/4x9	8-3/4x9	8-3/4x12
Southern pine	5x5-1/2	6-3/4x6-7/8	6-3/4x9-5/8	8-1/2x9-5/8	8-1/2x11

*Other sizes are available. Contact the local supplier or manufacturer for additional information.

Member Lay-up and Design Stresses. Since compression parallel to grain stresses are distributed uniformly over the cross-section of an axially loaded member, glued laminated timber columns are typically manufactured using a single grade of lumber throughout the depth of the member. Examples of lay-up combinations and some of the associated design stresses for single grade glued laminated timber members are shown in Table 4.37.

Two distinct values are provided for F_b and F_v, depending on which axis the load is applied to, i.e., parallel to the wide or to the narrow face of the member. If a column is going to be loaded as a combined axial and bending member, it may be preferable to specify a bending member lay-up such as a 24F-V8 DF or 24F-V5 SP combination. Such members use a graded lumber lay-up throughout the depth of the member and are more efficient for resisting high bending stresses.

For a complete listing of available glued laminated timber layup combinations for both members primarily loaded axially or as bending members, refer to Tables 4.3 and 4.4.

Column Design Equations. Until the promulgation of the 1991 NDS, wood columns were designed based on a methodology that required classifying the member as a short, intermediate, or long column. This required a trial-and-error solution

TABLE 4.37 Column Lay-up Designations

Species and Layup Combination	Lam Grade	F_c	E	$F_{bx\text{-}x}$	$F_{by\text{-}y}$	$F_{vx\text{-}x}$	$F_{vy\text{-}y}$
DF – No. 2	L2	1900	1.7×10^6	1700	1800	240	210
SP – No. 47	N2M	1900	1.4×10^6	1400	1750	270	235

*All stress values are in psi and assume 4 or more laminations (up to 15 inches) without special tension laminations. Numerous other species and layup combinations are available. See EWS Data File Y117 for more information.

when it was not obvious which classification applied for a specific design situation, and many designers considered it to be a cumbersome procedure.

Based on extensive research conducted at the USDA Forest Products Laboratory and other research institutions, the 1991 NDS was revised to reflect the use of a single column design formula regardless of the length-to-depth (l/d) ratio previously used to classify columns as short, intermediate, or long. This is shown as Eq. (4.18) for a member subjected to *concentric axial loads* only.

$$F'_c = F^*_c \left\{ \frac{1 + (F_{cE}/F^*_c)}{2c} - \sqrt{\left[\frac{1 + (F_{cE}/F^*_c)}{2c}\right]^2 - \frac{(F_{cE}/F^*_c)}{c}} \right\} \qquad (4.18)$$

where F'_c = allowable compression parallel to grain design value
F^*_c = tabulated compression parallel to grain design value adjusted for service conditions (moisture, temperature, load duration) and size effect when applicable
F_{cE} = critical buckling design value
c = 0.8 for sawn lumber, 0.85 for round timber poles and piles, 0.9 for glued laminated timber and structural composite lumber

The critical buckling design value is determined by the well-known Euler column formula:

$$F_{cE} = \frac{K_{cE}E'}{(L_e/d)^2} \qquad (4.19)$$

where E' = allowable modulus of elasticity adjusted for service conditions (moisture, temperature)
d = least unbraced dimension of column
L_e = effective column length based on unbraced length and end fixity conditions
K_{cE} = 0.300 for products with the coefficient of variation of MOE, COV_E = 0.25, such as visually graded lumber and round timber poles and piles; 0.384 for products with COV_E = 0.15, such as machine-evaluated lumber (MEL); 0.418 for products with $COV_E \leq 0.11$, such as glued laminated timber, machine stress-rated (MSR) lumber, and structural composite lumber

The solution of this equation, which determines the allowable compression parallel to grain stress, is based on the physical dimensions of the column, the published material properties such as E and F_c, and several constants. The two constants, c and K_{cE}, are material-dependent, with higher values assigned to wood products with lower variability such as glued laminated timber, thus resulting in higher column capacities.

Through the laminating process, naturally occurring strength-reducing characteristics in the lumber are randomly distributed throughout the member, resulting in lower variability in mechanical properties for glued laminated timber as compared to sawn lumber products. For example, the typical coefficient of variation for the modulus of elasticity of glued laminated timber is about 10%, which is equal to or lower than other comparable wood products. Based on the relative homogeneity of glued laminated timber and its low variability, the values of c and K_{cE} for glued laminated timber have been established as 0.9 and 0.418, respectively.

Column Design Tables. Tables 4.38A–4.38F have been generated to provide column capacities for two typical glued laminated timber lay-up combinations for an eccentric condition of load application. These tables are summarized as follows:

No. 2 DF Tables 4.38A–4.38C (eccentric loading)

No. 47 SP Tables 4.38D–4.38F (eccentric loading)

All tables have been truncated at an L/d ratio of 50.

For most applications, the No. 2 DF and No. 47 SP combinations will result in the most cost-efficient columns. These permit the use of all L2 laminations for the No. 2 DF and all No. 2 medium grain laminations for the No. 47 SP combinations.

For those applications requiring greater capacities, the use of a No. 5 DF (all L1 laminations) or a No. 50 SP (all No. 1 dense laminations) are recommended. For additional tables of glulam columns, see EWS publication Y240A, *Design of Structural Glued Laminated Timber Columns*.[12]

Since wood columns are typically not truly loaded concentrically, Tables 4.38A–4.38F are provided based on the conservative assumption that the load is applied with an eccentricity of ⅙ of the least dimension of the column. This degree of eccentricity is considered to be representative of many actual in-service column installations such as an end column supporting a beam. As such, it provides a conservative solution based on an allowance for some degree of field framing inconsistencies. It is recommended that these tables be used for those applications where it is desirable to use a simple tabular solution for preliminary design sizing.

For applications with greater degrees of eccentricities or side loads, the designer is referred to the NDS[6] for equations that account for these conditions of loading.

As with the use of all design tables, it is recommended that the advice of a design professional be obtained to verify the capacity and applicability of any column size provided in Tables 4.38A–4.38F.

Where higher capacities are required and it can be ensured that the loads will be applied concentrically, the column may be designed in accordance with Eq. (4.18), as shown in the following example.

Column Design Example Determine the size of a glued laminated timber column required to support a 45 kip axial floor load (DOL = 1.0) applied concentrically. Assume the length of the column is 15 ft and that it is in a dry use service condition. Use a Douglas fir combination No. 2. Assume the column is unbraced and that the end conditions are pinned.

Tabulated allowable stresses (see Table 4.4):

$$F_{c//} = 1900 \text{ psi}$$

$$E = 1{,}700{,}000 \text{ psi}$$

Adjusted allowable stresses:

$$F_c^* = 1900 \times 1.0 = 1900 \text{ psi}$$

$$E' = E = 1{,}700{,}000 \text{ psi}$$

Try a 6¾ in. × 7½ in. section: net area = $6.75 \times 7.5 = 50.62 \text{ in.}^2$

Determine effective length $(L_e) = 15(12) \times 1.0 = 180$ in.

Determine slenderness ratio $= L_e/d = 180/6.75 = 26.67 < 50$

TABLE 4.38A Glulam Column Design Tables
ECCENTRIC END LOADS FOR DOUGLAS-FIR COMBINATION NO. 2 GLULAM COLUMNS
Allowable axial loads (pounds). Side loads and bracket loads are not permitted. End loads are limited to a maximum eccentricity of either 1/6 column width or 1/6 column depth.

Effective Column Length (ft)	Lamination Net Width = 5-1/8 in.								
	Net Depth = 6 in. (4 lams)			Net Depth = 7-1/2 in. (5 lams)			Net Depth = 9 in. (6 lams)		
	Load Duration Factor			Load Duration Factor			Load Duration Factor		
	1.00	1.15	1.25	1.00	1.15	1.25	1.00	1.15	1.25
8	24,349	26,269	27,389	30,437	32,837	34,236	36,524	39,404	41,084
9	21,771	23,206	24,036	27,214	29,008	30,045	32,657	34,810	36,054
10	19,370	20,464	21,096	24,213	25,580	26,370	29,055	30,696	31,644
11	17,232	18,087	18,580	21,541	22,609	23,225	25,849	27,130	27,870
12	15,370	16,051	16,444	19,212	20,064	20,555	23,054	24,077	24,666
13	13,759	14,312	14,631	17,199	17,890	18,288	20,639	21,468	21,946
14	12,369	12,824	13,086	15,461	16,030	16,357	18,553	19,236	19,628
15	11,166	11,545	11,763	13,958	14,431	14,703	16,749	17,317	17,644
16	10,122	10,441	10,624	12,652	13,051	13,280	15,183	15,661	15,935

17	9,212	9,483	9,638	11,515	11,853	12,047	13,818	14,224	14,457
18	8,415	8,647	8,780	10,519	10,809	10,974	12,623	12,970	13,169
19	7,714	7,914	8,029	9,643	9,893	10,036	11,572	11,872	12,043
20	7,096	7,269	7,369	8,870	9,087	9,211	10,643	10,904	11,053
21	6,547	6,699	6,785	8,184	8,373	8,481	9,820	10,048	10,178

Notes:

(1) The tabulated allowable loads apply to glulam members made with all L2 laminations (Combination 2) without special tension laminations.

(2) Applicable service conditions = dry.

(3) The tabulated allowable loads are based on axially loaded columns subjected to a maximum eccentricity of either 1/6 column width or 1/6 column depth, whichever is worse. For side loads, other eccentric end loads, or other combined axial and flexural loads, see 1997 NDS.

(4) The column is assumed to be unbraced, except at the column ends, and the effective column length is equal to the actual column length.

(5) Design properties for normal load duration and dry-use service conditions:
Compression parallel to grain (F_c) = 1,900 psi for 4 or more lams, or 1,600 psi for 2 or 3 lams.
Modulus of elasticity (E) = 1.7×10^6 psi.
Flexural stress when loaded parallel to wide faces of lamination (F_{by}) = 1,800 psi for 4 or more lams, or 1,600 psi for 3 lams.
Flexural stress when loaded perpendicular to wide faces of lamination (F_{bx}) = 1,700 psi for 2 lams to 15 in. deep without special tension laminations.
Volume factor for F_{bx} is in accordance with 1997 NDS. Size factor for F_{by} is $(12/d)^{0.111}$, where d is equal to the lamination width in inches.

(6) This table is for preliminary design use only. Final design should include a complete analysis, including the bearing capability of the foundation supporting the column.

TABLE 4.38B Glulam Column Design Tables

ECCENTRIC END LOADS FOR DOUGLAS-FIR COMBINATION NO. 2 GLULAM COLUMNS

Allowable axial loads (pounds). Side loads and bracket loads are not permitted. End loads are limited to a maximum eccentricity of either 1/6 column width or 1/6 column depth.

Effective Column Length (ft)	Lamination Net Width = 6-3/4 in.								
	Net Depth = 7-1/2 in. (5 lams)			Net Depth = 9 in. (6 lams)			Net Depth = 10-1/2 in. (7 lams)		
	Load Duration Factor			Load Duration Factor			Load Duration Factor		
	1.00	**1.15**	**1.25**	**1.00**	**1.15**	**1.25**	**1.00**	**1.15**	**1.25**
8	46,419	51,735	55,055	57,007	63,204	67,033	66,509	73,738	78,206
9	43,906	48,470	51,252	53,421	58,620	61,752	62,324	68,390	72,044
10	41,184	44,902	46,984	49,641	53,883	56,381	57,915	62,863	65,778
11	38,162	41,002	42,653	45,794	49,203	51,184	53,426	57,403	59,714
12	35,027	37,314	38,637	42,032	44,777	46,365	49,037	52,240	54,092
13	32,072	33,936	35,013	38,486	40,723	42,016	44,900	47,510	49,019
14	29,356	30,897	31,788	35,228	37,077	38,145	41,099	43,256	44,503
15	26,898	28,188	28,933	32,278	33,826	34,720	37,658	39,464	40,507
16	24,690	25,782	26,412	29,628	30,938	31,694	34,566	36,094	36,976
17	22,712	23,644	24,182	27,254	28,373	29,019	31,797	33,102	33,855
18	20,942	21,745	22,207	25,130	26,094	26,648	29,318	30,442	31,090

19	19,356	20,052	20,453	23,227	24,062	24,543	27,098	28,073	28,634
20	17,933	18,541	18,890	21,519	22,249	22,668	25,106	25,957	26,446
21	16,653	17,187	17,493	19,984	20,624	20,992	23,314	24,061	24,490
22	15,500	15,971	16,241	18,600	19,165	19,489	21,700	22,359	22,737
23	14,458	14,876	15,115	17,349	17,851	18,138	20,241	20,826	21,161
24	13,514	13,886	14,099	16,217	16,663	16,919	18,920	19,441	19,739

Notes:

(1) The tabulated allowable loads apply to glulam members made with all L2 laminations (Combination 2) without special tension laminations.

(2) Applicable service conditions = dry.

(3) The tabulated allowable loads are based on axially loaded columns subjected to a maximum eccentricity of either 1/6 column width or 1/6 column depth, whichever is worse. For side loads, other eccentric end loads, or other combined axial and flexural loads, see 1997 NDS.

(4) The column is assumed to be unbraced, except at the column ends, and the effective column length is equal to the actual column length.

(5) Design properties for normal load duration and dry-use service conditions:
Compression parallel to grain (F_c) = 1,900 psi for 4 or more lams, or 1,600 psi for 2 or 3 lams.
Modulus of elasticity (E) = 1.7×10^6 psi.
Flexural stress when loaded parallel to wide faces of lamination (F_{by}) = 1,800 psi for 4 or more lams, or 1,600 psi for 3 lams.
Flexural stress when loaded perpendicular to wide faces of lamination (F_{bx}) = 1,700 psi for 2 lams to 15 in. deep without special tension laminations.
Volume factor for F_{bx} is in accordance with 1997 NDS. Size factor for F_{by} is $(12/d)^{0.111}$, where d is equal to the lamination width in inches.

(6) This table is for preliminary design use only. Final design should include a complete analysis, including the bearing capability of the foundation supporting the column.

TABLE 4.38C Glulam Column Design Tables

ECCENTRIC END LOADS FOR DOUGLAS-FIR COMBINATION NO. 2 GLULAM COLUMNS

Allowable axial loads (pounds). Side loads and bracket loads are not permitted. End loads are limited to a maximum eccentricity of either 1/6 column width or 1/6 column depth.

Effective Column Length (ft)	Lamination Net Width = 8-3/4 in.								
	Net Depth = 9 in. (6 lams)			Net Depth = 10-1/2 in. (7 lams)			Net Depth = 12 in. (8 lams)		
	Load Duration Factor			Load Duration Factor			Load Duration Factor		
	1.00	1.15	1.25	1.00	1.15	1.25	1.00	1.15	1.25
8	77,152	86,969	93,273	92,611	104,825	112,726	107,117	120,632	129,293
9	74,473	83,480	89,193	90,045	101,094	107,917	103,206	115,536	123,334
10	71,555	79,699	84,789	86,575	96,266	102,297	98,943	110,018	116,911
11	68,426	75,672	80,122	82,582	91,135	96,363	94,379	104,155	110,129
12	65,121	71,458	75,279	78,377	85,795	90,240	89,573	98,051	103,132
13	61,687	67,146	70,385	74,027	80,361	84,096	84,602	91,841	96,110
14	58,187	62,847	65,581	69,620	74,983	78,112	79,565	85,695	89,271
15	54,702	58,674	60,991	65,261	69,799	72,433	74,583	79,770	82,780
16	51,311	54,711	56,689	61,047	64,908	67,144	69,767	74,181	76,736
17	48,075	51,005	52,707	57,051	60,361	62,277	65,201	68,984	71,174
18	44,948	47,502	48,986	53,312	56,174	57,829	60,928	64,198	66,091

19	42,045	44,279	45,576	49,844	52,336	53,777	56,964	59,813	61,460
20	39,367	41,334	42,475	46,644	48,830	50,092	53,307	55,805	57,248
21	36,904	38,645	39,654	43,700	45,627	46,740	49,942	52,146	53,418
22	34,640	36,189	37,086	40,994	42,703	43,689	46,851	48,804	49,931
23	32,560	33,945	34,745	38,509	40,032	40,909	44,011	45,751	46,754
24	30,648	31,891	32,609	36,226	37,588	38,372	41,401	42,958	43,854

Notes:

(1) The tabulated allowable loads apply to glulam members made with all L2 laminations (Combination 2) without special tension laminations.

(2) Applicable service conditions = dry.

(3) The tabulated allowable loads are based on axially loaded columns subjected to a maximum eccentricity of either 1/6 column width or 1/6 column depth, whichever is worse. For side loads, other eccentric end loads, or other combined axial and flexural loads, see 1997 NDS.

(4) The column is assumed to be unbraced, except at the column ends, and the effective column length is equal to the actual column length.

(5) Design properties for normal load duration and dry-use service conditions:
Compression parallel to grain (F_c) = 1,900 psi for 4 or more lams, or 1,600 psi for 2 or 3 lams.
Modulus of elasticity (E) = 1.7×10^6 psi.
Flexural stress when loaded parallel to wide faces of lamination (F_{by}) = 1,800 psi for 4 or more lams, or 1,600 psi for 3 lams.
Flexural stress when loaded perpendicular to wide faces of lamination (F_{bx}) = 1,700 psi for 2 lams to 15 in. deep without special tension laminations.
Volume factor for F_{bx} is in accordance with 1997 NDS. Size factor for F_{by} is $(12/d)^{0.111}$, where d is equal to the lamination width in inches.

(6) This table is for preliminary design use only. Final design should include a complete analysis, including the bearing capability of the foundation supporting the column.

TABLE 4.38D Glulam Column Design Tables
ECCENTRIC END LOADS FOR SOUTHERN PINE COMBINATION NO. 47 GLULAM COLUMNS
Allowable axial loads (pounds). Side loads and bracket loads are not permitted. End loads are limited to a maximum eccentricity of either 1/6 column width or 1/6 column depth.

Effective Column Length (ft)	Lamination Net Width = 5 in.								
	Net Depth = 5-1/2 in. (4 lams)			Net Depth = 6-7/8 in. (5 lams)			Net Depth = 8-1/4 in. (6 lams)		
	Load Duration Factor			Load Duration Factor			Load Duration Factor		
	1.00	1.15	1.25	1.00	1.15	1.25	1.00	1.15	1.25
8	17,783	19,146	19,948	23,985	25,571	26,489	28,782	30,685	31,787
9	15,858	16,901	17,512	21,036	22,210	22,888	25,243	26,652	27,466
10	14,114	14,927	15,401	18,458	19,353	19,870	22,150	23,224	23,844
11	12,579	13,224	13,599	16,256	16,956	17,359	19,508	20,347	20,831
12	11,244	11,764	12,066	14,388	14,945	15,266	17,265	17,934	18,319
13	10,089	10,513	10,758	12,802	13,253	13,513	15,362	15,904	16,215
14	9,088	9,438	9,627	11,450	11,821	12,034	13,740	14,185	14,440
15	8,219	8,481	8,622	10,293	10,601	10,778	12,352	12,721	12,933
16	7,438	7,645	7,763	9,297	9,556	9,704	11,157	11,467	11,645

17	6,748	6,923	7,023	8,435	8,654	8,779	10,122	10,385	10,535
18	6,147	6,297	6,383	7,684	7,872	7,979	9,221	9,446	9,574
19	5,622	5,751	5,825	7,028	7,189	7,281	8,433	8,627	8,737
20	5,160	5,272	5,336	6,450	6,590	6,670	7,740	7,908	8,004

Notes:

(1) The tabulated allowable loads apply to glulam members made with all N2M laminations (Combination 47) without special tension laminations.

(2) Applicable service conditions = dry.

(3) The tabulated allowable loads are based on axially loaded columns subjected to a maximum eccentricity of either 1/6 column width or 1/6 column depth, whichever is worse. For side loads, other eccentric end loads, or other combined axial and flexural loads, see 1997 NDS.

(4) The column is assumed to be unbraced, except at the column ends, and the effective column length is equal to the actual column length.

(5) Design properties for normal load duration and dry-use service conditions:
Compression parallel to grain (F_c) = 1,900 psi for 4 or more lams, or 1,150 psi for 2 or 3 lams.
Modulus of elasticity (E) = 1.4×10^6 psi.
Flexural stress when loaded parallel to wide faces of lamination (F_{by}) = 1,750 psi for 4 or more lams, or 1,550 psi for 3 lams.
Flexural stress when loaded perpendicular to wide faces of lamination (F_{bx}) = 1,400 psi for 2 lams to 15 in. deep without special tension laminations.
Volume factor for F_{bx} is in accordance with 1997 NDS. Size factor for F_{by} is $(12/d)^{0.111}$, where d is equal to the lamination width in inches.

(6) This table is for preliminary design use only. Final design should include a complete analysis, including the bearing capability of the foundation supporting the column.

TABLE 4.38E Glulam Column Design Tables
ECCENTRIC END LOADS FOR SOUTHERN PINE COMBINATION NO. 47 GLULAM COLUMNS
Allowable axial loads (pounds). Side loads and bracket loads are not permitted. End loads are limited to a maximum eccentricity of either 1/6 column width or 1/6 column depth.

Effective Column Length (ft)	Lamination Net Width = 6-3/4 in.								
	Net Depth = 6-7/8 in. (5 lams)			Net Depth = 8-1/4 in. (6 lams)			Net Depth = 9-5/8 in. (7 lams)		
	Load Duration Factor			Load Duration Factor			Load Duration Factor		
	1.00	1.15	1.25	1.00	1.15	1.25	1.00	1.15	1.25
8	35,774	39,505	41,796	45,733	50,861	54,038	55,427	61,905	65,938
9	33,301	36,391	38,245	43,086	47,409	50,018	52,597	57,387	60,096
10	30,759	33,271	34,757	40,208	43,755	45,857	48,352	51,969	54,074
11	28,251	30,291	31,489	37,230	40,118	41,595	44,014	46,872	48,528
12	25,872	27,544	28,524	34,253	36,217	37,354	39,961	42,254	43,580
13	23,677	25,065	25,875	31,098	32,700	33,627	36,281	38,150	39,231
14	21,686	22,849	23,527	28,277	29,602	30,368	32,990	34,536	35,429
15	19,893	20,878	21,450	25,773	26,882	27,522	30,068	31,362	32,109
16	18,286	19,126	19,613	23,553	24,491	25,032	27,479	28,573	29,204
17	16,847	17,569	17,987	21,587	22,387	22,848	25,185	26,119	26,656
18	15,557	16,182	16,543	19,841	20,530	20,926	23,148	23,951	24,413

19	14,399	14,943	15,257	18,288	18,884	19,227	21,336	22,032	22,431
20	13,359	13,835	14,110	16,903	17,422	17,721	19,720	20,326	20,674
21	12,422	12,841	13,082	15,663	16,118	16,380	18,273	18,805	19,110
22	11,575	11,946	12,159	14,550	14,952	15,182	16,975	17,444	17,712
23	10,809	11,138	11,327	13,548	13,904	14,108	15,806	16,221	16,459
24	10,114	10,408	10,576	12,644	12,961	13,142	14,751	15,121	15,332

Notes:

(1) The tabulated allowable loads apply to glulam members made with all N2M laminations (Combination 47) without special tension laminations.

(2) Applicable service conditions = dry.

(3) The tabulated allowable loads are based on axially loaded columns subjected to a maximum eccentricity of either 1/6 column width or 1/6 column depth, whichever is worse. For side loads, other eccentric end loads, or other combined axial and flexural loads, see 1997 NDS.

(4) The column is assumed to be unbraced, except at the column ends, and the effective column length is equal to the actual column length.

(5) Design properties for normal load duration and dry-use service conditions:
Compression parallel to grain (F_c) = 1,900 psi for 4 or more lams, or 1,150 psi for 2 or 3 lams.
Modulus of elasticity (E) = 1.4×10^6 psi.
Flexural stress when loaded parallel to wide faces of lamination (F_{by}) = 1,750 psi for 4 or more lams, or 1,550 psi for 3 lams.
Flexural stress when loaded perpendicular to wide faces of lamination (F_{bx}) = 1,400 psi for 2 lams to 15 in. deep without special tension laminations.
Volume factor for F_{bx} is in accordance with 1997 NDS. Size factor for F_{by} is $(12/d)^{0.111}$, where d is equal to the lamination width in inches.

(6) This table is for preliminary design use only. Final design should include a complete analysis, including the bearing capability of the foundation supporting the column.

TABLE 4.38F Glulam Column Design Tables
ECCENTRIC END LOADS FOR SOUTHERN PINE COMBINATION NO. 47 GLULAM COLUMNS
Allowable axial loads (pounds). Side loads and bracket loads are not permitted. End loads are limited to a maximum eccentricity of either 1/6 column width or 1/6 column depth.

Effective Column Length (ft)	Lamination Net Width = 8-1/2 in.								
	Net Depth = 8-1/4 in. (6 lams)			Net Depth = 9-5/8 in. (7 lams)			Net Depth = 11 in. (8 lams)		
	Load Duration Factor			Load Duration Factor			Load Duration Factor		
	1.00	1.15	1.25	1.00	1.15	1.25	1.00	1.15	1.25
8	58,811	65,794	70,212	71,516	80,466	86,187	83,914	94,750	101,717
9	56,008	62,197	66,050	68,804	76,927	82,043	81,234	91,197	97,510
10	53,030	58,411	61,697	65,839	73,070	77,541	78,224	87,198	92,778
11	49,925	54,519	57,274	62,644	68,949	72,772	74,878	82,775	87,580
12	46,763	50,638	52,935	59,262	64,666	67,889	71,222	77,545	80,942
13	43,626	46,890	48,811	55,778	60,369	63,080	66,784	71,587	74,380
14	40,599	43,362	44,983	52,297	56,199	58,490	61,972	65,972	68,292
15	37,742	40,099	41,478	48,919	52,252	54,203	57,426	60,790	62,739
16	35,086	37,112	38,295	45,708	48,574	50,248	53,207	56,064	57,718
17	32,639	34,393	35,415	42,700	45,180	46,550	49,332	51,782	53,200
18	30,396	31,923	32,813	39,880	41,926	42,997	45,797	47,915	49,139

19	28,331	29,670	30,448	37,258	38,871	39,802	42,580	44,424	45,488
20	26,439	27,619	28,303	34,699	36,112	36,927	39,656	41,271	42,203
21	24,715	25,760	26,365	32,374	33,618	34,335	36,998	38,421	39,240
22	23,142	24,071	24,609	30,258	31,360	31,994	34,580	35,840	36,564
23	21,705	22,536	23,015	28,330	29,310	29,874	32,377	33,498	34,142
24	20,391	21,136	21,566	26,571	27,447	27,950	30,367	31,368	31,943

Notes:

(1) The tabulated allowable loads apply to glulam members made with all N2M laminations (Combination 47) without special tension laminations.

(2) Applicable service conditions = dry.

(3) The tabulated allowable loads are based on axially loaded columns subjected to a maximum eccentricity of either 1/6 column width or 1/6 column depth, whichever is worse. For side loads, other eccentric end loads, or other combined axial and flexural loads, see 1997 NDS.

(4) The column is assumed to be unbraced, except at the column ends, and the effective column length is equal to the actual column length.

(5) Design properties for normal load duration and dry-use service conditions:
Compression parallel to grain (F_c) = 1,900 psi for 4 or more lams, or 1,150 psi for 2 or 3 lams.
Modulus of elasticity (E) = 1.4×10^6 psi.
Flexural stress when loaded parallel to wide faces of lamination (F_{by}) = 1,750 psi for 4 or more lams, or 1,550 psi for 3 lams.
Flexural stress when loaded perpendicular to wide faces of lamination (F_{bx}) = 1,400 psi for 2 lams to 15 in. deep without special tension laminations.
Volume factor for F_{bx} is in accordance with 1997 NDS. Size factor for F_{by} is $(12/d)^{0.111}$, where d is equal to the lamination width in inches.

(6) This table is for preliminary design use only. Final design should include a complete analysis, including the bearing capability of the foundation supporting the column.

Determine allowable compression parallel to grain design value using Eq. (4.18):

$$F_{cE} = \frac{(K_{cE}E')}{(L_e/d)^2} = \frac{(0.418 \times 1{,}700{,}000)}{(26.67)^2} = 999 \text{ psi}$$

$$F_{cE}/F_c^* = \frac{999}{1900} = 0.526$$

$$F_c' = 1900 \left\{ \frac{1 + 0.526}{2 \times 0.9} - \sqrt{\left[\frac{1 + 0.526}{2 \times 0.9}\right]^2 - \frac{0.526}{0.9}} \right\} = 914.5 \text{ psi}$$

Determine allowable axial load $= F_c' \times A = 914.5 \times 50.62 = 46.3$ kips > 45 kips.
Use a 6¾ in. × 7½ in. No. 2 Douglas fir glued laminated timber combination.

4.7.3 Substitution of Glued Laminated Timber Beams for Steel or Sawn Lumber

Glued laminated timber beams of equal or greater strength and stiffness can often be substituted for sawn lumber or steel beams. Refer to APA Data File S570, *Substitution of Glulam Beams for Steel or Solid-Sawn Lumber,*[11] for substitution tables and additional information.

4.8 *FIELD HANDLING AND INSTALLATION CONSIDERATIONS*

Glued laminated timber beams must be stored properly and handled with care to ensure optimum performance. When they leave the mill, glued laminated timber beams are often protected with sealants, primers, or wrappings as specified by the designer. Care must be taken during loading, unloading, and transporting as well as in the yard and on the job site to protect these members from damage.

4.8.1 Sealants, Primers, and Wrappings

Sealants applied to the ends of beams help guard against moisture penetration and excessive end grain checking. A coat of sealant should be field applied to the ends of beams (see Fig. 4.34) if they are trimmed to length, dapped, or otherwise cut after leaving the mill.

Surface sealants, which can be applied to the top, bottom, and sides of beams, resist dirt and moisture intrusion and help control checking and grain raising. Specify a paintable penetrating sealant if beams are to be stained or given a natural finish. A primer coat also helps protect beams from moisture and soiling and provides a paintable surface for subsequent finishing, if specified. See Chapter 9 for a further discussion of paints and stains.

Water-resistant wrappings are often specified to protect beams from moisture, soiling, and surface scratches during transit and job-site storage. Because exposure to sunlight can discolor beams, opaque wrappings are recommended. Beams can be wrapped individually, by bundle, or by load tarping. In applications where appearance is especially important, individual wrapping should be left intact until

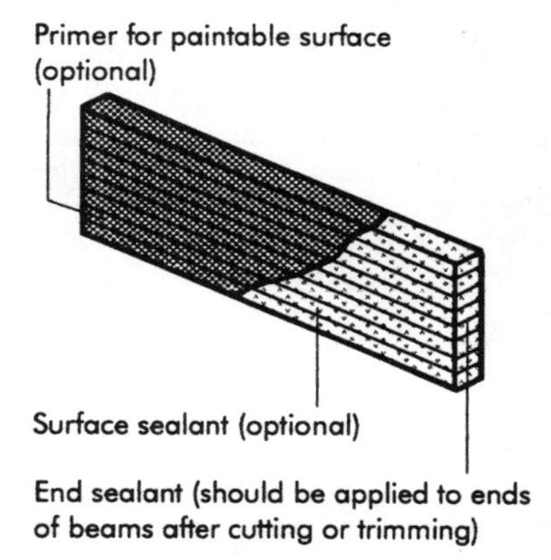

FIGURE 4.34 Glulam sealant at the end of the beam.

installation to minimize exposure to job-site conditions. Individual wrapping should be completely removed during the erection process to minimize uneven surface bleaching due to exposure to sunlight.

4.8.2 Loading and Unloading

Glued laminated timber beams are commonly loaded and unloaded with forklifts. For greater stability and handling safety, place the sides of the beams, rather than the bottoms, flat on the forks (see Fig. 4.35). Carrying extremely long beams on their sides, however, can cause them to flex excessively. To control flex in these cases, use two or more forklifts, lifting in unison. If a crane with slings or chokers is used to load or unload beams, provide adequate blocking at all beam edges between the sling and the members to protect corners and edges. Only fabric slings should be used to lift glued laminated timber members. Using spreader bars can reduce the likelihood of damage when lifting long beams.

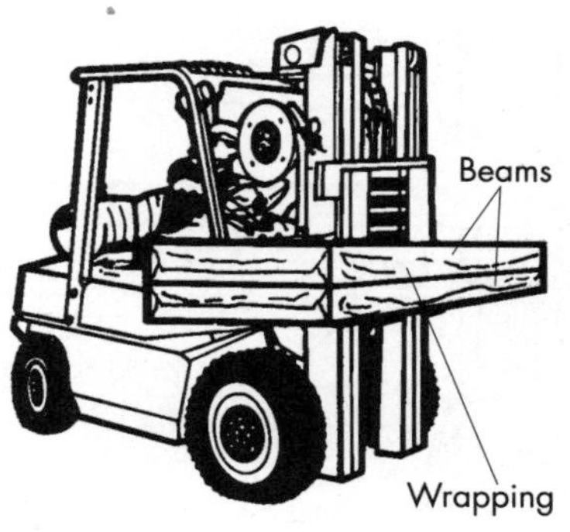

FIGURE 4.35 Glulam handling.

4.8.3 Storage in the Yard

A level, well-drained covered storage site is recommended. Keep beams off the ground using lumber blocking, skids or rack systems as shown in Figs. 4.36 and 4.37. If the beams are wrapped, the wrapping should be left in place to protect them from moisture, soiling, sunlight, and scratches. For long-term storage, cut slits in the bottom of the wrapping to allow ventilation and drainage of any entrapped moisture (see Fig. 4.36). Proper ventilation and drainage will reduce the likelihood of water damage, staining and the start of decay.

4.8.4 Transportation

Stack beams on lumber blocking or skids when loading them on rail cars or trucks. Beams can rest on their sides or bottoms. Secure the load with straps to keep it from shifting. Protect beam edges with softeners or wood blocking when strapping down the load.

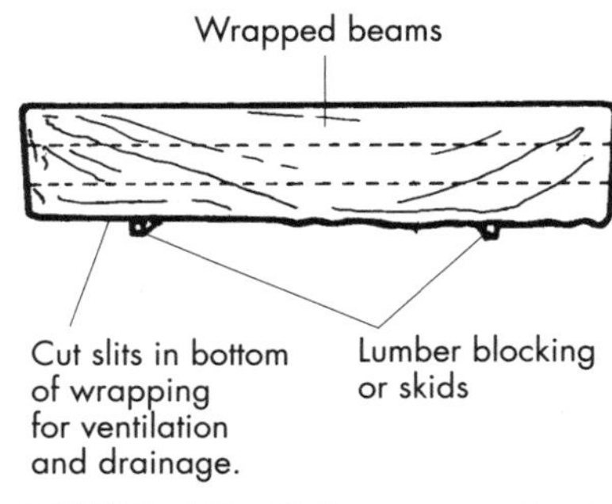

FIGURE 4.36 Glulam storage in the yard.

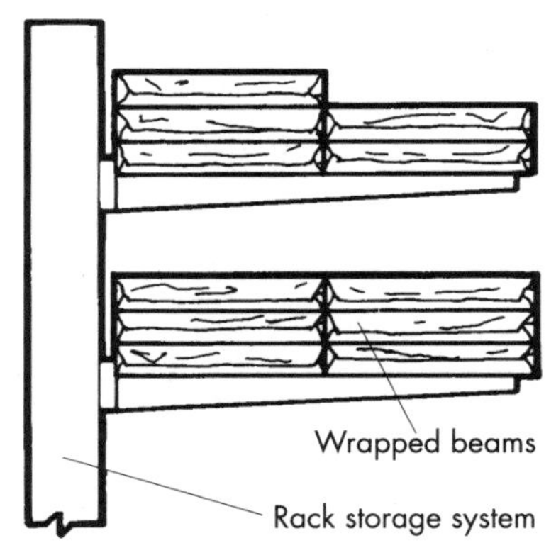

FIGURE 4.37 Glulam storage on a rack.

4.8.5 Storage at the Job Site

If possible, store glued laminated timber under cover to protect the beams from moisture, soiling, and sunlight. Place the beams on spaced lumber blocking on level, well-drained ground (see Fig. 4.36). Seal ends of beams immediately after trimming or cutting. Once beams are installed, allow them to season gradually and adjust to the temperature and moisture conditions of the structure. Do not expose glued laminated timber members to rapid changes in moisture and temperature such as may occur from temporary heating units. Such exposure may result in excessive surface checking.

4.9 SPECIFYING GLUED LAMINATED TIMBER

4.9.1 Common Lay-up Combinations

While the use of a wide variety of species and grades results in optimizing the use of the lumber resource by the manufacturer, the multiplicity of layup combinations as tabulated in Tables 4.2 and 4.3 can be confusing to the design professional.

To simplify the selection process, the lay-up combinations typically available have been highlighted in Tables 4.2 and 4.3. For bending members, these are 24F-V4 (unbalanced) and 24F-V8 (balanced) combinations for Douglas fir, 24F-V3 (unbalanced) and 24F-V5 (balanced) combinations for southern pine, and the 24F-1.8E lay-up (unbalanced), which permits the use of several different species.

By selecting one of these highlighted combinations, the designer will be identifying glued laminated timber products that have sufficiently high design properties to satisfy most design situations and are typically available in most major market areas in the United States. Other lay-up combinations are available on a regional basis, and the designer should verify availability of any combination for a given geographic area by contacting local suppliers or the glued laminated timber manufacturers.

4.9.2 Specific End-Use Lay-up Combinations

It is important to note that certain lay-up combinations in Tables 4.2 and 4.3 have been developed for specific end-use applications. Several examples of these are as follows:

- The *20F-V12* (unbalanced) and *20F-V13* (balanced) combinations use Alaska yellow cedar. These are intended for applications exposed to the elements or high-humidity conditions where the use of the heartwood of a naturally durable species is preferred instead of a pressure preservative-treated glued laminated timber.
- The *24F-1.8E* lay-up is a general-purpose lay-up combination intended primarily for stock beams used in residential construction. This lay-up permits the use of a variety of species and is suitable for virtually any simple-span beam application.
- The *26F-E/DF1* and *26F-E/DF1M1* combinations were developed for use in combination with prefabricated wood I-joists and are often referred to as I-joist depth compatible (IJC) lay-ups.

- The *28F-E/SP1, 30F-E/SP1, 30F-E2M2/SP,* and *30F-E2M3/SP* lay-ups are southern pine combinations that are intended for applications where high bending stress and MOE properties are specified.

4.9.3 Specifying by Stresses

When the designer or end user is uncertain as to the availability or applicability of a specific lay-up combination, the most efficient way to specify glued laminated timber is to provide the manufacturer or supplier with the required stresses to satisfy a given design. For example, assume a simple-span beam design requires the following allowable stresses to carry the in-service design loads:

$$F_b = 2250 \text{ psi}$$

$$F_v = 150 \text{ psi}$$

$$F_{c\perp} = 500 \text{ psi}$$

$$\text{MOE} = 1.65 \times 10^6 \text{ psi}$$

If the designer provides the manufacturer or supplier with these required stresses, a number of lay-up combinations satisfying these stress requirements could then be supplied, depending on availability. This will often result in the lowest-cost option being supplied while still satisfying all design requirements. Note that the specific gravity of the wood species affects the connection strength, which in turn may govern the diaphragm design. Therefore, it may be necessary to specify the minimum specific gravity of wood species when the connection strength of the structure is critical.

4.9.4 Member Sizes

In addition to specifying the allowable design stresses, it is also necessary to specify the size of member required. While glued laminated timber can be manufactured in virtually any cross-sectional size and length required, it is important to understand that since glued laminated timber is manufactured using dimension lumber, certain widths and depths become de facto standards which should be specified whenever possible. Table 4.39 provides typical net finished widths for glued laminated timber.

The depths of glued laminated timber are typically specified in multiples of 1½ in. for western species and 1⅜ in. for southern pine. Thus, a 10-lamination member

TABLE 4.39 Typical Net Finished Glulam Widths

Nominal width	3	4*	6*	8	10	12
Western species	2½	3⅛	5⅛	6¾	8¾	10¾
Southern pine	2½	3	5	6¾	8½	10½

*For the 4 in. and 6 in. nominal widths, glulam may also be available in 3½ in. and 5½ in. widths respectively. These "full-width" members correspond to the dimensions of 2 × 4 and 2 × 6 framing lumber and are supplied with "hit or miss" surfacing which is only acceptable for concealed applications. For additional information on the appearance characteristics of glulam, see EWS, Technical Note Y110, *Glued Laminated Timber Appearance Classifications for Construction Applications.*

using western species will have a net depth of 15 in., while a 10-lamination southern pine member will have a net depth of 13¾ in. Other thicknesses of laminations may be specified but these will require a custom order. An example would be the use of ¾ in. thick laminations to produce members with a tight radius of curvature such as occurs in most arch members.

When used in conjunction with I-joists, glued laminated timber may be supplied in I-joist compatible (IJC) depths. For residential construction, these are 9½, 11⅞, 14, and 16 in. Section properties for these depths are shown in Tables 4.23 and 4.29 for 3½ and 5½ in. net widths.

4.9.5 Glulam Specification Guide

To specify a glulam member properly, use the Specification Guide given in this section. While all possible design considerations cannot be covered by a general specification of this type, most of the common specification concerns are incorporated.

Typical Glued Laminated Timber Specification

General

1. Structural glued laminated timber shall be furnished as shown on the plans and in accordance with the following specifications.
2. Shop drawings and details shall be furnished and approved before fabrication is commenced.
3. The (manufacturer) (seller) (general contractor) shall furnish connection steel and hardware for joining structural glued laminated timber members to each other and to their supports, exclusive of anchorage embedded in masonry or concrete, setting plates, and items field-welded to structural steel. Steel connections shall be finished with one coat of rust-inhibiting paint.

Manufacture

1. Materials, manufacture, and quality assurance. Structural glued laminated timber shall be in conformance with ANSI Standard A190.1, *American National Standard for Structural Glued Laminated Timber,* or other code-approved design, manufacturing, and/or quality assurance procedures.
2. End-use application. Structural glued laminated timber members shall be manufactured for the following structural uses as applicable:

   Simple-span bending member—B

   Continuous or Cantilevered span bending member—CB

   Compression member—C

   Tension member—T

3A. Design values. Structural glued laminated timber shall provide design values as shown for normal load duration and dry-use condition.* See Table 4.40.

* Dry service condition—moisture content of the member will be below 16% in service; wet service condition—moisture content of the member will be at or above 16% in service. When structural glued laminated timber members are to be preservative treated, wet-use adhesives must be specified.

TABLE 4.40 Specific Required Design Stresses

Application	Structural use (identification symbol)	Design stress	psi
Bending member loaded perpendicular to wide face of the lamination	Simple-span member (*B*)	Bending, F_b	______
	Continuous or cantilevered-span member (*CB*)	Horizontal shear, F_v	______
		Compression (perpendicular to grain), $F_{c\perp}$	______
		Top lamination	
		Bottom lamination	______
		Modulus of elasticity, *E*	______
Concentric[a] Axially Loaded Member	Tension member (*T*)	Tension parallel to grain, F_t	______
		Modulus of elasticity, *E*	______
	Compression member (*C*)	End grain bearing	______
		Compression (parallel to grain), F_c	______
		Modulus of elasticity, *E*	______

[a] For members subjected to combined bending and axial loading, additional design values for all other applicable stresses such as but not limited to Fb_x, Fb_y, Fv_x, Fv_y must also be specified.

3B. Lamination combination number. An alternative to specifying the required design stresses is to specify a specific laminating combination symbol if known.**

4. Appearance classification. Members shall be (framing) (industrial) (architectural) (premium) classification in accordance with industry recommendations.†

5. Laminating adhesives. Adhesives used in the manufacture of structural glued laminated timber shall meet requirements for (wet-use) (dry-use) service conditions.*

6. Camber (when applicable). Structural glued laminated timber (shall) (shall not) be manufactured with a built-in camber. Camber radius shall be (1600) (2000) (3500) or (other) feet or a specific amount of camber may be specified in inches.

7. Preservative treatment (when applicable). Members shall be pressure-treated in accordance with American Wood Preservers Association (AWPA) Standard C28 with (creosote, or creosote/coal tar solution) (pentachlorophenol in oil) (pentachlorophenol in light solvent) (waterborne) preservatives as required for (soil contact) (above ground) exposure.‡

** Laminating combination should be based on design requirements and section capacities published in Engineered Wood Systems, AITC, or manufacturer's brochures. National Evaluation Report 486 and ICBO ES Report ER-5714 provide a tabulation of laminating combinations available from EWS member manufacturers.

† Appearance classifications are described in Section 4.6.1.

‡ When pentachlorophenol in light solvent or waterborne preservative treatments are specified for protection against decay or insect attack, individual laminations usually are treated prior to manufacturing structural glued laminated timber members. These treatments are not available from all manufacturers, and the designer should verify availability prior to specification. Where paintable surfaces are required, specify pentachlorophenol in light solvent or a waterborne preservative. Wood treated with creosote, creosote/coal tar solution, or pentachlorophenol in oil should not be used in contact with materials subject to staining.

8. Fire resistance (when applicable). Members shall be sized and manufactured for one-hour fire resistance.§

9. Protective Sealers and Finishes. Unless otherwise specified, sealer shall be applied to the ends of all members. Surfaces of members shall be (not sealed) (sealed with primer/sealer coating) (other).¶

10. Trademarks. Members shall be marked with the APA EWS or AITC trademark indicating conformance with the manufacturing, quality assurance and marking provisions of ANSI Standard A190.1

11. Certificates (when applicable). A certificate of conformance may be provided by the (manufacturer) (seller) to indicate conformance with ANSI Standard A190.1.

12. Protection for shipment. Members shall be (not wrapped) (load wrapped) (bundle wrapped) (individually wrapped) with a water-resistant covering for shipment.

As indicated under item 3A of manufacture, design stresses may be specified. Table 4.40 is a guide for specifying design stresses for structural glued laminated timber used for bending members such as purlins, beams, or girders or for axially loaded members such as columns or truss chords.

4.10 REFERENCES

1. ANSI A190.1, *American National Standard for Wood Products—Structural Glued Laminated Timber,* American National Standards Institute, New York, 1992.
2. ASTM D3737, *Standard Test Method for Establishing Stresses for Structural Glued Laminated Timber,* in *Annual Book of ASTM Standards,* American Society for Testing and Materials, Philadelphia, 1998.
3. APA EWS Technical Note S580, *Preservative Treatment of Glulam Beams,* APA—The Engineered Wood Association, Tacoma, WA, 1995.
4. APA EWS Technical Note Y260, *Dimensional Changes in Structural Glued Laminated Timber,* APA—The Engineered Wood Association, Tacoma, WA, 1998.
5. APA EWS Technical Note Y110, *Glued Laminated Timber Appearance Classifications for Construction Applications,* APA—The Engineered Wood Association, Tacoma, WA, 1998.
6. *National Design Specification for Wood Construction,* American Forest and Paper Association, Washington, DC, 1997.
7. Hooley, R. F. and Madsen, B. *Lateral Stability of Glued Laminated Beams,* Journal of the Structural Division, Proceedings of the American Society of Civil Engineering, vol. 90, No. ST3, June, 1964.
8. Hooley and Duval, *Lateral Buckling of Simply Supported Glued Laminated Beams.*

§When structural glued laminated timber with one-hour fire resistance is specified, minimum size limitations and additional lamination requirements are applicable. Supporting steel connectors and fasteners also must be protected to achieve a one-hour fire rating. Cover connectors or fasteners with fire-rated (type X) gypsum wallboard or sheathing, or 1½ in. wood, to provide the needed protection.

¶Specify a penetrating sealer when the finish will be natural or when a semitransparent stain is to be used. Primer/sealer coatings have a higher solids content and provide greater moisture protection and are suitable for use with opaque or solid-color finishes.

9. Cowan, W. C., *Shear Stress in Two Wood Beams over Wood Block Supports,* Report No. 2249, Forest Products Laboratory, USDA, Madison, WI, 1962.
10. APA EWS Technical Note R465, *Checking in Glued Laminated Timber,* APA—The Engineered Wood Association, Tacoma, WA, 1999.
11. Faherty, K. F., and T. G. Williamson, *Wood Engineering and Construction Handbook,* 3d ed., McGraw-Hill, New York, 1999.
12. APA EWS Data File Y24A, *Design of Structural Glued Laminated Timber Columns,* APA—The Engineered Wood Association, Tacoma, WA, 1999.
13. APA EWS Technical Note S570, *Substitution of Glulam Beams for Steel or Solid-Sawn Lumber,* APA—The Engineered Wood Association, Tacoma, WA, 1999.

CHAPTER FIVE
PREFABRICATED WOOD I-JOISTS AND ENGINEERED RIM BOARD

Edward Keith, P.E.
Senior Engineer, TSD

5.1 INTRODUCTION

While relatively new to the construction industry, when compared to products such as lumber, plywood, or glued laminated timber, both prefabricated wood I-joists and engineered rim board products are rapidly becoming the products of choice by quality- and environmentally-conscious builders alike. Both of these engineered wood products are discussed in detail in this chapter, starting with I-joists and then engineered rim board in Section 5.12.

5.1.1 The Development of Prefabricated Wood I-joists and Rim Board

Originally commercialized by the Trus Joist Corporation (now a Weyerhaeuser Company) in the 1960s, engineered wood I-joists owe their beginning, at least in part, to a publication developed by the Douglas Fir Plywood Association (a precursor to APA—The Engineered Wood Association) in 1959 entitled DFPA Specification BB-8, *Design of Plywood Beams.*[1] This specification, later published as Plywood Design Specification Supplement 2, *Design and Fabrication of Glued Plywood-Lumber Beams,*[2] outlined the original design procedures that ultimately provided the basis for current design recommendations.

The first universally recognized standard for wood I-joists was ASTM D5055, *Standard Specification for Establishing and Monitoring Structural Capacities of Prefabricated Wood I-Joists.*[3] This consensus standard provides guidelines for the evaluation of mechanical properties, physical properties, and quality of wood I-joists and is the current common testing standard for I-joists. However, since ASTM D5055 does not specify required levels of performance, individual manufacturers of I-joists generally have their own proprietary company standards that govern the everyday production practice for their products. The common sizes and design properties for I-joists are dictated by the market and the major I-joist manufacturers have similar product offerings.

Under the current building code rules and procedures, I-joist manufacturers can gain code recognition through evaluation reports provided by various code agencies, such as ICBO, BOCA, and SBCCI. For example, ICBO Evaluation Services AC14,[4] Acceptance Criteria for Prefabricated Wood I-Joists, provides guidelines on implementing performance features of the Uniform Building Code (UBC). Although ASTM D5055 and ICBO AC14 provide guidance for developing proprietary design values, no standard performance levels or grades are presented in the documents.

The lack of a code-recognized standard that recognizes performance levels has resulted in individual manufacturing standards being promulgated by the manufacturers covering their respective proprietary I-joist products. Each standard has the potential for having differing installation details, allowable spans, web penetration requirements, allowable stresses, etc. Some manufacturers feel that this lack of standardization has slowed the rate of acceptance of these products within the market place. As the history of other building materials such as plywood and oriented strand board (OSB) has shown, some degree of standardization of the industry is inevitable. Also inevitable is that along with standardization will come greater manufacturing efficiencies and greater use in construction.

To fill this need for standard performance levels, APA, in conjunction with several I-joist manufacturers, is developing performance-based standards for performance-rated wood I-joist products. The first such APA performance standard is for the use of wood I-joists in residential floors, designated as PRI-400.[5] It should be noted that this is a voluntary standard and not all I-joist manufacturers have chosen to produce PRI-400 products.

Since APA has promulgated the PRI-400 standard, much of the information presented in this chapter will be based on this standard. However, this does not preclude the use of any wood I-joists to achieve similar intended functions. Also manufacturers who do manufacture the PRI-400 products may also manufacture a proprietary series of I-joists for which they have obtained individual ES or NES code reports.

While clearly a good idea from the start for a number of structural reasons that will be discussed in detail in this chapter, the increased use of engineered wood products such as wood I-joists and engineered wood rim board in construction will have a positive impact on the environment, from the standpoint of reducing demand for products from older-growth forests. Historically, residential floors have been framed with 2 × 10s (38 × 235 mm) and 2 × 12s (38 × 305 mm). These sizes had to be milled from trees that were at least 18 in. (460 mm) in diameter, necessitating the use of older-growth trees. Engineered wood I-joists and rim board products are both made out of a number of engineered wood components, all currently being made economically out of second- and third generation plantation forests. No longer requiring the log sizes only found in older forests, these engineered wood products also permit the use of fast-growing species for which there was no commercial value just a few years ago. Engineered wood products such as I-joists and rim board have led the way in the early green building movement. Figure 5.1 illustrates the use of wood I-joists and structural wood panel rim board.

5.2 PREFABRICATED WOOD I-JOISTS

5.2.1 Growth of the Industry

Figure 5.2 shows market trends for North American (United States and Canada) I-joists from 1980 through 2000. The source of the data can be found in the legend

FIGURE 5.1 Engineered wood products used in a residential floor system.

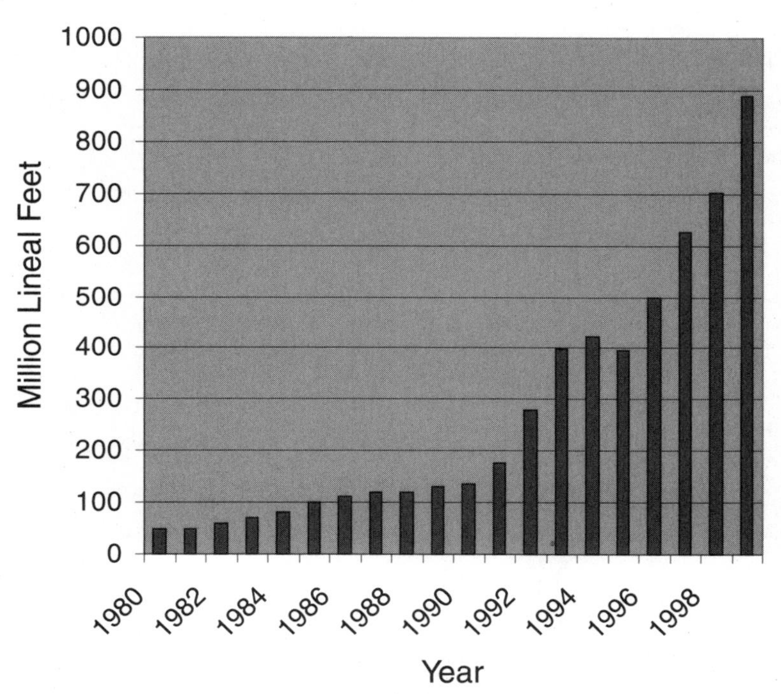

FIGURE 5.2 Engineered wood I-joists market growth. *(Market information based on APA.)*[6,7]

to the graph.[6,7] The residential market has been the driver in the United States and Canada, accounting for about 90% of the volume increase in the last 4 years. Remodeling and nonresidential construction uses are also increasing, and these markets will provide for even more market volume growth in the future. As shown by Fig. 5.2, the total U.S. and Canadian I-joist production was approximately 890 million lineal feet (271 million lineal meters) in 2000.

As engineered wood I-joists only represent about ⅓ of the raised floor joist market in single- and multifamily residential construction and only a negligible amount of the wood roof framing market, it can be seen that tremendous domestic market potential remains to be tapped.

5.2.2 What Is an APA Performance Rated I-Joist?

The APA Performance Rated I-joist (PRI) is an I-shaped engineered wood structural member designed for use in residential floor and roof construction (see Fig. 5.3). PRI I-joist products are manufactured under the rigorous quality assurance standards of APA—The Engineered Wood Association. Other I-joist products manufactured in accordance with other proprietary code evaluation reports may fulfill the same purpose.

Performance Rated I-joists are identified by their net depth followed by a designation such as "PRI-30" that relates to the joist design properties. These designations will be covered later in detail. In order to be classified as a PRI, the joist is limited to a $L/480$ live load maximum deflection (where L = span) for glued-nailed residential floor applications, a criterion that provides superior floor performance.

PRIs are manufactured to strict tolerances with the following characteristics:

- Flanges are either sawn lumber or structural composite lumber, typically LVL. The top flange is of the same type and grade of material as the bottom flange. The net flange size depends on the material used.

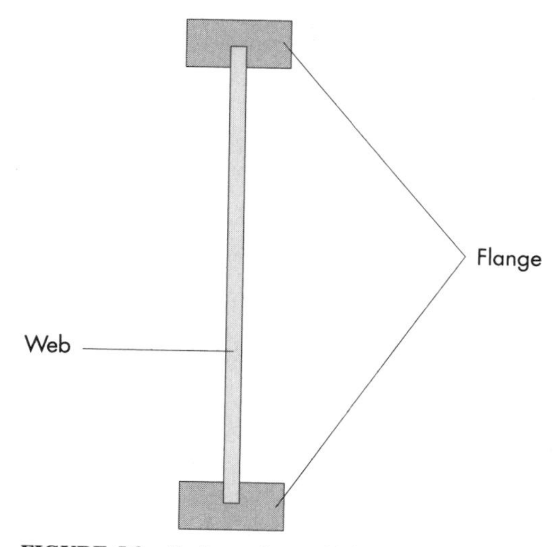

FIGURE 5.3 Engineered wood I-joist.

- Webs consist of wood structural panels, which can be plywood or OSB. All panels are classified as Exposure 1 or Exterior and are typically ⅜ in. (9.5 mm) in thickness.
- All PRIs are assembled using exterior-type adhesives per ASTM D2559.[8]
- APA PRIs are available in four depths as shown in Table 5.1.
- While PRIs of the same depth may be manufactured with various flange widths depending on the product designator, flange width is an important design consideration when specifying hangers. Unless the designer is very specific on his plans, he may not know what the actual width of the I-joist installed will be. Often designers and builders will insist that the I-joist supplier provide the hangers as well.
- Most mills supply I-joists to distributors and dealers in lengths up to 60 ft (18.30 m). These are then cut to frequently used lengths such as 16–36 ft (4.90–11 m). Check local supplier for availability.

It should be noted that many manufacturers produce I-joists for the commercial building market that are beyond the scope of APA Performance Rated I-joists. These I-joists are typically manufactured in depths of 18–30 in. or deeper in 2 in. depth

TABLE 5.1 Designations for APA Performance Rated I-Joists

Net depth	Joist designation
9½ in.	PRI-20
	PRI-30
	PRI-40
	PRI-50
	PRI-60
11⅞ in.	PRI-20
	PRI-30
	PRI-40
	PRI-50
	PRI-60
	PRI-70
	PRI-80
	PRI-90
14 in.	PRI-40
	PRI-50
	PRI-60
	PRI-70
	PRI-80
	PRI-90
16 in.	PRI-40
	PRI-50
	PRI-60
	PRI-70
	PRI-80
	PRI-90

For SI: 1 in. = 25.4 mm.

increments. Since no industry standard exists for these products, manufacturers must obtain building code Evaluation Service Reports prior to marketing. However, while these products are deeper than the PRIs, many of the design and construction philosophies presented in this chapter for PRIs are also applicable to other I-joists.

5.2.3 I-Joist Manufacturing Process

Wood I-joists are manufactured out of a number of different flange and web materials. As such, the manufacturing processes vary slightly to accommodate the material differences. In general, however, I-joists are manufactured in one of two basic methods: in fixed lengths or in continuous lines.

The fixed-length method gets its name from the fact that the flange stock—usually LVL for this method—arrives at the assembly point in finite lengths, usually around 60–65 ft (18.3–19.9 m) long. A wedge-shaped groove is machined into the flange material. The geometry of this groove is essential to the manufacturing process because its wedge shape provides the clamping pressure at the web-to-flange joint that is required by the adhesive to provide a good glue bond. The adhesives used are fully waterproof and are required by the standard to meet the requirements of ASTM D2559. After the flanges are initially pressed on to the web, the completed I-joists are carefully moved to the adhesive curing station of the manufacturing process. Radio frequency, microwave, or simply storing in a hot environment are three of the curing methods used, depending on the adhesive system used and physical plant layout. Once the adhesive cure has been accomplished, the joists are trademarked, bundled together and wrapped for shipment to the distributors (see Fig. 5.4).

The continuous-line method is most often used with sawn lumber flanges. In order to manufacture long-length I-joists, the flanges must be made in equally long lengths. When sawn lumber flanges are used, shorter lengths of lumber are finger-jointed together to form long lengths in a continuous process. Because the end joint is a structural joint, it does not matter where it occurs along the length of the I-joist. After end jointing, the joint moves through a radio frequency or microwave curing station to cure the adhesive in the joint. At the next station in the process, the groove is machined in the continuous flange. With two parallel flange lines operating simultaneously, the output of these lines is joined by the station that applies the adhesive to the grooves and inserts the web elements. These elements are initially pressed together as described previously, and the continuous I-joist is cut to usable lengths—as in the fixed length process, about 60–65 ft (18.3–19.9 m). The final curing of the flange-to-web joints can occur either before or after the cut-to-size operation depending on the curing method used. The I-joists are then trademarked, bundled, and wrapped for shipment to the distributors (see Fig. 5.5).

5.3 MOISTURE PERFORMANCE OF ENGINEERED WOOD VS. LUMBER

All engineered wood composites have many characteristics in common. They are stronger, more dimensionally stable, more homogeneous, better utilize available natural resources and, are typically more builder-friendly than sawn lumber. From a proper design and detailing point of view, there is another common characteristic

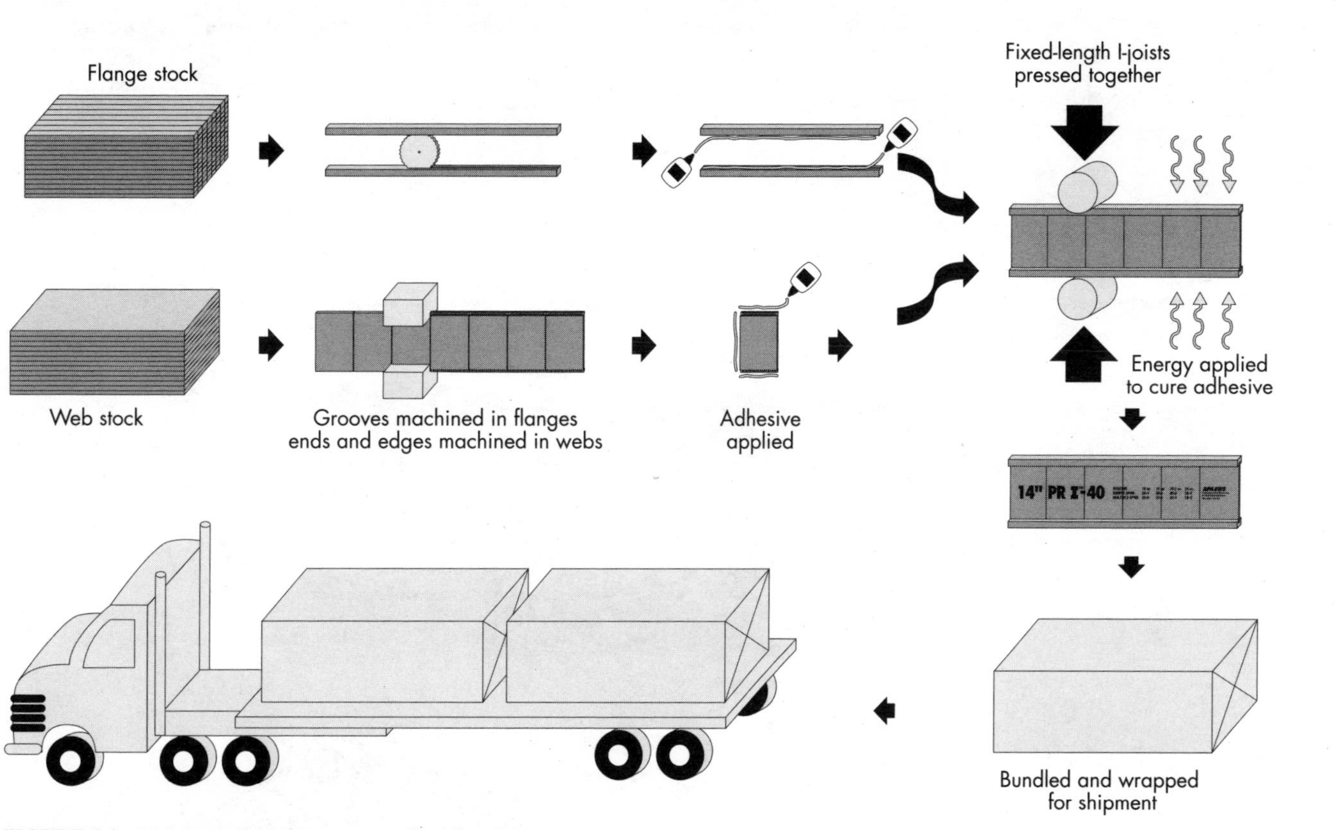

FIGURE 5.4 I-joist manufacturing process—fixed-length process.

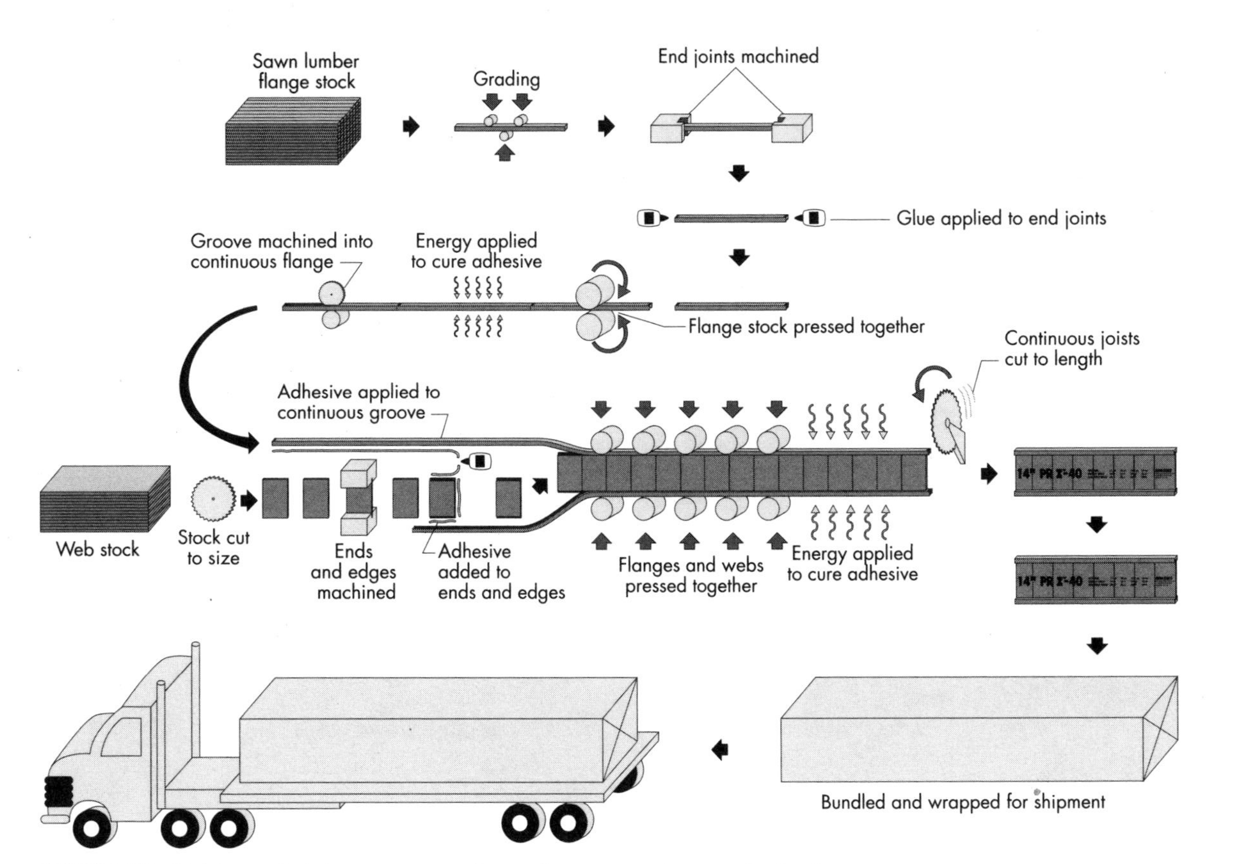

FIGURE 5.5 I-joist manufacturing process—continuous-line process using sawn lumber flanges.

that it is important to understand. This characteristic is that all engineered wood components are manufactured in a relatively dry state. The moisture content of engineered wood products at the time of manufacture ranges from approximately 4–12%. During the manufacturing process, the wood-based resource must be dried to these levels to ensure that a good glue bond is developed. A range of values is given because some adhesive systems used in some products have different moisture requirements.

It is also important to realize that these are not average moisture contents as traditionally measured. If a certain adhesive system requires a maximum 6% moisture content to develop an adequate glue bond, then every piece must meet that maximum during fabrication. A traditional average where 50% are above the maximum and 50% are below just doesn't work. Only those pieces at or below the maximum will ever get to the marketplace.

Traditional dry lumber, on the other hand, is dried to a much higher moisture content, typically 19%, although some lumber is dried to 16%. Because of natural variability, the range of moisture content of the lumber pieces in a given bundle may vary widely. A given lumber element may even have moisture gradients along the length or across the width.

In service, however, such as in a residential structure, after four to eight months of drying, all wood elements will reach an equilibrium moisture content of 6–10%, depending on the season and location of the structure. Because the engineered wood products are very close to this normal equilibrium moisture content as manufactured, and because they are typically shipped in a waterproof protective wrapping, they take on little or no additional moisture during this period. As such, their dimensions vary imperceptibly during this period. The sawn lumber, however, dries down during this period through a relatively large range of moisture content. Along with drying comes an equally significant shrinkage. As Fig. 5.6 shows, a 14 in. (337 mm) deep sawn lumber element can shrink as much as ¾ in. (19 mm) in its depth as it cycles from the as-dried to in-service equilibrium moisture content. This difference in behavior between solid-sawn lumber and engineered wood can lead to structural failure if the designer is not careful.

5.3.1 I-joists and Rim Board Used Together in an Engineered Wood System

APA EWS I-joists and APA EWS Rim Board products (discussed in Section 5.12) are made in 9½, 11⅞, 14, and 16 in. (241, 302, 356, and 406 mm) net depths. It is no accident that these sizes are not compatible with, and are larger than, traditional lumber net depths for 2 × 10s, 2 × 12s, 2 × 14s, and 2 × 16s (38 × 241, 38 × 302, 38 × 356, and 38 × 406 mms). There are many applications in roofing systems and especially residential floors, where other elements are used in conjunction with the I-joists for the express purpose of transferring load through the floor system without overloading the floor joists. Some examples of these other elements are blocking panels over an interior bearing wall and rim or starter joists. In these cases, the vertical load from the structure above the plane of the floor is transferred through the floor into the structure/foundation below by way of direct bearing on the blocking panels and rim or starter joist.

Because the load is transferred in direct bearing, it is essential that the blocking panels and rim or starter joist be the same height as the floor joist. Solid-sawn lumber cannot be used in applications like these because of the very likely potential for shrinkage. Shrinkage by as little as ⅛ in. (3 mm) can be enough to transfer the

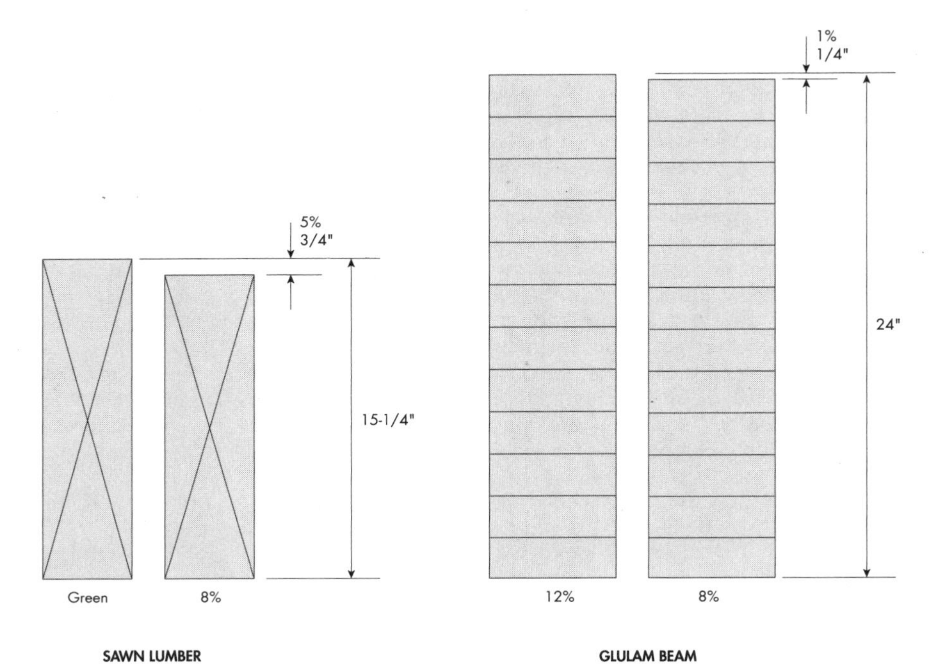

FIGURE 5.6 Shrinkage of glulam compared with sawn lumber.

vertical loads from the walls above directly to the floor joists, thus inducing possible bearing or reaction overload conditions at these locations. The solution to the problem is to use engineered wood products for these applications. They are manufactured in the correct depths and have the same dimensional stability properties (see Fig. 5.7).

While the previous discussion concerns vertical loads, the same is true of lateral loads such as those caused by wind and seismic events. The small gap between the floor sheathing above and the sawn lumber rim joist or blocking panel below re-

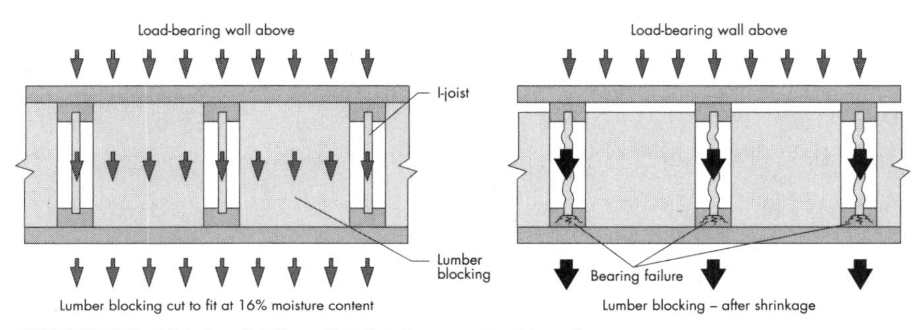

FIGURE 5.7 Effects of differential shrinkage on load transfer.

sulting from shrinkage of the lumber members can have a small but negative impact on the performance of the structure during the design event. Even greater however, will be the impact on the deformation of the structure caused by the potential slip at this location under design lateral loads. While not necessarily life threatening, these greater deformations can result in increased damage to the nonstructural components of the building, such as drywall, windows and doors, cabinets, and interior and exterior finishes. If the deformations are excessive, they can cause the structure to be irreparable.

Every application where solid-sawn lumber is used in conjunction with engineered wood must be looked at very carefully with respect to the different moisture states of materials at the time of construction. The safest alternative is to not mix engineered wood with solid-sawn lumber in any situation where load sharing might be an issue.

5.4 ENGINEERED WOOD I-JOIST STANDARD

APA—The Engineered Wood Association, formerly the American Plywood Association, and before that the Douglas Fir Plywood Association, has long been a leader in the development of engineered wood standards. From the development of the now nationally recognized plywood standard in the 1950s (see Chapter 2) to the new standard discussed in Section 5.4.1, APA has developed, or has been instrumental in the development of most of the glued engineered wood standards in use in the United States today. The APA PRI-I-joist Standard discussed below, is one of the most recent standards developed for the engineered wood industry.

5.4.1 I-Joist Standard

The APA I-joist trademark signifies that the I-joists are manufactured to strict quality standards by a member of Engineered Wood Systems (EWS), a related corporation of APA. PRIs are manufactured in conformance with PRI-400, *Performance Standard for APA EWS I-Joists.* A copy of this standard can obtained by contacting APA—The Engineered Wood Association. Member manufacturers are committed to a rigorous program of quality verification and testing, which ensures predictable product performance, regardless of the manufacturer.

PRI-400 brings product standardization while providing for a multitude of design and construction situations. The standard provides design information for numerous types and sizes of I-joists. Specifiers, designers, and builders can select and use I-joists from various APA EWS member manufacturers, using just one set of design and installation criteria. Because PRIs can be selected based on their allowable span for glued uniformly loaded residential floors, it is easy to incorporate them into any design. Because all APA EWS I-joists share a common set of installation and fastening details, the generation of design drawings is greatly simplified.

5.4.2 Identifying I-Joists

All three of the major model building codes as well as the new International Building Code require engineered wood products to be trademarked by an approved third-party inspection agency before they can be used in code-conforming construction. One purpose of these regulations is to allow for easy identification of the product. The information on these stamps is beneficial to the engineer, contractor, and building official and ultimately to the consumer. While the format of the trademark stamp may vary depending on the quality assurance agency representing the mill, they all contain the same general information. An example of a trademark stamp and an explanation of the information it contains are shown below.

I-Joist Trademarking and Identification. The I-joist trademark stamp shown in Fig. 5.8 and illustrated in detail in Fig. 5.9 below contains information valuable to the specifier, builder, building inspector, and future remodelers. In addition to the actual depth of the joist, the trademark contains information on the joists' specific designation, and may contain the allowable spans for both simple and multiple conditions for I-joist spacings of 12, 16, 19.2, and 24 in. (305, 406, 488, and 610 mm) on center (o.c.).

5.5 I-JOISTS DESIGN PROPERTIES

For those applications not covered specifically by this Handbook, Table 5.2 is provided. Table 5.2 contains the allowable design capacities for APA EWS performance rated I-joists. Similar design properties are also available from manufacturers for non-PRI-400 products. The values listed are single-joist, normal-duration design values. Duration of load adjustments may be applied to all of the values in the table

FIGURE 5.8 Field installation of I-joist and rim board showing APA stamps.

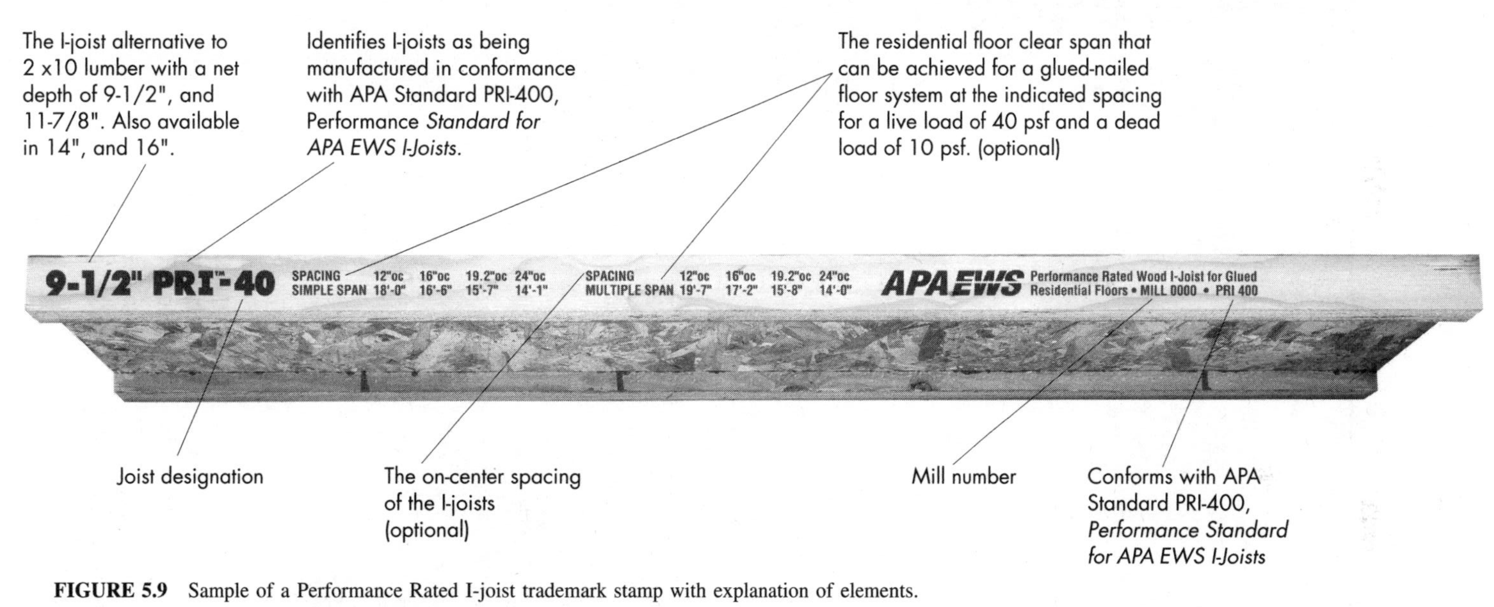

FIGURE 5.9 Sample of a Performance Rated I-joist trademark stamp with explanation of elements.

TABLE 5.2 Design Capacities for APA EWS Performance Rated I-Joists[a]

Depth	Joist designation	EI[b] 10^6 lbf-in.^{2}	M[c] Nonrepetitive lbf-ft	M[c] Repetitive lbf-ft	V[d] lbf	IR[e] lbf	ER[f] lbf	K[g] 10^6 lbf
9½″	PRI-20	145	2180	2265	1120	1700	830	4.94
	PRI-30	161	2800	2910	1120	1905	945	4.94
	PRI-40	193	2355	2520	1120	2160	1080	4.94
	PRI-50	186	3290	3420	1120	2040	1015	4.94
	PRI-60	231	3245	3470	1120	2160	1080	4.94
11⅞″	PRI-20	253	2910	3025	1420	1700	830	6.18
	PRI-30	280	3715	3860	1420	1905	945	6.18
	PRI-40	330	3145	3365	1420	2500	1200	6.18
	PRI-50	322	4375	4550	1420	2040	1015	6.18
	PRI-60	396	4335	4635	1420	2500	1200	6.18
	PRI-70	420	5600	5820	1420	2335	1160	6.18
	PRI-80	547	6130	6555	1420	2760	1280	6.18
	PRI-90	604	7770	8080	1925	3355	1400	6.18
14″	PRI-40	482	3860	4130	1710	2500	1200	7.28
	PRI-50	480	5350	5560	1710	2040	1015	7.28
	PRI-60	584	5320	5690	1710	2500	1200	7.28
	PRI-70	613	7120	7405	1710	2335	1160	7.28
	PRI-80	802	7525	8050	1710	3020	1280	7.28
	PRI-90	881	9535	9915	2125	3355	1400	7.28
16″	PRI-40	657	4535	4850	1970	2500	1200	8.32
	PRI-50	663	6270	6520	1970	2040	1015	8.32
	PRI-60	799	6250	6685	1970	2500	1200	8.32
	PRI-70	841	8350	8680	1970	2335	1160	8.32
	PRI-80	1092	8845	9460	1970	3020	1280	8.32
	PRI-90	1192	11,205	11,650	2330	3355	1400	8.32

For SI: 1 lbf = 4.45 kN, 1 lbf ft. = 1.356 N m, 1 lbf in.^{2} = 0.00287 N m^2, 1 in. = 25.4 mm.

[a] The tabulated values are design values for normal duration of load. All values, except for EI and K, are permitted to be adjusted for other load durations as permitted by the code for solid sawn lumber.

[b] Bending stiffness (EI) of the I-joist.

[c] Moment capacity (M) of a single I-joist. When I-joists are in contact or spaced not more than 24 in. on-center, are not less than three in number, and are joined by floor, roof, or other load-distributing elements adequate to support the design load, repetitive moment shall be permitted for use in design.

[d] Shear capacity (V) of the I-joist.

[e] Intermediate reaction (IR) of the I-joist with a minimum bearing length of 3½ in. without bearing stiffeners.

[f] End reaction (ER) of the I-joist with a minimum bearing length of 1¾ in. without bearing stiffeners. For a bearing length of 4 in. (5 in. for 14 in. and 16 in. PRI-50 joists), the end reaction may be set equal to the tabulated shear value. Interpolation of the end reaction between 1¾ and 4 in. (5 in. for 14 in. and 16 in. PRI-50s) bearing is permitted. For end-reaction values over 1550 plf, bearing stiffeners are required with the exception of PRI-90, which requires bearing stiffeners when end reaction values exceed 1,885 lbf.

[g] Coefficient of shear deflection (K). For calculating uniform load and center-point load deflections of the I-joist in a simple-span application, use Eqs. (1) and (2).

Uniform load:
$$\delta = \frac{5\omega\ell^4}{384EI} + \frac{\omega\ell^2}{K} \tag{1}$$

Centerpoint load:
$$\delta = \frac{P\ell^3}{48EI} + \frac{2P\ell}{K} \tag{2}$$

where δ = calculated deflection (in.)
P = concentrated load (lbf)
EI = bending stiffness of the I-joist (lbf-in.^{2})
K = coefficient of shear deflection (lbf)
ω = uniform load (lbf/in.)
ℓ = design span (in.)

except the *EI* value and the *K* value. Values for moment (*M*) capacity in the table are given for both repetitive and nonrepetitive applications.

There are a several significant differences in designing for I-joists compared to designing for solid-sawn lumber. The first is in calculating deflection. In calculating deflection of solid-sawn lumber beams, the deflection due to shear deformation is minimal and is accounted for by adjusting the published design MOE values to include shear deformation effects. With I-joists, however, the deflection due to shear can be a significant contributor to the overall deflection of the member and, as such, must be calculated separately. Footnote (*g*) of Table 5.2 provides the equations necessary to compute the total deflection for uniformly loaded or center point-loaded single spans.

The second significant difference in designing for I-joists vs. solid sawn lumber joists is related to the necessity to consider the intermediate and end reactions in I-joist design. Because of the narrow cross section of the I-joist web, the web exerts a potential splitting force in the flange of the joist. This is a design consideration foreign to solid lumber design, but may be critical in I-joist design. Notice that the assigned values are different for an intermediate reaction (IR) and an exterior reaction (ER). This is because the intermediate reaction capacity presupposes a minimum bearing length of 3½ in. (89 mm), where an exterior reaction is based on a minimum required bearing length of 1¾ in. (44 mm). If a designer were considering a cantilever situation where the support adjacent to the cantilever was fully supporting the I-joist over an area at least 3½ in. (89 mm) long, the intermediate reaction (IR) value would be appropriate at this location.

5.6 END-USE APPLICATIONS—FLOOR CONSTRUCTION

5.6.1 Residential Floors—Background

All PRIs are identified by an allowable span designated for uniformly loaded residential floor construction at various I-joist spacings. Therefore, the I-joist required for a specific application is easily determined by selecting the span needed and then choosing the I-joist that meets the span, spacing, and loading criteria for that application. Tables 5.4 and 5.5 are provided for simple-span and multiple-span applications, respectively. These two tables are based on normal code required uniform load criteria for residential construction—40 lb/ft^2 (1.9 kN/m^2) for live load and 10 lb/ft^2 (0.48 kN/m^2) for dead load. The APA PRI trademark may contain the spans listed on these two tables.

For those applications where stone or ceramic tile is desired for a finish floor surface, similar tables are provided based on a live load of 40 lb/ft^2 (1.9 kN/m^2) and 20 lb/ft^2 (0.96 kN/m^2) for dead load. Again two tables are provided, Table 5.6 for simple-span applications and Table 5.7 for multiple-span applications. These tables would also be applicable for the case of using a lightweight concrete or gypsum topping.

Discussion of L/480—Deflection Criteria for Floors. The maximum spans listed in Tables 5.4–5.7 are all based on a live-load deflection criteria of $L/480$ (where L = span) for glued-nailed residential floor applications. This deflection criteria, 33% stiffer than the traditional code-recognized deflection criteria of $L/360$, was selected during the development of APA's I-joist performance standard, PRI-400, to provide superior floor performance under live load applications. A detailed discussion of this decision by APA is provided in Section 5.8.7.

The Glued-Nailed Floor System. In order to achieve the allowable spans that are shown in Tables 5.4 through 5.7, floor sheathing must be field glued-nailed to the I-joist flanges as shown in Fig. 5.10. Spans must be reduced by 1 foot (0.3048 meters) when the sheathing is nailed only.

In addition to achieving greater span, by gluing the wood structural panels to the I-joists:

- Floor stiffness is increased appreciably, particularly when tongue-and-groove (T&G) joints in the floor panels are glued.
- The possibility of floor squeaks, excessive deflection, vibration, bounce, and nail-popping is greatly reduced.
- The number of cracks that can leak airborne noise is also reduced, thereby increasing the acoustical performance of the floor-ceiling assembly.

The Glued-Nailed Floor System. Panels recommended for glued floor construction are T&G APA-Rated Sturd-I-Floor for single-floor construction, and APA Rated Sheathing for the subfloor when used with a separate underlayment or with structural finish flooring. An additional layer of underlayment, or veneer-faced Sturd-I-Floor with "sanded face" should be applied in areas to be finished with resilient floor coverings such as tile, linoleum, vinyl or fully adhered carpet. In both

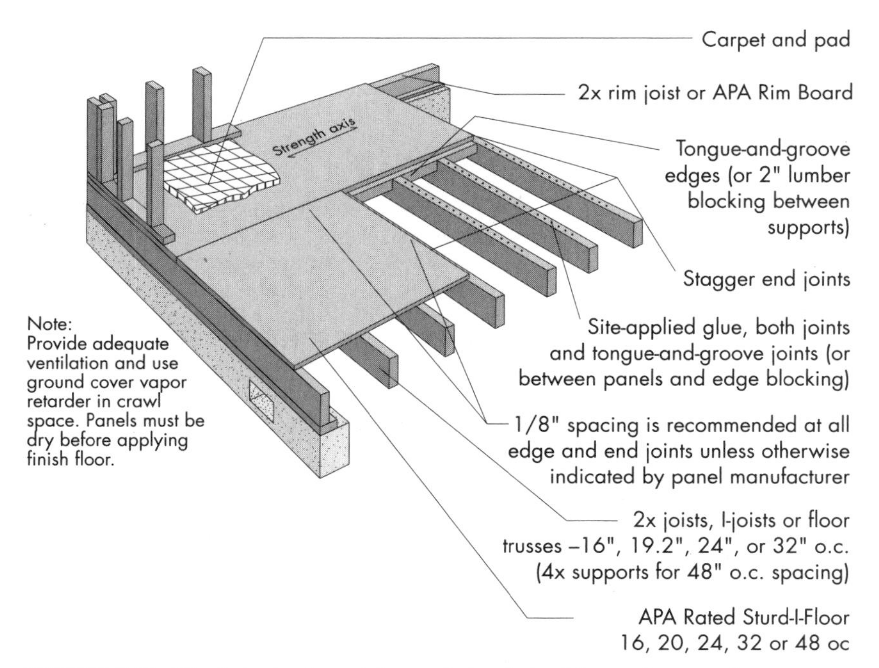

FIGURE 5.10 Wood structural panel floor nailed and glued for maximum performance.

cases, Sturd-I-Floor and subflooring panels should be installed continuous over two or more spans with the long dimension or strength axis across the I-joists. Other non-APA trademarked panels complying with either PS-1[9] or PS-2[10] may also be used.

Tongue-and-groove panels are highly recommended for single-floor construction. Before each panel is placed, a line of glue is applied to the I-joist with a caulking gun. The panel's T&G joint should also be glued, although less heavily to avoid squeeze-out. If square-edge panels are used, edges must be supported between I-joists with 2 × 4 (38 × 89 mm) blocking. Glue panels to blocking to minimize squeaks. Blocking is not required under structural finish flooring, such as wood strip flooring, or if a separate underlayment is installed.

Important Note: Wood structural panels must be glued with an adhesive meeting APA Specification AFG-01[11] *or ASTM D3498*[12] *to achieve the allowable spans shown in Tables 5.4–5.7. If the floor structure is built using OSB panels with sealed surfaces and edges, only solvent-based glues should be used. Always follow the specific application recommendations of the adhesive manufacturer.*

Application. For best results, follow these application procedures:

1. Wipe any mud, dirt, water, or ice from I-joist flanges before gluing.

2. Snap a chalk line across the I-joists 4 ft in from the wall for panel edge alignment and as a boundary for spreading glue.

3. Spread only enough glue to lay one or two panels at a time, or follow specific recommendations from the glue manufacturer.

4. Lay the first panel with tongue side to the wall, and nail in place. This protects the tongue of the next panel from damage when tapped into place with a block and sledgehammer.

5. Apply a continuous line of glue (about ¼ in. [6 mm] diameter) to the top flange of a single I-joist. Apply glue in a serpentine pattern on wide areas, such as with double I-joists.

6. Apply two lines of glue on I-joists where panel ends butt, to ensure proper gluing of each end.

7. After the first row of panels is in place, spread glue in the groove of one or two panels at a time before laying the next row. Glue line may be continuous or spaced, but avoid squeeze-out by applying a thinner line (⅛ in. [3 mm]) than used on I-joist flanges.

8. Tap the second row of panels into place, using a block to protect groove edges.

9. Stagger end joints in each succeeding row of panels. A ⅛ in. (3 mm) space between all end joints and ⅛ in. (6 mm) space at all edges, including T&G edges, is recommended. (Use a spacer tool or an 8d common (3.3 × 64 mm) nail to ensure accurate and consistent spacing.)

10. *Complete all nailing of each panel before glue sets.* Check the manufacturer's recommendations for allowable cure time. (Warm weather accelerates glue setting.) Use 6d ring- or screw-shank (3 × 51 mm) nails for panels ¾ in. (19 mm) thick or less, and 8d ring- or screw-shank (3 × 64 mm) nails for thicker panels. Space nails per Table 5.3. Closer nail spacing may be required by some codes, or for diaphragm construction. The finished deck can be walked on right away and will carry construction loads without damage to the glue bond.

TABLE 5.3 APA Rated Sturd-I-Floor/Single Floor Fastening Schedules for PRIs[a]

Span rating (maximum joist spacing) (in.)	Panel thickness[b] (in.)	Fastening: glued-nailed[c] (in.)		
			Maximum spacing (in.)	
		Nail size and type	Supported panel edges	Intermediate supports
16	23/32[d]	6d ring- or screw-shank[e] (3 × 51 mm)	12	12
20	23/32[d]	6d ring- or screw-shank[e] (3 × 51 mm)	12	12
24	23/32, 3/4	6d ring- or screw-shank[e] (3 × 51 mm)	12	12
	7/8	8d ring- or screw-shank[e] (3 × 64 mm)	6	12

For SI: 1 in. = 25.4 mm.

[a] Special conditions may impose heavy traffic and concentrated loads that require construction in excess of the minimums shown.

[b] Panels in a given thickness may be manufactured in more than one allowable span. Panels with an allowable span greater than the actual joist spacing may be substituted for panels of the same thickness with a allowable span matching the actual joist spacing.

[c] Use only adhesives conforming to APA Specification AFG-01 or ASTM D3498, applied in accordance with the manufacturer's recommendations. If OSB panels with sealed surfaces and edges are to be used, use only solvent-based glues; check with panel manufacturer.

[d] Recommended minimum thickness for use with I-joists.

[e] 8d common (3.3 × 64 mm) nails may be substituted if ring- or screw-shank nails are not available.

5.6.2 I-Joist Allowable Span Tables—Residential Floor

I-joist allowable span tables for residential applications are provided in Tables 5.4 to 5.7.

Design Assumptions. The floor span tables provided for both simple and multiple span are based on the following assumptions:

- Span for calculation purposes equals clear span + 0.25 ft (76 mm) (Spans given in Tables 5.4–5.7 equal the vertical projection measured between inside faces of supports.)
- Allowable spans are calculated based on the use of glued-nailed construction utilizing adhesives meeting the requirements of APA Specification AFG-01 or ASTM D3498.
- Bending capacities are adjusted for repetitive member stresses (1.04 for composite flanges and 1.07 for sawn lumber) as is applicable to all wood products when multiple members are spaced 24 in. (610 mm) on center or less.
- Bending stiffness and coefficient of shear deflection and other design properties are as listed in Table 5.2.

Joist Identification. All APA PRIs are identified by an allowable span designated for uniformly loaded residential floor construction at various I-joist spacings. Therefore, the specific I-joist needed for a given application is easily determined by

TABLE 5.4 Allowable Floor Spans for APA PRI I-Joists (simple-span only), Based on a Live Load of 40 psf and a Dead Load of 10 psf

Depth	I-joist designation	Simple-span floor I-joist spacing 12 in. o.c.	16 in. o.c.	19.2 in. o.c.	24 in. o.c.	32 in. o.c.	48 in. o.c.
9½″	PRI-20	16 ft, 7 in.	15 ft, 2 in.	14 ft, 4 in.	13 ft, 4 in.	11 ft, 3 in.	8 ft, 2 in.
	PRI-30	17 ft, 1 in.	15 ft, 8 in.	14 ft, 10 in.	13 ft, 11 in.	12 ft, 10 in.	9 ft, 4 in.
	PRI-40	18 ft, 0 in.	16 ft, 6 in.	15 ft, 7 in.	14 ft, 1 in.	11 ft, 9 in.	9 ft, 7 in.
	PRI-50	17 ft, 10 in.	16 ft, 4 in.	15 ft, 15 in.	14 ft, 5 in.	13 ft, 5 in.	10 ft, 0 in.
	PRI-60	19 ft, 0 in.	17 ft, 4 in.	16 ft, 4 in.	15 ft, 4 in.	13 ft, 10 in.	10 ft, 8 in.
11⅞″	PRI-20	19 ft, 11 in.	18 ft, 2 in.	17 ft, 2 in.	15 ft, 5 in.	12 ft, 4 in.	8 ft, 2 in.
	PRI-30	20 ft, 6 in.	18 ft, 9 in.	17 ft, 9 in.	16 ft, 7 in.	14 ft, 0 in.	9 ft, 4 in.
	PRI-40	21 ft, 6 in.	19 ft, 7 in.	18 ft, 2 in.	16 ft, 3 in.	13 ft, 7 in.	11 ft, 1 in.
	PRI-50	21 ft, 4 in.	19 ft, 6 in.	18 ft, 5 in.	17 ft, 3 in.	15 ft, 1 in.	10 ft, 0 in.
	PRI-60	22 ft, 8 in.	20 ft, 8 in.	19 ft, 6 in.	18 ft, 3 in.	16 ft, 0 in.	11 ft, 10 in.
	PRI-70	23 ft, 0 in.	21 ft, 0 in.	19 ft, 10 in.	18 ft, 7 in.	17 ft, 3 in.	11 ft, 5 in.
	PRI-80	24 ft, 11 in.	22 ft, 8 in.	21 ft, 4 in.	19 ft, 11 in.	18 ft, 6 in.	12 ft, 8 in.
	PRI-90	25 ft, 8 in.	23 ft, 4 in.	22 ft, 0 in.	20 ft, 6 in.	19 ft, 0 in.	13 ft, 10 in.
14″	PRI-40	24 ft, 4 in.	22 ft, 1 in.	20 ft, 2 in.	18 ft, 0 in.	15 ft, 1 in.	11 ft, 10 in.
	PRI-50	24 ft, 4 in.	22 ft, 3 in.	21 ft, 0 in.	19 ft, 8 in.	15 ft, 1 in.	10 ft, 0 in.
	PRI-60	25 ft, 9 in.	23 ft, 6 in.	22 ft, 2 in.	20 ft, 9 in.	17 ft, 9 in.	11 ft, 10 in.
	PRI-70	26 ft, 1 in.	23 ft, 10 in.	22 ft, 6 in.	21 ft, 0 in.	17 ft, 3 in.	11 ft, 5 in.
	PRI-80	28 ft, 3 in.	25 ft, 9 in.	24 ft, 3 in.	22 ft, 8 in.	19 ft, 1 in.	12 ft, 8 in.
	PRI-90	29 ft, 1 in.	26 ft, 6 in.	24 ft, 11 in.	23 ft, 3 in.	20 ft, 10 in.	13 ft, 10 in.
16″	PRI-40	27 ft, 0 in.	24 ft, 0 in.	21 ft, 11 in.	19 ft, 7 in.	16 ft, 4 in.	11 ft, 10 in.
	PRI-50	27 ft, 0 in.	24 ft, 8 in.	23 ft, 4 in.	20 ft, 2 in.	15 ft, 1 in.	10 ft, 0 in.
	PRI-60	28 ft, 7 in.	26 ft, 1 in.	24 ft, 7 in.	23 ft, 0 in.	17 ft, 10 in.	11 ft, 10 in.
	PRI-70	29 ft, 0 in.	26 ft, 5 in.	24 ft, 11 in.	23 ft, 1 in.	17 ft, 3 in.	11 ft, 5 in.
	PRI-80	31 ft, 4 in.	28 ft, 6 in.	26 ft, 11 in.	25 ft, 1 in.	19 ft, 1 in.	12 ft, 8 in.
	PRI-90	32 ft, 2 in.	29 ft, 3 in.	27 ft, 7 in.	25 ft, 9 in.	20 ft, 10 in.	13 ft, 10 in.

For SI: 1 in. = 25.4 mm, 1 ft = 304.8 mm.

1. Allowable clear span applicable to simple-span residential floor construction with a design dead load of 10 psf (0.48 kN/m²) and live load of 40 psf (1.9 kN/m²). The live load deflection is limited to span/480.

2. The tabulated spans are based on a composite floor with glued-nailed sheathing meeting the requirements for APA rated sheathing or APA rated Sturd-I-Floor conforming to PRP-108[13], PS 1, or PS 2 with a minimum thickness of 19⁄32 in. (15 mm) (40⁄20 or 20 oc) for a joist spacing of 19.2 in. (488 mm) or less, or 23⁄32 in. (18.3 mm) (48⁄24 or 24 oc) for a joist spacing of 24 in. (610 mm). Adhesive must meet APA Specification AFG-01 or ASTM D3498. Spans must be reduced 12 in. (305 mm) when the floor sheathing is nailed only.

3. Minimum bearing length must be 1¾ in. (44.5 mm) for the end bearings.

4. Bearing stiffeners are not required when APA PRI joists are used with the spans and spacings given in this table, except as required by hanger manufacturers.

selecting the span needed and then choosing the I-joist that meets the span, spacing, and loading criteria.

Tables 5.4 and 5.5 may be used when it is known that a simple- or a multiple-span application is involved. These tables indicate the allowable clear span for various joist spacings under typical residential floor loads (40 lb/ft² [1.9 kN/m²] live load and 10 lb/ft² [0.48 kN/m²] dead load) for glued-nailed systems. For loading conditions with 40 lb/ft² (1.9 kN/m²) live load and 20 lb/ft² (0.96 kN/m²) dead load, Tables 5.6 and 5.7 are provided.

TABLE 5.5 Allowable Floor Spans for APA PRI I-Joists (multiple-span only), Based on a Live Load of 40 psf and a Dead Load of 10 psf

Depth	I-Joist designation	Multiple-span floor I-joist spacing					
		12 in. o.c.	16 in. o.c.	19.2 in. o.c.	24 in. o.c.	32 in. o.c.	48 in. o.c.
9½″	PRI-20	18 ft, 1 in.	16 ft, 3 in.	14 ft, 10 in.	13 ft, 3 in.	10 ft, 0 in.	6 ft, 7 in.
	PRI-30	18 ft, 8 in.	17 ft, 1 in.	16 ft, 1 in.	15 ft, 0 in.	11 ft, 3 in.	7 ft, 5 in.
	PRI-40	19 ft, 7 in.	17 ft, 2 in.	15 ft, 8 in.	14 ft, 0 in.	11 ft, 8 in.	8 ft, 5 in.
	PRI-50	19 ft, 5 in.	17 ft, 9 in.	16 ft, 9 in.	15 ft, 8 in.	12 ft, 0 in.	7 ft, 11 in.
	PRI-60	20 ft, 8 in.	18 ft, 10 in.	17 ft, 9 in.	16 ft, 5 in.	12 ft, 9 in.	8 ft, 5 in.
11⅞″	PRI-20	21 ft, 8 in.	18 ft, 10 in.	16 ft, 9 in.	13 ft, 5 in.	10 ft, 0 in.	6 ft, 7 in.
	PRI-30	22 ft, 4 in.	20 ft, 5 in.	18 ft, 10 in.	15 ft, 0 in.	11 ft, 3 in.	7 ft, 5 in.
	PRI-40	23 ft, 0 in.	19 ft, 11 in.	18 ft, 2 in.	16 ft, 2 in.	13 ft, 6 in.	9 ft, 9 in.
	PRI-50	23 ft, 3 in.	21 ft, 3 in.	20 ft, 0 in.	16 ft, 1 in.	12 ft, 0 in.	7 ft, 11 in.
	PRI-60	24 ft, 8 in.	22 ft, 6 in.	21 ft, 2 in.	19 ft, 1 in.	14 ft, 9 in.	9 ft, 9 in.
	PRI-70	25 ft, 1 in.	22 ft, 11 in.	21 ft, 7 in.	18 ft, 6 in.	13 ft, 9 in.	9 ft, 1 in.
	PRI-80	27 ft, 1 in.	24 ft, 8 in.	23 ft, 3 in.	21 ft, 8 in.	16 ft, 4 in.	10 ft, 10 in.
	PRI-90	27 ft, 11 in.	25 ft, 5 in.	23 ft, 11 in.	22 ft, 3 in.	19 ft, 11 in.	13 ft, 2 in.
14″	PRI-40	25 ft, 6 in.	22 ft, 1 in.	20 ft, 1 in.	18 ft, 0 in.	14 ft, 9 in.	9 ft, 9 in.
	PRI-50	26 ft, 6 in.	24 ft, 2 in.	20 ft, 2 in.	16 ft, 1 in.	12 ft, 0 in.	7 ft, 11 in.
	PRI-60	28 ft, 1 in.	25 ft, 7 in.	23 ft, 8 in.	19 ft, 9 in.	14 ft, 9 in.	9 ft, 9 in.
	PRI-70	28 ft, 6 in.	25 ft, 11 in.	23 ft, 2 in.	18 ft, 6 in.	13 ft, 9 in.	9 ft, 1 in.
	PRI-80	30 ft, 10 in.	28 ft, 0 in.	26 ft, 5 in.	23 ft, 11 in.	17 ft, 11 in.	11 ft, 10 in.
	PRI-90	31 ft, 8 in.	28 ft, 10 in.	27 ft, 1 in.	25 ft, 3 in.	19 ft, 11 in.	13 ft, 2 in.
14″	PRI-40	27 ft, 8 in.	23 ft, 11 in.	21 ft, 10 in.	19 ft, 6 in.	14 ft, 9 in.	9 ft, 9 in.
	PRI-50	29 ft, 6 in.	24 ft, 3 in.	20 ft, 2 in.	16 ft, 1 in.	12 ft, 0 in.	7 ft, 11 in.
	PRI-60	31 ft, 2 in.	28 ft, 1 in.	24 ft, 9 in.	19 ft, 9 in.	14 ft, 9 in.	9 ft, 9 in.
	PRI-70	31 ft, 7 in.	27 ft, 10 in.	23 ft, 2 in.	18 ft, 6 in.	13 ft, 9 in.	9 ft, 1 in.
	PRI-80	34 ft, 2 in.	31 ft, 1 in.	29 ft, 3 in.	23 ft, 11 in.	17 ft, 11 in.	11 ft, 10 in.
	PRI-90	35 ft, 1 in.	31 ft, 11 in.	30 ft, 0 in.	26 ft, 7 in.	19 ft, 11 in.	13 ft, 2 in.

For SI: 1 in. = 25.4 mm, 1 ft = 304.8 mm.

1. Allowable clear spans applicable to multiple-span residential floor construction with a design dead load of 10 psf (0.48 kN/m²) and live load of 40 psf (1.9 kN/m²). The end spans must be 40% or more of the adjacent span. The live load deflection is limited to span/480.

2. The tabulated spans are based on a composite floor with glued-nailed sheathing meeting the requirements for APA rated sheathing or APA rated Sturd-I-Floor conforming to PRP-108, PS 1, or PS 2 with a minimum thickness of 19⁄32 in. (15 mm) (40⁄20 or 20 oc) for a joist spacing of 19.2 in. (488 mm) or less, or 23⁄32 in. (18.3 mm) (48⁄24 or 24 oc) for a joist spacing of 24 in. (610 mm). Adhesive must meet APA Specification AFG-01 or ASTM D3498. Spans must be reduced 12 in. (305 mm) when the floor sheathing is nailed only.

3. Minimum bearing length must be 1 ¾ in. (44.5 mm) for the end bearing, and 3½ (89 mm) in. for the intermediate bearing.

4. Bearing stiffeners are not required when APA PRI joists are used with the spans and spacings given in this table, except as required by hanger manufacturers.

While any of the PRIs shown in the allowable span tables in this chapter may be available in a specific market area, availability of any specific PRI product should be verified.

Example To illustrate the selection of a performance rated I-Joist product, assume a normal residential floor load (40 lb/ft² live load and 10 lb/ft² dead load) with a design simple span of 16 ft, 1 in. For architectural reasons, the joist depth is limited to 11⅞ in., joist spacing to 19.2 in. on center, and a simple span is desired. From the 11⅞″ portions of Table 5.4, looking down the 19.2 in. o.c. spacing column, it can be seen that *any* joist designation will work.

TABLE 5.6 Allowable Floor Spans for APA PRI I-Joists (simple-span only), Based on a Live Load of 40 psf and a Dead Load of 20 psf

Depth	I-Joist designation	Simple-span floor I-joist spacing					
		12 in. o.c.	16 in. o.c.	19.2 in. o.c.	24 in. o.c.	32 in. o.c.	48 in. o.c.
9½″	PRI-20	16 ft, 7 in.	14 ft, 11 in.	13 ft, 7 in.	12 ft, 2 in.	10 ft, 3 in.	6 ft, 9 in.
	PRI-30	17 ft, 1 in.	15 ft, 8 in.	14 ft, 10 in.	13 ft, 9 in.	11 ft, 8 in.	7 ft, 9 in.
	PRI-40	18 ft, 0 in.	15 ft, 9 in.	14 ft, 4 in.	12 ft, 10 in.	10 ft, 8 in.	8 ft, 9 in.
	PRI-50	17 ft, 10 in.	16 ft, 4 in.	15 ft, 5 in.	14 ft, 5 in.	12 ft, 7 in.	8 ft, 4 in.
	PRI-60	19 ft, 0 in.	17 ft, 4 in.	16 ft, 4 in.	15 ft, 1 in.	12 ft, 7 in.	8 ft, 10 in.
11⅞″	PRI-20	19 ft, 11 in.	17 ft, 3 in.	15 ft, 9 in.	13 ft, 8 in.	10 ft, 3 in.	6 ft, 9 in.
	PRI-30	20 ft, 6 in.	18 ft, 9 in.	17 ft, 9 in.	15 ft, 7 in.	11 ft, 8 in.	7 ft, 9 in.
	PRI-40	21 ft, 0 in.	18 ft, 2 in.	16 ft, 7 in.	14 ft, 10 in.	12 ft, 5 in.	9 ft, 10 in.
	PRI-50	21 ft, 4 in.	19 ft, 6 in.	18 ft, 5 in.	16 ft, 9 in.	12 ft, 7 in.	8 ft, 4 in.
	PRI-60	22 ft, 8 in.	20 ft, 8 in.	19 ft, 6 in.	17 ft, 5 in.	14 ft, 7 in.	9 ft, 10 in.
	PRI-70	23 ft, 0 in.	21 ft, 0 in.	19 ft, 10 in.	18 ft, 7 in.	14 ft, 4 in.	9 ft, 6 in.
	PRI-80	24 ft, 11 in.	22 ft, 8 in.	21 ft, 4 in.	19 ft, 11 in.	15 ft, 10 in.	10 ft, 6 in.
	PRI-90	25 ft, 8 in.	23 ft, 4 in.	22 ft, 0 in.	20 ft, 6 in.	17 ft, 4 in.	11 ft, 6 in.
14″	PRI-40	23 ft, 4 in.	20 ft, 2 in.	18 ft, 5 in.	16 ft, 5 in.	13 ft, 9 in.	9 ft, 10 in.
	PRI-50	24 ft, 4 in.	22 ft, 3 in.	21 ft, 0 in.	16 ft, 9 in.	12 ft, 7 in.	8 ft, 4 in.
	PRI-60	25 ft, 9 in.	23 ft, 6 in.	21 ft, 8 in.	19 ft, 4 in.	14 ft, 10 in.	9 ft, 10 in.
	PRI-70	26 ft, 1 in.	23 ft, 10 in.	22 ft, 6 in.	19 ft, 2 in.	14 ft, 4 in.	9 ft, 6 in.
	PRI-80	28 ft, 3 in.	25 ft, 9 in.	24 ft, 3 in.	21 ft, 2 in.	15 ft, 10 in.	10 ft, 6 in.
	PRI-90	29 ft, 1 in.	26 ft, 6 in.	24 ft, 11 in.	23 ft, 2 in.	17 ft, 4 in.	11 ft, 6 in.
16″	PRI-40	25 ft, 3 in.	21 ft, 11 in.	20 ft, 0 in.	17 ft, 10 in.	14 ft, 10 in.	9 ft, 10 in.
	PRI-50	27 ft, 0 in.	24 ft, 8 in.	21 ft, 0 in.	16 ft, 9 in.	12 ft, 7 in.	8 ft, 4 in.
	PRI-60	28 ft, 7 in.	25 ft, 9 in.	23 ft, 6 in.	19 ft, 10 in.	14 ft, 10 in.	9 ft, 10 in.
	PRI-70	29 ft, 0 in.	26 ft, 5 in.	24 ft, 0 in.	19 ft, 2 in.	14 ft, 4 in.	9 ft, 6 in.
	PRI-80	31 ft, 4 in.	28 ft, 6 in.	26 ft, 6 in.	21 ft, 2 in.	15 ft, 10 in.	10 ft, 6 in.
	PRI-90	32 ft, 2 in.	29 ft, 3 in.	27 ft, 7 in.	23 ft, 2 in.	17 ft, 4 in.	11 ft, 6 in.

For SI: 1 in. = 25.4 mm, 1 ft = 304.8 mm.

1. Allowable clear span applicable to simple-span residential floor construction with a design dead load of 20 psf (0.96 kN/m²) and live load of 40 psf (1.9 kN/m²). The live load deflection is limited to span/480.

2. The tabulated spans are based on a composite floor with glued-nailed sheathing meeting the requirements for APA rated sheathing or APA rated Sturd-I-Floor conforming to PRP-108, PS 1, or PS 2 with a minimum thickness of 19⁄32 in. (15 mm) (40⁄20 or 20 oc) for a joist spacing of 19.2 in. (488 mm) or less, or 23⁄32 in. (18.3 mm) (48⁄24 or 24 oc) for a joist spacing of 24 in. (610 mm). Adhesive must meet APA Specification AFG-01 or ASTM D3498. Spans must be reduced 12 in. (305 mm) when the floor sheathing is nailed only.

3. Minimum bearing length shall be 1¾ in. (44.5 mm) for the end bearings.

t4. Bearing stiffeners are not required when APA PRI joists are used with the spans and spacings given in this table, except as required by hanger manufacturers.

If the owners later decided that they wanted ceramic tile on the first floor, the designer would use Table 5.6 to check his design. Again from the 11⅞″ portions of Table 5.6, looking down the 19.2 in. o.c. spacing column, it can be seen that any joist designation will work *except* for the PRI-20. However, if the spacing of the I-joists were changed to 16 in. on center, then *any* joist designation would work for the floor under the ceramic tile installation.

Substituting APA Performance Rated I-Joists for Other APA PRIs

Uniformly Loaded Applications. The most common type of joist substitution is to substitute one PRI for another having the same depth. For PRIs that have been selected based on the use of the allowable span tables for uniformly loaded appli-

TABLE 5.7 Allowable Floor Spans for APA PRI I-Joists (multiple-span only), Based on a Live Load of 40 psf and a Dead Load of 20 psf

Depth	I-Joist designation	Multiple-span floor I-joist spacing					
		12 in. o.c.	16 in. o.c.	19.2 in. o.c.	24 in. o.c.	32 in. o.c.	48 in. o.c.
9½″	PRI-20	17 ft, 2 in.	14 ft, 10 in.	13 ft, 6 in.	11 ft, 1 in.	8 ft, 3 in.	5 ft, 5 in.
	PRI-30	18 ft, 8 in.	16 ft, 10 in.	15 ft, 4 in.	12 ft, 6 in.	9 ft, 4 in.	6 ft, 2 in.
	PRI-40	18 ft, 1 in.	15 ft, 8 in.	14 ft, 3 in.	12 ft, 9 in.	10 ft, 7 in.	7 ft, 0 in.
	PRI-50	19 ft, 5 in.	17 ft, 9 in.	16 ft, 8 in.	13 ft, 5 in.	10 ft, 0 in.	6 ft, 7 in.
	PRI-60	20 ft, 8 in.	18 ft, 5 in.	16 ft, 9 in.	14 ft, 2 in.	10 ft, 7 in.	7 ft, 0 in.
11⅞″	PRI-20	19 ft, 10 in.	16 ft, 9 in.	13 ft, 11 in.	11 ft, 1 in.	8 ft, 3 in.	5 ft, 5 in.
	PRI-30	22 ft, 4 in.	18 ft, 10 in.	15 ft, 8 in.	12 ft, 6 in.	9 ft, 4 in.	6 ft, 2 in.
	PRI-40	21 ft, 0 in.	18 ft, 2 in.	16 ft, 6 in.	14 ft, 9 in.	12 ft, 3 in.	8 ft, 1 in.
	PRI-50	23 ft, 3 in.	20 ft, 2 in.	16 ft, 9 in.	13 ft, 5 in.	10 ft, 0 in.	6 ft, 7 in.
	PRI-60	24 ft, 8 in.	21 ft, 4 in.	19 ft, 5 in.	16 ft, 5 in.	12 ft, 3 in.	8 ft, 1 in.
	PRI-70	25 ft, 1 in.	22 ft, 11 in.	19 ft, 3 in.	15 ft, 4 in.	11 ft, 5 in.	7 ft, 7 in.
	PRI-80	27 ft, 1 in.	24 ft, 8 in.	22 ft, 9 in.	18 ft, 2 in.	13 ft, 7 in.	9 ft, 0 in.
	PRI-90	27 ft, 11 in.	25 ft, 5 in.	23 ft, 11 in.	22 ft, 2 in.	16 ft, 7 in.	11 ft, 0 in.
14″	PRI-40	23 ft, 3 in.	20 ft, 1 in.	18 ft, 4 in.	16 ft, 4 in.	12 ft, 3 in.	8 ft, 1 in.
	PRI-50	26 ft, 6 in.	20 ft, 2 in.	16 ft, 9 in.	13 ft, 5 in.	10 ft, 0 in.	6 ft, 7 in.
	PRI-60	27 ft, 4 in.	23 ft, 8 in.	20 ft, 7 in.	16 ft, 5 in.	12 ft, 3 in.	8 ft, 1 in.
	PRI-70	28 ft, 6 in.	23 ft, 2 in.	19 ft, 3 in.	15 ft, 4 in.	11 ft, 5 in.	7 ft, 7 in.
	PRI-80	30 ft, 10 in.	28 ft, 0 in.	24 ft, 11 in.	19 ft, 11 in.	14 ft, 11 in.	9 ft, 10 in.
	PRI-90	31 ft, 8 in.	28 ft, 10 in.	27 ft, 1 in.	22 ft, 2 in.	16 ft, 7 in.	11 ft, 0 in.
16″	PRI-40	25 ft, 3 in.	21 ft, 10 in.	19 ft, 11 in.	16 ft, 5 in.	12 ft, 3 in.	8 ft, 1 in.
	PRI-50	27 ft, 0 in.	20 ft, 2 in.	16 ft, 9 in.	13 ft, 5 in.	10 ft, 0 in.	6 ft, 7 in.
	PRI-60	29 ft, 8 in.	24 ft, 9 in.	20 ft, 7 in.	16 ft, 5 in.	12 ft, 3 in.	8 ft, 1 in.
	PRI-70	30 ft, 11 in.	23 ft, 2 in.	19 ft, 3 in.	15 ft, 4 in.	11 ft, 5 in.	7 ft, 7 in.
	PRI-80	34 ft, 2 in.	30 ft, 0 in.	24 ft, 11 in.	19 ft, 11 in.	14 ft, 11 in.	9 ft, 10 in.
	PRI-90	35 ft, 1 in.	31 ft, 11 in.	27 ft, 9 in.	22 ft, 2 in.	16 ft, 7 in.	11 ft, 0 in.

For SI: 1 in. = 25.4 mm, 1 ft = 304.8 mm.

1. Allowable clear spans applicable to multiple-span residential floor construction with a design dead load of 20 psf (0.96 kN/m^2) and live load of 40 psf (1.9 kN/m^2). The end spans shall be 40% or more of the adjacent span. The live load deflection is limited to span/480.

2. The tabulated spans are based on a composite floor with glued-nailed sheathing meeting the requirements for APA rated sheathing or APA rated Sturd-I-Floor conforming to PRP-108, PS 1, or PS 2 with a minimum thickness of 19⁄32 in. (15 mm) (40⁄20 or 20 oc) for a joist spacing of 19.2 in. (488 mm) or less, or 23⁄32 in. (18.3 mm) (48⁄24 or 24 oc) for a joist spacing of 24 in. (610 mm). Adhesive must meet APA Specification AFG-01 or ASTM D3498. Spans must be reduced 12 in. (305 mm) when the floor sheathing is nailed only.

3. Minimum bearing length shall be 1¾ in. (44.5 mm) for the end bearing, and 3½ in. (89 mm) for the intermediate bearing.

4. Bearing stiffeners are not required when APA PRI joists are used with the spans and spacings given in this table, except as required by hanger manufacturers.

cations, this is a simple substitution. All that is needed is to select a PRI with an equivalent or greater allowable span from a given series classification.

Example Referring to Table 5.4, it can be seen that a 9½″ PRI-60 can be substituted for a 9½″ PRI-40 for all on-center spacings listed. Note also that a 9½″ PRI-50 is NOT a suitable substitution for a 9½″ PRI-40; its allowable spans are not as great.

If a designer wants to give the builder the greatest possible latitude in selecting PRI I-joists, they can design for the minimum allowable spans for each depth and then just specify PRI by depth, e.g., "9½″ PRI."

However, it is important to note that mills manufacture specific products based on flange resource availability and manufacturing limitations. This means that while the depths of a given series will be constant, not all series may be available in a given geographic area. Since flange width may vary for a given series, it may be best practice for a designer not to specify the hanger width but instead to specify "a flange-width-compatible hanger."

Example All 9½″ PRIs have a depth of 9½ in. regardless of flange type—the flange width may vary. If metal hangers are required, it is important to verify the appropriate hangers when a subsitution is being considered.

As an option, the designer can design the application for the minimum allowable span or section properties for the desired I-joist depth and then specify the depth and a flange-width-compatible hanger. For example: "Floor joists use 9⁄12″ APA PRIs with flange-width-compatible hangers. . . . "

Nonuniformly Loaded Applications. In some situations, a PRI may be designed to support nonuniform loads such as from a load-bearing wall above or from a concentrated point load. In these cases, a direct substitution based on the allowable span tables may not be possible. Because of variations in the design properties and physical dimensions of I-joists, each design property given in Table 5.2 must be compared to verify a candidate for product substitution.

Example Assume a substitution of a 9½″ PRI-60 for a 9½″ PRI-50 joist is desired. As indicated by Tables 5.4 or 5.5, the 9½″ PRI-60 joist has greater span capabilities than a 9½″ PRI-50 joist for all spacings listed and could be substituted on a uniform load design basis. However, as noted by comparing the design properties in Table 5.2, the moment capacity for the PRI-60 joist is lower than the value for the PRI-50. Thus, if the design is based on other than uniform load conditions, it would need to be verified that the design moment does not exceed the capacity of the 9½″ PRI-60 before the substitution can be made.

PRI Joists Substituted for Sawn Lumber

Uniformly Loaded Applications. Substituting PRIs for sawn lumber in uniform load applications is relatively simple. Table 5.8 illustrates several span comparisons of PRIs with sawn lumber joists for typical residential uniform loading. For the sawn lumber joists, a live-load deflection criterion of $L/360$ applies and the joists are assumed to be simple span. For the PRI joists, the live-load deflection criterion is $L/480$ and this is applicable to simple or multiple span applications.

Example Assume 9½″ PRIs are to be substituted for 2 × 10 sawn lumber joists spaced 16 in. on center. From Table 5.8, it can be seen that the maximum span for nay of the sawn lumber species tabulated is 16 ft, 1 in. for southern pine. A 9½″ PRI-30 could be directly substituted for any of the lumber joists, other than the southern pine joists, at the same on-center spacing when used in either a simple or multiple span. A 9½″ PRI-40 could be substituted for any of the lumber species shown at this for either simple or multiple spans.

Example For the same conditions as the example above but assuming the PRIs are used in a multiple-span-only application, the 9½″ PRI-30 joists could be spaced at 19.2 in. on center and substituted for all of the lumber species listed. This results in fewer joists to be handled and installed while still maintaining the superior deflection performance characteristics of a PRI.

TABLE 5.8 Comparison of PRIs with Uniformly Loaded Sawn Lumber[a]

2 × 10 sawn lumber No. 2, spaced 16 in. o.c. Live load deflection = $L/360$	9½″ PRIs Live load deflection = $L/480$			
			Maximum	
Species/maximum Simple span	I-joist designation	Spacing	Simple span	Multiple span
SPF (south)[b]	9½″ PRI-20	16 in. o.c.	15 ft, 2 in.	16 ft, 3 in.
14 ft, 3 in.	9½″ PRI-30	16 in. o.c.	15 ft, 8 in.	17 ft, 1 in.
Hem-fir[b]	9½″ PRI-40	16 in. o.c.	16 ft, 6 in.	17 ft, 2 in.
15 ft, 2 in.	9½″ PRI-50	16 in. o.c.	16 ft, 4 in.	17 ft, 9 in.
Douglas fir-larch[b]	9½″ PRI-60	16 in. o.c.	17 ft, 4 in.	18 ft, 10 in.
15 ft, 5 in.	9½″ PRI-20	19.2 in. o.c	14 ft, 4 in.	14 ft, 10 in.
Southern pine[c]	9½″ PRI-30	19.2 in. o.c.	14 ft, 10 in.	16 ft, 1 in.
16 ft, 1 in.	9½″ PRI-40	19.2 in. o.c	15 ft, 7 in.	15 ft, 8 in.
	9½″ PRI-50	19.2 in. o.c.	15 ft, 5 in.	16 ft, 9 in.
	9½″ PRI-60	19.2 in. o.c	16 ft, 4 in.	17 ft, 9 in.

2 × 12 sawn lumber No. 2, spaced 16 in. o.c. Live load deflection = $L/360$	11⅞″ PRIs Live load deflection = $L/480$			
			Maximum	
Species/maximum Simple span	I-joist designation	Spacing	Simple span	Multiple span
SPF (south)[b]	11⅞″ PRI-20	16 in. o.c.	18 ft, 2 in.	18 ft, 10 in.
16 ft, 6 in.	11⅞″ PRI-30	16 in. o.c.	18 ft, 9 in.	20 ft, 5 in.
Hem-fir[b]	11⅞″ PRI-40	16 in. o.c.	19 ft, 7 in.	19 ft, 11 in.
17 ft, 7 in.	11⅞″ PRI-50	16 in. o.c.	19 ft, 6 in.	21 ft, 3 in.
Douglas fir-larch[b]	11⅞″ PRI-60	16 in. o.c.	20 ft, 8 in.	22 ft, 6 in.
17 ft, 10 in.	11⅞″ PRI-70	16 in. o.c.	21 ft, 4 in.	23 ft, 2 in.
Southern pine[c]	11⅞″ PRI-80	16 in. o.c.	22 ft, 8 in.	24 ft, 8 in.
18 ft, 10 in.	11⅞″ PRI-90	16 in. o.c.	23 ft, 4 in.	25 ft, 5 in.
	11⅞″ PRI-20	19.2 in. o.c	14 ft, 4 in.	14 ft, 10 in.
	11⅞″ PRI-30	19.2 in. o.c.	17 ft, 9 in.	18 ft, 10 in.
	11⅞″ PRI-40	19.2 in. o.c	18 ft, 2 in.	18 ft, 2 in.
	11⅞″ PRI-50	19.2 in. o.c.	18 ft, 5 in.	20 ft, 0 in.
	11⅞″ PRI-60	19.2 in. o.c	19 ft, 6 in.	21 ft, 2 in.
	11⅞″ PRI-70	19.2 in. o.c	20 ft, 1 in.	21 ft, 10 in.
	11⅞″ PRI-80	19.2 in. o.c.	21 ft, 4 in.	23 ft, 3 in.
	11⅞″ PRI-90	19.2 in. o.c	22 ft, 0 in.	23 ft, 11 in.

For SI: 1 in. = 25.4 mm, 1 ft = 304.8 mm.
[a] Uniform live load = 40 psf (1.9 kN/m^2). Uniform dead load = 10 psf (0.48 kN/m^2).
[b] Western Lumber Use Manual—Base Values for Dimension Lumber.
[c] Southern Pine Use Guide—Empirical Design Values for Dimension Lumber.

Similar comparisons could be made for the other joists shown in Table 5.8, or for any PRI tabulated in Tables 5.4 or 5.5 based on uniformly loaded residential floors.

Nonuniformly Loaded Applications. As with the substitution of PRIs for one another, substituting a PRI for a sawn lumber joist when the joist selection is based on nonuniform loading conditions is more complicated than just comparing span capabilities. In these cases, an engineering analysis may be required before making this substitution. Table 5.2 is provided for such applications.

5.7 I-JOIST INSTALLATION DETAILS—FLOORS

5.7.1 Installing Performance Rated I-Joists—Floors

Because they are simple to install, I-joists provide many benefits to the designer and contractor alike. However, as with any construction material, it is essential to follow proper installation procedures.

I-joists are to be installed as shown in Fig. 5.11 and in accordance with the following guidelines, which show methods recommended for most applications found in common situations of residential floor construction.

General Installation Guidelines for All I-Joist Products

- I-joists must be plumb and anchored securely to supports before floor sheathing is attached.
- Supports for multiple-span joists must be level.
- To minimize settlement when using hangers, I-joists should be firmly seated in the hanger bottoms.
- Leave a 1/16 in. (1.5 mm) gap between I-joist ends and headers.
- Except for cutting to length, I-joist flanges should *never* be cut, drilled, or notched.
- I-joists must be protected from the weather prior to installation.
- I-joists must not be used in places where they will be permanently exposed to weather or will reach moisture content greater than 16%, such as in swimming pool or hot tub areas.
- They must not be installed where they will remain in direct contact with concrete or masonry.

Typical PRI Floor Framing and Construction Details and Installation Notes

1. Installation of APA PRIs must be as shown in Fig. 5.11.
2. Concentrated loads should only be applied to the top surface of the top flange. At no time should concentrated loads be suspended from the bottom flange with the exception of light loads (ceiling fans, light fixtures, etc.).
3. End bearing length must be at least 1¾ in. (44.5 mm). For multiple-span joists, intermediate bearing length must be at least 3½ in. (89 mm).
4. Ends of floor joists must be restrained to prevent rollover. Use performance rated rim board or I-joist blocking panels.

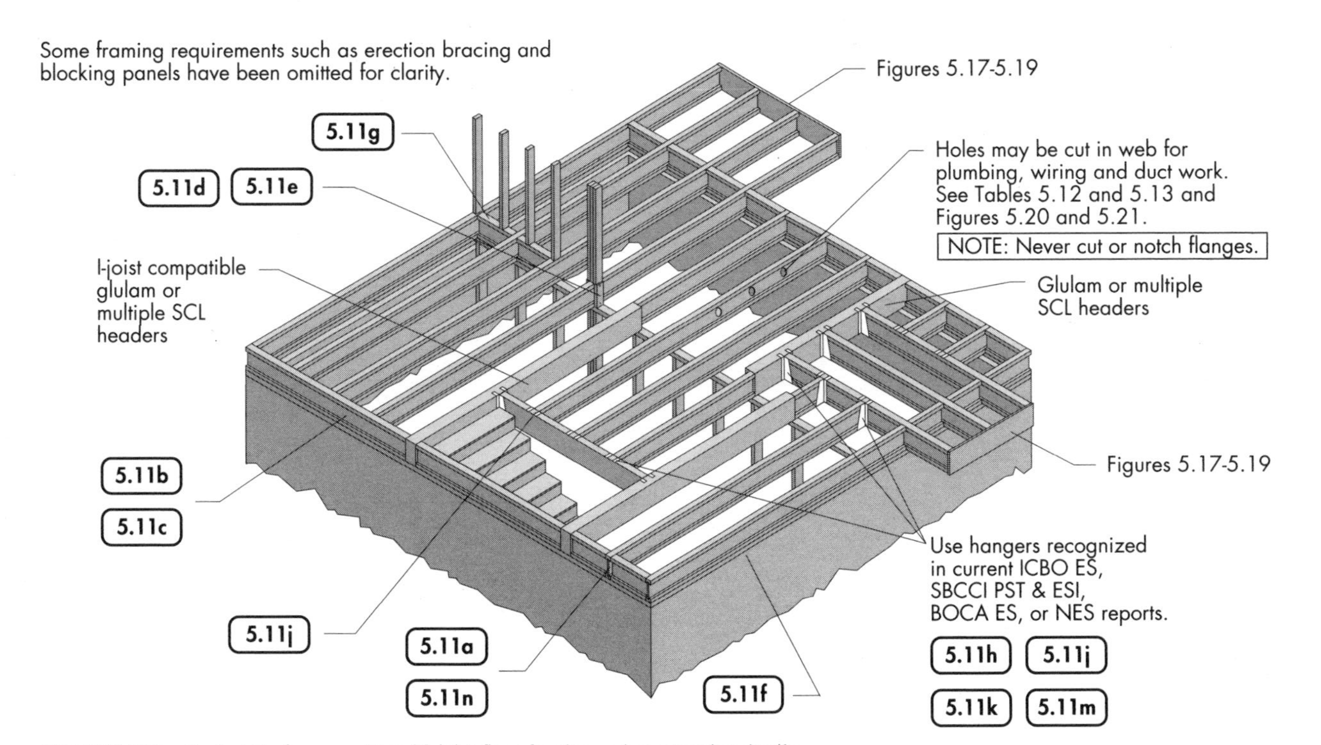

FIGURE 5.11 Typical Performance Rated I-joist floor framing and construction details.

All nails shown in the details 5.11a through 5.11m are assumed to be common nails unless otherwise noted. 10d box nails may be substituted for 8d common shown in details. Inividual components not shown to scale for clarity.

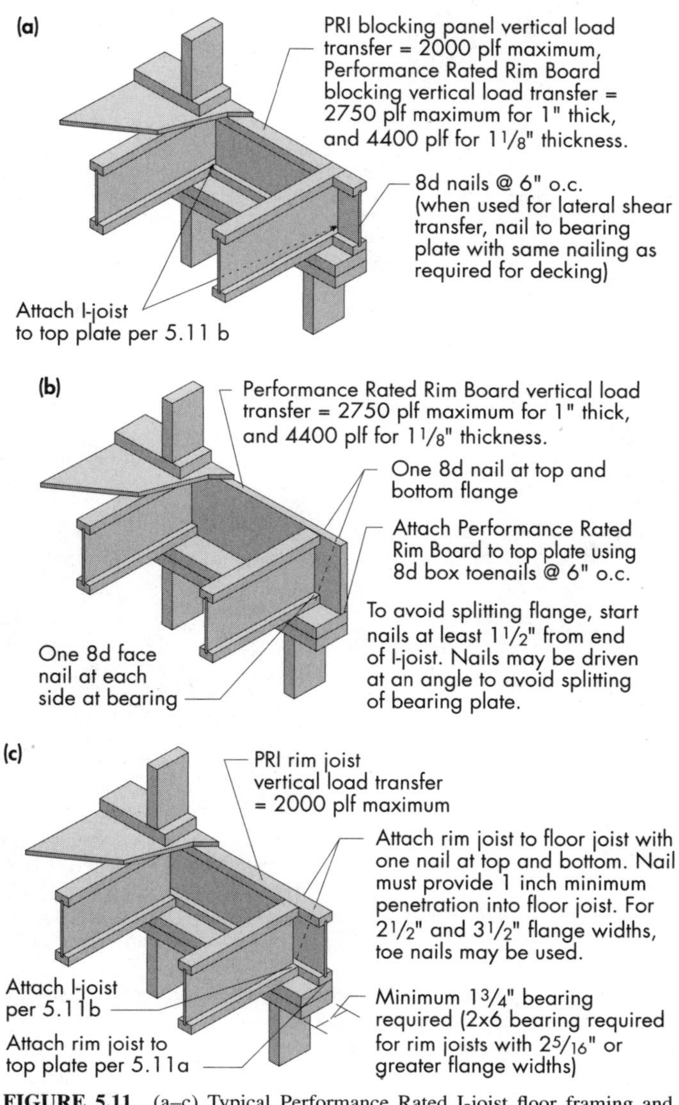

FIGURE 5.11 (a–c) Typical Performance Rated I-joist floor framing and construction details. (*Continued*)

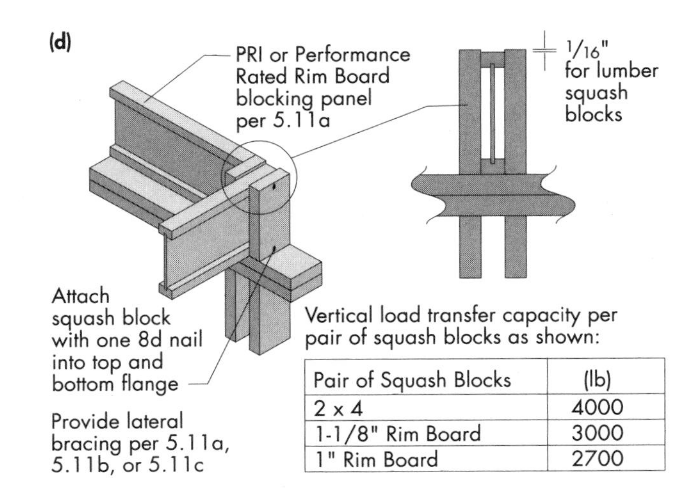

Pair of Squash Blocks	(lb)
2 x 4	4000
1-1/8" Rim Board	3000
1" Rim Board	2700

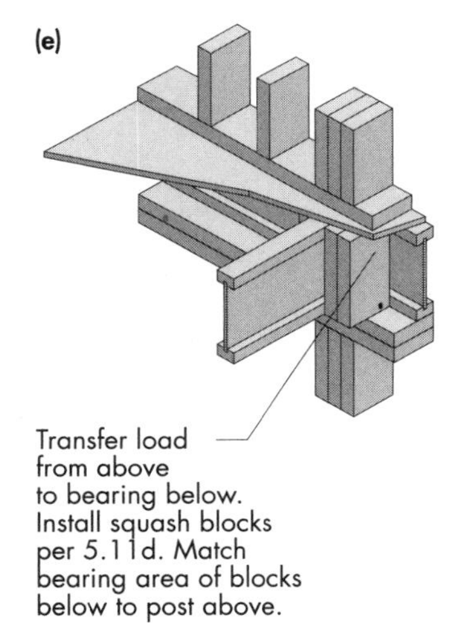

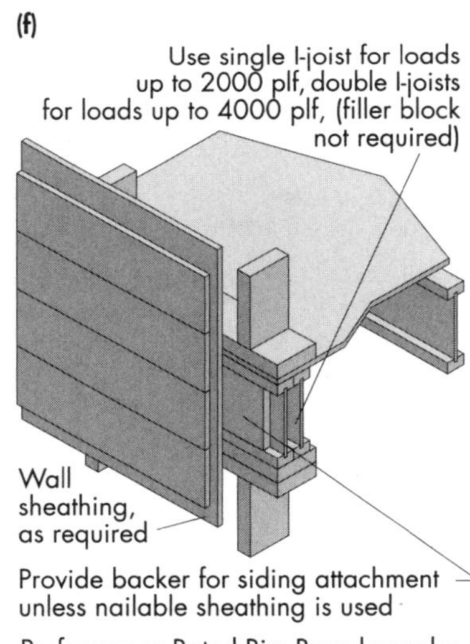

FIGURE 5.11 (d–f) Typical Performance Rated I-joist floor framing and construction details. (*Continued*)

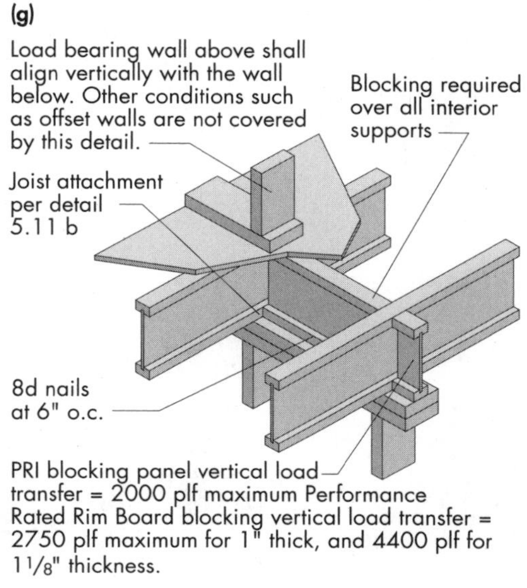

FIGURE 5.11 (g) Typical Performance Rated I-joist floor framing and construction details. (*Continued*)

5. I-joists installed beneath bearing walls perpendicular to the joists must have full depth blocking panels, performance rated rim board, or squash blocks (cripple blocks) to transfer gravity loads from above the floor system to the wall or foundation below.

6. For I-joists installed beneath bearing walls parallel to the joists, the maximum allowable vertical load using a single I-joist is 2000 plf (2.94 kN/m) and 4000 plf (5.9 kN/m) if double I-joists are used.

7. Continuous lateral support of the I-joist's compression flange is required to prevent rotation and buckling. In simple-span uses, lateral support of the top flange is normally supplied by the floor sheathing. In multiple-span or cantilever applications, bracing of the I-joist's bottom flange is also required at interior supports of multiple-span joists and at the end support next to the cantilever extension. The ends of all cantilever extensions must be laterally braced as shown in Figs. 5.17, 5.18, and 5.19.

8. Nails installed perpendicular to the wide face of the flange must be spaced in accordance with the applicable building code requirements or approved building plans but should not be closer than 3 in. (75 mm) o.c. per row (4 in. [100 mm] o.c. for I-joists with composite flanges 1½ in. [38 mm] wide) for 6 or 8d common (2.9 × 61 mm or 3.3 × 64 mm) nails. If more than one row of nails is used (not permitted for I-joists with composite flanges 1½ in. [38 mm] wide), the rows must be offset at least ½ in. (12.5 mm). Nails installed parallel to the wide face of the veneers in LVL flanges must not be spaced closer than 3 in. (75 mm) o.c. for 8d common (3.3 × 64 mm) nails, and 4 in. (100 mm) o.c. for 10d common (3.8 × 76 mm) nails.

9. Figure 5.11 details on the following pages show only I-joist-specific fastener requirements. For other fastener requirements, see the applicable building code.

(h)

Backer block (use if hanger load exceeds 250 lbs.) Before installing a backer block to a double I-joist, drive 3 additional 10d nails through the webs and filler block where the backer block will fit. Clinch. Install backer tight to top flange. Use twelve 10d nails, clinched when possible. Maximum capacity for hanger for this detail = 1280 lb.

BACKER BLOCKS (Blocks must be long enough to permit required nailing without splitting)

Flange Width	Material Thickness Required*	Minimum Depth**
$1^1/_2$"	$^{19}/_{32}$"	$5^1/_2$"
$1^3/_4$"	$^{23}/_{32}$"	$5^1/_2$"
$2^5/_{16}$"	1"	$7^1/_4$"
$2^1/_2$"	1"	$5^1/_2$"
$3^1/_2$"	$1^1/_2$"	$7^1/_4$"

* Minimum grade for backer block material shall be Utility grade SPF (south) or better for solid sawn lumber and Rated Sheathing grade for wood structural panels.

** For face-mount hangers use net joist depth minus $3^1/_4$" for joists with $1^1/_2$" thick flanges. For $1^5/_{16}$" thick flanges use net depth minus $2^7/_8$".

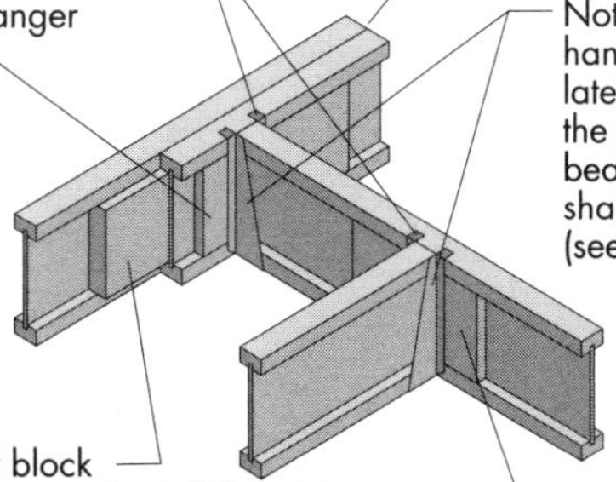

For hanger capacity see hanger manufacturer's recommendations. Verify double I-joist capacity to support concentrated loads.

FIGURE 5.11 (h) Typical Performance Rated I-joist floor framing and construction details. (*Continued*)

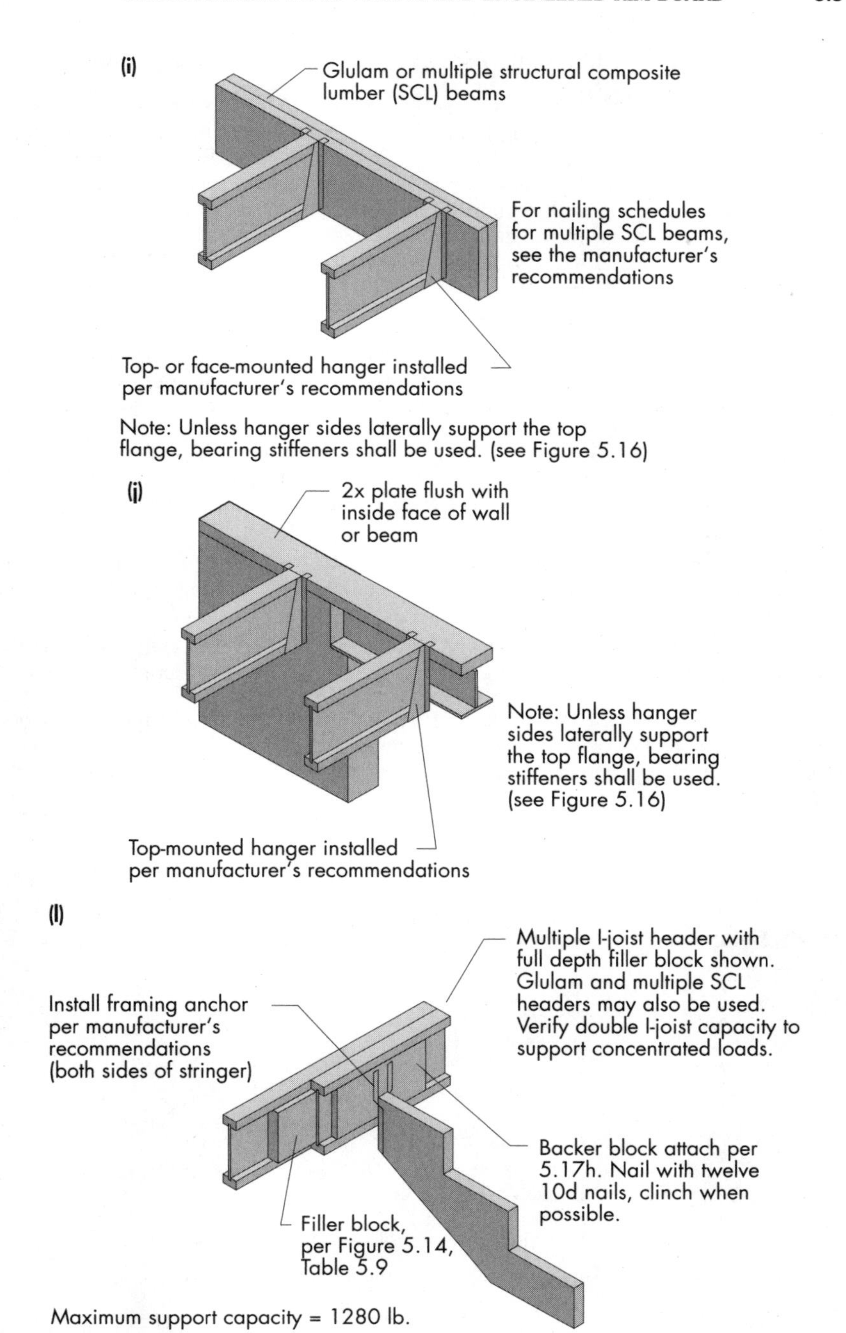

FIGURE 5.11 (i,j,l) Typical Performance Rated I-joist floor framing and construction details. (*Continued*)

5.7.2 Other Framing Elements Essential to Engineered Wood Floor Construction

When designing or working with I-joists for the first time, the user will be introduced to a number of new and sometimes esoteric terms. While some of these terms almost exclusive to I-joists are intuitive and therefore easy to grasp, there are a few that are a little difficult to understand in terms of their actual function. The following sections are an introduction to these terms.

Blocking Panels

Definition/Function. As far as the installation of I-joists is concerned, a blocking panel is a rectangular piece of engineered wood (I-joist, rim board, or I-joist compatible LVL) that is placed between adjacent joists at various locations described below (see Fig. 5.12). Blocking has three major functions:

1. To provide lateral support to the floor joists—to prevent them from physically rolling over due to lateral loads. This is analogous to preventing overturning in a shear wall. This is accomplished by the rectangular shape and stiffness of the blocking panel.
2. To provide a means of transferring shear/lateral loads from the walls above to the floor/foundation below. This is accomplished by nailing into the top and bottom flanges.
3. To provide a means of transferring vertical loads from the wall above to the foundation/floor below. The blocking is used in bearing to accomplish this.

Physical Description. Blocking is often site-fabricated out of engineered wood materials on hand. It is essential that engineered wood materials be used because the shrinkage that can be anticipated with the use of sawn lumber would make the

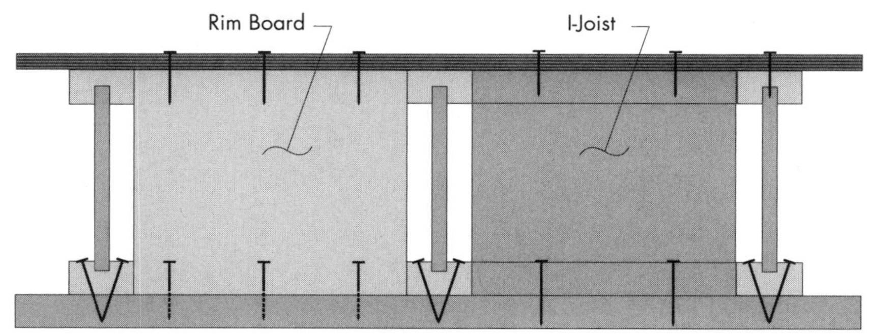

• Shape and rigidity of blocking panel prevent roll-over of floor or roof joists.
• Vertical capacity of blocking panel transfers vertical loads.
• Fastening at top and base of blocking panel transfer shear.

FIGURE 5.12 Engineered wood blocking panel used in I-joist floor system.

blocking unable to perform the vertical load transfer function and could seriously impede its ability to transfer shear/lateral loads.

Fabricate the blocking panels from engineered wood products of I-joist-compatible sizes and cut to fit snugly between the floor joists.

Use Recommendations.

1. Blocking panels are required at the ends of floor joists not otherwise restrained from overturning by band joist or rim board and at all other supports.
2. Blocking panels are required between floor joists supporting load-bearing walls running perpendicular to the joists.
3. Blocking panels are required between floor joists at the interior support adjacent to a *nonload-bearing* cantilever in the floor system.
4. For a *load-bearing* cantilever, blocking panels are required between floor joists at the interior support adjacent to the cantilever as well as for 4 ft along the support on either side of the cantilevered joists.

Squash Blocks/Cripple Blocks

Definition/Function. A squash block is a block of wood that is installed adjacent to an I-joist to prevent the I-joist from carrying a point load that it is not designed to support (see Fig. 5.13).

When acting as a bending member, in addition to checking bending strength, shear strength, and deflection, the I-joist designer must also consider the exterior reaction (ER) and intermediate reaction (IR) as possible design controls. This is the force in the upward direction at the support locations that balances or resists the downward force on the I-joist—the live and dead loads. The reason for needing to consider each of these reaction points is the propensity for the thin web to knife through the bottom flange of the I-joist, as a web-bearing failure. In some cases, the span of the I-joists will be limited by this design check. Even when not limited

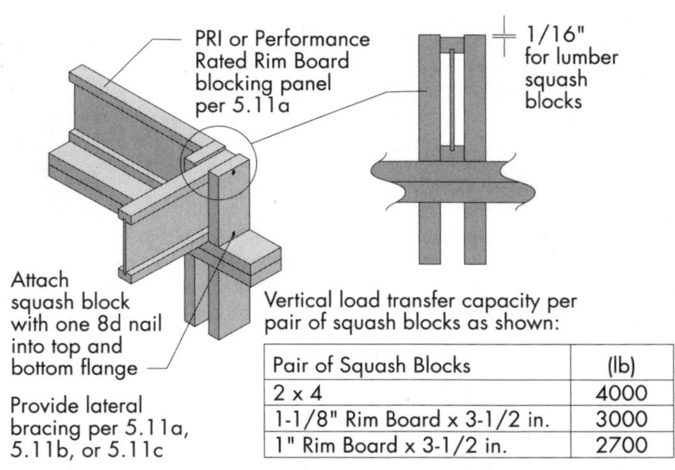

Pair of Squash Blocks	(lb)
2 x 4	4000
1-1/8" Rim Board x 3-1/2 in.	3000
1" Rim Board x 3-1/2 in.	2700

FIGURE 5.13 Installation of squash/cripple blocks to transfer vertical loads around floor joists.

by the IR or ER, when an I-joist is at or near its maximum span, there is little excess IR or ER capacity left within the joist to support other than the uniform load as the majority is used up by the floor loads.

The problem is that in conventional platform construction there are situations where loads from above are transferred down through the floor into the wall or foundation below. This occurs where load-bearing walls fall on floors or where posts supporting headers are located within the walls. Beneath these wall and point loads, the I-joists usually do not have enough IR or ER capacity remaining to safely transfer these loads without exceeding the design web bearing capacity of the I-joist floor joists. The solution is to place extra load-carrying members in line with these loads and insure that *they* carry the load and not the I-joists. For line loads like load-bearing walls, blocking panels (described in above under Blocking Panels) are normally used as described below to transfer these additional loads around the web of the floor joist. In the case of point loads, squash blocks are more often specified.

Physical Description. A squash block is a 2 × 4 (38 × 89 mm) or 2 × 6 (38 × 240 mm) lumber block that is oriented with the grain of the wood running parallel to the vertical axis of the web of the joist. The squash block is cut just slightly longer than the I-joist is deep, usually 1⁄16 in. (1.5 mm) longer. This is done to ensure that the block will pick up the vertical load and not the I-joist. The grain is oriented parallel to the vertical axis to minimize the impact of shrinkage by the lumber block.

The minimum grade of the squash blocks is utility grade spruce-pine-fir (SPF) (south).

In addition to lumber blocks, performance rated rim board may be cut into blocks and used for such applications. Because the rim board material is already cut to a compatible size for use with the I-joist, its use as a squash block eliminates the real danger associated with the inadvertent use of undersized squash blocks.

Use Recommendations.

1. Each lumber squash block has a capacity of at least 2000 lbf (8.9 kN). Because squash blocks are usually placed in pairs—to minimize load eccentricity—at least 4000 lbf (17.78 kN) is appropriate for a pair of blocks with full bearing on the top/sole plates. Often the designer will simply match the area of the squash blocks with that of the posts above.

   A pair of 1 1⁄8 in. rim board squash blocks has a total capacity of 3000 lb (13.34 kN), and a pair of 1 in. rim board squash blocks has a total capacity of 2700 lb (12.0 kN)—the capacity is reduced in direct proportion to the rim board thickness. Again, the above numbers assume full bearing on the top/sole plate.
2. Lumber squash blocks are to be cut 1⁄16 in. (1.5 mm) longer than the depth of the joist to ensure that the squash block carries the load. (Remember that the I-joist is "busy" carrying the floor load and may have little capacity remaining for the additional wall load.)
3. Squash blocks are installed with the wide side of the block flush with the edges of the I-joist flanges. When possible, they will be fully seated on the top/sole plate below. They will be attached to the top and bottom flange of the I-joist with one 8d (3.3 × 64 mm) common nail at each location. The extra 1⁄16 in. (1.5-mm) of the lumber squash block is to stick up above the surface of the top flange of the I-joist.
4. The use of squash blocks in lieu of blocking for the entire length of a load-bearing wall is not recommended. They could be used, however, to transfer

vertical loads in an occasional joist space to allow for the passage of a duct. The model building codes, however, require blocking under load-bearing walls to provide lateral stability and prevent rollover of the joists, as well as to transfer vertical load. While the squash blocks can transfer the vertical loads, they have *no* ability to provide lateral stability. Leaving out an occasional blocking panel and putting in squash block could be justified from an engineering perspective.

Multiple-I-Joist Construction. Some applications require that multiple I-joists be used to achieve the necessary span. Multiple joists may be required to frame openings, support concentrated loads, or support any other loads, which would exceed the capacity of a single I-joist. In these instances, an engineering evaluation may be required. Figure 5.14 illustrates the construction details required to assemble two or more I-joists, and Table 5.9 provides detailed requirements for installing the required filler blocks covered in below under Filler Blocks.

When Multiple-I-joist Construction Is Required.

1. Double I-joists are required along the foundation or lower support wall line when the vertical load from above is between 2000 plf (2.94 kN/m) and 4000 plf (5.89 kN/m) and no other means of supporting the vertical load is provided.
2. When reinforcing the floor for cantilevers caused by vertical wall offsets as shown in Fig. 5.19, double I-joists are one reinforcing option available.
3. Double I-joists are used under interior braced walls or shearwalls running parallel to the floor joists where required by code.
4. Double I-joists are used when determined by engineering calculation.

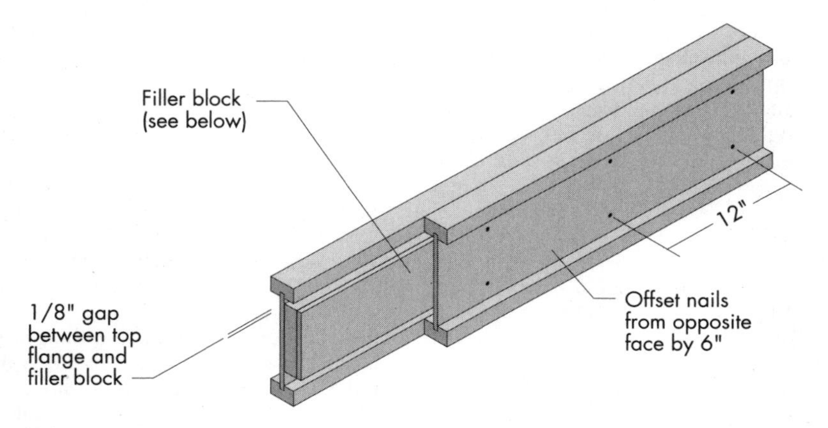

Notes:

1. Support back of I-joist web during nailing to prevent damage to web/flange connection.
2. Leave a 1/8-inch gap between top of filler block and bottom of top I-joist flange.
3. Filler block is required between joists for full length of span.
4. Nail joists together with two rows of 10d nails at 12 inches o.c. (clinched when possible) on each side of the double I-joist. Total of 4 nails per foot required. If nails can be clinched, only 2 nails per foot are required.

FIGURE 5.14 Multiple I-joist construction showing filler blocks and attachment details.

TABLE 5.9 Filler Block Requirements for Multiple I-Joist Construction

Flange width (in.)	I-Joist depth	Filler block size
1½	9½″	1⅛ × 6 in. high
	11⅞″	1⅛ × 8 in. high
1¾	9½″	1⅜ × 6 in.
	11⅞″	1⅜ × 8 in.
	14″	1⅜ × 10 in.
	16″	1⅜ × 12 in.
2⅚	11⅞″	2 × 8 in.
	14″	2 × 10 in.
	16″	2 × 12 in.
2½	9½″	2⅛ × 6 in.
	11⅞″	2⅛ × 8 in.
	14″	2⅛ × 10 in.
	16″	2⅛ × 12 in.
3½	11⅞″	3 × 8 in.
	14″	3 × 10 in.
	16″	3 × 12 in.

For SI: 1 in. = 25.4 mm.

Filler Blocks

Definition/Function. Filler blocks are used to fill the rectangular space between a pair or more of I-joists designed to act as a single bending member. The purpose of the blocks is to transfer load from one I-joist to the next to achieve load sharing. This is accomplished by forcing each of the joists to deflect the same amount under the applied load. When used between two joists subjected to a line load running the full length of the joists, the filler blocks must also be placed the full length of the double I-joists. Filler blocks do not, however, have to be continuous—they can be made up of shorter lengths of lumber and/or wood structural panels (see Fig. 5.14) provided the attachment is designed to transfer the point/noncontinuous load.

Physical Description. Filler blocks are made up of lumber, rim board, or wood structural panel (WSP) materials on hand—whatever it takes to meet the size requirements of Table 5.9. The minimum grade of WSP would be Rated Sheathing; minimum lumber grade is utility grade SPF (south) or better. Any Performance Rated Rim Board product would also work satisfactorily.

The depth of the filler block should equal the distance between the flanges of the joist minus ⅛ in. (3 mm). This gap is placed between the filler block and the top flange. There is nothing scientific about the ⅛ in. (3 mm) gap. Ideally, the carpenter would cut the blocks so they would just fit between the flanges. Rather than too tight a fit being risked, and possible damage to one or both of the flange-to-web joints, a slightly loose fit is recommended.

In a similar manner, the thickness of the filler block is also important. Too thick is better than too thin. If the filler block is too thick, the result is a small gap between the flanges, which will not cause a problem. Too thin can cause problems when the nailing schedule shown in Fig. 5.14 is attempted. Notice that the nails are placed near the top and bottom of the filler block. This puts them very close to the flanges of the joist. If there is a gap between the web of the joist and filler block, the mechanics of driving a nail will attempt to close up that gap. This can do one of two things. The first is that it can damage the web or the web-to-joist glue bond. If repeated every 12 in. (3.5 mm), this can cause a failure of the I-joist. The second is that it can cause the flange of the I-joist to rotate, making for an uneven surface and/or reducing the capacity of the I-joist due to the induced eccentric loading.

Use recommendations.

1. Follow the attachment recommendations of Fig. 5.14.
2. Size filler blocking in accordance with Table 5.9.
3. Filler blocks are *not* required where double I-joists occur under a load-bearing wall and are supported by the foundation or wall below as in the case of a double starter joist. In this case, the I-joists are not being used as bending members, so load sharing between joists is not required.

Backer Block Construction

Definition/Function. Backer blocks (shown in Fig. 5.15) are used to fill the rectangular space between the outside edge of the I-joist flange and the web of the I-joist. A backer block is very similar to a half-thickness filler block. The backer block does not run the full length of the I-joist. The purpose of the backer blocks is to provide a flat, flush surface by which surface- or top-mounted hangers or other structural elements can be attached to I-joists, and thus it is only as long as it needs to be to transfer these loads without splitting.

With *top-mounted hangers,* the backer block prevents rotation on the lower portion of the hanger by filling the void between the hanger and the web. If the top-mounted hanger does not rely on any attachment into the side of the I-joist supporting it, the backer block is really only providing bearing and an additional backer block on the other side of the I-joist web is not required. *Backer blocks are NOT required if load on top-mounted hanger is less than 250 lbf (1.11 kN).*

With *face-mounted hangers,* the backer block provides anchorage for the hanger nails. It also provides a means whereby the hanger load is transferred to the web of the I-joist. For such applications, backer blocks on both sides of the web are almost always required.

Physical Description. See Physical Description for Filler Blocks (above) for description of acceptable materials.

The depth of the backer block should equal the distance between the flanges of the joist minus ⅛ in. (3 mm). *Backer blocks should always be installed tight to the top flange.* This is to minimize the possibility of inadvertent loading of the bottom flange in a direction perpendicular to the web-to-flange joint.

The thickness of the backer block can be critical. If the backer block extends out beyond the edge of the flange when a *top-mounted* hanger was being used, this could cause insufficient bearing on the horizontal fold of the hanger or provide insufficient space for the required nails. This may not be so critical for the *face-mounted* hanger, depending on the nailing pattern of the hanger. If the backer block is too thin, it could cause the hangers to be installed with a slight rotation toward the web. This can have a detrimental impact on the bearing surface area and/or ultimate capacity of the hanger. A plus or minus tolerance of ⅛ in. (3 mm) is typically recommended. For tolerance verification and further information please contact the hanger manufacturer.

As mentioned previously, the backer block doesn't have to run the full length of the I-joist. The backer block should be long enough to fully support the flanges

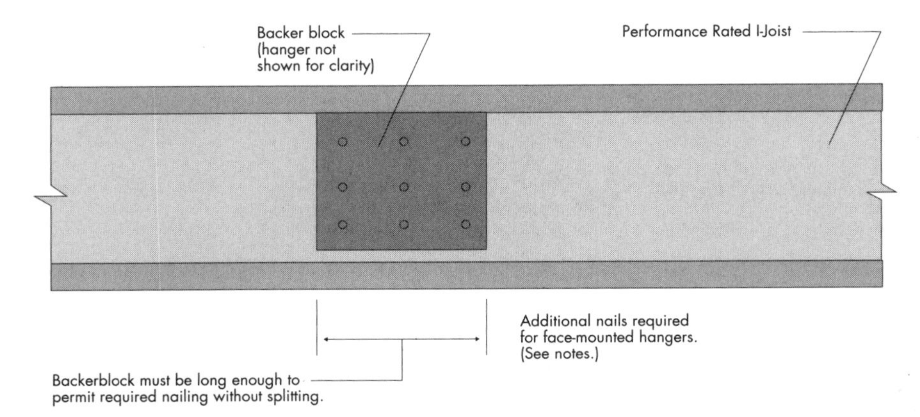

Notes:

a) For face-mounted hangers, backer blocks required on both sides of I-joist. If double I-joists are used, filler block between acts as second backer block.

b) Before installing a backer block to a double I-joist, drive 3 additional 10d nails through the webs and filler block where the backer block will fit. Clinch. Install backer tight to top flange. Use twelve 10d nails, clinched when possible. Maximum capacity for hanger for this detail = 1280 lb.

BACKER BLOCKS (Blocks must be long enough to permit required nailing without splitting)

Flange Width	Material Thickness Required*	Minimum Depth**
1-1/2"	19/32"	5-1/2"
1-3/4"	23/32"	5-1/2"
2-5/16"	1"	7-1/4"
2-1/2"	1"	5-1/2"
3-1/2"	1-1/2"	7-1/4"

* Minimum grade for backer block material shall be Utility grade SPF (south) or better for solid sawn lumber and Rated Sheathing grade for wood structural panels.

** For face-mount hanger use net joist depth minus 3-1/4 in. for joists with 1-1/2 in. thick flanges. For 1-5/16 in. thick flanges use net depth minus 2-7/8 in.

FIGURE 5.15 Backer block installation and connection details.

of the hanger and the required nailing without splitting the block. The hanger width plus 4–6 in. (100–150 mm) will be sufficient if wood structural panel blocking is used. If the backer block is to be made out of lumber, then additional length may be required to prevent splitting.

Use Recommendations.

1. For top-mounted hangers, backer blocks are *normally* required on one side of the I-joist web only. *Exceptions:*
   - When the load on the top-mounted hanger exceeds 1000 lbf (4.45 kN), the second backer block is required to act as a web stiffener. See above under Backer Block Construction.

- When the top-mounted hanger requires additional face nails to reach full capacity, the second backer block will be required. (See Recommendation 3 for face-mounted hangers below.)
- When the load on the top-mounted hanger does not exceed 250 lbf (1.11 kN), a backer block is not required.

2. For top-mounted hangers the backer block will be mounted tight to the top flange. (The load is applied to the I-joist through the top flange. To prevent knife-through of the top flange by the web, the joint between the backer block and the top flange must be tight.)

For face-mounted hangers backer blocks are required on both sides of a single I-joist. This is to allow sufficient nail penetration in the main member (the block-joist-block assembly) to develop the nail capacity. It also allows for a pseudo-double-shear connection at the I-joist web. As with the top-mounted hanger, the backer block should be mounted tight against the top flange.

Because the purpose of the backer block is to transfer the load from the hanger into the web of the I-joist, and because the web of the I-joist in only ⅜ in. (10 mm) thick, additional fasteners (fasteners other than those supplied with the hanger) are required. These are the fasteners that attach the backer block to the web of the I-joist. Approximately one additional fastener for each 120 lbf (0.53 kN) of load on the hanger is required. In addition, the fasteners that attach the hanger to the backer block must be long enough to penetrate the web and into the backer block on the far side.

3. When face-mounted hangers are used, care must be exercised in selecting the material used for the backer blocks. The fastening requirements specified by most hanger manufacturers are established based on a specified species or specific gravity (G) of the material receiving the fastener. If the hanger installations are based on a G of 0.50, then only OSB, Structural I plywood, or Douglas fir or southern yellow pine (SYP) lumber may be used for the backer blocks.

4. When a piece of lumber is being nailed parallel to and up against the side of an I-joist, as seen in Fig. 5.18 for a non-load-bearing cantilever, the backer block is only required on the lumber-side of the I-joist when the attachment nails can be clinched.

5. When installing the backer block to multiple I-joists, drive three additional 10d (3.8×79 mm) common nails, through the webs and filler block where the backer block will fit, clinch if possible. The blocks should be fitted tight against the top flange for both top- and face-mounted hangers).

Web Stiffeners

Definition/Function. A web stiffener, as shown in Fig. 5.16, is a wood block that is used to reinforce the web of an I-joist at locations where:

1. The webs of the I-joist are in jeopardy of buckling out of plane. This may occur in the deeper I-joist when they exceed their maximum exterior reaction (ER) capacity.
2. The webs of the I-joist are in jeopardy of knifing through the I-joist flanges. This can occur at any I-joist depth when the design reaction capacities are exceeded.
3. The I-joist is supported in a hanger and the sides of the hanger do not extend up to the top flange. With the top flange unsupported by the hanger sides, the joist may collapse to the side, putting a twist in the flange of the joist. The web

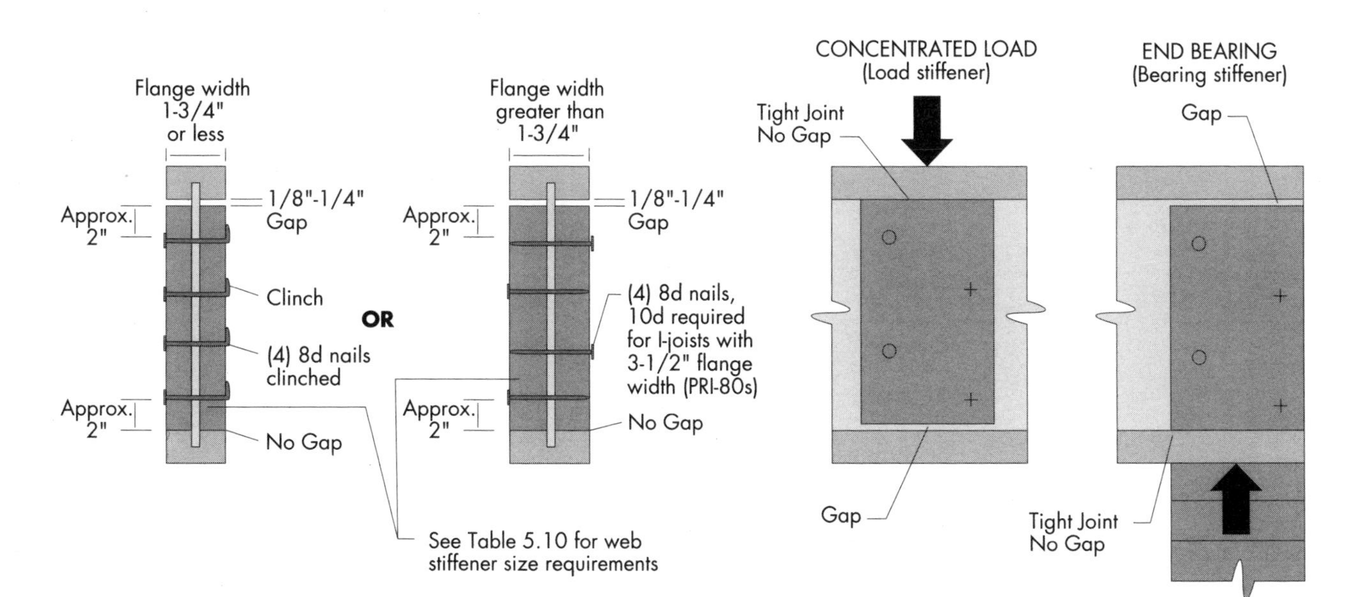

FIGURE 5.16 Performance rated I-joist web stiffener requirements.

stiffener supports the I-joist along a vertical axis as it was designed. (In this application the web stiffener acts very much like a backer block.)

There are two kinds of web stiffeners: *bearing stiffeners* and *load stiffeners.* They are differentiated by where the gap between the slightly undersized stiffener and the top or bottom flange is located.

Bearing stiffeners are located at the reactions, both intermediate and exterior, when required. I-joists subjected to the normal residential uniform loads and installed in accordance with the allowable spans either printed on the joist or located in Tables 5.4–5.7 *do not* need bearing stiffeners at any support.

Load stiffeners are located between supports where significant point loads are applied to the top flange of an I-joist.

Physical Description. Web stiffener blocks may be made up of lumber, rim board, or wood structural panels (WSP). Minimum grade of WSP would be Rated Sheathing; minimum lumber grade—utility grade SPF (south) or better. Any Performance Rated Rim Board product would also work satisfactorily.

Ideally, the depth of the web stiffener should equal the distance between the flanges of the joist minus ⅛–¼ in. (3–6 mm). For bearing stiffeners, this gap is placed between the stiffener and the bottom of the top flange. For load stiffeners, the gap is located at the bottom of the stiffener.

Recommendations. The following recommendations are applicable to APA PRI-400 Performance Rated I-joists. For other I-joists, contact the manufacturer for specific web stiffener recommendations.

1. A *bearing stiffener* is required in all engineered applications with design end reactions greater than 1550 lbf (6.9 kN). (For PRI-90 a stiffener is required when end-reactions exceed 1,885 lbf.) The gap between the stiffener and the flange is at the top.
2. A *load stiffener* is required at locations where a concentrated load greater than 1500 lbf (6.67 kN) is applied to the I-joist's top flange between supports, or in the case of a cantilever, anywhere between the cantilever tip and the support. In these applications *only,* the gap between the load stiffener and the flange is at the bottom.
3. A *bearing stiffener* is required when the I-joist is supported in a hanger and the sides of the hanger do not extend up to, and support, the top flange. The gap between the stiffener and flange is at the top.
4. When using APA Performance Rated I-Joists for residential construction at their design uniform load and span, *bearing stiffeners* are not required at support points.
5. Web stiffeners are to be sized in accordance with Table 5.10.

TABLE 5.10 Web Stiffener Size Required

I-joist flange width (in.)	Web stiffener size—each side of web
1½	15/32 in. × 2 5/16 in. minimum width
1¾	19/32 in. × 2 5/16 in. minimum width
2 5/16	1 in. × 2 5/16 in. minimum width
2½	1 in. × 2 5/16 in. minimum width
3½	1½ in. × 2 5/16 in. minimum width

For SI: 1 in. = 25.4 mm.

5.8 SPECIAL DESIGN CONSIDERATIONS—FLOORS

5.8.1 Non-Load-Bearing Floor Cantilever

Discussion. Non-load-bearing cantilevers or balconies may be constructed using either continuous APA PRIs (Fig. 5.17) or by adding lumber extensions (Fig. 5.18) to the I-joist. Continuous I-joist cantilevers are limited to ¼ the adjacent span when supporting uniform loads only. For applications supporting concentrated loads at the end of the cantilever, such as a wall, see Fig. 5.19.

Unless otherwise engineered, cantilevers are limited to a maximum of 4 ft when supporting uniform loads only. Blocking is required at the cantilever support as shown.

Uniform floor load must not exceed 40 psf (1.92 kN/m^2) live load and 10 psf (0.48 kN/m^2) dead load. The balcony load must not exceed 60 psf (2.87 kN/m^2) live load and 10 psf (0.48 kN/m^2) dead load.

5.8.2 Cantilever Details for Vertical Building Offset (Concentrated Wall Load)

Discussion. I-joists may also be used in cantilever applications supporting a concentrated load applied to the end of the cantilever, such as with a vertical building offset. For all cantilever-end concentrated load applications, the cantilever is limited

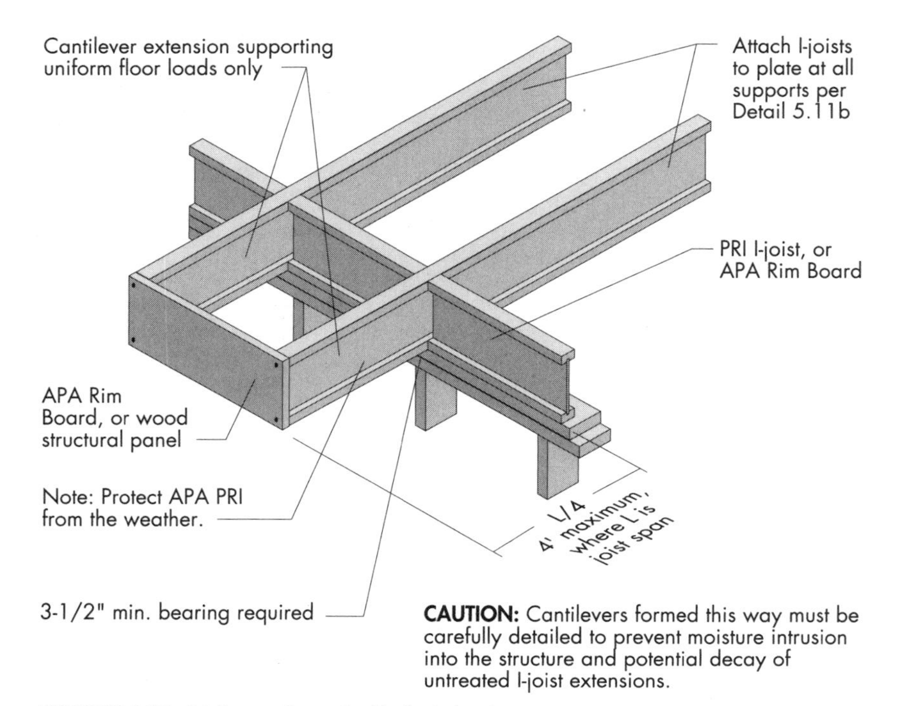

FIGURE 5.17 I-joist cantilever details for balconies.

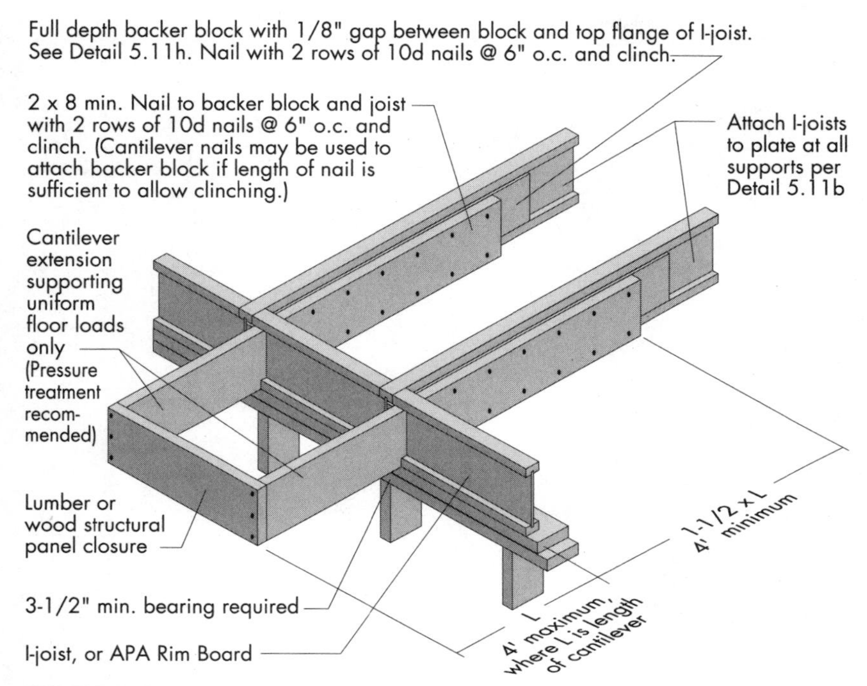

FIGURE 5.18 Lumber cantilever details for balconies.

to 2 ft (610 mm) maximum. In addition, blocking is required along the cantilever support and for 4 ft (1.2 m) on each side of the cantilever area. For 40 psf (1.92 kN/m^2) live loads and 10 psf (0.48 kN/m^2) dead loads, I-joists at their maximum allowable floor spans subject to the roof loads and layout (see Table 5.11) and as illustrated in Fig. 5.19, three ways of reinforcing are allowed in load-bearing cantilever applications: reinforcing sheathing applied to one side of the I-joist (method 1), reinforcing sheathing applied to both sides of the joist (method 2), or the use of double I-joists (alternative method 2).

Example The building designer wants to use a 9½ in. deep I-joist for a second story floor with a 2 ft offset. The roof is a preengineered truss with a 28 ft span, and the total load on the roof is 45 psf (30 psf live load + 15 psf dead load).

Referring to Table 5.11, for this example, no reinforcement is needed if the joists are spaced 12 or 16 in. on center. If the joists are spaced 19.2 in. on center, any of the reinforced cantilever details may be used. If the floor joists are placed 24 in. on center, a deeper joist or closer spacing must be used.

5.8.3 Cutting Holes in the Webs of I-Joists

Discussion. One of the benefits of using I-joists in floor construction is that holes may be cut in the joist webs to accommodate electrical wiring, plumbing lines, and other mechanical systems, therefore minimizing the depth of the floor system. See below under Calculation of Maximum-Size Hole at a Given Location for an ex-

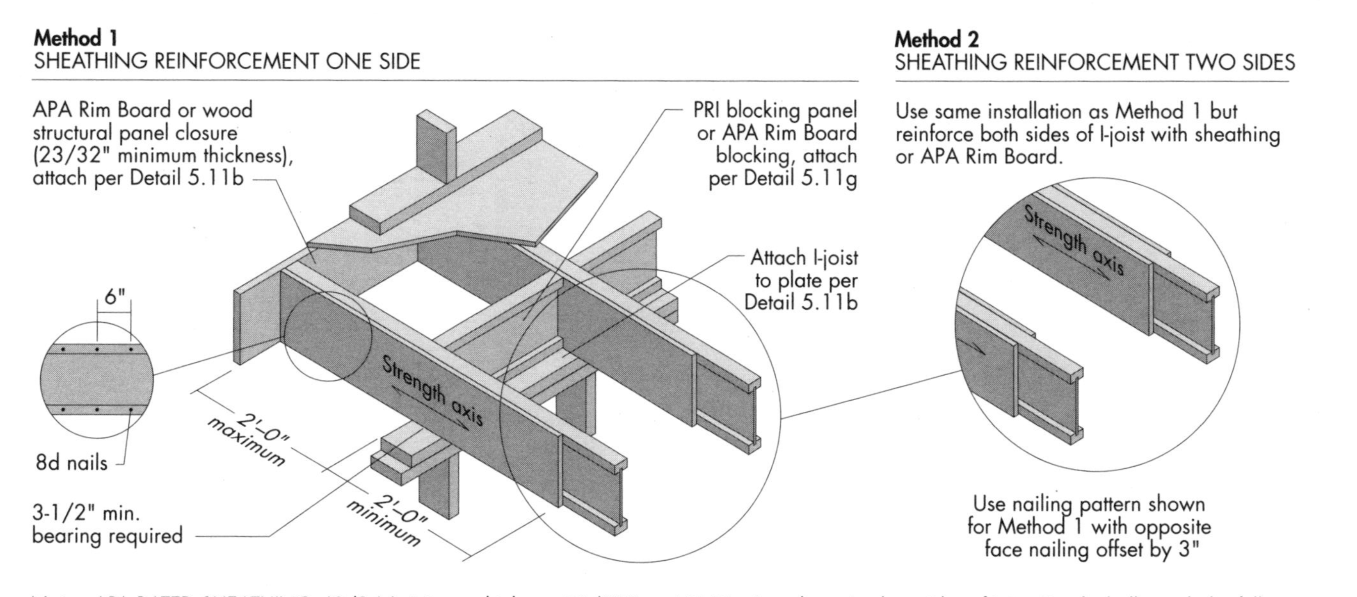

FIGURE 5.19 Methods 1 and 2: Cantilever reinforcement details for vertical building offset.

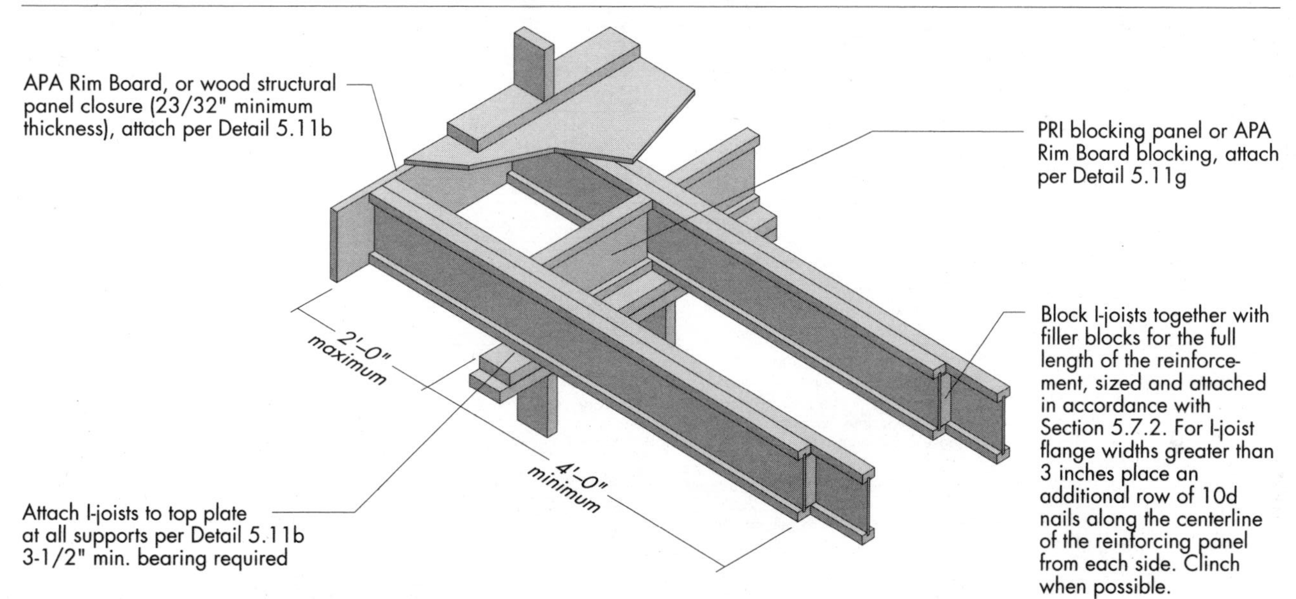

FIGURE 5.19 Alternate Method 2: Cantilever reinforcement details for vertical building offset. (*Continued*)

TABLE 5.11 PRI Cantilever Reinforcement Required

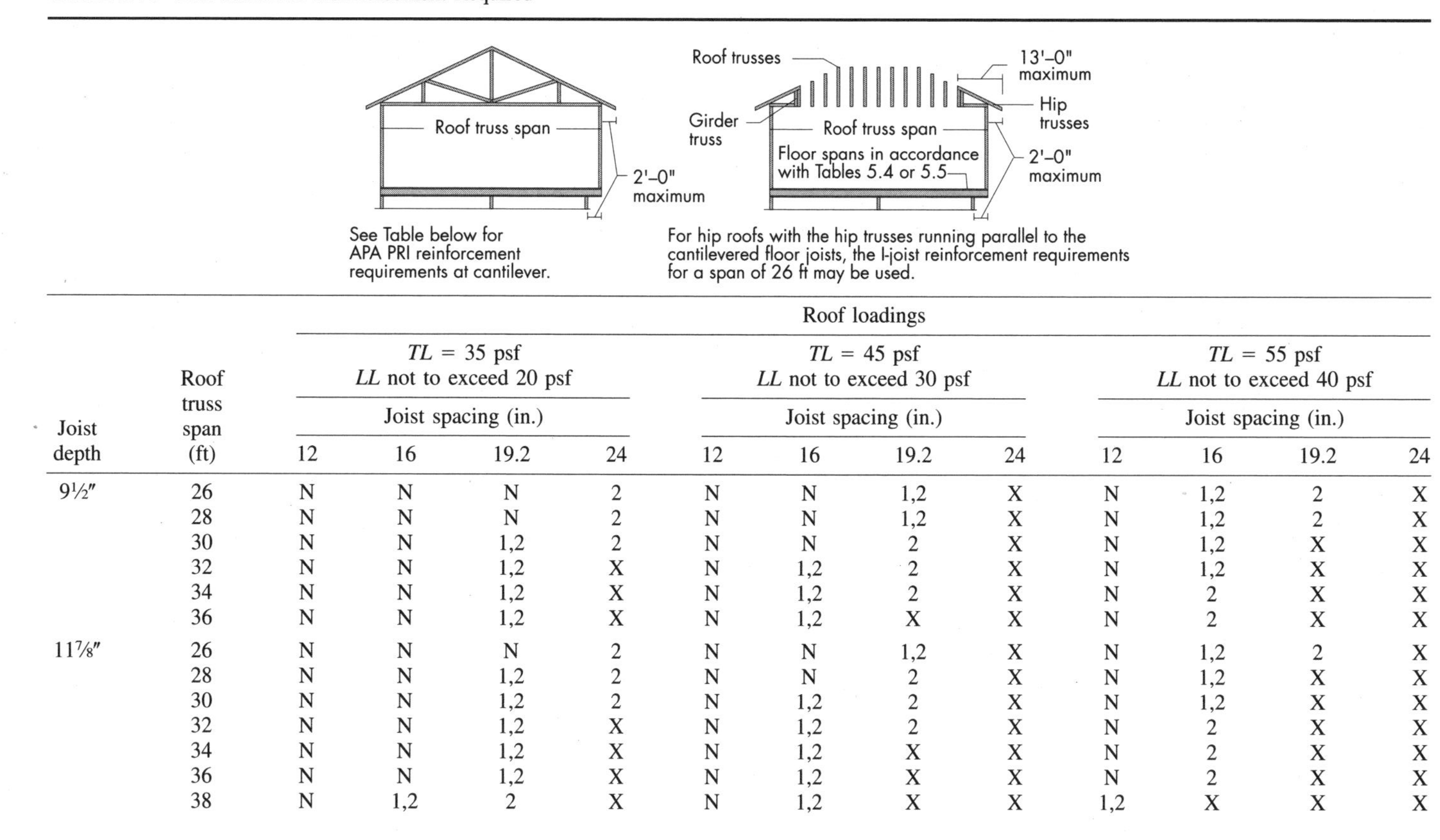

		Roof loadings											
		TL = 35 psf *LL* not to exceed 20 psf				*TL* = 45 psf *LL* not to exceed 30 psf				*TL* = 55 psf *LL* not to exceed 40 psf			
	Roof truss	Joist spacing (in.)				Joist spacing (in.)				Joist spacing (in.)			
Joist depth	span (ft)	12	16	19.2	24	12	16	19.2	24	12	16	19.2	24
9½″	26	N	N	N	2	N	N	1,2	X	N	1,2	2	X
	28	N	N	N	2	N	N	1,2	X	N	1,2	2	X
	30	N	N	1,2	2	N	N	2	X	N	1,2	X	X
	32	N	N	1,2	X	N	1,2	2	X	N	1,2	X	X
	34	N	N	1,2	X	N	1,2	2	X	N	2	X	X
	36	N	N	1,2	X	N	1,2	X	X	N	2	X	X
11⅞″	26	N	N	N	2	N	N	1,2	X	N	1,2	2	X
	28	N	N	1,2	2	N	N	2	X	N	1,2	X	X
	30	N	N	1,2	2	N	1,2	2	X	N	1,2	X	X
	32	N	N	1,2	X	N	1,2	2	X	N	2	X	X
	34	N	N	1,2	X	N	1,2	X	X	N	2	X	X
	36	N	N	1,2	X	N	1,2	X	X	N	2	X	X
	38	N	1,2	2	X	N	1,2	X	X	1,2	X	X	X

14″	26	N	N	N	1,2	N	N	N	2	N	N	1,2	X
	28	N	N	N	1,2	N	N	1,2	X	N	N	2	X
	30	N	N	N	2	N	N	1,2	X	N	1,2	2	X
	32	N	N	N	2	N	N	1,2	X	N	1,2	2	X
	34	N	N	N	2	N	N	1,2	X	N	1,2	X	X
	36	N	N	1,2	2	N	N	2	X	N	1,2	X	X
	38	N	N	1,2	X	N	1,2	2	X	N	1,2	X	X
	40	N	N	1,2	X	N	1,2	2	X	N	2,2	X	X
16″	26	N	N	N	1,2	N	N	1,2	2	N	N	1,2	X
	28	N	N	N	1,2	N	N	1,2	X	N	1,2	2	X
	30	N	N	N	2	N	N	1,2	X	N	1,2	2	X
	32	N	N	N	2	N	N	1,2	X	N	1,2	2	X
	34	N	N	1,2	2	N	N	2	X	N	1,2	X	X
	36	N	N	1,2	2	N	1,2	2	X	N	1,2	X	X
	38	N	N	1,2	X	N	1,2	2	X	N	2	X	X
	40	N	N	1,2	X	N	1,2	2	X	N	2	X	X
	42	N	N	1,2	X	N	1,2	X	X	N	2	X	X

For SI: 1 in. = 25.4 mm, 1 ft = 304.8 mm.

1. N = no reinforcement required. 1 = PRIs reinforced with $^{23}/_{32}$ in. (18.3 mm) wood structural panel on one side only; 2 = PRIs reinforced with $^{23}/_{32}$ in. (18.3 mm) wood structural panels on both sides or double I-joists; X = Try a deeper joist or closer spacing.

2. Maximum load must be: 15 psf (0.72 kN/m^2) roof dead load, 50 psf (2.4 kN/m^2) floor total load, and 80 plf (0.12 kN/m) wall load. Wall load is based on 3 ft, 0 in. (0.91 m) maximum width window or door opening. For larger openings or multiple 3 ft, 0 in. (0.91 m) width openings spaced less than 6 ft, 0 in. (1.82 m) on-center, additional joists between the openings cripple studs may be required.

3. Table applies to joists 12–24 in. (305–619 mm) o.c. Use 12 in. o.c. (305 mm) requirements for lesser I-joist spacings.

ample calculation for determining the allowable hole size for any specific application.

Rules for Cutting holes in PRI Joists (see Figs. 5.20 and 5.21)

1. The distance between the inside edge of the support and the centerline of any hole must not be less than that shown in Table 5.12. (The minimum distance from the centerline of the hole to the inside edge of the support is the *lesser* of two times the hole diameter or 12 in. plus ½ of the hole diameter. This criterion is built into the table but must be considered when calculating hole locations.)
2. I-joist top and bottom flanges must NEVER be cut, notched, or otherwise modified.
3. Whenever possible, field-cut holes should be centered on the middle depth of the web.
4. The maximum-size hole that can be cut into an I-joist web must equal the clear distance between the flanges of the I-joist minus ¼ in. (6 mm). A minimum of ⅛ in. (3 mm) should always be maintained between the top or bottom of the hole and the adjacent I-joist flange.
5. The sides of square holes or longest sides of rectangular holes should not exceed ¾ of the diameter of the maximum round hole permitted at that location.

Never drill, cut or notch the flange, or over-cut the web.

Holes in webs should be cut with a sharp saw.

For rectangular holes, avoid over cutting the corners, as this can cause unnecessary stress concentrations. Slightly rounding the corners is recommended. Starting the rectangular hole by drilling a 1" diameter hole in each of the 4 corners and then making the cuts between the holes is another good method to minimize damage to I-joist.

FIGURE 5.20 Rules for cutting and notching PRI I-joists.

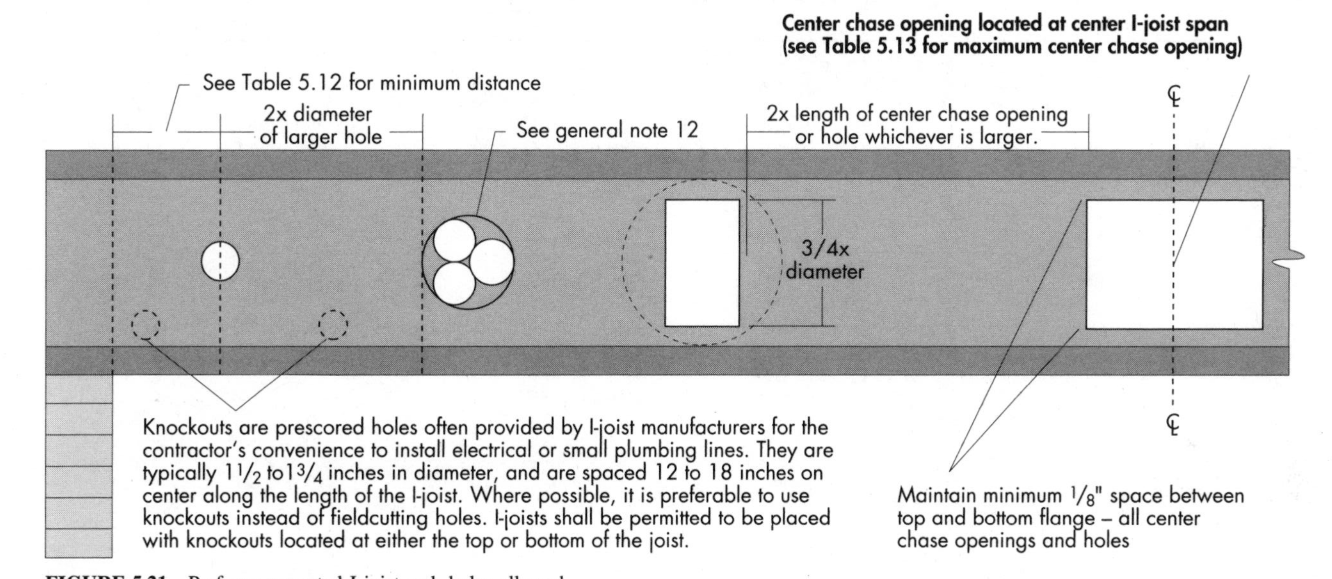

FIGURE 5.21 Performance rated I-joist web holes allowed.

TABLE 5.12 Location of Circular Holes in APA PRI I-Joists

Joist depth	Joist designation	Minimum distance from inside face of any support to center of hole (ft-in.) Round hole diameter (in.)														
		2	3	4	5	6	$6\frac{1}{4}$	7	8	$8\frac{5}{8}$	9	10	$10\frac{3}{4}$	11	12	$12\frac{3}{4}$
$9\frac{1}{2}$″	PRI-20	0-6	1-0	2-0	3-6	4-6	5-0									
	PRI-30	1-0	2-0	3-0	4-6	5-6	6-0									
	PRI-40	1-0	2-0	3-0	4-0	5-0	5-6									
	PRI-50	1-6	2-6	4-0	5-0	6-6	7-0									
	PRI-60	2-0	3-0	4-0	5-0	6-6	7-0									
$11\frac{7}{8}$″	PRI-20	0-6	0-6	1-0	1-0	2-0	2-6	3-6	5-0	6-0						
	PRI-30	0-6	0-6	1-0	2-0	3-0	3-6	4-6	6-6	7-6						
	PRI-40	0-6	0-6	1-6	2-6	3-6	4-0	4-6	6-0	7-0						
	PRI-50	0-6	0-6	1-0	2-6	4-0	4-6	5-6	7-0	8-0						
	PRI-60	0-6	1-6	3-0	4-0	5-0	5-6	6-6	8-0	9-0						
	PRI-70	0-6	1-6	3-0	4-6	6-0	6-0	7-6	9-0	9-5						
	PRI-80	2-0	3-6	4-6	6-0	7-0	7-6	8-6	10-0	11-0						
	PRI-90	1-0	2-0	3-6	5-0	6-0	6-6	7-6	9-0	9-6						
14″	PRI-40	0-6	1-0	2-0	3-0	4-0	4-0	4-6	5-6	6-0	6-6	8-0	9-6			
	PRI-50	0-6	0-6	1-0	1-0	1-0	1-6	2-6	4-6	6-0	6-6	8-6	10-0			
	PRI-60	0-6	1-0	1-0	1-6	3-0	3-6	4-6	6-0	7-0	7-6	9-0	10-6			
	PRI-70	0-6	0-6	1-0	1-0	2-6	3-0	4-0	6-0	7-0	8-0	10-0	11-6			
	PRI-80	0-6	2-0	3-0	4-6	5-6	6-0	7-0	8-6	9-6	10-0	11-6	13-0			
	PRI-90	0-6	0-6	1-6	3-0	4-6	5-0	6-0	7-6	8-6	9-6	11-0	12-0			
16″	PRI-40	0-6	0-6	1-0	1-0	2-0	2-0	3-0	4-0	4-6	5-0	6-0	7-0	7-0	9-0	10-6
	PRI-50	0-6	0-6	1-0	1-0	1-0	1-6	1-6	1-6	2-0	2-6	5-0	6-6	7-0	9-0	10-6
	PRI-60	0-6	0-6	1-0	1-0	1-0	1-6	2-0	3-6	4-6	5-0	6-6	8-0	8-6	10-6	12-0
	PRI-70	0-6	0-6	1-0	1-0	1-0	1-6	1-6	3-0	4-6	5-0	7-0	8-6	9-6	11-6	13-0
	PRI-80	0-6	0-6	1-0	2-0	3-6	4-0	5-0	6-6	7-6	8-0	10-0	11-0	11-6	13-6	15-0
	PRI-90	0-6	0-6	1-0	1-0	2-6	3-0	4-0	5-6	6-6	7-6	9-6	10-6	11-0	13-0	14-6

For SI: 1 in. = 25.4 mm. 1 ft = 304.8 mm.

1. Above tables may be used for I-joist spacing of 24 in. on-center or less and are valid for maximum spans based on 10 psf (0.48 kN/m^2) dead load + 40 psf (1.92 kN/m^2) live load as well as those based on 20 psf (1.2 kN/m^2) dead load + 40 psf (1.92 kN/m^2) live load.

2. Hole location distance is measured from inside face of any supports to center of hole.

3. Distances in this chart are based on uniformly loaded joists that are placed at or less than the maximum span for the loads given in note 1.

4. Table 5.12 is based on the maximum span for the load cases listed in note 1 above for each I-joist designation. The value in the table is based on the combination that results in the greatest minimum distance.

6. Where more than one hole is necessary, the distance between adjacent hole edges must exceed twice the diameter of the largest round hole or twice the size of the largest square hole (or twice the length of the longest side of the longest rectangular hole) and each hole must be sized and located in compliance with the requirements of Table 5.12. When a center chase opening is being placed within the web of an I-joist in conjunction with additional holes, the edge of the holes must not be placed any closed to the edge of the center chase opening than two times the length of the center chase opening. All center chase openings must be sized in accordance with Table 5.13.
7. A knockout is not considered a hole, may be utilized anywhere it occurs, and may be ignored for purposes of calculating minimum distances between holes.
8. One and one-half in. holes must be permitted anywhere in a cantilevered section of a PRI Joist. Holes of greater size may be permitted subject to verification.
9. A 1½ in. (38 mm) hole can be placed anywhere in the web provided that it meets the requirements of item 6 above.
10. All holes must be cut in a workmanlike manner in accordance with the restrictions listed above.
11. No span should contain more than three holes. A group of round holes may be permitted if they meet the requirements for a single round hole circumscribed around them. This group of holes will be considered a single hole in counting the maximum number of holes permitted in a single span.

Example A residential housing designer wants to provide guidance on his house plans to show the contractor where 6 in. diameter holes may be placed in the web of the floor joists. In his design he has specified 11⅞ in. PRI-70 floor joists that are spaced at 16 in. on center and are used in both simple and multiple span applications at spans of 16 ft, 3 in. measured between inside faces of supports.

From Table 5.12, an 11⅞ in. PRI-70 with a maximum hole size of 6 in. yields a required *minimum* end-distance of 6 ft. That means that a 6 in. diameter hole may be placed in a 4 ft, 3 in. long area centered in the middle of any 16 ft, 3 in. [16 ft, 3 in. $-$ (2 $\times$ 6 ft, 0 in.) = 4 ft, 3 in.]. If the designer reduces the diameter of the hole to 5 in., the area in the center of the span expands to 7 ft, 3 in. long [16 ft, 3 in. $-$ (2 $\times$ 4 ft, 6 in.) = 7 ft, 3 in.]. See Example figure below.

Note: The designer's notes and drawings should reflect the information provided above under Rules for Cutting Holes in PRI Joists.

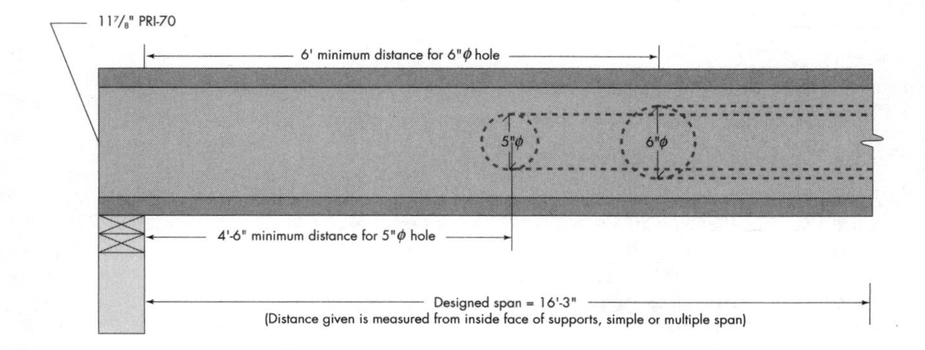

TABLE 5.13 Maximum Size of Rectangular Duct/Center Chase Openings That May Be Placed at Center of Span in Webs of PRI I-joists

I-joist designation	Joist spacing (in.)	Center chase opening length (in.)						
		8	10	12	14	16	18	20
PRI-20	12	X	X	X				
	16	X	X					
	19.2	X	X					
	24	X						
PRI-30	12	X	X	X				
	16	X						
	19.2							
	24							
PRI-40	12	X	X	X	X	X		
	16	X	X	X	X			
	19.2	X	X	X				
	24	X	X					
PRI-50	12	X	X					
	16	X						
	19.2	X						
	24	X						
PRI-60	12	X	X	X	X			
	16	X	X					
	19.2	X						
	24	X						
PRI-70	12	X	X	X	X			
	16	X	X					
	19.2	X	X					
	24	X	X					
PRI-80	12	X	X	X	X	X	X	
	16	X	X	X	X			
	19.2	X	X					
	24	X	X					
PRI-90	12	X	X	X	X	X	X	
	16	X	X	X	X			
	19.2	X	X					
	24							

For SI: 1 in. = 25.4 mm.

5.8.4 Center Chase Openings

Discussion. A center chase opening is a large rectangular hole that is often required within the web of an I-joist. These openings are used to provide passage for ventilation ducts through the floor joists and *must be located centered on the center of the I-joist span.* While rectangular holes can be cut in the webs of I-joists using Section 5.8.3, the size of rectangular holes generated by this method is often insufficient for this use.

Table 5.13 and the rules discussed in below under Rules for Cutting Center Chase Openings in PRI Joists have been generated for these applications. Based on the depth and spacing of the PRI Joists, this table tells the user the largest size center chase opening that may be placed *at center span only.*

Rules for Cutting Center Chase Openings in PRI Joists (see Figs. 5.20 and 5.21)

1. The maximum length of center chase openings cut in a PRI joist must be as shown in Table 5.13.
2. The center chase opening must ALWAYS be centered at the center of the I-joist span.
3. I-joist top and bottom flanges must NEVER be cut, notched, or otherwise modified.
4. The maximum depth of the center chase opening must equal the clear distance between the flanges of the I-joist minus ¼ in. (6 mm). A minimum of ⅛ in. (3 mm) should always be maintained between the top or bottom of the center chase opening and the adjacent I-joist flange.
5. When a center chase opening is being placed within the web of an I-joist in conjunction with additional holes, the edge of the holes must not be placed any closed to the edge of the center chase opening than two times the length of the center chase opening. All holes and center chase openings must be sized in accordance with Tables 5.12 and 5.13.
6. A knockout is not considered a hole and may be utilized wherever it occurs and may be ignored for purposes of calculating minimum distances between holes and center chase openings.
7. All center chase openings must be cut in a workmanlike manner in accordance with the restrictions listed above.

Example A residential housing designer wants to provide guidance on his house plans to show the contractor where an air conditioning center chase opening may be placed through an opening in the web of the floor joists. In his design, he has specified 11⅞ in. PRI-70 floor joists that are spaced at 16 in. on center and are used in both simple and multiple span applications at spans of 16 ft, 3 in. measured between inside faces of supports.

From Table 5.13, given a PRI-70, a 10 in. wide center chase opening can be placed at midspan of either the simple or multiple spans as long as the on-center spacing is 19.2 in. or less.

Note: The designer's notes on plans should reflect the information provided above under Rules for Cutting Center Chase Openings in PRI Joists.

5.8.5 Shear Transfer at Engineered Wood Floors

Discussion. Over the last couple of decades the use of engineered wood products for the fabrication of lightweight wood floor systems has increased in popularity. Performance Rated I-Joists, glued laminated beams, Performance Rated Rim Board, and laminated veneer lumber (LVL) are used as a framing system and feature precision, high strength, and superior quality.

As the use of these products expands beyond conventional construction, it is inevitable that questions will arise concerning their use in engineered applications. One such question concerns the use of APA I-joists and APA Performance Rated Rim Board around the perimeter of the structure in engineered applications where the wood floors are designed as diaphragms and the walls above are designed as shear walls.

In such applications, the engineered wood product forming the rim joist/rim board around a floor system is subject to a number of loads not normally considered in conventional, nonengineered applications. These additional engineering considerations include:

- Diaphragm perimeter nailing
- Transfer of shear wall forces from the walls above into the foundation/wall framing below
- Shear transfer of diaphragm loads to the foundation/wall framing below

Such design forces result in attachment requirements that often exceed conventional fastening schedules found in the model building codes. The challenge to the designer is to detail these critical connections such that all of the applied loads are transferred through the connection in an economical and practical way.

Mixing Engineered Wood Products and Sawn Lumber—Preventing Incompatibilities. Mixing engineered wood products and sawn lumber in a roof or floor system should *never* be done without a careful analysis of the potential consequences.

The use of sawn lumber blocking or rim boards in conjunction with wood I-joists in a floor system is a typical example of when these products might be incorrectly used together. In such situations blocking and rim board materials are used, at least in part, to assist the wood I-joist in distributing vertical loads through the floor system into the structure below. As the building materials in the structure reach equilibrium moisture conditions with their surroundings, sawn lumber used as blocking and rim boards may shrink while the I-joists do not. As a result, lumber components are not available to carry the applied vertical load that they were designed to carry, thus overloading the I-joists, as illustrated in Fig. 5.7.

There are applications, such as diaphragm blocking, squash blocks, and backer or filler blocking, where sawn lumber is acceptable for use in conjunction with engineered wood products.

Shear Capacity of APA Engineered Wood Products Used as Rim Joists

APA Rim Board (and Rim Board Plus). Each of the major model building codes are consistent in prescriptively requiring the rim joist to be attached to structural framing below with 8d (3.3 × 64 mm) common toe nails spaced at 6 in. (150 mm) on center. While this attachment schedule is sufficient to develop 180–200 plf (0.26–0.29 kN/m) of rim board required for the conventional construction provisions of the model building codes, it falls short of developing the internal shear capacity of the APA rim board products, which is 2300 plf (3.38 kN/m) for 1 in. (25 mm) thick product and 3000 plf (4.41 kN/m) for 1⅛ in. (29 mm) thick board. Spacing toenails closer than 6 in. (150 mm) on center to increase the load capacity of the rim board assembly must be viewed with extreme caution. Closer spacing may cause splitting in some rim board products, and an increased load capacity may not actually be achieved using this technique.

Code-required minimum nailing into the edge of a rim board product to anchor the perimeter of the floor diaphragm and other framing from above can only develop 180–200 plf (0.26–0.29 kN/m), depending on the type of rim board used. This is sufficient capacity to meet the conventional construction requirements of the codes.

Framing anchors or blocking as shown in Figs. 5.22 and 5.23, along with the information in Tables 5.14–5.18, may be used to develop the additional capacity required.

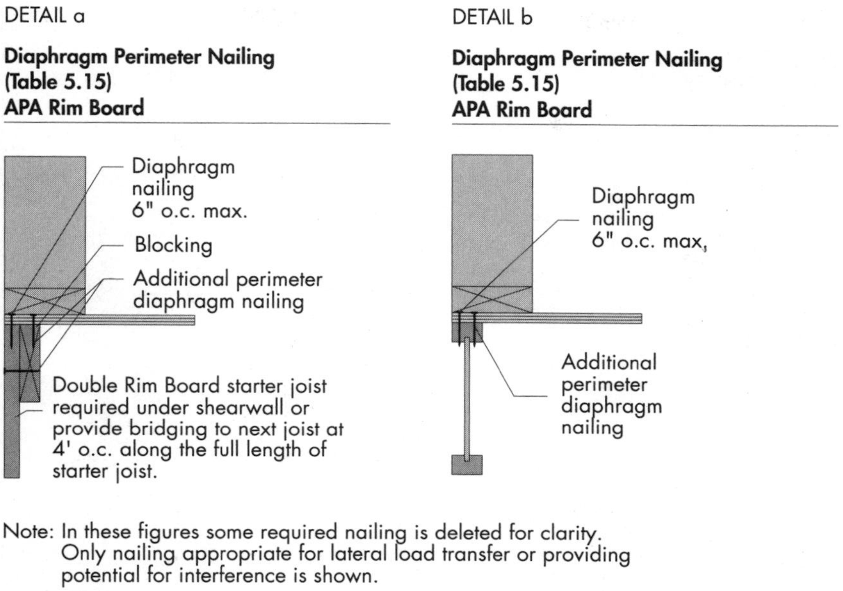

FIGURE 5.22 (a,b) Diaphragm perimeter nailing into engineered wood rim boards.

APA PRI I-Joists. Similar to rim board products, PRI I-joists when used as rim joists have considerably more capacity than the code-required nailing—8d common (3.3 × 64 mm) toenails at 6 in. (150 mm) on center—will develop. By virtue of their flanges, however, I-joists are considerably easier to attach to structural framing both above and below the joist.

On I-joists with a 1½ in. (38 mm) wide flange there is sufficient room to place a double row of nails in both the top and bottom flanges. I-joists with flanges 2⁵⁄₁₆ in. (59 mm) and wider can easily accommodate four staggered rows of nails. The full lateral load capacity of the I-joist (1000 plf [1.5 kN/m]) may be achieved in most cases. Table 5.14 is provided to aid the designer in selecting, for various flange widths and PRI series, the appropriate nailing schedules required to achieve the desired design load. (Note that the 1000 plf [1.5 kN/m] capacity mentioned above and shown in Table 5.14 below is a factored load and, as such, already includes a 1.6 adjustment for duration of load.)

Diaphragm Perimeter Nailing. In conventional nonengineered construction applications, the floor or roof diaphragm is prescriptively described in the model building codes. In these applications, the floor sheathing—normally wood structural panel sheathing—is attached to the floor perimeter framing with 8d (3.3 × 64 mm) common nails at 6 in. (150 mm) on center. In engineered applications, the design loads and geometry of the structure may dictate a diaphragm perimeter-nailing schedule of 4, 2½, or even 2 in. (100, 64, or 51 mm) on center. The performance of some engineered wood products, such as engineered wood rim boards, can be adversely impacted by these closer nail spacing schedules. It is important in such applications to develop design details to accommodate these loads and their corresponding close nailing schedules.

DETAIL a

Diaphragm Perimeter Nailing (Panel-to-Lumber, Table 5.15; Lumber-to-Lumber, Table 5.16) – APA Rim Board

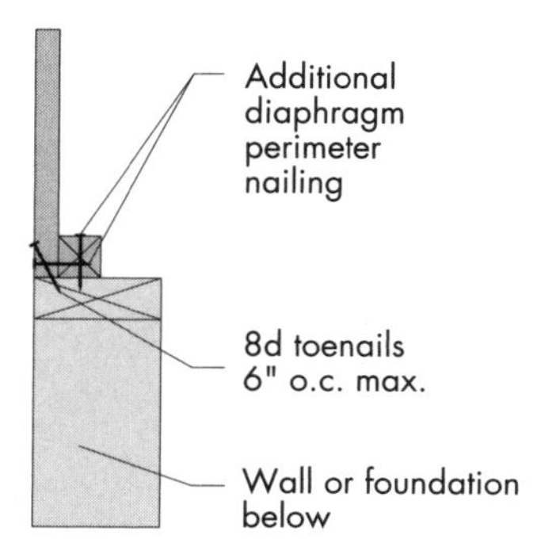

DETAIL b

Diaphragm Perimeter Nailing (Panel-to-Lumber, Table 5.16) – APA I-Joist Rim Board

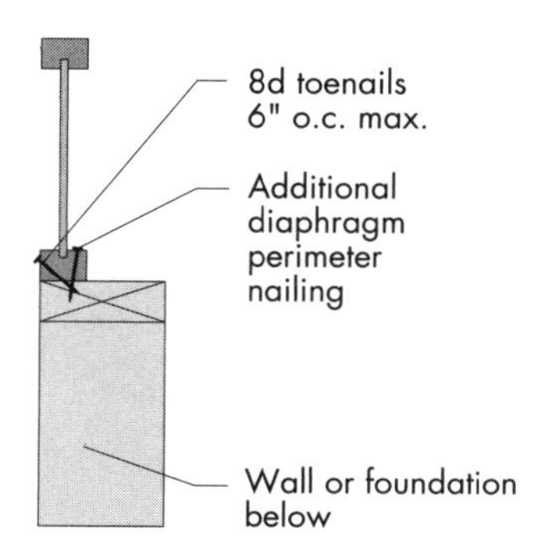

DETAIL c

Diaphragm Perimeter Nailing Using Framing Anchors (Table 5.17) – APA Rim Board

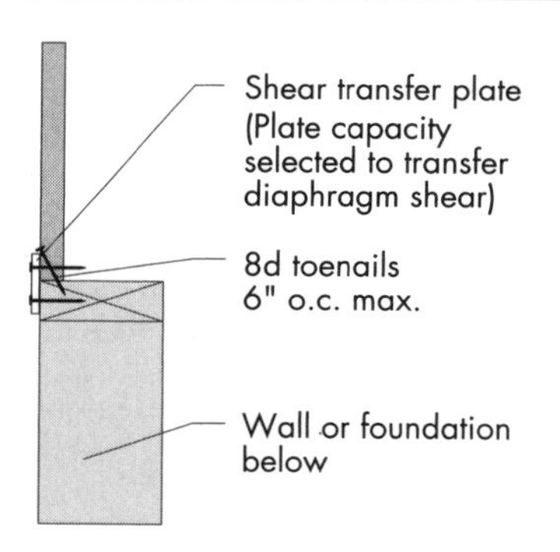

DETAIL d

Diaphragm Perimeter Nailing Using Framing Anchors (Table 5.17) – APA I-Joist Rim Board

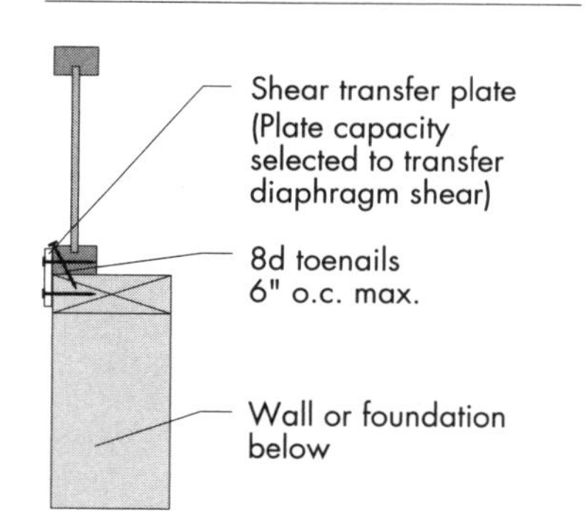

Note: Place shear transfer plates between toe nails to prevent splitting of framing.

Note: In these figures some required nailing is deleted for clarity. Only nailing appropriate for lateral load transfer or providing potential for interference is shown.

FIGURE 5.23 (a–d) Shear transfer of diaphragm loads to the foundation/wall framing below.

TABLE 5.14 Lateral Capacity of Face-Nailed Performance Rated I-Joist Flanges

Flange width in.	Joist designation: PRI series	Nail size	Maximum flange nailing in web (no. of rows at on-center spacing (S))	Total no. of nails per foot	Maximum capacity (plf), Flange specific gravity 0.42	0.46	0.49
1½	9½″ PRI-20,	8d common,	2 rows at 12 in.	2	248	256	264
	9½″ PRI-30,	10d box, or	2 rows at 6 in.	4	496	512	528
	11⅞″ PRI-20,	12d box	2 rows at 4 in.	6	736	768	784
	11⅞″ PRI-30		2 rows at 3 in.	8	984	1000	1000
or	or						
1¾	9½″ PRI-50,	10 d common,	2 rows at 12 in.	2	272	280	288
	11-⅞″ PRI-50,	12d common,	2 rows at 6 in.	4	536	560	576
	14″ PRI-50, 16″ PRI-50	or 16d sinker	2 rows at 4 in.	6	808	840	864
		16d common	2 rows at 12 in.	2	384	400	408
			2 rows at 16 in.	4	768	800	816
2$^{5}/_{16}$	14″ PRI-70, 16 ft PRI-70	8d common,	2 rows at 12 in.	2	248	256	264
or	or	10d box, or	2 rows at 6 in.	4	496	512	528
	9½″ PRI-40,	12d box	2 rows at 4 in.	6	736	768	784
	9½″ PRI-60,		2 rows at 3 in.	8	984	1000	1000
	11⅞″ PRI-40,						
	11⅞″ PRI-60,	10d common,	2 rows at 12 in.	2	272	280	288
2½	14″ PRI-40, 14″ PRI-60	12d common,	2 rows at 6 in.	4	536	560	576
	16″ PRI-40, 16″ PRI-60	or 16d sinker	2 rows at 4 in.	6	808	840	864
or	or		4 rows at 6 in.	8	1000	1000	1000
	11⅞″ PRI-80,	16d common	2 rows at 12 in.	2	384	400	408
3½	14″ PRI-80, 16″ PRI-80		2 rows at 6 in.	4	768	800	816

TABLE 5.14 Lateral Capacity of Face-Nailed Performance Rated I-Joist Flanges (*Continued*)

For SI: 1 in. = 25.4 mm, 1 ft = 304.8 mm, 1 lbf = 4.45 kN

Values given above include a 1.6 duration of load adjustment for high wind and seismic design (subject to local code variations).

The above values are based on the assumption that the nailing does not cause excessive splitting of the flange.

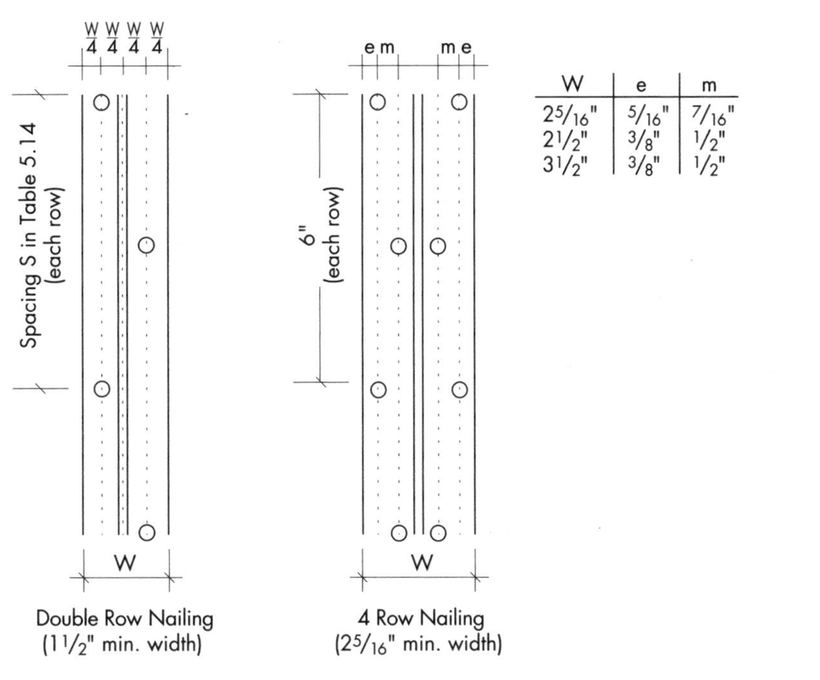

W	e	m
$2^5/_{16}$"	$^5/_{16}$"	$^7/_{16}$"
$2^1/_2$"	$^3/_8$"	$^1/_2$"
$3^1/_2$"	$^3/_8$"	$^1/_2$"

TABLE 5.15 Face-Nail Design Capacities for Attachment of Wood Structural Panels (lbf/nail) (Use for transfer-of-shear nailing shown in Figs. 5.22a or b, 5.23a, and 5.24–5.28)

Nail size (length × diameter)	Specific gravity of main member (G)	Thickness of wood structural panel side member (in.)				
		5/16	3/8	7/16	15/32	19/32
8d (2½ × 0.131 in.)	$G \geq 0.50$	110	114	117	118	128
	$0.50 > G \geq 0.46$	107	109	112	115	125
	$0.46 > G \geq 0.42$	102	104	107	110	120
10d (3 × 0.148 in.)	$G \geq 0.50$	133	136	139	141	152
	$0.50 > G \geq 0.46$	128	131	134	136	147
	$0.46 > G \geq 0.42$	123	125	128	131	141

For SI: 1 in. = 25.4 mm, 1 lbf = 4.45 kN,

1. Nail penetration into the main member of 1⅜ in. (35 mm) for 8d (3.3 × 64 mm) common nails and 1½ in. (38 mm) for 10d (3.8 × 76 mm) common nails is required to use the values listed above.

2. The main member is the member that receives the point of the fastener. The side member is the member that supports the head of the fastener.

3. Values given above include 1.6 duration of load adjustment for high wind and seismic design (subject to local code variations).

4. Main member OSB values are based on Douglas fir-larch species.

5. Main member Structural I plywood values are based on Douglas fir-larch species. Main member plywood Rated Sheathing values are based on plywood with an effective specific gravity of 0.42.

6. Side member wood structural panel values are appropriate for all grades of plywood and OSB.

7. Above design capacities are based on the 1997 edition of the *National Design Specification for Wood Construction* (NDS), except as noted in note 1.

8. Specific gravity (G) of common framing members:

Species	G
Douglas fir-larch	0.50
Hem-fir	0.43
Englemann spruce-Lodgepole pine (MSR 1650f and higher)	0.46
Southern pine	0.55
Spruce-pine-fir	0.42
Spruce-pine-fir (E of 2,000,000 and greater MSR and MEL)	0.50
Structural I plywood	≥0.50
OSB	≥0.50
Plywood Rated Sheathing	≥0.42

9. When the main member is an LVL I-joist flange, contact I-joist supplier for appropriate specific gravity.

TABLE 5.16 Face-Nail Design Capacities for I-Joist Flanges and Lumber Framing (lbf/nail) (Use for transfer-of-shear nailing shown in Figs. 5.23a, 5.23b, and 5.26)

Nail size (length × diameter)	Specific gravity of side member (G)	Specific gravity of main member (G)	Thickness of side member (in.)				
			1	1⅛	1¼	1⅜	1½
8d (2½ × 0.131 in.)	$G \geq 0.50$	$G \geq 0.50$	149	136	123	112	99
		$0.50 > G \geq 0.46$	142	131	118	107	94
		$0.46 > G \geq 0.42$	136	125	114	102	88
	$0.50 > G \geq 0.46$	$G \geq 0.50$	142	131	118	107	94
		$0.50 > G \geq 0.46$	138	126	115	102	91
		$0.46 > G \geq 0.42$	131	120	109	99	86
	$0.46 > G \geq 0.42$	$G \geq 0.50$	133	125	114	99	88
		$0.50 > G \geq 0.46$	130	120	109	94	83
		$0.46 > G \geq 0.42$	126	115	106	173	158
10d (3 × 0.148 in.)	$G > 0.50$	$G \geq 0.50$	189	189	186	173	158
		$0.50 > G \geq 0.46$	181	181	178	165	154
		$0.46 > G \geq 0.42$	173	173	170	157	146
	$0.50 > G \geq 0.46$	$G \geq 0.50$	181	181	178	165	154
		$0.50 > G \geq 0.46$	174	174	171	160	147
		$0.46 > G \geq 0.42$	166	166	165	152	141
	$0.46 > G \geq 0.42$	$G \geq 0.50$	162	173	170	157	146
		$0.50 > G \geq 0.46$	158	166	165	152	141
		$0.46 > G \geq 0.42$	154	170	157	146	134

For SI: 1 in. = 25.4 mm, 1 lbf = 4.45 kN.

1. Nail penetration into the main member of 1½ in. (38 mm) for 8d (3.3 × 64 mm) common nails and 1⅝ in. (41 mm) for 10d (3.8 × 76 mm) common nails is required to use the values listed above.

2. The main member is the member that receives the point of the fastener. The side member is the member that supports the head of the fastener.

3. Values given above include 1.6 duration of load adjustment for high wind and seismic design (subject to local code variations).

4. Above design capacities are based on the 1997 edition of the *National Design Specification for Wood Construction* (NDS), except as noted in note 1.

5. Specific gravity (G) of common framing members:

Species	G
Douglas fir-larch	0.50
Hem-fir	0.43
Englemann spruce-Lodgepole pine (MSR 1650f and higher)	0.46
Southern pine	0.55
Spruce-pine-fir	0.42
Spruce-pine-fir (E of 2,000,000 and greater MSR and MEL)	0.50
Structural I plywood	≥0.50
OSB	≥0.50
Plywood Rated Sheathing	≥0.42

6. When the main member is an LVL I-joist flange, contact I-joist supplier for appropriate specific gravity.

TABLE 5.17 Design Capacities of Nailed Connections with Metal Side Plates (lbf/nail) (Use for transfer-of-shear nailing shown in Figs. 5.23c, 5.23d, and 5.24, and 5.25, in accordance with manufacturer's recommendations)

Nail size (length × diameter)	Thickness/gage of metal side plates	Specific gravity of main member (G)	Thickness of lumber main member (in.)				
			1	1⅛	1¼	1⅜	1½
8d (2½ × 0.131 in.)	3/64 in (18 gage)	$G \geq 0.50$	94	106	118	130	141
		$0.50 > G \geq 0.46$	88	99	110	122	133
		$0.46 > G \geq 0.42$	82	91	102	112	122
	1/16 in. (16 gage)	$G \geq 0.50$	96	107	120	131	144
		$0.50 > G \geq 0.46$	90	101	112	123	134
		$0.46 > G \geq 0.42$	83	93	104	114	123
	5/64 in. (14 gage)	$G \geq 0.50$	98	109	122	134	146
		$0.50 > G \geq 0.46$	91	102	114	125	138
		$0.46 > G \geq 0.42$	85	94	106	117	126
10d (3 × 0.148 in.)	3/64 in. (18 gage)	$G \geq 0.50$	101	114	126	139	152
		$0.50 > G \geq 0.46$	94	106	118	130	142
		$0.46 > G \geq 0.42$	86	98	109	120	131
	1/16 in. (16 gage)	$G \geq 0.50$	102	115	128	141	154
		$0.50 > G \geq 0.46$	96	107	120	131	142
		$0.46 > G \geq 0.42$	88	99	110	122	133
	5/64 in. (14 gage)	$G \geq 0.50$	104	117	130	142	155
		$0.50 > G \geq 0.46$	98	109	122	133	146
		$0.46 > G \geq 0.42$	90	101	112	123	134

For SI: 1 in. = 25.4 mm, 1 lbf = 4.45 kN.

1. The dowel bearing strength of steel = 45 ksi (310 MPa).
2. The main member is the member that receives the point of the fastener. The side plate is the member that supports the head of the fastener.
3. Values given above include 1.6 duration of load adjustment for high wind and seismic design (subject to local code variations).
4. Main member OSB values are based on Douglas fir-larch species.
5. Main member Structural I plywood values are based on Douglas fir-larch species. Main member plywood Rated Sheathing values are based on plywood with an effective specific gravity of 0.42.
6. Above design capacities are based on the 1997 edition of the *National Design Specification for Wood Construction* (NDS).
7. Specific gravity (G) of common framing members:

Species	G
Douglas fir-larch	0.50
Hem-fir	0.43
Englemann spruce-Lodgepole pine (MSR 1650f and higher)	0.46
Southern pine	0.55
Spruce-pine-fir	0.42
Spruce-pine-fir (E of 2,000,000 and greater MSR and MEL)	0.50
Structural I plywood	≥0.50
OSB	≥0.50
Plywood Rated Sheathing	≥0.42

6. When the main member is an LVL I-joist flange, contact I-joist supplier for appropriate specific gravity.

TABLE 5.18 Recommended Design Shear for Staples Sheet Metal Blocking (lbf per staple)[a]

APA panel grade	Gage: Staple	Gage: Sheet metal[b]	Minimum panel thickness (in.): 5/16	3/8	15/32	19/32	23/32
APA Structural I rated sheathing	16	26	16	24	36	51	51
	16	24	16	24	36	51	51
	14[c]	24	–	–	50	75	75
APA rated sheathing	16	26	14	22	32	47	51
	16	24	14	22	32	47	51
	14[c]	24	–	–	45	68	75

For SI: 1 in. = 25.4 mm, 1 lbf = 4.45 kN.
[a] Based on normal duration of load.
[b] Strips 3 in. wide.
[c] Fourteen-gage staples through 26-gage metal strips not recommended.

Figure 5.22a and b illustrate methods that may be used to accommodate 4, 2½, or 2 in. (100, 64, or 51 mm) on-center diaphragm perimeter nailing requirements. (See Table 5.15 for nail capacities.)

Shear Transfer of Diaphragm Loads to the Foundation/Wall Framing Below In cases where either Fig. 5.22a or b is used to transfer higher diaphragm shear values into the diaphragm perimeter framing (the rim board), it is essential that the attachment schedule at the base of this framing be adjusted to accommodate the additional load into the foundation or framing below. The minimum nailing recommendations published in the model building codes for rim board to framing connections are insufficient to transfer the additional diaphragm loads that precipitated the use of 4, 2½, or even 2 in. (100, 64, or 51 mm) on-center diaphragm perimeter nailing schedules. Care must be taken when installing these additional fasteners to prevent splitting of the framing members. In addition, nail and lumber specifications must provide for a minimum depth of penetration that allows full connection capacity.

Figure 5.23a–d is provided to give the designer some examples of methods used to accommodate these loads. (See Tables 5.15–5.17 for nail capacities.)

Transfer of Shear Wall Forces from the Walls Above into the Foundation/Wall Framing Below. In engineered construction, lateral loads are transferred from the roof and floor diaphragms, through the shear walls, and eventually down into the foundation. Because most wood structures today are platform framed (i.e., the interior and exterior walls sit on the floor below), special detailing is required to transfer the forces from shear walls above to the walls or foundation below. As previously discussed, it is not always possible to transfer these forces from a shear wall above directly to the rim board below because of the possibility of splitting the framing member forming the rim board.

For this reason, various methods have been developed to safely transfer forces around this critical connection. Figures 5.24 and 5.25 follow the same pattern, in that they provide for all shear panel edges to occur over and be attached to common framing. Note that prior to making connections into the side of LVL flanges in

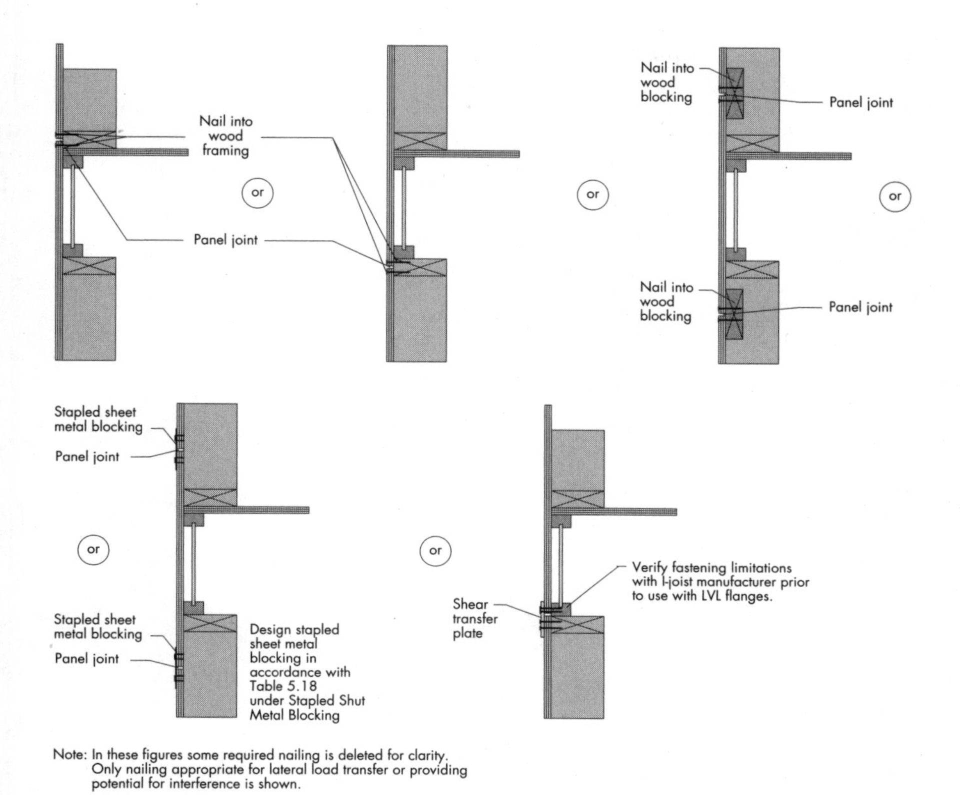

FIGURE 5.24 Transfer of shear wall forces between floors—Performance Rated I-joists.

engineered wood I-joists, the I-joist manufacturers should be contacted for fastener limitations. If connections are made at the web of the I-joist or at an APA rim board, backer blocking should be attached to ensure minimum nail penetration into framing (8d [3.3 × 64 mm] common shear nailing requires 1⅜ in. [35 mm] while 10d [3.8 × 76 mm] common nailing requires 1½ in. [38 mm]). (See Tables 5.15 and 5.16 for nail capacities.)

Similar attachment must be provided in those areas when shear wall shear transfer nailing is not accommodated by the fastener details described in Figs. 5.24 and 5.25. Figure 5.26 shows the transfer of shear wall forces around the floor diaphragm-to-rim board connection and directly into the rim board itself. At the bottom of the rim board additional nailing is required to transfer the shear wall forces into the foundation below. In Figs. 5.27 and 5.28 these shear wall forces are shown being transferred directly into the sill plate. (See Tables 5.15 and 5.16 for nail capacities.)

Framing Anchors. In addition to the direct attachment methods, framing anchors may also be used to transfer forces between the various elements of the structural frame. It is important to install all framing anchors in accordance with manufac-

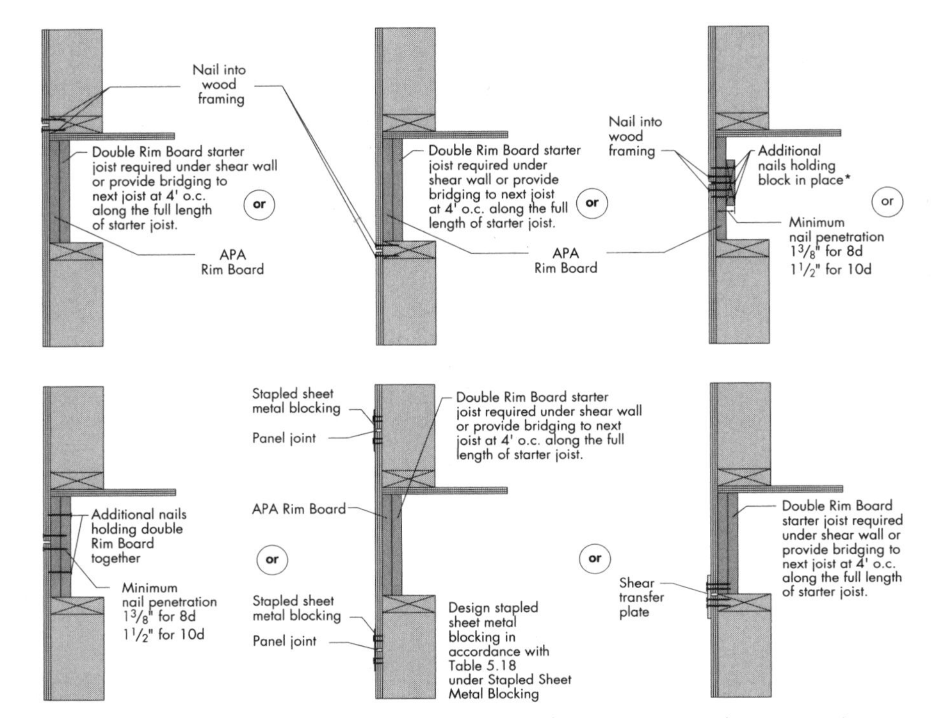

FIGURE 5.25 Transfer of shear wall forces between floors—Performance Rated rim board.

turer's recommendations. If it is necessary to nail into an LVL I-joist flange parallel to the glue lines when installing framing anchors, check with the I-joist manufacturer for nailing limitations. (See Table 5.17 for nail capacities.)

Fastener Design Capacities. Tables 5.15–5.17 are provided to assist the design professional in the proper detailing of these important connections. The capacities presented in the following tables are based on the European Yield Method as presented in the 1997 edition of the National Design Specification for Wood Construction (NDS) except that nail penetration into the main member of 1⅜ in. (35 mm) for 8d (3.3 × 64 mm) common nails and 1½ in. (38 mm) for 10d (3.8 × 79 mm) common nails was assumed. While this penetration is less than that normally used for such calculations, it allows the use of 1⅜ in. (35 mm) and 1½ in. (38 mm) thick main members as shown in Figs. 5.22–5.28.

Stapled Sheet Metal Blocking. Recommended design shear values for stapled galvanized steel sheet metal blocking are given in Table 5.18. Panel edges between framing must be supported by tongue-and-groove joints or panel clips. Recommendations are also applicable to two-layer systems where edge joints of the top layer are staggered from those of the bottom layer.

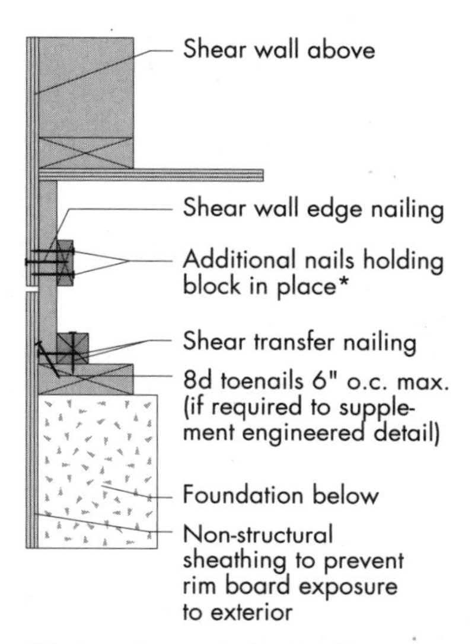

*Engineering analysis using European Yield Method (*1991 National Design Specification*, American Forest and Paper Association) may prove additional blocking unnecessary. If shear wall occurs over starter joist, double rim board may take place of additional blocking.

Note: In these figures some required nailing is deleted for clarity. Only nailing appropriate for lateral load transfer or providing potential for interferance is shown.

FIGURE 5.26 Transfer of shear forces at foundation—performance rated rim board.

Performance is sensitive to staple overdriving, particularly when using 26-gage sheet metal strips. For this reason, it is recommended that full inspection of workmanship be considered when sheet metal blocking is used. Staples should be driven so that the staple crown is flush with the top surface of the metal strip. Install staples with crowns oriented perpendicular to the plywood face grain or panel major axis (see Fig. 5.29).

Example A design professional is tasked with designing a two-story, platform-framed wood structure with a tile roof. As the structure is located in an area of high seismicity (Zone 4) and because of the mass of the roof, it is determined that the shear walls sitting on the second floor have a design requirement of 380 plf. The shear load along the edge of the second floor diaphragm parallel to the shear wall in question is 175 plf in the direction under consideration.

Since the capacity of the Performance Rated rim board with the minimum code-required nailing is 180 plf, no additional fastening is required at the perimeter of the diaphragm along this edge.

The design professional has selected 7/16 in. OSB for use as wall sheathing. Because all of the capacity at the floor sheathing-to-rim board connection available to transfer the diaphragm shear has been effectively utilized, the designer recognizes that it is not

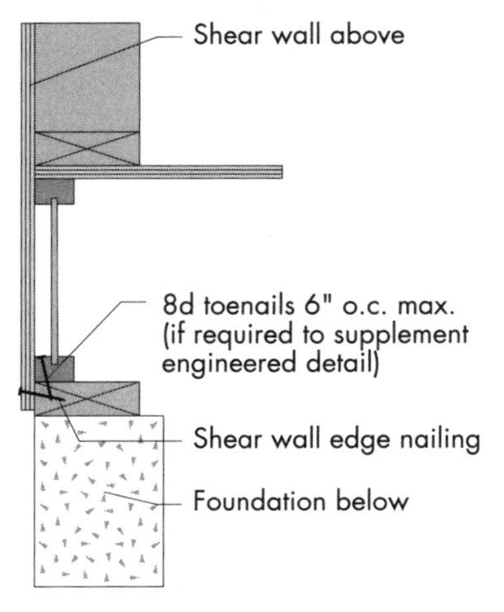

FIGURE 5.27 Transfer of shear forces at foundation—Performance Rated I-joist.

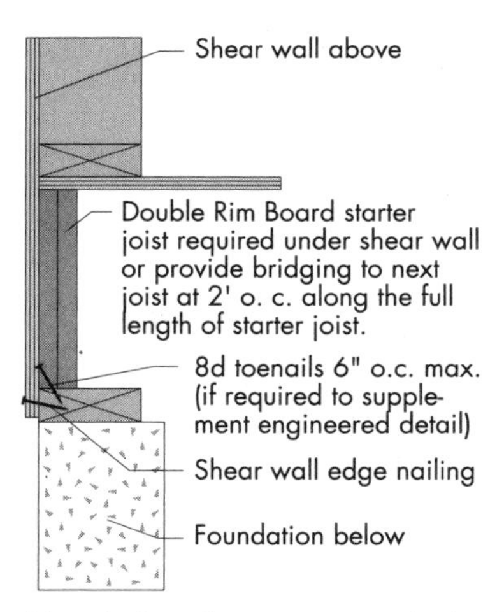

FIGURE 5.28 Transfer of shear forces at foundation—double rim board.

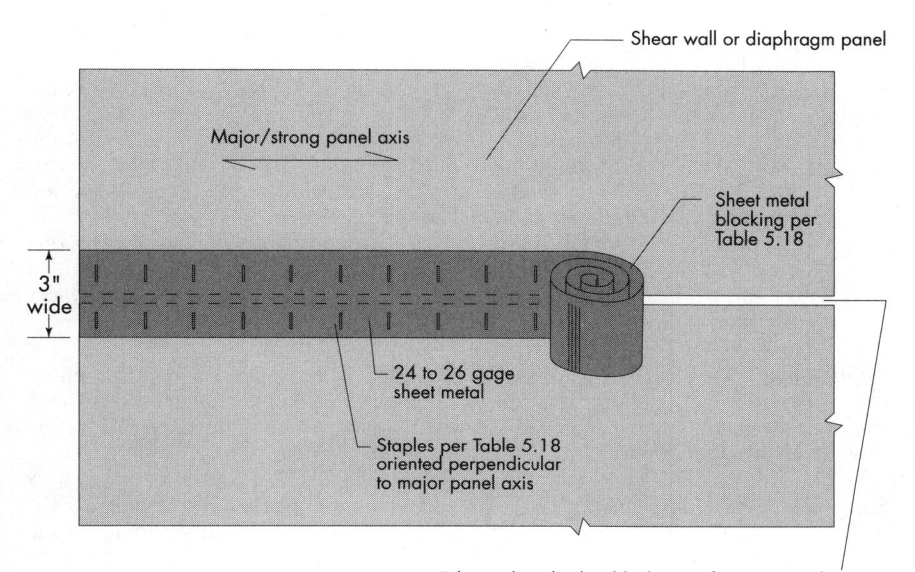

FIGURE 5.29 Sheet-metal blocking used for shear transfer in shear walls and diaphragms.

possible to apply any additional load to that connection. As such, it is decided to use the wall sheathing to transfer the shear wall forces around the rim board utilizing one of the methods shown in Fig. 5.25.

The fourth option in Fig. 5.25 is selected. The shear walls are being attached with 8d nails. The designer decides to use this same-size nail to transfer the shear stresses between floors as follows:

- With 7⁄16 in. thick APA OSB wall sheathing (a side member 7⁄16 in. thick) and a main member made up of two layers of 1⅛ in. APA OSB rim board (a depth of penetration of 1⅜ in. is required to develop the nail capacities), an adjusted single nail capacity of 117 lb/nail can be found in Table 5.15.
- Number of fasteners required per foot = 300/117 = 3.25 fasteners/ft.
- Distance between fasteners = 12/3.25 = 3.69 in. *Use 3 in. (75 mm) on center.*

Example Had it been decided to use the first option in Fig. 5.25 and the bottom plate had been spruce-pine-fir, the calculations would have been as follows:

- With 7⁄16 in. thick OSB wall sheathing (a side member 7⁄16 in. thick) and a spruce-pine-fir main member, an adjusted single nail capacity of 107 lb/nail can be found in Table 5.16.
- Number of fasteners required per foot = 380/107 = 3.55 fasteners/ft.
- Distance between fasteners = 12/3.55 = 3.38 in. *Use 3 in. (75 mm) on center.*

5.8.6 Framing Stairwell Openings in Floors

Discussion. When designing an I-joist floor for a residential structure, the designer is often faced with detailing an unsupported stairwell opening in the floor. The following information simplifies the selection of trimmers and headers, provides guidance on the appropriate detailing for their use, and quantifies hanger capacity requirements for I-joist-to-header and header-to-trimmer intersections.

These recommendations are based on the use of APA PRI joists used in either single or multiple maximum allowable spans for residential applications per Tables 5.4 and 5.5. The information is based on a total load of 50 psf (2.4 kN/m^2) for the floor and stair areas. The information provided is appropriate for stairwell openings, 10.5–12 ft (3.2–3.7 m) in length and 48 in. (1.2 m) in width, whose long dimension is either running parallel **or** perpendicular to the joist span, as shown in Fig. 5.30. *When these recommendations are followed, it is unnecessary to support the stairwell opening from below with vertical framing members.* It is also assumed that there is a non-load-bearing partition load of 64 plf (0.9 kN/m) along one header and one trimmer.

Limitations. The stair stringers may be attached to the header/trimmer at either end of the stairwell opening. For stairwells parallel or perpendicular to the I-joist spans, the opening may be placed anywhere in the floor without regard to the support of the floor framing.

Stairwells Parallel to I-Joist Span. The most common method for placing a stairwell in a wood-framed floor is to run the long axis of the opening parallel to the span of the I-joist. This generally requires smaller headers and trimmers than the perpendicular orientation.

Table 5.19 is a guide for determining the suggested (minimum capacity) I-joist requirement or the capacity of other engineered wood members required to frame the headers and trimmers seen in Fig. 5.30a.

Caution: In situations where the stairwell runs parallel to the floor joists, and the floor joists are installed over two or more spans, the header supporting the continuous floor joists may be subjected to uplift loads caused by the floor joists it supports. Cutting the interrupted joists at the center support will eliminate this uplift load. If this method is selected, the designer will have to ensure that the maximum allowable simple span for the I-joists is not exceeded. An alternative method would be to leave the floor joists continuous over the interior support and design the header and hangers for the resulting uplift load.

Stairwell Design and Detailing. The strength and stiffness requirements for headers and trimmers required for stairwells located in APA PRI I-joist floors in residential structures is shown in Table 5.19. The header and trimmer requirements can be satisfied by the use of single or double I-joists, or engineered wood headers or trimmers in I-joist-compatible depths (IJCs). Glulam beams or laminated veneer lumber are recommended (see Tables 5.21–5.23). Double I-joist construction is discussed in detail in above under Filler Blocks along with the proper sizing of backer blocks for use with hangers. The use of more than two I-joists is not recommended due to the difficulty in adequately connecting the I-joists together.

Headers. The stairwell header may be made up of a single I-joist as specified in Table 5.19, although some designers may prefer to use double I-joists for framing. Also provided are minimum shear, moment, and stiffness (*EI*) capacities for IJC engineered wood products that may be used in lieu of the I-joist. If another I-joist is going to be substituted for the suggested joist specified, all of the design capacities of the desired I-joist must be checked against those for the suggested I-joist.

Install backer blocks (see Section 5.7.2 under Backer Block Construction) behind any face- or top-mounted hangers attached to an I-joist or double I-joist header.

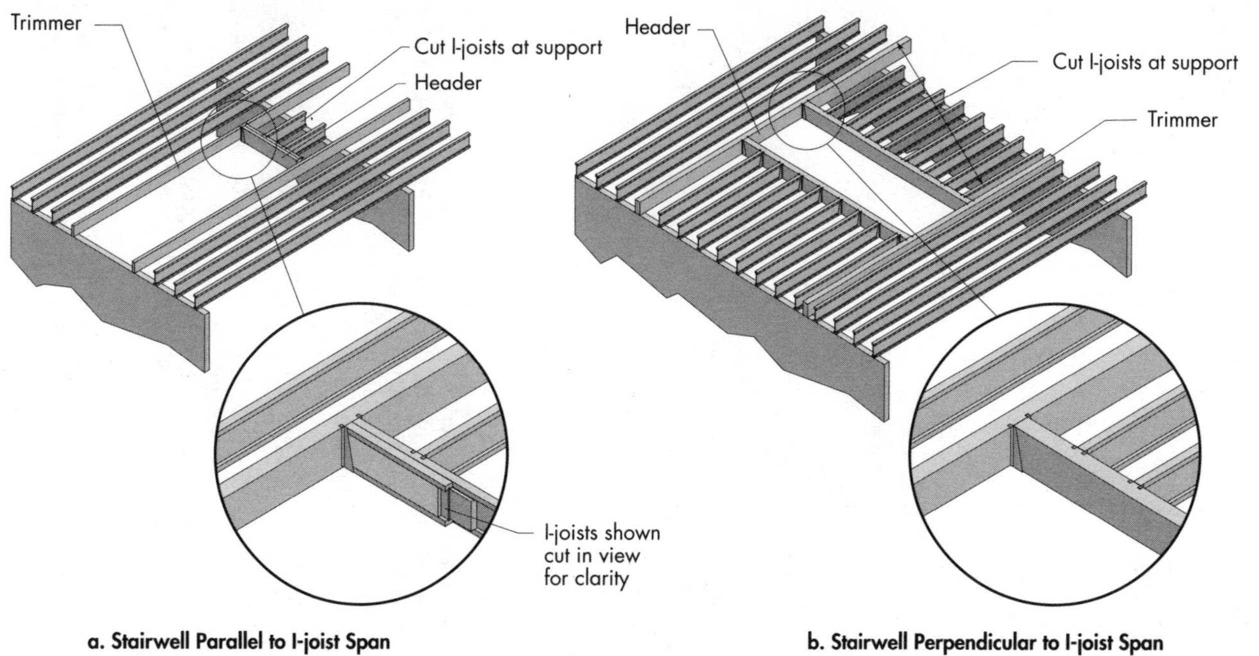

Terms:	Header:	Framing member that supports the cut ends of I-joists.
	Trimmer:	Framing member that supports the ends of the header.

FIGURE 5.30 (a,b) Stairwell openings in Performance Rated I-joist floors.

TABLE 5.19 Required Headers and Trimmers for Stairwell Openings Parallel to PRI Floor Joists

Maximum I-joist span (ft)	Header requirements: Suggested I-joist	Alternative IJC beam: Moment (lbf-ft)	Shear[2] (lbf)	*EI* (10^6 lb-in.^{2})	Joist to header hanger requirement
14	(1) 9½″ PRI-30	875	875	15	Type A
16	(1) 9½″ PRI-50	975	975	17	Type A
18	(1) 9½″ PRI-40	1075	1075	19	Type A
20	(1) 11⅞″ PRI-40	1175	1175	20	Type A
22	(1) 11⅞″ PRI-80	1275	1275	22	Type A

Maximum I-joist span (ft)	Trimmer requirements: Suggested I-joist	Alternative IJC beam: Moment (lbf-ft)	Shear[2] (lbf)	*EI* (10^6 lb-in.^{2})	Header to trimmer hanger requirements
14	(2 ea) 9-½″ PRI-50	6125	1900	270	Type A
16	(2 ea) 11-⅞″ PRI-50	7600	2050	390	Type A
18	(2 ea) 11-⅞ PRI-70	9225	2200	530	Type A
20	(2 ea) 11-⅞ PRI-70	11,000	2350	700	Type A
22	(2 ea) 11-⅞ PRI-90	12,925	2450	920	Type A

For SI: 1 in. = 25.4 mm, 1 lbf = 4.45 kN, 1 ft-lbf = 1.356 Nm.
Note: Type A—Top- or face-mounted framing anchor—1450 lbf (6450 kN).
Header lengths not to exceed 48″.

The headers are attached to the trimmers with metal hangers as shown in Fig. 5.30a. Select hangers that are sized to fit the dimensions of the I-joist, double I-joist, or engineered wood headers and that have a minimum capacity of 1450 lbf (6450 kN) for all applications covered in Table 5.19. (See Header to trimmer hanger requirement in Table 5.19.) Fourteen hundred fifty pounds (6450 kN) is the largest total hanger capacity required at each end of the header. As shown also in Table 5.19, under Joist to header hanger requirement, hangers with the same 1450 lb (6450 kN) capacity may also be used where the floor I-joists intersect the header.

Trimmers. As well as headers, Table 5.19 also gives the required moment, shear, and stiffness capacities for use in selecting IJC engineered wood products.

For headers and trimmers alike, Table 5.21–5.23 contain capacities for a number of common IJC LVL and glulam products to assist the designer in selecting the proper size when the engineered wood option is used.

Stairwells Perpendicular to I-Joist Span. Often the floor plan or architectural details of the building are such that it is not possible to orient the stairwell axis parallel to the I-joist span. Trimmers are placed parallel to the I-joist span and support the headers by way of metal hangers. In this case, the headers are up to 12 ft (3.7 m) long and support the cut ends of the floor joists also via metal hangers. This relationship can be seen in Fig. 5.30b. In addition to the header load, the trimmers are designed to carry the concentrated loads of the stair stringers.

Caution: Because the trimers intersect the span of the floor joists over a large length—up to 12 ft (3.7 m)—in cases where the floor joists are used continuously over multiple spans, special design consideration must be given to the adjacent clear span to ensure adequate floor performance. To eliminate design problems and allow maximum flexibility in locating the stairwell, consider limiting the maximum allowable spans for continuous floors containing stairwells perpendicular to I-joist spans to those given for single-span floors.

Upward thrust acting on the header adjacent to a center support can be eliminated by cutting the I-joists at the center of the support, thus providing two simple spans where the I-joists are interrupted by the headers.

Stairwell Design and Detailing. For a given maximum I-joist clear span, Table 5.20 gives the required moment, shear, and stiffness capacities for both the headers and trimmers. These values are provided for use in selecting IJC engineered wood headers and trimmers. Also provided are the joist-to-header and header-to-trimmer hanger requirements for these applications.

When the stairwell is framed perpendicular to the I-joist span, the trimmers run parallel to the I-joist span and are placed up to 12 ft (3.7 m) apart. The headers form the long-side boundaries of the stairwell, carry the weight of the cut floor joist, and are connected to the trimmers with metal hangers. The required capacity of these hangers is given in Table 5.20 under Header to trimmer hanger requirement.

TABLE 5.20 Required Headers and Trimmers for Stairwell Openings Parallel to PRI Floor Joists

I-joist span (ft)	Header requirements: Suggested I-joist	Alternative IJC beam: Moment (lbf-ft)	Shear[2] (lbf)	*EI* (10^6 lb-in.^{2})	Joist to header hanger requirement
14	(2) 9½″ PRI-30	5400	1800	232	Type A
16	(2) 9½″ PRI-60	6300	2100	278	Type A
18	(2) 11⅞″ PRI-60	7200	2400	325	Type A
20	(2) 11⅞″ PRI-90	8100	2700	371	Type A
22	Use alternative IJC beam	9000	3000	419	Type A

I-joist span (ft)	Trimmer requirements: Suggested I-joist	Alternative IJC beam: Moment (lbf-ft)	Shear (lbf)	*EI* (10^6 lb-in.^{2})	Header to trimmer hanger requirements
14	Use alternative IJC beam	11,200	3900	562	Type B
16		14,600	4400	818	Type B
18		18,400	4800	1152	Type B
20		22,600	5300	1548	2700 lbf
22		27,200	5700	2021	3000 lbf

For SI: 1 in. = 25.4 mm, 1 lbf = 4.45 kN, 1 ft-lbf = 1.356 Nm.

Note: Type A—Top- or face-mounted framing anchor—1450 lbf (6450 kN), Type B—Top- or face-mounted framing anchor—2500 lbf (11,120 kN).

The required capacities of the metal hangers supporting the floor joists where they intersect the headers is also given in Table 5.20 under Joist to header hanger requirement.

Headers. Because of the increased load on the headers when the stairwells are located perpendicular to the floor joist, it is often impractical to use double I-joists for these applications. For this reason, Table 5.20 provides the suggested double I-joist, where appropriate, as well as the required allowable bending, shear, and stiffness capacities for the header for these spans. IJC engineered wood products such as LVL or glulam beams are encouraged for these applications because of their size compatibility, superior performance, ease of use, and low moisture content. See Tables 5.21–5.23 for moment and shear capacities for some commonly available LVL and glulam members in I-joist-compatible depths. If other I-joists are going to be substituted for the suggested joists, all of the design properties for the desired I-joist must be checked against those for the suggested I-joist.

Trimmers. Most double I-joists do not have the capacity to be used as trimmers in situations where the stairwell is placed perpendicular to the floor joists. As such, in the Trimmer requirements section of Table 5.20 only IJC requirements for moment, shear, and stiffness are given. Again, Tables 5.21–5.23 are provided for substituting IJC glulam and LVL beams.

Example Stairwell Perpendicular to I-Joist Span

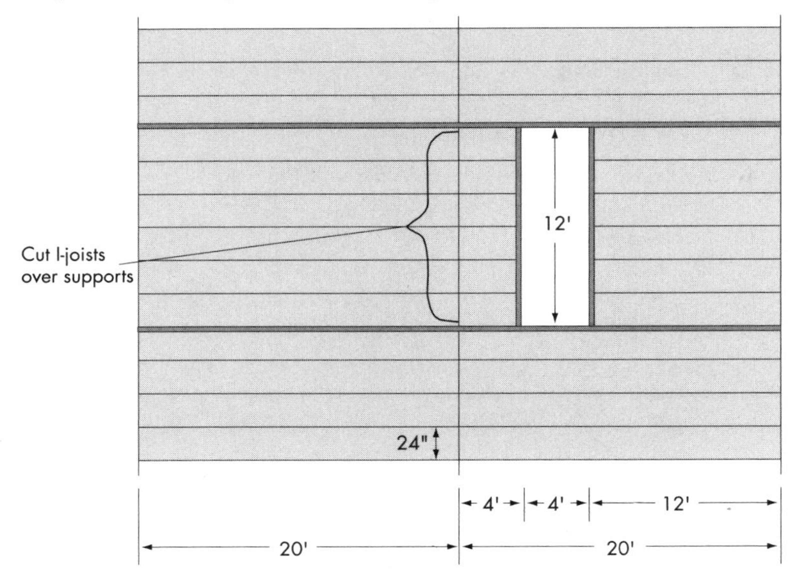

Given: A designer's plan for a residential floor called for the use of a 14″ PRI-80 over two 20-foot spans. These joists are to be placed at 24 in. on center and placed continuous over the intermediate support. A stairwell whose long axis is oriented perpendicular to the span of the I-joists is to be placed in the floor as shown above.

Stairwell Design

Check of floor design: Because the stairwell is to be placed perpendicular to the I-joist span, the designer checks both the simple and multiple maximum spans for a

TABLE 5.21 APA PRL-501 LVL Capacity Table

Depth (in.)	LVL thickness (in.)											
	1¾			3½ (2 pcs. 1¾″ LVL)			5¼ (3 pcs. 1¾″ LVL)			7 (4 pcs. 1¾″ LVL)		
	Moment (lbf-ft)	Shear (lbf)	*EI* (10^6 lb/in.^{2})	Moment (lbf-ft)	Shear (lbf)	*EI* (10^6 lb/in.^{2})	Moment (lbf-ft)	Shear (lbf)	*EI* (10^6 lb/in.^{2})	Moment (lbf-ft)	Shear (lbf)	*EI* (10^6 lb/in.^{2})
9½	6361	3159	250	12,723	6318	500	19,084	9477	750	25,446	16,635	1000
11⅞	9940	3949	488	19,879	7897	977	29,819	11,845	1465	39,758	15,793	1954
14	13,552	4655	800	27,103	9310	1601	40,655	13,965	2401	54,206	18,620	3201
16	17,407	5320	1195	34,814	10,640	2389	52,221	15,960	3584	69,627	21,280	4779

LVL capacities based on a minimum:

E = 2,000,000 psi

F_b = 2900 psi × $(12/d)^{0.125}$

F_v = 285 psi

TABLE 5.22 Douglas-Fir Glulam I-Joist-Compatible Beam Capacity Tables

Glulam beam depth (in.)	Glulam beam thickness (in.)								
	3½			5½			7		
	Moment (lbf-ft)	Shear (lbf)	EI (10^6 lb/in.^{2})	Moment (lbf-ft)	Shear (lbf)	EI (10^6 lb/in.^{2})	Moment (lbf-ft)	Shear (lbf)	EI (10^6 lb/in.^{2})
9½	10,529	5320	450	16,546	8360	707	21,058	10,640	900
11⅞	16,452	6650	879	25,853	10,450	1382	32,904	13,300	1758
14	22,867	7840	1441	35,933	12,320	2264	45,733	15,680	2881
16	29,967	8960	2150	46,933	14,080	3397	59,733	17,920	4301

Glulam capacities based on a minimum:

E = 1,800,000 psi

F_b = 2400 psi × volume factor at maximum I-joist span (16 in. oc)

F_v = 240 psi

TABLE 5.23 Southern Pine Glulam I-Joist-Compatible Beam Capacity Tables

Glulam beam depth (in.)	Glulam beam thickness (in.) 3½			5½			7		
	Moment (lbf-ft)	Shear (lbf)	EI (10^6 lb/in.^{2})	Moment (lbf-ft)	Shear (lbf)	EI (10^6 lb/in.^{2})	Moment (lbf-ft)	Shear (lbf)	EI (10^6 lb/in.^{2})
9½	13,161	5985	525	20,682	9405	825	26,323	11,970	1050
11⅞	20,565	7481	1026	32,316	11,756	1612	41,130	14,963	2051
14	28,583	8820	1681	44,917	13,860	2641	57,167	17,640	3361
16	37,333	10,080	2509	58,667	15,840	3942	74,667	20,160	5018

Glulam capacities based on a minimum:
$E = 2{,}100{,}000$ psi
$F_b = 3000$ psi × volume factor at maximum I-joist span (16 in. o.c.)
$F_v = 270$ psi

14″ PRI-80 floor joist to ensure that the required span is less. (For maximum allowable spans, see Tables 5.4 and 5.5.)

The maximum allowable simple span for a 14″ PRI-80 at 24 in. on center is 22 ft, 8 in. 23 ft, 11 in. 22 ft, 8 in > 20 ft—OK.

The maximum allowable multiple span for a 14″ PRI-80 at 24 in. on center is 23 ft, 11 in. 23 ft, 11 in. > 20 ft—OK.

Design header and trimmer: From Table 5.20, for an I-joist span of 20 ft, the maximum moment, shear, and stiffness requirements for an IJC header are 8100 lbf-ft, 2700 lbf, and 371 × 10^6 lb-in.^{2}, respectively. For trimmers, these values are 22,600 lbf-ft, 5300 lb, and 1548 × 10^6 lb-in.^{2}

LVL solution: See Table 5.21.

Select header: Try an LVL header 1¾ in. wide × 14 in. deep.

Moment capacity available = 13,552 lbf-ft, 8100 lbf-ft required—OK

Shear capacity available = 4655 lbf, 2700 lbf required—OK

Stiffness capacity available = 800 × 10^6 lb-in.^{2}, 371 × 10^6 lb-in.^{2} required—OK

Select trimmer: Try a three-piece LVL trimmer 5¼ in. wide × 14 in. deep.

Moment capacity available = 40,655 lbf-ft, 22,600 lbf-ft required → OK

Shear capacity available = 13,965 lbf, 5300 lbf required—OK

Stiffness capacity available = 2401 × 10^6 lb-in.^{2}, 1548 × 10^6 lb-in.^{2} required → OK

Note: The three piece LVL trimmer must be adequately fastened together. See Chapter 6 for recommended fastening.

Required capacities for hanger selection:

Floor joist to header capacity required—from Table 5.20—Type A—1450 lb

Header to trimmer hanger capacity required—from Table 5.20—2700 lb

Glulam solution: See Table 5.22 for a Douglas Fir Glulam Solution

Select header: Try a Douglas fir 24F, glulam header 3½ in. wide × 14 in. deep.

Moment capacity available = 22,867 lbf-ft, 8100 lbf-ft required—OK

Shear capacity available = 7840 lbf, 2700 lb required—OK

Stiffness capacity available = 1441 × 10^6 lb-in.^{2}, 371 × 10^6 lb-in.^{2} required—OK

Select trimmer: Try a Douglas fir 24F, glulam trimmer 5½ in. wide × 14 in. deep.

Moment capacity available = 35,933 lbf-ft, 22,600 lbf-ft required—OK

Shear capacity available = 12,320 lbf, 5300 lbf required—OK

Stiffness capacity available = 2264 × 10^6 lb/in.^{2}, 1548 × 10^6 lb-in.^{2} required—OK

Required capacities for hanger selection:

Floor joist to header hanger capacity required—from Table 5.20—Type A—1450 lbf

Header to trimmer hanger capacity required—from Table 5.20—2700 lbf

5.8.7 Floor Vibration

Traditionally, floor vibration has not been an issue with a well-designed and constructed floor. The model code-required serviceability deflection requirements of span/360 for live load and span/240 for total load have long served to keep code-conforming floors stiff enough to minimize vibration-related problems. These deflection requirements are over 100 years old and were based on the use of traditional lumber framing and prevailing architectural norms. Spans in traditional lumber-framed structure seldom exceeded 14–16 ft (4.3–4.9 m).

With the advent of engineered wood products, however, designers are no longer limited by the capacities and lengths of traditional lumber structural elements. Spans that were not possible with lumber just a few years ago are now common with engineered wood products. The traditional deflection limits may no longer be appropriate for the longer spans made possible by engineered wood products. For this reason, APA, in developing PRI-400, *Performance Standard for APA EWS I-joists,* has voluntarily adopted a live-load deflection criterion of span/480 that is 33% stiffer than that required in the current model building codes. This deflection criterion was selected to increase the floor stiffness because vibration loads are caused by transient or live loads, most often by people moving about the floor itself.

By increasing the stiffness of the floor, i.e., using a span/480 requirement instead of the more traditional span/360, the vibrations caused by a thundering herd of youngsters can be more easily tolerated. Designing the ideal floor is not, however, an exact science. Because one of the benefits of a wood floor is its ability to cushion footfalls, it is not desirable to make every floor overly stiff. As usual, a one-size solution does not fit all. APA's selection of span/480 as a serviceability requirement is a compromise. It provides a substantial decrease in floor vibration with a minimal cost penalty without making the floor so stiff that comfort is compromised.

Researchers have proposed a number of additional methods that can be used to reduce floor vibration even further. These methods include:

- Gluing the wood structural panel floor to the supporting framing members
- Gluing wood structural panels or gypsum board to the bottom of the floor joists
- Decreasing the floor-joist spacing
- Using full-depth blocking at regular intervals between all of the floor joists over the entire floor
- Adding concrete topping over the floor sheathing
- Gluing lumber to the bottom of every floor joist

By far the most practical and economical way to further increase the stiffness of a floor when using PRI joists is to select the most economical I-joist from the allowable span tables and then maintain the same joist designation but upgrade to the next net depth.

Example If a 9½″ PRI-40 is selected from any of Tables 5.4–5.7 for a given application, specifying an 11⅞″ PRI-40 will provide an increase in stiffness of over 70% for just a few cents per lineal foot.

5.9 END-USE APPLICATIONS—ROOF CONSTRUCTION

5.9.1 Engineered Wood Roofs—Background

As mentioned previously, APA PRI-400 Performance Rated I-joists may be identified by an allowable span designated for uniformly loaded residential floor construction at various I-joist spacings appearing on the trademark. These spans, however, are not applicable to roof framing construction. In many cases, the dead load associated with roof framing is greater than that used for calculating floor loads. In addition, roof live loads are of a shorter duration and consequently, many design properties may be increased due to this short-term load duration.

Their long lengths, light weight, and exceptional strength and stiffness characteristics make I-joists ideal for many rafter situations. The fact that they are available in depths far in excess of conventional lumber framing allows the use of inexpensive batt insulation while still leaving sufficient room for the required ventilation air space beneath the roof sheathing.

When placed with the factory-installed knockouts removed and oriented adjacent to the top flange, these openings aid in providing ventilation even to difficult areas such as where skylights, air-handling equipment, ducts, or roof-access hatches interrupt the ventilation path between a pair of joists. The ventilation air can flow around such obstacles by moving horizontally though the knockout holes into adjacent, unobstructed cavities. This is discussed further in Section 5.10.2 under Providing Roof Ventilation in I-Joist Rafter Roof Systems and Proper Positioning of Knockouts to Provide Ventilation of Roof Deck.

While it is possible to design an APA Performance Rated I-joist for almost any application using the design properties information in Table 5.2, for most common uniformly loaded roof applications allowable spans are provided in Tables 5.24–5.31.

5.9.2 I-Joist Allowable Span Tables—Roof

Design Assumptions. The roof span tables provided are based on the following assumptions:

- *Deflection criteria* (measured perpendicular to roof member):
  - Live load—$L/240$
  - Total load—$L/180$.
- *Loading criteria:*
  - *Wind load:*
    - Wind load was not considered in the development of these tables.
  - *Snow load:*
    - Duration of load adjustment factor (DOL)—1.15
    - Dead load—15 psf (0.72 kN/m^2)
    - Snow load—20 psf (0.96 kN/m^2), 25 psf (1.2 kN/m^2), 30 psf (1.44 kN/m^2), 40 psf (1.92 kN/m^2), 50 psf (2.4 kN/m^2) (based on vertical projection of roof framing).*
  - *Construction load:*
    - Duration of load adjustment factor (DOL)—1.25
    - Dead load—10 psf (0.48 kN/m^2), 15 psf (0.72 kN/m^2), and 20 psf (0.96 kN/m^2)

---

* Snow loads are equal to the actual snow load present on a flat roof. In ASCE 7-98 this would be the P_f load.

TABLE 5.24a Allowable Roof Spans for PRI I-Joists: Simple-Span (snow load = 20, dead load = 15 psf)

		Slope of 1/4:12 to 4:12			Slope of >4:12 to 8:12			Slope of >8:12 to 12:12		
Depth	Series	16 in. o.c.	19.2 in. o.c.	24 in. o.c.	16 in. o.c.	19.2 in. o.c.	24 in. o.c.	16 in. o.c.	19.2 in. o.c.	24 in. o.c.
9½″	PRI-20	19 ft, 11 in.	18 ft, 9 in.	16 ft, 10 in.	18 ft, 9 in.	17 ft, 7 in.	16 ft, 3 in.	17 ft, 3 in.	16 ft, 3 in.	15 ft, 0 in.
	PRI-30	20 ft, 8 in.	19 ft, 5 in.	17 ft, 11 in.	19 ft, 5 in.	18 ft, 3 in.	16 ft, 10 in.	17 ft, 11 in.	16 ft, 10 in.	15 ft, 7 in.
	PRI-40	21 ft, 9 in.	19 ft, 10 in.	17 ft, 9 in.	20 ft, 7 in.	19 ft, 3 in.	17 ft, 3 in.	19 ft, 0 in.	17 ft, 10 in.	16 ft, 6 in.
	PRI-50	21 ft, 8 in.	20 ft, 4 in.	18 ft, 10 in.	20 ft, 4 in.	19 ft, 1 in.	17 ft, 8 in.	18 ft, 9 in.	17 ft, 8 in.	16 ft, 4 in.
	PRI-60	23 ft, 4 in.	21 ft, 11 in.	20 ft, 3 in.	21 ft, 11 in.	20 ft, 7 in.	19 ft, 0 in.	20 ft, 2 in.	19 ft, 0 in.	17 ft, 7 in.
11⅞″	PRI-20	23 ft, 11 in.	21 ft, 10 in.	19 ft, 6 in.	22 ft, 7 in.	21 ft, 2 in.	18 ft, 11 in.	20 ft, 10 in.	19 ft, 7 in.	18 ft, 2 in.
	PRI-30	24 ft, 11 in.	23 ft, 5 in.	21 ft, 8 in.	23 ft, 5 in.	22 ft, 0 in.	20 ft, 4 in.	21 ft, 7 in.	20 ft, 3 in.	18 ft, 9 in.
	PRI-40	25 ft, 3 in.	23 ft, 0 in.	20 ft, 6 in.	24 ft, 6 in.	22 ft, 4 in.	19 ft, 11 in.	22 ft, 10 in.	21 ft, 5 in.	19 ft, 2 in.
	PRI-50	26 ft, 1 in.	24 ft, 6 in.	22 ft, 8 in.	24 ft, 6 in.	23 ft, 0 in.	21 ft, 4 in.	22 ft, 7 in.	21 ft, 3 in.	19 ft, 8 in.
	PRI-60	27 ft, 11 in.	26 ft, 3 in.	24 ft, 2 in.	26 ft, 3 in.	24 ft, 8 in.	22 ft, 10 in.	24 ft, 3 in.	22 ft, 9 in.	21 ft, 1 in.
	PRI-70	28 ft, 6 in.	26 ft, 9 in.	24 ft, 9 in.	26 ft, 9 in.	25 ft, 2 in.	23 ft, 3 in.	24 ft, 9 in.	23 ft, 3 in.	21 ft, 6 in.
	PRI-80	31 ft, 1 in.	29 ft, 3 in.	27 ft, 0 in.	29 ft, 3 in.	27 ft, 5 in.	25 ft, 5 in.	27 ft, 0 in.	25 ft, 4 in.	23 ft, 6 in.
	PRI-90	32 ft, 2 in.	30 ft, 2 in.	27 ft, 11 in.	30 ft, 3 in.	28 ft, 4 in.	26 ft, 3 in.	27 ft, 11 in.	26 ft, 2 in.	24 ft, 3 in.
14″	PRI-40	28 ft, 0 in.	25 ft, 6 in.	22 ft, 9 in.	27 ft, 2 in.	24 ft, 9 in.	22 ft, 1 in.	25 ft, 11 in.	23 ft, 9 in.	21 ft, 3 in.
	PRI-50	29 ft, 10 in.	28 ft, 1 in.	26 ft, 0 in.	28 ft, 1 in.	26 ft, 4 in.	24 ft, 5 in.	25 ft, 11 in.	24 ft, 4 in.	22 ft, 6 in.
	PRI-60	31 ft, 10 in.	29 ft, 11 in.	26 ft, 9 in.	29 ft, 11 in.	28 ft, 1 in.	26 ft, 0 in.	27 ft, 8 in.	25 ft, 11 in.	24 ft, 0 in.
	PRI-70	32 ft, 5 in.	30 ft, 5 in.	28 ft, 2 in.	30 ft, 5 in.	28 ft, 7 in.	26 ft, 5 in.	28 ft, 1 in.	26 ft, 5 in.	24 ft, 5 in.
	PRI-80	35 ft, 5 in.	33 ft, 3 in.	30 ft, 9 in.	33 ft, 3 in.	31 ft, 3 in.	28 ft, 11 in.	30 ft, 8 in.	28 ft, 10 in.	26 ft, 9 in.
	PRI-90	36 ft, 6 in.	34 ft, 4 in.	31 ft, 9 in.	34 ft, 4 in.	32 ft, 3 in.	29 ft, 10 in.	31 ft, 8 in.	29 ft, 9 in.	27 ft, 7 in.
16″	PRI-40	30 ft, 4 in.	27 ft, 8 in.	24 ft, 9 in.	29 ft, 5 in.	26 ft, 10 in.	24 ft, 0 in.	28 ft, 3 in.	25 ft, 9 in.	23 ft, 0 in.
	PRI-50	33 ft, 4 in.	31 ft, 3 in.	28 ft, 8 in.	31 ft, 3 in.	29 ft, 4 in.	27 ft, 2 in.	28 ft, 10 in.	27 ft, 1 in.	25 ft, 1 in.
	PRI-60	35 ft, 5 in.	32 ft, 6 in.	29 ft, 1 in.	33 ft, 3 in.	31 ft, 3 in.	28 ft, 2 in.	30 ft, 8 in.	28 ft, 10 in.	26 ft, 9 in.
	PRI-70	36 ft, 0 in.	33 ft, 10 in.	31 ft, 4 in.	33 ft, 10 in.	31 ft, 9 in.	29 ft, 5 in.	31 ft, 3 in.	29 ft, 4 in.	27 ft, 2 in.
	PRI-80	39 ft, 3 in.	36 ft, 11 in.	34 ft, 2 in.	36 ft, 11 in.	34 ft, 8 in.	32 ft, 1 in.	34 ft, 1 in.	32 ft, 0 in.	29 ft, 8 in.
	PRI-90	40 ft, 5 in.	38 ft, 0 in.	35 ft, 2 in.	38 ft, 0 in.	35 ft, 8 in.	33 ft, 0 in.	35 ft, 1 in.	32 ft, 11 in.	30 ft, 6 in.

For SI: 1 in. = 25.4 mm, 1 ft = 304.8 mm.

1. Allowable clear span applicable to simple-span roof construction with a design dead load of 15 psf (0.72 kN/m^2) and a snow load of 20 psf (0.96 kN/m^2). The snow load deflection is limited to span/240 and total load deflection to span/180.

2. Spans are based on a duration of load (DOL) factor of 1.15.

3. Minimum bearing length must be 1¾ in. (44.5 mm) for the end bearings and 3½ in. (89 mm) on end bearing adjacent to cantilever.

4. Bearing stiffeners are not required when I-joists are used with the spans and spacings given in this table, except as required by hanger manufacturers.

5. Spans include a cantilever of up to 2 ft on one end of the I-joist.

TABLE 5.24b Allowable Roof Spans for PRI I-Joists: Multiple-Span Only (flat roof (¼–12); snow load = 20 psf, dead load = 15 psf)

Depth	Designation	16 in. o.c.	19.2 in. o.c.	24 in. o.c.
9½″	PRI-20	20 ft, 11 in.	19 ft, 1 in.	17 ft, 1 in.
	PRI-30	23 ft, 3 in.	21 ft, 8 in.	19 ft, 4 in.
	PRI-40	22 ft, 1 in.	20 ft, 2 in.	18 ft, 0 in.
	PRI-50	24 ft, 4 in.	22 ft, 11 in.	21 ft, 0 in.
	PRI-60	25 ft, 11 in.	23 ft, 8 in.	21 ft, 2 in.
11⅞″	PRI-20	24 ft, 2 in.	22 ft, 1 in.	19 ft, 9 in.
	PRI-30	27 ft, 5 in.	25 ft, 0 in.	22 ft, 4 in.
	PRI-40	25 ft, 6 in.	23 ft, 4 in.	20 ft, 10 in.
	PRI-50	29 ft, 4 in.	27 ft, 1 in.	24 ft, 3 in.
	PRI-60	30 ft, 0 in.	27 ft, 5 in.	24 ft, 6 in.
	PRI-70	32 ft, 0 in.	30 ft, 1 in.	27 ft, 5 in.
	PRI-80	34 ft, 11 in.	32 ft, 7 in.	29 ft, 2 in.
14″	PRI-40	28 ft, 4 in.	25 ft, 10 in.	23 ft, 1 in.
	PRI-50	32 ft, 11 in.	30 ft, 0 in.	26 ft, 7 in.
	PRI-60	33 ft, 3 in.	30 ft, 4 in.	27 ft, 2 in.
	PRI-70	36 ft, 4 in.	34 ft, 2 in.	30 ft, 6 in.
	PRI-80	39 ft, 7 in.	36 ft, 2 in.	32 ft, 4 in.
16″	PRI-40	30 ft, 9 in.	28 ft, 0 in.	25 ft, 0 in.
	PRI-50	35 ft, 8 in.	32 ft, 6 in.	26 ft, 7 in.
	PRI-60	36 ft, 1 in.	32 ft, 11 in.	29 ft, 5 in.
	PRI-70	40 ft, 5 in.	37 ft, 7 in.	30 ft, 6 in.
	PRI-80	43 ft, 0 in.	39 ft, 3 in.	35 ft, 1 in.

For SI: 1 in. = 25.4 mm, 1 ft = 304.8 mm.

1. Allowable clear span applicable to multiple-span roof construction with a design dead load of 15 psf (0.72 kN/m^2) and a roof SNOW LOAD of 20 psf (1.2 kN/m^2). The end spans must be 40% or more of the adjacent span. The snow load deflection is limited to span/240 and total load deflection to span/180.

2. Spans are based on a duration of load (DOL) factor of 1.15.

3. Minimum bearing length must be 1¾ in. (44.5 mm) for the end bearings and 3½ in. (89 mm) for the intermediate bearing.

4. Bearing stiffeners are not required when APA PRI joists are used with the spans and spacings given in this table, except as required by hanger manufacturers.

TABLE 5.25a Allowable Roof Spans for PRI I-Joists: Simple-Span (snow load = 25 psf, dead load = 15 psf)

		Slope of 1/4:12 to 4:12			Slope of >4:12 to 8:12			Slope of >8:12 to 12:12		
Depth	Series	16 in. o.c.	19.2 in. o.c.	24 in. o.c.	16 in. o.c.	19.2 in. o.c.	24 in. o.c.	16 in. o.c.	19.2 in. o.c.	24 in. o.c.
9½″	PRI-20	19 ft, 1 in.	17 ft, 7 in.	15 ft, 9 in.	17 ft, 11 in.	16 ft, 10 in.	15 ft, 4 in.	15 ft, 6 in.	14 ft, 7 in.	13 ft, 4 in.
	PRI-30	19 ft, 9 in.	18 ft, 7 in.	17 ft, 2 in.	18 ft, 7 in.	17 ft, 5 in.	16 ft, 2 in.	16 ft, 1 in.	15 ft, 1 in.	14 ft, 0 in.
	PRI-40	20 ft, 5 in.	18 ft, 7 in.	16 ft, 7 in.	19 ft, 9 in.	18 ft, 1 in.	16 ft, 2 in.	17 ft, 1 in.	15 ft, 10 in.	14 ft, 1 in.
	PRI-50	20 ft, 9 in.	19 ft, 5 in.	18 ft, 0 in.	19 ft, 6 in.	18 ft, 4 in.	16 ft, 11 in.	16 ft, 11 in.	15 ft, 10 in.	14 ft, 8 in.
	PRI-60	22 ft, 3 in.	20 ft, 11 in.	19 ft, 4 in.	21 ft, 0 in.	19 ft, 8 in.	18 ft, 3 in.	18 ft, 2 in.	17 ft, 0 in.	15 ft, 9 in.
11⅞″	PRI-20	22 ft, 4 in.	20 ft, 5 in.	18 ft, 3 in.	21 ft, 8 in.	19 ft, 10 in.	17 ft, 9 in.	18 ft, 9 in.	17 ft, 4 in.	15 ft, 6 in.
	PRI-30	23 ft, 10 in.	22 ft, 4 in.	20 ft, 7 in.	22 ft, 5 in.	21 ft, 1 in.	19 ft, 6 in.	19 ft, 5 in.	18 ft, 3 in.	16 ft, 10 in.
	PRI-40	23 ft, 7 in.	21 ft, 6 in.	19 ft, 3 in.	23 ft, 0 in.	20 ft, 11 in.	18 ft, 9 in.	20 ft, 1 in.	18 ft, 4 in.	16 ft, 4 in.
	PRI-50	24 ft, 11 in.	23 ft, 5 in.	21 ft, 8 in.	23 ft, 6 in.	22 ft, 1 in.	20 ft, 5 in.	20 ft, 4 in.	19 ft, 1 in.	17 ft, 8 in.
	PRI-60	26 ft, 9 in.	25 ft, 1 in.	22 ft, 7 in.	25 ft, 2 in.	23 ft, 8 in.	21 ft, 10 in.	21 ft, 9 in.	20 ft, 5 in.	18 ft, 11 in.
	PRI-70	27 ft, 3 in.	25 ft, 7 in.	23 ft, 8 in.	25 ft, 8 in.	24 ft, 1 in.	22 ft, 4 in.	22 ft, 3 in.	20 ft, 10 in.	19 ft, 4 in.
	PRI-80	29 ft, 9 in.	27 ft, 11 in.	25 ft, 10 in.	28 ft, 0 in.	26 ft, 4 in.	24 ft, 4 in.	24 ft, 3 in.	22 ft, 9 in.	21 ft, 1 in.
	PRI-90	30 ft, 9 in.	28 ft, 10 in.	26 ft, 8 in.	28 ft, 11 in.	27 ft, 2 in.	25 ft, 2 in.	25 ft, 1 in.	23 ft, 6 in.	21 ft, 9 in.
14″	PRI-40	26 ft, 2 in.	23 ft, 10 in.	21 ft, 4 in.	25 ft, 6 in.	23 ft, 3 in.	20 ft, 9 in.	22 ft, 3 in.	20 ft, 4 in.	18 ft, 2 in.
	PRI-50	28 ft, 7 in.	26 ft, 10 in.	24 ft, 10 in.	26 ft, 11 in.	25 ft, 3 in.	23 ft, 4 in.	23 ft, 3 in.	21 ft, 10 in.	20 ft, 3 in.
	PRI-60	30 ft, 6 in.	28 ft, 1 in.	25 ft, 1 in.	28 ft, 8 in.	26 ft, 11 in.	24 ft, 5 in.	24 ft, 10 in.	23 ft, 4 in.	21 ft, 4 in.
	PRI-70	31 ft, 0 in.	29 ft, 1 in.	26 ft, 11 in.	29 ft, 2 in.	27 ft, 5 in.	25 ft, 4 in.	25 ft, 3 in.	23 ft, 8 in.	21 ft, 11 in.
	PRI-80	29 ft, 9 in.	27 ft, 11 in.	25 ft, 10 in.	31 ft, 11 in.	29 ft, 11 in.	27 ft, 8 in.	27 ft, 7 in.	25 ft, 11 in.	24 ft, 0 in.
	PRI-90	34 ft, 11 in.	32 ft, 9 in.	30 ft, 4 in.	32 ft, 11 in.	30 ft, 11 in.	28 ft, 7 in.	28 ft, 6 in.	26 ft, 9 in.	24 ft, 9 in.
16″	PRI-40	28 ft, 5 in.	25 ft, 11 in.	23 ft, 2 in.	27 ft, 8 in.	25 ft, 3 in.	22 ft, 6 in.	24 ft, 2 in.	22 ft, 0 in.	19 ft, 8 in.
	PRI-50	31 ft, 10 in.	29 ft, 11 in.	26 ft, 10 in.	30 ft, 0 in.	28 ft, 2 in.	26 ft, 1 in.	25 ft, 11 in.	24 ft, 4 in.	22 ft, 7 in.
	PRI-60	33 ft, 5 in.	30 ft, 5 in.	27 ft, 2 in.	31 ft, 11 in.	29 ft, 8 in.	26 ft, 6 in.	27 ft, 7 in.	25 ft, 11 in.	23 ft, 2 in.
	PRI-70	34 ft, 5 in.	32 ft, 4 in.	29 ft, 11 in.	32 ft, 5 in.	30 ft, 6 in.	28 ft, 2 in.	28 ft, 1 in.	26 ft, 4 in.	24 ft, 5 in.
	PRI-80	37 ft, 7 in.	35 ft, 3 in.	32 ft, 5 in.	35 ft, 4 in.	33 ft, 3 in.	30 ft, 9 in.	30 ft, 8 in.	28 ft, 9 in.	26 ft, 7 in.
	PRI-90	38 ft, 8 in.	36 ft, 4 in.	33 ft, 7 in.	36 ft, 5 in.	34 ft, 2 in.	31 ft, 8 in.	31 ft, 6 in.	29 ft, 7 in.	27 ft, 5 in.

For SI: 1 in. = 25.4 mm, 1 ft = 304.8 mm.

1. Allowable clear span applicable to simple-span roof construction with a design dead load of 15 psf (0.72 kN/m^2) and a roof snow load of 25 psf (1.20 kN/m^2). The snow load deflection is limited to span/240 and total load deflection to span/180.

2. Spans are based on a duration of load (DOL) factor of 1.15.

3. Minimum bearing length must be 1¾ in. (44.5 mm) for the end bearings and 3½ in. (89 mm) on end bearing adjacent to cantilever.

4. Bearing stiffeners are not required when I-joists are used with the spans and spacings given in this table, except as required by hanger manufacturers.

5. Spans include a cantilever of up to 2 ft on one end of the I-joist.

TABLE 5.25b Allowable Roof Spans for PRI I-Joists: Multiple-Span Only (flat roof (¼–12); snow load = 25 psf, dead load = 15 psf)

Depth	Designation	16 in. o.c.	19.2 in. o.c.	24 in. o.c.
9½″	PRI-20	19 ft, 7 in.	17 ft, 10 in.	15 ft, 11 in.
	PRI-30	22 ft, 2 in.	20 ft, 3 in.	18 ft, 1 in.
	PRI-40	20 ft, 8 in.	18 ft, 10 in.	16 ft, 10 in.
	PRI-50	23 ft, 3 in.	21 ft, 10 in.	19 ft, 7 in.
	PRI-60	24 ft, 3 in.	22 ft, 1 in.	19 ft, 9 in.
11⅞″	PRI-20	22 ft, 8 in.	20 ft, 8 in.	18 ft, 5 in.
	PRI-30	25 ft, 7 in.	23 ft, 4 in.	20 ft, 10 in.
	PRI-40	23 ft, 10 in.	21 ft, 9 in.	19 ft, 5 in.
	PRI-50	27 ft, 10 in.	25 ft, 4 in.	22 ft, 8 in.
	PRI-60	28 ft, 1 in.	25 ft, 7 in.	22 ft, 11 in.
	PRI-70	30 ft, 7 in.	28 ft, 8 in.	25 ft, 8 in.
	PRI-80	33 ft, 4 in.	30 ft, 6 in.	27 ft, 3 in.
14″	PRI-40	26 ft, 6 in.	24 ft, 2 in.	21 ft, 7 in.
	PRI-50	30 ft, 9 in.	28 ft, 1 in.	23 ft, 3 in.
	PRI-60	31 ft, 1 in.	28 ft, 5 in.	25 ft, 4 in.
	PRI-70	34 ft, 9 in.	32 ft, 5 in.	26 ft, 8 in.
	PRI-80	37 ft, 1 in.	33 ft, 10 in.	30 ft, 3 in.
16″	PRI-40	28 ft, 9 in.	26 ft, 2 in.	23 ft, 5 in.
	PRI-50	33 ft, 4 in.	29 ft, 1 in.	23 ft, 3 in.
	PRI-60	33 ft, 9 in.	30 ft, 9 in.	27 ft, 6 in.
	PRI-70	38 ft, 6 in.	33 ft, 4 in.	26 ft, 8 in.
	PRI-80	40 ft, 2 in.	36 ft, 8 in.	32 ft, 9 in.

For SI: 1 in. = 25.4 mm, 1 ft = 304.8 mm.

1. Allowable clear span applicable to multiple-span roof construction with a design dead load of 15 psf (0.72 kN/m²) and a roof snow load of 25 psf (1.2 kN/m²). The end spans must be 40% or more of the adjacent span. The snow load deflection is limited to span/240 and total load deflection to span/180.

2. Spans are based on a duration of load (DOL) factor of 1.15.

3. Minimum bearing length must be 1¾ in. (44.5 mm) for the end bearings and 3½ in. (89 mm) for the intermediate bearing.

4. Bearing stiffeners are not required when APA PRI joists are used with the spans and spacings given in this table, except as required by hanger manufacturers.

TABLE 5.26a Allowable Roof Spans for PRI I-Joists: Simple-Span (snow load = 30 psf, dead load = 15 psf)

Depth	Series	Slope of 1/4:12 to 4:12			Slope of >4:12 to 8:12			Slope of >8:12 to 12:12		
		16 in. o.c.	19.2 in. o.c.	24 in. o.c.	16 in. o.c.	19.2 in. o.c.	24 in. o.c.	16 in. o.c.	19.2 in. o.c.	24 in. o.c.
9½″	PRI-20	18 ft, 3 in.	16 ft, 7 in.	14 ft, 10 in.	17 ft, 3 in.	16 ft, 3 in.	14 ft, 6 in.	16 ft, 0 in.	15 ft, 1 in.	13 ft, 11 in.
	PRI-30	19 ft, 0 in.	17 ft, 10 in.	16 ft, 6 in.	17 ft, 11 in.	16 ft, 10 in.	15 ft, 7 in.	16 ft, 7 in.	15 ft, 7 in.	14 ft, 5 in.
	PRI-40	19 ft, 3 in.	17 ft, 6 in.	15 ft, 8 in.	18 ft, 9 in.	17 ft, 1 in.	15 ft, 3 in.	17 ft, 8 in.	16 ft, 7 in.	14 ft, 10 in.
	PRI-50	19 ft, 11 in.	18 ft, 8 in.	17 ft, 3 in.	18 ft, 9 in.	17 ft, 8 in.	16 ft, 4 in.	17 ft, 5 in.	16 ft, 4 in.	15 ft, 2 in.
	PRI-60	21 ft, 5 in.	20 ft, 1 in.	18 ft, 5 in.	20 ft, 2 in.	18 ft, 11 in.	17 ft, 6 in.	18 ft, 9 in.	17 ft, 7 in.	16 ft, 3 in.
11⅞″	PRI-20	21 ft, 1 in.	19 ft, 3 in.	17 ft, 2 in.	20 ft, 7 in.	18 ft, 9 in.	16 ft, 9 in.	19 ft, 4 in.	18 ft, 2 in.	16 ft, 3 in.
	PRI-30	22 ft, 11 in.	21 ft, 6 in.	19 ft, 5 in.	21 ft, 7 in.	20 ft, 3 in.	18 ft, 9 in.	20 ft, 0 in.	18 ft, 10 in.	17 ft, 5 in.
	PRI-40	22 ft, 3 in.	20 ft, 4 in.	18 ft, 2 in.	21 ft, 9 in.	19 ft, 10 in.	17 ft, 8 in.	21 ft, 1 in.	19 ft, 2 in.	17 ft, 2 in.
	PRI-50	24 ft, 0 in.	22 ft, 6 in.	20 ft, 10 in.	22 ft, 7 in.	21 ft, 3 in.	19 ft, 8 in.	21 ft, 0 in.	19 ft, 9 in.	18 ft, 3 in.
	PRI-60	25 ft, 8 in.	23 ft, 11 in.	21 ft, 4 in.	24 ft, 3 in.	22 ft, 9 in.	20 ft, 10 in.	22 ft, 6 in.	21 ft, 1 in.	19 ft, 7 in.
	PRI-70	26 ft, 2 in.	24 ft, 7 in.	22 ft, 9 in.	24 ft, 8 in.	23 ft, 2 in.	21 ft, 6 in.	22 ft, 11 in.	21 ft, 6 in.	19 ft, 11 in.
	PRI-80	28 ft, 7 in.	26 ft, 10 in.	24 ft, 10 in.	27 ft, 0 in.	25 ft, 4 in.	23 ft, 5 in.	25 ft, 1 in.	23 ft, 6 in.	21 ft, 9 in.
	PRI-90	29 ft, 6 in.	27 ft, 8 in.	25 ft, 7 in.	27 ft, 10 in.	26 ft, 2 in.	24 ft, 2 in.	25 ft, 11 in.	24 ft, 4 in.	22 ft, 6 in.
14″	PRI-40	24 ft, 8 in.	22 ft, 6 in.	20 ft, 1 in.	24 ft, 1 in.	22 ft, 0 in.	19 ft, 8 in.	23 ft, 4 in.	21 ft, 3 in.	19 ft, 0 in.
	PRI-50	27 ft, 5 in.	25 ft, 9 in.	23 ft, 5 in.	25 ft, 11 in.	24 ft, 4 in.	22 ft, 6 in.	24 ft, 0 in.	22 ft, 7 in.	20 ft, 11 in.
	PRI-60	29 ft, 0 in.	26 ft, 6 in.	23 ft, 8 in.	27 ft, 7 in.	25 ft, 10 in.	23 ft, 1 in.	25 ft, 8 in.	24 ft, 1 in.	22 ft, 4 in.
	PRI-70	29 ft, 9 in.	27 ft, 11 in.	25 ft, 10 in.	28 ft, 1 in.	26 ft, 4 in.	24 ft, 5 in.	26 ft, 1 in.	24 ft, 6 in.	22 ft, 8 in.
	PRI-80	32 ft, 6 in.	30 ft, 6 in.	28 ft, 2 in.	30 ft, 8 in.	28 ft, 10 in.	26 ft, 8 in.	28 ft, 6 in.	26 ft, 9 in.	24 ft, 9 in.
	PRI-90	33 ft, 6 in.	31 ft, 6 in.	29 ft, 1 in.	31 ft, 8 in.	29 ft, 9 in.	27 ft, 6 in.	29 ft, 5 in.	27 ft, 7 in.	25 ft, 7 in.
16″	PRI-40	26 ft, 9 in.	24 ft, 5 in.	21 ft, 10 in.	26 ft, 2 in.	23 ft, 10 in.	21 ft, 4 in.	25 ft, 4 in.	23 ft, 1 in.	20 ft, 8 in.
	PRI-50	30 ft, 7 in.	28 ft, 4 in.	25 ft, 4 in.	28 ft, 10 in.	27 ft, 1 in.	24 ft, 9 in.	26 ft, 9 in.	25 ft, 2 in.	23 ft, 4 in.
	PRI-60	31 ft, 6 in.	28 ft, 9 in.	25 ft, 8 in.	30 ft, 8 in.	28 ft, 1 in.	25 ft, 1 in.	28 ft, 6 in.	26 ft, 9 in.	24 ft, 3 in.
	PRI-70	33 ft, 1 in.	31 ft, 1 in.	28 ft, 9 in.	31 ft, 3 in.	29 ft, 4 in.	27 ft, 2 in.	29 ft, 0 in.	27 ft, 3 in.	25 ft, 3 in.
	PRI-80	36 ft, 1 in.	33 ft, 10 in.	30 ft, 7 in.	34 ft, 1 in.	32 ft, 0 in.	29 ft, 7 in.	31 ft, 8 in.	29 ft, 8 in.	27 ft, 6 in.
	PRI-90	37 ft, 2 in.	34 ft, 10 in.	32 ft, 3 in.	35 ft, 1 in.	32 ft, 11 in.	30 ft, 6 in.	32 ft, 7 in.	30 ft, 7 in.	28 ft, 4 in.

For SI: 1 in. = 25.4 mm, 1 ft = 304.8 mm.

1. Allowable clear span applicable to simple-span roof construction with a design dead load of 15 psf (0.72 kN/m^2) and a roof snow load of 30 psf (1.44 kN/m^2). The snow load deflection is limited to span/240 and total load deflection to span/180.

2. Spans are based on a duration of load (DOL) factor of 1.15.

3. Minimum bearing length must be 1¾ in. (44.5 mm) for the end bearings and 3½ in. (89 mm) on end bearing adjacent to cantilever.

4. Bearing stiffeners are not required when I-joists are used with the spans and spacings given in this table, except as required by hanger manufacturers.

5. Spans include a cantilever of up to 2 ft on one end of the I-joist.

TABLE 5.26b Allowable Roof Spans for PRI I-Joists: Multiple-Span Only (flat roof (¼–12); snow load = 30 psf, dead load = 15 psf)

Depth	Designation	16 in. o.c.	19.2 in. o.c.	24 in. o.c.
9½″	PRI-20	18 ft, 5 in.	16 ft, 10 in.	15 ft, 0 in.
	PRI-30	20 ft, 11 in.	19 ft, 1 in.	17 ft, 0 in.
	PRI-40	19 ft, 5 in.	17 ft, 9 in.	15 ft, 10 in.
	PRI-50	22 ft, 4 in.	20 ft, 8 in.	18 ft, 6 in.
	PRI-60	22 ft, 10 in.	20 ft, 10 in.	18 ft, 7 in.
11⅞″	PRI-20	21 ft, 4 in.	19 ft, 5 in.	17 ft, 2 in.
	PRI-30	24 ft, 1 in.	22 ft, 0 in.	19 ft, 3 in.
	PRI-40	22 ft, 6 in.	20 ft, 6 in.	18 ft, 4 in.
	PRI-50	26 ft, 2 in.	23 ft, 11 in.	20 ft, 8 in.
	PRI-60	26 ft, 5 in.	24 ft, 2 in.	21 ft, 7 in.
	PRI-70	29 ft, 4 in.	27 ft, 1 in.	23 ft, 8 in.
	PRI-80	31 ft, 6 in.	28 ft, 9 in.	25 ft, 8 in.
14″	PRI-40	24 ft, 11 in.	22 ft, 9 in.	20 ft, 4 in.
	PRI-50	29 ft, 0 in.	25 ft, 10 in.	20 ft, 8 in.
	PRI-60	29 ft, 4 in.	26 ft, 9 in.	23 ft, 11 in.
	PRI-70	33 ft, 4 in.	29 ft, 7 in.	23 ft, 8 in.
	PRI-80	34 ft, 11 in.	31 ft, 10 in.	28 ft, 6 in.
16″	PRI-40	27 ft, 1 in.	24 ft, 8 in.	22 ft, 1 in.
	PRI-50	31 ft, 1 in.	25 ft, 10 in.	20 ft, 8 in.
	PRI-60	31 ft, 10 in.	29 ft, 0 in.	25 ft, 4 in.
	PRI-70	35 ft, 7 in.	29 ft, 7 in.	23 ft, 8 in.
	PRI-80	37 ft, 11 in.	34 ft, 7 in.	30 ft, 8 in.

For SI: 1 in. = 25.4 mm, 1 ft = 304.8 mm.

1. Allowable clear span applicable to multiple-span roof construction with a design dead load of 15 psf (0.72 kN/m^2) and a roof snow load of 30 psf (1.44 kN/m^2). The end spans must be 40% or more of the adjacent span. The snow load deflection is limited to span/240 and total load deflection to span/180.

2. Spans are based on a duration of load (DOL) factor of 1.15.

3. Minimum bearing length must be 1¾ in. (44.5 mm) for the end bearings and 3½ in. (89 mm) for the intermediate bearing..

4. Bearing stiffeners are not required when APA PRI joists are used with the spans and spacings given in this table, except as required by hanger manufacturers.

TABLE 5.27a Allowable Roof Spans for PRI I-Joists. Simple-Span (snow load = 40 psf, dead load = 15 psf)

Depth	Series	Slope of 1/4:12 to 4:12			Slope of >4:12 to 8:12			Slope of >8:12 to 12:12		
		16 in. o.c.	19.2 in. o.c.	24 in. o.c.	16 in. o.c.	19.2 in. o.c.	24 in. o.c.	16 in. o.c.	19.2 in. o.c.	24 in. o.c.
9½″	PRI-20	16 ft, 6 in.	15 ft, 0 in.	13 ft, 5 in.	16 ft, 2 in.	14 ft, 9 in.	13 ft, 2 in.	15 ft, 1 in.	14 ft, 2 in.	12 ft, 10 in.
	PRI-30	17 ft, 9 in.	16 ft, 7 in.	15 ft, 3 in.	16 ft, 9 in.	15 ft, 9 in.	14 ft, 7 in.	15 ft, 7 in.	14 ft, 8 in.	13 ft, 7 in.
	PRI-40	17 ft, 5 in.	15 ft, 10 in.	14 ft, 2 in.	17 ft, 1 in.	15 ft, 7 in.	13 ft, 11 in.	16 ft, 7 in.	15 ft, 2 in.	13 ft, 6 in.
	PRI-50	18 ft, 7 in.	17 ft, 5 in.	16 ft, 1 in.	17 ft, 7 in.	16 ft, 6 in.	15 ft, 3 in.	16 ft, 5 in.	15 ft, 5 in.	14 ft, 3 in.
	PRI-60	20 ft, 0 in.	18 ft, 8 in.	16 ft, 8 in.	18 ft, 11 in.	17 ft, 9 in.	16 ft, 4 in.	17 ft, 7 in.	16 ft, 6 in.	15 ft, 4 in.
11⅞″	PRI-20	19 ft, 1 in.	17 ft, 5 in.	15 ft, 7 in.	18 ft, 9 in.	17 ft, 1 in.	15 ft, 3 in.	18 ft, 3 in.	16 ft, 7 in.	14 ft, 10 in.
	PRI-30	21 ft, 4 in.	19 ft, 8 in.	17 ft, 7 in.	20 ft, 3 in.	19 ft, 0 in.	17 ft, 3 in.	18 ft, 10 in.	17 ft, 8 in.	16 ft, 4 in.
	PRI-40	20 ft, 2 in.	18 ft, 4 in.	16 ft, 5 in.	19 ft, 9 in.	18 ft, 0 in.	16 ft, 1 in.	19 ft, 3 in.	17 ft, 6 in.	15 ft, 8 in.
	PRI-50	22 ft, 5 in.	21 ft, 0 in.	19 ft, 1 in.	21 ft, 2 in.	19 ft, 11 in.	18 ft, 5 in.	19 ft, 9 in.	18 ft, 6 in.	17 ft, 2 in.
	PRI-60	23 ft, 8 in.	21 ft, 7 in.	19 ft, 4 in.	22 ft, 8 in.	21 ft, 2 in.	18 ft, 11 in.	21 ft, 2 in.	19 ft, 10 in.	18 ft, 4 in.
	PRI-70	24 ft, 5 in.	22 ft, 11 in.	21 ft, 2 in.	23 ft, 2 in.	21 ft, 9 in.	20 ft, 1 in.	21 ft, 7 in.	20 ft, 3 in.	18 ft, 9 in.
	PRI-80	26 ft, 8 in.	25 ft, 0 in.	23 ft, 0 in.	25 ft, 3 in.	23 ft, 8 in.	21 ft, 11 in.	23 ft, 7 in.	22 ft, 1 in.	20 ft, 5 in.
	PRI-90	27 ft, 7 in.	25 ft, 10 in.	23 ft, 11 in.	26 ft, 1 in.	24 ft, 6 in.	22 ft, 8 in.	24 ft, 4 in.	22 ft, 10 in.	21 ft, 2 in.
14″	PRI-40	22 ft, 4 in.	20 ft, 5 in.	18 ft, 2 in.	21 ft, 11 in.	20 ft, 0 in.	17 ft, 10 in.	21 ft, 4 in.	19 ft, 5 in.	17 ft, 4 in.
	PRI-50	25 ft, 7 in.	23 ft, 8 in.	21 ft, 2 in.	24 ft, 3 in.	22 ft, 9 in.	20 ft, 9 in.	22 ft, 7 in.	21 ft, 3 in.	19 ft, 8 in.
	PRI-60	26 ft, 3 in.	24 ft, 0 in.	21 ft, 5 in.	25 ft, 9 in.	23 ft, 6 in.	21 ft, 0 in.	24 ft, 1 in.	22 ft, 8 in.	20 ft, 5 in.
	PRI-70	27 ft, 9 in.	26 ft, 1 in.	24 ft, 1 in.	26 ft, 4 in.	24 ft, 8 in.	22 ft, 10 in.	24 ft, 6 in.	23 ft, 0 in.	21 ft, 4 in.
	PRI-80	30 ft, 4 in.	28 ft, 6 in.	25 ft, 6 in.	28 ft, 9 in.	27 ft, 0 in.	24 ft, 11 in.	26 ft, 10 in.	25 ft, 2 in.	23 ft, 3 in.
	PRI-90	31 ft, 4 in.	29 ft, 5 in.	27 ft, 2 in.	29 ft, 8 in.	27 ft, 10 in.	25 ft, 9 in.	27 ft, 8 in.	25 ft, 11 in.	24 ft, 0 in.
16″	PRI-40	24 ft, 3 in.	22 ft, 1 in.	19 ft, 9 in.	23 ft, 9 in.	21 ft, 8 in.	19 ft, 4 in.	23 ft, 2 in.	21 ft, 1 in.	18 ft, 10 in.
	PRI-50	28 ft, 2 in.	25 ft, 8 in.	21 ft, 10 in.	27 ft, 0 in.	25 ft, 2 in.	22 ft, 6 in.	25 ft, 2 in.	23 ft, 8 in.	21 ft, 11 in.
	PRI-60	28 ft, 6 in.	26 ft, 0 in.	23 ft, 3 in.	27 ft, 11 in.	25 ft, 6 in.	22 ft, 9 in.	26 ft, 10 in.	24 ft, 10 in.	22 ft, 2 in.
	PRI-70	30 ft, 11 in.	29 ft, 0 in.	24 ft, 11 in.	29 ft, 3 in.	27 ft, 6 in.	25 ft, 5 in.	27 ft, 3 in.	25 ft, 7 in.	23 ft, 8 in.
	PRI-80	33 ft, 8 in.	31 ft, 0 in.	27 ft, 7 in.	31 ft, 11 in.	29 ft, 11 in.	27 ft, 2 in.	29 ft, 9 in.	27 ft, 11 in.	25 ft, 10 in.
	PRI-90	34 ft, 8 in.	32 ft, 7 in.	30 ft, 1 in.	32 ft, 10 in.	30 ft, 10 in.	28 ft, 6 in.	30 ft, 7 in.	28 ft, 9 in.	26 ft, 7 in.

For SI: 1 in. = 25.4 mm, 1 ft = 304.8 mm.

1. Allowable clear span applicable to simple-span roof construction with a design dead load of 15 psf (0.72 kN/m^2) and a roof snow load of 40 psf (1.92 kN/m^2). The snow load deflection is limited to span/240 and total load deflection to span/180.

2. Spans are based on a duration of load (DOL) factor of 1.15.

3. Minimum bearing length must be 1¾ in. (44.5 mm) for the end bearings and 3½ in. (89 mm) on end bearing adjacent to cantilever.

4. Bearing stiffeners are not required when I-joists are used with the spans and spacings given in this table, except as required by hanger manufacturers.

5. Spans include a cantilever of up to 2 ft on one end of the I-joist.

TABLE 5.27b Allowable Roof Spans for PRI I-Joists: Multiple-Span Only (flat roof (¼–12); snow load = 40 psf, dead load = 15 psf)

Depth	Designation	16 in. o.c.	19.2 in. o.c.	24 in. o.c.
9½″	PRI-20	16 ft, 8 in.	15 ft, 2 in.	13 ft, 7 in.
	PRI-30	18 ft, 11 in.	17 ft, 3 in.	15 ft, 5 in.
	PRI-40	17 ft, 7 in.	16 ft, 0 in.	14 ft, 4 in.
	PRI-50	20 ft, 6 in.	18 ft, 8 in.	16 ft, 8 in.
	PRI-60	20 ft, 8 in.	18 ft, 10 in.	16 ft, 10 in.
11⅞″	PRI-20	19 ft, 3 in.	17 ft, 7 in.	14 ft, 0 in.
	PRI-30	21 ft, 10 in.	19 ft, 8 in.	15 ft, 9 in.
	PRI-40	20 ft, 4 in.	18 ft, 6 in.	16 ft, 7 in.
	PRI-50	23 ft, 8 in.	21 ft, 1 in.	16 ft, 10 in.
	PRI-60	23 ft, 11 in.	21 ft, 10 in.	19 ft, 6 in.
	PRI-70	26 ft, 10 in.	24 ft, 2 in.	19 ft, 4 in.
	PRI-80	28 ft, 6 in.	26 ft, 0 in.	22 ft, 10 in.
14″	PRI-40	22 ft, 7 in.	20 ft, 7 in.	18 ft, 4 in.
	PRI-50	25 ft, 4 in.	21 ft, 1 in.	16 ft, 10 in.
	PRI-60	26 ft, 6 in.	24 ft, 2 in.	20 ft, 8 in.
	PRI-70	29 ft, 1 in.	24 ft, 2 in.	19 ft, 4 in.
	PRI-80	31 ft, 7 in.	28 ft, 10 in.	25 ft, 0 in.
16″	PRI-40	24 ft, 5 in.	22 ft, 4 in.	19 ft, 11 in.
	PRI-50	25 ft, 4 in.	21 ft, 1 in.	16 ft, 10 in.
	PRI-60	28 ft, 9 in.	25 ft, 11 in.	20 ft, 8 in.
	PRI-70	29 ft, 1 in.	24 ft, 2 in.	19 ft, 4 in.
	PRI-80	34 ft, 3 in.	31 ft, 3 in.	25 ft, 0 in.

For SI: 1 in. = 25.4 mm, 1 ft = 304.8 mm.

1. Allowable clear span applicable to multiple-span roof construction with a design dead load of 15 psf (0.72 kN/m^2) and a roof snow load of 40 psf (1.92 kN/m^2). The end spans must be 40% or more of the adjacent span. The snow load deflection is limited to span/240 and total load deflection to span/180.

2. Spans are based on a duration of load (DOL) factor of 1.15.

3. Minimum bearing length must be 1¾ in. (44.5 mm) for the end bearings and 3½ in. (89 mm) for the intermediate bearing.

4. Bearing stiffeners are not required when APA PRI joists are used with the spans and spacings given in this table, except as required by hanger manufacturers.

TABLE 5.28a Allowable Roof Spans for PRI I-Joists. Simple-Span (snow load = 50 psf, dead load = 15 psf)

Depth	Series	Slope of ¼:12 to 4:12			Slope of >4:12 to 8:12			Slope of >8:12 to 12:12		
		16 in. o.c.	19.2 in. o.c.	24 in. o.c.	16 in. o.c.	19.2 in. o.c.	24 in. o.c.	16 in. o.c.	19.2 in. o.c.	24 in. o.c.
9½″	PRI-20	15 ft, 2 in.	13 ft, 10 in.	12 ft, 4 in.	14 ft, 11 in.	13 ft, 7 in.	12 ft, 2 in.	14 ft, 4 in.	13 ft, 3 in.	11 ft, 10 in.
	PRI-30	16 ft, 8 in.	15 ft, 7 in.	14 ft, 0 in.	15 ft, 10 in.	14 ft, 11 in.	13 ft, 9 in.	14 ft, 10 in.	13 ft, 11 in.	12 ft, 10 in.
	PRI-40	16 ft, 0 in.	14 ft, 7 in.	13 ft, 0 in.	15 ft, 9 in.	14 ft, 4 in.	12 ft, 10 in.	15 ft, 5 in.	14 ft, 0 in.	12 ft, 6 in.
	PRI-50	17 ft, 6 in.	16 ft, 5 in.	15 ft, 2 in.	16 ft, 8 in.	15 ft, 7 in.	14 ft, 5 in.	15 ft, 7 in.	14 ft, 7 in.	13 ft, 6 in.
	PRI-60	18 ft, 9 in.	17 ft, 2 in.	15 ft, 4 in.	17 ft, 11 in.	16 ft, 9 in.	15 ft, 1 in.	16 ft, 9 in.	15 ft, 8 in.	14 ft, 6 in.
11⅞″	PRI-20	17 ft, 7 in.	16 ft, 0 in.	14 ft, 4 in.	17 ft, 3 in.	15 ft, 9 in.	14 ft, 1 in.	16 ft, 10 in.	15 ft, 5 in.	13 ft, 9 in.
	PRI-30	19 ft, 11 in.	18 ft, 1 in.	16 ft, 2 in.	19 ft, 2 in.	17 ft, 10 in.	15 ft, 11 in.	17 ft, 11 in.	16 ft, 9 in.	15 ft, 6 in.
	PRI-40	18 ft, 6 in.	16 ft, 11 in.	15 ft, 1 in.	18 ft, 3 in.	16 ft, 7 in.	14 ft, 10 in.	17 ft, 10 in.	16 ft, 3 in.	14 ft, 6 in.
	PRI-50	21 ft, 0 in.	19 ft, 8 in.	17 ft, 7 in.	20 ft, 1 in.	18 ft, 10 in.	17 ft, 3 in.	18 ft, 9 in.	17 ft, 7 in.	16 ft, 3 in.
	PRI-60	21 ft, 10 in.	19 ft, 11 in.	17 ft, 9 in.	21 ft, 5 in.	19 ft, 7 in.	17 ft, 6 in.	20 ft, 1 in.	18 ft, 10 in.	17 ft, 1 in.
	PRI-70	23 ft, 0 in.	21 ft, 7 in.	19 ft, 11 in.	21 ft, 11 in.	20 ft, 7 in.	19 ft, 0 in.	20 ft, 6 in.	19 ft, 3 in.	17 ft, 9 in.
	PRI-80	25 ft, 1 in.	23 ft, 6 in.	21 ft, 2 in.	23 ft, 11 in.	22 ft, 5 in.	20 ft, 9 in.	22 ft, 4 in.	21 ft, 0 in.	19 ft, 5 in.
	PRI-90	25 ft, 11 in.	24 ft, 3 in.	21 ft, 4 in.	24 ft, 8 in.	23 ft, 2 in.	21 ft, 5 in.	23 ft, 1 in.	21 ft, 8 in.	20 ft, 1 in.
14″	PRI-40	20 ft, 7 in.	18 ft, 9 in.	16 ft, 9 in.	20 ft, 3 in.	18 ft, 5 in.	16 ft, 6 in.	19 ft, 9 in.	18 ft, 0 in.	16 ft, 1 in.
	PRI-50	23 ft, 11 in.	21 ft, 10 in.	18 ft, 5 in.	22 ft, 11 in.	21 ft, 5 in.	19 ft, 2 in.	21 ft, 5 in.	20 ft, 2 in.	18 ft, 8 in.
	PRI-60	24 ft, 2 in.	22 ft, 1 in.	19 ft, 8 in.	23 ft, 9 in.	21 ft, 8 in.	19 ft, 4 in.	22 ft, 11 in.	21 ft, 2 in.	18 ft, 11 in.
	PRI-70	26 ft, 1 in.	24 ft, 6 in.	21 ft, 1 in.	24 ft, 11 in.	23 ft, 4 in.	21 ft, 7 in.	23 ft, 3 in.	21 ft, 10 in.	20 ft, 2 in.
	PRI-80	28 ft, 6 in.	26 ft, 3 in.	23 ft, 4 in.	27 ft, 2 in.	25 ft, 6 in.	23 ft, 1 in.	25 ft, 5 in.	23 ft, 10 in.	22 ft, 1 in.
	PRI-90	29 ft, 5 in.	27 ft, 7 in.	25 ft, 6 in.	28 ft, 1 in.	26 ft, 4 in.	24 ft, 4 in.	26 ft, 3 in.	24 ft, 8 in.	22 ft, 9 in.
16″	PRI-40	22 ft, 4 in.	20 ft, 4 in.	18 ft, 2 in.	21 ft, 11 in.	20 ft, 0 in.	17 ft, 10 in.	21 ft, 5 in.	19 ft, 7 in.	17 ft, 5 in.
	PRI-50	25 ft, 11 in.	23 ft, 1 in.	18 ft, 5 in.	25 ft, 6 in.	23 ft, 3 in.	20 ft, 4 in.	23 ft, 11 in.	22 ft, 5 in.	20 ft, 3 in.
	PRI-60	26 ft, 3 in.	23 ft, 11 in.	21 ft, 4 in.	25 ft, 10 in.	23 ft, 6 in.	21 ft, 0 in.	25 ft, 2 in.	23 ft, 0 in.	20 ft, 6 in.
	PRI-70	29 ft, 1 in.	26 ft, 5 in.	21 ft, 1 in.	27 ft, 8 in.	26 ft, 0 in.	23 ft, 4 in.	25 ft, 11 in.	24 ft, 4 in.	22 ft, 6 in.
	PRI-80	31 ft, 3 in.	28 ft, 6 in.	23 ft, 4 in.	30 ft, 2 in.	28 ft, 0 in.	25 ft, 1 in.	28 ft, 3 in.	26 ft, 6 in.	24 ft, 6 in.
	PRI-90	32 ft, 7 in.	30 ft, 7 in.	28 ft, 3 in.	31 ft, 1 in.	29 ft, 2 in.	27 ft, 0 in.	29 ft, 1 in.	27 ft, 3 in.	25 ft, 3 in.

For SI: 1 in. = 25.4 mm, 1 ft = 304.8 mm.

1. Allowable clear span applicable to simple-span roof construction with a design dead load of 15 psf (0.72 kN/m^2) and a roof snow load of 50 psf (2.4 kN/m^2). The snow load deflection is limited to span/240 and total load deflection to span/180.

2. Spans are based on a duration of load (DOL) factor of 1.15.

3. Minimum bearing length must be 1¾ in. (44.5 mm) for the end bearings and 3½ in. (89 mm) on end bearing adjacent to cantilever.

4. Bearing stiffeners are not required when I-joists are used with the spans and spacings given in this table, except as required by hanger manufacturers.

5. Spans include a cantilever of up to 2 ft on one end of the I-joist.

TABLE 5.28b Allowable Roof Spans for PRI I-Joists: Multiple-Span Only (flat roof (¼–12); snow load = 50 psf, dead load = 15 psf)

Depth	Designation	16 in. o.c.	19.2 in. o.c.	24 in. o.c.
9½″	PRI-20	15 ft, 4 in.	13 ft, 11 in.	11 ft, 10 in.
	PRI-30	17 ft, 4 in.	15 ft, 10 in.	13 ft, 3 in.
	PRI-40	16 ft, 2 in.	14 ft, 9 in.	13 ft, 2 in.
	PRI-50	18 ft, 10 in.	17 ft, 2 in.	14 ft, 3 in.
	PRI-60	19 ft, 0 in.	17 ft, 4 in.	15 ft, 1 in.
11⅞″	PRI-20	17 ft, 8 in.	14 ft, 10 in.	11 ft, 10 in.
	PRI-30	20 ft, 0 in.	16 ft, 8 in.	13 ft, 3 in.
	PRI-40	18 ft, 8 in.	17 ft, 0 in.	15 ft, 3 in.
	PRI-50	21 ft, 5 in.	17 ft, 10 in.	14 ft, 3 in.
	PRI-60	22 ft, 0 in.	20 ft, 0 in.	17 ft, 6 in.
	PRI-70	24 ft, 7 in.	20 ft, 5 in.	16 ft, 4 in.
	PRI-80	26 ft, 2 in.	23 ft, 10 in.	19 ft, 4 in.
14″	PRI-40	20 ft, 9 in.	18 ft, 11 in.	16 ft, 11 in.
	PRI-50	21 ft, 5 in.	17 ft, 10 in.	14 ft, 3 in.
	PRI-60	24 ft, 4 in.	21 ft, 11 in.	17 ft, 6 in.
	PRI-70	24 ft, 7 in.	20 ft, 5 in.	16 ft, 4 in.
	PRI-80	29 ft, 0 in.	26 ft, 6 in.	21 ft, 2 in.
16″	PRI-40	22 ft, 6 in.	20 ft, 6 in.	17 ft, 6 in.
	PRI-50	21 ft, 5 in.	17 ft, 10 in.	14 ft, 3 in.
	PRI-60	26 ft, 4 in.	21 ft, 11 in.	17 ft, 6 in.
	PRI-70	24 ft, 7 in.	20 ft, 5 in.	16 ft, 4 in.
	PRI-80	31 ft, 6 in.	26 ft, 6 in.	21 ft, 2 in.

For SI: 1 in. = 25.4 mm, 1 ft = 304.8 mm.

1. Allowable clear span applicable to multiple-span roof construction with a design dead load of 15 psf (0.72 kN/m^2) and a roof snow load of 50 psf (2.4 kN/m^2). The end spans must be 40% or more of the adjacent span. The snow load deflection is limited to span/240 and total load deflection to span/180.

2. Spans are based on a duration of load (DOL) factor of 1.15.

3. Minimum bearing length must be 1¾ in. (44.5 mm) for the end bearings and 3½ in. (89 mm) for the intermediate bearing.

4. Bearing stiffeners are not required when APA PRI joists are used with the spans and spacings given in this table, except as required by hanger manufacturers.

TABLE 5.29a Allowable Roof Spans for PRI I-Joists. Simple-Span (construction load = 20 psf, dead load = 10 psf)

Depth	Series	Slope of 1/4:12 to 4:12			Slope of >4:12 to 8:12			Slope of >8:12 to 12:12		
		16 in. o.c.	19.2 in. o.c.	24 in. o.c.	16 in. o.c.	19.2 in. o.c.	24 in. o.c.	16 in. o.c.	19.2 in. o.c.	24 in. o.c.
9½″	PRI-20	21 ft, 1 in.	19 ft, 10 in.	18 ft, 4 in.	19 ft, 11 in.	18 ft, 8 in.	17 ft, 3 in.	18 ft, 5 in.	17 ft, 4 in.	16 ft, 0 in.
	PRI-30	21 ft, 10 in.	20 ft, 6 in.	19 ft, 0 in.	20 ft, 7 in.	19 ft, 4 in.	17 ft, 11 in.	19 ft, 1 in.	17 ft, 11 in.	16 ft, 7 in.
	PRI-40	23 ft, 2 in.	21 ft, 9 in.	20 ft, 1 in.	21 ft, 10 in.	20 ft, 6 in.	19 ft, 0 in.	20 ft, 4 in.	19 ft, 1 in.	17 ft, 8 in.
	PRI-50	22 ft, 11 in.	21 ft, 6 in.	19 ft, 11 in.	21 ft, 7 in.	20 ft, 3 in.	18 ft, 9 in.	20 ft, 1 in.	18 ft, 10 in.	17 ft, 5 in.
	PRI-60	24 ft, 7 in.	23 ft, 1 in.	21 ft, 5 in.	23 ft, 3 in.	21 ft, 10 in.	20 ft, 2 in.	21 ft, 7 in.	20 ft, 3 in.	18 ft, 9 in.
11⅞″	PRI-20	25 ft, 5 in.	23 ft, 11 in.	22 ft, 0 in.	24 ft, 0 in.	22 ft, 6 in.	20 ft, 10 in.	22 ft, 3 in.	20 ft, 11 in.	19 ft, 4 in.
	PRI-30	26 ft, 4 in.	24 ft, 9 in.	22 ft, 11 in.	24 ft, 10 in.	23 ft, 4 in.	21 ft, 7 in.	23 ft, 0 in.	21 ft, 8 in.	20 ft, 0 in.
	PRI-40	27 ft, 10 in.	26 ft, 0 in.	23 ft, 3 in.	26 ft, 3 in.	24 ft, 7 in.	22 ft, 8 in.	24 ft, 4 in.	22 ft, 10 in.	21 ft, 2 in.
	PRI-50	27 ft, 7 in.	25 ft, 11 in.	24 ft, 0 in.	26 ft, 0 in.	24 ft, 5 in.	22 ft, 7 in.	24 ft, 1 in.	22 ft, 8 in.	21 ft, 0 in.
	PRI-60	29 ft, 6 in.	27 ft, 9 in.	25 ft, 8 in.	27 ft, 10 in.	26 ft, 2 in.	24 ft, 3 in.	25 ft, 10 in.	24 ft, 3 in.	22 ft, 6 in.
	PRI-70	30 ft, 1 in.	28 ft, 3 in.	26 ft, 2 in.	28 ft, 5 in.	26 ft, 8 in.	24 ft, 8 in.	26 ft, 4 in.	24 ft, 9 in.	22 ft, 11 in.
	PRI-80	32 ft, 11 in.	30 ft, 10 in.	28 ft, 7 in.	31 ft, 0 in.	29 ft, 2 in.	27 ft, 0 in.	28 ft, 9 in.	27 ft, 1 in.	25 ft, 1 in.
	PRI-90	34 ft, 0 in.	31 ft, 11 in.	29 ft, 6 in.	32 ft, 1 in.	30 ft, 1 in.	27 ft, 10 in.	29 ft, 9 in.	27 ft, 11 in.	25 ft, 11 in.
14″	PRI-40	31 ft, 7 in.	28 ft, 10 in.	25 ft, 9 in.	29 ft, 9 in.	28 ft, 0 in.	25 ft, 2 in.	27 ft, 8 in.	26 ft, 0 in.	24 ft, 1 in.
	PRI-50	31 ft, 6 in.	29 ft, 8 in.	27 ft, 5 in.	29 ft, 9 in.	27 ft, 11 in.	25 ft, 11 in.	27 ft, 7 in.	25 ft, 11 in.	24 ft, 0 in.
	PRI-60	33 ft, 8 in.	31 ft, 7 in.	29 ft, 3 in.	31 ft, 9 in.	29 ft, 10 in.	27 ft, 7 in.	29 ft, 6 in.	27 ft, 8 in.	25 ft, 8 in.
	PRI-70	34 ft, 3 in.	32 ft, 2 in.	29 ft, 9 in.	32 ft, 3 in.	30 ft, 4 in.	28 ft, 1 in.	29 ft, 11 in.	28 ft, 2 in.	26 ft, 1 in.
	PRI-80	37 ft, 5 in.	35 ft, 1 in.	32 ft, 6 in.	35 ft, 3 in.	33 ft, 2 in.	30 ft, 8 in.	32 ft, 9 in.	30 ft, 9 in.	28 ft, 6 in.
	PRI-90	38 ft, 7 in.	36 ft, 3 in.	33 ft, 6 in.	36 ft, 5 in.	34 ft, 2 in.	31 ft, 8 in.	33 ft, 9 in.	31 ft, 9 in.	29 ft, 5 in.
16″	PRI-40	34 ft, 3 in.	31 ft, 3 in.	27 ft, 11 in.	33 ft, 1 in.	30 ft, 6 in.	27 ft, 3 in.	30 ft, 8 in.	28 ft, 10 in.	26 ft, 5 in.
	PRI-50	35 ft, 2 in.	33 ft, 0 in.	30 ft, 7 in.	33 ft, 2 in.	31 ft, 2 in.	28 ft, 10 in.	30 ft, 9 in.	28 ft, 11 in.	26 ft, 9 in.
	PRI-60	37 ft, 5 in.	35 ft, 2 in.	32 ft, 6 in.	35 ft, 3 in.	33 ft, 2 in.	30 ft, 8 in.	32 ft, 9 in.	30 ft, 9 in.	28 ft, 6 in.
	PRI-70	38 ft, 1 in.	35 ft, 9 in.	33 ft, 1 in.	35 ft, 11 in.	33 ft, 9 in.	31 ft, 3 in.	33 ft, 4 in.	31 ft, 4 in.	29 ft, 0 in.
	PRI-80	41 ft, 6 in.	39 ft, 0 in.	36 ft, 1 in.	39 ft, 2 in.	36 ft, 9 in.	34 ft, 1 in.	36 ft, 4 in.	34 ft, 2 in.	31 ft, 8 in.
	PRI-90	42 ft, 9 in.	40 ft, 1 in.	37 ft, 2 in.	40 ft, 4 in.	37 ft, 10 in.	35 ft, 1 in.	37 ft, 5 in.	35 ft, 2 in.	32 ft, 7 in.

For SI: 1 in. = 25.4 mm, 1 ft = 304.8 mm.

1. Allowable clear span applicable to simple-span roof construction with a design dead load of 10 psf (0.48 kN/m^2) and a construction load of 20 psf (1.2 kN/m^2). The construction load deflection is limited to span/240 and total load deflection to span/180.

2. Spans are based on a duration of load (DOL) factor of 1.25.

3. Minimum bearing length must be 1¾ in. (44.5 mm) for the end bearings and 3½ in. (89 mm) on end bearing adjacent to cantilever.

4. Bearing stiffeners are not required when I-joists are used with the spans and spacings given in this table, except as required by hanger manufacturers.

5. Spans include a cantilever of up to 2 ft on one end of the I-joist.

TABLE 5.29b Allowable Roof Spans for PRI I-Joists: Multiple-Span Only (flat roof (¼–12); construction load = 20 psf, dead load = 10 psf)

Depth	Designation	16 in. o.c.	19.2 in. o.c.	24 in. o.c.
9½″	PRI-20	23 ft, 7 in.	21 ft, 6 in.	19 ft, 3 in.
	PRI-30	24 ft, 6 in.	23 ft, 0 in.	21 ft, 3 in.
	PRI-40	24 ft, 11 in.	22 ft, 8 in.	20 ft, 3 in.
	PRI-50	25 ft, 8 in.	24 ft, 2 in.	22 ft, 4 in.
	PRI-60	27 ft, 7 in.	25 ft, 11 in.	23 ft, 10 in.
11⅞″	PRI-20	27 ft, 3 in.	24 ft, 11 in.	22 ft, 3 in.
	PRI-30	29 ft, 6 in.	27 ft, 9 in.	25 ft, 2 in.
	PRI-40	28 ft, 9 in.	26 ft, 3 in.	23 ft, 6 in.
	PRI-50	30 ft, 11 in.	29 ft, 0 in.	26 ft, 11 in.
	PRI-60	33 ft, 1 in.	30 ft, 10 in.	27 ft, 7 in.
	PRI-70	33 ft, 9 in.	31 ft, 8 in.	29 ft, 4 in.
	PRI-80	36 ft, 10 in.	34 ft, 7 in.	32 ft, 0 in.
14″	PRI-40	31 ft, 11 in.	29 ft, 1 in.	26 ft, 0 in.
	PRI-50	35 ft, 4 in.	33 ft, 3 in.	30 ft, 3 in.
	PRI-60	37 ft, 6 in.	34 ft, 3 in.	30 ft, 7 in.
	PRI-70	38 ft, 4 in.	36 ft, 0 in.	33 ft, 4 in.
	PRI-80	41 ft, 11 in.	39 ft, 4 in.	36 ft, 5 in.
16″	PRI-40	34 ft, 7 in.	31 ft, 7 in.	28 ft, 3 in.
	PRI-50	39 ft, 5 in.	36 ft, 8 in.	32 ft, 9 in.
	PRI-60	40 ft, 8 in.	37 ft, 1 in.	33 ft, 2 in.
	PRI-70	42 ft, 8 in.	40 ft, 1 in.	37 ft, 1 in.
	PRI-80	46 ft, 6 in.	43 ft, 8 in.	39 ft, 6 in.

For SI: 1 in. = 25.4 mm, 1 ft = 304.8 mm.

1. Allowable clear span applicable to multiple-span roof construction with a design dead load of 10 psf (0.48 kN/m^2) and a construction load of 20 psf (1.2 kN/m^2). The end spans must be 40% or more of the adjacent span. The construction load deflection is limited to span/240 and total load deflection to span/180.

2. Spans are based on a duration of load (DOL) factor of 1.25.

3. Minimum bearing length must be 1¾ in. (44.5 mm) for the end bearings and 3½ in. (89 mm) for the intermediate bearing.

4. Bearing stiffeners are not required when APA PRI joists are used with the spans and spacings given in this table, except as required by hanger manufacturers.

TABLE 5.30a Allowable Roof Spans for PRI I-Joists. Simple-Span (construction load = 20 psf, dead load = 15 psf)

		Slope of ¼:12 to 4:12			Slope of >4:12 to 8:12			Slope of >8:12 to 12:12		
Depth	Series	16 in. o.c.	19.2 in. o.c.	24 in. o.c.	16 in. o.c.	19.2 in. o.c.	24 in. o.c.	16 in. o.c.	19.2 in. o.c.	24 in. o.c.
9½″	PRI-20	19 ft, 11 in.	18 ft, 9 in.	17 ft, 4 in.	18 ft, 9 in.	17 ft, 7 in.	16 ft, 3 in.	17 ft, 3 in.	16 ft, 3 in.	15 ft, 0 in.
	PRI-30	20 ft, 8 in.	19 ft, 5 in.	17 ft, 11 in.	19 ft, 5 in.	18 ft, 3 in.	16 ft, 10 in.	17 ft, 11 in.	16 ft, 10 in.	15 ft, 7 in.
	PRI-40	21 ft, 11 in.	20 ft, 7 in.	18 ft, 6 in.	20 ft, 7 in.	19 ft, 4 in.	17 ft, 11 in.	19 ft, 0 in.	17 ft, 10 in.	16 ft, 6 in.
	PRI-50	21 ft, 8 in.	20 ft, 4 in.	18 ft, 10 in.	20 ft, 4 in.	19 ft, 1 in.	17 ft, 8 in.	18 ft, 9 in.	17 ft, 8 in.	16 ft, 4 in.
	PRI-60	23 ft, 4 in.	21 ft, 11 in.	20 ft, 3 in.	21 ft, 11 in.	20 ft, 7 in.	19 ft, 0 in.	20 ft, 2 in.	19 ft, 0 in.	17 ft, 7 in.
11⅞″	PRI-20	24 ft, 1 in.	22 ft, 7 in.	20 ft, 4 in.	22 ft, 7 in.	21 ft, 3 in.	19 ft, 8 in.	20 ft, 10 in.	19 ft, 7 in.	18 ft, 2 in.
	PRI-30	24 ft, 11 in.	23 ft, 5 in.	21 ft, 8 in.	23 ft, 5 in.	22 ft, 0 in.	20 ft, 4 in.	21 ft, 7 in.	20 ft, 3 in.	18 ft, 9 in.
	PRI-40	26 ft, 4 in.	24 ft, 0 in.	21 ft, 5 in.	24 ft, 9 in.	23 ft, 2 in.	20 ft, 9 in.	22 ft, 10 in.	21 ft, 5 in.	19 ft, 10 in.
	PRI-50	26 ft, 1 in.	24 ft, 6 in.	22 ft, 8 in.	24 ft, 6 in.	23 ft, 0 in.	21 ft, 4 in.	22 ft, 7 in.	21 ft, 3 in.	19 ft, 8 in.
	PRI-60	27 ft, 11 in.	26 ft, 3 in.	24 ft, 4 in.	26 ft, 3 in.	24 ft, 8 in.	22 ft, 10 in.	24 ft, 3 in.	22 ft, 9 in.	21 ft, 1 in.
	PRI-70	28 ft, 6 in.	26 ft, 9 in.	24 ft, 9 in.	26 ft, 9 in.	25 ft, 2 in.	23 ft, 3 in.	24 ft, 9 in.	23 ft, 3 in.	21 ft, 6 in.
	PRI-80	31 ft, 1 in.	29 ft, 3 in.	27 ft, 0 in.	29 ft, 3 in.	27 ft, 5 in.	25 ft, 5 in.	27 ft, 0 in.	25 ft, 4 in.	23 ft, 6 in.
	PRI-90	32 ft, 2 in.	30 ft, 2 in.	27 ft, 11 in.	30 ft, 3 in.	28 ft, 4 in.	26 ft, 3 in.	27 ft, 11 in.	26 ft, 2 in.	24 ft, 3 in.
14″	PRI-40	29 ft, 2 in.	26 ft, 7 in.	23 ft, 9 in.	28 ft, 1 in.	25 ft, 10 in.	23 ft, 1 in.	25 ft, 11 in.	24 ft, 4 in.	22 ft, 2 in.
	PRI-50	29 ft, 10 in.	28 ft, 1 in.	26 ft, 0 in.	28 ft, 1 in.	26 ft, 4 in.	24 ft, 5 in.	25 ft, 11 in.	24 ft, 4 in.	22 ft, 6 in.
	PRI-60	31 ft, 10 in.	29 ft, 11 in.	27 ft, 8 in.	29 ft, 11 in.	28 ft, 1 in.	26 ft, 0 in.	27 ft, 8 in.	25 ft, 11 in.	24 ft, 0 in.
	PRI-70	32 ft, 5 in.	30 ft, 5 in.	28 ft, 2 in.	30 ft, 5 in.	28 ft, 7 in.	26 ft, 5 in.	28 ft, 1 in.	26 ft, 5 in.	24 ft, 5 in.
	PRI-80	35 ft, 5 in.	33 ft, 3 in.	30 ft, 9 in.	33 ft, 3 in.	31 ft, 3 in.	28 ft, 11 in.	30 ft, 8 in.	28 ft, 10 in.	26 ft, 9 in.
	PRI-90	36 ft, 6 in.	34 ft, 4 in.	31 ft, 9 in.	34 ft, 4 in.	32 ft, 3 in.	29 ft, 10 in.	31 ft, 8 in.	29 ft, 9 in.	27 ft, 7 in.
16″	PRI-40	31 ft, 8 in.	28 ft, 10 in.	25 ft, 9 in.	30 ft, 8 in.	28 ft, 0 in.	25 ft, 0 in.	28 ft, 9 in.	26 ft, 11 in.	24 ft, 0 in.
	PRI-50	33 ft, 4 in.	31 ft, 3 in.	28 ft, 11 in.	31 ft, 3 in.	29 ft, 4 in.	27 ft, 2 in.	28 ft, 10 in.	27 ft, 1 in.	25 ft, 1 in.
	PRI-60	35 ft, 5 in.	33 ft, 3 in.	30 ft, 4 in.	33 ft, 3 in.	31 ft, 3 in.	28 ft, 11 in.	30 ft, 8 in.	28 ft, 10 in.	26 ft, 9 in.
	PRI-70	36 ft, 0 in.	33 ft, 10 in.	31 ft, 4 in.	33 ft, 10 in.	31 ft, 9 in.	29 ft, 5 in.	31 ft, 3 in.	29 ft, 4 in.	27 ft, 2 in.
	PRI-80	39 ft, 3 in.	36 ft, 11 in.	34 ft, 2 in.	36 ft, 11 in.	34 ft, 8 in.	32 ft, 1 in.	34 ft, 1 in.	32 ft, 0 in.	29 ft, 8 in.
	PRI-90	40 ft, 5 in.	38 ft, 0 in.	35 ft, 2 in.	38 ft, 0 in.	35 ft, 8 in.	33 ft, 0 in.	35 ft, 1 in.	32 ft, 11 in.	30 ft, 6 in.

For SI: 1 in. = 25.4 mm, 1 ft = 304.8 mm.

1. Allowable clear span applicable to simple-span roof construction with a design dead load of 15 psf (0.72 kN/m^2) and a construction load of 20 psf (1.2 kN/m^2). The construction load deflection is limited to span/240 and total load deflection to span/180.

2. Spans are based on a duration of load (DOL) factor of 1.25.

3. Minimum bearing length must be 1¾ in. (44.5 mm) for the end bearings and 3½ in. (89 mm) on end bearing adjacent to cantilever.

4. Bearing stiffeners are not required when I-joists are used with the spans and spacings given in this table, except as required by hanger manufacturers.

5. Spans include a cantilever of up to 2 ft on one end of the I-joist.

TABLE 5.30b Allowable Roof Spans for PRI I-Joists: Multiple-Span Only (flat roof (¼–12); construction load = 20 psf, dead load = 15 psf)

Depth	Designation	16 in. o.c.	19.2 in. o.c.	24 in. o.c.
9½″	PRI-20	21 ft, 10 in.	19 ft, 11 in.	17 ft, 9 in.
	PRI-30	23 ft, 3 in.	21 ft, 10 in.	20 ft, 2 in.
	PRI-40	23 ft, 0 in.	21 ft, 0 in.	18 ft, 9 in.
	PRI-50	24 ft, 4 in.	22 ft, 11 in.	21 ft, 2 in.
	PRI-60	26 ft, 2 in.	24 ft, 7 in.	22 ft, 1 in.
11⅞″	PRI-20	25 ft, 3 in.	23 ft, 0 in.	20 ft, 7 in.
	PRI-30	28 ft, 0 in.	26 ft, 1 in.	23 ft, 3 in.
	PRI-40	26 ft, 8 in.	24 ft, 4 in.	21 ft, 8 in.
	PRI-50	29 ft, 4 in.	27 ft, 6 in.	25 ft, 3 in.
	PRI-60	31 ft, 4 in.	28 ft, 7 in.	25 ft, 6 in.
	PRI-70	32 ft, 0 in.	30 ft, 1 in.	27 ft, 10 in.
	PRI-80	34 ft, 11 in.	32 ft, 9 in.	30 ft, 4 in.
14″	PRI-40	29 ft, 6 in.	26 ft, 11 in.	24 ft, 1 in.
	PRI-50	33 ft, 6 in.	31 ft, 4 in.	28 ft, 0 in.
	PRI-60	34 ft, 8 in.	31 ft, 8 in.	28 ft, 4 in.
	PRI-70	36 ft, 4 in.	34 ft, 2 in.	31 ft, 7 in.
	PRI-80	39 ft, 9 in.	37 ft, 4 in.	33 ft, 8 in.
16″	PRI-40	32 ft, 0 in.	29 ft, 3 in.	26 ft, 1 in.
	PRI-50	37 ft, 2 in.	33 ft, 11 in.	28 ft, 11 in.
	PRI-60	37 ft, 8 in.	34 ft, 4 in.	30 ft, 8 in.
	PRI-70	40 ft, 5 in.	38 ft, 0 in.	33 ft, 2 in.
	PRI-80	44 ft, 1 in.	40 ft, 11 in.	36 ft, 7 in.

For SI: 1 in. = 25.4 mm, 1 ft = 304.8 mm.

1. Allowable clear span applicable to multiple-span roof construction with a design dead load of 15 psf (0.72 kN/m²) and a construction load of 20 psf (1.2 kN/m²). The end spans must be 40% or more of the adjacent span. The construction load deflection is limited to span/240 and total load deflection to span/180.

2. Spans are based on a duration of load (DOL) factor of 1.25.

3. Minimum bearing length must be 1¾ in. (44.5 mm) for the end bearings and 3½ in. (89 mm) for the intermediate bearing.

4. Bearing stiffeners are not required when APA PRI joists are used with the spans and spacings given in this table, except as required by hanger manufacturers.

TABLE 5.31a Allowable Roof Spans for PRI I-Joists. Simple-Span (construction load = 20 psf, dead load = 25 psf)

Depth	Series	Slope of ¼:12 to 4:12			Slope of >4:12 to 8:12			Slope of >8:12 to 12:12		
		16 in. o.c.	19.2 in. o.c.	24 in. o.c.	16 in. o.c.	19.2 in. o.c.	24 in. o.c.	16 in. o.c.	19.2 in. o.c.	24 in. o.c.
9½″	PRI-20	19 ft, 0 in.	17 ft, 10 in.	16 ft, 4 in.	17 ft, 10 in.	16 ft, 9 in.	15 ft, 6 in.	16 ft, 4 in.	15 ft, 4 in.	14 ft, 3 in.
	PRI-30	19 ft, 8 in.	18 ft, 6 in.	17 ft, 1 in.	18 ft, 5 in.	17 ft, 4 in.	16 ft, 0 in.	16 ft, 11 in.	15 ft, 11 in.	14 ft, 9 in.
	PRI-40	20 ft, 11 in.	19 ft, 4 in.	17 ft, 3 in.	19 ft, 7 in.	18 ft, 5 in.	16 ft, 8 in.	18 ft, 0 in.	16 ft, 11 in.	15 ft, 8 in.
	PRI-50	20 ft, 8 in.	19 ft, 5 in.	17 ft, 11 in.	19 ft, 4 in.	18 ft, 2 in.	16 ft, 10 in.	17 ft, 9 in.	16 ft, 8 in.	15 ft, 6 in.
	PRI-60	22 ft, 3 in.	20 ft, 10 in.	19 ft, 4 in.	20 ft, 10 in.	19 ft, 6 in.	18 ft, 1 in.	19 ft, 2 in.	18 ft, 0 in.	16 ft, 7 in.
11⅞″	PRI-20	23 ft, 0 in.	21 ft, 2 in.	18 ft, 11 in.	21 ft, 6 in.	20 ft, 2 in.	18 ft, 3 in.	19 ft, 9 in.	18 ft, 7 in.	17 ft, 2 in.
	PRI-30	23 ft, 9 in.	22 ft, 4 in.	20 ft, 8 in.	22 ft, 3 in.	20 ft, 11 in.	19 ft, 4 in.	20 ft, 5 in.	19 ft, 2 in.	17 ft, 9 in.
	PRI-40	24 ft, 6 in.	22 ft, 5 in.	20 ft, 0 in.	23 ft, 6 in.	21 ft, 7 in.	19 ft, 4 in.	21 ft, 7 in.	20 ft, 3 in.	18 ft, 5 in.
	PRI-50	24 ft, 11 in.	23 ft, 4 in.	21 ft, 8 in.	23 ft, 4 in.	21 ft, 11 in.	20 ft, 3 in.	21 ft, 5 in.	20 ft, 1 in.	18 ft, 7 in.
	PRI-60	26 ft, 8 in.	25 ft, 0 in.	23 ft, 2 in.	25 ft, 0 in.	23 ft, 5 in.	21 ft, 8 in.	22 ft, 11 in.	21 ft, 7 in.	19 ft, 11 in.
	PRI-70	27 ft, 2 in.	25 ft, 6 in.	23 ft, 7 in.	25 ft, 5 in.	23 ft, 11 in.	22 ft, 1 in.	23 ft, 5 in.	22 ft, 0 in.	20 ft, 4 in.
	PRI-80	29 ft, 8 in.	27 ft, 10 in.	25 ft, 9 in.	27 ft, 10 in.	26 ft, 1 in.	24 ft, 2 in.	25 ft, 7 in.	24 ft, 0 in.	22 ft, 3 in.
	PRI-90	34 ft, 0 in.	31 ft, 11 in.	29 ft, 6 in.	32 ft, 1 in.	30 ft, 1 in.	27 ft, 10 in.	29 ft, 9 in.	27 ft, 11 in.	25 ft, 11 in.
14″	PRI-40	27 ft, 3 in.	24 ft, 10 in.	22 ft, 2 in.	26 ft, 3 in.	24 ft, 0 in.	21 ft, 5 in.	24 ft, 7 in.	22 ft, 10 in.	20 ft, 5 in.
	PRI-50	28 ft, 6 in.	26 ft, 9 in.	24 ft, 9 in.	26 ft, 8 in.	25 ft, 1 in.	23 ft, 2 in.	24 ft, 6 in.	23 ft, 0 in.	21 ft, 4 in.
	PRI-60	30 ft, 5 in.	28 ft, 6 in.	26 ft, 1 in.	28 ft, 6 in.	26 ft, 9 in.	24 ft, 9 in.	26 ft, 2 in.	24 ft, 7 in.	22 ft, 9 in.
	PRI-70	30 ft, 11 in.	29 ft, 0 in.	26 ft, 10 in.	28 ft, 11 in.	27 ft, 2 in.	25 ft, 2 in.	26 ft, 7 in.	25 ft, 0 in.	23 ft, 2 in.
	PRI-80	33 ft, 9 in.	31 ft, 8 in.	29 ft, 4 in.	31 ft, 7 in.	29 ft, 8 in.	27 ft, 6 in.	29 ft, 1 in.	27 ft, 4 in.	25 ft, 3 in.
	PRI-90	38 ft, 7 in.	36 ft, 3 in.	33 ft, 6 in.	36 ft, 5 in.	34 ft, 2 in.	31 ft, 8 in.	33 ft, 9 in.	31 ft, 9 in.	29 ft, 5 in.
16″	PRI-40	29 ft, 6 in.	26 ft, 11 in.	24 ft, 1 in.	28 ft, 6 in.	26 ft, 0 in.	23 ft, 3 in.	27 ft, 2 in.	24 ft, 10 in.	22 ft, 2 in.
	PRI-50	31 ft, 9 in.	29 ft, 10 in.	27 ft, 7 in.	29 ft, 9 in.	27 ft, 11 in.	25 ft, 10 in.	27 ft, 4 in.	25 ft, 8 in.	23 ft, 9 in.
	PRI-60	33 ft, 9 in.	31 ft, 8 in.	28 ft, 3 in.	31 ft, 8 in.	29 ft, 9 in.	27 ft, 4 in.	29 ft, 1 in.	27 ft, 4 in.	25 ft, 4 in.
	PRI-70	34 ft, 4 in.	32 ft, 3 in.	29 ft, 10 in.	32 ft, 2 in.	30 ft, 3 in.	28 ft, 0 in.	29 ft, 7 in.	27 ft, 10 in.	25 ft, 9 in.
	PRI-80	37 ft, 6 in.	35 ft, 2 in.	32 ft, 7 in.	35 ft, 1 in.	32 ft, 11 in.	30 ft, 6 in.	32 ft, 3 in.	30 ft, 4 in.	28 ft, 1 in.
	PRI-90	42 ft, 9 in.	40 ft, 1 in.	37 ft, 2 in.	40 ft, 4 in.	37 ft, 10 in.	35 ft, 1 in.	37 ft, 5 in.	35 ft, 2 in.	32 ft, 7 in.

For SI: 1 in. = 25.4 mm, 1 ft = 304.8 mm.

1. Allowable clear span applicable to simple-span roof construction with a design dead load of 20 psf (1.2 kN/m^2) and a construction load of 20 psf (1.2 kN/m^2). The construction load deflection is limited to span/240 and total load deflection to span/180.

2. Spans are based on a duration of load (DOL) factor of 1.25.

3. Minimum bearing length must be 1¾ in. (44.5 mm) for the end bearings and 3½ in. (89 mm) on end bearing adjacent to cantilever.

4. Bearing stiffeners are not required when I-joists are used with the spans and spacings given in this table, except as required by hanger manufacturers.

5. Spans include a cantilever of up to 2 ft on one end of the I-joist.

TABLE 5.31b Allowable Roof Spans for PRI I-Joists: Multiple-Span Only (flat roof (¼–12); construction load = 20 psf, dead load = 20 psf)

Depth	Designation	16 in. o.c.	19.2 in. o.c.	24 in. o.c.
9½″	PRI-20	20 ft, 5 in.	18 ft, 7 in.	16 ft, 7 in.
	PRI-30	22 ft, 2 in.	20 ft, 10 in.	18 ft, 10 in.
	PRI-40	21 ft, 6 in.	19 ft, 7 in.	17 ft, 6 in.
	PRI-50	23 ft, 3 in.	21 ft, 10 in.	20 ft, 2 in.
	PRI-60	25 ft, 0 in.	23 ft, 1 in.	20 ft, 7 in.
11⅞″	PRI-20	23 ft, 7 in.	21 ft, 6 in.	19 ft, 3 in.
	PRI-30	26 ft, 8 in.	24 ft, 4 in.	21 ft, 9 in.
	PRI-40	24 ft, 11 in.	22 ft, 9 in.	20 ft, 3 in.
	PRI-50	28 ft, 0 in.	26 ft, 3 in.	23 ft, 8 in.
	PRI-60	29 ft, 3 in.	26 ft, 8 in.	23 ft, 10 in.
	PRI-70	30 ft, 7 in.	28 ft, 8 in.	26 ft, 6 in.
	PRI-80	33 ft, 4 in.	31 ft, 4 in.	28 ft, 5 in.
14″	PRI-40	27 ft, 7 in.	25 ft, 2 in.	22 ft, 6 in.
	PRI-50	32 ft, 0 in.	29 ft, 3 in.	25 ft, 3 in.
	PRI-60	32 ft, 5 in.	29 ft, 7 in.	26 ft, 5 in.
	PRI-70	34 ft, 9 in.	32 ft, 7 in.	29 ft, 0 in.
	PRI-80	37 ft, 11 in.	35 ft, 3 in.	31 ft, 6 in.
16″	PRI-40	29 ft, 11 in.	27 ft, 4 in.	24 ft, 5 in.
	PRI-50	34 ft, 9 in.	31 ft, 8 in.	25 ft, 3 in.
	PRI-60	35 ft, 2 in.	32 ft, 1 in.	28 ft, 8 in.
	PRI-70	38 ft, 8 in.	36 ft, 3 in.	29 ft, 0 in.
	PRI-80	41 ft, 11 in.	38 ft, 3 in.	34 ft, 2 in.

For SI: 1 in. = 25.4 mm, 1 ft = 304.8 mm.

1. Allowable clear span applicable to multiple-span roof construction with a design dead load of 20 psf (1.2 kN/m²) and a construction load of 20 psf (1.2 kN/m²). The end spans must be 40% or more of the adjacent span. The construction load deflection is limited to span/240 and total load deflection to span/180.

2. Spans are based on a duration of load (DOL) factor of 1.25.

3. Minimum bearing length must be 1¾ in. (44.5 mm) for the end bearings and 3½ in. (89 mm) for the intermediate bearing.

4. Bearing stiffeners are not required when APA PRI joists are used with the spans and spacings given in this table, except as required by hanger manufacturers.

- Construction load—20 psf (0.96 kN/m²) (based on vertical projection of roof framing)

- *Span:* Span for calculation purposes equals clear span + 0.25 ft (75 mm) (spans given in Tables 5.24–5.31 equal the vertical projection measured between inside faces of supports). In generating the multiple-span tables (Tables 5.24b–5.31b), the end span was assumed to be at least 40% of the adjacent span.
- *Cantilever:* For the simple-span pitched roof tables (Tables 5.24a–5.31a) *only*, a 2 ft (610 mm) cantilever is assumed on one end of the span.
- *Roof slopes:* The following roof slopes were analyzed:
  - >¼:12 to ≤4:12
  - >4:12 to ≤8:12
  - >8:12 to ≤12:12

- *Design properties:*
  - Bending capacities are adjusted for repetitive member and duration of load adjustment factor (DOL = 1.25 for construction loads and 1.15 for snow loads).
  - Shear capacities, exterior reactions, and interior reactions are adjusted for DOL (1.25 for construction loads and 1.15 for snow loads).
  - Intermediate reaction capacity is used to check exterior wall reaction adjacent to the 2 ft (610 mm) cantilever.
  - Bending stiffness and coefficient of shear deflection and other design properties are as listed in Table 5.2.

Roof Span Tables. The allowable span information given in Tables 5.24–5.31 is based on the horizontal distance between inside faces of supports. The *a* tables are single-span tables for pitched roofs and include a 2 ft (610 mm) overhang also measured horizontally (see Fig. 5.31). The *b* tables are for near-flat roofs (roofs with a slope of ¼:12 or less) and may be used for multiple spans assuming the shorter span is at least 40% of the longer.

The first set of tables, Tables 5.24–5.28, is tables for snow loads of 20 psf (0.96 kN/m^2), 25 psf (1.2 kN/m^2), 30 psf (1.44 kN/m^2), 40 psf (1.92 kN/m^2), and 50 psf (2.4 kN/m^2), generated using a duration-of-load adjustment factor of 1.15. All of the tables are based on a dead load of 15 psf (0.72 kN/m^2), which was used to include the weight of roofing, roof sheathing, insulation, and inside drywall.

Construction loads, sometimes called "sunshine loads," are covered in Tables 5.29–5.31. Construction loads are associated with a duration-of-load adjustment

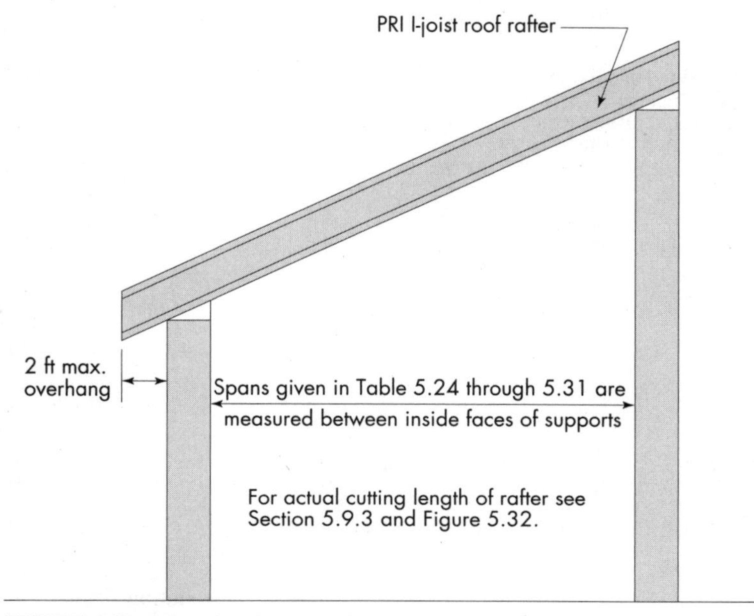

FIGURE 5.31 Measuring the span of a sloped roof.

factor of 1.25. These tables all assume a 20 psf (0.96 kN/m^2) construction load as is normally required by the model building codes, but have dead loads of 10 psf (0.48 kN/m^2), 15 psf (0.72 kN/m^2), and 20 psf (0.96 kN/m^2). These loads represent structures like carports, uninsulated roofs, and insulated roofs, respectively.

See Section 5.11.2 for an example calculation of these roof spans.

Example To illustrate the selection of an APA PRI product for a roof application, assume a roof design snow load of 40 psf and 15 psf dead load with a roof slope of 5:12 and horizontal simple span of 17 ft, 1 in. For architectural reasons, the joist depth is limited to 9½ in. and spacing to 16 in. on center. From the 9½ in. depths of Table 5.27a, looking down the 16 in. on center spacing column for a slope of 5:12, it can be seen that either a PRI-40, PRI-50, or PRI-60 may be used for this application.

5.9.3 Calculating True Cutting Length of Roof Rafters from Horizontal Dimensions

Discussion. A builder/designer is often faced with the problem of finding the true cut length of a roof joist for a given slope and horizontal dimension as well as determining the angle of the plumb cuts at the ends of the joist. While it is sometimes possible to do a full-size layout of the sloped element at the job site, it is considerably quicker and easier to calculate the true cut length, as illustrated in Fig. 5.32. One note of caution: the horizontal dimension used for this calculation is the horizontal dimension measured from beginning of joist to end of joist, including any overhangs, if present. Figure 5.32 and Table 5.32 are provided for making this calculation. *The horizontal dimension is NOT the clear span, or span measured from center of bearing to center of bearing, as sometimes used in this handbook.*

Example Assume that it is necessary to calculate the true length of a rafter for a given situation. In this case, the horizontal length of the rafter including the overhang (L_h) is 16.33 ft (measured off the plan view of the roof) and the rafter pitch is 7:12. A 9½ in. deep I-joist rafter is desired.

From Fig. 5.32 and Table 5.32, the true cutting length of the rafter is:

Horizontal span $= L_h = 16.33$ ft

$L_h \times \text{SF} + L_p =$ true cutting length

16.33 ft × 1.158 + 0.462 ft = 19.372 = 19 ft, 4½ in.

Therefore, cut joist length to <u>19 ft, 4½ in.</u>

(Note that the L_p can also be used to layout plumb cut at ends of I-joist.)

5.10 I-JOIST INSTALLATION DETAILS—ROOF

5.10.1 Installing Performance Rated I-Joists—Roofs

In addition to their use as floor joists, Performance Rated I-Joists are also well suited for roof applications. As with floor joists, specific installation details must be followed to achieve optimum performance.

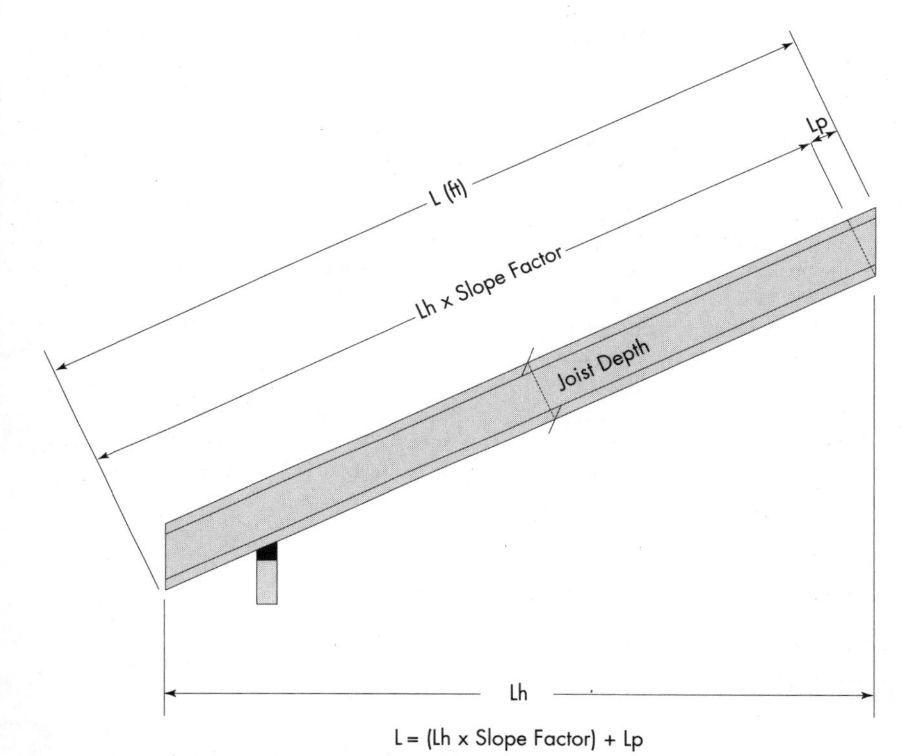

FIGURE 5.32 Calculating cut length of rafter (L) from horizontal dimension measured between outside faces of joists.

TABLE 5.32 Slope Factors and Length Increase for Plumb Cut

Roof slope	Slope factor (SF)	Length increase for plumb cut in (ft) (L_p)			
		Joist depth			
		9½″	11⅞″	14″	16″
2½:12	1.021	0.165	0.206	0.243	0.278
3:12	1.031	0.198	0.247	0.292	0.333
3½:12	1.042	0.231	0.289	0.340	0.389
4:12	1.054	0.264	0.330	0.389	0.444
4½:12	1.068	0.297	0.371	0.438	0.500
5:12	1.083	0.330	0.412	0.486	0.556
6:12	1.118	0.396	0.495	0.583	0.667
7:12	1.158	0.462	0.577	0.681	0.778
8:12	1.202	0.528	0.660	0.778	0.889
9:12	1.250	0.594	0.742	0.875	1.000
10:12	1.302	0.660	0.825	0.972	1.111
11:12	1.357	0.726	0.907	1.069	1.222
12:12	1.414	0.792	0.990	1.167	1.333

For SI: 1 in. = 25.4 mm, 1 ft = 304.8 mm.

Typical I-Joist Roof Framing Construction Details and Installation Notes

1. Installation of APA PRIs must be as shown in Figs. 5.33–5.35.
2. Except for cutting to length, or for providing birdsmouth bearings as detailed in Fig. 5.34, I-joist top or bottom flanges should NEVER be cut, drilled, or notched.
3. Performance rated I-joists are permitted to be birdsmouth cut at the lower end of the joist only. The birdsmouth cut must have full bearing and not overhang the inside face of the plate. Bearing/web stiffeners are required at the birdsmouth cut on both sides of the web.
4. When beveled bearing plates are used at I-joist supports, I-joist attachment to the bevel plate must be designed to transfer lateral thrust.
5. Concentrated loads should only be applied to the top surface of the top flange. At no time should concentrated loads be suspended from the bottom flange, with the exception of light loads (lighting fixtures, ceiling fans, etc.).
6. I-joists must be protected from the weather prior to installation.

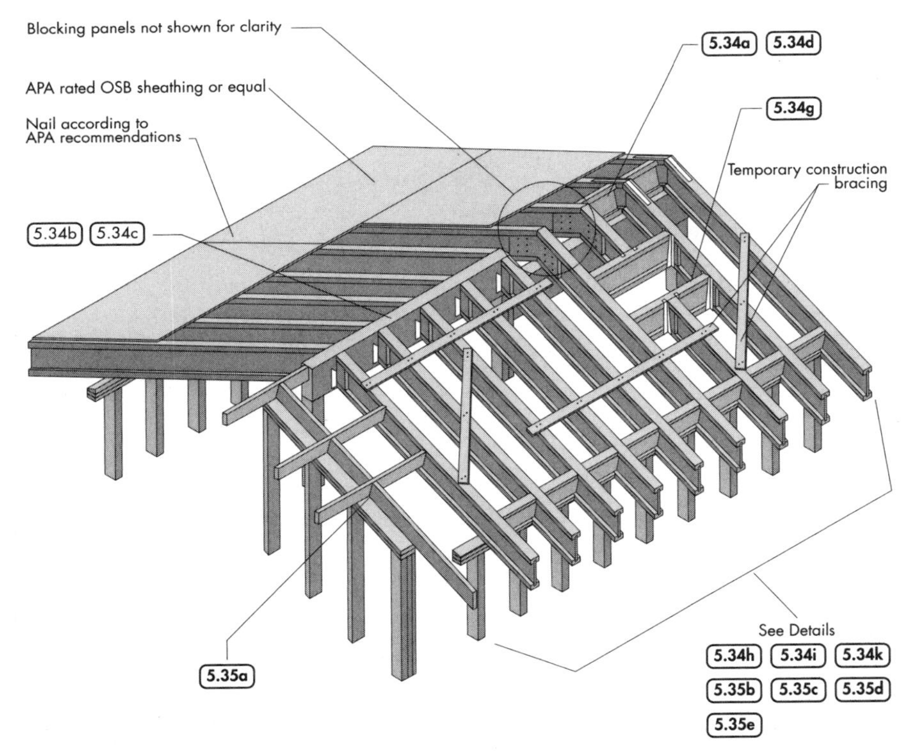

FIGURE 5.33 Typical Performance Rated I-joist roof framing.

All nails shown in the details 5.34a through 5.34l are assumed to be common nails unless otherwise noted. 10d box nails may be substituted for 8d common shown in details. Individual components not shown to scale for clarity. 12:12 Maximum roof slope.

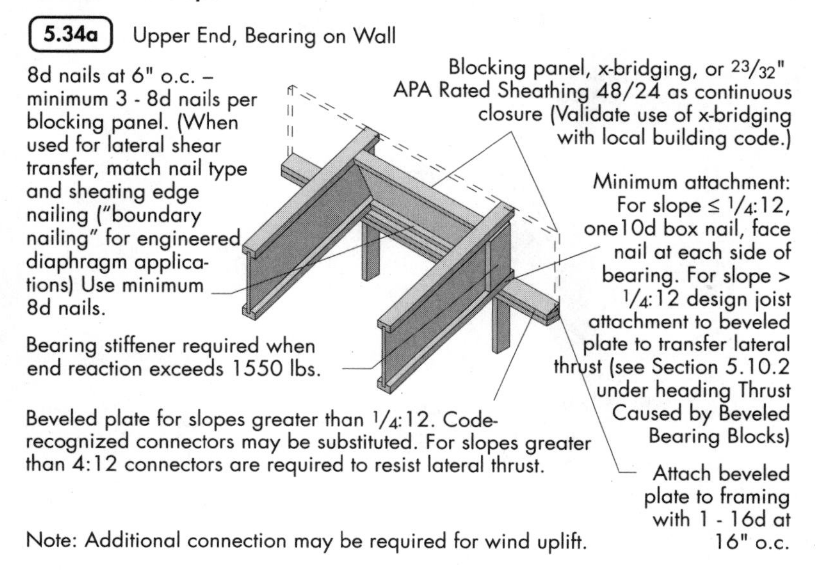

5.34b Peak Connection

For roof slopes > 1/4:12 and 12:12, provide a strap nailed in accordance with Section 5.10.2 under heading Thrust Caused by Beveled Bearing Blocks, (min. 3" nail spacing) wrapped around ridge beam

Ridge beam (Glulam or LVL)

Beveled bearing stiffener required each side

Adjustable Slope Hanger with a minimum unadjusted uplift capacity of 300 lb.

Note: Additional connection may be required for wind uplift.

FIGURE 5.34 (a,b) Typical Performance Rated I-joist roof framing and construction details.

5.34c I-Joist to Ridge Beam Connection

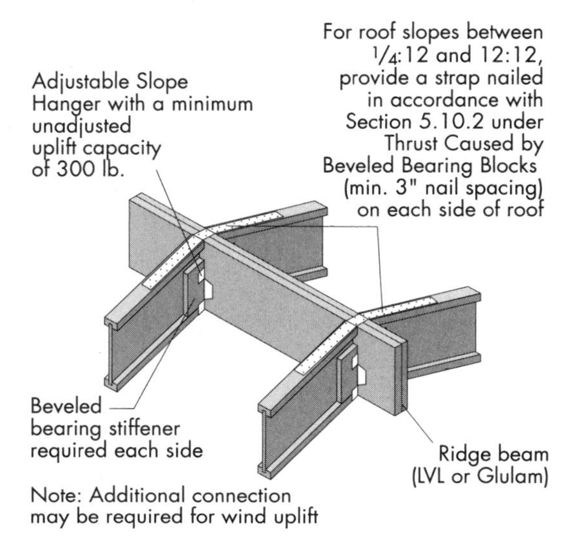

5.34d I-Joist Connection with Wood Structural Panel Gussets

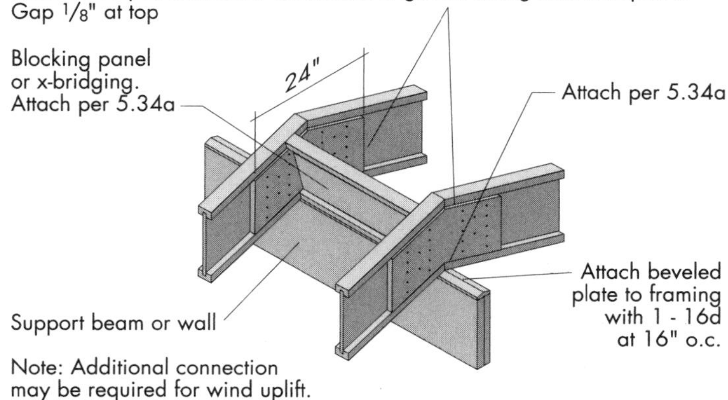

FIGURE 5.34 (c,d) Typical Performance Rated I-joist roof framing and construction details. (*Continued*)

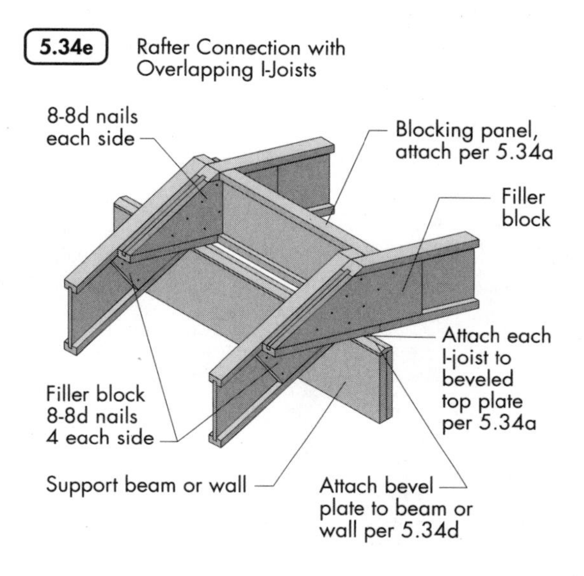

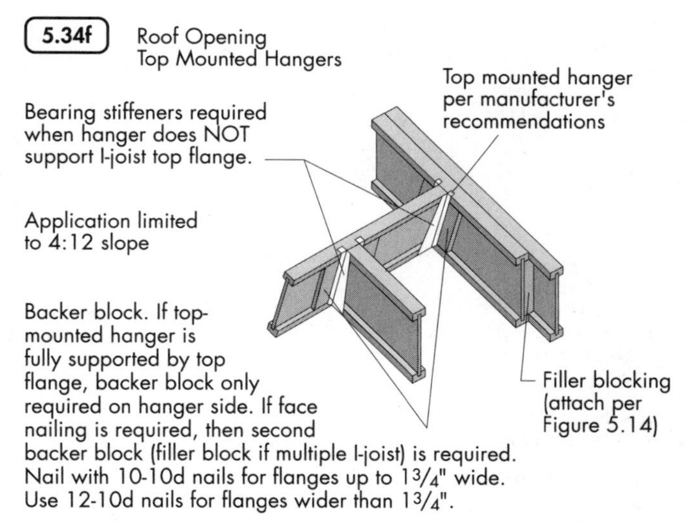

FIGURE 5.34 (e,f) Typical Performance Rated I-joist roof framing and construction details. (*Continued*)

5.34g Roof Opening, Face-Mounted Hangers

Backer block on both sides of web (or backer block and filler block, if multiple I-joists), nail with 12-10d nails clinch when possible.

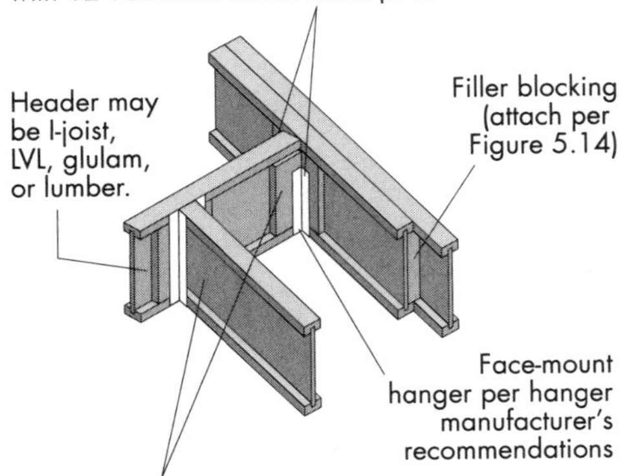

Bearing stiffeners required when hanger does not support I-joist top flange

5.34h Birdsmouth Cut & Bevel Cut Bearing Stiffener

Permitted on low end of I-joist only

1/8" gap at top

Bearing stiffeners required each side of I-joist. Bevel cut bearing stiffener to match roof slope (see Section 5.10.2 under Web Stiffeners in I-Joist Roof Rafters).

One 10d box nail, face nail at each side of bearing (face nail where flange is 7/8" to 1" thick)

4-8d nails (two each side) clinched when possible.

Birdsmouth cut shall bear fully and not overhang the inside face of plate

Note: Additional connection may be required for wind uplift.

FIGURE 5.34 (g,h) Typical Performance Rated I-joist roof framing and construction details. (*Continued*)

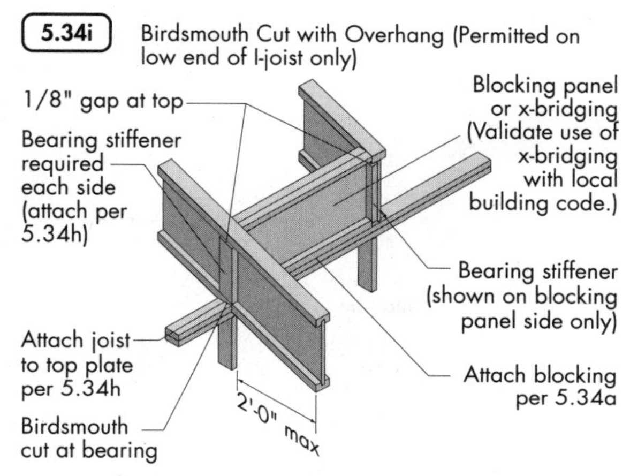

Note: Additional connection may be required for wind uplift.
Note: Outside corner of blocking panel may be trimmed if it interferes with roof sheathing. In such cases, position blocking panel on top plate to minimize trimming and still allow required nailing into top plate.

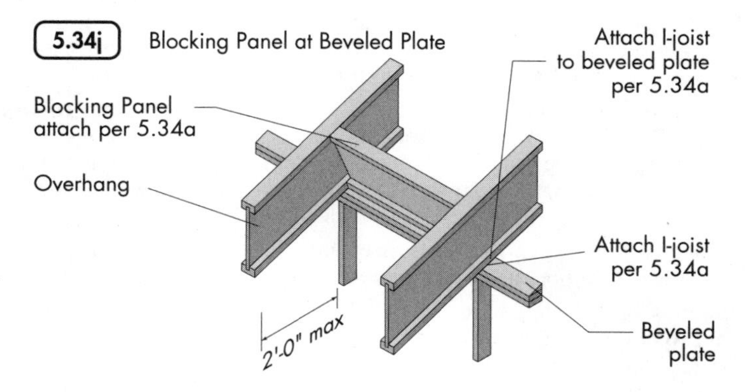

Note: Additional connection may be required for wind uplift.

FIGURE 5.34 (i,j) Typical Performance Rated I-joist roof framing and construction details. (*Continued*)

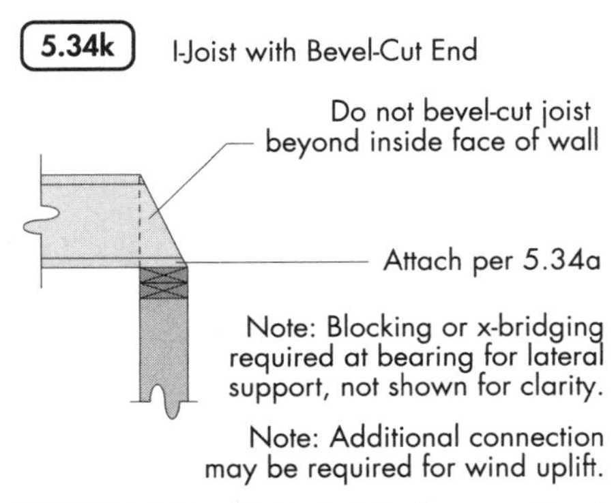

FIGURE 5.34 (k) Typical Performance Rated I-joist roof framing and construction details. (*Continued*)

7. I-joists must not be used in places where they will be permanently exposed to weather (overhangs are exceptionally vulnerable) or in areas where they will reach a moisture content greater than 16%, such as in swimming pool or hot-tub enclosures. They must not be installed where they will remain in direct contact with concrete or masonry.
8. End-bearing length must be at least 1¾ in. (44.5 mm). For continuous framing and roof framing with cantilevers, the intermediate support and end bearing adjacent to the cantilever both must be at least 3½ in. (89 mm). For multiple-span joists, intermediate bearing length must be at least 3½ in.
9. Ends of roof joists must be restrained at the bearing to prevent rollover. Performance Rated Rim Board or I-joist blocking panels are preferred. Cantilever-end blocking must be placed at the support adjacent to the cantilever, and ends of all cantilever extensions must be laterally braced by a fascia board or other similar method.
10. Performance Rated Rim Board, I-joist blocking panels, or other means of providing lateral support must be provided at all I-joist bearing points.
11. Continuous lateral support of the I-joist's compression flange is required to prevent rotation and buckling. In simple span roof applications, lateral support of the top flange is normally supplied by the roof sheathing. Bracing of the I-joist's bottom flange is also required at interior supports of multiple-span joists and at the end support next to an overhang. Lateral support of the entire bottom flange may be required in cases of load reversal such as those caused by high wind.
12. Nails installed perpendicular to the wide face of the flange must be spaced in accordance with the applicable building code requirements or approved building plans but should not be closer than 3 in. (75 mm) o.c. per row for I-joists with lumber flanges using 8d common (3.3 × 64 mm) nails. For I-joists with composite flanges 1½ in. (38 mm) wide, 8d common (3.3 × 64 mm) nails should be spaced no closer than 4 in. (100 mm) o.c.
13. Nails installed parallel to the wide face of the veneers in LVL flanges must not be spaced closer than 3 in. (75 mm) o.c. for 8d common (3.3 × 64 mm) nails,

All nails shown in the details 5.35a through 5.34e are assumed to be common nails unless otherwise noted. 10d box nails may be substituted for 8d common shown in details. Individual components not shown for clarity.

5.35a Outrigger

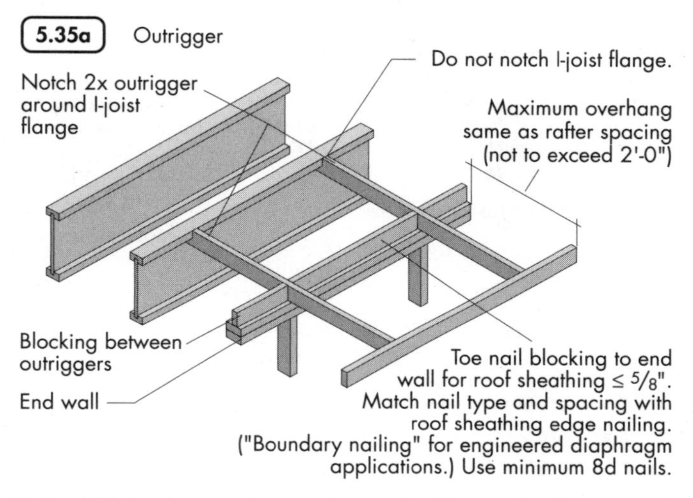

Note: Additional connection may be required for wind uplift.

5.35b I-Joist Overhang with Beveled Plate

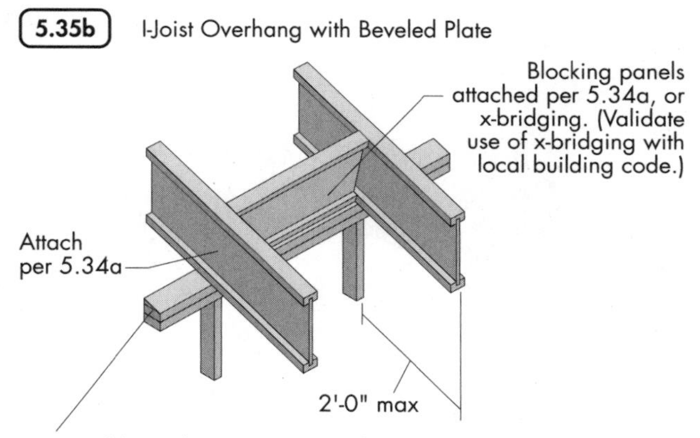

Note: Additional connection may be required for wind uplift.

FIGURE 5.35 (a,b) Typical Performance Rated I-joist roof overhang details.

5.35c Lumber Overhang with Beveled Plate

(Blocking panel or x-bridging not shown for clarity)

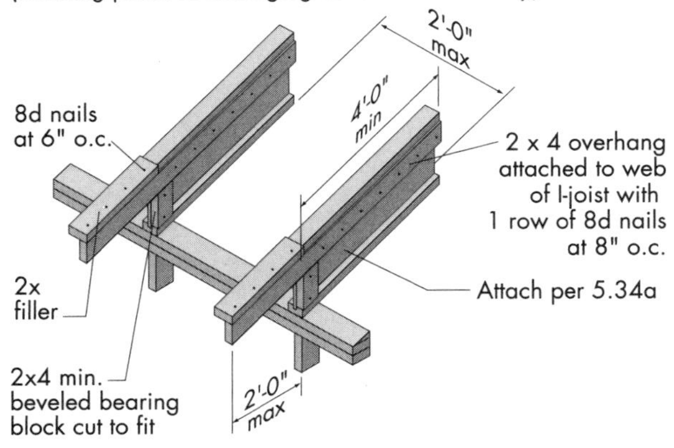

Note: Additional connection may be required for wind uplift.

Note: Lumber overhang shall be 2 x 4 Spruce-Pine-Fir #2 or better, or stronger species.

5.35d I-Joist Overhang for Fascia Support with Birdsmouth Cut

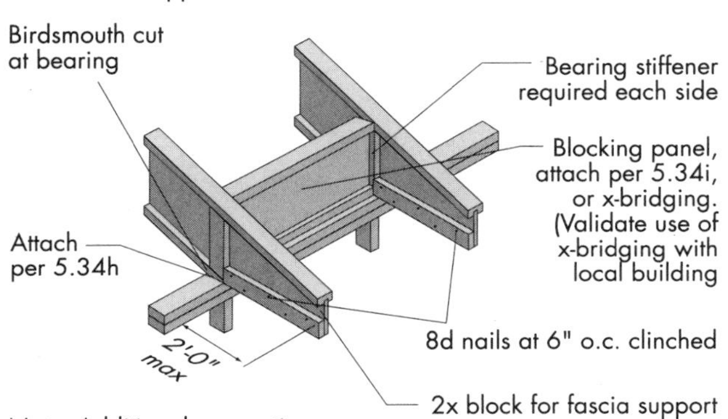

Note: Additional connection may be required for wind uplift.

FIGURE 5.35 (c,d) Typical Performance Rated I-joist roof overhang details. (*Continued*)

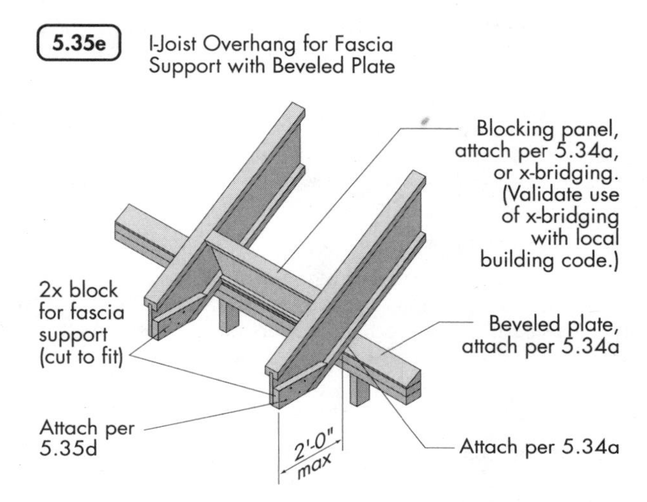

FIGURE 5.35 (e) Typical Performance Rated I-joist roof overhang details. (*Continued*)

and 4 in. (100 mm) o.c. for 10d common (3.8 × 76 mm) nails. If more than one row of nails is used (not permitted for I-joists with composite flanges 1½ in. [38 mm] wide), the rows must be offset at least ½ in. (12 mm).

14. Details in Figs. 5.34 and 5.35 show only I-joist-specific fastener requirements. For other fastener requirements, such as wind uplift requirements or other member attachment details, see the applicable building code. If I-joists are oriented so that the knockouts provided by the manufacturer are adjacent to the top flange, they may be removed to aid ventilation.

15. The top and bottom flanges of the I-joist must be kept within ½ in. (12.5 mm) of true alignment. The use of I-joist blocking panels or engineered wood rim board greatly simplifies this requirement.

16. All roof details are valid up to a 12:12 slope unless otherwise noted.

Typical I-Joist Roof Framing Temporary Bracing Notes

1. All engineered wood rim boards, blocking, connections, and temporary bracing must be installed before workers are allowed on the structure.

2. For temporary bracing, use lines of 1 × 4s (19 × 89 mm) nailed to each I-joist with two 8d (3.3 × 64 mm or 2.9 × 64 mm) nails. The lines should be parallel, about 8 ft (2.4 m) apart, and should have ends overlapped.

3. To prevent rollover of the entire roof system, brace each end and every 25 ft (8 m) of roof with blocking at ends or diagonal bracing. *Please note that in a roof system framed with parallel-chord rafters such as I-joists, the panel roof sheathing alone does not provide bracing for the roof framing! The blocking or bridging at the bearing points must be provided.*

4. The continuous 1 × 4 (19 × 89 mm) bracing must be attached to the braced bays.

5.10.2 Other Framing Considerations Essential to Engineered Wood Roof Construction

In addition to the installation details provided in Fig. 5.34, there are a number of other framing elements that must be properly designed/specified/installed to ensure the expected performance of the components and of the roof system as a whole. The most common of these elements are covered in the next several sections.

Roof Overhangs

Discussion. While often considered along with the roof installation details, the roof overhang details are presented separately in Fig. 5.35. This approach was selected because any number of overhang details may be used with any of the roof framing details presented in Fig. 5.34.

Ridge Beams

Discussion. A ridge beam, as shown in Fig. 5.36, is a horizontal load-bearing member of a roof frame that supports the upper ends of the roof rafters. A ridge beam differs from a ridge board (see below) in that a ridge beam is a *structural* bending member and, as such, must be supported by vertical load-carrying members, which must provide a continuous load path from the beam down to the foundation. The vertical load-carrying members are usually a set of multiple lumber studs embedded within the end, and sometimes interior walls. Glulam columns may also be used as the vertical load-carrying element. Because, the load path from the ridge beam must be continuous into the foundation, headers are required over doors and windows that interrupt that path. Care also must be taken at intermediate floors to ensure that the load is transferred from the column, through the floor system, and into the wall or foundation below. This is normally accomplished by the use of squash blocks between the floor sheathing and the framing/foundation below. See Detail 5.29e.

The rafters can bear on the top of the beam via a tapered bearing plate or a bevel cut on the top of the beam or can be supported at the sides of the beam by prefabricated joist hangers.

Caution: When I-joist rafters are being used, never use a birdsmouth cut at the ridge beam end of the rafter!

When the I-joists are supported along the sides of the beam, the ridge beam must be deep enough to support fully both flanges of the joist, as shown in Fig. 5.36. Both solid-sawn and engineered wood products such as laminated veneer lumber or glulam beams can be used as ridge beams.

Ridge Boards

Discussion. A ridge board, as shown in Fig. 5.37, is a horizontal member of a roof frame that is placed on edge at the ridge and into which the upper ends of the rafters are fastened. Unlike a ridge beam, discussed above, a ridge board is not a vertical load-carrying element and does not have to be supported at each end.

Because of the geometry of wood I-joists and their relatively deep depths, especially when cut at an angle, it can be difficult to size a ridge board properly for roof construction applications. The ridge board must be deep enough to support fully both flanges of the I-joist as shown in Fig. 5.37. Considering a 16 in. (406

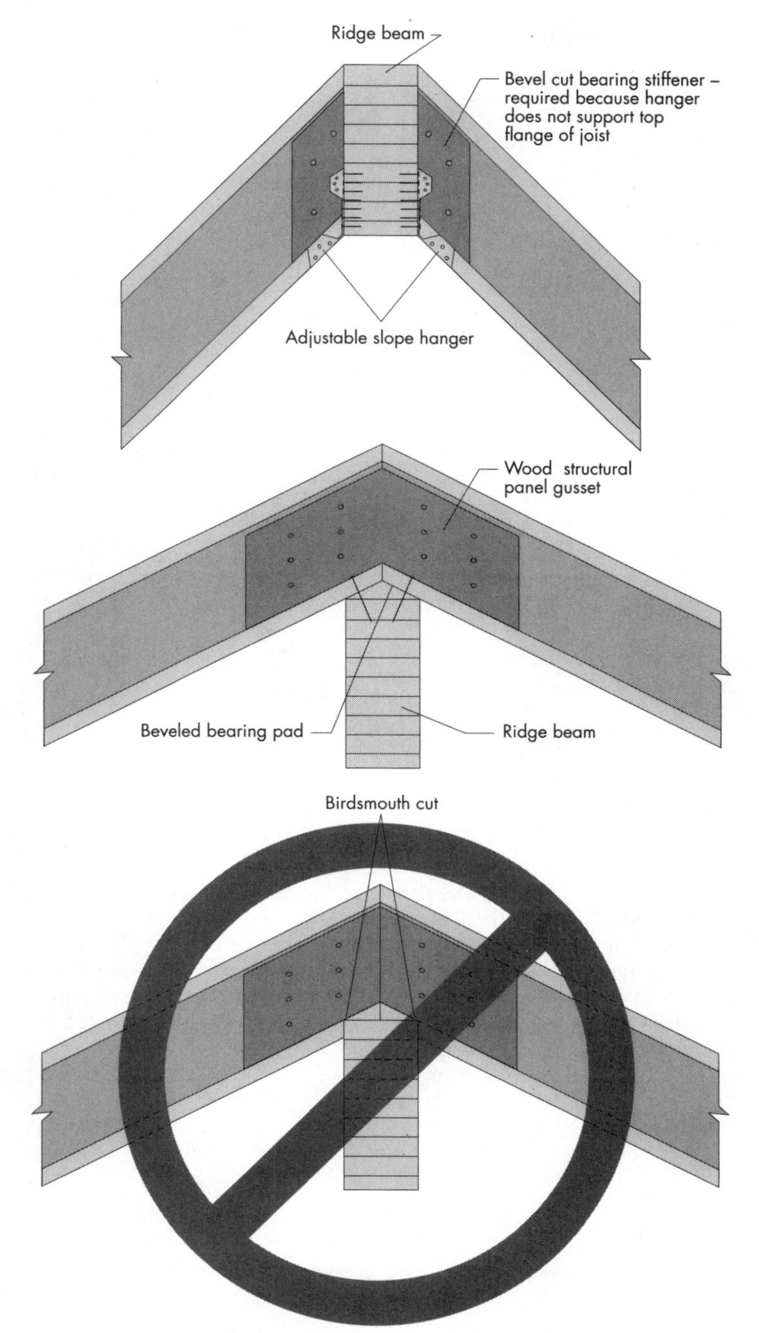

FIGURE 5.36 Typical I-joist ridge beam details.

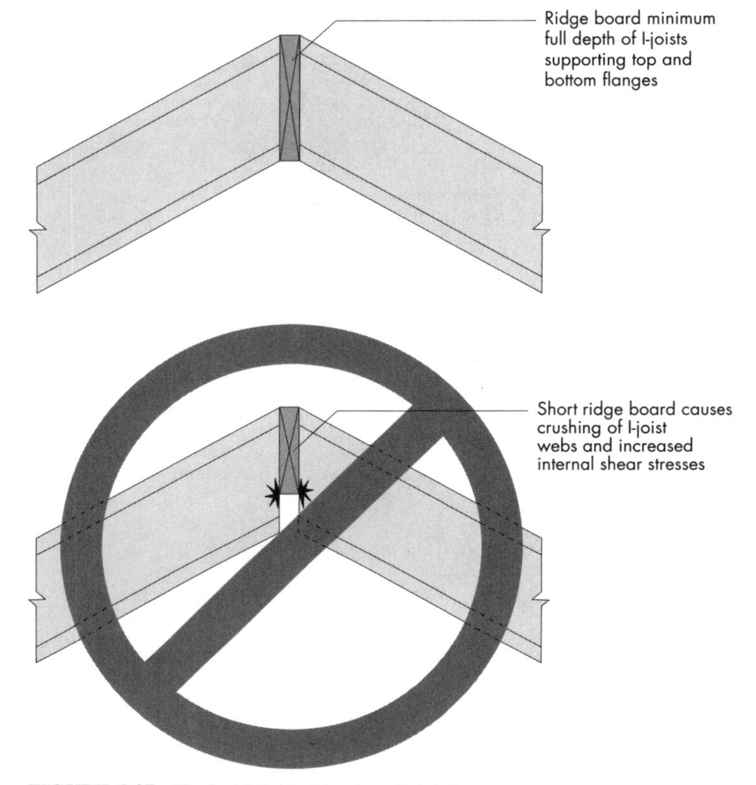

FIGURE 5.37 Typical I-joist ridge board details.

mm) deep I-joist and a 12:12 slope, the ridge board would have to be at least 22⅝ in. (464 mm) deep. If both of the flanges are not supported, additional shear stresses not accounted for in the design are generated in the web of the I-joist.

A possible solution to this problem is to use rim board products for the ridge board. These can be obtained in depths up to 24 in. (610 mm) and lengths up to 24 ft (7.3 m). While somewhat more expensive, LVL can also be used because it comes in similar depths and lengths up to 65 ft (19.8 m).

Design. A ridge board is a part of a relatively complicated load-carrying system that includes the ridge board, both rafters, and the collar tie. Each of these elements must be sized and interconnected properly to support the design loads. As such, these are indeterminate structures and the designs should always be conducted by a registered design professional.

Lateral Bracing and Diaphragm Shear Transfer Elements

Discussion. An essential part of the design, detailing, and fabrication of any wood-framed building is the requirement to adequately tie the various components of the structure's lateral-force-resisting elements together. (See Chapter 7 for a full discussion on this subject.) The one element of this load path that is most often found deficient in the field is the transfer of the roof diaphragm shear into the resisting shear walls. It is an unfortunate circumstance that this deficiency is normally noted only after a high wind or seismic event and the consequences of this deficiency are almost always very serious, typically resulting in structural collapse.

A common error made by designers that often leads to such deficiencies is to think of the roof diaphragm as existing only in two dimensions, as in a single plane—a mono-slope roof as one plane, a gable roof as two planes intersecting at the ridge, etc. While this analogy is convenient for discussion and sketching on paper, for design it is important to remember that the roof diaphragm assembly has a depth associated with it. The problem lies in the fact that the diaphragm shear is carried in the roof sheathing at the top of the diaphragm assembly, while the shear transfer from the diaphragm to the shear walls is carried out at the bottom of the framing.

In designing a roof with a lightweight wood truss bearing on the top chord at the shear wall location, the distance between the shear plane and the attachment point to the shear walls is usually only 4–6 in. (100–150 mm), the vertical dimension of the top chord of the truss. In small residential-sized structures, this short distance, along with the propensity of the truss to resist the overturning of the top chord due to its geometry, account for the apparently satisfactory performance of such connections even when improperly detailed and installed.

In using conventional lumber or I-joist rafter framing, however, the depth of the rafters at the supporting wall is significantly larger. This depth can be as much as 22.63 in. (572 mm) for a 16 in. (406 mm) deep I-joist rafter at a 12:12 slope. In addition, the rafter, being a linear element, does not have any inherent resistance to overturning, as does the top chord of a truss. The model building codes, recognizing this phenomenon, recommend bridging or blocking at the ends of the rafters to resist this overturning. Common practice is to use a piece of I-joist for blocking in these locations. This blocking is normally shown placed perpendicular to the top flange of the rafter, as seen in Figs. 5.34 and 5.35. While this may be beneficial for providing overturning resistance, only in the shallowest of roof slopes can this detail be used to provide the requisite shear transfer from the roof diaphragm to the top plate of the shear wall below (see Fig. 5.38).

In a similar manner, the use of sheet metal bridging, as often specified, limits the diaphragm shear capacity transfer to the lateral capacity of the two nails attaching the I-joist to the top plate. If toe nails are used to attach the I-joist rafters to the supporting walls, one on each side of the rafter, then only the capacity of one toe nail may be used because the other is being loaded in withdrawal. The performance of such a system can be greatly enhanced by the use of framing anchors to increase the lateral capacity of the I-joist-to-wall connection.

How can these connections be designed and detailed? There are a number of solutions that can be used to make this important connection.

I-Joist Blocking. For low-slope roofs, especially when a tapered bearing block is used, an I-joist blocking panel as illustrated in Fig. 5.38 may be used for this purpose. The flange at the top will provide adequate nailing surface for the roof diaphragm perimeter nailing. The flange at the bottom allows the use of face nailing to transfer the roof diaphragm shears into the resisting walls below. As the blocking panels are usually attached perpendicular to the roof slope, they make contact with the wall below at an angle to the vertical. As such, the blocking panels are most easily installed before the roof is sheathed because it will greatly improve the builders' ability to fasten the blocks to the top plate of the wall below.

If the rafters are birdsmouthed over the supporting wall, the distance between the support wall and the roof sheathing is reduced accordingly. For this reason, the

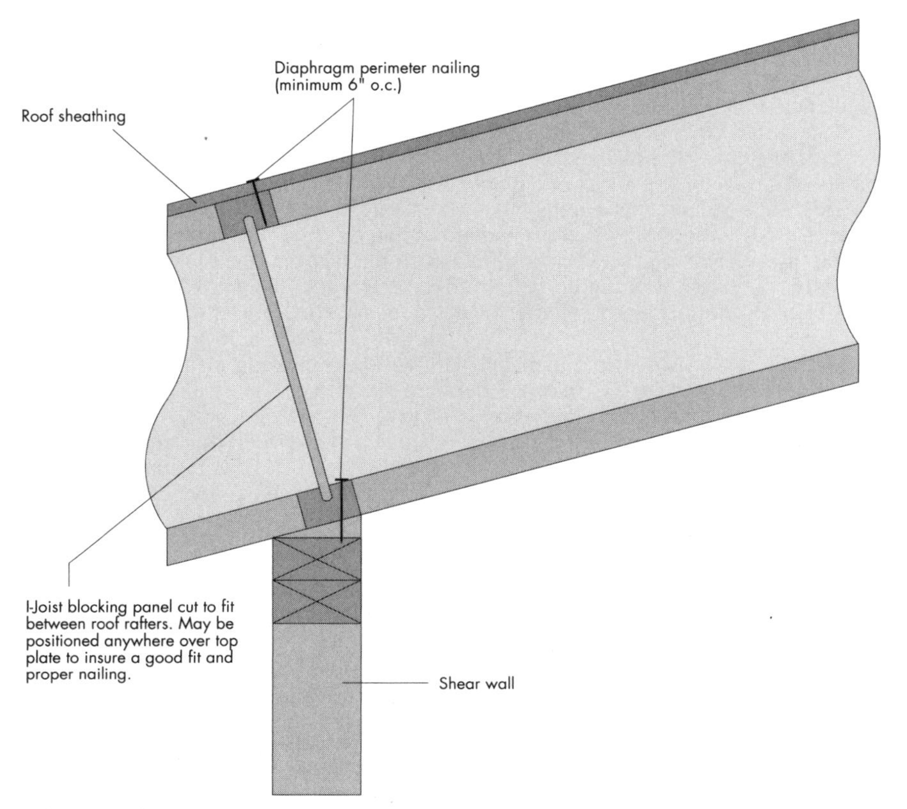

FIGURE 5.38 I-joist blocking panel in a low-slope roof.

roof slope range of a common I-joist blocking panel is extended considerably when a sloped bearing pad is used instead of a birdsmouth cut in the bottom flange of the joist.

Rim Board Blocking. For roofs of any slope, engineered wood rim board may be used for blocking. Because engineered wood rim board is made in depths of up to 24 in. (610 mm), it can be ripped to whatever depth is required to run from the underside of the sheathing to the top plate of the wall below. Since it must be ripped to fit in most applications, the upper edge of the rim board can be cut to match the slope of the roof. Figure 5.39 shows a rim board being used for such an application

Notice that the attachment of the blocking panel to the top plate is shown to be made with toe nails. Because toe nails have a capacity only ⅚ that of a face nail, it would appear necessary to adjust the nail spacing accordingly at this location. For example, if the required diaphragm perimeter nailing is 8d (3.3 × 64 mm) common nails at 6 in. (150 mm) on center, the nail spacing at the base of the blocking panel will have to be 8d (3.3 × 64 mm) common nails at 5 in. (125 mm) on center to compensate for the reduced capacity of the toe nails. This, however, is not usually the case. As the shear is transferred from the diaphragm into the shear wall at the bottom of the framing, all of the fasteners from the roof framing

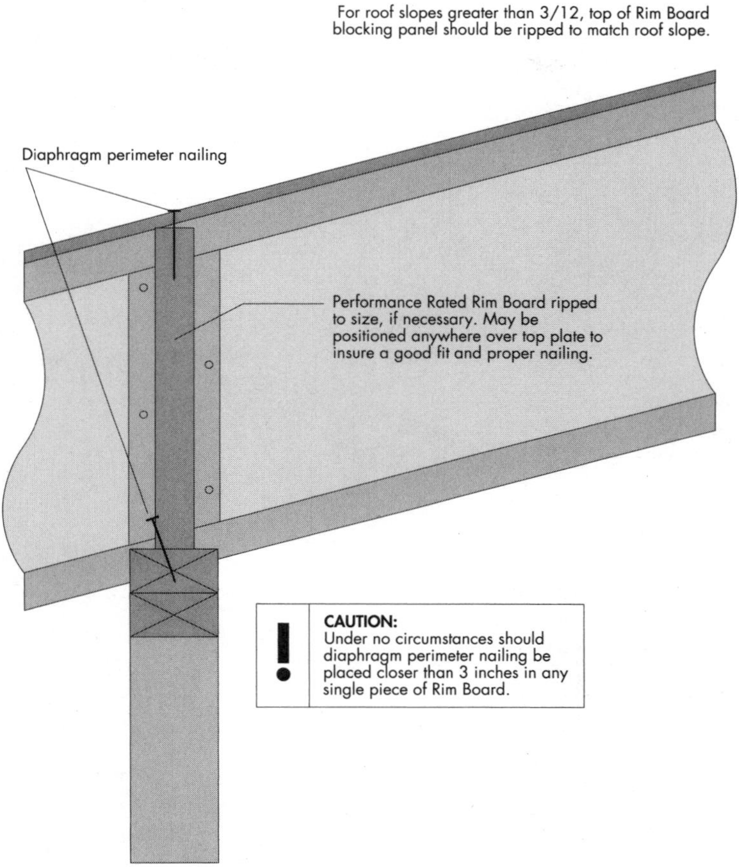

FIGURE 5.39 Performance Rated Rim board used as a blocking panel.

that penetrate the top plate may be used to assist in that transfer of shear. In addition to the fasteners at the bottom of the blocking panels, the fasteners that connect the I-joists to the top plate may also be considered when designing for this shear transfer.

There is a design limitation associated with the use of engineered wood rim board products for such applications. This limitation has to do with the spacing of the roof sheathing nails in the edge of the rim board. Engineered wood rim boards are made out of a number of different products including laminated veneer lumber, plywood, OSB, and glulam beams, and out of a number of different thicknesses. Consequently, the minimum nail spacing for various nail sizes may vary depending on the rim board product selected. *Under no circumstances should the nail spacing be less than 3 in. (75 mm) on center.*

Fabricated Blocking Panels. Another method for transferring the shear from the diaphragm perimeter into the shear walls below is through the use of prefabricated blocking panels as shown in Fig. 5.40. These are most cost-effective when

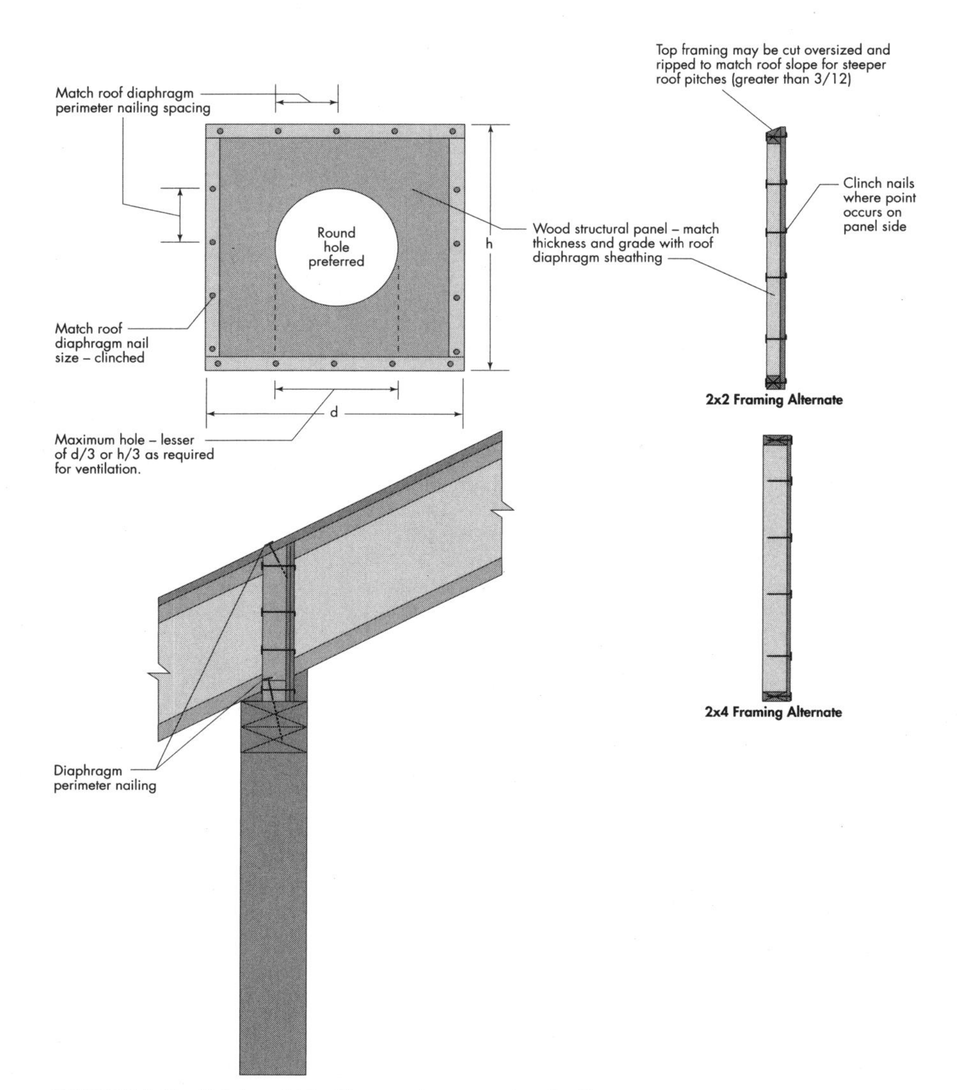

FIGURE 5.40 Fabricated blocking panel construction detail.

the distance between the top plate and the roof diaphragm is large and the rafters or trusses are spaced at 19.2–24 in. (488–610 mm) on center. As can be seen in Fig. 5.40, these blocking panels are little more than small shear walls that are placed in the roof framing system to transfer the shear from the diaphragm to the shear wall below. Because the roof diaphragm is already designed for the shear load acting on it, by borrowing its thickness and diaphragm perimeter-nailing requirement for the fabricated blocking panel, the designer effectively guarantees the capacity of the blocking panel. One advantage of this method is that the fabricated blocking panel does not have the capacity limitations that the rim board blocking panel does because of its edge nailing limitations. The blocking panel can be fabricated with sufficient nailing surface and panel thickness to match any roof diaphragm shear capacity.

Because of the exceptional shear-through-the-thickness values associated with the wood structural panels used in the blocking panel, it is possible to provide ventilation holes in the blocking panel. This is covered in more detail in Figs. 5.40 and 5.44 and discussed briefly below under Providing Roof Ventilation in I-Joist Rafter Roof Systems.

Recently, truss manufacturers have begun to provide blocking panels made out of 2 × (38 mm ×) framing with a single diagonal member. These are held together with pressed-in truss plates. These can be fabricated to fit, can be designed for shear transfer, and provide excellent ventilation of the roof cavity.

Metal Bridging or Cross Bridging. Metal bridging or cross bridging, as shown in Fig. 5.41, is often seen in such applications, being easy to install and economical and having the added benefit of providing ventilation to the roof deck. This practice should be used with caution by the designer because these systems have very limited ability to transfer shear from the roof diaphragm to the supporting structure

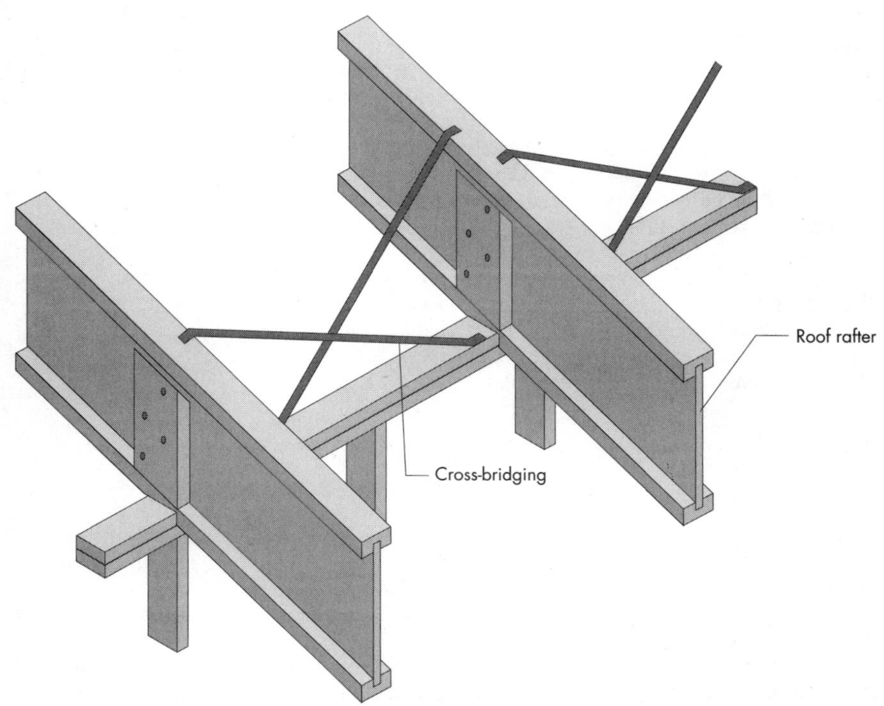

Caution: Cross-bridging has limited capacity to transfer shear forces from roof diaphragm into shear walls below. Consult local code before specifying or detailing.

FIGURE 5.41 Metal bridging or cross-bridging used as a blocking panel.

below. It is advised that the designer ensure that the system selected is allowed by the local building code and/or is covered by a current code evaluation report that include lateral load capacities for such applications.

Such a system *may* be adequate for small structures in parts of the county where seismic or high wind loads are not a consideration. Again, the designer is advised to proceed with caution.

Shear Transfer from Interior Shear Walls—Parallel and Perpendicular

Discussion. Often, in designing wood buildings for high wind or seismic events, it becomes advantageous to utilize interior shear walls. This usually occurs when the diaphragm length-to-width ratio of the structure exceeds three to one. Because the shear transfer load path in such instances is from the roof diaphragm to the interior shear wall, it is imperative that the connection from the diaphragm to the shear wall be properly made. It is also important to realize that the interior shear wall is resisting shear from both sides of the roof diaphragm and therefore may be loaded up to two times as high as the exterior shear walls. This increased capacity must be reflected by the nailing at the diaphragm and at every interelement connection from the roof sheathing to the foundation. It is not uncommon to see two-sided shear walls in these applications.

For parallel applications, as seen in Fig. 5.42, the most efficient way to transfer the shear is to construct the shear wall right up to the bottom of the roof or floor sheathing and transfer the diaphragm directly into the shear wall. The interior shear wall looks very much like a gable end wall or firewall. In multifamily housing the interior shear wall can be easily incorporated into the firewall, and very often is.

In those applications where the shear wall runs perpendicular to the roof framing, transfer of diaphragm shear into the interior shear wall can be a little difficult when continuous roof framing is desired. In this case, shear transfer blocking panels must be utilized to carry the shear from the diaphragm into the shear walls. This is illustrated in Fig. 5.43.

A cleaner alternative is to run the interior shear wall up to the roof diaphragm and support the rafters running perpendicular to the wall on the wall with hangers. The advantages of running the I-joist continuous over the interior support are lost, but the ease of providing shear transfer at this location, or designing a fire wall at this location may well provide sufficient compensation.

See Chapter 7 for further information on shear transfer.

Providing Roof Ventilation in I-joist Rafter Roof Systems

Discussion. Building designers are often faced with an apparent conflict between providing the required shear transfer at the point where the roof diaphragm passes over the resisting shear walls (see the previous two sections) and the requirement of providing roof ventilation at the same location. The only bracing method that lends itself easily to good ventilation is the metal bridging or cross-bridging method. As discussed previously, these methods provide little or no shear transfer and should never be used in locations of seismic activity or high wind.

The solution to providing ventilation is to use one of the other methods described above under Lateral Bracing and Diaphragm Shear Transfer Elements and carefully cut circular holes into the bracing elements to allow for ventilation. Figure 5.40 shows a fabricated blocking panel with an allowable circular hole cut in it with a diameter of up to ⅓ of the smallest dimension of the blocking panel. This is based on the shear-through-the-thickness capacity of the blocking panel and presupposes

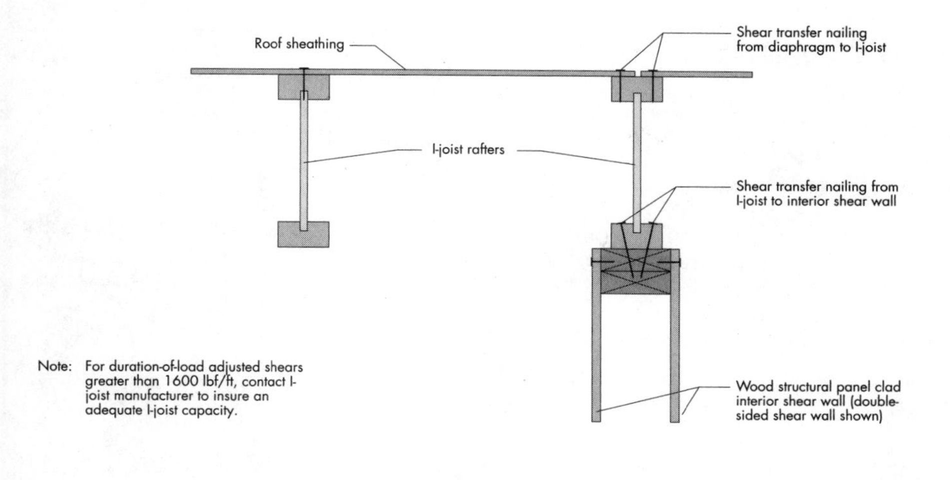

or

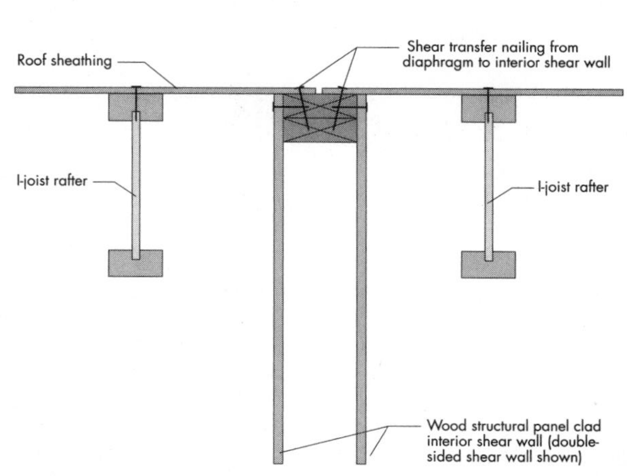

FIGURE 5.42 Shear transfer from roof diaphragm into interior shear wall—shear wall parallel to rafter.

that the blocking-panel sheathing is the same thickness and grade as is the roof diaphragm sheathing. As mentioned previously, some truss manufacturers have begun to sell blocking panels made out of 2 × (38 mm ×) framing with one diagonal member that can be designed for shear transfer and provide for excellent ventilation of the roof cavity.

While it is theoretically possible to place holes in I-joist blocking panels, the I-joist manufacturer should be contacted for specific details.

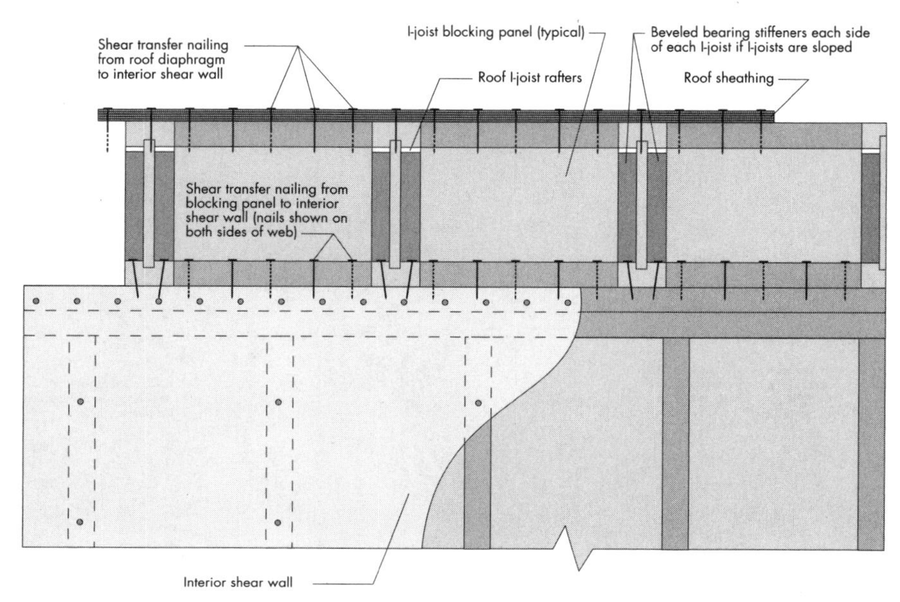

Note: For duration-of-load adjusted shears greater than 1600 lb/ft, contact I-joist manufacturer to insure adequate capacity. I-joist blocking panels should be adjusted for their true length when computing capacity available for shear transfer as follows:

$$L' = L - \left(\left(\frac{L \times 12}{S} + 1\right) \times \frac{w}{12}\right)$$

where:
L' = adjusted wall length (ft)
L = actual wall length (ft)
S = I-joist rafter/joist spacing (in.)
w = I-joist flange width (in.)

FIGURE 5.43 Shear transfer from roof diaphragm into interior shear wall—shear wall perpendicular to rafter.

By far the most popular type of blocking panels used for roof applications is the rim board blocking panel. Figure 5.44 in provides information on how to modify such blocking panels to allow roof ventilation. The method used depends on the required capacity of the roof diaphragm above. If the V-notch method is used, the number of fasteners that may be placed into the top of the blocking panel will be reduced and consequently its ability to transfer shear will be correspondingly reduced. For such applications the hole method may prove best. While round, rectangular, or square holes may be used in the allowable zone, be careful to avoid overcutting the corners of square or rectangular holes. A round hole cut with a hole saw is always the preferred method.

Proper Positioning of Knock-outs to Provide Ventilation of Roof Deck

Discussion. Performance Rated I-joists are typically supplied with prescored holes in the web of the joist for the convenience of the contractor. They are from 1⅜–1½ in. (35–38 mm) in diameter and are spaced 12–24 inches (300–600 mm) on center along the length of the I-joist, normally located close to one flange. In floor applications, the I-joists are typically oriented so the knock-outs are close to the bottom flange, and these are often used for small plumbing lines and electrical

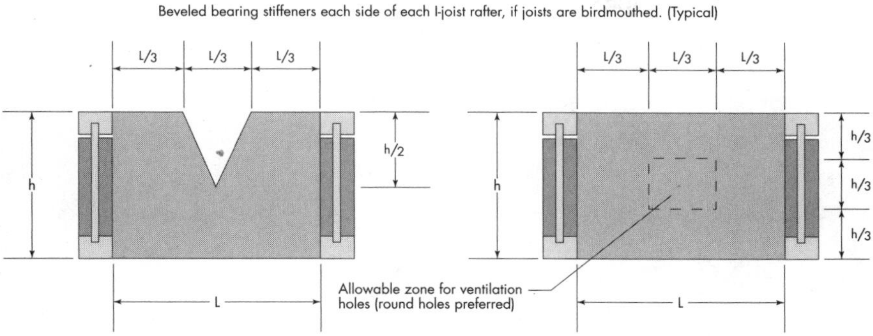

FIGURE 5.44 Performance Rated Rim Board shear blocking modified to provide ventilation—roof only.

wiring. In roof applications, however, it is desirable to orient the I-joists such that the knock-outs are adjacent to the top flange. Once removed, the openings resulting may be used to provide ventilation under the roof deck when the normal ventilation path is eliminated by framing around roof penetrations such as skylights, ducts, or access doors (see Fig. 5.45).

I-joist roof rafters can be sized to have sufficient depth to provide room for insulation as well as an air space above the insulation for ventilation.

Compression-Side Bracing for Continuous I-Joist Roof Framing

Discussion. For simply supported I-joists under gravity loads, the top flange of the beam is in compression. Fortunately, wood structural panel sheathing is usually attached to this flange to form the floor or roof deck. The sheathing provides full lateral support of the compression-side flange. In high-wind conditions, however, the resulting uplift can often overcome the dead and live loads on the roof and shift the compression zone to the often unsheathed/unsupported bottom flange.

In multiple-span continuous or cantilevered applications, portions of the bottom flange of the I-joist are put in compression under gravity loads. Gypsum sheathing provides adequate lateral support for the bottom flange in these cases when applied. In those circumstances where gypsum sheathing is not used, however, those portions of the bottom flange experiencing compression must be provided with lateral support in accordance with the requirements of the local building code.

For the various PRI designations, as a rule of thumb, the following will provide the required lateral support to the unsupported bottom flange of I-joists in the compression zone:

Provide lateral support at 32 in. (800 mm) on center throughout the unsupported compression zone. A lateral capacity of 150 lb (0.66 N) per connection is required at these locations.

For a more exact solution, Table 5.33 and Fig. 5.46 are provided.

Reinforcing Roof for Air-Handling Equipment—on Rooftop or Suspended from the Underside

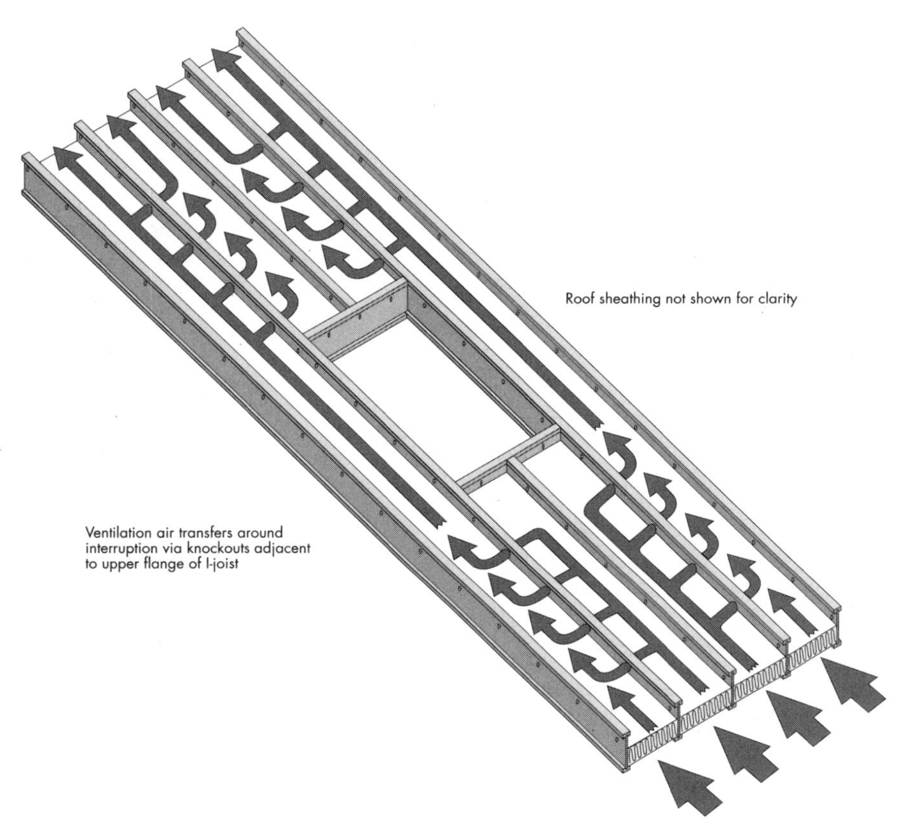

FIGURE 5.45 Positioning of knock-outs to provide ventilation of roof deck around penetrations.

TABLE 5.33 Lateral Connection Requirements for PRI Joists with Unsupported Compression Flanges

PRI designation	Lateral restraint spacing required—D_l (in.) Floor	Roof: Dead load	Roof: Wind load	Lateral connection capacity required (lbf)
PRI-20	32	34	36	55
PRI-30	32	33	35	71
PRI-40	56	62	65	63
PRI-50	34	37	39	87
PRI-60	52	57	61	87
PRI-70	45	48	52	116
PRI-80	72	79	85	123
PRI-90	74	77	81	156

For SI: 1 in. = 25.4 mm, 1 lbf = 4.45 N.

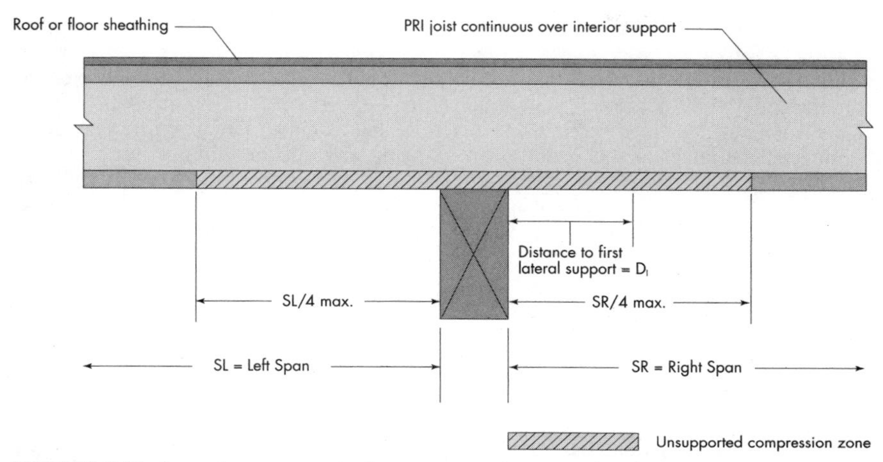

FIGURE 5.46 Lateral support required at unsupported compression flange.

Discussion. The weight of heaters and air conditioners and the associated air-handling equipment must be taken into consideration during the initial layout and design of the roof. In this respect, framing with I-joists, as shown in Fig. 5.47, is essentially the same as framing with sawn lumber. What differences arise are due primarily to the necessity to use filler blocks between the elements of multiple I-joist construction and the fact that performance rated I-Joists come with an allowable span rating.

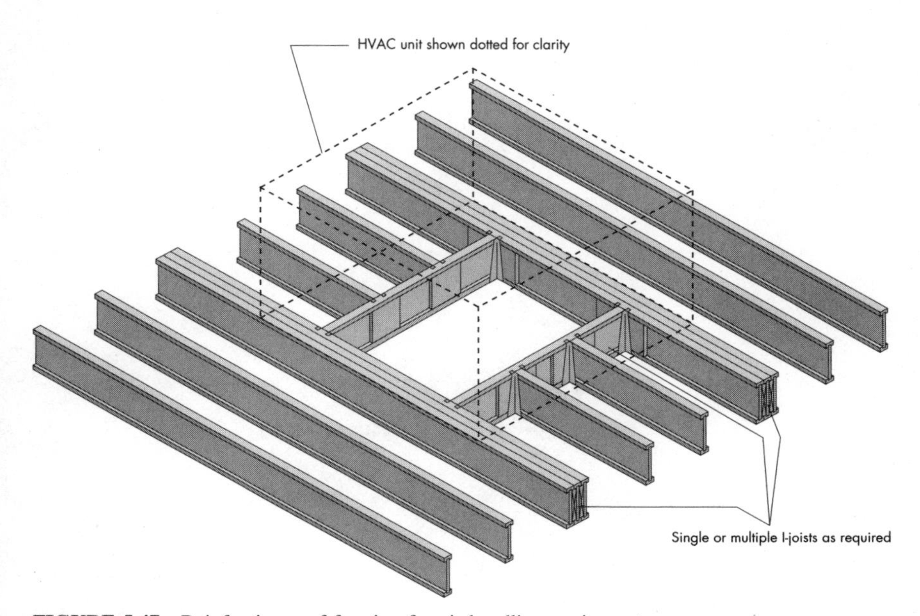

FIGURE 5.47 Reinforcing roof framing for air-handling equipment.

The use of filler blocks in multiple I-joists is covered in Section 5.7.2 under Multiple I-Joist Construction and Filler Blocks and is appropriate for roof framing applications as well as floor framing. While double I-joists may not be avoided directly underneath the heating, ventilation, and air conditioning unit itself, routing of the air-handling ducts perpendicular to the span of the floor framing can often minimize the impact of the additional weight on any one individual joist.

Allowable floor spans in the floor and roof span tables (Tables 5.4–5.7 and 5.24–5.31, respectively) are based on uniform load applications and would have to be adjusted for any concentrated loads applied. The design capacity values are provided in Table 5.2 to facilitate the necessary engineering.

Thrust Caused by Beveled Bearing Blocks

Discussion. While not an issue with birdsmouthed I-joists, the use of sloped or beveled bearing blocks, also known as bevel blocks, at the low end of a pitched roof induces an inward thrust at the top of the wall supporting the bevel block. Similarly, a beveled bearing block at the top of the pitched I-joist rafter produces an outward thrust. If the beveled bearing block is located over a ridge board and the roof is symmetrical about the ridge, the thrust from one side is canceled out by an equal thrust in the opposite direction from the other side, at least for the uniform load condition. The gusset and attachment details shown in Fig. 5.36 are usually sufficient to resist these loads at the top of the rafter.

The inward thrust at the lower end of the roof joist, however, must be considered when attaching the bevel blocks and attaching the I-joist rafters to the bevel blocks. This thrust is also an important consideration when sizing a variable-slope framing anchor for this application.

Calculating Thrust at Beveled Bearing Pads. The thrust at the bearing pad is a result of the vertical force—the end reaction of the roof rafter—being resolved into components parallel and perpendicular to the beveled surface of the bearing pad, as seen in Fig. 5.48. The result of these forces at the lower end of a roof rafter is to push the bearing block toward the inside of the building. At the upper end of the rafter, if not restrained, the force will act to push the wall toward the outside.

The task of the designer is to determine the magnitude of this inward or outward force and attach the I-joist to the beveled bearing block with sufficient capacity to resist these forces. The first step is to compute the unit load on the roof rafter.

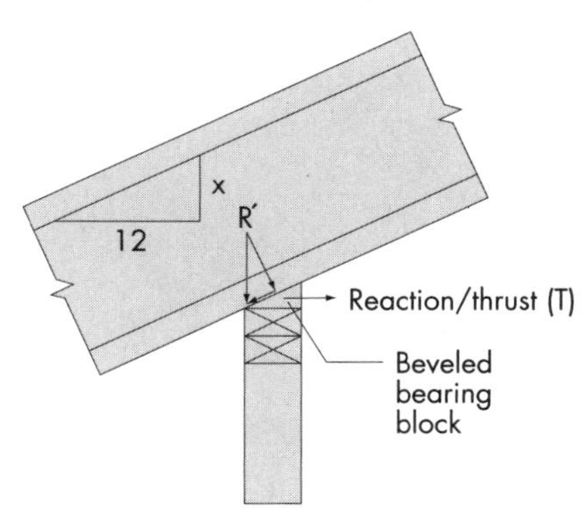

FIGURE 5.48 Calculating thrust at beveled bearing blocks.

Compute Unit Load on Horizontal Projection of the Rafter (w). Because the design live loads are traditionally given in terms of *horizontal* unit area, the building code-adjusted live loads need no further correction for slope length. The dead loads, however, reflect the actual weights of building and finishing materials, insulation, and roof covering and therefore have to be corrected for the actual slope length of the roof system.

Given: the slope of the roof is x in 12:

$$w = \text{live load} + \text{dead load}$$

where live load $= w_{ll}$

dead load $= \dfrac{w_{dl} \times (12^2 + x^2)^{1/2}}{12}$

x = vertical component of slope

Compute Unit Load on Beveled Bearing Pad (R). Using the horizontal distance between the supports and the unit load calculated above, the equation below can be used to calculate the vertical unit load acting on the beveled bearing pad.

$$R = \frac{w \times (L + a)^2}{2 \times L}$$

where R = vertical unit load on beveled plate (lbf/ft)
w = unit load on horizontal projection, from above (lbf/ft^2)
L = horizontal distance between supports (ft)
a = horizontal length of cantilever, if any (ft)

Compute Load on Beveled Bearing Pad per Rafter (R')

$$R' = \frac{R \times s}{12}$$

where R' = vertical load on beveled plate at roof joist (lbf/rafter)
R = vertical unit load perpendicular to the beveled plate, from above (lbf/ft)
s = distance between roof rafters (in.)

Compute Inward Thrust on Beveled Bearing Pad (T)

$$T = \frac{R' \times x}{(12^2 + x^2)^{1/2}}$$

where T = inward thrust on beveled bearing pad (lbf)
R' = vertical load on beveled plate at roof joist, from above (lbf)
x = vertical component of slope

Example Assume a 7:12 roof slope with a horizontal span of 16 ft plus a 2 ft overhang. The design snow load (w_{ll}) is 40 psf (based on the horizontal projection of the roof) and the dead load (w_{dl}) of the roof system (based on the actual slope length of the roof system) is 15 psf. The roof rafters are placed at 16 in. on center.

Compute Unit Load on Horizontal Projection of the Rafter (w)

Given: the slope of the roof is 7:12:

$$w = \text{live load} + \text{dead load}$$

$$w = 40 + \frac{15 \times (12^2 + 7^2)^{1/2}}{12}$$

$$w = 57.37 \text{ psf}$$

Compute Unit Load on Beveled Bearing Pad (R)

$$R = \frac{w \times (1 + a)^2}{2 \times l} = \frac{57.37 \times (16 + 2)^2}{2 \times 16} = 581 \text{ lbf/ft}$$

Compute Load on Beveled Bearing Pad per Rafter (R')

$$R' = \frac{R \times s}{12} = \frac{581 \times 16}{12} = 775 \text{ lbf/rafter}$$

Compute Inward Thrust on Beveled Bearing Pad (T)

$$T = \frac{R' \times x}{(12^2 + 7^2)^{1/2}} = \frac{775 \times 7}{13.89} = 391 \text{ lbf/rafter inward thrust}$$

Attachment of Sprinkler Piping to I-Joist Roof Framing

Discussion. The subject of attachment of sprinkler piping to I-joist roof framing falls into two categories. The first category is the main supply lines that can contribute greatly to the stress level of the roof framing system due to their weight. Because the I-joists carrying these support heavier systems, they will always require an engineered solution. The designer is reminded that when I-joists are doubled up to support such a system, filler blocks are required as described in Section 5.7.2 under Multiple I-Joist Construction and Filler Blocks. In addition, the load from the hangers supporting these main supply lines should always be supported from the top flange of the I-joists, as seen in Fig. 5.49.

The second category of sprinkler piping is the small diameter lightweight piping that feeds the sprinkler heads. This piping is normally counted as a part of the 15 psf (0.72 kN/m^2) dead load used in the roof design and is often suspended by wire from the roof-framing members. Figure 5.50 shows two optional details suitable for such applications.

Note: It may be necessary to pre drill holes in I-joist flanges prior to placing screw eyes to prevent splitting of the I-joist flange.

Blocking of I-Joist Roof Rafters

Discussion. Blocking panels are discussed at length in Section 5.7.2 under Blocking Panels, and the information presented there is appropriate for blocking of I-joist roof rafters given a near-flat or low sloped roof. Given any slope great enough that the blocking panel is not in contact with both the roof sheathing and the top plate of the supporting wall below, the capacity of the blocking panel to function fully to provide overturning resistance, shear transfer, and vertical load transfer is seriously reduced.

While the blocking panel is oriented perpendicular to the roof plane and is of the same depth as the roof joists, it will still act to prevent the rolling over of the I-joists. However, if the slope of the roof is such that the I-joist cannot be attached to both the roof sheathing and the wall below, it will not be able to transfer the

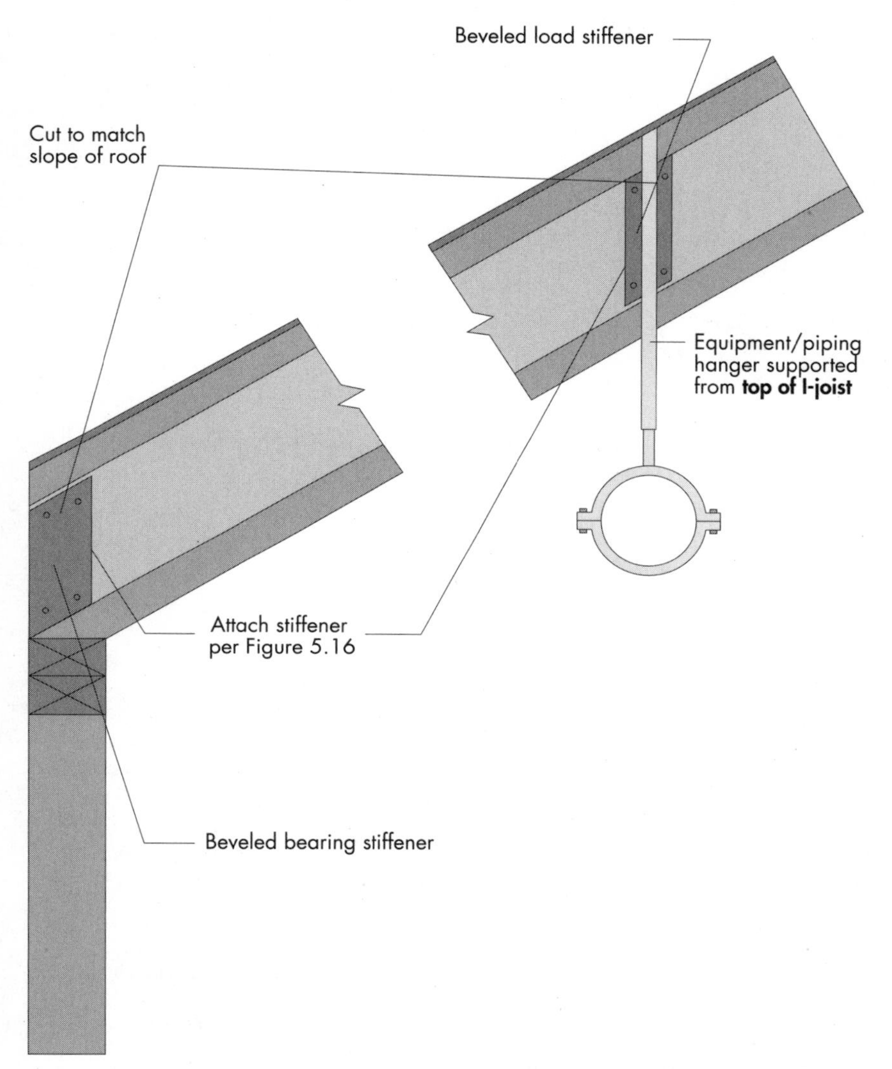

FIGURE 5.49 Attachment of heavy weight sprinkler piping (main supply lines) to I-joist framing.

roof diaphragm shear or vertical load into the wall below. While vertical load transfer is normally not a consideration in roof blocking, the shear transfer aspect can be very important. This is discussed more completely above under Lateral Bracing and Diaphragm Shear Transfer Elements and Shear Transfer from Interior Shear Walls—Parallel and Perpendicular.

As discussed above under Providing Roof Ventilation in I-Joist Rafter Roof Systems, building code-required roof ventilation can also be a challenge in situa-

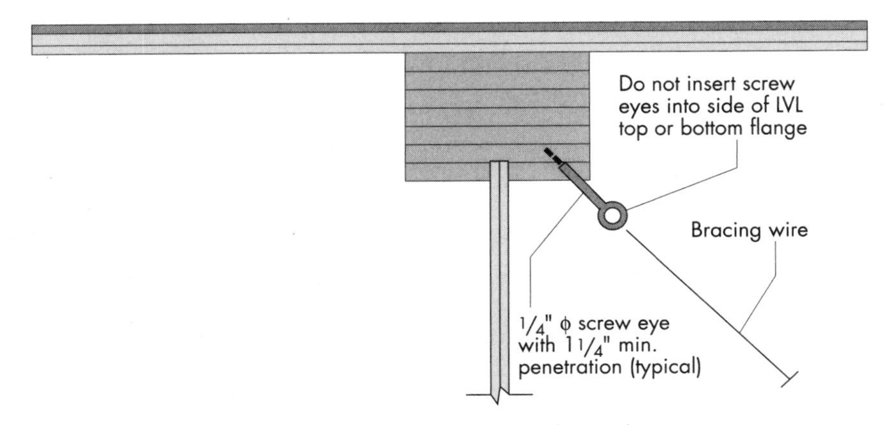

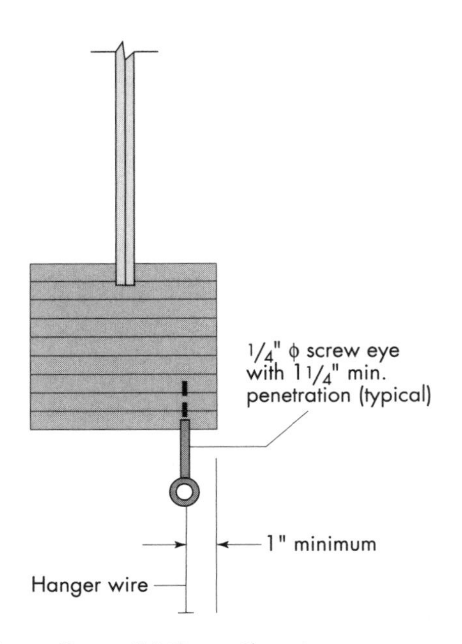

FIGURE 5.50 Attachment of low-weight sprinkler piping to I-joist roof framing.

tions where roof blocking and ventilation are both required. Probably the best solution is to cut rim board to fit and drill round holes in the center of each block as shown in Fig. 5.44 or to use a prefabricated blocking panel as described in Fig. 5.40.

Filler Blocking and Multiple I-joist Construction of I-Joist Roof Rafters

Discussion. The positioning of rooftop equipment or access penetrations will almost always require the use of double or multiple I-joist construction. For these applications the information found in Section 5.7.2 under Filler Blocks is applicable to roof framing applications.

Backer Blocking of I-Joist Roof Rafters

Discussion. As in floor construction, changes in I-joist roof framing direction will often require the use of framing anchors and along with these, backer blocking. The information specified in Section 5.7.2 under Backer Block Construction should also be applied to roof framing applications.

Web Stiffeners in I-Joist Roof Rafters

Discussion. Web stiffeners in roof framing applications serve the same functions as they do in floor application, and as such, the information presented in Section 5.7.2 under Web Stiffeners is appropriate, with one clarification. In pitched roofs, the web stiffeners must be cut and positioned such that they are in line with the direction of the load, i.e., vertical. Because the stiffeners must fit tight against the top or bottom flange, depending on whether they are load or bearing stiffeners, the top and bottom of the stiffener must be carefully cut to match the slope of the roof. As an alternative, if the geometry of the location allows, an oversized rectangular stiffener may be used, as seen in Fig. 5.51.

Flat Roof Construction

Discussion. Truly flat roofs—roofs with a slope of 0:12—should ONLY be considered when some other means of guaranteeing roof drainage is provided. They are not recommended for wood framed roofs, due to the propensity of the wood framing members to deflect under load. An example of a roof drainage system used over a flat roof is a system using tapered foam blocks that are placed over the top of the flat roof deck to provide insulation and, due to their taper, provide for drainage. These blocks also provide a foundation for the roofing system. One problem associated with such systems is that at a later date the whole insulation system may be removed and replaced with an unsloped system without the designer's knowledge or input.

Outside of the fact that it is almost impossible to design a truly flat roof, there is a valid structural reason for not designing flat roofs. This is to prevent a phenomenon known as *ponding,* which may occur when water collects in a depression on a flat roof. As water builds up in the depression, the load on the supporting members increases. This causes additional deflection, which permits more water to accumulate. Once initiated, this cycle can continue until a roof failure occurs. Ponding is more prevalent in areas where roofs are designed for low live-to-dead load ratios. Roofs designed for snow loads are typically stiffer and have greater ability to resist ponding than similar roofs designed only for relatively light construction loads.

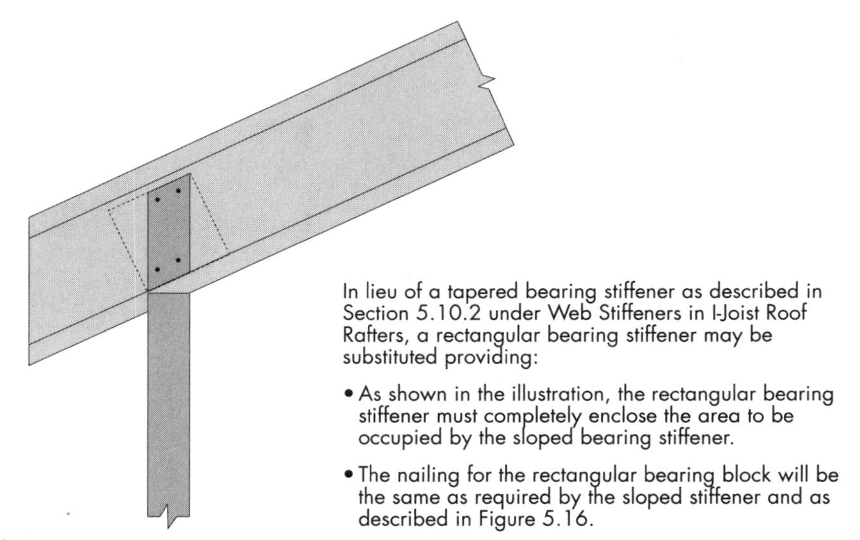

FIGURE 5.51 Beveled bearing stiffeners for pitched roof applications and a rectangular option where appropriate.

For the most part, the roof framing will be designed for drainage by the application of a slight slope leading to roof drains installed in the lowest portions of the roof. The minimum slope recommended is ¼ in. in 12 in. (¼:12) or about a 2% slope. This should be treated as a minimum only. In very large roofs where long spans and beam cambers—intentional or otherwise—are present, or where heavy lateral connection hardware is installed under the roof sheathing, the minimum slope may not be sufficient to ensure proper drainage. It is always good practice to provide as much slope as can be allowed by the geometry of the building. In any case, avoid 0 slope roofs.

5.10.3 SPECIAL DESIGN CONSIDERATIONS—ROOFS

Roof Access Openings

Discussion. It is often necessary to provide small openings in a near-flat-to-low-slope roof structures to provide access to the roof tops for inspection and maintenance. The details shown in Fig. 5.52 are provided to allow the framing of such an opening without design, subject to the following:

- The details in Fig. 5.52 are appropriate for single-span roofs with a pitch up to 4:12 (see Tables 5.24a–5.31a) and multiple-span near-flat roofs (see Tables 5.24b–5.31b). The opening may be up to 48 in. (1.2 m) wide, measured perpendicular to the span by 3 ft (0.9 m) parallel to the span. The framing shown in Fig. 5.52 is designed to support the dead and live loads upon which the allowable span tables are based. Figure 5.52 should NOT be used to support roof mounted air handling equipment without further engineering analysis.

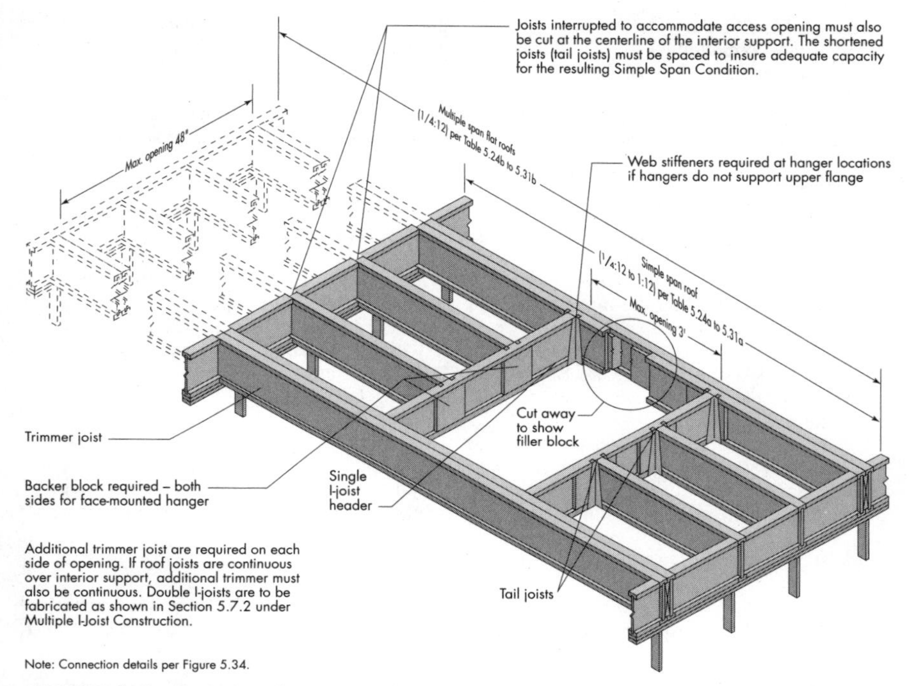

FIGURE 5.52 Typical roof access opening details.

- The details in Fig. 5.52 are also appropriate for residential floors with a maximum deal load of 10 psf (0.48 kN/m^2) and live load of 40 psf (1.92 kN/m^2) in which the I-joist floor joists meet the floor span requirements of Tables 5.4 and 5.5. Figure 5.52 should NOT be used to support stairs or permanent access ladders.
- Multiple openings between trimmer joists are permitted. Openings occurring on both sides of a trimmer are not covered by this detail and must be designed.
- Headers, trimmers, and tail joists must be of identical depth and designation as those selected for the roof/floor framing. In multiple-span near-flat roof or floor applications, the additional trimmers required at the sides of the opening shall extend full length across all spans.
- In multiple-span applications, tail joists shall be cut at centerline of adjacent intermediate support. Verify that both the tail joist span and the inline joist span adjacent to the tail joist are appropriate for simple-span applications. A closer spacing of the joists at this location may be required (see Example below).

Example A designer wants to place a roof access opening in a near-flat roof. The desired opening is 3 × 4 ft and the roof consists of 14″ PRI-70 I-joists spaced at 16 in. on-center spanning continuously over the interior load bearing support. This results in two spans at 26 ft long. The roof dead load is 15 psf and the design snow load on the roof is 30 psf. Check the existing spans and properly specify the framing around the opening.

Check Existing Roof Joist Spans. From Table 5.26b the allowable clear span for a two-span continuous 14″ PRI-70 spaced at 16 in. on-center is 33 ft, 4 in., which is greater than 26 ft. The design spans are OK.

Detail Opening. In accordance with Fig. 5.52, orient the 3 ft dimension on the opening parallel to the I-joist span and the 4 ft dimension perpendicular to the I-joist span.

1. One header is required.
2. Two trimmers, or one joist and one trimmer, are required on each side of the opening. These framing members must run full length across both spans.

Check the Simple-Span Capacity of the Interrupted I-Joists. Because all interrupted I-joists must be cut at the intermediate support and designed as simple-span members, their allowable simple span capacity must be checked. From Table 5.26a, for a slope between 0:12 and 4:12 and a joist spacing of 16 in. on-center, the allowable span for a 14 in. PRI-70 is 29 ft, 3 in. Because this is greater than the required span of 26 ft, the back span and any of the shorter spans running from the supports to the headers of the opening are OK.

Note that if the specified joists had not had the required simple span capacity, the spacing between the joists in line with the opening could have been reduced to increase their span capability. As an alternative, an I-joist designation with the appropriate single-span and multiple-span capacity could be specified for the entire roof system.

Special Roof Cantilever Considerations

Discussion. When engineered wood I-joists are used in cantilevered roof situations, a number of special considerations must be accounted for in the design and detailing.

1. *Weathering-in considerations:* The first consideration has to do with the potential for outside exposure in roof cantilevers. Engineered wood I-joists are fabricated out of very high-quality dried wood components and should not be exposed to exterior conditions for extended lengths of time. Whenever wood I-joists are used in roof or floor cantilevers, they must always be protected from ambient water by means of a weather barrier. Boxed soffits, lumber end closures, exterior sheathing, and APA rated sidings are examples of acceptable methods. Some methods that are not acceptable are wood, vinyl, or metal lap siding without a building paper backing and the use of paint only.
2. *Structural considerations:* Several structural considerations that warrant discussion.

    - *Lateral bracing:* While not usually a consideration in residential-type construction, where cantilevers seldom exceed 24 in. (610 mm), in commercial structures the designer may want to utilize the long available lengths of the wood I-joists for relatively long cantilevers. If this is the case, the designer is reminded that the cantilever situation is similar to a continuous I-joist over a center support in that the bottom flange of the joist is put in compression and as such requires lateral bracing similar to that described above under Compression-Side Braring for Continuous I-Joist Roof Framing. In general, finishing off the underside of the cantilevered I-joists with a wood structural panel sheathing—"boxing the soffit"—will provide all of the lateral support necessary. On the inside of the structure gypsum wallboard applied to the bottom side of the I-joists will provide the required lateral support. If the joists are to remain exposed on the bottom side of the backspan side of the cantilever, then blocking or bridging will have to be used for this purpose.

- *Diaphragm boundary nailing:* Also not usually a consideration in residential-type construction, properly defining the roof diaphragm boundary can be a source of confusion when cantilevers are present. The diaphragm boundary is not always the edge of the roof. Because the lateral support for the diaphragm is provided by the shear walls that support it, the diaphragm boundary is usually directly over the shear walls. The load from the diaphragm boundary is transferred to the shear walls by the blocking panels or other wall framing as discussed above under Lateral Bracing and Diaphragm Shear Transfer Elements and Shear Transfer from Interior Shear Walls—Parallel and Perpendicular at the shear wall location.
- *Balcony railings:* This can be a problem whenever I-joists are used in cantilevered situations, be it roofs or floors. Because only the ends of the flanges are available for attachment, many of the conventional methods used for the attachment of railings may not be acceptable. The ends of the I-joists may have to be reinforced with stiffeners to provide sufficient end-nailing area. If this is done, the stiffener-to-web attachment detail will have to be carefully designed to transfer the applied loads into the ⅜ in. (9.5 mm) thick web.

5.11 CALCULATION EXAMPLES

The following sections are examples of calculations used to develop the tables used in this chapter. The first example shown in Section 5.11.1 describes the methodology used to calculate the transformed section properties based on partial composite action resulting from the use of the glued floor system and the generation of floor spans using these properties. Section 5.11.2 illustrates the calculation of allowable spans for a pitched roof. Note that this example does not assume a glued roof, and therefore section transformation is not required. The final example in Section 5.11.3 described the methodology used to determine the maximum allowable hole size that may be cut into the web of an I-joist.

5.11.1 Calculation Example for the Effective Floor Stiffness and Allowable Span of Glued-Floor System

Assumptions. Steps 1–6 are used to demonstrate the calculation procedure used in determining the effective floor *EI* for 14″ PRI-40 at a 19.2 in. (488 mm) spacing. The effective floor stiffness calculation is based on the premise that the floor system meets the following requirements:

- The floor panels must meet the requirements for APA Rated Sheathing, APA Rated Sturd-I-Floor, PS 1, or PS 2 with a minimum thickness of $^{19}/_{32}$ in. (15.1 mm) (40/20 or 20 oc) for a joist spacing of 19.2 in. (488 mm) or less, or $^{23}/_{32}$ in. (18.3 mm) (48/24 or 24 oc) for a joist spacing of 24 in. (610 mm).
- Mechanical properties of the floor I-joists must meet the requirements for APA EWS Performance Rated I-Joists as set forth in APA PRI-400.
- The floor panels must be glued and nailed to the I-joists based on the nailing schedule required by the governing building code.

- Design values (EI and EA) for floor panels are in accordance with Chapter 2, and the lower value obtained from APA Rated Sheathing and APA Rated Sturd-I-Floor is used.
- The glued floor construction factor was set equal to 0.45.

Calculation of Effective Floor Stiffness (English units only) (see Fig. 5.53)

Step 1. 14″ PRI-40 I-Joist Design Properties (From Table 5.2.)

Net depth = 14 inches

Bending $EI = 482 \times 10^6$ lbf-in.^{2}

M (Repetitive Member) = 4130 lbf-ft

IR (3½ in. intermediate bearing) = 2500 lbf

Joist spacing = 19.2 in.

$K = 7.28 \times 10^6$ lbf

$V = 1710$ lbf

ER (1¾ in. end bearing) = 1200 lbf

Step 2. Flange Materials (1.5E SCL)

Width = 2½ in.

Depth = 1½ in.

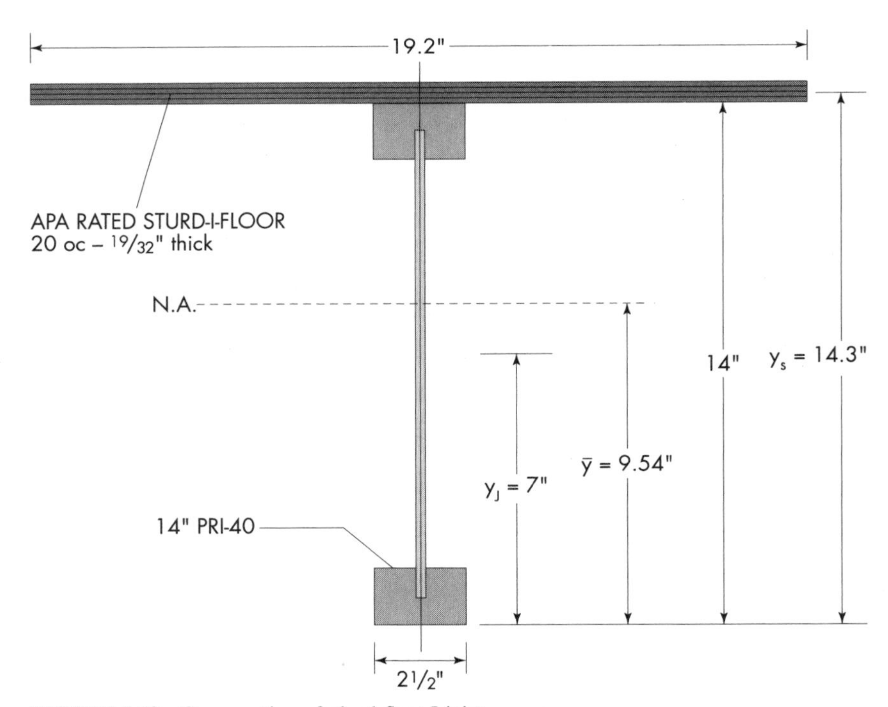

FIGURE 5.53 Cross section of glued floor I-joist.

Step 3. Web Materials (3/8 in. rated sheathing—OSB)
$EA_{\perp} = 2.9 \times 10^6$ lbf/ft
$A = 4.5$ in.^{2}/ft
Thickness = 3/8 in.

Step 4. I-Joist Sectional Properties
$EA_{\text{flange}} = 1.5 \times 10^6 \times (2 \times 2.5 \times 1.5) = 11.25 \times 10^6$ lbf
$EA_{\text{web}} = 2.9 \times 10^6 \times (14 - 2 \times 1.5)/12 = 2.66 \times 10^6$ lbf
$EA_{\text{joist}} = (11.55 + 2.66) \times 10^6 = 13.91 \times 10^6$ lbf

Step 5. Floor Sheathing (APA Rated Sturd-I-Floor 20 oc—OSB)**
Thickness = 19/32 in.
$EI = 13{,}000$ lbf-in.^{2}/ft $\times 19.2/12 = 20{,}800$ lbf-in.^{2}
$EA = 4.6 \times 10^6$ lbf/ft $\times 19.2/12 = 7.4 \times 10^6$ lbf

Step 6. Composite Floor

	EA (10^6 lbf)	y (in.)	EAy (10^6 lbf-in.)
I-joist	13.9	7.00	97.4
Sheathing	7.4	14.30	105.8
$\Sigma =$	21.3		203.2

Therefore, $\bar{y} = 203.2/21.3 = 9.54$ in.

	EA (10^6 lbf)	EI (10^6 lbf-in.^{2})	$(y - \bar{y})^2$	$EI + EA(y - \bar{y})^2$
I-joist	13.9	482	6.44	572
Sheathing	7.4	0.021	22.66	168
$\Sigma = EI_{\text{composite}} =$				740

Therefore,

$$EI_{\text{effective}} = 0.45 \times 740 \times 10^6 + 0.55 \times 482 \times 10^6$$
$$= \underline{598 \times 10^6 \text{ lbf-in.}^2}$$

Calculation of Allowable Floor Span (English units only)

Assumptions. The allowable spans given in Tables 5.4 and 5.5 are based on a design dead load of 10 psf and live load of 40 psf, as typically applicable to residential floor construction. The allowable spans were established based on the premise that the floor system meets the following requirements:

** This rating has a lower design EI than APA Rated Sheathing 40/20. In addition, OSB has a lower design EI than plywood and thus is conservative for calculation purposes.

- The floor system must contain no less than three I-joists.
- The coefficient of shear deflection, K, was set equal to 0.52×10^6 lbf/in. times the net I-joist depth in inches.
- Total load deflection limit (δ_{TL}) = span/240
- Live load deflection limit (δ_{LL}) = span/480

Based on the effective stiffness (EI) from the previous calculation, steps 1 through 6 are used to determine the allowable span for 14″ PRI-40 at the 19.2 in. floor joist spacing.

Step 1. Live Load Deflection

1. For simple-span applications,

$$\delta_{LL} = \frac{5\omega_{LL}\ell^4}{384EI} + \frac{\omega_{LL}\ell^2}{K} \leq \frac{\ell}{480}$$

where ω_{LL} = live load = 40 lbf/ft² × 19.2 in./12 in./ft = 64 lbf/ft
$EI = 598 \times 10^6$ lbf-in.²
$K = 7.28 \times 10^6$ lbf
ℓ = unknown span (from the center to center of supports)

Solving the above equation for ℓ yields 21 ft, 2 in.

2. For multiple-span applications,

$$\delta_{LL} = \frac{0.0099\omega_{LL}\ell^4}{EI} + \frac{1.028\omega_{LL}\ell^2}{K} \leq \frac{\ell}{480}$$

Solving the above equation for ℓ yields 23 ft, 1 in.

Step 2. Total Load Deflection

1. For simple-span applications,

$$\delta_{TL} = \frac{5\omega_{TL}\ell^4}{384EI} + \frac{\omega_{TL}\ell^2}{K} \leq \frac{\ell}{240}$$

where ω_{TL} = total load = 50 lbf/ft² × 19.2 in./12 in./ft = 80 lbf/ft
$EI = 598 \times 10^6$ lbf-in.²
$K = 7.28 \times 10^6$ lbf
ℓ = unknown span (from the center to center of supports)

Solving the above equation for ℓ yields 24 ft, 11 in.

2. For multiple-span applications,

$$\delta_{TL} = \frac{0.0099\omega_{TL}\ell^4}{EI} + \frac{1.028\omega_{TL}\ell^2}{K} \leq \frac{\ell}{240}$$

Solving the above equation for ℓ yields 27 ft, 3 in.

Step 3. Moment Capacity

For simple- or multiple-span applications,

$$M = \frac{\omega_{TL}\ell^2}{8}$$

where ω_{TL} = total load = 50 lbf/ft² × 19.2 in./12 in./ft = 80 lbf/ft
M (Repetitive Member) = 4130 lbf-ft
ℓ = unknown span (from the center to center of supports)

Solving the above equation for ℓ yields 20 ft, 4 in.

Step 4. Shear Capacity

1. For simple-span applications,

$$V = \frac{\omega_{TL}\ell}{2}$$

where ω_{TL} = total load = 50 lbf/ft² × 19.2 in./12 in./ft = 80 lbf/ft
V = 1,710 lbf
ℓ = unknown span (from the center to center of supports)

Solving the above equation for ℓ yields 42 ft, 9 in.

2. For multiple-span applications,

$$V = \frac{5\omega_{TL}\ell}{8}$$

Solving the above equation for ℓ yields 34 ft, 2 in.

Step 5. Intermediate Reaction Capacity

For multiple-span applications,

$$IR = \frac{5\omega_{TL}\ell}{4}$$

where ω_{TL} = total load = 50 lbf/ft² × 19.2 in./12 in./ft = 80 lbf/ft
IR = 2500 lbf
ℓ = unknown span (from the center to center of supports)

Solving the above equation for ℓ yields 25 ft, 0 in.

Step 6. End Reaction Capacity

1. For simple-span applications,

$$ER = \frac{\omega_{TL}\ell}{2}$$

where ω_{TL} = total load = 50 lbf/ft² × 19.2 in./12 in./ft = 80 lbf/ft
ER = 1200 lbf
ℓ = unknown span (from the center to center of supports)

Solving the above equation for ℓ yields 30 ft, 0 in.

2. For multiple-span applications,

$$ER = 0.45\omega_{LL}\ell + 0.40\omega_{DL}\ell$$

where ω_{LL} = live load = 40 lbf/ft^2 × 19.2 in./12 in./ft = 64 lbf/ft
ω_{DL} = Dead load = 10 lbf/ft^2 × 19.2 in./12 in./ft = 16 lbf/ft
ER = 1200 lbf
ℓ = unknown span (from the center to center of supports)

Solving the above equation for ℓ yields 34 ft, 1 in.

Therefore, the allowable span is 20 ft, 4 in. for simple-span applications (controlled by moment capacity in step 3) and 20 ft, 4 in. for multiple-span applications (also controlled by moment capacity in step 3). To calculate the clear span, the allowable span is reduced by 1¾ in. for simple-span applications and (1¾ + 3½)/2 or 2⅝ in. for multiple-span applications. Therefore, the published clear span is 20 ft, 2 in. for simple-span applications (see Table 5.4) and 20 ft, 1 in. for multiple-span applications (see Table 5.5).

5.11.2 Calculation Example for Allowable Roof Spans (English units only)

Assumptions. The allowable spans given in Tables 5.24–5.28 are based on the design dead load of 15 psf and snow loads of 20, 25, 30, 40, and 50 psf, as may be applicable to residential roof construction. Tables 5.27–5.31 are based on a 20 psf construction load with dead loads of 10, 15, and 20 psf. Because gluing of roof sheathing is not traditionally done, nor recommended by APA, it is not necessary to compute composite stiffness properties. As such, the I-joist design capacities used for roof design may be taken directly from Table 5.2. The allowable spans were established based on the premise that the roof system meets the following requirements:

- The roof system must contain no less than three I-joists.
- The coefficient of shear deflection, K, was set equal to 0.52 × 10^6 lbf/in. times the net I-joist depth in inches.
- Near-flat roof tables assume a two-span condition with a short-span length of at least 40% of the longer span.
- Pitched roof tables are based on worst-case conditions for simple or simple spans with two foot cantilever conditions.
- Total load deflection limit (δ_{TL}) = span/240.
- Live load deflection limit (δ_{LL}) = span/180.

Steps 1–5 are used to determine the allowable span for 14″ PRI-40 at the 19.2 in. roof rafter spacing. A snow load of 40 psf is assumed, as well as a dead load of 15 psf, with a roof pitch of 4:12. A duration of load adjustment of 1.15 applies to the snow load.

For calculating allowable span, the vertical live load is converted to a load acting perpendicular to the slope length of the roof rafter. (While ASCE 7-98 allows for a further reduction of snow loads based on some combinations of roof slope, temperature, and roughness, none was taken for the development of these spans.) The dead load, on the other hand, is an estimate of the weight of the actual materials and, as such, is assumed to act in a vertical direction with relation to the roof rafter.

The dead load and slope-adjusted live load are broken up into components acting perpendicular and parallel to the slope. The load component perpendicular to the slope was used with the slope length to calculate the allowable spans. As such, the allowable span calculations actually yield the slope length of the rafter. This length must be converted to the horizontal distance between the rafter supports, because this is how the roof allowable loads are presented in the tables. In addition, the published tables define the span as the distance between the faces of the supports, so a small distance must be subtracted from the calculated distance because the span as calculated represents the length measured from center bearing to center bearing. Three inches are subtracted from the calculated span to represent this distance.

Calculation of Allowable Roof Span

Step 1. Live Load Deflection

Simple span with no cantilever controls,

$$\delta_{LL} = \frac{5\omega_{LL}\ell^4}{384EI} + \frac{\omega_{LL}\ell^2}{K} \leq \frac{\ell}{240}$$

where ω_{LL} = snow load = 40 lbf/ft^2 × 19.2 in./12 in./ft = 64 lbf/ft
Sloped roof adjustment for snow loads = C_s
For a cold roof, 4:12 pitch, C_s = 1.0
Live load adjustment for roof slope—$64 \times 12/(4^2 + 12^2)^{1/2}$ = 60.72 lbf/ft
Adjusted normal to slope—$60.72 \times 12/(4^2 + 12^2)^{1/2}$ = 57.60 lbf/ft

ω_{LL} = 1.0 × 57.60 = 57.60 lbf/ft
EI = 482 × 10^6 lbf-in.^{2}
K = 7.28 × 10^6 lbf
ℓ = unknown slope length (from the center to center of supports)

Solving the above equation for ℓ yields 26.05 ft

Step 2. Total Load Deflection

Simple span with no cantilever controls,

$$\delta_{TL} = \frac{5\omega_{TL}\ell^4}{384EI} + \frac{\omega_{TL}\ell^2}{K} \leq \frac{\ell}{180}$$

where ω_{TL} = total load = dead load + adjusted live load
Dead load = 15 lbf/ft^2 × 19.2 in./12 in./ft = 24 lbf/ft

Live load adjustment for roof slope—$64 \times 12/(4^2 + 12^2)^{1/2}$ = 60.72 lbf/ft (see step 1 above)
Total load adjusted normal to slope—$(24 + 60.72) \times 12/(4^2 + 12^2)^{1/2}$ = 80.37 lbf/ft

ω_{TL} = 80.37 lbf/ft
EI = 482 × 10^6 lbf-in.^{2}
K = 7.28 × 10^6 lbf
ℓ = unknown slope length (from the center to center of supports)

Solving the above equation for ℓ yields 25.64 ft

Step 3. Moment Capacity

Simple span with no cantilever controls,

$$M = \frac{\omega_{TL}\ell^2}{8}$$

where ω_{TL} = Total load = dead load + adjusted live load
ω_{TL} = 84.72 lbf/ft (see Step 2 above)
M (repetitive member) = 4130 lbf-ft * 1.15 = 4750 lbf-ft
ℓ = unknown slope length (from the center to center of supports)

Solving the above equation for ℓ yields 21.74 ft

Step 4. Shear Capacity

Simple span with 2 ft cantilever controls,

$$V = \frac{\omega_{TL}(\ell^2 + a^2)}{2}$$

where ω_{TL} = total load = dead load + adjusted live load
ω_{TL} = 84.72 lbf/ft (see step 2 above)
V = 1710 × 1.15 = 1967 lbf
a = 2 ft
ℓ = unknown slope length (from the center to center of supports)

Solving the above equation for ℓ yields 48.85 ft

Step 5. Exterior Reaction Capacity

Simple span with 2 ft cantilever controls,

$$ER = \frac{\omega_{TL}(\ell + a)^2}{2\ell}$$

where ω_{TL} = total load = dead load + adjusted live load
ω_{TL} = 84.72 lbf/ft (see step 2 above)
ER = because of the 2 ft cantilever, the most conservative approach would be to use the tabular ER value adjusted for full bearing—3½ in.—in accordance with note 6 of Table 5.2. ((1710 − 1200) /(4 − 1.75)) × 1.75 + 1200)
ER = 1597 lbf × 1.15 = 1837 lbf
ℓ = unknown slope length (from the center to center of supports)

Solving the above equation for ℓ yields 34.34 ft

Therefore, the limiting slope length is 21.74 ft, based on moment (Step 3). To calculate the clear span, the limiting slope length must be converted to the horizontal allowable span. In addition, the calculated horizontal allowable span is reduced by 0.25 ft to convert to the allowable clear span.

Convert limiting slope length to allowable clear span as follows:

Slope length = 21.74 ft

Horizontal allowable span = $(21.18 \times 12/(4^2 + 12^2)^{1/2}) = 20.62$ ft

Allowable clear span = $20.62 - 0.25 = 20.37$ ft

Allowable clear span = 20 ft, 5 in (see Fig. 5.54)

5.11.3 Calculation Example for Proper Placement of Hole in an I-Joist Web

In developing Table 5.12, the assumption was made that the I-joist, at a section cut through the center of a hole, would have a shear capacity directly proportional to the amount of web remaining (the web is assumed to run the full depth of the joist). The shear load at any hole location would have to be less than or equal to the shear capacity of the remaining web at that location.

In developing Table 5.12, the shear load on the I-joist was calculated using the Wood I-Joist Manufacturers Association (WIJMA) load cases 1, 2, 3, and 5. All of these load cases are for two-span continuous applications and provide conservative

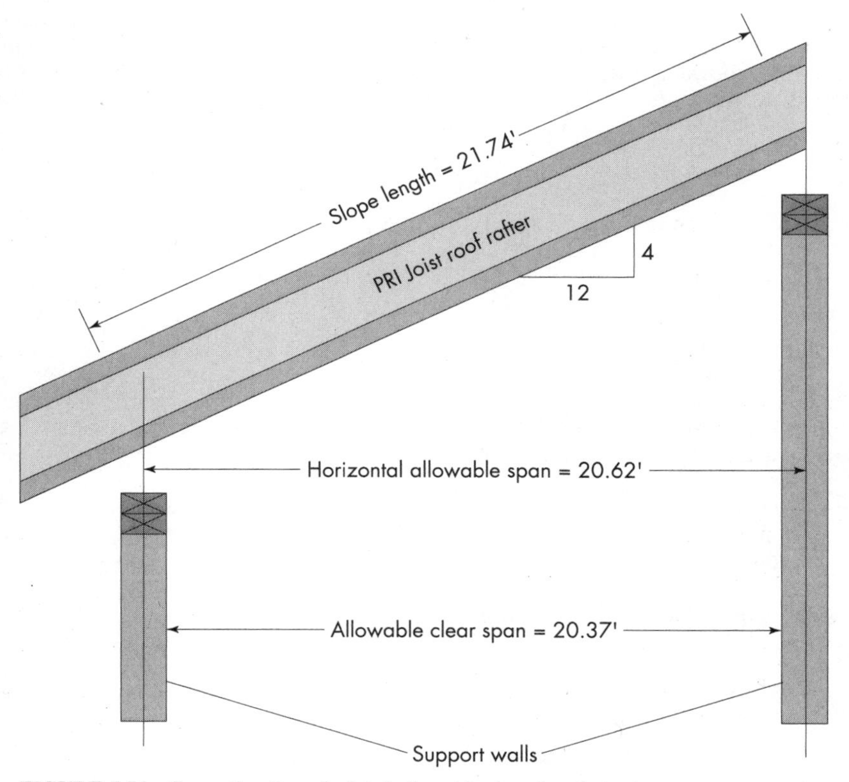

FIGURE 5.54 Converting the calculated allowable slope length to clear span.

results even when used for locating holes in single span applications. The shear load was calculated for the limiting I-joist designation for maximum spans at 12, 16, 19.2, and 24 in. (305, 406, 488, and 610 mm) on center I-joist spacings. The shear load was calculated at ½ ft (152 mm) increments from the inside support of the two-span case and used for both the multiple and single span conditions. This shear load was converted to an allowable hole diameter at each ½ ft (152 mm) increment in an Excel spreadsheet.

In addition to the allowable hole size calculated as described briefly above and in more detail below, there is an additional requirement that is built in to the recommendations found in Table 5.12. This requirement is for a minimum hole distance from the center of the hole to the inside face of the support equal to the lesser of two times the hole diameter or 12 in. (305 mm) plus ½ of the hole diameter.

Calculation Methodology (English units only). To calculate the allowable hole diameter for a given location in a given set of circumstances is relatively straightforward. Because the I-joist designation, span, floor loading, and desired hole location are usually known, the first step is to calculate the critical shear at the given location considering all of the load cases.

The maximum hole diameter permitted can then be determined by calculating the proportion of the I-joist's allowable shear capacity that was *not* used to resist the live and dead load-induced shear stress. The ratio of the shear capacity remaining to the allowable shear capacity is equal to the ratio of the maximum allowable hole diameter to the total depth of the I-joist. The general equation used for the allowable hole diameter at a given point is as follows:

$$(V_a - V_x)/V_a = d/D$$

where V_a = allowable shear capacity of the I-joist designation (lb)
V_x = load-induced shear at a point along the span of the beam (lb)
d = maximum hole diameter allowed in section (in.)
D = depth of the I-joist (in.)

While very straightforward in concept, the difficulty lies in the assumed load cases. The partial-span uniform load cases required make hand calculation of the shears for a multiple-span situation very time-consuming. If a multiple-span situation is contemplated, one of a number of single-element software design programs, such as the APA EWS WOODCAD program, is recommended to find the shear at any specific location.

Because the purpose of the design example is to illustrate the procedure for determining the allowable hole size at a specific point, a single span I-joist is used in the following example.

Calculation of Maximum-Size Hole at a Given Location

Given:

- An 18 ft simple-span situation with a floor live load of 40 psf and a dead load of 10 psf
- 11⅞″ PRI-70 floor joists specified spaced at 16 in. on-center
- Calculate the largest hole possible at a point 3 ft in from either support

Calculation:

1. Calculate the shear at a point 3 ft in from one end of the span under uniform load:

$$V_x = w \times (\ell/2 - x)$$

where V_x = shear at x ft in from the face of the support (lb)
w = uniform load on the beam (lb/ft)
ℓ = span of the I-joist measured between faces of supports (ft)
x = distance from face of support to hole location (ft)

$$V_x = 50 \times (18/2 - 3) = 300 \text{ lbf}$$

2. Determine the allowable shear for an 11⅞″ PRI-70:

From Table 5.2, V_a = 1420 lbf

3. Determine allowable hole size at a point 3 ft in from one end of the span:

$$(V_a - V_x)/V_a = d/D$$

where V_a = allowable shear capacity of the I-joist designation (lbf)
V_x = load-induced shear at a point along the span of the beam (lbf)
d = maximum hole diameter allowed in section (in.)
D = depth of the I-joist (in.)

$$(1420 - 300)/1420 = d/11\tfrac{7}{8}$$

$$\underline{d = 9.36 \text{ in.}}$$

4. Determine the maximum hole that may be placed in an 11⅞ in. I-joist:

$$D_{max} = D - (2 \times D_f) - 0.25$$

where D_{max} = maximum diameter hole allowed (in.) (see Section 5.8.3 under Rules for Cutting Holes in PRI Joists, item 4)
D = depth of the I-joist (in.)
D_f = depth or thickness of flange (in.)

$$D_{max} = 11.875 - (2 \times 1.5) \times 0.25 = \underline{8.625 \text{ in.}}$$

5. Determine lesser of step 3 and step 4:

Use 8.625 in. or <u>8⅝ in.</u>

6. Check minimum distance between 8⅝ in. hole and inside face of support: From Section 5.8.3, under Rules for Cutting Holes in PRI Joists, item 1, a minimum distance (rounded up to the nearest ½ ft) from the centerline of the hole to the inside face of the support is 1.5 ft. As the location of the hole is 3 ft from the support, the minimum distance requirement is met.

7. Conclusion:
<u>An 8⅝ in. hole may be placed at a point 3 ft from either support.</u>

5.12 ENGINEERED WOOD RIM BOARD

5.12.1 Growth of the Industry

Figure 5.55 shows end-use market growth for North American engineered wood rim board over 1998 and 1999, since this is a relatively new product. The source of the data can be found in the footnote of the graph. Since rim board products are used primarily in I-joist floor and roof applications, like I-joists, the residential market has accounted for over 90% of the increase noted. Again, as I-joist use increases in nonresidential and remodeling markets, the engineered wood rim board market is expected to track right along.

5.12.2 What Is a Performance Rated Rim Board?

A rim board is the wood component that fills the space between the sill plate and bottom plate of a wall or, in second floor construction, between the top plate and bottom plate of two wall sections. The rim board must match the depth of the framing members between floors or between the floor and foundation to function properly. In addition to supporting the wall loads, the rim board ties the floor joists together. It is an integral component in an engineered wood system because it transfers both lateral loads and vertical bearing forces.

While lumber has been the traditional product used for rim boards, it is not compatible with the new generation of wood I-joists used in today's floor and roof

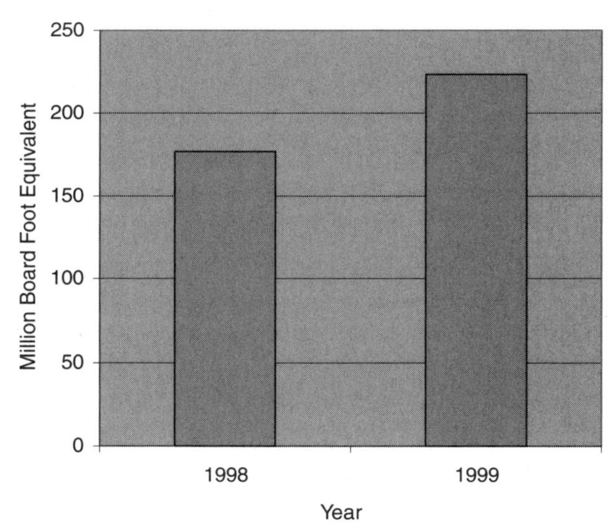

FIGURE 5.55 Engineered wood rim board market growth. 1999 Information extrapolated from 1998 rim board market information and I-joist growth information for 1999. (*APA*.)

construction. Lumber will typically have a different depth than I-joists and will also exhibit different shrinkage characteristics. See Section 5.3.1. With the increased use of wood I-joists, a demand for compatible engineered wood rim boards has resulted. To meet this need, APA promulgated a standard for Performance-Rated Rim Board designated as PRR-401.[14]

PRR-401 rim boards can be manufactured using plywood, oriented strand board (OSB), glued laminated timber (glulam), or laminated veneer lumber (LVL). APA-EWS performance rated rim boards are available in lengths up to 24 ft (7.3 m), depending on the product used for fabrication. Typical sizes are shown in Table 5.34.

As engineered wood products, Performance Rated Rim Boards have greater dimensional stability, higher strength, increased structural reliability, more consistent quality, and a lower tendency to check or split than sawn lumber. Because rim board products are designed to be used specifically with I-joists, they make an excellent economical source for blocking panels, squash blocks, load-bearing cantilever reinforcement, ridge boards, and other reinforcement applications. A typical application of an engineered wood rim board product is illustrated in Fig. 5.56.

5.12.3 Manufacture of Engineered Rim Boards

Because APA EWS Rim Board can be made out of several different materials, there are a number of different manufacturing processes that are utilized. Rim Board is generally available in lengths from 8–24 ft (2.4–7.3 m) long, depending on the material/manufacturing process. The maximum length is in many cases limited by the contractor's ability to handle the element in the field, as it is relatively flexible along the shallow axis.

OSB specifically trademarked as Rim Board is ripped (cut parallel to the strength axis) to I-joist-compatible sizes out of large sheets of OSB. Because these sheets can be manufactured in lengths up to 24 ft (7.3 m) along the strong axis, rim board up to this length is possible.

LVL is traditionally manufactured in a continuous line, that is, the length is governed only by handling and shipping restraints from the mill, to the distribution yard, to the job site. As such, because of its relatively narrow depth, LVL Rim Board is also usually limited to a maximum length of 24 ft (7.3 m) to minimize handling problems on the job site.

While this is uncommon, Rim Board can also be made out of resawn glulam beams. Glulam beams are made up of individual laminations of end-jointed lumber and can be manufactured in lengths in excess of 100 ft (30.5 m) long. Once the adhesive has set, the glulam is sent through a secondary cutting operation that rips the beam into planks of I-joist-compatible depths and the requisite thickness for

TABLE 5.34 APA Performance Rated Rim Boards—Standard Sizes and Dimensions

Property	Standard sizes
Thickness (in.)	1, 1⅛, 1¼, and 1½
Depth (in.)	9½, 11⅞, 14, 16, 18, 20, 22, 24
Length (ft)	8 to 24

For SI: 1 in. = 25.4 mm, 1 ft = 304.8 mm.

FIGURE 5.56 Performance Rated Rim Board.

APA EWS Rim Board. As mentioned for LVL, the actual length as a Rim Board is usually limited to a maximum of 24 ft (7.3 m) due to handling considerations.

APA EWS Rim Board can also be made by ripping plywood trademarked as Rim Board to the appropriate depth. Because plywood is typically manufactured in sheets 8 ft (2.8 m) long, Rim Board so manufactured is limited by the length of the panel.

5.12.4 Moisture Performance of Engineered Wood vs. Lumber

An understanding of the moisture performance of engineered wood products such as Rim Board in important to prevent misapplication in the field. This subject is covered in detail in Section 5.3.

5.12.5 Engineered Wood Rim Board Standard

As mentioned previously, APA—The Engineered Wood Association has long been a leader in the development of engineered wood standards. The APA EWS Rim Board Standard is one of the most recent standards developed for the engineered wood industry.

Glulam Rim Boards are a resawn grade of glued laminated timbers manufactured in accordance with the *Performance Standard for APA EWS Rim Boards* and ANSI A190.1.[15]

The *Performance Standard for APA EWS Rim Boards* meets or exceeds the requirements given in the *ICBO ES Acceptance Criteria for Wood-Based Rim Board—AC 124.*[16]

5.12.6 Identifying Rim Boards

While the three major model building codes as well as the new International Building Code do not require Rim Board to be trademarked by an approved third-party inspection agency before they can be used in code-conforming construction, APA provides this service to their members. The purpose of providing this trademarking service is to allow for easy identification of the product. The information on these stamps is beneficial to the engineer, contractor, building official, and ultimately the consumer. While the format of the trademark stamp may vary depending on the quality assurance agency representing the mill, they all contain the same kind of information. Examples of these trademark stamps and an explanation of the information they contain are shown in Fig. 5.57.

Rim Board Trademarking and Identification. Similar to I-joist trademark stamps, discussed in Section 5.4.2, the rim board trademark stamps shown in Fig. 5.58 also contain helpful information. The rim board grade is given in the trademark along with the thickness and references the standard, PRR-401.

FIGURE 5.57 Field installation of I-joist and rim board showing APA stamps.

FIGURE 5.58 Sample Performance Rated Rim Board trademark stamp with explanation of elements.

5.12.7 Rim Board Designations

There are currently two APA EWS Performance Rated Rim Board designations in use, Rim Board and Rim Board Plus. While products of both of these designations exceed the minimum requirements for residential rim boards as defined by ICBO Acceptance Criteria 124, the Rim Board Plus product has slightly higher design properties required for some engineered applications. Manufacturers may also supply proprietary rim board products with even higher design properties.

5.12.8 Rim Board Design Properties

When using Performance Rated Rim Board for applications other than satisfying the prescriptive requirements of the model building codes, it may be necessary to have allowable design capacities for these engineered applications. Table 5.35 is provided for such applications.

5.12.9 Rim Board Application Notes

Performance Rated Rim Boards Used as Starter Joists. Rim boards can be installed in a direction not only perpendicular, but also parallel to the framing as starter joists. Design capacities of the rim boards are the same for both orientations with the exception that the vertical load capacity (V) should be reduced as follows unless the rim board (starter joist) is doubled or blocking panels are installed no more than 24 in. (610 mm) o.c. between the rim board (starter joist) and adjacent floor joist:

1. For 1 in. (24 mm) rim boards, use the bearing load capacity of 450 lbf/ft (6.6 kN/m) when the rim board depth is not greater than 18 in. (460 mm) or 350 (5.1 kN/m) when the rim board depth is greater than 18 in. (460 mm) but does not exceed 24 in. (610 mm).
2. For 1⅛ in. (29-mm) rim boards, use the bearing load capacity of 850 lbf/ft (12.4 kN/m) when the rim board depth is not greater than 18 in. (460 mm) or 750 lbf/ft (10.9 kN/m) when the rim board depth is greater than 18 in. (460 mm) but does not exceed 24 in. (610 mm).

Note that some starter joists are installed at a non-load-bearing end wall. As a result, it is not always necessary to use double rim boards or blocking panels for starter joists, since the design load may be within that prescribed in 1 or 2.

TABLE 5.35 APA Performance Rated Rim Board Design Capacities[a]

		H (lbf/ft)	V (lbf/ft)		Z (lbf)	P (lbf)
		Depth (d) limitation (in.)				
Grade	t_{min} (in.)	$d \leq 24$	$d \leq 18$	$18 < d \leq 24$	$d \leq 24$	$d \leq 24$
Rim Board	1	180	2750	1650	300	3500
	1⅛	180	4400	3000	350	3500
Rim Board Plus	1	N/A[b]	N/A[b]	N/A[b]	N/A[b]	N/A[b]
	1⅛	200	5700	3500	350	3500

For SI: 1 in. = 25.4 mm, 1 ft = 304.8 mm, 1 lbf = 4.45 kN.

[a] These design values are applicable only to rim board applications in compliance with the connection requirements given in this document and should not be used in the design of a bending member, such as joist, header, r after, or ledger. The design values are applicable to the normal load duration (10 years) for wood products, except for the horizontal load transfer capacity, which is based on the short-term load duration (10 minutes). All values may be adjusted for other load durations in accordance with the applicable code.

where t_{min} = minimum thickness for design capacities listed.
H = horizontal (shear) load transfer capacity.
V = bearing (vertical) load capacity. See in Section 5.12.9, item 1, when used as starter joists.
Z = lateral resistance of a ½ in. (12 mm) diameter lag screw.
P = concentrated load capacity, which must be satisfied simultaneously with the bearing load capacity (V). See Section 5.12.9, item 5 for additional information.

[b] The minimum thickness for APA Rim Board Plus is 1⅛ in. (28.6 mm).

Performance Rated Rim Boards Applied over Opening. APA performance rated rim boards are not generally intended for use as headers. Instead, use glulam, I-joist, or LVL headers.

Caution: Performance Rated Rim Boards should not be designed or used as a bending member, such as a joist, header, rafter, or ledger, and should be installed with full bearing along the entire length. However, they may used to span standard window openings when supporting no more than one story.

Performance Rated Rim Boards Used as Fire-Blocking Panels. The minimum thickness of 1 in. (25 mm) for rim boards exceeds the minimum requirement of 23/32 in. (18 mm) published in the model building codes as long as the joints are backed by another rim board or a 23/32 in. (18 mm) wood structural panel.

Performance Rated Rim Boards Used in Applications where a High Lateral Load Transfer Capacity Is Required. When the applied lateral loads exceed the published horizontal load capacities of APA rim boards, add a commercially available specialty connector made by connector manufacturers between the rim board and framing or sole plate to transfer these loads. This type of connector is installed using face nailing into the rim board and has a typical lateral load capacity of 400 (1.78 kN) to 500 lbf (2.22 kN) per connector. See Section 5.8.5 for further information on shear transfer at engineered wood floors.

Performance Rated Rim Boards Subjected to a Combination of Uniform and Concentrated Vertical Loads. First, the applied concentrated load must not exceed the concentrated load capacity (P) of the rim board. Second, the applied concentrated load must be calculated as an equivalent uniform load based on the applied loading length increased by a 45° load distribution through decking and plate, as applicable. The equivalent uniform load must be added to the applied uniform load to determine the total applied uniform load, which must not exceed the bearing load capacity (V) of the rim board. If the total applied uniform load exceeds the bearing load capacity (V), use appropriate squash blocks, double rim boards, or a higher grade of Performance Rated Rim Board to carry the concentrated vertical load.

Example A mechanical unit distributes a weight of 3000 lbf for a distance of 12 in. along the top of a 1⅛ × 18 in. APA rim board through $^{23}/_{32}$ in. sheathing. In addition to the mechanical unit, the rim board carries a uniform veritcal load of 3000 lbf/ft.

Check:

1. Concentrated vertical load = 3000 lbf < 3500 lbf—OK.
2. Equivalent uniform load = 3,000/[12 + 2 × $^{23}/_{32}$)/12] = 2680 lbf/ft. Total uniform load = 2680 + 3000 = 5680 lbf/ft > 4400 lbf/ft—NG.

Use APA Performance Rated Rim Board Plus that has an allowable bearing (vertical) load capacity of 5700 lbf/ft (83.2 kN/m), or use double rim boards under the concentrated load area.

5.12.10 Rim Board Installation and Connection Details

A rim board is the wood component that fills the space between the sill plate and bottom plate of a wall or, in second-floor construction, between the top plate and second-floor floor sheathing of the two wall sections. The proper installation of the rim board is essential to the overall structural integrity of the structure. Because of its important structural function, the rim board must be properly installed (see Fig. 5.59).

Attachment of Floor Sheathing to APA Performance Rated Rim Board. Use 8d nails (box or common) (2.9 × 64 mm or 3.3 × 64 mm) at 6 in. (150 mm) o.c.

Caution: The horizontal load capacity is not necessarily increased with a decreased nail spacing. Under no circumstances should the nail spacing be less than 3 inches (75 mm).

The 16d (box or common) (3.4 × 89 mm or 4.1 × 89 mm) nails used to connect the bottom plate of a wall to the rim board through the sheathing do not reduce the horizontal load capacity of the rim board provided that the 8d (2.9 × 64 mm or 3.3 × 64 mm) nail spacing (sheathing-rim board) is 6 in. (150 mm) o.c. and the 16d spacing (3.4 × 89 mm or 4.1 × 89 mm) (bottom plate-sheathing-rim board) is in accordance with the prescriptive requirements of the applicable code.

Attachment of APA Performance Rated Rim Board to I-Joist. Use two 8d nails (box or common), one each into the top and bottom flanges. This is typical for rim board having a thickness up to 1⅛ in. (29 mm). A larger nail size may be required by the I-joist manufacturer or for thicker rim board products.

Note: Attempting to nail two pieces of rim board where they abut to a common floor joist can cause damage to the joist. Rim board butt joints should always be made between floor joists.

FIGURE 5.59 Performance Rated Rim Board used as a band joist.

Attachment of APA Performance Rated Rim Board to Sill Plate. Toe nail using 8d (box or common) (2.9 × 64 mm or 3.3 × 64 mm) at 6 in. (150 mm) o.c. Install toenails as shown in Fig. 5.60.

Attachment of APA Performance Rated Rim Board to Floor Framing. Attach rim board to floor framing members in accordance with the details shown in Fig. 5.61.

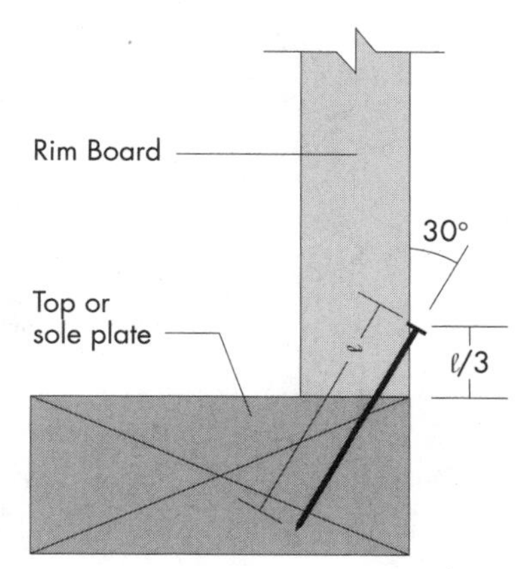

FIGURE 5.60 Toe nail connection at rim board.

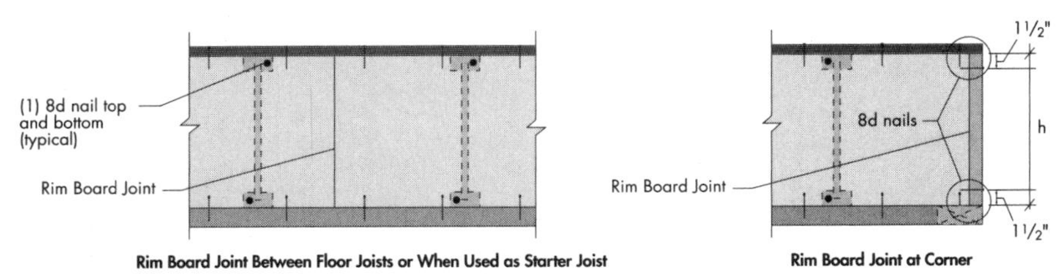

FIGURE 5.61 Rim-board to rim board attachment details.

Attachment of 2× Lumber Ledgers to APA Performance Rated Rim Board. Use ½ in. (12.5 mm) diameter lag screws with a minimum nominal length of 4 in. (100 mm) or ½ in. (12.5 mm)diameter through-bolts with washers and nuts to attach lumber ledgers to the rim board. In both cases, use a vertical and horizontal load design value of 350 lbf (1.56 kN) per fastener if the rim board thickness is 1⅛ in. (29 mm) or 300 lbf (1.33 kN) per fastener if the rim board thickness is 1 in. (25 mm) (see Fig. 5.62).

Caution: The lag screw should be inserted in a lead hole by turning with a wrench, not by driving with a hammer. Overtorquing can significantly reduce the lateral resistance of the lag screw and should therefore be avoided. See the 1997 National Design Specification for Wood Construction (NDS), published by the

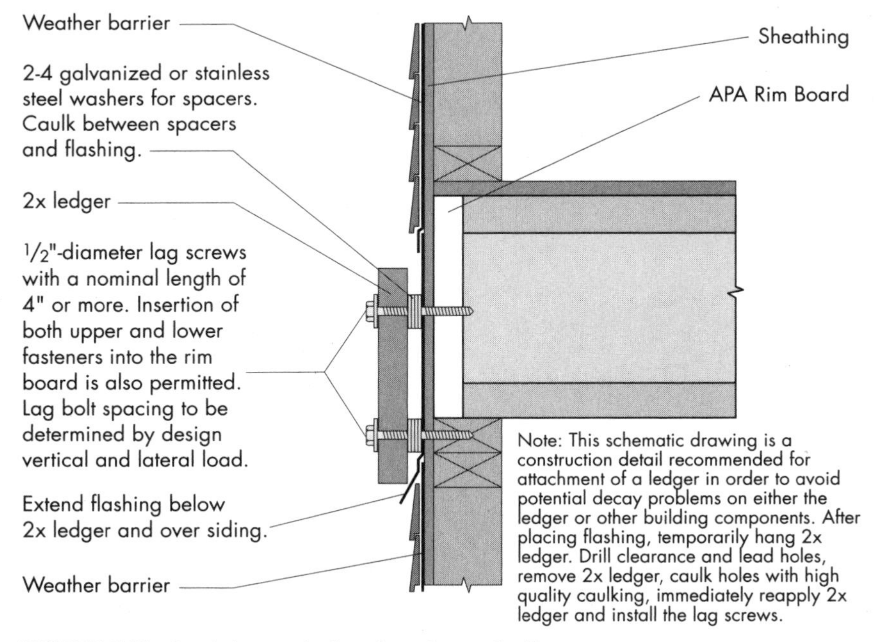

FIGURE 5.62 2 × ledger to rim board attachment detail.

American Forest and Paper Association, for the appropriate size of clearance and lead holes.

Lateral Resistance of Nails Applied to Faces of APA Performance Rated Rim Board. Calculate the lateral nail resistance based on the procedures given in the 1997 NDS and the following guidelines:

- If the APA Performance Rated Rim Board is made of OSB, use the dowel bearing strength equivalent to Douglas fir-larch.
- If the APA Performance Rated Rim Board is made of plywood, use the dowel bearing strength equivalent to Douglas fir-larch. The calculated lateral resistance should then be adjusted by a factor of 0.9.
- If the APA Performance Rated Rim Board is made of glulam, use the wood species of the lay-up combination to determine the dowel bearing strength.
- If the APA Performance Rated Rim Board is made of LVL, use the wood species of the veneer or use the equivalent species published in the manufacturer's code report to determine the dowel bearing strength.
- If the product information is unavailable, refer to the appropriate 1997 NDS dowel bearing design values applicable to spruce-pine-fir.

Example A design professional is assigned the task of designing a nailed connection using rim board. In this connection, the main member—the last member to receive the tip of the nail—is to be a Performance Rated Rim Board. As the nailed connection calculation is to be based on the European Yield Method as found in the 1997 National Design Specification (NDS), it will first be necessary for the designer to determine the dowel bearing strength of the rim board. In accordance with this section and Table 12A of the 1997 NDS, the following is determined:

Rim board material	Dowel bearing strength (lbf-in.^{2})
OSB	4650
Plywood	4650 (use 0.9 of final lateral resistance calculated)
Douglas fir-larch glulam	4650
Southern pine glulam	5550
Douglas fir-larch LVL	4650 (Check manufacturer's code report for "equivalent" species)
Southern pine LVL	5550 (Check manufacturer's code report for "equivalent" species)
Information unavailable—use spruce-pine-fir	3350

For SI: 1 lbf-in.^{2} = 0.00287 Nm^2

If the designer is unsure of what the makeup of the rim board will be, it would be conservative to base the calculations on spruce-pine-fir—3350 psi (9.61 Nm^2). Using these dowel bearing strengths, the connection can be designed for any other nailed joint.

5.13 STORAGE, HANDLING, AND SAFETY RECOMMENDATIONS FOR ENGINEERED WOOD PRODUCTS

The superior performance of Performance Rated I-joists, rim board, and other engineered wood products is at least partially due to the fact that they are delivered to the job site in a dry state. This minimizes the possibility of shrinkage, warping, cutting, crowning, and all the other realities that come with building with sawn lumber.

To ensure optimum performance and safe handling, rim board and I-joists must be stored and installed properly. The following guidelines will help protect joists from damage in storage, during shipment, and on the construction site and protect the installer from job-site injury.

5.13.1 Storage and Handling Guidelines

1. Store, stack, and handle I-joists and rim board vertically and level only.
2. Do not store I-joists or rim board in direct contact with the ground and/or flatwise.
3. Protect engineered wood from weather, and use stickers to separate bundles.
4. To protect engineered wood further from dirt and weather, do not open bundles until time of installation.
5. Take care not to damage I-joists or rim board with forklifts or cranes.
6. Do not twist or apply loads to the I-joist or rim board when horizontal.
7. Never use or try to repair a damaged engineered wood products.
8. When handling I-joists with a crane on the job site, take a few simple precautions to prevent damage to the I-joists and injury to the work crew.
   - Pick I-joists in bundles as shipped by the supplier (see Fig. 5.63).
   - Orient the bundles so that the webs of the I-joists are vertical (see Fig. 5.64).
   - Pick the bundles at the 5th points, using a spreader bar if necessary (see Fig. 5.65).

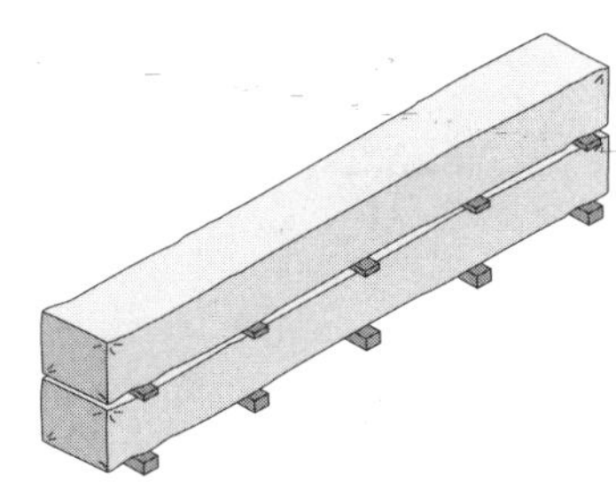

FIGURE 5.63 Proper stacking of bundles.

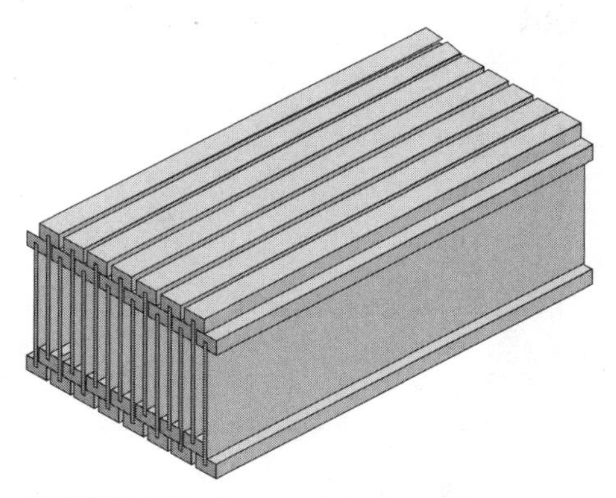

FIGURE 5.64 Proper orientation of bundles.

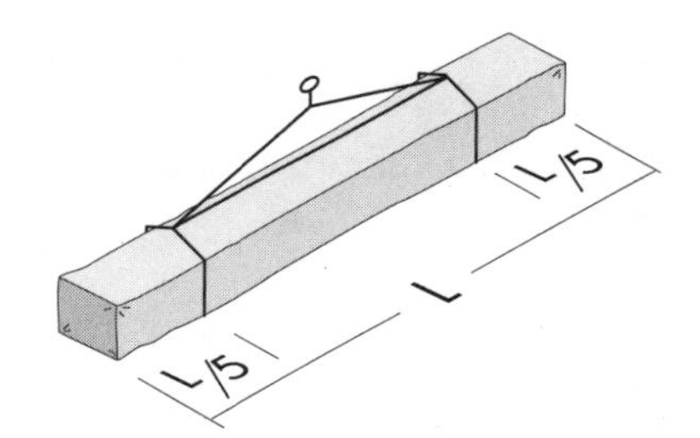

FIGURE 5.65 Proper lifting procedure for bundles.

5.13.2 Safety Tips (see Fig. 5.66)

1. I-joists are not stable until completely installed, and will not carry any load until fully braced and sheathed.
2. Do not allow workers to walk on I-joists until joists are fully installed and braced, or serious injuries can result.
3. Never stack building materials over unsheathed I-joists. Stack only over beams or walls.

5.13.3 Avoid Accidents During Construction by Following These Important Guidelines

1. Brace and nail each I-joist as it is installed, using hangers, blocking panels, rim board, and/or cross-bridging at joist ends. When I-joists are applied continuous over interior supports and a load-bearing wall is planned at that location, blocking will be required at the interior support.
2. When the building is completed, the floor sheathing will provide lateral support for the top flanges of the I-joists. Until this sheathing is applied, temporary

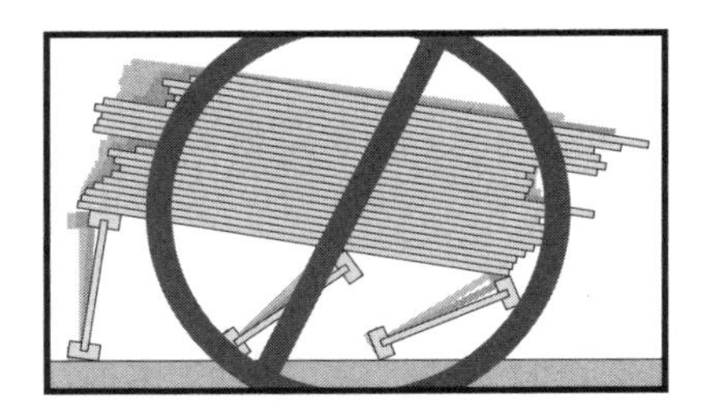

FIGURE 5.66 I-joist safety during construction.

bracing, often called struts or strong backs, or temporary sheathing must be applied to prevent I-joist rollover or buckling.

- Temporary bracing or struts must be 1 × 4 in. (19 × 89 mm) minimum, at least 8 ft (2.4 m) long, and spaced no more than 8 ft (2.4 m) on center, and must be secured with a minimum of two 8d (3.3 × 64 or 2.9 × 64 mm) nails fastened to the top surface of each I-joist. Nail bracing to a lateral restraint at the end of each bay. Lap ends of adjoining bracing over at least two I-joists.
- Alternatively, sheathing (temporary or permanent) can be nailed to the top flange of the first 4 ft (1.2 m) of I-joists at the end of the bay.

3. For cantilevered I-joists, brace top and bottom flanges and brace ends with closure panels, rim board, or cross-bridging.
4. Install and nail permanent sheathing to each I-joist before placing loads on the floor system. Then stack building materials over beams or walls only.
5. Never install a damaged I-joist.

Failure to follow applicable model building codes and Span Ratings, use allowable hole sizes and locations, or use web stiffeners when required can result in serious accidents.

5.14 GLOSSARY

Backer block: Blocks made from lumber or wood structural panels that are attached to the web of an I-joist to develop the design capacities of the fasteners used in the web.

Band joist: See *Rim board.*

Birds-mouth cut: A notch cut out of the flange at the bottom end of a sloped I-joist rafter to provide a parallel mating surface at the point of bearing. The cut surface should never extend out beyond the inside face of the bearing surface.

Blocking panel: Short-length, full-depth sections of wood I-joists or *APA EWS* Rim Board that are cut to fit between floor joists. Blocking panels prevent the floor joists from overturning and distribute vertical and lateral loads through the floor system into the structure below.

Center chase opening: A large rectangular hole that is placed within the web of an I-joist and is used to provide passage for ventilation ducts through the floor joists. Center chase openings sized in accordance with Table 5.13 are always located centered on the center of the I-joist span.

Cold roof: A cold roof is a roof with a $C_t > 1.0$ and shown in Table 7-3 of ASCE 7-98. For purposes of calculation in this publication, a structure with a cold, ventilated roof in which the thermal resistance factor (R-value) between the ventilated space and the heated space exceeds 25 F°*hr*sq ft/Btu (4.4 K*m^2/W) was assumed.

Cripple block: See Squash block.

Duct chase: See Center chase opening.

Duration of load (*DOL*): An adjustment to some allowable stress values for wood-based products based on the cumulative time the member is exposed to the design load. For loads of short duration such as snow, seismic or wind, an increase in certain allowable stresses is permitted. For loads of permanent duration—10 years or longer—certain allowable stresses must be reduced.

Filler block: Filler material made of lumber and/or wood structural panels placed between the webs of adjacent I-joists. The webs of the adjacent I-joists are nailed to the filler blocks to provide load sharing between the joists. Filler blocks are used the full length of the I-joist but they do not have to be made up of continuous members.

Flange: The top and bottom pieces of a wood I-joist, which are comprised of solid-sawn lumber or structural composite lumber.

Flat root: A roof with a 0:12 slope. Zero slope roofs are not recommended due to the potential for ponding.

"H" clips: A deformed or extruded metal clip with the cross-sectional shape of an upper case "H." When placed between adjacent panel edges they provide nominal edge support and assist in maintaining the recommended panel-edge spacing.

Header: A horizontal load-carrying member that spans between load-carrying members running in a direction perpendicular to the header. Where the header interrupts the span of these load-carrying members it is designed to support the ends of the members and distribute the load they are designed to carry into an alternative load path.

Load-bearing wall: A wall that transfers any load, vertical or lateral, in addition to its own weight into the structure below.

Low slope roof: A roof with a pitch of less than 4:12.

Multiple span: A situation where a horizontal load-carrying element is placed continuously over three or more supports.

Near-flat roofs: Roofs with a roof slope between ¼:12 to 1:12.

Performance Rated I-joist: A Performance Rated I-joist (PRI) is an I-joist that is developed primarily for residential markets and manufactured in accordance with the Performance Standard for APA EWS I-joists, PRI-400. PRI's can be identified by a table of allowable spans printed right on the I-joist in the APA trademark stamp. I-joists with the same depth and designation can be substituted for each other regardless of the manufacturer or how fabricated.

Ponding: Ponding is a phenomenon that may occur when water collects in a depression on a flat roof. As water builds up in the depression, the load on the supporting members increases. This causes additional deflection, which causes additional deflection, which permits more water to accumulate. Once initiated, this cycle can continue until a roof failure occurs. Ponding is prevented by building roofs with a slight slope (a *minimum* slope of ¼:12) to allow for drainage of water from the roof.

Ridge board: A ridge board is a horizontal member of a roof frame that is placed on edge at the ridge and into which the upper ends of the rafters are fastened.

Ridge beam: A ridge beam is a horizontal load-bearing member of a roof frame that supports the upper ends of the roof rafters. A ridge beam differs from a ridge board in that a ridge beam is a *structural* bending member and, as such, must be supported by vertical load-carrying members. These vertical load-carrying members must provide a continuous load path from the beam all the way down to the foundation.

Rim board: A glued engineered wood product that is attached continuously to the ends of the floor joists, preventing overturning of the joist. Rim boards also complete the "weathering-in" of the floor system and transfer both vertical and lateral loads.

Simple span: A situation where a horizontal load-carrying element spans between two supports.

Slope length: The true length of a non-horizontal member measured from centerline of bearing to centerline of bearing. The length used for designing pitched roof rafters.

Span: The horizontal distance between supports. In this chapter, I-joist span is measured from inside face to inside face of supports.

Squash block: Sawn lumber members that are nailed to each side of an I-joist at the location of concentrated loads up to 4000 lbf (17.8 kN) (per pair of squash blocks). The squach blocks are cut ¹⁄₁₆″ (1.5 mm) longer than the I-joist depth to insure that the load is transferred by the squash blocks around the I-joist to the support directly below. Rim Board product may be used for squash blocks providing the capacity per pair of blocks is reduced in proportion to the squash block thickness. Select a Rim Board product equal to the depth of the I-joist. Oversizing is not necessary.

Structural composite lumber (SCL): Lumber manufactured by parallel orientation of wood fibers of various geometries, bonded with a structural adhesive. Most commonly, this takes the form of laminated veneer lumber (LVL).

Tail joists: The short joists formed when a header interrupts the span of a floor or roof joist. These joists run parallel to the uncut joists from the header to the normal bearing location.

Trimmers: Additional load-carrying members placed adjacent to existing members to assist in carrying the load transferred by an intersecting header. When I-joists, load is shared by the adjacent members through the use of filler blocks.

Web: The vertical element of a wood I-joist that is made of plywood or OSB. The web joins the I-joist's top and bottom flanges and transmits shear forces.

Web stiffener: Wood structural panel or sawn lumber members placed between the flanges of I-joists in locations where increased reaction capacity (bearing stiffener) or concentrated load-carrying capacity (load stiffener) of the I-joist is required. Web stiffeners are cut 1/8″ (3 mm) less than the dimension between the flanges and placed tightly against the flange resisting the applied load. (Generally, against the bottom flange for bearing stiffeners and the top for load stiffeners.)

5.15 REFERENCES

1. Design Specification No. BB-8, *Design of Plywood Beams,* 1959, available through APA—The Engineered Wood Association, Tacoma, WA.
2. Plywood Design Specification, Supplement 2, *Design and Fabrication of Glued Plywood-Lumber Beams,* APA—The Engineered Wood Association, Tacoma, WA, 1998.
3. ASTM D5055, Standard Specification for Establishing and Monitoring Structural Capacities of Pre-Fabricated Wood I-Joists, ASTM, West Conshohocken, PA, 1998.
4. ICBO ES Acceptance Criteria for Prefabricated Wood I-Joists, AC 14, ICBO Evaluation Services Inc., Whitten, CA.
5. PRI-400, *Performance Standard for APA EWS I-Joists,* APA—The Engineered Wood Association, Tacoma, WA, 1999.
6. *Regional Production and Market Outlook for Structural Panels and Engineered Wood Products,* 1999–2003, APA—The Engineered Wood Association, Tacoma, WA, 1999.
7. *I-Joist and Rim Board Market Update,* APA—The Engineered Wood Association, Tacoma, WA, December 1999.
8. ASTM D2559, *Standard Specification for Adhesives for Structural Laminated Wood Products for Use under Exterior (Wet Use) Exposure Conditions,* ASTM, West Conshohocken, PA, 1999.
9. National Institute of Standards and Technology, Voluntary Product Standard PS1-95, *Construction and Industrial Plywood,* available through APA—The Engineered Wood Association, Tacoma, WA.
10. National Institute of Standards and Technology, Voluntary Product Standard PS2-92, *Performance Standard for Wood-Based Structural-Use Panels,* available through APA—The Engineered Wood Association, Tacoma, WA.
11. AFG-01, *Adhesives for Field-Gluing Plywood to Wood Framing,* 1984, available through APA—The Engineered Wood Association, Tacoma, WA.
12. ASTM D3498, *Standard Specification for Adhesives for Field-Gluing Plywood to Lumber Framing for Floor Systems,* ASTM, West Conshohocken, PA, 1999.
13. PRP-108, *Performance Standards and Policies for Structural-Use Panels,* APA—The Engineered Wood Association, Tacoma, WA, 1994.
14. PRR-401, *Performance Standard for APA EWS Rim Board,* APA—The Engineered Wood Association, Tacoma, WA.
15. ANSI A190.1, *American National Standard for Wood Products—Structural Glued Laminated Beams,* American National Standards Institute, Inc., New York, 1992.
16. *ICBO ES Acceptance Criteria for Wood-Based Rim Board—AC 124,* ICBO Evaluation Services, Inc., Whittier, CA.

CHAPTER SIX
STRUCTURAL COMPOSITE LUMBER

Zhaozhen Bao, PhD
Associate Scientist, TSD

6.1 INTRODUCTION

Structural composite lumber (SCL) is a generic engineered wood structural lumber product family that includes laminated veneer lumber (LVL), parallel strand lumber (PSL), laminated strand lumber (LSL), oriented strand lumber (OSL), and other wood composite lumber products with similar engineering and configuration features. LVL was first produced in the early 1970s and since then has been commercially available in the United States. PSL was introduced to market in the 1980s. SCL products are used in the same structural applications as sawn lumber and timber.

6.1.1 Laminated Veneer Lumber

Laminated veneer lumber (LVL) was the earliest type of SCL product commercially manufactured for the marketplace. It is now the most widely used structural composite lumber product in the residential housing market. LVL is produced by bonding layers of wood veneers in a large billet under proper temperature and pressure. Typically, LVL is produced in 4 ft wide billets with different lengths, and the billet is then sawn to different dimensions to meet the needs of the final applications. There are no limitations on wood species for LVL. Basically, any species that are used in manufacturing plywood can be used to produce LVL. The most common species for LVL are Douglas fir, southern pine, and SPF (spruce-pine-fir). Generally, LVL is used as headers and beams, chords for trusses, ridge beams in mobile homes, flanges in prefabricated I-joists, and scaffold planks. It is also used as columns, shear wall framing studs, and even structural members in upholstered furniture frames.

Description and General Features. LVL is a glued engineered wood composite lumber product manufactured by laminating wood veneers using exterior type adhesives (Fig. 6.1). Generally, the grain directions of wood veneers in LVL are all

FIGURE 6.1 Laminated veneer lumber of different depths; picture (photo) showing a shot of LVL.

parallel to the length direction of the billet, although LVL products with cross lamination are sometimes seen for meeting specific structural needs. Similar to the lay-up strategy in glued-laminated timber (glulam), veneers with higher grades are placed on the faces while lower-grade veneers are in the core. This specific lay-up configuration of the wood veneers effectively utilizes materials' strength and improves the strength and stiffness of the manufactured products in a desired way. Natural defects of wood such as knots, knotholes, and splits can be closely controlled, and their individual effects are virtually eliminated in LVL.

Desired width, depth, and length can be technically achieved by various manufacturing techniques, including side jointing and end jointing (butt, scarf, and finger jointing). Since the grade and quality of each individual layer of veneers can be closely controlled, the variations in product properties are lowered as compared to sawn lumber products, and therefore the properties and performance of final LVL products can be more confidently predicted than sawn lumber. LVL has improved mechanical properties and dimensional stability. It can offer a broader range in product width, depth, and length than lumber. Various wood species, even those considered low-grade or previously underutilized species, can be used to manufacture LVL.

Briefs on Manufacturing Process

Veneer. Veneers for manufacturing LVL are made in the same way as veneers for plywood manufacturing. The commonly used veneer thicknesses are $\frac{1}{10}$–$\frac{1}{8}$ in. Typically veneer thicknesses in LVL manufacturing do not exceed $\frac{1}{4}$ in. Similar to plywood manufacturing, fresh-peeled veneers are mechanically dried. Veneers are then trimmed, repaired, and sorted according to the number and sizes of defects such as knots, knotholes, and splits. In order to ensure the desired engineering properties of the finished LVL product, individual veneer is passed though a veneer grade tester for measuring moisture content, density, and E values. Stress wave propagation time, also called ultrasonic propagation time (UPT), is also used to sort veneers. UPT outputs correspond to veneer grades in accordance to plant manuals. According to the output UPT values, veneers are graded and sorted for LVL lay-ups. Other alternative grading systems may also be used at the manufacturers option.

Lay-up. Scarf and finger jointing techniques are used to end-joint pieces of veneer sheets to a desired length. End joints between layers are staggered along the length to minimize strength-reducing effects end jointing may have. In some manufacturing processes, the ends of veneer sheets are overlapped in lieu of end-jointing. Exterior-type adhesives, such as phenol-formaldehyde (PF) resorcinol base adhesives, are used in bonding the veneers together. Adhesives are spread onto the veneer surfaces and mats are laid according to the predetermined lay-up pattern. The mat sheets are then ready to be pressed in a hot press.

Pressing. The LVL mat sheets are sent into either a stationary or staging hot press or a continuous hot press. During hot pressing, the thermoset type adhesive is cured under heat and pressure, permanently bonding the plies of veneers together to form billets. The LVL billets are then ripped to specific widths (depths) and cut to given lengths. Figure 6.2 shows a typical LVL manufacturing operation.

6.1.2 Parallel Strand Lumber

Parallel strand lumber (PSL) is manufactured by glue-bonding wood strands to form a condensed billet in such a way that the wood fiber (grain) direction of the strands is primarily oriented parallel to the length of the member. The thickness (least dimension) of the strands usually is less than 0.25 in., and the average length of strands is about 150 times the thickness of the strands. One source of strands used in PSL can be clipped veneers from the process wastes in plywood or LVL plants, or full-size veneer sheets such as used in plywood and LVL manufacturing can be used. Presently, PSL is made primarily from Douglas fir, western hemlock, southern pine, and yellow poplar, although there are no restrictions on species. PSL is commonly used in structures as headers and beams, as well as columns and studs.

Description and General Features. PSL has greater strength properties than sawn lumber. Its strength is enhanced by increasing the amount of densification of the pressed billet rather than obtained through the optimized use of veneer species and grades in lay-up processing, as for LVL. Usually, a PSL billet is made in a cross section of 11×19 inches. Its larger cross section allows the use of PSL as a direct substitution for structural timber products without secondary gluing. Due to its high degree of strand alignments and increased densification, PSL possesses excellent strength and stiffness in the primary axis. On the minus side, however, PSL is heavier than the same-sized sawn or glue-laminated timber (glulam). Also, its adhesive is more abrasive to saws and drills.

Briefs on Manufacturing Process. The dimensions of veneers (strands) used in PSL are about ⅛ in. thick by ¾ in. wide by 24 in. long. The strands are coated with exterior type adhesive, commonly phenol-resorcinol formaldehyde resin. The strands are all oriented parallel to the length of the billet. Heat used for curing the adhesive is generated by microwave in the hot press. This makes it possible to cure the adhesive from the inside out. The cross section of PSL billet, therefore, can be

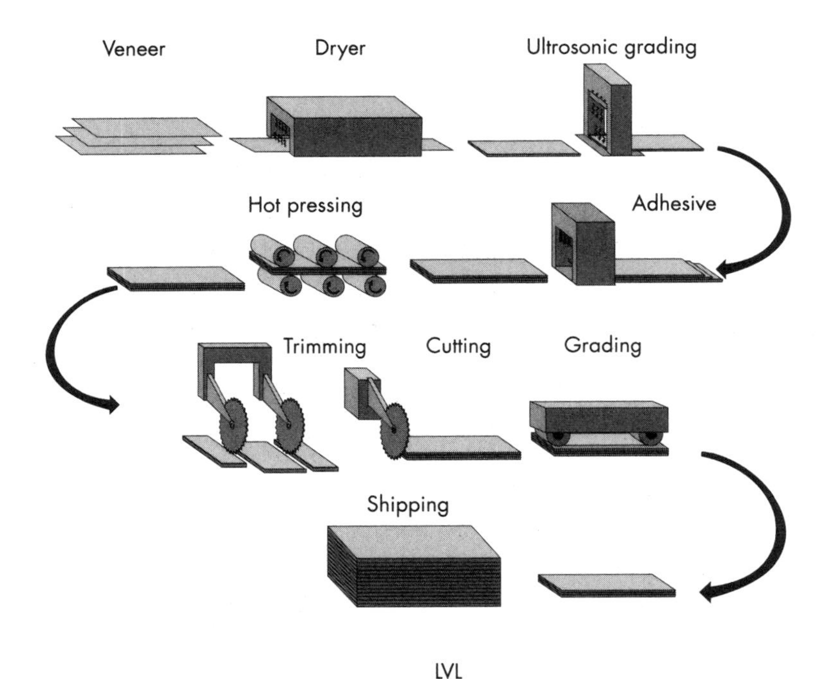

FIGURE 6.2 Typical manufacturing process of LVL using ultrasonic grading.

made up to 11 × 19 in., which is greater than that for LVL, where heat is typically transferred from the outside to the inside of the billet during the hot pressing. The continuous pressing operation allows a higher degree of densification to be achieved. Although there is no length limitation in the continuous pressing, typical lengths of PSL billets are up to 60 ft and are actually limited by handling restrictions. Billets are then resized to desired dimensions. If needed, larger cross-section dimensions can be achieved by secondary operations.

6.1.3 Laminated Strand Lumber and Oriented Strand Lumber

Laminated strand lumber (LSL) is an extension of the technology used to manufacture oriented strand board (OSB). Wood strands used in LSL are about 12 in. long, which is longer than the strands normally used in OSB (which are about 3–4 in. long). Among those SCL products, LSL perhaps is the most efficient in utilizing wood resources. It has no restrictions on raw materials. Small logs and crooked logs of many species, including aspen, yellow poplar, and other underutilized, fast-growing species, can be used in manufacturing LSL.[1] A higher degree of strand orientation in LSL is required than in OSB, and greater pressing pressures are needed in order to obtain increased densification.

Oriented strand lumber (OSL) is another type of laminated strand lumber (LSL) product and has a similar manufacturing process to that for LSL. The primary difference between them is that the length of strand used in OSL is shorter (up to

6 in.) than that used in LSL (approximately 12 in.). OSL has somewhat lower strength and stiffness values than LSL.

Description and General Features. Generally LSL has somewhat lower strength and stiffness properties than LVL and PSL. The lay-up operation in LSL mat forming processing resembles OSB mat forming, improving transverse strength and limiting cupping potential. Waterproof adhesives are used in LSL manufacturing. The unique steam injection pressing process achieves curing of the sprayed adhesive on strands and densification of the LSL mat, thus resulting in enhanced strength properties of the final product. LSL demonstrates excellent fastener-holding strength and mechanical connector performance. LSL in general is less dimensionally stable than LVL and PSL due to its greater densification. Since more wood substances are compacted to form the relatively dense LSL, it exhibits more thickness swell as compared to LVL and PSL as its moisture content changes.

Briefs on Manufacturing Process. Wood strands are first produced by a strander, and then the strands are screened out to eliminate unwanted sizes. The green strands are driven through a drum-type dryer. Exterior-type adhesives, wax, and other additives are blended in a blender and the liquid mixture is sprayed onto the strand surfaces as they tumble through the inside of a rotating drum. Strands are oriented through a forming machine with multiple forming heads. The mats are sent into a hot press. Curing of the resins and densification of the mat are achieved under proper temperature and pressures during hot pressing. Billets are then trimmed to the required sizes. With this technology, billets 8 ft wide and up to 5 in. thick and 48 ft long can be poduced. Billets are finally cut and ripped to desired dimensions for different end uses.

Figure 6.3a and b shows a typical manufacturing process of PSL and LSL respectively.

6.2 GROWTH OF INDUSTRY

It was reported in the 1940s that an LVL-type product was developed for high strength parts for aircraft structures using Sitka spruce veneer.[2] This was probably the earliest example of LVL. In the late 1960s, the U.S. and Canadian governments searched for methods to increase the raw material efficiency of lumber manufacturing since the sawn lumber industry conventionally renders half or more of logs into sawdust, wood slabs, and other types of residues. Technology of parallel lamination of rotary-peeled veneers was developed and the so-produced laminated veneer lumber product was termed LVL. LVL has improved strength properties, allows more economical and efficient material utilization, and utilizes affordable phenol-formaldehyde resins, and all of these factors combined to encourage manufacturers to enter into this industry.

Laminated veneer lumber (LVL) was introduced to the building industry by Trus Joist Corporation in 1970[3] and has been available since 1971.[4] Parallel strand lumber (PSL) was developed and patented by MacMillan Bloedel Ltd. around 1978 and became commercially available in the early 1990s.[3] Structural composite lumber is a growing segment of the engineered wood industry and is used extensively as a replacement for sawn lumber products. More production lines of LVL have been added to meet the increasing demand on LVL as headers, beams, and partic-

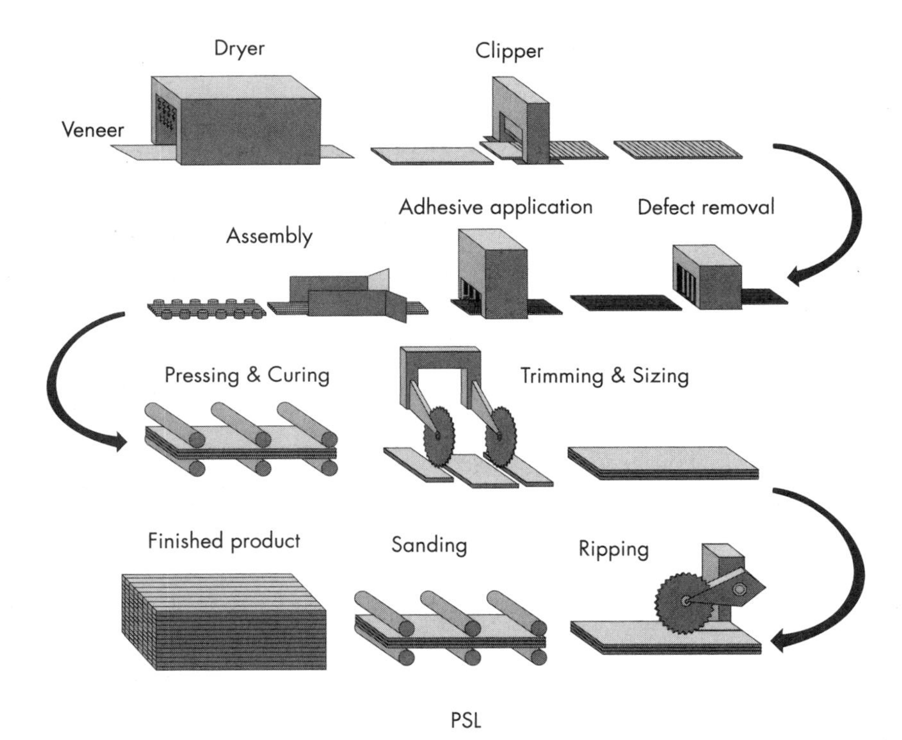

FIGURE 6.3a PSL manufacturing process.

ularly the flange materials of I-joists. According to APA—The Engineered Wood Association, the LVL production estimates (United States and Canada) for the year of 2003 will be 84.0×10^6 ft^3, as compared to a total production of 31.5×10^6 ft^3 in 1996. Table 6.1 lists the LVL production estimates for United States and Canada from 1996–2003. One of the newest members of the SCL family is LSL which was introduced into the marketplace by the Trus Joist Corporation in the 1990s.

6.3 STANDARDS

6.3.1 Standards

ASTM D5456, *Standard Specification for Evaluation of Structural Composite Lumber Products,*[5] provides guidelines for the evaluation of mechanical properties, physical properties, and quality of structural composite lumber products. Since ASTM D5456 is not a product standard for the SCL industry, individual manufacturers of SCL generally have their own proprietary manufacturing product standards that govern the everyday production practice for their products. The common grades and their design stresses for SCL, particularly for LVL, are dictated by the market, and the major LVL products have similar or comparable design values.

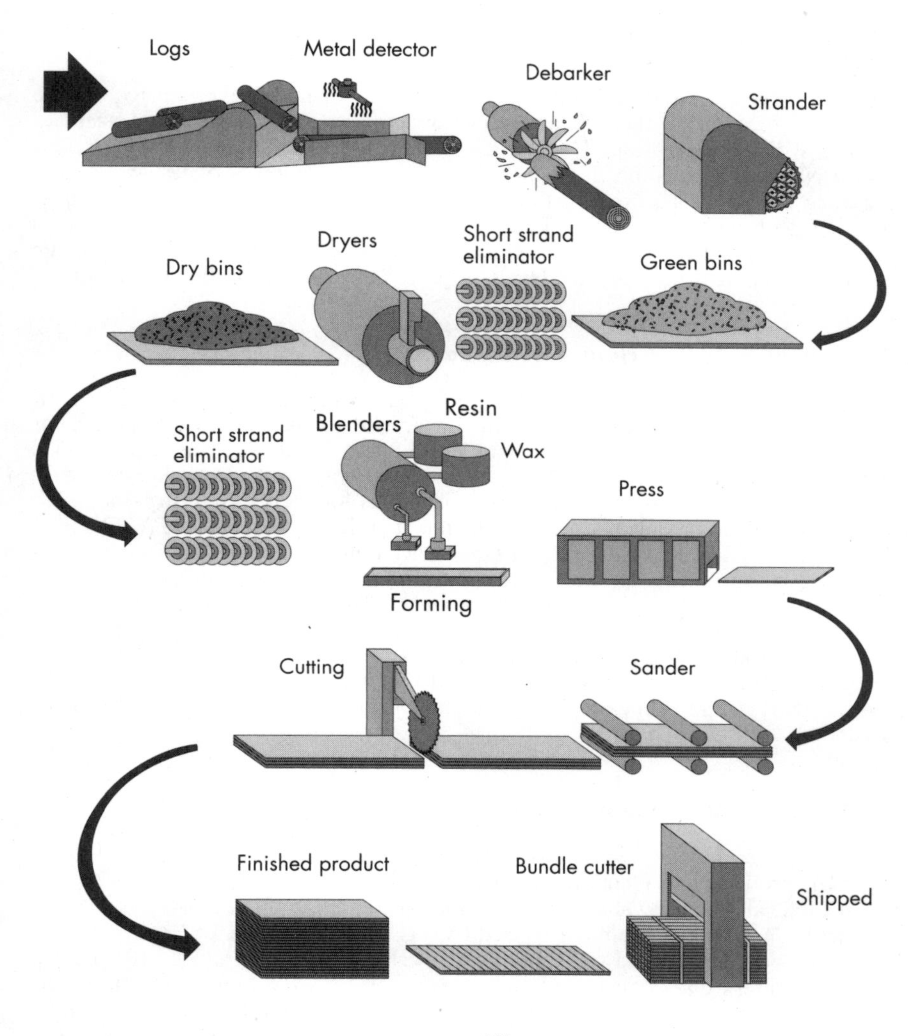

FIGURE 6.3b LSL manufacturing process.

TABLE 6.1 LVL Production Estimates—United States and Canada

	LVL production estimates—U.S. and Canada (million cubic feet)							
	1996	1997	1998	1999	2000	2001	2002	2003
United States			41.0	46.2	50.5	61.5	68.3	75.1
Canada[a]			1.7	3.8	7.5	8.5	8.7	8.9
Total North America	31.5[a]	37.7[a]	42.7	50.0	58.0	70.0	77.0	84.0

[a]1996–1997 only combined statistics available.

6.3.2 Code Recognition

Under the current building code jurisdictions, SCL manufacturers are required to gain code recognition through an evaluation process provided by code agencies such as ICBO (International Conference of Building Officials) Evaluation Service or the National Evaluation Service (NES). ICBO Evaluation Services AC47,[6] *Acceptance Criteria for Structural Composite Lumber,* provides guidelines on implementing performance features of the Uniform Building Code (UBC). Although ASTM D5456 and ICBO AC47 provide guidance for developing proprietary design values, no standard performance levels or grades are defined in the documents.

6.3.3 Industry LVL Performance Standard

Several LVL manufacturers felt that a performance standard covering major LVL products was needed to provide a common ground for performance requirements and acceptance criteria of existing products. Such a standard was also deemed important to provide an industry-wide uniform quality control and quality assurance system for the products. In order to meet the needs for such an industry-wide LVL standard, APA and its members have developed an LVL standard—APA PRL-501, *Performance Standard for APA EWS Laminated Veneer Lumber.*[7] The standard addresses grading, materials, tolerances, performance criteria, quality assurance requirements, and trademarking for LVL products. This standard has been approved by each of the three model code agencies and is intended to serve as an industry-wide LVL product standard and provide a direct avenue for the code approval and recognition for LVL products manufactured under this standard. The adoption of this standard is also intended to simplify the design and specification of LVL products.

6.3.4 Grades and Grading System

As with the machine stress-rated (MSR) grading system used by the lumber industry, SCL products are identified according to their assigned stress classes. Stress classes indicate the allowable designated modulus of elasticity, E, and design bending stress, F_b. The most commonly used LVL grades range from 1.5E (an allowable design $E = 1.5 \times 10^6$ psi) to 2.1E (an allowable design $E = 2.1 \times 10^6$ psi). In general, a stress class consists of two parts separated by a hyphen ("-"). To the left of the hyphen is the designated E value while to the right is the design bending stress (F_b), both of which are in the units of pounds per square inch (psi). For instance, a grade of 1.5E-2250F signifies LVL with an allowable E of 1.5×10^6 psi and F_b of 2250 psi. Table 6.2 is excerpted from APA PRL-501. It lists the major design properties for APA EWS performance rated LVL. Other grades of LVL products can be manufactured as special orders to meet special needs.

6.4 PHYSICAL PROPERTIES

6.4.1 Specific Gravity and Moisture Content

Specific gravity of SCL products is an important indicator of various physical and mechanical properties. As a wood-based laminated layered material, SCL's density

TABLE 6.2 Design Properties for APA EWS Performance Rated LVL[a]

APA EWS LVL stress classes	E^b 10^6 psi	$F_b{}^c$ psi	$F_t{}^d$ psi	$F_{c\parallel}{}^e$ psi	$F_v{}^f$ Edgewise psi	$F_{c\perp}{}^g$ Edgewise psi
1.5E-2250F	1.5	2250	1500	1950	220	575
1.8E-2600F	1.8	2600	1700	2400	285	700
1.9E-2600F	1.9	2600	1700	2550	285	700
2.0E-2900F	2.0	2900	1900	2750	285	750
2.1E-3100F	2.1	3100	2200	3000	285	850

[a] The tabulated values are design values for normal duration of load. All values, except E and $F_{c\perp}$, are permitted to be adjusted for other load durations as permitted by the code. The design stresses are limited to conditions in which the maximum moisture content is less than 16 percent.

[b] Bending modulus of elasticity (E), which is applicable in either edgewise or flatwise applications, includes shear deflections. For calculating uniform load and center-point load deflections of the LVL in a simple-span application, use Eqs. (1) and (2).

$$\text{Uniform load:} \qquad \delta = \frac{5\omega\ell^4}{384EI} \tag{1}$$

$$\text{Center-point load:} \qquad \delta = \frac{P\ell^3}{48EI} \tag{2}$$

where δ = calculated deflection (in.)
ω = uniform load (lbf/in.)
P = concentrated load (lbf)
ℓ = design span (in.)
I = moment of inertia of the LVL (in.^{4})

[c] Allowable bending stress (F_b) is applicable in either edgewise or flatwise applications. For depths of 3½ in. or deeper when loaded edgewise, the tabulated F_b value shall be modified by $(12/d)^{1/8}$, where d is the actual depth (in.). For depths less than 3½ in. when loaded edgewise, use the adjusted F_b for 3½ in. No adjustment on F_b is required for flatwise applications.

[d] Axial tension (F_t) of the LVL is based on a gage length of 4 ft. For specimens longer than 4 ft, the tabulated F_t value shall be adjusted by $(4/L)^{1/8}$, where L is the actual length (ft).

[e] Compression parallel to grain ($F_{c\parallel}$) of the LVL.

[f] Allowable shear stress (F_v) of the LVL when loaded edgewise.

[g] Allowable compressive stress perpendicular to grain ($F_{c\perp}$) of the LVL when loaded edgewise.

is affected greatly by the wood species from which it is made. However, due to the variables of species being used, the significant amounts of adhesives being added, and the densification achieved under the heat and pressure during hot pressing, the density of SCL products may be different from that of the original wood species. Since virtually all kinds of available wood species can be used to manufacture SCL products, their density falls in a wide range of 20–42 lb/ft³ (approximately 0.33–0.68 specific gravity, respectively). The equivalent specific gravity of an SCL product must be established during the development of mechanical properties because this will also affect fastener performance (see below under Equivalent Specific Gravity).

The moisture content (MC) of SCL at time of production will be around 8–10%. Moisture content of SCL starts changing as the conditions of the ambient air change. As the changes in moisture content in wood change the weight and the volume of the member, they can change the specific gravity of the member. In general, the change is not noticeable since the moisture changes of SCL are generally small under a protected environment.

6.4.2 Dimensional Stability

Wood is a hygroscopic material, as are SCL products, since they inevitably inherit this feature from the original wood. However, the dimensional changes in SCL are usually less than the original wood from which it is produced. This is attributed to the drying of individual veneers/strands and the subsequent hot pressing and drying, plus the addition of the exterior-type adhesives.

Since SCL products are recommended for use in a protected dry end-use condition where moisture content is below 16%, little change in moisture content will occur in such protected service conditions. The linear expansion and shrinkage of SCL under the typical in-service moisture changes will be minimal and generally not noticeable.

6.4.3 Durability

Like any wood-based product, SCL products may be subjected to wood degradation or deterioration during use associated with attacks of fungi and insect, temperature and UV, and, most commonly, water. However, since SCL products are recommended for dry-use conditions they are relatively unaffected by those attacks.

Also, it is generally recognized that SCL products have better durability than untreated wood members due to the relatively thorough drying process of wood elements, addition of the exterior-type adhesives, and hot pressing. Thus, SCL products can be expected to have a prolonged service life without significant or noticeable degradation in strength when properly installed and maintained.

6.5 MECHANICAL PROPERTIES

SCL products are primarily loaded in either of the two major directions: joist (edgewise) or plank (flatwise) orientation. Load applied to members can be from any of the three directions: (1) parallel to grain, (2) perpendicular to grain and parallel to glue-plane, or (3) perpendicular to grain and perpendicular to glue-plane. In order to define the three axes that determine the dimensions of SCL, a drawing is provided in Fig. 6.4. The three axes will be referenced throughout the following text.

6.5.1 Bending Properties

Bending probably is the most common loading situation for most SCL members. Examples of structural members used as bending members are joists, headers, beams, and floor girders. Bending properties of modulus of rupture (MOR) and modulus of elasticity (MOE) are the two most frequently used properties in assessing the strength and stiffness of a material.

Edgewise and Flatwise Bending. SCL members can be used as headers and beams, which are conventionally loaded on edge. They are also used as scaffold planks and deck boards, which are primarily loaded flatwise. It is typically observed that flatwise SCL bending specimens yield higher bending strength (MOR) and stiffness (MOE) values than the specimens tested on edge.

The explanation is the so-called I-beam effect. Here, the two outer surface layers (surface layers usually consist of several individual plies of veneer) of SCL are of

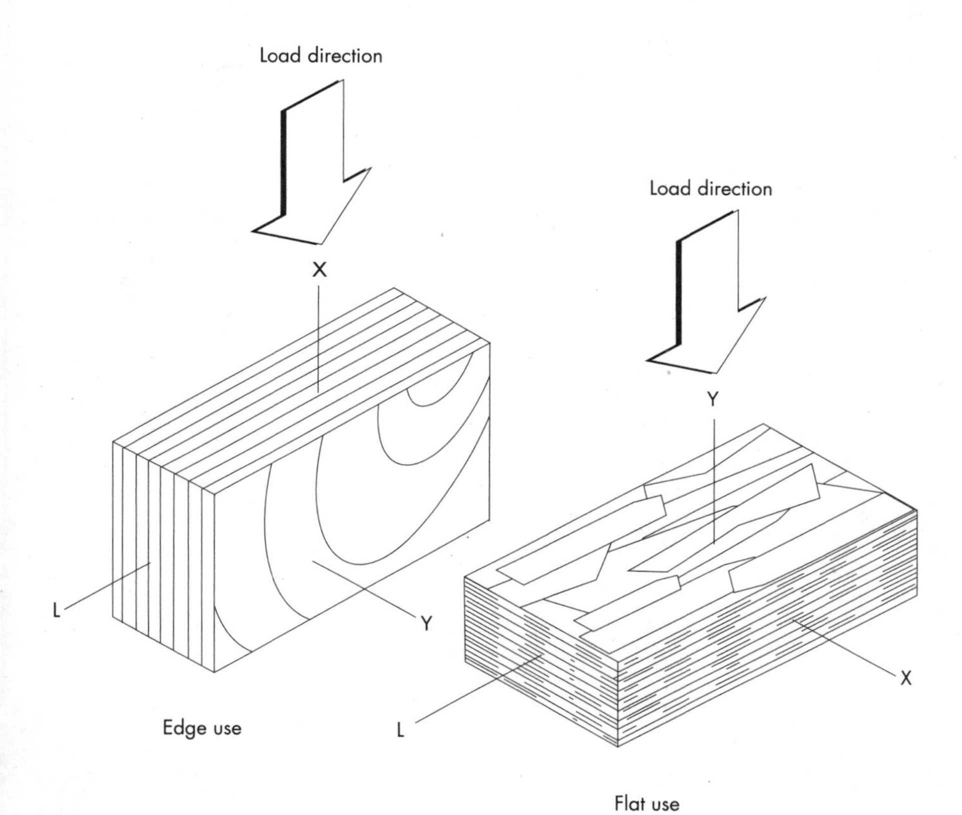

FIGURE 6.4 The three axes and the orientations for SCL.

higher grade veneer (or strands). In addition, the outer layers have a higher degree of densification relative to the center, resulting from the heat and pressure gradient in hot-pressing process.

This densified outer veneer layer concentrates relatively more wood into the outer region of the SCL. This gives the outer region a slightly higher apparent MOE. Treating the SCL as a composite, accounting for this slightly higher MOE causes the ratio of the $E_{densified}/E_{normal}$ to be slightly higher than 1.0. Calculating a transformed section using this ratio makes the outer, densified region slightly wider than the interior section of the SCL, analogous to I-beam flanges.

It is just this I-beam effect that more efficiently resists bending moment in the flatwise direction and therefore yields higher bending strength and stiffness than does the edgewise bending where the I-beam effect does not apply.

In the edgewise orientation, SCL beams and headers are trimmed to different depths ranging from 3½–24 inches to meet the requirements for different end uses.

Figure 6.5a and b shows typical edgewise bending and flatwise bending applications, respectively.

For edgewise application, commercially available LVL products are cut to some common standardized depths. They can be used as one single-piece member or glue- or nail-laminated with two or more pieces sidewise to form a wider bending

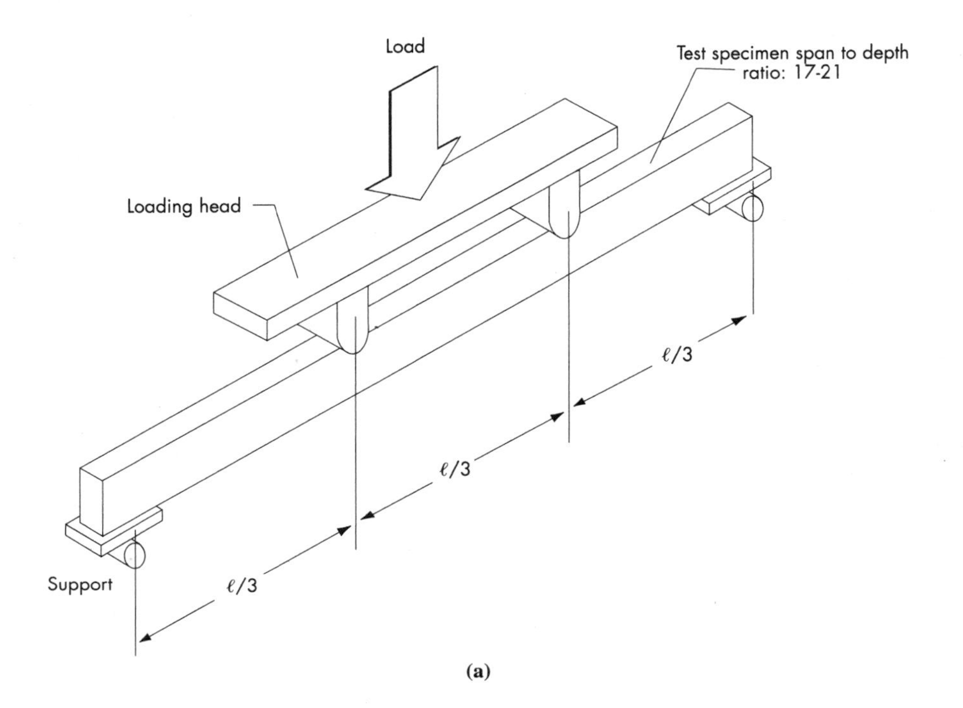

(a)

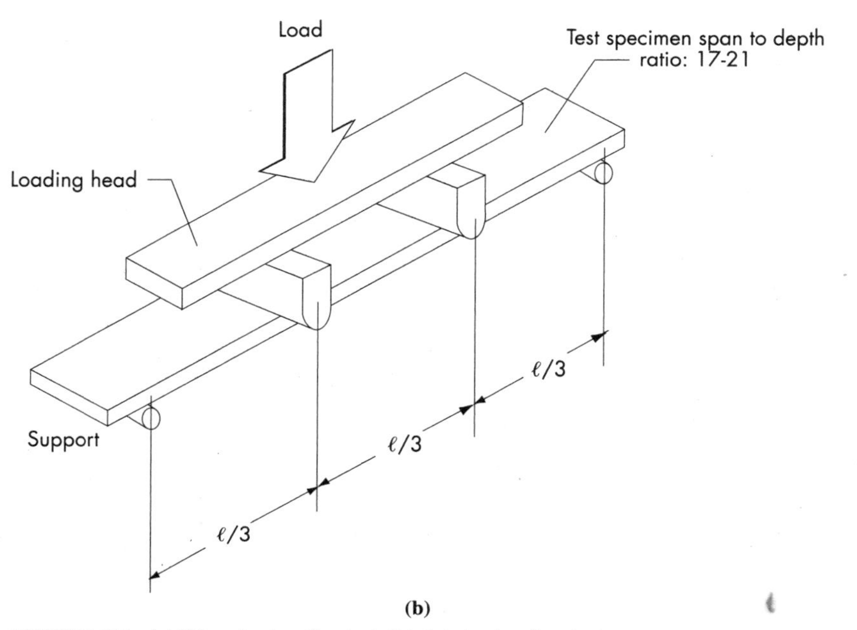

(b)

FIGURE 6.5 (a) Edgewise bending test, (b) flatwise bending test.

member. Lengths of members range from several ft to 80 ft. Table 6.3 gives some commonly used LVL depths.

Beam Size (Depth) Effect on Strength. Wood structural materials, including SCL, exhibit so-called size (or volume) effects. This means that members with different cross-sectional areas and lengths will exhibit different strengths when being tested. A theory termed "weakest link," based on the statistical theory of strength by Weibull,[8] assumes that failure of a specimen will occur when the stress in the specimen is the same as the stress that would cause the failure of the weakest element of volume if tested independently.[9] Statistically, specimens with smaller depth (same width) have lower probability of flaws in the high-stress portion than does the specimen with a deeper depth.[4] This theory provides a well-accepted explanation for why specimens with smaller depth always have higher bending strength (MOR) than deeper specimens. Numerous test results show that similar behavior of SCL occurs.

Bending size effect generally takes the following form:

$$K_d = \left(\frac{d_1}{d}\right)^{1/m} \left(\frac{L_1}{L}\right)^{1/m} \tag{6.1}$$

where K_d = bending size (depth) effect factor
d_1, L_1 = depth and length of base member (in.)
d, L = depth and length of member of other size (in.)
m = parameter determined based on test data

Equation (6.1) only considers the depth and length of members since test results indicate that increasing the width of SCL bending members does not result in strength reduction, at least within the limits given in Annex A1 of ASTM D5456. In general, a constant span/depth (L/d) ratio is used, and therefore Eq. (6.1) can be simplified as:

$$K_d = \left(\frac{d_1}{d}\right)^{2/m} \tag{6.2}$$

The size factor, K_d, is therefore actually a function of the depth of the specimens for a constant L/d ratio. That is why the size factor is also called a depth factor. According to ASTM D5456, a minimum of four different depths, including the base depth, should be included in determining the depth factor of SCL. Annex A1 of ASTM D5456 specifies the detailed procedures in deriving the depth factor. Typically, design values for SCL are published for a 12 in. depth as a base value. The tabulated design bending stress values, F_b, in Table 6.2 are for APA EWS performance rated LVL with a base depth of 12 in. As the depth factor is derived, the design bending stresses for SCL members with other than the base depth can be adjusted by:

$$F_b = F_{b12''} \times K_d \tag{6.3}$$

TABLE 6.3 Commonly Used Depths of LVL Beam Members

	Commonly used depths of LVL beam members (in.)								
Depth (in.)	1¾	3½	5½	9½	11⅞	14	16	18	24

where F_b = design bending stress for members other than the base depth (psi)
$F_{b12''}$ = design bending stress for the base depth (12 in.) (psi)
K_d = bending depth (size) effect factor

The exponent of the depth factor of LVL, $2/m$, depends primarily on veneer species and lay-ups. From the published values by code agencies (ICBO and NES) of industry-wide bending depth factors, K_d, this exponent ranges from $\frac{1}{5}$–$\frac{1}{11}$, with a majority having a value of or close to $\frac{1}{8}$. Table 6.2 is based on the use of an exponent of $\frac{1}{8}$ for the range of LVL products in PRL-501. Generally, the minimum depth that this bending depth factor applies to is 3½ in. except otherwise stated. For specimen depth less than 3½ in. the design stress for the 3½ in. depth is used.

6.5.2 Axial Tension Properties

SCL products are also used in structures as tension chords in a truss that receive axial forces (parallel to grain), or as members that encounter both tension and bending loads (headers and beams). The use of SCL as the flanges of I-joists is another example of a tension application. According to ASTM D5456, tension stress of SCL is determined in accordance with the principles in ASTM D198, *Standard Test Method of Static Tests of Lumber in Structural Sizes,*[10] or ASTM D4761, *Standard Test Method for Mechanical Properties of Lumber and Wood-Based Structural Material.*[11] Figure 6.6 illustrates the tension test setup.

Length (Size) Effect on Strength. Tensile strength parallel to grain of SCL is found to be affected by the length of the test specimens. A general trend is that as the specimen length increases, the tensile strength decreases for specimens of the same thickness. This phenomenon is similar to the bending depth effect and is also believed to be caused primarily by the weakest link theory. As the probability of encountering a weak spot in a longer member increases, the chances of failure increase.[12] An exponential relationship between specimens with the base length and

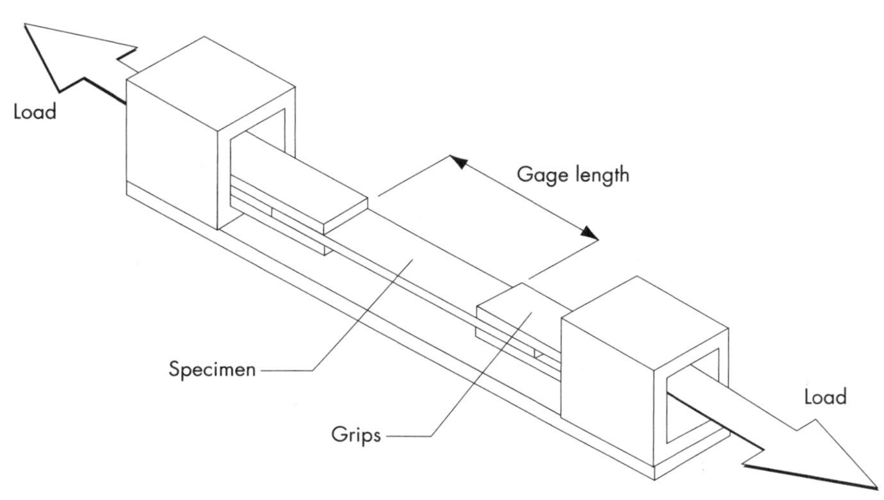

FIGURE 6.6 Tension test setup.

other lengths can be determined in a similar manner as for the bending size factor. The length used in deriving the length effect factor is the gage length, which is the net distance between the grips at each end. The tension length factor takes the following form:

$$K_L = \left(\frac{L_1}{L}\right)^{1/m} \tag{6.4}$$

where K_L = tension length (size) effect factor
L_1 = length (gage length) of base member (ft)
L = length (gage length) of member of other length, (ft)
m = parameter to be determined

According to ASTM D5456, a minimum of four different lengths, including the base length, should be involved to determine the length factor of SCL. Annex A1 of ASTM D5456 specifies the detailed procedures in determining the length factor. Table 6.2 gives the design tension stress for APA EWS performance rated LVL. Usually, design tension values for SCL are based on a gage length of 4 ft. For specimens with a gage length longer than 4 ft, the tabulated design tension stress, F_t, shall be adjusted by the length effect factor, K_L:

$$F_t = F_{t4'} \times K_L \tag{6.5}$$

where F_t = design tensile stress for members other than the base length (psi)
$F_{t4'}$ = design tensile stress for the base length (4 ft) (psi)
K_L = tension length (size) effect factor

6.5.3 Shear Properties

Shear capacity is another important property of SCL. Bending members are constantly under shear load, in addition to a flexural moment. In beam and header applications, as the span-to-depth ratio becomes small (less than 5), or a concentrated load is located close to supports, shear capacity (stress) will be critical in governing the design. Horizontal shear is determined in the two major directions, edgewise (joist orientation, shear in *L–Y* plane) and flatwise (plank orientation, shear in *L–X* plane) (see Fig. 6.4). For SCL, particularly for LVL, shear in the edgewise direction is always higher than shear in the flatwise direction. This is because in the edgewise direction shear load is resisted by each individual layer of veneer plus the glue lines. In the flatwise direction, however, maximum shear will most likely fall within a single layer or a glue line. The particular layer of veneer that resists the shear would probably be of a lower grade since the general principle of LVL layup is to position the higher-grade veneers at or adjacent to the two faces. Therefore, the shear in the *L–Y* plane (joist) is higher than the shear in the *L–X* plane (flat). The design shear stress, F_v, given in Table 6.2 is the shear in the edgewise direction for APA EWS performance rated LVL. Figure 6.7 shows the shear test setup.

6.5.4 Compression

SCL may also be used in applications where the primary stresses are compression. Depending on the directions of compression loads applied to the members, com-

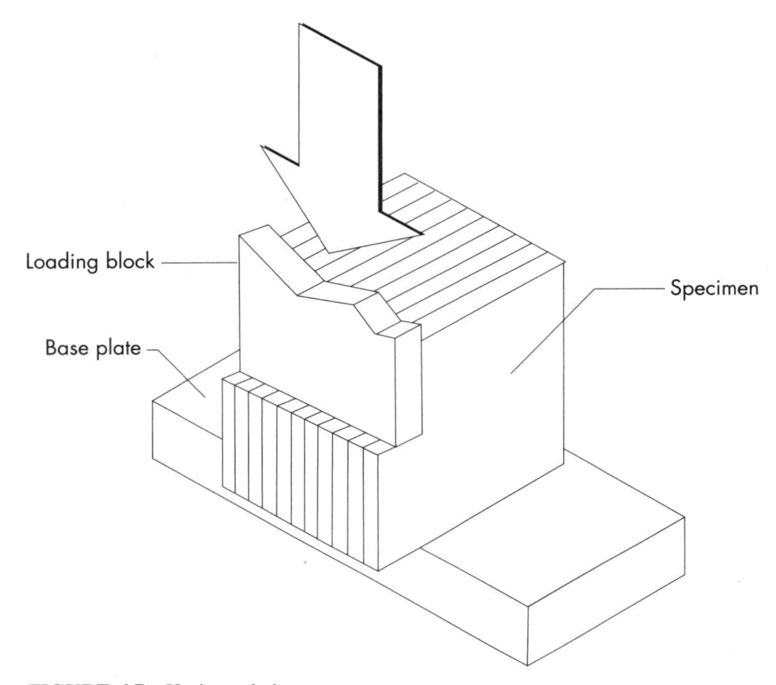

FIGURE 6.7 Horizontal shear test.

pression is categorized as compression parallel to grain and compression perpendicular to grain. Members supporting compression loads applied parallel to the grain are typically columns or compression chords of trusses or compression flanges of I-joists. When an SCL bending member is loaded, the bearing area under the supports is subjected to compression perpendicular to grain stress. The load is transmitted to the bearing area and causes deformation of the bearing area. Bearing stress are thus established based on a maximum permissible deformation.

Compression Parallel to Grain. Columns and studs are the most common structural members that are designed for withstanding compression parallel to grain. As an engineered structural material, SCL is also used as columns and studs in residential and nonresidential construction. Columns should be securely braced in order to prevent them from being laterally instable and to prevent buckling when axial compression load is applied. A proper slenderness ratio for columns is required. Slenderness ratio is defined as the ratio of the unbraced, or unsupported, length to the least radius of gyration r, in which $r = \sqrt{I/A}$, where I is the moment of inertia of the cross section about the weak axis and A is the area of the cross section. For rectangular wood columns, the slenderness ratio is expressed as the ratio of the unbraced column length to the least cross-sectional dimension in the plane of lateral support (see Fig. 6.8). The slenderness ratio is expressed as following.

$$S = \frac{L_e}{d} = \frac{K_e \cdot L}{d} \tag{6.6}$$

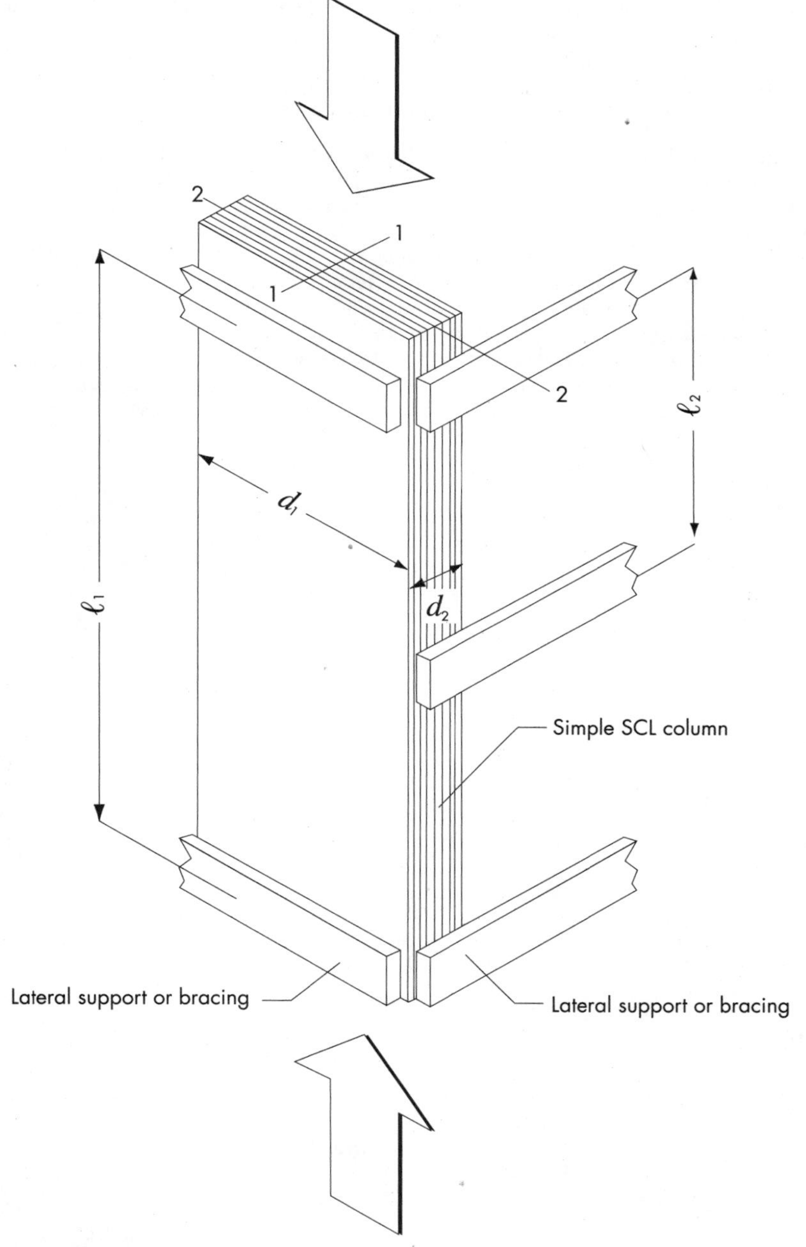

FIGURE 6.8 Simple solid column, showing d_1, d_2, l_1, l_2.

where S = slenderness ratio for solid wood column with rectangular cross sections
L_e = effective column length, $L_e = K_e L$
L = unbraced column length
K_e = buckling length coefficient for compressive members
d = cross-sectional dimension in the plane of lateral support

Figure 6.8 shows a typical column loading situation. Since the buckling length coefficient, K_e, is dependent on the member end restraint conditions, lower values of effective length is associated with more end fixity and less lateral translation while higher values will be associated with the less end fixity and more lateral translation.[13] Figure 6.9 shows some typical buckling shapes of columns with different end fixity conditions. Table 6.4 provides the values of buckling length coefficient, K_e, corresponding to each column end fixity condition in Fig. 6.9.

Column Design Equations. Prior to 1991, design procedures for solid wood columns required classifying the members as short, intermediate, or long column, based on slenderness ratios. The calculation procedures needed trial and error solutions and were considered cumbersome. Based on extensive research conducted

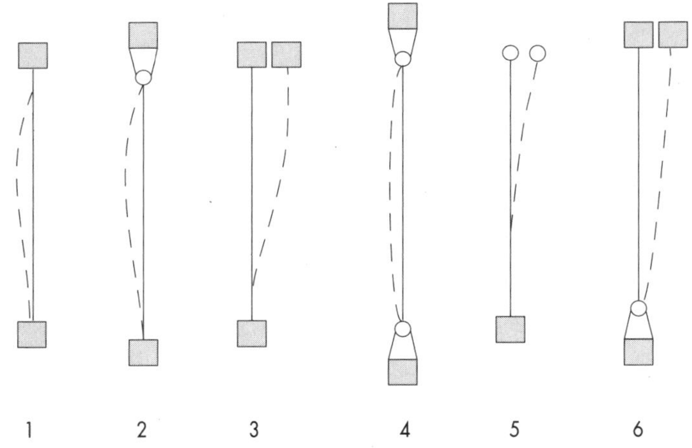

FIGURE 6.9 Typical buckling shapes of columns with different end fixity conditions.

TABLE 6.4 Buckling Length Coefficients, K_e

	Column number in Fig. 6.9					
	1	2	3	4	5	6
End fixity	Both ends fixed	One end fixed, one end pinned	One end fixed, one end translation free	Both ends pinned	One end fixed, one end free of translation and rotation	One end pinned, one end translation free
Theoretical K_e value	0.5	0.7	1.0	1.0	2.0	2.0
Recommended design K_e	0.65	0.80	1.2	1.0	2.1	2.4

at the USDA Forest Products Laboratory and other research institutions, the 1991 NDS was revised to reflect the use of a single-column formula that covers all range of slenderness ratios. Equation (4.17) provides the design basis.

Equation (4.17) establishes the adjusted design values for compression parallel to grain without requiring the initial classification of three column groups (short, intermediate, and long) and applying three different equations. The allowable compression parallel to grain stress is based on physical dimensions of the column, the published material properties such as E and F_c, and some constants. The two constants, c and K_{cE}, are material dependent with higher values assigned to wood products with lower variability such as SCL and glulam, thus resulting in higher column capacities.

Since the manufacturing process of structural composite lumber reduces or eliminates some strength-reducing characteristics that occur in natural wood, it therefore reduces the variability in material properties and increases the homogeneity of the products. A typical coefficient of variation (COV) for the modulus of elasticity (E) of LVL, for example, is less than 10%. This allows higher values of c and K_{cE} to be used in Eq. (4.17) and translates into greater compression parallel to grain design capacity.

Design compression stress parallel to grain for SCL is determined according to ASTM D5456 and D198. A slenderness ratio (L/r) < 17 of test specimens is required. Table 6.2 lists the design compression parallel to grain values for APA EWS performance rated LVL. As discussed above, depending on column physical dimensions, lateral supporting (bracing) conditions, and related allowable properties, adjusted design compression parallel to grain stress can be determined using Eq. (4.17).

Compression Perpendicular to Grain. Compression perpendicular to grain (bearing) measures the capacity of a member to carry bearing loads within a given deformation limit. According to ASTM D5456, a deformation limit of 0.04 in. is chosen for a test specimen of 1.5 in. thick in determining bearing stress of SCL. It is unlikely that a member will fail catastrophically under bearing stress even if the load has gone beyond the design load based on the 0.04 in. deformation limit. Overloaded bearing stress will cause more deformation in the bearing area and will be represented by a noticeable indentation along the supported areas. It is therefore reasonable that bearing is governed by a serviceability criterion. Sometimes a lower design bearing stress (bearing stress based on 0.02 in. deformation limit) is used in cases where a lower deformation level is desired or a close control of the deformation is required.

For SCL products, bearing stress in the edgewise direction is always higher than the bearing stress in the flatwise direction. For LVL in edgewise compression, each individual veneer aligned side by side resists the compression load. The higher-grade veneers may predominately contribute to the bearing resistance. On the other hand, in flatwise compression, load is transmitted through the layers of parallelly laid veneers, the layers with a lower grade may deform more, and the deformations of each layer will accumulate. Another reason is associated with the strength differences between the earlywood and latewood in a growth ring. In flatwise compression, load is applied and transmitted through the growth rings of each layer of veneers. The weaker earlywood cells may be compacted more easily than latewood. In edgewise compression, however, this growth ring effect does not apply, since load is applied parallel to growth rings of all veneers. In the case of PSL and LSL, the explanation for LVL may also be valid since the configurations of the PSL and LSL are similar to LVL. Figure 6.10 illustrates test setups for the compression perpendicular to grain.

The design values for edgewise compression perpendicular to grain for APA EWS performance rated LVL are listed in Table 6.2. Flatwise design values will be lower.

6.5.5 Fastener Properties

Numerous types of mechanical fasteners have been developed and are used with wood construction. They are discussed in detail in Chapter 7. One common functionality of all these different types of fasteners is to connect wood or structural composite lumber elements to meet required load-carrying performance. Fasteners should be properly selected and capable of transmitting shear or bearing load to designated parts or elements in a given structure. Among those commonly used with SCL are nails, screws, dowels, plates, and bolts. According to ASTM D5456, the nail withdrawal strength, nail dowel bearing strength, and bolt dowel bearing strength of SCL must be determined by test.

Nail Withdrawal. Nail withdrawal strength is a measure of the holding strength of nails when they encounter a force or load that tends to pull the nails from the nailed structural components. Nail withdrawal strength is determined in both face nailing (Y-direction) and edge nailing (X-direction). In the nail withdrawal test, an 8d common wire nail (with a 0.131 in. diameter and 2.5 in. length) is inserted into the specimens with a minimum penetration of 1.25 in. The nail is then pulled out and the result is converted to a lb/in. of penetration basis. In the case of LVL, in general, nail withdrawal strength in the face direction is expected to be higher than that in the edge direction. This is because in face nailing the nail penetrates across multiple layers of veneers and each individual layer contributes to the resistance to withdrawal. The layers of veneers with higher grade may supply more resistance owing to their higher specific gravity. On the other hand, however, nails inserted into the edge are only embedded in one or two veneers. Besides, the nails are usually inserted into the center portion of the edge, and that location is usually where the lower grade veneers are positioned. This explains why nail withdrawal strength of LVL in face and edge is different. Figure 6.11 illustrates the nail withdrawal test setup.

Dowel Bearing. Dowel bearing is a measure of load-deformation behavior of wood and SCL material that is laterally loaded by a fastener without bending of the fastener during loading. It actually tests the bearing resistance capacity of the fastened assemblies before excessive yielding failure of the fastener occurs. Loads on the fastened assemblies are usually a compressive load applied in a direction perpendicular to the axis of the dowels. Examples are those fasteners under lateral shear load in nailed or bolted connections.

Nail Dowel Bearing. Nail dowel bearing stress for wood and structural composite lumber is determined according to ASTM D5456. A 10d common wire nail (with a 0.148 in. diameter and 3 in. length) is used in the test. Nail dowel bearing tests can be conducted on both edge and face (Y and X) directions. It will be easier to identify the nailing direction as well as the loading direction if the following two-letter system is introduced. The first letter indicates the dowel insertion direction while the second shows the loading direction. For example, XL represents that the nail is inserted in the X direction and the load is applied along the L direction; XY indicates the nail is inserted in the X direction but the load is applied along the Y direction. This is illustrated in Fig. 6.12. In general, nail dowel bearing strength

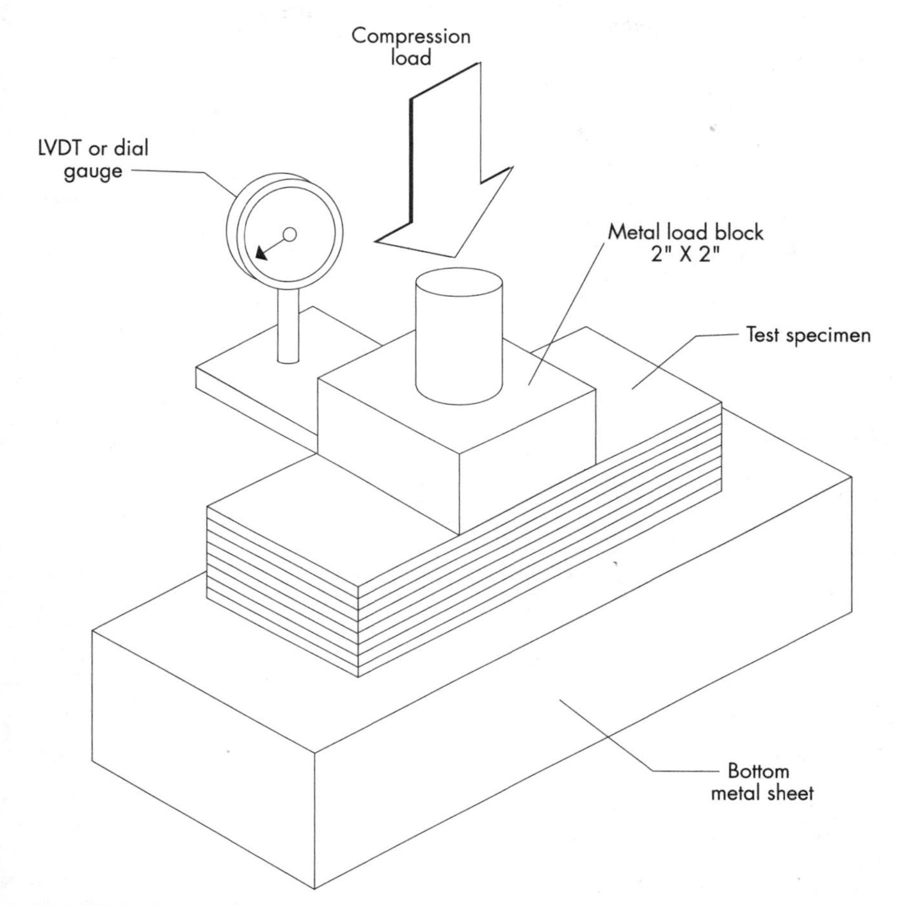

FIGURE 6.10 Compression perpendicular to grain.

in edge (*Y*) direction (including loading in *L* and *X* direction) is higher than nail dowel bearing strength in face (*X*) direction. In edge dowel bearing, the nail is supported on each individual layer of veneers through the thickness and the bearing load is withstood by all layers, as in shear through-the-thickness. In face dowel bearing, bearing load falls onto the surface layer (in the *XY* direction) or onto the center layer (or two adjacent layers) thickness (in the *XL* direction). In a similar fashion as in the compression perpendicular to grain, the dowel bearing strength on the edge (*Y* direction) is higher than that on the face (*X* direction).

Bolt Dowel Bearing. It is not unusual to connect pieces of SCL side by side to form a wider member or to fasten the end of a column into the column caps using bolts. When such connected structural members undergo applied loading, compression loads will be transferred through the bolt (dowel) to the members. Dowel bearing stress of the member must be known in order to complete a connection design. According to ASTM D5456, bolt dowel bearing stress for structural

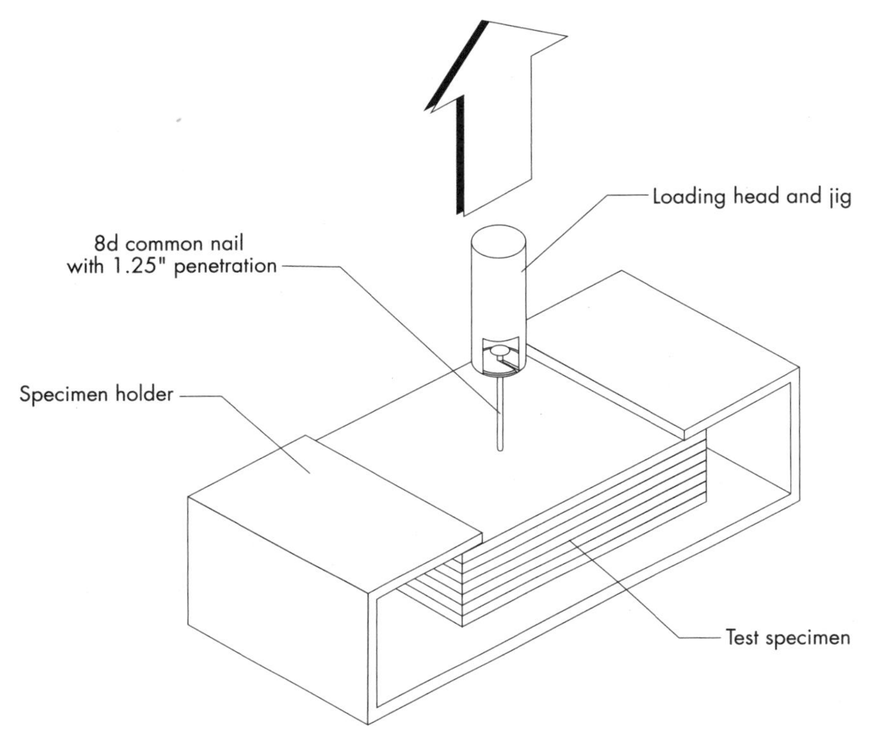

FIGURE 6.11 Nail withdrawal test setup.

composite lumber must be determined by test. Dowels with a diameter of ½ in. and ¾ in. are used in determining the dowel bearing stress. Tests of bolts installed in the *Y* direction and loaded in both *X* and *L* directions are usually required. Dowel bearing tests for bolts installed in the *X* direction are generally not required since fastening pieces of SCL edgewise is not a common practice.

In order to prevent the test specimens from splitting before the completion of the dowel bearing test, a full-hole dowel bearing test is recommended for SCL. The yield load is obtained by drawing an offset line parallel to the fitted straight line of the initial linear portion of the load-deformation curve. The horizontal distance between the fitted straight line and the offset line is 5% of the dowel diameter. The load at which the offset line intersects the load-deformation curve is the yield load and the dowel bearing stress is derived by dividing the yield load by the product of the diameter and the thickness of the specimens. Figure 6.13 shows dowel bearing test setups for both half-hole and full-hole tests.

Equivalent Specific Gravity. It is customary in the lumber industry to express the fastening capacity by a known species and their corresponding specific gravities. According to ASTM D5456, the connection properties of SCL should be expressed based on an equivalent specific gravity that indicates the SCL has a connection property equivalent to a known species or specific gravity. For both nail withdrawal

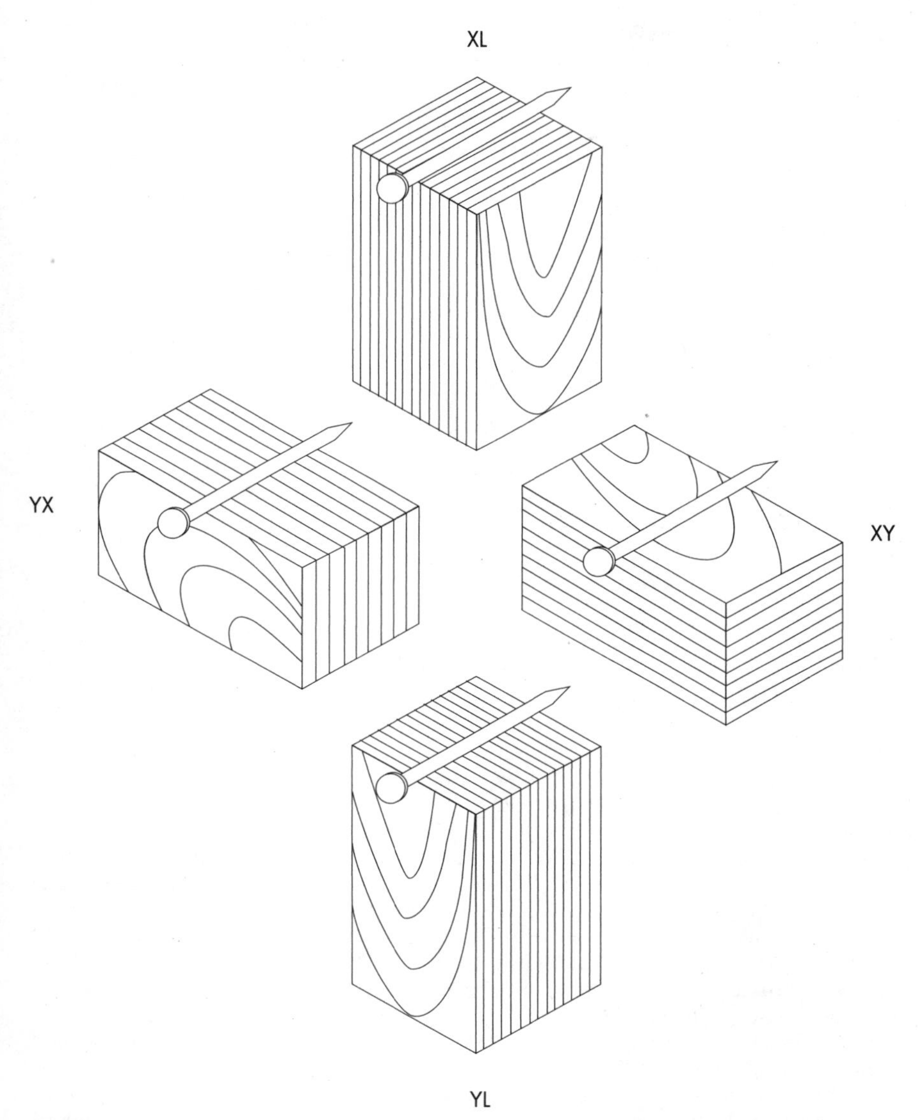

FIGURE 6.12 The directions of nail insertion and load applied in nail dowel bearing.

strength and dowel bearing stress (nail dowel bearing and bolt dowel bearing), actual test results are to be compared with the tabulated design values of each property in the corresponding tables in the National Design Specification (NDS).[13] For each test property, an equivalent specific gravity is determined by comparing the tabulated design value and the actual test average. The comparison is conducted

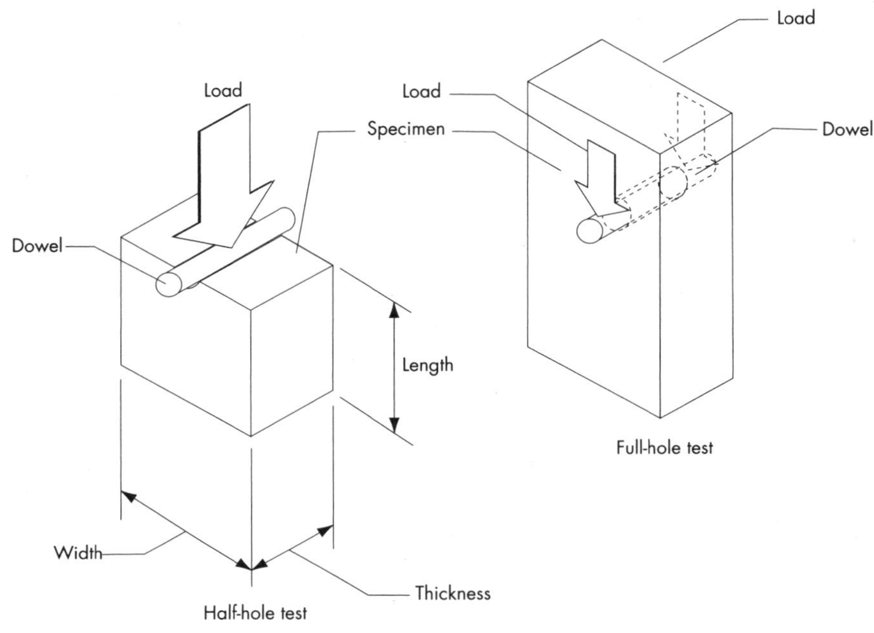

FIGURE 6.13 The dowel bearing test, half-hole and full-hole test.

such that the test average (divided by an appropriate load factor, if applicable) is greater than the tabulated design value and the specific gravity corresponding to this design value will be assigned to the SCL as the equivalent specific gravity.

For nail withdrawal strength, test results are normalized to a lb/in. of penetration basis. Table 12.2A in the NDS should be used for this purpose. For nail dowel bearing stress, Table 12A in the NDS is followed in determining the equivalent specific gravity of test specimens. It is not unusual that for the test results to fall between gaps of those tabulated design values in the tables. Equation (12.2-1) in the NDS provides a means of calculating specific gravity by the test results of nail withdrawal strength while the equation given in footnote 2 of Table 12A of the NDS shows the mathematical relationship between the specific gravity and nail dowel bearing stress. Applying the two equations will give a desirable accuracy in determining the equivalent specific gravities in cases where the differences between the test results and the nearest tabulated design values are significant. Table 8A of the NDS lists the design values of bolt dowel bearing and their corresponding species and specific gravities. Footnote 2 provides equations relating the specific gravity and bolt dowel bearing stress in parallel and perpendicular to grain directions, if they need to be used. The equivalent specific gravity of bolt dowel bearing of SCL is determined and assigned to specimens in a similar way for nail dowel bearing.

The recommended fastener design properties for APA EWS performance rated LVL, expressed in equivalent specific gravity, are given in Table 6.5.

TABLE 6.5 Fastener Design Properties for APA EWS Performance Rated LVL

APA EWS LVL Stress Class	Minimum equivalent specific gravity for fasteners[a]				
	Nails and wood screws			Bolts and lag screws	
		Lateral (dowel)		Lateral (dowel)	
	Withdrawal	Face[b]	Edge[c]	Face[b]	Edge[c]
1.5E-2250F	0.42	0.42	0.39	0.42	NA
1.8E-2600F	0.50	0.50	0.46	0.50	NA
1.9E-2600F	0.50	0.50	0.46	0.50	NA
2.0E-2900F	0.50	0.50	0.46	0.50	NA
2.1E-3100F	0.50	0.50	0.46	0.50	NA

[a] Established by qualification tests in accordance with Annex A2 of ASTM D5456.
[b] When the fastener is installed in the *Y* direction and loaded in either *L* or *X* direction (see Fig. 6.4).
[c] When the fastener is installed in the *X* direction and loaded in either *L* or *Y* direction (see Fig. 6.4).

6.6 END-USE APPLICATIONS

6.6.1 Major Applications

Structural composite lumber products are widely used as a framing material for residential construction in North America. In addition to use in housing, SCL is finding increasing use in light commercial construction.[14] Among those common uses of SCL, beams, headers, joists, rim boards, columns, and as flanges in prefabricated I-joists are the predominant applications in construction, in both residential and light and low-rise commercial construction projects.

Structural composite lumber's excellent performance in strength, stiffness, dimensional stability, universal availability, variety of available dimensions, and cost-saving attributes make it a viable and competitive alternative to other wood and nonwood products.

Figure 6.14 shows some major applications of SCL as members in construction.

6.6.2 Other Applications

SCL has also been used as planks in scaffolding due to its superior strength, high stiffness, low variability, high strength-to-weight ratio and dimensional stability. A significant percentage of scaffold plank used in the United States is now LVL.[1]

In some industrial applications, such as material handling, packaging, and transportation, SCL products are well suited for runners, stringers, and posts due to their excellent strength and durability. Being high temperature-treated in the manufacturing process, SCL is considered wood-harmful insects-free and can be used as packing materials without the fear of the infestation of harmful insects, which may be transmitted in some solid wood packing materials.

In recent years, upholstered furniture frame manufacturers have tried using LVL as the front rails and spring rails in upholstered furniture frames. Although the use in upholstered furniture frames is on a trial basis, it is believed that the use of LVL for some major beam members in upholstered furniture frames will increase since LVL possesses high strength and stiffness.

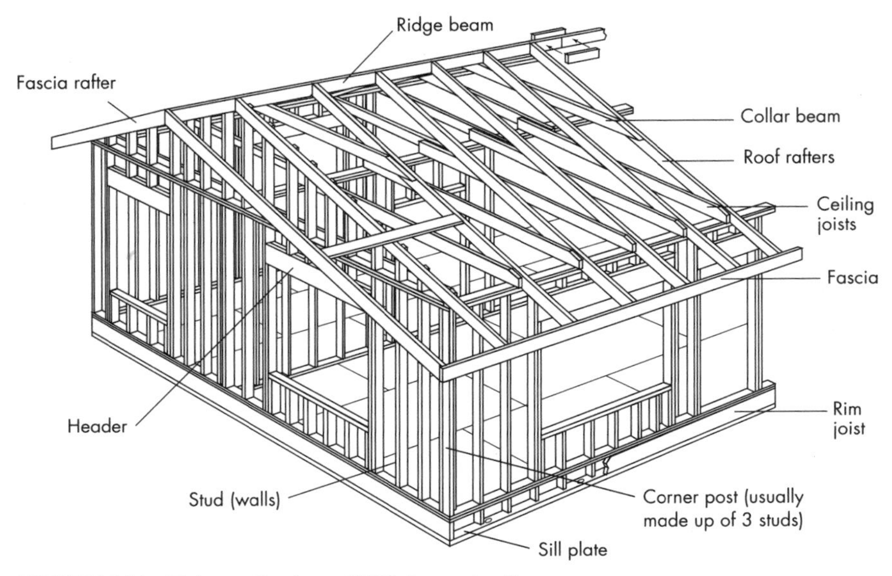

FIGURE 6.14 Major applications of SCL in construction.

6.7 SPECIAL DESIGN CONSIDERATIONS

6.7.1 Fire Performance

SCL has fire-resistive characteristics very similar to those of conventional wood lumber products. Wood is combustible. When wood burns, it forms a layer of char that is an insulator and covers unburned parts inside the char layer. This char layer segregates the fire and slows down the rate of combustion. As a result, wood members with large cross sections retain substantial proportions of their original load-carrying capacity during fire exposure.[4] SCL also has this charring ability, which tends to slow down the rate of combustion. This is discussed in more detail in Chapter 10.

Fire retardant agents can be added during manufacturing for some SCL products. Fire-treated SCL will have better fire resistance, although at a cost of losing some strength of the wood.[15] Fire resistance information on SCL products can usually be found in the evaluation reports issued by evaluation services of the model code agencies. Therefore, the code report should be reviewed in order to assess the fire performance of SCL products in design.

6.7.2 Moisture Content Effect on Major Mechanical Properties

Moisture content of wood changes as the conditions of ambient air change. In general, the changes in moisture contents range that are below the fiber saturation point, FSP, will cause changes in dimensions as well as in mechanical properties. Since SCL products are normally used in a protected dry-use environment in which moisture content is below 16%, all design values of SCL are published assuming a dry-use condition. Generally, for a covered structure, an equilibrium moisture content of approximately 7–11% is reached in a typical ambient temperature of about 70°F. If this dry-use condition can be maintained, no adjustment on design values is needed. When the moisture content of the service environment is above 16%, adjustment factors should be used to accommodate materials design capacities to a wet-use condition (MC $\geq$ 16%). Wet-use factors provided in building codes and product standards for sawn lumber and timber are generally considered appropriate for SCL products.

6.7.3 Creep of Structural Composite Lumber

Creep is a time-dependent behavior of materials and is the increase of deformation (or deflection) of a material under a constant load. As wood members are under sustained loading, they exhibit a time-dependent deformation that progresses gradually and persistently as times elapses. Creep is a function of the initial constant load, and it is also related to the time of loading exponentially. The deflection of wood bending members under long-term loading is oftentimes critical to their performance. In considering the time-dependent deflection (creep) under long-term loading, an adjustment factor is applied to the deflection due to the long-term component of the design load. It has been customary practice to use a creep deflection factor of 1.5 for glued laminated timber or seasoned timber and 2.0 for unseasoned lumber. There is no universally accepted creep deflection factor available for SCL at this time, as the creep behavior of SCL has not been studied thoroughly. It is believed that the creep performance of SCL is very similar to that of seasoned lumber or glulam.

The draft ASTM standard, *Standard Specification for Evaluation of Load Duration and Creep Effects of Wood and Wood-Based Products,* provides test methods and evaluation criteria guidance for determining creep properties of SCL.

Based on limited edgewise LVL bending creep test results conducted at APA in recently years, it appears that LVL does have satisfactory long-term bending creep performance.

6.8 DESIGN TABLES

In order to facilitate selections of SCL beam under given load and span conditions or determine the allowable load capacity of a given size beam under a known span, load-span tables are always helpful in obtaining a trial size. In using the load-span tables, a complete engineering verification conducted by a qualified engineer is always recommended before a final sizing decision is made.

The following load-span tables are generated based on the design values of APA EWS performance rated LVL. They are intended to be used to determine the allowable uniform load capacity for edgewise-loaded LVL beams, such as for joists, headers and beam applications. The design stress values for APA EWS performance rated LVL are based on stress classes established by APA EWS LVL manufacturers and represent a group of generic LVL products.

Tables 6.6–6.8 are load-span tables generated according to the allowable design stresses of Table 6.2. Refer to the following notes pertinent to the use of the tables.

TABLE 6.6 Allowable Uniform Loads (PLF) for APA EWS LVL—Floor (100%)
(Single-piece beam with 1¾ in. width [thick] loaded edgewise)

Span (ft)	Grade	1.5E-2250F				1.8E-2600F				1.9E-2600F				2.0E-2900F				2.1E-3100F			
	Depth (in.)	9½	11⅞	14	16	9½	11⅞	14	16	9½	11⅞	14	16	9½	11⅞	14	16	9½	11⅞	14	16
6	LL[a]																				
	TL[b]	1100	1511	1955	2458	1301	1959	2533	3185	1301	1959	2533	3185	1426	1959	2533	3185	1426	1959	2533	3185
	BRG[c]	4.2	5.8	7.4	9.4	4.1	6.1	7.9	9.9	4.1	6.1	7.9	9.9	4.4	6.1	7.9	9.9	4.4	6.1	7.9	9.9
8	LL	539				647				683				720				756			
	TL	632	961	1263	1534	730	1110	1513	1944	730	1110	1513	1944	815	1239	1637	1988	871	1307	1637	1988
	BRG	3.2	4.9	6.4	7.8	3.0	4.6	6.3	8.1	3.0	4.6	6.3	8.1	3.4	5.1	6.8	8.3	3.6	5.4	6.8	8.3
10	LL	274	538			329	646			348	682			366	719			385	755		
	TL	403	613	836	1074	466	709	966	1242	466	709	966	1242	520	791	1078	1386	556	846	1153	1444
	BRG	2.6	3.9	5.3	6.8	2.4	3.7	5.0	6.5	2.4	3.7	5.0	6.5	2.7	4.1	5.6	7.2	2.9	4.4	6.0	7.5
11	LL	205	403	663		246	484	796		260	511			274	539	885		288	566	929	
	TL	309	506	690	887	372	585	797	1025	384	585	797	1025	413	653	890	1144	434	698	952	1223
	BRG	2.2	3.5	4.8	6.2	2.1	3.3	4.6	5.9	2.2	3.3	4.6	5.9	2.4	3.7	5.1	6.5	2.5	4.0	5.4	7.0
12	LL	157	309	509		189	372	612		200	393	646		210	414	680		221	435	715	
	TL	237	424	579	744	285	491	669	860	301	491	669	860	318	548	747	960	334	586	799	1027
	BRG	1.8	3.2	4.4	5.7	1.8	3.1	4.2	5.4	1.9	3.1	4.2	5.4	2.0	3.4	4.7	6.0	2.1	3.7	5.0	6.4
13	LL	123	242	399	598	148	291	480	718	156	308	507		165	324	534	799	173	341	561	839
	TL	186	361	492	633	224	417	569	732	236	417	569	732	249	466	636	817	262	499	680	874
	BRG	1.5	3.0	4.1	5.2	1.5	2.8	3.8	4.9	1.6	2.8	3.8	4.9	1.7	3.1	4.3	5.5	1.8	3.4	4.6	5.9
14	LL	98	193	319	478	117	232	383	574	124	245	405	606	131	259	426	638	138	272	448	670
	TL	148	292	424	545	178	351	490	630	188	359	490	630	198	390	547	704	209	410	585	753
	BRG	1.5	2.6	3.8	4.8	1.5	2.6	3.6	4.6	1.5	2.6	3.6	4.6	1.5	2.8	4.0	5.1	1.5	3.0	4.3	5.5
15	LL	79	156	258	387	95	188	310	465	100	199	328	491	106	209	345	518	111	220	363	544
	TL	120	237	368	474	144	284	426	548	152	300	426	548	161	317	476	612	169	333	509	655
	BRG	1.5	2.3	3.5	4.5	1.5	2.2	3.3	4.3	1.5	2.3	3.3	4.3	1.5	2.5	3.7	4.8	1.5	2.6	4.0	5.1

16	LL	64	128	212	318	77	154	255	382	82	163	269	404	86	172	284	425	91	180	298	447
	TL	98	194	320	416	118	233	374	481	125	247	374	481	132	260	418	537	138	273	447	575
	BRG	1.5	2.0	3.3	4.2	1.5	1.9	3.1	4.0	1.5	2.1	3.1	4.0	1.5	2.2	3.5	4.5	1.5	2.3	3.7	4.8
17	LL	53	106	176	264	64	127	211	317	68	135	223	335	71	142	235	353	75	150	247	371
	TL	81	161	266	368	98	194	320	425	103	205	330	425	109	216	356	475	115	227	374	508
	BRG	1.5	1.7	2.9	4.0	1.5	1.7	2.8	3.8	1.5	1.8	2.9	3.8	1.5	1.9	3.1	4.2	1.5	2.0	3.3	4.5
18	LL	44	88	147	221	53	107	177	266	56	113	187	281	59	119	197	297	63	125	208	312
	TL	68	135	223	327	82	162	269	379	86	172	284	379	91	181	299	423	96	190	314	453
	BRG	1.5	1.5	2.6	3.7	1.5	1.5	2.5	3.5	1.5	1.6	2.7	3.5	1.5	1.7	2.8	4.0	1.5	1.8	2.9	4.2
19	LL	37	75	124	187	45	90	150	225	47	95	158	238	50	100	167	251	53	106	176	264
	TL	57	114	189	284	69	137	227	339	73	145	240	339	77	153	253	379	81	161	266	400
	BRG	1.5	1.5	2.3	3.4	1.5	1.5	2.2	3.3	1.5	1.5	2.4	3.3	1.5	1.5	2.5	3.7	1.5	1.6	2.6	3.9
20	LL	31	63	106	160	38	76	127	192	40	81	135	203	42	85	142	214	45	90	150	225
	TL	48	97	161	243	58	117	194	292	62	124	205	305	65	131	216	325	69	137	227	342
	BRG	1.5	1.5	2.0	3.1	1.5	1.5	2.0	3.0	1.5	1.5	2.1	3.2	1.5	1.5	2.2	3.4	1.5	1.5	2.4	3.5
22	LL	22	46	78	118	27	56	94	143	29	60	100	151	31	63	105	159	32	66	111	168
	TL	35	72	120	181	43	87	144	218	46	92	153	230	48	97	161	243	51	102	169	255
	BRG	1.5	1.5	1.7	2.5	1.5	1.5	1.6	2.5	1.5	1.5	1.7	2.6	1.5	1.5	1.8	2.8	1.5	1.5	1.9	2.9
24	LL		35	59	90	20	42	71	108	21	45	76	115	23	47	80	121	24	50	84	128
	TL		54	91	138	32	66	110	166	34	70	116	176	36	73	123	185	38	77	129	195
	BRG		1.5	1.5	2.1	1.5	1.5	1.5	2.1	1.5	1.5	1.5	2.2	1.5	1.5	1.5	2.3	1.5	1.5	1.6	2.4
26	LL		26	45	69		32	55	84		34	58	89		36	62	94		38	65	99
	TL		42	70	107		51	85	129		54	90	137		57	95	144		60	100	152
	BRG		1.5	1.5	1.8		1.5	1.5	1.7		1.5	1.5	1.8		1.5	1.5	1.9		1.5	1.5	2.1
30	LL			27	43			34	52		20	36	55		22	38	59		23	40	62
	TL			44	68			53	82		33	57	87		35	60	92		37	63	96
	BRG			1.5	1.5			1.5	1.5		1.5	1.5	1.5		1.5	1.5	1.5		1.5	1.5	1.5

[a] Maximum deflection = $L/360$ under live load, other deflection limits may apply.
[a] Maximum deflection = $L/240$ under total load, other deflection limits may apply.
[c] Required bearing length (inches), based on plate bearing of 450 psi for 1.5*E* and 550 psi for 1.8*E*, 1.9*E*, 2.0*E*, and 2.1*E* grades.

TABLE 6.7 Allowable Uniform Loads (PLF) for APA EWS LVL—Roof (115%, snow load) (Single-piece beam with 1¾ in. width [thick] loaded edgewise)

Span	Grade	1.5E-2250F				1.8E-2600F				1.9E-2600F				2.0E-2900F				2.1E-3100F			
(ft)	Depth (in.)	9½	11⅞	14	16	9½	11⅞	14	16	9½	11⅞	14	16	9½	11⅞	14	16	9½	11⅞	14	16
6	LL[a]																				
	TL[b]	1266	1738	2248	2826	1496	2253	2913	3663	1496	2253	2913	3663	1640	2253	2913	3663	1640	2253	2913	3663
	BRG[c]	4.8	6.6	8.6	10.8	4.7	7.0	9.1	11.4	4.7	7.0	9.1	11.4	5.1	7.0	9.1	11.4	5.1	7.0	9.1	11.4
8	LL																				
	TL	726	1105	1452	1764	839	1277	1740	2236	839	1277	1740	2236	937	1425	1883	2286	1002	1503	1883	2286
	BRG	3.7	5.6	7.4	9.0	3.5	5.3	7.2	9.3	3.5	5.3	7.2	9.3	3.9	5.9	7.8	9.5	4.2	6.2	7.8	9.5
10	LL																				
	TL	463	705	961	1235	536	815	1111	1428	536	815	1111	1428	598	910	1240	1594	639	973	1326	1661
	BRG	2.9	4.5	6.1	7.8	2.8	4.2	5.8	7.4	2.8	4.2	5.8	7.4	3.1	4.7	6.4	8.3	3.3	5.1	6.9	8.6
11	LL	309				372								413				434			
	TL	382	582	793	1020	442	673	917	1179	442	673	917	1179	493	751	1024	1316	528	803	1095	1407
	BRG	2.7	4.1	5.5	7.1	2.5	3.8	5.2	6.7	2.5	3.8	5.2	6.7	2.8	4.3	5.8	7.5	3.0	4.6	6.3	8.0
12	LL	237				285				301				318				334			
	TL	320	488	665	856	371	564	769	989	371	564	769	989	414	630	859	1104	443	674	919	1181
	BRG	2.4	3.7	5.1	6.5	2.3	3.5	4.8	6.2	2.3	3.5	4.8	6.2	2.6	3.9	5.4	6.9	2.8	4.2	5.7	7.4
13	LL	186				224				236				249				262			
	TL	272	415	566	728	315	480	655	842	315	480	655	842	352	536	731	940	377	573	782	1005
	BRG	2.2	3.4	4.7	6.0	2.1	3.2	4.4	5.7	2.1	3.2	4.4	5.7	2.4	3.6	4.9	6.3	2.5	3.9	5.3	6.8
14	LL	148	292			178	351			188				198	390			209	410		
	TL	199	357	487	627	271	413	563	725	271	413	563	725	303	461	629	809	279	494	673	866
	BRG	1.8	3.2	4.3	5.6	2.0	3.0	4.1	5.3	2.0	3.0	4.1	5.3	2.2	3.4	4.6	5.9	2.0	3.6	4.9	6.3
15	LL	120	237			144	284			152	300			161	317			169	333		
	TL	161	310	424	545	194	359	490	630	205	359	490	630	215	401	547	704	226	429	585	753
	BRG	1.5	3.0	4.0	5.2	1.5	2.8	3.8	4.9	1.6	2.8	3.8	4.9	1.7	3.1	4.3	5.5	1.8	3.3	4.6	5.9

16	LL	98	194	320		118	233			125	247			132	260			138	273		
	TL	132	272	372	478	159	315	430	553	168	315	430	553	177	352	480	618	186	377	514	661
	BRG	1.5	2.8	3.8	4.9	1.5	2.6	3.6	4.6	1.5	2.6	3.6	4.6	1.5	2.9	4.0	5.1	1.5	3.1	4.3	5.5
17	LL	81	161	266		98	194	320		103	205			109	216	356		115	227	374	
	TL	109	241	328	423	132	278	380	489	139	278	380	489	147	311	425	546	1544	333	454	585
	BRG	1.5	2.6	3.5	4.6	1.5	2.5	3.4	4.3	1.5	2.5	3.4	4.3	1.5	2.7	3.7	4.8	1.5	2.9	4.0	5.2
18	LL	68	135	223		82	162	269		86	172	284		91	181	299		96	190	314	
	TL	92	181	292	376	110	248	338	435	117	248	338	435	123	277	378	486	129	255	404	521
	BRG	1.5	2.1	3.3	4.3	1.5	2.3	3.2	4.1	1.5	2.3	3.2	4.1	1.5	2.6	3.5	4.5	1.5	2.4	3.8	4.9
19	LL	57	114	189	284	69	137	227		73	145	240		77	153	253		81	161	266	400
	TL	77	154	262	337	93	185	303	390	99	222	303	390	104	206	339	436	109	216	362	466
	BRG	1.5	1.9	3.2	4.1	1.5	1.8	3.0	3.8	1.5	2.2	3.0	3.8	1.5	2.0	3.3	4.3	1.5	2.1	3.6	4.6
20	LL	48	97	161	243	58	117	194	292	62	124	205		65	131	216	325	69	137	227	342
	TL	66	131	236	303	79	158	273	351	84	167	273	351	89	176	305	393	93	185	326	420
	BRG	1.5	1.7	3.0	3.9	1.5	1.6	2.8	3.6	1.5	1.7	2.8	3.6	1.5	1.8	3.2	4.1	1.5	1.9	3.4	4.4
22	LL	35	72	120	181	43	87	144	218	46	92	153	230	48	97	161	243	51	102	169	255
	TL	48	97	162	250	59	117	194	289	62	124	224	289	66	131	217	323	69	138	228	346
	BRG	1.5	1.5	2.3	3.5	1.5	1.5	2.2	3.3	1.5	1.5	2.6	3.3	1.5	1.5	2.5	3.7	1.5	1.6	2.6	4.0
24	LL	26	54	91	138	32	66	110	166	34	70	116	176	36	73	123	185	38	77	129	195
	TL	37	74	123	209	44	89	148	241	47	94	157	241	50	100	166	270	52	105	174	289
	BRG	1.5	1.5	1.9	3.2	1.5	1.5	1.9	3.0	1.5	1.5	2.0	3.0	1.5	1.5	2.1	3.4	1.5	1.5	2.2	3.6
26	LL	20	42	70	107	24	51	85	129	26	54	90	137	28	57	95	144	29	60	100	152
	TL	28	57	96	145	34	69	115	174	36	73	122	205	38	77	129	195	40	81	136	205
	BRG	1.5	1.5	1.6	2.4	1.5	1.5	1.6	2.4	1.5	1.5	1.7	2.8	1.5	1.5	1.7	2.6	1.5	1.5	1.8	2.8
30	LL		26	44	68		31	53	82		33	57	87		35	60	92		37	63	96
	TL		36	60	92		43	73	111		46	77	118		49	82	124		51	86	131
	BRG		1.5	1.5	1.8		1.5	1.5	1.7		1.5	1.5	1.8		1.5	1.5	1.9		1.5	1.5	2.0

[a] Maximum deflection = $L/240$ under live load, other deflection limits may apply.
[b] Maximum deflection = $L/180$ under total load, other deflection limits may apply.
[c] Required bearing length (inches), based on plate bearing of 450 psi for 1.5*E* and 550 psi for 1.8*E*, 1.9*E*, 2.0*E*, and 2.1*E* grades.

TABLE 6.8 Allowable Uniform Loads (PLF) for APA EWS LVL—Roof (125%, non-snow load) (Single-piece beam with 1¾ in. width [thick] loaded edgewise)

Span	Grade	1.5E-2250F				1.8E-2600F				1.9E-2600F				2.0E-2900F				2.1E-3100F			
(ft)	Depth (in.)	9½	11⅞	14	16	9½	11⅞	14	16	9½	11⅞	14	16	9½	11⅞	14	16	9½	11⅞	14	16
6	LL[a]																				
	TL[b]	1376	1889	2443	3072	1626	2449	3166	3981	1626	2449	3166	3981	1783	2449	3166	3981	1783	2449	3166	3981
	BRG[c]	5.2	7.2	9.3	11.7	5.1	7.6	9.9	12.4	5.1	7.6	9.9	12.4	5.6	7.6	9.9	12.4	5.6	7.6	9.9	12.4
8	LL																				
	TL	789	1201	1578	1917	912	1388	1891	2430	912	1388	1891	2430	1018	1549	2046	2485	1089	1633	2046	2485
	BRG	4.0	6.1	8.0	9.7	3.8	5.8	7.9	10.1	3.8	5.8	7.9	10.1	4.2	6.4	8.5	10.3	4.5	6.8	8.5	10.3
10	LL																				
	TL	504	766	1045	1343	582	886	1208	1552	582	886	1208	1552	650	989	1348	1732	695	1058	1441	1805
	BRG	3.2	4.9	6.6	8.5	3.0	4.6	6.3	8.1	3.0	4.6	6.3	8.1	3.4	5.1	7.0		3.6	5.5	7.5	9.4
11	LL	309				372								413				434			
	TL	415	632	862	1108	480	731	997	1281	480	731	997	1281	536	816	1113	1430	574	873	1190	1529
	BRG	2.9	4.4	6.0	7.7	2.7	4.2	5.7	7.3	2.7	4.2	5.7	7.3	3.1	4.7	6.4	8.2	3.3	5.0	6.8	8.7
12	LL	237				285				301				318				334			
	TL	348	530	723	930	403	613	836	1075	403	613	836	1075	450	685	934	1200	481	733	999	1284
	BRG	2.7	4.0	5.5	7.1	2.5	3.8	5.2	6.7	2.5	3.8	5.2	6.7	2.8	4.3	5.8	7.5	3.0	4.6	6.2	8.0
13	LL	186				224				236				249				262			
	TL	296	451	615	791	342	522	711	915	342	522	711	915	383	583	794	1021	409	623	850	1093
	BRG	2.4	3.7	5.1	6.5	2.3	3.5	4.8	6.2	2.3	3.5	4.8	6.2	2.6	3.9	5.4	6.9	2.8	4.2	5.7	7.4
14	LL	148	292			178	351			188				198	390			209	410		
	TL	199	388	530	681	295	449	612	788	295	449	612	788	329	501	684	880	279	536	732	941
	BRG	1.8	3.5	4.7	6.1	2.1	3.3	4.5	5.7	2.1	3.3	4.5	5.7	2.4	3.6	5.0	6.4	2.0	3.9	5.3	6.8
15	LL	120	237			144	284			152	300			161	317			169	333		
	TL	161	337	460	592	194	390	533	685	205	390	533	685	215	436	595	765	226	467	636	818
	BRG	1.5	3.2	4.4	5.6	1.5	3.0	4.1	5.3	1.6	3.0	4.1	5.3	1.7	3.4	4.6	6.0	1.8	3.6	5.0	6.4

16	LL	98	194	320		118	233			125	247			132	260			138	273		
	TL	132	296	404	520	159	342	467	601	168	342	467	601	177	382	522	671	186	409	558	718
	BRG	1.5	3.0	4.1	5.3	1.5	2.8	3.9	5.0	1.5	2.8	3.9	5.0	1.5	3.2	4.3	5.6	1.5	3.4	4.6	6.0
17	LL	81	161	266		98	194	320		103	205			109	216	356		115	227	374	
	TL	109	261	357	460	132	302	413	532	139	302	413	532	147	338	461	594	154	362	494	635
	BRG	1.5	2.8	3.9	5.0	1.5	2.7	3.6	4.7	1.5	2.7	3.6	4.7	1.5	3.0	4.1	5.2	1.5	3.2	4.4	5.6
18	LL	68	135	223		82	162	269		86	172	284		91	181	299		96	190	314	
	TL	92	181	318	409	110	269	368	473	117	269	368	473	123	301	411	529	129	255	440	566
	BRG	1.5	2.1	3.6	4.7	1.5	2.5	3.4	4.4	1.5	2.5	3.4	4.4	1.5	2.8	3.8	4.9	1.5	2.4	4.1	5.3
19	LL	57	114	189	284	69	137	227		73	145	240		77	153	253		81	161	266	400
	TL	77	154	284	366	93	185	329	424	99	241	329	424	104	206	368	474	109	216	394	507
	BRG	1.5	1.9	3.4	4.4	1.5	1.8	3.2	4.2	1.5	2.4	3.2	4.2	1.5	2.0	3.6	4.7	1.5	2.1	3.9	5.0
20	LL	48	97	161	243	58	117	194	292	62	124	205		65	131	216	325	69	137	227	342
	TL	66	131	256	330	79	158	296	382	84	167	296	382	89	176	331	427	93	185	355	457
	BRG	1.5	1.7	3.3	4.2	1.5	1.6	3.1	4.0	1.5	1.7	3.1	4.0	1.5	1.8	3.4	4.4	1.5	1.9	3.7	4.7
22	LL	35	72	120	181	43	87	144	218	46	92	153	230	48	97	161	243	51	102	169	255
	TL	48	97	162	271	59	117	194	314	62	124	244	314	66	131	217	351	69	138	228	376
	BRG	1.5	1.5	2.3	3.8	1.5	1.5	2.2	3.6	1.5	1.5	2.8	3.6	1.5	1.5	2.5	4.0	1.5	1.6	2.6	4.3
24	LL	26	54	91	138	32	66	110	166	34	70	116	176	36	73	123	185	38	77	129	195
	TL	37	74	123	227	44	89	148	262	47	94	157	262	50	100	166	294	52	105	174	315
	BRG	1.5	1.5	1.9	3.5	1.5	1.5	1.9	3.3	1.5	1.5	2.0	3.3	1.5	1.5	2.1	3.7	1.5	1.5	2.2	3.9
26	LL	20	42	70	107	24	51	85	129	26	54	90	137	28	57	95	144	29	60	100	152
	TL	28	57	96	145	34	69	115	174	36	73	122	222	38	77	129	195	40	81	136	205
	BRG	1.5	1.5	1.6	2.4	1.5	1.5	1.6	2.4	1.5	1.5	1.7	3.0	1.5	1.5	1.7	2.6	1.5	1.5	1.8	2.8
30	LL		26	44	68		31	53	82		33	57	87		35	60	92		37	63	96
	TL		36	60	92		43	73	111		46	77	118		49	82	124		51	86	131
	BRG		1.5	1.5	1.8		1.5	1.5	1.7		1.5	1.5	1.8		1.5	1.5	1.9		1.5	1.5	2.0

[a] Maximum deflection = $L/240$ under live load, other deflection limits may apply.
[b] Maximum deflection = $L/180$ under total load, other deflection limits may apply.
[c] Required bearing length (in.), based on plate bearing of 450 psi for 1.5*E* and 550 psi for 1.8*E*, 1.9*E*, 2.0*E*, and 2.1*E* grades.

6.8.1 Notes on Using Allowable Uniform Load Tables

1. Tables are based on uniform loads, simple spans and dry-use conditions.
2. Each cell is checked for the two deflection criteria ($L/360$ under live load, and $L/240$ under total load, in floor applications; $L/240$ under live load, and $L/180$ under total load, in roof applications) given in the table footnotes, bending, and shear. If both loads determined by the two deflection criteria are greater than the loads limited by bending and/or shear, then the smaller of either bending or shear strength will govern. The smallest load will control the design.
3. When no live load is shown, total load will control. The total load may be governed by either bending or shear, or by the applicable deflection criterion.
4. Duration of load factors have been applied using 1.00 for floor beams, 1.15 for snow load, and 1.25 for non-snow load.
5. Bearing across full width of beam is assumed. A plate bearing stress of 450 psi for 1.5E grade and 550 psi for the other grades is assumed.
6. Proper lateral supports at bearing area and along the beam length are assumed.
7. The weight of the beam has been included in the calculations.

6.9 DESIGN EXAMPLES

Following are some design examples to illustrate the general design procedures and familiarize readers with the common design concepts and considerations for SCL products.

6.9.1 Beam Design Examples

Beam Design Example 1. Determine an LVL floor beam size for the given design conditions. Use allowable design properties of materials from Table 6.2.

Given:

- Single-span LVL beam loaded edgewise, span $l = 6$ ft; beams spaced at 8 ft on center
- Floor live load = 125 psf (light storage)
- Dead load = 10 psf
- Allowable live load deflection = $L/360$
- Allowable total load deflection = $L/240$
- Use APA EWS performance rated LVL
- Normal duration of load (DOL = 1.00)
- Dry-use condition

Design:

- Assume a 1.8E-2600F LVL grade is used (Table 6.2):

$F_b = 2600$ psi
$E = 1.8 \times 10^6$ psi

$F_v = 285$ psi
$F_{c\perp} = 700$ psi
Live load, $LL = 125 \times 8 = 1000$ lb/ft
Dead load, $DL = 10 \times 8 = 80$ lb/ft
Total load, $TL = LL + DL = 1000 + 80 = 1080$ lb/ft

- From Table 6.6, select a trial size of a 9½ in. deep LVL beam:

$$I = 125 \text{ in.}^4$$

$$S = 26.3 \text{ in.}^3$$

$$F_b = 2600\ (12/9.5)^{1/8} = 2677 \text{ psi}$$

$$w_{\text{beam}} = 4.0 \text{ lb/ft (estimated)}$$

1. Check allowable live load for $L/360$ deflection limit:

$$\Delta = \frac{5 \cdot w \cdot l^4}{384 \cdot EI} \qquad \frac{l}{360} = \frac{5 \cdot w \cdot l^4}{384 \cdot EI}$$

$$w = \frac{384 \cdot 10^6 \cdot EI}{360 \cdot 5 \cdot l^3 \cdot 144} = \frac{384 \cdot 10^6 \cdot 1.8 \cdot 125}{360 \cdot 5 \cdot 6^3 \cdot 144} = 1543 \text{ lb/ft}$$

Net load = 1539 lb/ft (22.46 KN/m)

2. Check allowable total load for $L/240$ deflection limit:

$$\Delta = \frac{5 \cdot w \cdot l^4}{384 \cdot EI} \qquad \frac{l}{240} = \frac{5 \cdot w \cdot l^4}{384 \cdot EI}$$

$$w = \frac{384 \cdot 10^6 \cdot EI}{240 \cdot 5 \cdot l^3 \cdot 144} = \frac{384 \cdot 10^6 \cdot 1.8 \cdot 125}{240 \cdot 5 \cdot 6^3 \cdot 144} = 2315 \text{ lb/ft}$$

Net load = 2311 lb/ft (33.72 kN/m)

3. Check allowable total load for bending stress:

$$S \cdot F_b = \frac{w \cdot l^2}{8}$$

$$w = \frac{8 \cdot S \cdot F_b}{l^2 \cdot 12} = \frac{8 \cdot 26.3 \cdot 2677}{6^2 \cdot 12} = 1304 \text{ lb/ft}$$

Net load = 1300 lb/ft (18.97 kN/m)

4. Check allowable total load for shear:

$$V = \frac{w \cdot l}{2} - w \cdot d$$

$$w = \frac{2 \cdot V}{\left(l - \dfrac{2d}{12}\right)} = \frac{2 \cdot 3159}{\left(6 - \dfrac{2 \cdot 9.5}{12}\right)} = 1430 \text{ lb/ft}$$

Net load = 1426 lb/ft 20.81 kN/m)

Under the given span and load conditions, the two deflection criteria resulted in higher allowable loads than the allowable loads determined by the two strength limits. The actual loading capacity is governed by bending. The allowable load is therefore 1300 lb/ft, which is greater than the required load of 1080 lb/ft. The selected 9½ in. deep, 1.8E-2600F grade LVL will work under the given conditions. In Table 6.6, the allowable uniform load for a 9½ in. 1.8E-2600F LVL beam at a 6 ft span is found to be 1301 lb/ft. This matches the above design results.

Finally, check the required bearing length under the actual loads. If the calculated bearing length is less than 1.5 in., use 1.5 in. as the minimum bearing length.

$$L_{\text{brg}} = \frac{w \cdot l/2}{F_c \cdot b} = \frac{(1080 \cdot 6)/2}{550 \cdot 1.75} = 3.37 \text{ in. (86 mm)}$$

where L_{brg} = bearing length (in.)
w = actual total load, 1080 lb/ft, given
l = span, 6 ft, given
F_c = allowable plate bearing stress, 550 psi for 1.8E LVL
b = width of the beam, 1.75 in.

Beam Design Example 2. Select an LVL roof beam for the given design conditions. Use load-span tables for APA-EWS LVL to select a trial size.

Given:

- Single-span LVL roof beam loaded edgewise, span l = 10 ft. Beams are spaced at 8 ft on center
- Live load = 30 psf (snow)
- Dead load = 12 psf
- Allowable total load deflection = ½ in.
- Use APA EWS performance rated LVL
- Duration of load =1.15 (snow load)

Design:

Live load, $LL = 30 \times 8 = 240$ lb/ft
Dead load, $DL = 12 \times 8 = 96$ lb/ft
Total load, $TL = LL + DL = 240 + 96 = 336$ lb/ft

From Table 6.7, for a 10-ft span, select APA 1.5E-2250F, 1¾ in. × 9½ in. LVL. It has an allowable load of 463 lb/ft (6.76 kN/m), which is greater than the required 336 lb/ft under the given loading condition. Although all of the other grades with the same size meet this load requirement, the use of the 1.5*E* grade will be most cost effective and therefore it is selected.

- Check total load deflection and bearing length:
  1. Deflection under total load:

$$\Delta = \frac{5 \cdot w \cdot l^4}{384 \cdot EI}$$

$$\Delta = \frac{5 \cdot 336 \cdot 10^4 \cdot 12^3}{384 \cdot 1.5 \cdot 10^6 \cdot 125} = 0.40 \text{ in. } (10.16 \text{ mm}) \leq \tfrac{1}{2} \text{ in. } (12.7 \text{ mm})\text{—OK}$$

  2. Determine length of bearing:

$$L_{\text{brg}} = \frac{336(10)/2}{450(1.75)} = 2.13 \text{ in. } (54.2 \text{ mm})$$

6.9.2 Column Design Examples

Column Design Example 1, Simple Solid Column. Determine the size of an APA EWS performance rated LVL column required to support a 12,000 lb floor load (DOL = 1.0) applied concentrically under dry-use service condition. The total length of the column is 10 ft. The end restraints are each pinned joints. The column is laterally braced by 2 × 4 lumber chords that are nailed at the midheight (see Fig. 6.15).

Try a 1¾ in. × 12 in. APA EWS performance rated LVL, 2.0E-2900F. The allowable stresses (Table 6.2):

$$F_{c//} = 2750 \text{ psi}$$
$$E = 2{,}000{,}000 \text{ psi}$$

Given:

- $L_1 = 10$ ft, $d_1 = 12$ in.
- $L_2 = 58$ in., $d_2 = 1.75$ in.
- K_e, buckling length coefficient, 1.0, based on the end restraint conditions
- $c = 0.9$ for LVL
- Euler buckling coefficient = 0.418 for LVL

Design:

- Determine the adjusted allowable stresses:

$$F_c^* = 2750 \times 1.0 = 2750 \text{ psi}$$
$$E' = E = 2{,}000{,}000 \text{ psi}$$

- Determine the effective length and slenderness ratio:

$$L_{e1} = K_eL_1 = 1.0 \times 10 \times 12 = 120 \text{ in.}$$
$$L_{e2} = K_eL_2 = 1.0 \times 58 = 58 \text{ in.}$$
$$L_{e1}/d_1 = 120/12 = 10 < 50$$
$$L_{e2}/d_2 = 58/1.75 = 33.14 < 50$$

Since $(L_{e1}/d_1) < (L_{e2}/d_2)$, use the larger slenderness ratio (L_{e2}/d_2)

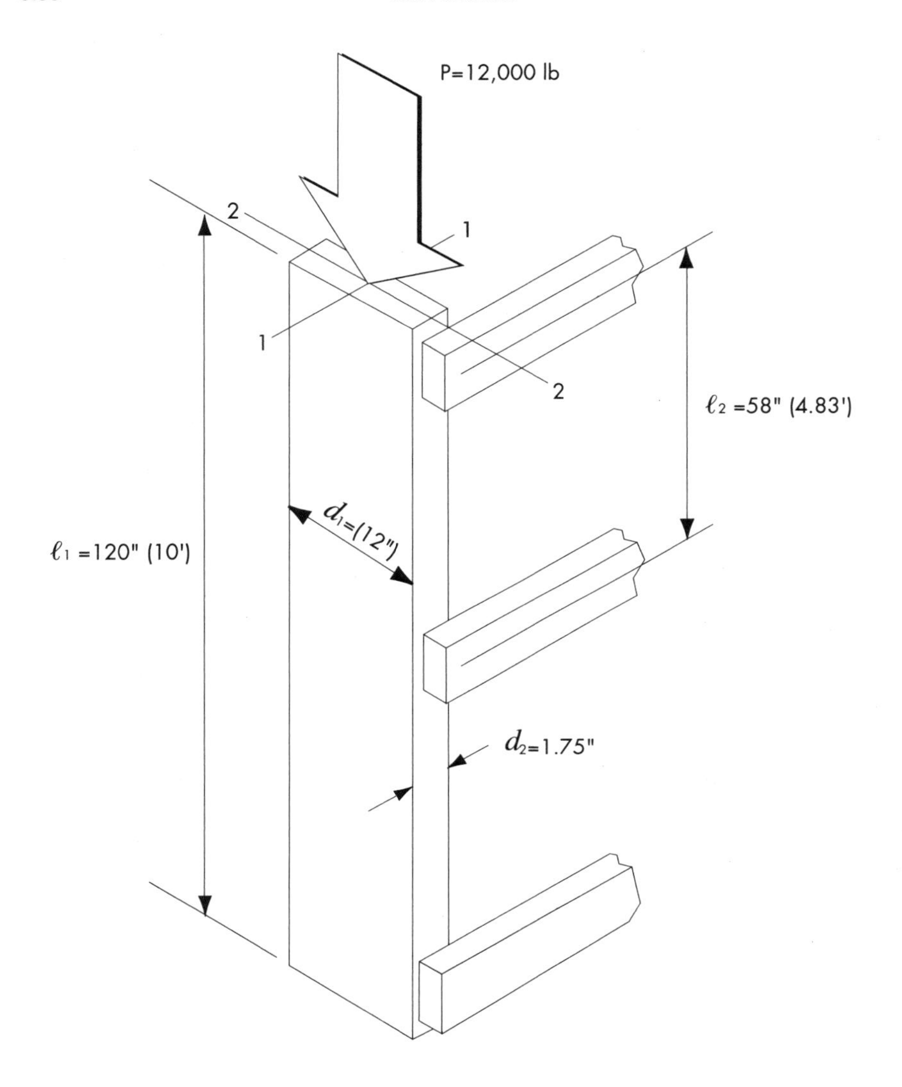

FIGURE 6.15 Column calculation simple column.

- Determine the allowable compression parallel to grain stress using Eq. 4.17:

$$F_{cE} = \frac{K_{cE} \cdot E'}{(L_e/d)^2} = \frac{0.418 \cdot 2{,}000{,}000}{(33.14)^2} = 761 \text{ psi}$$

$$F_{cE}/F_c^* = \frac{761}{2750} = 0.277$$

$$F_c' = F_c^* \left\{ \frac{1 + (F_{cE}/F_c^*)}{2c} - \sqrt{\left[\frac{1 + (F_{cE}/F_c^*)}{2c}\right]^2 - \frac{(F_{cE}/F_c^*)}{c}} \right\}$$

$$= 2750 \left\{ \frac{1 + (0.277)}{2 \cdot 0.9} - \sqrt{\left[\frac{1 + (0.277)}{2 \cdot 0.9}\right]^2 - \frac{(0.277)}{0.9}} \right\} = 735 \text{ psi}$$

- Determine allowable axial load:

$$P = F_c' \times A = 735 \times 1.75 \times 12 = 15{,}435 \text{ lb (68.66 kN)} > 12{,}000 \text{ lb—OK}$$

Column Design Example 2, Simple Solid Column, Combined Axial Compression and Bending Load. Determine the size of an APA EWS performance rated LVL column required to support a 24,000 lb floor load applied concentrically under dry-use service condition. The length of the column is 10 ft and the column is spaced 4 ft on center. The end restraints are pinned joints. The narrow face of the column receives wind load of 25 psf. Assume the edge of column is supported laterally throughout the length (see Fig. 6.16).

- Try a 1¾ in. × 8 in. APA EWS performance rated LVL, 2.0E-2900F, allowable stresses (Table 6.2):

$F_{c//} = 2750$ psi
$F_{b1} = 2900$ psi
$E = 2{,}000{,}000$ psi

Given:

- $L_1 = 10$ ft, $d_1 = 8$ in.
- $L_2 = 10$ ft, $d_2 = 1.75$ in.
- K_e, buckling length coefficient, 1.0, based on the two end restraint conditions
- $c = 0.9$ for LVL
- Euler buckling coefficient = 0.418 for LVL
- DOL = 1.6, based on wind load

Design:

- Determine the adjusted allowable stresses:

$F_c^* = 2750 \times 1.6 = 4400$ psi
$F_{b1}' = 2900 \times 1.6 = 4640$ psi
$E' = E = 2{,}000{,}000$ psi
$S_1 = bd^2/6 = 1.75 \times 8^2/6 = 18.667 \text{ in.}^3$

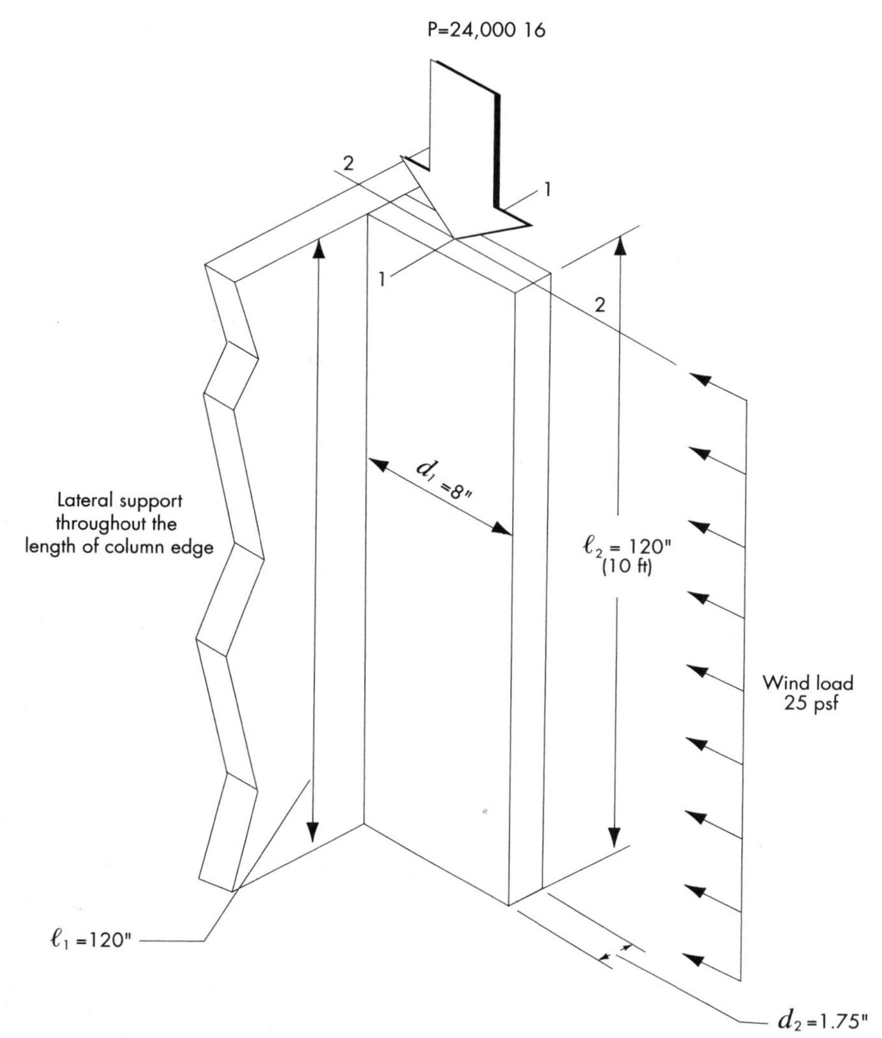

FIGURE 6.16 Column calculation simple column combined axial and bending loads.

Compression

- Determine the effective length and slenderness ratio:

$L_{e1} = K_e\, L_1 = 1.0 \times 10 \times 12 = 120$ in.
$L_{e1}/d_1 = 120/8 = 15 < 50$
$L_{e2}/d_2 = 0$, since the edge is fully supported

Use the slenderness ratio (L_{e1}/d_1) since the bending moment is applied about the 1–1 axis.

- Determine the allowable compression parallel to grain stress using Eq. (4.17):

$$F_{cE} = \frac{K_{cE} \cdot E'}{(L_e/d)^2} = \frac{0.418 \cdot 2{,}000{,}000}{(15)^2} = 3716 \text{ psi}$$

$$F_{cE}/F_c^* = \frac{3716}{4400} = 0.844$$

$$F_c' = F_c^* \left\{ \frac{1 + (F_{cE}/F_c^*)}{2c} - \sqrt{\left[\frac{1 + (F_{cE}/F_c^*)}{2c}\right]^2 - \frac{(F_{cE}/F_c^*)}{c}} \right\}$$

$$= 4400 \left\{ \frac{1 + (0.844)}{2 \cdot 0.9} - \sqrt{\left[\frac{1 + (0.844)}{2 \cdot 0.9}\right]^2 - \frac{(0.844}{0.9}} \right\} = 3037 \text{ psi}$$

- Determine actual compression load:

$$f_c = P/A = 24{,}000/(1.75 \times 8)$$

$$= 1714 \text{ psi } (11.82 \times 10^3 \text{ kN/m}^2) < 3037 \text{ psi—OK}$$

Bending

$$M_{1\ \max} = \frac{w \cdot l^2}{8} = \frac{25 \cdot 4 \cdot 10^2 \cdot 12}{8} = 15{,}000 \text{ lb-in.}$$

$$f_{b1} = \frac{M_{1\ \max}}{S_1} = \frac{15{,}000}{18.667} = 804 \text{ psi } (5.54 \times 10^3 \text{ kN/m}^2) < F_b' = 4640 \text{ psi—OK}$$

Combined Bending and Axial Compression. According to NDS 3.9.2, members subjected to a combination of bending about its edgewise axis (Fig. 6.16) and axial compression shall be so proportioned that:

$$\left(\frac{f_c}{F_c'}\right)^2 + \frac{f_{bx}}{F_{bx}'(1 - f_c/F_{cEx})} \le 1.0$$

$$\left(\frac{1714}{3037}\right)^2 + \frac{804}{4{,}640 \cdot (1 - 1714/3716)} = 0.319 + 0.322 = 0.641 < 1.0\text{—OK}$$

where f_c = actual (computed) compression stress (psi)
F_c' = allowable design value of compression parallel to grain including the adjustment for slenderness ratio (in this case, about 1–1 axis) (psi)
f_{bx} = actual (computed) edgewise bending stress (psi)
F_{bx}' = allowable edgewise design bending stress including all applicable adjustments (psi)
F_{cEx} = critical buckling design stress about 1–1 axis (psi) and is determined by Eq. (4.17) by using slenderness ratio of $(L_e/d)_x$

6.10 INSTALLATION DETAILS

6.10.1 General Installation Guidance

It is extremely important that correct and appropriate installation practices be followed for the purpose of safe use of SCL products. Since it is not possible to cover all possible installation details that have been used and proven to be safe and efficient in practice, this section provides some basics of recommended installations of SCL products. It is always helpful to the users to follow the manufacturer's guides on installations and check with local code requirements and recommendations. In general, the following guidelines should be used:

- SCL products should be used in a covered dry condition only. Although temporarily wetting due to the weather caused by construction delays is not a major concern, since the material will reach a final equilibrium condition as the construction is completed, it is recommended that proper covering be used to protect the materials from wetting whenever possible during construction.
- SCL products are allowed to be cut to different desired length. However, any other types of cutting, including notching, that are not recommended by manufacturers, to alter the shape of the SCL beams for fitting certain needs are not permitted since such cutting or notching may reduce the strength and may cause failure in service.
- Drilling holes other than recommended in manufacturer's guidelines is not permitted for SCL beams.
- Never try installing any damaged SCL in a structure.

6.10.2 Some Installation Details

Figures 6.17–6.24 illustrate some typical installation details for SCL products.

In beam-to-beam connections, various types of hangers can be selected to accommodate different applications. Figure 6.17 illustrates a typical beam-to-beam

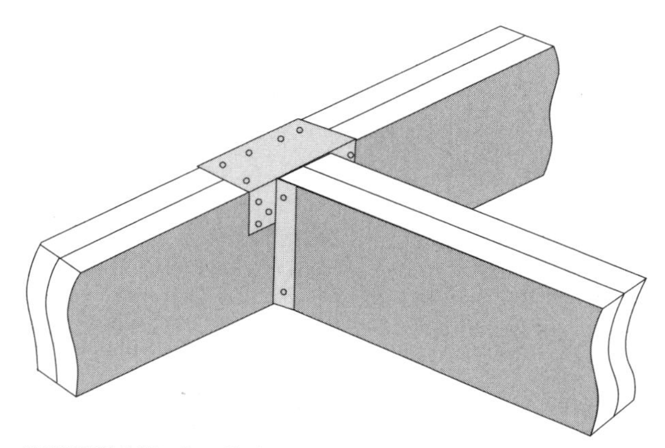

FIGURE 6.17 Installation detail 1: beam to beam connection.

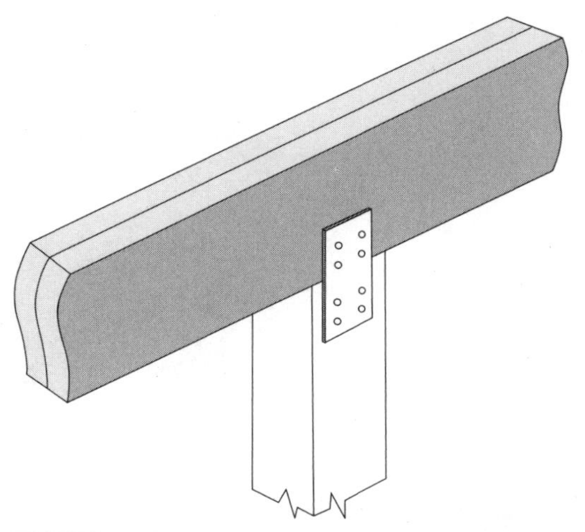

FIGURE 6.18 Installation detail 2: bearing on wood column.

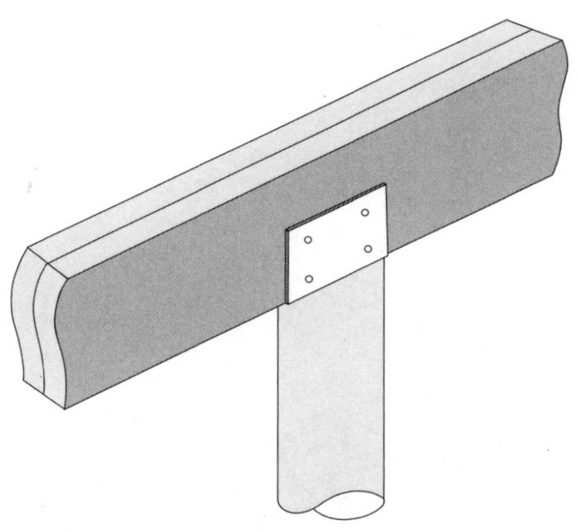

FIGURE 6.19 Installation detail 3: bearing on steel column.

connection. It is important to select a hanger that is appropriate for each application. Check bearing, fastening, and evenness of installations to ensure the full capacity of the SCL beams will be utilized.

In beam-to-column connection applications such as shown in Figures 6.18 and 6.19, verify the required bearing area and the ability of the supporting member to provide adequate strength and stability.

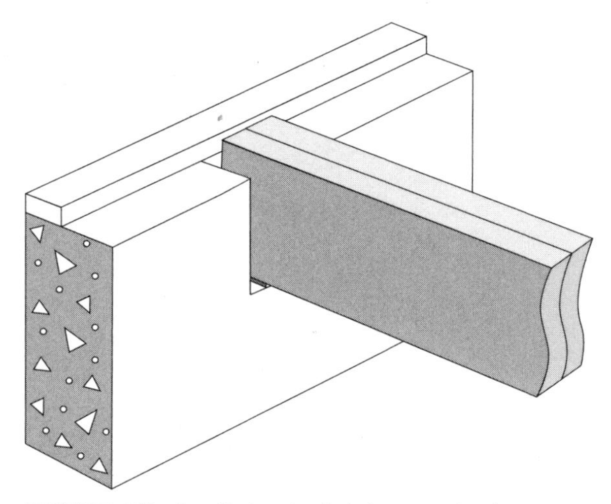

FIGURE 6.20 Installation detail 4: beam pocket in masonry wall.

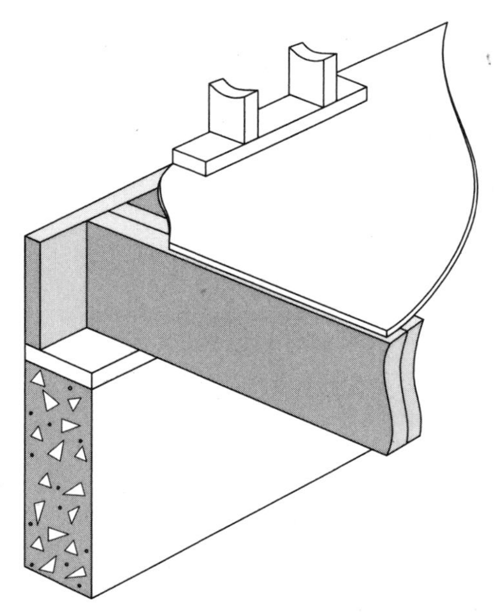

FIGURE 6.21 Installation detail 5: bearing on exterior wall (prevent direct contact with concrete).

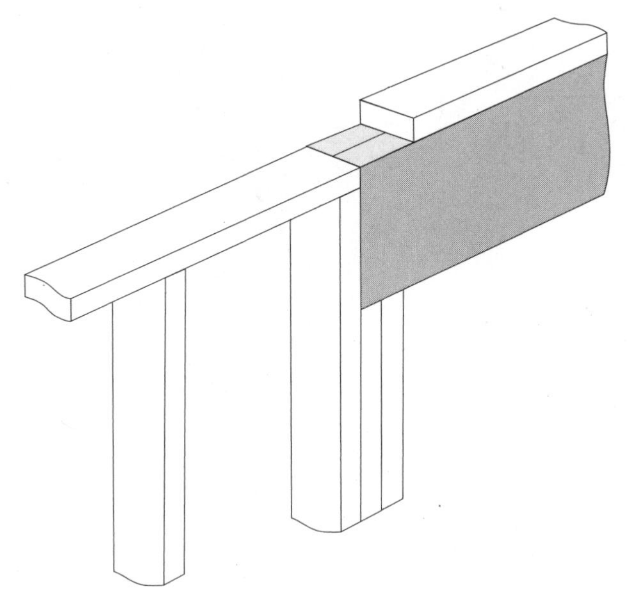

FIGURE 6.22 Installation detail 6: bearing for door or window header.

Prevent direct contact of SCL members with concrete in applications where SCL members are to be seated in a pocket in a masonry wall. Clearance should be maintained between the member and the masonry pocket to allow for ventilation as shown in Fig. 6.20. Consult local building code for other requirements.

When bearing on an exterior wall as shown in Fig. 6.21, prevent direct contact of SCL members with masonry or concrete by providing a wood or steel bearing plate between the SCL and the masonry wall. Consult local building code for requirements.

When SCL is used as a beam or header, as shown in Fig. 6.22, a minimum bearing length of 1½ in. with a bearing width across full width of the beam or header is required. Lateral supports are required at bearing points. Check with manufacturers' installation guidelines or Tables 6.6–6.8 for bearing length requirements.

Do not notch at bearing; do not drill or cut in ways that are not recommended by manufacturers. Such cutting may cause significant reduction in load-carrying capacity, particularly at critical locations of load or bearing points, and may result in unexpected sudden failure, as shown in Fig. 6.23.

LVL members can be spliced side by side to form a wider header or beam. Nailing or bolting two or more pieces is recommended for such secondary splicing. Usually 16d common nails and ½ in. diameter bolts are used for the splicing. Depending on the numbers of pieces and the depths of LVL to be joined, different connectors and fastening patterns can be used, as shown in Fig. 6.24.

Always follow the fastening recommendations provided by manufacturers. In general, the following shall be used as guidelines for the side-joining of LVL members.

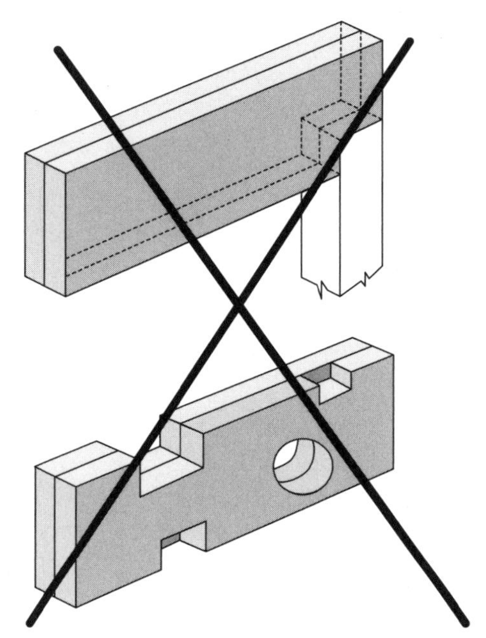

FIGURE 6.23 Installation detail 7: do not notch at bearing, or drill or cut beams not recommended by manufacturer.

1. For top and bottom rows, a minimum edge distance of 2 in. should be maintained.
2. For three-piece member splicing, nails should be driven from each side.
3. Bolt hole size shall be the same as the bolt diameter. Each bolt shall be fully extended through the full thickness of the members jointed. Washes shall be used under head and nut.
4. For a four-piece splice, regardless of depth of the members, bolts must be used.

6.11 TIPS ON FIELD HANDLING

Structural composite lumber (SCL) products are usually shipped in water-resistant wrapping that protects them from moisture, soiling, and surface scratches during transit and job-site storage. Care must be taken to protect the SCL in all transit periods, from the point where the product is delivered, to job-site handling and storage, to final installation to structures.

6.11.1 Loading and Unloading

The most commonly used equipment for handling the packages of SCL is a forklift. For safety considerations, always lift the SCL packages on the flat side and keep

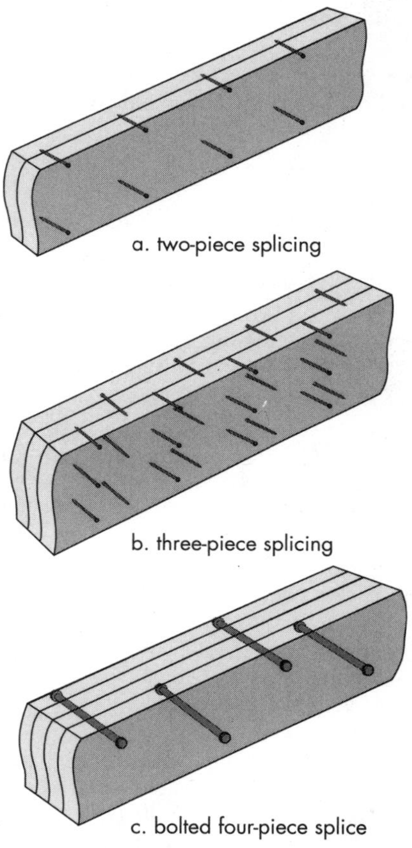

FIGURE 6.24 Installation detail 8: side splicing.

balanced loading all the time, particularly when the beams are long. When unloading, put blocks evenly at a reasonable length in order not to cause excessive deflection between two blocks. Excessive deflection during storage may cause some material damage.

6.11.2 Storage

SCL packages should be set on level, well-drained ground. Storing them in a covered area is recommended for long storage periods. Lumber bumpers or blocks should be used to keep SCL packages from direct contact with ground. For long-term storage, cut slits in the bottom of the wrapping to allow ventilation and drainage of any possible entrapped moisture. This will reduce the possibility of water damage, staining and the start of decay owing to the presence of entrapped water.

6.12 GLOSSARY AND INDEX

Veneer: A thin layer or sheet of wood. In LVL and PSL manufacturing, wood veneers are produced using peeling lathe and the veneer is also called rotary-cut veneer.

Butt joint: End joint formed by abutting the squared ends of two pieces of wood veneers.

Finger joint: An end joint made up of several meshing wedges or fingers of wood bonded together with adhesive.

Scarf joint: An end joint formed by tapping and beveling two pieces on end with adhesive.

Laminate: A product made by bonding together pieces of layers (lamination) of wood or to form a thicker wood composite product.

Lay-up: The process of loosely assembling the adhesive-coated veneers, to be pressed or clamped.

Adhesive: A substance that can be used to bonding veneers and other wood elements together by surface attachment. In SCL manufacturing hot-setting (or thermo-setting) type adhesives such as phenol formaldehyde- and phenol resorcinol-based adhesives are used.

Billet: A long, rectangular cross-sectional board of pressed laminated SCL off the production line. The thickness is the same as the laminated sheet but the width can be any desired size from 12–48 in. Typical length is less than 80 ft.

Stationary press: A hot press that has single or multiple layers of daylights. Metal plates exert load and transmit heat to the mat of wood composites and finally form condensed composite products.

Continuous press: A hot press that has only one opening and consists of a pair of closed stainless metal sheet bends. The gap (distance) between the two metal bends is gradually reduced to the final nominal thickness of the product. The pressure and the heat are adjustable along the length direction to meet the pressing schedule. As the mat is fed in, the two metal bends rotate at the same rate while maintaining the required pressure and heat at each stage during the pressing cycle, forming a continuous long sheet of wood structural lumber.

Edgewise load: Load that applies to SCL's narrow side, parallel to the X axis in Fig. 6.4. This loading is also frequently called "in the joist direction." Examples are beams and headers.

Flatwise load: Load that applies to SCL's wide side, parallel to the Y axis in Fig. 6.4. This loading is also frequently called "in the face direction." Examples are scaffolding planks and stair boards.

6.13 REFERENCES

1. *Engineered Wood Products: A Guide for Specifiers, Designers and Users,* Forest Research Foundation, Madison, WI, 1997.
2. "Glued Structural Members," in *Wood Handbook,* Forest Products Laboratory Madison, WI.
3. Wilson, A., and N. Malin, "Structural Engineered Wood," *Environmental Building News,* vol. 8, no. 11, pp. 11–17, 1999.

4. Faherty, K., and T. Williamson, *Wood Engineering and Construction Handbook,* 3d ed., McGraw-Hill, New York, 1999.
5. ASTM D5456-98, *Standard Specification for Evaluation of Structural Composite Lumber Products,* ASTM, West Conshohocken, PA, 1999.
6. AC47, *Acceptance Criteria for Structural Composite Lumber,* ICBO Evaluation Services, Inc, Whittier, CA, 1996.
7. APA PRL-501, *Performance Standard for APA EWS Laminated Veneer Lumber,* APA—The Engineered Wood Association, Tacoma, WA, 2000.
8. Welibull, W., "A Theory of the Strength of Materials," *Swedish Royal Inst. Eng. Res. Proc.,* Stockholm, Sweden, 1939.
9. Gurfinkel, G., *Wood Engineering,* Kendall/Hunt, Dubuque, IA, 1981.
10. ASTM D198-98, *Standard Test Method of Static Tests of Lumber in Structural Sizes,* ASTM, West Conshohocken, PA, 2000.
11. ASTM D4761-96, *Standard Test Method for Mechanical Properties of Lumber,* ASTM, West Conshohocken, PA, 2000.
12. Madsen, B., *Structural Behavior of Timber,* Timber Engineered Ltd., North Vancouver, BC, Canada, 1992.
13. *National Design Specification for Wood Construction,* American Forest and Paper Association, Washington, DC, 1999.
14. Structural Composite Lumber Guideline, *Allowable Stress Design (ADS),* American Forest and Paper Association, Washington, DC, 1999.
15. Bodig, J., and B. Jayne, *Mechanics of Wood and Wood Composites,* Van Nostrand Reinhold, New York, 1982.

CHAPTER 7
DESIGNING FOR LATERAL LOADS

Thomas D. Skaggs, Ph.D., P.E.
Senior Engineer

Zeno A. Martin, P.E.
Associate Engineer

7.1 INTRODUCTION

This chapter addresses the design of wood-framed structural-use panel-sheathed diaphragms and shear walls, which also may be referred to as vertical diaphragms. The supplement does not cover the method in which lateral loads are determined, nor cover prescriptive requirements for nonengineered wall and roof construction, as these topics are addressed in complete detail by the applicable model building codes.

7.2 TERMINOLOGY

Average story drift: is the average of the deformations of all lines of shear walls oriented parallel to the applied load. The deformation shall be calculated as a portion in each line of shear walls at the top of the shear wall (see Eq. [7.1]).

Base shear: the total reaction at the base of a wall parallel to the axis of the wall or structure due to an applied lateral load; a "sliding" force.

Blocked diaphragm: a diaphragm in which all panel edges occur over and are fastened to common framing; the additional fastening provides a load path to transfer shear at all panel edges, thus increasing the overall shear capacity and rigidity (stiffness) of the diaphragm.

Boundary element: diaphragm and shear wall boundary members to which sheathing transfers forces. Boundary elements include chords and drag struts at diaphragm and shear wall perimeters, interior openings, discontinuities, and re-entrant corners.

Box-type structure: when diaphragms and shear walls are used as the lateral force resisting system of a building, the structural system is called a box system.

Chord: the edge members of a diaphragm or a shear wall, typically the joists, ledgers, truss elements, double top plates, end posts, etc., that resist axial forces. Chords are oriented perpendicular to the applied lateral load.

Collector: a structural building component that distributes the diaphragm shear from one building element to another; typically served by the double top plate. Collectors are oriented parallel to the applied lateral load. Also called drag struts.

Diaphragm: a flat, or nearly flat, structural unit acting like a deep, thin beam. The term is usually applied to roofs and floors designed to withstand lateral loads. Diaphragms are commonly created by installing structural-use panels over roof or floor supports.

Diaphragm, blocked: a diaphragm in which adjacent sheathing edges are fastened to a common member for transferring shear. Examples to achieve blocked status are panels being fastened to common framing members, sheet metal fastened to adjacent panels typically with staples, or staples driven through the tongue-and-groove for 1⅛ in. wood structural panels.

Diaphragm, flexible: a diaphragm is flexible for the purpose of distribution of story shear when the lateral deformation of the diaphragm under the applied lateral load (see Eq. [7.2]) is greater than or equal to two times the average story drift. For analysis purposes, it can be assumed that a flexible diaphragm distributes story shear by tributary area into lines of shear walls oriented parallel to the applied lateral load. Traditionally, wood diaphragms are considered flexible.

Diaphragm, unblocked: a diaphragm in which only panel edges in one direction (i.e., the 4 ft wide panel ends) occur over and are fastened to common framing; the typical diaphragm for standard residential construction.

Drag strut: see *collector.*

Lateral load: horizontal forces that result from wind or seismic forces. Wind forces act on the side of the building and on sloped roofs. Seismic forces result from ground accelerations causing inertial forces to act on the structural mass.

Lateral stiffness: the slope of the load-displacement for a lateral force-resisting system.

Load path: the path taken by forces acting on a building. Loads are transferred by the elements in the building and by the connections between those elements into the foundation.

Overturning: occurs when a lateral force acts on a wall or structure and the wall is restrained from sliding; a "tip-over" or overturning force results.

Shear wall: a vertical, cantilevered diaphragm that is constructed to resist lateral shear loads by fastening structural-use panels over wood wall framing. This structural system transfers lateral forces from the top of the wall to the bottom of the wall, and eventually transfers the lateral loads to the foundation.

Shear wall segment: a portion of the shear wall that runs from the diaphragm above to the diaphragm/foundation below; also known as full-height segment. Shear wall segments occur between building wall discontinuities such as doors, windows or corners in the shear wall.

Structural-use panel: a structural panel product composed primarily of wood and meeting the requirements of USDOC PS-1, USDOC PS-2, or equivalent proprietary standard recognized by the code authority. Structural-use panels include all-veneer plywood, composite panels containing a combination of veneer and wood-based material, and matformed panels such as oriented strand board and waferboard.

Subdiaphragm: a portion of a larger wood diaphragm designed to transfer local forces to primary diaphragm collectors.

Tie-down (*hold-down*)*:* a device used to resist uplift of the chords of shear walls.

Wall bracing: a building element that resists lateral loads under low load situations; the configuration and connections are prescribed by the building codes for light-framed wood structures.

7.3 SHEAR WALLS

Shear walls and diaphragms are designed to transfer in-plane forces. When these two assemblies are used to resist lateral design forces of buildings, the structural system is sometimes referred to as a box system. The shear walls provide reactions for the roof and floor diaphragms, and transmit the forces into the foundation (see Fig. 7.1).

The structural design of buildings using diaphragms is a relatively simple, straightforward process if the designer keeps in mind the overall concept of structural diaphragm behavior. Actually, with ordinary good construction practice, any sheathed element in a building adds considerable strength to the structure. Thus, if the walls and roofs are sheathed with panels and are adequately tied together and tied to the foundation, many of the requirements of a diaphragm structure are met. This fact explains the good performance of structural-use panel sheathed buildings

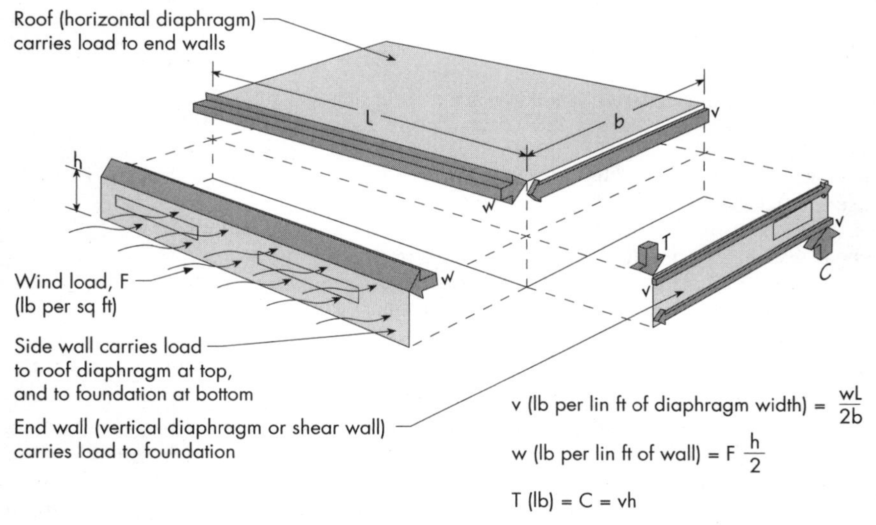

FIGURE 7.1 Distribution of lateral loads on buildings.

in hurricane and earthquake events even when they have not been engineered as diaphragms or shear walls per se.

Various textbooks and other resources provide in-depth coverage of this topic (see the References). For the purposes of this chapter, the following assumptions are made:

- These assemblies act as deep beams.
- In-plane shear resistance is provided by the structural-use panel-to-framing connections.
- Axial tension and compression resistance is provided by the chord members (analogous to an I-beam flange).
- Nailed assemblies, as shown in Tables 7.1, 7.2, and 7.7, exhibit ductile, energy-absorbing behavior.

While shear resistance of these assemblies can be computed by principles of engineering mechanics[1] it is recommended that designers use Tables 7.1, 7.2, and 7.7 for typical design purposes. In addition to eliminating labor-intensive calculations, these tables limit configurations to those that have proven to exhibit the aforementioned ductile behavior by demonstration via structural testing and years of successful in-use performance.

7.3.1 Shear-Wall Testing

Buildings are subjected to a variety of loads during their lifetime. All loads resisted by the structure must be transferred into the foundation. Gravitational loads (roof live load, snow load, dead load) react vertically on the structure and are typically transferred to the foundation through load-bearing walls. Wind and earthquakes apply lateral load to a structure. Lateral loads are transferred to the foundation through lateral force-resisting systems. For light-frame construction, the gravitational forces are typically resisted by nominal dimension lumber in the form of wall studs, and the lateral loads are commonly resisted by wood structural panel sheathed shear walls.

APA—The Engineered Wood Association has conducted research on the behavior of shear walls for almost 50 years. The first technical report was published in 1953. Two years later the Uniform Building Code (UBC) recognized shear-wall design values based on the APA tests. This recognition allowed shear walls to be used as lateral force-resisting systems in buildings designed per the UBC.

The intent of the following sections are to review the methods in which current shear-wall values are derived, discuss some of the questions raised about shear-wall performance based on monotonic and reversed cyclic load testing, and briefly discuss research efforts that are currently being conducted by APA.

Static Test Methods

ASTM E72. The current version of ASTM E72[2] covers standard tests of wall, floor and roof elements. Although ASTM E72 is often thought of as a racking test, the actual standard test method is much more broad-based. The racking test portion of ASTM E72 has the longest history of any of the standard test methods discussed in this chapter. Much of the data that have traditionally been used to support shear-wall design values have been developed from tests following ASTM E72. This particular test method is based on one-directional (also referred to as monotonic) load tests on 8 × 8 ft (2.4 × 2.4 m) shear walls. Figure 7.2 is a schematic diagram of a typical shear wall with studs spaced 24 in. (610 mm) o.c., based on ASTM E72 test methodology.

The ASTM E72 test method has several notable attributes. Since the standard test method does not specify any dead load on the top of the wall, some type of overturning restraint is required. The overturning forces in the wall are resisted by steel hold-down rods. Although it can be argued that the steel rods do not model the real world, the rods are designed not to represent real-world hold-down mechanisms but to model the effects of dead load incurred in an actual structure. Another notable feature of an ASTM E72 test wall is the stop at the end of the wall. This stop is intended to prevent lateral slippage of the wall assembly on the base of the test frame.

Typically, several incremental monotonic (unidirectional) loads are applied and removed before the wall is taken to failure. These loads allow the measurement of permanent wall set to be recorded at different load levels. The magnitude of the load cycles specified by this test standard is not related to the sheathing being tested. For example, low-strength and high-strength shear walls follow the same load cycle. The test method is partially intended to eliminate confounding factors of the test that may affect the overall results. The test method is intended primarily to evaluate panel shear resistance, including the sheathing attachments to framing, not the behavior of the structural assembly or system.

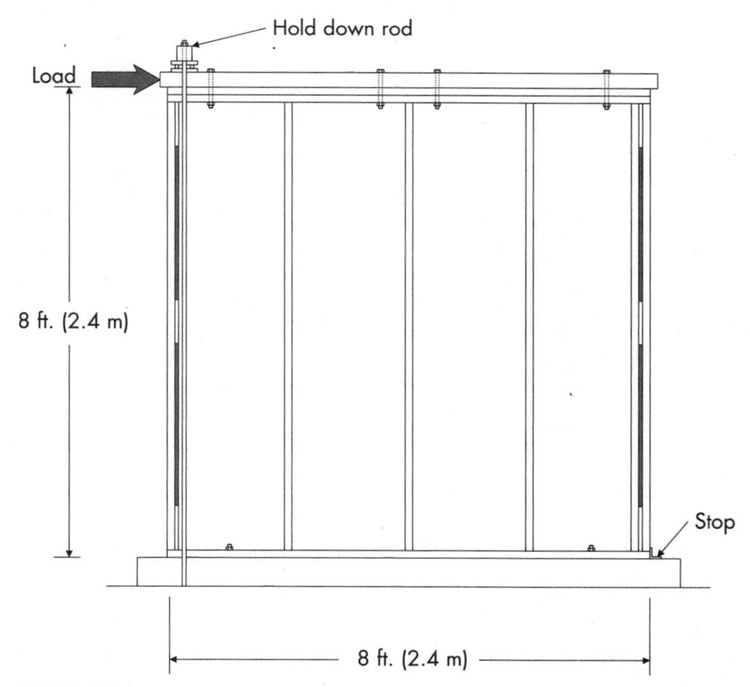

FIGURE 7.2 Schematic diagram of shear-wall test following the methods outlined by ASTM E72.[2]

The following observations can be made. The vast majority of shear wall tests conducted over the past 50 years follow ASTM E72. One of the problems with following this test method is that it specifies load increments, regardless of the wall sheathing material. ASTM E72 was originally intended for evaluating nominal wall bracing for residential applications that exhibited relatively low shear capacities. The load increments for shear walls that are designed for high loads may represent only a fraction of the design load.

Although ASTM E72 allows other loading patterns to be used, no suggestion is given on how the loads should be chosen. APA test methodology, covered in PRP-108,[3] deviates from ASTM E72 on this point, in that the loads are determined as a function of the design load. This ensures that the wall will be loaded to design load at least three times. The results from ASTM E72 tests are often used to verify the allowable design values published in the U.S. model building codes or the 2000 International Building Code (IBC).[4] A load factor (strength limit state/design shear load) for a test series typically averages about 3.0.

ASTM E564. Some designers and agencies have been critical of the ASTM E72 test method due to the use of the artificial hold-down mechanism (steel hold-down rods); therefore, an alternative ASTM standard was developed, ASTM E564.[5] The ASTM E564 test method is designed to evaluate the performance of shear wall assemblies instead of focusing only on the behavior of the sheathing. Figure 7.3 illustrates a schematic diagram of an ASTM E564 assembly test.

For ASTM E564, hold-down connectors are required to resist overturning forces instead of hold-down rods as used in ASTM E72. Unlike ASTM E72, ASTM E564 provides an option to apply vertical loads to simulate gravitational forces. Another

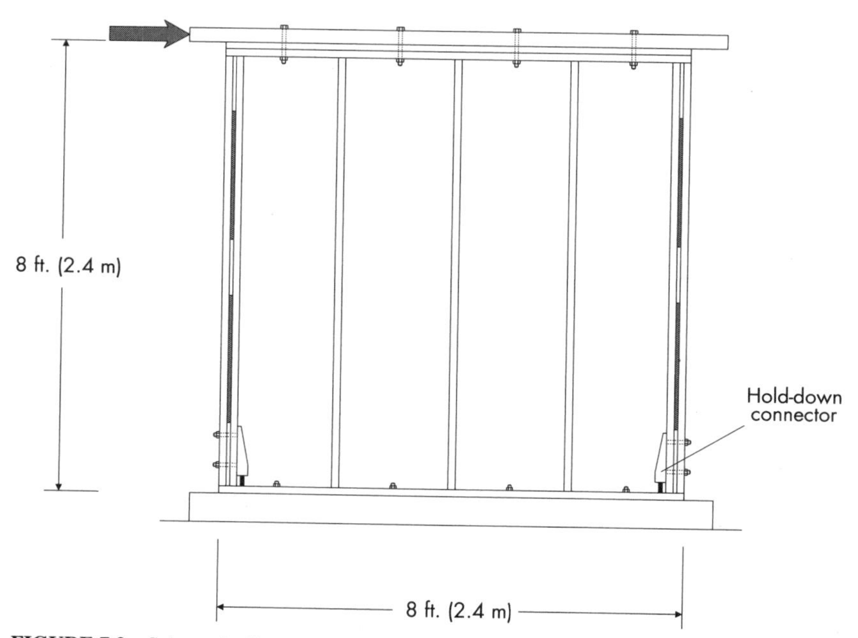

FIGURE 7.3 Schematic diagram of shear-wall tests following the methods outlined by ASTM E564.[5]

difference in the two test methods is that ASTM E564 does not utilize a stop at the sole plate; lateral slippage is prevented only by the sole plate bolts.

ASTM E564 test procedures also subject the shear walls to several monotonic load and unload cycles. In contrast to ASTM E72, the magnitude of the loads is based on the expected maximum shear capacity (strength limit state) of the walls. Although, some may view the ASTM E72 and E564 load tests as being cyclic, since the walls are loaded and unloaded several times, the methods do not model true behavior of earthquakes because: (1) the loads are only applied for a small number of cycles, (2) the load is not fully reversing (it only loads the wall monotonically in one direction) and (3) the monotonic tests are typically performed slowly, thus modeling static behavior. Such loading is considered applicable to wind loading conditions. Conversely, earthquake loads are typically fully reversing cyclic loads that continue for numerous cycles and cause dynamic loads on the structures.

Quasi-static Cyclic Test Methods

Structural Engineers Association of Southern California (SEAOSC) Test Method. Code officials, engineers, and researchers have questioned how well monotonic laboratory tests of shear walls relate to the behavior of full-size shear walls subjected to reversed cyclic loads. After the Northridge, California, earthquake (January 17, 1994), the City of Los Angeles adopted several building code changes that effectively reduced (by 25%) the allowable design loads for shear walls based on monotonic tests, until cyclic (reversed) load shear wall tests were conducted to confirm or modify previously recognized values. Since very few data were available from matched monotonically and cyclically tested shear walls, the reductions were based on experience of structural engineers in Southern California and were somewhat arbitrary. The intention of the interim reductions was to provide conservative design values until more informed design recommendations could be formed. One of the hurdles of cyclic testing was that there was no universally recognized protocol for cyclic load testing of shear walls. In 1994, the Structural Engineers Association of Southern California (SEAOSC) took on the task of developing a test standard for cyclic (reversed) loading of shear walls. The proposed standard was finalized in 1996,[6] with minor revisions in 1997.

The shear-wall test specimen used in the SEAOSC test method is similar to the wall used for ASTM E564 monotonic tests (Fig. 7.3). The overturning forces that are generated in the wall are resisted by the use of hold-down connectors. Load applied to the top of the wall is a fully reversing, sinusoidal load with decay cycles, as illustrated by Fig. 7.4.

The magnitude of the load cycles is based on the first major event (FME). By definition, the FME is the "first significant limit state to occur." In lay terms, FME represents the point in which permanent damage to the wall begins to occur. This point can also be thought of as the proportional limit or yield limit state of the wall. The load cycle that is used by the SEAOSC test method is based on a proposed sequential phased displacement (SPD) procedure.[7] The proposed procedure was developed by a joint U.S. and Japan Technical Coordinating Committee on Masonry Research and is commonly referred to as either the SPD or the TCCMAR procedure. The SPD procedure is designed to provide fully reversing displacements with progressively increasing displacements.

After FME occurs (after nine cycles in Fig. 7.4), a degradation cycle is included in the test cycle. This degradation cycle represents 100%, 75%, 50%, and finally 25% of the maximum displacement increment. The degradation cycle is included to analyze the effect of systems that develop slack. The area of the hysteresis loops

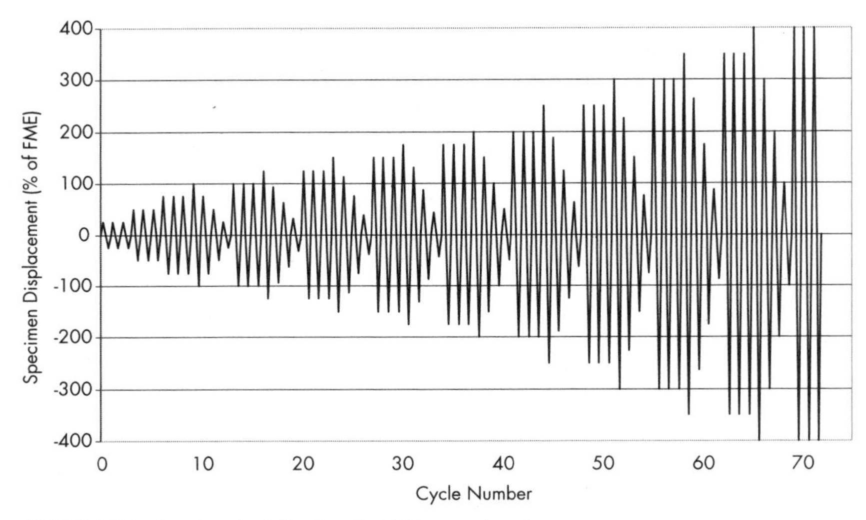

FIGURE 7.4 Load cycle following the SPD procedure.[6]

represent the energy dissipated by the shear wall. Observing the load vs. displacement hysteresis loops enables a slack system to be identified by the load essentially approaching zero during the degradation cycles. A slack system will lose the ability to dissipate energy at lower displacements. Slack systems may occur for some bolted connections (localized wood crushing causing loose joints) or some types of brittle systems; however, wood structural panel shear walls with mechanical fasteners, such as nails, generally do not behave as a slack system.

The next distinct portion of the load cycle is the stabilization load cycles. Each incremental displacement has three stabilization cycles. The intent of this cycle is for the strength and stiffness degradation to stabilize before the next incremental cycle. The stabilization cycle can be thought of as "at a given displacement, the wall incurs as much damage as it can"; then the next load cycle increment is initiated.

Although the original SPD proposed procedure was based on quasi-static loading, the SEAOSC procedure specifies that the tests will be conducted at a loading rate from 0.2–1.0 Hz. The specified load rate is at a lower (or at least on the low end) frequency than what is typically observed in an earthquake. There are several reasons to test at a lower frequency. First, it is believed that lower frequency tests will limit the inertial effects that might occur during the load tests. The second reason is that these types of tests are performed with the aid of an effective force-testing system. An effective force testing system simulates the dynamic loads of earthquakes by applying an effective force to the wall. This effective force is similar to forces generated by the mass of the structure and the ground acceleration. The effective force testing system is used in lieu of shake table testing. The lower frequencies are more attainable by typical effective force-testing systems.

In 1997, the International Conference of Building Officials Evaluation Service, Inc. (ICBO ES) adopted the SEAOSC test protocol by means of an acceptance criteria[8] for evaluation of prefabricated sheathed shear-wall elements. Although the acceptance criteria do not currently apply to site-built wood structural panel sheathed shear walls, it is believed that the recognition of the acceptance criteria sets a precedent for the evaluation for all light-frame lateral force resisting elements.

Also in 1997, the SEAOSC test method was presented to ASTM Committee E06 for recognition as a consensus test method. The SEAOSC test method follows the guidelines of ASTM format to expedite approval. During the time this chapter was being written, the test method was balloted several times. It is expected that the standard will be balloted at the main committee level soon.

In 1995, APA—The Engineered Wood Association installed a cyclic loading system that has the capability to test shear walls following the SEAOSC procedure. One of APA's objectives is to gain knowledge about shear-wall performance in the cyclic domain. An extensive multiyear test program has been initiated to answer many questions that have not been addressed in the cyclic test domain. A few of the tests programs are:

- Investigate the behavior of narrow shear walls subjected to cyclic loading and develop design recommendations for their design and construction.
- Analyze the behavior of walls that contain two dissimilar sheathing materials: for example, wood structural panel on one side, gypsum wallboard on the other side.
- Analyze the effect of sheathing orientation (vertical or horizontal).
- Analyze the effect of blocked and unblocked sheathing.
- Study alternative sheathing fastener systems.

The test programs are designed to study factors that have raised questions about the behavior of cyclically loaded shear walls. Due to the importance of resolving several of the initial issues, APA funded a series of tests at the University of California-Irvine (UCI). This test program consisted of eight different shear walls tested with fully reversed cyclic tests in accordance with the SEAOSC procedure (note that the test method was not finalized when these test were conducted, but the tests were conducted following a draft of the test method). However, no matching monotonic tests (either ASTM E72 or ASTM E564) were conducted on these shear walls. The preliminary findings[9] of these load tests indicate that the maximum shear loads tended to be lower than monotonic tested walls following ASTM E72 test procedures. Although this finding could be significant, the comparison is not based on matched specimens. For example, fastener fatigue failure, as related to cyclic load protocol, has a significant effect on the final results but is not a failure mode for monotonic testing.

Most of the UCI tests were conducted following the SPD test procedure. One test was conducted following a modified SPD procedure. This modified procedure eliminated the degradation cycles that occur after FME. Figure 7.5 is a representation of the load cycle followed for one of the eight tests.

Matched walls were tested following the SPD procedure (with degradation cycles and stabilization cycles) and the modified SPD procedure (stabilization cycles only). The two tests indicated that the degradation cycles did not significantly affect the results of wood structural panel shear-wall tests. The degradation cycles may not be necessary for future shear-wall tests, although other tests in progress with wood structural panels and other sheathing materials over steel framing indicate that degradation cycles may provide useful information.

One of the significant observations of the APA tests conducted at UCI was that sheathing nails around the perimeters exhibited fatigue. The majority of the load cycles in the SPD procedure exceed the FME displacement (yield limit state). Since wall damage is dependent on total number of cycles and the number of cycles and

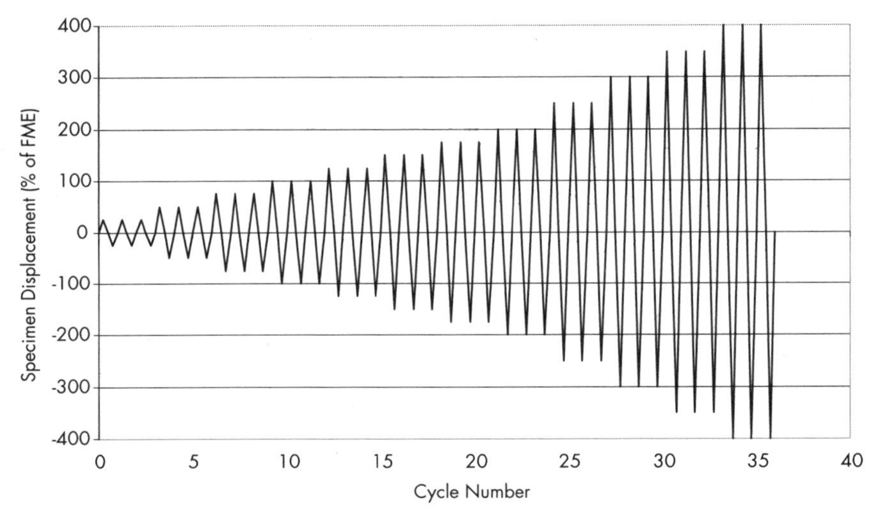

FIGURE 7.5 Simplified SPD load cycle as reported by Rose.[9]

the amount of displacements that exceed FME, it follows that the behavior of the wall will be dependent on the load history.

Since nail fatigue was not reported in observations of damaged shear walls after the Northridge, California, earthquake or other recent seismic events, Ficcadenti et al.[10] conducted cyclically loaded shear-wall tests following a modification of the SEAOSC procedure. Steel hold-down rods, similar to ASTM E72, were utilized. The hold-down rods were required on both ends of the test wall to resist overturning loads in both directions. The SPD load cycle was also modified by removing the stabilization cycles, but included the degradation cycles of the SPD procedure. Figure 7.6 illustrates the modification of the load cycle.

In these tests, failures due to nails pulling through the structural panel were observed. Smaller-sized nails and thinner panels were used in these tests. Some of the differences may be attributed to load history, and some may be attributed to the fastening conditions.

These walls also achieved maximum shear loads in excess of 3.0 times design load. The maximum shear capacity of the walls was achieved about 2 in. (50 mm) displacement. Conversely, several of the APA tests reached the maximum shear loads at around the 1 in. (25 mm) displacement cycles. Figure 7.7 is a typical hysteresis loop observed in one of the eight APA wall tests conducted at UCI. Note that the cycled shear capacity occurred at about 1 in. of displacement.

The maximum shear capacity of shear walls tested following the SEAOSC test procedure will clearly be affected by the fatigue failures of the nails. It appears unrealistic to assume that cyclically tested walls will achieve load factors that are as high as matching monotonically tested counterparts due to the difference in failure modes.

CUREe Protocol and Others. As previously discussed, the SPD loading protocol, in many researchers' opinions, is a very severe test cycle that causes a failure mode (nail fatigue) that is not typical of seismically loaded shear walls. Researchers realize the need and importance of cycle testing and its importance in aiding advanced shear wall modeling tools. However, it is also important to match real shear-

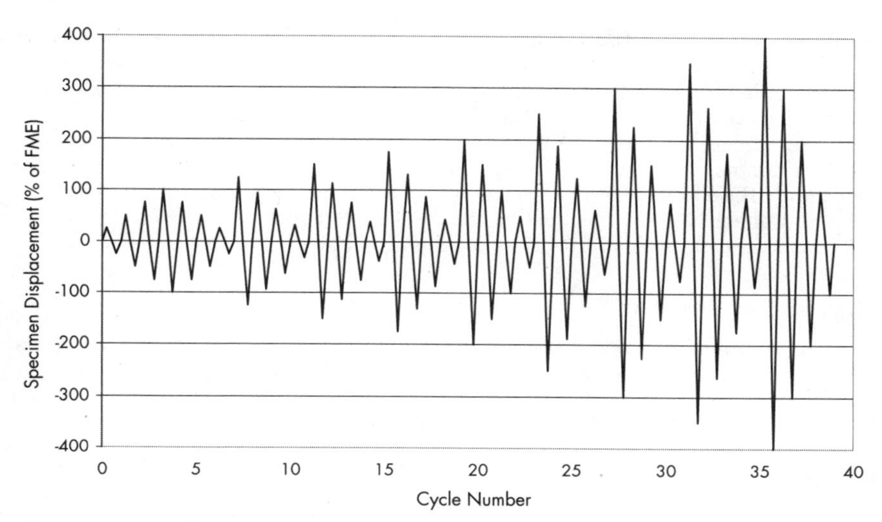

FIGURE 7.6 Modified SPD load cycle used by Ficcadenti et al.[10]

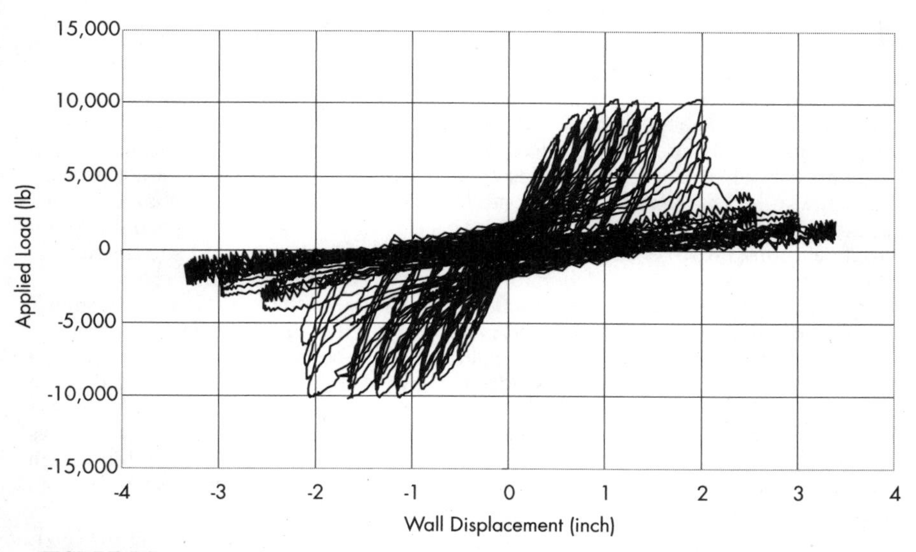

FIGURE 7.7 Load versus displacement response of the APA.

wall performance with lab-tested walls. Therefore, there has been a significant amount of research on the effects of load cycles. The authors' experience is mostly with North American researchers, but some work has been initiated on the international front and will eventually be finalized in an ISO standard.

In North America alone, the authors are aware of at least three different cyclic test regimes other than the ISO test cycle and the SPD loading protocol. Forintek Canada developed a test cycle that leads to more realistic failures (reported by Lam et al.[11] Forintek Research Reports available through their website and numerous conference proceedings).[12] He et al.[13] modified the Forintek procedure because the authors' opinion was that the Forintek load cycle was also too severe. Finally, one of the elements of the CUREe-Caltech Woodframe Project developed yet another cyclic test cycle. This load cycle and others was studied under another element of the Woodframe project (unpublished data, but will be available from the CUREe website).[14] Based on other researchers' conclusions and preliminary data from the CUREe-Caltech Woodframe Project, the SPD cycle appears to provide the lowest ultimate load capacity and lowest deflection at ultimate load.

Dynamic Test Methods

Pseudo-dynamic. Characterizing the performance of shear-wall assemblies though quasi-static cyclic testing provides a vast amount of data that can be very useful. Particularly, the data can be used to develop and refine computer-based models for more advanced design analysis of wood-framed buildings. However, an important concept should not be forgotten about the quasi-static testing. Cyclic testing only simulates the affects of an undefined seismic event in an undefined structure, which, in the authors' experience, is a difficult concept for many laypersons to understand. A very common question that laypersons ask when they visit the APA Laboratory and witness a cyclic test is "What magnitude earthquake did the cyclic test simulate?" The common answer to this question is that it depends on soil conditions, mass of the structure, plan of structure, and ground motion. Typically quasi-static cyclic testing, particularly the SPD cycle, subjects walls to many more displacement repetitions that would be expected from a seismic event.

Pseudo-dynamic testing is intended to simulate the effects of an assumed ground motion on an assumed structure. Karacabeyli and Ceccotti[15] summarize the results of their pseudo-dynamic tests. Based on previous either quasi-static cyclic or monotonic shear-wall testing, a nonlinear time history analysis can be conducted on an assumed building. Based on the time-history analysis, the deflection of an individual shear wall can be modeled. Once this deflection history is developed from the nonlinear time history analysis, a test shear wall can be constructed and tested with the deflection history under laboratory conditions. Furthermore, this testing can be conducted at a much slower testing rate, hence the name "pseudo-dynamic testing."

The advantage of this test method is that it actually simulates dynamic performance of shear walls, unlike the quasi-static tests. Another advantage is that the test equipment required to conduct pseudo-dynamic tests is much more common than what would be required to conduct a true dynamic test. The disadvantage of this method is that many assumptions are required to conduct this type of test, such as building geometry and ground motion. Thus, to obtain useful results for design inferences, a range of ground motions would need to be conducted on walls from a range of building types.

Pseudo-dynamic test methods can also be very effective in estimating the loads on a particular shear wall subjected to a particular event. Of course, detailed data must be known about this shear wall, so it is unlikely this method could be useful to simulate real seismic events. More likely, this method would be useful for simulating the effects of a laboratory type event, such as a shake table tests. One of the CUREe-Caltech Woodframe project elements will conduct pseudo-dynamic tests on a wall assembly similar to a wall in a full-sized structure that was subjected to shake table testing.

Shake Table Testing. As one can imagine, shake table testing has not been a terribly common occurrence for timber framed construction in North America. The reason for this lack of testing is several-fold: (1) the study of seismic design of wood-framed structures has traditionally not been viewed as academically rigorous by some universities, (2) historically, wood-framed structures have performed well in seismic events, which leads to (3) lack of funding for this type of research.

To the authors' knowledge, Dolan[16] was one of the pioneers in dynamic testing of wood framed elements. Dolan's tests were conducted on single wall elements, tested on a relatively small shake table at the University of British Columbia. Even though shake table testing had been conducted on single elements, shake table testing for full-sized structures has been limited in North America. An interesting point is that after the 1995 Kobe, Japan, earthquake, several full-sized structures were tested using the Kobe ground motion on very large shake tables in Japan. Although the results of these tests are interesting in a general sense, the construction techniques are somewhat different for North America and Japan.

To the authors' knowledge, the first full-sized wood-framed structures shake table test in North America was conducted in 2000 as part of the CUREe-Caltech Wood-framed project. The test structure was intended to be representative of a two-story residential structure. The findings of this testing are still being studied, but some important initial conclusions are that nonstructural elements such as gypsum wall-board and stucco provide significantly more stiffness than assumed. There is no doubt that the design methodology for wood-frame structures will be updated based on this more recent shake table testing. Finally, another element of the CUREe-Caltech Woodframe project is shake table testing of a multifamily dwelling with tuck-under parking commonly referred to as soft story construction. The historical performance of these types of structures has not been particularly good, given the torsional issue caused by the asymmetric soft story associated with the parking level. This testing will likely lead to retrofit recommendations for these types of buildings.

7.3.2 Shear-Wall Design

A shear wall behaves similarly to a horizontal diaphragm. In fact, a shear wall is simply a cantilevered diaphragm to which load is applied at the top of the wall and is transmitted out along the bottom of the wall. This creates a potential for overturning that must be accounted for and any overturning force is typically resisted by hold-downs, or tension ties, at each end of the shear-wall segments.

Table 7.1 and 7.2 present the tabulated values for structural-use panel sheathed wood frame shear walls for wind loading and seismic loading, respectively. The table applicable to wind loading lists design values that are 40% higher than the shear-wall tables intended for seismic loading. This 40% increase is to account for the better understanding and refinement of wind loading design. Furthermore, shear walls subjected to extreme wind loads have had a proven track record with excellent performance. Some model building codes have adopted the 40% increase. Thus, the designer should confirm that the increase is applicable under the local code.

Design Tables. Table 7.1 provides recommended shear (lb/ft) for structural-use panel shear walls with framing of Douglas fir, larch, or southern pine for seismic loading. Table 7.2 provides similar shear values for wind loading.

Other Design Considerations

Estimated Deflection. The deflection (Δ) of a blocked shear wall uniformly nailed throughout may be estimated by use of the following formula:

$$\Delta = \frac{8vh^3}{EAb} + \frac{vh}{G_v t_v} + 0.75\ he_n + \frac{h}{b}\ d_a \tag{7.1}$$

TABLE 7.1 Recommended Shear (lb/ft) for Structural-Use Panel Shear Walls with Framing of Douglas Fir, Larch, or Southern Pine[a] for Seismic Loading[b]

Panel grade	Minimum nominal panel thickness (in.)	Minimum nail penetration in framing (in.)	Panels applied directly to framing: Nail size (common or galvanized box)	Nail spacing at panel edges (in.): 6	4	3	2[e]	Panels applied over 1/2 in. or 5/8 in. gypsum sheathing: Nail size (common or galvanized box)	Nail spacing at panel edges (in.): 6	4	3	2[e]
	5/16	1 1/4	6d	200	300	380	510	8d	200	300	390	510
APA Structural I grades	3/8			230[d]	360[d]	460[d]	510[d]					
	7/16	1 3/8	8d	255[d]	395[d]	505[d]	670[d]	10d	280	430	550[f]	730
	15/32			280	430	550	730					
	15/32	1 1/2	10d	340	510	665[f]	870		–	–	–	–
APA Rated Sheathing; APA Rated Siding[g] and other APA grades except species group 5	5/16 or 1/4[c]			180	270	350	450		180	270	350	450
	3/8	1 1/4	6d	200	300	390	510	8d	200	300	390	510
	3/8			220[d]	320[d]	410[d]	530[d]					
	7/16	1 3/8	8d	240[d]	350[d]	450[d]	585[d]	10d	260	380	490[f]	640
	15/32			260	380	490	640					
	15/32			310	560	600[f]	770	–	–	–	–	–
	19/32	1 1/2	10d	340	510	665[f]	870	–	–	–	–	–
APA Rated Siding[g] and other APA grades except species group 5			Nail size (galvanized casing)					Nail size (galvanized casing)				
	5/16[c]	1 1/4	6d	140	210	275	360	8d	140	210	275	360
	3/8	1 3/8	8d	160	240	310	410	10d	160	240	310[f]	410

[a] For framing of other species: (1) find specific gravity for species of lumber in the AF&PA National Design Specification; (2) (a) for common or galvanized box nails, find shear value from table above for nail size for actual grade; (3) multiply value by the following adjustment factor: specific gravity adjustment factor = [1 – (0.5 – SG)], where SG = specific gravity of the framing. This adjustment shall not be greater than 1.

[b] All panel edges backed with 2 in. nominal or wider framing. Install panels either horizontally or vertically. Space nails maximum 6 in. o.c. along intermediate framing members for 3/8 in. and 7/16 in. panels installed on studs spaced 24 in. o.c. For other conditions and panel thickness, space nails maximum 12 in. o.c. on intermediate supports. Fasteners shall be located 3/8 in. from panel edges.

[c] 3/8 in. or APA Rated Siding 16 o.c. is minimum recommended when applied direct to framing as exterior siding.

[d] Shears may be increased to values shown for 15/32 in. sheathing with same nailing provided (1) studs are spaced a maximum of 16 in. o.c. or (2) if panels are applied with strength axis across supports.

[e] Framing at adjoining panel edges shall be 3 in. nominal or wider, and nails shall be staggered where nails are spaced 2 in. o.c. Check local code for variations of these requirements.

[f] Framing at adjoining panel edges shall be 3 in. nominal or wider, and nails shall be staggered where 10d nails having penetration into framing of more than 1½ in. are spaced 3 in. o.c. Check local code for variations of these requirements.

[g] Values apply to all-veneer plywood APA Rated Siding panels only. Other plywood siding panels may also qualify on a proprietary basis. Plywood siding 16 o.c. plywood may be 11/32 in., 3/8 in., or thicker. Thickness at point of nailing on panel edges governs shear values.

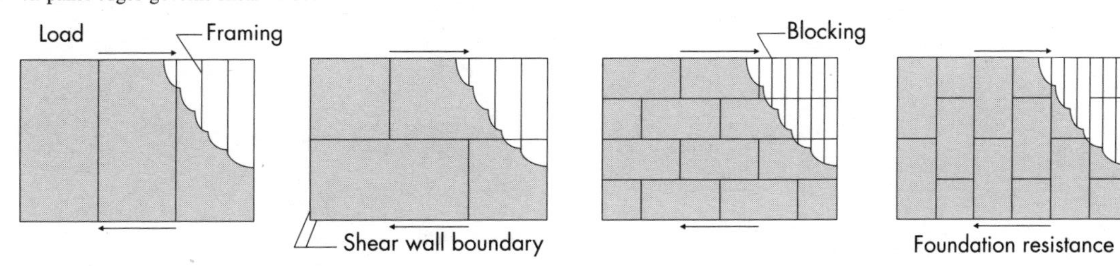

TABLE 7.2 Recommended Shear (lb/ft) for Structural-Use Panel Shear Walls with Framing of Douglas Fir, Larch, or Southern Pine for Wind Loading Only[b]

Panel grade	Minimum nominal panel thickness (in.)	Minimum nail penetration in framing (in.)	Panels applied directly to framing: Nail size (common or galvanized box)	Nail spacing at panel edges (in.): 6	4	3	2[e]	Panels applied over ½ in. or ⅝ in. gypsum sheathing: Nail size (common or galvanized box)	Nail spacing at panel edges (in.): 6	4	3	2[e]
APA Structural I grades	5/16	1¼	6d	280	420	545	715	8d	280	420	545	715
	3/8	1⅜	8d	320[d]	505[d]	645[d]	855[d]	10d	390	600	770[f]	1020
	7/16			355[d]	555[d]	705[d]	940[d]					
	15/32			390	600	770	1020					
	15/32	1½	10d	480	715	930[f]	1220		–	–	–	–
APA Rated Sheathing; APA Rated Siding[g] and other APA grades except species group 5	5/16 or 1/4[c]	1¼	6d	250	270	490	630	8d	250	380	490	630
	3/8			280	420	545	715		280	420	545	715
	3/8	1⅜	8d	310[d]	450[d]	575[d]	740[d]	10d	365	530	685[f]	895
	7/16			335[d]	490[d]	630[d]	820[d]					
	15/32			365	530	685	895					
	15/32	1½	10d	435	645	840[f]	1080	–	–	–	–	–
	19/32			475	715	930[f]	1220	–	–	–	–	–
APA Rated Siding[g] and other APA grades except species group 5			Nail size (galvanized casing)					Nail size (galvanized casing)				
	5/16[c]	1¼	6d	195	295	385	505	8d	195	295	385	505
	3/8	1⅜	8d	225	335	435	575	10d	225	335	435[f]	575

[a] For framing of other species: (1) find specific gravity for species of lumber in the AF&PA National Design Specification; (2) (a) for common or galvanized box nails, find shear value from table above for nail size for actual grade; (3) multiply value by the following adjustment factor: specific gravity adjustment factor = [1 – (0.5 – SG)], where SG = specific gravity of the framing. This adjustment shall not be greater than 1.

[b] All panel edges backed with 2 in. nominal or wider framing. Install panels either horizontally or vertically. Space nails maximum 6 in. o.c. along intermediate framing members for $\frac{3}{8}$ in. and $\frac{7}{16}$ in. panels installed on studs spaced 24 in. o.c. For other conditions and panel thickness, space nails maximum 12 in. o.c. on intermediate supports. Fasteners shall be located $\frac{3}{8}$ in. from panel edges.

[c] $\frac{3}{8}$ in. or APA Rated Siding 16 o.c. is minimum recommended when applied direct to framing as exterior siding.

[d] Shears may be increased to values shown for $\frac{15}{32}$ in. sheathing with same nailing provided (1) studs are spaced a maximum of 16 in. o.c. or (2) if panels are applied with strength axis across supports.

[e] Framing at adjoining panel edges shall be 3 in. nominal or wider, and nails shall be staggered where nails are spaced 2 in. o.c. Check local code for variations of these requirements.

[f] Framing at adjoining panel edges shall be 3 in. nominal or wider, and nails shall be staggered where 10d nails having penetration into framing of more than $1\frac{1}{2}$ in. are spaced 3 in. o.c. Check local code for variations of these requirements.

[g] Values apply to all-veneer plywood APA Rated Siding panels only. Other plywood siding panels may also qualify on a proprietary basis. Plywood siding 16 o.c. plywood may be $\frac{11}{32}$ in., $\frac{3}{8}$ in., or thicker. Thickness at point of nailing on panel edges governs shear values.

Notes: The recommended shears presented in Table 7.2 are based on the historically used design stresses for shear walls, which are listed in Table 7.1. Some building codes have allowed a 40% increase due to the better understanding of wind loading. The designer should confirm that the 40% increase is applicable under the local code.

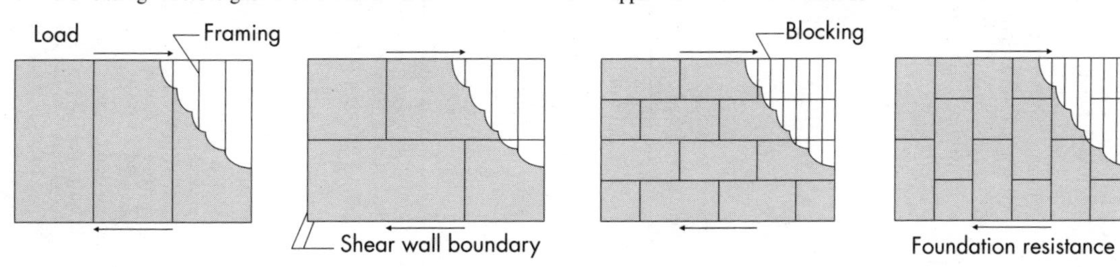

where Δ = calculated deflection (in.)
v = maximum shear due to design loads at the top of the wall (lb/ft)
h = wall height (ft)
E = elastic modulus of boundary element (vertical member at shear-wall boundary) (psi)
A = area of boundary element cross section (vertical member at shear wall boundary) (in.^{2})
b = wall width (ft)
$G_v t_v$ = rigidity through the thickness lb/ft (see Table 2.22). This value should be adjusted by the appropriate C_g factor listed in Table 2.22A.
e_n = nail deformation (in.) (see Table 7.3)
d_a = deflection due to anchorage details (rotation and slip at tie-down bolts). See Table 7.4 for range of hold-down deflections.

Overturning Moment. Overturning moments result from shear walls being loaded by horizontal forces. The overturning moments are resisted by force couples. The tension couple is typically achieved by a hold-down. Figure 7.8 and the equation below present a method for calculating overturning forces.

$$\text{Unit shear} = \frac{V}{L}$$

$$\text{Overturning force} = \text{chord force} = \frac{vh}{L}$$

TABLE 7.3 Fastener Slip Equations e_n (in.) for Use in Calculating Diaphragm and Shear-Wall Deflection Due to Nail Slip (Structural I)

Fastener	Minimum penetration (in.)	For maximum loads up to (lb)	Approximate slip, e_n (in.)[a,b]	
			Green/dry	Dry/dry
6d common nail	1¼	180	$(V_n/434)^{2.314}$	$(V_n/456)^{3.144}$
8d common nail	1⅜	220	$(V_n/857)^{1.869}$	$(V_n/616)^{3.018}$
10d common nail	1½	260	$(V_n/977)^{1.894}$	$(V_n/769)^{3.276}$
14-ga staple	1 to 2	140	$(V_n/902)^{1.464}$	$(V_n/596)^{1.999}$
14-ga staple	2	170	$(V_n/674)^{1.873}$	$(V_n/461)^{2.776}$

[a] Fabricated green/tested dry (seasoned); fabricated dry/tested dry. V_n = fastener load (lb/nail).
[b] Values based on Structural I plywood fastened to lumber with a specific gravity of 0.50 or greater. Increase slip by 20% when plywood is not Structural I.

TABLE 7.4 Approximate Deflection (in.) Due to Anchorage Detail d_a for Use in Shear-Wall Deflection

Hold-down type	Approximate range[a] of d_a (in.)
Bolted to post	0.15–0.25
Nailed to post	0.10–0.13
Lag screw to post	0.03–0.05

[a] These numbers are intended to represent an approximate range. A better estimate can be achieved by consulting the specific hold-down manufacturer.

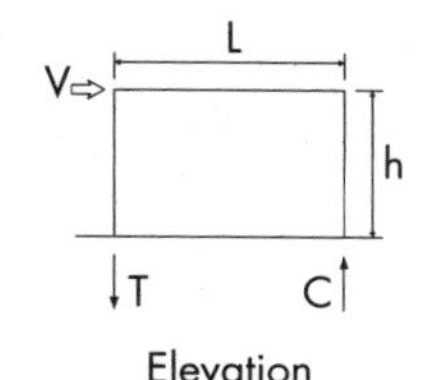

FIGURE 7.8 Overturning forces.

Examples

Example 7.1 Simple Shear-Wall Design for Wind Loading

Given:

- Commercial building
- Wind loading
- Wall requires ⅝ in. gypsum sheathing for one-hour fire rating
- Required shear-wall capacity 670 lb/ft (9,780 N/m).

Find:
Panel thickness, nail size and nailing schedule

solution Using Table 7.2, check the "Panels applied over ½ in. or ⅝ in. gypsum sheathing" area of table. Check "Rated Sheathing . . ." rows first since Structural I may not be readily available in all areas. From the table, note that 10d nails with 3 in. nail spacing at panel edges and 12 in. nail spacing at intermediate framing for a sheathing thickness of ⅜, 7/16, or 15/32 in. will provide a capacity of 685 lb/ft provided that the framing at adjoining panel edges is 3 in. nominal or wider. Because 685 lb/ft (10,000 N/m) is greater than 670 lb/ft (9780 N/m), this selection is OK for use.

Example 7.2 Simple Shear-Wall Design for Seismic Loading

Given:

- Residential building
- Seismic loading
- Typical wall sheathing thickness of 7/16 in.
- Typical nail size of 8d common
- Wall stud spacing of 24 in. o.c.
- Required shear-wall capacity is 435 lb/ft (6350 N/m).

Find:
Required nail spacing

solution Using Table 7.1, check the "Panels applied direct to framing" area of table. Check "Rated Sheathing . . ." rows first because Structural I may not be readily available in all areas. From the table, note that 7/16 in. structural-use panels with 8d nails spaced at 3 in. at the panel edges and 6 in. at intermediate framing (see footnote *b*) will provide a capacity of 450 lb/ft. Because 450 lb/ft (6570 N/m) is greater than 435 lb/ft (6350 N/m), this selection is OK for use.

Example 7.3 Shear-Wall Design (Traditional Segmented)

Design shear walls for the wall shown in Figure 7.9. The shear load on the wall from the diaphragm is 3 kips (13,344 N). The controlling load is assumed to be from wind pressures.

solution The total length of full-height segments is 10 ft (3.05 m). Note that 3.5:1 is the minimum shear wall aspect ratio for wind loading or for seismic loads in design categories A–C (per the 2000 IBC)[4] or seismic zones 1–3 (per the 1997 UBC). For seismic design category D–F (per the 2000 IBC) or seismic zone 4 (per the 1997 UBC) the maximum shear-wall aspect ratio is 2:1. Thus, the wall segments shown in this example are too narrow to meet the 2:1 aspect ratio criteria, but do meet the 3.5:1 criteria.

1. The unit shear is:

$$v = V/L = 3000/10 = 300 \text{ lb/ft } (4380 \text{ N/m})$$

2. Assuming the framing will be spruce-pine-fir (with specific gravity, SG = 0.42), the shear values from the capacity table Table 7.2 must be adjusted according to footnote *a*. The specific gravity adjustment factor (SGAF) is:

$$\text{SGAF} = 1 - (0.5 - \text{SG}) = 1 - (0.5 - 0.42) = 0.92$$

According to Section 2306.4.1 of the 2000 IBC, for wind loads the allowable shear capacities are permitted to be increased by 40%. The design shears in Table 7.2 have been adjusted in accordance with this section of the building code.

From the shear-wall design table, 7⁄16 in. (11.1 mm) wood structural panels with 8d common nails at 6 in. (152 mm) on supported edges will provide an allowable capacity, v_{allow} of:

$$v_{\text{allow}} = 365(0.92) = 335 \text{ lb/ft } (4890 \text{ N/m}) \geq 300 \text{ lb/ft } (4380 \text{ N/m})$$

Note the increase to 15⁄32 in. (11.9 mm) panels is taken assuming studs will be placed 16 in. (406 mm) o.c. and the panel will be oriented with the 8 ft (2.44 m) direction vertical.

3. The hold-downs must be located at the ends of each full-height segment as shown in Figure 7.9 and designed to resist, H:

$$H = vh = 300(8) = 2400 \text{ lb } (10{,}700 \text{ N})$$

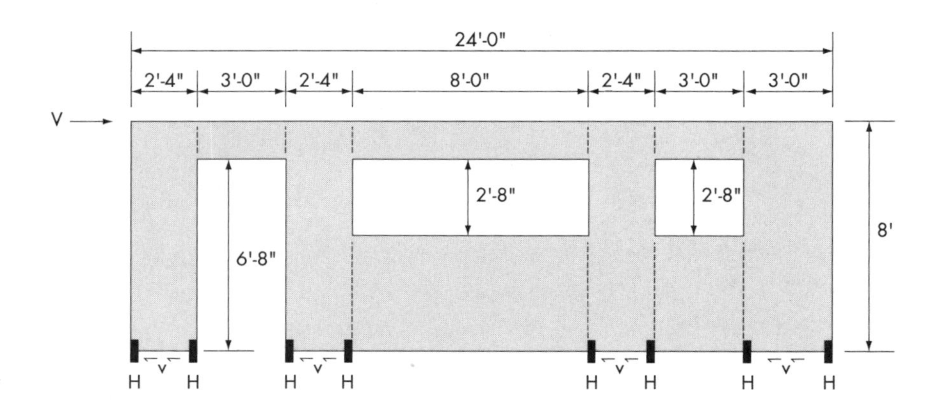

FIGURE 7.9 Building elevation for segmented shear-wall example.

Example 7.4 Shear-Wall Deflection

Calculate the deflection of the shear wall in Example 7.3.

solution The total shear-wall deflection will be considered to be a function of the deflection of the full-height segments. For this example a weighted average based on wall rigidities will be used to calculate the total shear-wall deflection. Wall rigidities will be assumed to be relative to wall length, assuming consistent framing and nailing patterns. Another approach to this problem would be to iterate using the load-deflection equation to converge on the displacement caused by the total load, assuming the wall segments all deflect an equal amount.

Shear-wall deflection analysis usually involves engineering judgment. In this example the walls are narrow, with an approximate aspect ratio of 3.5:1. The accuracy of the shear wall deflection equation at aspect ratios greater than 2:1 is questionable. However, in the absence of formal guidelines for narrow shear-wall deflection, the code equation will be used. Other aspects that would reasonably influence the accuracy of wall deflection calculations (by stiffening the wall) are the presence of sheathing above and below openings, and wall finish materials (such as siding, stucco, and gypsum). No formal guidelines currently exist for these aspects either, but there are several large research projects investigating shear wall behavior, and in the near future guidelines will likely be developed. In the meantime, an estimate will be made as follows.

The deflection of the 2.33 ft (0.71 m) wall segment is calculated with the following equation (using English units):

$$\Delta = \frac{8vh}{EAb} + \frac{vh}{Gt} + 0.75he_n + \frac{hd_a}{b}$$

where v = 300 lb/ft, unit load
h = 8 ft, wall height
E = 1,200,000 psi, for s-p-f studs—stud grade
A = 10.5 in.^{2}, for two 2 × 4 vertical end studs
$G_v t_v = Gt$ = 27000(3.1) lb/in., for 7⁄16 OSB (from Chapter 2)
b = 2.33 ft, wall width

Load per nail, v_{nail}:

$$v_{nail} = v/(12/S) = 300/(12/6) = 150 \text{ lb/nail (where } S = \text{nail spacing, in.)}$$

Nail slip, e_n (equation from Table 7.3 using English units):

$$e_n = 1.2(v_{nail}/857)^{1.869} = 1.2(150/857)^{1.869} = 0.0462 \text{ in.}$$

Hold-down slip, d_a = 0.11 in. (2.8 mm), from hold-down manufacturer's catalog

The deflection of each component is:

$$\Delta_{bending} = 8vh^3/(EAb) = 8(300)8^3/(1{,}200{,}000(10.5)2.33) = 0.042 \text{ in. (1.07 mm)}$$

$$\Delta_{shear} = vh/(Gt) = 300(8)/(27{,}000(3.1)) = 0.029 \text{ in. (0.74 mm)}$$

$$\Delta_{nail\ slip} = 0.75h(e_n) = 0.75(8)0.0462 = 0.277 \text{ in. (7.04 mm)}$$

$$\Delta_{hold\text{-}down} = h(d_a)/b = 8(0.11)/2.33 = 0.377 \text{ in. (9.58 mm)}$$

The total shear-wall deflection, Δ, is a summation of each component:

$$\Delta = 0.042 + 0.029 + 0.277 + 0.377 = 0.725 \text{ in. (18.4 mm)}$$

The deflection for the 3 ft wall segment is calculated (not shown) as 0.632 in. (16.05 mm). Table 7.5 summarizes the wall segment relative rigidities, load, and deflection.

The total wall deflection is calculated as a weighted average:

$$\Delta = 3(0.23)0.725 + 0.3(0.632) = 0.69 \text{ in. (17.53 mm)}$$

Shear-wall deflection is important for checking drift limitations and determining whether the diaphragm should be considered rigid or flexible.

TABLE 7.5 Relative Shear-Wall Rigidities for Example 7.4

Wall segment	Length	R^a	$R/\Sigma R$	V^b lb	v lb/ft	Δ (in.)
1	2.33	0.78	0.23	700	300	0.725
2	2.33	0.78	0.23	700	300	0.725
3	2.33	0.78	0.23	700	300	0.725
4	3	1.00	0.30	901	300	0.632
Σ	10.0	3.33		3000		

$^a R$ = relative rigidity based on wall length.
^{b}Shear distributed to wall segments in proportion to wall length.

7.3.3 Advanced Shear-Wall Topics

Principles of Mechanics. The 2000 International Building Code[4] allows principles of mechanics to be used to calculate the resistance of shear walls based on fastener strength and sheathing shear resistance per Section 2305.1.1. Tissell provides guidance in the principles of mechanics calculations, published in APA Research Report 154.[17] Tissell's approach is based on single-fastener design values adjusted by various empirically determined factors that account for such things as nail spacing, panel grade, framing size, stud spacing, and the potential for panel buckling. APA originally derived Table 7.1 based on this type of approach. Through the years, the load-duration factor, diaphragm factor, and nail design values have changed, and therefore following Tissell's approach will not provide perfect agreement with Table 7.1. However, in most cases the agreement is reasonable. When the individual fasters design values changed, APA chose not to change the shear-walls tables accordingly because, through testing, the design values have been demonstrated to be appropriate.

The principles of mechanics approach can be a very useful tool, especially when, for some reason or another, the shear wall is beyond the scope of Table 7.1 or 7.2. It should be noted that Tissell's approach does not consider crushing of the top and bottom plate, which, based on recent testing experience, may result in further empirical adjustments to account for this. Crushing of the top and bottom plate can be an issue for highly loaded shear walls; it does not appear to be a significant issue for walls within the scope of Table 7.1 or 7.2. APA recommends that of Table 7.1 or Table 7.2 be used in most cases over the principles of mechanics approach; however, sometimes a tabulated solution is not an option.

Shear Transfer Around Openings (pier method). Traditional shear walls only consider full-height segments of wood structural panels as being effective in resisting shear forces. Therefore, any realistic wall line that contains openings will consist of one or more segments of shear walls. Each full-height segment must also be detailed to resist overturning forces, which often requires tie-downs at each end of the full-height segments.

Another approach is to treat the entire wall as one unit, hence reducing the number of tie-downs. This method requires the shear forces to be transferred around the openings using the sheathing above and below openings, and some time may require additional metal strapping to supplement the strength of the sheathing. The 1999 SEAOC Bluebook[18] credits Ed Diekmann, S.E., with developing this approach circa 1982. The approach is similar to the "pier method" in masonry design, and appears in the 2000 International Building Code[4] in Section 2305.3.7.1 under the title *Force Transfer Around Openings.*

Two main advantages of this method are that it reduces the number of tie-downs requires for a wall line and allows the designer to use narrower piers since the aspect ratio limitations in the building code apply to the pier, as opposed to the full-height segments. The disadvantage of this method is that it can require a large amount of detailed calculations, especially if the wall becomes very complicated. It also requires special field detailing, extra blocking, extra nailing, and/or additional straps to achieve the field detailing.

Three good references for additional information are Ed Diekmann's shear-wall chapter in the Faherty and Williamson Handbook,[19] Duquette's *Timber Solutions Manual,*[20] and the SEAOC *Seismic Design Manual,* volume 2.[21]

Perforated Shear-Wall Approach. Intuitively, one could argue that there are two ways to design walls with openings. One method is to reinforce the openings, discussed in the previous subsection; the other option is to place a penalty on the shear wall in some form or fashion. The concept of providing an empirically developed penalty is the basis for the perforated shear-wall approach. Based on a significant amount of monotonic and cyclic testing in both the United States and Japan, an empirically developed penalty was developed that accounts for maximum wall opening size and percent wall sheathing (Table 7.6). The shear capacity adjustment factor is applied as a reduction factor to Tables 7.1 and 7.2. Technically, any wall with openings is classified as a perforated shear wall. *Perforated shear walls* can have several meanings, but in general the term implies the empirically developed approach.

TABLE 7.6 Shear Capacity Adjustment Factor for Use with the Perforated Shear-Wall Method

	Maximum opening height[a]				
Wall height	$H/3$	$H/2$	$2H/3$	$5H/6$	H
8 ft wall	2 ft, 8 in.	4 ft, 0 in.	5 ft, 4 in.	6 ft, 4 in.	8 ft, 0 in.
10 ft wall	3 ft, 4 in.	5 ft, 0 in.	6 ft, 8 in.	8 ft, 4 in.	10 ft, 0 in.
Percent full-height sheathing[b]	Effective shear capacity ratio				
0%	1.00	0.67	0.50	0.40	0.33
10%	1.00	0.69	0.53	0.43	0.36
20%	1.00	0.71	0.56	0.45	0.38
30%	1.00	0.74	0.59	0.49	0.42
40%	1.00	0.77	0.63	0.53	0.45
50%	1.00	0.80	0.67	0.57	0.50
60%	1.00	0.83	0.71	0.63	0.56
70%	1.00	0.87	0.77	0.69	0.63
80%	1.00	0.91	0.83	0.77	0.71
90%	1.00	0.95	0.91	0.87	0.83
100%	1.00	1.00	1.00	1.00	1.00

[a] H = the vertical dimension of the tallest opening in the shear wall. Where areas above and below an opening remain unsheathed, the height of the opening shall be defined as the height of the wall.

[b] The sum of the lengths of the shear segments that are sheathed full height and meet the aspect ratio requirements of the building code divided by the total length of the shear wall.

The main advantage of the perforated shear-wall approach is that it reduces the number of tie-downs, is simple to apply, and can lead to very minor penalties for small openings. The disadvantage of this system is that it can lead to very large penalties for large opening sizes, which essentially forces the designer to use the pier method or the traditional segmented approach or redefine the perforated wall to eliminate large openings (see Example 7.5 and 7.6 for this comparison). Another disadvantage is that currently there is no theoretical derivation of the empirical adjustment factors. Some engineers have difficulty accepting a so-called black box approach without an engineering mechanics backup. It is also the authors' experience that many engineers are comfortable using the method for wind design but argue that it is not appropriate for seismic design. As previously noted, a reasonable amount of cyclic testing has been conducted on this methodology, and the empirical adjustment factors have always proven to be conservative with respect to the ultimate strength of the tested walls.

This method was first introduced in the building codes in 1995. The Standard Building Code was the first to accept the methodology. The most current building code reference for this method is the 2000 International Building Code[4] in Section 2305.3.7.2 under the title *No Force Transfer Around Openings*. In addition, the 2000 NEHRP Provisions[22] provide the most current thinking on the subject, which is slightly more up to date than the 2000 IBC. The 2000 NEHRP Commentary[21] also provides a detailed design example for the perforated shear-wall method.

Example 7.5 Shear Wall Designed with Openings (Using the Perforated Shear-Wall Approach)

This example will show an empirical shear-wall design method that accounts for the effect of openings using the perforated shear wall method.

The same wall section as shown in Example 7.3 will be redesigned as a perforated shear wall to highlight the differences between the two methods. The entire wall, not just full-height segments, will be considered as the shear wall and the openings will be accounted for with a shear capacity adjustment factor (SCAF). Hold-downs are always required at the ends of shear walls. With this design method hold-downs are required only at the ends of the wall since the entire wall is treated as one shear wall with openings.

Design the shear walls for the wall shown in Figure 7.10. The shear load on the wall from the diaphragm, V, is 3 kips (13.34 kN) from wind pressures. The length of the perforated shear wall is defined by hold-down location, H, as shown in Fig. 7.10.

solution

1. The unit shear in the wall is 300 lb/ft (4380 N/m) (see Example 7.3)
2. Design variables:

   The specific gravity adjustment factor (SGAF) is 0.92 (from Example 7.3).

   The allowable shear capacities can be increased by 40% for wind loads, per section 2306.4.1.[4] Alternatively, Table 7.2 can be used which implements the 40% increase.

   The length of full-height segments is 10 ft (3.05 m). See note from Example 7.3 regarding the narrow full-height segments shown here. Two items are needed for finding the shear capacity adjustment factor (SCAF):

   a. percent full height sheathed

   b. maximum opening height

   The percent full-height sheathed is the length of the full-height segments divided by the total length of wall = 10/24 = 0.42%. The maximum opening height is 6 ft, 8 in. From Table 7.6 or Section 2305.3.7 of the 2000 IBC the SCAF is 0.53 (conservative using 40% without interpolation).

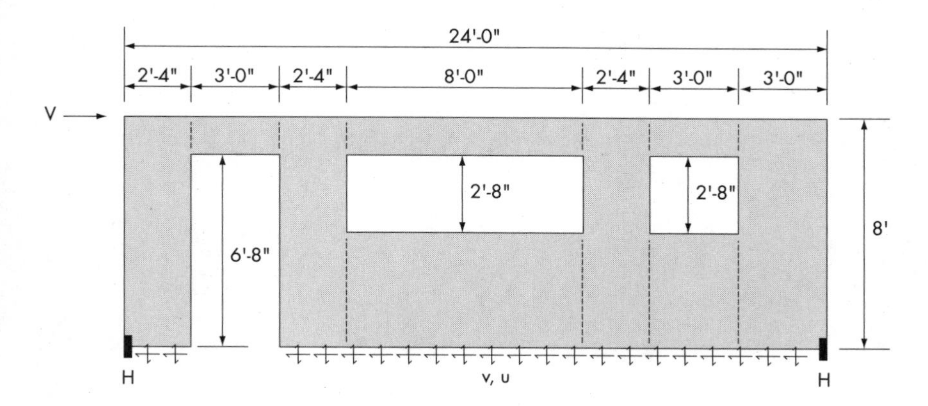

FIGURE 7.10 Building elevation for perforated shear-wall example.

3. From Table 7.2, the shear-wall design table for wind loading, 7⁄16 in. (11.1 mm) wood structural panels with 8d common nails spaced at 3 in. (76.2 mm) at panel supported edges will provide an adjusted allowable capacity, v_{allow} of:

$$v_{\text{allow}} = 630(0.92)(0.53) = 307 \text{ lb/ft } (4480 \text{ N/m}) \geq 300 \text{ lb/ft } (4380 \text{ N/m}) \quad \therefore \text{OK}$$

4. The hold-downs must be designed to resist the unadjusted allowable shear capacity = 450 lb/ft (note that the 40% increase in the wind resistance is reversed for this calculations, 630/1.4 = 450):

$$H = vh = 450(8) = 3600 \text{ lb } (16{,}000 \text{ N})$$

Additionally, uplift and shear connections must be provided along the base of the wall between hold-downs to resist the unadjusted allowable shear capacity (450 lb/ft), v and u shown in Fig. 7.10.

5. Provisions for calculating the total shear-wall deflection of a perforated shear wall state that the total deflection shall be based on the maximum deflection of any full height segment divided by the SCAF. Using the deflection equation from Example 7.4 with all terms the same but with 3 in. o.c. edge nailing, the deflection of the 2.33 ft wall segment becomes 0.524 in. The total perforated shear wall deflection is calculated as:

$$\Delta = 0.524/\text{SCAF} = 0.524/0.53 = 0.989 \text{ in. } (21.12 \text{ mm})$$

Example 7.6 Shear Wall Designed with Openings (Using the Perforated Shear-Wall Approach)

Repeat Example 7.4 but in this example the perforated shear wall will be defined with hold-downs as shown in Fig. 7.11 which eliminates the effect of the door at one end. The length of the perforated shear wall is 18.67 ft (5.7 m). The length of full-height segments is 7.67 ft (2.33 m).

solution

1. The unit shear in the wall is:

$$v = V/18.67 = 3000/18.67 = 392 \text{ lb/ft } (5720 \text{ N/m})$$

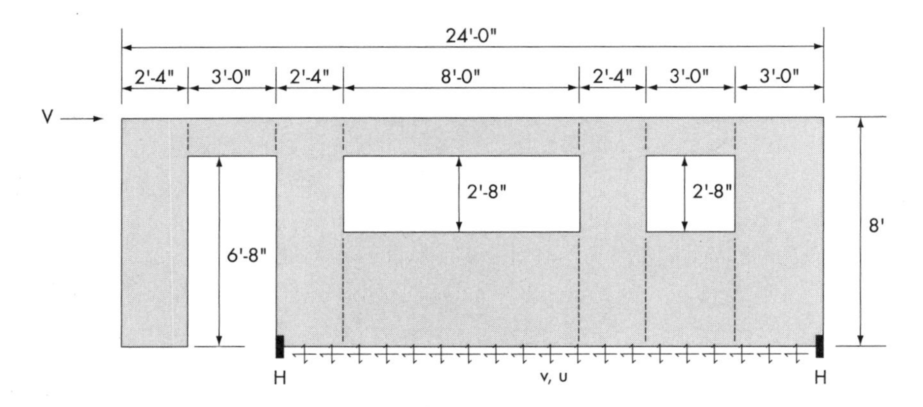

FIGURE 7.11 Building elevation for perforated shear wall example with alternate wall length definition.

2. Design variables:

   The specific gravity adjustment factor (SGAF) is still 0.92 (from Example 7.3).

   The allowable shear capacities can be increased by 40% for wind loads, per section 2306.4.1 (IBC).

   The percent full height sheathed is the length of the full-height segments divided by the total length of wall = 7.67/18.67 = 0.41%. The maximum opening height is now 2 ft, 8 in. From Table 7.6 or Section 2305.3.7, the SCAF is 1.0.

3. From Table 7.2, 7⁄16 in. (11.1 mm) wood structural panels with 8d common nails spaced 4 in. o.c. at supported panel edges will provide an adjusted allowable capacity, v_{allow} of:

$$v_{allow} = 490(0.92)(1.0) = 451 \text{ lb/ft } (6582 \text{ N/m}) \geq 392 \text{ lb/ft } (5720 \text{ N/m}) \quad \therefore \text{OK}$$

4. The hold-downs must be designed to resist the unadjusted allowable shear capacity = 350 (note that the 40% increase in the wind resistance is reversed for this calculations, 490/1.4 = 350):

$$H = vh = 350(8) = 2800 \text{ lb } (12{,}500 \text{ N})$$

   Additionally, uplift and shear connections must be provided along the base of the wall between hold downs to resist the unadjusted allowable shear capacity (350 lb /ft), v and u shown in Fig. 7.11.

5. As in Example 7.4, the deflection equation from Example 7.3 with all terms the same but with 4 in. o.c. edge nailing is used here. The deflection of the 2.33 ft wall segment becomes 0.578 in. The total perforated shear wall deflection is calculated as:

$$\Delta = 0.578/\text{SCAF} = 0.578/1.0 = 0.578 \text{ in. } (14.68 \text{ mm})$$

commentary This example and Example 7.4 show that by redefining the perforated shear wall to eliminate the door opening, fewer nails are required and smaller shear, uplift, and hold-down forces must be resisted and a smaller deflection results.

Portal Frame Approach. Rose and Skaggs[24] present an additional method for handling large openings in shear walls. The approach is to treat a wall system as a moment frame, or portal frame. The methodology is to extend the header material into the field of the shear wall, then to overlap the wood structural panel sheathing over the header, and nail the sheathing to the header with a heavy nailing grid. The nailing grid provides moment rotation for the portal frame and transfers the moment utilizing the sheathing. This method allows for narrow wall segments and can be a solution to narrow shear walls for garage door fronts.

Some manufacturers of prefabricated shear-wall elements present a similar solution using the proprietary components. For additional information on site-built portal frames, consult APA Research Report 154.[17]

7.4 DIAPHRAGMS

Wood structured panel diaphragms have been used extensively for roofs, floors, and walls in both new construction and rehabilitation of older buildings. A complete diaphragm analysis includes analyzing chord stresses, connections, and tie-downs.

A diaphragm acts in a manner analogous to a deep beam or girder, where the panels act as a web resisting shear while the diaphragm edge members perform the function of flanges resisting bending stresses. These edge members are commonly called chords in diaphragm design, and may be joists, ledgers, trusses, bond beams, studs, etc.

Due to the great depth of most diaphragms in the directions parallel to application of load and to their means of assembly, their behavior differs slightly from that of the usual relatively shallow beam. Shear stresses have been proven essentially uniform across the depth of the diaphragm rather than showing significant parabolic distribution as the web of a shallow beam. Similarly, chords in a diaphragm carry all flange stresses acting in simple tension and compression rather than sharing these stresses significantly with the web. As in any beam, consideration must be given to bearing stiffeners, continuity of webs and chords, and web buckling that is normally resisted by framing members.

Diaphragms vary considerably in load-carrying capacity, depending on whether they are blocked or unblocked (see Fig. 7.12). Blocking consists of lightweight nailers, usually 2 × 4s, framed between the joists or other primary structural supports for the specific purpose of connecting the edges of the panels. Systems that provide support framing at all panel edges, such as panelized roofs, are also considered blocked. The reason for blocking in diaphragms is to provide for connection of panels at all edges for better shear transfer. Unblocked diaphragm loads may be controlled by buckling of the unsupported panel edges, which would result in an ultimate capacity of the diaphragm being unaffected by increasing the nail schedule since more nailing will have little effect on the buckling performance of the panels. For the same nail spacing, design load on a blocked diaphragm can be designed as high as two times the design load of its unblocked counterpart.

7.4.1 Diaphragm Testing

Static Test Methods. One of the first investigations of wood structural panel diaphragms was conducted by Countryman in 1951.[25] This research was directed primarily at establishing the concept that plywood functioned as an efficient shear-

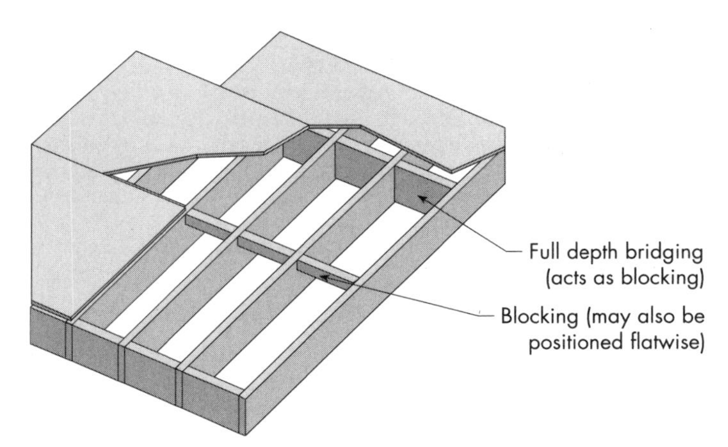

FIGURE 7.12 Blocking of adjacent panel edges.

resistant diaphragm and formed a basis for plywood diaphragm design. This testing, coupled with the shear-wall testing, was used to develop a diaphragm table that first appeared in the 1952 Uniform Building Code.

Three years later, additional research was conducted to expand the knowledge through investigations of the effect of panel layout, blocking, and orientation of the framing and plywood panel joints relative to the load.[26] The 1955 edition of the UBC began the tradition of footnoting the diaphragm table for further clarification.

In 1966, an additional 19 full-sized diaphragms were tested.[27] This research reflected changes in the manufacture of plywood due to the promulgation of U.S. Product Standard PS 1-66. Other construction variables were also evaluated, such as short plywood nails, panelized roof sections, and plywood over steel bar joists. The aforementioned reports are no longer in print, but fortunately Tissell and Elliot summarized these earlier diaphragm tests in Appendix A of APA Research Report 138.[1]

Research Report 138 also reports the original testing that was conducted in the late 1970s and early 1980s. The testing was intended to investigate diaphragms that provided higher resistance than the standard building code tables. Panelized construction, multilayers of panels, as well as multiple rows of fasteners were investigated during this series of tests. This testing is the original basis for APA's ICBO Evaluation Service Evaluation Report 1952.[28]

In addition to the research done at the facilities of APA, plywood diaphragms have been tested at other laboratories, particularly the Forest Products Laboratory at Oregon State University.

Interestingly enough, much of the historical diaphragm testing was conducted prior to the development of the standardized test method for conducting diaphragm tests, ASTM Standard E455.[29] The authors' have reviewed ASTM E455 as well as all of the APA reports prior to the first version of E455, and it appears that the monotonic testing conducted by APA was generally in accordance with ASTM E455.

Quasi-static Cyclic Test Methods. A significant amount of diaphragm testing has been conducted in the monotonic domain. However, testing diaphragms using cyclic procedures has been fairly limited. A large testing program funded by the National

Science Foundation was conducted in the early 1980s by a consortium of Southern California Engineering firms.[30] The authors of the report call the testing quasi-static cyclic. However, it appears that some type of nonlinear modeling was conducted to generate actual diaphragm deflections from seismic events. The scope of the testing was intended on developing mitigation techniques for unreinforced masonry buildings. The ABK research apparently has led to the development of monumental rehabilitation techniques in this area.

Dynamic Test Methods. Although the dynamic properties of shear walls are limited, the properties of diaphragms tested under dynamic loads are even more limited. The lack of testing can likely be attributed to three factors:

1. Diaphragms elements have typically performed well in past seismic events.
2. The problems related to diaphragms have been related to detailing of the diaphragm, especially anchorage issues.
3. Diaphragms are typically very large, hence expensive to test.

To the authors' knowledge, the only full-sized shake table tests in North America studying diaphragms are part of the CUREe-Caltech Woodframe project. As a follow-up to the shake table testing, individual element testing following monotonic and quasi-static cyclic testing is part of this project. The results of the CUREe-Caltech Woodframe project will undoubtedly improve our understanding of the behavior of wood diaphragms.

7.4.2 Diaphragm Design

Design Table. Table 7.7 presents the tabulated values for both blocked diaphragms and unblocked diaphragms for wind loading and seismic loading. As with shear walls, some building codes have implemented a 40% increase in diaphragm resistances. The designer should confirm that the increase is applicable under the governing code.

Estimated Deflection. Calculations for diaphragm deflection shall account for the usual bending and shear components as well as many other factors, such as nail deformation, that will contribute to the deflection.

The deflection (Δ) of a blocked structural-use panel diaphragm uniformly nailed throughout may be estimated by use of the following formula:

$$\Delta = \frac{5vL^3}{8EAb} + \frac{vL}{4G_vt_v} + 0.188Le_n + \frac{\sum(\Delta_cX)}{2b} \tag{7.2}$$

where Δ = the calculated deflection (in.)

v = maximum shear due to design loads in direction under consideration (lb/ft)

L = diaphragm length (ft)

E = elastic modulus of chords (psi)

A = area of chord cross section (in.^{2})

b = diaphragm width (ft)

G_vt_v = rigidity through the thickness (lb/ft) (see Table 2.22). This value should be adjusted by the appropriate C_g factor listed in Table 2.22A.

e_n = nail deformation (in.) (see Table 7.3)

$\Sigma(\Delta_cX)$ = sum of individual chord-splice slip values of the diaphragm, each multiplied by its distance to the nearest support

TABLE 7.7 Recommended Shear (lb/ft) for Horizontal Structural-Use Panel Diaphragms with Framing of Douglas Fir, Larch or Southern Pine[a] for Wind or Seismic loading[b]

Panel grade	Common nail size	Minimum nail penetration in framing (in.)	Minimum nominal panel thickness (in.)	Minimum nominal width of framing member (in.)	Blocked diaphragms: Nail spacing (in.) at diaphragm boundaries (all cases), at continuous panel edges parallel to load (cases 3 and 4), and at all panel edges (cases 5 and 6): 6 / Nail spacing (in.) at other panel edges (cases 1, 2, 3, and 4)[b]: 6	4 / 6	2½[c] / 4	2[c] / 3	Unblocked diaphragms: Nails spaced 6 in. max. at supported edges[b]: Case 1 (no unblocked edges or continuous joints parallel to load)	All other configurations (cases 2, 3, 4, 5, and 6)
APA Structural I grades	6d[e]	1¼	5/16	2	185	250	375	420	165	125
				3	210	280	420	475	185	140
	8d	1⅜	3/8	2	270	360	530	600	240	180
				3	300	400	600	675	265	200
	10d[d]	1½	15/32	2	320	425	640	730	285	215
				3	360	480	720	820	320	240
APA Rated Sheathing, APA Rated Sturd-I-floor and other APA grades except species group 5	6d[e]	1¼	5/16	2	170	225	335	380	150	110
				3	190	250	380	430	170	125
			3/8	2	185	250	375	420	165	125
				3	210	280	420	475	185	140
	8d	1⅜	3/8	2	240	320	480	545	215	160
				3	270	360	540	610	240	180
			7/16	2	255	340	505	575	230	170
				3	285	380	570	645	255	190
			15/32	2	270	360	530	600	240	180
				3	300	400	600	675	265	200
	10d[d]	1½	15/32	2	290	385	575	655	255	190
				3	325	430	650	735	290	215
			19/32	2	320	425	640	730	285	215
				3	360	480	720	820	320	240

[a]For framing of other species: (1) find specific gravity for species of lumber in the AF&PA National Design Specification; (2) (a) For common or galvanized box nails, find shear value from table above for nail size for actual grade; (3) Multiply value by the following adjustment factor: specific gravity adjustment factor = $[1 - (0.5 - SG)]$, where SG = specific gravity of the framing. This adjustment shall not be greater than 1.

[b]Space nails maximum 12 in. o.c. along intermediate framing members (6 in. o.c. when supports are spaced 48 in. o.c.). Fasteners shall be located 3/8 in. from panel edges.

[c]Framing at adjoining panel edges shall be 3 in. nominal or wider, and nails shall be staggered where nails are spaced 2 in. o.c. or 2½ in. o.c.

[d]Framing at adjoining panel edges shall be 3 in. nominal or wider, and nails shall be staggered where 10d nails having penetration into framing of more than 1⅝ in. are spaced 3 in. o.c.

[e]8d is recommended minimum for roofs due to negative pressures of high winds.

Notes: (1) Design for diaphragm stresses depends on direction of continuous panel joints with reference to load, not on direction of long dimension of sheet. Continuous framing may be in either direction for blocked diaphragms. (2) Some building codes have allowed a 40% increase due to the better understanding of wind loading. The designer should confirm that the 40% increase is applicable under the governing code.

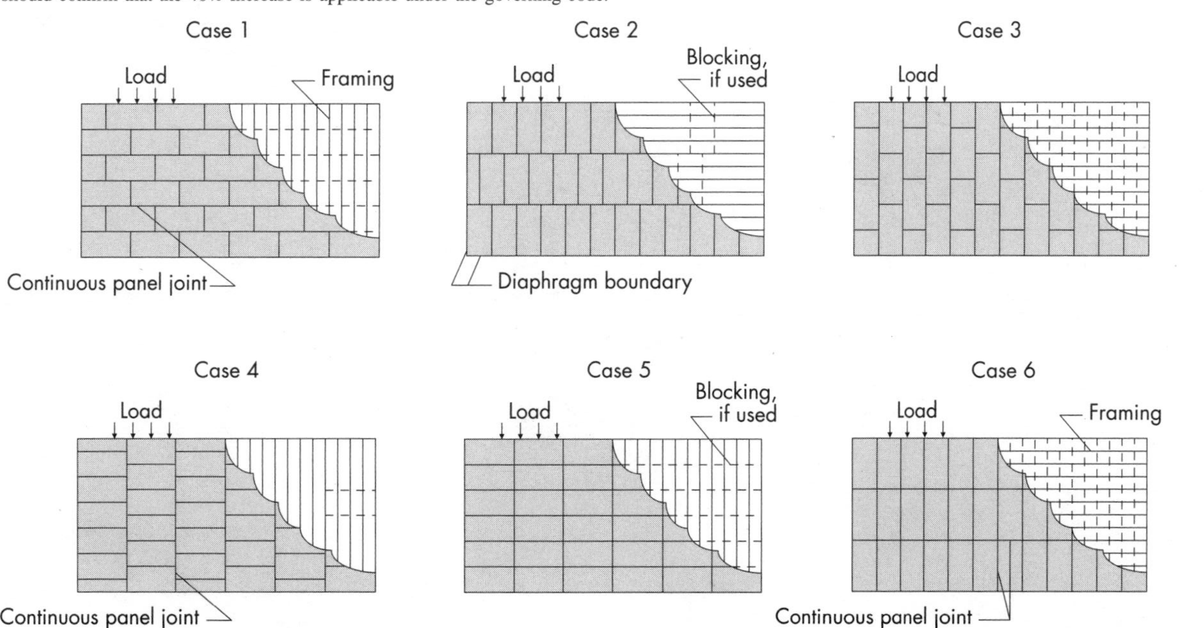

If not uniformly nailed, the constant 0.188 in the third term must be modified accordingly. The recommendation from Applied Technology Council 7[31] is to modify the constant 0.188 in proportion to the average load on each nail compared to the average load on each nail that would have been present had the constant nail schedule been maintained. Example 7.7 illustrates a simple way to modify the 0.188 factor.

Example 7.7 Determining the Nail Slip Modification Factor for Nonuniformly Nailed Diaphragms

Given:

- Commercial building with a floor plan of 150 × 50 ft
- Design for seismic loading, which generates a uniform load on the diaphragm of 200 lb/ft
- Panelized roof with long dimension of panels parallel to supports; therefore assume case 4 fully blocked diaphragm layout
- Nail size of 8d common (note that a more common nailing would be 10d common for this diaphragm, but 8d was used for illustration of the modification of the 0.188 factor)
- Assume ½ in. Structural I sheathing with 2 × _ subpurlin spacing of 24 in. oc.

Find:

The appropriate modification factor for the nail slip term

solution Based on the ATC-7 procedure, estimate the average load per nail.

From Fig. 7.13, the diaphragm can be broken into two nailing zones, a heavy-density nailing and a light-density nailing.

Choose heavy-density nail patterns for the 32 ft sections at each end of the diaphragm:

2 in. at the diaphragm boundary and the continuous panel edges parallel to the load

3 in. at the other panel edges

12 in. in the field of the panel

This results in a diaphragm shear design value of 600 lb/ft from Table 7.7 (OK)

Choose a light-density nail pattern for the interior 86 ft of the diaphragm:

4 in. at the diaphragm boundary and the continuous panel edges parallel to the load

6 in. at the other panel edges

12 in. in the field of the panel

This results in a diaphragm shear design value of 360 lb/ft from Table 7.7 (OK)

Based on above nailing pattern, the maximum load per nail for the high-density nailing zone can be estimated as 600 lb/ft/6 nails/foot = 100 lb/nail. For the lower-density nailing region located at 32 ft from the end of the diaphragm the maximum load per nail can be estimated as 344/3 = 115 lb/nail.

By inspection, if the diaphragm was uniformly nailed at the high-density nailing pattern, the average load per nail is 50 lb/nail.

The average load per nail for the staggered nailing is based on a weighted average using geometry to determine the area of the trapezoid and triangle.

$$\frac{\frac{1}{2}(100 + 57.3)32 + \frac{1}{2}(114.7)43}{75} = 66.5 \text{ lb/nail}$$

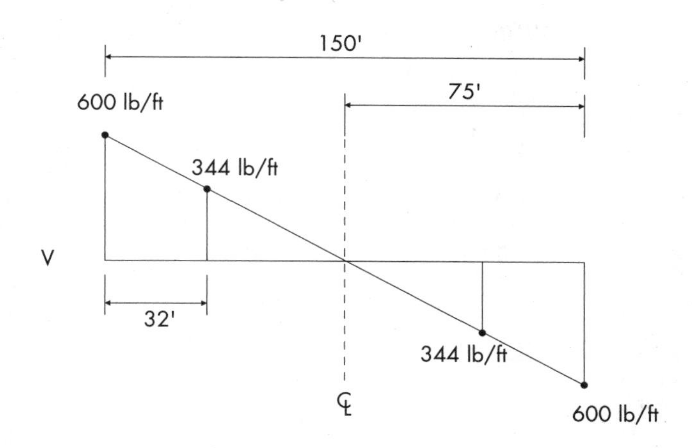

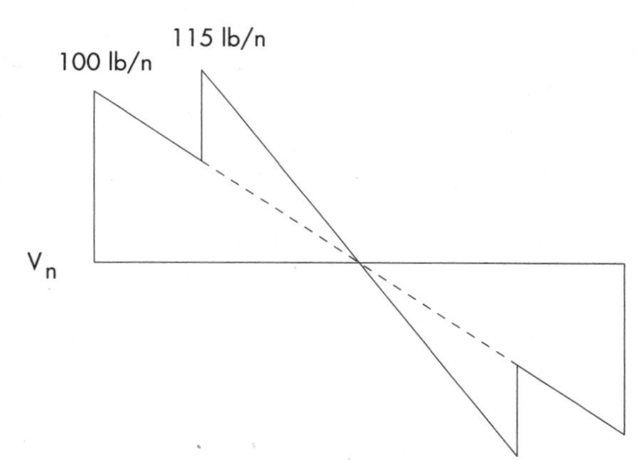

FIGURE 7.13 Diaphragm shear (top) and load per nail (bottom) along the length of the diaphragm.

The adjusted modifier for the nail slip term is 0.188 * 66.4/50 = 0.250. Therefore, the third term in Eq. (7.2) should be modified for this particular example as 0.250 Le_n. This term applies only to this particular example; a similar procedure can be implemented for other staggered nailing cases.

Another limitation of Eq. (7.2) is that it is applicable only for blocked diaphragms. There has been a need for estimating the deflection of unblocked diaphragms. To address this need, APA engineers reviewed the load-deflection data for blocked and unblocked diaphragms developed historically in full-scale diaphragm tests. Based on this review, the following guidance is offered regarding methods for determining the stiffness of unblocked diaphragms until more formal theoretical analysis tools are available (SEAOC Blue Book, §805.3.2):[18]

Limited testing of diaphragms[1,25,26,27] suggests that the deflection of an unblocked diaphragm at its tabulated allowable shear capacity will be about 2.5 times the calculated deflection of a blocked diaphragm of similar construction and dimensions, at the same shear capacity. If diaphragm framing is spaced more than 24 in. o.c., testing indicates a further increase in deflection of about 20% for unblocked diaphragms (e.g., to three times the deflection of a comparable blocked diaphragm). This relationship can be used to develop an estimate of the deflection of unblocked diaphragms.

Chord Forces. Blocked diaphragms are assumed to act like long deep beams. This model assumes that shear forces are accommodated by the structural-use panel web of the beam and that moment forces are carried by the tension or compression forces in the flanges, or chords of the beam. These chord forces are often assumed to be carried by the double top plate of the supporting perimeter walls. Given the magnitude of the forces involved in most light-framed wood construction projects, the double top plate has sufficient capacity to resist the tensile and compressive forces, assuming adequate detailing at the splice locations. A problem lies in wall lines that make a continuous diaphragm chord impossible.

Because shear walls are little more than blocked cantilevered diaphragms, they too develop chord forces and require chords. The chords in a shear wall are the double studs that are required at the end of each shear wall. Just as the chords need to be continuous in a diaphragm, the chords in a shear wall also need to maintain their continuity. This is accomplished by the tension ties (hold-downs) that are required at each end of each shear wall and between the chords of stacked shear walls to provide overturning restraint.

Figure 7.14 shows a plan view of a simple diaphragm loaded with a uniform load, w. Use of this figure and the equations below allows the calculation of diaphragm forces.

$$\text{Diaphragm reaction} = \frac{L_w}{2}$$

$$\text{Diaphragm unit shear} = \frac{L_w}{2L_2}$$

$$\text{Diaphragm moment} = \frac{wL^2}{8}$$

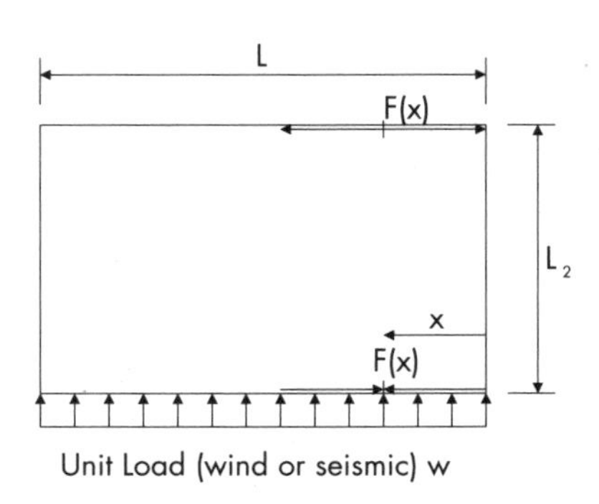

FIGURE 7.14 Diaphragm chord forces.

$$\text{Max. chord force} = \frac{wL^2}{8L_2}$$

$$\text{Chord force at point } x,\ F(x) = \frac{wLx}{2L_2} - \frac{wx^2}{2L_2}$$

Examples

Example 7.8 Simple Diaphragm Design for Wind Loading

Given:

- Residential roof diaphragm
- Wind loading
- Trussed roof
- Unblocked diaphragm required
- Required diaphragm capacity is 180 lb/ft (2630 N/m)
- Panel orientation unknown

Find:

Panel thickness, nail size and nailing schedule

solution Using Table 7.7, refer to the "Unblocked diaphragm" area of the table. Because panel orientation is unknown, use the "All other configurations . . ." column since these values will be conservative. Check "Rated Sheathing . . ." rows first since Structural I may not be readily available in all areas. Similarly, check only rows with 2 in. minimum nominal framing width because the framing is made up of trusses. From the table, note that 8d nails with $^{15}/_{32}$ in. sheathing over 2 × _ framing yields a capacity of 180 lb/ft with a minimal 6 and 12 in. nail spacing. This value may be increased by 40% to 250 lb/ft (3,680 N/m) depending on the governing building code. Because 180 lb/ft (2,630 N/m) is equal to the required 180 lb/ft (2630 N/m) capacity, this selection is OK for use.

Example 7.9 Simple Diaphragm Design for Seismic Loading

Given:

- Commercial roof diaphragm
- Seismic loading
- Trussed roof
- Required diaphragm Capacity 350 lb/ft (5110 N/m)
- Case 1 panel orientation

Find:

Panel thickness, nail size and nailing schedule

solution Using Table 7.7, refer to the "Unblocked diaphragm" area of the table first. Note that no solution is possible. Next, check the "Blocked diaphragm" area of the table. Check "Rated Sheathing . . ." rows first. Similarly, check only rows with 2 in. minimum nominal framing width because the framing is made up of trusses. From the table, note that 8d nails with $^{15}/_{32}$ in. sheathing over 2 × _ framing yields a capacity of 360 lb/ft. Nails must be placed 4 in. on center at all diaphragm boundaries and 6 in. on center at all other panel edges. Because 360 lb/ft (5250 N/m) is greater than 350 lb/ft (5110 N/m), this selection is OK for use.

Example 7.10 Design of Wood Structural Panel Roof Diaphragm—Blocked with Douglas Fir Framing. Assume seismic loading controls.

Design the diaphragm for the roof system shown in Fig. 7.15. The roof has purlins spaced 8 ft (2.44 m) o.c. with subpurlins spaced 2 ft (610 mm) o.c., thus achieving blocking at all panel edges. Load case 4 is appropriate for load in the N-S direction and case 2 for the E-W direction. The uniformly distributed load on the diaphragm is 516 lb/ft (7530 N/m) in the N-S direction and 206 lb/ft (3000 N/m) in the E-W direction. The building dimensions are 192 × 120 ft (58.5 × 36.6 m) as shown in Fig. 7.15.

solution

1. The maximum diaphragm shear is:

$$v_{\text{N-S}} = \frac{wl}{2B} = \frac{516(192)}{2(120)} = 413 \text{ lb/ft } (6030 \text{ N/m})$$

$$v_{\text{E-W}} = \frac{wl}{2B} = \frac{206(120)}{2(192)} = 64 \text{ lb/ft } (934 \text{ N/m})$$

2. Table 7.7, $^{15}\!/_{32}$ in. (11.9 mm) Rated Sheathing wood structural panels with 8d common nails spaced at 2.5 in. (64 mm) o.c. at the diaphragm boundary; 2.5 in. (64 mm) o.c. on all N-S panel edges and 4 in. (100 mm) o.c. at all E-W panel edges will provide an allowable capacity, v_{allow}, of:

$$v_{\text{allow}} = 530 \text{ lb/ft } (7730 \text{ N/m}) \geq 413 \text{ lb/ft } (6030 \text{ N/m}) \quad \therefore \text{OK}$$

Examining the shear in the diaphragm along the length, as shown in Fig. 7.16, provides an opportunity to reduce the nail density.

Table 7.8 summarizes the edge nail schedule requirements for different selected zones in the diaphragm. All field nailing should be at 12 in. (305 mm) o.c.

By inspection, the 6 in. (152 mm) o.c. nailing on all edges is adequate for E-W load (max. = 64 lb/ft (934 N/m)).

3. The maximum chord force, T or C, is obtained by resolving the maximum diaphragm moment into a couple (dividing the moment by the depth):

$$T_{\text{N-S}} = C_{\text{N-S}} = \frac{wl^2}{8B} = \frac{516(192)^2}{8(120)} = 19.8 \text{ kips } (88{,}100 \text{ N})$$

$$T_{\text{E-W}} = C_{\text{E-W}} = \frac{wl^2}{8B} = \frac{206(120)^2}{8(192)} = 1.92 \text{ kips } (85{,}400 \text{ N})$$

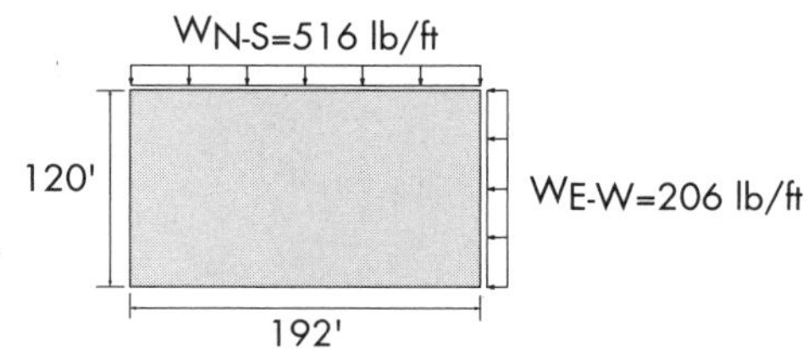

FIGURE 7.15 Building plan view for blocked diaphragm example.

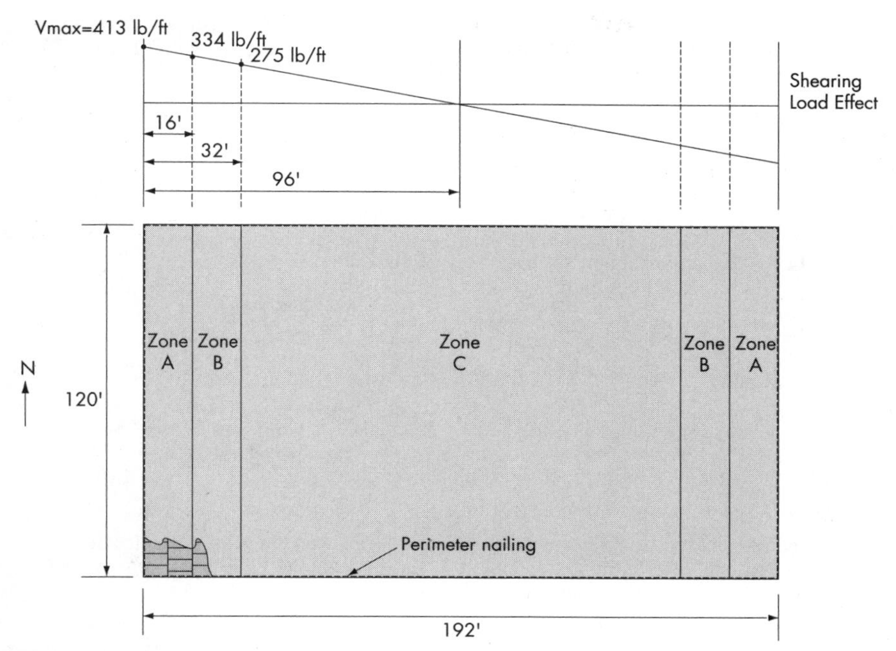

FIGURE 7.16 Panel layout and diaphragm shear distribution for blocked diaphragm example.

TABLE 7.8 Summary of Edge Nailing Requirements for Example 7.10

15/32 rated sheathing wood structural panel with 8d common nails							
N-S Load	Zone	Panel edge nailing (in.) Perimeter	N-S edges	E-W edges	Shear capacity	Shear applied	Status
Case 4	A	2.5	2.5[a]	4	530 >	413	OK
	B	4	4	6	360 >	334	OK
	C	6	6	6	270 ~	275	OK

[a]N-S purlins must be 2.5 in. or wider.
Note: 1 in. = 25.4 mm.

The chord force can be calculated at any distance along the length by using the moment equation as a function of length and then dividing the moment by the diaphragm depth. The ledger will carry the chord force, which is often either steel or wood. The ledger design is not shown in this example.

Example 7.11 Design of Wood Structural Panel Roof Diaphragm—Unblocked with Spruce-Pine-Fir Framing

Design the diaphragm for a roof system consisting of light-frame wood trusses spaced at 24 in. (610 mm) o.c. It is desired not to have blocking. Panel orientation is assumed as load case 1 for load in the N-S direction and case 3 for the E-W direction. The specific gravity of spruce-pine-fir is 0.42. A wind pressure of 19 psf (910 N/m²) is assumed to act uniformly in both the E-W and N-S directions as shown in Fig. 7.17. The building dimensions are 72 × 42 ft (21.9 × 12.8 m).

solution

1. The uniformly distributed load acting on the roof diaphragm by tributary area is:

$$w = 19\left(7 + \frac{12}{2}\right) = 247 \text{ lb/ft } (3600 \text{ N/m})$$

2. The maximum diaphragm shear in the roof diaphragm is:

$$v_{\text{N-S}} = \frac{wl}{2B} = \frac{247(72)}{32(42)} = 212 \text{ lb/ft } (3090 \text{ N/m})$$

$$v_{\text{E-W}} = \frac{wl}{2B} = \frac{247(42)}{2(72)} = 72 \text{ lb/ft } (1050 \text{ N/m})$$

3. Diaphragm nailing capacity. Since the framing will be spruce-pine-fir with SG = 0.42, the shear values from the capacity table must be adjusted according to footnote a of Table 7.7. The specific gravity adjustment factor (SGAF) is:

$$\text{SGAF} = 1 - (0.5 - \text{SG}) = 1 - (0.5 - 0.42) = 0.92$$

From the diaphragm design table, 7⁄16 in. (11.1 mm) wood structural panels with 8d common nails spaced at 6 in. (152 mm) on the supported edges will provide an adjusted allowable shear capacity, v_{allow}, of:

$$v_{\text{allow case 1}} = 230(0.92) = 212 \text{ lb/ft } (3090 \text{ N/m})$$

$$\geq 212 \text{ lb/ft } (3090 \text{ N/m}) \quad \therefore \text{ OK (for the N-S direction)}$$

$$v_{\text{allow case 2}} = 180(0.92) = 166 \text{ lb/ft } (2420 \text{ N/m})$$

$$\geq 72 \text{ lb/ft } (1050 \text{ N/m}) \quad \therefore \text{ OK (for the E-W direction)}$$

Note that these values could have been increased by 40%, depending on the governing building code.

4. The maximum chord force in tension and compression, T and C, is:

$$T_{\text{N-S}} = C_{\text{N-S}} = \frac{wl^2}{8B} = \frac{247(72)^2}{8(42)} = 381 \text{ lb/ft } (17{,}000 \text{ N})$$

$$T_{\text{E-W}} = C_{\text{E-W}} = \frac{wl^2}{8B} = \frac{247(72)^2}{8(42)} = 756 \text{ lb/ft } (3360 \text{ N})$$

A double top plate, spliced together, will resist the chord force along the length. The splice design is not shown in this example. Recall from Example 7.10 that, the

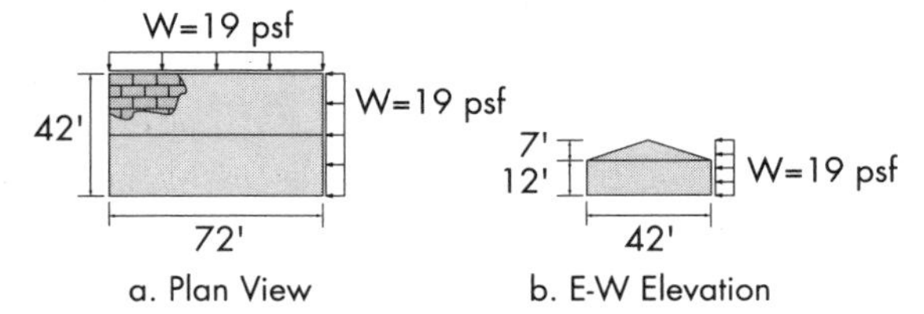

FIGURE 7.17 Building dimensions for unblocked diaphragm example.

chord force can be calculated at any distance along the length by using the moment equation as a function of length and then dividing the moment by the diaphragm depth.

Example 7.12 Calculate Deflection of an Unblocked Diaphragm
Calculate the deflection of the diaphragm designed in Example 7.11

solution Research by APA has shown that unblocked diaphragms deflect about 2.5 times that of blocked diaphragms, and for diaphragm framing spaced greater than 24 in. (610 mm) o.c. this difference increases to about 3. The deflection of a blocked diaphragm is calculated by the following equation (based on English units):

$$\Delta = \frac{5vL^3}{8EAb} + \frac{vL}{4Gt} + 0.188Le_n + \frac{\Sigma(\Delta_c X)}{2b}$$

where v = 212 lb/ft, max. unit shear in diaphragm
L = 8 ft, diaphragm length
E = 1,200,000 psi, for s-p-f studs—stud grade
A = 16.5 in.^{2}, for two, 2 × 6 vertical end studs
$G_v t_v = Gt$ = 27000(3.1) lb/in., for 7/16 OSB from Chapter 2
b = 42 ft, diaphragm depth
e_n = nail slip
Δ_c = chord splice slip
X = distance from chord splice to closest supporting shear wall

Load per nail, v_{nail}:

$v_{nail} = v/(12/S) = 212/(12/6) = 106$ lb/nail (where S = nail spacing in in.)

Nail slip, e_n, for the case assembled when the framing is green ($MC > 19\%$) and the 20% increase in nail slip is warranted since the panels are not Structural I. (Use equation from Table 7.3, based on English units):

$$e_n = 1.2(v_{nail}/857)^{1.869} = 1.2(106/857)^{1.869} = 0.024 \text{ in.}$$

Chord splice slip, Δ_c, will be assumed to be 0.03 in the tension chord splices and 0.005 in the compression chord splices. These values are based on a review of diaphragm tests.[1] APA test values for tension chord slip range from 0.011–0.156 in. for the different configurations tested, with a value of about 0.03 being an estimated average. Additionally, APA research shows that compression chord slip is about ⅙ of the tension chord slip. Selecting values for chord splice slip involve considerable engineering judgment. Alternate assumptions and techniques can be found in other sources,[21,32] but no values appear to be definitive since so many variables can be involved.

Chord splices will be located every 8 ft.

The deflection of each component is:

$\Delta_{bending} = 5vL^3/(8EAb) = 5(212)72^3/(1{,}200{,}000(16.5)42) = 0.060$ in. (1.52 mm)

$\Delta_{shear} = vL/(4Gt) = 212(72)/(4(27{,}000)3.1) = 0.046$ in. (1.17 mm)

$\Delta_{nail\ slip} = 0.188L(e_n) = 0.188(8)0.024 = 0.327$ in. (8.31 mm)

$\Delta_{chord\ splice} = \Sigma(\Delta_c X)/(2b)$

$\Delta_c X_{tension\ chord} = 2[0.03(8) + 0.03(16) + 0.03(24) + 0.03(32)] = 4.8$ in.*ft.

$\Delta_c X_{compression\ chord} = 2[0.005(8) + 0.005(16) + 0.005(24) + 0.005(32)] = 0.8$ in.*ft.

$\Delta_{chord\ splice} = (4.8 + 0.8)/(2(42)) = 0.067$ in. (1.70 mm)

The total deflection, Δ, is a summation of the terms above multiplied by 2.5 to account for the unblocked diaphragm construction:

$$\Delta = (0.060 + 0.046 + 0.327 + 0.067)2.5 = 1.25 \text{ in. } (31.8 \text{ mm})$$

7.4.3 Advanced Diaphragm Topics

Principles of Mechanics. The 2000 International Building Code[4] allows calculations, tests, or analogies drawn there from to estimate the resistance of diaphragms per Section 2305.2.1 provided that the permissible deflections of the attached distributing or resisting elements are not exceeded. Tissell and Elliot[1] provide guidance on principles of mechanics calculations approach, published in APA Research Report 138. This approach is similar to the shear-wall approach—based fundamentally on a single fastener design value adjusted by various empirically determined factors that account for such things as nail spacing, panel grade, framing width, and number of nails or fasteners. APA originally derived Table 7.7 based on this approach and confirmed by various testing programs. The high-load diaphragms that currently appear in ICBO Evaluation Service ER-1952[28] are also based on this type of approach. Through the years, the load duration factor, diaphragm factor and nail design values have changed, therefore, following the Tissell and Elliot approach will not provide perfect agreement with Table 7.7. However in most cases the agreement is reasonable. When the individual fasteners design values changed, APA chose not to change the diaphragm tables accordingly because, through testing, the design values have been demonstrated to be appropriate. The tables in ICBO Evaluation Service ER-1952 have been updated based on current recognized fastener design values.

The principles of mechanics approach can be a very useful tool, especially when, for some reason or another, the diaphragm is beyond the scope of Table 7.7. APA recommends that Table 7.7 be used in most cases over the principles of mechanics approach; however, sometimes a tabulated solution is not an option.

Diaphragms with Openings. One of the similarities between diaphragms and shear walls is that many of the diaphragms have openings. Skylights, stairwells, and elevator shafts are but a few of the types of large openings that can appear in diaphragms. Unlike shear walls, designing for openings in diaphragms is limited to a method that requires a moderate amount of calculations (i.e., there is no empirical reduction factor for diaphragms). This most common method treats the diaphragm opening similar to a Vierendeel truss and was apparently first formalized by ATC-7.[31] The currently available version of Tissell and Elliot[1] (last printed in 2000) illustrates a relatively easy design example. Diekmann's chapter in the Faherty and Williamson Handbook[19] also contains a design example for the design and detailing of openings in diaphragms.

7.5 ADVANCED TOPICS FOR LATERAL LOAD DESIGN

7.5.1 Drag Struts

As described in Section 7.2, the load path for a box-type structure is from the diaphragm into the shear walls running parallel to the direction of the load (i.e., the diaphragm loads are transferred to the shear walls that support it). Because the diaphragm acts like a long, deep beam, it loads each of the supporting shear walls evenly along the length of the walls. The problem lies with the fact that seldom is each shear wall solid throughout its full length. Typically a wall contains windows and doors.

The traditional model used to analyze shear walls only recognizes wall segments that run full height as shear wall segments. This means that at locations with windows or doors, a structural element is needed to distribute the diaphragm shear over the top of the opening and into the full-height segments adjacent to it. This element is called a drag strut or collector.

Fortunately, in residential construction the double top plates existing in most stud walls will serve as a drag strut. It may be necessary to detail the double top plate such that no splices occur in critical zones. Or it may be necessary to specify the use of a tension strap at butt joints to transfer these forces. The maximum force seen by drag struts is generally equal to the diaphragm design shear in the direction of the shear wall multiplied by the distance between the shear wall segments.

Drag struts are also used to tie different parts of irregularly shaped buildings together.

To simplify design, irregularly shaped buildings (such as L- or T-shaped) are typically divided into simple rectangles. When the structure is reassembled after the individual designs have been completed, drag struts are used to provide the necessary continuity between these individual segments to ensure that the building will act as a whole.

Figures 7.18, 7.19 and 7.20 and the generalized equations given below provide methods to calculate the drag strut forces.

Unit shear above opening $= V/L - L_0 = v_b$

Max. force in drag strut $=$ greater of

$$\frac{(V/L)L_0L_1}{(L - L_0)} = \frac{v_aL_0L_1}{(L - L_0)}$$

or

$$\frac{(V/L)L_0L_2}{(L - L_0)} = \frac{v_aL_0L_2}{(L - L_0)}$$

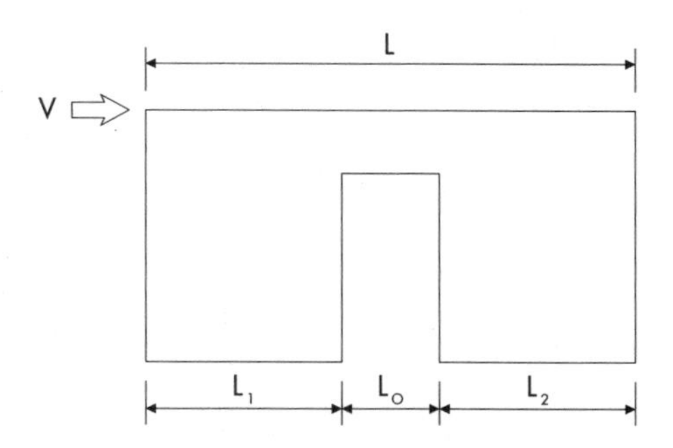

FIGURE 7.18 Shear-wall drag strut.

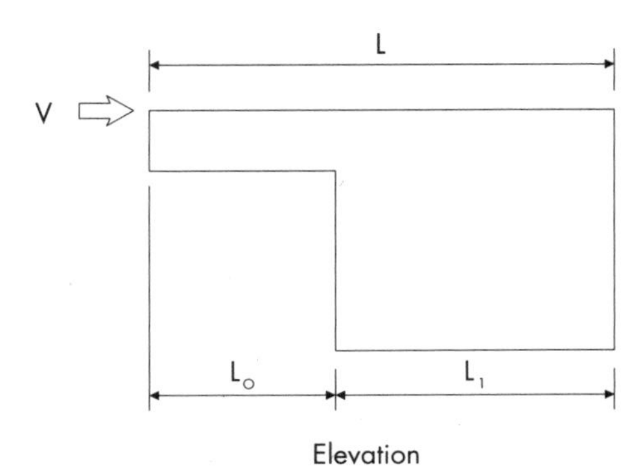

FIGURE 7.19 Shear-wall special case drag strut.

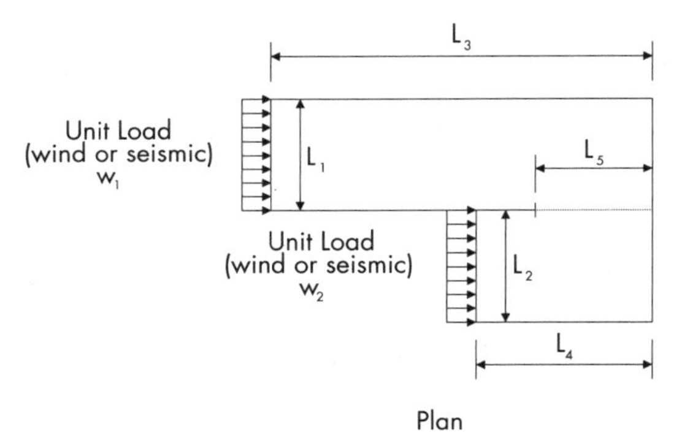

FIGURE 7.20 Diaphragm drag strut, L-shaped building.

Unit shear above opening $V_a = V/L$
Unit shear below opening $V_b = V/L_1$
Max. force in drag strut $= (V/L)L_0$
Force on drag strut $= ((w_1L_1/2L_3) + (w_2L_2/2L_4))L_5$

7.5.2 Subdiaphragm Concepts

Due to failures of several large flat-roof diaphragms after the San Fernando earthquake of 1971, a number of code changes were introduced into the Uniform Building Code that require continuous cross-ties. These and other code changes that require minimum attachment of concrete and masonry walls to wood diaphragms were proposed and enacted to improve the seismic performance of large flat-roofed structures using flexible structural-use panel diaphragms.

Since the enactment of these code changes, the subdiaphragm concept (also known as the minidiaphragm) has been recognized and extensively used to provide a method of meeting the wall attachment and continuous cross-tie code requirements while minimizing the number and length of ties required to achieve continuity between chords. A formal definition of a subdiaphragm can be found in the 1997 Uniform Building Code: "SUBDIAPHRAGM is a portion of a larger wood diaphragm designed to anchor and transfer local forces to primary diaphragm struts and the main diaphragm."

In practice, the subdiaphragm approach is used to concentrate and transfer local lateral forces to main structural members that support the roof vertical loads. The subdiaphragm approach is often an economical solution to code-required cross-ties for the following reasons:

- Boundary members are already present.
- Boundary members generally span the full length and width of the buildings with few connectors.
- Boundary members are large enough to accommodate loads easily.
- Boundary members are large enough to allow room for requisite connections.

In general, the larger the roof area, the greater the savings that can be achieved by using subdiaphragms.

Each subdiaphragm must meet all applicable diaphragm requirements provided in the applicable building code. As such, each subdiaphragm must have chords, continuous tension ties, and sufficient sheathing thickness and attachment to transfer the shear stresses generated within the diaphragm sheathing by the subdiaphragm. In addition, building codes may contain aspect ratios that are specific to subdiaphragms.

The subdiaphragm is actually the same structure as the main roof diaphragm; thus, the subdiaphragm utilizes the same roof sheathing to transfer shear stresses as the main diaphragm. As such, sheathing nailing and thickness requirements of the roof diaphragm may not be sufficient for the subdiaphragm requirements. In this case, the subdiaphragm requirements would control and dictate the roof sheathing and fastening requirements in the subdiaphragm locations. Fortunately, the portion of the main diaphragm that is utilized as a subdiaphragm is a choice left to the designer; thus, the dimensions of the subdiaphragm can be chosen to minimize potential discontinuities in sheathing thicknesses or nail schedules. Similarly, the roof diaphragm requirements may be more stringent than those for the subdiaphragm.

Example 7.13 Design of a Subdiaphragm

A common problem observed after large seismic events is roof-to-wall separation, particularly for high-mass walls such as concrete or masonry. In recent building codes more attention has been given to this critical connection with increased connection force requirements (2000 IBC Sec. 1620.2.1). Continuous ties from one main diaphragm chord to the other opposite chord are also required.

Subdiaphragms are useful for concentrating the forces and connections needed to provide a continuous cross-tie path from one diaphragm support to the other. In this example, the east and west wall are required to have a continuous cross-tie connection. This can be achieved in two ways: (1) by directly connecting all, or enough, of the subpurlins together from east to west so that continuity is achieved, which requires many connections; or (2) by using a subdiaphragm to concentrate the wall connection force into the main girders, which requires relatively fewer connections. Continuity between the north and south walls can be achieved by purlin connections.

Considering the diaphragm designed in Example 7.10, design a subdiaphragm for the main diaphragm to function as the connection of the roof diaphragm to the east-

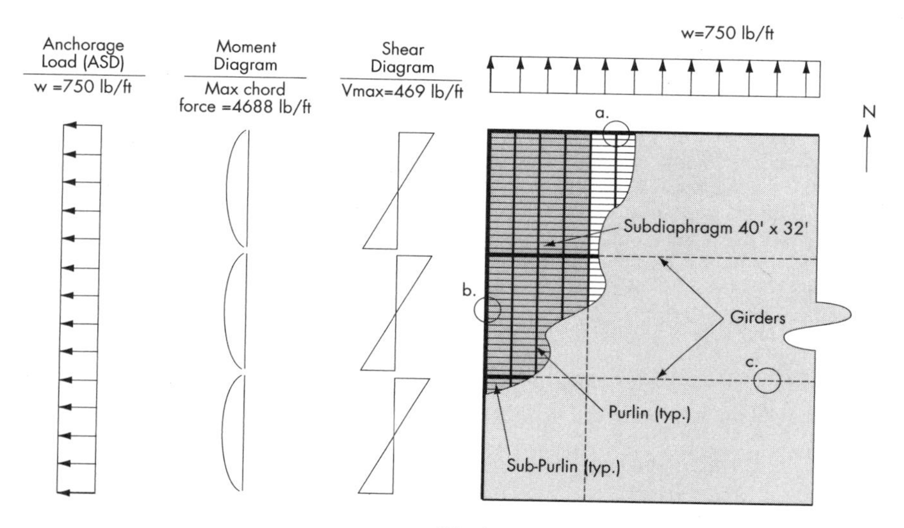

FIGURE 7.21 Plan view of subdiaphragm design example.

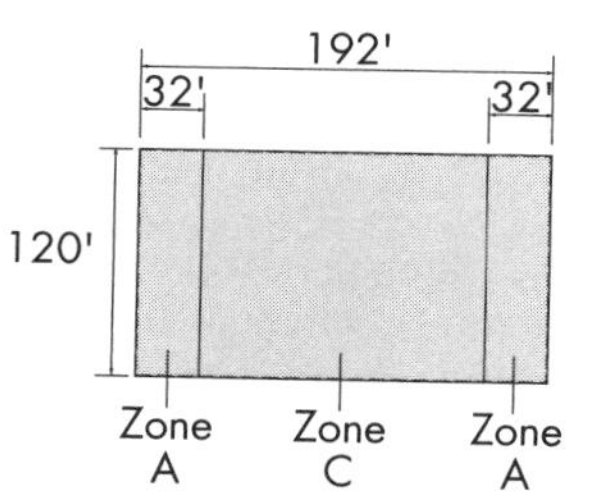

FIGURE 7.22 Zone layout to accommodate subdiaphragm nailing.

west walls. The design anchorage force, F_p, for the wall-to-diaphragm connection is 750 lb/ft (10,900 N/m) as shown in Fig. 7.21.

solution The maximum length-to-width ratio of the structural subdiaphragm is 2.5:1. Thus, the minimum subdiaphragm depth is 40 ft/2.5 = 16 ft (4.88 m)

1. After several iterations a subdiaphragm depth of 32 ft (9.75 m) was selected for load compatibility with the existing main diaphragm nailing pattern. The maximum subdiaphragm shear for this depth is:

$$v = \frac{wl}{2B} = \frac{750(40)}{2(32)} = 469 \text{ lb/ft (6,840 N/m)}$$

Zone A nailing, as shown in Example 7.10, Fig. 7.16, is adequate, though the area of Zone A must be increased to extend 32 ft (9.75 m) from the east and west walls as shown in Fig. 7.22.

2. The maximum chord force, T or C, in the subdiaphragm is:

$$T = C = \frac{wl^2}{8B} = \frac{750(40)^2}{8(32)} = 4688 \text{ lb } (20{,}900 \text{ N})$$

The steel channel ledger and purlin act as subdiaphragm chords. Their design is not shown.

3. The three general connection forces are:

- N-S walls (see location *a* in Fig. 7.21), for 8 ft (2.44 m) purlin spacing the anchorage force is:

$$F = 750(8) = 6000 \text{ lb } (26{,}700 \text{ N})$$

- E-W walls (see location *b* in Fig. 7.21), for 2 ft (610 mm) subpurlin spacing the anchorage force is:

$$F = 750(2) = 1{,}500 \text{ lb } (6{,}670 \text{ N})$$

- Main girder connection force is (see location *c* in Fig. 7.21):

$$F = 750(40) = 30{,}000 \text{ lb } (133.5 \text{ kN})$$

4. The final diaphragm sheathing design is shown in Fig. 7.22, and the zone nailing is specified in Example 7.10.

7.5.3 Rigid vs. Flexible Diaphragms

Historically, wood structural panel sheathed diaphragms have been considered to behave in a flexible manner. This assumption implies that the diaphragm behaves similar to simply supported beams spanning from shear wall to shear wall. Half of the reaction, thus shear force, is distributed to one shear wall, and the other half of the reaction is distributed to the other shear wall. The force transferred to each shear wall is based on the tributary area of each wall. This assumption makes the distribution of shear forces a fairly simple calculation.

Engineering theory points out the flaws in this simple design methodology, particularly when some walls in a system are very stiff relative to other walls. A more complicated assumption, that of a rigid diaphragm, is to distribute forces to shear walls based on the relative stiffness of each wall. Although this approach makes engineering sense, the reality is that determining shear-wall stiffness based on Eq. (7.1) is somewhat problematic. For the first iteration, an arbitrary load is applied to each line of shear walls to determine the relative stiffness of the lines of walls. Once the relative stiffness of the wall lines has been determined, the applied lateral load is distributed proportionally. The shear walls are redesigned, the lateral stiffness is recalculated, and the applied load is reapportioned. This is continued until convergence is satisfactorily achieved. Since the shear-wall deflection equation is nonlinear, convergent problems may arise for some numerical solutions.

The provision for determining rigid vs. flexible diaphragms was first included in the 1988 Uniform Building Code. The code definition of "flexible diaphragm" is when diaphragm deflection is two or more times the average shear wall deflection. When the diaphragm deflects less than two times the average story drift, the diaphragm is assumed to be rigid. Although this definition has been in the building code for several code cycles, it is the authors' experience that the issue was not considered until after the 1994 Northridge, California, earthquake. The effects of the earthquake generated significant discussion about the behavior of wood diaphragms, particularly in structures that have torsional irregularities—for example, tuck-under parking or soft-story construction. If a diaphragm is classified as rigid, then 5% accidental torsion must be added, regardless of plan symmetry.

The result of the greater attention to the code provision that defines rigid diaphragm ranges from the complexity of the design process significantly increasing to some engineers stating that all wood diaphragms are flexible. On the conservative end of this spectrum, some engineers conduct a rigid diaphragm analysis and a flexible diaphragm analysis and estimate a load envelope, thus sizing walls accordingly. In the authors' opinion, this concept has not been fully embraced by engineers outside of California, especially given that the analysis is at least twice as difficult and is often not justified. On the other end of the spectrum, it would appear that always classifying wood diaphragms as rigid (or for that matter flexible) could lead to problems, particularly when there are significant torsional irregularities.

It appears that the future building codes may allow exceptions for light-frame structures. The draft 2003 International Building Code will allow diaphragms using wood structural panels to be considered flexible. However, this approach requires that the simplified analysis procedure be used, which has a 20% increase in calculated design forces. Some designers may choose to implement the 20% increase in lieu of the more complicated force distribution procedure. Other designers may choose to sharpen their pencils and use the more complicated analysis procedures. Other design offices may have a rule of thumb that a certain class of wood-framed buildings should always be designed as a rigid or flexible diaphragm or both.

In the future, perhaps better computer analysis programs will provide improved force distribution on structures. Another approach would be to develop simple tables that implement force penalties based on the difference in shear wall rigidity as well as diaphragm aspect ratio. APA is currently investigating the second scenario.

7.6 REFERENCES

1. Tissell, J. R., and J. R. Elliot, *Plywood Diaphragms,* Research Report 138, American Plywood Association (now APA—The Engineered Wood Associtaion), Tacoma, WA, 1980.
2. ASTM E72-98, *Standard Test Methods of Conducting Strength Tests of Panels for Building Construction,* ASTM, West Conshohocken, PA, 2000.
3. APA—The Engineered Wood Association, *Performance Standards and Policies for Structural-Use-Panels,* Tacoma, WA, 1994.
4. 2000 International Building Code, International Code Council, Falls Church, VA.
5. ASTM E564-95, *Standard Method of Static Load Test for Shear Resistance of Framed Walls for Buildings,* ASTM, West Conshohocken, PA, 2000.
6. Structural Engineers Association of Southern California (SEAOSC), *Standard Method of Cyclic (Reversed) Load Test for Shear Resistance of Framed Walls for Buildings,* SEAOSC, Whittier, CA, 1996.
7. Porter, M. L., "Sequential Phase Displacement (SPD) Procedure for TCCMAR Testing," in *Proceedings, 3rd Meeting of the Joint Technical Coordination Committee on Masonry Research,* U. S.-Japan Coordinated Earthquake Research Program, Tomanu, Japan, 1987.
8. ICBO Evaluation Service, Prefabricated Wood Shear Panels, Acceptance Criteria AC-130, ICBO, Whittier, CA, 1997.
9. Rose, J. D., *Preliminary Testing of Wood Structural Panel Shear Walls Under Cyclic (Reversed) Loading,* APA Research Report 158, APA—The Engineered Wood Association, Tacoma, WA, 1996.

10. Ficcadenti, S. J., T. A. Castle, D. A. Sandercock, and R. K. Kazanjy, "Laboratory Testing to Investigate Pneumatically Driven Box Nails for the Edge Nailing of ⅜″ Thick Plywood Shear Walls," in *Proceedings, 64th SEAOC Annual Convention,* Indian Wells, CA, October 19–21, Structural Engineers Association of California, Sacramento, CA, 1995.
11. Lam, F., H. G. L. Prion, and H. Ming, "Lateral Resistance of Wood Shear Walls with Large Sheathing Panels," *Journal of Structural Engineering,* vol. 123, no. 12, pp. 1666–1673, 1997.
12. Forintek Canada Corporation, *http://www.forinteck.ca/eng/publications/publication.html.*
13. He, M., H. Magnusson, F. Lam, and H. G. L. Prion, "Cyclic Performance of Perforated Wood Shear Walls with Oversize OSB Panels," *Journal of Structural Engineering,* vol. 125, no. 1, pp. 10–18, 1999.
14. The Consortium of Universities for Research in Earthquake Engineering (CUREE), *http://www.curee.org/projects/woodframe project/woodframe project.html.*
15. Karacabeyli, E., and A. Ceccotti, "Nailed Wood-Frame Shear Walls for Seismic Loads: Test Results and Design Considerations," Paper Reference T207-6, in *Proceedings of Structural Engineers World Congress,* July 19–23, 1998, San Francisco, CA.
16. Dolan, J. D., "The Dynamic Response of Timber Shear Walls," Ph.D. thesis, The University of British Columbia, Vancouver, 1989.
17. Tissell, J. R., *Wood Structural Panel Shear Walls,* Research Report 154, APA—The Engineered Wood Association, Tacoma, WA, 1993.
18. Structural Engineers Association of California (SEAOC), *Recommended Lateral Force Requirements and Commentary* (SEAOC Blue Book), 7th ed., Seismology Committee, SEAOC, Sacramento, CA, 1999.
19. Faherty, K. F., and T. G. Williamson, eds., *Wood Engineering and Construction Handbook,* 3d ed., McGraw-Hill, New York, 1999.
20. Duquette, D. W., *Timber Solutions Manual,* Argulus, New York, 1997.
21. Structural Engineers Association of California (SEAOC), *Seismic Design Manual,* vol. 2, *Building Design Examples: Light Frame, Masonry and Tilt-up,* SEAOC, Sacramento, CA, 2000.
22. Federal Emergency Management Agency (FEMA), *NEHRP Recommended Provisions for Seismic Regulations for New Buildings and Other Structures,* Part 1, Provisions, FEMA 268, FEMA, Washington, DC, 2001.
23. Federal Emergency Management Agency (FEMA), NEHRP Recommended Provisions for Seismic Regulations for New Buildings and Other Structures, Part 2, Commentary, FEMA-269, Federal Emergency Management Agency, Washington, DC, 2001.
24. Rose, J. D., and T. D. Skaggs, *Wood Structural Panel Sheathing for Narrow Shear Walls and Wall Bracing,* Research Report 156, APA—The Engineered Wood Associtionn, Tacoma, WA, 2001.
25. Countryman, D., *Lateral Tests on Plywood Sheathed Diaphragms,* Laboratory Report 55, Douglas Fir Plywood Association (now APA—The Engineered Wood Association), Tacoma, WA, 1952 (no longer in print).
26. Countryman, D., *1954 Horizontal Plywood Diaphragm Tests,* Laboratory Report 63, Douglas Fir Plywood Association (now APA—The Engineered Wood Association), Tacoma, WA, 1955 (no longer in print).
27. Tissell, J. R., *1966 Horizontal Plywood Diaphragm Tests,* Laboratory Report 106, American Plywood Association (now APA—The Engineered Wood Association), Tacoma, WA, 1967 (no longer in print).
28. ICBO Evaluation Service, *303 Siding and High-Load Diaphragms for APA—The Engineered Wood Association,* Tacoma, WA, Evaluation Report ER-1952, ICBO, Whittier, CA, 1999.
29. ASTM E455-98, *Standard Test Methods for Static Load Testing of Framed Floor and Roof Diaphragm Construction for Buildings,* American Society for Testing and Materials, West Conshohocken, PA, 2000.

30. ABK-TR-03, *Methodology for Mitigation of Seismic Hazarads in Existing Unreinforced Masonry Buildings: Diaphragm Testing,* National Science Foundation, Washington, DC, 1981.

31. ATC-7, *Guidelines for the Design of Horizontal Wood Diaphragms,* Applied Technology Council, Redwood City, OR, 1981.

32. Breyer, D. E., K. J. Fridley, and K. E. Cobeen, *Design of Wood Structures—ASD,* 4th ed., McGraw-Hill, New York, 1998.

7.7 ADDITIONAL READING

ASCE 7-98, *Minimum Design Loads for Buildings and Other Structures,* American Society of Civil Engineers, New York, 1998.

Department of the Army, Navy and Air Force, *Seismic Design for Buildings,* TM 5-809-10, NAV FAC P-355, AFM 88-3 ("Tri-Services" Manual), U.S. Government Printing Office, Washington, DC, 1992, chap. 13.

Federal Housing Administration (FHA), *A Standard for Testing Sheathing Materials for Resistance to Racking,* Technical Circular No. 12, FHA, Underwriters Division, Washington, DC, October 5, 1949.

FEMA-273, *NEHRP Guidelines for the Seismic Rehabilitation of Buildings,* Federal Emergency Management Agency, Washington, DC, 1997.

FEMA-274, *NEHRP Commentary on the Guidelines for the Seismic Rehabilitation of Buildings,* Federal Emergency Management Agency, Washington, DC, 1997.

Forest Products Laboratory, *Wood: Engineering Design Concepts,* Materials Education Council, 110 Materials Research Laboratory, Pennsylvania State University, University Park, PA, 1986.

Load Path and Continuity in "Engineered" Wood-Frame Buildings, Seminary workbook, International Conference of Building Officials, Whittier, CA, 1998.

National Institute of Standards and Technology. *Voluntary Product Standard PS 1-95 for Construction and Industrial Plywood,* Office of Standards Services, Washington, DC, 1995.

National Institute of Standards and Technology, *Voluntary Product Standard PS 2-92, Performance Standard for Wood-Based Structural-Use Panels,* Office of Standards Services, Washington, DC, 1992.

Plywood Design Specification, Form Y510, 1997.

Seismic Retrofit Training for Building Contractors and Inspectors, Seminar workbook, Federal Emergency Management Agency, Washington, DC, 1998. Available at *http://www.abag.ca.gov/bayarea/eqmaps/fixit/training.html.*

Stalnaker, J. J., and E. C. Harris, *Structural Design in Wood,* Van Nostrand Reinhold, New York, 1997.

Timber Construction Manual, John Wiley & Sons, New York, 1994.

UBC Earthquake Regulations (1997) Overview and Perspective, Seminar 109, International Conference of Building Officials, Whittier, CA, 1998.

Western Wood Products Association, *Western Woods Use Book,* Western Wood Products Association, Portland, OR, 1996.

Williams, A. *Seismic Design of Buildings and Bridges,* 2d ed., Engineering Press, Austin, TX, 1998.

CHAPTER EIGHT
MECHANICAL FASTENERS AND CONNECTIONS*

Zeno A. Martin, P.E.
Associate Engineer, TSD

8.1 INTRODUCTION

Everyone understands the concept that a chain is only as strong as its weakest link, and connections are the critical link between elements of a structure. Connections maintain load path continuity and provide structural integrity. When the importance of proper connection details is overlooked, structural failure can occur due to this weak link. Properly designed and detailed connections are what hold a structure together, and the designer needs to understand some fundamental principles associated with connections for wood structures:

- Since the consequence of connection failure is severe, the design values have relatively high built-in factors of safety and relatively low probabilities of failure under design loads.
- Fasteners and connectors for wood have continued to improve over the years, resulting in more reliable and accurate design guidelines—such as those presented in this chapter.
- Strength or serviceability failures usually occur due to poor design or construction practices—basically, not following design guidelines.

The guidelines in this chapter provide safe and reliable connections. Furthermore, this chapter addresses common issues to help avoid potential problems such as how to specify nails correctly and accurately and also how to remedy common field problems such as overdriven fasteners. These guidelines mostly follow those of the *National Design Specification for Wood Construction* (*NDS*),[1] with additional information on fastening engineered wood products from APA—The Engineered Wood Association (APA)[2] recommendations and other sources as referenced.

*The author gratefully acknowledges Keith Faherty for his permission to use excerpts of his previous work on this topic as published in the McGraw-Hill Wood Engineering and Construction Handbook, 3rd edition, 1999.

8.1.1 Types of Fasteners

Many types of fasteners and connectors are available for use in attaching one wood structural unit to another. Some of the common types include nails, screws, lag screws, bolts, power-driven fasteners (special nails and staples), and thin-gauge steel plate connectors. Other connector types are split rings, shear plates, timber rivets, dowels, drift pins, and metal plate connectors. Some of these are shown in Fig. 8.1.

- Nails are generally used when loads are light. They are used for light-frame construction, diaphragms, and shear walls. It has been reported that the average home has a many as 70,000 nails.[3] For increased installation efficiency, many nails are now installed with pneumatic nailers.
- Screws are not as frequently used as nails or lag screws in wood construction. But they are more satisfactory than nails under vibratory or withdrawal loads, since they have less tendency to work loose.
- Lag screws, bolts, and timber connectors are used for large elements and in light-frame construction when loads of relatively large magnitude must be transferred. Bolts are less efficient than split rings or shear plates; however, they are often adequate. Lag screws are used when bolts are undesirable, when the member is too thick, or when one face of the member is not accessible for the installation of washers and nuts. Lag screws can be used in conjunction with split rings and shear plates and are especially effective for large magnitude withdrawal loads.
- Split rings and shear plates are used for connections in heavy timber construction. They also may be used in wood trusses when spans are relatively long and the trusses are widely spaced, causing high connection forces.
- Thin-gauge steel plate connectors are used extensively to join wood structural elements, fasten one subassembly to another, and anchor a structure to its foundation. Many different types of proprietary steel plate connectors are commercially available, including joist and purlin hangers, beam seats, column caps, strap ties, framing anchors, seismic and hurricane anchors, mud sills, hold-downs, column bases, and many others. Toothed sheet-metal plates are used extensively in the construction of light wood trusses, trussed rafters, and trussed floor joists. One or more of these types of structural components are used in approximately 90% of all residential structures, and they are also widely used in commercial and institutional buildings.
- Corrosion-resistant fasteners are available and used when the wood elements are in high moisture environments and/or have been chemically treated to resist decay or fire.

8.2 CONNECTION DESIGN

For many fasteners, nominal design values are defined by a table or equation in the *National Design Specification for Wood Construction* (NDS)[1] and are based on a certain set of assumed end-use conditions such as normal load duration (10-year), dry condition of use, no sustained exposure to elevated temperatures, and others. The word "nominal" is used in this chapter and in the NDS to describe a fastener design value of preassumed end-use conditions. After actual anticipated end-use conditions are accounted for by applying all appropriate adjustment factors, an

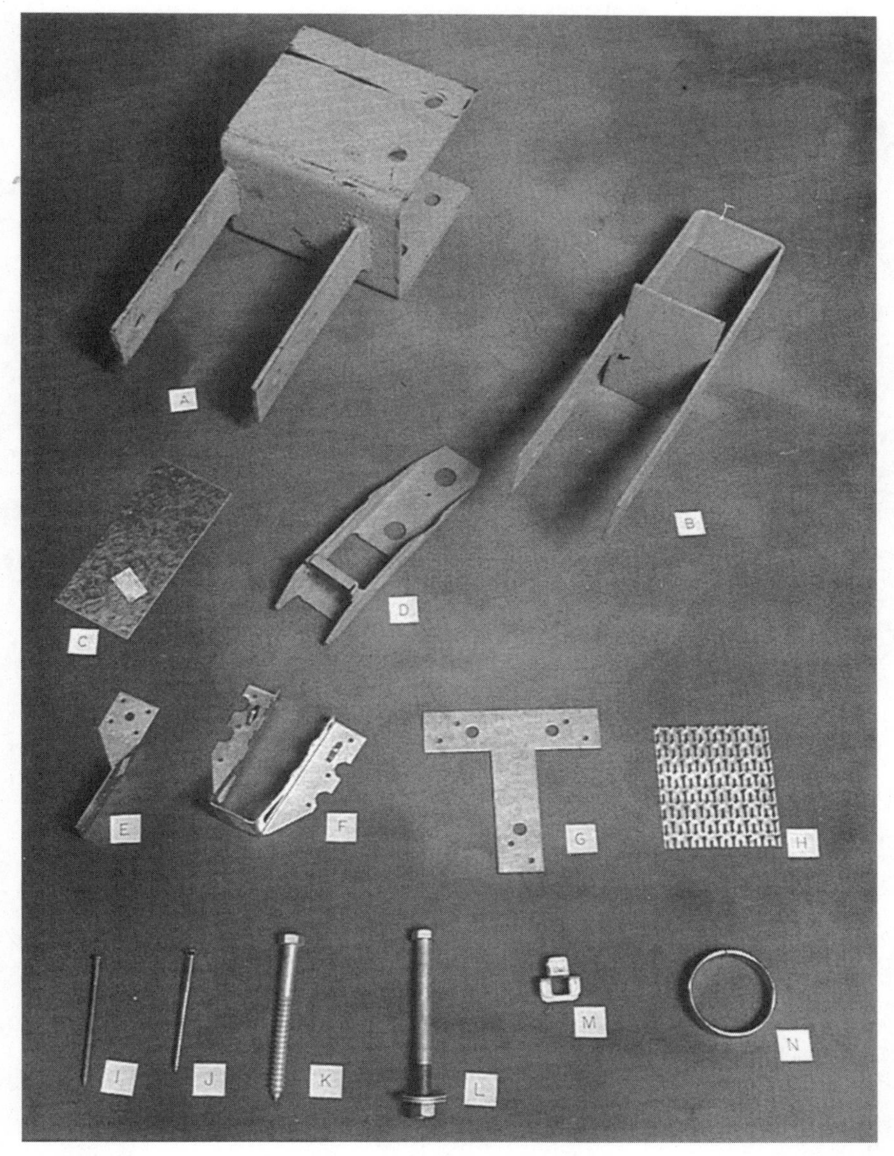

FIGURE 8.1 Various fasteners and connectors: (*A*) column cap, (*B*) column base, (*C*) truss plate, (*D*) hold-down, (*E*) framing anchor, (*F*) joist hanger, (*G*) T-strap tie, (*H*) toothed sheet-steel plate, (*I*) nail, (*J*) screw, (*K*) lag bolt, (*M*) plywood clip, (*N*) split ring.

allowable design value is obtained that can be equated to the design load requirements.

Provisions are usually given for calculating nominal design values for various types of single fasteners. When a connection has multiple fasteners of similar size and type, the total connection strength is a summation of the strength for each individual fastener with the appropriate adjustment to account for group action.

For nails, spikes, bolts, lag screws, and wood screws, the lateral load design values can be calculated by yield limit equations that account for the different potential yield modes—a mechanics-based approach often called the European yield model. Withdrawal load design values for these same fastener types can be calculated from empirical equations. For other fastener types, such as split rings, shear plates, dowels, drift pins, and timber rivets, the lateral and withdrawal design values are generally obtained from empirically based tables.

8.2.1 Yield Model

In the early 1990s, use of a fastener yield model, frequently referred to as the European yield model (EYM) for determining lateral load capacity was adopted in the United States. Research has shown that this model is fairly accurate to somewhat conservative in predicting actual fastener behavior.[4] Figure 8.2 shows the EYM failure modes. Mechanics-based equations predict yield for each mode, but due to the soft conversion process in NDS's transition from the old empirical equations to the EYM, these EYM equations have different constants for different fastener types. For a given connection, each possible failure mode must be considered and the lowest capacity of the possible failure modes is taken as the design load for the connection. Since several or more equations must be considered, solutions with a computer, using a spreadsheet or other equation-solving software, can be time saving.

8.2.2 Adjustment Factors

Adjustment factors are used to account for end-use conditions that differ from those assumed for nominal design values. Generally, nominal design values are based on single fasteners having: a normal load duration (10 years of constant or cumulative loading), dry wood at the time of connection fabrication and in use, no sustained exposure to temperatures exceeding 100°F, the fastener bearing perpendicular to the longitudinal axis of the wood grain, and adequate fastener penetration and spacing. When multiple fasteners are used in a connection, an adjustment is necessary to account for the group action. Although limited to certain connection types, several other adjustment factors are applicable, such as the metal side plate factor for split rings, shear plates, and timber rivets, the diaphragm factor for use in a mechanics-based method for diaphragm design, and the toe-nail factor for toe-nailed connections.

Allowable design values are determined by multiplying the nominal values by appropriate adjustment factors. Table 8.1 specifies which adjustment factors must be considered for different connections and load types, where Z, P, and Q refer to lateral loads in the three primary axes directions and W refers to withdrawal loads.

Load-Duration Factor, C_D. Wood-bearing (crushing) strength is dependent upon the duration of loading, due to damage accumulation in the wood material, and the same load-duration factors are applied to wood connections as are applied to other

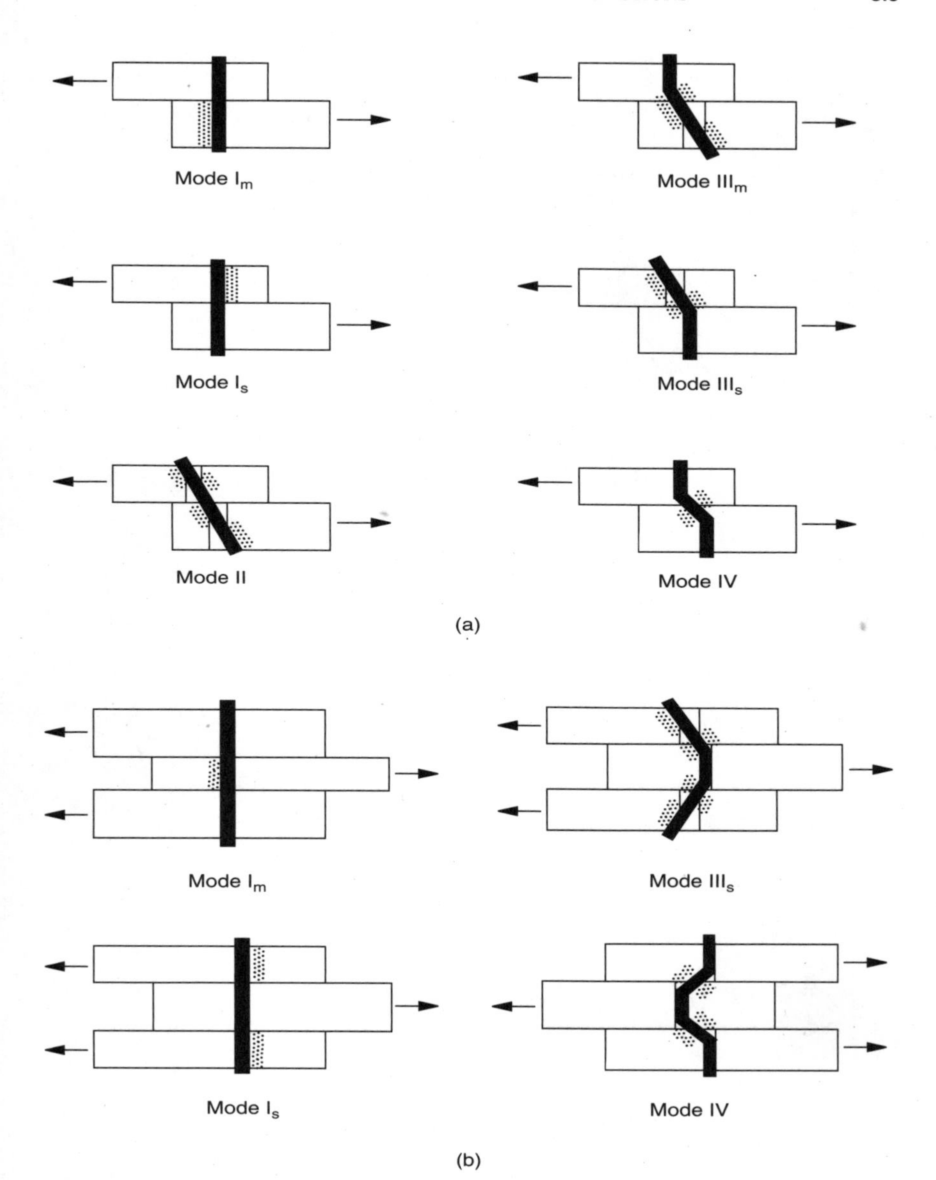

FIGURE 8.2 Single-faster connection yield modes: (*a*) single shear connections, (*b*) double shear connections.

TABLE 8.1 Applicability of Adjustment Factors for Connection

		Load-duration factor*	Wet-service factor†	Temperature factor	Group action factor	Geometry factor‡	Penetration-depth factor‡	End-grain factor‡	Metal side plate factor‡	Diaphragm factor‡	Toe-nail factor‡
Bolts	$Z' = Z$	C_D	C_M	C_t	C_g	C_Δ	•	•	•	•	•
Lag screws	$W' = W$	C_D	C_M	C_t	•	•	•	C_{eg}	•	•	•
	$Z' = Z$	C_D	C_M	C_t	C_g	C_Δ	C_d	C_{eg}	•	•	•
Split-ring and shear-plate connectors	$P' = P$	C_D	C_M	C_t	C_g	C_Δ	C_d	•	C_{st}	•	•
	$Q' = Q$	C_D	C_M	C_t	C_g	C_Δ	C_d	•	•	•	•
Wood screws	$W' = W$	C_D	C_M	C_t	•	•	•	•	•	•	•
	$Z' = Z$	C_D	C_M	C_t	•	•	C_d	C_{eg}	•	•	•
Nails and spikes	$W' = W$	C_D	C_M	C_t	•	•	•	•	•	•	C_{tn}
	$Z' = Z$	C_D	C_M	C_t	•	•	C_d	C_{eg}	•	C_{di}	C_{tn}
Metal plate connectors	$Z' = Z$	C_D	C_M	C_t	•	•	•	•	•	•	•
Drift bolts and drift pins	$W' = W$	C_D	C_M	C_t	•	•	•	C_{eg}	•	•	•
	$Z' = Z$	C_D	C_M	C_t	C_g	C_Δ	C_d	C_{eg}	•	•	•
Spike grids	$Z' = Z$	C_D	C_M	C_t	•	C_Δ	•	•	•	•	•
Timber rivets	$P' = P$	C_D¶	C_M	C_t	•	•	•	•	C_{st}***	•	•
	$Q' = Q$	C_D¶	C_M	C_t	•	C_Δ**	•	•	C_{st}***	•	•

*The load-duration factor C_D shall not exceed 1.6 for connections.

†The wet service factor C_M shall not apply to toe nails loaded in withdrawal.

‡Specific information concerning geometry factors C_Δ, penetration depth factors C_d, end-grain factors C_{eg}, metal side plate factors C_{st}, diaphragm factors C_{di}, and toenail factors C_{tn} is provided in Chapters 8, 9, 10, 11, 12 and 14 of the NDS.[1]

¶The load-duration factor, C_D, is only applied when wood capacity, P_w or Q_w, controls. See section on Timber Rivets.

**The geometry factor, C_Δ, is only applied when wood capacity, Q_w, controls. See section on Timber Rivets.

***The metal side plate factor, C_{st}, is only applied when rivet capacity (P_r, Q_r) controls. See section on Timber Rivets.

Source: This table is from AF&PA's *National Design Specification® for Wood Construction.*

wood elements, with two exceptions. The first is that the impact load-duration factor (C_D = 2.0) is not allowed for connections. The second is when the capacity of a connection is controlled by the strength of a material other than wood, such as metal, concrete, or masonry. The allowable strength of these materials shall not be adjusted for load duration because they do not exhibit wood's load duration behavior.

The 1997 Uniform Building Code[5] (UBC) has more conservative load-duration factors for wind and earthquake loadings for all members, including connections, than does the NDS. For wind and earthquake loads, the UBC specifies C_D = 1.33, except that C_D = 1.6 can be used for nailed and bolted connections exhibiting mode III and IV behavior. The NDS and 2000 International Building Code[6] (IBC) permit C_D = 1.6 for wind and earthquake loads regardless of failure mode. Breyer et al.[7] report that C_D = 1.6 for modes III and IV in the UBC is permitted due to the greater energy dissipation attributes of these more ductile failure modes. They also report that 99% of the tabulated nail connection design values in the NDS are governed by modes III and IV, so the UBC and NDS are essentially in agreement regarding wind and seismic load duration factors for nailed connections. Reference 8 provides tables giving both the controlling failure mode and the tabulated nominal design values for nailed and bolted connections, as in the NDS.

In this text, NDS provisions are followed, but the above discussion serves as a reminder to designers to verify local code acceptance of the appropriate load-duration factor.

The load-duration factors are given in Chapter 1.

Wet Service Factor, C_M. The moisture content in the wood is considered both at the time of fabrication of the connection and as it will be in service. Dry use describes sawn wood elements with moisture contents of 19% or less and engineered wood products such as glulam and LVL with moisture contents of 16% or less. Most covered structures will remain continuously dry for the service life of the structure, and the issue of dry at fabrication pertains mostly to sawn lumber, which often does have higher moisture contents after the harvested wood is cut. However, most engineered wood products require a low moisture content for good adhesion in their manufacture, and thus engineered wood elements will almost always be dry at time of fabrication. Furthermore, many engineered wood products, such as I-joists, are manufactured for specific end-use applications and are intended to remain dry at all times.

Nominal design values must be multiplied by the appropriate wet service factors specified in Table 8.2.

Temperature Factor, C_t. The temperature factor can usually be set equal to one, but for uncommon applications where the wood members and their connections will experience sustained exposure to elevated temperatures (above 100°F), the nominal design values are to be multiplied by the appropriate temperature factors specified in Table 8.3.

Group Action Factor, C_g. Tests have shown that when more than two bolts, lag screws, or metal connectors (such as shear plates, split rings, and drift bolts and pins) are placed in a row, the distribution of force to the fasteners in the row is not uniform. The behavior of a group of fasteners is as follows:[9]

1. As the number of fasteners in a row increases, the allowable load per fastener decreases.

TABLE 8.2 Wet Service Factors for Connections

Fastener type	Moisture content: At time of fabrication	Moisture content: In-service	Load: Lateral	Load: Withdrawal
Shear plates and split rings[a]	≤19%	≤19%	1.0	–
	>19%	≤19%	0.8	–
	any	>19%	0.7	–
Metal connector plates[b]	≤19%	≤19%	1.0	–
	>19%	≤19%	0.8	–
	any	>19%	0.8	–
Bolts, drift pins, and drift bolts	≤19%	≤19%	1.0	–
	>19%	≤19%	0.4[c]	–
	any	>19%	0.7	–
Lag screws and wood screws	≤19%	≤19%	1.0	1.0
	>19%	≤19%	0.4[c]	1.0
	any	>19%	0.7	0.7
Nails and spikes	≤19%	≤19%	1.0	1.0
	>19%	≤19%	0.7	0.25
	≤19%	>19%	0.7	0.25
	>19%	>19%	0.7	1.0
Threaded hardened nails	any	any	1.0	1.0
Timber rivets	≤19%	≤19%	1.0	–
	>19%	≤19%	0.9	–
	≤19%	>19%	0.8	–
	>19%	>19%	0.8	–

[a] For split ring or shear plate connectors, moisture content limitations apply to a depth of ¾ in. below the surface of the wood.

[b] For more information on metal connector plates see NDS.[1]

[c] C_m = 1.0 for wood screws. For bolt and lag screw connections with: (1) one fastener only, (2) two or more fasteners placed in a single row parallel to grain, or (3) fasteners placed in two or more rows parallel to grain with separate splice plates for each row, C_m = 1.0.

Source: This table is from AF&PA's *National Design Specification® for Wood Construction.*

TABLE 8.3 Temperature Factors, C_t, for Connections

In-service moisture conditions[a]	C_t: $T \leq$ 100°F	C_t: 100°F $< T \leq$ 125°F	C_t: 125°F $< T \leq$ 150°F
Dry (1)	1.0	0.8	0.7
Wet	1.0	0.7	0.5

[a] Dry in-service for wood is defined as a moisture content of 19% or less and the wood remains under continuously dry conditions such as in most covered structures.

Source: This table is from AF&PA's *National Design Specification® for Wood Construction.*

2. If a connection contains more than two fasteners in a row:
   - An uneven distribution of fastener load occurs.
   - The two end fasteners carry a greater load than the interior fasteners, and with six or more fasteners in a row, the two end fasteners carry over 50% of the load.
   - The elastic strength of the connection will not increase significantly if additional fasteners (more than six) are added to a row.
3. Small misalignment of bolt holes may cause large shifts in bolt loads.
4. Ultimate strength tests show that:
   - A slight redistribution of the load from the more heavily loaded end bolts to the less heavily loaded interior bolts occurs when wood crushing under the bolt is the mode of failure.
   - A partial specimen failure occurs before substantial redistribution takes place if final failure is in shear.

Therefore, it is necessary in the design of a connection with more than two fasteners in a row to include a group action factor C_g.

$$C_g = \left[\frac{m(1 - m^{2n})}{n[(1 + R_{EA}m^n)(1 + m) - 1 + m^{2n}]}\right]\left(\frac{1 + R_{EA}}{1 - m}\right) \tag{8.1}$$

where n = the number of fasteners in a row
R_{EA} = the lesser of E_sA_s/E_mA_m or E_mA_m/E_sA_s
E_m = modulus of elasticity of main member (psi)
E_s = modulus of elasticity of side member (psi)
A_m = gross cross-sectional area of main member (in.^{2})
A_s = gross cross-sectional area of side member, (in.^{2})
$m = u - \sqrt{u^2 - 1}$
$u = 1 + \gamma \frac{s}{2}\left(\frac{1}{E_mA_m} + \frac{1}{E_sA_s}\right)$
s = center-to-center spacing between adjacent fasteners in a row (in.)
γ = load/slip modulus for connection, lb/in.
γ = (180,000)($D^{1.5}$) for bolts or lag screws in wood-to-wood connections
γ = (270,000)($D^{1.5}$) for bolts or lag screws in wood-to-metal connections
D = diameter of bolt or lag screw (in.)

Tables 8.4 and 8.5 are based on Eq. (8.1).

Effective Cross-Sectional Areas. In order to obtain the adjustment factor, C_g, from tabular data provided in Table 8.4 and Table 8.5, the effective cross-sectional areas must first be determined as follows:

1. *For load applied parallel to grain:* Use the gross cross-sectional area of the member.
2. *For load applied perpendicular to grain:* The effective cross-sectional area is the product of the member thickness and overall width of the fastener group, see Example 8.2. If only one row of fasteners is used, use the minimum parallel-to-grain spacing of the fasteners as the width of the fastener group.
3. *For load applied at an angle to grain:* No guidance is provided by NDS; however, based on engineering judgment, appropriate cross-sectional areas can be determined.

TABLE 8.4 Group Action Factors, C_g, for Bolt or Lag Screw Connections with Wood Side Members* for $D = 1$ in., $s = 4$ in., $E = 1{,}400{,}000$ psi

		Number of fasteners in a row										
A_s/A_m†	A_s,† in.^{2}	2	3	4	5	6	7	8	9	10	11	12
0.5	5	0.98	0.92	0.84	0.75	0.68	0.61	0.55	0.50	0.45	0.41	0.38
	12	0.99	0.96	0.92	0.87	0.81	0.76	0.70	0.65	0.61	0.57	0.53
	20	0.99	0.98	0.95	0.91	0.87	0.83	0.78	0.74	0.70	0.66	0.62
	28	1.00	0.98	0.96	0.93	0.90	0.87	0.83	0.79	0.76	0.72	0.69
	40	1.00	0.99	0.97	0.95	0.93	0.90	0.87	0.84	0.81	0.78	0.75
	64	1.00	0.99	0.98	0.97	0.95	0.93	0.91	0.89	0.87	0.84	0.82
1	5	1.00	0.97	0.91	0.85	0.78	0.71	0.64	0.59	0.54	0.49	0.45
	12	1.00	0.99	0.96	0.93	0.88	0.84	0.79	0.74	0.70	0.65	0.61
	20	1.00	0.99	0.98	0.95	0.92	0.89	0.86	0.82	0.78	0.75	0.71
	28	1.00	0.99	0.98	0.97	0.94	0.92	0.89	0.86	0.83	0.80	0.77
	40	1.00	1.00	0.99	0.98	0.96	0.94	0.92	0.90	0.87	0.85	0.82
	64	1.00	1.00	0.99	0.98	0.97	0.96	0.95	0.93	0.91	0.90	0.88

*Tabulated group action factors (C_g) are conservative for $D < 1$ in., $s < 4$ in., or $E > 1{,}400{,}000$ psi.
†When $A_s/A_m > 1.0$, use A_m/A_s and use A_m instead of A_s.
Source: This table is from AF&PA's *National Design Specification® for Wood Construction.*

Row of Fasteners. The following are considered to be a row of fasteners:

1. Two or more bolts of the same diameter loaded in single or multiple shear
2. Two or more connector units or lag bolts of the same type and size loaded in single shear
3. Adjacent staggered rows of fasteners that are spaced apart less than ¼ the spacing between the fasteners in a row (see Fig. 8.3a)

Examples of staggered fasteners are shown in Fig. 8.3.

Example 8.1 Group of Fasteners Loaded Parallel to Grain

Determine the group action factor for the bolted butt joint shown in Fig. 8.4.

solution

A_m (3 × 6) = 2.5 × 5.5 = 13.75 in.^{2} (8871 mm^2)

A_s (two 2 × 6s) = 2 × (1.5 × 5.5) = 16.50 in.^{2} (10,645 mm^2)

Since $A_s/A_m > 1.0$, consider $A_m/A_s = 13.75/16.5 = 0.833$

Interpolating between $A_m/A_s = 0.5$ and 1.0 for $A_m = 12$ in.^{2} from Table 8.4 gives

$C_g = 0.87 + (0.93 - 0.87)(0.833 - 0.5)/(1 - 0.50) = 0.91$ (conservative since based on $A_m = 12$ in.^{2})

Note: double interpolation from the tables or Eq. (8.1) can be used to obtain a less conservative exact value for C_g.

TABLE 8.5 Group Action Factors, C_g, for Bolt and Lag Screw Connections with Steel Side Plates* for $D = 1$ in, $s = 4$ in, $E_{wood} = 1{,}4000{,}000$ psi, $E_{steel} = 30{,}000{,}000$ psi

		Number of fasteners in a row										
A_m/A_s	A_m, in.^{2}	2	3	4	5	6	7	8	9	10	11	12
12	5	0.97	0.89	0.80	0.70	0.62	0.55	0.49	0.44	0.40	0.37	0.34
	8	0.98	0.93	0.85	0.77	0.70	0.63	0.57	0.52	0.47	0.43	0.40
	16	0.99	0.96	0.92	0.86	0.80	0.75	0.69	0.64	0.60	0.55	0.52
	24	0.99	0.97	0.94	0.90	0.85	0.81	0.76	0.71	0.67	0.63	0.59
	40	1.00	0.98	0.96	0.94	0.90	0.87	0.83	0.79	0.76	0.72	0.69
	64	1.00	0.99	0.98	0.96	0.94	0.91	0.88	0.86	0.83	0.80	0.77
	120	1.00	0.99	0.99	0.98	0.96	0.95	0.93	0.91	0.90	0.87	0.85
	200	1.00	1.00	0.99	0.99	0.98	0.97	0.96	0.95	0.93	0.92	0.90
18	5	0.99	0.93	0.85	0.76	0.68	0.61	0.54	0.49	0.44	0.41	0.37
	8	0.99	0.95	0.90	0.83	0.75	0.69	0.62	0.57	0.52	0.48	0.44
	16	1.00	0.98	0.94	0.90	0.85	0.79	0.74	0.69	0.65	0.60	0.56
	24	1.00	0.98	0.96	0.93	0.89	0.85	0.80	0.76	0.72	0.68	0.64
	40	1.00	0.99	0.97	0.95	0.93	0.90	0.87	0.83	0.80	0.77	0.73
	64	1.00	0.99	0.98	0.97	0.95	0.93	0.91	0.89	0.86	0.83	0.81
	120	1.00	1.00	0.99	0.98	0.97	0.96	0.95	0.93	0.92	0.90	0.88
	200	1.00	1.00	0.99	0.99	0.98	0.98	0.97	0.96	0.95	0.94	0.92
24	40	1.00	0.99	0.97	0.95	0.93	0.89	0.86	0.83	0.79	0.76	0.72
	64	1.00	0.99	0.98	0.97	0.95	0.93	0.91	0.88	0.85	0.83	0.80
	120	1.00	1.00	0.99	0.98	0.97	0.96	0.95	0.93	0.91	0.90	0.88
	200	1.00	1.00	0.99	0.99	0.98	0.98	0.97	0.96	0.95	0.93	0.92
30	40	1.00	0.98	0.96	0.93	0.89	0.85	0.81	0.77	0.73	0.69	0.65
	64	1.00	0.99	0.97	0.95	0.93	0.90	0.87	0.83	0.80	0.77	0.73
	120	1.00	0.99	0.99	0.97	0.96	0.94	0.92	0.90	0.88	0.85	0.83
	200	1.00	1.00	0.99	0.98	0.97	0.96	0.95	0.94	0.92	0.90	0.89
35	40	0.99	0.97	0.94	0.91	0.86	0.82	0.77	0.73	0.68	0.64	0.60
	64	1.00	0.98	0.96	0.94	0.91	0.87	0.84	0.80	0.76	0.73	0.69
	120	1.00	0.99	0.98	0.97	0.95	0.92	0.90	0.88	0.85	0.82	0.79
	200	1.00	0.99	0.99	0.98	0.97	0.95	0.94	0.92	0.90	0.88	0.86
42	40	0.99	0.97	0.93	0.88	0.83	0.78	0.73	0.68	0.63	0.59	0.55
	64	0.99	0.98	0.95	0.92	0.88	0.84	0.80	0.76	0.72	0.68	0.64
	120	1.00	0.99	0.97	0.95	0.93	0.90	0.88	0.85	0.81	0.78	0.75
	200	1.00	0.99	0.98	0.97	0.96	0.94	0.92	0.90	0.88	0.85	0.83
50	40	0.99	0.96	0.91	0.85	0.79	0.74	0.68	0.63	0.58	0.54	0.51
	64	0.99	0.97	0.94	0.90	0.85	0.81	0.76	0.72	0.67	0.63	0.59
	120	1.00	0.98	0.97	0.94	0.91	0.88	0.85	0.81	0.78	0.74	0.71
	200	1.00	0.99	0.98	0.96	0.95	0.92	0.90	0.87	0.85	0.82	0.79

*Tabulated group action factors (C_g) are conservative for $D < 1$ in. or $s < 4$ in.
Source: This table is from AF&PA's *National Design Specification® for Wood Construction.*

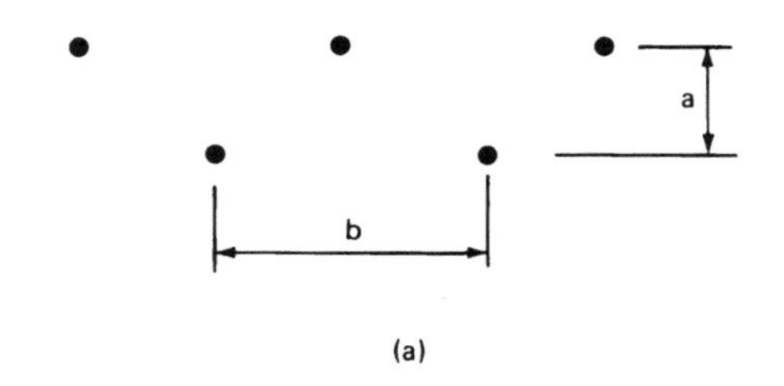

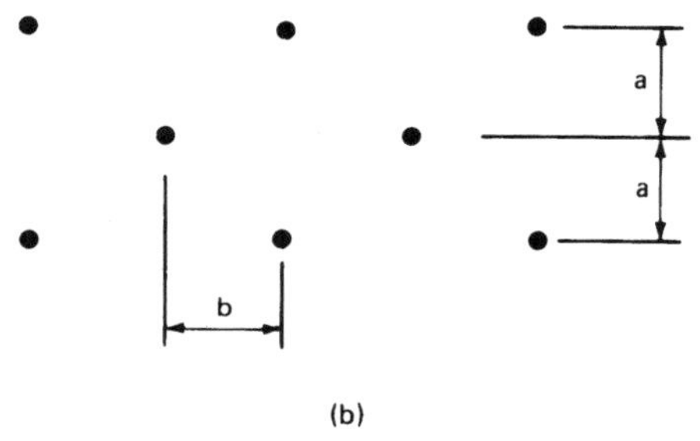

FIGURE 8.3 Staggered fasteners: (*a*) if $a < b/4$, single row; if $a > b/4$, two rows, (*b*) if $a < b/4$, top two lines act as one row and bottom lines acts as a second row; if $a > b/4$, consider as three rows.

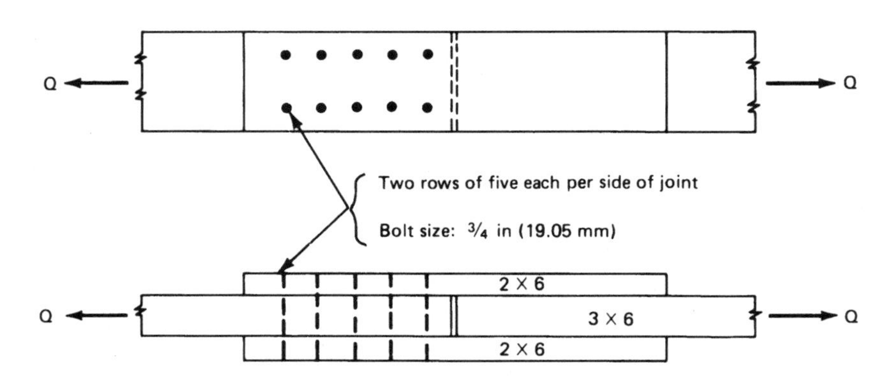

FIGURE 8.4 Bolted butt joint of fasteners loaded parallel to grain; see Example 8.1.

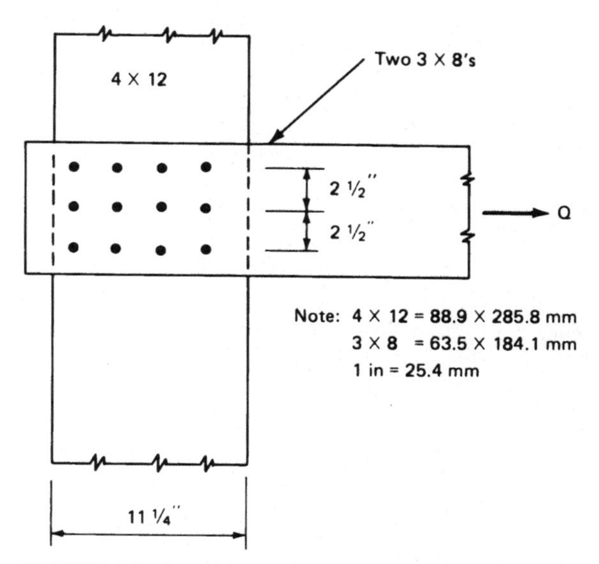

FIGURE 8.5 Bolted connection fasteners loaded perpendicular to grain; see Example 8.2.

Example 8.2 Group of Fasteners Loaded Perpendicular to Grain
Determine the group action factor for the bolted connection in Fig. 8.5.

solution An effective area for the 4 × 12 in. member is:

$$A_m = 2.5 \times 2 \times 3.5 = 17.5 \text{ in.}^2 \text{ (11290 mm}^2\text{)}$$

where 2.5 × 2 = overall width of fastener group and 3.5 = thickness of the main member.

$$A_s = \text{(two 3} \times \text{8's)} = 2 \times 18.125 = 36.25 \text{ in.}^2 \text{ (23387 mm}^2\text{)}$$

$$A_m/A_s = 17.5/36.25 = 0.48 \text{ (round to 0.5 for simplicity)}$$

From Table 8.4, $C_g = 0.96$ (conservatively based on $A_s = 28$ in.²)

Note: extrapolation and interpolation from the tables or Eq. (8.1) can be used to obtain a more accurate, less conservative value.

Geometry Factor, C_Δ. Nominal design values for bolts, lag screws, and other connectors (such as shear plates, split rings, rivets, and drift bolts and pins) require that specific end and edge distances are maintained. The NDS provides two edge and end distance requirement levels, one is for full design values, and the other is the minimum distance permitted. When the actual distances used are less than those required for full design values but greater than the minimum permitted, the geometry factor can be applied to account for this. Three geometry factors are considered:

- $C_{\Delta s}$ = spacing factor
- $C_{\Delta e}$ = edge distance factor
- $C_{\Delta n}$ = end distance factor

The smallest factor for any fastener in a group is applied to all fasteners in that group.

The minimum edge and end distances, spacing, and geometry factors are the same for lag screws and bolts for equal bolt and lag screw shank diameters. These are given in Section 8.3.6 under Placement.

Penetration Depth Factor, C_d. The penetration depth factor adjusts the design value for fastener penetration less than that required for full design values but greater than the required minimum. The method for determining penetration depth factors for the different fastener types is described in their respective sections (see Section 8.3.1 for nails, 8.3.4 for wood screws, and 8.3.5 for lag screws).

End Grain Factor, C_{eg}. End grain factors simply account for fastener penetration into end grain. These factors for the different connection types are described in the fastener type section (see Section 8.3.1 for nails, 8.3.4 for wood screws, and 8.3.5 for lag screws). The use of fasteners in end grain is typically not recommended.

Metal Side Plate Factor, C_{st}. The metal side plate factor is only applicable to timber rivets and split ring and shear plate connectors. Values for this adjustment factor can be found in the NDS.

Diaphragm Factor, C_{di}. Although uncommon, a mechanics-based approach can be used as an alternative to building code-accepted tables for designing horizontal and vertical diaphragms (vertical diaphragms are also known as shear walls).[10] The diaphragm factor is used to adjust the nominal lateral design value for a fastener and is applicable only for attachment of the wood structural panel to framing members when used in a mechanics-based diaphragm design. The diaphragm factor is 1.1.

Toe-Nail Factor, C_{tn}. The nominal withdrawal design values for toe-nailed connections are to be adjusted by the toe-nail factor of 0.67 and the nominal lateral design values by 0.83. According to the NDS, the wet service factor is not to be used for toe nails loaded in withdrawal. The toe-nail factor is discussed further in Section 8.3.1 under Toe Nails.

8.2.3 Angle to Grain Loading

Wood is not as strong or as stiff perpendicular to its longitudinal grain as it is parallel. The Hankinson formula Eq. (8.2) can be used to approximate elastic properties, such as bearing strength, in directions other than parallel or perpendicular to the grain of the wood:

$$F_{e\theta} = \frac{F_{e\parallel} F_{e\perp}}{F_{e\parallel} \sin^2\theta + F_{e\perp} \cos^2\theta} \tag{8.2}$$

where $F_{e\theta}$ = dowel bearing strength at inclination θ with direction of grain
$F_{e\parallel}$ = dowel bearing strength parallel to grain

$F_{e\perp}$ = dowel bearing strength perpendicular to grain
θ = angle between direction of grain and direction of desired property (ranges from 0° for parallel to grain and 90° for perpendicular to grain)

See Chapter 1 for further discussion of the Hankinson formula.

8.3 TYPES OF MECHANICAL FASTENERS AND CONNECTORS

8.3.1 Nails

Introduction. Formerly, the common wire nail was typically the most frequently used fastener in wood construction and still may be for hammer-driven nails due to their relatively thicker shank diameter, which will better resist bending. However, with the advent of power-driven nails and foreign manufacturing, the common wire nail is now one of many types of nails in wide use, including those with coated or deformed shanks. Along with the increased use of power-driven nailing has come increased confusion about nail nomenclature in regard to pennyweight, type, and sizes as older terminology may no longer be applicable.

This section is aimed more at the traditional nail user where many years have contributed to the current design recommendations and vast amounts of data exist. Section 8.3.3 is devoted to the topic of power driven fasteners. An attempt will be made in both in this section and in that on power-driven fasteners to explain the differences between traditional nail nomenclature and actual power-driven nail sizes.

Many types of nails have been developed, such as those with deformed shanks and coated surfaces. Some of these are shown in Fig. 8.6.

The diameters and lengths for the nails traditionally used in construction are shown in Table 8.6.

Loading of Nails

1. Lateral or withdrawal loading results from a force applied perpendicular or parallel, respectively, to the axis of the nail, as shown in Fig. 8.7.
2. Nails should be laterally loaded and not subjected to withdrawal forces, if possible.
3. End-grain withdrawal (nails driven into wood parallel with the longitudinal fibers) is not recommended.

Design Values. Design values for nails can be referenced from tables, or they can be calculated. When appropriate tables are not available, an empirical equation can be used to calculate withdrawal resistance (as presented in Section 8.3.1 under Withdrawal Resistance) and the EYM equations can be used to calculate lateral resistance (as presented in Section 8.3.1 under Lateral Resistance).

When more than one nail is used in a connection, the total design value for the connection loaded laterally or in withdrawal is the sum of the design values for a single nail.

Prebored Holes. Nominal design values for nails are appropriate whether holes are predrilled or not. When a predrilled hole is desired to prevent splitting of wood, the diameter of the hole should not exceed:

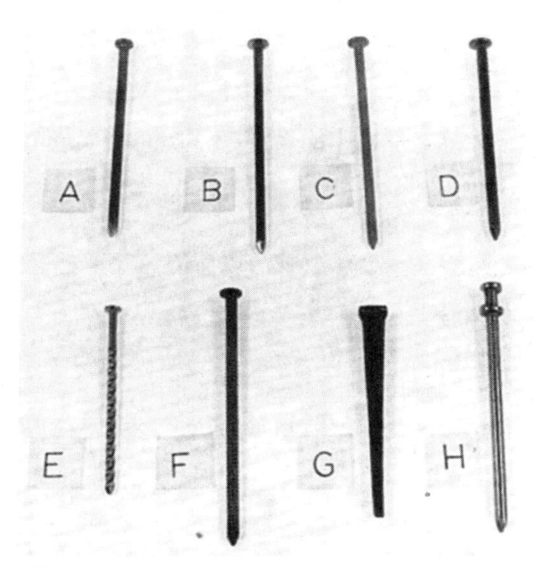

FIGURE 8.6 Different nail types: (*A*) common wire nail, (*B*) box nail, (*C*) zinc-coated, (*D*) cement-coated, (*E*) helically threaded, (*F*) annularly threaded, (*G*) cut nail, (*H*) double-headed construction nail.

- 90% of the nail diameter for wood with a specific gravity greater than 0.6
- 75% of the nail diameter for wood with a specific gravity of 0.6 or less

See Table 8.7 for specific gravity values of commonly used wood species.

Placement. It has often been the practice in the past to let the placement of nails be determined on the job site, with the primary objective being to avoid splitting of the wood. However, in present-day design, this may not be satisfactory. While some guidance is available on the placement of nails in wood, the recommendations are not consistent. The following recommendations have been obtained from the literature (note: d = shank diameter of nail and * refers to distance in direction of stress):

1. *Wood Handbook*[11]

   End distance* = 15d
   End distance = 12d
   Edge distance = 10d

2. *Uniform Building Code*[5]

   End distance* = ½ required penetration of nail
   Edge distance* = ½ required penetration of nail
   Spacing in row* = required penetration of nail

3. *Timber Design and Construction Handbook*[12] *softwoods:*

   End distance = ½ nail length
   Edge distance = ¼ nail length
   Spacing = ½ nail length

4. *Wood Technology in the Design of Structures*[13]

Without prebored holes:

End distance = 20d
Edge distance = 5d
Perpendicular-to-grain spacing = 10d
Parallel-to-grain spacing = 20d

With prebored holes:

End distance = 10d
Edge distance = 5d
Perpendicular-to-grain spacing = 3d
Parallel-to-grain spacing = 10d

The recommended values for nail placement as given in the four references listed above show considerable variation. For example, the recommended values for end distance can vary from approximately 5d to 20d. Minimum spacing as well as end and edge distances are necessary to prevent the wood from splitting as the nail is driven. The density, moisture content, straight grain vs. interlocking grain, and tangential vs. radial grain orientation of the wood and the diameter of the shank and type of point of the nail all have an effect on the splitting of the wood as the nails are driven.

Nail Nomenclature: Pennyweight and Specifying Nails. As can be seen in Table 8.6, a particular pennyweight (e.g., 10d) can have a different diameter and/or length, depending on its type (box, common, or sinker). Both pennyweight *and* type (e.g., 16d *and* common) describe a particular nail size according to the standard ASTM F 1667 (the old standard still sometimes referenced is FF-N-105). Thus, pennyweight refers only to a range of diameters and lengths, not to a specific size, but to include both the pennyweight *and* the type would be to describe a particular length and diameter.

Perhaps in the past pennyweight was used to specify a particular nail, but today this alone is insufficient. With the advent of power-driven fasteners and increased import and nondomestic manufacturing, nails are more frequently being defined by their length and shank diameter. Confusion arises because along with nail dimensions printed on most power-driven nails is a meaningless pennyweight label put there simply because many people "know pennyweight" (mistakenly think pennyweight describes a size). Different manufcturers, or even the same one, may sell the same pennyweight nails, but they may have different dimensions.

To avoid confusion, the specifier should list both the pennyweight *and* type for those going by traditional nomenclature and the length and shank diameter for those using pneumatic nailers (e.g., 8d common, 2½ × 0.131 in.). A common error to avoid is to simply specify a pennyweight, such as 8d, because 8d alone doesn't describe any particular nail size. In fact, the range of actual nail sizes available as 8d, for example, would surprise many. Further discussion of power-driven fasteners continues in Section 8.3.3.

Withdrawal Resistance. Withdrawal loading results from a force or force component, applied parallel to the axis of the nail, as shown in Fig. 8.7. The equation given in the NDS to obtain nominal withdrawal design values is:

$$W = 1380G^{5/2}D \tag{8.3}$$

TABLE 8.6 Typical Nail Dimensions

Type	Dimension[a]	Pennyweight[b]									
		6d	8d	10d	12d	16d	20d	30d	40d	50d	60d
Common	L	2 in.	2½ in.	3 in.	3¼ in.	3½ in.	4 in.	4½ in.	5 in.	5½ in.	6 in.
	D	0.113 in.	0.131 in.	0.148 in.	0.148 in.	0.162 in.	0.192 in.	0.207 in.	0.225 in.	0.244 in.	0.263 in.
	H	0.266 in.	0.281 in.	0.312 in.	0.312 in.	0.344 in.	0.406 in.	0.438 in.	0.469 in.	0.5 in.	0.531 in.
Box	L	2 in.	2½ in.	3 in.	3¼ in.	3½ in.	4 in.	4½ in.	5 in.	–	–
	D	0.099 in.	0.113 in.	0.128 in.	0.128 in.	0.135 in.	0.148 in.	0.148 in.	0.162 in.	–	–
	H	0.266 in.	0.297 in.	0.312 in.	0.312 in.	0.344 in.	0.375 in.	0.375 in.	0.406 in.	–	–
Sinker	L	1⅞ in.	2⅜ in.	2⅞ in.	3⅛ in.	3¼ in.	3¾ in.	4¼ in.	4¾ in.	–	5¾ in.
	D	0.092 in.	0.113 in.	0.12 in.	0.135 in.	0.148 in.	0.177 in.	0.192 in.	0.207 in.	–	0.244 in.
	H	0.234 in.	0.266 in.	0.281 in.	0.312 in.	0.344 in.	0.375 in.	0.406 in.	0.438 in.	–	0.5 in.

[a] L = nail length, D = shank diameter, H = head diameter.
[b] Tolerances specified in ASTM F 1667. See ASTM F 1667 for other nail types.
Source: This table is from AF&PA's *National Design Specification® for Wood Construction.*

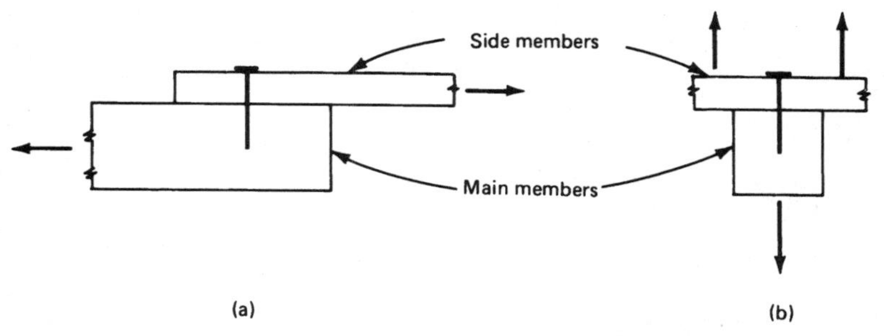

FIGURE 8.7 Type of loading on nails: (*a*) lateral loading; (*b*) withdrawal loading.

TABLE 8.7 Specific Gravity of Some Commonly Used Wood Species

Species of wood	Specific gravity, G
Aspen	0.39
Coast sitka spruce	0.39
Douglas fir-larch	0.50
Douglas fir-larch (north)	0.49
Douglas fir-south	0.46
Eastern hemlock	0.41
Eastern softwoods	0.36
Eastern spruce	0.41
Hem-fir	0.43
Hem-fir (north)	0.46
Mixed southern pine	0.51
Mountain hemlock	0.47
Ponderosa pine	0.43
Redwood, close grain	0.44
Redwood, open grain	0.37
Sitka spruce	0.43
Southern pine	0.55
Spruce-pine-fir	0.42
Spruce-pine-fir (south)	0.36
Western hemlock	0.47
Western hemlock (north)	0.46
Western white pine	0.40

where W = nominal design value per inch of penetration in member holding point of nail
G = specific gravity of wood
D = diameter of nail

Values obtained using Eq. (8.3) are given in Table 8.8. However, an adjustment (factor of 0.192/0.177 to account for shank diameter differences) has been made to the common wire nail values to arrive at values for the threaded hardened nails (see NDS commentary).

TABLE 8.8 Nail and Spike Withdrawal Design Values (*W*)*
(are in pounds per inch of penetration into side grain of main member)

Specific gravity	Common wire nails, box nails and common wire spikes, diameter *D*, in.															Threaded nails, wire diameter *D*, in.				
G	0.099	0.113	0.128	0.131	0.135	0.148	0.162	0.192	0.207	0.225	0.244	0.263	0.283	0.312	0.375	0.12	0.135	0.148	0.177	0.207
0.55	31	35	40	41	42	46	50	59	64	70	76	81	88	97	116	41	46	50	59	70
0.50	24	28	31	32	33	36	40	47	50	55	60	64	59	76	91	32	36	40	47	55
0.43	17	19	21	22	23	25	27	32	35	38	41	44	47	52	63	22	25	27	32	38
0.42	16	18	20	21	21	23	26	30	33	35	38	41	45	49	59	21	23	26	30	35

* Tabulated withdrawal design values (*W*) for nail or spike connections shall be multiplied by all applicable adjustment factors.

Note: 1 in. = 25.4 mm; 1 lb = 4.45 N.

Source: This table is from AF&PA's *National Design Specification® for Wood Construction.*

Equation (8.3) and Table 8.8 provide design values per inch of penetration for:

1. Nails driven into wood without splitting
2. Seasoned wood, which remains dry, or unseasoned wood, which remains wet
3. Normal load duration
4. Nails driven into the side grain
5. Normal temperature

Despite the similar (about 10% different) design values for smooth-shank and threaded-shank nails in the NDS, studies have shown that threaded shank nails have significantly higher withdrawal capacity than comparable smooth-shank nails.[14,15] The purpose of the thread (or other deformation) is to increase the friction between the shank of the nail and the wood fiber. The result of this increased friction is a nail with increased withdrawal resistance, and Eq. (8.3), being a function of specific gravity and shank diameter only, doesn't account for this.

Common wire nails or spikes are not to be loaded in withdrawal from the end grain of wood.

If the end-use conditions of the nominal design values are not expected for the connection under consideration, then appropriate adjustments as described in Section 8.2.2 must be made.

Example 8.3 Withdrawal Loading of Nails

Four 20d common nails are driven into a Douglas fir member (Fig. 8.8). The wood is unseasoned ($MC > 19\%$) but will season with use ($MC \leq 19\%$). Determine the allowable withdrawal dead load.

solution

Specific gravity of Douglas fir = 0.50 (see Table 8.7)

Length of nail = 4 in. (102 mm) (see Table 8.6)

Penetration of nail = 4 − 0.5 = 3.5 in. (88.9 mm)

Load-duration factor, $C_D = 0.9$ (see Chapter 1)

Wet service factor, $C_M = 0.25$ (see Table 8.2)

Nominal strength per nail per inch of penetration: $W = 47$ lb/in. (8.23 N/mm) (see Table 8.8)

W' (allowable) = $W \times$ penetration $\times C_D \times C_M \times$ number of nails

W' (allowable) = $47 \times 3.5 \times 0.9 \times 0.25 \times 4 = 148$ lb (658 N)

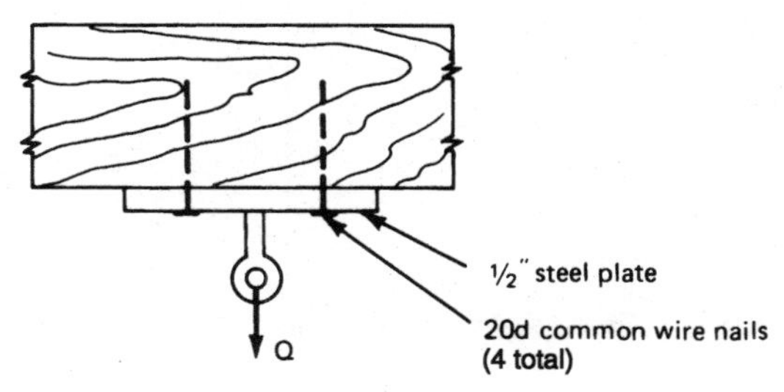

FIGURE 8.8 Example 8.3.

Example 8.4 Nail Withdrawal from Glulam or LVL

Repeat Example 8.3, but instead of solid-sawn lumber the wood member is to be glulam (Douglas fir) or 1.8E LVL. The following principles are equally appropriate for either. Engineered wood products, such as glulam or LVL, will always arrive on site dry since their manufacture requires the use of dry wood components. Therefore, $C_m = 1.0$. The withdrawal strength of the nail in glulam is taken as that for Douglas fir (specific gravity = 0.50), from which the glulam is made. The withdrawal strength of the 1.8E LVL is given an equivalent specific gravity of 0.50 (as presented in Section 8.4.4, Example 8.15). So in this example the two engineered wood products have the same design withdrawal capacities.

$$W' \text{ (allowable)} = W \times \text{penetration} \times C_D \times C_M \times \text{number of nails}$$

$$W' \text{ (allowable)} = 47 \times 3.5 \times 0.9 \times 1.0 \times 4 = 592 \text{ lb (2633 N)}$$

Note the significant difference in design values between Example 8.3 and Example 8.4 due to the wood being dry at the time of connection assembly. Solid-sawn wood, particularly that with larger dimensions, frequently has moisture contents greater than 19% since it can take months for solid wood to lose its moisture when not kiln dried.

Lateral Resistance. Lateral loading results from a force or force component applied perpendicular to the axis of the nail or spike, as shown in Fig. 8.7. A typical load-slip curve for an 8d common nail is shown in Fig. 8.9.

The ultimate lateral load capacity for a nail in single shear (two members) is dependent on several material and dimensional properties of the connection. These include the thickness of the two joined members, the dowel bearing strength of the wood (crushing), and the diameter and yield strength of the nail.

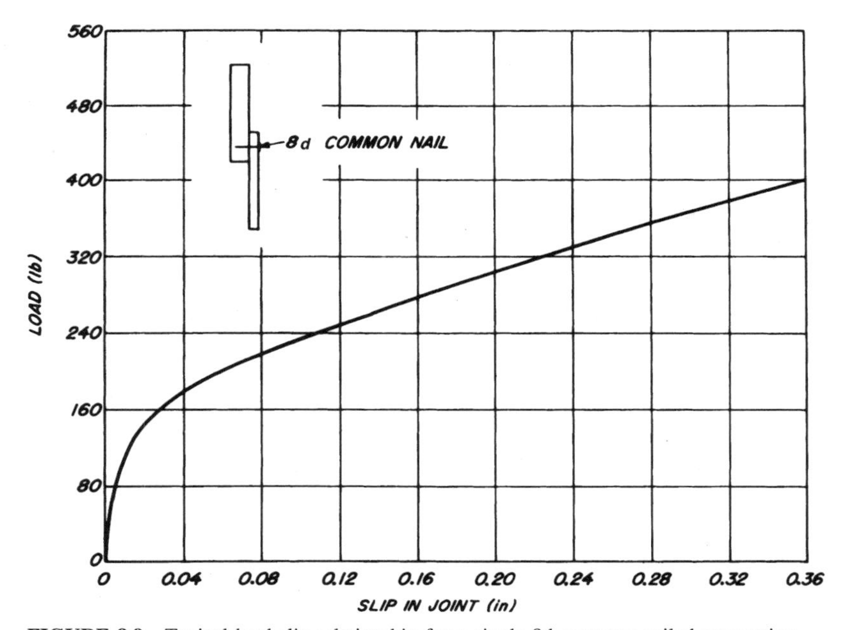

FIGURE 8.9 Typical load-slip relationship for a single 8d common nailed connection.

The European yield model (EYM) provides four possible failure modes for a two-member connection in single shear. The model assumes that the bearing capacity is reached when either the wood crushes under the nail or one or two plastic hinges are formed in the nail. Shown in Table 8.9 are the four yield modes and their respective equations. The nominal nail or spike lateral design value, Z, shall be the lowest value of the four possible failure modes.

Nominal lateral design values have been obtained using the yield model equations and are given in Tables 8.10 and 8.11a and b. These values are given for different nail types and sizes, which meet the minimum size given in Table 8.6 before the application of any protective finish or coating.

If a connection does not qualify under the geometry and material properties given in the design tables, then the yield equations may be used to obtain a lateral

TABLE 8.9 Yield Modes for Single-Shear Nail Connections

	Yield mode
$Z = \dfrac{Dt_sF_{es}}{K_D}$	Mode I_s
$Z = \dfrac{k_2DpF_{em}}{K_D(1 + 2R_e)}$	Mode III_m
$Z = \dfrac{k_3Dt_sF_{em}}{K_D(2 + R_e)}$	Mode III_s
$Z = \dfrac{D^2}{K_D}\sqrt{\dfrac{2F_{em}F_{yb}}{3(1 + R_e)}}$	Mode IV

$$\text{where } k_2 = -1 + \sqrt{2(1 + R_e) + \frac{2F_{yb}(1 + 2R_e)D^2}{3F_{em}p^2}}$$

$$k_3 = -1 + \sqrt{\frac{2(1 + R_e)}{R_e} + \frac{2F_{yb}(2 + R_e)D^2}{3F_{em}t_s^2}}$$

$R_e = F_{em}F_{es}$.
p = penetration, main member (member holding point), in.
t_s = thickness side member or $L/3$ for toe-nailed connections, in.
F_{em} = dowel bearing strength, main member, lb/in.^2
F_{es} = dowel bearing strength, side member, lb/in.^2
D = nail or spike diameter, in (use root diameter for threaded nails)
F_{yb} = bending yield strength of nail or spike, psi
K_D = 2.2 for $D \leq 0.17$ in., $K_D = 10D + 0.5$ for 0.17 in. $< D <$ 0.25 in.

TABLE 8.10 Nail Lateral Load Design Values for Single-Shear Connections with Both Members of Identical Species[a]

Side member thickness, in.	Nail diameter, in.	Pennyweight			G = 0.55 Southern pine	G = 0.5 Douglas fir larch	G = 0.43 Hem-fir	G = 0.42 Spruce-pine-fir
		Common	Box	Sinker				
½	0.099	–	6d	7d	55	48	39	38
	0.113	6d	8d	8d	67	59	49	47
	0.120	–	–	10d	74	65	54	53
	0.128	–	10d, 12d	–	82	73	61	59
	0.131	8d	–	–	85	76	63	61
	0.135		16d	12d	89	79	66	65
	0.148	10d, 12d	20d	16d	101	90	75	73
	0.162	16d	40d	–	117	105	89	87
	0.177	–	–	20d	127	114	96	94
	0.192	20d	–	30d	137	124	105	103
	0.207	30d	–	40d	148	134	115	112
	0.225	40d	–	–	162	147	126	123
	0.244	50d	–	60d	166	151	130	127
¾	0.099	–	6d	7d	61	55	48	47
	0.113	6d	8d	8d	79	72	58	57
	0.120	–	–	10d	89	80	64	62
	0.128	–	10d, 12d	–	101	87	70	68
	0.131	8d	–	–	104	90	73	70
	0.135	–	8d	12d	108	94	76	74
	0.148	10d, 12d	20d	16d	121	105	85	83
	0.162	16d	40d	–	138	121	99	96
	0.177	–	–	20d	148	130	106	103
	0.192	20d	–	30d	157	138	114	111
	0.207	30d	–	40d	166	147	122	119
	0.225	40d	–	–	178	158	132	129
	0.244	50d	–	60d	182	162	136	132
1	0.099	–	6d	7d	61	55	48	47
	0.113	6d	8d	8d	79	72	63	61
	0.120	–	–	10d	89	81	71	69
	0.128	–	–	–	101	93	80	79
	0.131	8d	10d, 12d	–	106	97	84	82
	0.135	–	–	12d	113	103	89	86
	0.148	10d, 12d	16d	16d	128	118	99	96
	0.162	16d	20d	–	154	141	113	109
	0.177	–	40d	20d	168	151	121	117
	0.192	20d	–	30d	183	159	128	124
	0.207	30d	–	40d	192	167	135	131
	0.225	40d	–	–	202	172	144	140
	0.244	50d	–	60d	207	181	148	143
1¼	0.099	–	6d[d]	7d[d]	61	55	48	47
	0.113	6d[d]	8d[d]	–	79	72	63	51
	0.120	–	–	10d	89	81	71	69
	0.128	–	10d, 12d	–	101	93	80	79
	0.131	8d[d]	–	–	106	97	84	82
	0.135	–	16d	12d	113	103	89	88
	0.148	10d, 12d	20d	16d	128	118	102	100
	0.162	16d	40d	–	154	141	122	120
	0.177	–	–	20d	168	154	133	130
	0.192	20d	–	30d	185	170	145	140
	0.207	30d	–	40d	203	186	152	147

TABLE 8.10 Nail Lateral Load Design Values for Single-Shear Connections with Both Members of Identical Species[a] (*Continued*)

Side member thickness, in.	Nail diameter, in.	Pennyweight			$G = 0.55$ Southern pine	$G = 0.5$ Douglas fir larch	$G = 0.43$ Hem-fir	$G = 0.42$ Spruce-pine-fir
		Common	Box	Sinker				
1¼	0.225	40d	–	–	224	200	160	15
	0.244	50d	–	60d	230	204	163	158
	0.207	30d	–	40d	203	186	152	147
	0.225	40d	–	–	224	200	160	155
	0.244	50d	–	60d	230	204	163	158
1½	0.099	–	6d[d]	7d[d]	61	55	48	47
	0.113	6d[d]	8d[d]	6d[d]	79	72	63	61
	0.120	–	–	10d	89	81	71	69
	0.128	–	10d, 12d	–	101	93	80	79
	0.131	8d[d]	–	–	106	97	84	82
	0.135	–	16d	12d	113	103	89	88
	0.148	10d, 12d	20d	16d	128	118	102	100
	0.162	16d	40d	–	154	141	122	120
	0.177	–	–	20d	168	154	133	130
	0.192	20d	–	30d	155	170	147	144
	0.207	30d	–	40d	203	186	161	158
	0.225	40d	–	–	224	205	178	172
	0.244	50d	–	60d	230	211	181	175

[a] Tabulated lateral design values (Z) shall be multiplied by all applicable adjustment factors.

[b] Tabulated lateral design values are for common wire, box, and sinker nails inserted in side grain with nails perpendicular to wood fibers; minimum nail penetration, p, into the main member equal to $10D$; and nail bending yield strengths (F_{yb}):

F_{yb} = 100,000 psi for 0.099 in. $D \leq 0.142$ in.

F_{yb} = 90,000 psi for 0.142 in. $< D \leq 0.177$ in.

F_{yb} = 80,000 psi for 0.177 in. $< D \leq 0.244$ in.

[c] When $6D \leq p \leq 10D$, tabulated lateral design values (Z) shall be multiplied by $p/10D$.

[d] Nail length is insufficient to provide $10D$ penetration. Tabulated lateral design values (Z) shall be adjusted per footnote.[c]

Source: This table is from AF&PA's *National Design Specification® for Wood Construction.*

design value. Dowel bearing strengths, F_{em} and F_{es}, for lumber are given in Table 8.12, and for wood structural panels (plywood and OSB) in Table 8.42. Yield values for the different nail types are given in the design table footnotes.

The load-carrying capacity of most fasteners is affected considerably by the large difference in the properties of wood parallel and perpendicular to the wood grain. However, the lateral load capacity for nails is approximately the same regardless of the direction of bearing and grain orientation.[16] There is some indication that for the larger-diameter spikes the parallel and perpendicular to grain proportional-limit values may not be the same,[17] but in the NDS the lateral load-carrying capacity of nails is not affected by grain orientation.

The nominal lateral load design values given the design tables or by the EYM equations are for:

1. Nails driven into wood without splitting
2. Seasoned wood, which remains dry in service

TABLE 8.11a Nail Lateral Load Design Values for Single-Shear Connections with ASTM A653, Grade 33 Steel Side Plates[a]

Side member thickness, in.	Nail diameter, in.	Pennyweight			G = 0.55 Southern pine	G = 0.5 Douglas fir larch	G = 0.43 Hem-fir	G = 0.42 Spruce-pine-fir
		Common	Box	Sinker				
0.036 (20 gauge)	0.099		6d	7d	59	54	48	47
	0.113	6d	8d	8d	76	70	62	60
	0.120	–	–	10d, 12d	86	79	69	68
	0.128	–	10d	–	97	90	79	77
	0.131	8d	–	–	102	94	82	81
	0.135	–	16d	12d	108	100	87	86
	0.148	10d, 12d	20d	16d	123	114	100	98
0.048 (18 gauge)	0.099	–	6d	7d	60	55	49	48
	0.113	6d	8d	8d	77	71	63	61
	0.120	–	–	10d	87	80	70	69
	0.128	–	10d, 12d	–	96	91	80	78
	0.131	8d	–	–	103	95	83	82
	0.135	–	16d	12d	109	101	88	87
	0.148	10d, 12d	20d	16d	124	115	101	99
	0.162	16d	40d	–	148	137	120	118
	0.177	–	–	20d	171	158	138	136
	0.192	20d	–	30d	178	164	144	141
	0.207	30d	–	40d	195	179	157	154
0.060 (16 gauge)	0.099	–	6d	7d	62	57	51	50
	0.113	6d	8d	8d	79	73	64	63
	0.12	6d	–	10d	88	82	72	71
	0.128	–	10d, 12d	–	100	92	81	80
	0.131	8d	–	–	104	97	85	83
	0.135	–	16d	12d	111	102	90	88
	0.148	10d	20d	16d	126	116	102	100
	0.162	16d	40d	–	150	138	121	119
	0.177	–	–	20d	172	159	140	137
	0.192	20d	–	30d	179	165	145	142
	0.207	30d	–	40d	195	180	158	155
	0.225	40d	–	–	215	199	174	171
	0.244	50d	–	60d	221	204	179	176
0.075 (14 gauge)	0.099	–	6d	7d	65	60	53	52
	0.113	69	8d	8d	82	76	87	66
	0.120	–	–	10d	91	85	75	73
	0.128	–	10d, 12d	–	103	95	84	82
	0.131	8d	–	–	107	99	88	86
	0.135	–	16d	12d	113	105	93	91
	0.148	10d, 12d	20d	16d	129	119	105	103
	0.162	16d	40d	–	152	141	124	122
	0.177	–	–	20d	175	162	142	139
	0.192	20d	–	30d	182	168	148	145
	0.207	30d	–	40d	198	183	161	157
	0.225	40d	–	–	217	201	176	173
	0.244	50d	–	60d	223	206	181	178
0.105 (12 gauge)	0.099	–	6d	7d	73	68	60	59
	0.113	6d	8d	8d	90	84	74	73
	0.12	–	–	10d	100	93	82	80
	0.128	–	10d, 12d	–	111	103	91	90
	0.131	8d	–	–	116	107	95	93
	0.135	–	16d	12d	122	113	100	98

TABLE 8.11a Nail Lateral Load Design Values for Single-Shear Connections with ASTM A653, Grade 33 Steel Side Plates[a] (*Continued*)

Side member thickness, in.	Nail diameter, in.	Pennyweight			$G = 0.55$ Southern pine	$G = 0.5$ Douglas fir larch	$G = 0.43$ Hem-fir	$G = 0.42$ Spruce-pine-fir
		Common	Box	Sinker				
0.105 (12 gauge)	0.148	10d, 12d	20d	16d	137	127	113	110
	0.162	16d	40d		161	149	132	129
	0.177	–	–	20d	183	169	149	147
	0.192	20d	–	30d	189	175	155	152
	0.207	30d	–	40d	205	190	167	164
	0.225	40d	–	–	223	207	182	179
	0.244	50d	–	60d	230	212	187	183

[a]Tabulated lateral design values (Z) shall be multiplied by all applicable adjustment factors.

[b]Tabulated lateral design values are for common wire, box, and sinker nails inserted in side grain with nails perpendicular to wood fibers; minimum nail penetration, p, into the main member equal to $10D$; dowel bearing strengths (F_e) of 61,850 psi for ASTM A653, Grade 33 steel and nail bending yield strengths (F_{yb}):

F_{yb} = 100,000 psi for 0.099 in. $D \leq 0.142$ in.

F_{yb} = 90,000 psi for 0.142 in. $< D \leq 0.177$ in.

F_{yb} = 80,000 psi for 0.177 in. $< D \leq 0.244$ in.

[c]When $6D \leq p \leq 10D$, tabulated lateral design values (Z) shall be multiplied by $p/10D$.

Source: This table is from AF&PA's *National Design Specification® for Wood Construction.*

3. Normal load duration
4. A nail or spike in single shear
5. Full penetration
6. Nail inserted in side grain
7. Normal temperature
8. Wood or steel side plates

If the end-use conditions assumed for nominal design value tables are not expected for the connection under consideration, then appropriate general adjustments as described in Section 8.2.2 must be made. Specific adjustments are described as follows.

Double Shear. Lateral design values, Z, for nails or spikes in double shear (three-member connections) in wood-to-wood connections are two times the least value of Z determined for each shear plane, provided $t_m > 6D$, where t_m is the thickness of main (center) member and D is the shank diameter of nail. In addition, the penetration depth factor is applied based on the penetration of the nail in the member containing the point of the nail.

When clinching a nail in a double-shear connection, an exception in the NDS states that the lateral design values Z shall be 1.75 times the least value of Z determined for each shear plane provided:

- Nail penetration $\geq 10D$ ($C_d = 1.0$).
- The nails to be clinched are 12d or smaller and extend at least $3D$ beyond the side member.
- The side members are at least ⅜ in. (9.5 mm) thick.

TABLE 8.11b Nail Lateral Load Design Values for Single-Shear Connections with ASTM, Grade 33 Steel Side Plates[a]

Side member thickness, in.	Nail diameter, in.	Pennyweight			$G = 0.55$ Southern pine	$G = 0.5$ Douglas-fir larch	$G = 0.43$ Hem-fir	$G = 0.42$ Spruce-pine-fir
		Common	Box	Sinker				
0.120 (11 gauge)	0.113	6d	8d	8d	95	89	79	77
	0.120	–	–	10d	105	97	86	85
	0.128	–	10d, 12d	–	116	108	96	94
	0.131	8d	–	–	121	112	99	97
	0.135	–	16d	12d	127	118	104	102
	0.148	10d, 12d	20d	16d	143	133	117	115
	0.162	16d	40d		166	154	137	134
	0.177	–	–	20d	188	174	154	151
	0.192	20d	–	30d	195	181	159	156
	0.207	30d	–	40d	210	194	172	168
	0.225	40d	–	–	228	211	186	183
	0.244	50d	–	60d	234	217	191	187
0.134 (10 gauge)	0.099	–	6d	7d	82	76	66	65
	0.113	6d	8d	8d	100	93	83	81
	0.120	–	–	10d	110	102	91	89
	0.128	–	10d, 12d	–	122	113	100	96
	0.131	8d	–	–	126	117	104	102
	0.135	–	16d	12d	132	123	109	107
	0.148	10d, 12d	20d	16d	148	138	122	120
	0.162	–	40d	–	172	160	142	139
	0.177	–	–	20d	194	180	159	156
	0.192	20d	–	30d	200	186	164	161
	0.207	30d	–	40d	215	199	176	173
	0.225	40d	–	–	233	216	191	187
	0.244	50d	–	60d	239	221	195	192
0.179 (7 gauge)	0.099	–	6d	7d	82	76	66	65
	0.113	6d	8d	8d	107	99	86	84
	0.120	–	–	10d	121	111	97	95
	0.126	–	10d, 12d	–	137	126	111	106
	0.131	8d	–	–	144	132	116	114
	0.135	–	16d	12d	152	141	123	121
	0.148	10d, 12d	20d	16d	170	158	140	137
	0.162	16d	40d	–	194	180	160	157
	0.177	–	–	20d	215	200	178	174
	0.192	20d	–	30d	222	206	183	179
	0.207	30d	–	40d	236	219	194	190
	0.225	40d	–		252	234	207	203
	0.244	50d	–	60d	258	240	212	208
0.239 (3 gauge)	0.099	–	6d	7d	82	76	66	65
	0.113	6d	8d	8d	101	99	87	84
	0.12	–	–	10d	121	111	97	95
	0.128	–	10d, 12d	–	137	126	111	108
	0.131	8d	–	–	144	132	116	114
	0.135	–	16d	12d	153	141	123	121
	0.148	10d, 12d	20d	16d	174	160	140	137
	0.162	16d	40d	–	209	192	168	165
	0.177	–	–	20d	241	222	195	191
	0.192	20d	–	30d	251	231	202	198
	0.207	30d	–	40d	270	251	222	217
	0.225	40d	–	–	285	265	235	231
	0.244	50d	–	60d	291	271	240	236

TABLE 8.11b Nail Lateral Load Design Values for Single-Shear Connections with ASTM, Grade 33 Steel Side Plates[a] (*Continued*)

[a]Tabulated lateral design values (Z) shall be multiplied by all applicable adjustment factors.
[b]Tabulated lateral design values are for common wire, box, and sinker nails inserted in side grain with nails perpendicular to wood fibers; minimum nail penetration, p, into the main member equal to $10D$; dowel bearing strengths (F_e) of 61,850 psi for ASTM A653, Grade 33 steel and nail bending yield strengths (F_{yb}):

F_{yb} = 100,000 psi for 0.099 in. $D \leq 0.142$ in.

F_{yb} = 90,000 psi for 0.142 in. $< D \leq 0.177$ in.

F_{yb} = 80,000 psi for 0.177 in. $< D \leq 0.244$ in.

[c]When $6D \leq p \leq 10D$, tabulated lateral design values (Z) shall be multiplied by $p/10D$.
Source: This table is from AF&PA's *National Design Specification® for Wood Construction.*

TABLE 8.12 Dowel Bearing Strength for Nail and Wood Screw Connections

Species combination	Specific gravity, G^a	Dowel bearing strength, F_e, psi[b]
Southern pine	0.55	5550
Douglas fir-larch	0.50	4650
Hem-fir	0.43	3500
Spruce-pine-fir	0.42	3350

Note: 1 lb/in.^{2} = 6.895 kPa.
[a]Specific gravity based on weight and volume when oven dry.
[b]$F_e = 16{,}6000G^{1.84}$, tabulated design values are rounded to nearest 50 psi.
Source: This table is from AF&PA's *National Design Specification® for Wood Construction.*

Length of Penetration. The design values given in the tables apply only when the depth of penetration, p, of the nail or spike into the member holding the point is 10 times the nail or spike diameter, D ($p = 10D$). When penetration is less than $10D$, the design value must be adjusted by C_d, where $C_d = p/10D < 1.0$, but the minimum penetration permitted is $6D$.

Lateral Resistance in End Grain. The design value in lateral resistance for a nail or spike into the end grain (parallel to wood fibers) is ⅔ (C_{eg} 0.67) of the design value given in the tables or equations, which are for the lateral load strength when the nail or spike is driven into the side grain.

Variation of Specific Gravities in Joined Members. When the joined members are sawn lumber of different species groups, an appropriate design value, Z, can be calculated using the four yield model equations previously given. The dowel bearing strengths corresponding to the species, or material, for the main and side members are to be used.

Example 8.5 Lateral Load on Nail

Determine the number and size of common nails needed to transfer a 900 lb (4.0 kN) (dead plus snow) load from the diagonal to the vertical member shown in Fig. 8.10. The wood is seasoned ($MC \leq 19\%$) Douglas fir that will remain dry ($MC \leq 19\%$).

solution

Try 16d common nails; L = 3.5 in. (88.9 mm), D = 0.162 in. (4.12 mm) (from Table 8.6)

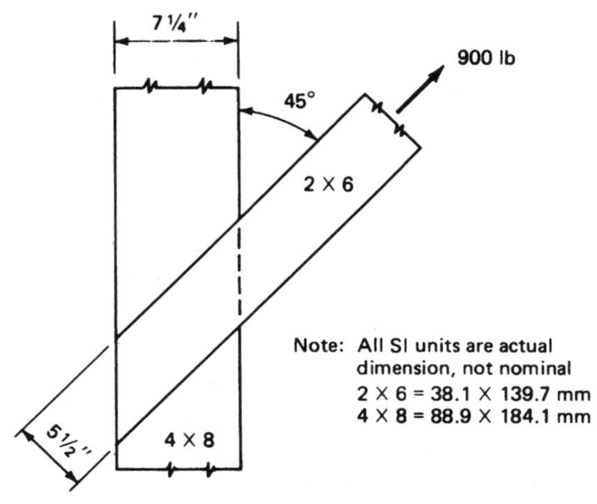

FIGURE 8.10 Example 8.5.

Penetration needed to use full lateral load design value $= 10D = 1.62$ in. (41.2 mm)

Actual penetration $= 3.50 - 1.50 = 2.00$ in. (50.8 mm) $\therefore$ use full nominal design value

Z (nominal) $= 141$ lb (627 N) (from Table 8.10)

Z' (allowable) $= Z \times C_D \times C_M \times C_d \times C_t$

$$C_d = 2.00/1.94 = 1.03 > 1.0 \therefore \text{use } C_d = 1.0$$

Z' (allowable) $= 141 \times 1.15 \times 1.0 \times 1.0 \times 1.0 = 162$ lb (721 N)

Number of nails required $= 900/162 = 5.6 \therefore$ use six 16d common nails

Example 8.6 Nails, Lateral Load, Steel Side Plates

Steel gusset plates are used to transfer the load Q through the joint shown in Fig. 8.11. If 12 8d common nails are used per side as shown, what is the allowable load assuming only dead load is carried? The wood is unseasoned ($MC > 19\%$) spruce-pine-fir that will be used in a dry location ($MC \leq 19\%$).

solution

For an 8d common nail, $L = 2.50$ in. (63.5 mm), $D = 0.131$ in. (3.33 mm) (see Table 8.6)

Penetration needed to use full design value $= 10D = 1.31$ in. (33.3 mm)

Actual penetration $= 2.50 - 0.06 = 2.44$ in. > 1.31 in. (62 mm $>$ 33.3 mm) $\therefore$ the full design value can be used

Z (nominal) $= 83$ lb (369 N) (from Table 8.11a)

Z' (allowable) $= Z \times C_D \times C_M \times C_t$

Z' (allowable) $= 83 \times 0.90 \times 0.70 \times 1.0 = 52$ lb (231 N) per nail

$$Q = 52 \times 12 = 624 \text{ lb (2777 N)}$$

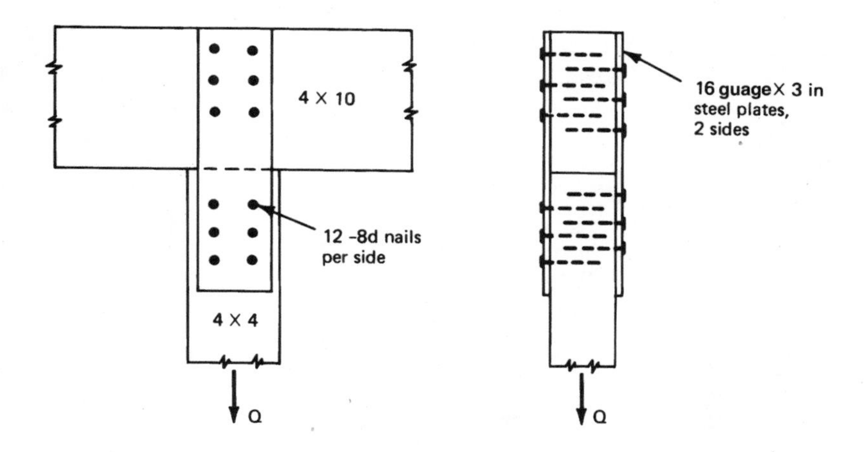

Note: 4 X 4 = 88.9 X 88.9 mm
4 X 10 = 88.9 X 235.0 mm

FIGURE 8.11 Example 8.6 and 8.7.

Example 8.7 Nails, LVL, Lateral Load, Steel Side Plates

Repeat Example 8.6, but instead of the solid-sawn 4 × 10, the member is to be two pieces of 1¾ × 9½, 1.8E LVL, and instead of an unseasoned 4 × 4, a kiln-dried solid sawn member will be used.

$C_m = 1.0$ for dry wood

The equivalent specific gravity of the1.8E LVL is given as 0.50 (as presented in Section 8.4.4, Example 8.15). Since the spruce-pine-fir has a lower bearing strength (specific gravity = 0.42) than the LVL, nails into the 4 × 4 will control the connection design strength. Then,

$$Q \text{ (allowable)} = 12 \times Z' = 12 \times 83 \times 0.90 \times 1.0 \times 1.0 = 896 \text{ lb (3987 N)}$$

If the 4 × 4 were to be Douglas fir instead of spruce-pine-fir, then the individual fastener capacity would become 97 instead of 83 for both the solid-sawn and LVL nailed members (see Table 8.11a). Then,

$$Q \text{ (allowable)} = 12 \times Z' = 12 \times 97 \times 0.90 \times 1.0 \times 1.0 = 1048 \text{ lb (4662 N)}$$

Example 8.8 Clinched Nails, Lateral Load, Plywood Side Plates

A tension splice is constructed of seasoned ($MC \leq 19\%$) southern pine lumber and southern pine plywood (19⁄32 in., with a 40/20 span rating) that will remain dry. Determine the number, size, and placement of common nails needed to transmit the 1800 lb (8.01 kN) load (dead plus wind) through the connection shown in Fig. 8.12.

Note: as discussed in Section 8.4.2, under Bearing Strength, the bearing strength of the plywood is dependent on the species used in the plies, and recommendations are given for when the species making up the plywood is known or unknown. In this example, it is assumed that all plies will be southern pine.

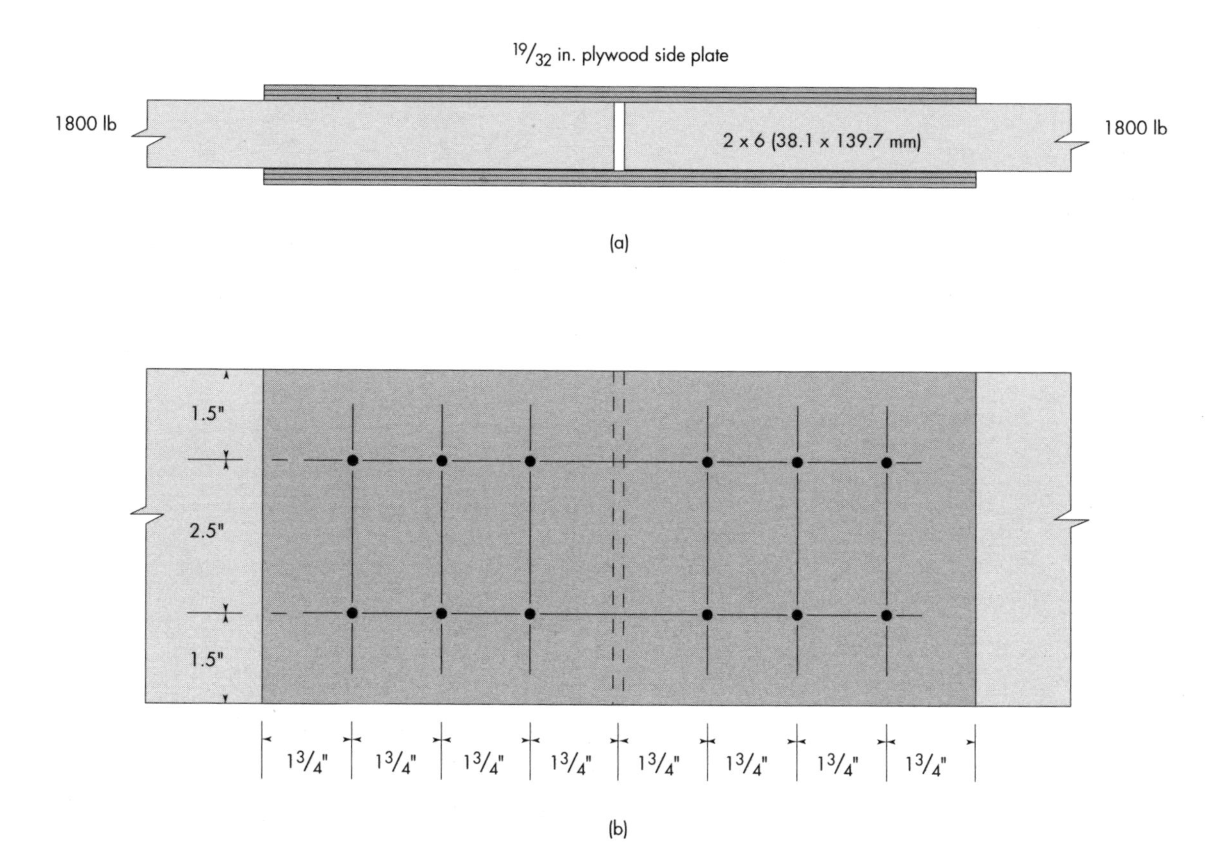

FIGURE 8.12 Example 8.8.

solution

Try 10d common nails; length = 3.0 in. (76.2 mm), D = 0.148 in. (3.76 mm); clinch if possible.

Protruding length required for clinching = $3D = 3 \times 0.148 = 0.444$ in. (11.3 mm)

Actual protruding length = 3.00 −(1.5 + 0.5 + 0.5) = 0.50 in. > 0.444 in. (12.7 mm > 11.3 mm) ∴ clinching may be used.

Z (nominal) = 107 lb/nail from EYM calculations not shown. Mode III controls. A conservative alternative approach would be to use the value for a ½ in. wood side plate from Table 8.10, which is 101 lb/nail.

Z' (allowable) = $Z \times C_D \times C_M \times C_t \times C_d \times$ clinching adjustment (for double shear).

Z' (allowable) = 107 × 1.60 × 1.0 × 1.0 × 1.0 × 1.75 = 300 lb (1335 N) per nail.

Number of nails required = 1800/300 = 6.

Using the conservative alternate approach of 101 lb/nail, seven nails would be required.

Check the tensile capacity of the plywood side plates:

F_tA = 2900 lb/ft (42.3 kN/m) width along strength axis of plywood from Chapter 2 for a 40/20 rated wood structural panel.

Plywood adjustments are discussed in Chapter 2. The nominal tensile capacity F_tA must be adjusted by:

- C_s = 0.5 for plywood width less than 8 in. to account for a potential localized defect.
- 2.0 to account for the two side plates.
- C_g = 1.0 (not to be confused with the group action factor of this chapter) for three- or four-ply plywood. If five-ply were to be used, then C_g = 1.3 would be permitted.
- C_D = 1.6 for wind load duration.

The allowable tensile capacity is:

$$F_tA' = 2900 \times 0.5 \times 2 \times 1.0 \times 1.6$$
$$= 2126 \text{ lb} > 1800 \text{ lb } (9457 \text{ N} > 8007 \text{ N}) \therefore \text{ ok}$$

Toe Nails. Toe nails are often used to fasten studs and joists to plates, although they should be avoided when possible since it is very difficult to provide a good quality toe-nail connection. A toe nail is driven as shown in Fig. 8.13. All end

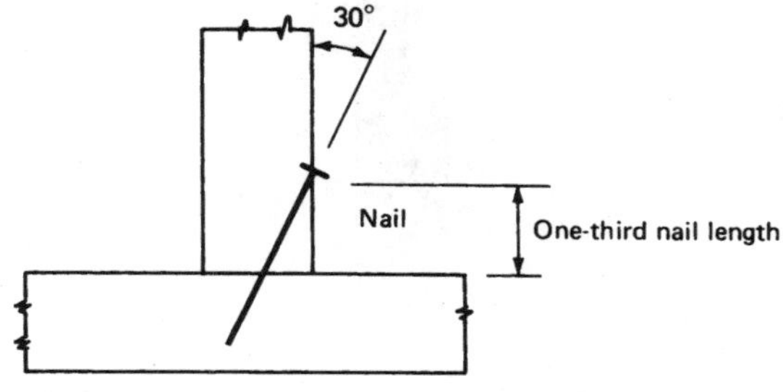

FIGURE 8.13 Toe nail.

distance recommendations are violated with a toe nail, and from the author's experience toe-nailing frequently splits the side member (usually the stud or blocking). Additionally, toe nails are limited in their directional resistance as shown in Fig. 8.14 and cannot be relied upon to provide load capacity in all directions. Without further detail, these reasons alone are enough to avoid engineering or specifying toe-nailed connections even though they are permitted by code. Significantly stronger connections can be made using simple proprietary connectors.[23,24]

Withdrawal Loading. The design value for withdrawal loading at a toe nail should not exceed two-thirds of the side-grain withdrawal value.

Lateral Loading. The design value for a toe nail subjected to a lateral load should not exceed 5⁄6 of that permitted for a nail driven in side grain and laterally loaded.

Adjustments—Toe Nails. All adjustment factors that apply to lateral and withdrawal values for side grain also apply to toe nails (see Section 8.2.2), except that the wet-service factor C_M does not apply to toe nails loaded in withdrawal, per the NDS.

Slant Nailing. Although the terms *slant nailing* and *toe-nailing* might be used interchangeably, there are design implications for when the toe-nail adjustment factor is appropriate. Slant nailing is commonly used to join two decking members, or joists or rafters to wall plates (such as a rim board or rim joist to a top plate).

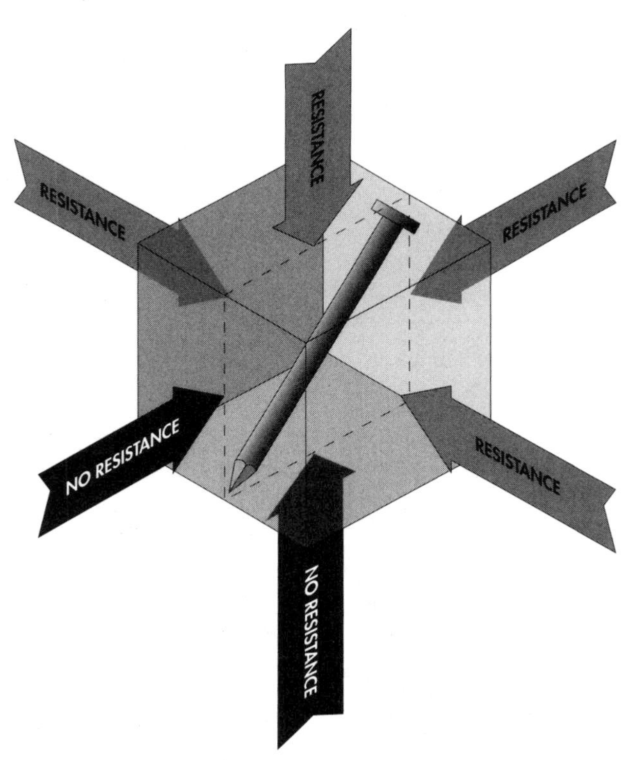

FIGURE 8.14 Toe nail limitations.

A distinction is drawn because toe-nail adjustment factors are appropriate only when the longitudinal axes of the two members being joined intersect,[13] as shown in Fig. 8.13. In other words,[7] in a toe-nail connection the nail goes through the end grain of the side member and the slant nail does not. *The toe-nail adjustment is only appropriate for toe nails and need not be applied to slant nails.*

Other Nails. Other nail types not addressed in this chapter exist, but little or no structural advantage is gained by using these nails under normal loading and use conditions, although deformed shank nails have been shown to have significantly greater withdrawal strength (as discussed in Section 8.3.1 under Withdrawal Resistance). Often power-driven nails will have name types listed on their box if the power-driven nail size corresponds to a particular pennyweight and type (e.g., 0.113 × 2⅜ in. = 8d cooler).

Appropriate design values can be obtained for many other nails from NER-272,[18] which was specifically developed to accommodate a wide range of types (power-driven fasteners in particular), basing the design values on the nail dimensions only. In addition, lateral load design values for nonstandard nails can be obtained by using the EYM equations given above under Lateral Resistance, based on the properties of the nails.

Generally, the adjustment factors given in Section 8.2.2 apply to all nails.

8.3.2 Staples

Staples are usually U-shaped wire fasteners with two same-size pointed or pointless legs connected by a common crown. They are designed to be driven by manual strike, pneumatic, electric, or spring tools and to hold two or more pieces together.

If hammer-driven into wood or wood-based materials, these staples have to be relatively stout to prevent buckling during driving. If tool-driven, the staples can be relatively slender, since they are driven at a rapid rate while laterally supported by the tool's guide body. Tool-driven staples are usually provided with flats along their legs to facilitate tight collating into strips, as shown in Fig. 8.15. They are often coated with polymers in order to decrease the staple's driving resistance.

Table 8.13 gives the sizes of collated and cohered heavy-wire steel staples that are typically available. Allowable loads for staples can be reasonably taken to be

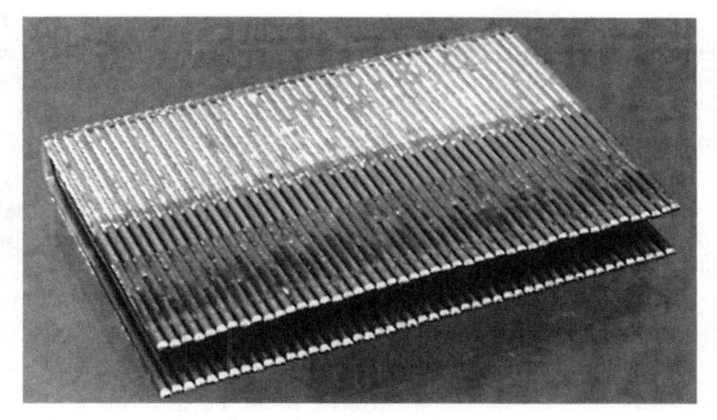

FIGURE 8.15 Staples.

TABLE 8.13 Available Sizes of Standard, Collected, and Cohered Heavy-Wire Steel Staples[a]

	Crown width, in.													
	0.18	0.25	0.38	0.44	0.44	0.50	0.75	0.88	0.94	1.00	1.00	1.38	1.56	2.12
Leg length, in. / Wire sizes:	18	18	18 16 14	16 15	14	16 15 14	16 14	16 14	16 14	16 14	10	12	12	10
1.00	x	x		x	x	x	x	x	x	x				
1.12	x	x	x	x	x	x	x	x	x	x				
1.25	x	x	x	x	x	x	x	x	x	x				
1.38			x	x	x	x	x	x	x	x				
1.50	x	x	x	x	x	x	x	x	x	x				
1.62			x	x	x	x	x	x				x	x	x
1.75			x	x	x	x	x	x			x			
1.88				x	x	x	x	x						
2.00				x	x	x	x	x			x			
2.25				x	x	x								
2.50				x	x	x								
2.75					x	x								
3.00					x	x								
3.25					x	x								
3.50					x									

Note: 1 in. = 25.4 mm.
[a] Outside crown width is given.

equal to twice the value of a nail with a shank diameter equal to that of one of the staple's legs. See Section 8.3.3 under Design Values for design values for staples.

8.3.3 Power-Driven Fasteners

Power-driven fasteners, generally nails and staples, are an acceptable substitute for hand-driven fasteners provided they are of equivalent size and quality. Several things should be considered when using power-driven fasteners:

1. *Nail nomenclature:* Power-driven fasteners are usually defined by their shank diameter and length. A pennyweight, which means nothing alone, is also sometimes added to the labeling. On some power-driven nail packages, labels include the shank diameter, length, and corresponding pennyweight and type (e.g., 0.162 × 3½ = 16d common). See Section 8.3.1 under Nail Nomenclature for a discussion on nail nomenclature and specifying nails.
2. *Contact:* Lightweight, pneumatic nail or staple tools do not always provide sufficient force against the surface when fastening wood structural panels to framing to ensure that the panel is tight against the framing. This can cause problems with fastener pop or squeaks when fastening floor sheathing or underlayment. Normally the force required to eliminate gaps beneath underlayment or floor sheathing can be provided by the operator if he stands on the panel or applies hand pressure adjacent to where the fastener is being driven.
3. *Thin galvanizing:* Power-driven fasteners must be smooth and uniform in diameter to avoid jamming the feed mechanism of the nail or staple tool. The

usual electro-galvanizing, which provides the required smoothness, is too thin to provide long-term corrosion resistance. Where a high degree of corrosion resistance is required, it will usually be necessary to use fasteners of a material that does not corrode, such as aluminum or stainless steel.

4. *Overdriving:* Some nail or staple tools commonly used for building construction may exert sufficient force to countersink the fastener, which is an undesirable condition when installing wood structural panels for roof, wall, or floor applications (some effects of overdriven fasteners are discussed in Section 8.4.2 under Effect of Overdriven Fasteners).

Adjusting the air pressure is an unpredictable way of controlling the depth to which the fastener is driven. Since material densities vary, it is impossible to find an air pressure setting that will consistently drive the fasteners flush with the panel surface.

The solution to overdriving fasteners is to use nail or staple tools that have a depth-control adjustment feature, which permits driving fasteners so that their head or crown is flush with the panel surface. Alternatively, some tool manufacturers offer shorter driver blades that can be installed as a modification in their tools by the manufacturer's service representative to accomplish this objective. For further information, contact the manufacturers through their local or regional distributors.

Specifying Power-Driven Fasteners. It is recommended that the designer, when knowing that power-driven fasteners will be used, specify a certain fastener size (length and diameter) only, or the size, pennyweight and type (e.g., 0.131 diameter × 2½ in. length = 8d common). If pennyweight and type alone are specified, there's a chance that the type will be overlooked and a box labeled with the pennyweight will be used, resulting in an unknown nail size actually being used. See further discussion of nail nomenclature and specifying in Section 8.3.1 under Nail Nomenclature.

Design Values (NER-272). A comprehensive source of design information for power-driven staples and nails is National Evaluation Report 272.[18] The same model for connection withdrawal strength is used for nails and staples. A different model is used for nails and staples for determining lateral load capacity. Appropriate adjustment factors as discussed in Section 8.2.2 must be applied to the nominal design values to obtain the allowable.

8.3.4 Wood Screws

Introduction. The general procedure for the design of connections using wood screws is very similar to that used for nails, but there are some notable differences in construction practices. Screws are inserted by turning, and lead holes may be drilled.

The common types of wood screws and their principal parts are shown in Fig. 8.16. Screws are designated by gauge (diameter of shank) and their overall length. For example, a No. 9, 2 in. (50.8 mm) wood screw would have a shank diameter of 0.177 in. (4.50 mm) and a total length of 2.0 in. (50.8 mm). Table 8.14 gives some commonly available screw sizes. It is adequate for design purposes to assume that ⅔ of the screw length is threaded. If possible, the structural design should avoid using wood screws in withdrawal from end grain.

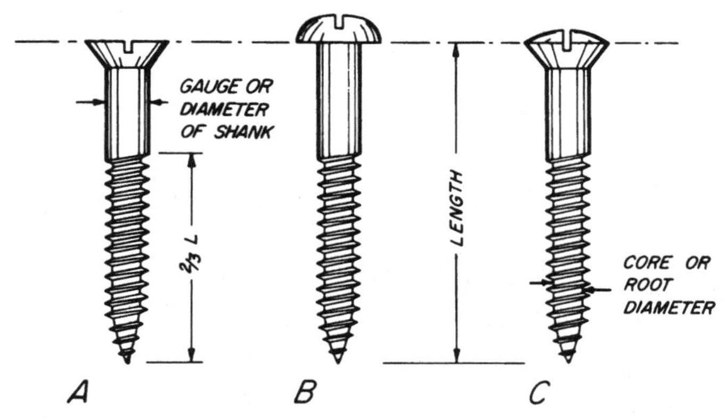

FIGURE 8.16 Common types of wood screws: (*A*) flathead, (*B*) roundhead, (*C*) ovalhead.

TABLE 8.14 Some Lengths and Gauges of Available Wood Screws

Length, in.	Gauge limits	Length, in.	Gauge limits
½	2 to 8	2	8 to 20
⅝	3 to 10	2¼	9 to 20
¾	4 to 11	2½	12 to 20
⅞	6 to 12	2¾	14 to 20
1	6 to 14	3	16 to 20
1¼	7 to 17	3½	18 to 24
1½	6 to 18	4	18 to 24
1¾	8 to 20		

Note: 1 in. = 25.4 mm.

Withdrawal Resistance. The equation used to obtain the nominal withdrawal design value given in the NDS is:

$$W = 2850G^2D \tag{8.4}$$

where W = maximum withdrawal load (lb)
G = specific gravity of wood
D = shank diameter of screw (in.)

Withdrawal tests on wood screws have shown that the ultimate load for a screw inserted into the side grain of seasoned wood may be estimated from the same equation with 2850 replaced by 15,700.

Table 8.15, derived from Eq. (8.4), gives nominal withdrawal design values for wood screws. The nominal design values are for

1. Seasoned wood, which remains dry in service
2. Normal load duration (10 year)

TABLE 8.15 Cut Thread or Rolled Thread Wood Screw Withdrawal Design Values (W)[a]

Specific gravity, G	Wood screw gauge										
	6g	7g	8g	9g	10g	12g	14g	16g	18g	20g	24g
0.55	119	130	141	152	163	186	208	231	253	275	320
0.50	98	107	117	126	135	154	172	191	209	228	264
0.43	73	79	86	93	100	114	127	141	155	168	196
0.42	69	76	82	89	95	108	121	134	147	161	187

Note: 1 in. = 25.4 mm; 1 lb = 4.45 N.

[a] Tabulated withdrawal design values (W) for wood screw connections shall be multiplied by all applicable adjustment factors; W, in pounds per inch of thread penetration into side grain of main member. Thread length is approximately two-thirds the total wood screw length.

Source: This table is from AF&PA's *National Design Specification® for Wood Construction.*

3. Screw turned into side grain
4. Normal temperature (<100°F)

If the end-use conditions assumed for the nominal design values are not satisfied for the connection under consideration, then appropriate adjustments as described in Section 8.2.2 must be made.

The wood screw used should have a tensile strength at net (root) section greater than the applied load as failure should occur in the wood rather than in the metal.

- *Number of wood screws:* The design value when more than one wood screw is used is equal to the sum of the design values permitted for each screw.
- *Lead holes:* Lead holes should be drilled according to Table 8.16.
- *Penetration:* The effective penetration to be used is the length of the threaded portion of the screw in the member receiving the point.
- *End grain:* Wood screws should not be loaded in withdrawal from the end grain of the wood.

Example 8.9 Withdrawal Load, Wood Screws

Determine the size and number of wood screws needed to support a 900 lb (3.56 kN) load for two months (Fig. 8.17). The wood is seasoned hem-fir that will remain dry.

solution

Try No. 16 × 2½ in. wood screws

Specific gravity of wood = 0.43 (see Table 8.7)

TABLE 8.16 Lead Holes for Wood Screws Loaded in Withdrawal

Specific gravity, G	Percent of root diameter
$G > 0.6$	90
$0.5 \leq G \leq 0.6$	70
$G \leq 0.5$	No lead hole required

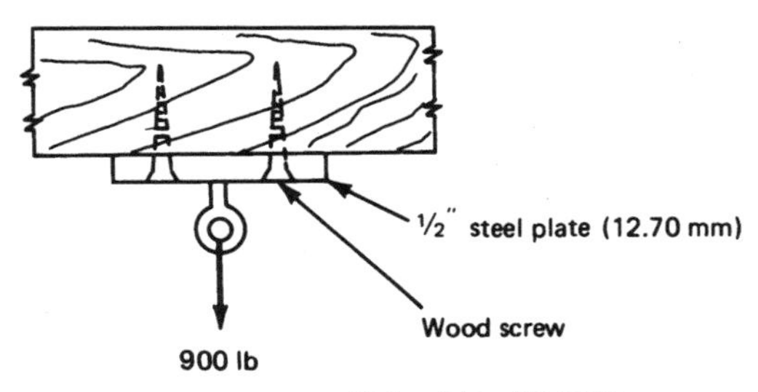

FIGURE 8.17 Example 8.9.

Penetration = 2.50 − 0.50 = 2.00 in. (50.8 mm)

Threaded length (approximately) = ⅔ × screw length = ⅔ × 2.50 = 1.67 in. < 2.00 in ∵ use 1.67 in. (42.4 mm)

W (nominal) = 141 lb/in of threaded length (Table 8.15)

W' (allowable) = 141 × C_D × C_M × C_t × threaded length

W' (allowable) = 141 × 1.15 × 1.0 × 1.0 × 1.67 = 271 lb (1206 N)

Number required = 900/271 = 3.32 ∴ Use four no. 16 × 2½ in. wood screws (6.8 × 63.5 mm)

Lateral Resistance. The ultimate lateral load capacity of a wood screw in single shear (two members) is dependent on several material and dimensional properties of the connection. These include the thickness of the two joined members, the bearing (crushing) strength of the wood, and the diameter and yield strength of the wood screw.

The European yield model (EYM) provides three possible failure modes for a two-member wood screw connection in single shear. The model assumes that the bearing capacity is reached when either the wood crushes under the wood screw or one or two plastic hinges are formed in the wood screw. The three yield modes and corresponding equations that provide the nominal lateral load value for each yield mode are given in Table 8.17. The controlling value for a particular connection is the lowest value obtained from the three equations given in Table 8.17.

Lateral load design values have been obtained using the yield model equations and are given in Table 8.18 for wood-to-wood connections and in Table 8.19 for wood-to-metal connections.

If a connection does not qualify under the geometry and material properties given in Table 8.18 or Table 8.19, then the three yield model equations may be used to obtain a lateral design value. Table 8.12 provides the dowel bearing strengths, F_{em} and F_{es}, needed in solving these equations, and the footnotes of Tables 8.18 and 8.19 provide the yield strengths of the screws. These equations have a built-in factor of safety about equal to that of nailed connections.

The nominal lateral load design values from Tables 8.18 and 8.19 and the yield equations are for:

1. Seasoned wood, which remains dry in service
2. Normal load duration (10 years)
3. A screw in single shear

TABLE 8.17 Yield Modes for Single-Shear Screw Connections

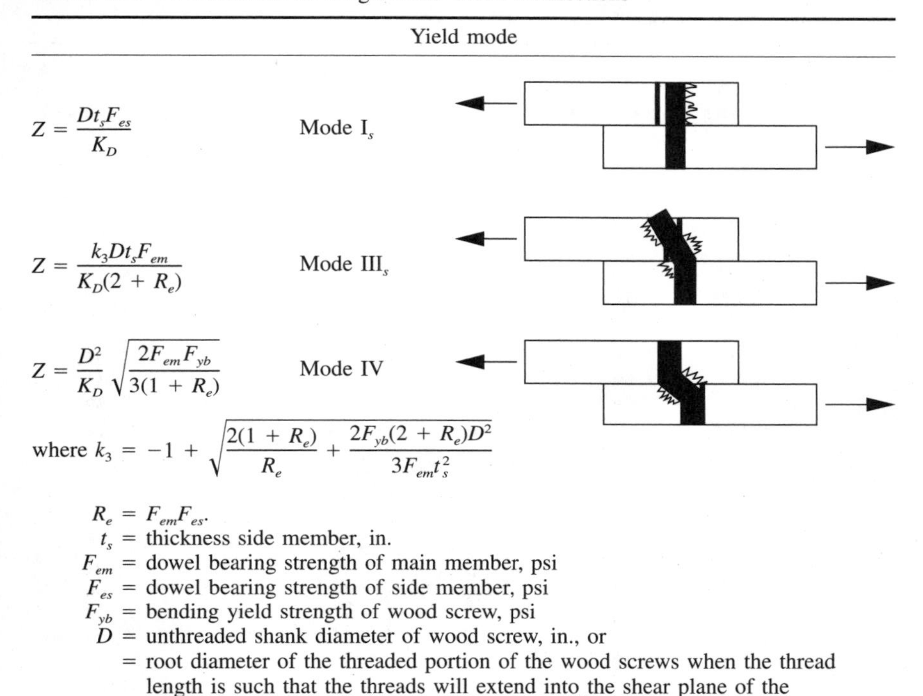

	Yield mode
$Z = \dfrac{Dt_sF_{es}}{K_D}$	Mode I_s
$Z = \dfrac{k_3Dt_sF_{em}}{K_D(2 + R_e)}$	Mode III_s
$Z = \dfrac{D^2}{K_D}\sqrt{\dfrac{2F_{em}F_{yb}}{3(1 + R_e)}}$	Mode IV

where $k_3 = -1 + \sqrt{\dfrac{2(1 + R_e)}{R_e} + \dfrac{2F_{yb}(2 + R_e)D^2}{3F_{em}t_s^2}}$

$R_e = F_{em}F_{es}$.
t_s = thickness side member, in.
F_{em} = dowel bearing strength of main member, psi
F_{es} = dowel bearing strength of side member, psi
F_{yb} = bending yield strength of wood screw, psi
D = unthreaded shank diameter of wood screw, in., or
= root diameter of the threaded portion of the wood screws when the thread length is such that the threads will extend into the shear plane of the connection, in.
K_D = 2.2 for $D \leq 0.17$ in., $K_D = 10D + 0.5$ for 0.17 in. $< D < 0.25$ in.

4. Full penetration ($10D$)
5. Screw inserted in side grain
6. Normal temperature ($<100°F$)

If these conditions are not satisfied for the connection under consideration, then appropriate general adjustments as described in Section 8.2.2 must be made. Specific adjustments for wood screws are as follows:

- *Number of wood screws:* The design value when more than one wood screw is used is equal to the sum of the design values permitted for each screw.
- *Penetration:* For lateral resistance, the penetration of the screw into the main member should be approximately ten times the shank diameter. If the penetration is less than ten diameters, the design value should be reduced in proportion to its reduced penetration using the equation $C_d = p/10D \leq 1.0$. The minimum penetration should not be less than six shank diameters.
- *Lead holes:* Lead holes for laterally loaded connections should be drilled according to Table 8.20.
- *Angle of load to grain:* The design values for wood screws apply to any angle of load to grain for wood or steel side plates.

TABLE 8.18 Wood Screw Design Values (Z)[a] for Single-Shear (Two-Member) Connections with Both Members of Identical Species

Side member thickness, t_s, in.	Wood screw diameter, D, in.	Wood screw number	$G = 0.55$ Southern pine, Z, lb	$G = 0.50$ Douglas fir-larch, Z, lb	$G = 0.43$ Hem-fir, Z, lb	$G = 0.42$ Spruce-pine-fir, Z, lb
½	0.138	6	67	59	48	47
	0.151	7	73	65	53	52
	0.164	8	82	72	60	59
	0.177	9	93	83	69	68
	0.190	10	101	89	75	73
	0.216	12	122	110	93	90
	0.242	14	133	119	101	99
⅝	0.138	6	75	66	53	51
	0.151	7	82	72	58	56
	0.164	8	91	80	65	63
	0.177	9	103	90	74	72
	0.190	10	110	97	80	77
	0.216	12	132	117	97	94
	0.242	14	142	126	105	102
¾	0.138	6	78	72	58	56
	0.151	7	87	79	63	61
	0.164	8	100	88	70	68
	0.177	9	113	99	80	77
	0.190	10	122	106	86	83
	0.216	12	144	126	103	100
	0.242	14	154	135	111	108
1	0.138	6	78	72	62	61
	0.151	7	87	79	69	67
	0.164	8	100	91	79	78
	0.177	9	117	107	93	90
	0.190	10	127	116	100	97
	0.216	12	160	147	118	114
	0.242	14	178	157	126	122
1¼	0.138	6	78	72	62	61
	0.151	7	87	79	69	67
	0.164	8	100	91	79	78
	0.177	9	117	107	93	91
	0.190	10	127	116	101	99
	0.216	12	160	147	127	125
	0.242	14	177	162	141	138
1½	0.138	6	78	72	62	61
	0.151	7	87	79	69	67
	0.164	8	100	91	79	78
	0.177	9	117	107	93	91
	0.190	10	127	116	101	99
	0.216	12	160	147	127	125
	0.242	14	177	162	141	138

TABLE 8.18 Wood Screw Design Values $(Z)^a$ for Single-Shear (Two-Member) Connections with Both Members of Identical Species (*Continued*)

Note: 1 in. = 25.4 mm; 1 lb = 4.45 N.

[a] Tabulated lateral design values (Z) for wood screw connections shall be multiplied by all applicable adjustment factors. Tabulated lateral design values (Z) are for rolled thread wood screws inserted in side grain with nail axis perpendicular to wood fibers; minimum screw penetration, p, into the main member equal to $10D$; and screw bending yield strengths (F_{yb}):

F_{yb} = 100,000 psi for $0.099'' \leq D \leq 0.142''$

F_{yb} = 90,000 psi for $0.142'' < D \leq 0.177''$

F_{yb} = 80,000 psi for $0.177'' < D \leq 0.244''$

When $6D \leq p < 10D$, tabulated lateral design values (Z) shall be multiplied by $p/10D$.

Source: This table is from AF&PA's *National Design Specification® for Wood Construction.*

- *End grain:* For wood screws inserted into the end grain and laterally loaded, the end grain adjustment factor, C_{eg}, is 0.67.

Combined Lateral and Withdrawal Loads. When a wood screw is inserted perpendicular to the wood fibers and is subjected to both lateral and withdrawal loadings, the equation given in Section 8.3.5 under Combined Lateral and Withdrawal Loads may be used to obtain a design value.

Example 8.10 Lateral Load on Wood Screw

Determine the allowable wind load Q for the joint shown (Fig. 8.18). The wood is seasoned southern pine that will be exposed to the weather.

solution

Wood specific gravity = 0.55 (see Table 8.7).

For no. 12 × 2½ in. wood screws: L = 2.5 in. (63.5 mm) and D = 0.216 in. (5.49 mm) (see Table 8.19)

Penetration needed to use full design value = $10D$ = 2.16 in. (54.9 mm)

Minimum penetration permitted = $6D$ = 1.30 in. (32.9 mm)

Actual penetration = 2.5 − 0.105 = 2.4 in. > $10D$ ∴ $C_d = 1.0$

Z (nominal) = 168 lb (see Table 8.19)

Z' (allowable) = $168 \times C_D \times C_M \times C_d$

Z' (allowable) = 168 × 1.6 × 0.7 × 1.0 × 1.0 = 188 lb (903 N)

Q = 188 × 8 = 1504 lb (6696 N)

8.3.5 Lag Screws

Introduction. Lag screws (also called lag bolts) are typically used for convenience or where bolts are impossible to install or undesirable to use. The size of a lag screw is designated by its shank diameter, D_s, and nominal length, L (see Fig. 8.19). Table 8.21 provides typical data on lag screw sizes and dimensions. The strength of a lag screw connection is dependent on factors similar to those of both the wood screw and the bolt.

Lag screws are turned into prebored holes. The recommended diameter of the hole is dependent on the density of the wood and the diameter of the shank of the

TABLE 8.19 Wood Screw Design Values (Z)[a] for Single-Shear (Two-Member) Connections with ASTM A653, Grade 33 Steel Side Plate

Steel side plate	Wood screw diameter, D, in.	Wood screw gauge, g	$G = 0.55$ Southern pine, Z, lb	$G = 0.50$ Douglas fir-larch, Z, lb	$G = 0.43$ Hem-fir, Z, lb	$G = 0.42$ Spruce-pine-fir, Z, lb
0.036	0.138	6g	75	70	61	60
(20 gage)	0.151	7g	83	77	68	66
	0.164	8g	96	89	78	76
0.048	0.138	6g	76	71	62	61
(18 gage)	0.151	7g	84	78	69	67
	0.164	8g	97	90	79	77
0.060	0.138	6g	78	72	64	63
(16 gage)	0.151	7g	86	80	70	69
	0.164	8g	99	91	80	79
	0.177	9g	115	107	94	92
	0.190	10g	124	115	101	99
0.075	0.138	6g	81	75	67	65
(14 gage)	0.151	7g	90	83	73	72
	0.164	8g	102	94	83	82
	0.177	9g	118	110	97	95
	0.190	10g	128	118	104	102
	0.216	12g	159	147	129	127
	0.242	14g	174	161	142	139
0.105	0.138	6g	90	83	74	72
(12 gage)	0.151	7g	98	91	81	79
	0.164	8g	111	103	91	89
	0.177	9g	127	118	104	102
	0.190	10g	137	127	112	110
	0.216	12g	168	156	138	135
	0.242	14g	183	169	150	147
0.120	0.138	6g	95	88	78	77
(11 gage)	0.151	7g	104	96	85	84
	0.164	8g	116	108	96	94
	0.177	9g	132	123	109	107
	0.190	10g	143	133	117	115
	0.216	12g	174	161	143	140
	0.242	14g	188	175	154	151
0.134	0.138	6g	100	93	82	81
(10 gage)	0.151	7g	109	101	90	88
	0.164	8g	122	113	100	98
	0.177	9g	138	128	114	112
	0.190	10g	149	138	122	120
	0.216	12g	180	167	148	145
	0.242	14g	194	180	160	156

TABLE 8.19 Wood Screw Design Values (Z)[a] for Single-Shear (Two-Member) Connections with ASTM A653, Grade 33 Steel Side Plate (*Continued*)

Steel side plate	Wood screw diameter, D, in.	Wood screw gauge, g	$G = 0.55$ Southern pine, Z, lb	$G = 0.50$ Douglas fir-larch, Z, lb	$G = 0.43$ Hem-fir, Z, lb	$G = 0.42$ Spruce-pine-fir, Z, lb
0.179 (7 gage)	0.138	6g	106	98	85	84
	0.151	7g	117	108	95	93
	0.164	8g	135	125	109	107
	0.177	9g	159	147	128	126
	0.190	10g	171	158	139	136
	0.216	12g	203	189	167	164
	0.242	14g	217	202	179	175
0.239 (3 gage)	0.138	6g	106	98	85	84
	0.151	7g	117	108	95	93
	0.164	8g	135	125	109	107
	0.177	9g	159	147	128	126
	0.190	10g	172	158	139	136
	0.216	12g	217	200	175	172
	0.242	14g	240	221	194	190

Note: 1 in. = 25.4 mm; 1 lb = 4.45 N.

[a]Tabulated lateral design values (Z) for wood screw connections shall be multiplied by all applicable adjustment factors. Tabulated lateral design values (Z) are for rolled thread wood screws inserted in side grain with screw axis perpendicular to wood fibers; minimum screw penetration, p, into the main member equal to $10D$; dowel bearing strengths (F_e) of 61,850 psi for ASTM A653, Grade 33 steel and screw bending yield strengths (F_{yb}):

F_{yb} = 100,000 psi for $0.099'' \leq D \leq 0.142''$

F_{yb} = 90,000 psi for $0.142'' < D \leq 0.177''$

F_{yb} = 80,000 psi for $0.177'' < D \leq 0.244''$

When $6D \leq p < 10D$, tabulated lateral design values (Z) shall be multiplied by $p/10D$.

Source: This table is from AF&PA's *National Design Specification® for Wood Construction.*

TABLE 8.20 Lead Holes for Wood Screws Loaded Laterally

Specific gravity, G	Part of screw	Percent of root diameter
$G > 0.6$	Shank	100
	Thread	100
$G \leq 0.6$	Shank	87.5
	Thread	87.5

lag screw. The total length of the hole drilled should be equal to the length of the lag screw. Table 8.22 gives the required values for both diameter and length of hole. When loaded primarily in withdrawal, ⅜ in. (9.5 mm) and smaller-diameter lag screws may be inserted into a member of specific gravity, $G \leq 0.5$ without a lead or clearance hole provided that spacings, end distances, and edge distances are adequate to prevent splitting.

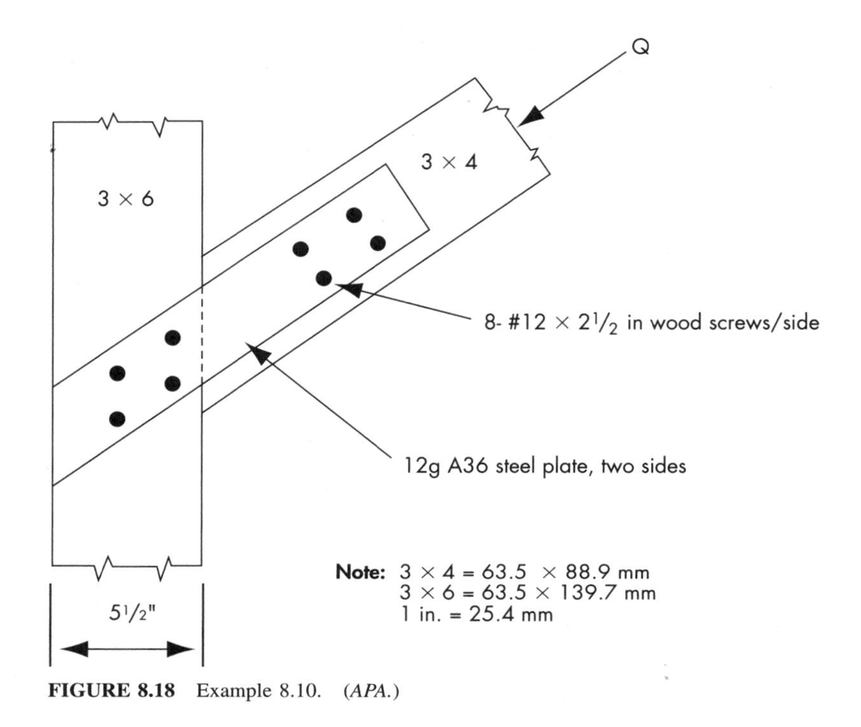

FIGURE 8.18 Example 8.10. (*APA*.)

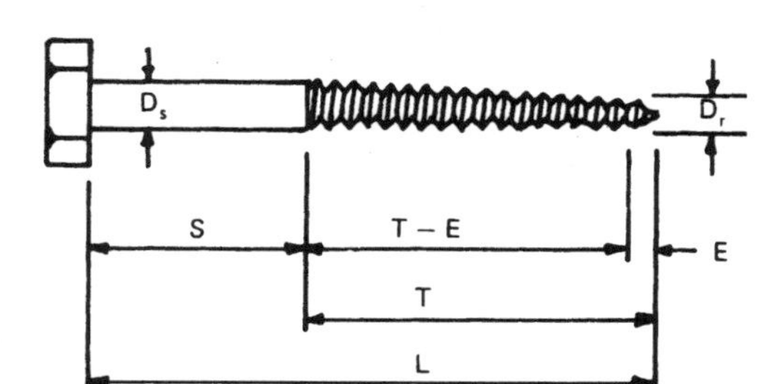

FIGURE 8.19 Lag screw dimensions. *Note:* All terms are defined in Table 8.21.

The threaded portion of the screw is to be turned into its lead hole and not driven in by a hammer. Soap or other lubricant is recommended to facilitate insertion and to prevent damage to the connection.

Withdrawal Resistance. The nominal withdrawal design value for a lag screw inserted into side grain with the lag screw axis perpendicular to wood grain can be calculated by the empirical equation

$$W = 1800G^{3/2}D_s^{3/4} \tag{8.5}$$

TABLE 8.21 Typical Dimensions of Standard Lag Screws for Wood[1]
All dimensions in inches.

D = diameter

D_r = root diameter

S = unthreaded shank length

T = thread length[2]

E = lenth of tapered tip

N = number of threads/inch

Length L		Diameter, D										
		1/4	5/16	3/8	7/16	1/2	5/8	3/4	7/8	1	1 1/8	1 1/4
	D_r	0.173	0.277	0.265	0.328	0.371	0.471	0.579	0.683	0.780	0.887	1.012
	E	5/32	3/16	7/32	9/32	5/16	13/32	1/2	13/32	11/16	25/32	7/8
	H	11/64	7/32	1/4	19/64	11/32	27/64	1/2	37/64	43/64	3/4	27/32
	F	7/16	1/2	9/16	5/8	3/4	15/16	1 1/8	1 5/16	1 1/2	1 11/16	1 7/8
	N	10	9	7	7	6	5	4 1/2	4	3 1/2	3 1/4	3 1/4
1	*S*	1/4	1/4	1/4	1/4	1/4						
	T	3/4	3/4	3/4	3/4	3/4						
	T-E	19/32	9/16	17/32	15/32	7/16						
1 1/2	*S*	1/4	1/4	1/4	1/4	1/4						
	T	1 1/4	1 1/4	1 1/4	1 1/4	1 1/4						
	T-E	1 3/32	1 1/16	1 1/32	31/32	15/16						
2	*S*	1/2	1/2	1/2	1/2	1/2	1/2					
	T	1 1/2	1 1/2	1 1/2	1 1/2	1 1/2	1 1/2					
	T-E	1 11/32	1 5/16	1 9/32	1 7/32	1 3/16	1 3/32					
2 1/2	*S*	3/4	3/4	3/4	3/4	3/4	3/4					
	T	1 3/4	1 3/4	1 3/4	1 3/4	1 3/4	1 3/4					
	T-E	1 19/32	1 9/16	1 17/32	1 15/32	1 7/16	1 11/32					
3	*S*	1	1	1	1	1	1	1	1	1		
	T	2	2	2	2	2	2	2	2	2		
	T-E	1 27/32	1 13/16	1 25/32	1 23/32	1 11/16	1 5/8	1 1/2	1 13/32	1 5/16		

TABLE 8.21 Typical Dimensions of Standard Lag Screws for Wood
All dimensions in inches. (Continued)

Length L		Diameter, D										
		1/4	5/16	3/8	7/16	1/2	5/8	3/4	7/8	1	1 1/8	1 1/4
4	*S*	1 1/2	1 1/2	1 1/2	1 1/2	1 1/2	1 1/2	1 1/2	1 1/2	1 1/2	1 1/2	1 1/2
	T	2 1/2	2 1/2	2 1/2	2 1/2	2 1/2	2 1/2	2 1/2	2 1/2	2 1/2	2 1/2	2 1/2
	T-E	2 11/32	2 5/16	2 9/32	2 7/32	2 3/16	2 1/8	2	1 29/32	1 13/16	1 23/32	1 5/8
5	*S*	2	2	2	2	2	2	2	2	2	2	2
	T	3	3	3	3	3	3	3	3	3	3	3
	T-E	2 27/32	2 13/16	2 25/32	2 23/32	2 11/16	2 19/32	2 1/2	2 13/32	2 5/16	2 7/32	2 1/8
6	*S*	2 1/2	2 1/2	2 1/2	2 1/2	2 1/2	2 1/2	2 1/2	2 1/2	2 1/2	2 1/2	2 1/2
	T	3 1/2	3 1/2	3 1/2	3 1/2	3 1/2	3 1/2	3 1/2	3 1/2	3 1/2	3 1/2	3 1/2
	T-E	3 11/32	3 5/16	3 9/32	3 7/32	3 3/16	3 3/32	3	2 29/32	2 13/16	2 23/32	2 5/8
7	*S*	3	3	3	3	3	3	3	3	3	3	3
	T	4	4	4	4	4	4	4	4	4	4	4
	T-E	3 27/32	3 13/16	3 25/32	3 23/32	3 11/16	3 19/32	3 1/2	3 13/32	3 5/16	3 7/32	3 1/8
8	*S*	3 1/2	3 1/2	3 1/2	3 1/2	3 1/2	3 1/2	3 1/2	3 1/2	3 1/2	3 1/2	3 1/2
	T	4 1/2	4 1/2	4 1/2	4 1/2	4 1/2	4 1/2	4 1/2	4 1/2	4 1/2	4 1/2	4 1/2
	T-E	4 11/32	4 5/16	4 9/32	4 7/32	4 3/16	4 3/32	4	3 29/32	3 13/16	3 23/32	3 5/8
9	*S*	4	4	4	4	4	4	4	4	4	4	4
	T	5	5	5	5	5	5	5	5	5	5	5
	T-E	4 27/32	4 13/16	4 25/32	4 23/32	4 11/16	4 19/32	4 1/2	4 13/32	4 5/16	4 7/32	4 1/8
10	*S*	4 1/2	4 1/2	4 1/2	4 1/2	4 1/2	4 1/2	4 1/2	4 1/2	4 1/2	4 1/2	4 1/2
	T	5 1/2	5 1/2	5 1/2	5 1/2	5 1/2	5 1/2	5 1/2	5 1/2	5 1/2	5 1/2	5 1/2
	T-E	5 11/32	5 5/16	5 9/32	5 7/32	5 3/16	5 3/32	5	4 29/32	4 13/16	4 23/32	4 5/8
11	*S*	5	5	5	5	5	5	5	5	5	5	5
	T	6	6	6	6	6	6	6	6	6	6	6
	T-E	5 27/32	5 13/16	5 25/32	5 23/32	5 11/16	5 19/32	5 1/2	5 13/32	5 5/16	5 7/32	5 1/8
12	*S*	6	6	6	6	6	6	6	6	6	6	6
	T	6	6	6	6	6	6	6	6	6	6	6
	T-E	5 27/32	5 13/16	5 25/32	5 23/32	5 11/16	5 19/32	5 1/2	5 13/32	5 5/16	5 7/32	5 1/8

1. Tolerances specified in ANSI B18.2.1 Full body diameter lag screw is shown. For reduced body diameter lag screws, the unthreaded shank diameter may be reduced to approximately the root diameter, D_r.
2. Thread length (T) for intermediate lag screw lengths (L) is 6″ or ½ the lag screw length plus 0.5″, whichever is

TABLE 8.22 Recommended Diameter and Length of Holes for Lag Screws

Part of hole	Length of hole[a]	Specific gravity of wood, G	Diameter of hole[b]
Shank	S	All	D_s
Threaded	$T - E$	$G > 0.6$	65–85% of D_s
		$0.5 < G \leq 0.6$	60–75% of D_s
		$G \leq 0.5$	40–70% of D_s

[a] See Fig. 8.19 for dimensions of lag screw.
[b] Larger percentages for larger diameter of lag screws.

where W = maximum nominal withdrawal load (lb/in.)
G = specific gravity of wood
D_s = shank diameter (in.)

Tabulated values, derived from Eq. (8.5), are given in Table 8.23. The average ultimate withdrawal strength can be estimated as 4.5 times the nominal design value.

Lag screws loaded in withdrawal can, if long enough, develop loads exceeding the yield strength of the steel. This should be prevented, and the load-duration factors are not appropriate for adjusting the steel strength for this failure mode (see Section 8.2.2 under Load Duration Factor).

Equation (8.5) and Table 8.23, give nominal design values for

1. Seasoned wood that remains dry
2. Normal load duration (10 years)
3. Withdrawal from side grain

If these conditions are not satisfied for the connection under consideration, then appropriate adjustments as described in Section 8.2.2 must be made. Specific lag screw adjustments are as follows:

- *End grain:* If possible, the design should avoid withdrawal from the end grain of the wood. When this condition cannot be avoided, the design value in withdrawal from the grain should be multiplied by an end-grain factor, C_{eg} = 0.75.

TABLE 8.23 Lag Screw Withdrawal Design Values (W)[a]

Specific gravity, G	Lag screw unthreaded shank diameter, D, in.										
	1/4	5/16	3/8	7/16	1/2	5/8	3/4	7/8	1	1 1/8	1 1/4
0.55	260	307	352	395	437	516	592	664	734	802	868
0.50	225	266	305	342	378	447	513	576	636	695	752
0.43	179	212	243	273	302	357	409	459	508	554	600
0.42	173	205	235	264	291	344	395	443	490	535	579

Note: 1 in. = 25.4 mm; 1 lb = 4.45 N.
[a] Tabulated withdrawal design values (W) for lag screw connections shall be multiplied by all applicable adjustment factors; W, in pounds per inch of thread penetration into side grain of main member. Length of thread penetration in main member shall not include the length of the tapered tip.
Source: This table is from AF&PA's *National Design Specification® for Wood Construction.*

- *Placement:* The minimum spacing and end and edge distances for lag screws loaded in withdrawal and not laterally loaded should be: spacing = $4D$, end distance = $4D$, and edge distance = $1.5D$.

Lateral Resistance. The ultimate lateral load capacity of a lag screw in single shear (two members) is dependent on several material and dimensional properties of the connection. These include the thickness of the two joined members, the bearing strength of the wood, and the diameters (shank and root) and yield strength of the lag screw.

The European yield model (EYM) provides three possible failure modes for a two-member lag screw connection in single shear. The model assumes that the capacity is reached when either the wood crushes under the lag screw or one or two plastic hinges is formed in the lag screw. Shown in Table 8.24 are the three yield modes and the corresponding equations that provide the nominal lateral load

TABLE 8.24 Yield Modes for Single-Shear Lag Screw Connections

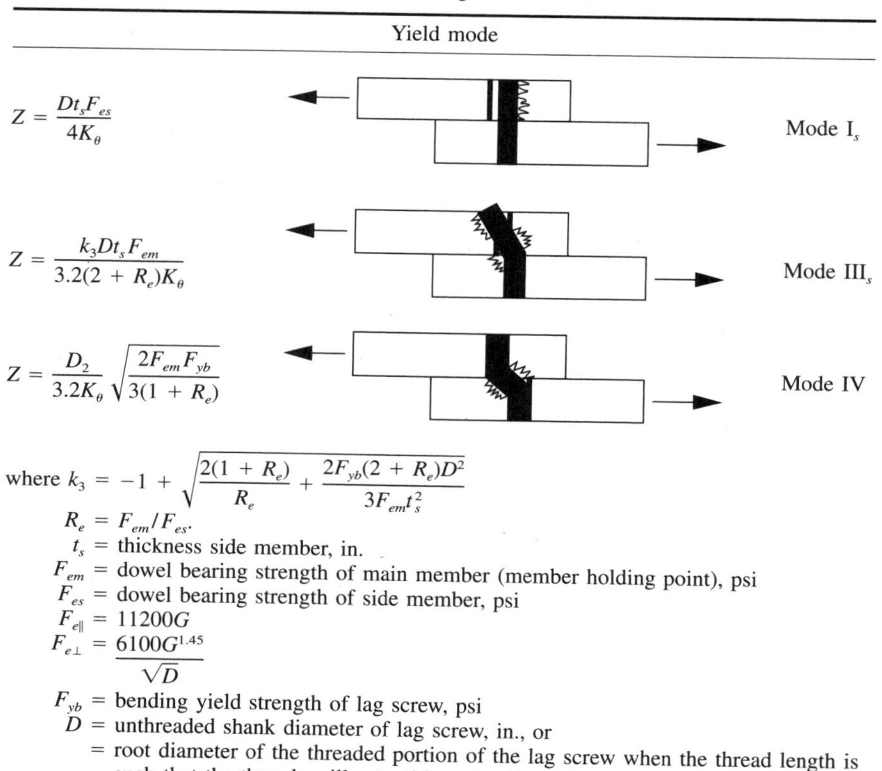

	Yield mode	
$Z = \dfrac{Dt_sF_{es}}{4K_\theta}$		Mode I_s
$Z = \dfrac{k_3Dt_sF_{em}}{3.2(2 + R_e)K_\theta}$		Mode III_s
$Z = \dfrac{D_2}{3.2K_\theta}\sqrt{\dfrac{2F_{em}F_{yb}}{3(1 + R_e)}}$		Mode IV

where $k_3 = -1 + \sqrt{\dfrac{2(1 + R_e)}{R_e} + \dfrac{2F_{yb}(2 + R_e)D^2}{3F_{em}t_s^2}}$

$R_e = F_{em}/F_{es}$.

t_s = thickness side member, in.

F_{em} = dowel bearing strength of main member (member holding point), psi

F_{es} = dowel bearing strength of side member, psi

$F_{e\parallel} = 11200G$

$F_{e\perp} = \dfrac{6100G^{1.45}}{\sqrt{D}}$

F_{yb} = bending yield strength of lag screw, psi

D = unthreaded shank diameter of lag screw, in., or

= root diameter of the threaded portion of the lag screw when the thread length is such that the threads will extend into the shear plane of the conneciton, in.

$K_\theta = 1 + (\theta_{max}/360°)$

θ_{max} = maximum angle of load to grain ($0° \leq \theta \leq 90°$) for any member in a connection.

design value for each yield mode. The controlling value for a particular connection is the lowest value obtained from the three equations.

Lateral design values have been calculated using the yield model equations and are shown in Table 8.25 for wood-to-wood connections and in Table 8.26 for wood-to-metal connections. Design values are given in both tables for loading applied parallel to grain ($Z_{\parallel}$) and perpendicular to grain ($Z_{\perp}$).

If a connection does not qualify under the geometry and material properties given in Table 8.25 or Table 8.26, then the yield model equations may be used to obtain a lateral design value. Table 8.27 gives dowel bearing strengths needed for the EYM equations, and the footnotes to Tables 8.25 and 8.26 provide lag screw yield strengths. These equations will provide a design value of about ⅕ the ultimate capacity of the lag screw.

For other angles of loading, the allowable load may be computed from values parallel and perpendicular to the grain by using the Hankinson formula, given in Section 8.2.3.

The nominal design values from tables or the yield equations are based on:

1. Seasoned wood that remains dry in service
2. Normal load duration (10 years)
3. One lag screw, in single shear
4. Lag screw laterally loaded parallel or perpendicular to grain

If these end-use conditions assumed are not expected for the connection under consideration, then appropriate general adjustments as described in Section 8.2.2 must be made, and specific adjustments for lag screws subjected to lateral load are described as follows.

- *Penetration depth factor:* Lateral load design values for lag screws are based on a penetration (not including length of tapered tip) into the main member of eight times the shank diameter of the lag screw, $p = 8D$. The minimum lag screw penetration for a reduced design value is four times the diameter of the lag screw shank; when $4D \leq p \leq 8D$, the design value should be multiplied by C_d, where $C_d = p/8D \leq 1.0$.
- *Lag screw in end grain:* When the loads act perpendicular to the grain and the lag screw is inserted parallel to the fibers (i.e., in the end grain of the member), design values for lateral resistance should be multiplied by the end-grain factor, $C_{eg} = 0.67$.

Placement. The recommended spacing, end distances, edge distances, and net section for lag screws are the same as those given for a bolt diameter equal to the lag screw shank diameter. These are given in Section 8.3.6 under Placement.

Combined Lateral and Withdrawal Loads. When a lag screw is subjected to a combined lateral and withdrawal loading (see Fig. 8.20), it is recommended that the allowable design value be obtained from the equation

$$Z'_{\alpha} = \frac{(Q'\,p)Z'}{W'\,p\cos^2\alpha + Z'\sin^2\alpha} \tag{8.6}$$

where α = angle between the wood surface and the direction of applied load
p = length of thread penetration in main member

TABLE 8.25 Lag Screw Design Values (Z) for Single-Shear (Two-Member) Connection with Both Members of Identical Species[1,2,3]

Side member thickness, t_s, in.	Lag screw diameter, D, in.	$G = 0.55$ Southern pine				$G = 0.50$ Douglas fir-larch				$G = 0.43$ Hem-fir				$G = 0.42$ Spruce-pine-fir			
		$Z_{\parallel}$ lbs.	$Z_{s\perp}$ lbs.	$Z_{m\perp}$ lbs.	$Z_{\perp}$ lbs.	$Z_{\parallel}$ lbs.	$Z_{s\perp}$ lbs.	$Z_{m\perp}$ lbs.	$Z_{\perp}$ lbs.	$Z_{\parallel}$ lbs.	$Z_{s\perp}$ lbs.	$Z_{m\perp}$ lbs.	$Z_{\perp}$ lbs.	$Z_{\parallel}$ lbs.	$Z_{s\perp}$ lbs.	$Z_{m\perp}$ lbs.	$Z_{\perp}$ lbs.
1/2	1/4	130	90	100	90	120	90	90	80	110	80	80	70	110	80	80	70
	5/16	190	130	140	130	180	110	130	110	170	90	120	90	160	90	120	90
	3/8	160	110	110	100	150	100	110	90	140	80	100	80	130	80	90	80
5/8	1/4	140	100	110	100	130	90	100	90	120	80	90	80	110	80	90	70
	5/16	200	140	150	130	190	130	140	120	170	110	120	110	170	110	120	110
	3/8	170	110	120	100	160	100	110	100	140	90	100	80	140	90	100	80
3/4	1/4	150	110	120	110	140	100	110	100	130	90	100	80	120	80	90	80
	5/16	210	150	160	140	200	140	150	130	180	120	130	110	180	120	130	110
	3/8	180	120	130	110	170	110	120	100	150	100	110	90	150	90	110	90
1	1/4	160	120	120	120	150	120	120	110	140	100	110	90	140	100	100	90
	5/16	240	160	180	150	220	150	170	140	200	130	150	120	200	130	140	120
	3/8	210	130	150	120	200	120	140	110	170	100	120	100	170	100	120	90
1 1/4	1/4	160	120	120	120	150	120	120	110	140	110	110	100	140	100	100	100
	5/16	270	180	200	170	250	160	190	150	220	140	160	130	220	140	160	130
	3/8	210	150	150	140	200	140	140	130	190	120	130	110	180	110	130	110
1 1/2	1/4	160	120	120	120	150	120	120	110	140	110	110	100	140	100	100	100
	5/16	270	200	200	180	250	180	190	170	230	150	170	140	230	150	170	140
	3/8	210	150	150	140	200	140	140	130	190	130	130	120	180	130	130	110
	7/16	320	220	230	200	310	200	210	180	290	170	190	150	280	160	190	150
	1/2	410	250	290	230	390	220	270	200	350	190	240	180	350	190	240	170
	5/8	600	340	420	310	560	310	380	280	500	280	340	240	490	270	330	240
	3/4	830	470	560	410	770	440	510	380	700	360	450	330	690	350	440	330
	7/8	1080	560	710	540	1020	490	660	490	930	390	580	390	910	380	570	380
	1	1360	600	870	600	1290	530	810	530	1180	420	720	420	1160	410	710	410

2½	¼	160	120	120	120	150	120	120	110	140	110	110	100	140	100	100	100
	5⁄16	270	200	200	180	250	190	190	170	230	170	170	150	230	170	170	150
	⅜	210	150	150	140	200	140	140	130	190	130	130	120	180	130	130	110
	7⁄16	320	230	230	210	310	210	210	190	290	190	190	170	280	190	190	170
	½	410	290	290	250	390	270	270	240	360	240	240	210	360	240	240	210
	⅝	670	430	440	390	640	390	420	350	590	330	380	290	580	320	370	290
	¾	1010	550	650	490	960	500	610	450	890	430	550	380	880	420	540	370
	⅞	1370	690	880	600	1280	630	830	550	1130	550	730	470	1110	540	710	460
	1	1660	830	1080	720	1550	770	990	660	1380	680	870	580	1360	670	850	570

1. Tabulated lateral design values (Z) shall be multiplied by all applicable adjustment factors.

2. Tabulated lateral design values (Z) are for "reduced diameter body" lag screws inserted in side grain with nail axis perpendicular to wood fibers; minimum screw penetration, p, into the main member equal to $8D$; screw bending yield strengths (F_{yb}):

F_{yb} = 70,000 psi for D = ¼″

F_{yb} = 60,000 psi for D = 5⁄16″

F_{yb} = 45,000 psi for D > ⅜″

3. When $4D \le p < 8D$, tabulated lateral design values (Z) shall be multiplied by $p/8D$.

Source: This table is from AF&PA's *National Design Specification® for Wood Construction.*

TABLE 8.26 Lag Screw Design Values (Z) for Single-Shear (Two-Member) Connections with ¼ in. ASTM A36 Steel Side Plate or ASTM A653, Grade 33 Steel Side Plate (for $t_s \leq$ ¼ in.)[1,2,3]

Steel side plate, t_s, in.	Lag screw diameter, D, in.	$G = 0.55$ Southern pine		$G = 0.50$ Douglas fir-larch		$G = 0.43$ Hem-fir		$G = 0.42$ Spruce-pine-fir	
		$Z_{\parallel}$, lb	$Z_{\perp}$, lb	$Z_{\parallel}$, lb	$Z_{\perp}$, lb	$Z_{\parallel}$, lb	$Z_{\perp}$, lb	$Z_{\parallel}$, lb	$Z_{\perp}$, lb
0.075 (14 gage)	¼	200	150	190	140	180	120	180	120
	5/16	240	170	230	160	210	140	210	140
	3/8	250	170	240	160	220	140	220	140
0.105 (12 gage)	¼	210	160	210	150	190	140	190	130
	5/16	250	180	240	170	230	150	230	150
	3/8	260	180	250	170	240	150	230	150
0.120 (11 gage)	¼	230	170	220	160	200	140	200	140
	5/16	260	190	250	170	240	160	230	160
	3/8	270	180	260	170	240	160	240	160
0.134 (10 gage)	¼	240	180	230	170	210	150	210	150
	5/16	270	190	260	180	240	160	240	160
	3/8	280	190	270	180	250	160	250	160
0.179 (7 gage)	¼	260	190	250	180	230	160	230	160
	5/16	310	220	290	210	280	190	270	180
	3/8	320	220	310	200	290	190	280	180
0.239 (3 gage)	¼	260	190	250	180	230	160	230	160
	5/16	330	230	310	210	290	190	290	190
	3/8	340	230	320	210	300	190	300	190
	7/16	480	320	470	300	440	270	430	270
	½	590	380	560	350	530	320	530	320
	5/8	850	520	820	490	770	440	760	440
	¾	1190	690	1150	660	1070	590	1070	580
	7/8	1600	900	1540	840	1440	760	1420	750
	1	2050	1110	1970	1040	1840	940	1820	930
¼	¼	260	190	250	180	240	160	230	160
	5/16	330	230	320	220	290	190	290	190
	3/8	340	230	330	210	300	190	300	190
	7/16	510	330	490	310	450	280	450	270
	½	660	410	630	380	580	340	580	340
	5/8	950	580	920	540	860	490	850	490
	¾	1300	760	1250	710	1170	640	1160	630
	7/8	1710	960	1650	900	1540	810	1520	800
	1	2170	1170	2080	1100	1940	990	1930	980

1. Tabulated lateral design values (Z) shall be multiplied by all applicable adjustment factors.

2. Tabulated lateral design values (Z) are for "reduced body diameter" lag screws inserted in side grain with screw axis perpendicular to wood fibers; minimum screw penetration, p, into the main member equal to $8D$; a dowel bearing strengths (F_e) of 61,850 psi for ASTM A653. Grade 33 steel and 87,000 psi for ASTM A36 steel and screw bending yield strengths (F_{yb}):

F_{yb} = 70,000 psi for D = ¼″

F_{yb} = 60,000 psi for D = 5/16″

F_{yb} = 45,000 psi for $D \geq$ 3/8″

3. When $4D \leq p < 8D$, tabulated lateral design values (Z) shall be multiplied by $p/8D$.

Source: This table is from AF&PA's *National Design Specification® for Wood Construction*.

TABLE 8.27 Dowel Bearing Strength for Lag Screw Connections

Species combination	Specific gravity, G^a	Dowel bearing strength, psi[b]									
		$F_{e\parallel}$	$F_{e\perp}$ $D = 1/4$ in.	$F_{e\perp}$ $D = 5/16$ in.	$F_{e\perp}$ $D = 3/8$ in.	$F_{e\perp}$ $D = 7/16$ in.	$F_{e\perp}$ $D = 1/2$ in.	$F_{e\perp}$ $D = 5/8$ in.	$F_{e\perp}$ $D = 3/4$ in.	$F_{e\perp}$ $D = 7/8$ in.	$F_{e\perp}$ $D = 1$ in.
Southern pine	0.55	6150	5150	4600	4200	3900	3650	3250	2950	2750	2550
Douglas fir-larch	0.50	5600	4450	4000	3650	3400	3150	2800	2600	2400	2250
Hem-fir	0.43	4800	3600	3200	2950	2700	2550	2250	2050	1900	1800
Spruce-pine-fir	0.42	4700	3450	3100	2850	2600	2450	2200	2000	1850	1750

Note: 1 lb/in.^{2} = 6.895 kPa.

[a] Specific gravity based on weight and volume when oven dry.

[b] $F_{e\parallel} = 11200G$; $F_{e\perp} = 6100G^{1.45}/\sqrt{D}$; tabulated values are rounded to the nearest 50 psi.

Source: This table is from AF&PA's *National Design Specification® for Wood Consruction.*

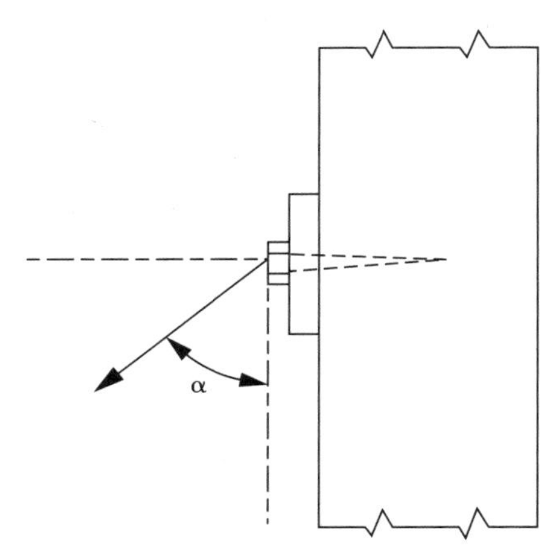

FIGURE 8.20 Combined lateral and withdrawal loading.

Z' = allowable lateral design load
W' = allowable withdrawal design load

Example 8.11 Lag Screws Loaded at an Angle to the Grain

Determine the allowable load, Q (dead load plus earthquake load), for the connection shown in Fig. 8.21. The wood is no. 1 Douglas fir surface green (MC > 19%) and is used in a dry-service condition (MC < 19%).

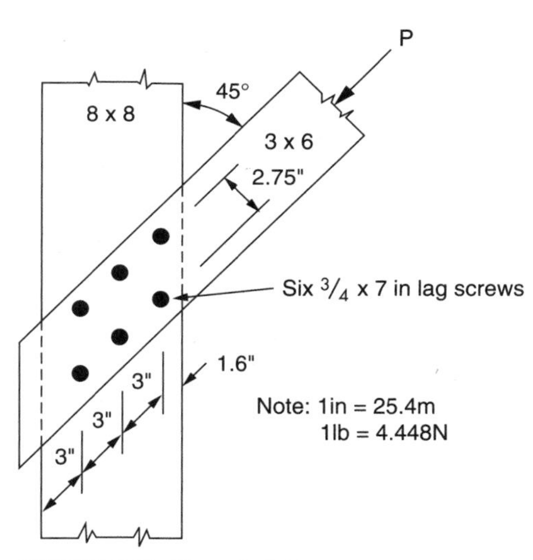

FIGURE 8.21 Example 8.11.

solution

Specific gravity = 0.50 (see Table 8.7)

$Z_{\parallel}$ = 960 lb, $Z_{M\perp}$ = 610 lb (4539 N) (from Table 8.25)

Check penetration of lag screw:

$p = 7.0 - 2.5 = 4.5$ in (114 mm)

$C_d = p/8D = 4.5/(8 \times 3/4) = 0.75$

$Z_{45°} = (960 \times 610)/(960 \times \sin^2 45 + 610 \cos^2 45) = 746$ lb (3318 N) (see Section 8.2.3)

Group reduction factor:*

$A_m = 3.25 \times 4.50 = 14.62$ in.^{2} (9432 mm^2)

$A_s = 2.5 \times 5.5 = 13.75$ in.^{2} (8871 mm^2)

$A_m/A_s = 14.62/13.75 > 1.0 \therefore$ use $A_s/A_m = 13.75/14.62 = 0.94$ and use A_m = 14.62 in.^{2} (9432 mm^2)

$C_g = 0.99$ by interpolation from Table 8.4

$C_m = 0.4$ (see Table 8.2)

Allowable load:

$Z'_{45°} = Z_{45°} \times$ No. of fasteners $\times C_g \times C_d \times C_D \times C_m = 746 \times 6 \times 0.99 \times 0.75 \times 1.6 \times 0.4 = 2127$ lb (9461 N)

8.3.6 Bolts

Introduction. Although bolts do not have as great a load-carrying capacity as timber connectors such as shear plates and split rings, described in Section 8.3.7, in most cases they provide more than adequate strength. The bolts most typically used conform to ANSI/ASME Standard B18.2.1-1981, *Square and Hex Bolts and Screws* (*Inch Series*). Standard bolts range in size from ¼–1½ in. (6.35–38.1 mm) in diameter and from 1–11½ in. (25.4–292 mm) in length, as shown in Table 8.28.

Steel bolts can be electroplated or galvanized. Special bolts for a variety of purposes may be made of aluminum, bronze, stainless steel, or even specialized wood products. The use of a high-strength bolt may or may not increase the allowable design value, as wood crushing can limit the connection strength.

Bolts are placed in predrilled holes that are 1/32–1/16 in. (0.79–1.59 mm) larger than the bolt diameter, with the larger value applying to the larger-diameter bolts. Shrinkage of the bolt hole can be about 5–8% of bolt diameter across the grain; therefore, the larger hole size will minimize the potential for splitting of the wood if it dries to reach an equilibrium moisture content. Larger holes also facilitate field installation.

*The group reduction factor C_g is partially based on the stiffness of the connected members. Therefore, since the load is not parallel or perpendicular to the grain, it is appropriate to use an equivalent width slightly larger than the spacing between rows and a thickness equal to the penetration of the lag screw into the main member for the effective area of the main member. The use of a wider width is justified based on the fact that the wood when loaded at an angle to the grain (not including 90°) has a greater stiffness than when loaded perpendicular to the grain.

TABLE 8.28 Some Available Sizes of Standard Regular-Stock Steel Bolts

Diameter, in.	Lengths, in.[a]
1/4	1 to 8
5/6	1 to 8 and 8 1/2 to 10
3/8	1 to 8 and 8 1/2 to 11 1/2
7/16	1 to 8 and 8 1/2 to 11 1/2
1/2	1 to 8 and 8 1/2 to 11 1/2
5/8	1 to 8 and 8 1/2 to 11 1/2
3/4	1 to 8 and 8 1/2 to 11 1/2
7/8	1 1/4 to 8 and 8 1/2 to 11 1/2
1	1 1/2 to 8 and 8 1/2 to 11 1/2
1 1/8	1 1/2 to 8 and 8 1/2 to 11 1/2
1 1/4	1 1/2 to 8 and 8 1/2 to 11 1/2

Note: 1 in. = 25.4 mm.

[a] Bolt lengths vary by 1/4 in. from 1 to 8 in. and by 1/2 in. from 8 1/2 to 11 1/2 in.

Washers are to be used with all bolts when the head or the nut will bear against the wood to prevent the head or nut from being drawn into the wood when the bolt is tightened. It is recommended that all nuts be tight (snug) at the time of installation.

If bolt holes are not properly aligned, a considerable shift in the distribution of load to the bolts may occur, and excessive deformations may result. However, it has been found that normal fabrication tolerances have been sufficient for bolted structures to perform in an acceptable manner even if exposed to vibrations, shocks, and earthquakes.

Loads on Bolts. Bolts are generally subjected to lateral loading, where the load is applied perpendicular to the axis of the bolt. Loads can be applied either parallel, perpendicular, or at an angle to the grain of the wood members being joined.

The ultimate load capacity of a bolted connection, in single, double, or multiple shear, is dependent on several material and dimensional properties and the number of bolt shear planes.

The European yield model (EYM) provides six possible modes of failure for bolted connections in single shear and four for double shear. Failures occur when either the wood crushes under the bolt or when one, two, or four plastic hinges are formed in the bolt.

The controlling value for a two-member (single-shear) connection is the least value of Z obtained from the six equations given in Table 8.29.

When a member is loaded at an angle to the grain, the dowel bearing strength shall be determined as discussed in Section 8.2.3.

The controlling value for a three-member (double-shear) connection is the least value of Z obtained by solving the four equations given in Table 8.30.

A typical load-deformation curve is shown in Fig. 8.22 for a connection with a 1/2 in. (12.7 mm) bolt and three members. The two splice plates are steel. As can be seen, the first part of the curve is linear, and little slip takes place in the connection. At the proportional limit, slip begins to increase rapidly with load, the curve finally flattens out, and slip becomes very large for little or no load increase.

TABLE 8.29 Yield Modes for Single-Shear Bolted Connections

Equation	Yield mode
$Z = \dfrac{Dt_m F_{em}}{4K_\theta}$	Mode I_m
$Z = \dfrac{Dt_s F_{es}}{4K_\theta}$	Mode I_s
$Z = \dfrac{k_1 Dt_s F_{es}}{3.6K_\theta}$	Mode II
$Z = \dfrac{k_2 Dt_m F_{em}}{3.2(1 + 2R_e)K_\theta}$	Mode III_m
$Z = \dfrac{k_3 Dt_s F_{em}}{3.2(2 + R_e)K_\theta}$	Mode III_s
$Z = \dfrac{D^2}{3.2K_\theta}\sqrt{\dfrac{2F_{em}F_{yb}}{3(1 + R_e)}}$	Mode IV

where $k_1 = \dfrac{\sqrt{R_e + 2R_e^2(1 + R_t + R_t^2) + R_t^2 R_e^3} - R_e(1 + R_t)}{(1 + R_e)}$

$$k_2 = -1 + \sqrt{2(1 + R_e) + \frac{2F_{yb}(1 + 2R_e)D^2}{3F_{em}t_m^2}}$$

$$k_3 = -1 + \sqrt{\frac{2(1 + R_e)}{R_e} + \frac{2F_{yb}(2 + R_e)D^2}{3F_{em}t_s^2}}$$

$R_e = F_{em}/F_{es}$
$R_t = t_m/t_s$
t_m = thickness of main (thicker) member, in.
t_s = thickness of side (thinner) member, in.
F_{em} = dowel bearing strength of main (thicker) member, psi (see table 8.39)
F_{es} = dowel bearing strength of side (thinner) member, psi (see table 8.39)
$F_{e\parallel}$ = $11200G$, dowel bearing strength parallel to grain, lb
$F_{e\perp} = \dfrac{6100G^{1.45}}{\sqrt{D}}$, dowel bearing strength perpendicular to grain, lb
F_{yb} = bending yield strength of bolt, psi
D = nominal bolt diameter, in.
$K_\theta = 1 + (\theta_{max}/360°)$
θ_{max} = maximum angle of load to grain ($0° \leq \theta \leq 90°$) for any member in a connection.

Quality of Bolt Hole. The bearing strength of wood under a bolt is affected to a considerable extent by the size and quality of the hole into which the bolt is inserted. If too large a hole is drilled, the bearing strength under the bolt will not be uniform, and if the hole is too small, the wood will split when the bolt is driven in or as it dries to the equilibrium moisture content.

A smooth hole will provide greater bearing strength than a rough-cut hole, as shown in Fig. 8.23. Also, deformations due to load will increase with an increase

TABLE 8.30 Yield Modes for Double-Shear Bolted Connections

Equation	Yield mode
$Z = \dfrac{Dt_m F_{em}}{4K_\theta}$	Mode I_m
$Z = \dfrac{Dt_s F_{es}}{2K_\theta}$	Mode I_s
$Z = \dfrac{k_3 Dt_s F_{em}}{1.6(2 + R_e)K_\theta}$	Mode III_s
$Z = \dfrac{D^2}{1.6K_\theta}\sqrt{\dfrac{2F_{em}F_{yb}}{3(1 + R_e)}}$	Mode IV

where $k_3 = -1 + \sqrt{\dfrac{2(1 + R_e)}{R_e} + \dfrac{2F_{yb}(2 + R_e)D^2}{3F_{em}t_s^2}}$

$R_e = F_{em}/F_{es}$
t_m = thickness of main (center) member, in.
t_s = thickness of side member, in.
F_{em} = dowel bearing strength of main (center) member, psi
F_{es} = dowel bearing strength of side members, psi
F_{yb} = bending yield strength of bolt, psi
D = nominal bolt diameter, in.
$K_\theta = 1 + (\theta_{max}/360°)$

in the roughness of the bolt hole surface. Rough surfaces are usually due to using dull drill bits, fast rates of feed, and a slow drill revolution.

Nominal Design Values. Nominal design values for bolts laterally loaded are given in Tables 8.31 and 8.32 for both parallel and perpendicular to grain for single-shear (two-member) connections (see Fig. 8.24). Separate values are given for loading of side or main members perpendicular to grain including the value $Z_\perp$ that applies when both members are loaded perpendicular to the grain of the wood. Table 8.31 provides nominal design values for sawn lumber when both members are of identical species. Table 8.32 gives nominal design values when a ¼ in. (6.35 mm) A36 steel side plate is used with a sawn wood member. Nominal design values when glued-laminated timber members are used in conjunction with a sawn lumber side member or a ¼ in. (6.35 mm) A36 steel side member are given in Tables 8.33 and 8.34, respectively.

Tables 8.35 and 8.36 provide nominal bolt design values for double-shear connections when the load is applied parallel or perpendicular to grain of the side or main member (see Fig. 8.24). Table 8.35 is for main and side members of identical species, whereas Table 8.36 is for a double-shear connection with side members of ¼ in. (6.35 mm) A36 steel. For nominal design values when the main member is glued-laminated timber and the side members are sawn lumber or ¼ in. (6.35 mm) A36 steel side plates, see Tables 8.37 and 8.38, respectively.

If a connection does not qualify under the geometry and material properties given in the tables, then the EYM equations may be used to obtain the nominal

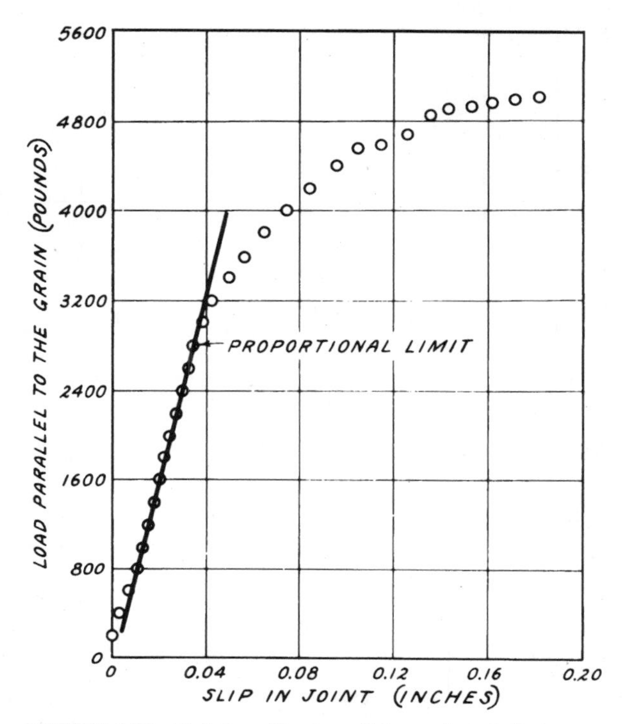

FIGURE 8.22 Variation of load parallel to grain and slip in connection.

lateral design value. Table 8.39 provides dowel bearing strengths, F_{em} and F_{es}. For angles of load between 0° (parallel to grain) and 90° (perpendicular to grain) the dowel bearing strength can be calculated from the Hankinson formula (given in Section 8.2.3). The bolt yield strength in the tables is assumed to be 45 ksi (310 MPa), corresponding to an A307 bolt.

If the wood members being joined are of different species, then the nominal design value should be selected based on the species with the lowest dowel bearing strength. This will usually be the species with the lowest specific gravity (see Table 8.39).

Design values for large-diameter bolts, more than 1 in. (25 mm), may or may not be justified based on their behavior in actual structures and the results of research conducted. The designer is cautioned before selecting large-diameter bolts as fasteners to check current industry design recommendations.

Wood-to-Concrete Connections. A wood-to-concrete connection is composed of a single wood member that is attached by bolts embedded in the concrete or masonry. Design values for wood-to-concrete connections are given in the NDS. Wood in direct contact with concrete or masonry that is in contact with soil grade must be preservative treated because these materials can wick moisture into the wood. Further discussion of wood-concrete connection details can be seen in Chapter 13.

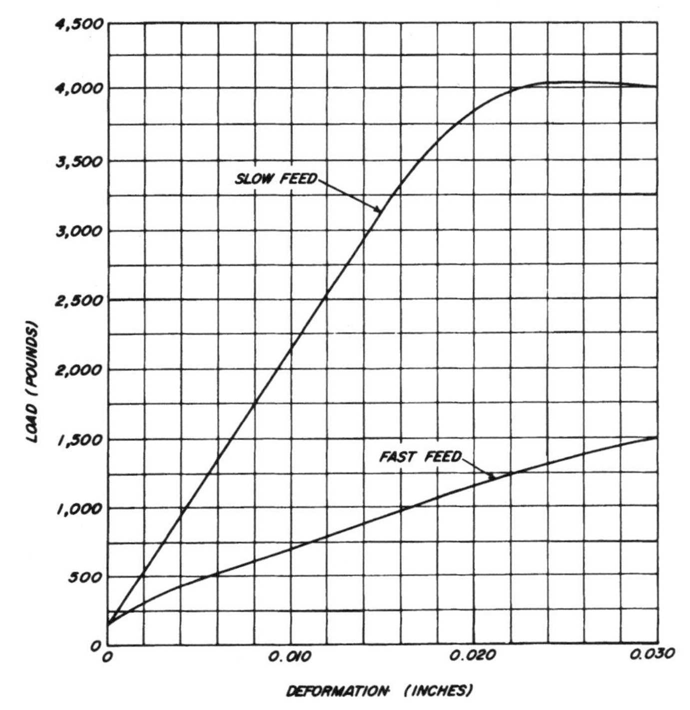

FIGURE 8.23 Effects of surface condition of bolt holes, resulting from a past feed rate and a slow feed rate of drill bit, on load deformation of bolted connections.

Placement. In the placement of bolts, it is necessary to consider the following:

1. End distance: tension or compression
2. Edge distance: loaded and unloaded
3. Spacing in a row
4. Spacing between rows

Since it may be necessary to determine the placement of bolts in both the parallel and perpendicular to grain directions, and it may even be necessary in some cases to consider a load applied at an angle to grain, the calculation of all dimensions can be a considerable task. The NDS gives values for spacing when loads are applied either parallel or perpendicular to the grain. No spacing values are given for placement of the bolt when loads are applied at an angle to grain, although "uniform stress in main members and a uniform distribution of load to all bolts requires that the gravity axis of the members shall pass through the center of resistance of the bolt group."[1] Figure 8.25 shows the NDS requirements. Most spacing values are based on a multiple of the bolt diameter *D* and are given from the center of the bolt hole.

TABLE 8.31 Bolt Design Values (Z) for Single-Shear (Two-Member) Connections[a] for Sawn Lumber with Both Members of Identical Species

Thickness			$G = 0.55$ Southern pine				$G = 0.50$ Douglas fir-larch				$G = 0.43$ Hem-fir				$G = 0.42$ Spruce-pine-fir			
Main member, t_m, in.	Side member, t_s, in.	Bolt diameter, D, in.	$Z_{\parallel}$, lb	$Z_{s\perp}$, lb	$Z_{m\perp}$, lb	$Z_{\perp}$, lb	$Z_{\parallel}$, lb	$Z_{s\perp}$, lb	$Z_{m\perp}$, lb	$Z_{\perp}$, lb	$Z_{\parallel}$, lb	$Z_{s\perp}$, lb	$Z_{m\perp}$, lb	$Z_{\perp}$, lb	$Z_{\parallel}$, lb	$Z_{s\perp}$, lb	$Z_{m\perp}$, lb	$Z_{\perp}$, lb
1½	1½	½	530	330	330	250	480	300	300	220	410	250	250	180	410	240	240	170
		⅝	660	400	400	280	600	360	360	240	520	300	300	190	510	290	290	190
		¾	800	460	460	310	720	420	420	270	620	350	350	210	610	340	340	210
		⅞	930	520	520	330	850	470	470	290	720	390	390	230	710	380	380	220
		1	1060	580	580	350	970	530	530	310	830	440	440	250	810	430	430	240
2½	1½	½	660	400	420	350	610	370	370	310	550	320	310	250	540	320	300	240
		⅝	930	560	490	390	850	520	430	340	730	420	360	270	710	410	350	270
		¾	1120	660	560	430	1020	590	500	380	870	460	410	300	850	450	400	290
		⅞	1300	720	620	470	1190	630	550	410	1020	500	450	320	1000	490	440	310
		1	1490	770	680	490	1360	680	610	440	1160	540	500	350	1140	530	490	340
3	1½	½	660	400	470	360	610	370	420	330	550	320	350	290	540	320	330	280
		⅝	940	560	550	460	880	520	480	400	790	420	400	320	780	410	390	310
		¾	1270	660	620	500	1190	590	550	440	1020	460	450	350	1000	450	440	340
		⅞	1520	720	690	550	1390	630	610	480	1190	500	500	380	1160	490	490	370
		1	1740	770	750	580	1590	680	670	510	1360	540	550	410	1330	530	540	400
3½	1½	½	660	400	470	360	610	370	430	330	550	320	380	290	540	320	370	280
		⅝	940	560	620	500	880	520	540	460	790	420	440	370	780	410	430	360
		¾	1270	660	690	580	1200	590	610	510	1100	460	500	400	1080	450	480	390
		⅞	1680	720	770	630	1590	630	680	550	1370	500	550	430	1340	490	540	420
		1	2010	770	830	670	1830	680	740	590	1570	540	600	470	1530	530	590	460
	3½	½	750	520	520	460	720	490	490	430	660	440	440	390	660	430	430	380
		⅜	1170	780	780	650	1120	700	700	560	1040	600	600	450	1020	590	590	440
		¾	1690	960	960	710	1610	870	870	630	1450	740	740	500	1420	730	730	480
		⅞	2170	1160	1160	780	1970	1060	1060	680	1690	910	910	540	1660	890	890	520
		1	2480	1360	1360	820	2260	1230	1230	720	1930	1030	1030	580	1890	1000	1000	560

TABLE 8.31 Bolt Design Values (Z) for Single-Shear (Two-Member) Connections[a] for Sawn Lumber with Both Members of Identical Species (*Continued*)

Thickness: Main member, t_m, in.	Thickness: Side member, t_s, in.	Bolt diameter, D, in.	G = 0.55 Southern pine: $Z_{\parallel}$, lb	$Z_{s\perp}$, lb	$Z_{m\perp}$, lb	$Z_{\perp}$, lb	G = 0.50 Douglas fir-larch: $Z_{\parallel}$, lb	$Z_{s\perp}$, lb	$Z_{m\perp}$, lb	$Z_{\perp}$, lb	G = 0.43 Hem-fir: $Z_{\parallel}$, lb	$Z_{s\perp}$, lb	$Z_{m\perp}$, lb	$Z_{\perp}$, lb	G = 0.42 Spruce-pine-fir: $Z_{\parallel}$, lb	$Z_{s\perp}$, lb	$Z_{m\perp}$, lb	$Z_{\perp}$, lb
4½	1½	⅝	940	560	640	500	880	520	590	460	790	420	530	410	780	410	520	400
		¾	1270	660	840	660	1200	590	750	590	1100	460	600	460	1080	450	590	450
		⅞	1680	720	930	720	1590	630	820	630	1460	500	660	500	1440	490	640	490
		1	2150	770	1000	770	2050	680	890	680	1800	540	720	540	1760	530	710	530
	3½	⅝	1170	780	780	680	1120	700	730	630	1040	600	660	520	1020	590	650	510
		¾	1690	960	1090	830	1610	870	1000	730	1490	740	840	570	1480	730	820	560
		⅞	2300	1160	1300	900	2190	1060	1160	780	1950	920	960	620	1920	910	940	600
		1	2870	1390	1440	950	2610	1290	1290	840	2240	1140	1070	670	2190	1120	1050	650
5½	1½	⅝	940	560	640	500	880	520	590	460	790	420	530	410	780	410	520	400
		¾	1270	660	850	660	1200	590	790	590	1100	460	700	460	1080	450	690	450
		⅞	1680	720	1090	720	1590	630	980	630	1460	500	780	500	1440	490	760	490
		1	2150	770	1190	770	2050	680	1060	680	1800	540	860	540	1760	530	830	530
	3½	⅝	1170	780	780	680	1120	700	730	630	1040	600	660	530	1020	590	650	520
		¾	1690	960	1090	850	1610	870	1030	780	1490	740	920	650	1480	730	900	640
		⅞	2300	1160	1410	1020	2190	1060	1260	910	1950	920	1030	720	1920	910	1010	700
		1	2870	1390	1550	1100	2660	1290	1390	970	2370	1140	1150	780	2330	1120	1120	760
7½	1½	⅝	940	560	640	500	880	520	590	460	790	420	530	410	780	410	520	400
		¾	1270	660	850	660	1200	590	790	590	1100	460	700	460	1080	450	690	450
		⅞	1680	720	1090	720	1590	630	1010	630	1460	500	900	500	1440	490	890	490
		1	2150	770	1350	770	2050	680	1270	680	1800	540	1130	540	1760	530	1110	530
	3½	⅝	1170	780	780	680	1120	700	730	630	1040	600	660	530	1020	590	650	520
		¾	1690	960	1090	850	1610	870	1030	780	1490	740	920	650	1480	730	910	640
		⅞	2300	1160	1450	1020	2190	1060	1360	930	1950	920	1210	790	1920	910	1180	780
		1	2870	1390	1830	1210	2660	1290	1630	1110	2370	1140	1340	970	2330	1120	1300	950

Note: 1 in. = 25.4 mm; 1 lb = 4.45 N.

[a] Tabulated lateral design values (Z) for bolted connections shall be multiplied by all applicable adjustment factors. Tabulated lateral design values (Z) are for "full diameter" bolts with a bending yield strength (F_{yb}) of 45,000 psi.

Source: This table is from AF&PA's *National Design Specification® for Wood Construction.*

TABLE 8.32 Bolt Design Values (Z) for Single-Shear (Two-Member) Connections for Sawn Lumber with ¼ in. ASTM A36 Steel Side Plate[1,2]

Thickness			G = 0.55 Southern pine		G = 0.50 Douglas fir-larch		G = 0.43 Hem-fir		G = 0.43 Spruce-pine-fir	
Main member, t_m, in.	Steel side plate, t_s, in.	Bolt diameter, D, in.	$Z_{\parallel}$, lb	$Z_{\perp}$, lb	$Z_{\parallel}$, lb	$Z_{\perp}$, lb	$Z_{\parallel}$, lb	$Z_{\perp}$, lb	$Z_{\parallel}$, lb	$Z_{\perp}$, lb
1½	¼	½	620	350	580	310	520	280	510	270
		⅝	780	400	730	360	650	320	640	320
		¾	940	450	870	420	780	360	770	360
		⅞	1090	510	1020	470	910	410	900	400
		1	1250	550	1170	510	1040	450	1030	450
2½	¼	½	860	470	830	410	740	350	720	340
		⅝	1150	530	1050	470	920	400	910	390
		¾	1370	590	1270	530	1110	450	1090	440
		⅞	1600	650	1480	590	1290	490	1270	480
		1	1830	700	1690	640	1480	540	1450	530
3	¼	½	860	540	830	470	770	400	770	380
		⅝	1260	610	1210	540	1070	450	1050	440
		¾	1610	670	1480	610	1290	500	1270	490
		⅞	1880	740	1730	660	1500	550	1480	540
		1	2150	790	1980	720	1720	600	1690	590
3½	¼	½	860	550	830	510	770	450	770	430
		⅝	1260	690	1210	610	1130	500	1120	490
		⅞	1740	760	1670	680	1480	560	1450	540
		⅞	2170	840	1990	740	1720	610	1690	590
		1	2480	890	2270	800	1970	660	1930	650
4½	¼	⅝	1260	760	1210	710	1130	610	1120	600
		¾	1740	950	1670	840	1560	680	1550	660
		⅞	2320	1030	2220	910	2070	740	2050	720
		1	2980	1100	2860	980	2490	810	2440	790
5½	¼	⅝	1260	760	1210	710	1130	640	1120	630
		¾	1740	1000	1670	940	1560	810	1550	790
		⅞	2320	1240	2220	1090	2070	880	2050	860
		1	2980	1320	2860	1170	2670	950	2640	930
7½	¼	⅝	1260	760	1210	710	1130	640	1120	630
		¾	1740	1000	1670	940	1560	850	1550	840
		⅞	2320	1280	2220	1210	2070	1080	2050	1070
		1	2980	1590	2860	1500	2670	1270	2640	1230
9½	¼	¾	1740	1000	1670	940	1560	850	1550	840
		⅞	2320	1280	2220	1210	2070	1080	2050	1070
		1	2980	1590	2880	1500	2670	1350	2640	1330
11½	¼	⅞	2320	1280	2220	1210	2070	1080	2050	1070
		1	2980	1590	2860	1500	2670	1350	2640	1330
13½	¼	1	2980	1590	2860	1500	2670	1350	2640	1330

1. Tabulated lateral design values (Z) for bolted connections shall be multiplied by all applicable adjustment factors.

2. Tabulated lateral design values (Z) are for "full diameter" bolts with bending yield strength (F_{yb}) of 45,000 psi and dowel bearing strength, (F_e) of 87,000 psi for ASTM A36 steel.

Source: This table is from AF&PA's *National Design Specification® for Wood Construction.*

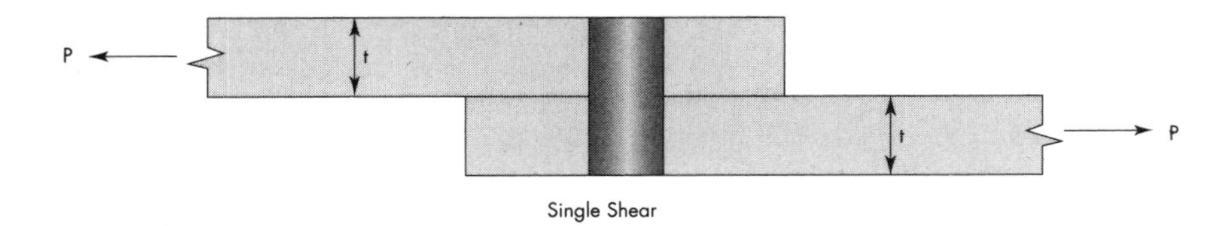

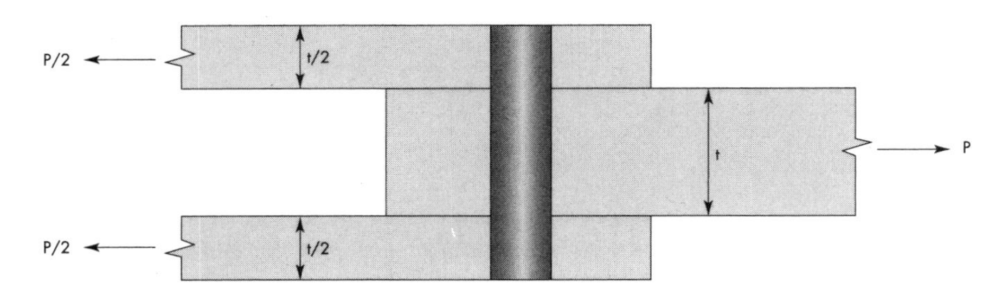

FIGURE 8.24 Bolt in single and double shear.

The l/D ratio is defined as the length, l, of the bolt in the member divided by the bolt diameter, D. The length of the bolt is the lesser of the thickness of the main member or the thickness of the side member(s). For example:

1. For a two-member connection with t_s = 1.5 in. (38.1 mm), t_m = 2.5 in. (63.5 mm) and D = 0.75 in. (19.1 mm),

$$l_s/D = 1.5/0.75 = \underline{2.0 \text{ controls}}$$

$$l_m/D = 2.5/0.75 = 3.33$$

2. For a three-member connection with t_s = 2 × 1.5 in. (76.2 mm), t_m = 2.5 in. (63.5 mm), and D = 0.75 in. (19.1 mm),

$$l_s/D = 2 \times 1.5/0.75 = 4.0$$

$$l_m/D = 2.5/0.75 = \underline{3.33 \text{ controls}}$$

The NDS permits end distances and spacings for bolts to be reduced if the design load applied to the connection is less than its allowable design capacity. The NDS requirements are given in Tables 8.40 and 8.41. However, it is recommended that the spacings given in Fig. 8.25 be maintained when the use of these spacings will not significantly influence the overall cost of the connection.

If an end distance or spacing is reduced, the nominal design value should be adjusted by the geometry factor C_Δ as defined in Tables 8.40 and 8.41 and discussed in Section 8.2.2.

On occasion, designers select relatively large diameter washers. Care must be taken that adequate spacing is provided between the bolts to prevent an overlap of washers.

TABLE 8.33 Bolt Design Values for Single-Shear Connections of Glulam Main Member with Lumber Side Member of Identical Species[a]

Thickness: Main member, t_m, in.	Thickness: Side member, t_s, in.	Bolt diameter, D, in.	G = 0.55 Southern pine: $Z_{\parallel}$, lb	$Z_{s\perp}$, lb	$Z_{m\perp}$, lb	$Z_{\perp}$, lb	G = 0.50 Douglas-fir-larch: $Z_{\parallel}$, lb	$Z_{s\perp}$, lb	$Z_{m\perp}$, lb	$Z_{\perp}$, lb	G = 0.43 Hem-fir: $Z_{\parallel}$, lb	$Z_{s\perp}$, lb	$Z_{m\perp}$, lb	$Z_{\perp}$, lb	G = 0.52 Spruce-pine-fir: $Z_{\parallel}$, lb	$Z_{s\perp}$, lb	$Z_{m\perp}$, lb	$Z_{\perp}$, lb
2½	1½	½	–	–	–	–	610	370	370	310	550	320	310	250	540	320	300	240
		⅝	–	–	–	–	850	520	430	340	730	420	360	270	710	410	350	270
		¾	–	–	–	–	1020	590	500	380	870	460	410	300	850	450	400	290
		⅞	–	–	–	–	1190	630	550	410	1020	500	450	320	1000	490	440	310
		1	–	–	–	–	1360	680	610	440	1160	540	500	350	1140	530	490	340
3	1½	½	660	400	470	360	–	–	–	–	–	–	–	–	–	–	–	–
		⅝	940	560	550	460	–	–	–	–	–	–	–	–	–	–	–	–
		¾	1270	660	620	500	–	–	–	–	–	–	–	–	–	–	–	–
		⅞	1520	720	690	550	–	–	–	–	–	–	–	–	–	–	–	–
		1	1740	770	750	580	–	–	–	–	–	–	–	–	–	–	–	–
3⅛	1½	½	–	–	–	–	610	370	430	330	550	320	360	290	540	320	340	280
		⅝	–	–	–	–	880	520	500	410	790	420	410	330	780	410	400	320
		¾	–	–	–	–	1200	590	570	460	1060	460	460	360	1040	450	450	350
		⅞	–	–	–	–	1440	630	630	490	1230	500	510	390	1210	490	500	380
		1	–	–	–	–	1640	680	690	530	1410	540	560	420	1380	530	550	410
5	1½	⅝	940	560	640	500	–	–	–	–	–	–	–	–	–	–	–	–
		¾	1270	660	850	660	–	–	–	–	–	–	–	–	–	–	–	–
		⅞	1680	720	1020	720	–	–	–	–	–	–	–	–	–	–	–	–
		1	2150	770	1100	770	–	–	–	–	–	–	–	–	–	–	–	–
							–	–	–	–	–	–	–	–	–	–	–	–
5⅛	1½	⅝	–	–	–	–	880	520	590	460	790	420	530	410	780	410	520	400
		¾	–	–	–	–	1200	590	790	590	1100	460	670	460	1080	450	660	450
		⅞	–	–	–	–	1590	630	920	630	1460	500	740	500	1440	490	720	490
		1	–	–	–	–	2050	680	990	680	1800	540	810	540	1760	530	780	530
6¾	1½	⅝	940	560	640	500	880	520	590	460	790	420	530	410	780	410	520	400
		¾	1270	660	850	660	1200	590	790	590	1100	460	670	460	1080	450	660	450
		⅞	1680	720	1090	720	1590	630	1010	630	1460	500	900	500	1440	490	890	490
		1	2150	770	1350	770	2050	680	1270	680	1800	540	1030	540	1760	530	1000	530

Note: 1 in. = 25.4 mm; 1 lb = 4.45 N.

[a]Tabulated lateral design values (Z) for bolted connections shall be multiplied by all applicable adjustment factors. Tabulated lateral design values (Z) are for full-diameter bolts with a bending yield strength F_{yb}) of 45,000 psi.

Source: This table is from AF&PA's *National Design Specification® for Wood Construction.*

TABLE 8.34 Bolt Design Values for Single-Shear Connections of Glulam Main Member with ¼ in. ASTM A36 Steel Side Plate[1,2]

Thickness: Main member, t_m, in.	Thickness: Steel side plate, t_s, in.	Bolt diameter, D, in.	$G = 0.55$ Southern pine: $Z_{\parallel}$, lb	$G = 0.55$ Southern pine: $Z_{\perp}$, lb	$G = 0.50$ Douglas-fir-larch: $Z_{\parallel}$, lb	$G = 0.50$ Douglas-fir-larch: $Z_{\perp}$, lb	$G = 0.43$ Hem-fir: $Z_{\parallel}$, lb	$G = 0.43$ Hem-fir: $Z_{\perp}$, lb	$G = 0.42$ Spruce-pine-fir: $Z_{\parallel}$, lb	$G = 0.42$ Spruce-pine-fir: $Z_{\perp}$, lb
2½	¼	½	–	–	830	410	740	350	720	340
		⅝	–	–	1050	470	920	400	910	390
		¾	–	–	1270	530	1110	450	1090	440
		⅞	–	–	1480	590	1290	490	1270	480
		1	–	–	1690	640	1480	540	1450	530
3	¼	½	860	540	–	–	–	–	–	–
		⅝	1260	610	–	–	–	–	–	–
		¾	1610	670	–	–	–	–	–	–
		⅞	1880	740	–	–	–	–	–	–
		1	2150	790	–	–	–	–	–	–
3⅛	¼	½	–	–	830	490	770	410	770	400
		⅝	–	–	1210	550	1110	460	1090	450
		¾	–	–	1540	620	1340	510	1310	500
		⅞	–	–	1790	680	1560	560	1530	550
		1	–	–	2050	740	1780	610	1750	600
5	¼	⅝	1260	760	–	–	–	–	–	–
		¾	1740	1000	–	–	–	–	–	–
		⅞	2320	1140	–	–	–	–	–	–
		1	2980	1210	–	–	–	–	–	–
5⅛	¼	⅝	–	–	1210	710	1130	640	1120	630
		¾	–	–	1670	940	1560	760	1550	740
		⅞	–	–	2220	1020	2070	830	2050	810
		1	–	–	2860	1100	2670	900	2640	880
6¾	¼	⅝	1260	760	1210	710	1130	640	1120	630
		¾	1740	1000	1670	940	1560	850	1550	840
		⅞	2320	1280	2220	1210	2070	1060	2050	1030
		1	2980	1590	2860	1420	2670	1150	2640	1120
8½	¼	¾	1740	1000	–	–	–	–	–	–
		⅞	2320	1280	–	–	–	–	–	–
		1	2980	1590	–	–	–	–	–	–
8¾	¼	¾	–	–	1670	940	1560	850	1550	840
		⅞	–	–	2220	1210	2070	1080	2050	1070
		1	–	–	2860	1500	2670	1350	2640	1330
10½	¼	⅞	2320	1280	–	–	–	–	–	–
		1	2980	1590	–	–	–	–	–	–
10¾	¼	⅞	–	–	2220	1210	2070	1080	2050	1070
		1	–	–	2860	1500	2670	1350	2640	1330
12¼	¼	⅞	–	–	2220	1210	2070	1080	2050	1070
		1	–	–	2860	1500	2670	1350	2640	1330
14¼	¼	1	–	–	2860	1500	2670	1350	2640	1330

1. Tabulated lateral design values (Z) for bolted connections shall be multiplied by all applicable adjustment factors.
2. Tabulated lateral design values (Z) are for "full diameter" bolts with bending yield strength (F_{yb}) of 45,000 psi and dowel bearing strength, (F_e) of 87,000 psi for ASTM A36 steel.

Source: This table is from AF&PA's *National Design Specification® for Wood Construction.*

TABLE 8.35 Bolt Design Values (Z) for Double-Shear (Three-Member) Connections[a] for Sawn Lumber with Both Members of Identical Species

Thickness: Main member, t_m, in.	Thickness: Side member, t_s, in.	Bolt diameter, D, in.	$G = 0.55$ Southern pine: $Z_{\parallel}$, lb	$Z_{s\perp}$, lb	$Z_{m\perp}$, lb	$G = 0.50$ Douglas fir-larch: $Z_{\parallel}$, lb	$Z_{s\perp}$, lb	$Z_{m\perp}$, lb	$G = 0.43$ Hem-fir: $Z_{\parallel}$, lb	$Z_{s\perp}$, lb	$Z_{m\perp}$, lb	$G = 0.42$ Spruce-pine-fir: $Z_{\parallel}$, lb	$Z_{s\perp}$, lb	$Z_{m\perp}$, lb
1½	1½	½	1150	800	550	1050	730	470	900	650	380	880	640	370
		⅝	1440	1130	610	1310	1040	530	1130	840	420	1100	830	410
		¾	1730	1330	660	1580	1170	590	1350	920	460	1320	900	450
		⅞	2020	1440	720	1840	1260	630	1580	1000	500	1540	970	490
		1	2310	1530	770	2100	1350	680	1800	1080	540	1760	1050	530
2½	1½	½	1320	800	910	1230	743	790	1100	650	640	1080	z640	610
		⅝	1870	1130	1020	1760	1040	880	1590	840	700	1570	830	690
		¾	2550	1330	1110	2400	1170	980	2190	920	770	2160	900	750
		⅞	3360	1440	1200	3060	1260	1050	2630	1000	830	2570	970	810
		1	3840	1530	1280	3500	1350	1130	3000	1080	900	2940	1050	880
3	1½	½	1320	800	940	1230	730	860	1100	650	760	1080	640	740
		⅝	1870	1130	1220	1760	1040	1050	1590	840	840	1570	830	830
		¾	2550	1330	1330	2400	1170	1170	2190	920	920	2160	900	900
		⅞	3360	1440	1440	3180	1260	1260	2920	1000	1000	2880	970	970
		1	4310	1530	1530	4090	1350	1350	3600	1080	1080	3530	1050	1050
3½	1½	½	1320	800	940	1230	730	860	1100	650	760	1080	640	740
		⅝	1870	1130	1290	1760	1040	1190	1590	840	980	1570	830	960
		¾	2550	1330	1550	2400	1170	1370	2190	920	1080	2160	900	1050
		⅞	3360	1440	1680	3180	1260	1470	2920	1000	1160	2880	970	1130
		1	4310	1530	1790	4090	1350	1580	3600	1080	1260	3530	1050	1230
	3½	½	1500	1040	1040	1430	970	970	1330	880	880	1310	870	860
		⅝	2340	1560	1420	2240	1410	1230	2070	1190	980	2050	1170	960
		¾	3380	1910	1550	3220	1750	1370	2980	1490	1080	2950	1460	1050
		⅞	4600	2330	1680	4290	2130	1470	3680	1840	1160	3600	1810	1130
		1	5380	2780	1790	4900	2580	1580	4200	2280	1260	4110	2240	1230

TABLE 8.35 Bolt Design Values (Z) for Double-Shear (Three-Member) Connections[a] for Sawn Lumber with Both Members of Identical Species (*Continued*)

Thickness: Main member, t_m, in.	Thickness: Side member, t_s, in.	Bolt diameter, D, in.	$G = 0.55$ Southern pine: $Z_{\parallel}$, lb	$Z_{s\perp}$, lb	$Z_{m\perp}$, lb	$G = 0.50$ Douglas fir-larch: $Z_{\parallel}$, lb	$Z_{s\perp}$, lb	$Z_{m\perp}$, lb	$G = 0.43$ Hem-fir: $Z_{\parallel}$, lb	$Z_{s\perp}$, lb	$Z_{m\perp}$, lb	$G = 0.42$ Spruce-pine-fir: $Z_{\parallel}$, lb	$Z_{s\perp}$, lb	$Z_{m\perp}$, lb
4½	1½	⅝	1870	1130	1290	1760	1040	1190	1590	840	1050	1570	830	1040
		¾	2550	1330	1690	2400	1170	1580	2190	920	1380	2160	900	1350
		⅞	3360	1440	2170	3180	1260	1890	2920	1000	1500	2880	970	1460
		1	4310	1530	2300	4090	1350	2030	3600	1080	1620	3530	1050	1580
	3½	⅝	2340	1560	1560	2240	1410	1460	2070	1190	1270	2050	1170	1240
		¾	3380	1910	1990	3220	1750	1760	2980	1490	1380	2950	1460	1350
		⅞	4600	2330	2170	4390	2130	1890	3900	1840	1500	3840	1810	1460
		1	5740	2780	2300	5330	2580	2030	4730	2280	1620	4660	2240	1580
5½	1½	⅝	1870	1130	1290	1760	1040	1190	1590	840	1050	1570	830	1040
		¾	2550	1330	1690	2400	1170	1580	2190	920	1400	2160	900	1380
		⅞	3360	1440	2170	3180	1260	2030	2920	1000	1800	2880	970	1780
		1	4310	1530	2700	4090	1350	2480	3600	1080	1980	3530	1050	1930
	3½	⅝	2340	1560	1560	2240	1410	1460	2070	1190	1320	2050	1170	1310
		¾	3380	1910	2180	3220	1750	2050	2980	1490	1690	2950	1460	1650
		⅞	4600	2330	2650	4390	2130	2310	3900	1840	1830	3840	1810	1780
		1	5740	2780	2810	5330	2580	2480	4730	2280	1980	4660	2240	1930
7½	1½	⅝	1870	1130	1290	1760	1040	1190	1590	840	1050	1570	830	1040
		¾	2550	1330	1690	2400	1170	1580	2190	920	1400	2160	900	1380
		⅞	3360	1440	2170	3180	1260	2030	2920	1000	1800	2880	970	1780
		1	4310	1530	2700	4090	1350	2530	3600	1080	2270	3530	1050	2240
	3½	⅜	2340	1560	1560	2240	1410	1460	2070	1190	1320	2050	1170	1310
		¾	3380	1910	2180	3220	1750	2050	2980	1490	1850	2950	1460	1820
		⅞	4600	2330	2890	4390	2130	2720	3900	1840	2450	3840	1810	2420
		1	5740	2780	3680	5330	2580	3380	4730	2280	2700	4660	2240	2630

Note: 1 in. = 25.4 mm; 1 lb = 4.45 N.

[a] Tabulated lateral design values (Z) for bolted connections shall be multiplied by all applicable adjustment factors. Tabulated lateral design values (Z) are for full-diameter bolts with a bending yield strength (F_{yb}) of 45,000 psi.

TABLE 8.36 Bolt Design Vlaues (Z) for Double-Shear (Three-Member) Connections for Sawn Lumber with ¼ in. ASTM A36 Steel Side Plates[1,2]

Thickness			G = 0.55 Southern pine		G = 0.50 Douglas fir-larch		G = 0.43 Hem-fir		G = 0.42 Spruce-pine-fir	
Main member, t_m, in.	Steel side plate, t_s, in.	Bolt diameter, D, in.	$Z_\parallel$, lb	$Z_\perp$, lb	$Z_\parallel$, lb	$Z_\perp$, lb	$Z_\parallel$, lb	$Z_\perp$, lb	$Z_\parallel$, lb	$Z_\perp$, lb
1½	¼	½	1150	550	1050	470	900	380	880	370
		⅝	1440	610	1310	530	1130	420	1100	410
		¾	1730	660	1580	590	1350	460	1320	450
		⅞	2020	720	1840	630	1580	500	1540	490
		1	2310	770	2100	680	1800	540	1760	530
2½	¼	½	1720	910	1650	790	1500	640	1470	610
		⅝	2400	1020	2190	880	1880	700	1840	690
		¾	2880	1110	2630	980	2250	770	2200	750
		⅞	3360	1200	3060	1050	2630	830	2570	810
		1	3840	1280	3500	1130	3000	900	2940	880
3	¼	½	1720	1100	1650	950	1540	770	1530	740
		⅝	2510	1220	2410	1050	2250	840	2200	830
		¾	3460	1330	3150	1170	2700	920	2640	900
		⅞	4040	1440	3680	1260	3150	1000	3080	970
		1	4610	1530	4200	1350	3600	1080	3530	1050
3½	¼	½	1720	1100	1650	1030	1540	890	1530	860
		⅝	2510	1420	2410	1230	2260	980	2230	960
		¾	3480	1550	3340	1370	3120	1080	3080	1050
		⅞	4630	1680	4290	1470	3680	1160	3600	1130
		1	5380	1790	4900	1580	4200	1260	4110	1230
4½	¼	⅝	2510	1510	2410	1420	2260	1270	2230	1240
		¾	3480	1990	3340	1760	3120	1380	3090	1350
		⅞	4630	2170	4440	1890	4150	1500	4110	1460
		1	5960	2300	5720	2030	5330	1620	5280	1580
5½	¼	⅝	2510	1510	2410	1420	2260	1280	2230	1270
		¾	3480	2000	3340	1890	3120	1690	3090	1650
		⅞	4630	2570	4440	2310	4150	1830	4110	1780
		1	5960	2810	5720	2480	5330	1980	5280	1930
7½	¼	⅝	2510	1510	2410	1420	2260	1280	2230	1270
		¾	3480	2000	3340	1890	3120	1690	3090	1670
		⅞	4630	2570	4440	2410	4150	2160	4110	2130
		1	5960	3180	5720	3000	5330	2700	5280	2630
9½	¼	¾	3480	2000	3340	1890	3120	1690	3090	1670
		⅞	4630	2570	4440	2410	4150	2160	4110	2130
		1	5960	3180	5720	3000	5330	2700	5280	2660
11½	¼	⅞	4630	2570	4440	2410	4150	2160	4110	2130
		1	5960	3180	5720	3000	5330	2700	5280	2660
13½	¼	1	5960	3180	5720	3000	5330	2700	5280	2660

1. Tabulated lateral design values (Z) for bolted connections shall be multiplied by all applicable adjustment factors.

2. Tabulated lateral design values (Z) are for "full diameter" bolts with bending yield strength (F_{yb}) of 45,000 psi and a dowel bearing strength (F_e) of 87,000 psi for ASTM A36 steel.

Source: This table is from AF&PA's *National Design Specification® for Wood Construction.*

TABLE 8.37 Bolt Design Values for Double-Shear Connections of Glulam Main Member with Sawn Lumber Side Members[a]

Thickness: Main member, t_m, in.	Thickness: Side member, t_s, in.	Bolt diameter, D, in.	$G = 0.55$ Southern pine			$G = 0.50$ Douglas-fir-larch			$G = 0.43$ Hem-fir			$G = 0.52$ Spruce-pine-fir		
			$Z_{\parallel}$, lb	$Z_{s\perp}$, lb	$Z_{m\perp}$, lb	$Z_{\parallel}$, lb	$Z_{s\perp}$, lb	$Z_{m\perp}$, lb	$Z_{\parallel}$, lb	$Z_{s\perp}$, lb	$Z_{m\perp}$, lb	$Z_{\parallel}$, lb	$Z_{s\perp}$, lb	$Z_{m\perp}$, lb
2½	1½	½	–	–	–	1230	730	790	1100	650	640	1080	640	610
		⅝	–	–	–	1760	1040	880	1590	840	7-00	1570	830	690
		¾	–	–	–	2400	1170	980	2190	920	770	2160	900	750
		⅞	–	–	–	3060	1260	1050	2630	1000	830	2570	970	810
		1	–	–	–	3500	1350	1130	3000	1080	900	2940	1050	880
3	1½	½	1320	800	940	–	–	–	–	–	–	–	–	–
		⅝	1870	1130	1220	–	–	–	–	–	–	–	–	–
		¾	2550	1330	1330	–	–	–	–	–	–	–	–	–
		⅞	3360	1440	1440	–	–	–	–	–	–	–	–	–
		1	4310	1530	1530	–	–	–	–	–	–	–	–	–
3⅛	1½	½	–	–	–	1230	730	860	1100	650	760	1080	640	740
		⅝	–	–	–	1760	1040	1090	1590	840	880	1570	830	860
		¾	–	–	–	2400	1170	1220	2190	920	960	2160	900	940
		⅞	–	–	–	3180	1260	1310	2920	1000	1040	2880	970	1010
		1	–	–	–	4090	1350	1410	3600	1080	1130	3530	1050	1090
5	1½	⅝	1870	1130	1290	–	–	–	–	–	–	–	–	–
		¾	2550	1330	1690	–	–	–	–	–	–	–	–	–
		⅞	3360	1440	2170	–	–	–	–	–	–	–	–	–
		1	4310	1530	2550	–	–	–	–	–	–	–	–	–
5⅛	1½	⅝	–	–	–	1760	1040	1190	1590	840	1050	1570	830	1040
		¾	–	–	–	2400	1170	1580	2190	920	1400	2160	900	1380
		⅞	–	–	–	3180	1260	2030	2920	1000	1700	2880	970	1660
		1	–	–	–	4090	1350	2310	3600	1080	1850	3530	1050	1790
6¾	1½	⅝	1870	1130	1290	1760	1040	1190	1590	840	1050	1570	830	1040
		¾	2550	1330	1690	2400	1170	1580	2190	920	1400	2160	900	1380
		⅞	3360	1440	2170	3180	1260	2030	2920	1000	1800	2880	970	1780
		1	4310	1530	2700	4090	1350	2530	3600	1080	2270	3530	1050	2240

Note: 1 in. = 25.4 mm; 1 lb = 4.45 N.

[a] Tabulated lateral design values (Z) for bolted connections shall be multiplied by all applicable adjustment factors. Tabulated lateral design values (Z) are for full-diameter bolts with a bending yield strength (F_{yb}) of 45,000 psi.

Source: This table is from AF&PA's *National Design Specification® for Wood Construction.*

TABLE 8.38 Bolt Design Values for Double-Shear Connections of Glulam Main Member with ¼ in. ASTM A36 Steel Plate[a]

Thickness												
Main member, t_m, in.	Steel side plate, t_s, in.	Bolt diameter, D, in.	$G = 0.55$ Southern pine		$G = 0.50$ Douglas-fir-larch		$G = 0.43$ Hem-fir		$G = 0.42$ Spruce-pine-fir			
			$Z_{\parallel}$, lb	$Z_{\perp}$, lb	$Z_{\parallel}$, lb	$Z_{\perp}$, lb	$Z_{\parallel}$, lb	$Z_{\perp}$, lb	$Z_{\parallel}$, lb	$Z_{\perp}$, lb		
2½	¼	½	–	–	1650	790	1500	640	1470	610		
		⅝	–	–	2190	880	1880	700	1840	690		
		¾	–	–	2630	980	2250	770	2200	750		
		⅞	–	–	3060	1050	2630	830	2570	810		
		1	–	–	3500	1130	3000	900	2940	880		
3	¼	½	1720	1100	–	–	–	–	–	–		
		⅝	2510	1220	–	–	–	–	–	–		
		¾	3460	1330	–	–	–	–	–	–		
		⅞	4040	1440	–	–	–	–	–	–		
		1	4610	1530	–	–	–	–	–	–		
3⅛	¼	½	–	–	1650	980	1540	800	1530	770		
		⅝	–	–	2410	1090	2260	880	2230	860		
		¾	–	–	3280	1220	2810	960	2750	940		
		⅞	–	–	3830	1310	3280	1040	3210	1010		
		1	–	–	4380	1410	3750	1130	3670	1090		
5	¼	⅝	2510	1510	–	–	–	–	–	–		
		¾	3480	2000	–	–	–	–	–	–		
		⅞	4630	2410	–	–	–	–	–	–		
		1	5960	2550	–	–	–	–	–	–		
5⅛	¼	⅝	–	–	2410	1420	2260	1280	2230	1270		
		¾	–	–	3340	1890	3120	1580	3090	1540		
		⅞	–	–	4440	2150	4150	1700	4110	1660		
		1	–	–	5720	2310	5330	1850	5280	1790		
6¾	¼	⅝	2510	1510	2410	1420	2260	1280	2230	1270		
		¾	3480	2000	3340	1890	3120	1690	3090	1670		
		⅞	4630	2570	4440	2410	4150	2160	4110	2130		
		1	5960	3180	5720	3000	5330	2430	5280	2360		
8½	¼	¾	3480	2000	–	–	–	–	–	–		
		⅞	4630	2570	–	–	–	–	–	–		
		1	5960	3180	–	–	–	–	–	–		
8¾	¼	¾	–	–	3340	1890	3120	1690	3090	1670		
		⅞	–	–	4440	2410	4150	2160	4110	2130		
		1	–	–	5720	3000	5330	2700	5280	2660		
10½	¼	⅞	4630	2570	–	–	–	–	–	–		
		1	5960	3180	–	–	–	–	–	–		
10¾	¼	⅞	–	–	4440	2410	4150	2160	4110	2130		
		1	–	–	5720	3000	5330	2700	5280	2660		
12¼	¼	⅞	–	–	4440	2410	4150	2160	4110	2130		
		1	–	–	5720	3000	5330	2700	5280	2660		
14¼	¼	1	–	–	5720	3000	5330	2700	5280	2660		

1. Tabulated lateral design values (Z) for bolted connections shall be multiplied by all applicable adjustment factors.

2. Tabulated lateral design values (Z) are for "full-diameter" bolts with bending yield strength (F_{yb}) of 45,000 psi.

Source: This table is from AF&PA's *National Design Specification® for Wood Construction.*

TABLE 8.39 Dowel Bearing Strength for Bolted Connections

Species combination	Specific gravity, G^a	Dowel bearing strength, psi[b]					
		$F_{e\parallel}$	$F_{e\perp}$ $D = \frac{1}{2}$ in.	$F_{e\perp}$ $D = \frac{5}{8}$ in.	$F_{e\perp}$ $D = \frac{3}{4}$ in.	$F_{e\perp}$ $D = \frac{7}{8}$ in.	$F_{e\perp}$ $D = 1$ in.
Southern pine	0.55	6150	3650	3250	2950	2750	2550
Douglas fir-larch	0.50	5600	3150	2800	2600	2400	2250
Hem-fir	0.43	4800	2550	2250	2050	1900	1800
Spruce-pine-fir	0.42	4700	2540	2200	2000	1850	1750

Note: 1 lb/in.^{2} = 6.895 kPa.
[a] Specific gravity based on weight and volume when oven dry.
[b] $F_{e\parallel} = 11200G$; $F_{e\perp} = 6100G^{1.45}/\sqrt{D}$; tabulated values are rounded to the nearest 50 psi.
Source: This table is from AF&PA's *National Design Specification® for Wood Construction.*

Net Section. Members must be checked for load-carrying capacity at the critical net section of the connections. The net section is the gross cross-sectional area of the member reduced to account for bolt holes. In tension and compression members, the required net section, in square inches, is determined by dividing the total load transferred through the critical net section by the allowable design stress for the species and grade of material used. When the bolts are staggered, adjacent bolts or lag screws should be considered as being placed in the critical section unless the bolts in a row are spaced at least eight diameters apart.

Adjustments to Allowable Values. After the nominal value has been selected from the appropriate table, or calculated, general adjustments must be made to account for anticipated end-use conditions, as discussed in Section 8.2.2. The geometry adjustment factor, C_Δ, for bolts is given in Tables 8.40 and 8.41.

Example 8.12 Bolt, Parallel to Grain Load, Placement

Determine the size, number, and placement of bolts needed to transfer the 7500 lb (33.4 kN) load (dead load plus snow load) through the butt joint shown in Fig. 8.26. Wood is seasoned no. 1 Douglas fir (MC $\leq$ 19%), which will remain dry in service (MC $\leq$ 19%).

solution

Size of member:

A (required) $= P/F_t = 7500/(675 \times 1.15 \times 1.2^*) = 8.05$ in.^{2} (5194 mm^2)

Try a 2 × 8 in., $A = 10.875$ in.^{2} (7016 mm^2), for both main member and side plates

Try $\frac{5}{8}$ in. (15.9 mm) bolts

Z (nominal) $= 1310$ lb per bolt (Table 8.35)

Z' (allowable) $= Z \times C_D \times C_M = 1310 \times 1.15 \times 1.0 = 1506$ lb (5830 N)

Number required $= 7500/1506 = 4.98$

Try six $\frac{5}{8}$ in. bolts, two rows of three bolts

*Assuming a nominal 2 × 8 is used, 1.2 is the size factor from Table 4a of the NDS.

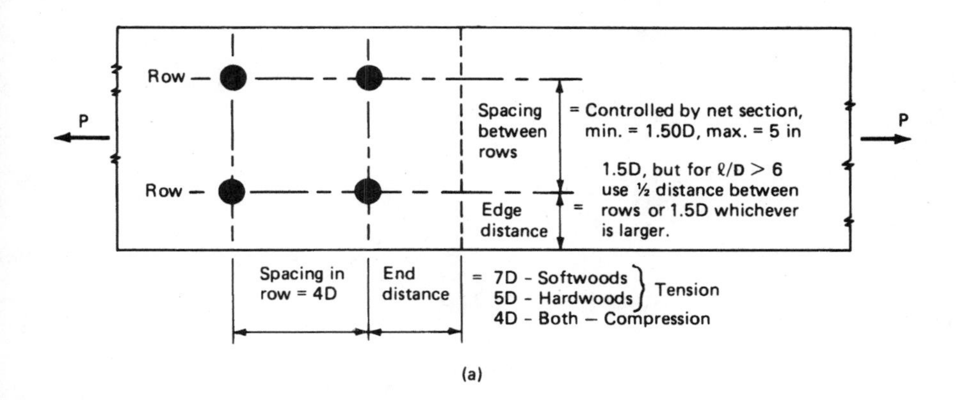

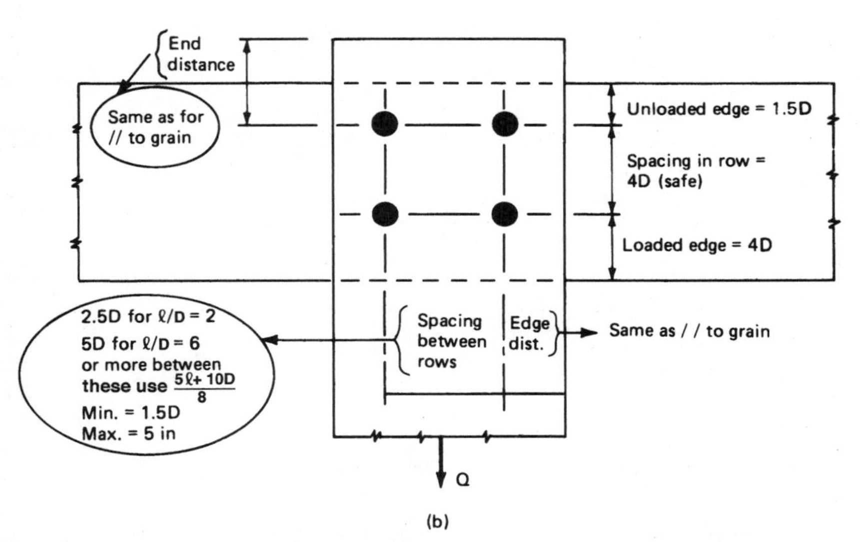

FIGURE 8.25 Placement of bolts: (*a*) Parallel to grain, (*b*) perpedicular to grain. *Note:* All values are minimum unless otherwise stated.

Group reduction:

$A_m = 1.5 \times 7.25 = 10.875$ in.^{2} (7016 mm^2)

$A_s = 2 \times 10.875 = 21.75$ in.^{2} (14032 mm^2)

$A_s/A_m > 1 \therefore$ use $A_m/A_s = 10.875/21.75 = 0.5$ and use A_m in place of A_s; Thus,

$C_g = 0.96 - (12 - 10.875) \times ((0.96 - 0.92)/(12 - 5)) = 0.95$ (from Table 8.4)

$Q = 1506 \times 0.95 \times 6 = 8584$ lb (38.2 kN) > 7500 lb (33.4 kN) $\therefore$ OK

Example 8.13 Bolt, Perpendicular to Grain Loading, Placement

Determine the size, number, and placement of bolts needed to transfer the 6000 lb (26.7 kN) (two-month duration) load through the connection shown in Fig. 8.27. All wood is seasoned (MC ≤ 19%) and will remain dry (MC ≤ 19%).

TABLE 8.40 End Distance for Bolts

Direction of loading	Minimum end distance: For reduced design value	Minimum end distance: For full design value
⊥ to grain	$2D$	$4D$
∥ to grain,		
• Compression	$2D$	$4D$
• Tension		
Softwoods	$3.5D$	$7D$
Hardwoods	$2.5D$	$5D$

$$C_\Delta = \frac{\text{actual end distance}}{\text{full design value end distance}}$$

TABLE 8.41 Spacing for Bolts in Row

Direction of loading	Minimum spacing: For reduced design value	Minimum spacing: For full design value
∥ to grain	$3D$	$4D$
⊥ to grain	$3D$	Required spacing for attached member(s)

$$C_\Delta = \frac{\text{actual spacing}}{\text{full design value spacing}}$$

solution

Try ¾ in. bolts:

$Z_{m\perp}$ (nominal) = 1580 lb (7031 N) (see Table 8.37)
$Z_{m\perp}$ (allowable) = 1580 lb × 1.15 = 1817 lb (8086 N)
Number required = 6000 lb/ 1817 lb = 3.3 ∴ use 4 with two rows of two bolts

Note: No group reduction factor required for two or fewer in a row.

Net section (2 × 6):

A (net)* = (1.5 × 5.5 × 2) − (2 × 0.78125 × 2) = 13.375 in.^{2} (8629 mm^2)
A (required) = $6000/F_t$ = 6000/(675 × 1.3 × 1.15) = 6.21 in.^{2} (4010 mm^2) < 13.375 in^2 (8629 mm^2) ∴ ok

Note: Net section applies only to member loaded parallel to grain.

Spacing in row:

Percent full load capacity of bolts = 3.30/4 × 100 = 83%

*Area removed by bolt hole is 1/32 in. larger than bolt diameter.

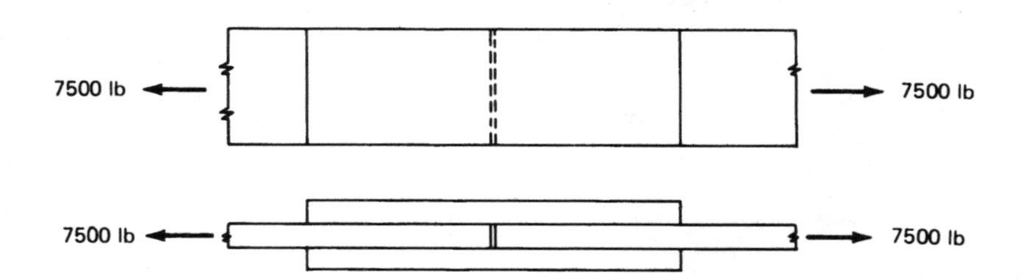

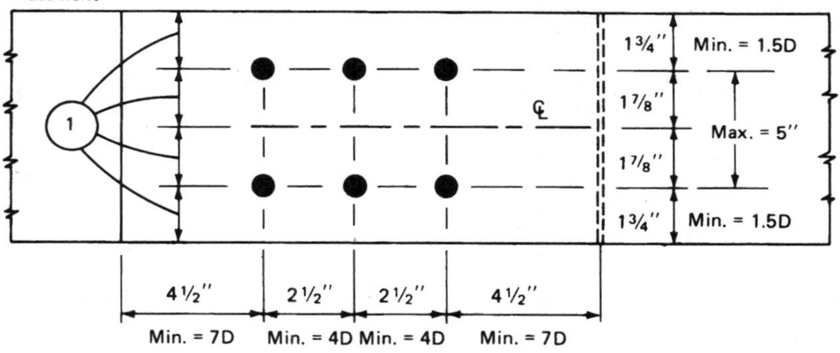

Notes: (1) As closely as possible make these values all equal for uniform distribution of stress across the section.
(2) 1 in = 25.4 mm

FIGURE 8.26 Example 8.12.

$4D \times 0.83 = 2.47$ in (62.7 mm) (spacing in row)

Spacing between rows:

For l/D, use lesser of l_m/D or l_s/D:

$$l_m/D = 5.125/0.75 = 6.83$$

$$l_s/D = 3.0/0.75 = 4.0 \text{ (controls)}$$

Minimum spacing between rows is given by the equation $l_{\min} = (5l + 10D)/8$ (see Fig. 8.25) when $2 < l/D < 6$; thus,

$$l_{\min} = (5 \times 3 + 10 \times 0.75)/8 = 2.81 \text{ in. (71.4 mm)}$$

Required width $= 1.125 + 2.81 + 1.125 = 5.06$ in. < 5.5 in. (128.5 mm < 139.7 mm) $\therefore$ 2×6 will provide adequate spacing

Use two 2×6's (38.1×139.7 mm) for vertical members

Example 8.14 Bolt, Metal Side Plate, Analysis

Determine the allowable load, Q (dead load + construction load), for the connection shown in Fig. 8.28 using a load duration factor, C_D, of 1.25. The connection will remain dry. All bolt spacings, ends, and edge distances are assumed to be adequate, as is the capacity of the glulam members.

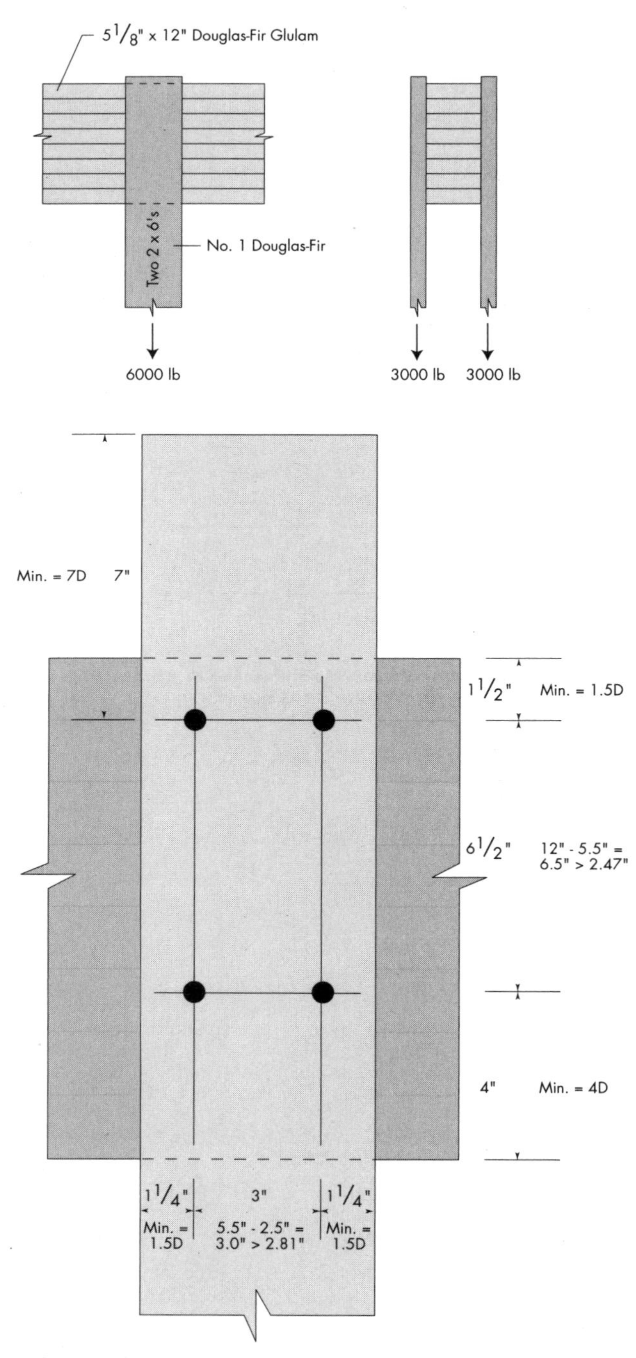

FIGURE 8.27 Example 8.13.

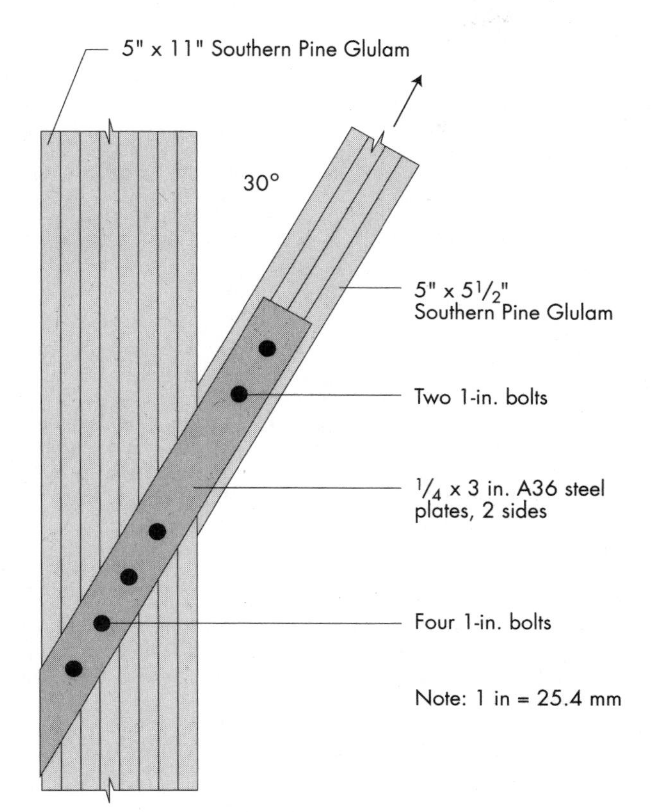

FIGURE 8.28 Example 8.14.

solution

Analysis of diagonal member:

Z = 5960 lb per bolt (see Table 8.38)

Z' (allowable) = $Z \times C_D \times$ number of bolts = 5960 × 1.25 × 2 = 14,900 lb (66,278 N)

Analysis of vertical member:

$Z_{\parallel}$ = 5960 lb (see Table 8.38)

$Z_{\perp}$ = 2550 lb (see Table 8.38)

$\theta = 30°$

$Z_{30°} = (Z_{\parallel} \times Z_{\perp})/(Z_{\parallel} \sin^2\theta + Z_{\perp}\cos^2\theta) = (5960 \times 2550)/(5960 \times \sin^2 30 + 2550 \times \cos^2 30)$ = 4467 lb (19,869 N) (see Section 8.2.3)

Group reduction: $A_m = 5 \times 6 = 30$ in.^{2} (19354 mm^2), assumed width equal to minimum spacing for bolts in a row, ($4D$ = 4 in.) plus some additional distance (2 in.) to account for load not acting perpendicular to grain (see footnote to Example 8.11 and bolt placement chart)

$A_s = 0.25 \times 3.0 \times 2 = 1.5$ in^2 (968 mm^2) (steel side plates)

$A_m/A_s = 30/1.5 = 20.0$

$C_g = 0.96$ (a conservative estimate from Table 8.5, based on $A_m/A_s = 18$ and $A_m = 24$, interpolation or using Eq. (8.1) would yield a more accurate and advantageous value)

$Z_{30°}'$ (allowable) $= Z_{30°} \times C_D \times C_g \times$ no. of bolts $= 4467 \times 1.25 \times 0.96 \times 4 =$ 21,442 lb (95,377 N)

Therefore, Q (max) = 14,900 lb (66,278 N) and is controlled by the two bolts in the diagonal member.

8.3.7 Other Connectors

Shear Plates and Split Rings

Shear Plates. Shear plates are capable of transferring large shear forces. They are used to attach beams to columns through steel straps, in the fabrication of heavy timber trusses by using steel splice plates, and quite extensively for connections between glulam members of timber structures. Shear plates are also used in pairs to make wood-to-wood connections.

Shear plates require that a groove be cut and a hole drilled using a special tool. The shear plate does not provide a wedge fit; thus, a shear-plate connection will allow greater slip than a split-ring connection and is less likely to cause a wood member to split as a result of shrinkage due to seasoning of the wood in service. The bolt or lag screw in a shear-plate connection serves a dual function of clamping the connection together and assisting in transferring the load. See the NDS[1] for further design information on shear plates.

Split Rings. Split rings are also capable of transferring large shear forces. They may be used in virtually any wood-to-wood connection requiring the transfer of high shear forces and were traditionally used in the fabrication of heavy timber trusses. In the past, they also were successfully used on occasion for light wood trusses.

Split rings require that a groove be cut by a special tool in both faces of the mating pieces of wood. At the same time the groove is cut, a hole is bored for the bolt. As the name implies, a split ring has a split, which allows it to expand as it is placed in the groove, thus allowing it to fit tightly in the groove. Therefore, very little slip will take place as load is applied. Because of the extra effort needed to install split rings properly and the potential shrinkage problems, most designers prefer to use shear plates. See the NDS[1] for further design information on split rings.

The bolt in a split-ring connection serves merely to clamp the two pieces of wood together, thereby keeping the split ring in place. The bolt does not assist in the transfer of load as do the bolts in a shear-plate connection.

Solid Rings. Solid ceramic rings are used in place of steel split-ring connectors for special-purpose structures where environmental conditions are such that the use of nonmetallic and/or corrosion-resistant fasteners is required. These rings are placed in a groove precut with a special tool. The rings are located and used like split-ring connectors. They are designed for shear load transfer in wood-to-wood connections.

Tests conducted by private industry have shown that solid ceramic rings perform as well under load as steel split rings. Therefore, the allowable design value for a solid ceramic ring may be taken to be the equivalent to that of a steel split ring of equal diameter.

Design Values: Shear Plates and Split Rings. Other sources for design provisions for shear plate and split ring connectors can be found in a booklet prepared by Timber Engineering Company (TECO)[19] and the *Timber Construction Manual,* published by AITC.[21]

Timber Rivets. The glulam rivet has been available in Canada for several decades. It was included in the 1997 NDS[1] and called the timber rivet. The name was changed to reflect its applicability to any wood member, not just glulam. The timber rivet is manufactured from galvanized-hardened steel and has a rectangular shank of ¼ × ⅛ in. (6.35 × 3.18 mm) with rounded corners, a ¼ in. (6.35 mm) long wedge point and a 11⁄32 × 7⁄32 in. (8.73 × 5.56 mm) upset head. It is generally used with A36 steel side plates, at least ⅛ in. (3.18 mm) thick. The rivet is available in lengths of 1½–3½ in. (38.1 mm–88.9 mm) and must have a minimum ultimate tensile strength of 145,000 psi (1000 MPa). The rivet is driven through 17⁄64–9⁄32 in. (6.75–88.9 mm) holes, which provides a wedge fit.

Nominal capacities of rivet connections can range from 2,000–129,000 lb (8.9–574 N), depending on how many rows and rivets are used and the thickness of the wood member. Design values for parallel, perpendicular, and at an angle of load to grain can be obtained from the NDS. The NDS also contains provisions for timber rivet penetration, placement, and design value adjustment factors.

Dowels, Drift Pins, and Drift Bolts. Dowels, drift pins, and drift bolts are hammer-driven or pressed into tight-fitting or slightly undersized holes. They are designed primarily to transmit shear loads from member to member. Drift pins are also used to align members. Dowels, drift pins, and drift bolts may be made of metal, wood, or high-strength plastics.

The performance of these fasteners depends to a considerable extent on their bending stiffness and shear resistance as well as on their bearing areas and the bearing resistance of the fastened members. Because of their tight fit, multiples of these fasteners in a connection, designed to transmit shear loads, act like multiple lag screws.

So that a tight fit of the fastener is maintained in the hole, the shrinkage and swelling of the fastener, if any, and that of the surrounding wood must be similar. Thus, the moisture content of the wood surrounding the fastener should be similar to that of the fastener at the time of installation if the fastener is made of wood.

Design Values for Dowels, Drift Pins, and Drift Bolts. Connections assembled with dowels, drift pins, and drift bolts are designed in a manner similar to that used for bolts. For a drift pin or drift bolt an appropriate design value can be obtained by using 75% of the design value for a common bolt of the same diameter and length in main member. If however, the drift pin or drift bolt is used in a manner similar to a lag screw, the design value should be taken to be that of a lag screw with equivalent shank diameter. It is recommended that some additional penetration of the pin into members be provided to offset the lack of the washer, head, and nut of a common bolt. Samsalone and White[20] give design values for dowels placed in the end grain of glue-laminated timber for monotonic, repeated, and sustained loads. Other references also give design information for these types of connections.[11,21]

Metal Plate Connectors. Truss plates are flat steel plates with protruding teeth. The teeth are typically forced into the wood by a press or roller, and one plate is installed in each face of a wood member. Their greatest use is as gusset plates in prefabricated wood roof and floor trusses.

Plates are typically proprietary devices, and design values are not readily available. The manufacturer of the metal gusset plate obtains approval of their design values from building code agencies through Evaluation Service reports. Generally, the design for trusses using metal plate connectors is carried out by the plate manufacturer, the truss manufacturer, or specialized consultants. Usually the designer can obtain truss plans by contacting the truss manufacturer.

The sizing of metal plate connectors is typically based on an allowable load for the connector in terms of pounds per square inch. Allowable values, although not typically available, are in the range of 80–200 psi (552–1379 kPa), with the lower value applying to lower-density wood and the higher values to southern pine or Douglas fir. Angle of load to grain has some effect on allowable loads of metal-plate connectors, and adjustments for load duration and fire-retardant treatment also need to be made.

A recent innovative use of truss plate connectors is to interconnect glulam components to create a rigid frame. Further information on this can be obtained from APA.[22]

Joist Hangers and Framing Anchors. Joist hangers and framing anchors are designed to reduce difficulties and uncertainties of nailing and to provide considerably stronger shear connections than are practically feasible otherwise. These connectors also serve to help locate and seat wood members accurately.

Catalogs are available[23,24] that show the configuration and give all dimensions of joist hangers, framing anchors, and other light-gauge connectors. These catalogs provide design loads for the connectors and the values given have received approval by the model codes through issuance of Evaluation Service reports.

8.3.8 Choice of Fasteners

Structural Considerations

Balanced Deformations. It is not considered good practice to intermix sizes or types of fasteners in a connection, because the difference in slippage (stiffness) of each can cause an overload on the stiffest connectors before the less stiff fasteners receive significant load.

Thickness of Lumber. Most structural lumber is 1½ in. (38.1 mm) or more in thickness; therefore, any connector or fastener could be used. However, nails, screws, and staples are less commonly used in lumber greater than 1½ in. (38.1 mm) thick. For heavy timber construction, lag screws, bolts, split rings, shear plates, and special-purpose timber connectors such as beam hangers and post caps are recommended.

Type of Load. Vibratory loads, or those that cause a stress reversal, may influence the choice of fastener or connector type. Bolts have less of a tendency to wear or work loose than do nails, screws, or staples, and shear plates and split rings will typically perform better than bolts.

Size and Type. It is generally better to use the same size and type of fastener or connector throughout a structural subunit in order to maintain uniform magnitudes of slippage at all connections per unit of force.

Cost and Construction Considerations

Number of Fasteners and Connectors. The smaller the number of fasteners and connectors, in general, the lower the cost.

Uniform Size and Type of Fastener or Connector. Maintaining a single size and type will reduce overall cost since it will reduce labor cost. It also reduces the chance of error.

Labor Cost. Labor is a large part of the total cost of many structures. It may range from approximately 25–50% or more of the total cost.

Assembly and Erection. Flexible structural components must be properly supported and braced during construction.

Connection Design: General Considerations

Shrinkage and Swelling. Wood shrinks as it dries and swells when it gains moisture in the direction perpendicular to the longitudinal grain. Little shrinkage or swelling takes place in wood parallel to grain (generally much less than 1%, and this is considered negligible). However, in the perpendicular-to-grain direction, commonly referred to as across the grain, the magnitude of shrinkage or swelling will be many times that of parallel to grain and can be significant. Therefore, in the design of connections it is necessary that proper consideration be given to movements caused by shrinkage and swelling of wood members. Chapter 1 provides a detailed discussion of shrinkage and swelling of wood with changes in moisture content.

Tension Perpendicular to Grain. The ultimate strength of wood in tension perpendicular to grain is extremely variable such that values are not published. In addition, translation of applied loads into induced stresses is a subject of debate in engineering circles. For these and other reasons, avoid loading wood in tension perpendicular to grain.

Moisture Content. Wood, when maintained at low moisture contents, will function satisfactorily for many years. However, if it stays wet and is untreated, its useful life is limited due to potential damage by decay-causing organisms (see Chapter 9 for further discussion). The expected in-service moisture content needs to be given due consideration in connection design.

Typical Connections. Some of the beam connections typically used in glulam construction are shown in Chapter 4. A good source of up-to-date information for many commonly used connections is the catalogs of connector manufacturers.[23,24]

Eccentric Connections. If the resultant of all loads applied to a connection does not pass through the center of gravity of the fastener (or connector) group, then an eccentricity is introduced. Each fastener in the connection is then subjected to two force components. One component is a direct force due to the resultant force, and the second component results from the moment caused by the eccentricity of load. The force components may be added vectorially to obtain the total force acting on each fastener. Any connection involving eccentricity must be carefully analyzed, and the designer should avoid the use of eccentric connections whenever possible.

8.4 SPECIAL FASTENING AND CONNECTION ISSUES

8.4.1 Mixing Engineered Wood Products and Sawn Lumber

Mixing engineered wood products and sawn lumber in a roof or floor system should not be done without a careful analysis of the potential consequences.

Engineered wood products are manufactured at very low moisture contents (5% or lower for wood structural panel products) and to a high degree of dimensional accuracy. Sawn lumber products, on the other hand, unless kiln dried, may be supplied at moisture contents in excess of 19%. When used together on a job site, the engineered wood products have a tendency to expand slightly as they absorb moisture, while the sawn lumber products are subject to shrinkage as they dry. In situations where lumber and engineered wood products are used together in floor or roof systems, this differential shrinkage can lead to situations that must be considered by the design professional.

The use of sawn lumber blocking or rim boards in conjunction with wood I-joists in a floor system is a classic example. In such situations blocking and rim board materials are used, at least in part, to assist the wood I-joist in distributing vertical loads through the floor system into the structure below. As the building materials in the structure reach equilibrium moisture content with their surroundings, sawn lumber blocking and rim board shrink while the I-joists do not. As a result, lumber components are not available to carry the applied vertical load that they were designed to carry, thus overloading the I-joists.

This is one of the reasons that engineered wood products such as I-joists are manufactured in depths different from nominal sawn lumber products. A nominal 2 × 10 (1.5 × 9.25 in. [38.1 × 235 mm]) may not easily be used to block a 9½ in. (241 mm) deep I-joist. Sawn lumber is not effective for such applications and should never be used without careful consideration. There are applications, such as diaphragm blocking, squash blocks, and backer or filler blocking where sawn lumber is acceptable for use in conjunction with engineered wood products. See Chapter 5 for further information on I-joists.

8.4.2 Fastening Wood Structural Panels

Bearing Strength. An important component of the European yield model (EYM) is the bearing strength of the elements being connected. This section presents APA recommendations for bearing strength of plywood and oriented strand board (OSB) wood structural panels for use in the EYM equations or with tabular results obtained using EYM methodology.

Douglas fir-larch has a listed specific gravity of 0.50.[1] Due to the species limitations prescribed in Voluntary Product Standard PS 1, plywood trademarked structural I or marine grade can also be taken as having a specific gravity of 0.50. Plywood not identified as structural I or marine grade can be taken as having a specific gravity of 0.42, unless the species of plies is known, in which case the specific gravity listed for the actual species may be used.

Dowel bearing strength of OSB has been measured in a limited number of tests conducted by APA. Average results exceeded 6000 psi (41.4 MPa). However, until further study is completed, it is APA's conservative recommendation that the dowel bearing value for Douglas fir lumber, which is tabulated in the NDS as 4650 psi (32 MPa), be used for OSB panels that bear the APA trademark. This corresponds to a specific gravity of 0.50. Table 8.42 summarizes these recommendations.

Effect of Overdriven Fasteners. The following is an APA suggested guideline for determining if overdriven fasteners affect the shear capacity of diaphragm or shear wall construction.[25]

1. If all fasteners around the perimeter of panels appear to be overdriven by the same amount—about 1⁄16 in. (1.6 mm)—and it appears that panels have been wetted during construction, it can be assumed that the fastener embedment is due to panel thickness swelling. This can be verified by measuring the thickness of panels where fasteners appear to be overdriven and comparing to measurements where panels have been protected from the weather, or to the original nominal panel thickness, which is part of the APA trademark. In this case no reduction in shear capacity need be taken.
2. If no more than 20% of the fasteners around the perimeter of panels are overdriven by up to 1⁄8 in. (3.2 mm), no reduction in shear capacity needs to be taken.
3. If more than 20% of the fasteners around the perimeter of panels are overdriven, or if any are overdriven by more than 1⁄8 in. (3.2 mm), additional fasteners shall be driven to maintain the required shear capacity. For every two fasteners overdriven, one additional fastener shall be driven. If nails were used in the original installation and are spaced too close to allow the placement of additional nails, then approved staples must be used for the additional fasteners required.

Another consideration that should not be overlooked when judging overdriven fasteners is the minimum nominal panel thickness required for the design shear. If design shear for the construction requires a 15⁄32 in. (11.9 mm) minimum nominal panel thickness and the actual sheathing is 19⁄32 in. (15.1 mm) with all fasteners overdriven 1⁄8 in. (3.2 mm), the net result is a 15⁄32 in. (11.9 mm) panel, which meets the design shear requirements.

Do not forget to consider the maximum nailing recommendations when adding additional fasteners. The recommendations prevent the splitting of the framing. If the additional nails violate the minimum spacing requirements, use staples and ignore the original nails.

TABLE 8.42 Dowel Bearing Strength of Wood Structural Panels

Wood structural panel	Specific gravity, G	Dowel bearing strength, F_e
Plywood		
Structural I, marine	0.50	4650 psi
Other grades[a]	0.42	3350 psi
Oriented strand board		
All grades	0.50	4650 psi

[a] Use $G = 0.42$ when species of the plies is not known. When species of the plies is known, specific gravity listed for the actual species and the corresponding dowel bearing strength may be used, or the weighted average may be used for mixed species.

8.4.3 Corrosion-Resistant Fasteners for Construction

Corrosion resistance of fasteners is of primary importance for construction applications that are exposed to the weather or other high moisture conditions, particularly in certain cases where the wood is treated.

Exterior Siding. APA has historically recommended corrosion-resistant fasteners for exterior wood structural panel or lap siding, an application that is obviously subject to weathering and frequent high moisture conditions. Although corrosion of fasteners can ultimately affect structural performance, a more immediate consideration for siding is to prevent unsightly staining.

Based on outdoor exposure and accelerated laboratory tests, APA has determined that greater-than-minimum corrosion resistance is required for fasteners in order to preserve the desired appearance of exterior siding when exposed to weather. For this reason, hand-driven nails should be hot-dipped or hot-tumbled galvanized steel, or stainless steel. Although aluminum nails also perform satisfactorily, they are not as easily driven by hand. Aluminum and stainless steel nails are available, however, for use with power-driven (pneumatic) fastener tools.

APA tests also show that electrically or mechanically galvanized steel nails appear satisfactory when plating thickness meets or exceeds requirements for ASTM A641, class 3 coatings and are further protected by a yellow chromate coating. However, extensive in-service experience with such fasteners is lacking. Even high-quality galvanized fasteners described above may sometimes react under wet conditions with the natural extractives of some wood species, and may cause staining if left unfinished. Such staining can be minimized if the siding is finished in accordance with APA recommendations or if the roof overhang protects the siding from direct exposure to moisture and weathering. However, for best performance in such cases, use of aluminum or stainless steel fasteners should be considered.

Permanent Wood Foundations. Accelerated laboratory corrosion tests show that fasteners other than stainless steel can corrode in preservative-treated material at moisture contents that may occur in wood foundation components below grade. Plywood and lumber in wood foundation walls must remain attached to provide shear-wall strength to resist earth, wind, and seismic loads, and also to restrain the lumber studs so they do not buckle under applied vertical loading. Since plywood-to-lumber fasteners must have long-term durability, only stainless steel fasteners are recommended for attaching plywood to lumber below grade. Above grade, where the material normally is subject to drying, hot-dipped or hot-tumbled galvanized steel, stainless steel, silicon bronze, or copper fasteners may be used to attach plywood to the framing, since the average moisture content in the lumber and plywood will not be sufficient to cause corrosion.

The lumber-to-lumber connections above grade will be in a dry environment, and hot-dipped or hot-tumbled galvanized nails may be used. Lumber-to-lumber connections below grade only carry loads during fabrication and erection. After the final structure is completed, the plywood-to-lumber fasteners hold the lumber components together. Consequently, the lumber-to-lumber fasteners below grade may be hot-dipped or hot-tumbled galvanized steel nails. It is recognized that these fasteners have less corrosion resistance than stainless steel fasteners, but tests have shown that the strength of shear-wall assemblies is reduced very little even when certain lumber-to-lumber fasteners are completely removed.

Lumber knee walls are sometimes used to support brick veneer. In those cases, the studs are toe-nailed to the wood foundation wall and are directly exposed to the soil. Lumber-to-lumber fasteners in brick knee wall assemblies, therefore, must be stainless steel since they must remain intact throughout the life of the structure.

For complete construction details and recommendations for the Permanent Wood Foundation (PWF) system, contact the Southern Forest Products Association.[26]

Fire-Retardant-Treated Plywood. Fire-retardant treating of panels involves a secondary production process involving impregnation of proprietary chemical formulations by treatment companies. Chemicals used in fire-retardant-treated (FRT) plywood are typically inorganic or organic salts that may become corrosive when used in high moisture conditions. The hygroscopic nature and corrosion characteristics may vary between treatments. Therefore, APA recommends that fastener recommendations be obtained from the company providing the treatment and redrying service.

8.4.4 LVL (Laminated Veneer Lumber)

The bearing strength of LVL can be obtained from Table 8.7 using the appropriate equivalent specific gravity for LVL as given in Chapter 6. Once the equivalent bearing strength is obtained, the connection design procedure is the same as if the LVL were solid-sawn lumber. The LVL has different equivalent specific gravities for fasteners applied parallel or perpendicular to the plies.

Example 8.15 LVL Equivalent Specific Gravity and Bearing Strength

Determine the equivalent specific gravity and bearing strength for APA EWS LVL stress class 1.8E-2600F for nail fasteners loaded in withdrawal and nails loaded laterally.

solution From Chapter 6 the equivalent specific gravity is 0.50 for withdrawal and face lateral loads and 0.46 for edge lateral loads.

- The withdrawal capacity for nails in the LVL can be determined from Eq. (8.3) or Table 8.8 for specific gravity of 0.50.
- The edge bearing strength is 4000 psi (27.6 MPa) for specific gravity equal to 0.46 from the footnote equation in Table 8.12.
- The face bearing strength is 4650 psi (32.1 MPa) for specific gravity of 0.50 from Table 8.12.

8.4.5 I-Joists, Rim Board, and Glulam

Connection details for I-joists and rim board can be found in Chapter 5, and for glulam in Chapter 4.

8.5 REFERENCES

1. *National Design Specification for Wood Construction,* rev. ed., American Forest and Paper Association, Washington, DC, 1997.
2. APA—The Engineered Wood Association, 7011 S. 19th St., P.O. Box 11700, Tacoma, WA 98411.

3. Chow, P., J. D. McNatt, S. J. Lambrechts, and G. Z. Gertner, "Direct Withdrawal and Head Pull-Through Performance of Nails and Staples in Structural Wood Based Panel Materials," *Forest Products Journal,* vol. 38, no. 6, pp. 19–25, 1988.
4. McLain, T. E., and S. Thangjitham, "Bolted Wood Connection Yield Model," *Journal of Structural Engineering, ASCE,* vol. 109, no. 8, pp. 1820–1835, August 1983.
5. Uniform Building Code, International Conference of Building Officials, Whittier, CA, 1991.
6. International Building Code, International Code Council, Falls Church, VA, 2000.
7. Breyer, D. E., K. J. Fridley, and K. E. Cobeen, *Design of Wood Structures,* 4th ed., McGraw-Hill, New York, 1999.
8. Tucker, B. J., D. G. Pollock, K. J. Fridley, and J. J. Peters, "Governing Yield Modes for Common Bolted and Nailed Wood Connections," *Practice Periodical on Structural Design and Construction,* vol. 5, no. 1, February 2000.
9. Cramer, C. O., Load Distribution in Multiple-Bolt Tension Connections," *Journal of the the Structural Division, ASCE,* vol. 94, no. ST5, pp. 1101–1117, May 1968.
10. Tissel, J. R., and J. R. Elliot, *Research Report 138—Plywood Diaphragms,* APA—The Engineered Wood Association Tacoma, WA, 1997.
11. *Wood Handbook: Wood as an Engineering Material,* Agricultural Handbook no. 72, rev. August 1974, Forest Products Laboratory, U.S. Department of Agriculture, Washington, DC, 1987.
12. Timber Engineering Co., *Timber Design and Construction Handbook,* McGraw-Hill, New York, 1956.
13. Hoyle, R. J., and F. E. Woeste, *Wood Technology in the Design of Structures,* 5th ed., Iowa State University Press, Ames, IA, 1989.
14. *Roof Sheathing Fastening Schedules for Wind Uplift,* APA Report T92-28, APA—The Engineered Wood Association, Tacoma, WA, March 1993.
15. Rammer, D. R., S. G. Winistorfer, and D. A. Bender, "Withdrawal Strength of Threaded Nails," *Journal of Structural Engineering, ASCE,* vol. 127, no. 4, pp. 442–443, April 2001.
16. Scholten, J. A., *Strength of Wood Connections Made with Nails, Staples, or Screws,* Research Note FPL 0100, U.S. Department of Agriculture, Washington, DC, March 1965.
17. *Structural Wood Research: State-of-the-Art and Research Needs,* American Society of Civil Engineers, New York, 1984.
18. National Evalution Report NER-272, National Evalution Service, Inc., Falls Church, VA, 1997.
19. *Design Manual for TECO Timber Connector Construction,* Publ. No. 109, Timber Engineering Co., Colliers, WV, 1973.
20. Samsalone, M. J., and R. N. White, *End-Grain Dowel Connections in Laminated Timber: Monotonic, Repeated, and Sustained Load Behavior*, Research Report no. 84-6, Cornell University, Ithaca, NY, March 1984.
21. American Institute of Timber Construction (AITC), *Timber Construction Manual,* 4th ed., AITC, Englewood, CO, 1994.
22. A510, *Glulam Frame Building,* APA—The Engineered Wood Association, Tacoma, WA.
23. Simpson Strong-Tie Co. Inc., 4637 Chabot Dr., #200, Pleasonton, CA 94588, tel. (925) 460-9912.
24. USP Lumber Connectors, 703 Rogers Dr., Montgomery, MN 56069-1324, tel. (328) 328-5934.
25. Andreason, K. R., and J. R. Tissell, T94-9, *Effects of Overdriven Nails in Shear Walls,* APA—The Engineered Wood Association, Tacoma, WA, 1994.
26. Southern Forest Products Association, P.O. Box 641700, Kenner, LA 70064-1700.

CHAPTER NINE
TREATMENTS AND FINISHES FOR WOOD

Richard Carlson
Senior Scientist, TSD

9.1 INTRODUCTION

Wood exposed to weathering must be protected from the elements in order to maintain its appearance and integrity. Exposure to sunlight and moisture can damage wood surfaces. Additionally, unprotected wood exposed to extended periods of high moisture is subject to decay and attack by wood-destroying organisms, which can undermine its integrity and structural adequacy.

Finishes such as paints and stains help protect wood surfaces from ultraviolet light and rapid fluctuations in moisture content. Although many finishes contain additives to provide some degree of protection from surface molds and fungi such as mildew, these additives provide little protection from wood-destroying fungi or insects. Various types of chemical treatments are available, however, which can provide protection from these organisms. The type and extent of treatment needed will vary depending on the severity of exposure and associated risk factors. Additionally, not all treated wood products are paintable. Therefore, if a product is to be finished, it is important to specify the use of a paintable wood preservative.

Finally, awareness is increasing of environmental and health concerns associated with the use of some chemicals for treating wood. The U.S. Environmental Protection Agency (EPA) requires the registration of all preservatives used in products manufactured for treating wood. They also publish guidelines for the handling and disposal of some types of treated wood. Further guidance on this subject is available from the American Wood-Preservers' Association.

9.2 PRESERVATIVE TREATMENTS

Wood may be preservatively treated through various processes such as brush-applied or dip applications, addition of preservatives into the product during manufacture, or pressure treatment after manufacture. To select the most suitable method, one must consider various factors such as code requirements (if applicable), the severity of exposure, level of protection required, cost, availability, and performance expectations.

9.2.1 Dips or Surface Applications

Nonpressure treatments such as water-repellent preservatives or other preservatives are available that can be applied by brush or dip treatment. The latter method is generally more effective when practical. These types of treatments provide limited protection in lower decay risk applications or in nonstructural situations. They are topical and provide little penetration on wood surfaces, although they are readily absorbed by the end grain. If brushed on, they should be liberally applied and worked into any open voids. Dip treatments are most effective when the product is immersed for several minutes. Follow the manufacturer's recommendations. Surface treatments are not as effective as pressure-preservative treatments in protecting wood. Not all water repellents contain preservatives and many are not paintable. Therefore, it is important to specify a product suitable for the intended end use. Other types of treatments, such as copper naphthenate and copper-8-quinolinolate, are also available for surface applications. Copper-8-quinolinolate may be used in accordance with FDA and/or USDA regulations where food contact with the treated wood is possible.

9.2.2 Blended Treatments

There are currently no existing treating standards for OSB, LVL, or I-joists. However, some manufacturers have evaluated the use of preservatives incorporated into the furnish of OSB. Furnish may be defined as wood-based material, such as flakes or strands, including applied resin, wax, and other additives, as the primary constituent of wood-based panels such as waferboard or oriented strand board (OSB). These blended preservatives are intended to provide protection from wood-destroying organisms. Check with the manufacturers for additional information. Other products, such as LVL, may be treatable with pressure processes or brush-on applications or dip treatments, subject to agreement between buyer and seller. The suitability of any of these processes will depend on many factors, such as developing technology, the effect of the treating process and/or treatment chemicals on the durability of the glue bond, the effectiveness and penetration of the preservative, and the level of decay hazard.

9.2.3 Pressure-Treated Plywood

Most construction applications don't require preservative-treated wood. But in some uses, and in certain climates, treated wood is recommended or required by local building codes to protect against decay, fungi, termites, carpenter ants, or other wood-destroying insects, or from marine borers in seawater exposure. Preservative-treated plywood is impregnated with preservatives by a pressure process. The resulting deep penetration of preservative provides protection against decay and insect attack. See below, Standards and Use Recommendations, for a summary list of AWPA standards for plywood.

Pressure-preservative-treated plywood is used in a variety of applications:

- Wood foundations
- Bulkheads
- Retaining walls
- Swimming pools
- Highway noise barriers
- Irrigation structures
- Cooling towers
- Electrical transformer vaults
- Decks, docks, piers, and floats
- Tanks
- Liquid manure storage tanks
- Planters
- Food transport, processing, and storage
- Roof, wall, and floor sheathing in tropical regions or applications where resistance to termites or decay is required

Preservative-treated plywood maintains stiffness and strength, thermal properties, workability, light weight, and economy.

Wood preservatives and wood treated with these chemicals should be used and handled appropriately (see below, Standards and Use Recommendations and Precautions for Use and Handling). Always follow the recommendations provided by the preservative manufacturer or wood-treating company.

Acceptances and Availability. Preservative treatments are recognized in model building codes, federal and military specifications, and American Association of State Highway and Transportation Officials (AASHTO) specifications.

Preservative-treated plywood is generally available in most metropolitan areas. For availability, contact local building materials suppliers or consult the Yellow Pages of the telephone directory under "Lumber Treating."

Some panel grades, thicknesses, span ratings, or species may be difficult to obtain in some areas. Check with your supplier for availability or include an alternative panel in the specifications.

Organic Preservatives. Organic preservatives used for preservative treating of plywood include creosote, pentachlorophenol (penta), and copper-8-quinolinolate.

Creosote is a coal tar product that is dissolved in a distilled solution or petroleum oil. It is an effective preservative in commercial, industrial, or marine applications when there is severe exposure to decay or insect attack, or marine borers in saltwater environments. Creosote-treated plywood has an odor and a dark, oiled surface appearance and therefore is not recommended where painting is required.

Penta is commonly dissolved in light petroleum oil or solvent or in a petroleum solvent/water solution. It is suitable as a preservative for ground contact or aboveground uses. Plywood treated with penta has an odor, and oil-borne penta has an oiled surface appearance and thus is not recommended for painting. However, plywood treated with solventborne or waterborne penta, including penta in LPG, has an oil-free surface and natural wood appearance and should be specified where staining or painting is desired. *Note: Penta dissolved in methylene chloride should not be used to treat plywood containing synthetic repairs in face veneers, since the solvent can damage the repairs.*

Plywood treated with copper-8-quinolinolate preservative can be used in applications where food is harvested, transported, or stored. The preservative is dissolved in liquid petroleum gas or light hydrocarbon solvents so that the treated plywood has a clean surface and is free of solvent odor. Check with the company providing the treatment service regarding applicable FDA and USDA acceptances.

Inorganic Preservatives. Inorganic preservatives are the most popular and commonly available types used for preservative treating of plywood. They include leach-resistant waterborne arsenical preservatives such as CCA, ACA, ACZA, and ACQ-B, which are highly effective in preventing decay and attack by termites, carpenter ants, and marine organisms (see Table 9.1 for chemical names of common waterborne preservatives). These preservatives are forced into plywood under pressure and become insoluble or fixed in the wood cells after impregnation and drying.

Waterborne preservatives are recommended where clean, odorless, and paintable products are required. Wood treated with such preservatives may be used inside residences or commercial and industrial buildings, providing sawdust and construction debris are cleaned up and disposed of after construction.

Note: Treatment of tongue-and-groove (T&G) panels with waterborne preservatives may result in difficulty in mating T&G edges. Using square-edge panels or milling T&G edges after treatment should be considered.

Standards and Use Recommendations. Table 9.2 lists applicable American Wood-Preservers' Association (AWPA) preservative-treating standards for plywood, according to the intended end use. Table 9.3 summarizes plywood applications and recommended preservative treatments for these uses, in accordance with AWPA Standards.

Preservative Penetration and Retention. AWPA standards specify preservative retention in terms of pounds of retained preservative per cubic foot of wood. Specified retention levels vary according to the type of preservative and the severity of exposure (see Table 9.3). For preservative-treated plywood, AWPA standards require that all veneers must be penetrated by preservative. The net retention of preservative in plywood is high—up to 25 lb/ft^3 in some treatments—because of plywood's high ratio of surface area to volume. Therefore, preservative-treated plywood may be cut in the field without loss of preservative protection at cut edges.

TABLE 9.1 Chemical Names of Common Waterborne Preservatives

ACA	Ammoniacal copper arsenate[a]
ACZA	Ammoniacal copper zinc arsenate[a]
CCA	Chromated copper arsenate[a] (types A, B, and C)
ACC	Acid copper chromate
ACQ-B	Ammoniacal copper quat (type B)[a]
ACQ-D	Amine copper quat (type D)
CC	Ammoniacal copper citrate
SBX	Borate oxide

[a] These preservatives are highly leach-resistant.

TABLE 9.2 Preservative-Treating Standards for Plywood[a]

		Preservative types			
Use	AWPA standard	Creosote	Pentachlorophenol	Copper-8-quinolinolate	Waterborne preservatives
General	C9	X	X	X	X
Highway	C14	X			X
Commercial-residential construction	C15			X	X
Farm	C16	X	X	X	X
Marine	C18	X	X		X
Wood foundation	C22				X
Food handling, storage, transportation	C29			X	
Cooling tower	C30	X			X

[a] For government and military procurement, Federal Specification TT-W-571 should be cited when specifying preservative-treated wood.

Fasteners. Hot-dipped or hot-tumbled galvanized steel, stainless steel, silicon bronze, or copper fasteners are recommended for use in preservative-treated wood. Only stainless steel fasteners are recommended for attaching preservative-treated plywood to lumber below grade in wood foundations.

Gluability. Plywood can be glued when treated with most of the waterborne preservatives and with some penta treatments using light petroleum oil or solvent, liquid petroleum gas, or petroleum solvent/water as a carrier. Consult treating firms or their trade associations for specific recommendations.

Structural Properties. Preservative treatments applied under AWPA standards do not affect plywood strength and stiffness. See APA *Plywood Design Specification*[1] (Form Y510) or APA Technical Note N375[2] for allowable working stresses and section properties or design capacities for plywood.

Finishing Recommendations. Waterborne preservatives are readily paintable when dried after treatment. Finishing recommendations are generally the same as those for untreated plywood. Some slight surface degradation is possible in sanded plywood after drying because of surface checking and/or discoloration. For this reason, Medium Density Overlay plywood gives best results where treated paintable surfaces are required. Unsanded grades of plywood do not require further finishing but can be finished with two-coat acrylic paints or opaque stains. Stain finishes or two-coat acrylic paints (stain-blocking primer and topcoat) are recommended for textured plywood.

Painting plywood treated with creosote or oil-borne preservatives such as pentachlorophenol is not recommended. Painting can be done only with difficulty and

TABLE 9.3 Recommended Treatments

Exposure	Typical application[a]	Minimum preservative treatment[b] (lb/ft^3 by assay)			
		Creosote	Pentachloro-phenol[d]	Copper-8-quinolinolate[f]	Waterborne preservatives[c]
Contact with sea water, exposed to marine borer attack.	Pontoons, wharf bulkheads, scows, floats and flood gates, etc.	Creosote: full-cell 2.50	Not recommended	Not recommended	ACZA 2.50 ACA 2.50 CCA 2.50 (Type A, B or C)
Contact with ground, chemicals, continuous moisture, or high humidity.	Permanent trench and tunnel lining, retaining walls, skirting for post and pier or pole-type foundations, snow sheds, floats, irrigation structures, tanks, linings for wet process industries, poultry dropping trays, septic tanks, some chemical storage tanks, industrial sewers, and smelter roofs.	Creosote: empty-cell 10.0	Penta: empty-cell 0.50	Not recommended	ACZA 0.40 ACA 0.40 ACC 0.40 CCA 0.40 (Type A, B or C) ACQ 0.40 (Type B or D) CC 0.40
	Permanent wood foundation system[e]	Not recommended	Not recommended	Not recommended	ACA, 0.60 ACZA, CCA (Type A, B or C) or ACQ (Type B)

Above-ground uses where plywood is subject to insect infestation or fungus attack.	Under these exposure conditions, protection may be advisable for subflooring over unexcavated areas or shallow crawl spaces; sheathing and other uses such as fences, exterior siding, exposed structural units such as stressed-skin panels and box beams, reservoir roofs, splash boards in pole-type buildings.	Creosote: empty-cell 8.0	Penta: empty-cell 0.40	Copper-8-quinolinolate: empty-cell 0.20	ACZA 0.25 ACA 0.25 ACC 0.25 CCA 0.25 (Type A, B or C) ACQ 0.25 (Type B or D) CC 0.25 SBX[g] 0.17
	Interior linings of refrigerators and box cars; food processing plants and warehouses; greenhouses and milk processing facilities; fruit, vegetable, and grain harvesting; transport and storage containers requiring low human toxicity fungicide.	Not recommended	Not recommended	Copper-8-quinolinolate: empty-cell 0.20	Not recommended

[a] See applicable EPA-approved Consumer Information Sheet (CIS) for precautions involving uses and handling of creosote, pentachlorophenol or inorganic arsenical pressure-treated wood.

[b] Recommended minimums from the American Wood-Preservers' Association Standards.

[c] Based on dry preservative per cubic foot, oxide basis, full-cell treatment.

[d] Oil, gas or other solvent-borne.

[e] Plywood marked PS 1 or PS 2, or APA Standard PRP-108 is required for the Permanent Wood Foundation system.

[f] Volatile petroleum (AWPA P9, Type B) or light hydrocarbon (AWPA P9, Type C) solvents only.

[g] For dry above-ground applications not subjected to liquid water or Formosan termites.

requires extensive care using an aluminum base paint. Paintable pentachlorophenol treatments are available. (See discussion under Organic Preservatives.)

For certain interior applications in commercial, industrial, or farm buildings, creosote- or pentachlorophenol-treated wood may be used if exposed surfaces are sealed by painting with two coats of urethane or epoxy paint or shellac (varnish may also be used for pentachlorophenol-treated wood). For guidelines on use precautions in these cases, refer to Table 9.4 and the EPA-approved Consumer Information Sheet (CIS) for the applicable preservative treatment.

Precautions for Use and Handling. The chemical formulations used for preservative treatment of plywood are registered with the EPA, which has approved guidelines for the use of pressure-treated wood to ensure safe handling and avoid envi-

TABLE 9.4 Use Precautions for Pressure-Treated Wood[a]

	Organic preservatives		Inorganic preservatives
Application	Creosote	Pentachlorophenol	Arsenicals
1. Skin contact applications.	OK[b]	OK[b]	OK
2. Residential interiors.	No	No[c]	OK
3. For industrial and farm buildings, interior components which are in ground contact and subject to decay or insect attack. Also see application 5.	OK[b]	OK[b]	OK
4. Laminated beams for commercial or industrial buildings	No	OK[b]	OK
5. Interiors of farm buildings when animals can crib (bite) or lick the treated wood.	No	No	OK
6. Agricultural farrowing or brooding facilities.	No	No	OK
7. Applications where preservative may become component of food or animal feed, such as structures or containers for storing silage or food.	No	No	No
8. Cutting boards or countertops for preparing food.	No	No	No
9. Decks, patios and walkways if surface is visibly clean and free from residues.	OK	OK	OK
10. Portions of beehives which may come into contact with honey.	No	No	No
11. Applications where treated wood can come into direct or indirect contact with drinking water for public or animal consumption.	No[d]	No[d]	No[d]

[a] Based on EPA-approved Consumer Information Sheets.
[b] Must be painted with recommended sealer (two coats).
[c] Except for glued laminated beams or for building components which are in ground contact and are subject to decay or insect infestation and where two coats of an appropriate sealer are applied.
[d] OK for incidental contact such as bridges or docks.

ronmental or health hazards. The use precautions for creosote, pentachlorophenol, and inorganic arsenical preservative-treated wood are published in EPA-approved Consumer Information Sheets (CISs) for these treatments, available from treaters, and are briefly summarized in Table 9.4.

Handling Tips

- Treated wood should not be burned in open fires or in stoves, fireplaces, or residential boilers.
- Treated wood from commercial or industrial use (e.g., construction sites) may be burned only in commercial or industrial incinerators or boilers in accordance with state and federal regulations.
- Avoid frequent or prolonged inhalation of sawdust from treated wood. When sawing and machining treated wood, wear a dust mask. Whenever possible, these operations should be performed outdoors to avoid indoor accumulations of air-borne sawdust from treated wood.
- When power sawing and machining, wear goggles to protect eyes from flying particles.
- Avoid frequent or prolonged skin contact with pentachlorophenol- or creosote-treated wood; when handling wood treated with these chemicals, wear long-sleeved shirts and long pants and use gloves.
- After working with treated wood, and before eating, drinking, or using tobacco products, wash exposed skin areas thoroughly.
- If preservatives or sawdust accumulates on clothes, launder before reuse. Wash work clothes separately from other household clothing.

9.2.4 Preservative Treatment of Glued Laminated Timber

Structural glued laminated timbers (glulams) bearing the APA EWS trademark are produced by members of Engineered Wood Systems (EWS), a related corporation of APA—The Engineered Wood Association. These glulams are manufactured to conform with AITC/ANSI Standard A190.1, *American National Standard for Structural Glued Laminated Timber*.[3]

Although glulams do not require preservative treatment for most uses, certain applications may present environmental conditions conducive to decay or insect or marine borer attack. Conditions that favor such attack are the presence of moisture (20% or greater moisture content of the wood) accompanied by temperatures ranging from 50–90°F. Decay progresses more slowly at temperatures outside this range and virtually ceases at temperatures below 35 or above 100°F.

These hazards are typically controlled through recognized design principles and construction techniques such as use of overhangs, flashings, ventilation, and proper joint connection details such as shown in Chapter 12. Elimination of potential decay hazards through effective design detailing is the preferred method of controlling decay. When hazards of decay or insect attack cannot be avoided, glulams must be pressure-preservative treated or a naturally durable species must be used. Examples of construction where such hazards may exist include direct exposure to weather, ground contact (including direct contact with concrete foundations and footings), contact with fresh water or seawater, and exposure to excessive condensation.

Outdoor uses of preservative-treated glulams include bridges, utility structures, marine applications, highway noise barriers, and decks. Indoor uses that may require pressure treatment include environments subject to high humidity or condensation, such as indoor swimming pools or greenhouses, where moisture content of the wood may exceed 20%. Indoor applications such as post and beam construction in some farm buildings may also involve ground contact and thus require preservative treatment for those members in contact with the ground. Table 9.5 lists common preservatives used to pressure-treat glulams and provides a summary of relevant considerations.

Applicable Standards. Applicable standards for preservative treatment of glulams include American Wood-Preservers' Association Standards C28[4] (and all other standards referenced therein) and M4.[5] Related specifications include American Forest and Paper Association National Design Specification;[6] American National Standards Institute, Inc. ANSI A190.1; Federal Specification TT-W-571,[7] and American Association of State Highway and Transportation Officials (AASHTO) (*Standard Specification for Preservative Pressure Treatment Process for Timber*).[8]

Preservatives. Pressure preservative treatments listed in American Wood Preservers' Association (AWPA) Standard C28 for glulams include creosote, pentachlorophenol, copper naphthenate, and water-borne inorganic arsenicals.

Other treatments and processes specified should be agreed to by purchaser, seller, and governing code body. Required retention and penetration levels depend on end use and exposure according to AWPA or other applicable specifications.

Organic Preservatives. Organic preservatives listed in AWPA Standard C28 include creosote and pentachlorophenol and are the primary treatments used in glulams manufactured from western species. See Section 9.2.3 under Organic Preservatives for a discussion of these organic preservatives.

Inorganic Preservatives. Inorganic preservatives are waterborne treatments such as ammoniacal copper arsenate (ACA) and chromated copper arsenate (CCA). While these are not recommended for use with western species, they may be used to treat southern pine laminations prior to gluing of the finished product. See Section 9.2.3 under Inorganic Preservatives for a further discussion of these inorganic preservatives.

Treatment Recommendations. Glulams are available in custom and stock sizes. Stock sizes are typically cut to length at a distribution center or on the job site. Most glulams to be pressure-treated will be custom sizes and should be ordered to exact dimensions, when possible, to avoid field cuts, which must be retreated. In addition, all fabrication, cutting, and predrilling of holes for fasteners is recommended prior to pressure treating.

Glulams may be treated after gluing, or the individual laminations may be treated prior to gluing, depending on the wood species and treatment specified. Treatments such as creosote or pentachlorophenol (penta) in oil are typically only specified for treatment of the finished member. Penta in light hydrocarbon solvents may be specified for the laminations prior to gluing or for the finished member. Waterborne salt treatments may be specified for laminations prior to gluing. However, glulam manufactured using pretreated laminations is not available from all glulam manufacturers and for use with all species and availability should be verified prior to specifying.

The use of waterborne preservatives for the treatment of finished glulam members is not recommended. The only waterborne preservative listed in American Wood-Preservers' Association Standard C28 for use after gluing is ACZA, and it is limited to coastal region Douglas-fir, western hemlock, and hem-fir. It is impor-

TABLE 9.5 Treatment Type Characteristics

	Creosote	Pentachlorophenol in oils	Pentachlorophenol in light solvents	Waterborne preservatives
Suitable applications	Saltwater or fresh water applications, wood block floor, bridges, towers, ground contact	Fresh water, ground contact, aboveground uses including docks, bridges, towers, and beams		Saltwater or fresh water applications and ground contact. May be used indoors provided sawdust and construction debris are cleaned up and removed.
Appearance	Dark, oily, odor	Oily, may be blotchy, may have odor	Varies from natural appearance of wood to some darkening of wood.	Green to brown, depending on chemicals used and exposure to light.
Paintability	Not paintable	Not practical	Can be finished with water repellent or oil-based semitransparent stain	Can be stained or painted when surface is dry and prepared in accordance with coating manufacturer's recommendations.
Comments	Should not be used in residential interiors. May be used in industrial interiors when two coats of effective sealer are applied.	May be used in residential, industrial, or commercial interior when two coats of effective sealer are applied.		May develop greenish discoloration of finish. Stain-blocking primer will help to minimize discoloration.

tant to note that waterborne types of treatments may cause dimensional changes such as warping and twisting or lead to excessive checking, splitting, or raised grain due to the posttreatment drying process resulting in a finished product having an unacceptable appearance. They may also result in discoloration of the wood, possible raised grain, excessive checking, or warping of the member.

Glulams that are to be preservative treated must be bonded together with wet-use adhesives conforming to AITC/ANSI A190.1. Table 9.6 provides a summary of treatment recommendations.

Species. Softwood species listed in AWPA Standard C28 for preservative treatment include coastal Douglas-fir, western hemlock, hem-fir, and southern pine. Listed hardwood species include red oak, red maple, and yellow poplar when treated after gluing. The most commonly available West Coast species are Douglas-fir and hem-fir. Other species of glulams may also be available for pressure treatment, subject to agreement by the seller and purchaser and approval by the governing code body.

Incising is recommended for Douglas-fir, western hemlock, hem-fir, red maple, and yellow poplar. Such incising is normally performed after gluing of the finished glulam. If laminating lumber is to be incised prior to gluing, the mating faces to be glued should not be incised. Incising is not considered to have a detrimental effect on the strength of glulam. However, the effects of incising on appearance should be considered when ordering glulams where aesthetics are important. Lack of incising, if specified, may cause difficulties in meeting the specified treatment retention and penetration levels and should only be considered with caution.

Retention and Penetration Levels. Retention and penetration levels are specified in AWPA Standards in pounds of retained preservative per cubic foot of wood and depth of penetration in inches. Specified retention and penetration levels vary according to the type of preservative and the level of exposure. Table 9.7 lists standards referenced in AWPA Standard C28 for specified retention and penetration levels according to the intended end use.

Field Cuts. It is recommended that all fabrication, trimming, and boring of glulams be performed prior to pressure treating. If there is any field fabrication or surface damage to the glulams, all cuts, holes, or damaged areas must be field-treated to protect the exposed wood material. Copper naphthenate may be used to reseal exposed areas of glulams treated with creosote or pentachlorophenol. It may leave a greenish coloration. Field treatments should be applied to saturation by dipping, brushing, spraying, soaking, or coating in accordance with AWPA Standard M4.

Fasteners. Fasteners used to connect preservative-treated glulams should be corrosion resistant to withstand the effects of the high moisture environment to which these members are typically exposed.

Corrosion of fasteners is influenced by the amount of moisture present, temperature, wood pH, extractives, chemicals in the treatment, and environmental factors such as chlorine, salt, and pollutants. Oil-borne treatments are generally not corrosive whereas the waterborne arsenical treatments can be highly corrosive, depending upon environmental conditions. Hot-dipped galvanized connectors are typically adequate, but other materials, such as stainless steel or monel, may be required in certain applications.

TABLE 9.6 APA-EWS Recommended Treatments for Preservative Treatment of Glulam

Treatment type	Western species		Southern pine		Hardwoods	
	Glulam treated prior to laminating	Glulam treated after laminating	Glulam treated prior to laminating	Glulam treated after laminating	Glulam treated prior to laminating	Glulam treated after laminating
Creosote	No	Yes	No	Yes	No	Yes
Oil-borne penta	No[b]	Yes	No[b]	Yes	No	No
Copper naphthenate	No	Yes	No	Yes	No	No
Cu-8-Q[1]	No	Yes	Yes	Yes	No	No
CCA	No	No[c]	Yes	No[c]	No	No
ACZA	No	No[d]	Yes	No[c]	No	No
ACA	No	No[c]	Yes	No[c]	No	No
ACC	No	No[c]	Yes	No[c]	No	No

[a] For aboveground use only.
[b] Exception when penta in light hydrocarbon solvents is used.
[c] Not specifically provided in AWPA Standard C28. If specified for treatment after laminating, these treatments can result in discoloration, excessive checking, raised grain, twisting, and warping.
[d] Although not recommended, this treatment is permitted by AWPA Standard C28.

TABLE 9.7 Preservative Retention and Penetration Specifications

Use	AWPA standard
General	C28
Highway	C14
Farms	C16
Marine	C18
Commercial-residential construction	C15

Structural Properties. Most building codes generally recognize design values for glulams as specified in the latest edition of the National Design Specification (NDS). Although the NDS does not specify reductions in the dry design values for glulams preservatively treated according to AWPA Standards, it does specify that wet-use design values shall be used whenever the moisture content in service is 16% or more.

Use and Handling Precautions. The EPA requires registration of pesticides used in pressure treatments. They have approved use and handling precautions for treated wood as published in Consumer Information Sheets. These sheets also list recommended sealers for treated wood used in certain indoor applications. These sheets are available from treaters and should accompany each shipment of treated wood. They can also be obtained from the American Wood Preservers Institute or APA—The Engineered Wood Association. Use precautions are summarized in Table 9.4 and appropriate sealers are listed in Table 9.8. Also see handling tips in Section 9.2.3 under Handling Tips. See Consumer Information Sheets for complete information.

Finishing. Creosote or pentachlorophenol in oil is not paintable on a practical basis. Pentachlorophenol in light solvents can be finished with natural finishes such as a clear water repellent or an oil-based semitransparent stain. Clear film-forming finishes such as lacquers, varnishes, or urethanes are not recommended for glulams used outdoors because they have a short service life and require extensive surface preparation prior to refinishing.

Glulams that have been treated with waterborne preservatives such as CCA can be finished with clear water repellents, oil-based semitransparent stains, or film-forming finishes such as solid-color stains or paint systems. However, the treatment may leave a green or brown color on the glulam surface that can affect the color and appearance of the finish.

If an opaque coating is desired, the most durable finish is a top-quality paint system such as a stain-blocking acrylic latex primer followed by two all-acrylic

TABLE 9.8 EPA Recommended Sealers for Treated Glulams

Creosote	Urethane, epoxy, shellac. Coal tar pitch or coal tar pitch emulsions suitable for wood block flooring.
Pentachlorophenol	Urethane, shellac, latex epoxy enamel, varnish.

latex topcoats, preferably from the same manufacturer. A stain-blocking primer may also be required under light-colored acrylic latex solid-color stains to help minimize discoloration of the finish. Always follow the coating manufacturer's recommendations.

If treated wood is to be used indoors, follow the EPA recommendations for appropriate sealing of the wood.

9.3 FIRE-RETARDANT FINISHES AND TREATMENTS

Structural wood panels such as plywood, oriented strand board (OSB), and composite panels have a Class III (or C) flame spread rating and are permitted for use in most occupancies. In instances where building codes require additional protection from fire hazards, certain finishes and treatments are available to provide enhanced flame spread performance. Details on the basics of fire protection are provided in Chapter 10. This section of this chapter will focus on available finishing and treatment options, which include fire-retardant paints, fire-resistive overlays, and fire-retardant-treated plywood.

9.3.1 Fire-Retardant Paints

Fire-retardant (FR) paints can be used on plywood for nonstructural interior finish applications such as wall and ceiling paneling to reduce flame spread ratings to 25 or less (Class I or A) or from 26 to 75 (Class II or B), depending on the paint selected. They can also be used on other engineered wood products, such as glulam and I-joists for flame spread reduction. FR paints are tested per ASTM E-84[9] for 10 minutes, as compared to 30 minutes for FRT plywood.

There are two broad types of fire-retardant paints. One is a pigmented type designed as a masking finish similar to ordinary paint systems. Most of these paints typically intumesce or swell to a thick insulating layer when exposed to rapidly rising temperatures. These types of paints are designed for their fire-retardant properties and may not provide the full range of colors and decorative effects of conventional finishing systems. Generally, they are flat or nongloss types, although semigloss formulations are also available. Some are more durable under washing than others, and most are designed for interior use only. A few companies manufacture fire-retardant paint with proprietary topcoat finishes designed for exterior use.

The second type of fire-retardant paint is a clear finish, similar in appearance to a lacquer or clear varnish, that is used for high appearance paneling such as fine hardwoods. Clear fire-retardant finishes may develop a characteristic bloom or milkiness when subjected to high humidity or direct exposure to the weather during construction.

9.3.2 Fire-Resistive Overlaid Panels

These panels have proprietary overlays and are tested for fire performance equivalent to fire-retardant-treated wood as defined in model building codes. They have a maximum flame spread rating of 25 tested with the flame font 10½ ft from the center of the burner. They must develop no progressive combustion as determined in a 30-minute test according to ASTM E84.

9.3.3 Pressure Impregnated

In any projected use of fire-retardant treatments (FRT), thorough investigation should first determine that it is the best overall solution in view of long-term insurance costs and adequate fire protection at lowest construction cost. FRT wood products are more expensive than untreated wood products, which in most cases perform satisfactorily in regard to both life safety and protection of property.

FRT is not recommended for use with glulam (see Chapter 4), structural composite lumber (see Chapter 6) or prefabricated wood I-joists (see Chapter 5) due to unknown effects of the treatment on strength and glue bond durability. Such deleterious effects may compromise the integrity of structural members.

Fire-retardant treating is a secondary process to plywood manufacturing. FRT wood is pressure-impregnated with chemicals in water solution to inhibit combustion. This qualifies it for lower flame spread (at least as low as gypsum wallboard) and smoke generation ratings and reduces its fire hazard classification. When it is identified as such by a code-recognized testing agency label, it is rated on a parity with noncombustible constructions by many insurance rating bureaus.

Precisely defined, FRT plywood has been impregnated with fire-retardant chemicals in accordance with American Wood-Preservers' Association Standard AWPA C27.[10] When tested for 30 minutes under ASTM Standard E-84 (the tunnel test), it has a flame spread not over 25 and shows no evidence of significant progressive combustion. The use of fire-retardant-treated wood products is primarily dictated by building code and insurance agencies.

Function of FRT. The function of fire retardants is to reduce the flame spread of a material. Most commercial fire retardants involve salts that become acidic under elevated temperatures. Fire-retardant chemicals directly alter the thermal degradation of wood such that the thermal decomposition of the wood occurs at temperatures lower than ignition point. The chemicals increase the production of char and reduce the production of volatile combustion vapors.

Design recommendations for FRT wood products recognize that the treatments reduce the strength of wood products due to the treatment and redrying. Prior to 1985, design guides for lumber and plywood included stress adjustments based on past research. Subsequent field problems related to corrosion of fasteners and gross degradation of fire-retardant-treated structural wood panels led APA to remove sress adjustments in the *Plywood Design Specification*. A note was added to the Specification indicating that the provider of the treatment and redrying service should be contacted for information regarding the effect of the treatment. Currently, structural design information is made available only from companies that manufacture fire-retardant treatments.

When It Is Used. Fire-retardant-treated wood is used in applications where fire safety is a concern. The use of FRT plywood is affected by those model codes that are adopted for a geographical area of the United States. One of the most common applications for FRT plywood is for roof sheathing as an alternative to a parapet or noncombustible deck in townhouse roofs. This alternative was first accepted by code in 1979 when adopted by the Building Officials and Code Administrators

International (BOCA) for the National Building Code. Details on code provisions and standards are provided in Chapter 10.

Field History. Commercial fire-retardant treatment of wood dates back to 1895, when the U.S. Navy published specifications for the use of FRT wood in ship construction. In 1899, New York City passed a requirement that wood trim used above the 12th floor of high-rise buildings must be fire-retardant treated. In the 1920s the use of ammonium phosphates and borax became common for treating cellulosic materials such as wood.[11] Shortages of metal during World War II increased the demand for nonflammable building materials. During this period the demand for FRT wood soared.

Since World War II, the primary market for FRT wood has been in multifamily, commercial, and industrial buildings where the use is primarily driven by regulatory considerations. The demand for FRT plywood has generally depended on performance and the cost of the product compared to alternative materials.

Fire-retardant treating of wood products improves fire resistance. However, performance problems with FRT products were reported in the 1980s related to side effects of the chemicals such as fastener corrosion and strength reductions of wood. Recent developments in phosphate chemistry and flame-retardant research have led to new formulations.[12]

Corrosion of Fasteners. The salts used in commercial fire-retardant treatments have a natural affinity for water. This hygroscopic nature can lead to corrosion of metal fasteners when FRT wood is exposed to high relative humidities. In such cases, FRT wood can become saturated with water such that the treatment chemicals drip in a solution from the wood.[13]

When the U.S. Navy started using FRT wood in the late 1800s, they encountered problems due to the hygroscopic and corrosive nature of the treatments. The corrosive effect the chemicals had on metal fasteners was a factor in their discontinuation of FRT wood use in 1902.[13]

Hygroscopicity and corrosion problems from the use of FRT wood continued to be reported in the early 1980s. Concern over the hygroscopic and corrosive nature of these treatments led to development of standards for less hygroscopic treatments. The modification in AWPA standards to reflect new treatments was drafted in 1982. Three major treatment companies, Hoover, Koppers, and Osmose, all introduced Interior Type A (low-hygroscopic) treatments in 1981 or 1982.

Strength. In 1986, an increasing trend of strength-related complaints concerning interior-type FRT plywood was noticed by APA. While it had been recognized that overdrying after treating may lead to strength loss and brittleness, these new reported concerns generally arose after installation. The plywood was usually reported charred in appearance and brittle with low strength. In many instances, replacement of the FRT plywood was deemed necessary due to low strength and/or related sagging of the panels.

Generally, the performance problems that were reported were related to low hygroscopic Interior Type A formulations. Subsequent research by the treating companies led to reformulated treatments to address these problems. Treating companies now submit data on FRT products to building code agencies to obtain an evaluation report. Because particular formulations represent proprietary information of the treater not subject to public disclosure, specific information such as design recommendations concerning a treated panel product must be obtained from the treater.

Treating Standards. Fire-retardant treating practices are covered by standards written by the American Wood-Preservers' Association (AWPA). The AWPA is an association of treatment companies, chemical suppliers, wood industry representatives, government and regulatory agencies, and academia. Standards written by AWPA address acceptable treating practices and performance levels for fire-retardant-treated wood products. Standards developed by American Society of Testing and Materials (ASTM) are also applicable to aspects of FRT plywood. The standards written by the ASTM are consensus standards developed by committees made up of a balanced membership of producers, users and general interest members. The following AWPA and ASTM standards may be consulted for specific details on FRT treatment and analysis of fire-retardant-treated wood.

AWPA

- A2, *Standard Methods for Analysis of Waterborne Preservatives and Fire-Retardant Formulations*[14]
- A3, *Standard Methods for Determining Penetration of Preservatives and Fire Retardants*[15]
- A26, *Standard Method for Analysis of Fire Retardant Solutions and Wood by Titration*[16]
- C20, *Structural Lumber-Fire-Retardant Treatment by Pressure Process*[17]
- C27, *Plywood-Fire-Retardant Treatment by Pressure Process*[10]
- E6, *Standard Method for Determining the Equilibrium Moisture Content of Treated Wood*[18]
- P17, *Fire Retardant Formulations*[19]

ASTM

- D2898, *Test Methods for Accelerated Weathering of Fire-Retardant-Treated Wood for Fire Testing*[20]
- D3201, *Test Method for Hygroscopic Properties of Fire-Retardant Wood and Wood-Base Products*[21]
- D4502, *Test Method for Heat and Moisture Resistance of Wood-Adhesive Joints*[22]
- D5516, *Test Methods for Evaluating the Mechanical Properties of Fire-Retardant Treated Softwood Plywood Exposed to Elevated Temperatures*[23]
- D5664, *Test Method for Evaluating the Effects of Fire-Retardant Treatments and Elevated Temperatures on Strength Properties of Fire-Retardant Treated Lumber*[24]
- D6305, *Practice for Calculating Bending Strength Design Adjustment Factors for Fire-Retardant-Treated Plywood Wood Sheathing*[25]

9.4 FINISHING UNTREATED WOOD PRODUCTS

Finishing recommendations provided in this chapter are intended as general guidelines and may or may not be applicable for all wood-based products. For specific recommendations, check with the product manufacturer and paint suppliers or manufacturers. The finish is the final touch on any building, so it's important to do it right the first time.

For best performance and appearance, wood must be properly finished. The primary functions of a finish are to protect the wood from the weather and enhance its appearance. To select the best finish for a particular application, weigh the aesthetic considerations against durability and maintenance requirements.

Some finishes, such as semitransparent stains, accent the natural beauty of wood. These types of finishes, however, generally require frequent maintenance. Other finishes, such as acrylic latex house paints, are less natural in appearance but typically offer superior durability.

Another consideration is the type of product to be finished and its surface characteristics. For example, the use of semitransparent stains is not appropriate for sidings that contain numerous surface repairs.

9.4.1 Common Performance Problems

There are many performance problems that can develop during the service life of wood products exposed to outdoor weathering. This discussion does not attempt to address all of them but instead focuses on some of the more common problems. Additional discussion is also provided in the sections dealing with specific products.

Mildew Discoloration of Wood Siding. Mildew is a living organism, a fungus, which is normally observed as a black or dull grey deposit on exterior or interior surfaces. It can easily be mistaken for dirt. Mildew develops from microscopic spores, which are carried in the air to a surface where they feed and grow. Fungi are plants that contain no green chlorophyll and thus cannot make their own food. They must live on material that possesses organic matter. Mildew will grow on almost any surface where conditions are favorable. For example, mildew will grow on metal, feeding not on the metal itself, but on a thin film of organic material that has accumulated on the metal. In addition to a food source, warmth, moisture, and a shady location all encourage the growth of mildew. Mildew is a surface phenomenon that is a problem principally because of its appearance. It does not affect the strength or other important properties of wood. Naturally, light-colored finishes show mildew more readily than dark-colored ones. Mildew may grow on painted sidings, but proper construction practices, maintenance, and the use of finishes containing mildewcides will help minimize its growth and often eliminate it.

Black spotted areas on siding are quite frequently thought to be dirt, when in reality they are mildew. Identification is important because the treatment is different than for dirt. A test for mildew can be made by applying a drop of 5% sodium hypochlorite solution (common household bleaching solution) to the dark area. Mildew will usually bleach within one or two minutes. Dark areas that do not bleach are probably dirt. *Note: It is important to use fresh bleach solution because, after standing six months or longer, bleach deteriorates and may no longer be sufficiently potent.*

Mildew can be removed with commercial mildew removers, following manufacturer's directions, or with a solution of one part household bleach (5% sodium hypochlorite) mixed in three parts by volume of warm water. On some finishes a milky film may develop from the bleaching treatment. Thorough rinsing of the entire area with clean water will help alleviate this condition.

When using bleach, avoid breathing the vapors and contact with the skin and eyes. Wear rubber gloves to protect hands and use goggles to prevent eye damage. Shrubs and other plant life should also be protected. Children and pets should be kept away from these products.

Many mildew problems require only the bleach treatment mentioned above. However, if the finish has deteriorated to the point where a new finish is required, mildew must be killed and the surface cleaned before refinishing. Otherwise it may continue to grow up through a newly applied finish.

It is important to use finishes that contain ingredients (mildewcides) that inhibit the future growth of mildew on the finish. It is also important to note that mildewcides are generally not permanent and will gradually lose their effectiveness. Thus, if conditions favorable to the growth of mildew continue to exist, in time mildew may reappear on the new finish. If this occurs, remove the mildew following the previously described procedure.

For mildew-prone areas, some paint manufacturers provide "special mildewcides" that can be added to a finish for improved mildew protection of the finish. However, certain mildewcide additives can adversely affect performance of some paint formulations. Check with the paint manufacturer before adding anything to a paint. Mildewcide additives must be thoroughly dispersed in the finish in order to be effective. Therefore, if the paint manufacturer recommends a mildewcide additive, have it added by your paint supplier using his mixing equipment.

Staining of Finishes from Water-Soluble Wood Extractives. Many wood species, such as redwood, western red cedar, and Douglas-fir, contain natural chemical compounds known as extractives. These extractives have no significant effect on strength properties of the wood or on gluelines of plywood. Some extractives are colored in various shades of yellow, red, and brown. Others are colorless. Among other properties, extractives help instill the natural color and beauty of wood.

The concentration of extractives may vary greatly between species, between trees, and within a tree. For instance, the wood just inside the inner bark of a tree (sapwood) will normally contain few or no extractives, while the extractive content of the inner part of a tree (heartwood) can be very high. Since the face veneers may come from different areas of a tree, it is not unusual for plywood to have considerable variation in extractive content.

Extractives are soluble in various solvents, such as water, alcohol, and benzene. When a wood substrate containing water-soluble extractives is exposed to sufficient moisture, the extractives may dissolve and migrate to the surface of the wood. This phenomenon can lead to discoloration (i.e., extractive staining) of finishes, especially light-colored finishes.

The moisture required to dissolve extractives may develop from the application of water-thinned finishes in conjunction with other sources of moisture already present in the environment or the wood. In this case, the discoloration may develop during or soon after application of the finish.

Extractive staining may also develop months after the finish is applied. For example, if water-soluble extractives are present in sufficient quantity but not enough moisture was present at the time of finishing to dissolve them, the extractives may still be dissolved by subsequent moisture from heavy dew, rain, or condensation. The moisture can wet and penetrate permeable water-thinned or oil-based finishes. It then dissolves the extractives and allows them to leach out through the finish.

The intensity of extractive staining is frequently associated with the color of the wood. Darker-colored woods often stain more heavily than lighter-colored woods. However, its occurrence is difficult to predict. It may occur on one panel but not on another. Or it may even occur only on certain portions of a panel face.

A good test to determine if extractives can be removed from a finished surface is to scrub the finish with a mild detergent solution using a soft bristle brush, then rinse thoroughly with clean water. If this approach is effective, all discolored surfaces may be cleaned in a similar manner.

In some instances, the extractives will wash away during weathering and virtually disappear. This is normally the case if the staining is not heavy and it occurs during the beginning of the rainy season. When staining occurs at the end of the rainy season or during the summer, ultraviolet light and air may chemically alter the extractives so that they are no longer water-soluble. If this occurs, refinishing may be necessary.

Whether the wood surface is new or has been previously finished, liberal application of a paintable water repellent to all exposed edges and ends prior to priming and/or topcoating will help to minimize the ingress of moisture into the wood.

Extractive staining is most frequently associated with light-colored one-coat finishes. Discoloration of finishes from extractives is not as apparent when earth tones or dark colors are used. The extractives may still migrate to the surface of the finish, but they will not be as noticeable because they will blend in with the finish color. Finishing with these colors is generally the simplest solution to avoid discoloration from extractives.

The best method for retarding extractive staining is to use a primer formulated to prevent the extractives from reaching the surface of the finish coat. One way of accomplishing this is to use a primer that forms a physical barrier (film) that blocks the migration of the extractives. High-quality latex or oil-based primers may be so formulated. Another method is to use a stain-blocking latex primer that is formulated to react chemically with the extractives. This type of primer is often discolored by the water-soluble extractives. However, if the primer is effective, the extractives should not discolor the topcoat.

To test the effectiveness of a primer in preventing extractive staining, select a small area of the primed surface that is most heavily discolored by extractives. Make sure that the primer is dry. Apply the topcoat to this area and allow it to dry. If discoloration of the topcoat occurs, another prime coat may be required. If no discoloration of the test area is observed, proceed with application of the topcoat over the previously primed surface. As discussed earlier, extractive staining may occur soon after application of a finish, or months later. If discoloration from extractives occurs over a previously finished surface, the solutions are similar to the procedures outlined above.

If using a darker color or earth-tone is satisfactory, select a small test area that is the most heavily discolored and apply the finish. If the appearance is satisfactory, proceed with application of the new finish. If a lighter color is preferred, apply a stain-blocking primer to the test area prior to application of the topcoat.

For best overall protection, durability, and general finish performance on plywood, one to two coats of acrylic latex primer formulated to prevent extractives from reaching the topcoat are recommended (consult finish manufacturer's recommendations). The topcoat should be compatible with the primer, preferably an all-acrylic latex formulation from the same manufacturer.

Flaking or Peeling Finish. Most film-forming finishes eventually fail by flaking or peeling. This type of failure results from loss of finish adhesion to the wood surface and is manifested by actual detachment of film fragments. While such failures may eventually be expected, premature flaking or peeling indicates that other problems may exist. Early loss of adhesion can be due to poor surface preparation, improper finish application, moisture problems, low-quality finishes, or selection of the wrong finish for the surface to be painted. To correct the problem, the cause must be identified and eliminated.

Loose or poorly bonded finish must be removed prior to recoating or the failure will soon repeat. Flaking or peeling finishes can be removed in a number of ways such as scraping, sanding, brushing or pressure washing. Small areas such as trim are often cleaned of loose, deteriorated finish using a paint remover followed by thorough rinsing. A heat gun, in combination with scraping, can also be effective.

Smooth surfaces can often be prepared by scraping and then sanding to a feathered edge in areas where the finish is still well bonded. Textured surfaces may be cleaned with a garden hose followed by scrubbing with a stiff, nonmetallic bristle brush. More severe cases of flaking or peeling may require pressure washing. Because pressure washers can severely damage wood surfaces if improperly used, this procedure is best left in the hands of an experienced professional.

Once the surface is properly cleaned and prepared, it should be refinished with a top-quality coating according to the manufacturer's instructions. The most durable finishes for wood exposed outdoors are top-quality house paints such as an acrylic latex system composed of a stain-blocking acrylic latex primer and an all-acrylic latex topcoat. However, an oil-based primer may be required for wood products with high extractive content, such as cedar or redwood lumber. Two topcoats can significantly extend the life of the finish.

9.4.2 APA Rated Siding Products

APA Rated Siding is a family of wood-based products that includes a wide variety of sidings such as plywood, overlaid oriented strand board, and composite materials. Most APA plywood siding products are classified as APA 303 Siding. Although the surface characteristics of APA Rated Sidings may vary, APA 303 Sidings can be manufactured in 13 different face grades, as shown in Table 9.9. They are also available in a wide variety of surface textures and groove patterns.

Siding surface textures and patterns range from a rough, rustic look to one of clean, smooth elegance. Textures such as rough-sawn (Fig. 9.1), brushed, and

TABLE 9.9 APA 303 Siding Face Grades

Class	Face grade[a]	Patches	
		Wood	Synthetic
Special Series 303	303-OC (clear)	Not permitted	Not permitted
	303-OL (overlaid, e.g., MDO siding)	Not applicable for overlays	
	303-NR (natural rustic)	Not permitted	Not permitted
	303-SR (synthetic rustic)	Not permitted	Permitted as natural-defect shape only
303-6	303-6-W	Limit 6	Not permitted
	303-6-S	Not permitted	Limit 6
	303-6-S/W	Limit 6—any combination	
303-18	303-18-W	Limit 18	Not permitted
	303-18-S	Not permitted	Limit 18
	303-18-S/W	Limit 18—any combination	
303-30	303-30-W	Limit 30	Not permitted
	303-30-S	Not permitted	Limit 30
	303-30-S/W	Limit 30—any combination	

[a] W—wood patch, S—synthetic patch, S/W—both wood and synthetic patches, number indicates maximum number of patches allowed.

FIGURE 9.1 Rough-sawn. Manufactured with a slight rough-sawn texture running across panel. Available without grooves, or with grooves of various styles, and in lap sidings, as well as in panel form. Generally available in 11⁄32, 3⁄8, 15⁄32, 1⁄2, 19⁄32, and 5⁄8 in. thicknesses. Rough sawn also available in APA Texture 1-11®, reverse board-and-batten, channel groove, and V-groove. Available in Douglas-fir, cedar, southern pine, and other species.

smooth or texture-embossed overlay are available with or without grooves. Varying the surface texture and the depth, width, and interval of grooves makes a wide variety of surface patterns possible. Typical patterns are illustrated on the following pages (Figs. 9.1 to 9.8).

Plywood APA 303 Siding trademarks carry a face grade designation designed to help builders, contractors, architects, and other users and specifiers select the appropriate siding appearance for each project. The four basic APA 303 Siding appearance classifications are Special Series 303, 303-6, 303-18 and 303-30. Each class is further divided into face grades according to categories of repair and appearance characteristics, as shown in Table 9.9. Limitations on grade characteristics are based on 4 × 8 ft panel size. Limits on other sizes vary in proportion.

Lap siding may be manufactured with square or beveled edges, in widths up to 12 in., and in lengths to 16 ft. Because it is manufactured in long lengths, APA lap siding is simple and easy to install. APA lap siding possesses the clean, horizontal lines of traditional lap siding, yet offers excellent dimensional stability and split resistance to help it maintain a consistent, even appearance.

Finishing APA Rated Siding. Whether finishing a new structure or refinishing an older one, a quality stain or paint can keep the wood panel or lap siding looking good. And care in applying the proper finish can make a considerable difference in the length of its life. Sidings are commonly used for both interior and exterior applications. The recommendations provided here are specifically for exterior fin-

FIGURE 9.2 APA Texture 1-11. APA 303 Siding panel with shiplapped edges and parallel grooves ¼ in. deep, ⅜ in. wide; grooves 4 or 8 in. o.c. are standard. Other spacings sometimes available are 2, 6, and 12 in. o.c., check local availability. T1-11 is available in $^{19}/_{32}$ and ⅝ in. thicknesses. Also available with scratch-sanded, unsanded, overlaid, rough-sawn, brushed, and other surfaces. Available in Douglas-fir, cedar, southern pine, and other species.

ishes. When used inside, APA panel sidings can usually be finished with any type of finish that is suggested by the finish manufacturer for indoor use on paneling.

Wood-based siding used outdoors should always be finished to protect it from the weathering process and to help maintain its appearance. Weathering erodes and roughens unfinished wood. Face-checking (separations between fibers parallel to the grain of the face veneer) can be expected on non-overlaid sidings that are exposed to weathering, even when finished. These checks usually blend with the textured surface of veneered sidings. They do not affect the siding's integrity or the performance of a properly applied finish. Refinishing usually obscures the checks, but if a check-free surface is desired, use an overlaid siding. Overlaid siding is available with smooth or embossed surfaces (Fig. 9.5).

Finish Selection. Different finishes give varying degrees of protection, so the type of finish, its quality and quantity, and the application method must be considered in selecting and planning the finishing or refinishing job. Oil-based, semitransparent stains may be used on certain veneer-faced sidings. Solid-color stains may be used on most APA Rated Sidings and generally provide better protection. In general, however, APA siding products perform best with all-acrylic latex paint systems.

Sanded plywood is not an APA Rated Siding material and is therefore not recommended for siding applications. However, it is frequently used for soffits and other miscellaneous exterior uses. Sanded plywood for exterior use should be finished only with all-acrylic latex house paints, composed of a stain-blocking primer and companion topcoat.

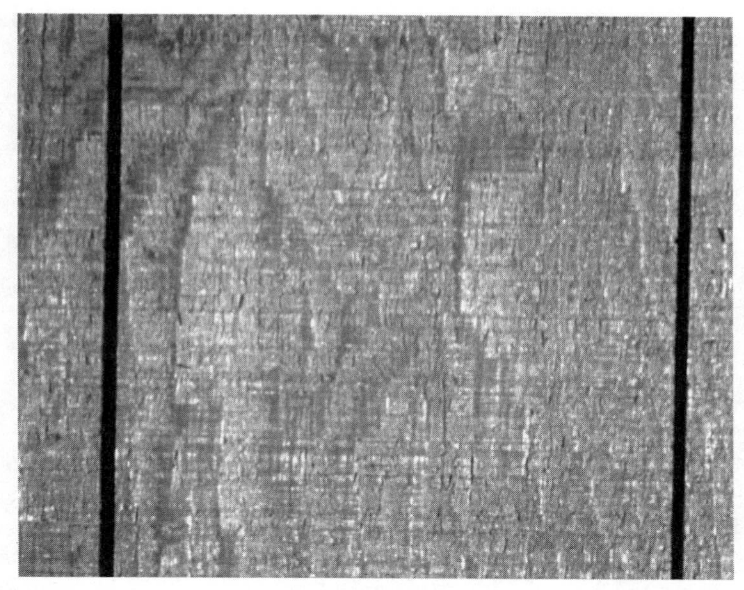

FIGURE 9.3 Kerfed Rough-sawn. Rough-sawn surface with narrow grooves providing a distinctive effect. Long edges shiplapped for continuous pattern. Grooves are typically 4 in. o.c. Also available with grooves in multiples of 2 in. o.c. Generally available in 11⁄32, 3⁄8, 15⁄32, 1⁄2, 19⁄32, and 5⁄8 in. thicknesses. Depth of kerf grooves varies with panel thickness.

Film-forming clear finishes, such as varnish, are not recommended on APA Rated Sidings for exterior use, since they tend to fail quickly by flaking and extensive surface preparation is required prior to refinishing.

Clear water-repellent treatments, including those with preservatives, are not considered satisfactory as the sole finish for APA Rated Sidings, since they do not protect the substrate from solar radiation. However, they are sometimes applied to bare wood as a treatment prior to painting and offer added protection from moisture. Allow the treatment to dry completely before applying the primer.

Many finishes contain additives to protect against mildew and other fungi. Some paint manufacturers provide mildewcide additives that can be added to a finish for additional protection in mildew-prone areas. However, certain mildewcide additives can adversely affect performance of some paint formulations. Therefore, check with the paint manufacturer before adding anything to a finish system. Mildewcides must be thoroughly dispersed in a finish to be effective and should be added by your paint supplier using his mixing equipment. This helps to ensure proper dispersion and maximum effectiveness of the mildewcide.

High-quality stains fit well with the architectural intent of textured sidings. They add color and beauty to the siding, showing off its rustic, rough texture. Stains are manufactured in two categories: semitransparent and solid-color.

Semitransparent stains are available in a variety of hues and provide color and some protection against weathering but still allow the wood grain and other natural characteristics to show. This type of stain will also show color differences in the wood itself, or between the wood and any repairs. The amount of pigmentation in semitransparent stains is closely associated with the amount of protection provided

FIGURE 9.4 Reverse Board-and-Batten. Deep, wide grooves cut into brushed, rough sawn, scratch-sanded or other textured surfaces. Grooves about ¼ in. deep, 1–1½ in. wide, spaced 8 or 12 in. o.c. with panel thickness of 19⁄32 and ⅝ in. Provides deep, sharp shadow lines. Long edges shiplapped for continuous pattern. Available in cedar, Douglas-fir, southern pine, and other species.

to the substrate. Only oil-based semitransparent stains are recommended. Properly applied, they will weather to a surface that is easily refinished without much preparation. Bleaching finishes (stains or oils) are not recommended for APA Rated Sidings. For a natural weathered appearance, use a gray oil-based semitransparent stain. Stains should be applied in accordance with the manufacturer's directions.

As shown in Table 9.10, semitransparent stains are recommended for plywood APA 303 Siding with face grades of 303-0C (clear, no patches), 303-NR (natural rustic, no patches but permits open knotholes), and 303-6-W (up to six wood patches per panel). Other APA Rated Siding face grades should not be finished with semitransparent stains unless specifically recommended by the siding manufacturer.

Brushed plywood, often referred to as abraded plywood, is textured at the mill by removing essentially all the soft grain from the surface (see Fig. 9.6). It should be finished according to the plywood manufacturer's recommendations, or with an oil-based semitransparent stain.

Solid-color stains are highly pigmented opaque stains that cover the wood's natural color but allow its texture to show. They also tend to obscure panel characteristics such as knots and repairs.

The natural water-soluble extractives in some woods (e.g., cedar, Douglas-fir) may bleed through lighter-colored latex stains and cause discoloration unless a stain-blocking primer is used. Darker-colored stains may generally be used without a primer. Follow manufacturer's recommendations.

FIGURE 9.5 APA Lap Siding. Rough-sawn, smooth, overlaid, or embossed surfaces, with square or beveled edges. Available in 3⁄8, 7⁄16, and 19⁄32 in. thicknesses.

FIGURE 9.6 Brushed. Brushed or relief-grain textures accent the natural grain pattern to create striking surfaces. Generally available in 11⁄32, 3⁄8, 15⁄32, 1⁄2, 19⁄32, and 5⁄8 in. thicknesses. Available in Douglas-fir, cedar, and other species.

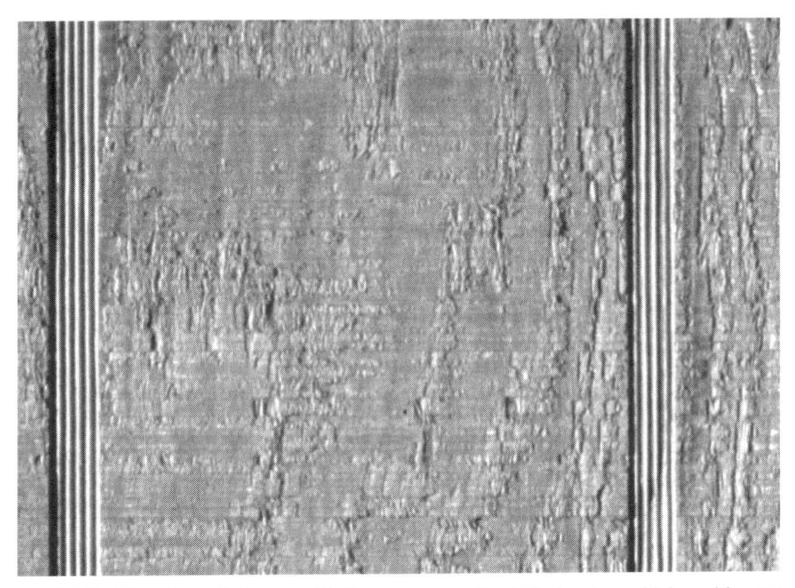

FIGURE 9.7 Channel Groove. Shallow grooves about $\frac{1}{16}$ in. deep, $\frac{3}{8}$ in. wide, cut into faces of panels, 4 or 8 in. o.c. Other groove spacings available. Shiplapped for continuous patterns. Generally available in surface patterns and textures similar to Texture 1-11 and in $\frac{11}{32}$, $\frac{3}{8}$, $\frac{15}{32}$, and $\frac{1}{2}$ in. thicknesses. Available in Douglas-fir, cedar, southern pine, and other species.

Solid-color stains, which are available in either latex- or oil-based formulations, may be used on all APA 303 Sidings except brushed plywood. Latex-based solid-color stains should be used on grades 303-NR or 303-SR. For best performance, use a stain-blocking, acrylic latex primer followed by a compatible topcoat of all-acrylic latex stain.

This is essential when light-colored latex finishes are used on wood containing water-soluble extractives. However, if a primer is not used, two topcoats of an all-acrylic latex stain will provide greater protection and depth of color and generally extend the life of the finish, as compared to only one coat. This is especially important on the south and west sides of buildings where the effects of the weather are the most severe. Always follow the finish manufacturer's recommendations.

Solid-color stains are suitable for most APA Rated Sidings, although some manufacturers recommend only acrylic latex formulations. Correctly applied, quality opaque stains are durable and provide good protection to the panel. Never use thick, low-gloss finishes such as typical shake and shingle stains or paints on APA Rated Sidings. They often form a weak, brittle film on the surface and lose their bond quickly, resulting in a surface that is very difficult to prepare for refinishing.

Top-quality acrylic latex house paint systems are recommended for all APA Rated Sidings, except brushed plywood. House paints require at least two coats, a primer and a topcoat. Primers are formulated specifically for controlled penetration, optimum bonding to the substrate, and minimizing extractive staining. Some house paint systems use an oil-alkyd primer followed by a latex topcoat. Other systems use up to two coats of a stain-blocking acrylic latex primer followed by an acrylic

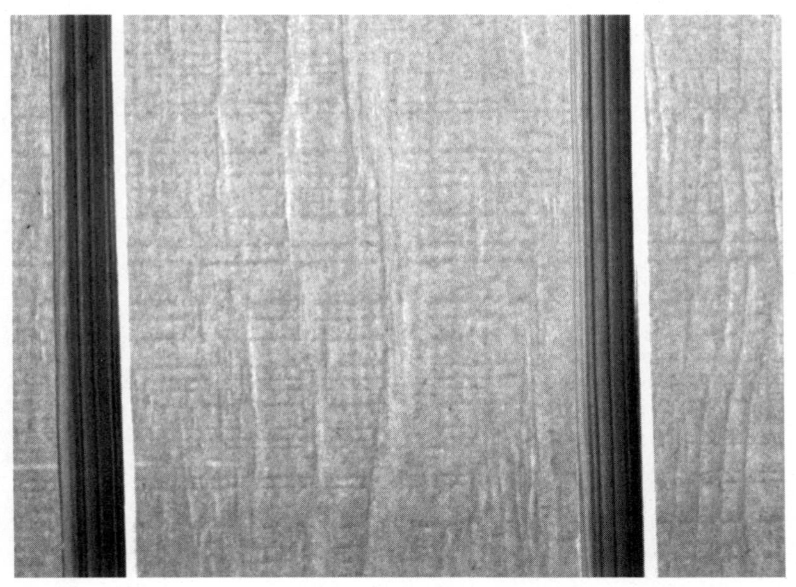

FIGURE 9.8 Overlaid Siding. Overlaid siding, such as Medium Density Overlay (MDO) and overlaid oriented strand board, is available without grooving; with V-grooves (spaced 6 or 8 in. o.c. usually standard); or in T1-11 or reverse board-and-batten grooving as illustrated above. Overlaid panel siding available in $^{11}/_{32}$, $^{3}/_{8}$, $^{7}/_{16}$, $^{15}/_{32}$, $^{1}/_{2}$, $^{19}/_{32}$, and $^{5}/_{8}$ in. thicknesses; also in lap siding. Overlaid siding is available factory-primed. Siding is overlaid on one side and available with texture-embossed or smooth surface.

latex topcoat. These latter systems are effective in reducing face-checking and generally offer superior performance. In any case, select companion products designed to be used together and preferably from the same manufacturer. Two topcoats will provide significant improvement in the life and performance of the finish.

House paint systems with oil or oil-alkyd topcoats are *not* recommended, as they tend to crack prematurely, permitting ingress of moisture and flaking of the finish.

Care and Preparation. Siding should be stored and handled with care to avoid exposure to water or weather before it is finished. Storage in a cool, dry place out of the sunlight and weather is best. If left outdoors, any straps on the siding bundles should be loosened or cut and the stacks covered. The covering should allow good air circulation between the siding and the cover itself to prevent condensation and mold growth.

The first coat of finish should be applied as soon as possible. Weathering of unprotected sidings can cause surface damage within four weeks and adversely affect finish performance.[26] For best performance, apply finishes within two weeks of siding installation.

Edge Sealing. Edge sealers retard the movement of moisture, making the siding edges less susceptible to sudden changes in weather. Some sidings are edge sealed by the manufacturer. Sidings that are not factory sealed should be sealed at the job site to minimize moisture fluctuations. Horizontal edges, especially lower drip

TABLE 9.10 Recommended Finishes

Finishes	Siding product					
	303 -OC, -OL, -NR, -SR	303 -6-W, -6-S, -6 S/W	303 -18-W, -18-S, -18 S/W	303 -30-W, -30-S, -30 S/W	COM-PLY	Overlaid
Semitransparent stain oil	-OC[b] -OL (not recommended) -NR[b] -SR[c]	-6-W[b] -6-S[c] -6 S/W[c]	[c]	[c]	Not recommended	Not recommended
Solid-color stain oil or latex[a]	-OC[b] -OL[d] -NR[e] -SR[e]	[b]	[b]	[b]	[d]	[d]
House paint acrylic latex Minimum 1 primer plus 1 topcoat	[b]	[b]	[b]	[b]	[b]	[b]

[a] Except for overlaid panels, use a stain-blocking primer with light-colored latex stains, since the wood extractives may cause a discoloration of the finish.

[b] Recommended with provisions given in text.

[c] These face grades should not be finished with semitransparent stains unless specifically recommended by the siding manufacturer.

[d] Some siding manufacturers recommend only acrylic latex formulations. Consult the manufacturer's recommendations.

[e] Only latex formulations are recommended.

edges, should be treated with special care because of the greater exposure to wetting from rain and water. Whether the siding is factory sealed or sealed at the job site, any sealed edges cut during construction should be resealed.

A liberal application of a good water-repellent preservative compatible with the final finish may be used for edge sealing if a stain is to be used on the siding's face. If the siding will be painted, use the same exterior house paint primer that will be used on the face. Edge sealing is easiest when the panels are in a stack.

Application Methods. The application method is as important as the finish material itself. It is poor economy to buy a quality stain or paint and then cause it to fail prematurely because of improper application.

Apply finishes only to clean surfaces under good weather conditions. Remove dirt and loose wood fibers with a stiff nonmetallic bristle brush. Final brushing should be along the grain to erase any marks. If the siding is very dry, application and performance of a latex finish is improved if the surface is dampened first. The surface should *not* be moistened when using oil-based products, however.

Use only top-quality tools and equipment in preparing and finishing all surfaces. Check with paint suppliers for recommendations on application equipment.

Do not paint or stain in the rain, when temperatures are low, or in direct sunlight when the panel is hot. Minimum temperatures are commonly 50°F (10°C) for latex and 40°F (5°C) for oil-based systems. Check with the coating manufacturer for specific application recommendations.

For best performance and a uniform appearance, be sure that the first coat of the finish is worked well into the siding surface. It's best to apply the finish with a brush (Fig. 9.9). If a spray is used, the finish should be applied liberally and then back-brushed or back-rolled while it is still wet. *Spraying alone does not work the finish into a textured wood surface.* Back-brushing or back-rolling also helps to even out spray patterns, giving a more uniform appearance. If spray is fogged onto the siding, too little is applied and it adheres only to the outer surface of loose dust and fibers. These quickly erode in the natural weathering process, causing premature failure of the finish. Long-napped paint rollers may also be used, but it is important that the proper amount of finish be applied and that the finish be worked well into the wood surface.

A second coat may be applied using any of these methods. However, back-brushing is not necessary.

Regardless of the method of application, follow the coverage rate recommended by the finish manufacturer. For textured plywood, up to twice as much finish is needed as for sanded plywood. Be sure to finish complete sections at one time in order to avoid lap marks.

Finishes can be factory-applied using machines that help to work them into the wood surface. The finish is applied under controlled conditions to ensure optimum spread rate and uniformity. This technique also allows the panels to be protected from the weather before delivery to the job site. Some APA Rated Siding products are available preprimed from the siding manufacturer.

Summary

- Protect the siding on the job site.
- Seal the edges of all panels, especially lower drip edges.
- Brush the surface with a stiff bristle brush and apply finish to clean surfaces in good weather. Surfaces must be dry for oil finishes but may be dampened for latex.

FIGURE 9.9 Brush application of finish.

- Apply the first coat of finish as soon as possible, preferably within two weeks of installation of the siding.
- Work the finish into the surface, especially on rough-sawn plywood.
- Select a high-quality stain or paint suitable for the siding grade.
- Use top-quality tools and equipment.
- Always follow the manufacturer's instructions.
- Do not apply finishes to hot or cold surfaces.

Testing of new finish formulations continues and recommendations are being revised as new information becomes available. Be sure to use the latest APA recommendations.

Refinishing and Maintenance. Any finish will begin to show its age after prolonged exposure to sun and weather. The weathering process can gradually erode the finish or cause it to become brittle and crack, lose its adhesion, fade, or mildew. Finishes vary with regard to their weathering characteristics. A high-quality finish, properly applied, will give the best protection and the longest life.

Where the siding is not protected by a finish, or the finish has weathered away, sunlight and moisture will erode and roughen the surface. For continuing satisfactory performance, it is important to maintain the finish. It is also important, however, not to refinish too often, especially when oil-based solid-color stains are used. If too many coats of an oil-based finish are applied, the finish becomes too brittle and will fail prematurely.

Finishes should be renewed when areas of unprotected bare wood first become visible or when obvious deterioration of the finish, such as peeling or flaking, is noticed. These types of finish failures usually occur first on the south and west sides of the building.

Thorough surface preparation is essential. Any finish applied over dirt, mildew, chalk, loose finish, or weathered wood that has not been cleaned of debris will not last long.

Mildew can develop on almost any type of finish, especially in warm, humid conditions or in areas with poor air circulation. It usually begins as dark spots that look like surface dirt. Unlike dirt, it disappears when household bleach is applied. In severe cases it may cause a uniform gray or black discoloration on large areas. Mildew must be killed and removed before refinishing or it may continue to grow through the newly applied finish. Follow the procedures discussed in Section 9.4.1 under Mildew Discoloration of Wood Siding.

Remove chalk and dirt by washing the siding surface with a household detergent solution. Rinse well to remove the detergent. A stiff nonmetallic brush can be used for rapid, thorough cleaning. Avoid wire brushes since broken-off metal fragments can cause staining. When loose wood fibers are present due to weathering, brush first across the grain to remove the fibers and then along the grain to remove any brush marks.

If the finish shows evidence of flaking or peeling, remove loosened particles with a stiff nonmetallic bristle brush. Wetting the surface prior to brushing sometimes helps. For more stubborn cases, a pressure washer can be used. This process should be left to a professional skilled in its use because pressure washers can damage wood surfaces and force water into the wall cavity around window casings and siding joints if improperly used. A water-rinsable type of paint remover can also be used to remove opaque finishes if deterioration is advanced.

When removing loose finish from overlaid sidings, be careful not to damage the overlaid surface. A stiff bristle brush or water-rinsable paint remover normally works best on overlaid sidings. After removing any loosened finish, be sure to wash the surface thoroughly prior to refinishing.

The suitability of a cleaned surface for refinishing is easily checked with adhesive tape. Apply the tape with finger pressure to a small area of cleaned and repainted surface. If the tape does not remove the finish when it is removed by a quick pull, the surface is satisfactory for repainting.

The New Finish. The clean siding is refinished just as if it were new (refer to the finishing section). However, one coat is satisfactory except where opaque finishes are applied over bare wood. Bare wood areas should be spot-primed. Also, when making color changes, two coats may be necessary to hide the old color. The selection of the new finish will depend upon the appearance desired as well as the siding's condition and the appearance of the old finish. Because weathered wood surfaces often require more finish than new wood, allow for more material when refinishing. Liberal application will give longer life and greater protection.

Refinishing helps to obscure face-checking that may occur on the veneer surfaces during weathering. The new finish should be thoroughly worked into the face-checks with a brush. If the finish is spray-applied, back-brush the finish into the face-checks. Face-checking is least noticeable when flat earth-tone colors are used.

Stains and paints differ in appearance and performance. Always select a top-quality product formulated for wood and apply it correctly to help assure lasting performance and beauty. And finally, selection of high-quality equipment will help ensure proper application of the finish. Check with the paint supplier for recommendations.

9.4.3 Sanded Plywood

Sanded plywood is a structural plywood panel with face and back plies that are sanded smooth in the manufacturing process. The panels have three or more cross-laminated layers of wood veneer, each layer consisting of one or more plies. The high grade of face and back veneers and the smooth sanded surface make sanded plywood the preferred panel for a variety of applications where appearance is important. And because sanded plywood is a structural panel, it also offers the advantages of high strength and stiffness, dimensional stability, impact resistance, fastener-holding ability, and workability.

Grade Designations. Most sanded plywood grades are identified by the veneer grade used on the face and back of the panel, such as, A-C, B-D. These veneer grades are defined in Voluntary Product Standard PS 1 for Construction and Industrial Plywood. Veneer grades define veneer quality according to natural unrepaired growth characteristics and allowable number and size of repairs permitted during manufacture (see Table 2.2). Veneer grades in descending order of quality are N, A, B, C-Plugged, C, and D. The minimum grade of veneer permitted in Exterior plywood is C. Use of D-grade veneer for sanded grades is limited to backs and inner plies of Exposure 1 or Interior panels.

Sanded plywood has B-grade or better veneer on one or both sides. Panels with B-grade or better veneer on both sides usually carry the APA trademark on the panel edge. Otherwise, the trademark is stamped on the back of the panel. Sanded plywood is produced in three basic exposure glue bond classifications: Exterior, Exposure 1, and Interior. See Chapter 2 for a discussion of glue bond classifications, species groups, and panel size and thickness.

Interior Finishing Recommendations. A wide variety of finishes is available for sanded plywood used in interior applications. The most common are described below. Always use finishes formulated for wood and follow the finish manufacturer's recommendations for best results. For interior (and exterior) applications, little or no sanding of the mill-sanded plywood surface is advised before application of the finish, to avoid uneven highlighting of hard and soft grain areas on the surface of the panels.

Various clear finishes and oils can be used on sanded plywood to provide the ever-popular real wood appearance. For the most natural effect, use two coats of a clear penetrating sealer. This type of finish resists soiling and allows easy cleaning. Some sealers can be tinted or used with light stains to add color and to produce a variety of attractive effects. Other clear finishes can also be used. Many finish manufacturers recommend that a sealer be used before applying a film-forming clear finish such as varnish.

Repairs and grain irregularities in sanded plywood can be pleasantly subdued by color toning. Tones of light gray, brown, or tan go well with wood colors and provide the best masking. Two-color toning techniques are recommended. The easiest method uses a heavy-bodied nonpenetrating sealer containing nonhiding pigments, and companion stains for color. Tint a small amount of the sealer with stains until the desired tone is obtained on a panel sample. Then mix the same proportions of stain and sealer in sufficient quantity for the entire job and apply by brush or spray. After drying and light sanding, apply a coat of clear finish to give the desired luster.

Where more control of the panel color differences is desired, begin by whitening the surface with pigmented resin sealer or diluted interior white undercoat. Wipe off before becoming tacky to display the grain desired. Then apply a clear resin sealer, allow to dry, and sand lightly. Next, apply a light stain, pigmented sealer, or tinted undercoat and wipe to the desired color depth. After drying and light sanding, apply a coat of satin varnish or brushing lacquer to provide luster and durability.

Semitransparent stains are highly recommended where both color and show-through of the grain and natural wood characteristics are desired. When light colors are used, only oil-based semitransparent stains are recommended. These help prevent discoloration of the finish caused by natural water-soluble compounds (called extractives) in the wood.

Solid-color stains and paints are opaque and mask repairs and wood grain patterns. Paints typically provide a smoother surface than solid-color stains. Paints are available in either oil-based or water-thinned (latex). Both normally require two coats, a primer or undercoat and a topcoat. The oil-based and darker colored latex solid-color stains often require only one coat. However, lighter-colored latex stains usually require a stain-resistant undercoat to prevent discoloration of the finish by extractives.

Paints are available in a full range of gloss levels, including flat, semigloss, and gloss. The flat finishes are generally more difficult to clean when soiled.

Exterior Finishing Recommendations. Sanded plywood is not recommended as an exterior siding on most buildings. However, it is frequently used for soffits and miscellaneous other exterior uses. For these applications, acrylic latex house paints are recommended.

House paints require at least two coats, a primer and a topcoat. Primers are formulated specifically for controlled penetration, optimum bonding to the substrate, and minimal extractive staining. Some acrylic latex systems use oil or oil-alkyd primer followed by the acrylic latex topcoat. Other systems use one or two coats of a stain-blocking acrylic latex primer and generally offer superior performance. In any case, select companion products that are designed to be used together, and preferably from the same manufacturer. Two topcoats will provide significant improvement in the life and performance of the finish.

All edges of plywood panels used for exterior applications should receive edge protection to minimize the effects of moisture absorption. Use the same exterior house paint primer for the edges that will be used on the face.

9.4.4 Overlaid Plywood

HDO plywood is manufactured with a thermosetting resin-impregnated fiber surface bonded to both sides under heat and pressure. It is the more rugged of the overlaid panels and ideal for such punishing applications as concrete forming and industrial tanks. HDO brings to the job all the proven advantages of plywood's large size, high strength, light weight, dimensional stability, and racking resistance. The tough resin overlay withstands severe exposure without further finishing. It also resists abrasion, moisture penetration, and deterioration from many common chemicals and solvents.

As required by Voluntary Product Standard PS 1, the minimum HDO overlay thickness before pressing is 0.012 in. The overlay weight is not less than 60 lb/ 1000 ft^2 of panel surface. HDO plywood is bonded with 100% waterproof glue and has inner ply construction of C- or C-Plugged grade veneer. Face veneers are B-grade or better.

HDO usually comes in a natural, semiopaque color. The overlay gives a soft wood-tone appearance to the panel surface. Other colors, such as black, brown, or olive drab, are also available.

MDO plywood is produced with a resin-treated fiber overlay with just the right tooth for rapid, even paint application. It is a preferred panel, therefore, for structural siding, exterior color accent panels, soffits, and other applications where long-lasting paint or coating performance is required.

Like HDO, Medium Density Overlay plywood is an Exterior-type panel manufactured only with 100% waterproof adhesive. Regular MDO is produced with B-grade face and back veneers and C-grade inner plies. Panels with B-grade veneers throughout or C-grade backs for siding can also be manufactured.

The MDO overlay surface may be specified on the face only or on both the face and back. The overlay is smooth and generally opaque, although it may show some evidence of the underlying wood grain. Siding panels with a texture-embossed surface and grooved panels with either smooth or textured overlays are also available. Most manufacturers produce MDO with a wood-tone surface color, although some supply their own identifying brand colors. Some also offer factory-primed and textured MDO, particularly for painted signs and residential siding applications.

Both HDO and MDO are easy to work using ordinary shop and carpentry tools. The overlays provide high resistance to edge splitting and slivering. They are tightly bonded and overlay separation is not a problem, even at high machine speeds. Both panels can be produced with nonskid surfaces. Both can be pressure-treated with preservatives. And both are produced in all standard sizes and thicknesses. Extra-long panels, including 9- and 10-ft siding panels, can be special ordered from some mills. While HDO is best suited for some applications and MDO for others, either panel may be used for a broad range of jobs. Table 9.11 lists some common uses for which either HDO or MDO offers high performance and low maintenance.

TABLE 9.11 Common Uses of APA Overlaid Panels

Painted signs
Concrete forming
Siding
Soffits and fascias
Cabinets and built-ins
Industrial tanks and vats
Counter tops
Truck and trailer linings
Highway signs
Agricultural bins

Finishing. MDO is an ideal base for paint and is designed to be exposed to the weather when finished. Although it performs perfectly well without further finishing in applications where it is not exposed to the weather, MDO should always be face-primed and top-coated with a compatible solid-color stain or house paint if used outdoors or subjected to wet, humid conditions. If solid-color stain is desired, some panel manufacturers recommend only acrylic-latex formulations. Check the panel manufacturer's recommendations. Some producers of MDO offer panels with a preprimed surface. HDO is designed to be used without further finishing, although it too is an excellent base for conventional paints after a light surface roughening.

Like any finish material, HDO and MDO should be stored in a cool, dry place out of the sun and protected from heaters or highly humid conditions, which frequently exist at construction sites. Be sure panels are dry when finish is applied and that the specific application recommendations of the paint manufacturer are followed.

Panels intended for exterior exposure should be edge-sealed as soon as possible. Edge sealing is not permanent, nor does it necessarily make the edges moisture proof. It does, however, minimize sudden changes in moisture content due to weather cycles. Panel edges may be sealed with one or two heavy coats of top-quality exterior house paint primer formulated for wood. Edges are most easily sealed while panels are in a stack.

To ensure a good paint or reflective sheeting bond, HDO is prepared by one of the following simple surface conditioning treatments. One method is scuff-sanding with fine grit sandpaper, which slightly roughens the surface and provides better tooth for the paint. Scuff-sanding also helps remove any surface contaminants. Panel surfaces should then be wiped clean to remove all dust.

The surface of HDO can also be conditioned for painting by thoroughly scrubbing with a nylon abrasive pad saturated in VM&P naptha or similar solvent. The liquid solvent should then be wiped off with a dry cloth to remove completely any surface contaminants. Panels should be exposed to good air circulation at least overnight to ensure complete evaporation of all solvent from the overlay. If stacked, panels should be separated with stickers. The time required to permit complete evaporation will depend upon the temperature and air movement through the stack.

Only paint products formulated for wood should be used to finish overlaid plywood. Primer and finish materials produced by the same manufacturer and formulated as companion products should be specified to ensure good adhesion between successive paint coats. Allow each coat to dry before applying the next, but complete as soon as practical to obtain good adhesion between coats. Follow the manufacturer's instructions carefully for best results. Conventional, high-quality exterior house paints as well as sign and bulletin paints perform well on both HDO and MDO. Best finish durability can be expected when using a top-quality acrylic latex house paint system composed of primer and topcoat. Hard, brittle finishes and clear finishes should be avoided. Both air drying and baking finish systems may be used.

Oil-based finishes should be allowed to erode before repainting to avoid a thick paint buildup. Overly thick oil-based films tend to become brittle and fail within themselves.

9.5 REFERENCES

1. Form Y510, *Plywood Design Specification,* APA—The Engineered Wood Association, Tacoma, WA.

2. Form N375, *Design Capacities of APA Performance Rated Structural-Use Panels,* APA—The engineered Wood Association, Tacoma, WA.
3. ANSI A190.1, *American National Standard for Wood Products—Structural Glued Laminated Timber,* American National Standards Institute, Inc., NY.
4. AWPA C28, *Standard for Preservative Treatment of Structural Glued Laminated Members and Laminations before Gluing of Southern Pine, Pacific Coast Douglas-Fir, Hemfir and Western Hemlock by Pressure Processes,* American Wood-Preservers' Association, Granbury, TX.
5. AWPA M4, *Standard for the Care of Preservative-Treated Wood Products,* American Wood-Preservers' Association, Granbury, TX.
6. *National Design Specification for Wood Construction,* American Forest and Paper Association, Washington, DC.
7. Federal Specification TT-W-571, *Wood Preservation Treating Practices,* U.S. Federal Supply Service (USFSS), Washington, DC.
8. American Association of State Highway and Transportation Officials, *Standard Specifications for Transportation Materials and Methods of Sampling and Testing,* Washington, DC.
9. ASTM E84, *Test Method for Surface Burning Characteristics of Building Materials,* American Society for Testing and Materials, Philadelphia, PA.
10. AWPA C27, *Plywood-Fire-Retardant Treatment by Pressure Process,* American Wood Preservers' Association, Granbury, TX.
11. Levan, S. L., "Chemistry of Fire Retardancy," in *The Chemistry of Solid Wood,* R. M. Rowell, ed., American Chemical Society, Washington, DC, 1984.
12. LeVan, S. L., and J. E. Winandy, "Effects of Fire-Retardant Treatments on Wood Strength: A Review," *Wood and Fiber Science,* vol. 21, no. 1, pp. 113–131, 1990.
13. Mader, H. J., "Bridging the Generation Gap in Fire-Retardant-Treated Wood," *Construction Specifier,* February 1984.
14. AWPA A2, *Standard Methods for Analysis of Waterborne Preservatives and Fire-Retardant Formulations,* American Wood-Preservers' Association, Granbury, TX.
15. AWPA A3, *Standard Methods for Determining Penetration of Preservatives and Fire Retardants,* AWPA A3, American Wood-Preservers' Association, Granbury, TX.
16. AWPA A26, *Standard Method for Analysis of Fire Retardant Solutions and Wood by Titration,* American Wood-Preservers' Association, Granbury, TX.
17. AWPA C20, *Structural Lumber-Fire-Treatment by Pressure Process,* American Wood-Preservers' Association, Granbury, TX.
18. AWPA E6, *Standard Method for Determining the Equilibrium Moisture Content of Treated Wood,* American Wood-Preservers' Association, Granbury, TX.
19. AWPA P17, *Fire Retardant Formulations,* American Wood-Preservers' Association, Granbury, TX.
20. ASTM D2898, *Test Methods for Accelerated Weathering of Fire-Retardant-Treated Wood for Fire Testing,* American Society for Testing and Materials, Philadelphia, PA.
21. ASTM D3201, *Test Method for Hygroscopic Properties of Fire-Retardant-Treated Wood and Wood-Base Products,* American Society for Testing and Materials, Philadelphia, PA.
22. ASTM D4502, *Test Method for Heat and Moisture Resistance of Wood-Adhesive Joints,* American Society for Testing and Materials, Philadelphia, PA.
23. ASTM D5516, *Test Methods for Evaluating the Effects of Mechanical Properties of Fire-Retardant Treated Softwood Exposed to Elevated Temperature,* American Society for Testing and Materials, Philadelphia, PA.
24. ASTM D5664, *Test Method for Evaluating the Effects of Fire -Retardant Treatements and Elevated Temperatures on Strength Properties of Fire-Retardant Treated Lumber,* American Society for Testing and Materials, Philadelphia, PA.
25. ASTM D6305, *Practice for Calculating Bending Strength Design Adjustment Factors for Fire—Retardant-Treated Plywood Wood Sheathing,* American Society for Testing and Materials, Philadelphia, PA.
26. Williams, R. S., P. L. Plantinga, and W. C. Feist, "Photodegredation of Wood Affects Paint Adhesion," *Forest Products Journal,* vol. 40, no. 1, pp. 45–49, 1990.

CHAPTER TEN
FIRE- AND NOISE-RATED SYSTEMS

John D. Rose
Retired Senior Engineer, TSD

10.1 INTRODUCTION TO FIRE-RATED SYSTEMS

When designing or building for fire protection, it is important to recognize that fireproof buildings do not exist. Building contents are a critical factor, and almost all contents can burn. Smoke and heat thus generated can cause extensive damage and loss of life long before the building itself begins to burn, regardless of the type of construction.

After studying residential fires involving combustible contents, the USDA Forest Products Laboratory (FPL) concluded that "wall and ceiling materials, whether combustible or noncombustible, had little or no effect on the time or temperature of the critical point"—the point at which human life is untenable. In the FPL studies, the critical point was reached in four to seven minutes. Other tests have shown that untenable conditions can occur in as little as two minutes.

So-called fireproof building materials do not guarantee safety for occupants or property. A classic demonstration of this was the 1953 fire in a General Motors manufacturing plant in Livonia, Michigan. The plant was considered completely noncombustible, yet was a complete loss due to the collapse of unprotected metal construction.

Another example was the 1967 disaster at McCormick Place, Chicago's exhibition hall. All of its structural members, including interior nonbearing walls, were noncombustible. Yet a fire that began in the contents spread with such heat that the entire ceiling fell as steel beams, girders and trusses buckled and collapsed.

The type of construction is, of course, important. To protect the occupants—always the first concern—as well as to safeguard property, a prompt detection and alarm system and the accessibility of numerous exits are vital. Also of importance are the type of contents and furnishings, interior finishes, degree of sprinkler protection, and availability of adequate fire-fighting equipment.

With proper construction in conformance with model building code regulations and with recognition of the above factors, fire-safe buildings can be designed with combustible or noncombustible materials. This puts wood framed systems in proper perspective. Fire- and noise-rated wood floor and roof ceiling systems and wall systems are being increasingly used for multifamily residential and nonresidential building construction, for low-rise (three stories and less) and medium-rise (four to six stories) buildings.

10.1.1 Basics of Fire Protection

In order to evaluate fire safety of a structure, building authorities consider many factors, including flame spread and fire-resistance ratings.

Flame Spread and Smoke Indexes. Flame spread relates to potential for spread of fire along the surfaces of the wall and ceiling within a room. It is measured by the flame travel along the surface of materials used for interior finish, such as walls, ceilings, partitions, paint, and wallpaper. Not considered in codes are such nonstructural materials as drapes and furnishings, though these may often be primary fuel sources. Flame spread is a property of the surface material, not the structure, when fire has started.

The recognized flame spread test is the tunnel test, American Society for Testing and Materials (ASTM) Test Method E84.[1] A test sample of material, 20 in. wide and 25 feet long, is installed as ceiling of a test chamber and exposed to a gas flame at one end. The distance of flame spread along the surface of the test sample is measured during a 10 minute test duration. Flame spread index is calculated as the area under a flame spread distance—time curve, divided by comparable areas for standard materials (inorganic reinforced cement board and red oak), and multiplied by 100. The flame spread index is 0 for inorganic reinforced cement board and 100 for red oak.

Another property measured in the ASTM E84 test is the opacity of the smoke generated by the burning material during a 10 minute test exposure. A photelectric cell, installed in the test chamber exhaust vent pipe, measures the light absorption (opacity) due to smoke generated during the test, which is compared to the amount of smoke generated by standard materials (inorganic reinforced cement board and red oak). Smoke developed index is calculated as the area under the light absorption-time curve, divided by the comparable area for standard materials, and multiplied by 100. The smoke developed index is 0 for inorganic reinforced cement board and 100 for red oak.

Materials with the lowest flame spread index (0–25) are classified as Class A (or I). Such materials are permitted for areas where fire hazard is most severe, for example vertical exit ways of unsprinklered buildings for public assembly.

Materials with a flame spread index from 26–75 are Class B (or II) are permitted in areas of intermediate severity, such as corridors providing exit way access in business and industrial buildings. For exit ways and for most interiors where Class A or Class B flame spread performance is required, fire-retardant-treated plywood (which falls in Class A) is permitted.

Materials with a flame spread index from 76–200 are Class C (or III). Wood structural panels such as plywood, oriented strand board (OSB), and composite panels (veneer faces with structural wood core) generally fall in this class and are permitted in rooms of most occupancies. (Exceptions: hospitals, or institutions where occupants are restrained.)

Table 10.1 shows flame spread and smoke developed index values of some commonly used construction materials. Table 10.2 shows typical flame spread requirements, as specified in the International Building Code (IBC). The Flame spread index for untreated wood structural panels falls within Class C, but ratings vary,

TABLE 10.1 Interior Finish Classifications

Interior finish or flame spread classification	Flame spread index	Smoke developed index
Class A (or I)	0–25	
Class B (or II)	26–75	450 max.
Class C (or III)	76–200	
Material	Flame spread index	Smoke developed index
Inorganic reinforced cement board	0	0
Fire-retardant-treated construction plywood	0–25	0–80
Fire-retardant-coated construction plywood	0–45	0–200
Fire-retardant-treated lumber	0–25	10–360
Red oak lumber	100	100
Wood structural panels	76–200	25–270

TABLE 10.2 Typical Flame Spread Classification Requirements for Interior Finish Based on the 2000 International Building Code

	Unsprinklered		
Group	Vertical exits and exit passageways[a,b]	Exit access corridors and other exitways	Rooms and enclosed spaces[c]
A-1 & A-2	A	A[d]	B[e]
A-3,[f] A-4, A-5	A	A[d]	C
B, E, M, R-1, R-4	A	B	C
F	B	C	C
H	A	A	B
I-1	A	B	B
I-2	A	A	B
I-3	A	A	B
I-4	A	A	B
R-2	B	B	C
R-3	C	C	C
S	B	B	C
U		No restrictions	

[a]Class C interior finish materials shall be permitted for wainscoting or paneling of not more than 1000 ft^2 of applied surface area in the grade lobby where applied directly to a noncombustible base or over furring strips applied to a noncombustible base and fireblocked.

[b]In vertical exits of buildings less than three stories in height of other than Group I-3, Class B interior finish for unsprinklered buildings shall be permitted.

[c]Requirements for rooms and enclosed spaces shall be based upon spaces enclosed by partitions. Where a fire-resistance rating is required for structural elements, the enclosing partitions shall extend from the floor to the ceiling. Partitions that do not comply with this shall be considered enclosing spaces and the rooms or spaces on both sides shall be considered one. In determining the applicable requirements for rooms and enclosed spaces, the specific occupancy thereof shall be the governing factor regardless of the group classification of the building or structure.

[d]Lobby areas in A-1, A-2, and A-3 occupancies shall not be less than Class B materials.

[e]Class C interior finish materials shall be permitted in places of assembly with an occupant load of 300 persons or less.

[f]For churches and places of worship, wood used for ornamental purposes, trusses, paneling, or chancel furnishing shall be permitted.

depending on species, thickness, and glue type. In general, for plywood, panels with exterior adhesives perform better than those with interior adhesives; thick panels better than thin; and low density species better than heavier species.[2,3]

Fire-Resistance Ratings. Though codes are concerned with how fast fire can spread on a room's surface, they are even more specific about fire resistance: the measure of containment of fire within a room or building. It is defined as protection against fire penetrating a wall, floor, or roof, either directly or through a high rate of heat transfer that might cause combustible materials to be ignited on the side of the wall or floor away from the actual fire. Thus, it is a property of an assembly of several materials, including fastenings, and of the workmanship.

Fire-resistive construction provides time to discover a fire, restrict or suppress it before it spreads, and evacuate the building if necessary.

The standard test for measuring fire resistance is ASTM E119.[4] Ratings of assemblies are determined by test procedures somewhat simulating actual fire conditions. Floor-ceilings and roof-ceilings are tested flat while loaded to their full allowable stress. Walls are tested vertically, either as bearing walls under full or limited axial load or as nonbearing walls under no load. The resistance rating is expressed in hours or minutes before some limiting condition is reached (flame passage or heat transmission on the unexposed surface, or structural collapse). It approximates the time an assembly would be expected to withstand actual fire conditions.

A one-hour rating, for example, is taken to mean that an assembly similar to that tested will not collapse, nor transmit flame or a high temperature, while supporting its full load, for at least one hour after the fire commences.

Thermal Barrier Index. When foam plastic insulation is used in building construction, an approved thermal barrier material is required to serve as a protective membrane to separate the interior of the building from the insulation, which is often highly flammable when exposed to fire conditions. The fire performance of the thermal barrier is evaluated by a special standard fire testing method (ICBO),[5] with fire exposure for 15 minutes on the bottom surface of a 3 × 3 ft horizontal test specimen of the thermal barrier material. The thermal barrier material is backed with noncombustible ½ in. thick calcium-silicate board for the test, to standardize testing conditions. Temperature rise on the unexposed surface is limited to an average of 250°F. A product meeting these requirements is defined as a thermal barrier membrane and is classified as having a thermal barrier index of 15, which is the minimum specified in the codes for applications where foam plastic insulation is used. Thermal barrier membranes are not required by codes for certain applications of foam plastic insulation, such as:

- When used as roof insulation that is separated from the interior of the building by minimum nominal 7⁄16 in. wood structural panels
- When used as a substrate for fire-classified roofing
- For insulation in attics or underfloor crawl spaces with limited access, where insulation is protected against ignition by minimum nominal ¼ in. wood structural panels or other prescriptive materials specified in the codes
- For insulation in cooler and freezer walls when the foam plastic insulation meets specified requirements

As for thermal resistance in fires: because of its superior insulating qualities, wood structural panels may be expected to develop a finish resistance (based on time to develop an average temperature rise of 250° on the back of the panel) of approximately 20 or more minutes per inch of thickness when subjected to heat and flame based on the ASTM E119 time-temperature curve. Pressure treatment with fire-retardant chemicals does not materially affect the finish resistance, though coating with fire retardant paints may be somewhat more effective.

Fire-Retardant Treated Wood and Fire-Retardant Coatings. Fire-retardant-treated (FRT) wood or plywood is pressure-impregnated with fire-retardant chemicals in water solution in accordance with American Wood-Preservers' Association Standards AWPA C27 (plywood) or C20 (lumber), to inhibit combustion and retard flame spread under fire exposure conditions. However, no treatment processes or standards have been developed for fire-retardant treatment of other wood structural panels (oriented strand board, or com-ply) or engineered wood composite framing members such as structural glued laminated timber (glulam), I-joists, laminated veneer lumber, parallel strand lumber, or oriented strand lumber.

When tested for 30 minutes under ASTM Standard E84, FRT wood and plywood have a flame spread index of 25 or less and show no evidence of significant progressive combustion. Also, there is a maximum limit on the flame travel during the test. FRT wood reduces its fire hazard classification and qualifies it for lower flame spread (at least as low as gypsum wallboard) and smoke index ratings.

Fire-retardant (FR) coatings can be used on wood and wood structural panels for nonstructural interior finish applications such as wall and ceiling paneling to reduce flame spread ratings to 25 or less (Class A) or from 26–75 (Class B), depending on the coating selected. FR coatings are tested per ASTME 84 for 10 minutes, as compared to 30 minutes for FRT wood. FR coatings can be factory- or field-applied as interior finish coats over new or existing wood surfaces; some FR coatings are available with proprietary topcoat finishes for exterior use. FR coatings are available as opaque or clear finishes.

10.1.2 Model Building Code Provisions (2000 International Building Code)

In the past, four model building codes have been used in the United States. These are the Standard Building Code (primarily used in the South); Uniform Building Code (primarily used in the Midwest and West); National Building Code (widely used in the Northeast); and the One- and Two-Family Dwelling Code. Most of the regional and state codes in the country are similar to or adaptations of these codes. Building code provisions have the authority of law (unlike insurance requirements, which are optional).

Beginning in the year 2000, two new national codes were promulgated to replace the model codes. These were the International Building Code (IBC)[6] and the International Residential Code for One and Two Family Dwellings (IRC).[7] Adoption of these codes by local and state jurisdictions will occur over a period of years with the model codes remaining in effect in the interim.

In addition, the National Fire Protection Association (NFPA) will be publishing an alternative national building code in 2002. Therefore, designers are cautioned to check their local area for the applicable code.

Types of Construction. The International Building Code standardized the types of construction prescribed in the previous model building codes, as shown in Table 10.3. Construction-type classifications are based on fire-resistance ratings of structural elements. Of the three types of wood construction, Type IV (Heavy Timber) construction is permitted for multistory buildings (up to four stories, or five stories

TABLE 10.3 Typical Types of Construction Permitting Wood Systems Based on 2000 International Building Code

Non-wood systems	
Types I and II construction	Noncombustible structural building elements. Includes subtypes A and B with specific fire-resistive ratings required for each building element. These construction types permit untreated or fire-retardant-treated wood for certain building elements such as partitions, roof framing, decking, etc. Heavy timber construction permitted for roof framing and decking, where one-hour or less fire-resistance rating is required. See code for specifics.
Wood systems	
Type III construction	Noncombustible exterior walls, interior building elements of light framing with protected or unprotected wood members. Includes sub-types A and B with specific fire-resistive ratings required for each building element.
Type IV contruction (heavy timber)	Noncombustible exterior walls, interior building elements of heavy timber wood members without concealed spaces.
Type V Construction	Structural elements, exterior and interior walls of light framing with protected or unprotected wood members throughout. Includes subtypes A and B with specific fire-resistive ratings required for each building element.

for certain occupancies) such as educational, religious, manufacturing, warehouse, supermarket; and permits the largest areas. The next largest areas are permitted for Type III construction, commonly used for commercial or public buildings up to three or four stories high, or five stories for some occupancies. Finally, Type V construction is used in 80% of all residential and many commercial, institutional, industrial and assembly buildings.

If the building requires a larger area than is permitted for the type of construction selected, the designer has several choices, including breaking up the area with fire walls, adding sprinklers, increasing property line setbacks, and specifying a more fire-resistant construction. (See Building Area Increases.)

In most cases, conventional wood-frame construction with wood structural panel sheathing and regular gypsum wallboard interior finish provides ample fire safety and is completely acceptable for one- and two-family residential applications.

Certain building applications, such as multifamily residential construction, and nonresidential construction, require additional protection. In these cases, the designer's options include protected construction or Heavy Timber construction. For certain applications, fire-retardant-treated wood is permitted for construction.

Protected Construction. Protected construction consists of conventional wood-framed assemblies, such as floor-ceiling or wall, with a fire-resistive material added to give primary protection to the wood framing. The material may be fire-resistive gypsum wallboard, plaster, or acoustical tile. The fire-resistive material, in conjunction with wood structural panel sheathing, prevents flame passage and temperature rise while reinforcing framing against collapse under load. Tables 10.4A and 10.4B are examples of typical fire-resistive requirements in model building codes. Fire-rated floor-ceiling and wall assemblies have been developed for one- and two-hour ratings using wood systems, for building applications in the United States. In Canada, fire-rated assemblies for 45-minute and 1½-hour ratings are permitted by the National Building Code of Canada.

Heavy Timber Construction. Heavy Timber construction provides fire protection through use of noncombustible exterior walls in conjunction with interior structural elements of large, solid wood members, including solid lumber girders, columns, and floor and roof decking, glued laminated wood, and engineered wood framing, installed without concealed spaces. See Table 10.5 for code definitions of minimum sizes of members for Heavy Timber construction.

The requirements for Heavy Timber construction in model building codes do not constitute one-hour fire resistance. The terminology is descriptive of early eastern U.S. textile mills, where it was known as mill construction, plank-on-timber, or slow-burning. Although outside surfaces of wood members may char during exposure to fire, the surface char layer acts as insulation. The strength and size of

TABLE 10.4A Fire-Resistance Rating Requirements for Building Elements (Hours) Based on 2000 International Building Code

Building element	Type III		Type IV	Type V	
	A[c]	B	HT	A[c]	B
Structural frame[a]					
Including columns, girders, trusses	1	0	HT	1	0
Bearing walls					
Exterior[e]	2	2	2	1	0
Interior	1	0	1/HT	1	0
Nonbearing walls and partitions					
Exterior			See Table 602		
Interior[d]			See Section 602		
Floor construction	1	0	HT	1	0
Including supporting beams and joists					
Roof construction					
Including supporting beams and joists	1[c]	0	HT	1[b]	0

[a] The structural frame shall be considered to be the columns and the girders, beams, trusses, and spandrels having direct connections to the columns and bracing members designed to carry gravity loads. The members of floor or roof panels that have no connection to the columns shall be considered secondary members and not a part of the structural frame.

[b] 1. Except in factory-industrial (F-I), hazardous (H), mercantile (M) and moderate hazard storage (S-1) occupancies, fire protection of structural members is not required, including protection of roof framing and decking where every part of the roof construction is 20 ft or more above any floor immediately below. Fire-retardant-treated wood members are allowed to be used for such unprotected members.

2. In all occupancies, heavy timber is allowed where a one-hour or less fire-resistance rating is required.

[c] An approved automatic sprinkler system shall be allowed to be substituted for one-hour fire-resistance-rated construction, provided such system is not otherwise required by other provisions of the code or used for an allowable area increase or an allowable height increase. The one-hour substitution for the fire resistance of exterior walls is not permitted.

[d] For interior nonbearing partitions in Type IV construction.

[e] Not less than the fire-resistance rating based on fire separation distance (see Table 10.4B).

TABLE 10.4B Typical Fire-Resistance Rating Requirements for Exterior Walls Based on Fire Separation Distance[a] Based on 2000 International Building Code

Fire separation distance (ft)	Type of construction	Group H	Group F-1, M, S-1	Group A,B,E, F-2, I, R,[b] S-2, U
<5[c]	All	3	2	1
<10	III, IV, V	2	1	1
<30	II-B, V-B	1	0	0
	Others	1	1	1
≥30	All	0	0	0

[a] Load-bearing exterior walls shall also comply with the fire-resistance rating requirements of Table 10.4A.

[b] Group R-3 and Group U, when used as accessory to Group R-3, shall not be required to have a fire-resistance rating where fire separation distance is 3 ft or more.

[c] See Section 503.2 of 2000 IBC for party walls.

TABLE 10.5 Dimensions of Components for Heavy Timber Construction

Heavy Timber construction is defined in the 2000 International Building Code by the following minimum sizes for the various members or portions of a building:

	in., nominal
Columns	
Supporting floor loads	8 × 8
Supporting roof and ceiling loads only	6 × 8
Floor framing	
Beams and girders	6 wide × 10 deep
Arches and trusses	8 in any dimension
Roof framing—not supporting floor loads	
Arches springing from grade	6 × 8 lower half 6 × 6 upper half
Arches, trusses, other framing springing from top of walls, etc.	4 × 6
Floor (covered with nominal 1 in. flooring, or ½ in. wood structural panels, or other approved surfacing)	
Splined or tongue-and-groove planks	3
Planks set on edge	4
Roof decks	
Splined or tongue-and-groove planks	2
Planks set on edge	3
Tongue-and-groove wood structural panels	1⅛

wood members are such that they continue to support its load, so the chance of building collapse is greatly diminished.

Based on comparative fire tests, 1⅛ in. thick wood structural panels with tongue-and-groove edges are accepted as an alternative to nominal 2 in. thick planks (or laminated planks at least 3 in. wide and set on edge) for Heavy Timber roof decks. See Fig. 10.1. Oriented strand board (OSB) wood structural panels, having a min-

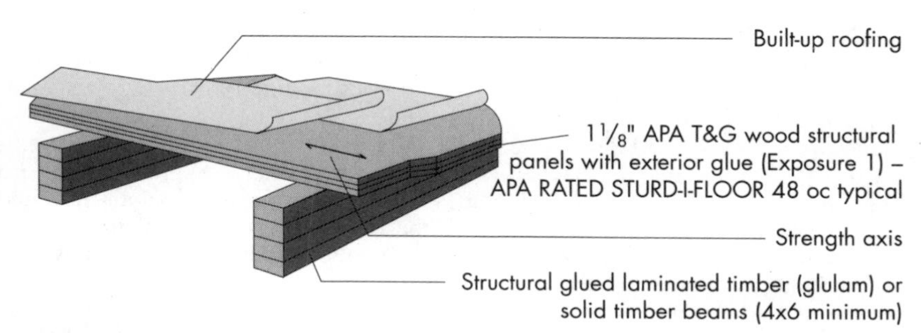

FIGURE 10.1 Heavy timber roof construction.

imum nominal thickness of 1³⁄₃₂ in. and tongue-and-groove edges, also are recognized as an alternative to 1⅛ in. wood structural panels for Heavy Timber roof decks in the Uniform Building Code.

This code recognition can simplify roof construction practices while providing fire protection. Performance of Heavy Timber construction is superior to most unprotected "noncombustible" (metal) structures, under fire conditions. There are no concealed spaces where fire can spread. Firefighting is simpler and safer. Firefighters who have had long experience with wood's structural integrity under fire conditions can more accurately predict how long wood will carry its load than they can with other materials, enabling them to stay on or in the building to combat the fire.

Codes also permit ¹⁵⁄₃₂ or ½ in. wood structural panels over nominal 3 in. planks for Heavy Timber floors. These provisions allow structural design of the building to resist wind or seismic loading, based on utilizing the wood structural panel diaphragm capacity for floor and roof decks.

Guidelines for structural evaluation of the effect of damage to Heavy Timber members and glulam after a fire are available.[8,9]

Fire Retardant Treated (FRT) Construction. Fire-retardant-treated (FRT) wood or plywood is permitted for certain applications in the model building codes. FRT wood reduces its fire hazard classification and qualifies it for lower flame spread (at least as low as gypsum wallboard) and smoke index ratings. When FRT wood is identified by a code-recognized testing agency label, it is rated on a parity with noncombustible construction by many fire insurance rating bureaus.

Span ratings for wood structural panels, and load capacities for wood framing and plywood wood structural panels are based on untreated materials and may not apply following fire-retardant treatment. Structural performance characteristics and use recommendations for FRT wood and plywood should be obtained from the company providing the treatment and redrying service. For structural applications, use only FRT wood that has recognition through building code evaluation service reports.

When considering use of FRT wood, first determine through thorough investigation that it is the best overall solution. Required fire protection at lowest construction cost, and cost of annual building insurance, should be taken into consideration. FRT wood is more expensive than untreated wood, which in most cases can be used in structural floor, wall, and roof assemblies to meet fire-resistive requirements in regard to both life safety and protection of property.

Calculated Fire Resistance. The International Building Code (as well as the three major model building codes) permits calculation as an alternative to prescriptive or tested assemblies for one-hour fire-rated wood-framed floors, roofs, and load-bearing and nonbearing walls. The codes provide tables of assigned times for components, which have been developed empirically from extensive studies of assemblies tested with nominal 2 in. wood framing in accordance with ASTM Standard E119. End-point criteria in the standard also were considered. A one-hour fire-rated assembly can be determined by combining the individual component times of the assembly in accordance with the method and limitations in the codes, thereby providing additional choices for the designer.

Methods also provide for determining the required size of exposed timber beams and columns with a minimun nominal dimension of 6 in., including structural glued laminated timber, for fire-resistance ratings up to one hour.[6,10] (See Chapter 4 for a further discussion.)

Building Occupancy, Area, and Height Limitations. All buildings must meet code requirements with respect to permissible heights and floor areas. These requirements are based on certain characteristics of the building, including the fire zone, type of occupancy, construction materials and systems, setbacks from property lines, exits, and automatic extinguishing systems.

Fire Zones. Some cities have established one or more geographic fire zones (or fire limits), which restrict type of use or occupancy, percentage of lot coverage, and type of construction permitted. The purpose is to make fire protection easier by concentrating in one area those buildings of similar fire hazard. Usually, wood-frame building construction is not permitted in central fire zones, where congestion and closeness of other buildings would make fire spread most likely and fire fighting most difficult.

Occupancy. Codes traditionally have specified use or occupancy classifications, including assembly, business, educational, factory and industrial, hazardous, institutional, mercantile, residential, storage, and utility or miscellaneous. Within occupancy classifications, codes also consider whether manufacturing or storage is of potentially explosive or dangerous materials; whether the residents are elderly, disabled, or confined, etc. Unprotected wood construction is not permitted in specific high-hazard or medical institutional occupancies.

Setbacks. Codes traditionally have recognized the advantage of large open areas around buildings, to make fire fighting easier and prevent fire spread. When buildings have more than 25% of their perimeter on a 20 ft minimum open space, or face on a street with a 20 foot minimum width, the International Building Code permits a building with larger area than when buildings are closer to property lines or other buildings.

Exits. The number and type of exits required depend on occupant load and travel distance to exits. All exit assemblies are classified by fire-resistance ratings, and except for certain high-hazard or institutional occupancies, protected wood construction is usually permitted. The maximum distance to an exit is 200 ft in accordance with the International Building Code, for most occupancy classifications without sprinklers including business or residential buildings. Previous codes differed somewhat in these provisions.

Sprinklers. Sprinklers are another option that can be used to increase building height and/or area, expanding the options for using wood systems in large multi-family residential and non-residential buildings. With sprinkler protection, code requirements for flame spread and fire-resistance ratings may be relaxed. It may be possible to add another story or increase building area. Reduced annual insurance premiums for buildings and contents mean that sprinklers generally will pay for themselves within a few years, depending on the value of the building and its contents. The difference in insurance rates between sprinklered wood and sprinklered unprotected steel buildings is usually minimal.

Building Area Increases. Building codes place limitations on the height and area of a building according to compliance with certain established criteria that are based on safety of occupants and fire services. These criteria include occupancies, types of construction, and location within fire zones.

Light-frame wood construction is often the best choice from the standpoint of cost and simplicity. However, the basic allowable areas may limit the size of the building. It is to the designer's advantage to utilize permitted construction features that allow increases in allowable building area, in order to take advantage of the economy and versatility of wood framed construction systems. The following suggestions, and the data in Table 10.6, help to identify these features.

- *One-hour fire resistance:* Codes allow the area of a wood-frame building to be increased when one-hour fire resistance is provided for all structural elements in the building, including beams and columns, floors, walls, and roofs.
- *Automatic sprinkler protection:* Codes have provisions which allow building areas to be increased when an automatic sprinkler system is installed throughout the building. For example, under the 2000 IBC, a 200% increase is permitted for one- and two-story buildings. An additional benefit is the likelihood of substantially lower insurance rates with sprinklers. Sprinkler systems can be connected to a central alarm system for additional protection.
- *Building separation:* Basic area increases are allowed if there are large open areas on two or more sides of a building. Under the 2000 IBC, a 150% increase is allowed if all sides face toward public streets.
- *Unlimited areas:* In some codes, provisions are made for the construction of unlimited area buildings for industrial, storage, or business uses. Generally, there must be large areas of open space surrounding the building, and the building must be completely sprinklered.
- *Fire walls:* The equivalent effect of area increases can be achieved by introduction of properly constructed fire walls. In effect, two contiguous buildings are erected but are separated by a rated wall or partition, with all openings protected.

The 2000 IBC provides a simple formula that can be used to determine area increases for the values shown in Table 10.6. This is given as:

$$A_a = A_t + \left[\frac{A_t\, l_f}{100}\right] + \left[\frac{A_t\, l_s}{100}\right] \tag{10.1}$$

where A_a = adjusted allowable area (ft^2)
A_t = area per Table 10.6
l_f = area increase due to open frontage
l_s = area increase due to sprinkler protection

TABLE 10.6 Typical Allowable Height and Building Areas Based on 2000 International Building Code. Height limitations shown as stories and feet above grade plane; area limitations as determined by the definition of "area building," per floor

		Type of construction				
	HGT (ft)	Type III		Type IV	Type V	
Group	HGT (story)	65	55	65	50	40
A-1	S	3	2	3	2	1
	A	14,000	8,500	15,000	11,500	5,500
A-2	S	3	2	3	2	1
	A	14,000	9,500	15,000	11,500	6,000
A-3	S	3	2	3	2	1
	A	14,000	9,500	15,000	11,500	6,000
A-4	S	3	2	3	2	1
	A	14,000	9,500	15,000	11,500	6,000
A-5	S	UL	UL	UL	UL	UL
	A	UL	UL	UL	UL	UL
B	S	5	4	5	3	2
	A	28,500	19,000	36,000	18,000	9,000
E	S	3	2	3	1	1
	A	23,500	14,500	25,500	18,500	9,500
F-1	S	3	2	4	2	1
	A	19,000	12,000	33,500	14,000	8,500
F-2	S	4	3	5	3	2
	A	28,500	18,000	50,500	21,000	13,000
H-1	S	1	1	1	1	NP
	A	9,500	7,000	10,500	7,500	NP
H-2	S	2	1	2	1	1
	A	9,500	7,000	10,500	7,500	3,000
H-3	S	4	2	4	2	1
	A	17,500	13,000	25,500	10,000	5,000
H-4	S	5	3	5	3	2
	A	28,500	17,500	36,000	18,000	6,500
H-5	S	3	3	3	3	2
	A	28,500	19,000	36,000	18,000	9,000
I-1	S	4	3	4	3	2
	A	16,500	10,000	18,000	10,500	4,500
I-2	S	1	NP	1	1	NP
	A	12,000	NP	12,000	9,500	NP
I-3	S	2	1	2	2	1
	A	10,500	7,500	12,000	7,500	5,000
I-4	S	3	2	3	1	1
	A	23,500	13,000	25,500	18,500	9,000
M	S	4	4	4	3	1
	A	18,500	12,500	20,500	14,000	9,000
R-1	S	4	4	4	3	2
	A	24,000	16,000	20,500	12,000	7,000
R-2	S	4	4	4	3	2
	A	24,000	16,000	20,500	12,000	7,000

TABLE 10.6 Typical Allowable Height and Building Areas Based on 2000 International Building Code. Height limitations shown as stories and feet above grade plane, area limitations as determined by the definition of "area building," per floor (*Continued*)

		Type of construction				
	HGT (ft)	Type III		Type IV	Type V	
Group	HGT (story)	65	55	65	50	40
R-3	S	4	4	4	3	3
	A	UL	UL	UL	UL	UL
R-4	S	4	4	4	3	2
	A	24,000	16,000	20,500	12,000	7,000
S-1	S	3	3	4	3	1
	A	26,000	17,500	25,500	14,000	9,000
S-2	S	4	4	5	4	2
	A	39,000	26,000	38,500	21,000	13,500
U	S	3	2	4	9	1
	A	14,000	8,500	18,000	9,000	5,500

UL = unlimited.

l_f may be determined by the following equation:

$$l_f = 100 \left[\frac{F}{P} - 0.25 \right] \frac{W}{30} \tag{10.2}$$

where l_f = area increase due to frontage (percent)
F = building perimeter fronting on public way or open space
P = perimeter of entire building
W = minimum width of public way or open space
$W/30$ not to exceed 1.0

The sprinkler increase is given as follows:

- Multistory buildings: $I_s = 200\%$
- Single-story buildings: $I_s = 300\%$
- Sprinkler system must be in accordance with applicable sections of the 2000 IBC

10.1.3 Fire-Rated Systems

Roof and floor systems that are accepted by building codes, while providing maximum strength and economy, include numerous fire-rated constructions involving wood structural panels over a variety of support systems including solid lumber joists and engineered wood framing components such as I-joists and open web trusses. Popular fire-rated wood systems that are economical and achieve good fire protection are protected floor- and roof-ceiling systems and wall systems. Heavy timber construction also is used for applications where exposed large wood members are desired for aesthetic reasons.

Protected Floor and Roof Ceiling Systems. There are numerous fire-rated assemblies combining wood framing and wood structural panels with fire-resistive gypsum wallboard that are especially suitable for nonresidential and multifamily residential buildings. They include one-hour rated and several two-hour rated protected wood-frame floor-ceiling and roof-ceiling systems.

In these assemblies, materials such as gypsum wallboard, plaster, and acoustical tile provide primary fire protection. The panel floor or roof acts to prevent flame passage and temperature rise, as well as to reinforce wood framing members against collapse under load after the effectiveness of the ceiling has been lost.

Because these systems contain wood and possibly other combustible materials, they are designated as combustible constructions. At present, codes don't permit them in so-called noncombustible (Type I or II) structures, even though their tested performance meets performance requirements that are the same as for assemblies classified as noncombustible.

Full-scale tests on assemblies representative of intended floor- or roof-ceiling construction are conducted by fire testing laboratories that are recognized by building code evaluation services. Many of these laboratories publish a listing directory that is updated annually or periodically, describing the materials and construction of the tested assemblies, and the fire endurance rating obtained in the test. One such directory is the Underwriters Laboratory Inc. (UL) Fire Resistance Directory;[11] others are listed under Additional Reading at the end of this chapter. Over 40 wood floor-ceiling (or roof-ceiling) systems using wood structural panels are listed in the UL directory. Listed designs in fire directories of recognized fire testing laboratories are used by architects and designers for building design and construction, and are accepted by building code officials for fire-rated construction.

Selected examples of fire-rated floor-ceiling construction options with lumber joists and engineered wood framing (wood I-joists or trusses) are shown in Fig. 10.2.

Many systems consist of a double layer floor of plywood subfloor and underlayment ($^{15}/_{32}$ in. and $^{19}/_{32}$ in., respectively). For fire-rated roof-ceiling assemblies, codes permit omission of the top layer of plywood. More recently, fire-rated floor-ceiling assemblies have been developed, which have a single-layer floor system of $^{19}/_{32}$ in. or thicker wood structural panels for combination subfloor-underlayment; these systems are more frequently specified than those having a two-layer floor.

Any finish flooring material may be used over the underlayment or single-layer floor. Many systems permit lightweight concrete or gypsum concrete floor topping, applied over the single-layer floor or over the subfloor in lieu of the top plywood underlayment layer, for one- and two-hour floor-ceiling assemblies. Other systems using a single-layer $^{23}/_{32}$ in. wood structural panel floor with wood I-joists or trusses also are shown in this figure.

Based on comparative tests, wood structural panels such as oriented strand board (OSB) or composite panels containing veneer faces and a wood structural core (COM-PLY®) may be used in tested plywood floor- or roof-ceiling systems without jeopardizing fire-resistance ratings. In double-layer wood systems, minimum $^{7}/_{16}$ in. oriented strand board may be used in lieu of a $^{15}/_{32}$ in. plywood subfloor. Other substitutions are based on equivalent panel thickness. For all assemblies, specified thicknesses and dimensions for sheathing and framing components are minimums, and thicker or larger materials may be used. For framing members, spacing may be reduced without affecting the fire rating.

As previously discussed, model building codes also include procedures for calculating fire resistance of one-hour fire-rated floor- or roof-ceiling systems, using a component additive method with standard building components. The time-

Some rated assemblies incorporate proprietary products. When designing and specifying, check the Underwriters Laboratories Fire Resistance Directory for complete details on a particular assembly. A change in details may affect fire resistance of the assembly.

1. Two-layer floor systems with joists.[a] For details, see U.L. Design Nos. L001, L003, L004, L005, L006, L201, L202, L206, L208 (1-1/2 hr), L209, L210, L211 (2 hr), L212, L501, L502, L503, L505 (2 hr), L511 (2 hr), L512, L514, L515, L516, L519, L522, L523, L525, L526, L533, L535, L536 (2 hr), L537, L541 (2 hr) and L545. Also see U.L. Designs No. L524 and L527 (1-1/2 hr single layer) with steel joists spaced 24" o.c., and L521 with wood trusses spaced 24" o.c.

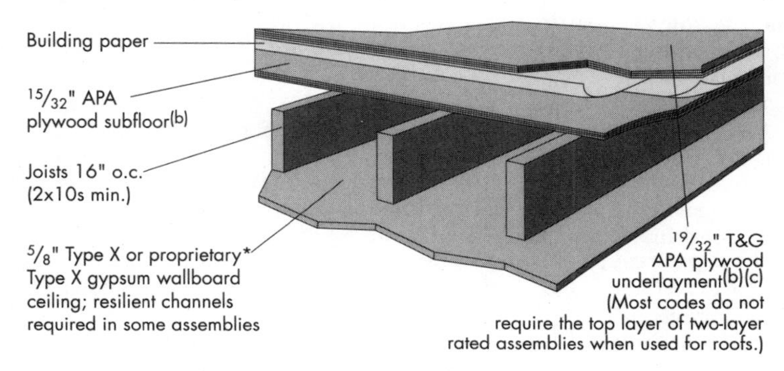

2. Single-layer floor systems with wood I-joists or trusses. For details, see U.L. Design Nos. L528, L529, L534, L542 and L544 (shown). Also see U.L. Design No. L513 for single-layer floor system with lumber joists spaced 24" o.c.

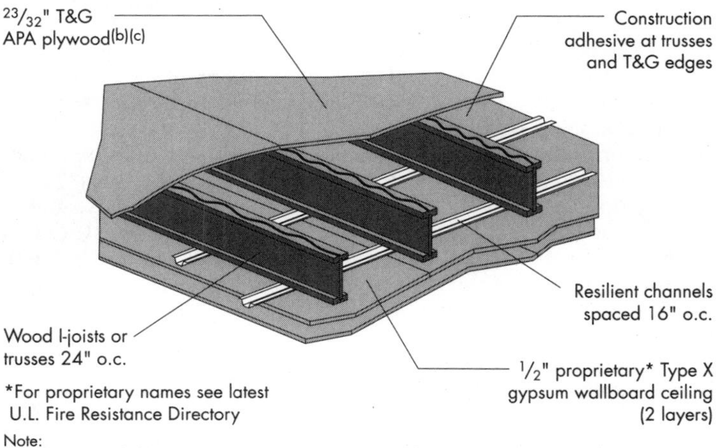

*For proprietary names see latest U.L. Fire Resistance Directory

Note:

(a) Substitution of 1 1/8" APA RATED STURD-I-FLOOR 48 oc for the combination of subfloor, paper, and underlayment is often allowed. Check with local Building Official.

(b) Tests have shown that substitution of OSB or composite APA RATED SHEATHING subfloor and APA RATED STURD-I-FLOOR underlayment for the plywood panels in rated assemblies will not jeopardize fire-resistance ratings. Substitution is based on equivalent panel thickness, except that 7/16" OSB subfloor panels may be used in place of 15/32" plywood subfloor panels in two-layer assemblies. OSB panels are listed as alternates to plywood subflooring or finish flooring in U.L. Design Nos. L501, L503, L505 (2hr), L508, L511 (2hr), L513, L514, L516, L521, L526, L528, L529, L532 (1-1/2 hr), L539, L540, L543, L544, L546, L548, L550, L551 and L552.

(c) Lightweight concrete or gypsum concrete floor topping permitted over single-layer floor or as alternate to plywood underlayment in many assemblies (check details).

FIGURE 10.2 One-hour fire-rated combustible floor-ceiling or (roof-ceiling) assemblies.

assigned values for these components are based on analysis of full-scale fire tests conducted with nominal 2 in. framing members; thus, values should not be applied to framing systems using lightweight engineered wood framing members. However, fire-rated floor-ceiling or roof-ceiling systems that use proprietary engineered wood framing members, are recognized by model building code evaluation services, under evaluation reports issued to individual manufacturers or trade associations.

Wall Systems. Options for fire-rated wall systems for light-frame wood construction include conventional wood-framed walls with wood structural panel sheathing or siding, commonly in conjunction with fire-resistive (Type X) gypsum wallboard interior finish

Protected Exterior Walls. Protected light-frame construction, with fire-resistive (Type X) gypsum wallboard interior finish, is rated by codes between ordinary unprotected and Heavy Timber construction in terms of fire performance.

Under 2000 IBC provisions, when separation between buildings is greater than 5 feet, the fire-resistive rating for the wall applies only to fire exposure from the interior side of the wall. Fire-resistive (Type X) gypsum wallboard sheathing is not required on the exterior side of the wall. Figure 10.3 illustrates a one-hour fire-rated wall construction based on UL Design No. U356 that is applicable under these code provisions.

In populated regions subject to urban wildfire hazards, such as the western and southwestern United States, or where buildings are separated by 5 ft or less, exterior walls must be rated for fire exposure from the exterior side as well as the interior side to comply with requirements of the 2000 IBC and the Urban-Wildland Interface Code (IFCI). In such applications, fire-resistive (Type X) gypsum wallboard sheathing is required on the exterior side of the wall, either under or over wood structural panel sheathing, over which an approved exterior finish is applied. Wood structural panel siding also can be installed over fire-resistive gypsum wallboard sheathing. Alternatively, noncombustible exterior finishes such as brick veneer or portland cement plaster (stucco) can be applied over wood structural panel sheathing on the exterior side of the wall to provide one-hour fire endurance when subjected to fire exposure on the exterior side of the wall.

An example of protected wall construction, rated for fire exposure on both sides, is shown in Fig. 10.4. Such walls consist of minimum 2 × 4 studs, spaced 16 in. or 24 in. o.c. with wood structural panel siding over ⅝ in. Type X gypsum sheathing on the exterior side, and ⅝ in. Type X or proprietary Type X (Type C or G) gypsum wallboard on the interior side. Another option permitted in the 2000 IBC uses stucco over wood structural panel sheathing. Protected wall constructions qualify for the same ratings if other materials are added; e.g., siding may be attached to the outside of a fire-rated wall to add shear-wall value without impairing the rating. Also, wood structural panel sheathing can be added between the gypsum wallboard fire-resistive membrane and wood framing, on either the interior or exterior side of the wall.

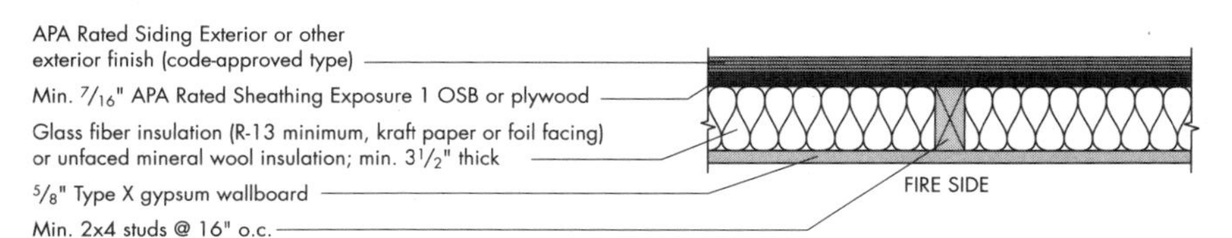

FIGURE 10.3 Load-bearing exterior wall system.

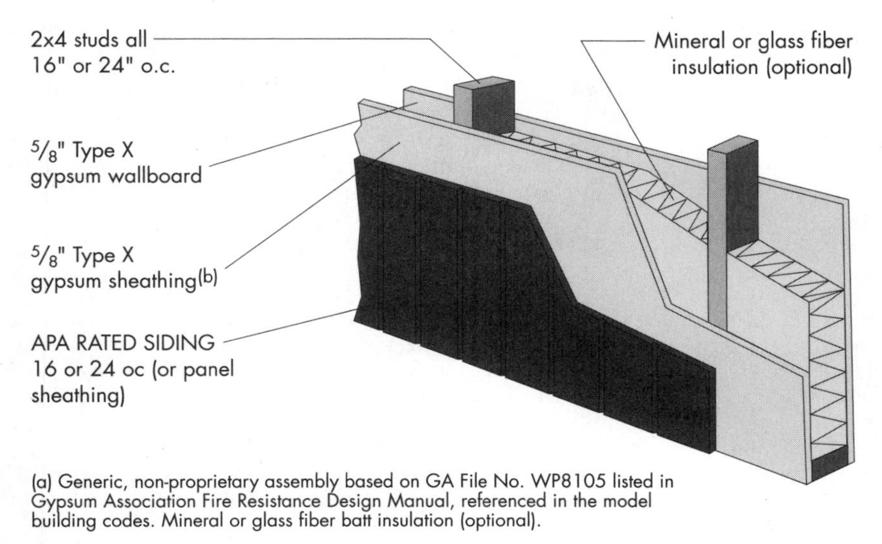

(a) Generic, non-proprietary assembly based on GA File No. WP8105 listed in Gypsum Association Fire Resistance Design Manual, referenced in the model building codes. Mineral or glass fiber batt insulation (optional).

(b) Exterior layer of gypsum sheathing not required under the National and Standard Building Codes, or the International Building Code, when separation is greater than 5 feet. Check local provisions. See U.L. Design U356 in U.L. Fire Resistance Directory.

FIGURE 10.4 One-hour fire-rated exterior load-bearing wall assembly.

2000 IBC provisions also permit determination of one-hour fire-rated wall systems using procedures for calculated fire resistance of components.

Interior Walls and Partitions. Building code regulations require a flame spread index of 200 maximum on materials used for interior finishes on exposed surfaces (in areas other than certain exit ways and corridors, as noted earlier). Flame spread index ratings for wood structural panels generally fall within Class A (FSI 76-200) and thus are within the range of acceptable materials.

For interior areas requiring lower flame spread ratings, fire-retardant-treated plywood paneling is acceptable. Panels identified by labeling from a recognized fire testing laboratory or quality assurance agency indicating conformance to a Class A flame spread rating, are accepted by code officials. Fire retardant paints, properly applied, also may be used to reduce the flame spread rating to Class A or B, and are often recognized by building officials.

In single-family residential use, Class C flame spread rating is acceptable for interior wall and ceiling finishes. Softwood plywood paneling is well within the acceptable range, and has been used for interior as well as exterior walls, where plywood is used to resist wind or seismic loading on the structure.

Roofing Systems. Fire-resistance ratings of roofing materials (built-up or single-ply roofing; prepared roofing such as asphalt composition shingles; or wood shingles or shakes) are classified as Class A, B, or C in descending order of fire protection afforded. Their use is prescribed by building codes, and also affects insurance rates. The standard test for measuring the fire characteristics of roof coverings is ASTM E108.[12] This standard specifies fire testing of roofing over untreated plywood as a roof deck substrate. However, fire classification of roofing over other types of wood structural panels, such as oriented strand board for roof deck substrates, is permitted if roofing fire tests are conducted over the intended roof deck substrate materials, by recognized fire testing laboratories. Listing directories published by recognized fire testing laboratories should be consulted for individual roofing specifications and minimum requirements for the roof deck substrate, for built-up or single-ply roofing membranes, spray-applied foam insulation and roof coating systems, or prepared roof covering materials such as shingles, shakes, cement tile, and metal roofing panels.

Structural Glued Laminated Timber (Glulam). Procedures are available in the model building codes to calculate the size of sawn timber and glulam beams required for projects in which one-hour fire resistance is required.[13] A structural member's fire resistance is measured by the time it can support its design load during a fire. An exposed beam or column sized for a minimum one-hour fire resistance will support its full design load for at least one hour during standard fire test conditions that simulate an actual fire. As with all other structural framing, final specifications of members designed to have one-hour fire resistance should be carefully checked by a professional engineer or architect to ensure compliance with all local building codes.

The following is a brief discussion of the fire performance of Heavy Timber members such as glulam. See Section 4.6.3 for a detailed discussion of the fire performance of glulam.

Beams. Charring of wood surfaces during a fire places a premium on cross-sectional area. Charring weakens a wood member's cross-section slowly because of the insulating characteristics of the developing char layer. Glulam beams with a minimum width of 5⅛ in. (nominal 6 in.) can be adapted to a one-hour fire rating in accordance with procedures recognized by the model building codes, although the beam bending capacity is reduced.

For 6¾ in. and 8¾ in. widths, there is a minimum depth at and above which all members with these widths can be adapted at 100% of the allowable design load for a one-hour fire rating. The minimum depth increases when the design calls for the beam to be exposed on four rather than three sides. See Table 4.21A.

To adapt beams whose dimensions qualify for one-hour fire rating, the basic lay-up must be modified as shown in Fig. 4.13. One center core lamination adjacent to the tension laminations is replaced by an additional tension lamination to augment the beam bending capacity provided by the tension laminations.

Fire resistance for glulam beams also may be provided by covering the exposed surfaces with two layers of fire-resistive ⅝ in. Type X gypsum wallboard. This modification has been demonstrated to provide one-hour fire resistance for beams having a minimum size of 4½ in. wide by 9½ in. deep (nominal 5 × 10 in. beam).[10] With such gypsum wallboard protection, a standard glulam beam or Heavy Timber member can be rated for one-hour fire resistance. A glulam beam, meeting the above lay-up requirements for a one-hour fire rating, can be upgraded to two-hour fire resistance when also protected with gypsum wallboard in accordance with the recognized fire-rated design.

Columns. Columns are produced with a single grade of laminations throughout, and therefore need no special lay-up to qualify for a one-hour fire rating. For glulam beams having 8¾ in. and 10¾ in. widths, columns meeting the minimum size standard satisfy the one-hour fire rating requirement at 100% of the allowable design load.

However, column length plays a significant role in determining minimum size for one-hour ratings. The column size needed for a one-hour fire rating is determined by calculating the l/d and then using the appropriate minimum dimensions in Tables 4.21B and 4.21C.

l = column length (in.)

d = column least dimension in (in.)

If l/d is less than or equal to 11, the minimum required size is smaller than when l/d is greater than 11.

Metal Connectors for One-Hour Rating. In structures using one-hour fire-rated timbers, all supporting metal connectors and fasteners must be protected to achieve a one-hour fire rating. A 1½ in. covering of wood, fire-resistive Type X gypsum wallboard, or any coating approved for a one-hour rating provides the needed protection.

10.1.4 Insurance Considerations

Fire Insurance Considerations. Compliance with building insurance provisions is voluntary. The means of doing so is not specified or even suggested by the rating bureau. This puts a special responsibility on the designer to determine the best combination of economical construction and sufficient protection to qualify for low premiums. If possible, the designer also should determine during preliminary planning just what kind of rating can be counted on, while there is still time to adjust such factors as setbacks, materials, building size, etc.

Most agencies will cooperate with a design professional who wishes to do this kind of research. However, rating bureaus are constrained to give out rate information only to, or by authority of, the building owner.

Insurance companies, through assessments against their premium income, have for many years supported state and regional rating bureaus throughout the country. Most of these have been consolidated into the Insurance Services Office (ISO).

The intent of the ISO is to develop a nationwide rate schedule, which will straighten out much of the confusion due to the multiplicity of rates throughout the United States. However, until a genuinely universal rate schedule is developed, it will still be necessary for architects to be familiar with local situations and be alert to changes.

There are two kinds of rates: class rates (generally for one- and two-family dwellings, small multifamily residential buildings, and garages), and schedule, or specific rates (all other types of buildings). The former, with many shared physical characteristics, may be rated according to their position in established classes based on type of construction, occupancy, and public fire protection. Only major differences in these characteristics would be reason for rate distinction.

In the latter category, however, each building is considered a separate case. It is rated after consideration of five factors: type of construction; effectiveness of public fire protection; private fire protection in the building; occupancy (there are well over 100 occupancy classes, according to how the building is used, whether the building is occupied, etc.); and exposure from other properties nearby.

Awareness is increasing that building content, and not structure, is the most critical element with respect to fire performance (based on experience with fires in high-rise steel and concrete buildings). More emphasis is now being placed on warning and sprinkler systems and less on "fireproof" structural materials.

Furthermore, the potential for collapse of unprotected steel-framed structures is being recognized, compared with wood construction. Insurance rates for wood sys-

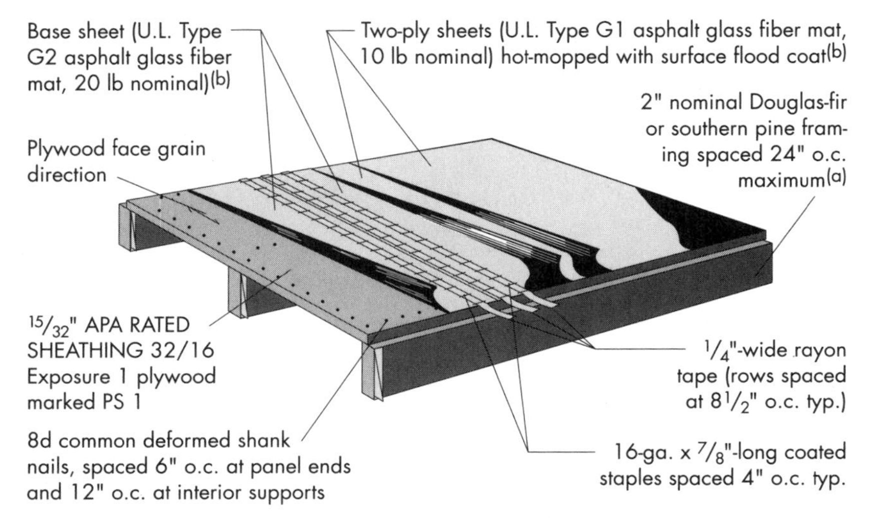

(a) Design in accordance with local building code requirements for roof loads and anchorage. All framing must have 2" nominal or greater width for plywood deck nailing.
(b) Install roofing base and ply sheets with roll direction parallel to plywood face grain directions.

FIGURE 10.5 Wind resistive roof assembly—U.L. class 90.

tems with sprinklers or fire-resistive ceiling systems are generally competitive with rates for unprotected and protected steel systems for most occupancies.

Extended Coverage Considerations. Wind uplift resistance of the building's roof system also determines the extended coverage endorsement (ECE) insurance rates. Wood roof deck and roofing systems have been evaluated by code-recognized testing laboratories to provide maximum protection against fire and wind uplift, in addition to complying with building and fire code requirements, to qualify for low insurance rates. Figure 10.5 illustrates an example of a UL Class 90 wind resistive roof assembly.

Although insurance rates for wood buildings may be higher than for noncombustible buildings, construction cost savings for wood buildings may result in a substantially reduced combined annual cost for the building and insurance.

Building owners may be able to save costs over the long term with wood construction, in spite of possibly higher annual insurance rates than for steel, concrete, or masonry construction. Savings realized from lower construction cost of wood buildings are often greater than the extra insurance cost. Interest on funds needed to build a more expensive building may exceed savings from lower annual insurance rates on the building and contents. Thus, spending more money on higher-cost construction in order to secure a lower insurance rate may result in an annual loss, rather than a financial gain, for years to come. If there is only a slight insurance premium saving when other construction systems are used, it may take so long to repay the additional construction cost, plus interest, that the higher-priced construction will represent a poor building investment compared with wood systems.

10.2 INTRODUCTION TO NOISE-RATED SYSTEMS

Noise originating outside a building is controlled largely by the location of the site, by the location of rooms relative to the exterior of the building, and by landscaping. Background noise of this type is an important consideration because it helps mask intermittent intruding sounds. For example, an intruding noise which would be intolerable in a quiet suburb might go unnoticed in a busy city, where traffic noise might mask noises from an adjoining room and the background noise itself might not seem unpleasant. A sudden, obtrusive noise, as distinguished from a general level of background noise, is disturbing in any context.

Because of the reduced level of background noise, suburban office/apartment/condominium buildings usually require higher levels of sound insulation than those in a busier urban environment.

10.2.1 Types of Noise

Noise between occupied units and within a unit is controlled through construction methods and materials that interrupt sound transmission paths. Such noises are transmitted through airborne paths, by impact such as foot traffic, and by structural vibrations.

- *Airborne noises,* such as traffic, voices, and television, penetrate through walls, doors and other structural elements. Open windows, cracks around doors, heating and ventilating ducts, and other imperfectly sealed openings may also leak airborne noise. Wall and floor-ceiling constructions are rated for airborne noise transmission.
- *Impact sounds* are produced by falling objects, foot traffic, and mechanical impacts. Floor constructions are rated for impact noise reduction to minimize annoying and critical impact sounds that are transmitted through the floor, as well as for airborne noise transmission.
- *Structural vibrations* are set up from the vibrations of mechanical apparatus such as heating/ventilating/air conditioning equipment and plumbing fixtures. Unless plumbing is properly isolated (as by acoustically designed hangers), annoying sounds can be transmitted throughout the entire structure.

10.2.2 Building Design Considerations

Proper design and layout of a building can do much to eliminate noise problems. Consideration should be given to such points as location and orientation of the building, landscaping, segregation of quiet areas, and offsetting of entrance doors.

Good construction can minimize sound problems, but all details must be carefully accomplished. Sound leaks can be sealed with nonporous, permanently resilient materials, such as acoustical caulking materials and acoustically designed gaskets and weather stripping. Piping penetrations can be wrapped or caulked. Airborne and impact noise can be controlled through properly designed and constructed wall and floor-ceiling assemblies.

Use of light-frame wood construction systems challenges designers to insulate against noise rather than simply relying on the massiveness of heavy walls and floors. Excellent levels of noise control can be achieved with good acoustical design in wood-frame structures surfaced with wood structural panels in combination with other materials, such as thermal or acoustical insulation, gypsum wallboard, and accessories such as steel resilient or furring channels. Sound control can be achieved by applying floor and wall materials over isolated air spaces that absorb sound. Addition of steel resilient channels significantly reduces sound transmission by isolating membrane surfaces such as gypsum wallboard from framing. Acoustically rated constructions shown in this chapter are based on laboratory- or field-tested acoustical performance since design procedures are not available.

Wood structural panels are excellent for this type of construction. Large panel size reduces the number of joints and cracks that can leak airborne noise. Wood structural panels also provide a good substrate for resilient floor coverings that reduce impact noise. A variety of popular acoustically rated wall and floor-ceiling assemblies are described in the following sections.

10.2.3 Coordination with Fire-Rated Systems

For multifamily residential and nonresidential buildings, it is often necessary that the wall or floor-ceiling assembly be rated for both acoustical and fire performance. In these situations, it is important that the same construction features have been included in assemblies evaluated for both fire and acoustical tests. For example, consider a construction in which sound-absorbing mineral wool acoustical insulation was used for an acoustically tested assembly but was omitted from a similar assembly that was evaluated for fire-resistance. In this situation, such insulation may affect the endurance of the fire-rated assembly. Similarly, substitution of other framing or sheathing materials, or another type of insulation such as glass fiber thermal insulation having similar or different thickness or density, may affect the acoustical rating of the assembly, as well as fire endurance. Guidelines for such material substitutions and additions are published in the Gypsum Association *Fire Resistance Design Manual*.[14]

10.2.4 Noise Measurement

Sound Transmission Class (STC). The ability of walls and floors to reduce airborne noise is the significance of STC numbers as illustrated in Table 10.7. In comparing rated constructions, it is noted that 3 db is the smallest difference that the human ear can clearly detect. Thus, differences of 1 or 2 points may be considered negligible. Also note that even this general comparison is valid only with respect to a given level of background noise. The STC value can be determined in accordance with ASTM E336[18] and ASTM E90.[15]

Impact Insulation Class (IIC). In addition to being rated for airborne sound transmission, floors are also rated by impact insulation class (IIC), formerly referred to as impact noise rating (INR). IIC values rate the capacity of floor assemblies to control impact noise such as footfalls or items dropping on floor surfaces. The IIC value is determined in accordance with ASTM E492[17] and ASTM E989.[10] INR ratings can be approximately converted (± 2 db) to IIC ratings by the (algebraic) addition of 51 db.

TABLE 10.7 Relation of Airborne Noise to Sound Transmission Class

STC ratings
25 Normal speech can be understood quite clearly.
30 Loud speech can be understood fairly well.
35 Loud speech audible but not intelligible.
42 Loud speech audible as a murmur.
45 Must strain to hear loud speech.
48 Some loud speech barely audible.
50 Loud speech not audible

The best way to reduce impact noise is to cover a floor with a resilient surfacing material such as carpet and padding. Where a hard surface finish flooring is used, a resiliently mounted ceiling system is effective, as is insulation fiberboard or other sound-absorbing composites that are installed between the subfloor and the underlayment.

The 2000 IBC specifies minimum STC 50 and IIC 50 values for floor-ceiling assemblies and STC 50 values for wall assemblies that are located between adjacent dwelling units or dwelling units and adjacent public areas such as corridors or stairs, or service areas. If the sound transmission values are determined by field testing, a minimum STC and IIC value of 45 is permitted. Higher values would provide a greater level of satisfaction to building occupants.

10.2.5 Flanking Paths

Acoustical ratings do not reflect the effect of noise that bypasses, or flanks, the specific construction. Flanking can increase noise transmission significantly. For example, a heating duct in a wall partition that has an STC 48 rating could reduce the STC rating for the assembly to about 30 if the duct is not properly sound-isolated.

Other significant flanking paths occur through back-to-back electrical and plumbing outlets, and joist spaces over fire separation walls or party walls and wall partitions. All flanking paths should be taken into account and eliminated if at all possible. Otherwise sound reduction by floor and wall construction may prove ineffective.

Field tests demonstrate that site-built construction can closely approximate sound insulation values of laboratory-tested assemblies, providing components are installed carefully and subject to supervision. Improper installation, on the other hand, can destroy the sound-insulating values of the best designs.

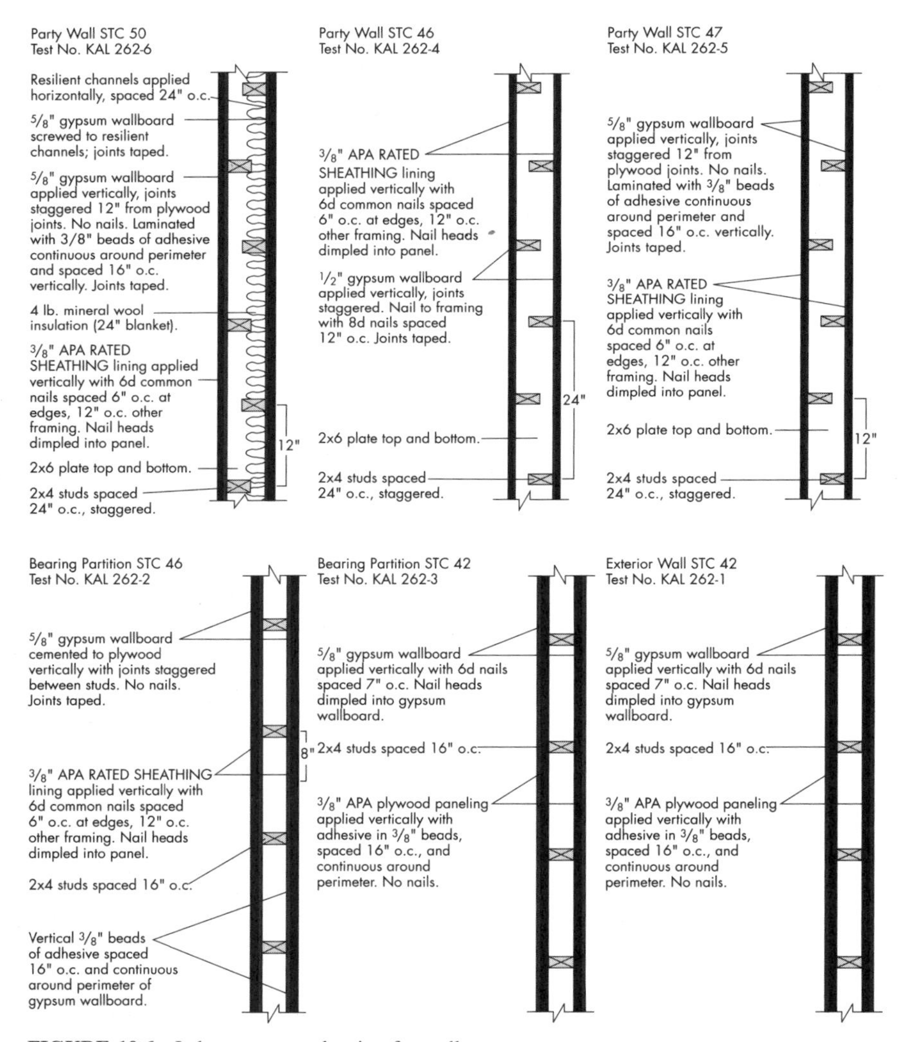

FIGURE 10.6 Laboratory sound rating for walls.

10.2.6 Laboratory-Tested Acoustical Ratings for Wall and Floor-Ceiling Assemblies

Wall and floor-ceiling assemblies described in this chapter have been tested by recognized acoustical laboratories and rated for sound transmission according to standard test methods.

Acoustical ratings for wood-framed wall constructions are shown in Fig. 10.6. This figure shows constructions for party walls, interior bearing walls, and an exterior wall. Walls and partitions using wood structural panels nailed directly to the framing can be used to develop racking resistance, which is often important in multifamily residential and nonresidential buildings requiring shear walls to resist wind or earthquake forces.

Floor-ceiling assemblies shown in Tables 10.8 and 10.9 provide transmission and impact ratings for popular basic constructions. Acoustical tests on similar floor-ceiling constructions indicated that floors with wood I-joists had equal or better sound transmission ratings (STC and IIC) than similar floors constructed with nominal 2 in. solid lumber joists (5).

Sound transmission and impact insulation class ratings shown for these constructions are well within the range of acceptable acoustical ratings for multifamily and nonresidential buildings. They should apply to site-built construction, provided that recognized precautions are taken for preventing flanking noise and sound leaks, and provided the construction conforms to the assembly that has been tested. Quality of workmanship and material are important factors that affect performance to rated levels. When interpreting these figures, it may be possible to modify some construction details without detracting from sound-insulating properties. Minor modifications of these floor-ceiling and wall assemblies to conform to requirements for a one-hour fire-resistance rating should be carefully evaluated for their potential effect on fire and acoustical ratings. In general, proposed modifications to construction details for fire- and acoustically rated assemblies should be coordinated with local building department officials before proceeding with final design and construction. Guidelines for such material substitutions and additions are published in the Gypsum Association *Fire Resistance Design Manual.*[14]

1. Species and grade of wood structural panels and framing can be changed.
2. Thickness of wall studs can be nominal 3 in. or 4 in., and depths of floor joists can be nominal 8–12 in.
3. OSB and composite panels may be substituted for plywood on a thickness-for-thickness basis.
4. Acoustical ratings obtained with assemblies containing ½, ⅝, or ¾ in. wood structural panels are unchanged if panel thicknesses are 15⁄32, 19⁄32, or 23⁄32 in., respectively. Thicker-than-tested wood structural panels or gypsum wallboard can be substituted.
5. Glass fiber or mineral wool insulation can be substituted.
6. Resilient channels can be used on one or both sides of a wall.

Major modifications in thickness, or cumulative changes, can alter a rated system's acoustical ratings. While assemblies were tested using plywood, other wood structural panels (oriented strand board [OSB] and COM-PLY® panels) may be substituted on a thickness-for-thickness basis. Because of their substantially similar strength and stiffness properties and slightly higher density, use of such wood struc-

TABLE 10.8 Laboratory Ratings for Floors: Conventional Wood Joist Floor with Resilient Finish Floor

Test number and sponsor	Finish floor	Deck[a]	Gypsum wallboard ceiling	Insulation	STC	IIC	Weight (lbs/sq ft)
Case 1 NBS-728A NBS	1/8 in. vinyl asbestos tile on 1/2 in. plywood underlayment	5/8 in. rated sheathing subfloor on 2 × joists at 16 in. o.c.	1/2 in. nailed to joists	None	37	34	9.0
Case 2 NBS-728B NBS	1/4 in. foam rubber pad and 3/8 in. nylon carpet on 1/2 in. plywood underlayment	5/8 in. rated sheathing subfloor on 2 × joists at 16 in. o.c.	1/2 in. nailed to joists	None	37	56	approx. 9.5
Case 3 KAL-2241 & 2 APA	0.075 vinyl sheet on 3/8 in. plywood underlayment	5/8 in. rated sheathing subfloor on 2 × joists at 16 in. o.c.	5/8 in. screwed to resilient channels	3 in. glass fiber	46	46	8.9
Case 4 KAL-224-32 & 33 APA	1/16 in. vinyl sheet	19/32 in. T&G Sturd-I-Floor on 2 × joists at 16 in. o.c.	5/8 in. screwed to resilient channels	3 in. glass fiber	48	45	7.8
Case 5 KAL-224-3 & 4 APA	44 oz gropoint carpet and 40 oz hair pad	19/32 in. T&G Sturd-I-Floor on 2 × joists at 16 in. o.c.	5/8 in. screwed to resilient channels	3 in. glass fiber	48	69	8.6
Case 6 CK 6512-8 USG	44 oz gropoint carpet and 40 oz hair pad, on 25/32 in. oak	1/2 in. rated sheathing subfloor on 2 × joists at 16 in. o.c.	1/2 in. screwed to resilient channels	3 in. mineral wool	50	71	9.5
Case 7 G&H-APA-1ST APA	Vinyl tile	19/32 in. T&G Sturd-I-Floor glued to 2 × joists at 16 in. o.c.	5/8 in. screwed to resilient channels	1 in. mineral wool stapled to side of joists and bottom of subfloor	51	51	approx. 9.4
	Carpet and pad					74	10.2

[a] While the listed assemblies were tested using plywood, it is accepted that other wood structural panels, OSB and COM-PLY®, may be substituted on a thickness for thickness basis based on their substantially similar strength and stiffness properties and slightly higher density without compromising the STC or IIC ratings of the tested assemblies.

TABLE 10.8 Laboratory Ratings for Floors: Conventional Wood Joist Floor with Resilient Finish Floor (*Continued*)

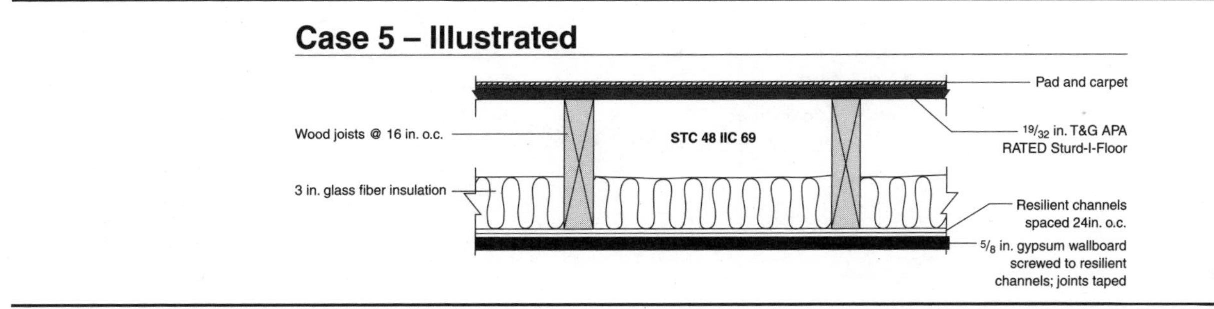

TABLE 10.9 Laboratory Ratings for Floors: Lightweight and Gypsum Concrete over Conventional Wood Joist Floor

Test number and sponsor	Finish floor	Deck[a]	Gypsum wallboard ceiling[b]	Insulation	STC[c]	IIC	Weight (psf)
Case 1 G&H-USDA-1ST USDA	None	1½ in. of 100 pcf cellular concrete over ⅝ in. rated sheathing subfloor on 2 × joists at 16 in. o.c.	⅝ in. nailed to joists	None	48		21.0
	Carpet and pad					68	
Case 2 G&H-USDA-3ST USDA	None	1½ in. of 100 pcf cellular concrete over ⅝ in. rated sheathing subfloor on 2 × joists at 16 in. o.c.	⅝ in. screwed to resilient channels	None	55		20.0
	Carpet and pad					67	
Case 3 G&H-USDA-2ST USDA	None	1½ in. of 100 pcf cellular concrete over ⅝ in. rated sheathing subfloor on 2 × joists at 16 in. o.c.	⅝ in. screwed to resilient channels	3 in. glass fiber	58		20.5
	Carpet and pad					67	
Case 4 G&H-USDA-5ST USDA	None	1½ in. of 100 pcf cellular concrete over ⅝ in. rated sheathing subfloor on 2 × joists at 16 in. o.c.	⅝ in. mineral-fiber acoustical tile on suspension system hung from resilient channels	3 in. glass fiber	59		22.2
Case 5 G&H-USDA-6ST USDA	None	1½ in. of 100 pcf cellular concrete over ⅝ in. rated sheathing subfloor on 2 × joists at 16 in. o.c.	⅝ in. mineral-fiber acoustical tile on suspension system hung from joists	3 in. glass fiber	58		22.1
Case 6 G&H-USDA-9ST USDA	0.075 vinyl sheet	1½ in. of 100 pcf cellular concrete on ½ in. sound board over ⅝ in. rated sheathing subfloor on 2 × joists at 16 in. o.c.	⅝ in. screwed to resilient channels	3 in. glass fiber	59	52	22.0
							22.4
	Carpet and pad					72	
Case 7 KAL-224-29 & 30 APA	44 oz carpet and 40 oz. pad	1⅝ in. of 75 pcf perlite/sand concrete over ⅝ in. rated sheathing subfloor on 2 × joists at 16 in. o.c.	⅝ in. nailed to joists	None	47	66	18.4
Case 8 KAL-224-31 & 34 APA	0.075 vinyl sheet	1⅝ in. of 75 pcf perlite/sand concrete over ⅝ in. rated sheathing subfloor on 2 × joists at 16 in. o.c.	⅝ in. screwed to resilient channels	3 in. glass fiber	50	47	17.9

Case 9 KAL-224- 27 & 28 APA	44 oz carpet and 40 oz pad	1⅝ in. of 75 pcf perlite/sand concrete over ⅝ in. rated sheathing subfloor on 2 × joists at 16 in. o.c.	⅝ in. screwed to resilient channels	3 in. glass fiber	53	74	18.5
Case 10 KAL-736- 12 & 13 ISU	5/16 in. wood block set in mastic	1⅝ in. of 60 pcf lightweight concrete on ½ in. insulation board over ½ in. rated sheathing subfloor on 2 × joists at 16 in. o.c.	⅝ in. screwed to resilient channels	3 in. batt	54	53	18.8
Case 11 BBN 670602 & 670601 USG	None	⅝ in. gypsum concrete over ⅝ in. rated sheathing subfloor on 2 × joists at 16 in. o.c.	½ in. screwed to resilient channels	3 in. mineral wool	56	54	14.5
Case 12 R-TL 81-16 R-IN 81-1 & -2	None	¾ in. of 111 pcf gypsum concrete over ⅝ in. rated sheathing subfloor on 2 × joists at 16 in. o.c.	½ in. Type X screwed to resilient channels	3½ in. glass fiber	60	55	12.9
GC	0.10 in. cushioned vinyl 0.09 in. vinyl sheet					49	
Case 13 R-TL 81-17	None	¾ in. of 111 pcf gypsum concrete over ⅝ in. rated sheathing subfloor on 2 × joists at 16 in. o.c.	½ in. Type X screwed to resilient channels	None	58		12.7
R-IN 81-3						50	
GC	0.10 in. cushioned vinyl						
Case 14 R-TL 81-19	None Carpet and pad	¾ in.[d] of 111 pcf gypsum concrete over ⅝ in. rated sheathing subfloor on 2 × joists at 16 in. o.c.	½ in. Type X screwed to joists	None	50		12.6
R-IN 81-6 GC						56	
Case 15 IAL 5-761-	None	¾ in. of 105 pcf gypsum concrete over ⅝ in. rated sheathing subfloor on 2 × joists at 16 in. o.c.	⅝ in. screwed to resilient channels	2½ in. glass fiber	60*	79	15.0
1 & 2 GC	72 oz carpet and 46 oz pad						

TABLE 10.9 Laboratory Ratings for Floors: Lightweight and Gypsum Concrete over Conventional Wood Joist Floor (*Continued*)

Test number & sponsor	Finish floor	Deck[a]	Gypsum wallboard ceiling[b]	Insulation	STC[c]	IIC	Weight (lbs/sq ft)
Case 16 IAL 5-761- 3 & 4 GC	None 72 oz carpet and 48 oz pad	3/4 in. of 105 pcf gypsum concrete over 5/8 in. rated sheathing subfloor on 2 × joists at 16 in. o.c.	5/8 in. screwed to resilient channels	None	55	 75	15.0
Case 17 IAL 6-019-1 & 6-035 GC	None 60 oz carpet and 24 oz pad	1 in. of 111.5 pcf gypsum concrete on 1/2 in. of 24.9 pcf sheathing over 5/8 in. rated sheathing subfloor on 2 × joists at 16 in. o.c.	1/2 in. Type X screwed to joists	None	50	 75	16.0
Case 18 R-IN 81 -11, -12, -13 & -14	0.36 in. foam-back parquet 0.31 in. regular parquet 1/8 in. VA tile 0.09 in. vinyl sheet	1 3/8 in. of 106 pcf gypsum concrete over 5/8 in. rated sheathing subfloor on 2 × joists at 16 in. o.c.	1/2 in. Type X screwed to resilient channels	3 1/2 in. glass fiber		55 51 51 51	18.2
Case 19 R-IN 81-8 & -10 GC	39 oz glued carpet	1 3/8 in. of 106 pcf gypsum concrete over 5/8 in. rated sheathing subfloor on 2 × joists at 16 in. o.c.	1/2 in. Type X screwed to joists 1/2 in. Type X screwed to resilient channels	None		46 51	17.9
Case 20 G&H CA-6MT CCA-7MT CCA	0.063 in. vinyl-asbestos tile Carpet and pad	1 1/2 in. of 100 pcf cellular concrete over 5/8 in. rated sheathing subfloor on 2 ×10 joists at 16 in. o.c.	5/8 in. fire-rated nailed to joists	None	49 48	33 63	22.3 22.0
Case 21 G&H CCA-8MT CCA-9MT CCA	0.063 in. vinyl-asbestos tile Carpet and pad	1 1/2 in. of 100 pcf cellular concrete over 5/8 in. rated sheathing subfloor on 2 × 10 joists at 16 in. o.c.	5/8 in. fire-rated screwed to resilient furring over 5/8 in. fire-rated nailed to joists	None	55 54	43 63	24.8 25.5
Case 22 G&H CCA-10MT CCA-11MT CCA	0.063 in. vinyl-asbestos tile Carpet and pad	1 1/2 in. of 100 pcf cellular concrete over 5/8 in. rated sheathing subfloor on 2 × 10 joists at 16 in. o.c.	5/8 in. fire-rated screwed to resilient furring screwed to joists	None	58	37 73	22.4 23.1

Case 23 G&H CCA-12MT CCA-13MT	0.063 in. vinyl-asbestos tile Carpet and pad	1½ in. of 100 pcf cellular concrete over ⅝ in. rated sheathing subfloor on 2 × 10 joists at 16 in. o.c.	⅝ in. fire-rated screwed to resilient furring screwed to joists	3½ in. glass fiber	61	46	22.6
CCA						79	23.3
Case 24 G&H CCA-14MT CCA-15MT	0.063 in. vinyl-asbestos tile Carpet and pad	1½ in. of 100 pcf cellular concrete over ⅝ in. rated sheathing subfloor on 2 × 10 joists at 16 in. o.c.	½ in. fire-rated screwed to resilient furring screwed to joists	3½ in. glass fiber	60	47	22.0
CCA						73	22.7
Case 25 G&H CCA-16MT CCA-17MT	.063 in. vinyl-asbestos tile Carpet and pad	1½ in. of 100 pcf cellular concrete over ⅝ in. rated sheathing subfloor on 2 × 10 joists at 16 in. o.c.	½ in. fire-rated screwed to resilient furring screwed to joists	None	56	37	21.8
CCA						70	22.5

[a] While the listed assemblies were tested using plywood, it is accepted that other wood structural panels, OSB and COM-PLY®, may be substituted on a thickness for thickness basis based on their substantially similar strength and stiffness properties and slightly higher density without compromising the STC or IIC ratings of the tested assemblies.

[b] Except Cases 4 and 5.

[c] Asterisk (*) indicates values are for field sound transmission class (FSTC).

[d] Manufacturer recommends 1 in. thickness.

Case 9 – Illustrated

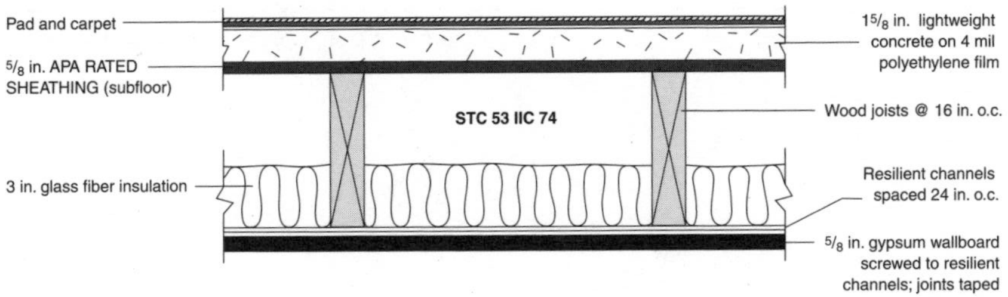

Pad and carpet

5/8 in. APA RATED SHEATHING (subfloor)

3 in. glass fiber insulation

STC 53 IIC 74

15/8 in. lightweight concrete on 4 mil polyethylene film

Wood joists @ 16 in. o.c.

Resilient channels spaced 24 in. o.c.

5/8 in. gypsum wallboard screwed to resilient channels; joints taped

TABLE 10.10 Field Tests of Sound Insulation Compared with Laboratory Tests[a]

Details of basic construction	Laboratory tests			Field tests		
	STC	IIC	Test ref.	FSTC	IIC	Field modifications or conditions
Conventional wood joists Subfloor, insulation, ceiling on resilient	46	46	KAL224-1 & -2	46	44	1⅛ in. T&G plywood subfloor
channels. Tile flooring.				51	41	½ in. fiberboard over ¾ in. plywood
Conventional wood joists; lightweight	47	66	KAL224-29 & -30	47		Possible perimeter leaks
concrete topping				44		
over plywood subfloor	48	68	G&H-USDA-1ST	49	63	
Ceiling nailed; no insulation.			G&H-LCR-1MT	48	66	
Carpet flooring.				48	71	
Ceiling on resilient channels; no insulation. Carpet flooring.	55	67	G&H-USDA-3ST	52	74	
1½ in. glass fiber. Carpet flooring.	58	67	G&H-USDA-2ST	56	75	Corrected a leak through fireplace
½ in. sound board under concrete; 3 in. insulation. Carpet flooring.	59	72	G&H-USDA-9ST	57	77	
½ in. sound board under ceiling channels; 3 in. insulation.	56		G&H-USDA-4ST	57	78	Carpet flooring
Long-span joists Particleboard on ¾ in. plywood; no insulation; ceiling nailed. Carpet flooring.	48	62	R-TL 70-48 R-IN 70-7	45	70	

Stressed-skin panels Lightweight concrete over factory-glued panels;[b] ceiling on resilient channels.				56		
Floor-ceiling assembly, as achieved by stacking modular units Plywood floor on upper unit. No "roof," but 3 in. insulation and nailed gypsum wallboard ceiling on lower unit.	53 51	45 80	KAL-224-12 & -13[c] KAL-224-14 & -15[d]	49	42	No finish floor. 3/8 in. plywood on 1/2 in. sound board "roof" on top of joists of lower unit.
Party wall Achieved by assembling modular units with 1 in. space between. Each wall with 5/8 in. gypsum wallboard, 2 × 3 studs spaced 16 in. o.c., insulation, and 1/2 in. sound deadening board.	49			47		
Same, but with glued plywood instead of the sound deadening board.	50					

[a] Floors carpeted except as noted.
[b] Stressed-skin panels had 5/8 in. top skin, 2 × 8 stringers, 2 × 4 T-flanges on bottom, no bottom skin.
[c] Hardwood flooring (25/32 in.) over 1/2 in. subfloor.
[d] Carpet and pad over 1 1/8 in. subfloor.

tural panel products in lieu of plywood should not compromise the STC or IIC of the tested systems.

Some assemblies contain proprietary products, so test sponsors should be contacted for additional details.

Test Sponsors

APA	APA—The Engineered Wood Association, Tacoma, Washington
USDA	USDA Forest Service, Wood Construction Research, Seattle, Washington
ISU	Iowa State University, and U.S. Forest Service, Division of Forest Economics and Marketing Research, Washington, DC
NBS	National Bureau of Standards, Washington, DC (now National Institute of Standards and Technology, Gaithersburg, Maryland)
USG	United States Gypsum Company, Chicago, Illinois
W	Weyerhaeuser Company, Dierks Division, Hot Springs, Arkansas
WWPA	Western Wood Products Association, Portland, Oregon
GC	Gyp-Crete Corporation, Hamel, Minnesota (now Maxxon Corp.)
CCA	Cellular Concrete Association (inactive)
GFP	Greenwood Forest Products, Lake Oswego, Oregon

10.2.7 Field-Tested Acoustical Ratings for Walls and Floor-Ceiling Assemblies

Field acoustical tests performed by the USDA Forest Service in apartment buildings in Seattle, Washington, demonstrated that field STC (FSTC) values can closely approach laboratory values if careful installation practices are followed. Comparison of field-measured insulation of wall and floor-ceiling constructions with laboratory tests of similar wall and floor assemblies revealed that the sound transmission predicted by laboratory tests can be closely approximated in the field, unless oversights in construction contribute to sound leaks or flanking. However, these ratings can be achieved only with determined effort at every stage in the design and construction process. Table 10.10 shows results of field acoustical tests.

For field-tested walls that compare with laboratory tests of comparable constructions, the average difference between predicted and actual performance was 3½ points. When corrections to sound flanking and leaking paths were made, the average difference reduced to 2½ points, which is less than can be perceived by the human ear.

Values were even closer for floors. Average FSTC values determined for field-tested floor-ceiling assemblies were only one point lower than the average of STC values based on laboratory tests of related assemblies. Comparisons between impact insulation class (IIC) values measured in the laboratory and in the field were more limited than those of airborne sound transmission values, but they indicate that laboratory and field impact noise ratings were of the same general magnitude. Field STC values also were measured for party wall and floor-ceiling assemblies in a stacked two-story manufactured (modular) motel building. Plywood sheathing was used in both wall and floor-ceiling construction, primarily for resistance to racking forces during transportation and erection.

10.3 REFERENCES

1. ASTM E84, *Standard Test Method for Surface Burning Characteristics of Building Materials,* American Society for Testing and Materials, West Conshohocken, PA.

2. APA Form Y380, *Fire Hazard Classification of PS-1 Plywood,* Research Report 128, APA—The Engineered Wood Association, Tacoma, WA.
3. Forest Products Society, *Wood Handbook: Wood as an Engineering Material*, Madison, WI, 1999.
4. ASTM E119, *Standard Test Methods for Fire Tests of Building Construction and Materials,* American Society for Testing and Materials, West Conshohocken, PA.
5. Uniform Building Code Standard 26-2, *Test Method for the Evaluation of Thermal Barriers,* International Conference of Building Officials, Whittier, CA, 1997.
6. International Building Code—2000, International Code Council, Inc.; available from International Council of Building Officials, Whittier, CA; Building Officials and Code Administrators International, Inc., Country Club Hills, IL; Southern Building Code Congress International, Inc., Birmingham, AL.
7. International Residential Code for One- and Two-Family Dwellings—2000, International Code Council, Inc.; available from International Council of Building Officials, Whittier, CA; Building Officials and Code Administrators International, Inc., Country Club Hills, IL; Southern Building Code Congress International, Inc., Birmingham, AL.
8. Keenan, F. J., and A. T. Quaile, "Evaluation: Rehabilitation of Fire-Damaged Members—Axially Loaded Members; Glulam Bending Members; Solid Sawn Timbers," in *Evaluation, Maintenance and Upgrading of Wood Structures,* ed. A. D. Freas, American Society of Civil Engineers, New York, 1982, 172–178.
9. Freas, A. D., "Designing to Avoid Problems: Fire—Heavy Timber and Glued-Laminated Constructions; Ordinary and Light-Frame Construction," in *Evaluation, Maintenance and Upgrading of Wood Structures,* ed. A. D. Freas, American Society of Civil Engineers, New York, 1982, 172–178.
10. ASTM E989, *Standard Classification for Determination of Impact Insulation Class* (*IIC*), American Society for Testing and Materials, West Conshohocken, PA.
11. Underwriters Laboratories Inc., *Fire Resistance Directory,* Northbrook, IL (annual).
12. ASTM E108, *Standard Test Methods for Fire Tests of Roof Coverings,* American Society for Testing and Materials, West Conshohocken, PA.
13. APA Form EWS Y245, APA Technical Note: Calculating Fire Resistance of Glulam Beams and Columns, APA—The Engineered Wood Association, Tacoma, WA.
14. Gypsum Association, *Fire Resistance Design Manual,* Publication 600-2000, Washington, DC, 2000.
15. ASTM E90, *Standard Test Method for Laboratory Measurement of Airborne Sound Transmission Loss of Building Partitions and Elements,* American Society for Testing and Materials, West Conshohocken, PA.
16. ASTM E413, *Classification for Rating Sound Insulation,* American Society for Testing and Materials, West Conshohocken, PA.
17. ASTM E492, *Standard Test Method for Laboratory Measurement of Impact Sound Transmission Through Floor-Ceiling Assemblies Using the Tapping Machine,* American Society for Testing and Materials, West Conshohocken, PA.
18. ASTM E336, *Standard Test Method for Measurement of Airborne Sound Insulation in Buildings,* American Society for Testing and Materials, West Conshohocken, PA.

10.4 ADDITIONAL READING

APA Form E30, *APA Design/Construction Guide: Residential and Commercial,* APA—The Engineered Wood Association, Tacoma, WA.

APA Form A310, *APA Design/Construction Guide: Nonresidential Roof Systems,* APA—The Engineered Wood Association, Tacoma, WA.

APA Form W305, *APA Design Construction Guide: Fire-Rated Systems,* APA—The Engineered Wood Association, Tacoma, WA.

APA Form W460, *APA Design Construction Guide: Noise-Rated Systems,* APA—The Engineered Wood Association, Tacoma, WA.

APA Form EWS X440, *APA/EWS Product Guide: Glulam,* APA—The Engineered Wood Association, Tacoma, WA.

APA Form X710, *APA Design/Construction Guide: I-Joists for Residential Floors,* APA—The Engineered Wood Association, Tacoma, WA.

Canadian Wood Council, *Wood and Fire Safety,* Ottawa, ON, 1991.

DuPree, R. B., *Catalog of STC and IIC Ratings for Wall and Floor/Ceiling Assemblies,* 1988; available from State of California, Department of Health Services, Office of Noise Control, Berkeley, CA.

Factory Mutual Research Corporation, *Approval Guide,* Norwood, MA (annual).

Grantham, J. B., and T. B. Heebink, "Field/Laboratory STC Ratings of Wood-Framed Partitions," *Sound and Vibration,* October 1971, pp. 12–18.

ICBO Evaluation Service, Inc., *303 Siding and High-Load Diaphragms*, Evaluation Report No. 1952, Whittier, CA.

Intertek Testing Services/Warnock Hersey, Inc., *Directory of Listed Products Book,* Middleton, WI (annual).

NAHB Research Foundation, *Acoustical Manual—Apartment and Home Construction,* NAHB Study 21021, 1971.

National Fire Protection Association, *Fire Protection Handbook,* 18th ed., Quincy, MA, 1997.

National Building Code of Canada, National Research Council of Canada, Institute for Research in Construction, Ottawa, ON, 1995.

Omega Point Laboratories, Inc., *Directory of Listed Building Products, Materials and Assemblies,* Elmendorf, TX.

Puchovsky, M. T., *Automatic Sprinkler Systems Handbook,* 8th ed., National Fire Protection Association, Quincy, MA, 1999.

Underwriters Laboratories Inc., *Building Materials Directory,* Northbrook, IL (annual).

Underwriters Laboratories Inc., *Roofing Materials and Systems Directory,* Northbrook, IL (annual).

Underwriters Laboratories of Canada, *List of Equipment and Materials,* vol. 2, *Building Materials;* vol. 3, *Fire Resistance Ratings,* Scarborough, ON, 1993.

Urban-Wildland Interface Code, International Fire Code Institute, Whittier, CA, 1997; available from International Conference of Building Officials, Whittier, CA.

Yerges, L. F., *Sound, Noise, and Vibration Control,* Van Nostrand Reinhold, New York, 1969.

CHAPTER ELEVEN
FIBER-REINFORCED POLYMER (FRP)-WOOD HYBRID COMPOSITES

Sheldon Q. Shi, Ph.D.
Associate Scientist

11.1 INTRODUCTION

Fiber-reinforced polymer (FRP)-wood hybrid composite materials are high-performance advanced engineered wood composites resulting from a combination of wood with fiber-reinforced polymers as reinforcements. The composites may have the advantages of both wood and FRP. Wood is a versatile structural material that can be processed into different forms of wood elements, such as lumber, veneers, flakes, strands, or fibers. These elements can then be combined with adhesives to form a variety of structural wood composites, including glued laminated timber (glulam), structure composite lumber (SCL), I-joists, and structural panel products. Wood has high performance-to-cost and strength-to-weight ratios. FRPs, sometimes called advanced fiber composites (AFCs), are also versatile materials consisting of both synthetic fibers and polymers. Because of FRP's higher strength and stiffness properties compared to wood materials, they can be used as reinforcement of conventional wood composites. With a suitable design for the material configuration, the reinforced engineered wood composites exhibit better performance and may show many potential advantages over traditional engineered wood composites. The following benefits are identified as possible enhancements to engineered wood composites that can be achieved by introducing FRP:

- Increase strength and/or stiffness.
- Reduce the variability in mechanical properties, which allows for higher design values.
- Allow using lower-grade and/or fast-growing species in construction products.
- Reduce the size and weight of the structural members.
- Increase the product ductility, serviceability, and fatigue performance.
- Enhance product durability and dimensional stability.

The reinforcement of structural wood products has been studied for more than 40 years. In the earlier stages of the research, the focus was mainly on using metallic reinforcement, including steel bars, prestressed stranded cables, and bonded steel and aluminum plates.[1–9] The major problem for the metallic reinforcement was the incompatibility between the wood and the reinforcing materials. For example, wood beams reinforced with bonded aluminum sheets had metal-wood bond delamination when the moisture content changed a few percent.[9] Metal is an elastic material, while wood is a viscoelastic material (a material exhibiting both viscous and elastic properties). Also, the hygro-expansion and stiffness behaviors between the wood and reinforcement materials are so different that separation at the glue line or tension failure in the wood near the glue line may occur.

Unlike the traditional metallic reinforcement, fiber-reinforced polymers could be a better reinforced material for structural wood products. Both wood and FRP are viscoelastic materials. Also, there are some similarities in material processing (e.g., resin curing process) between the two materials. Therefore, the incompatibility problems between the wood and the reinforcing FRP are minimized if they are designed properly. Another advantage of FRP over metallic reinforcement is that the FRP materials can be more easily incorporated into the manufacturing process used to produce the wood composites.

The development of FRP-reinforced wood composite materials may significantly increase the potential of expanding the use of engineered wood composites in building construction, including residential, commercial, and nonbuilding construction application. For the FRP itself, it is also an excellent material for retrofitting damaged wood members, which will extend the service life of wood material. FRP-reinforced wood composites have a far greater strength-to-weight ratio than either concrete or steel. From the material and technology standpoint, FRP-reinforced wood composites show remarkable strength in terms of properties and performance. Considerable attention has been focused by the wood industry and different research institutions on this new class of engineered wood material. A state-of-the-art research center, the Advanced Engineered Wood Composites (AEWC) center, has been established at the University of Maine and is supported by several funding agencies and industries. The major research focuses of the AEWC center are on FRP-wood hybrid composites. A commercial FRP-wood composite product, FiRP glulam (using pultruded FRP as reinforcement), has been patented and an Evaluation Report (ER-5100) was issued by the International Conference of Building Officials (ICBO) in September of 1995 on this new reinforced wood product.[10] The American Society for Testing and Materials (ASTM) is drafting a standard for establishing and monitoring structure capacities of fiber-reinforced glulam. The American Institute of Timber Construction (AITC) has established a task committee to develop a reinforcement supplement to the American National Standard Institute (ANSI) glulam standard A190.1. FRP-reinforced wood composites promise to revolutionize the structural wood and wood composites industry.

The objectives of this chapter are: (1) to introduce background information on the fiber-reinforced polymers and their fabrication processes; (2) to discuss the techniques of creating advanced FRP-wood hybrid composites; and (3) to show examples of how FRP-wood hybrid composites have been used in construction applications.

11.2 FIBER-REINFORCED POLYMERS

Fiber-reinforced polymers encompass a wide variety of composite materials with a polymer resin matrix that is reinforced (combined) with fibers in one or more directions. FRP composite properties are directional with the best mechanical properties in the direction of fiber placement. Compared to wood, FRPs have a much higher tensile strength and stiffness. With their high strength and stiffness, the fibers carry the loads imposed on the composite, while the resin matrix distributes the load across all the fibers in the structure. The combination of reinforced fibers and resin matrix is more useful than the individual components. By aligning fibers in one direction in a thin plate or shell, called lamina, layer, or ply, the maximum strength and stiffness of the unidirectional lamina can be obtained. If the fibers are randomly oriented, the same properties in every direction on the plane of the lamina are achieved. The properties of the FRP material are not just predicted by simply summing the properties of its components. The combination of the fibers and resin matrix is the complementary nature of the components. Most polymer resins are weak in tensile strength but are extremely tough and malleable, while the thin fibers have high tensile strength but are susceptible to damage. The following sections detail the comparisons of the different reinforcement fibers and resin matrices that are commonly used in the FRP processing. The different FRP fabrication processes are also discussed.

11.2.1 Reinforcement Fibers

When a material is shaped into a form of fiber, its strength and stiffness are usually much higher than that of the bulk because of the preferential orientation of molecules along the fiber direction and reduced number of defects presented in a fiber. For example, the tensile strength of bulk E-glass is only $0.22–0.84 \times 10^6$ psi (1.5–5.8 GPa) vs. 10.48×10^6 psi (72.3 GPa) for those in fiber form.[11]

The reinforcement fibers are typically 3–20 μm (1 μm = 0.00004 in.) in diameter, similar to human hair. They can be in the form of continuous filaments or discontinuous fibers (chopped fibers). Continuous fibers are long fibers that usually attain maximum strength and stiffness due to their controlled anisotropy and low number and size of surface defects with the load carried mostly by the fibers oriented along the load direction. Continuous filaments are supplied in bundles, such as strands, rovings, or yarns. A strand is a bundle of more than one continuous filament. A roving is a collection of parallel continuous strands forming a cylindrical element. A yarn is a collection of filaments or strands that are twisted together. Rovings are the most common forms of fibers that can be chopped, woven, or processed to create secondary fiber forms for composite manufacturing, such as woven fabrics, knitted fabrics, braided fabrics, and mats. Mats formed either by chopped strand or continuous strand are nonwoven fabrics that provide equal strength in all directions. Continuous-strand mats are formed by swirling continuous strands of fiber onto a moving belt and are finished with a chemical binder that holds the fiber in place.

Discontinuous fibers (short or chopped fibers) are cut from rovings into about 1.5–2.5 in. (2.81–6.35 cm). The length-to-diameter ratio (L/D) is called the aspect ratio or slenderness ratio. The aspect ratio of the discontinuous fiber significantly affects the properties of the short fiber reinforced composites. Chopped strand mats contain randomly distributed discontinuous fibers and held together with a resin. FRP composites made with these chopped fibers arranged randomly have nearly isotropic properties in the plane of the laminate. Short fiber composites usually have lower strength than continuous fiber composites and do not reduce the creep of polymer matrices as effectively as continuous reinforcement. However, short fiber composites find their good application in molded products where the short fibers can be adapted to the product contours more easily.

There are three main types of fiber reinforcements used in polymer matrix: glass fibers, carbon/graphite fibers, and synthetic polymer fibers (such as kevlar and aramid). The majority of the fibers used in the composites industry are glass. The basic building blocks for these fibers are carbon, silicon, oxygen, and nitrogen, each of which is characterized by strong covalent interatomic bonds, low density, thermal stability, and relative abundance in nature. Depending on the fiber type, filament diameter, sizing chemistry, and fiber form, a wide range of properties and performance can be achieved. Table 11.1 shows the major properties of some commonly used reinforced fibers.[11,12]

Glass Fibers. Glass fibers were commercialized in the 1930s. The basis of nearly all the commercial glasses is formed by silica, SiO_2 (about 55–72%). Other constituents of the glass compositions include aluminum oxide, Al_2O_3 and magnesium oxide, MgO. Silica does not melt, but begins to decompose at a temperature of 3,632°F (2,000°C). Using silica as a glass is perfectly suitable for many industrial applications. Glass fibers exhibit many advantages including hardness, corrosion resistance, inertness, light weight, flexibility, and inexpensiveness. However, it needs a high temperature for processing. All glass fibers have similar stiffness but different strength values and different resistance to environmental degradation. The commonly used glass fibers are E-glass (E for electrical), S-glass (S for strength), and C-glass (C for corrosion). Other types of glass fibers include D-glass (D for dielectric) and A-glass or AR-glass (AR for alkaline resistant). E-glass is the most commonly used glass fiber because it is the most economical for composites, offering sufficient strength at a low cost. E-glass is an excellent electrical insulator and is designed for better resistance to water and mild chemical concentrations. S-glass has the highest strength for uses in high-performance applications, such as the aerospace industry, where high specific strength and stiffness are important. However, S-glass cost three to four times more than E-glass, which limits its application. C-glass is usually used for corrosion-resistant applications since it has a much-improved durability upon exposure to acid and alkalis compared to E-glass. D-glass is used for electrical applications such as the core reinforcement of high voltage ceramic insulators. A-glass is used only in a few minor applications.

Carbon/Graphite Fibers. Carbon fibers, also called graphite fibers, are strong, lightweight, and chemically resistant. The beginning of the modern carbon fiber production was in the 1960s. Generally, carbon fibers are produced using the following three types of raw materials or precursors: polyacrylonitrile (PAN), pitch, and rayon $(C_6H_{10}O_5)_n$. Carbon fibers can be produced in three different ways: from gas, liquid, or solid raw materials. Gas-produced carbon fibers use hydrocarbons or organic compounds of transition metals; liquid produced fibers are asphaltic, high-viscosity pitches or bitumens; and solid-formed fibers use polymer fibers such as PAN and rayon. For the three major types of carbon fibers, rayon has the lowest yield (about 25%) and lowest initial modulus. However, the major advantage of rayon fibers is that they possess superior qualities when used as the reinforcement in metal matrix composites. They are slightly dense compared to PAN and pitched fibers.

Carbon fibers made from PAN precursors are much stronger than those made from rayon. They also have a better electrical conductivity. There are two major advantages to use pitch as a precursor for carbon fibers: higher yield and faster production rates. However, pitched fibers are more brittle than those from PAN, and they have a higher density, causing lower specific properties. Pitch fibers are less expensive but have lower strength than PAN fiber. The maximum operating

TABLE 11.1 Properties of the Commonly Used Fibers in FRP

Fibers	Diameter (μm)	Density (g/cm^3)	Tensile modulus (GPa)	Tensile strength (GPa)	Elongation (%)	Coefficient of thermal expansion (10^{-6}/°C)	Thermal conductivity (W/m/°C)	Specific heat (J/kg/K)
E-glass	8 to 14	2.54	72.4	3.45	1.8 to 3.2	5.0	1.3	840
C-glass	–	2.49	68.9	3.16	4.8	7.2	–	780
S-glass	10	2.49	85.5	4.59	5.7	5.6	–	940
D-glass	–	2.14	55.0	2.50	4.7	3.1	–	–
PAN carbon	7 to 10	1.67 to 1.90	228 to 517	1.72 to 2.93	0.3 to 1.0	−0.1 to −1.0	20 to 140	925 to 950
Pitch carbon	10 to 11	2.02	345	1.72	0.4 to 0.9	−0.9 to −1.6	–	–
Rayon carbon	6.5	1.53 to 1.66	41 to 393	0.62 to 2.20	1.5 to 2.5	–	38	–
Kevlar-29	12	1.44	62	2.76	3 to 4	−2	–	–
Kevlar-49	12	1.48	131	2.80 to 3.79	2.2 to 2.8	−2	0.04 to 0.5	1,420
Kevlar-149	–	1.47	179	3.62	1.9	–	–	–
Spectra 900	38	0.97	117	2.58	4 to 5	–	–	–

Note: The values in this table are only for general products. The properties for special formulated or treated fibers can have different property values.

temperature of carbon fibers can be from 600–1000°F (315–537°C). Carbon fibers are much stronger and stiffer than glass fibers. The stress corrosion (static fatigue) phenomenon is less marked. Carbon fibers are also good electrical conductors. The major limitation factor for the application of carbon fibers is the cost. To be competitive, with the advanced composites moving into new application areas, the cost of carbon fibers is decreasing while the demand for them is increasing. The material suppliers are optimistic that the cost of carbon fiber composites can be reduced over the next 10 years.[13]

For carbon fibers used as reinforcement in composite materials, the fibers must go through several processing steps to ensure compatibility with matrix resin system. The first step involves oxidation or chemical treatment of the fiber surface to introduce functional groups (OH, NH_2, COOH, etc.) capable of interacting with matrix resin. The second step involves sizing or coating the oxidized fiber with a coupling agent, and/or resin precursor.

Polymer Fibers. Polymer fibers, sometimes called organic fibers, are made by a process of aligning the polymer chains along the axis of the fiber. They can also exhibit very high strength and stiffness, good chemical resistance, and low density if a suitable process is used. The best-known polymer fibers are the aramid fibers, first commercialized by DuPont in 1971 under the trade name Kevlar. Kevlar fiber is an aromatic polyamide called poly(paraphenylene terephthalamide) in which the aromatic rings make the fiber fairly rigid. The Kevlar fibers usually exhibit high specific strength and stiffness or high specific toughness. They also have a high thermal stability, low creep, and good chemical resistance. Kevlar fibers have been successfully used in different structural applications, including advanced composites, rubber reinforcement, ballistic protection, friction products, and ropes/cables. Spectra is another polymer fiber made from oriented polyethylene. The advantage for the Spectra fibers is that they have a good chemical resistance and low density. However, their maximum operating temperature is relatively low (212°F, or 100°C). There are some other polymeric fibers including aromatic co-polyesters and aromatic heterocyclic polymers that have very limited commercial applications.

Other Fibers. Some other fibers, such as boron fibers and silicon carbide (SiC) fibers, can also be used in fiber-reinforced polymer composites. Boron fibers, produced by chemical vapor deposition on a tungsten wire, commonly have high stiffness, high strength, and low density. Because of the low production rate, boron fibers are among the most costly of the fibers presently made. SiC fibers, characterized by high stiffness, high strength, and higher temperature capacity, are also as high-cost as boron fibers. They have been used as reinforcement in both metal and polymeric resin matrices. SiC fibers also exhibit high stiffness and strength as well as high temperature capability.

11.2.2 Polymer Matrices

The polymer matrix generally accounts for 30–40% of a FRP composite material. The purposes of the matrix material are to hold the fibers together and maintain the fiber orientation, transfer the load between fibers during the FRP composite application, and carry transverse and interlaminar shear stresses within the FRP composites. The polymer matrix also protects the fibers from the environment and mechanical abrasion. The FRP creep property is controlled by the polymer matrix. The rationale for choosing the polymer matrix will depend on the cost, properties, and processing. A polymer matrix falls into two categories: thermoplastics and thermosets. Raw thermoplastic resins can be heated and cooled repeatedly to change their state from liquid to solid and vice versa, while a thermoset resin cannot return to its original state.

Each type of resin offers benefits for particular applications. Thermoplastics will not undergo polymerization during the storage and therefore have an unlimited shelf life. They are easy to handle, easy to repair by welding, solvent bonding, etc., and recyclable (post-formable). However, thermoplastics usually have poor melt flow characteristics, which make them more difficult to process. They are also prone to creep. Thermosets have a low resin viscosity, which will be of benefit to the fiber wet-out during the processing. They have excellent thermal stability after polymerization and are chemical-resistant and creep-resistant. The disadvantages of thermosets are that they are nonrecyclable via standard techniques, brittle, and not post-formable.

Thermoplastics such as polyethylene, polystyrene, polypropylene, and thermoplastic polyesters have been used in the manufacture of wood fiber/polymer composites (WPCs), also called plastic lumber. This product uses wood cellulosic fibers as reinforcements, with the thermoplastics as the matrix. The majority of resins used in the fiber-reinforced polymer processing are thermosets because of their low melting viscosity, good fiber impregnation, fairly low processing temperatures, and low cost. In this chapter, the main focus is on the commonly used thermosets used in the FRP applications. The most common thermosets are polyester, vinyl ester, epoxy, and phenolic resins. The general property information of these thermosetting resins is outlined in Table 11.2.[11,12,14]

Polyester. Polyesters are the most widely used class of thermosets in the construction market. They have a relatively low price, ease of handling, and a good balance of mechanical, electrical, and chemical properties. The unsaturated polyester resin has a low viscosity and can be dissolved in a reactive monomer, such as styrene, divinyle benzene, or methyl methacrylate. These diluents are usually used during the impregnation to reduce the viscosity and increase the degree of cross-linking after cure. Cross-linking reaction between the unsaturated polymer and unsaturated monomer can occur by the addition of heat and a free radical initiator (e.g., organic peroxide), and the low-viscosity solution is converted into a three-dimensional thermosetting polymer. Cross-linking can also be obtained using peroxides and suitable activators at room temperature. The ratio of saturated to unsaturated components controls the degree of cross-linking and thus rigidity of the product.

Polyester can be used in several fabrication processes, including hand lay-up, compression molding, resin transfer molding, and injection molding. Glass fibers are the most common reinforcements for polyester matrices. Polyester can be formulated for use in many outdoor applications (UV resistance, durability, color retention, and resistance to fiber erosion) since it has some degree of resistance to burning, improves impact and abrasion resistance and surface appearance of the final product, and has resistance to chemical attack.

Vinyl Ester. Vinyl ester offers a transition in mechanical properties and cost between the easily processed polyesters and higher-performance epoxy resins which are described in the following paragraph. Vinyl esters are synthesized from an unsaturated carboxylic acid (usually methacrylic acid) and an epoxy resin. Typical commercial resins have only terminal unsaturation (rather than inside the chain), pendant hydroxyl groups, and no carboxyl or hydroxyl end groups, so they are less susceptible to chemical attack. Compared to polyesters, vinyl ester resins shrink

TABLE 11.2 Properties of the Commonly Used Resin Matrices in FRP

Resin matrix	Density (g/cm^3)	Tensile modulus (GPa)	Tensile strength (MPa)	Compressive strength (MPa)	Elongation (%)	Coefficient of thermal expansion ($10^{-6}/°C$)	Thermal conductivity (W/m/°C)	Shrinkage on curing (%)	Specific heat (J/kg/K)	Glass transition temperature T_g (°C)
Polyester	1.10 to 1.50	1.2 to 4.5	40 to 90	90 to 250	2 to 5	60 to 200	0.2	4 to 12	–	50 to 110
Vinyl ester	1.15	3.0 to 4.0	65 to 90	127	1 to 5	53	–	1 to 6	–	100 to 150
Epoxy	1.1 to 1.4	2 to 6	35 to 130	100 to 200	1 to 8.5	45 to 70	0.1 to 0.2	1.5	1250 to 1800	50 to 250
Phenolic	1.25 to 1.4	–	55	–	1.8	–	–	1.1	–	–

Note: The values in this table are only for general products. The properties of special formulated resin matrix can have different property values.

less and absorb less water and are more chemically resistant. Different vinyl ester resins are available for applications.

Epoxy. Epoxy resins, first developed in the 1940s, are widely used in applications such as honeycomb structures, airframe and missile application, and tooling because of their versatility, high mechanical properties, high corrosion and chemical resistance, and good dimensional stability. Many epoxy resins are based upon the reaction of phenols with epichlorohydrin having an oxirane ring as their reactive moiety. Compared to polyester, epoxy resins shrink less and have higher strength/stiffness at moderate temperatures. They also cure slowly and are quite brittle after they are fully cured. However, they can be toughened with additives, including the addition of thermoplastics or multifunctional epoxides. Epoxy resins typically are twice the cost of vinyl esters.

Phenolics. Phenolic resins are the predominately used adhesive system for the wood composite industry. Therefore, as a reinforcement of wood composites, FRPs manufactured using phenolic resin should be more compatible with the wood composite materials. Phenolic resins are usually dimensionally stable to temperature. They have excellent physical and mechanical durability. They also have a good adhesive property, low smoke production, and low flammability. Phenolic resins are usually used in sheet molded compound (SMC), pultrusion, and filament winding. Processing phenolics is more complicated than processing other thermosets because of water being released during the curing process. Phenolic resins are usually modified using an elastomer or resorcinol. Phenol-resorcinol formaldehyde (PRF) resins are very popular as a resin matrix for FRP and as a binder in many other applications.

Polyurethane. Polyurethanes appear in a variety of forms used as coating, elastomer, foam, or adhesive. However, they are all based on the exothermic reaction of an organic polyisocyanate with a polyol. As a coating material in exterior or interior finishes, polyurethanes are tough, flexible, chemical-resistant, and fast-curing. As an adhesive, polyurethane bond usually has good impact resistance, fast curing, and good bond to different surfaces. One formulation of polyurethane adhesive, polymeric diphenylemethane diisocyanate (PMDI), has been widely used in the wood structural panel industry. Similar to phenolic resin, polyurethane is also dimensionally stable and has excellent physical and mechanical durability.

11.2.3 Fillers and Additives

Filler is the least expensive major ingredient in FRP components. The major purposes of adding fillers are to reduce the cost and improve the performance. By using the fillers properly, the properties of the FRP, including dimensional stability, water resistance, weathering, surface smoothness, stiffness, flame/smoke suppression, and temperature resistance, can be improved. The inorganic fillers are being used increasingly. The most common inorganic fillers are calcium carbonate, kaolin, alumina trihydrate, and calcium sulfate. Other commonly used fillers used in FRP are silica, talc, and mica.

Additives are also used in FRP process. The additives may increase the material cost, but they can enhance the FRP processability by modifying the properties and performance of the materials. It has been shown that the additives can enhance the fire resistance, emission control, air-release capability, viscosity control, electrical conductivity, and toughness. For example, antioxidants can help to inhibit polymer oxidation and degradation. Plasticizers help to improve processing characteristics and give a wider range of physical and mechanical properties. Heat or ultraviolet (UV) stabilizers are used to prevent either polymer degradation or surface and physical property changes due to UV radiation. Colorants are often used in the FRP manufacturing to provide color to the products. In the polyester processing, an organic peroxide such as methylethylketone peroxide (MEKP) is typically used as a catalyst or initiator for room temperature-cured process, and benzoyl peroxide is used for heat-cured molding.

11.2.4 Fabrication of FRP

FRP processing methods can be separated into two groups: opened-mold processes and closed-mold processes. In the opened-mold processes, open-contact molding for forming a new product is in one-sided molds. Wet lay-up (or hand lay-up) and open-mold spray-up are usually used in the opened-mold processes. Closed-mold processes are to transfer the liquid resin from an external source into a dry preform that has been placed in a two-sided matched closed mold. Closed-mold processes include filament winding, pultrusion, compression molding, vacuum bag molding, resin transfer molding, and extrusion. The choice of processing type depends on the type of matrix and fibers used in the FRP manufacturing, the temperature required for forming the products and curing the resin matrix, production rate, quality and performance of the final products, and cost-effectiveness of the process. For example, continuous fibers are primarily used in compression molding, resin transfer molding, and pultrusion application. Chopped strand mat is usually used in the hand lay-up process, continuous laminating, and some closed-molding applications. Following are brief discussions of each processing method.

Hand Lay-up. Hand lay-up, also called wet lay-up, is the simplest, lowest-cost, and most widely used process of FRP manufacturing. Figure 11.1 illustrates the hand lay-up process. In this process, the mold is first treated with the mold release, such as wax, polyvinyl alcohol, silicones, and release papers. The choice of release agent depends on the type of material to be molded. After the release agent is cured, a gel coat, such as polyester, mineral-filled, and pigmented layer, is then applied to the mold before the reinforcement to produce a good surface appearance of the FRP products. Precut continuous strand fiber in the forms of mat, woven roving, or fabric is manually placed in the mold. Catalyzed resin with a viscosity of 1000–1500 centerpoise is applied to the mat. Serrated hand rollers are used to compact the material against the mold for removing the entrapped air. Curing is usually accomplished at room temperature, and the final molded part is removed by pulling the molded product from the mold. The hand lay-up can be partially automated by the spray-up process. The viscosity of the resin in the spray-up process is in the range from 500–1000 centerpoise. Simple wet-preg machines can also be used to introduce resin in controlled amount into the woven fabrics which are then laid on the mold. Typical fiber volume is 15% with the spray-up and 25% with the hand lay-up. The advantages of the hand lay-up and spray-up processes are (1) minimal equipment investment, (2) easy operation, (3) design flexibility allowing larger parts and complex items to be produced, and (4) low void content of the composites. The disadvantages of these processes are (1) labor intensive, (2) low production rate, (3) high emission of volatile organic compounds (VOCs), (4)

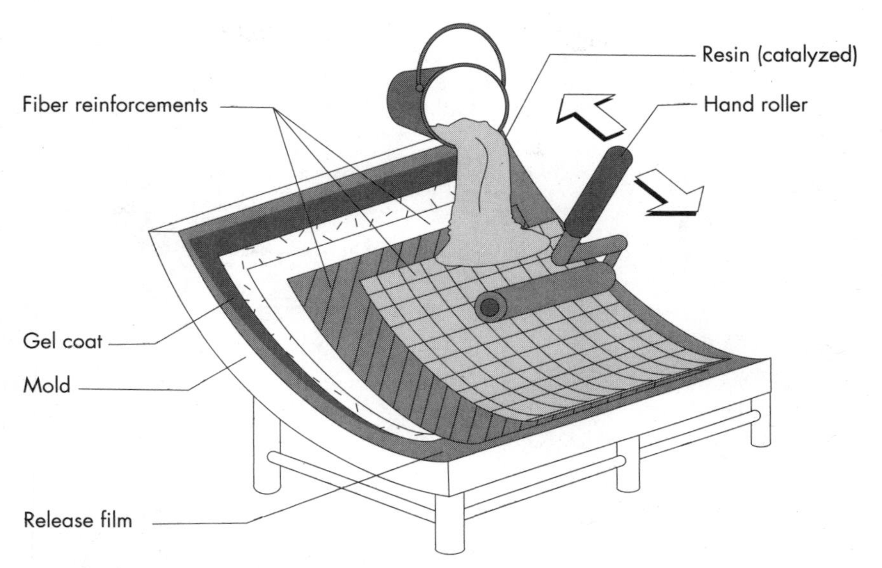

FIGURE 11.1 Hand lay-up process.

only one smooth surface, (5) long curing time, and (6) difficulty of maintaining high quality and product uniformity.

Prepreg Lay-up. Prepreg is a method wherein the fiber reinforcement is impregnated with a predetermined amount of uniformly distributed resin and is partially cured or thickened to attain optimum handling characteristics prior to molding. Most prepregs are made from epoxy resin systems. Thermoplastic matrices can also be used. The reinforcement in the prepreg lay-up process includes glass, carbon, and aramid fibers. The pregregs are usually supplied in rolls and are cut, oriented, and stacked with layers until the desired thickness is reached. An autoclave or vacuum is usually required to provide heat and pressure to assist in consolidating and curing parts laminated with prepregs. Prepreg lay-up methods simplify the manufacturing process of FRP. Fiber distribution is more uniform compared to hand lay-up, and a high fiber volume fraction can be obtained. The major disadvantage of the prepreg lay-up method is the added cost of making prepreg and curing equipment. Like the hand lay-up method, the prepreg lay-up method is a slow and labor-intensive.

Bag Molding. In the bag molding process, fiber and matrix systems are enclosed in a membrane. A uniform pressure, with the aid of a flexible diaphragm or bag, is applied to the laminate to improve the consolidation of the fibers and removal of the excess resin, air, and volatiles from the matrix. The pressure can be exerted either by evacuating the air inside the envelope (vacuum bag molding) or by applying external air pressure (pressure bag molding). Figure 11.2 shows an example of the vacuum bagging. Autoclave molding is used when pressure and high temperature are required. A release agent should be used on both sides of the laminate for easy separation from the mold. A breather layer is usually added to help dis-

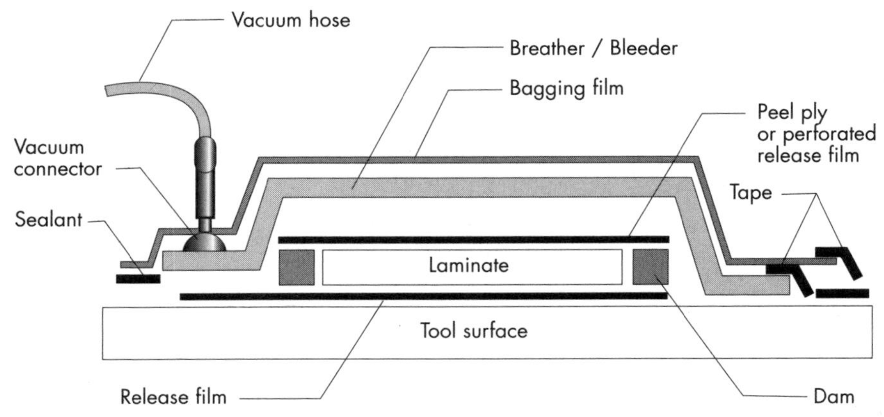

FIGURE 11.2 Bag molding process.

tributing the vacuum and channels the volatiles and excess resin to the vacuum port. By use of the bag molding procedure, a variety of products can be produced. This is an economical method for producing high-quality products with complex shapes.

Compression Molding. Compression molding uses matched male and female metal dies to form the mold. Fiber and resin in wet and prepreg form are placed in open mold. Heat and pressure are used to cure the fibers and resin by closing the male and female halves of the mold in a hydraulic press. After the material is cured, the pressure is released and the part is removed from the mold. Figure 11.3 shows an example of the compression molding process. Preheaters and preformers can be used for the compression molding process to shorten the molding cycles and reduce the entrapped air in the final products. Fiber types used in the compression molding can be continuous fibers, woven fabric, prepreg, or chopped fibers.

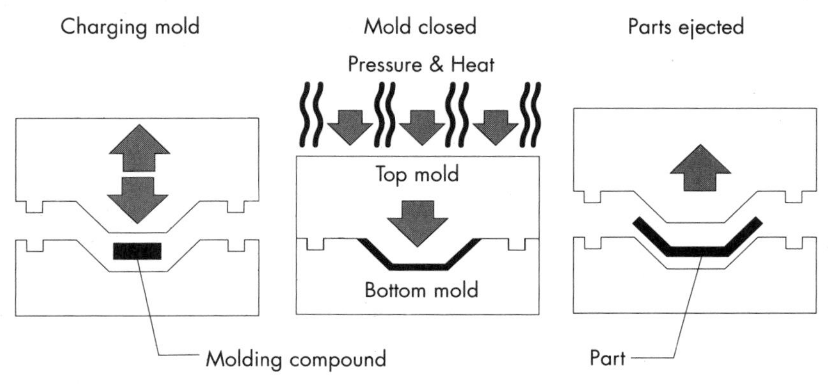

FIGURE 11.3 Compression molding process.

A combination of different fiber forms and types can also be used in one product. Thermoset and some thermoplastic can be used in the compression molding process. The products made from compression molding are not suitable for primary structures because it does not allow a high content of continuous fibers. However, many compression molded products are used in secondary structures.

Pultrusion. Pultrusion is a continuous, automated, cost effective process for high-volume production of continuous-length FRP with constant cross sections. In the pultrusion process, the roving is pulled by a continuous pulling device (e.g., reciprocating pullers) through a resin bath or an injection chamber, where the fibers are saturated with the resin mixture (wetted out). As the wet fibers are pulled through the heated steel forming dies, where the resin is hardened or cured, a rigid profile is formed that corresponds to the shape of the die. Usually, the cured profile shrinks and separates from the walls of the dies and is then pulled into a saw system for dimensioning. For certain applications, a radio frequency (RF) wave generator unit is used to preheat the composites before they enter the die. The RF heater is usually positioned between the resin bath and the preformer. Figure 11.4 shows a simple pultrusion line. The main fiber form used in the putrusion process is roving and continuous strand mat. Different resin systems, including polyesters, vinyl esters, epoxies, and phenolics, have proven to be successful for the pultrusion method. Because pultrusion is a continuous process, the production rate is relatively fast. However, there are some limitations on the pultrusion process:

1. A minimum length of roving or continuous fibers must be used.
2. Fiber volume fraction cannot exceed a certain value.
3. The cross section of the product has to be constant, and the thickness of pultruded walls is limited when standard conduction heaters are used, because of the curing limitation and interlaminar cracking.
4. The void content is difficult to control on thick section.

Resin Transfer Molding (RTM). Resin transfer molding is a closed-mold process in which reinforcement material is placed between two matching mold surfaces.

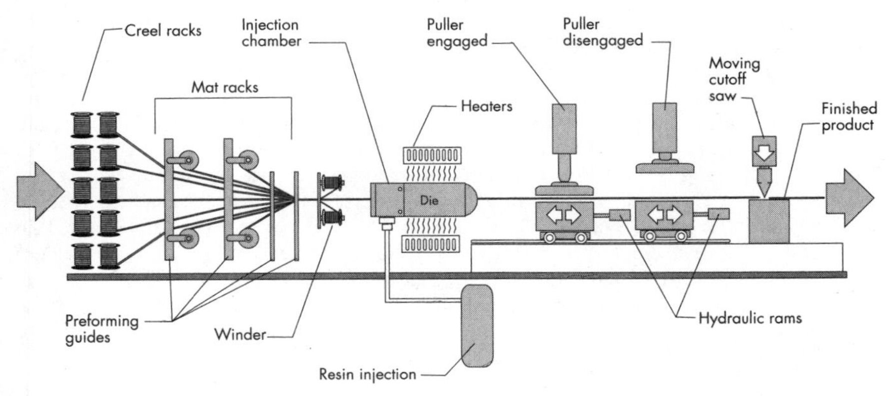

FIGURE 11.4 Pultrusion process.

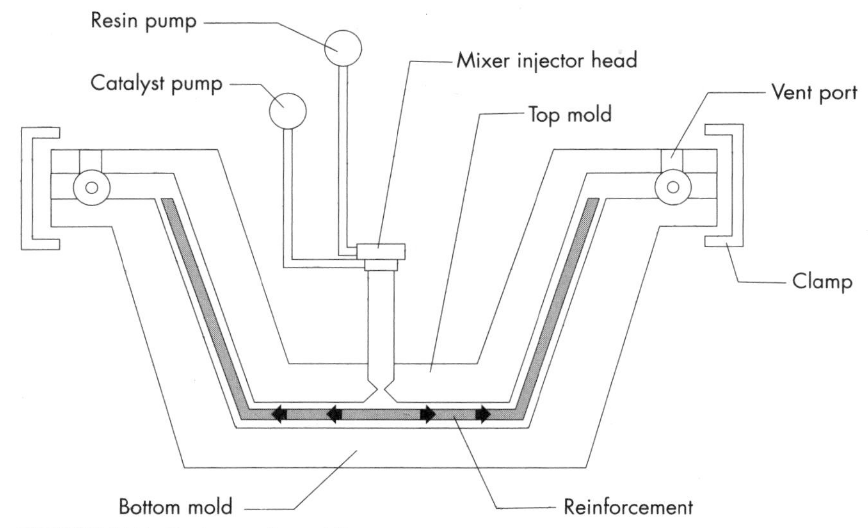

FIGURE 11.5 Resin transfer molding process.

One mold surface is male and the other is female. The matching mold set is then closed and clamped and a low-viscosity thermoset resin is injected under pressure into the mold cavity through a port or series of ports within the mold. The resin is injected to fill all voids within the mold set and thus penetrates and wets out all surfaces of the reinforcing materials. The part is typically cured with heat. Figure 11.5 shows a diagram of the resin transfer molding. A variety of resin systems can be used in the RTM including polyester, vinyl ester, epoxy, and phenolic. The advantages of the RTM include low emissions, high-quality surface, fast process, and tighter dimensional tolerances.

***Vacuum-Assisted Resin Transfer Molding* (*VARTM*).** Unlike resin transfer molding, vacuum-assisted resin transfer molding requires only one tool side in conjunction with a flexible bag. Figure 11.6 shows a diagram of the VARTM process. The reinforcing fabrics made from the dry fibers are placed into a mold. Vacuum is applied to the outlet of the mold and the resin is drawn into the mold by vacuum. A vacuum bag is used to apply pressure to the open surface. While pulling a vacuum, a resin is infused to saturate the fibers until the composite is fully cured. Controlled bagging of preforms and repeatable resin infusions enable the production of specifically designed, high-quality precision parts with consistent dimensional accuracy. The major differences between VARTM and the traditional RTM are that the injection of resin in combination with a vacuum is used for the VARTM process and a single open mold can be used to accomplish the fabrication of the parts by applying a vacuum bag. Since vacuum is used in the VARTM process, the resin flow is enhanced and the void formation is reduced. SCRIMP (Seemann Composite Resin Infusion Molding Process) is a typical VARTM process invented by Bill Seemann in the late 1980s that can be used to infuse thick laminates with the same high-quality results as a simple ⅛ in. (3.17 mm) laminate and provides close control over the resin and reinforcement mix.

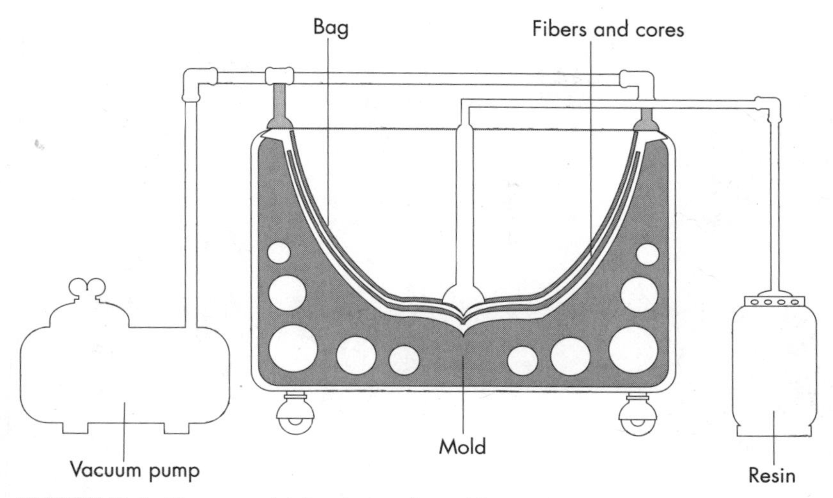

FIGURE 11.6 Vacuum-assisted resin transfer molding process.

Filament Winding. Filament winding is an automated, high-volume process for manufacturing cylindrical-shape products such as pipe, shaft, and pressure vessels. Figure 11.7 shows the typical filament winding process. In the filament winding, the winding machine pulls dry fibers from supply racks through a resin bath and winds the wet fibers around a mandrel, which can either be removable or in situ. The mandrel rotates continuously while the delivery eye moves back and forth. The angle fiber orientation can be varied from 5–90° by adjusting the rotational speed of the mandrel and the linear speed of the delivery eye. The winding angle used in the FRP depends on the strength-performance requirements. The major advantages of the filament winding process are: (1) a high production rate can be obtained, (2) the strength in different directions can be controlled, and (3) different sizes of the products can be produced.

11.2.5 Compatibility of Fibers and Polymer Matrix

The mechanical properties of fiber-reinforced polymer composites are highly dependent on good load transfer from the fibers to the matrix material, which in turn is significantly impacted by the interface between the fiber and matrix. The compatibility between fibers and polymer matrix is the critical step for the manufacture of the fiber-reinforced polymers. Most fiber-reinforced polymer composites fail because of inadequate bonding at the interface between reinforcement and matrix resin. This failure is often due to poor wettability and a lack of chemical compatibility and reactivity between the fibers and matrix. The successful development of fiber-reinforcement polymer composites is dictated by the quality of the fiber-matrix interface. To enhance interfacial bonding between the fibers and resin matrix, the fiber filaments are usually covered or treated by a number of agents to produce chemical change of the surface. These treatments or processes are referred to as sizing.

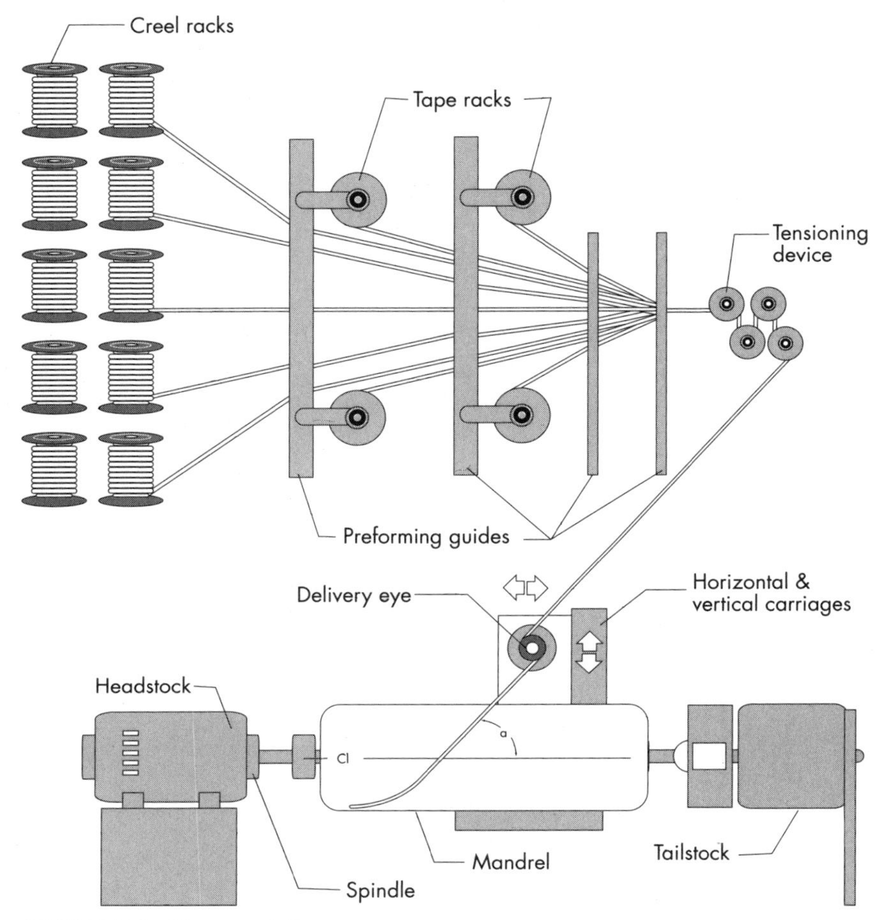

FIGURE 11.7 Filament winding process.

There are many different fiber treatments involved in sizing. Lubricant and antistatic agents are usually used to apply to the fiber filaments immediately after their formation to protect the fibers from breakage as they pass over guide rollers and winders during the processing. Coupling agents are used to improve the resin wet-out on the fibers during the resin impregnation and increase the adhesion between fibers and matrix so that it can enhance the overall strength of the final FRP composites. Also, fiber sizing can reduce the manufacturing time because of the fast resin wet-out on the fibers. Sizing chemistry varies dramatically from one application to another. The same fiber may require a different sizing, depending on the type of polymer that is being used as matrix. The sizing also serves to protect the fiber from moisture attack during the service life of the FRP composites.

The compatibility of fibers and resin matrix can be evaluated according to the interfacial bonding strength and resin wettability on the fiber surfaces. The surface energy of the fibers and resin system is directly related to the resin wettability on the fibers. Contact angle analysis is usually used to evaluate the wettability of a liquid on a solid surface. Interfacial bonding strength can be simply evaluated using interlaminated shear tests. At a micromechanics level, a microbond test or single fiber pull-out test can be employed for the evaluation of the interfacial bonding strength.

11.3 FRP STANDARDS

Several professional organizations have been involved in publishing codes, standards, test methods, and specifications for FRP composites. These organizations include the American Petroleum Institute (API), the American Society of Mechanical Engineers (ASME), the American Water Works Association (AWWA), and Underwriters Laboratories (UL). The American Society of Civil Engineers (ASCE) has a technical committee, Structural Composites and Plastics (SCAP) Committee, to address the design and implementation of FRP composites. The American Association of State Highway and Transportation Officials (AASHTO) Committee on Bridges and Structures has a Technical Committee on FRP composites developing design guidelines for the use of FRP composites in bridge applications. The American Society for Testing and Materials (ASTM) has published recognized industry test methods for FRP composites.[15] Several ASTM committees are working on test methods for the use of FRP repair materials and pultruded structural profiles. For example, ASTM Committee D20.18.01 (FRP material for concrete) is reviewing the test standards for FRP rebar and repair materials. ASTM committee D20.18.02 is developing the test methods for FRP pultruded profiles and shapes. ASTM committee D30.30.01 (composites for civil engineering) is reviewing possible standards for FRP composites products used in construction.

Among the ASTM standards, there are several test methods available for the evaluation of different properties of FRP materials. The major three mechanical properties of FRP materials are tensile strength, compressive strength, and interlaminar shear. In addition, fatigue, void content, and durability may also be important in their applications to reinforce wood composites.

Tensile strength of FRP materials is the most important for the reinforcement design of wood members. Three tensile test methods in the ASTM standards can be used for the tensile tests of FRP materials:[15] ASTM D638, *Standard Test Method for Tensile Properties of Plastic;* ASTM D3039, *Tensile Properties of Polymer Matrix Composite Materials;* and ASTM D5083, *Standard Test Method for Tensile Properties of Reinforced Thermosetting Plastics Using Straight-Sided Specimens.* ASTM D638 is primarily used for the testing of low-modulus fiber composites. The tests are often applied to FRP composites with laminate construction consisting of several layers of fiber mat material. ASTM D3039 is usually used for high-modulus fiber composites such as carbon or aramid construction. ASTM D5083 was designed for quality control (QC) testing. Unlike those used in ASTM D638 and ASTM D3039, the test specimens used in the ASTM D5083 are simple, straight-sided rectangular coupons without tabs. Since there are no tabs in the specimens for ASTM D5083, failure will likely happen in the grips, which can significantly affect the results, and lower test tensile strength and a greater variation are usually obtained compared to those from ASTM D638 and ASTM D3039.

ASTM D3410, *Standard Test Method for Compressive Properties of Unidirectional or Crossply Fiber-Resin Composites,* and ASTM D695, *Standard Test Method for Compressive Properties of Rigid Plastics,* have been developed for the testing of the compressive property of FRP.[15] ASTM D3410 may be more suited to the FRP materials. In ASTM D3410, three configurations for the compression testing are discussed: the Celanese fixture, the Illinois Institute of Technology Research Institute (IITRI) fixture, and the four-point-bending fixture. The IITRI fixture is the most widely accepted for the compression tests. ASTM D695 is designed specifically for low-modulus reinforced or unreinforced plastics.

Shear strength testing is used to evaluate the bonding property between the resin matrix and reinforced fibers. Many ASTM standards are available to evaluate the shear strength of FRP materials. ASTM D2344, *Apparent Interlaminar Shear Strength of Parallel Fiber Composites by Short-Beam Method,* describes the method to determine only the apparent shear strength and not the true interlaminar shear strength. Therefore, the shear results obtained from ASTM D2344 is very useful for quality control purposes but cannot be used to calculate design criteria. Another shear test method can be found in ASTM D5379, *Standard Test Method for Shear Properties of Composite Materials by the V-Notched Beam Method.* The shear values obtained from the method described in ASTM D5379 are 75% less than those obtained from ASTM D2344.[16]

Two standards are used in industry to evaluate the fiber-matrix compatibility, which largely depends on the sizing of the fibers:[15] ASTM D3914, *Standard Test Method for In-Plant Shear Strength of Pultruded Glass-Reinforced Plastic Rod,* and ASTM D4475, *Test Method for Apparent Horizontal Shear Strength of Pultruded Reinforced Plastic Rods by the Short-Beam Method.*

Void content of FRP is directly related to its strength. The voids and pores in the FRP become defects that effectively reduce the strength of the composites. Therefore, the less the void content, the higher the quality of the FRP materials. The measurement of the void content can follow ASTM D2734, *Void Content of Reinforced Plastics.*[15]

Fatigue is one of the critical properties for the materials used in many structural applications where the loading is cyclical, such as for bridges. The standards related to the evaluation of fatigue properties of FRP are:[15] ASTM D3479, *Tension-Tension Fatigue of Polymer Matrix Composite Material;* ASTM E739, *Practice for Statistical Analysis of Linear or Linearized Stress-Life* (*S-N*) *and Strain-Life* (*e-N*) *Fatigue data;* and ASTM E1049, *Practice for Cycle Counting in Fatigue Analysis.*

FRP can be used in cases where it is in direct contact with or is exposed to water or saltwater. Therefore, the resistances of the FRP material to ocean and reagent water can be important. Following are standards related to testing on durability:[15]

ASTM D1141, *Standard Specification for Ocean Water*

ASTM D1193, *Standard Specification for Reagent Water*

ASTM D2247, *Standard Practice for Testing Water Resistance of Coatings in 100% Relative Humidity*

ASTM G53, *Standard Practice for Operating Light-Exposure Apparatus* (*Carbon-Arc Type*) *with and without Water for Exposure of Non-metallic Materials*

ASTM G23, *Standard Practice for Operating Light and Water Exposure Apparatus* (*Fluorescent UV—condensation type*) *for Exposure of Non-metallic Materials*

11.4 APPLICATIONS OF FRP AS REINFORCEMENTS IN ENGINEERING WOOD COMPOSITES

There are many potential ways to reinforce engineered wood and wood composites using FRP materials. The potential benefits for the FRP-reinforced engineered wood composites were described in the Introduction to this chapter. Research and trial evaluations are ongoing looking for the best ways to incorporate FRP reinforcement into wood composites. Questions that need to be answered are:

1. What engineered wood products are suitable for reinforcement with FRP materials?
2. How can they be reinforced?
3. Is it economically feasible?

During recent years, several reinforcing techniques have been proposed and patented and many new technologies on the FRP-reinforced engineered wood composites have been released. Following is a discussion of the different potentials and technologies of FRP-reinforced wood composites.

11.4.1 Reinforcement for Glued Laminated Timber (Glulam)

FRP reinforcement of glulam may have a significant impact on the glulam industry. Glulam is an engineered wood composite manufactured by bonding laminations of dimension lumber together to form a large structural member. The design and manufacture of glulam enables efficient use of forest resources. Because the stresses across the depth of the cross section of the beam are different, lower-strength wood laminations can be positioned in the zones subjected to lower design stresses while the high-strength materials are positioned in those zones subjected to higher stresses. Since the FRP materials have a much higher tension strength than wood, positioning layer(s) of high-strength FRP composites in the high tensile stress area enables the glulam beam to be reinforced and the overall strength and performance of the glulam product increased.

Glulam is often used in beam applications that are subject to bending moment. It has been shown that the tension side of the glulam beam will control failure. Therefore, the optimum reinforcement is achieved by positioning the FRP layer(s) on the tension side of the glulam so that it can force plastic behavior of the wood in compression at failure load, and in turn increase the ultimate bending strength of the beam. Figure 11.8 shows a typical configuration of FRP-reinforced glulam. The outmost lamination of the glulam, called a bumper layer, is used to protect the FRP material from surface damage during handling and installation. It also acts to protect the FRP material from potential fire hazards.

Much research has been conducted on FRP-reinforced glulam during the last decade. Davalos et al. (1992) discuss the response of small yellow-poplar glulam beams reinforced on the tension side with glass/vinyl ester FRP.[17] Tingley and Leichti (1993) conducted research on glulam made from lower-grade ponderosa pine reinforced in the tension zone with pultruded Kevlar and carbon FRP.[18] Sonti et al. (1995) discuss yellow poplar glulam reinforced with pultruded glass/vinyl ester FRP in tension or in both tension and compression.[19] Dailey et al. (1995)

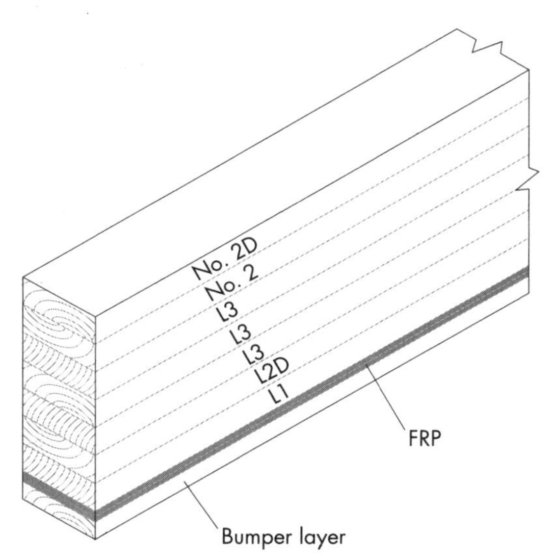

FIGURE 11.8 A design configuration of FRP-reinforced glulam.

studied glulam reinforced in the tension zone with pultruded glass/resorcinol-modified phenolic FRP sheets.[20] Researchers at the AEWC Research Center of the University of Maine have been conducting research on FRP-reinforced glulam using Douglas fir and eastern hemlock and are developing a probabilistic computer model, ReLAM, to predict the statistical properties of the strength and stiffness of a population of reinforced glulams.[21]

All research to date has shown that the strength of the glulam is increased significantly using the FRP reinforcement and that stiffness can also be increased, but to a lesser degree. Depending on the percentage of the reinforcements and material lay-up, the percentage of the property increase can be up to 120% for bending strength or modulus of rupture (MOR), and 20% for stiffness or modulus of elasticity (MOE). Research has shown that the creep behavior of FRP-reinforced wood is primarily dominated by the creep of wood.[22] It is also shown by Dagher et al. (1998) that despite the increased loading for FRP-reinforced glulam, there is no increase in relative creep compared to conventional glulam.[23]

The processing techniques of incorporating FRP into glulam can directly affect the product performance and overall cost of the glulam member. One simple method is to bond one or more layers of FRP sheets (typically ⅛ in. (3.17 mm) or less in thickness) to the wood laminations using adhesives. This process does not require any significant process changes for the conventional glulam manufacturing process. Adhesives used must be capable of bonding the FRP to the wood lamination. Currently, the most common FRP materials used for glulam reinforcements are manufactured through a pultrusion process, although the use of other processes such as wet prepregs has shown promise.

The cost issue is the major obstacle to marketing of FRP-reinforced glulam. Since the FRP materials are much more expensive than wood laminates, the overall cost of the FRP-reinforced glulam may be more expensive than that of the conventional glulam. With respect to reducing the overall cost of FRP-reinforced glulam, several concepts need to be considered. One is reducing the component cost in the composites. The other is reducing the processing cost of the composites.

From the material standpoint, FRP reinforcement allows the use of lower-grade wood materials in the glulam beam configuration. This use of lower-cost wood laminates in the FRP-reinforced glulam could reduce the overall cost of the beams. Efforts should also be made to reduce the cost of FRP materials used in wood composites. Using lower-cost adhesives to bond the FRP and wood together may also contribute to reducing the overall cost.

Another effective solution to reducing the overall cost of the FRP-reinforced glulam may be from the fabrication standpoint. There are many similarities in the manufacturing processing techniques between FRP and wood composites. In using thermosetting resins, the curing mechanisms for the resin matrix in FRP and the adhesives in glulam and other wood composites are similar. Therefore, directly using the reinforced fibers or fabrics in the glulam and other wood composites may reduce the cost significantly compared to using pultruded FRP as the reinforcement. Research is ongoing in this direction, and many new technologies should evolve in the future.

FRP-reinforced glulam has been used in many different applications. Following are some application examples of FRP-reinforced glulam developed by University of Maine, AEWC center.

Figure 11.9 shows a 44 ft (13.4 m) long, 16 ft (4.9 m) wide bridge constructed in 1997 at West Seboeis, Maine. This is a FRP-reinforced glulam girder bridge with a glulam deck. The design live load is 172,000 lb (765 kN). The glulams were made from red pine. The FRP used as reinforcement in the glulam has a tensile strength of 120,000 psi (827,000 kPa) and a tensile modulus of 6.6×10^6 psi (45.5 GPa).

Figure 11.10 shows another example of the application of FRP-reinforced glulam, Bar Harbor Yacht Club Pier at Bar Harbor, Maine. This pedestrian pier is 124 ft (37.8 m) long and 5 ft (1.5 m) wide. The design live load of this pier is 85 psf (4.1 kPa). The glulams are made from red maple. The FRP used in glulam has the same tensile strength and modulus as that used in the West Seboeis Bridge. This FRP-reinforced glulam pier costs 25% less than steel construction.

Figure 11.11 shows an FRP stress-laminated deck located in Milbridge, Maine, constructed in 1996. Usually steel stressing bars are used to post-tension the lon-

FIGURE 11.9 FRP-reinforced glulam used in West Seboeis Stream Bridge. (*Courtesy of University of Maine.*)

FIGURE 11.10 FRP-reinforced glulam used in Bar Harbor Yacht Club Pier. (*Courtesy of University of Maine.*)

FIGURE 11.11 FRP-reinforced glulam used in Milbridge bridge. (*Courtesy of University of Maine.*)

gitudinal wood laminations in short-span bridge construction. Glass fiber-reinforced polymer (GFRP) tendons were used in this bridge as an alternative stressing system. This bridge is 16 ft (4.9 m) long and 24 ft (7.3 m) wide and made from eastern hemlock. The GFRP tendons have a tensile strength of 26,000 lb/tendon (116 kN/tendon) and a tensile modulus of 6.0×10^6 psi (41.4 GPa). From the field report,[24] after 2.5 years of service the bridge deck retained 86% of its initial prestress and performed satisfactorily without being restressed. In contrast, the steel threadbars of a similar bridge located in a similar environment had to be retensioned twice during the first two months. The bars lost 67% of their prestress in 27 months following the second restressing.

Figures 11.12 and 11.13 show an example of the use of FRP-reinforced glulam used in a commercial building application. Figure 11.12 is a close-up showing the FRP material positioned between the outmost and second lamination on the tension side. Typically the percentage of FRP used in this type of application is between 0.5% and 1.5%. Figure 11.13 shows the FRP-reinforced glulam in the completed roof structure.

FIGURE 11.12 FRP-reinforced glulam used in a commercial building application.

FIGURE 11.13 FRP-reinforced glulam used in a completed roof structure.

11.4.2 Reinforcement for Structural Wood Panels

Structural wood panels, such as plywood and oriented strand board (OSB), are used extensively in residential and commercial building application as wall sheathing, roof sheathing, and flooring. They are also used in numerous nonbuilding applications such as shipping containers, railroad cars, truck liners, concrete forms, pallets, and bins. Most of these panel applications require good flexural properties such as high moment capacity and stiffness (*EI*). Their bending strength and stiffness can be improved by overlaying layers of FRP sheets on the surfaces of structural panels. Figure 11.14 shows a reinforcement configuration in a research report published by APA in 1972.[25] This is a sandwich panel combining fiberglass-reinforced polymer and plywood. The research results showed that in considering the properties parallel to face grain, the overlay systems contributed from 14–70% of the stiffness and from 22–88% of the strength, depending upon the relative thickness of the plywood and the overlay reinforcement and the weight and type of glass reinforcement.

Reinforcing structural wood panels by overlaying FRP sheet can be applied to different types of panels. Bulleit (1985) studied overlaying FRP on flakeboard and particleboard.[26] However, since the whole surface area is covered by the more expensive FRP sheet, the reinforced structural panel will be much more expensive

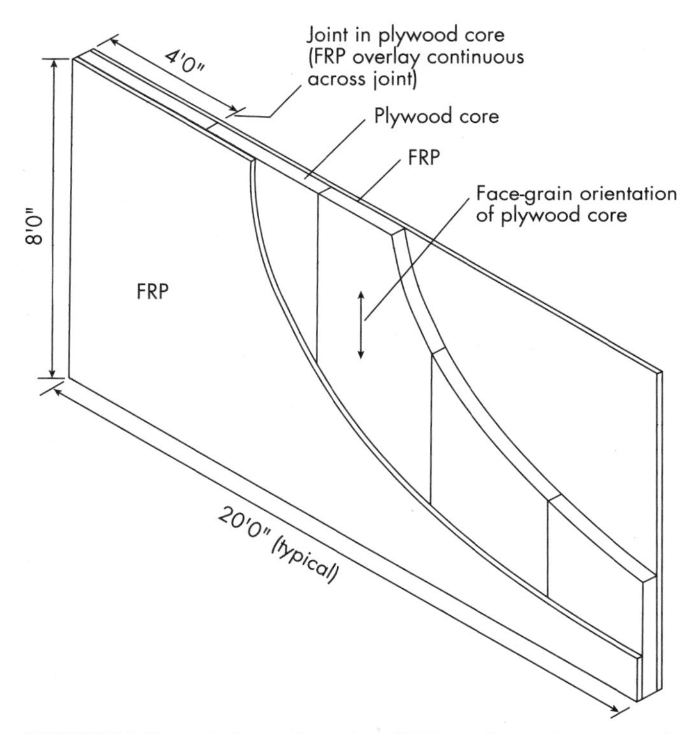

FIGURE 11.14 A design configuration of FRP-reinforced plywood, sandwich panel.

than conventional unreinforced wood panels. This higher cost may limit its applications.

FRPs can also be positioned in the core layers of the panels to improve the shear capacity. This reinforcement technique can be easily employed for plywood by incorporating an FRP sheet into the veneer lay-up. For OSB, it may require changes in the production process in order to position the reinforcement in the core. Another option for enhancing the properties of OSB may be to introduce glass fibers into the manufacturing process.

Considering the reinforcement cost, it may be more effective to apply the FRP reinforcing material to the structural panels during the end-use application. For example, in shear wall applications, since high fastener capacity around the edge of the panel is required in order to provide maximum shear resistance, it may be feasible to put FRP reinforcement around the edge of the panels to increase the fastener/panel strength so that the performance of the entire shear wall can be improved. In addition to increasing the structural performance of wood panels, the introduction of FRP composites may result in enhancing their dimensional stability, durability, and fire resistance.

11.4.3 Reinforcement for Laminated Veneer Lumber (LVL) and I-joist

Wood I-joists are highly engineered wood components used extensively in the floor framing of residential construction and in roof and commercial building applications. The failure modes of I-joists subjected to bending could be tension on the bottom flange, compression on the top flange, shear on the web or web-web joints, web-flange joints, or web buckling. Because they are highly engineered, the failure modes of wood I-joists may be any of these modes. Therefore, the benefit of the FRP reinforcement for enhancing the performance of I-joists may not be as obvious as those in glulam or wood structural panels. In an I-joist application, the maximum tension and compression stresses will occur in the flanges. Therefore, increasing the tensile strength of LVL flange could improve the moment capacity of the I-joists. One study evaluated using FRP to reinforce the LVL used as flanges in wood I-joists. It showed that the addition of two layers of 26 oz (737 g) unidirectional E-glass fabric to a 1 in. (25.4 mm) thick southern pine LVL could increase the mean tensile strength by 24%.[27]

Also, applying a layer of FRP in the LVL flange may prevent web-to-flange knifing through the thickness of the flange, which can control allowable reaction values. FRPs can also be used to reinforce the web in I-joists to improve shear capacity and stiffness. Figure 11.15 shows a conceptual example of a design configuration with FRP reinforcements of both LVL flanges and webs.

Research on FRP-reinforced I-joists also shows that the FRP reinforcement allows lower grades of sawn lumber to be used in the flange. Tingley (1999) reports that using 0.33% reinforcement by cross-section of flange yielded a reduction in the grade requirement from a 2,400 psi (16.6 mPa) tensile material to a 1,200 psi (8.3 mPa) tensile material, with a 25% increase in design capacity.[28]

In I-joist applications, high nailing strength of the top flanges is usually required in order to achieve composite structural action with the wood structural panel flooring system. Therefore, applying a FRP layer on the surface of the I-joist top flange may offer the potential of increasing the nailing capacity and thus improving the structural performance of the floor system. However, further evaluations are necessary for all of these reinforcement configurations to achieve market acceptance.

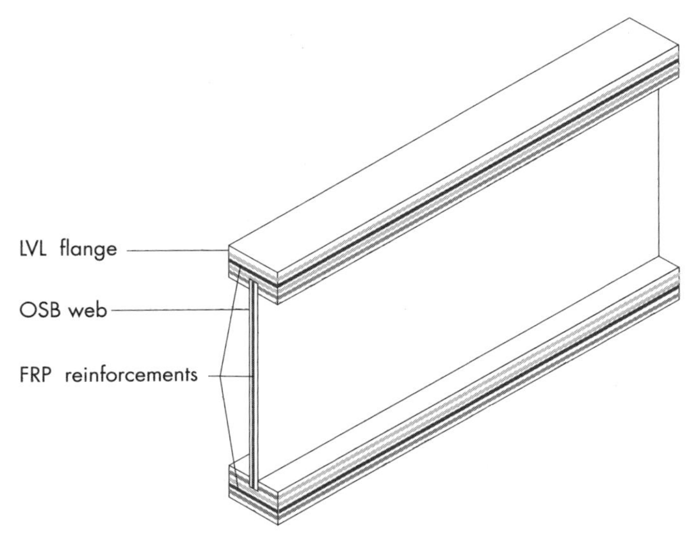

FIGURE 11.15 A design configuration of FRP-reinforced I-joist.

11.4.4 Reinforcement for Other Engineered Wood Products

FRP reinforcements could also apply to many other engineered wood products to improve the performance during their applications. Some research work has been conducted on the reinforcement of sawn lumber using FRP materials. Plevris and Triantafillou (1992) studied the effect of reinforcing fir wood with carbon/epoxy FRP.[29] The effect of reinforced European beech lumber with carbon/epoxy FRP is discussed by Triantafillou and Plevris (1992).[30] Abdel-Magid et al. (1994) studied nominal 2 × 4 in. (50.8 × 101.6 mm) hemlock beams reinforced in tension with carbon/epoxy and Kevlar/epoxy FRP.[31] All these research studies show a significant increase in strength properties for the FRP reinforced lumber, but at significantly higher cost.

Research has shown that FRP can be successfully used as reinforcement on truck trailer structural flooring or truck deck and other similar hardwood transport containers.[32] Traditionally, a relatively expensive high-quality laminated white oak with high density is used for truck decking because of its high strength requirement. It has been observed that many failures occur near the hook joints of the truck decking planks. Therefore, with the application of a thin layer of FRP to the bottom deck surface, the truck deck can be reinforced and a lower-quality species can be used. The reinforced truck deck will also have a lighter weight than the normally used high-quality white oak material. Taking into account the initial material cost and the continuing energy cost associated with the transport of heavier flooring, the overall cost of the FRP reinforced truck deck could be reduced compared to traditional wood materials. Also, the structural performance can be improved by the FRP reinforcement.

FRP is also a good reinforcement for utility pole applications. A research project is ongoing at the University of Minnesota on incorporating fiber reinforcements in hollow utility poles. Basically, the reinforced fibers are applied and wrapped around the stave cylinder and covered by layer(s) of wood veneers.[33] The FRP-reinforced utility pole could have higher strength and lighter weight due to the hollow configuration. It also has the potential to reduce costs.

Slave pallets, or open-deck pallets, also have a potential to be reinforced using FRPs. Pallet products usually require high strength, stiffness, and durability. Slave pallets are usually made with thicker wood panel components (i.e., 2 in. (50.8 mm) or greater). Surface smoothness and high strength-to-weight ratio also benefit this application. FRP reinforcement of slave pallets could result in a thinner pallet, which can save storage space, provide a higher strength-to-weight ratio, and a smoother surface. Similar benefits may also enhance open-deck pallets used in material handling system. FRP reinforcement may also find applications in concrete forming, containers and bins, and other industrial wood products. But research is needed to determine what type of FRPs best fit those end uses and establish value-added cost benefits.

11.4.5 Retrofitting Wood Structures

Rehabilitation of existing materials and structures has been regarded as one of the major areas for FRP composites applications. FRPs have been successfully used to retrofit concrete.[34,35,36] Structural wood members can be damaged by poor maintenance practice and surface degradation due to overloads, pests, or weather conditions over years of use. These structural members include timber bridges, utility poles, waterfront pilings, and other structural wood components. Damages could cause severe strength loss, requiring replacement of the structural members after a certain service time. Since replacement is very expensive, there is a demand to find an effective alternative. In bridge applications, due to the demand to increase the allowable gross vehicle weight, the flexural and shear strengthening of timber bridges to increase their load-carrying capacity is necessary. Instead of replacing the existing structures and members, applying reinforcements to the existing structures could be a more efficient solution. FRP reinforcements save time and money by allowing continuing use of the existing structure and have the added benefits of being a lightweight, noncorrosive, and high-strength material. There are many ways to implement the rehabilitation using FRP. One way is to attach the FRP composites either directly to the existing wood or to a wood laminate, which is then bonded to the wood.

Epoxy injection is another way to accomplish repairs.[37] This involves injecting epoxy into the interior of the damaged area of the wood members. The epoxy injection method is good for the repair of decay or splits in truss joints, glulam members, and solid-sawn timbers. The McGraw-Hill *Handbook of Wood Engineering and Construction* has a chapter devoted to the use of epoxy to repair wood components.[38]

The effectiveness of retrofitting using FRP materials can be shown in the following example. A warehouse, built in 1954, in the Washington, DC, area had severe cracking on its timber structural members, and there were concerns for its possible collapse. The cost of replacing the members was estimated at over $1.2 million. However, epoxy repair methodology saved over $600,000 on this warehouse.[37]

Older utility poles can be restored to their original strength by reinforcing with FRP. This is a cost-effective alternative to pole replacement. It helps extend the life of a pole by restoring bending strength without bonding to the pole while including effective preservative treatment. FRP materials are stronger than those used in other systems, providing superior performance when exposed to moisture, heat, cold, and freeze-thaw cycles. To satisfy the demand for a visually appealing restoration, a wood grain design on its outermost layer can be incorporated. Osmose Inc. developed the FiberWrap II system to provide utilities with an additional option to using steel trusses when restoring poles. This system consists of effective preservative treatments combined with a custom-fit fiberglass repair applied in the field. After the pole is treated internally and externally, several layers of high-strength fiberglass are wrapped onto the pole and saturated with a polyester resin.

11.5 FUTURE RESEARCH

It has been shown in the previous discussions that there are many potential opportunities to combine FRP composites with engineered wood materials. However, the technologies are not mature for many of the applications. Economic viability is a critical issue for FRP-reinforced wood composites. Cost effectiveness must be considered when a new technology is developed. Since there are many similarities in processing between FRP and wood composites, it would be more feasible and efficient to combine the two processes of FRP and wood composites than to combine the two materials. Future research may focus on the processing techniques of FRP reinforcement and on optimizing the manufacturing process to reduce the cost and improve the efficiency of the manufacturing.

Research is also needed to extend the applications of FRP in wood products. There are many opportunities to combine FRP and wood, resulting in a high-performance product. However, it is necessary to investigate these potential applications further. From the theoretical point of view, interfacial bonding between the wood and FRP is a key aspect of the properties of the FRP-reinforced engineering wood products. Research work is important in the development of higher-strength and lower-cost FRP that is compatible with wood materials. Long-term durability of the bonding between FRP and wood is another important issue for FRP-reinforced wood composites.

11.6 REFERENCES

1. Boomsliter, G. P., "Timber Beams Reinforced with Spiral Drive Dowels," *West Virginia University Bulletin,* October 1984.
2. Bulleit, W. M., "Reinforcement of Wood Materials: A Review," *Wood and Fiber Science,* vol. 16, no. 3 pp. 391–397, 1984.
3. Bulleit, W. M., L. B. Sandberg, and G. J. Wood, "Steel-Reinforced Glue Laminated Timber," *Journal of Structural Engineering, ASCE,* vol. 115, no. 2, pp. 433–444, 1989.
4. Coleman, G. E., and H. T. Hurst, "Timber Structures Reinforced with Light Gage Steel," *Forest Products Journal,* vol. 24, no. 7, pp. 45–53, 1974.
5. Hoyle. R. J., "Steel-Reinforced Wood Beam Design," *Forest Products Journal,* vol. 25 no. 4, pp. 17–23, 1975.
6. Krueger, G. P., "Ultimate Strength Design of Reinforced Timber," *Wood Science,* vol. 6, no. 2, pp. 175–186, 1973.
7. Lantos, G., "The Flexural Behavior of Steel Reinforced Laminated Timber Beam," *Wood Science,* vol. 2, no. 3, pp. 136–143, 1970.
8. O'Brien, M. E., Reinforced Laminated Beam, United States Patent 5026593, 1991.

9. Sliker, A., "Reinforced Wood Laminated Beams," *Forest Products Journal,* vol. 12, no. 1, pp. 91–96, 1962.
10. ICBO Evaluation Services, Inc., *Fiber-Reinforced Plastic (FiRP) Reinforced Glued-Laminated Wood Beams,* Report No. 5100, ICBO, Whittier, CA, 1995.
11. Barbero, E. J., *Introduction to Composite Materials Design,* Taylor & Francis, Philadelphia, 1998.
12. Hyer, M. W., *Stress Analysis of Fiber-Reinforced Composite Materials,* WCB/McGraw-Hill, New York, 1998.
13. Wrigley, A., "Interest Growing in Carbon Fiber Composites for Auto," *American Metal Market,* vol. 103, no. 26, p. 7, 1995.
14. Elias, H. G., *An Introduction to Plastics,* VCH, New York, 1993.
15. American Society of Testing and Materials (ASTM), *Annual Book of ASTM Standards,* ASTM, West Conshohocken, PA, 2000.
16. Tingley, A. T., and R. Dandu, *Critical Review of Test Methods to Determine Mechanical Properties of FRP Used to Reinforce Glulam Beams,* Research Paper, at Wood Science & Technology Institute (N.S.) Ltd. Corvallis, OR, 2000.
17. Davalos, J. F., H. A. Salim, and U. Munipalle, "Glulam-GFRP Composite Beams Reinforced with E-glass/Polyester Pultruded Composites," in *Proceedings of 10th Structures Congress,* ASCE, San Antonio, TX, 1992, pp. 47–50.
18. Tingley, D. A., and R. J. Leichti, "Reinforced Glulam: Improved Wood Utilization and Product Performance," Paper presented at Technical Forum, Globalization of Wood: Supply, Products, and Markets, Forest Products Society, Portland, OR, 1993.
19. Sonio, S., J. F. Davalos, R. Hernandez, R. C. Moody, and Y. Kim, "Laminated Wood Beams Reinforced with Pultruded Fiber-Reinforced Plastic," in *Proceedings of Composites Institute's 50th Annual Conference & Expo '95,* Composites Institute of the Society of the Plastics Industry, Inc., Cincinnati, OH, January 30–February 1, 1995, Session 10-B, pp. 1–5.
20. Dailey, T. H., Jr., R. A. Allison, J. Minneci, and R. L. Bender, "Hybrid Composites: Efficient Utilization of Resources by Performance Enhancement of Traditional Engineered Composites with Pultruded Sheets," in *Proceedings of Composites Institute's 50th Annual Conference & Expo '95,* Composites Institute of the Society of the Plastics Industry, Inc., Cincinnati, OH, January 30–February 1, 1995, Session 5-C, pp. 1–4.
21. Dagher, H. J., S. Shaler, J. Poulin, B. Abdel-Magid, W. Tjoelker, and B.Yeh, "Ultimate Strength of FRP-Reinforced Glulam Beams Made with Douglas-Fir and Eastern Hemlock," in *Proceedings ICE,* January 19–22, 1998.
22. Plevris, N., and T. Triantafillou, "Creep Behavior of FRP-Reinforced Wood Member," *Journal of Structural Engineering, ASCE,* vol. 12, no. 2, pp. 174–186, 1995.
23. Dagher, H. J., J. Breton, and S. Shaler, "Creep Behavior of FRP-Reinforced Glulam Beams," *Proceedings ICE,* January 19–22, 1998.
24. Dagher, H. J., and F. M. Altimore, *Field Performance of a GFRP Stress-Laminated Deck Located in Milbridge, Maine,* Research Report No. AEWC 00-02, Advanced Engineered Wood Composites Center, University of Maine, 2000.
25. American Plywood Association (APA), *Basic Panel Properties Plywood Overlaid with Fiberglass-Reinforced Plastic,* Research Report 119, Part 1, APA, Tacoma, WA, 1972.
26. Bulleit, W. M., "Elastic Analysis of Surface Reinforced Particleboard," *Forest Products Journal,* vol. 35, no. 5, pp. 61–68, 1985.
27. Dagher, H. J., S. M. Shaler, and C. Lowry, "FRP Reinforced LVL and Structural I-Joists," First International Conference on Advanced Engineered Wood Composites, Bar Harbor, ME, July 5–8, 1999.
28. Tingley, D., "High-Strength Fiber Reinforced Plastic as Reinforcement for Wood Flange I-Beams," First International Conference on Advanced Engineered Wood Composites, Bar Harbor, ME, July 5–8, 1999.

29. Plevris, N., and T. Triantafillou, "FRP-Reinforced Wood as Structural Material," *Journal of Materials, in Civil Engineering, ASCE,* vol. 4, no. 3, pp. 300–317, 1993.
30. Triantafillou, T., and N. Plevris, "Prestressed FRP Sheets as External Reinforcement of Wood Members," *Journal of Structural Engineering, ASCE,* vol. 118, no. 5, pp. 1270–1284, 1992.
31. Abdel-Magid, B., H. J. Dagher, and T. Kimball, "The Effect of Composite Reinforcement on Structural Wood," in *Proceedings of ASCE 1994 Materials Engineering Conference, Infrastructure: New Materials and Methods for Repair,* San Diego, CA, November 14–16, 1994.
32. Martin, Z. A., J. K. Stith, and D. A. Tingley, "Strength and Stiffness Performance of FRP Reinforced White Oak," in *Proceedings of World Conference on Timber Engineering,* Whistler, BC, 2000.
33. Erickson, R., University of Minnesota, St. Paul, MN, private communication, 2000.
34. Nanni, A., *Fiber Reinforced Plastic (GDRP) Reinforcement for Concrete Structures: Properties and Applications,* Developments in Civil Engineering, 42, Elsevier, Amsterdam, p. 450, 1993.
35. Tighiouart, B., B. Benmokrane, and D. Gao, "Investigation of Bond in Concrete Member with Fiber Reinforced Polymer (FRP) Bars," *Construction and Building Materials Journal,* vol. 12, pp. 453–462, 1998.
36. Theriault, M., B. Benmokrane, and O. Chaallal, "Cracking Behavior of Beams Reinforced with FRP Rebars," in *Proceedings of First International Conference on Composites in Infrastructures,* Tucson, AZ, H. Saadatmanesh and M. Ehsani, eds., pp. 374–378, 1996.
37. Avent, R., "Structure Repair of Timber Using Epoxies," *Structure,* Summer, pp. 39–45, 2000.
38. Faherty, K. F., and T. G. Williamson, eds., *Wood Engineering and Construction Handbook,* 3d ed., McGraw-Hill, New York, 1999.

CHAPTER TWELVE
DESIGNING AND DETAILING FOR PERMANENCE

Ed Keith, P.E.
Senior Engineer, TSD

The majority of the single- and multifamily homes built in the United States are constructed almost entirely out of wood or use a large percentage of wood products in addition to their primary construction material. Many light commercial buildings also make extensive use of wood construction due to its ease of installation, structural integrity, code compliance, and environmental benefits. Wood has an almost unlimited useful life span if protected from ultraviolet (UV) light, moisture, fire, and insect attack. Whereas most fires occurring in modern wood structures are the result of human error and cannot be totally prevented by good design and construction techniques, ultraviolet degradation, moisture problems, and insect attack can be virtually eliminated by proper design, construction, and routine maintenance.

Every element of the structure must be protected from each of the hazards mentioned above. Traditional framing and construction techniques do a good job of providing protection for most building elements. Many elements of traditional construction that we take for granted are there for the express purpose of providing the permanence that we have come to expect in wood structures. Roof overhangs, for example, keep rainwater and direct sunlight off the walls of the structure. The roof overhangs also minimize the amount of water that enters the ground adjacent to basement or crawl spaces.

However, designers often are unaware of the real purpose of such traditional building elements and treat them as mere architectural features. Once they are perceived as an architectural feature, they may be eliminated at the whim of the designer. In the case of the example given above, once the roof overhang is removed to save money or alter the look of the structure, the protection it would have provided is also eliminated.

Some of the consequences of the seemingly insignificant act of reducing or removing the roof overhang are:

- The siding has less protection from UV rays.
- The siding has greater exposure to direct rain.
- The siding undergoes greater extremes of wetting and drying.

- The useful life of the finish on the wood elements of the exterior is reduced.
- The flashing around the windows must be more precisely installed to prevent leaks.
- The basement or crawl spaces are more susceptible to water intrusion.

Another example of unintended consequences resulting from a simple design change is the decision to replace a trussed roof with a conventionally framed vaulted ceiling. A conventionally framed vaulted ceiling is traditionally formed by using 2 × lumber framing supported at one end by the walls of the structure and supported at the other by a beam at the peak. The roof sheathing is attached to the top of the rafters with the drywall attached to the bottom. Insulation is normally placed between the rafters. The problem occurs with trying to provide adequate ventilation between the top of the insulation and the bottom of the sheathing. While these types of roofs have traditionally performed well, given the amounts of insulation required in most parts of the United States today, it is very difficult to create an air space to provide for the passage of air unless the rafters are specifically oversized for this application.

In addition, because each space between the rafters is not connected to any other space, each space must be ventilated independently. The use of skylights interrupts these air passages, if any. The use of ceiling penetrations allows both air and water vapor to be drawn up into the poorly ventilated spaces between the rafters. Any leaks in the roofing due to flashing errors, wind-driven rain, or the formation of ice dams can result in the roof sheathing and lumber framing getting wet. The lack of ventilation will keep the wood from drying out and could result in the decay of the roof framing system. Thus, the designer must take special care in the design of this type of roof to prevent these problems.

Of course, any architectural feature can be successfully designed and detailed once the principles for designing for permanence are understood. In the case of the vaulted ceiling, the use of a scissors truss can impart the same effect without any of the unintended consequences. Wood I-joists could also be used, as described in Chapter 5, since they can be purchased deep enough to provide room for insulation and allow sufficient space for ventilation. The knockouts provided in the I-joist can also be used to provide ventilation above the insulation and around roof penetrations such as skylights. See Fig. 5.45.

There are numerous factors that impact the long-term performance or permanence of wood building components, ranging from ultraviolet degradation due to exposure to the sunlight to predation of the wood structural elements by insects or fungi.

The next sections will cover a number of these factors in practical terms, with specific recommendations for the designing and detailing of roof, walls, and foundations for permanence.

12.1 FACTORS CAUSING DEGENERATION OF WOOD STRUCTURES

There are several factors that can significantly shorten the life of a wood structure, including weathering, decay, insect attack, and excessive moisture intrusion.

12.1.1 Weathering

Natural weathering of wood and finishes involves a complex process that is difficult to duplicate in laboratory tests. Without protective treatment, all wood products exposed to the outdoor environment are subject to a number of physical changes. The most obvious, common, and benign is the change of color caused by ultraviolet light. Light woods become darker, and dark woods become lighter, both converging on the color gray. Hardwoods change color more slowly than soft woods. The gray color, made up of partially degraded cellulose fiber and microorganisms, does not extend very far into the surface of the wood and if left unaltered actually protects the wood from further degradation. The wind, rain, and wind-borne sand and debris continually abrade this fragile coating, allowing more degradation to occur. The erosion rate is about ¼ in. per 100 years.

Unprotected outdoor exposure causes other, more damaging alterations to wood products that must be carefully considered when designing the structure, such as splits and checks in exposed wood products. These splits and checks can cause finishes to fail, permit insects and decay organisms to get beneath preservative treatments, and adversely impact the strength of the wood element.

Environmental factors causing degradation to sidings and other exposed wood products, glued engineered or solid sawn, include solar radiation, heat and cold, water, normal air contaminants, and wind. The importance of a particular weathering factor varies with the geographical location in which the product is located. For example, solar radiation is an important factor in degrading organic polymers in finishes and is usually more severe in the southern part of the United States. Water and temperature variations can also assist in the degradation of finishes and are more likely to occur in the northern states.

12.1.2 Fungi

As evidenced by buildings worldwide, wood construction can provide centuries of service life. However, as a natural, organic material, wood is susceptible to degradation by organisms under certain conditions. The most common organisms that must be taken into consideration during building design are fungi.

Fungi are low forms of plant life that derive their nutrition by using other organic materials as food rather than producing it themselves, as green plants do. For practical purposes, fungi can be separated into decay fungi and nondecay fungi. Decay fungi are probably the most significant organisms responsible for degradation of wood. Degradation by fungi is commonly referred to as rot, decay, brown rot, or dry rot. Nondecay fungi include stains and molds.

Decay Fungi. Decay fungi are spread by microscopic spores, which are produced by the fruiting bodies of fungi. Spores are always present in the atmosphere but need proper conditions to begin their growth. Under suitable conditions, the fungi spores grow into thread-like hyphae that spread throughout the host material and may ultimately produce fruiting bodies, which produce spores for further propagation.

As with all organisms, certain threshold conditions are necessary for survival and propagation of decay fungi. These conditions can be categorized into temperature, food, oxygen, and moisture. In most wood structures, decay can be best eliminated by controlling the moisture content of the wood.

Moisture Control. Decay growth in wood requires prolonged conditions where wood moisture content is in excess of 20–25%. The moisture content of wood components is a function of:

- *Humidity:* Eventually, wood will equilibrate to approximately 6–12 percent moisture content during service in most geographical locations if un-wetted by rain or condensation. The exact moisture content at equilibrium is primarily a function of relative humidity. The time it takes for wood to equilibrate is a function of member size and can be substantial for large sawn timbers and glued-laminated timbers.
- *Direct wetting:* Exposure to direct wetting leads to elevated surface moisture content over the short term and high moisture content throughout the entire wood member if exposure is prolonged. As an example, due to the large exposed surface area, wood structural panels continuously exposed to several days of rain wetting may lead to a moisture content of 50% or more.
- *Condensation:* Most wood components used in construction are ultimately protected from direct exposure to weather. However, some components may be subject to wetting from condensation (see Section 12.1.4).
- *Climate:* Wood products exposed to weather vary considerably in moisture content as they are always seeking equilibrium with changing humidity or undergoing moisture cycling from wetting due to rain or snow. Climate conditions affect decay potential of wood used above ground and exposed to weather. In areas of high rainfall or high humidity, the moisture content may be elevated. For such cases, good design and construction practices combined with the use of preservative-treated or naturally durable woods will minimize risk of decay and help assure good performance. As the exterior exposure conditions and moisture content are a function of climate, a decay probability map can be developed as shown in Fig. 12.1.

Decay Control in Wood Construction. The primary method of preventing decay fungi in wood construction involves keeping the wood below the threshold moisture content needed for decay. The following discussion provides an overview of proper design, storage, construction, and maintenance details that minimize the potential of reaching this moisture level.

Floors. Since floors are enclosed by the building envelope, they are generally at low risk of decay except in circumstances where they are over a damp soil crawl space or where plumbing leaks lead to localized wet spots. Adherence to the following provisions will help ensure good performance.

- *Crawl space ventilation:* Model building codes require a ratio of 1 ft^2 of net free ventilation for every 150 ft^2 of floor area. The ventilation requirements can be reduced to 1 ft^2 for every 1,500 square feet of floor area when a vapor retarder ground cover is placed over exposed soil in a crawl space. See Section 12.1.4 under Ventilation Requirements for more information.
- *Distance between grade and nearest untreated wood:* Codes typically require a distance of at least 6 in. between the grade and nearest untreated wood. All wood in contact with the ground or below grade should be preservative-treated.
- *Treated sill plates:* Wood members in contact with concrete foundations should be preservative-treated or of naturally durable species.

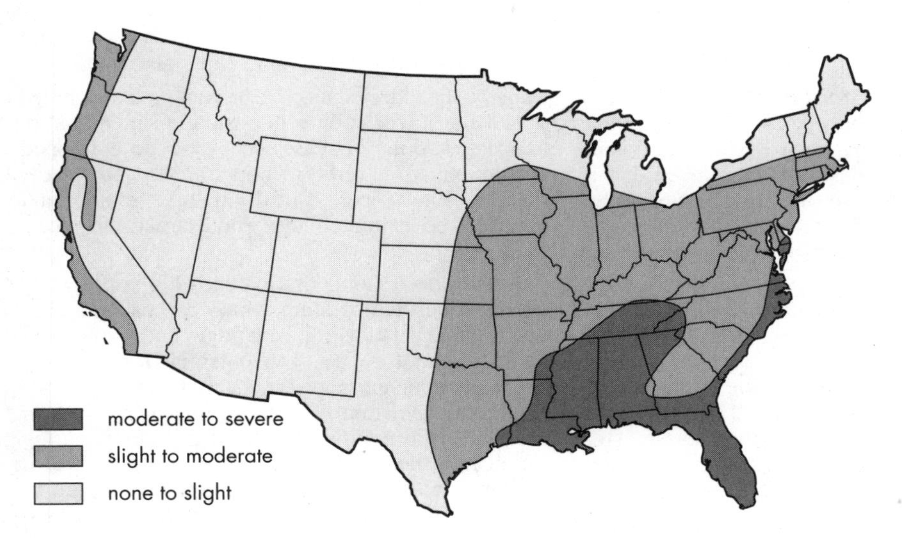

FIGURE 12.1 Decay probability map. (*Source:* International Residential Code for One- and Two-Family Dwelling.)[1]

Roofs. Wood roof members may be exposed to moisture from leaks, from moisture introduced at the time of construction, and from moisture generated from condensation. Attention to design and construction details can significantly reduce these moisture hazards as noted below.

- *Attic area ventilation/vapor retarders:* Model building codes typically require 1 ft^2 of net free ventilation for every 150 ft^2 of attic area. This provision can be reduced to 1 ft^2 for every 300 ft^2 when a ceiling vapor retarder is used. The same reduction applies for sloped (pitched) roofs when at least 50% of the required vent area is located in the upper portion of the space to be ventilated and is at least 3 ft above eave vents. See Section 12.1.4 under Ventilation Requirements more information.
- *Low-slope roofs:* It is often impractical to ventilate low-slope roofs since they generally do not contain ventable attic space. Experience has shown that low-slope roofs in commercial buildings can perform adequately even with minimal ventilation. These unique cases demand attention to other details to minimize entrapment or accumulation of moisture in the roof cavity. Attention to the following provisions minimizes decay hazard in low slope roofs.

  Penetrations in flat or low-slope roofs such as skylights, roof accesses, and HVAC duct openings pose special problems because they form isolated pockets within the roof framing. Often the installed vents do not draw from these pockets and can result in unventilated areas where moisture can build up. When using I-joists in the roof framing, the factory-installed knockouts can be removed from the joists forming these isolated pockets. The joists can then be installed with the knockouts on the top and sized such that the knockouts provide a path to an adjacent vented space. See Chapter 5 for more information on I-joists.

- *Moisture content of wood components:* The limited size of the roof cavity (located between the roofing membrane and the lower ceiling or insulation membrane), especially in conventionally constructed roofs, increases sensitivity to entrapped moisture introduced during construction. It is therefore important to specify the use of air- or kiln-dried lumber and to allow a period of drying after roofing and prior to installation of the insulation and ceiling if the wood structural panels were wetted during construction.
- *Moisture accumulation due to condensation:* Condensation in roofs is primarily dependent upon two factors: the roof deck temperature which can cause a condensation surface on the back side if it drops below the dew point, and the amount of moisture vapor accumulation in the roof cavity. The potential for moisture accumulation in the roof cavity depends upon entrapped moisture discussed above and interior humidity conditions. Interior moisture sources include such things as human occupancy, moisture generated from building use such as manufacturing operations, and moisture infiltration into the building. In some cases, additional venting to the outside of the building and use of vapor retarders are needed to avoid accumulated moisture that can lead to condensation. (See Section 12.1.4 for further information.)
- *After installation:* Once installed, protect wood structural panels as soon as possible with roofing felt or finish roofing material. If panels were wetted prior to roofing installation, allow some time for drying prior to installing the insulation and/or vapor retarder under the roof deck. Handheld moisture content meters are available for making a quick assessment of moisture conditions of the wood components. Target moisture content is 19% or lower.

Walls. Wood components used in walls may be subjected to moisture generated from condensation or moisture intrusion. Proper design, construction, and maintenance prevent these causes of moisture problems. See Section 12.2.2 for more detailed information.

- *Vapor retarders:* Condensation may occur in the inside surface of exterior wall sheathing in cold winter climates when moist air comes in contact with a cooler surface. Such moisture can come from sources inside the house such as cooking, clothes dryers, and showers. Installing exhaust fans over cooking stoves and in high-humidity areas such as bathrooms or laundry rooms can help vent excess moisture to the outside. Clothes dryers should also exhaust to the outside.

  Additionally, a vapor retarder should be installed on the warm side of the wall. Failure to protect the wall cavity from water vapor can result in condensation and elevated moisture content of the wall sheathing and studs. See Section 12.1.4 for more information on condensation.
- *Moisture intrusion:* Performance problems with walls can arise when the weather resistance of the exterior finish system degrades and allows moisture intrusion. Virtually any exterior finish system, whether it is wood siding, vinyl siding, stucco or proprietary systems such as exterior insulated finish systems (EIFS), can lose its moisture resistance, resulting in wall moisture problems. Often these moisture problems can be attributed to installation shortcomings such as lack of building paper and inadequate flashing around doors and windows. Wood structural panel sheathing as well as wall framing require protection from exposure to permanent moisture when used in wall systems.

Decay Control in Exposed Applications. Exposed products, such as siding, are often fully exposed to weather and thus have increased susceptibility to elevated moisture conditions. Although siding products will often experience moisture contents above the threshold value needed to support decay on an intermittent basis, wood-based siding products have a good history of performance due to the fact that they dry below this threshold value before decay can take hold. Proper architectural detailing, use of flashing and caulking, and adherence to the manufacturer's installation recommendations are essential for proper performance. For example, if trim is improperly installed around siding, it may trap moisture and/or reduce the drying ability of the siding. This can lead to long-term moisture accumulation that causes decay.

Exposed end grain of wood products warrants special consideration such as flashings or other means of protection, since the high capillarity of end grain increases water absorption. Use sealants and protective flashing on the end grain of wood members exposed to the elements.

Preservative Treatment. Preservative-treated wood products, which are pressure-impregnated in accordance with standards of the American Wood-Preservers' Association,[2] should be specified for applications that involve high decay hazard. See Chapter 9 for further information.

Mildew/Mold Fungi. Mildew/mold fungi, also known as nondecay fungi, are low forms of plant life similar to decay fungi. The major difference is that mildew and mold fungi do not cause the structural degradation of wood products.

Mold and mildew can be found both indoors and outdoors. *Mold* and *mildew* are terms commonly used interchangeably, although mold is often applied to black, blue, green, and red fungal growths and mildew to whitish growths. The presence of high concentrations of mold and mildew in buildings is an indication of high-moisture conditions that may be detrimental to the structure. Mold and mildew may also cause health problems. This section provides basic information on mold and mildew and methods to minimize the high moisture conditions that can lead to high levels of mold growth in structures. In this section, the term *mold* will be used to cover both mold and mildew unless otherwise stated.

Effect of Mold and Mildew Growth on Wood Components. As organic materials, mold and mildew can readily grow on wood if suitable moisture is present. Mold can grow on wood if exposed to water or prolonged humidity in excess of 70%.

When mold or mildew occurs on wood products during construction, it should be cleaned as specified below and then allowed to dry before being closed in.

When mold or mildew occurs on surfaces of wood products in a finished structure, the products should be similarly cleaned. In addition, the source of excessive moisture must be determined and rectified.

Control of mold and mildew in wood structures. Since mold and mildew require high-moisture conditions, proper moisture design, construction, and maintenance of structures are necessary to maintain moisture levels below the threshold for mold growth.

There are many sources of occupancy moisture that can lead to elevated interior humidity such that mold can grow. Following is a short list of interior moisture loads that can be anticipated in a residential structure:

- Shower — 0.5 pint per 5 minute shower
- Clothes dryer — 4.7 to 6.2 pints per load if vented indoors

• Cooking dinner	1.2 pints (plus 1.6 pints if gas cooking) per family of four
• Dishwashing by hand	0.7 pints per family of four
• House plants	0.9 pints per 6 plants

Cleaning Mold and Mildew. Mold or mildew is often mistaken as dirt. A simple test for mold or mildew is to apply a few drops of 5% solution of household bleach. (It is important to use fresh bleach since bleach deteriorates in potency when older than six months.)

Mold or mildew will usually bleach within one to two minutes. Areas that don't bleach are probably dirt.

Mold and mildew can be removed with commercial mold/mildew removers, following the manufacturer's directions, or with a solution of one part household bleach (5% sodium hypochlorite) mixed in three parts by volume warm water. When using bleach, avoid breathing the vapors and contact with skin and eyes. Children and pets should be kept away from these products.

Floods. Flooding represents an extremely high mold and mildew hazard level. Mold will start growing within 48 hours after floodwaters recede. Because of the high levels that may be encountered, flood-damaged structures present an extreme risk for mold. The American Red Cross Publication 4477, *Repairing your Flooded Home,*[3] and Institute of Inspection, Cleaning and Restoration Certification Standard S500-94, *Standard and Reference Guide for Professional Water Damage Restoration*[4] provide guidance on dealing with flood reclamation.

Health Issues with Mold and Mildew. Excessive mold and mildew growth can pose a potential health risk. The health aspects of molds are beyond the scope of this publication. The American Lung Association[5] is a source of information on health aspects of mold.

12.1.3 Termite Protection for Wood-Framed Construction

Termites occur in every state of the United States except Alaska. The presence or abundance of termites is determined by their environmental requirements such as temperature, humidity, soil moisture, and availability of food. Termite damage can be controlled with proper building practices and preventative measures.

Termite Species. Based on their habitat and mode of attack, termites found in the United States can be grouped in three classes: subterranean termites, drywood termites, and dampwood termites. For more information see Chapters 1 and 9.

Termite Protection. Techniques for termite protection involve prevention of access to wood or moisture required for termite existence.

Job-Site Sanitation. Houses built on land cleared of trees and brush are probably in the midst of subterranean termite colonies in those geographic areas where subterranean termites are known to exist. In these areas job-site sanitation is critical. Proper job-site cleanup includes removal or burning of all debris, lumber, logs, limbs and stumps. The presence of buried wood attracts termites and can lead to infestation of the house. Lumber scraps should be removed from the site prior to enclosing with the wood or concrete floor.

Construction. Where termites are prevalent, the best protection is to build using techniques that prevent their gaining access to the building. Foundations may be constructed using the Permanent Wood Foundation (PWF), poured concrete or masonry block with a poured concrete cap through which the termites cannot penetrate. Crawl space and attic vents must be screened to prevent access of winged termites during mating season.

Required minimum clearances between the ground surface and any untreated wood in the building are presented in Table 12.1. Lesser clearances are also acceptable provided such wood is pressure preservative-treated.

With less than 18 in. of clearance under floor framing or less than 12 in. under floor girders, the shallow under-floor space is generally inaccessible for inspection. In such cases any wood that is at or below the level of the floor sheathing (including the floor sheathing itself, the floor framing, girders, posts, rim joists and blocking, and PWF, if used) must be pressure preservative-treated.

Proper ventilation and use of vapor barriers on the ground in the crawl space will help prevent the moist conditions that subterranean and dampwood termites favor.

Soil Treatment/Wood Treatment. In regions where a termite hazard exists, treat the soil outside of foundation walls, along the inside of crawl space foundation walls, under basement floors or slabs, and at other points of ground contact with termiticides. For under-floor plenum heating/cooling systems use only termiticides that have been approved for plenum applications when treating soil inside the plenum.

If soil treatment is not used in termite hazard regions, preservative-treated wood should be considered for the subfloor sheathing, floor framing, and supports. The foundation walls and underside of the floor structure should be inspected periodically for evidence of termite infestation, especially if untreated materials are used in the floor sheathing, framing, and supports.

Plywood should be treated in accordance with American Wood-Preservers' Association Standards C9 (*Plywood Preservative Treatment by Pressure Process*)[6] and C15 (*Wood for Commercial-Residential Construction, Preservative Treatment by Pressure Processes*)[7] or the American Wood Preservers Bureau FDN Standard (for Permanent Wood Foundation)[8] or equivalent code-approved preservative-treating and quality control requirements and should be marked by an approved inspection agency certified to inspect preservative-treated wood, indicating compliance with these requirements. All such treated wood should be dried to moisture content of 19% (18% for plywood) or less after treatment to minimize subsequent shrinkage.

TABLE 12.1 Minimum Clearance for Untreated Wood to Grade (based on 2000 International Building Code,[9] Section 2304.11)

Outside grade	
• To framing and sheathing	8 in.
• To wood siding	6 in.
Inside grade (crawl space)	
• To floor joists or sheathing	18 in.
• To floor girders	12 in.

12.1.4 Reduction of Moisture through Condensation Control

Whenever moist air comes into contact with a cooler surface, condensation is likely to occur. The cool surface may be the underside of roof sheathing or the inside of exterior wall sheathing in winter, or the underside of a subfloor in summer when the building is air-conditioned.

The only requirements for condensation are moist air and a cool surface. In the winter, the moisture content of the indoor air (usually measured as relative humidity or vapor pressure) is important, as is the temperature of the surface on which this moisture could condense. The amount of moisture in the air outdoors is also sometimes a factor.

Condensation can be controlled in three ways: (1) reduce the amount of moisture initially in the air; (2) prevent the moisture from reaching a cold surface by introducing a vapor retarder; or (3) carry it away by ventilation.

The Language of Condensation Control. Water stays in the air as vapor as long as the temperature of the air and the amount of water vapor are such that the air can hold it. The amount of water in the air, relative to the amount that the air can hold, is called relative humidity. Warm air can hold more water vapor than cold air. Thus, as air with a given amount of water vapor in it is cooled, the relative humidity will rise until a temperature known as the dew point is reached. At this point relative humidity becomes 100%, and some of the moisture will condense as dew. If moist air contacts a surface at or below its dew point temperature, condensation will occur on that surface.

Water vapor in the air produces vapor pressure, which is a measure of moisture vapor concentration. Air with high vapor pressure tries to escape to or seek equilibrium with air of lower vapor pressure. The vapor can escape either with a flow of air through cracks or openings in the building shell or without it by direct penetration of building materials if a differential vapor pressure exists across the material.

Vapor permeance is a measure of the ease with which vapor can penetrate solid building materials. Materials with low permeance are rated as vapor barriers or, more properly, vapor retarders.

Changes in construction due to energy-saving features have tended to increase moisture levels within today's homes. Washers, dryers, cooking, showers, indoor steam rooms, and swimming pools are sources of water vapor within houses. In older houses, air infiltration around doors and windows, and often directly through cracks in the walls, more or less automatically eliminated condensation. With the tighter, energy-efficient houses being built today, control of condensation must be planned.

Condensation Control. The first step in the control of condensation involves reducing excess moisture inside the home.

- Vent clothes dryers to the outside and not into the attic or crawl space.
- Install range hoods over cooking stoves and operate them when any appreciable amount of steam is being generated.
- Install exhaust fans in bathrooms and vent them to the outside, not to the attic (consider wiring the fan so that it goes on automatically with the bathroom light).

Methods of moisture control vary with location in the house. For attics and enclosed cathedral ceilings, the simplest form of control involves ventilation. A ceiling vapor retarder is recommended in conjunction with ventilation for cathedral ceilings. With today's ever-increasing amounts of insulation and tighter construction, a ceiling vapor retarder may not be as necessary for attics when adequate ventilation is provided. Its omission would allow vapor to travel more easily through the ceiling and out through the attic vents. It is important, however, to seal or avoid penetrations for electrical ceiling fixtures, which can allow mass movement of moist air into the roof cavity or attic.

For walls, ventilation is impractical, and condensation control will generally take the form of vapor retarders. Vapor retarders in walls and at other locations should always be on or nearest the winter warm side in order to block vapor before it reaches a portion of the construction with a temperature below the dew point. (In hot, humid climates, a wall vapor retarder is sometimes omitted or even placed on the exterior wall. Check local practice in these areas.) If vapor is allowed to penetrate a wall and temperature reaches the dew point within the wall, the vapor may condense and cause trouble.

Wood floors are seldom so cool as to cause surface condensation of vapor from within the house. Structural panel floors bonded with exterior adhesives have sufficiently low vapor permeance (1 perm or less) to prevent excessive indoor moisture from escaping into the crawl space when penetrations or openings are adequately sealed. This is particularly important when insulation is applied to the under-floor area.

Use a vapor-retarder ground cover to prevent introduction of moisture from the ground beneath the house to the crawl space or interior. This is easy in crawl-space houses, where a layer of 6-mil polyethylene over the ground in the crawl space is usually all that is required. It is more difficult in basement houses, where vapor retarders should be installed under basement floors and outside foundation walls.

Condensation in the crawl space is unlikely in winter when a ground cover is used and adequate drainage is provided around the foundation to prevent moisture accumulation. Thus, foundation vents may be closed during the winter for energy savings. Closure of vents in winter for energy saving is particularly effective when foundation walls, rather than the underside of floors, are insulated. This technique is also more effective than floor insulation for preventing summer condensation, particularly when the building is air-conditioned.

In modern basement houses, ventilation is usually inherent with forced-air heating systems. Ventilation and air movement should be given separate consideration when heating systems are used that do not provide air circulation, such as baseboard heaters.

Ventilation Requirements. Minimum ventilation requirements are usually covered in building codes. The requirements in Table 12.2 are based on the 2000 International Residential Code (IRC)[1] and may be used as a guide for residential construction.

The required net free area of vents can be found by multiplying the value in the third column of the table by the appropriate floor or attic area of the building. Note that these are minimum code requirements, which have been found to be adequate under most normal residential circumstances. However, ventilation in excess of these minimums may be necessary when unplanned moisture is introduced by venting an appliance, such as a dryer, into the space (which is not recommended), or by misdirected surface or rainwater. Care should also be taken to provide adequate

TABLE 12.2 Minimum Ventilation Requirements—2000 International Residential Code

Location	Construction	Natural ventilation net free area opening as proportion of floor or attic area
Attic and structural spaces	No vapor retarder	$L/150$
	Vapor retarder in ceiling	$L/300$
	At least 50% and not more than 80% of required vent area in upper portion of space to be ventilated at least 3 ft above eave or cornice vents	$L/300$
Crawl space	No vapor retarder	$L/150$
	Vapor-retarder/ground cover and one vent within 3 ft of each corner	$L/1500$

extra attic vent area when moisture-laden air is introduced to the attic by whole-house fans. In such cases, attic vent area should be increased in accordance with manufacturer's recommendations. Attic ventilation strategy should also consider location of vents to minimize dead air spaces.

Ventilation in excess of minimums may be necessary in high-occupancy structures or in structures that contain moisture-producing activities, such as commercial kitchens or laundry facilities. It is traditionally the responsibility of the building design professional to determine the amount and location of ventilation to ensure satisfactory performance.

Ventilation Checklist. It is sometimes necessary to inspect an existing building for adequate ventilation where there are signs of unusual moisture. When checking for ventilation, be sure to note the following information:

1. Area of floor and attic to be ventilated
2. Presence of ground cover and ceiling vapor retarder
3. Signs of moisture accumulation, including decay, water stains, blistered paint, water standing in crawl space, rusty fasteners, or mold growth
4. Quantity, size, type, location, and condition of roof and foundation vents. Measure vents to be certain of size and check vent manufacturer's data for their net free area.

   The free ventilation area published by vent manufacturers varies slightly, so any calculations regarding vent area provided would be approximate. In some cases, the net free area may be marked on the vent. When manufacturer's data are not available, Table 12.3 may be used to estimate net free vent area.

   A vent may actually have zero free area and thus may be ineffective, either permanently or intermittently. Examples include closed foundation vents, covered roof ventilators, inoperative power vents, and eave vents that are clogged or blocked by insulation or paint.
5. Compare actual ventilation with minimum requirements, as shown in the following example.

TABLE 12.3 Net Free Area Guidelines for Vents and Screens

Ventilator type	Area (in.^{2})	Net free area (in.^{2})
	Roof	
Screened jacks, button caps	Vent pipe area $\left(\frac{\pi d^2}{4}\right.$, where d = pipe diameter, in.$\left.\right)$	Area × 0.6
Ridge	—	18 × lineal ft
	Gable or foundation (Louvered and screened)	
Rectangular	Height × width	Area × 0.44
Triangular	½ height × width	Area × 0.44
Soffit	Length × width	Area × 0.3
	Screens	
1/16 in. mesh	Height × width	Area × 0.5
1/8 in. mesh	Height × width	Area × 0.8
1/16 in. mesh with louvers	Height × width	Area × 0.33
1/8 in. mesh with louvers	Height × width	Area × 0.44

Example A 30 × 45 ft house has a vapor-retarder ground cover in the crawl space. There are four louver-type foundation vents (46 in.^{2} free area per vent). Triangular gable vents (155 in.^{2} free area per vent) are at the top of the gables at each end of the house for natural attic ventilation. Does this meet the code minimum ventilation criteria?

1. Determine Minimum Required Ventilation Area

Ratio for attic with at least 50% of vent area in upper half of space to be vented is 1/300. Note, however, that in this case 100% of vent area is in the upper half of the attic, reducing effectiveness of the ventilation and requiring that a ratio of 1/150 be used.

$$\text{Attic vent area required} = \text{horizontal projection of roof area} \times \text{ratio}$$
$$= (30 \times 45 \text{ ft}) \times 1/150$$
$$= 9.0 \text{ ft.}^2 = 1296 \text{ in.}^2$$

Ratio for crawl space with vapor-retarder ground cover is 1/1500.

$$\text{Crawl space vent area required} = \text{floor area times ratio}$$
$$= (30 \times 45 \text{ ft}) \times 1/1500$$
$$= 0.9 \text{ ft.}^2 = 130 \text{ in.}^2$$

2. Determine Total Net Free Area of Vents

$$\text{Attic vent area} = 155 \text{ in.}^2/\text{vent} \times 2 \text{ vents}$$
$$= 310 \text{ in.}^2 \text{ (free area of attic vents is less than that required)}$$

Crawl space vent area = 46 in.^{2}/vent × 4 vents

= 184 in.^{2} (free area of crawl space vents is more than that required)

3. Conclusion

Crawl space vents meet minimum ventilation requirements, but attic vents are less than ¼ of the required area. Additional attic venting is required, and could be accomplished by using larger gable vents or adding vents along eaves. Eave vents are recommended.

12.2 BUILDING DESIGN AND DETAILING FOR PERMANENCE

In Section 12.1 the factors leading to the premature degradation of wood structures were discussed. The long-term performance of the entire structure is predicated upon the control of these degrading agents through the proper design and detailing of the entire building envelope.

The building envelope provides a number of different environments in which the inhabitants, contents, and structural and nonstructural components exist. While the primary purpose of a structure is to provide an appropriate environment for the inhabitants and contents of the structure, a secondary but equally important purpose is to protect the structure itself from environments that will cause its deterioration. This section will concentrate on providing a building envelope suitable for protecting the building components. This is done with the knowledge that providing a suitable environment for the structural elements results in an excellent starting point for the protection of a structure's contents and inhabitants. Both must be successfully accomplished.

For purposes of this chapter, the building envelope has been divided into 3 components; the roof, walls, and the foundation. These three components essentially define the boundaries of the building envelope and are discussed below.

12.2.1 The Roof System

One of the most fundamental components of the building envelope is the roof system. This part of the building envelope provides essentially the first line of defense against the greatest source of moisture loading on the building: rainfall. In addition, properly sized roof overhangs protect the siding of the structure from all but wind-driven rain and, when equipped with gutters properly designed to discharge rainfall away from the foundation of the building, also alleviate many of the subsurface moisture problems often faced by building owners. While properly designed roofs and roof overhangs can also contribute significantly to the longevity of the exterior wall finishing system and provide a positive impact on energy savings by reducing solar load, the emphasis of this section will be on the prevention of damage due to water infiltration.

Roof Systems—Low-Slope and Pitched Roofs. As far as waterproofing is concerned, roofing systems fall easily into two categories; the first is the near-flat or low-slope roof, and the second is the pitched roof. Each of the roof types uses different waterproofing methods to keep water away from the interior. Pitched roof systems rely on the force of gravity to direct the flow of water downward and outward. These systems rely on a series of overlapping elements—roofing felts, shingles, tiles, and flashing details—to provide the system to accomplish this redirection of rainfall. The pitch of the roof provides the gravity and the detailing of the roofing system provides the redirection.

When the size or style of the building dictates the use of a near-flat or low-slope roof, such as for industrial or light commercial buildings, a different technology is used to provide waterproofing. In low-slope roofing systems water is kept outside of the building envelope by providing a waterproof barrier over the entire roof system and around every penetration in that roof system. In this case, the force of gravity actually works against the waterproofing system. Instead of providing the redirecting force to channel the water away from the inside of the building envelope, the force of gravity provides the motive force to drive the water into every imperfection in the roof waterproofing system. Because the roof is near flat, or of very little slope, the chances for standing water are very high. Moderately high winds can force water to pile up in areas that would normally drain adequately. Once standing water is present, even minor defects can cause major water leaks.

Roof Slopes. In a pitched roof the force of gravity is large enough to overcome both the force of wind-driven rain and the surface friction of the roofing material to permit the rainwater to be directed away from the interior of the building envelop. As the roof pitch gets steeper, the impact of gravity acting over the roof increases and it becomes easier to weatherproof/waterproof the roof. On the other hand, as the roof slope decreases, the effectiveness of such a system also decreases. There is no "magic slope" where conventional steep-slope waterproofing becomes ineffective and near-flat or low-slope roof waterproofing must be used. This point is impacted by the roof slope, the wind speed, the surface friction of the roofing material, and other, more esoteric factors, such as the shape of adjacent roof surfaces and the general topography around the building and the direction of the prevailing wind.

The roofing industry, however, has decided on a slope of greater than 3 in. in 12 in. (3:12) for the cutoff between a steeply pitched roof and a low-slope roof system. This is an arbitrary point, and a designer or builder may be well served to contact the local building department for guidance when dealing with a specific roof at or near this cutoff point. The local building official will know whether local conditions demand special considerations.

Low-Slope Roofs. Roof slope equal to or less than 3:12 is a low-slope roof. Outside of the fact that it is almost impossible to construct a truly flat roof, there is a valid structural reason for not designing one. This is to prevent a phenomenon known as *ponding,* which may occur when water collects in a depression on a flat roof. As water builds up in the depression, the load on the supporting members increases. This causes additional deflection, which permits more water to accumulate. Once initiated, this cycle can continue until a roof failure occurs. In fact, an initial depression is not even necessary to initiate ponding. Wind can cause water to build up in one area of the roof and cause the initial deflection. While melting snow is another potential culprit, ponding is more prevalent in areas where roofs are designed for low live-to-dead-load ratios. Roofs designed for snow loads are typically stiffer and have greater ability to resist ponding than similar roofs designed only for relatively light construction loads.

Roof framing should be designed for drainage by the application of a slight slope leading to roof drains installed in the lowest portions of the roof. The minimum slope recommended is ¼ in. in 12 in. (¼:12), or about a 2% slope. This should be treated as a minimum only. It is always good practice to provide as much slope as can be allowed by the geometry of the building.

Truly flat structural framing should be considered *only* when some other means of guaranteeing roof drainage is provided. An example of such a system is the use of foam blocks that are placed over the top of the flat roof deck to provide insulation and, due to their taper, provide for drainage. These blocks also provide a foundation for the roofing system. One problem associated with such systems is that during remodeling at a later date the whole insulation system may be removed and replaced with an unsloped system.

The Roofing System. The proper detailing and installation of each of the elements discussed below is essential to the construction of a sound watertight roofing system. Rather than listing a number of rules and regulations for the installation of each component that must be memorized, this section will discuss each of the elements and their interrelationship with the other elements and provide illustrations, with the hope that an understanding of the system will make the required details obvious to the reader. Because specific information may be required in certain applications, this section provides a list of informational sources for the various roofing-material types.

The roofing system, whether pitched or low-slope, is made up of a number of different components: roof sheathing, underlayment, roofing material, roof intersections, flashing, and ventilation details. Each of these components must be correctly installed, in the proper order, and with the proper relationship to each other in order to make the system work as planned. These components are discussed below and illustrated in Fig. 12.2.

Roof Sheathing. Roof sheathing forms the structural base or foundation for the roof system. It attaches to the roof framing and provides the nail base for the other components of the roofing system. In addition to supporting the roofing system, the roof sheathing is an important part of the building's structural frame, transferring water, snow, wind, earthquake, equipment, and construction loads into the structural frame below. This is true for low-slope roofs as well as pitched roofs. The most common types of roof sheathing used in residential and wood-framed light commercial structures are wood structural panels such as APA Rated Sheathing and Structural I Rated Sheathing.

A number of recent hurricanes have emphasized to the design, building, and code-enforcement communities the importance of designing the roof sheathing for all of the potential loads that may occur. In hurricane country, for example, the roof uplift loads can be far greater than the downward-acting traditional roof design loads, and the proper nailing of the roof sheathing is essential to resist the resulting withdrawal loads for the satisfactory performance of the roof system as well as the structural performance of the building itself.

Since the roof sheathing is installed with small gaps around the perimeter of each panel (to facilitate panel expansion), the sheathing itself will provide protection only against gross amounts of rainwater. If uncovered during a storm, a serious level of leakage can be anticipated. Waterproofing is not its job. Its purpose, as far as this discussion is concerned, is to provide the foundation for the rest of the roof system.

Installation Requirements. Chapter 2 gives the minimum nailing recommendations for attaching wood structural panel sheathing for a roof deck. In areas of high wind or seismic activity, additional nailing may be required. Panels should be installed in accordance with the recommendations presented in Chapter 2. Chapter

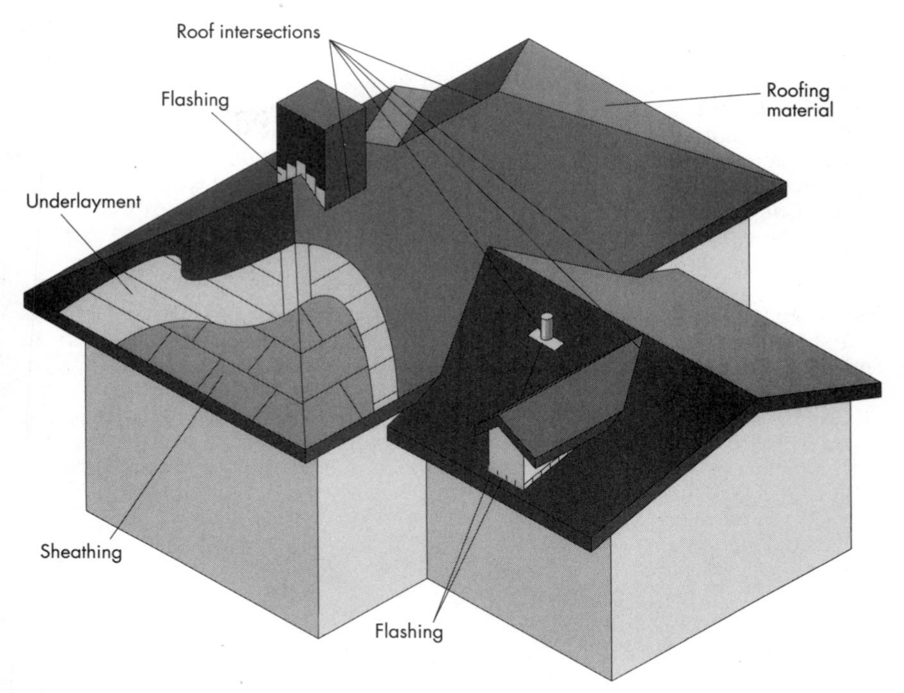

FIGURE 12.2 Roof system components.

8 also provides information on mechanical fasteners used to attach the sheathing to the framing.

Roofing Underlayment. Roofing underlayment, often consisting of building paper or felt, is really the first weatherproofing layer for a pitched roof. Properly installed underlayment is placed from the bottom of the pitched roof to the top, such that the upper layer overlaps the lower layer. The underlayment provides a path for any water that leaks through the roofing materials along the top of the paper to the edge of the roof while protecting the wood-based roof sheathing from this leakage. While wood structural panel roof sheathing is made with waterproof adhesive and has some ability to absorb and dissipate small amounts of water, it must be protected from prolonged exposure to high moisture contents in order to prevent the start and propagation of decay organisms.

In high-wind areas where the roofing materials can be blown off during a storm, the underlayment is often attached to the roof sheathing to prevent it from being blown off along with the roofing material. Small circles—about 2 in. in diameter—of thin sheet metal called "tin tabs" are often used in high-wind areas to secure the underlayment in case the roofing material is lost to the storm. The underlayment is nailed or stapled down through the tin tabs to increase the pull-off resistance of the paper.

While more common because of first cost, such mechanical devices are not as effective as a fully adhered system where the underlayment is overlapped and adhered to the roof with a waterproof adhesive system. The advantage of a fully adhered system is that it can easily resist tear-off by not allowing wind pressure to build up on its backside.

On low-slope roofs, the underlayment, if used, can form a number of different functions, depending on the type of roofing applied over it. Unless used as a part of the roofing material, almost none of these functions are water protection-related. In some systems the underlayment is attached mechanically to the roof sheathing and the roofing material is adhered to the underlayment. In this case it has the mechanical function of holding down the roofing material. If a leak forms in the roofing, the underlayment provides minimal leak protection because of the fastener penetrations.

In those roofing systems where the first layer is adhered to the roof deck, the distinction between the underlayment and the roofing material is blurred. In some roof systems the underlayment provides a slip surface between the roof sheathing and the roofing to enhance the performance of the roofing under thermal loads. There are single-ply, ballasted roof systems where no underlayment is used.

Installation Requirements. For sloped roofs underlayments can be installed in a number of different ways, depending on the roofing material used. See Fig. 12.3a–c for examples of proper underlayment installation. Note that the underlayment is

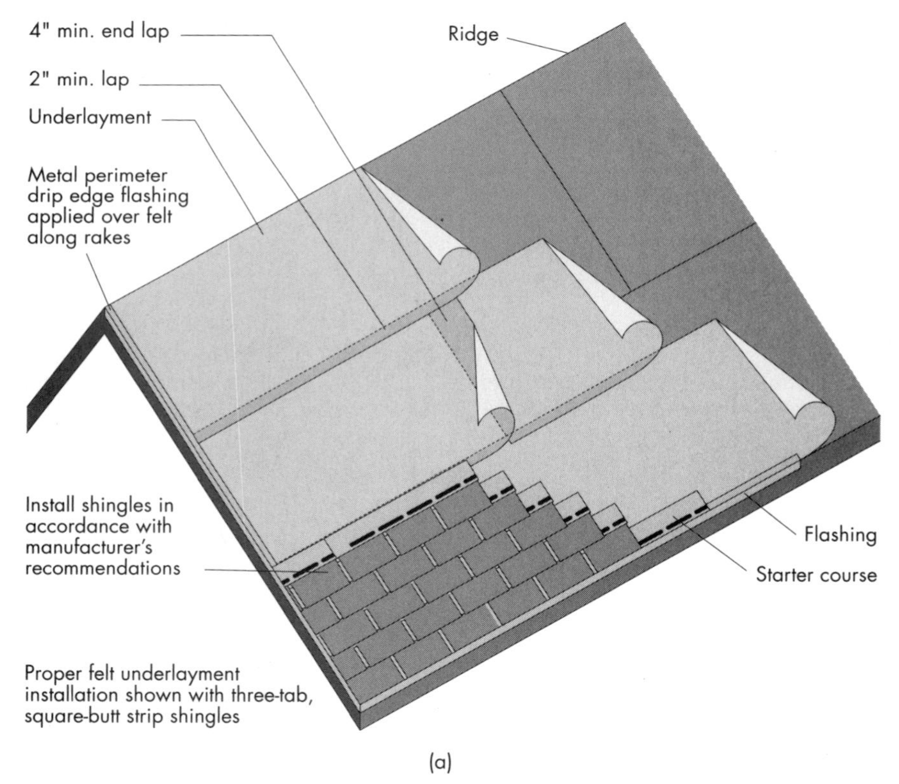

FIGURE 12.3 (*a*) Typical single-layer underlayment installation for steep-slope roofs.

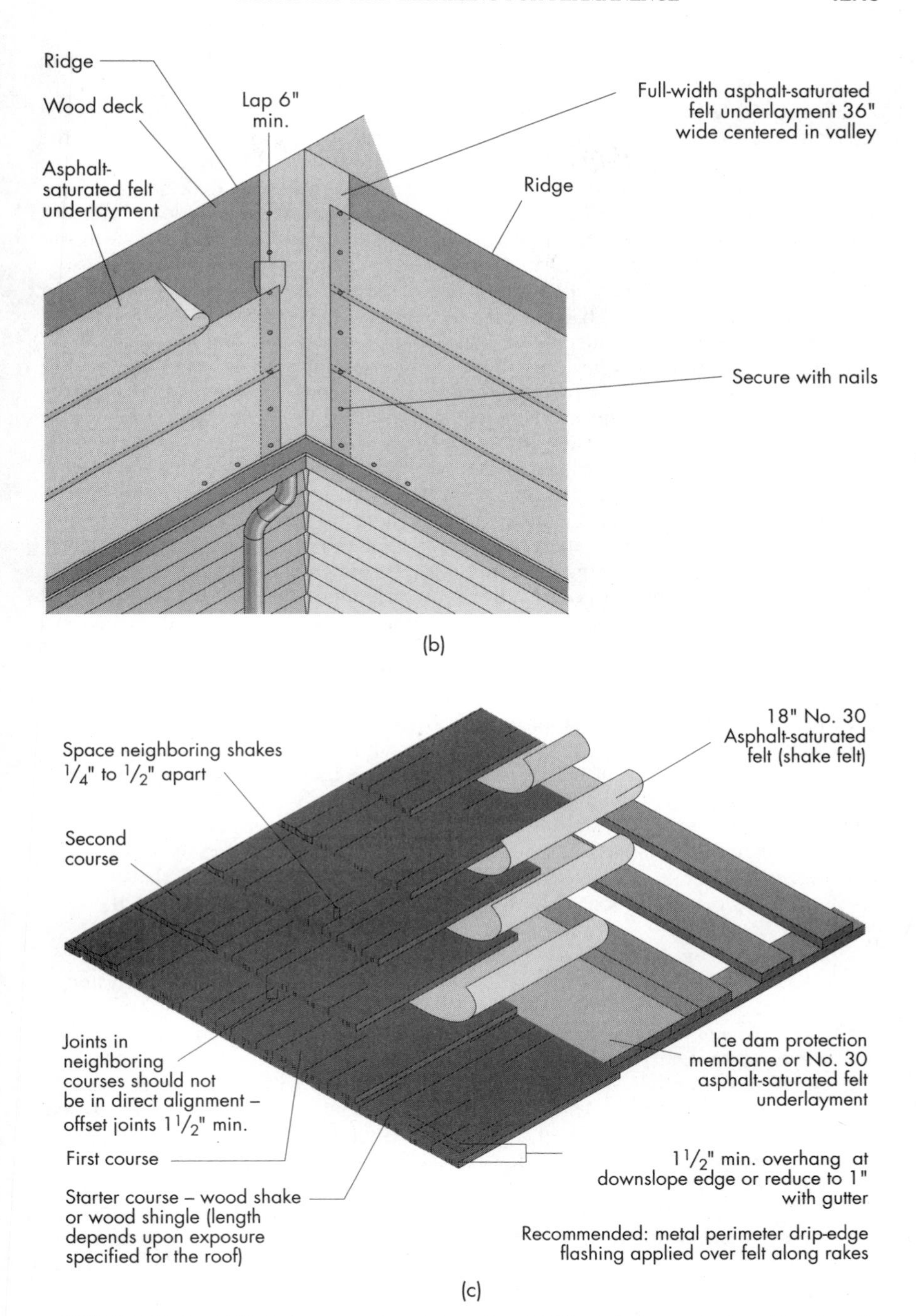

FIGURE 12.3 (*b*) Underlayment centered in valley. (*c*) Wood shake application. (*Continued*)

always installed in such a way as to channel the water out and down, away from the wood structural panel sheathing below.

Roofing Materials. The roofing material is the material that can be seen on the finished roof. The purpose of the roofing material, or roofing, is to provide the primary waterproof barrier for the structure. Because of the hostile nature of the roof surface, subjected to extremes of heat and cold, rain, snow, hail, flying debris, ultraviolet light, and foot traffic of maintenance personnel, the roofing material must have additional durability-related properties in addition to those required for keeping water out of the structure's envelope.

For pitched roofs, almost all roofing materials rely on some form of shingling to provide the weatherproof barrier. Like the underlayment, these roofs are installed from the bottom up, with successive layers overlapping, both vertically and horizontally. The roofing material most commonly used for pitched roofs are asphalt shingles, but other materials include slate, clay and concrete tiles, wood shingles and shakes, and metal shingles. The exceptions alluded to above are standing-seam and corrugated metal roofs. These roofs are often made up of single-piece elements that are full-length from the ridge to the overhang. Adjacent panels are connected to one another with a folded standing seam (the seam is elevated above the surface of the roof) or by overlapping adjacent corrugations.

Low-slope roofs utilize a staggering number of proprietary and nonproprietary roof systems, ranging from single- to multiple-ply; adhered, mechanically anchored, or ballasted; hot-mopped or cold-applied (solvent-, urethane-, or epoxy-based) systems; rolled on or poured on; vented or unvented: and any combination thereof. Volume 1 of the *NRCA Roofing and Waterproofing Manual*[10] contains over 1000 pages on the proper installation of such roofing types, including over 250 pages of flashing details for this vast array of roofing systems.

Roof Intersections. The majority of roof leaks occur in locations where the plane of the roof is interrupted by a ridge, another roof intersecting at an angle, a wall, or a penetration. Even the simplest of rooflines have dozens of potential leak sites due to chimneys, skylights, ridges and valleys, utility vent stacks, kitchen and bathroom ventilation fans, and code-required roof ventilation penetrations. Other than those caused by natural disasters, it is the improper execution of the required detailing around these discontinuities that is the cause of most roof leaks. Conversely, the proper execution of these details is a very important part of the roof system.

Proper Detailing of Roof Intersections. Figures 12.4a–d illustrate examples of proper roof ridge detailing for asphalt, slate, tile, and wood shake roofing systems, respectively.

Figures 12.5a–e illustrate typical valley intersection details. Figure 12.5a shows an example of an open metal valley flashing suitable for all roofing types, and Fig. 12.5b shows closed mitered valley flashing for utilization with flat roofing materials such as slate or flat tile. A closed mitered valley should not be used with a wood roof if leaves or needles can be an impediment to rapid water runoff. Figures 12.5c and d are examples of common valley details for asphalt shingles. Valley intersection details for corrugated metal roofing are shown in Fig. 12.5e.

Figures 12.6a–d deal with hip roof intersections. Figure 12.6a is provided to illustrate a typical example for flat roof products such as slate or low profile tile. Fig. 12.6b shows common hip details for asphalt shingles, and Fig. 12.6c illustrates a roof hip made with high-profile tile. Figure 12.6e is an example of a hip ridge detail designed for corrugated metal roofing.

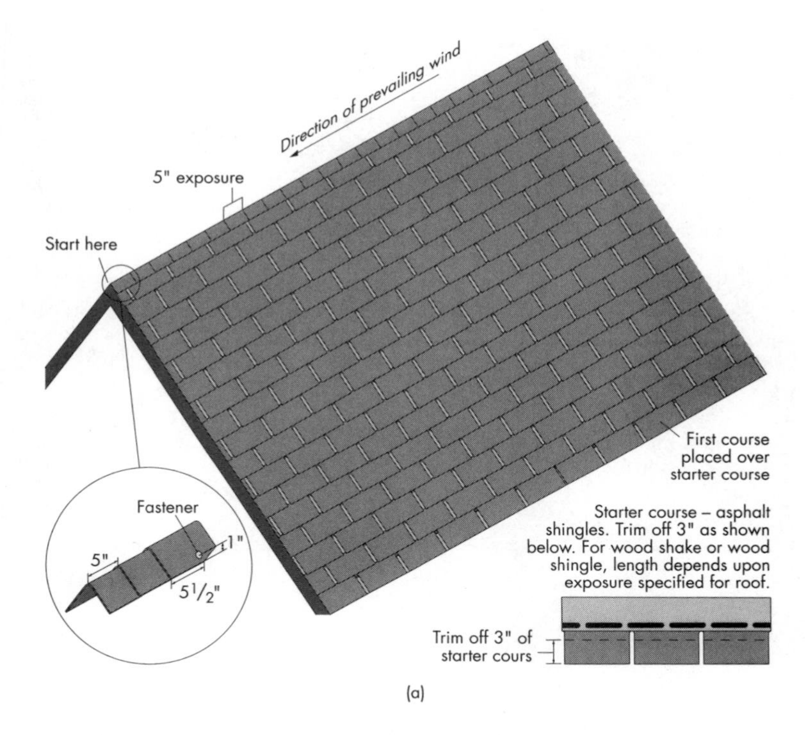

(a)

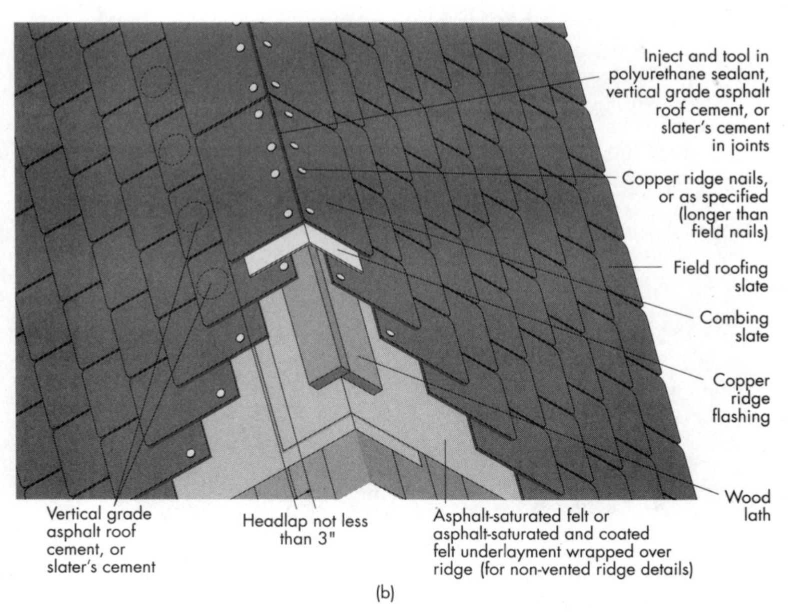

(b)

FIGURE 12.4 (*a*) Shingle roof ridge details (starter course detailed as shown). (*b*) Slate roof ridge details.

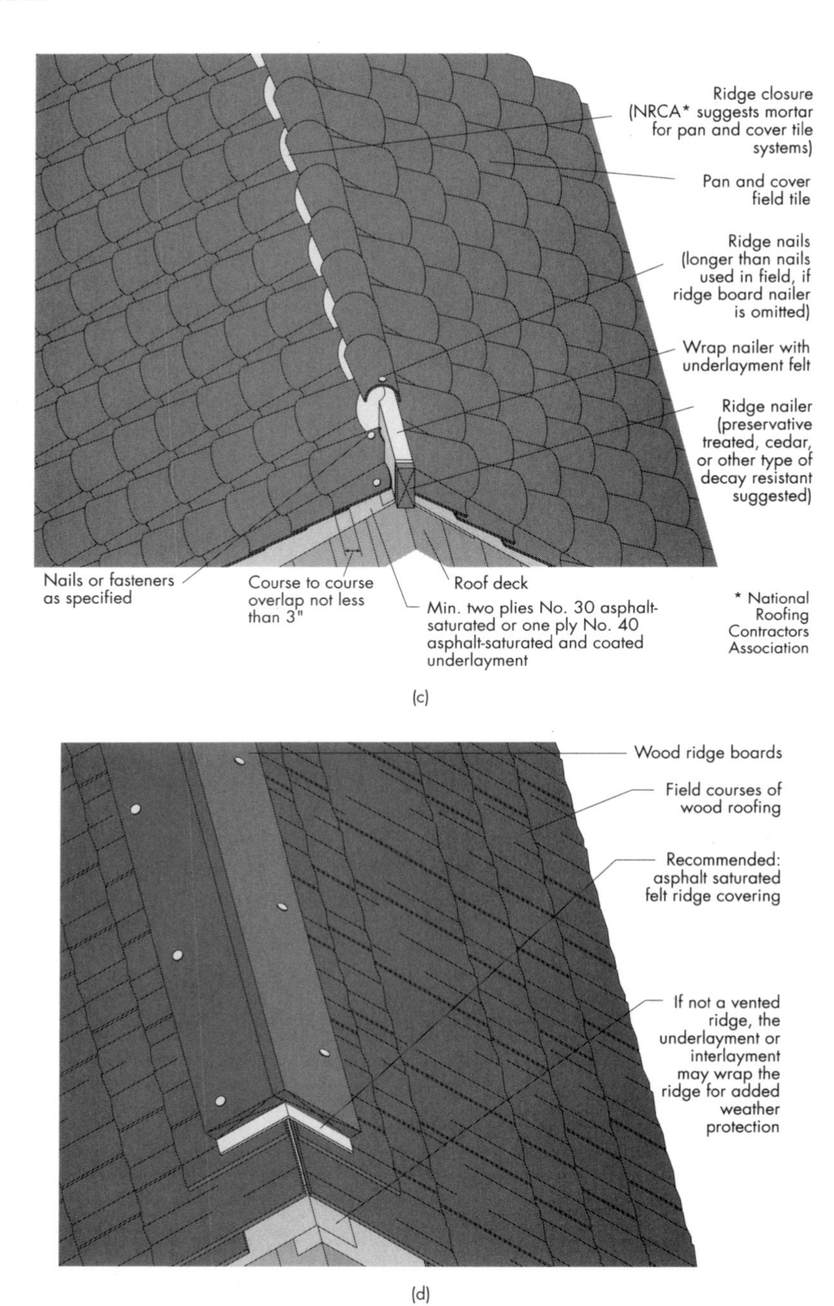

FIGURE 12.4 (*c*) Pan and cover tile roof ridge detail. (*b*) Wood ridge detail for use with shake or shingle-type roof. (*Continued*)

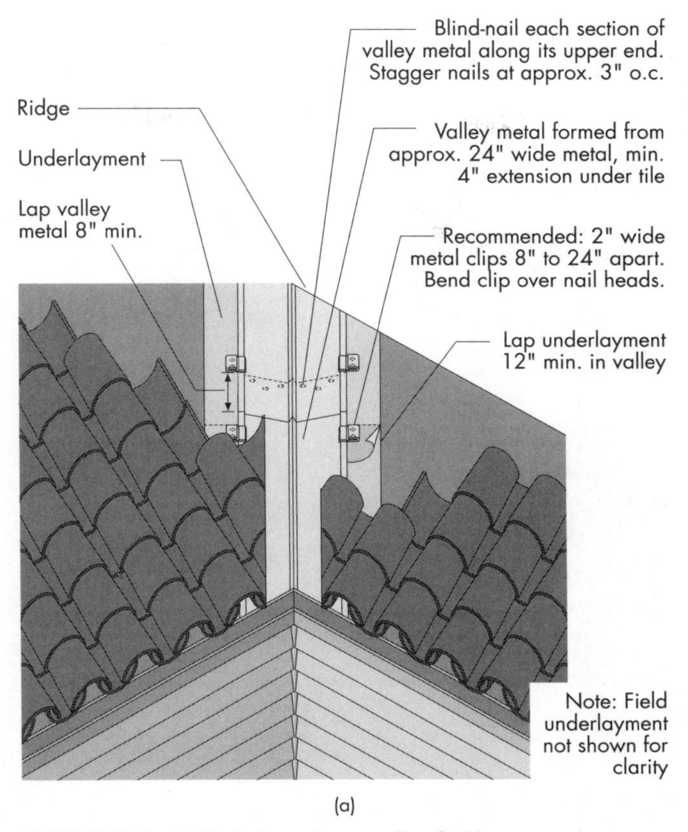

FIGURE 12.5 (*a*) Typical metal open valley flashing.

Figure 12.7 provides the most common detail used for eaves and rakes—the use of drip-edge material.

Flashing Details. Flashing is made up of thin sheets of corrosion-resistant material used in conjunction with the other elements of the roof system to prevent leaks around roof intersections and penetrations discussed above. Flashing is normally made up of galvanized steel, copper aluminum, lead, or vinyl. Often small roof penetrations such as vent stacks utilize flanged rubber boots in lieu of more conventional flashing because of the circular shape of the penetration.

In a pitched roof, regardless of the application or the type of flashing used, the principle used is always the same. The purpose of the flashing is to direct the flow of the water that leaks into the intersection down and away from the interior of the structure to the topside of the roofing material. In every case shown, the top edge of the flashing passes underneath the underlayment, the upper pieces of flashing pass over the lower pieces, and the lower edge of the flashing passes over the top of the roofing material. In such a manner, the flashing never directs the flow of water to the bottom side of the roofing underlayment, never putting it in contact with the wood structural panel sheathing.

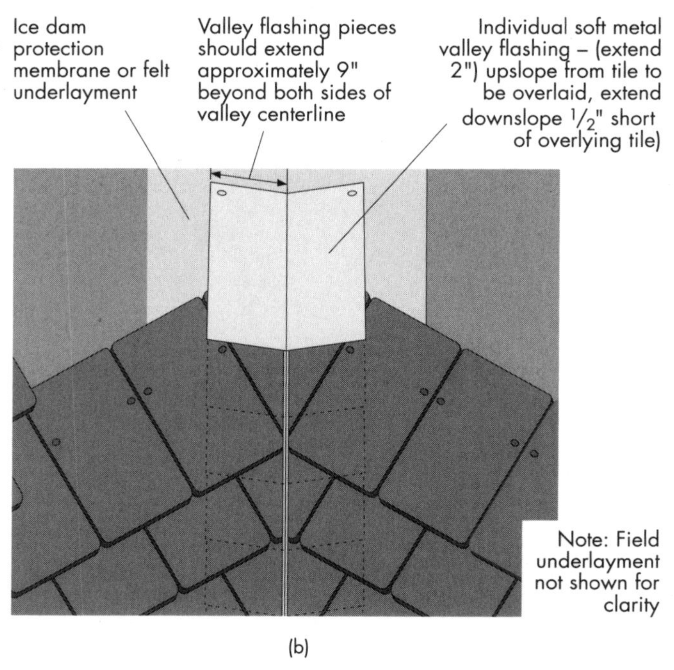

FIGURE 12.5 (*b*) Closed mitered valley with interwoven metal valley flashing—shown with flat tile roofing. (*Continued*)

Proper Flashing Installation Details. Some typical flashing details are covered in Figs. 12.8–12.11.

Fig. 12.8a is an illustration of a common vent stack penetration utilizing a preformed rubber of soft metal flashing designed specifically for that use. While the illustration shows a high-profile tile being used, the general procedure for properly installing the vent pipe flashing is the same for all roofing material types. Fig. 12.8b shows a similar stack penetration in a corrugated metal roof.

A series of illustrations is presented in Fig. 12.9 showing the steps necessary to flash around a masonry chimney. Many of the steps shown are common to other applications in steeply pitched roof applications.

Figures 12.10a and 12.10b are provided to illustrate the flashing details around a skylight or other similar applications. An example of a typical installation for low-profile roofing such as flat tile, slate, asphalt shingles, or wood shingles is provided, as well as one showing such an installation with high-profile clay tile.

The flashing required in intersections between roofing elements and vertical walls immediately adjacent are shown in Figs. 12.11a–c.

Ventilation. The purpose of each component of the roof system discussed so far has been to keep water from penetrating the building envelope. It can be anticipated that at some time during the life of the roof system some leakage will occur, whether by deterioration of the roofing material, wind-driven rain, ice dams, or other causes. A little leakage onto the wood structural panel sheathing in most cases can be tolerated because of the code-required ventilation underneath the roof sheathing.

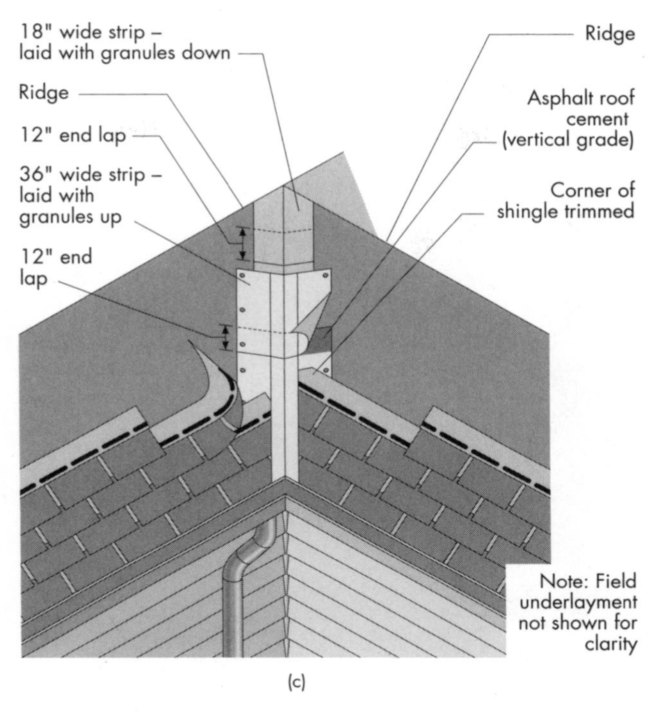

(c)

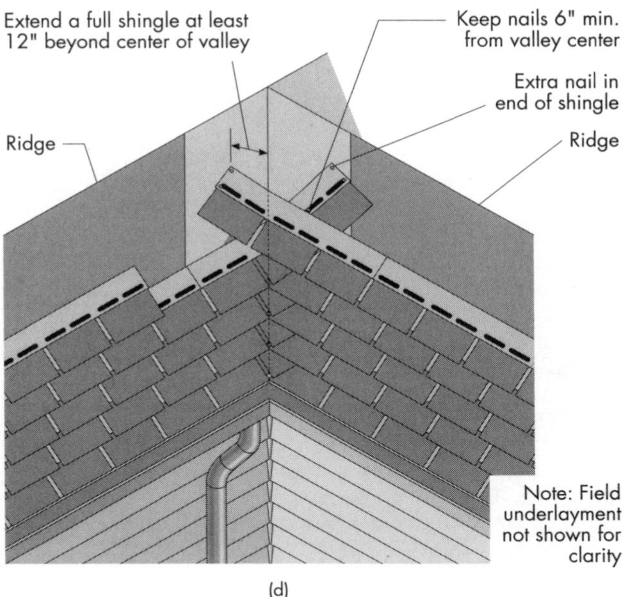

(d)

FIGURE 12.5 (*c*) Use of rolled roofing material for open valley construction. (*d*) Woven valley. (*Continued*)

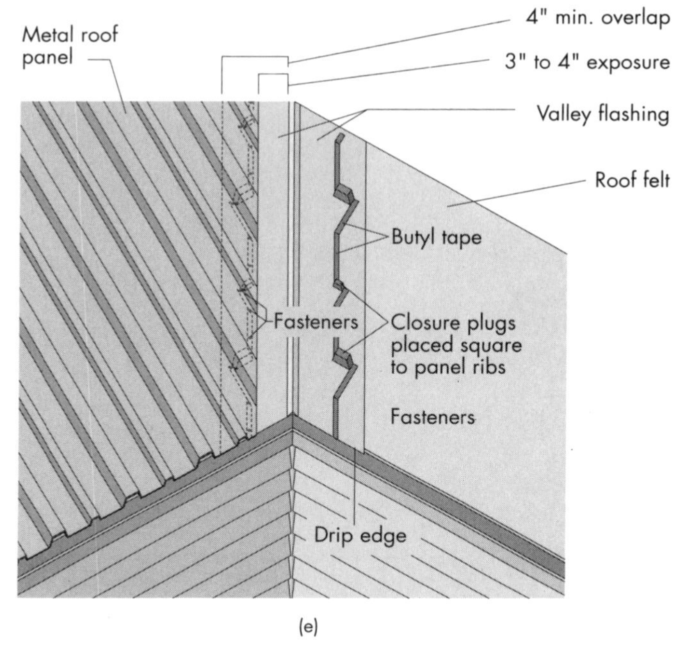

(e)

FIGURE 12.5 (*e*) Sealing details at metal roofing open valley. (*Continued*)

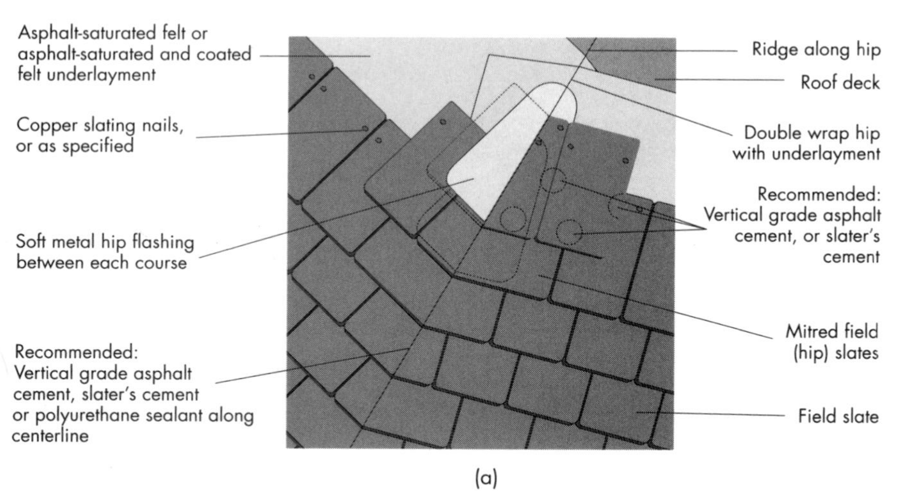

(a)

FIGURE 12.6 (*a*) Mitered hip shown with slate roofing and interwoven hip flashing.

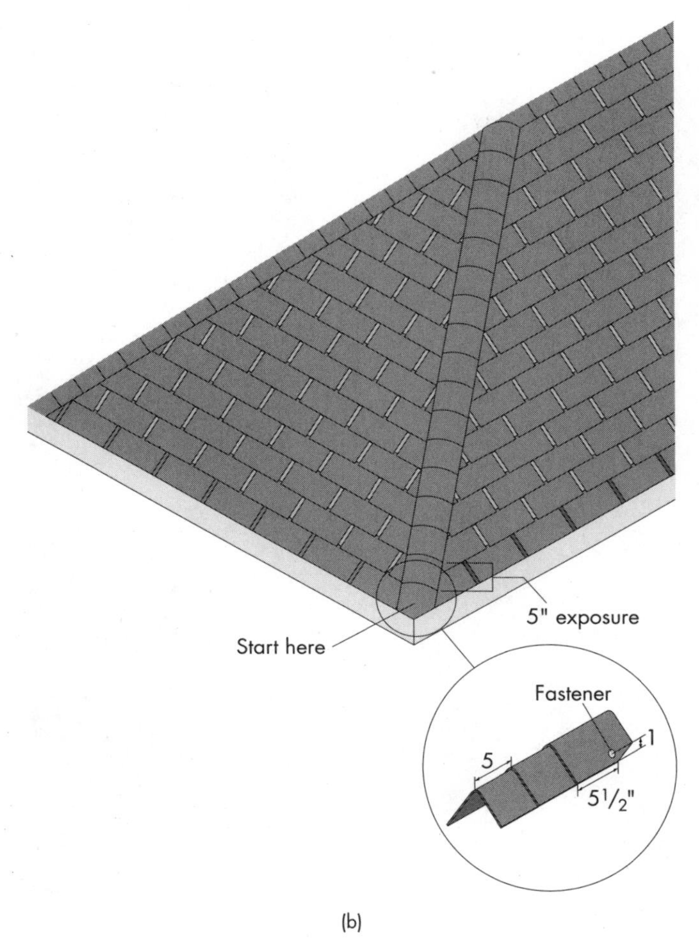

FIGURE 12.6 (*b*) Asphalt roof hip detail. (*Continued*)

The required ventilation serves two purposes. One is to provide a passage for airflow over the lower surface of the sheathing to assist in the drying out of the panel, which may have gotten wet from leakage of the roofing system or from condensation. As mentioned before, wood structural panels are manufactured with a fully waterproof adhesive and can tolerate the kind of wetting and drying associated with very minor roof leaks. The problems occur when the leakage is great enough that the required ventilation is insufficient to dry out the roof sheathing between wettings. In such cases, wood products are susceptible to mold and decay.

The second benefit of this ventilation air is to lower the temperature of the roof deck in the summer. High roof deck temperatures have been found to adversely impact the useful life of some types of fiberglass asphalt shingles. The flow of air

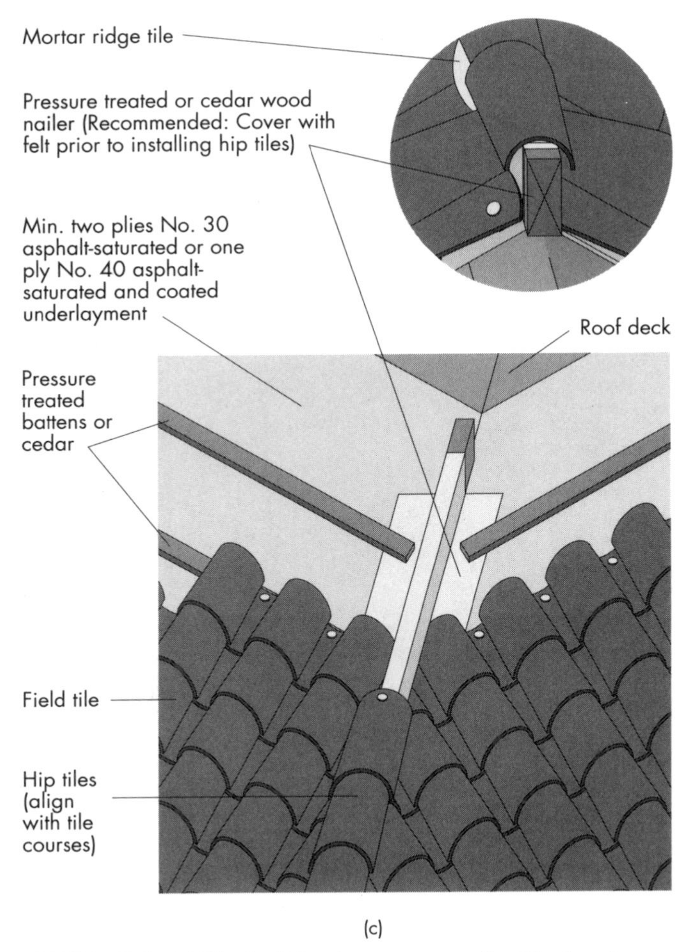

FIGURE 12.6 (*c*) Hip detail for clay or concrete tile. (*Continued*)

on the underside of the sheathing can reduce the temperature of the shingles by 20–30°.

Figure 12.12 is an illustration of roof ventilation in a sloped roof.

Special Considerations

Ice Dams. Ice dams are caused when natural heat losses through the roof cause the snow to melt. The meltwater flows downward until it hits the roof overhang and then refreezes because this area of the roof is at ambient winter temperature. Given the right set of circumstances, this layer of ice in the roofing material can get thicker and thicker. Eventually the meltwater will pond up behind the ice dam

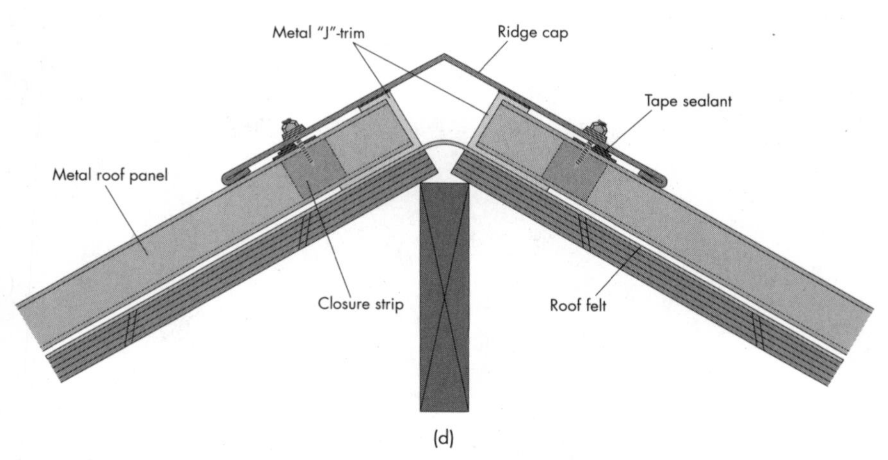

FIGURE 12.6 (*d*) Sealing the hip of a metal roof. (*Continued*)

and start backing up the roof slope. As it does this it moves up under the shingles and, if improperly applied, even the underlayment. This can saturate the wood structural panel sheathing and cause leaks into the habitable space. Figure 12.13a shows an example of proper detailing to mitigate the ice dam problem through the use of a double layer of roofing felt underlayment and, Fig. 12.13b shows the use of self-adhering underlayment for the same application. Resistance heaters are installed in some locations where the problem is especially severe.

High-Wind Considerations. While high-wind considerations were mentioned previously, some elaboration is warranted to help ensure good performance of the roof system. A number of special factors must be considered in designing the roof system of a structure in a high-wind area:

- The impact on ponding caused by high winds should be considered when designing a low-slope roof. As previously discussed, high winds can cause water to build up in areas of a low-slope roof that would normally have adequate drainage. Designing low-slope roofs with a slope greater than the minimum 2% is recommended.
- Impact on roof slope. In high-wind areas care should be taken in designing pitched roofs near the minimum slope of 3:12. The minimum slope for a pitched roof is based on the ability of the slope to channel water down and away from the interior of the structure. One of the forces that must be opposed by the gravity forces on the water is the wind force. On the windward side of the roof, the wind tries to force the water up under the shingles. This is opposed by the gravity force on the water due to the slope of the roof. As the slope increases, the roof becomes more resistant to wind-driven rain. Avoid roof slopes near the minimum for a pitched roof in high-wind areas.
- Proper selection of roofing material and underlayment. Asphalt shingles, in general, are designed for 65 mph winds. If asphalt shingles are used in high-wind areas, their loss during a windstorm must be anticipated. As such, to prevent damage to the structure and its contents when asphalt shingles are used, the underlayment must be selected and attached to provide the requisite weather-

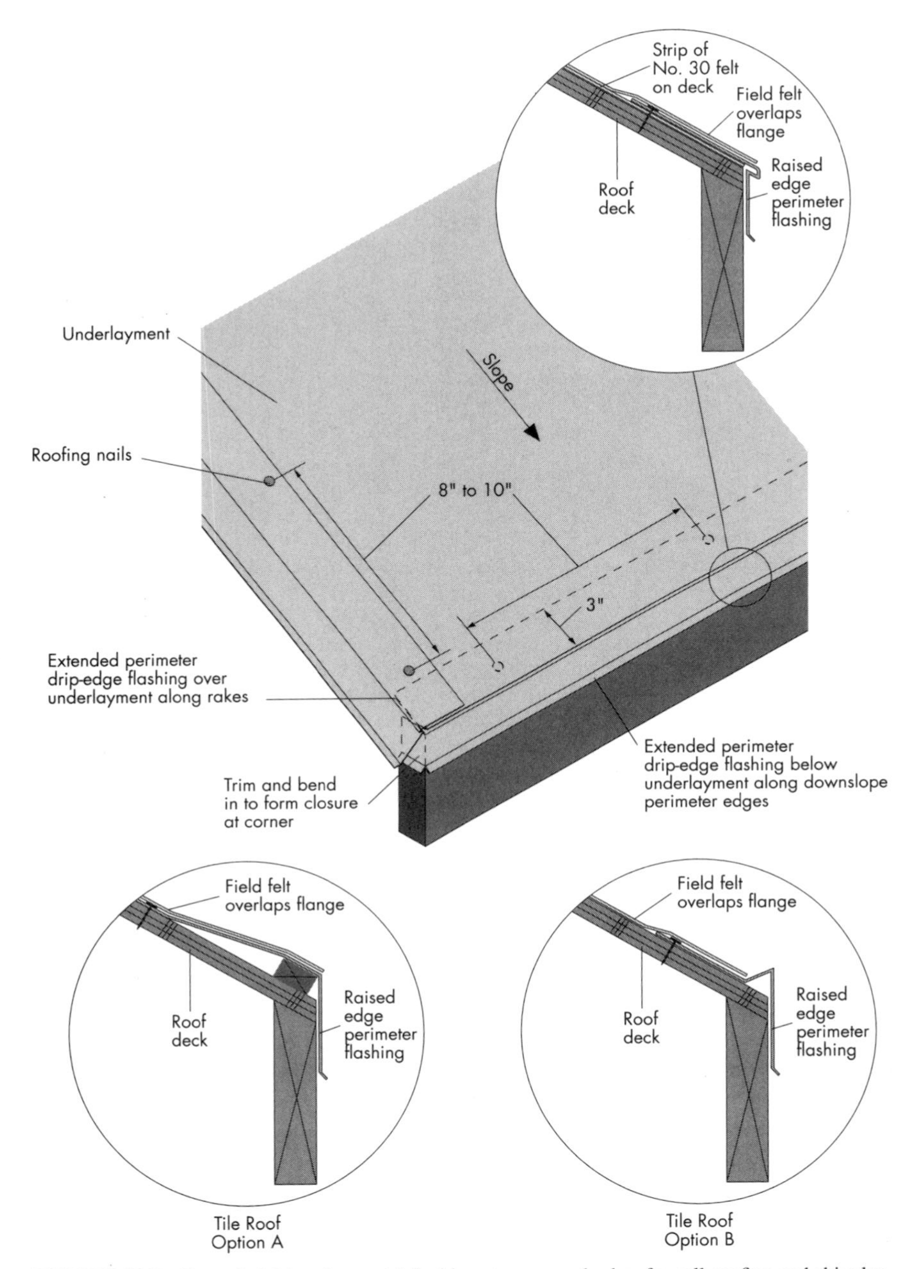

FIGURE 12.7 Extended drip-edge metal flashing at eaves and rakes for roll roofing and shingles. Options A and B are shown for tile roof.

STEP 1

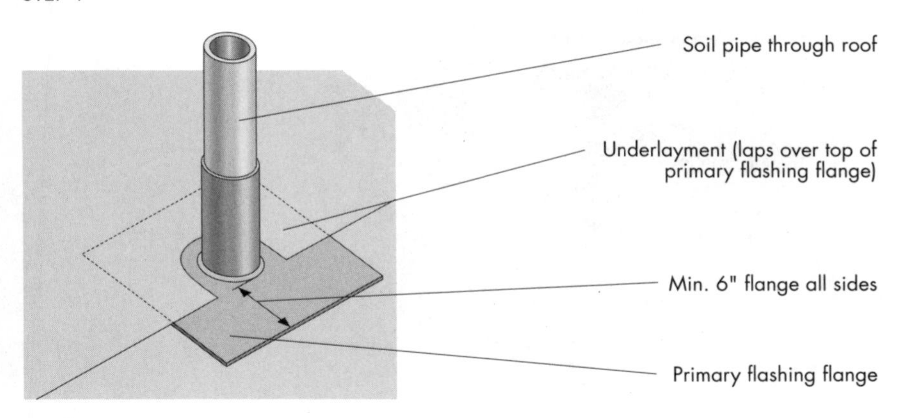

STEP 2

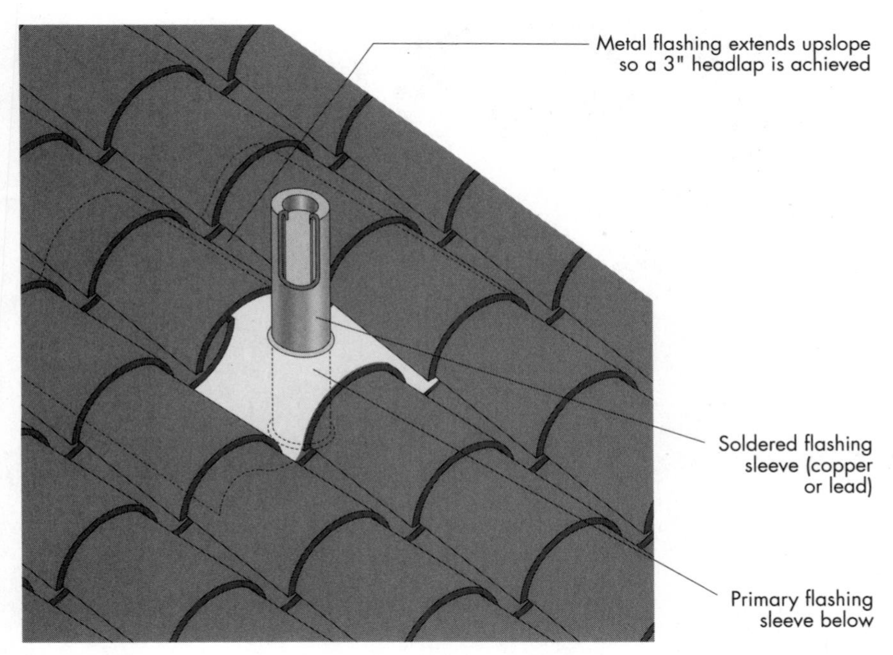

FIGURE 12.8 Two-stage flashing for sealing plumbing vent stack with pan and cover tile roof.

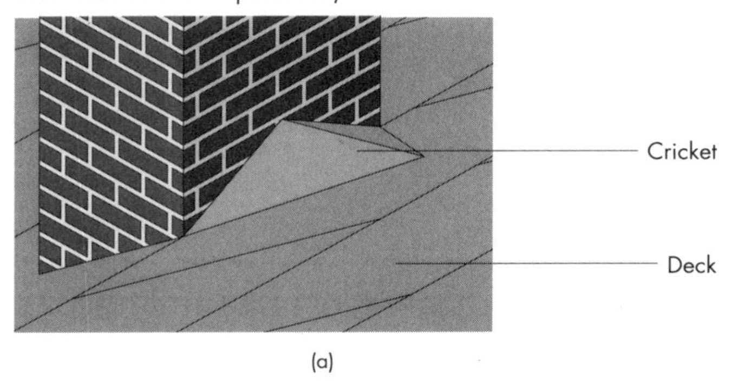

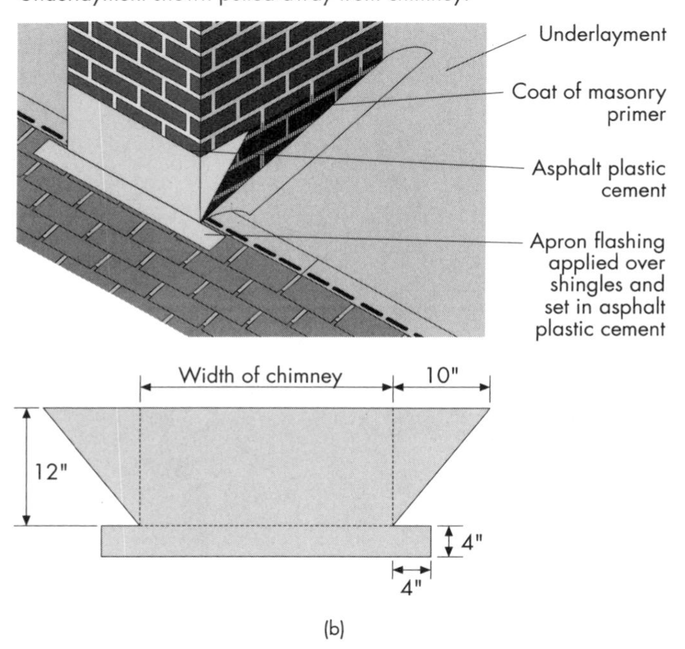

FIGURE 12.9 (*a*) Chimney flashing—Step 1. (*b*) Chimney flashing—Step 2.

Interlace step flashing with shingles. Set step flashing in asphalt plastic cement.

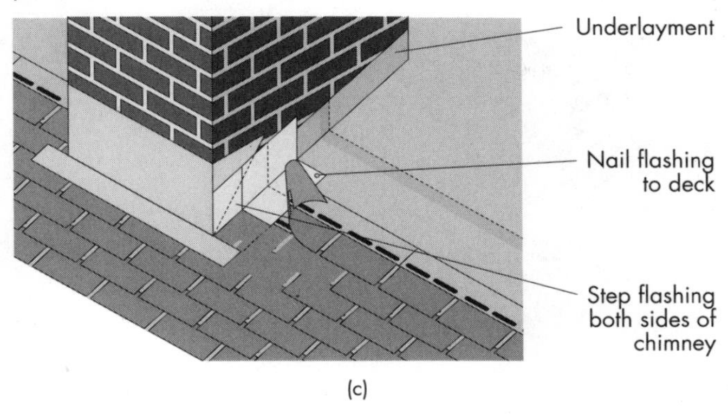

(c)

Extend step flashing up chimney and around corner. Nail corner flashing to deck and cricket.

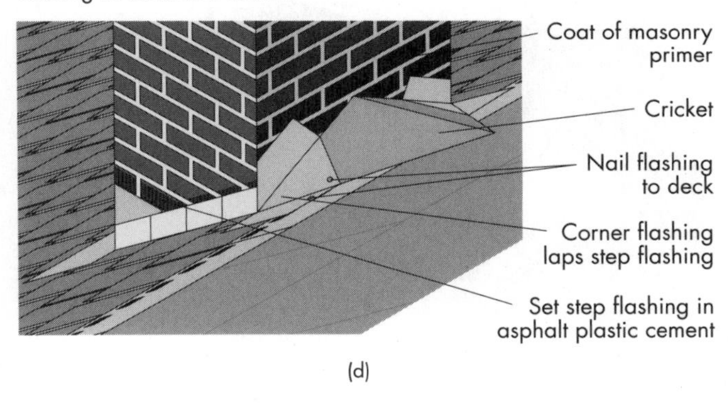

(d)

Place preformed cricket flashing over cricket and corner flashing. Set cricket flashing in asphalt plastic cement.

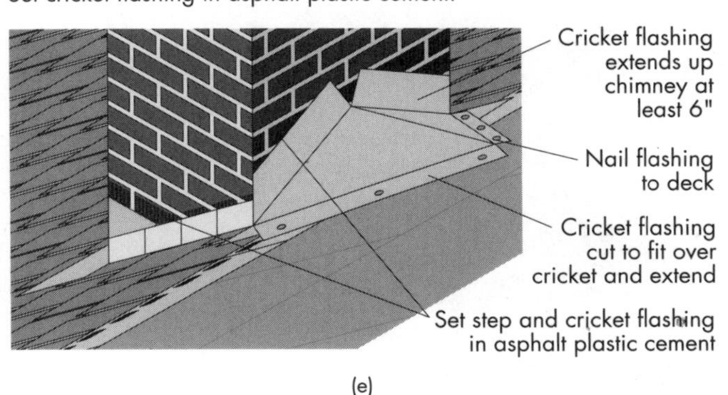

(e)

FIGURE 12.9 (*c*) Chimney flashing—Step 3. (*d*) Chimney flashing—Step 4. (*e*) Chimney flashing—Step 5. (*Continued*)

Flashing strip cut to contour of ridge in cricket. Size to extend up chimney at least 6".

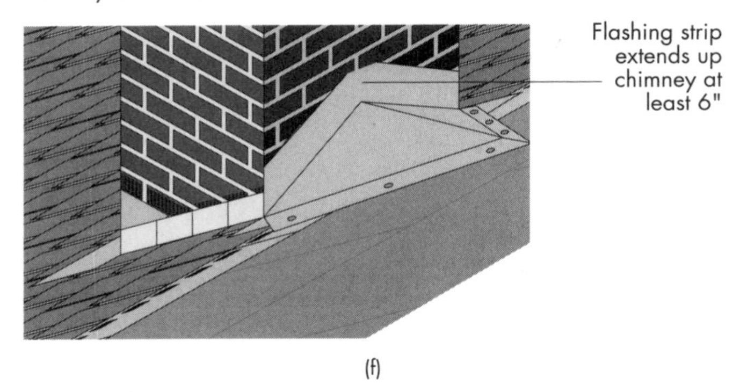

Place counter flashing over step and cricket flashing. Shingle remainder of roof.

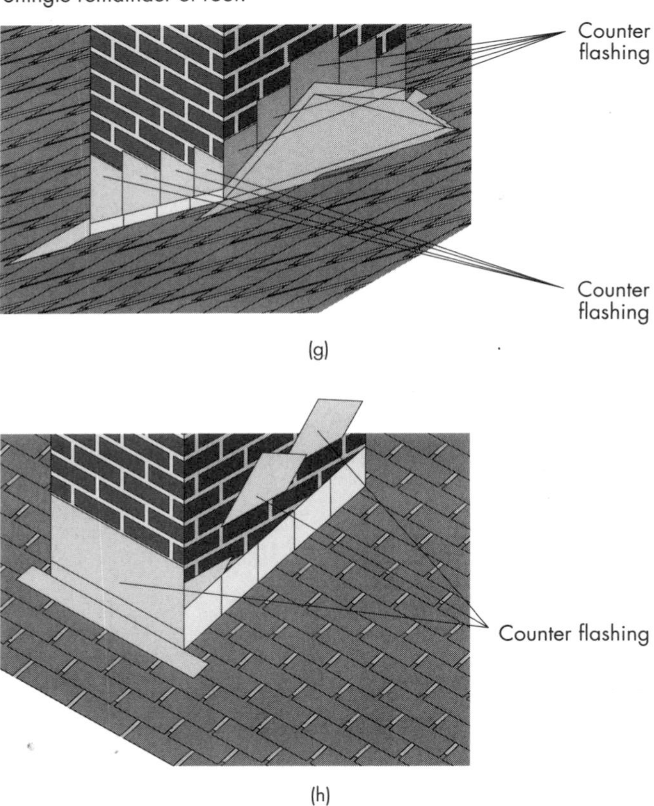

FIGURE 12.9 (*f*) Chimney flashing—Step 6. (*g*) Chimney flashing—Step 7. (*h*) Counter flashing installation on slope. (*Continued*)

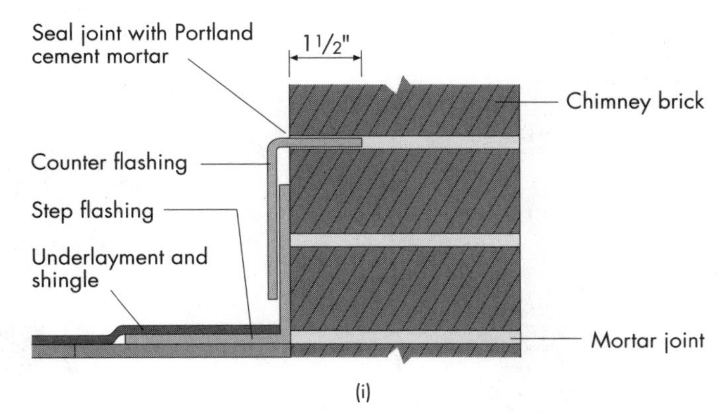

FIGURE 12.9 (*i*) Counter flashing details. (*Continued*)

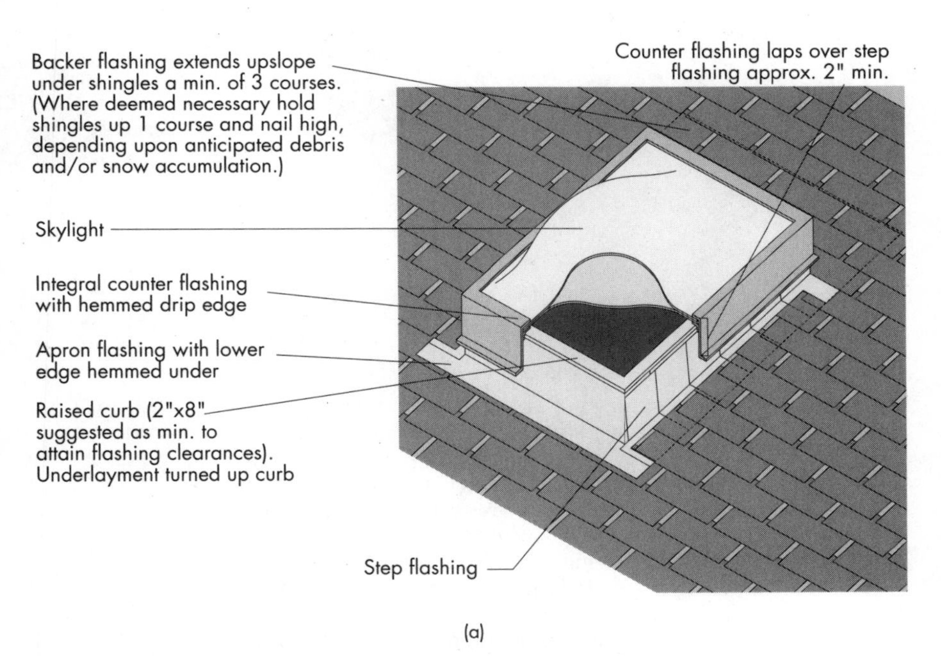

FIGURE 12.10 (*a*) Flashing around skylight with shingle-type flat roofing.

proofing for when the shingles blow off. The practice of relying on the shingle attachment to hold down the underlayment is not an option. Fully adhered underlayments or even a layer of hot-mopped asphalt roofing may be used. Obviously, this is not a consideration when other roofing materials that are designed and properly attached for the design wind load are used.

- Often the actual shape of the roof can be a factor in maintaining the weathertight integrity of the roof. The use of hip roofs as opposed to gable-end roofs

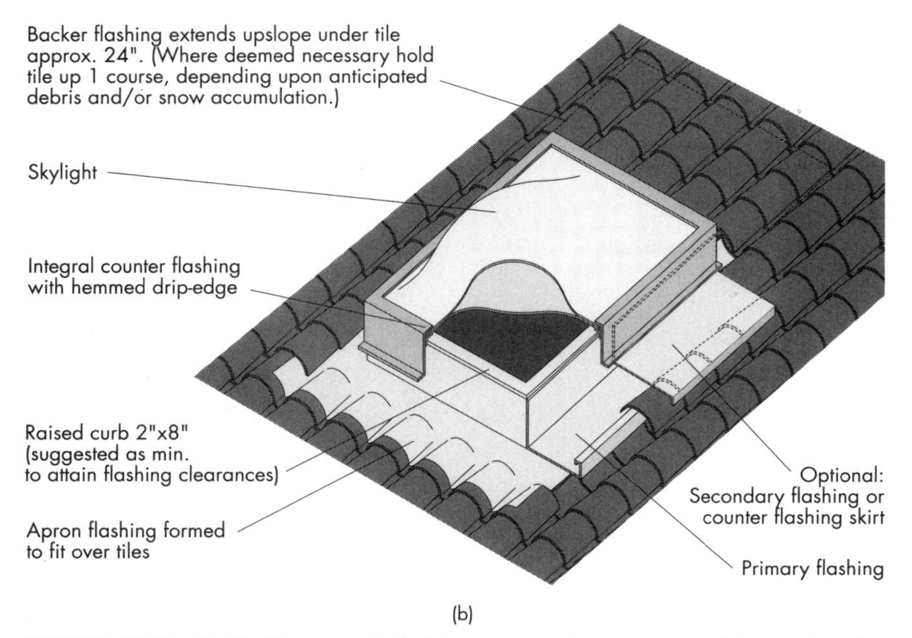

FIGURE 12.10 (*b*) Flashing around skylight—cover and pan concreter/clay roofing tile. (*Continued*)

can provide better bracing for the exterior walls and reduce wind uplift pressures on the roofing material. Better structural integrity and better performance of the roofing materials result with the use of hip roofs in high-wind areas.

- Another reason to use hip roofs in high-wind areas is because of the poor performance of gable-end vents in wind-driven rain situations. At hurricane wind speeds large quantities of rain can be driven in through gable end vents. While this water is transient in nature and poses little long-term threat to the structure from a decay perspective, it can cause a tremendous amount of damage to the contents and potentially to the occupants of the structure. The rainwater soaks the insulation while weakening the drywall. Within a short period of time the ceiling falls, destroying most of the contents of the room below. The use of hip roofs requires the use of ridge vents, or small surface-mounted roof vents, which have better resistance to water infiltration during high winds.
- As the sheathing forms the foundation for the whole roof system it is imperative that it be sufficiently attached to resist the kind of loads that can be found in high wind situations. APA has published nailing schedules for roof sheathing attachment in high-wind areas. See Chapter 7 for further information.

Cathedral Ceilings. In the past cathedral ceilings have been the source of some problems, due primarily to the difficulty in maintaining a workable ventilation path behind the sheathing as required by the building code while maintaining locally required levels of insulation. Difficulties arose from trying to get joists of suitable depths and the inability to maintain ventilation around roof deck penetrations such

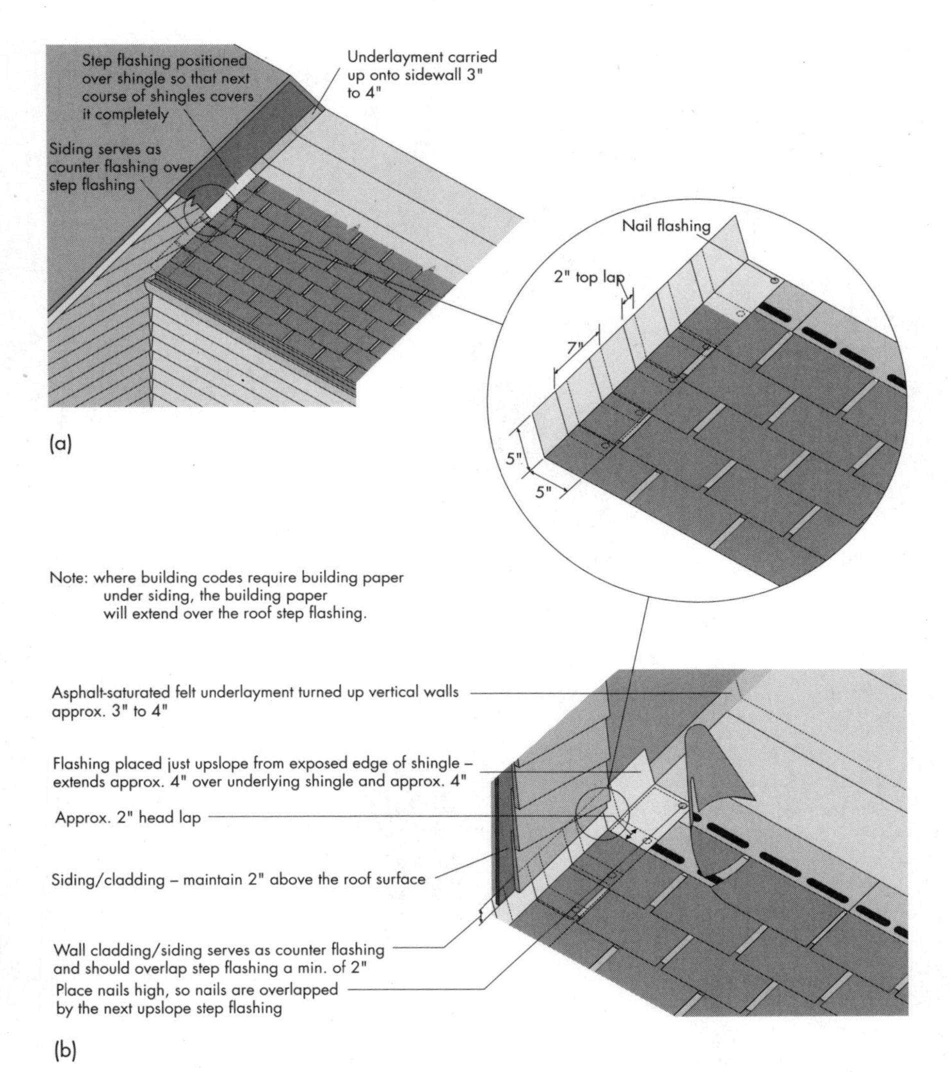

FIGURE 12.11 (*a*) A shingle-type roof at a sloped wall-to-roof intersection. (*b*) Close-up of flashing detail.

as skylights. There have been many documented cases of structural failures and nonstructural water damage because of these difficulties. Penetrations in the ceiling for the installation of canned lighting, ceiling fans, and other electrical equipment have also provided a path for air infiltration into this poorly ventilated, enclosed area.

With the introduction of engineered wood I-joists, most of the ventilation problems can be easily solved. I-joists are readily available in depths that easily permit required insulation levels and allow sufficient space for ventilation above the in-

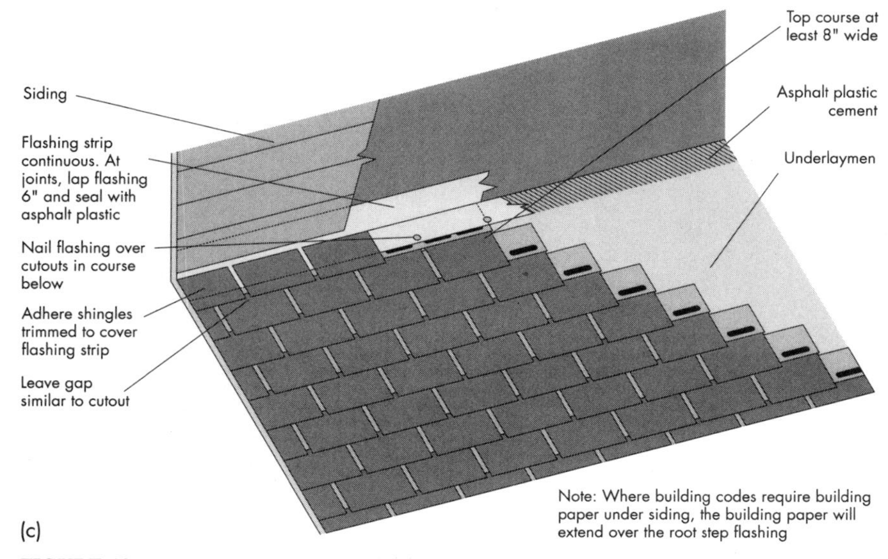

FIGURE 12.11 (*c*) A shingle-type roof at a horizontal wall-to-roof intersection. (*Continued*)

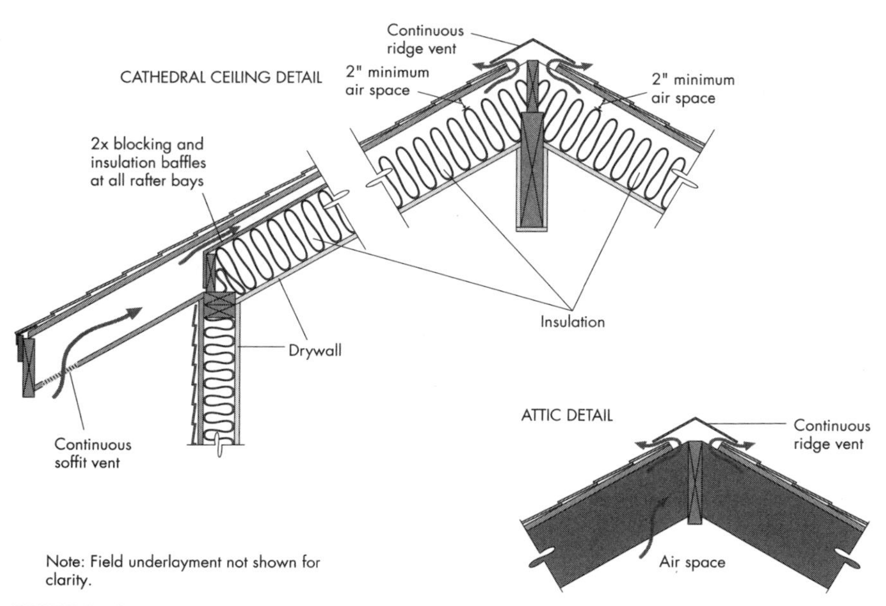

FIGURE 12.12 Code-required roof ventilation—at cathedral ceiling and attic—utilizing continuous ridge vents.

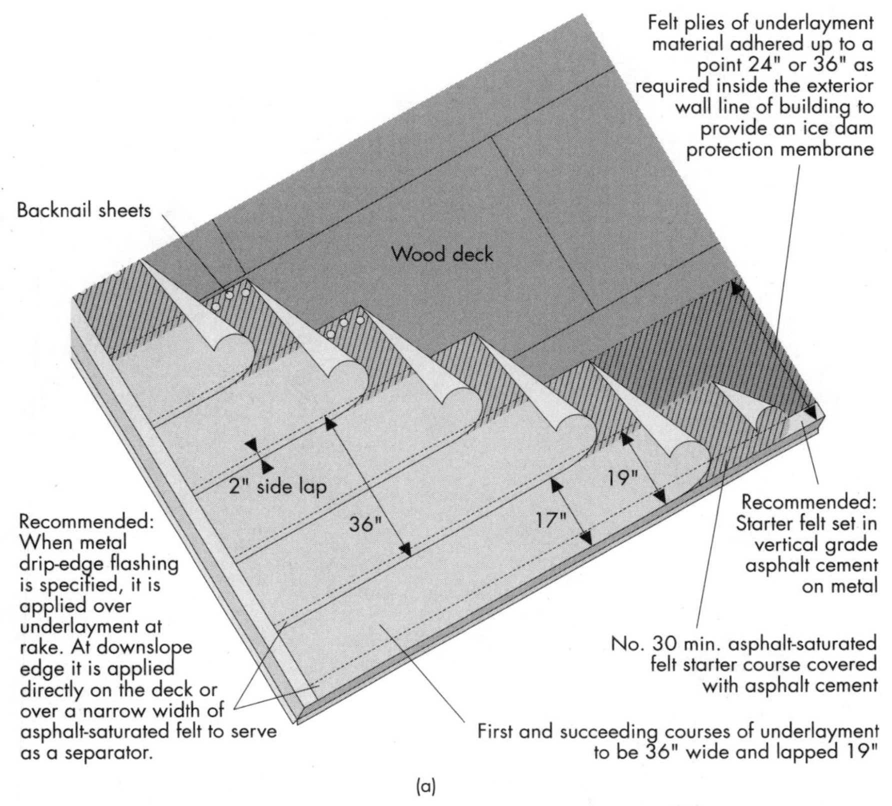

FIGURE 12.13 (*a*) Sealed asphalt-saturated felt underlayment used in low slope roofs or used in areas of wind-driven rain, snow or where ice dams are prevalent.

sulation. If roof penetrations are going to interrupt the flow of ventilation air, the prescored holes in the I-joist webs can be removed, the I-joists properly sized and positioned so the holes are adjacent to the top flange, and the ventilation restored around the penetration.

Because the ventilation requirements can only provide minimum protection against moisture in the roof cavity, it is a good idea to limit the use of ceiling penetrations that can cause air leaks. In addition, vapor-barrier paint may be used on the ceiling to minimize vapor transmission into the cavity.

Structural Insulated Panel (SIPs) Roof Systems. SIPs show great promise for roof applications. They are strong and lightweight and can easily accommodate the spans found in residential and light commercial roof systems. Chapter 3 provides detailed information on the design of SIPs. The very nature of their construction—the use of closed cell foam—virtually eliminates the possibility of air infiltration and greatly reduces the transmission of vapor into the system. The inability of the system to provide ventilation on the bottom side of the roof sheathing is bypassed by use of National Evaluation Reports (NER)[11] and model building code Engineering Reports (ES Reports).[12,13,14] With properly installed flashing details and the

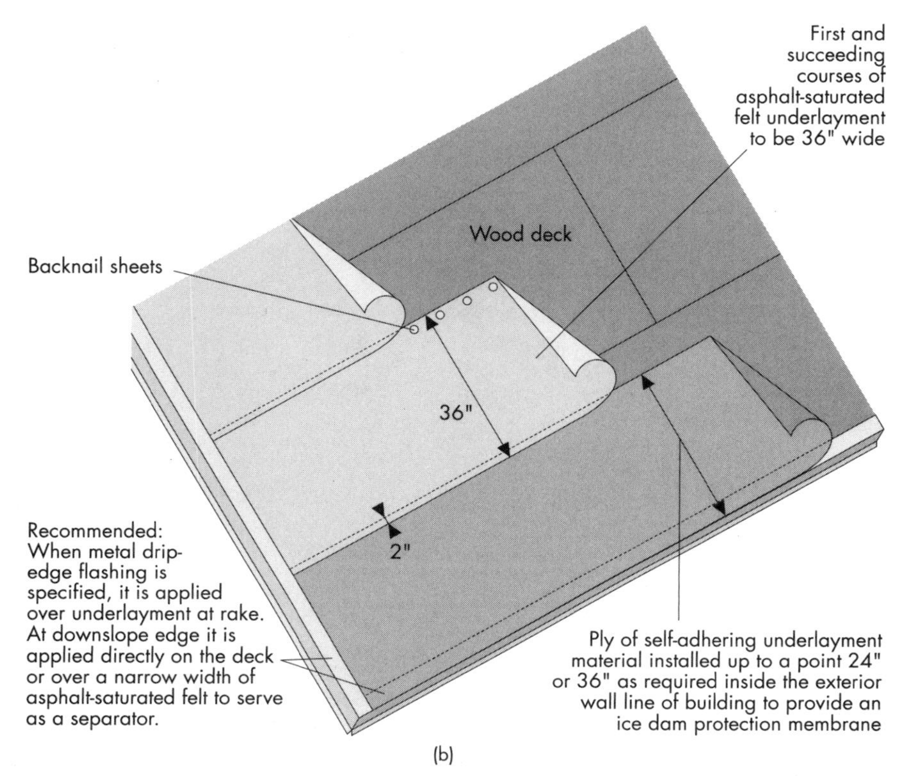

FIGURE 12.13 (*b*) Asphalt-saturated felt underlayment used in conjunction with an ice dam protecting membrane. (*Continued*)

other elements of the roofing system, use of SIPs greatly minimizes the required special finishes and sealing details that plague the builder where cathedral ceilings are concerned.

12.2.2 The Wall System

The exterior walls of the structure are also essential components of the building envelope. Aside from providing security and performing the essential function of supporting the rest of the building, the exterior walls form an integral part of the structure's water and weatherproofing system. Details in wall design and construction are important in preventing damaging moisture buildup, whether the moisture originates from outside or inside the building.

While the roof forms the first line of defense, as discussed in Section 12.2.1, the walls, by their very nature, are the next line of defense and represent the most complex components in the building envelope to design, detail, and construct for the purpose of maintaining a sound structure and a healthy environment for the occupants within the structure. Equally important is maintaining the health of the structure itself by controlling the environment within the elements of the structure, i.e., designing the structure for permanence.

Designing the structure for permanence is made difficult because in wood structures walls are:

- Usually discontinuous in the vertical and horizontal planes with multiple joints
- Made up of numerous relatively small elements, each relying on perfect installation to ensure proper performance
- Interrupted by a number of large penetrations such as windows and doors
- Interrupted by an enormous number of small penetrations such as nail holes, electrical wire and plumbing access holes, and roof and foundation vents
- Subject to differential pressures due to exterior wind conditions, stack effect, and ventilation imbalances
- Subject to differential moisture conditions resulting in vapor pressure differentials such as hot and moist inside, cold and dry outside, or vice versa
- Subject to wind-driven rain and/or inadequately protected by roof overhangs
- Subject to ultraviolet degradation, freeze-thawing, hot and cold cycles, and wet and dry cycles
- Designed to rely heavily on the use of sheet metal flashing and proper construction sequencing to provide waterproofing
- Placed close to the ground, hidden behind shrubs, and often subject to regular wettings by landscaping sprinklers
- Often placed in environments only marginally suited for wood construction—e.g., the use of untreated wood in tropical or termite-infested areas

Some and often all of the above factors make the design, fabrication, and installation of wood walls a very important aspect of designing for permanence. This section will cover these subjects in an integrated manner.

As discussed below, there are three primary sources of water in residential construction. From greatest to least, they are leakage, infiltration, and vapor transmission. Design, fabrication, and installation of wall systems for permanence with respect to these three moisture sources will be discussed in this section.

What will not be covered in this section is proper installation of wall covering treatments, which is covered in detail for wood products in Chapter 9 of this Handbook and in many other publications for wood and other construction materials.

Sources of Water in Wood Construction. The three primary sources of water in wood construction are:

- *Leakage,* by an order of magnitude, is the greatest contributor to water damage to wood structures. Caused by improperly installed flashing and roof construction, even the smallest error can introduce tens of gallons of water into the structural shell of the building over a relatively short period of time. Every year millions of dollars of damage are caused by water leakage. After Hurricane Andrew in 1992, the water damage done to thousands of homes caused by the loss of roofing materials and roof sheathing made up a very large percentage of the over 30 billion dollars worth of damage. It far exceeded the structural damage caused by the storm.
- *Infiltration* is the transportation of water into the structural system of a building by differential air pressure. The differential pressure between the inside and the

outside of a structure draws moisture-laden air into the cavities between the inside and outside walls and between the ceiling and roof surfaces. The differential pressure can be caused by the stack effect (warm air raises pulling in cold air from outside), improperly vented heating equipment, unbalanced ventilation systems, or high wind. When one takes into account that the total of all the little gaps in an average home envelope add up to over 1 ft^2, it can be seen that even a slight differential pressure can cause a good deal of air movement. Air infiltration by itself is not necessarily a bad thing. The problem arises when warm, moist air infiltrates a wall or roof cavity and impinges on a cold surface. If that surface is cold enough to lower the temperature of the air to below its dew point, water will condense on the cool surface. The warm, moist air can come from the outside (an air-conditioned home in Miami in the summer) or from the inside (a home in Wisconsin in the winter). The water always ends up in the same place—inside the wall or roof cavity. Fortunately, the ventilation required for the roof cavity has a proven track record of good performance except in the most extreme situations, such as when leaks are present. In walls, the use of an air barrier is the most effective way to control infiltration.

- *Vapor transmission* has become the superstar of moisture control problems. Unfortunately, it hardly deserves this distinction and distracts attention from the real culprits, leakage and, to a lesser extent, infiltration. Until a decade or so ago, no distinction was made between vapor transmission and air infiltration. Vapor transmission is the passage of molecular water through an obstacle, driven by a differential partial pressure for water across that obstacle. Put plainly, vapor transmission is the passage of moisture from wet to dry. In the most extreme cases that can be examined, the quantities of water are very small. The proper placement of a vapor retarder—on the warm side of the wall—will normally provide all of the protection required. Note that sloppy installation and penetrations in the vapor retarder for switch boxes, plumbing penetrations, and around windows and doors do not have an appreciable impact on the performance of the vapor barrier. They do, however, cause problems due to air infiltration at these points. Most of the damage associated with improperly installed vapor retarders is really caused by air infiltration problems.

Keeping Water out of a Wood Structure. There are numerous reasons why water must be kept out of the structural system. A few of the most important are:

- *Decay:* Protected wood in service in a residence in the United States will have an equilibrium moisture content ranging from 8–12%—a little higher in hot, humid climates and a little lower in cold, dry climates. At this moisture content range, wood can last almost indefinitely. When the equilibrium moisture content of wood rises above approximately 20%, decay organisms can thrive. Decay organisms use the wood as a food supply, seriously reducing its strength and durability over time. Wet wood also attracts a large number of wood-eating insects. One of the major functions of the building envelope is to keep excessive moisture away from the unprotected wood.
- *Insect infestation:* With a couple of notable exceptions, most insect species that consume wood are more apt to do so if the wood is wet than dry. In fact, as the wood gets wetter, it becomes a more viable source of food for a greater range of insect species. Aside from the obvious task of keeping wood-eating insects physically away from the wood structural components, the building envelope contrib-

utes most greatly to the minimization of insect infestation by keeping the wood dry.

- *Reduces effectiveness of insulation:* Aside from the structural problems that can result from the condensation of moisture within the wall cavity, the dampening of the insulation can reduce its insulative properties, causing the owner's heating and cooling bills to increase. In certain kinds of cellulose insulation, it is thought that continuous cycles of wetting and drying can cause fire-retardant chemicals placed in the insulation to leach out.
- *Reduces strength and stiffness of wood structural members:* Wood, in any form, loses a significant part of its strength and stiffness when exposed to moisture contents greater than 19% for prolonged periods of time. This is a temporary condition and not related to decay. When the moisture content drops back down to the safe range (below 16%), the loss in strength and stiffness is recovered. Certainly, wood subjected to high moisture conditions for long periods of time is also subject to a permanent loss of strength and stiffness through decay.
- *Mold and mildew production:* Moist conditions at the surface of building materials can promote the growth of molds and mildews. While in the short term these conditions cause little damage to the wood itself, they can be indicative of a high moisture condition that can cause serious structural damage to wood materials through wood decay. The spore production of certain species of molds and mildews has also been linked to potentially serious health risks in some individuals.
- *Reduces life of exterior finishes:* The durability of exterior finishes of wood elements can be seriously degraded by wet/dry cycling of wood members.
- *Comfort and health of occupants:* The comfort level of the inside environment of the building is directly related to how well the building envelope functions. Sick building syndrome, mold on inside walls, condensation on windows, decaying sills, insect infestations, allergic reactions, and unpleasant odors are just a few of the consequences of the failure of the building envelope to control water.

Water Leakage in Wood Wall Construction. By a huge margin, the largest contributor to damage to wood building construction is bulk water leakage through the envelope of the structure. This can be caused by a number of factors, including improper flashing details, improper installation of water/weatherproofing barriers, and poorly designed or executed wall intersections and penetrations. Wood structures have the ability to absorb, distribute, and dissipate small amounts of water, especially from intermittent sources. The problems arise when there are design or construction errors that channel large amounts of water into the wall cavities at a rate that exceeds the ability of the structure to absorb and eliminate the water. Some of the elements of the building envelope that can influence bulk water leakage are described below.

Flashing. Flashing is typically made up of thin sheets of corrosion-resistant material used in conjunction with the other elements of the building system to prevent leaks around wall intersections, window and door openings, and penetrations. Flashing is usually made up of galvanized steel, copper, aluminum, lead, or vinyl. At small wall penetrations such as exhaust vents, custom flashing is often used in lieu of conventional flashing because of the irregular shapes associated with some of these penetrations.

Flashing directs the flow of water down and away from the interior of the structure to the outside of the wall covering. In every example shown, the top edge of the flashing passes underneath the weather- or water-resistive barrier, the upper pieces of flashing pass over the lower pieces, and the lower edge of the flashing always laps over the top of the wall weather- or water-resistive barrier below. This order of installation is often called *shingle lapping*. In such a manner, the flashing never directs the flow of water to the interior side of the underlayment, never putting it in contact with the wood framing of the structure. Each flashing detail is part of a whole water-/weather-proofing system that is continuously redirecting water flow down and away from the interior of the structure.

Figures 12.14–12.21 illustrate examples of typical flashing details for both wood and masonry veneer walls and walls with stucco exterior finish. Flashing details for the exterior insulation finish system (EIFS) are not provided, as they are typically proprietary in nature. The finish system manufacturer should be contacted for applicable flashing installation details.

Figures 12.14–12.17 show flashing details commonly seen in wood wall construction around window and door openings and at the horizontal joints of plywood panel siding.

Similar information is provided for masonry veneer walls and walls with stucco exterior finish in Figs. 12.18–12.21.

Weather-Resistive Barriers. Long-term durability of wood framing, sheathing, and siding is enhanced when wall sheathing or framing is covered by a protective weather- or water-resistive barrier (building paper), or an approved alternative such as an air infiltration barrier or housewrap, as described below under Air Infiltration in Wood Wall Construction and Vapor Transmission in Wood Wall Construction.

For purposes of this discussion and the illustrations in this chapter, all references to weather-resistive barriers will include the use of air barriers. These products have a number of very beneficial properties when used in construction, as will be considered below. At this time, only their use to channel the flow of bulk water leakage will be described.

Weather-resistive barrier materials provide a second line of defense for the building envelope against the intrusion of bulk water through the exterior siding. The materials and labor costs associated with these products provide relatively inexpensive insurance to protect the structural components and reduce the risk of deterioration caused by unanticipated water leaks. However, techniques for installation of the weather-resistive barrier over flashing at penetrations and at corners or intersections with other surfaces are important to ensure satisfactory performance and minimum maintenance. Careless or improper installation of the weather-resistive barrier may result in water leaks behind the exterior wall covering or into the wall cavity itself. Over prolonged periods, these can lead to decay and deterioration, as well as increased risk of insect infestation in some regions, requiring costly structural and aesthetic repairs. Proper installation of the weather-resistive barriers at flashing, corners, and intersections is particularly important.

The basic principle is simple—provide a continuous barrier that sheds moisture down and away from the plane of the structural wall surface. This is accomplished by overlapping successive layers of weather-resistive barrier paper over the exterior walls of the structure. This applied surface is used in conjunction with properly applied flashing to direct bulk leakage away from the interior of the structural frame. The principle also extends to weather-proofing details at wall penetrations, such as windows and doors, plumbing hose bibbs, electrical boxes, wall-mounted air conditioners, vents for appliances, and at junctures with horizontal surfaces such as exterior decks and cantilevered balconies. In these cases, details of attaching components, sequence of installation of the weather-resistive barrier and flashing, and other components, such as windows or doors, are of special importance to ensure

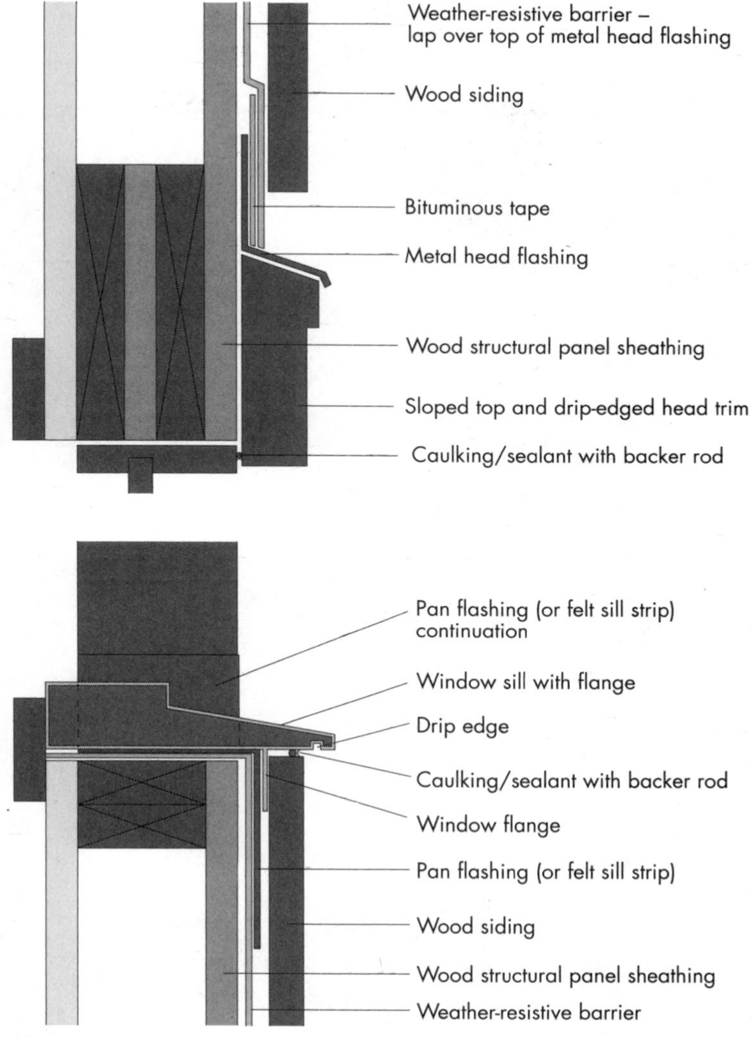

FIGURE 12.14 Cross section of window showing integration of structure's weather-resistive system in a wall with wood siding.

that water leaks are not directed to the wall sheathing, onto structural elements below, or into the wall cavity.

Figures 12.14–12.21 illustrate how the weather-resistive barrier materials are placed with respect to the flashing to channel any water running down the inside face of the barrier over the flashing and outside of the envelope. The relationship between proper weather-resistive barrier and flashing installation can be seen in the majority of the remaining figures in this chapter.

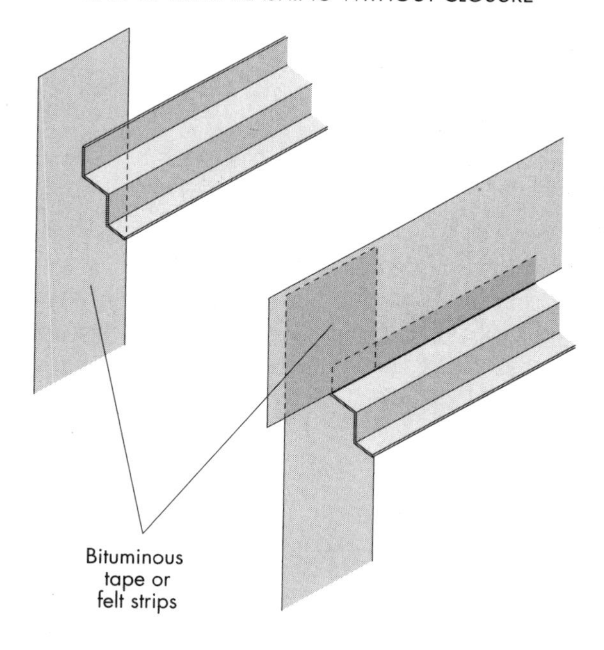

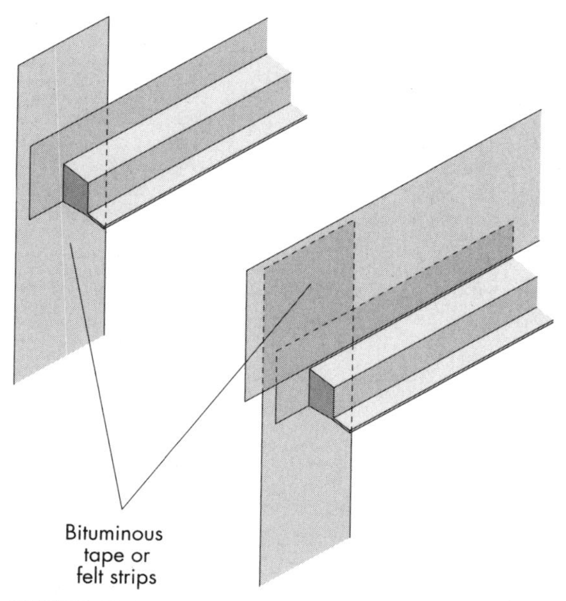

FIGURE 12.15 Proper use of bituminous tape or felt strips to seal head flashing over windows and doors.

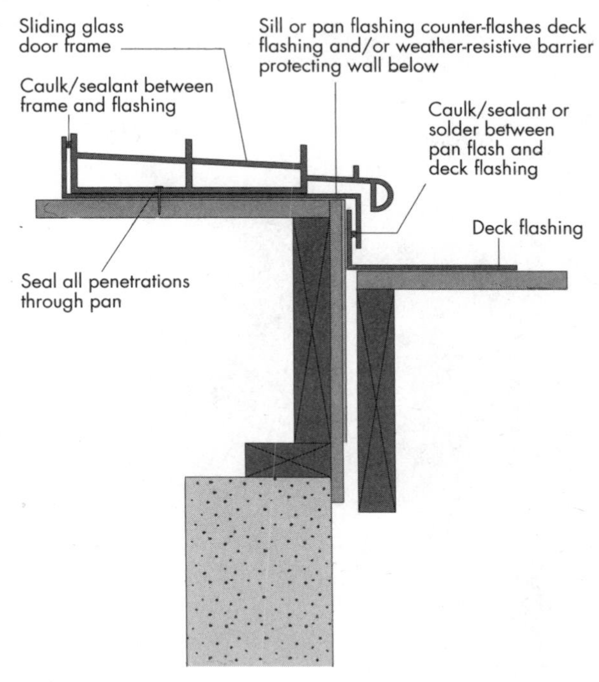

FIGURE 12.16 Sill flashing at sliding glass door.

Building Code Requirements for Weather-Resistive Barriers The importance of weather-resistive barriers can be evidenced by their requirement in all of the major model building codes, as seen below:

- 2000 International Building Code[9] (ICC), Sections 1404.2, Water-Resistive Barrier, and 1405.3, Flashing
- 2000 International Residential Code[1] (ICC), Sections R703.2, Weather-Resistant Sheathing Paper, and R703.8, Flashing; also Sections R703.5.1, Application for wood shingle or shake exterior cladding; R703.7.5, Flashing for masonry veneer exterior finish; and R703.9.1, Weather-Resistive Barrier, and R703.9.2, Flashing for Exterior Insulation Finish Systems (EIFS)
- 1999 National Building Code[15] (BOCA, Inc.), Sections 1404.3, Weather Protection, and 1406.3.10, Flashings
- 1999 Standard Building Code (SBCCI),*[16] Sections 2303.3, Moisture Protection, and 1403.1.4, Flashing
- 1997 Uniform Building Code (ICBO),*[17] Sections 1402.1, Weather-Resistive Barriers, and 1402.2, Flashing and Counterflashing

*These codes contain exceptions that permit omission of the weather-resistive barrier when the exterior cladding or wall finish consists of approved weather-proof panels or when walls are sheathed with water-repellent panel sheathing. Wood structural panel sheathing such as plywood and oriented strand board is recognized by the codes as water-repellent sheathing. However, applicable code provisions could be superseded by International Building Code requirements when locally adopted. The International Building Code requires building paper or an approved air barrier in addition to the use of wood structural panels.

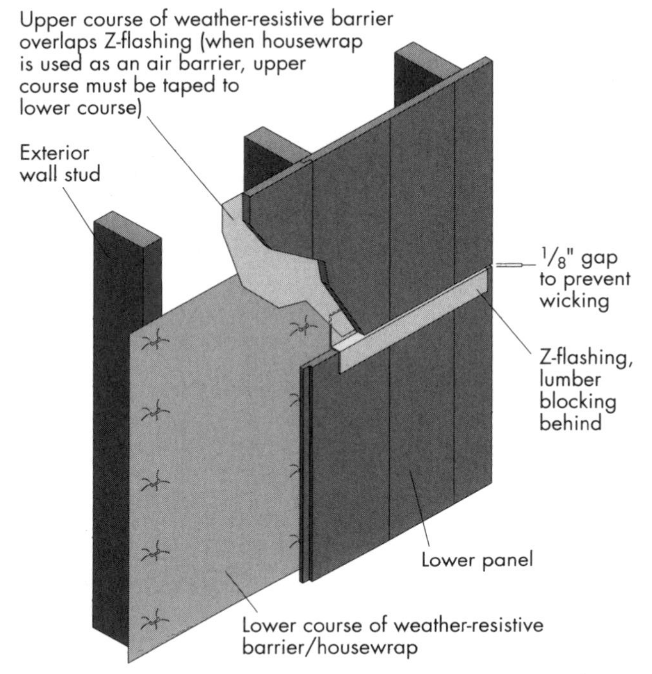

FIGURE 12.17 Proper installation of Z-flashing in a single wall system with APA rated siding.

Wall Intersections and Penetrations. While proper design, detailing, and execution of construction details relating to the building envelope as defined in this chapter are fairly straightforward, the trades often have difficulty adapting the same principles to unusual situations that can occur in modern construction. Such details include deck-to-wall intersections, wall-to-roof intersections, gutter-to-roof or wall intersections, and the installation of windows.

In addition, the more common penetrations, such as windows, doors, and ventilation openings, often have different flashing/waterproofing details, depending on the type of exterior siding used.

It is incumbent upon the designer to carefully detail in the drawings the flashing and waterproofing details that are appropriate for these unusual situations. The builder, on his part, must carefully follow the details provided on the drawings.

Figures 12.11a–c and 12.22 and 12.23 illustrate typical wall intersection details. Fig. 12.24 is an example of the detailing required at the intersection of an outside deck and an exterior wall. Fig. 12.25 shows a very common wall penetration detail.

Caulking. Elastomeric exterior sealants called caulks are a popular component of the waterproofing system used in modern structures. They are used to seal up the cracks between individual elements of the buildings exterior finish to keep wind and water from penetrating the skin of the structure. In this respect, the caulking

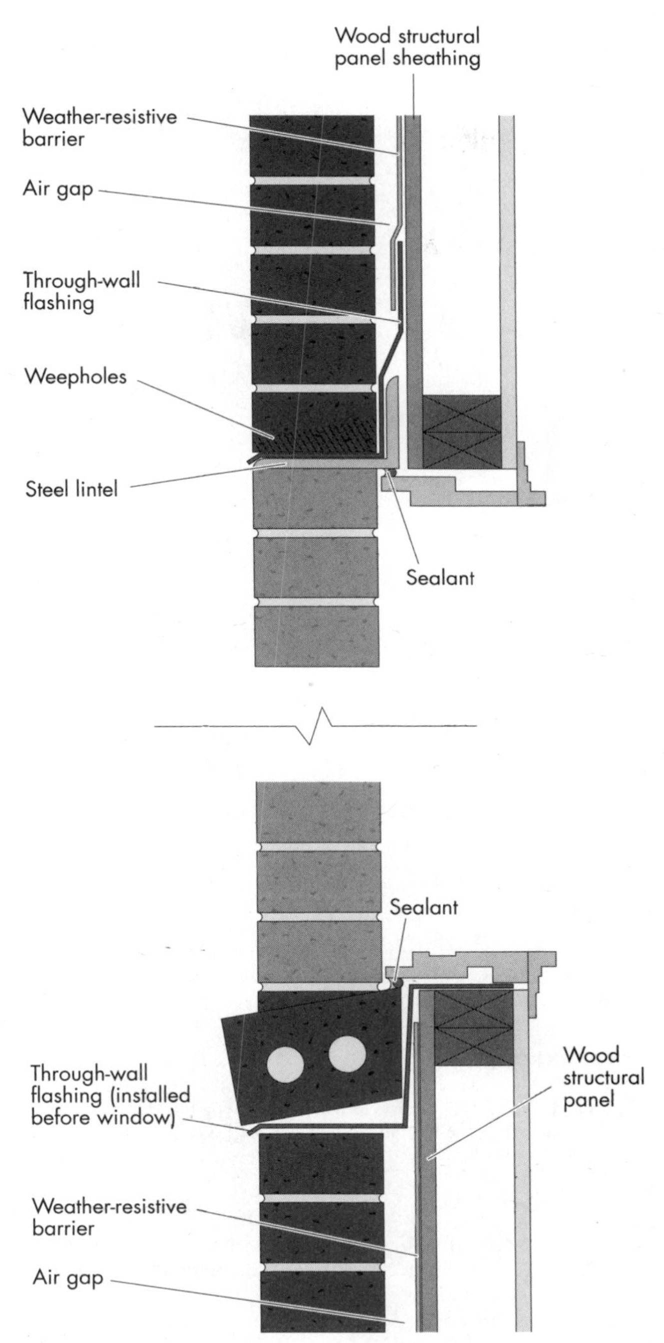

FIGURE 12.18 Cross section of window showing integration of structure's weather-resistive system in a wall with brick veneer.

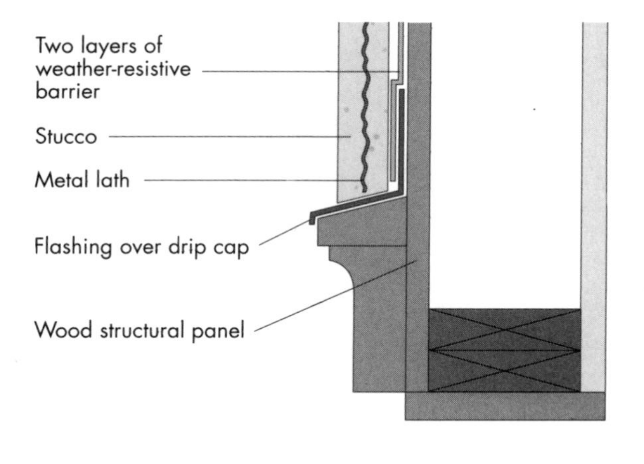

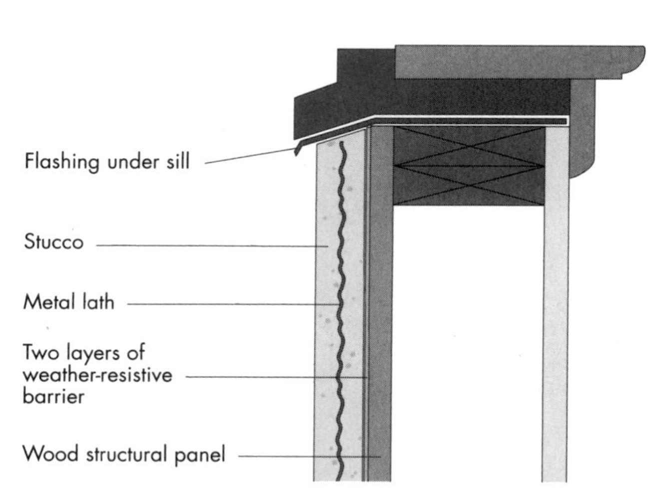

FIGURE 12.19 Cross section of window showing integration of structure's weather-resistive system in a wall with Portland cement stucco exterior wall covering.

provides a part of the walls' first line of defense against water intrusion. The waterproofing performance of modern structures often depends on many hundreds of feet of caulked joints.

Unlike flashing, properly applied building paper or air barriers, or even intelligent building design, CAULKS ARE NOT PERMANENT! They have a finite life

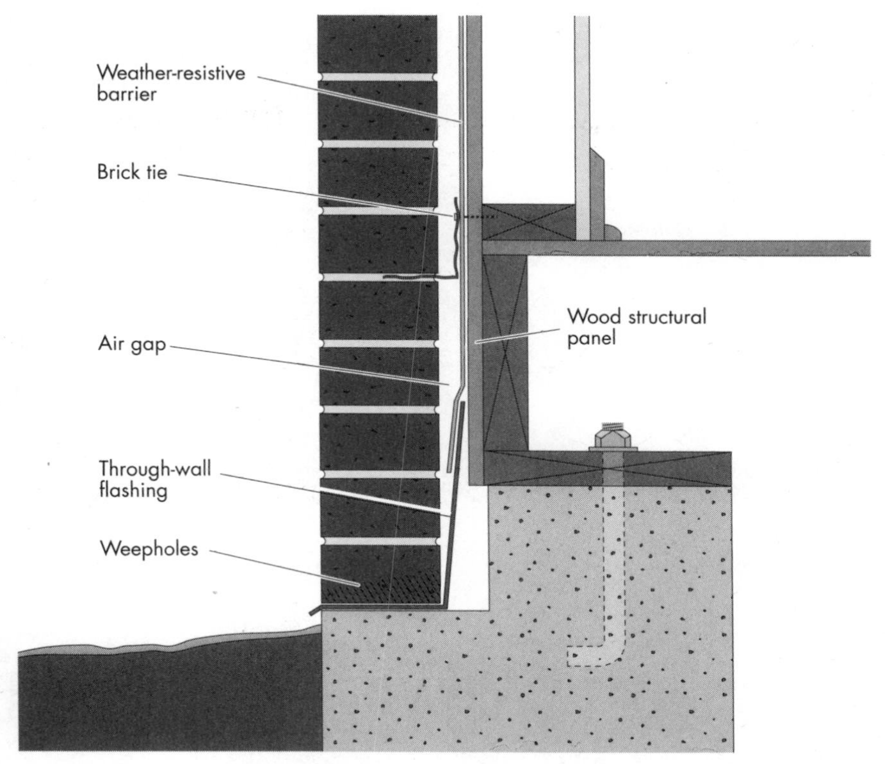

FIGURE 12.20 Flashing and weather-resistive barrier at brick ledge.

and must be replaced on a periodic basis. The actual life of a caulked joint depends on several things, including the quality of the caulking material, shape of the caulked joint, amount of differential movement between the adjacent elements, surface preparation, and exposure. Because every caulked joint in a given building has a different set of the above variables, it is safe to say that after a new building is a couple of years old, there will always be a caulked joint that needs recaulking.

As a result, a caulked joint should never be the sole form of waterproofing at a given location. Intelligent building design, back-up methods of waterproofing such as building paper or house wrap, and proper placement of flashing should always be used in conjunction with caulked joints.

The effectiveness and useful life of a caulked joint can be greatly improved through proper application. In addition to ensuring a clean surface to facilitate the adhesion of the caulked joint, the cross-sectional profile of the joint is also very important. A good caulked joint is one that maximizes the surface area between the caulk and the surface to which it is applied to enhance adhesion. In addition, a good caulked joint has a smaller cross section between the contact surfaces than it has at the mating surface. Examples are provided in Fig. 12.26. The smaller cross-section in the middle of the caulked joint allows differential movement to be taken up within the caulked joint and not to be concentrated at the mating surface

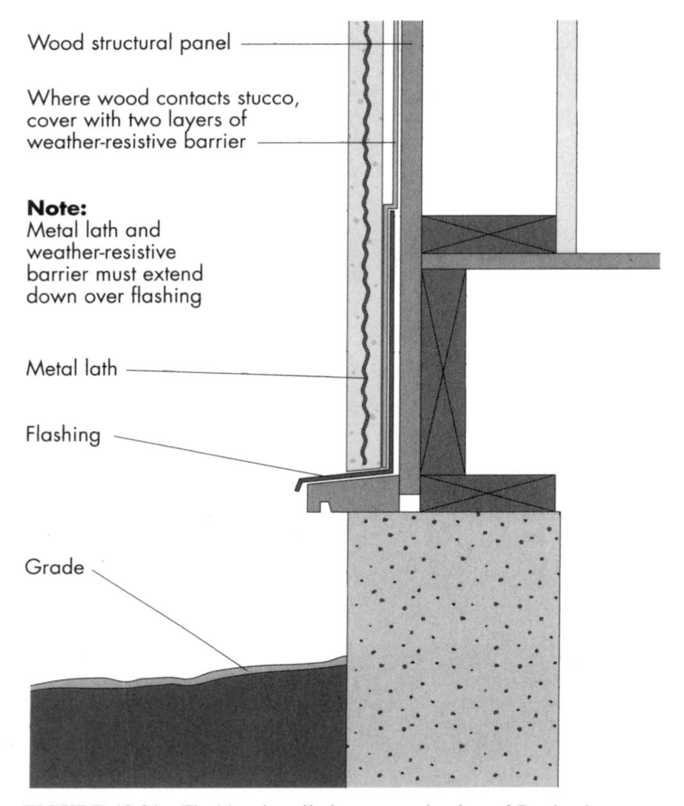

FIGURE 12.21 Flashing installation at termination of Portland cement stucco wall covering.

between the caulk and the surface to which it is applied. While the caulk is flexible, the joint between the caulk and the mating surface is not. If the stress is concentrated at the joint, it will cause premature failure of the joint.

Air Infiltration in Wood Wall Construction. Second to leakage by bulk water, the largest contributor to high moisture conditions in wall systems is air infiltration. Even the small differential pressures associated across a given wall can cause a relatively large volume of moisture-laden air to leak in to or out of the structure. The significance of this detail can be put in perspective when considering that the term *air changes per hour* is used to describe air leakage in structures.

Example A "tight" house is traditionally allowed about 0.4 air changes per hour. In a simple, single-story, 2000 ft^2 house this allows 6400 ft^3 of potentially moisture-laden air to leak in or out of the building every hour. Every 2-½ hours every bit of inside air that has been heated, cooled, dehumidified, or humidified is lost. Completely ignoring the energy issues, the significance of this leakage as far as moisture control is concerned is evident.

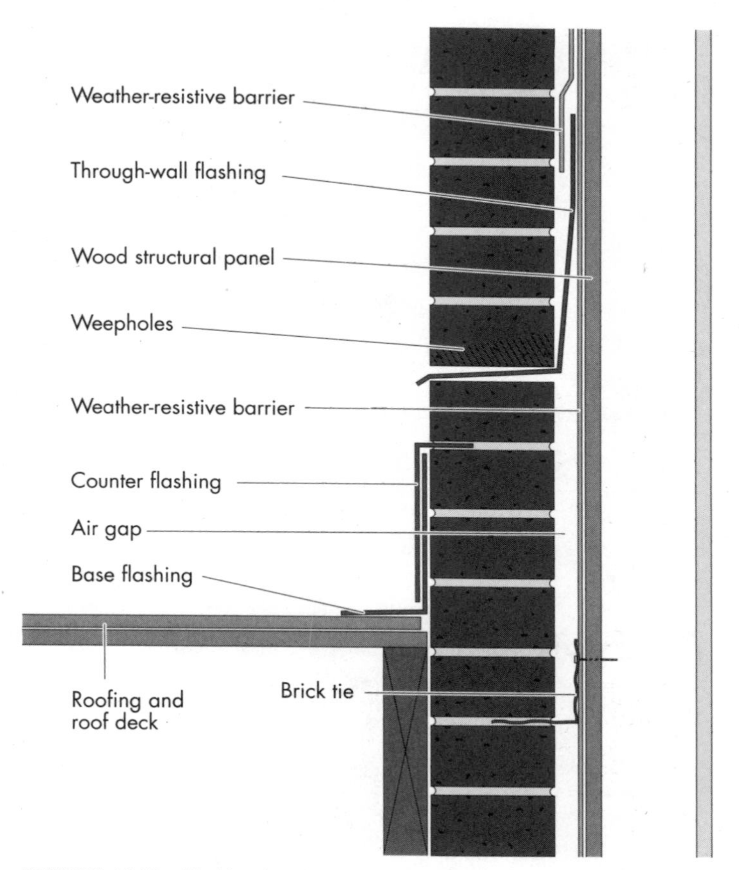

FIGURE 12.22 Flashing installation at brick veneer-to-roof intersection.

If there is a significant drop in temperature in the air as it passes through an insulated wall to the area of low pressure, there is the chance that the air will fall below the *dew point,* which is the temperature at which the air can no longer hold a part of the moisture as vapor within it. Wherever this point is reached, the excess vapor in the air settles out as condensation. If the dew point happens to be within the wall cavity, there may be a moisture problem as the building materials absorb this moisture and the moisture content of the building materials increases. The moist air can come from the inside or the outside depending on the pressure differential across the wall. In an area of hot, moist climate and air-conditioned spaces, the concern would be with leakage from the outside to the inside. In cold, dry climates the moisture laden inside air leaking out could cause the problem.

Air-Infiltration Barrier (*House Wrap*). An air-infiltration barrier blocks the flow of moisture-laden air into the wall cavity due to differential air pressures that exist across the wall. This differential air pressure can be the result of an unbalanced ventilation system, the stack effect caused by hot air rising within the structure, the

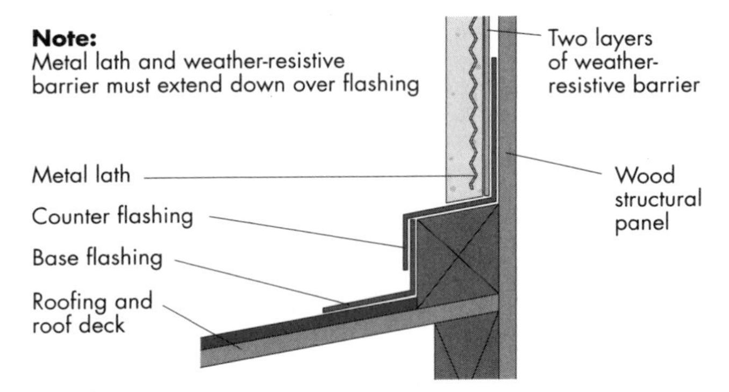

FIGURE 12.23 Flashing installation at Portland cement stucco-to-roof intersection.

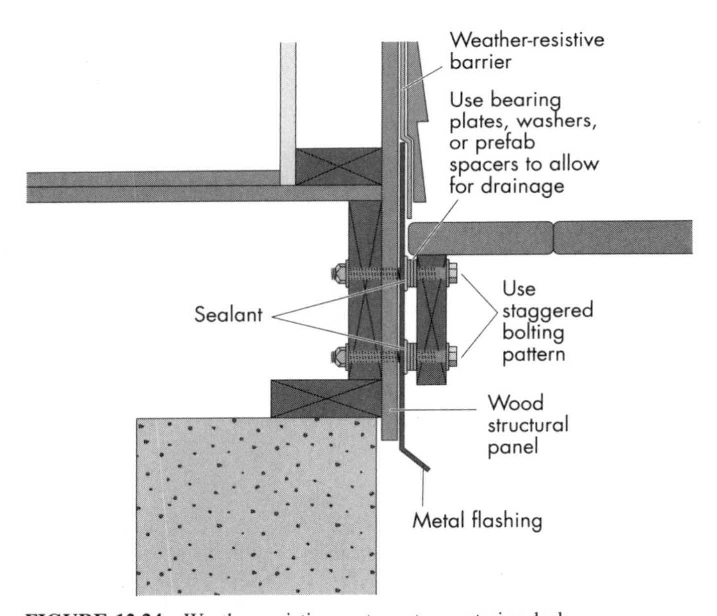

FIGURE 12.24 Weather-resistive system at an exterior deck.

use of unvented heating appliances, or exterior high-wind events. The actual differential pressure does not have to be very large to cause a significant amount of air leakage in one direction or another. As discussed previously, the problem with air leakage occurs when warm, moisture-laden air comes in contact with a cool surface and causes condensation of the moisture. If these conditions persist for a significant length of time and at such a level that the wood structure of the building cannot accommodate the additional moisture, the moisture buildup can cause a considerable amount of damage to the structure and degrade the living conditions within the structure.

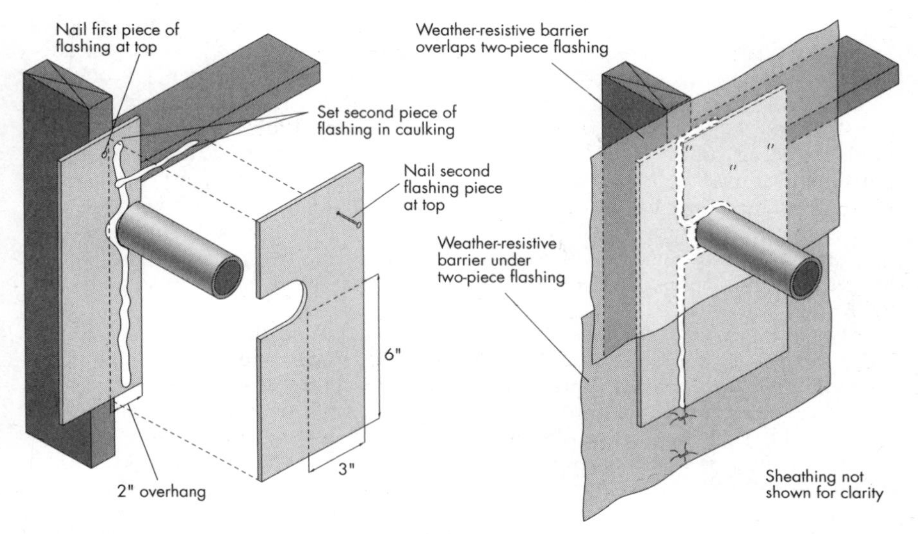

FIGURE 12.25 Integration of structure's weather-resistive system at a typical wall penetration.

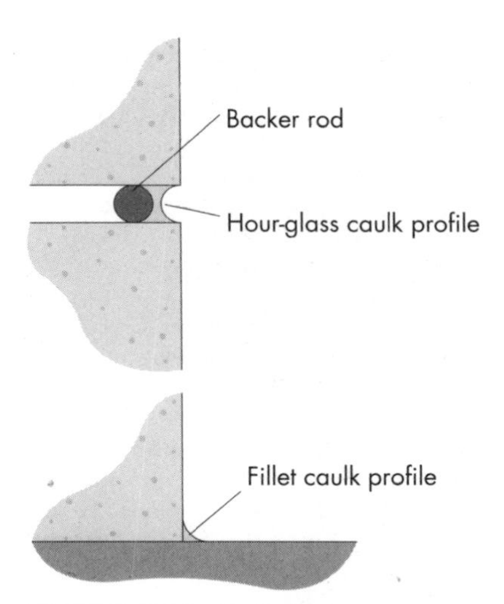

FIGURE 12.26 Caulked joint geometry.

It does not matter where the air barrier is placed, on the inside or outside surface of the wall. A good air-infiltration barrier on either surface will work equally well to prevent the flow of air. In a cold climate that requires a warm-side vapor retarder, the vapor retarder can act as the air barrier as well if properly applied and sealed.

The problem with using the inside vapor retarder as the air barrier as well lies in effectively sealing all of the penetrations and seams in the vapor retarder. Areas around plumbing and electrical wiring penetrations, joints in the vapor retarder at the top and bottom of interior and exterior walls, and discontinuities between floors and at foundation sill plates are all areas that are extremely difficult to seal adequately to ensure that the vapor retarder acts as an effective air barrier. The obvious solution is to seal the outside surface of the building because there are fewer penetrations and they are larger, easier to get at and easier to deal with. As the major model building codes allow house wraps to be used in lieu of building paper for most applications, the use of these products is on the rise. Note that when used in lieu of building paper for use as a weather-resistive barrier, the house wraps do not have to have all edges, seams, penetrations, rips, and tears taped. Air-infiltration barriers act as traditional weather-resistant barrier if applied like a building paper. To get the full benefits of an air barrier, however, they must be sealed as described below.

The new products developed for this application have the added benefit that while they block liquid water and restrict the flow of moisture-laden air, they also allow the passage of water vapor. The vapor permeability of these products as far as water vapor is concerned promotes the drying out of any moisture that may get trapped within the wall cavity. When they are used in conjunction with an inside vapor retarder, it is essential to provide a path for water vapor to escape to prevent the buildup of moisture within the wall cavity. A vapor-permeable house wrap does just this.

These air-infiltration barriers come in rolls up to 9 ft wide, allowing the builder to wrap the barrier all the way around the structure during construction. This is the origin of the term *house wrap*. The large size speeds up installation and minimizes the number of seams to be sealed. Because the air barrier must be airtight, all of the splits, seams, penetrations, and damaged areas must be repaired using a special adhesive-backed seam tape. This is one respect in which an air barrier differs from a vapor retarder. In a vapor retarder, seams, splits, and tears are of little significance. In an air barrier, they all add up.

Figures 12.27 and 12.28 show general installation techniques for proper application of an air barrier. Fig. 12.29 shows the proper sealing of the air barrier around openings such as windows and doors. Details are shown for installation of the air barrier both before and after installation of windows.

Vapor Transmission in Wood Wall Construction. The third and generally the smallest contributor to moisture in wood wall systems is vapor transmission. Vapor transmission is the introduction of molecular water into the structural system of a building caused by a differential water vapor pressure across the structure's envelope. As in the case of moisture-laden air inside the structure, vapor from the interior of the structure can permeate through the interior wall finish of exterior walls and condense on framing and sheathing surfaces in the wall cavity in cold weather conditions. To minimize this potential, an effective interior (warm side) vapor retarder is typically recommended beneath the interior wall finish for most regions.

An exception is that omission of the warm side vapor retarder in exterior walls may be considered in a normally moderate fringe temperature region in the southern and southeastern United States. In warm, humid regions close to the Gulf of Mexico and in Hawaii and the Caribbean regions, where air conditioning is prevalent, the vapor retarder should be installed on the exterior side of the wall. This will prevent the transmission of moisture vapor from the warm, humid exterior air into the wall cavity, which will result in increased potential for condensation on the cooler interior wall surface.

When the warm side is determined to be the inside wall, the vapor retarder can be a kraft paper or foil/kraft paper facing on the wall insulation. The effectiveness

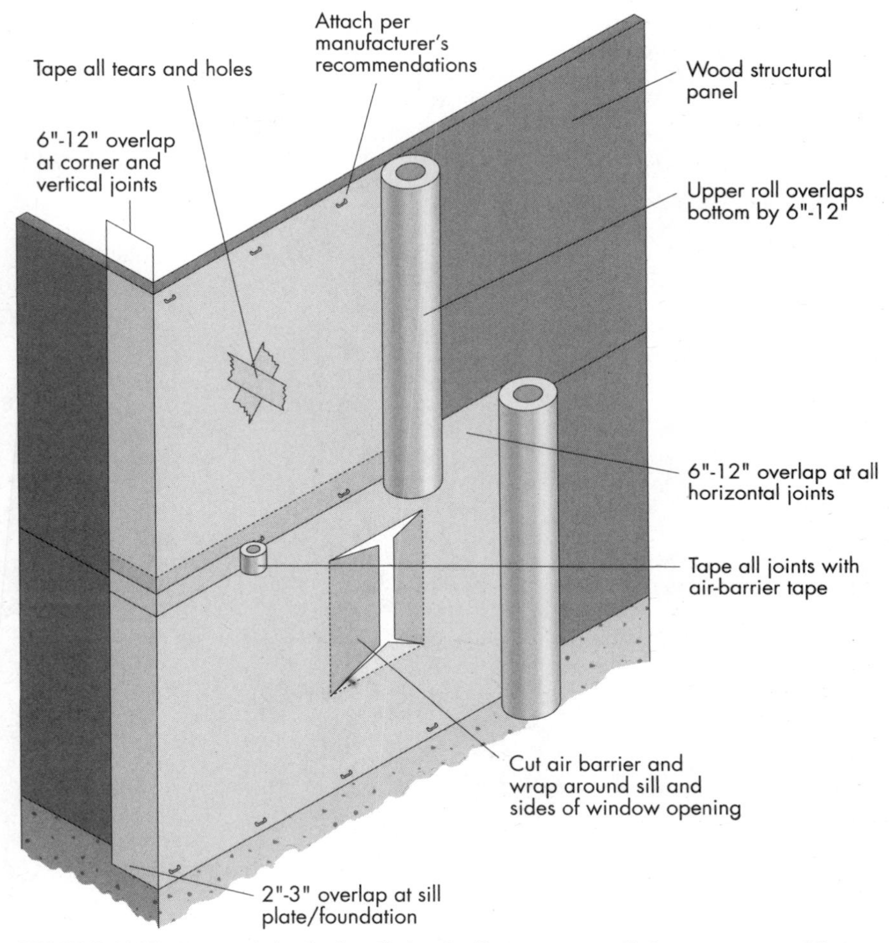

FIGURE 12.27 Proper air barrier installation details (two story wall shown—not to scale).

of this vapor retarder depends on how carefully the insulation is installed. The most effective installation technique is to cut the insulation batt length slightly oversized so that it can be friction-fit to avoid gaps at the top and bottom wall plates. Also, the installation tabs of the insulation facing should be lapped and stapled onto the nailing surface of the studs, rather than to the sides of the studs, to seal the insulation facing against air and moisture leakage and minimize gaps between the insulation and studs.

Alternatively, a continuous vapor retarder can be installed by using a separate layer of 4-mil polyethylene sheeting stapled over the interior side of the wall framing. In this case, unfaced insulation without an integral vapor retarder facing may be used and friction-fit to fill the stud cavities without gaps.

While polyethylene sheeting makes a very good vapor retarder, it is relatively difficult to install. In most cases, the use of polyethylene is not necessary, even in

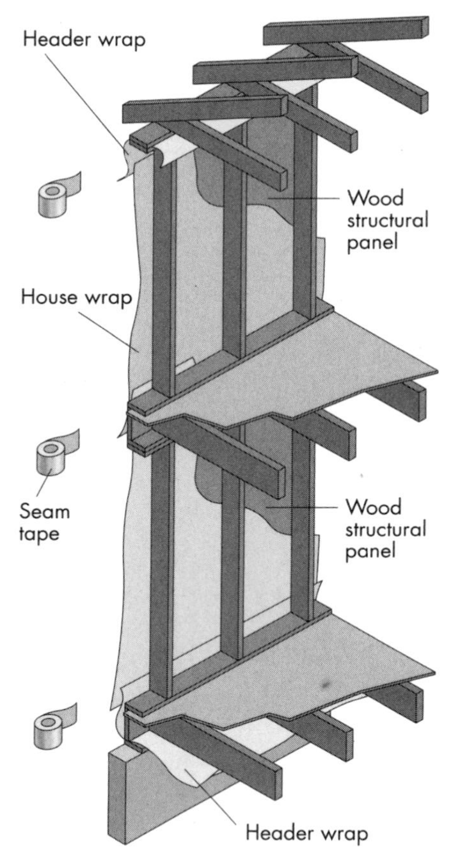

FIGURE 12.28 Air barrier installation details using header wrap.

very cold regions. Ordinary interior latex paint applied over drywall can provide sufficient vapor retardant properties.

Which Is the Warm Side of the Wall? For many years, designers assumed that the warm side of the structure was the inside surface of an exterior wall. For that reason, systems such as the airtight drywall system and the use of vapor retarders such as polyethylene behind the drywall along with gasketed electrical receptacles were promoted throughout the United States. Much of this early research was done in Canada and in areas of relatively cold climate in the United States. Is the warm side always the inside surface of the exterior wall? People living in Hawaii or Florida might say no. In many parts of the country, air conditioning is more often used than heating. In areas like this, the application of a vapor retarder beneath the drywall on the inside surface of the exterior wall can cause the very problem that the vapor retarder is installed to eliminate—condensation of water vapor within the wall cavity.

If placed on the inside surface of the inside wall in a structure that is predominantly air conditioned, warm, moist air within the wall cavity will condense on

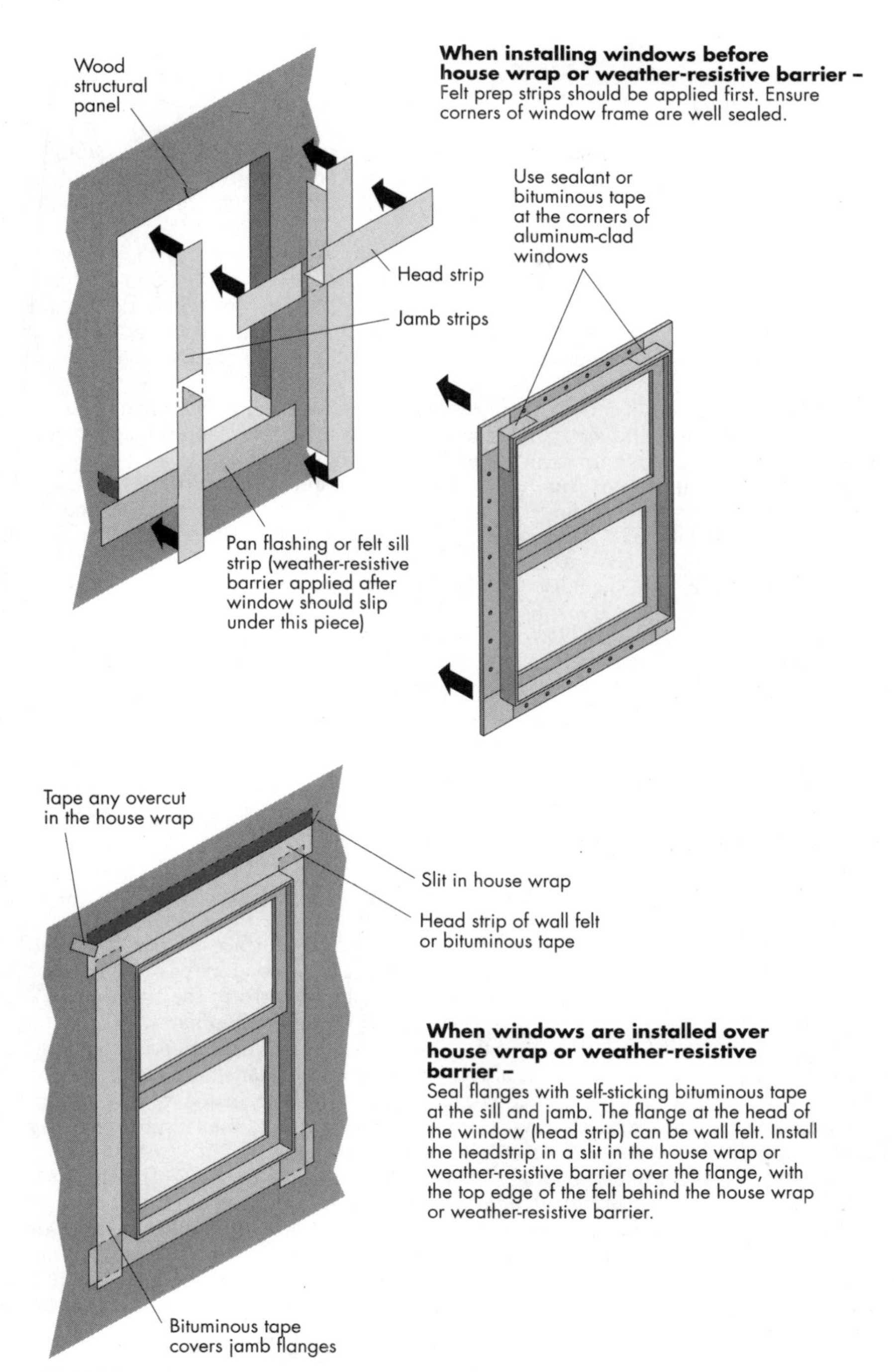

FIGURE 12.29 Window installation details.

the vapor retarder, within the wall cavity. Because the outside of the structure is the source of the moisture, it is unlikely that the wall will dry out in that direction. The vapor retarder on the inside surface will prevent drying out to the inside of the structure. If this is the case, the potential for a severe moisture problem is high.

What this means is that if a structure is in an area of predominant heating—northern tier states—put the vapor retarder on the inside surface of the exterior wall. If the structure is in an area where air conditioning predominates, the warm side is the outside, and that is where the vapor retarder should go.

The problem is what to do about the rest of the country (most of the country). There is more than a little controversy associated with this issue. Some experts advocate a very careful study of degree-days to determine on which side of the fence a given structure falls. Other experts argue that the use of a vapor retarder on any side of the wall, except in the most extreme climates, is inadvisable because for at least a portion of the year it is in the wrong location.

Lately, a consensus has been forming that it is best to leave the vapor retarder out completely unless the structure is in one of the most extreme climates, where the question "Which side is warm?" can be easily answered.

So, what is the impact of leaving out the vapor retarder? Very little if the design and construction of the building have eliminated the bulk water leakage and air infiltration. As mentioned, if the other two water sources are eliminated and mechanical ventilation is present in the moisture-generating areas of the structure (kitchen, bathrooms, and laundry rooms), the contribution to the moisture load on the structure due to vapor transmission will be minimal.

Weather-Resistive Barrier (Building Paper). Weather-resistive barriers applied on the exterior side of the wall must be permeable to permit water-vapor transmission, to avoid water or water vapor from being trapped in the wall. It must also provide resistance to absorption or passage of "free" water from unintentional water leakage. The traditional material for these purposes has been *building paper*.

The basic types of barrier materials are kraft waterproof building paper complying with Federal Specification UU-B-790a,[18] Grade D (water-vapor permeable), or No. 15 asphalt-saturated felt complying with ASTM D 226,[19] Type 1. These types of barrier materials are supplied in roll form (typically 3–4 ft wide), which are applied horizontally on the wall surface, starting at the base of the wall. Subsequent courses of building paper should be overlapped (shingle lapped) 2 in. or more over lower courses. End joints of building paper should be lapped 6 in. or more.

Building paper is traditionally added to the outside surface of the structural frame. Because it has some finite capacity to act as a vapor retarder (1 perm for 15 lb asphalt felt, and 0.3 perms for kraft paper), it can perform the function as a warm side vapor retarder in areas where air conditioning predominates. When it is used in areas where the inside wall of the structure is the warm wall, it is important that the inside vapor retarder have a greater perm rating than the building paper. This will ensure that any water vapor escaping through the inside vapor retarder will be easily able to escape through the outside retarder, thus preventing any buildup of moisture within the walls.

The proper installation of building paper is illustrated in Figs. 12.30 and 12.31.

Capillary Suction. *Capillary suction* is a term used to describe the flow of water through a material via small openings within the matrix of the material or between adjacent elements. In wood construction this term usually describes the wicking of moisture into wood when it is in contact with the ground or concrete in contact with the ground. The building codes try to prevent this in general by preventing untreated wood from coming in contact with grade. This is accomplished by requiring specific clearance distances between untreated wood and grade. See Table 12.1.

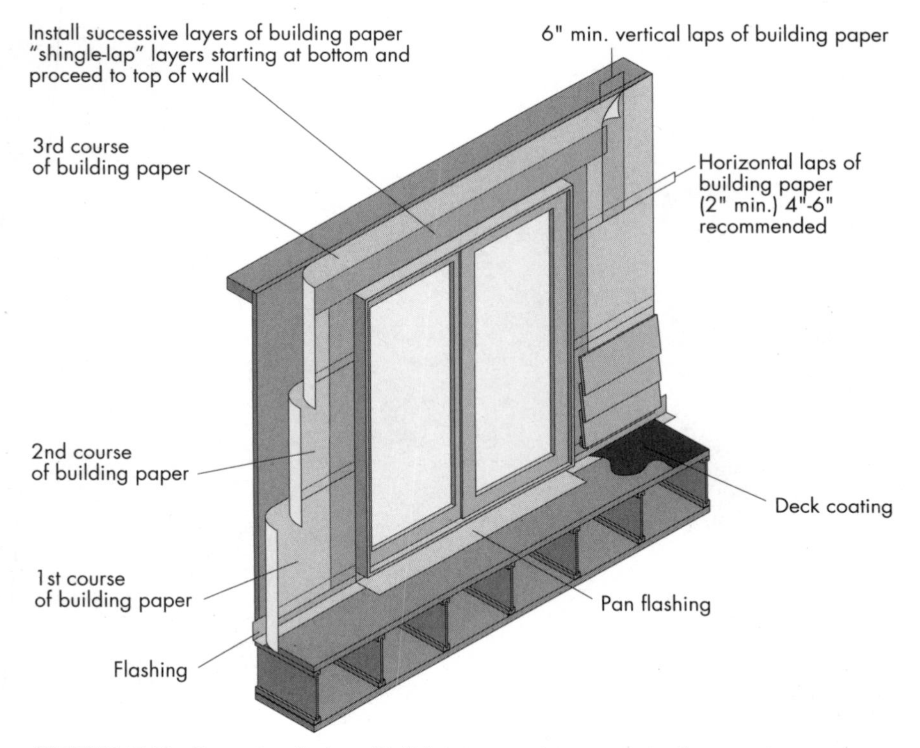

FIGURE 12.30 Proper installation of building paper—shown around a door opening.

Capillary suction can also cause water to be drawn up behind metal flashing that is in contact with other building materials. This can be prevented or minimized by using the proper installation sequence of flashing and other waterproofing materials.

Capillary suction is not all bad. In fact, it is capillary suction that makes wood construction so forgiving where occasional small leaks are concerned. It is capillary suction that carries this water to other portions of the building, increasing the surface of the affected area. This decreases the moisture content of the wood in the area of the leak, and the increased surface area increases the drying rate of the structure. Problems associated with leakage occur when the leak rate is equal to or greater than the capacity of the wood within the structure to dissipate the leakage.

Special Considerations

Wind-Driven Rain—Rain-Screen Wall. The entire exterior finish, weather-resistive barrier, and flashing system in wood construction are designed primarily to rely on gravity to keep bulk water out of the building envelope. When wind-driven rain is present, most of these safeguards can be compromised to one extent or another. If wind-driven rain is an infrequent occurrence, the forgiving nature of wood construction can often account for the occasional influx of water into the building system. The moisture in the affected areas will be removed through capillary suction, and the entire building frame will dry through vapor transmission.

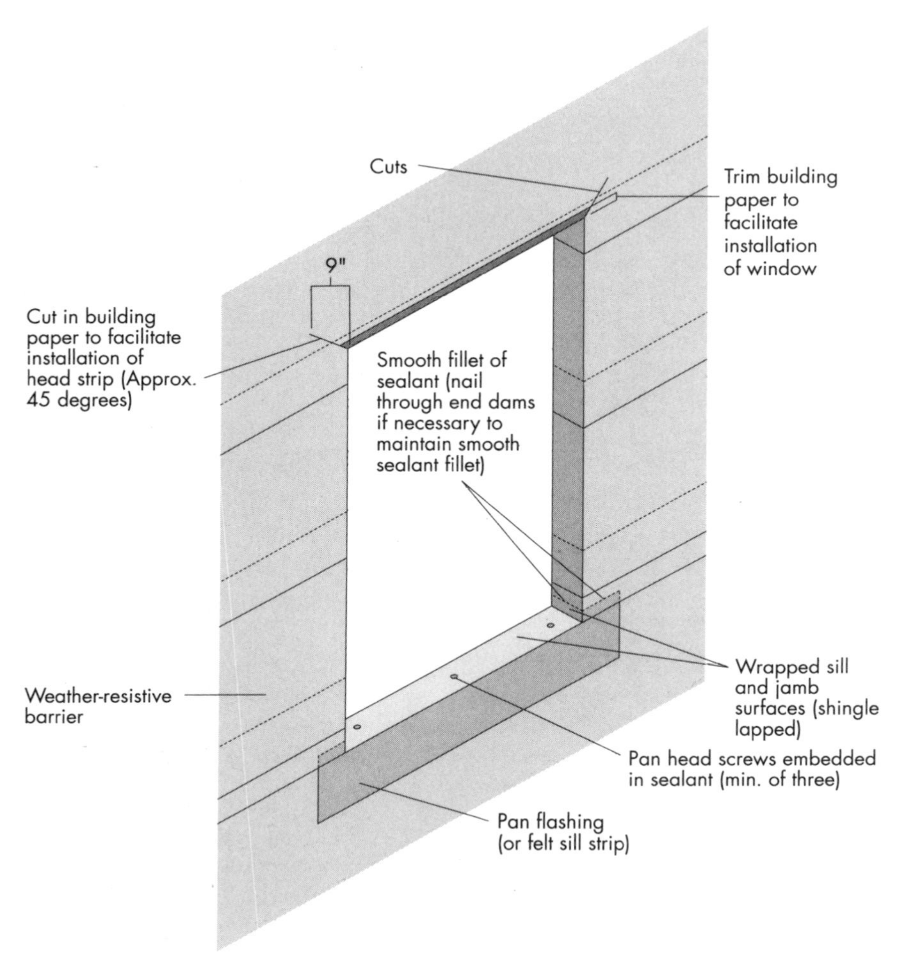

FIGURE 12.31 Proper installation of building paper—shown around a window opening.

If, however, the wind-driven rain is a frequent occurrence, the amount of water introduced into the structure could very easily outstrip the structure's ability to get rid of it. When this is the case, double-wall construction—also known as a rain-screen wall—is often recommended. Double-wall construction consists of building the exterior finish system out and away from the weather-resistive barrier, separating the first line of defense from the second line of defense with an air space. This separation is made with the use of pressure-treated lumber spacers that are installed vertically and carefully detailed around openings and penetrations to allow drainage of any water that makes it through the exterior finish. This space, ¾–1 in., is open at the bottom to promote drainage and closed at the top to allow the air space to equalize with the exterior air pressure. The opening at the bottom has a pest screen. This system is often used with an interior air barrier to allow the air pressure in the wall cavity to equalize with that in the air space behind the cladding. This eliminates the driving force behind water leakage into the wall. Fig. 12.32 illustrates this system.

Wood Clearance to Exterior Grade. The model building codes require that untreated wood framing and sheathing used for the building structure be located 8 in. or more above the exterior finish grade (6 in. or more to the bottom edge of exterior wood siding). This provides a margin to minimize backsplash from rain and allows for inspection of the foundation for signs of termite infestation.

Wood that has less than these clearances should be preservative-treated for aboveground contact in accordance with applicable standards of the American Wood-Preservers' Association.

Proper Installation of Siding Products. Because the siding product forms the primary water/weather barrier for the structure, it is important that it be properly applied. Some possible sources of information on the proper installation of siding products are the manufacturer, supplier, and trade associations representing the product types. It is important to consider these details early in the planning process because many of the specific flashing requirements and weather-resistive barrier details are dependent on the type of exterior skin the building will have.

Structural Insulated Panel (SIPs) Wall Systems. SIPs show great promise for wall applications. SIPs are discussed in detail in Chapter 3. They are strong, lightweight, and can easily accommodate the kind of vertical loads found in residential and light commercial applications. The very nature of their construction—the use of closed cell foam—virtually eliminates the possibility of air infiltration and greatly reduces the transmission of vapor into the system. Because of this, two of the three primary sources of water intrusion into the walls of wood structures can be eliminated. The only source left is bulk water leakage.

The inherent lack of framing in SIPs reduces the ability to distribute and disperse water that results from a small intermittent leak. For this reason, proper layering of building paper and proper installation of flashing is crucial to the permanence of such structures.

12.2.3 The Foundation System

The foundation is not only critical to the structural well-being of any building, it is also the most difficult of the structural systems to repair if a failure occurs. One of the major contributors to foundation problems is a lack of moisture control. Excessive moisture in the form of liquid water or excessive dampness can cause habitability problems and even structural damage. When it comes to liquid water and dampness control, details are critical. Repairs are often expensive or impractical, so it is vital to do it correctly the first time. It is recommended that a geotechnical survey be done prior to construction or prior to making extensive modifications to the building drainage systems. It is also recommended that a design professional be consulted to ensure the best chance of achieving success.

Potential Sources of Liquid Water and Dampness. There are several common causes of water intrusion into floors, crawl spaces, and basements:

1. Rainwater can flow on top of the ground, into the side of the foundation, and then into the crawl space or basement. This water can be from rain falling on the ground or paved areas near the home or from rain falling onto the roof and then being deposited too close to the foundation walls.

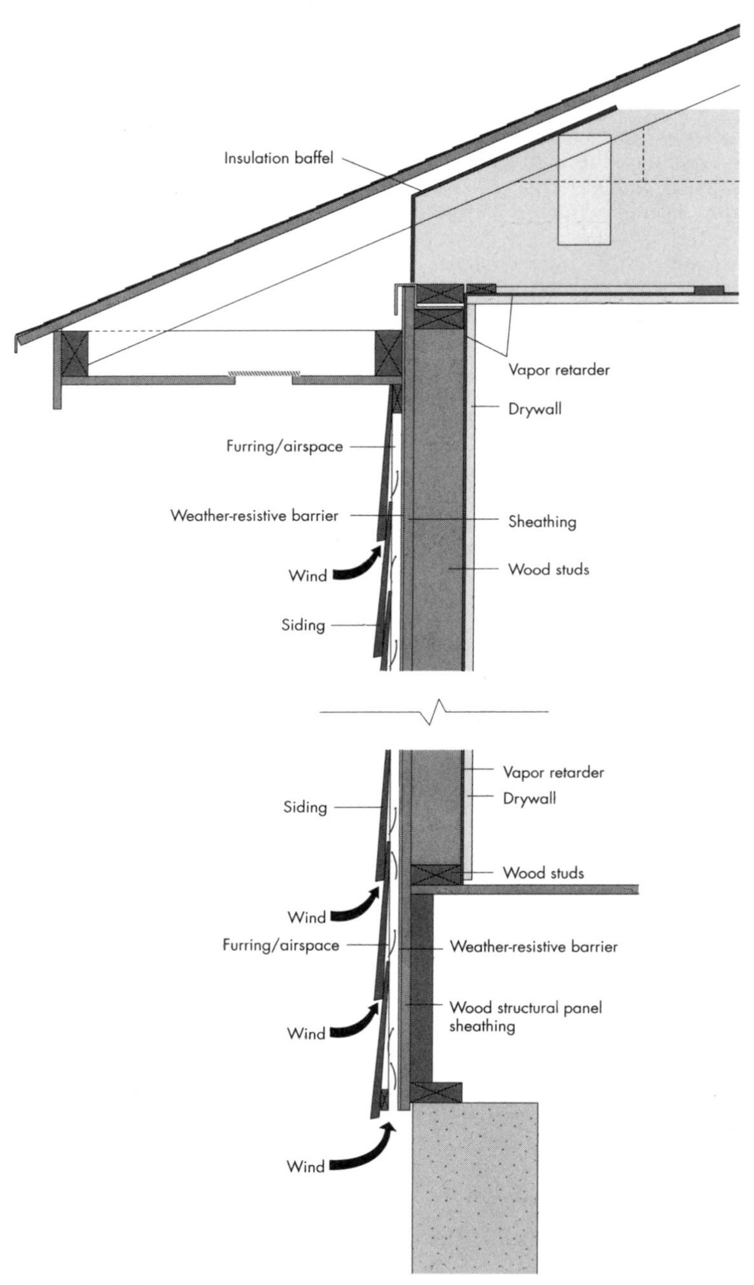

FIGURE 12.32 Rain-screen wall details.

2. Rainwater can enter the ground and flow through the soil into the area under the structure.
3. Natural springs sometimes emerge under homes. They may appear only seasonally.
4. Groundwater level can rise and fall seasonally. Sometimes it can actually rise all the way to the surface of the ground. See Fig. 12.33.
5. Groundwater below the level of the floor or foundation can wick upward through the soil by capillary action and cause dampness in basements, crawl spaces, and slab floors. In very fine soils or clays, this capillary rise can be as much as 8 ft above the groundwater table.
6. New concrete will usually be damp to the touch for several weeks or even months. This is because new concrete contains excess water and the dampness occurs as this water evaporates. This type of dampness is not a long-term problem.
7. During construction, landscaping, or remodeling or just over time, existing footing drain (perimeter drain) systems can become clogged with dirt or tree roots, crushed, broken, or severed. Subsequent water backup may cause dampness or even flooding.
8. Leaking water lines or sewers may cause water to enter the space beneath a home.

Foundation Types. Most wood structures are built on one of three types of foundations: slab-on-grade, crawl space, and full basement. Pressure-preservative-treated wood, masonry, post-and-pier, and pilings are occasionally used. Only the concrete foundations are addressed in this chapter. For information on pressure-preservative-treated wood foundations, referred to as Permanent Wood Foundations, contact the Southern Forest Association at P.O. Box 641700, Kenner, LA 70064-1700, phone (504) 443-4464.

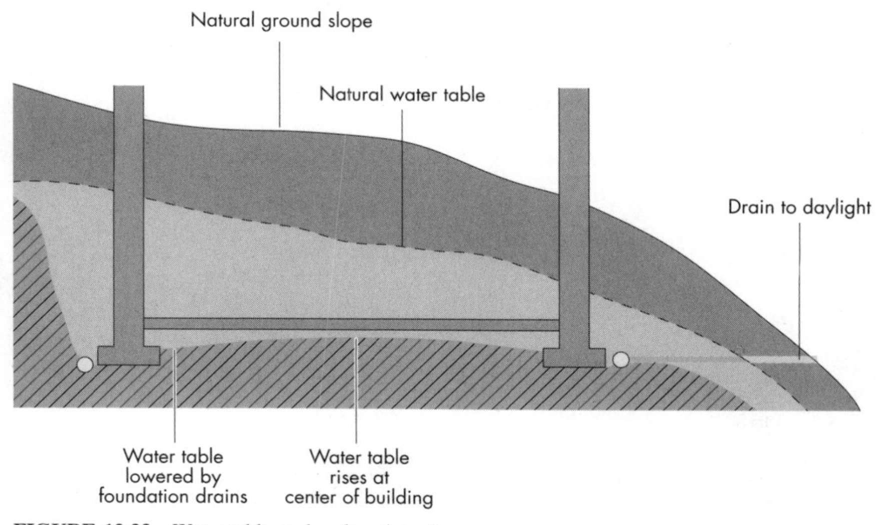

FIGURE 12.33 Water table under sloped grade.

Slab-on-Grade Foundation. This is probably the simplest type to build because the floor and footing can be poured as one unit. The floor is typically at, or only slightly above, ground level. Walls of any height can be formed on top of the slab-on-grade foundation system. Fig. 12.34 is provided as an illustration.

Crawl Space Foundation. This type of foundation is typically used to permit the construction of a wood-framed framed floor system above the ground. The crawl space under the floor provides access to wiring, plumbing, and heating ducts from under the floor. The foundation is composed of a separate footing plus concrete or masonry walls that may be only a few inches to several feet high. The unfinished dirt floor of this type of foundation system may be at or below ground level, as shown in Fig. 12.35.

Full Basement Foundation. This is actually a variation of the crawl space foundation but with full-height walls and a concrete floor. The basement walls may be below grade (ground level) on all sides. On hillsides, one or more sides of the basement may be above or partially above grade. See Fig. 12.36.

Preventive Measures before and during Construction of Foundations. Attention to the details of proper water control systems during construction can prevent water problems later. This section outlines the basics of good construction practices, which will minimize the chance of water problems later. Many of these suggestions, which are relatively common in commercial construction, are often overlooked in residential construction. Communication between the contractor and the owner is important. Consultation with a registered engineer, along with a geotechnical survey, are recommended.

Slab Floors. If fine soils or clays are present and the water table may rise to within 10 ft of the surface, the ground should be specially prepared to receive a slab floor.

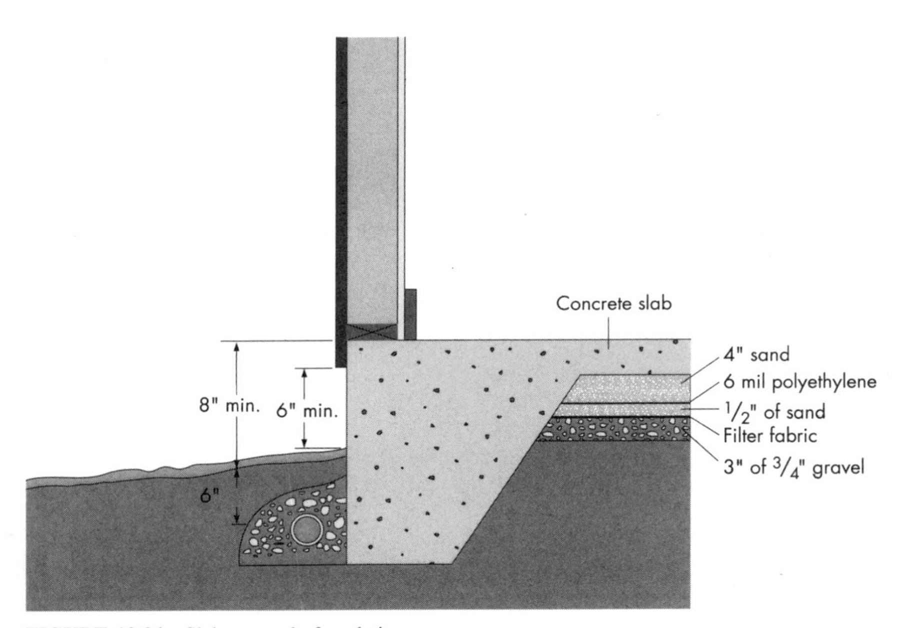

FIGURE 12.34 Slab-on-grade foundation.

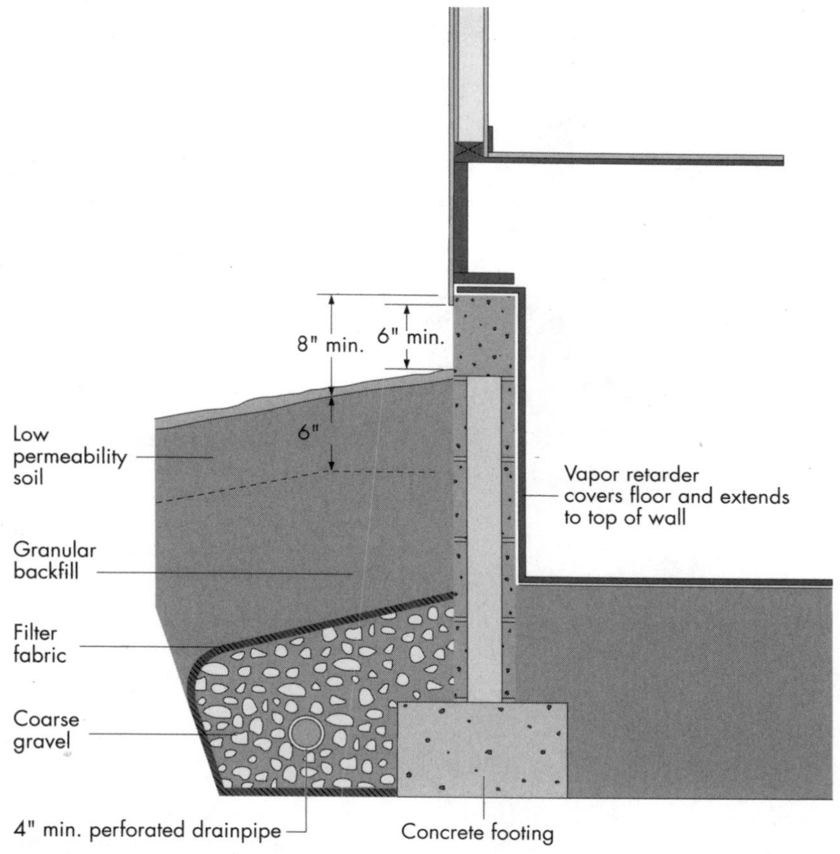

FIGURE 12.35 Concrete masonry crawl space foundation.

1. Install a drainage and capillary break of 3 in. of ¾ in. single-size coarse aggregate over the ground. This gravel should be compacted (Fig. 12.34).
2. Over the gravel, place a layer of geotechnical fabric.
3. Place ½ in. (15 mm) of compactable sand over the geotechnical fabric and compact well.
4. Place a layer of polyethylene (or reinforced polyethylene for greater puncture resistance) over the ½ in. (15 mm) of compacted sand.
5. Place 4 in. (100 mm) of compacted sand over the polyethylene. The sand should be sufficiently compacted to prevent a fully loaded premix concrete truck from indenting the fill more than ½ in. (15 mm).
6. Pour the slab over the sand.
7. Note that excess water in concrete evaporates and leaves microscopic holes through which water can move. To minimize concrete porosity, use a low water-cement ratio concrete mix with a high cement content and a superplasticizer additive.

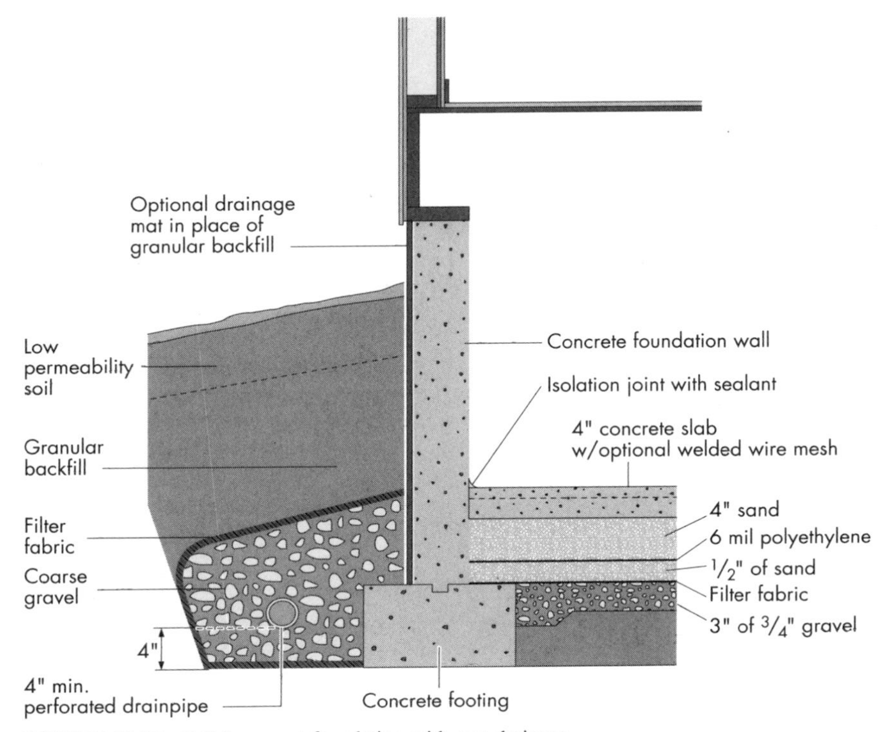

FIGURE 12.36 Full basement foundation with mat drainage.

Footing Drain. A footing drain (Figs. 12.34–12.36) should be installed around the perimeter of the foundation and discharged to a suitable location downhill from the home, into a drywell or into a storm sewer system. Install as follows:

1. Place a geotextile (filter) fabric on the bottom of the excavation.
2. Cover with 4 in. (100 mm) of one-size, clean, ¾ in. (20 mm) crushed stone or gravel.
3. Place 4 in. minimum diameter (100 mm) perforated drainpipe over the gravel if the soil is clay. Use a 6 in. (150 mm) perforated pipe if the soil is sandy and there is a lot of water to redirect. Orient the perforations down. The drainpipe may be laid level along the footing.
4. At the point where the footing drain leaves the perimeter of the structure, connect the perforated footing drain to an unperforated drainpipe of the same diameter and run to a downhill area, away from the structure, to a drywell or storm sewer or to a suitable location for an aboveground discharge.
5. Cover the perforated drainpipe with about 6 in. (150 mm) of clean gravel and cover the gravel with the geotextile (filter) fabric.

Slope the ground surface away from the building and foundation. Make sure that all water from downspouts is discharged away from the building into a drywell or storm sewer or suitable ground surface location downhill from the structure.

Crawl Space Foundations. The footing drain and gutter/downspout discharge systems should be installed the same way as for the slab floor described above. Through-wall connections for water, sewer and electrical should be minimized and thoroughly sealed. See Fig. 12.37. Again, make sure that the ground slopes away from the footings.

Crawl space ventilation should be provided to facilitate natural air circulation. There should be at least four ventilation openings, with at least one on each wall around the perimeter. Ventilation openings should be as high on the foundation walls as possible, and the total area should be evenly distributed among the walls.

The required total size of the openings may be calculated by the formula

$$a = \left(\frac{A}{150}\right) \tag{12.1}$$

where a = net area of all vents (ft^2)
A = area of crawl space (ft^2)

This formula assumes the use of ¼ in. (6 mm) mesh hardware cloth. The required area should be doubled for 1⁄16 in. (1.5 mm) mesh screen.

There should be a 6 mil (0.15 mm) minimum thickness polyethylene ground cover over bare ground under the structure to prevent ground moisture from escaping into the crawl space. Overlap the edges at least 6 in. (150 mm) and follow manufacturer's recommendations to seal all edges with tape or adhesive. Tape polyethylene to walls and around all objects protruding through the polyethylene. When the polyethylene is installed as recommended, the ventilation area may be reduced to 10% of the area calculated by the above equation.

Basements. The same steps that were taken for the slab foundation floors also apply to basement floors. Concrete vibrators should be used to prevent the formation of voids and cold joints in the walls. A liberal use of steel reinforcing bars in the basement walls will help minimize cracking and increase structural strength. Use low water-to-cement ratio concrete with high cement content. Superplasticizers will facilitate the workability of this concrete mix.

In addition, the below-grade portion of walls should be coated with asphaltic mastic or other suitable material. Depending on local groundwater conditions, a

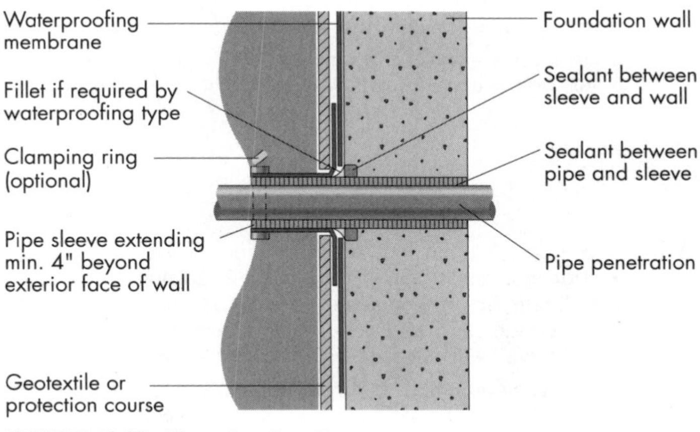

FIGURE 12.37 Through-wall sealing.

waterproof membrane should also be applied over the wall exterior. The following steps will help ensure a water-free interior:

1. Minimize through-wall penetrations for water, sewer, and electrical connections and seal all such penetrations to prevent water leakage. See Fig. 12.37. Seal around all joints and penetrations for pipes and conduits.
2. Coat the walls with a suitable damp proofing such as asphaltic mastic or other waterproofing material. Follow manufacturer's directions. This will be the backup system to help seal any existing or future minor cracks that may admit water.
3. When the natural water level is high up on the wall, the sealer should be covered with a waterproof membrane capable of withstanding the water pressure without leaking, even if a footing drain has been installed. When such a system is used, the components should all come from the same manufacturer to better ensure compatibility of components.
4. Attach a drainage mat to the wall, over the waterproof membrane or sealer, to provide a free pathway for water to follow easily to the footing drain. Gravel can also be used to create a drainage pathway to the footing drain, but care should be taken to not tear the waterproof membrane when placing the gravel.
5. Place a geotextile (filter) fabric on the bottom of the excavation and cover with 4 in. (100 mm) of one-size clean ¾ in. (20 mm) crushed stone or gravel.
6. Place 4 in. minimum diameter (100 mm) perforated drainpipe over the gravel if the soil is clay. Use a 6 in. (150 mm) perforated pipe if the soil is sandy and there is a lot of water to redirect. Orient the perforations down. The drainpipe may be laid level along the footing.
7. At the point where the footing drain leaves the perimeter of the building, connect the perforated footing drain to an unperforated drainpipe of the same diameter and run to a downhill area, away from the structure, to a drywell or storm sewer or suitable location for an aboveground discharge.

   When soil, adjacent structures, or ground elevations prevent gravity drainage of the water, a sump must be installed and the water pumped to a drywell or storm sewer or discharged downhill from the building. A licensed plumber may be required to attach to a storm sewer.
8. Cover the perforated drainpipe with about 6 in. (150 mm) of clean ¾ in. (20 mm) gravel and cover the gravel with the geotextile (filter) fabric.
9. Backfill the excavation. Compact the soil carefully in layers as backfill is added. Avoid overcompaction because that may damage the wall structurally.

12.3 GLOSSARY

Air infiltration barrier: A nonwoven synthetic fiber fabric that is used traditionally as the exterior-most substrate of the building envelope. Properly installed, with all seams and penetrations adequately taped, an air-infiltration barrier blocks the flow of moisture-laden air into the wall cavity due to differential air pressures that exist across the wall but allows the passage of water vapor.

Building envelope: The combination of roofing, waterproofing, dampproofing, and flashing systems that act together as a barrier to protect interior spaces from water and weather intrusion. These systems envelop a building from top to bottom, from below grade to roof.

Building Paper: See *Kraft paper.*

Capillary break: A physical impediment to capillary action. Air gaps, flashing, treated wood, plastic sheeting, and metal plates are all used to provide a capillary break in modern building construction.

Capillary suction: The upward-wicking motion of liquid water and vapor through small voids in soils and building materials. This action is natural and is caused by the surface tension of water.

Caulking: An elastomeric joint-sealing material appropriate for joints that exhibit little or no movement. In both interior and exterior applications, caulked joints are not permanent and require maintenance.

Counter flashing: Flashing that is surface-mounted or placed directly into walls with a portion exposed to flash various building elements, including roof flashing, waterproofing materials, building protrusions, and mechanical equipment.

Damp proofing: A mastic material applied to the outside surface of an underground structure to minimize the intrusion of moisture vapor into the structure.

Decay: Degradation of wood due to the presence of wood-inhabiting fungi. Fungi will only decay wood with moisture content above fiber saturation.

Dew point: The temperature at which a sample of humid air becomes saturated and the water vapor begins to condense to liquid water.

Equilibrium moisture content: That moisture content at which wood is neither gaining nor losing moisture.

Exterior insulated finish systems (EIFS): A number of proprietary finishing systems consisting of synthetic stucco over foam sheathing. These systems rely heavily on proper application of flashing systems and the presence of drainage layers to keep wood framing system dry.

Flashing: Thin sheets of corrosion-resistant metal that are formed and placed in such a way as to direct the flow of water in the intended direction. Flashing is used in modern construction to keep rainwater and snow melt away from the interior of the building envelope.

Geotextile: Also known as filter fabric. A porous, durable textile designed for use at or below grade, used to prevent the migration of soil fines by subsurface water movement.

Head flashing: Flashing installed above window head just below the adjacent facing material that the window abuts.

Hose bibb: A plumbing penetration in the building envelope used to accommodate an external water hose.

House wrap: Common name for *Air infiltration barrier,* described above. Called a house wrap because it is available in 9 ft rolls, allowing the builder to effectively wrap a single-story structure with minimal seams.

Ice dams: A buildup of a lens of ice under snow on a roof due to thawing and refreezing of the snow cover from the bottom side. This is a common occurrence on roof overhangs. As the snow over the ice dam melts, the resulting meltwater can build up behind the dam. This causes the meltwater to back up under the shingles of the roof and allow water to enter the building envelope.

Infiltration: The introduction of moisture-laden air into the structural system of a building caused by differential air pressure acting across small defects in the structure's envelope.

Kraft paper: A waterproof, water vapor-permeable building paper complying with Federal Specification UU-B-790a, Grade D.

Leakage: The introduction of bulk water into the structural system of a building due to improper design, detailing, construction, or maintenance of the structure's envelope.

Mold: A surface growth of fungus on damp or decaying matter. On wood these fungi form when the wood is above fiber saturation. While molds have minimal impact on the strength of the material themselves, they are an indicator of moisture conditions that can lead to more severe forms of decay.

Mildew: A relatively benign microscopic stain fungi that grows on wet or damp surfaces with little exposure to sunlight or air movement. Mildew can be removed by spraying with a mixture of one part household bleach and three parts water. The staining should disappear within two minutes of application.

No. 15 asphalt-saturated felt: A waterproof, water vapor-permeable building paper complying with ASTM D226, Type 1.

Rain-screen wall: A double-layer wall that provides an air space behind the exterior wall cladding to provide a place for the equalization of the exterior air pressure and a drainage plane for water driven through the exterior cladding. This system is often used with an interior air barrier to allow the air pressure in the interior of the wall to equalize with that in the air space behind the cladding. This will eliminate the driving force behind water leakage into the wall.

Roofing material: The material that can be seen on the finished roof. This material provides the final layer of protection against wind, rain, snow, and the sun.

Roof sheathing: A layer of wood structural panel sheathing that is placed over the roof framing to provide structural support for the applied roof and lateral loads. It also acts as the foundation for the attachment of the rest of the roofing system.

Shingle lapping: The placement of building elements such that the lower edge of the element on top always laps over the upper edge of the element below. In a similar manner, the vertical edges of overlapping elements are always offset from each other except at corners and breaks in the surface. The purpose of shingle lapping is to direct water away from the interior of the structure as it flows down under the force of gravity. The name comes from the manner in which roof or wall shingles are correctly installed.

Structural insulated panels: A structural composite panel made up of a solid foam core bonded to wood structural panel faces with a waterproof structural adhesive.

Superplasticizer: A chemical additive to concrete mixes that increased the workability of the concrete with a lower water-to-cement ratio.

Tin tabs: Small circles, about 2 in. in diameter, of thin sheet metal stapled through the underlayment into the roof sheathing below to provide blow-off protection of the underlayment in case the roofing material is lost in a storm.

Underlayment, roofing: A layer of material such as roofing felt that is placed over the roof sheathing to provide the first waterproof layer of protection to the structure and its contents.

Ultraviolet light (UV): A high-frequency component of sunlight outside the visible range that contributes significantly to the deterioration of unprotected wood.

UV degradation: The damage caused to unprotected wood by exposure to sunlight.

Vapor transmission: The introduction of molecular water into the structural system of a building caused by a differential water vapor pressure across the structure's envelope.

Water vapor pressure: The pressure of water vapor at a given temperature; the component of atmospheric pressure contributed by the presence of water vapor.

Water-resistive barrier: A barrier applied over the structural frame of a structure to keep exterior moisture and water away from the structural frame and interior of the structure. Traditionally, these barriers have the ability to pass water vapor to aid the structure in drying out when exposed to excessive moisture. Water-resistive barriers can be made up of wood structural panels, building felt, building paper, or air-infiltration barrier materials.

Weather-resistive barriers: See *Water-resistive barrier.*

12.4 REFERENCES

1. International Residential Code for One- and Two-Family Dwellings (IRC). International Code Council, Falls Church, VA, 2001.
2. American Wood-Preservers' Association, P.O. Box 5690, Granbury, TX 76049-0690.
3. American Red Cross Publication 4477, *Repairing your Flooded Home,* FEMA Publications, Washington, DC, 1992.
4. Standard S500, *Standard and Reference Guide for Professional Water Damage Restoration,* Institute of Inspection, Cleaning, and Restoration Certification, Vancouver, WA, 1994.
5. American Lung Association, 1740 Broadway, New York, NY 10019.
6. Standard C9, *Plywood Preservative Treatment by Pressure Process,* American Wood-Preservers' Association, Granbury, TX, 2000.
7. Standard C15, *Wood for Commercial-Residential Construction, Preservative Treatment by Pressure Processes,* American Wood-Preservers' Association, Granbury, TX, 2000.
8. FDN Standard (for Permanent Wood Foundation), American Wood Preservers Bureau (defunct).
9. International Building Code, International Code Council, Falls Church, VA, 2000.
10. National Roofing Contractors Association (NRCA), *NRCA Roofing and Waterproofing—The Manual,* Rosemont, IL, 1996.
11. National Evaluation Service, Inc.: contact through Ref. 11 or 12 below.
12. International Congress of Building Officials—Evaluation Services (ICBO ES), 5360 South Workman Mill Rd., Whittier, CA 90601.
13. Building Officials and Code Administrators, International—Evaluation Services (BOCA ES), 4051 West Flossmoor Road, Country Club Hills, IL 60478-5795.
14. SBCCI Public Safety Testing and Evaluation Services Inc. (SBCCI PST&ESI), 900 Montclair Road, Suite A, Birmingham, AL 35213-1206.
15. National Building Code, Building Officials and Code Administrators, International (BOCA), Country Club Hills, IL, 1999.
16. Standard Building Code, Southern Building Code Congress, International Inc. (SBCCI), Birmingham, AL, 1999.

17. Uniform Building Code, International Congress of Building Officials (ICBO), Whittier, CA, 1997.
18. Federal Specification UU-B-790, *Paper: Building, Waterproof,* U.S. Department of Commerce, Washington, DC.
19. ASTM D226, *Standard Specification for Asphalt-Saturated Organic Felt Used in Roofing and Waterproofing,* American Society for Testing and Materials, West Conshohocken, PA, 1997.

APPENDIX A
COMMON METRIC CONVERSION FACTORS

Quantity	Multiply	By	To obtain	
Length	inch	25.400 1	millimeter	mm
	inch	0.025 400 1	meter	m
	foot	0.304 800	meter	m
	yard	0.914 400	meter	m
	mile (U.S. Statute)	1.609 347	kilometer	km
	mile (U.S. Statute)	5,280	feet	ft
	mile (U.S. Statute)	1,760	yards	yd
	mile (U.S. Statute)	0.868 36	mile (nautical)	mi (naut)
	mile (nautical)	1.151 6	mile (U.S. Statute)	mi
	millimeter	$39.370\,079 \times 10^{-3}$	inch	in.
	meter	3.280 840	foot	ft
	meter	1.093 613	yard	yd
	kilometer	0.621 370	mile	mi
Area	square inch	$0.645\,160 \times 10^{3}$	square millimeter	mm^2
	square foot	0.092 903	square meter	m^2
	square yard	0.836 127	square meter	m^2
	square mile (U.S. Statute)	2.589 998	square kilometer	km^2
	acre	$4.046\,873 \times 10^{3}$	square meter	m^2
	acre	0.404 687	hectare	ha
	acre	43,560	square feet	ft^2
	acre	4,840	square yards	yd^2
	acre	4046.876	square meters	m^2
	$inch^2$/foot of width	2.116 7	$millimeter^2$/millimeter of width	mm^2/mm
	$inch^2$/foot of width	2,116.666 9	$millimeter^2$/meter of width	mm^2/m
	square millimeter	$1.550\,003 \times 10^{-3}$	square inch	$in.^2$
	square meter	10.763 910	square foot	ft^2
	square meter	1.195 990	square yard	yd^2
	square kilometer	0.386 101	square mile	mi^2
	square meter	$0.247\,104 \times 10^{-3}$	acre	ac

	hectare	2.471 044	acre	ac
	millimeter^2/millimeter of width	0.472 4	inch^2/foot of width	in.^{2}/ft
	millimeter^2/meter of width	$4.724\ 41 \times 10^{-4}$	inch^2/foot of width	in.^{2}/ft
Volume	cubic foot	$28.316\ 85 \times 10^{-3}$	cubic meter	m^3
	cubic foot	7.480 6	gallon	gal.
	cubic yard	0.764 555	cubic meter	m^3
	gallon (U.S. liquid)	0.133 68	cubic feet	ft^3
	gallon (U.S. liquid)	3.785 412	liter	l
	quart (U.S. liquid)	0.946 353	liter	l
	cubic millimeter	$61.023\ 759 \times 10^{-6}$	cubic inch	in.^{3}
	cubic meter	35.314 662	cubic foot	ft^3
	cubic meter	1.307 951	cubic yard	yd^3
	liter	0.264 172	gallon (U.S. liquid)	gal
	liter	1.056 688	quart (U.S. liquid)	qt
Force	ounce	0.278 014	newton	N
	pound	4.448 222	newton	N
	pound	0.453 592	kilograms	kg
	pound per foot	14.594	newtons per meter	N/m
	ton, short	2,000	pounds	lb
	ton, long	2,240	pounds	lb
	tonne (metric ton)	2,204.62	pounds	lb
	grain	$1.428\ 57 \times 10^{-4}$	pounds (avoirdupois)	lb (av.)
	grain	0.064 798 91	gram	g
	newton	3.596 942	ounce	oz
	newton	0.224 809	pound	lb
	newtons per meter	0.068 5	pounds per foot	lb/ft
	pounds	0.000 5	ton, short	tn sh.
	pounds	0.000 446 43	ton, long	tn l.
	pounds	0.000 453 593	tonne	metric ton.
	pounds	7,000	grains	gr

Quantity	Multiply	By	To obtain	
Mass	ounce (avoirdupois)	28.349 52	gram	g
	pound (avoirdupois)	0.453 592	kilogram	kg
	ton, short	$0.907\ 185 \times 10^{3}$	kilogram	kg
	gram	$35.273\ 966 \times 10^{-3}$	ounce (avoirdupois)	oz av.
	gram	15.432 32	grains	gr
	kilogram	2.204 622	pound (avoirdupois)	lb av.
	kilogram	$1.102\ 311 \times 10^{-3}$	ton, short	ton, tn sh.
	slug	14.594	kilograms	kg
	slug	32.174	pounds	lb
Angle	degree	$17.453\ 29 \times 10^{-3}$	radian	rad
	radian	57.295 788	degree	
Temperature	degree Fahrenheit	(°F − 32)/1.8	degree Celsius	°C
	degree Fahrenheit	(°F + 459.67)/1.8	degree Kelvin	K
	degree Fahrenheit	(°F + 459.67)/1.8	degree Rankine (absolute)	°R
	degree Celsius	(1.8 × °C) + 32	degree Fahrenheit	°F
	degree Celsius	°C + 273.15	degree Kelvin	K
	degree Celsius	(1.8 × °C) + 491.67	degree Rankine (absolute)	°R
	degree Kelvin	(1.8 × K) − 459.67	degree Fahrenheit	°F
	degree Kelvin	K − 273.15	degree Celsius	°C
	degree Kelvin	1.8 × K	degree Rankine (absolute)	°R
	degree Rankine (absolute)	°R − 459.67	degree Fahrenheit	°F
	degree Rankine (absolute)	(°R − 491.67)/1.8	degree Celsius	°C
	degree Rankine (absolute)	°R/1.8	degree Kelvin	K
Pressure, stress	pounds per square inch	6.894 757	kilopascal	kPa
	pounds per square inch	$6.894\ 757 \times 10^{-3}$	megapascals	MPa
	pounds per square inch	$6.894\ 757 \times 10^{-6}$	gigapascal	GPa
	pounds per square inch	6,894.757	newton/meter^2	N/m^2
	pounds per square inch	6.894 757	kilonewton/meter^2	kN/m^2

pounds per square inch	$6.894\,757 \times 10^{-3}$	newton/millimeter^2	N/mm^2
pounds per square inch	0.070 306 9	kilogram/centimeter^2	kg/cm^2
pounds per square inch	703.069 0	kilograms/meter^2	kg/m^2
pounds per square foot	47.880 3	newton/meter^2	N/m^2 (Pa)
pounds per square foot	$47.880\,3 \times 10^{-3}$	kilonewton/meter^2	N/m^2 (kPa)
pounds per square foot	$47.880\,3 \times 10^{-6}$	newton/millimeter^2	N/mm^2
pounds per square foot	4.882 4	kilogram/meter^2	kg/m^2
pounds per square foot	$4.882\,4 \times 10^{-6}$	gigagrams/meter^2	Gg/m^2
pounds/1000 feet^2	0.488 24	kilogram/100 meter^2	kg/100 m^2
pascal	$1.450\,4 \times 10^{-4}$	pounds/inch^2	psi
pascal	1.000 0	newton/meter^2	N/m^2
kilopascal	101.971 6	kilogram/meter^2	kg/m^2
kilopascal	0.145 038	pound per square inch	psi
kilopascal	0.334 562	foot of water (at 39.2°F)	ft H_2O
kilopascal	0.295 301	inch of mercury (32°F)	in. Hg
megapascal	101,971.6	kilogram/meter^2	kg/m^2
megapascals	145.038	pounds/inch^2	psi
gigapascal	145,039.815	pounds/inch^2	psi
gigapascal	$1.019\,716 \times 10^{8}$	kilogram/meter^2	kg/m^2
newton/meter^2 (Pa)	1.000 0	Pascal	Pa
newton/meter^2 (Pa)	0.101 971 6	kilogram/meter^2	kg/m^2
newton/millimeter^2	20,885.4	pounds/foot^2	psf
newton/millimeter^2	145.036 778	pounds/inch^2	psi
newton/meter^2	0.000 145 377 8	pounds/inch^2	psi
newton/meter^2	0.020 885 44	pounds/foot^2	psf
newton/millimeter^2	145.037 78	pounds/inch^2	psi
kilonewton/meter^2	0.145 037 78	pounds/inch^2	psi
kilonewton/meter^2	20.885 4	pounds/foot^2	psf
kilogram/meter^2	9.804 5	newton/meter^2	N/m^2
kilogram/centimeter^2	14.223 35	pounds/inch^2	psi
kilogram/meter^2	$9.806\,7 \times 10^{-9}$	gigapascal	GPa
kilogram/meter^2	0.009 806 7	kilopascal	kPa
kilogram/meter^2	$9.806\,7 \times 10^{-6}$	megapascal	MPa

Quantity	Multiply	By	To obtain	
Pressure, stress (*Continued*)	kilograms/meter^2	0.001 422 355	pounds/inch^2	psi
	kilogram/meter^2	0.204 816	pounds/foot^2	psf
	gigagrams/meter^2	$2.048\ 16 \times 10^{-7}$	pounds/foot^2	psf
	kilogram/100 meter^2	0.204 816	pounds/1000 feet^2	lb/1000 ft^2
	foot of water (39.2°F)	62.43	lb/ft^2	psf
	foot of water (39.2°F)	2.988 98	kilopascal	kPa
	inch of mercury (32°F)	3.386 38	kilopascal	kPa
	meter of water	9,806	newton/meter^2	N/m^2
	pound/square foot	0.016 02	feet of water (39.2°F)	ft H_2O
Density	pounds/foot^3	16.018 45	kilogram/meter^3	kg/m^3
	pounds/foot^3	0.016 018 465 25	gigagrams/meter^3	Gg/m^3
	kilogram/meter^3	0.062428	pounds/foot^3	lb/ft^3
	gigagrams/meter^3	62.428 795 132	pounds/foot^3	lb/ft^3
Moment of inertia (I)	inch^4	$4.162\ 374 \times 10^{-7}$	meter^4	m^4
	meter^4	2,402,475	inch^4	in^4
Moment of inertia per unit width (I/unit)	inch^4/foot of width	1,365.56	millimeter^4/millimeter of width	mm^4/mm
	inch^4/foot of width	1,365.5	10^3 millimeter^4/meter of width	10^3 mm^4/m
	inch^4/foot of width	$1.365\ 61 \times 10^{-6}$	meter^4/meter of width	m^4/m
	millimeter^4/millimeter of width	0.000 732 284 9	inch^4/foot of width	in.^{4}/ft
	10^3 millimeter^4/meter of width	0.000 732 284 9	inch^4/foot of width	in.^{4}/ft
	meter^4/meter of width	732,274	inch^4/foot of width	in.^{4}/ft
Section modulus	inches^3	$1.638\ 73 \times 10^4$	millimeter^3	mm^3
	millimeter^3	$6.102\ 30 \times 10^{-5}$	inches^3	in.^{3}
Moment section modulus per unit width (S/unit)	inch^3/foot of width	53,763.3	millimeter^3/meter of width	mm^3/m
	inch^3/foot of width	53.7633	millimeter^3/millimeter of width	mm^3/mm
	millimeter^3/meter of width	$1.860\ 0 \times 10^{-5}$	inch^3/foot of width	in.^{3}/ft
	millimeter^3/millimeter of width	$1.860\ 0 \times 10^{-2}$	inch^3/foot of width	in.^{3}/ft

Bending moment ($F_b S$)	pound-inch	0.112 985	newton-meter	N-m
	pound-foot	1.355 818	newton-meter	N-m
	pound-inch	1.129 848 × 10^{-4}	kilonewton-meter	kN-m
	newton-meter	8.850 748	pound-inch	lb-in.
	newton-meter	0.737 562	pound-foot	lb-ft
	kilonewton-meter	124,795.5	pound-inch	lb-in.
Bending strength per unit width ($F_b S$/unit)	pound-inch/foot of width	370.685 0	newton-millimeter/meter of width	N-mm/m
	pound-inch/foot of width	3.706 865 × 10^{-4}	kilonewton-meter/meter of width	kNm/m
	pound-inch/foot of width	0.370 685	newton-millimeter/millimeter of width	N-mm/mm
	pound-inch/foot of width	0.037 799 33	kilogram-centimeter/centimeter of width	kg-cm/cm
	newton-millimeter/meter of width	0.002 697 7	pound-inch/foot of width	lb-in./ft
	kilonewton-meter/meter of width	2,697.707	pound-inch/foot of width	lb-in./ft
	newton-millimeter/millimeter of width	2.697 707	pound-inch/foot of width	lb-in./ft
	kilogram centimeter/centimeter of width	26.455 49	pound-inch/foot of width	lb-in./ft
Stiffness (EI)	pound-inch^2	2.869 84 × 10^3	newton-millimeter^2	N-mm^2
	pound-inch^2	2.869 837 5	kilonewton-millimeter^2	kN-mm^2
	pound-inch^2	2.869 837 5 × 10^{-6}	kilonewton-meter^2	kN-m^2
	pound-foot^2	0.413 253 3	newton-meter^2	N-m^2
	newton-millimeter^2	3.484 515 × 10^{-4}	pound-inch^2	lb-ft^2
	kilonewton-millimeter^2	3.484 54 × 10^5	pound-inch^2	lb-ft^2
	kilonewton-meter^2	345,811.472 9	pound-inch^2	lb-ft^2
	newton-meter^2	2.419 823	pound-foot^2	lb-ft^2
Stiffness capacity per unit width (EI/unit)	pound-inch^2/foot of width	9,415.400 601	newton-millimeter^2/meter of width	N-mm^2/m
	pound-inch^2/foot of width	9.415 400 601	newton-millimeter^2/millimeter of width	N-mm^2/mm
	pound-inch^2/foot of width	0.009 415 400 601	kilonewton-millimeter^2/millimeter of width	kN-mm^2/mm
	pound-inch^2/foot of width	9.415 403 234	kilonewton-millimeter^2/meter of width	kN-mm^2/m
	pound-inch^2/foot of width	9.415 4 × 10^{-6}	kilonewton-meter^2/meter of width	kN-m^2/m
	pound-inch^2/foot of width	0.096 010 3	kilogram-centimeter^2/centimeter of width	kg-cm^2/cm
	pound-inch^2/foot of width	9.601 3 × 10^{-4}	kilogram-meter^2/meter of width	kg-m^2/m
	newton-millimeter^2/meter of width	1.062 089 7 × 10^{-4}	pound-inch^2/foot of width	lb-in.^{2}/ft
	newton-millimeter^2/millimeter of width	0.106 208 97	pound-inch^2/foot of width	lb-in.^{2}/ft

Quantity	Multiply	By	To obtain	
Stiffness capacity per unit width (EI/unit) (*Continued*)	kilonewton-millimeter^2/millimeter of width	106.208 97	pound-inch^2/foot of width	lb-in.^{2}/ft
	kilonewton-millimeter^2/meter of width	0.106 208 97	pound-inch^2/foot of width	lb-in.^{2}/ft
	kilonewton-meter^2/meter of width	106,208.97	pound-inch^2/foot of width	lb-in.^{2}/ft
	kilogram-centimeter^2/centimeter of width	10.415 548	pound-inch^2/foot of width	lb-in.^{2}/ft
	kilogram-meter^2/meter of width	1041.554 8	pound-inch^2/foot of width	lb-in.^{2}/ft
Axial stiffness capacity per unit width (EA/unit)	pounds/foot of width	$1.459\,390\,4 \times 10^{-5}$	kilonewtons/millimeter of width	kN/mm
	pounds/foot of width	0.014 593 904	newtons/millimeter of width	N/mm
	pounds/foot of width	14.593 9	newtons/meter of width	N/m
	kilonewtons/millimeter of width	68,521.76	pound/foot of width	lb/ft
	newtons/millimeter of width	68.5217 6	pound/foot of width	lb/ft
	newtons/meter of width	0.068 521 76	pound/foot of width	lb/ft
Rolling shear constant per unit width [(lb/Q)/unit]	inches^2/foot	2.117	millimeters^2/millimeter of width	mm^2/mm
	inches^2/foot	2.117	10^3 millimeter^2/meter of width	10^3 mm^2/m
	mm^2/mm	0.472 4	inches^2/foot	in.^{2}/ft
	10^3 mm^2/m	0.472 4	inches^2/foot	in.^{2}/ft
Rolling shear capacity per unit width [F_s(lb/Q)/unit]	pound/foot of width	14.593 904	newton/meter of width	N/m
	pound/foot of width	0.014 593 904	kilonewton/meter of width	kN/m
	pound/foot of width	0.014 593 904	newton/millimeter of width	N/mm
	pound/foot of width	0.014 881 63	kilogram/centimeter of width	kg/cm
	newton/meter of width	0.068 521 760 87	pound/foot of width	lb/ft
	kilonewton/meter of width	68.521 76	pound/foot of width	lb/ft
	newton/millimeter of width	65.521 76	pound/foot of width	lb/ft
	kilogram/centimeter of width	67.196 952	pound/foot of width	lb/ft
Axial tension capacity per unit width (F_tA/unit)	pounds/foot of width	0.014 593 904	newtons/millimeter	N/mm
	pounds/foot of width	0.0148 816	kilograms/centimeter of width	kg/cm
	newtons/millimeter of width	68.521 76	pounds/foot of width	lb/ft
	kilograms/centimeter of width	67.196 95	pounds/foot of width	lb/ft

Compression capacity per unit width (F_cA/unit)	pounds/foot of width	0.014 593 904	newtons/millimeter of width	N/mm
	pounds/foot of width	0.014 881 6	kilograms/centimeter of width	kg/cm
	newtons/millimeter of width	68.521 76	pounds/foot of width	lb/ft
	kilograms/centimeter of width	67.196 95	pounds/foot of width	lb/ft
Panel shear-through-the-thickness capacity per unit length (F_vT_v/unit length)	pounds/inch of length	0.175 126 85	newtons/millimeter of length	N/mm
	pounds/inch of length	0.178 579 53	kilograms/centimeter of length	kg/cm
	newtons/millimeter of length	5.710 146 7	pounds/inch of length	lb/in.
	kilograms/centimeter of length	5.599 746	pounds/inch of length	lb/in.
Panel Area (3/8-inch basis)	1/4-inch-thick-panel area	0.666 7	square feet-3/8-inch basis	$ft^2_{(3/8)}$
	3/8-inch-thick-panel area	1.000 0	square feet-3/8-inch basis	$ft^2_{(3/8)}$
	7/16-inch-thick-panel area	1.166 7	square feet-3/8-inch basis	$ft^2_{(3/8)}$
	15/32-inch-thick-panel area	1.250 0	square feet-3/8-inch basis	$ft^2_{(3/8)}$
	1/2-inch-thick-panel area	1.333 3	square feet-3/8-inch basis	$ft^2_{(3/8)}$
	19/32 inch-thick-panel area	1.583 3	square feet-3/8-inch basis	$ft^2_{(3/8)}$
	5/8-inch-thick-panel area	1.666 7	square feet-3/8-inch basis	$ft^2_{(3/8)}$
	11/16-inch-thick-panel area	1.833 3	square feet-3/8-inch basis	$ft^2_{(3/8)}$
	23/32-inch-thick-panel area	1.916 7	square feet-3/8-inch basis	$ft^2_{(3/8)}$
	3/4-inch-thick-panel area	2.000 0	square feet-3/8-inch basis	$ft^2_{(3/8)}$
	7/8-inch-thick-panel area	2.333 3	square feet-3/8-inch basis	$ft^2_{(3/8)}$
	1-inch-thick-panel area	2.666 7	square feet-3/8-inch basis	$ft^2_{(3/8)}$
	1 1/8-inch-thick-panel area	3.000 0	square feet-3/8-inch basis	$ft^2_{(3/8)}$
Panel volume	3/8-inch-basis square feet	$8.849\ 015\ 625 \times 10^{-4}$	cubic meters	m^3
Board feet	volume in inches cubed	0.006 944 4	board feet	bd. ft
Cord	cubic feet	0.007 812 5	cords	cd
	cords	128	cubic feet	ft^3

Quantity	Multiply	By	To obtain	
Permeance	perm (gr/h-ft^2-in. Hg)	0.659 043	grams/meter^2/day/millimeter of mercury	g/m^2/d/mm Hg
	g/m^2/d/mm Hg	1.517 35	grains/hour-square feet-inches of mercury	perm (gr/h-ft^2-in. Hg)
Permeability	perm inch (gr/h-ft^2-in. Hg-in.)	0.259 465	grams/square meter/day/millimeter of mercury/centimeter thickness	g/m^2/d/mm Hg/cm
	g/m^2/d/mm Hg/cm	3.854 08	grains/hour-square feet-inch of mercury inch of thickness	perm inch (gr/h-ft^2-in. Hg-in.)
Water vapor diffusion	gr/h-ft^2	16.739 752	grams/meter^2/day	g/m^2/d
	g/m^2/d	0.059 738	grains/hour/square foot	gr/h-ft^2
Thermal rating	(°F ft^2/Btu/h)	0.176	m^2 K/W	R
	m^2 K/W	5.678	(°F ft^2/Btu/h)	R

INDEX

ABOUT APA—THE ENGINEERED WOOD ASSOCIATION

APA—The Engineered Wood Association is a nonprofit trade association representing manufacturers of glued engineered wood composite products. Founded almost 70 years ago as the Douglas Fir Plywood Association (later the American Plywood Association) and now representing approximately 70% of the wood structural panel manufacturers in North America, APA is involved in conducting product and systems research, providing product use and application education, supporting market development, developing technical standards, and providing third party quality assurance for their member companies. You can learn more about APA and their members at their web site, http://www.apawood.org

ABOUT THE EDITOR

Thomas G. Williamson, P.E., is the Director of the Technical Services Division of APA, overseeing a staff of 20 scientists, engineers, technicians and support personnel who are responsible for the technical and research activities of the association. Mr. Williamson is also Executive Vice President of Engineered Wood Systems (EWS), a related corporation of APA. EWS represents manufacturers of glued laminated timber (glulam), laminated veneer lumber (LVL), prefabricated wood I-joists and other related glued engineered wood composites. Mr. Williamson is co-editor in chief of the McGraw-Hill *Wood Engineering and Construction Handbook* and is co-author of the chapter on Wood Construction of the McGraw-Hill *Building Design and Construction Handbook.*